ANNUAL REVIEW OF GENETICS

ANNUAL REVIEW OF GENETICS

VOLUME 35, 2001

ALLAN CAMPBELL, *Editor*
Stanford University, Stanford

WYATT W. ANDERSON, *Associate Editor*
University of Georgia, Athens

ELIZABETH W. JONES, *Associate Editor*
Carnegie Mellon University, Pittsburgh

www.AnnualReviews.org   science@AnnualReviews.org   650-493-4400

ANNUAL REVIEWS
4139 El Camino Way • P.O. BOX 10139 • Palo Alto, California 94303-0139

ANNUAL REVIEWS
Palo Alto, California, USA

International Standard Serial Number: 0066-4197
International Standard Book Number: 0-8243-1235-X
Library of Congress Catalog Card Number: 63-8847

TYPESET BY TECHBOOKS, FAIRFAX, VA
PRINTED AND BOUND IN THE UNITED STATES OF AMERICA

Annual Review of Genetics
Volume 35, 2001

CONTENTS

ERRATA
   An online log of corrections to *Annual Review of Genetics*
   chapters may be found at http://genet.AnnualReviews.org/errata.shtml

RELATED ARTICLES

ANNUAL REVIEWS is a nonprofit scientific publisher established to promote the advancement of the sciences. Beginning in 1932 with the *Annual Review of Biochemistry*, the Company has pursued as its principal function the publication of high-quality, reasonably priced *Annual Review* volumes. The volumes are organized by Editors and Editorial Committees who invite qualified authors to contribute critical articles reviewing significant developments within each major discipline. The Editor-in-Chief invites those interested in serving as future Editorial Committee members to communicate directly with him. Annual Reviews is administered by a Board of Directors, whose members serve without compensation.

Annu. Rev. Genet. 2001. 35:1–29

HYPOVIRUSES AND CHESTNUT BLIGHT: Exploiting Viruses to Understand and Modulate Fungal Pathogenesis

Angus L. Dawe and Donald L. Nuss

Center for Agricultural Biotechnology, University of Maryland Biotechnology Institute, College Park, Maryland 20742-4450; dawe@umbi.umd.edu, nuss@umbi.umd.edu

Key Words   biological control, hypovirulence, mycovirus, signal transduction, viral determinants

■ **Abstract**   Fungal viruses are considered unconventional because they lack an extracellular route of infection and persistently infect their hosts, often in the absence of apparent symptoms. Because mycoviruses are limited to intracellular modes of transmission, they can be considered as intrinsic fungal genetic elements. Such long-term genetic interactions, even involving apparently asymptomatic mycoviruses, are likely to have an impact on fungal ecology and evolution. One of the clearest examples supporting this view is the phenomenon of hypovirulence (virulence attenuation) observed for strains of the chestnut blight fungus, *Cryphonectria parasitica*, harboring members of the virus family Hypoviridae. The goal of this chapter is to document recent advances in hypovirus molecular genetics and to provide examples of how that progress is leading to the identification of virus-encoded determinants responsible for altering fungal host phenotype, insights into essential and dispensable elements of hypovirus replication, revelations concerning the role of G-protein signaling in fungal pathogenesis, and new avenues for enhancing biological control potential.

CONTENTS

INTRODUCTION

Fungal viruses are considered unconventional because they lack an extracellular route of infection and persistently infect their hosts, often in the absence of apparent symptoms. Given the mysteries surrounding the origin of viruses, however, it is conceivable that the introduction of an extracellular phase into the life cycle, rather than its elimination, represents the more recent development in viral evolution. Because mycoviruses are limited to intracellular modes of transmission, they can be considered as intrinsic fungal genetic elements. Such long-term genetic interactions, even involving apparently asymptomatic mycoviruses, are likely to have an impact on fungal ecology and evolution. One of the clearest examples supporting this view is the phenomenon of hypovirulence (virulence attenuation) observed for strains of the chestnut blight fungus, *Cryphonectria parasitica*, harboring members of the virus family Hypoviridae. The recent application of molecular genetic approaches to hypovirus research has provided the means for engineering these viral genetic elements for both fundamental and practical applications. The goal of this chapter is to document these recent advances and provide examples of how that progress is leading to the identification of virus-encoded determinants responsible for altering fungal host phenotype, insights into essential and dispensable elements of hypovirus replication, revelations concerning the role of G-protein signaling in fungal pathogenesis, and new avenues for enhancing biological control potential. The engineering of hypoviruses for purposes of controlling and understanding fungal pathogenesis has been compared to the use of animal viruses in contemporary gene therapy. The goal in both cases is to stably alter host phenotype in a predictable and efficacious manner.

DISCOVERY OF HYPOVIRULENCE AND HYPOVIRUSES

Discovery and Early Descriptive Studies

Chestnut blight, its origins, progression, and devastating consequences have been chronicled in considerable detail in several excellent reviews (3, 49, 75, 80, 82, 104). Discussions of the discovery of hypovirulence and hypoviruses are limited in this section to those points considered necessary to prepare the reader for the main focus of the review. A time-line of important milestones in hypovirus research is presented in Figure 1.

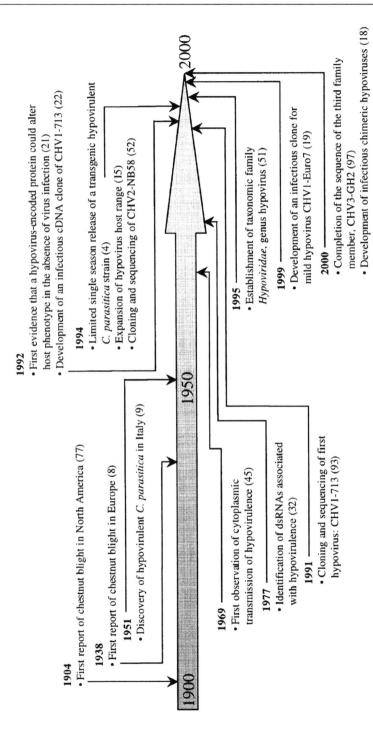

Figure 1 Time line of important milestones in hypovirus research.

1904
• First report of chestnut blight in North America (77)

1938
• First report of chestnut blight in Europe (8)

1951
• Discovery of hypovirulent *C. parasitica* in Italy (9)

1969
• First observation of cytoplasmic transmission of hypovirulence (45)

1977
• Identification of dsRNAs associated with hypovirulence (32)

1991
• Cloning and sequencing of first hypovirus: CHV1-713 (93)

1992
• First evidence that a hypovirus-encoded protein could alter host phenotype in the absence of virus infection (21)
• Development of an infectious cDNA clone of CHV1-713 (22)

1994
• Limited single season release of a transgenic hypovirulent *C. parasitica* strain (4)
• Expansion of hypovirus host range (15)
• Cloning and sequencing of CHV2-NB58 (52)

1995
• Establishment of taxonomic family *Hypoviridae*, genus hypovirus (51)

1999
• Development of an infectious clone for mild hypovirus CHV1-Euro7 (19)

2000
• Completion of the sequence of the third family member, CHV3-GH2 (97)
• Development of infectious chimeric hypoviruses (18)

While the chestnut blight epidemic is generally associated with devastation of the American chestnut *Castanea dentata* in North America beginning in 1904 (77), chestnut blight also appeared in Europe in the 1930s (8), initially causing a similar level of destruction of the European chestnut, *Castanea sativa*. However, within 15 years into the European epidemic, healing trees were observed in Italy (9), leading to the isolation, in the 1960s, of hypovirulent *C. parasitica* strains from apparently healing cankers by the French mycologist Jean Grente (44). Subsequent studies by Grente and co-workers (45) correlated the hypovirulence phenotype with the presence of cytoplasmic factors that were transmissible via hyphal anastomosis (fusion of hyphae). This finding established the basis of a biological control strategy and set the stage for a large-scale government-sponsored biological control program in France to treat chestnut orchards. The reader is referred to Heiniger & Rigling (49) for excellent accounts of reduced disease severity in European forest ecosystems and orchards as a result of the natural spread or application of hypovirulence. With very few exceptions, primarily in the state of Michigan (39), hypovirulence, either natural or introduced, has not significantly altered the severity of chestnut blight in North America. Possible explanations for geographic differences in hypovirulence efficacy are discussed in a subsequent section.

While practical application of hypovirulence for biological control was being implemented in Europe, the molecular basis of the phenomenon was being investigated in the United States. Day et al. (32) at the Connecticut Agricultural Experiment Station provided the first insight into the physical nature of the hypovirulence agent by showing that several European hypovirulent *C. parasitica* strains harbored double-stranded (ds) RNA. Subsequent studies by this group demonstrated a correlation between dsRNA transmission and conversion to the hypovirulence phenotype (5). Repeated efforts to identify virus-like particles in hypovirulent strains, a common observation for dsRNA-containing fungi (12), gave negative results (33, 47, 79, 105). Rather, these dsRNAs, and a related RNA polymerase activity (37), were consistently found associated with pleiomorphic membrane vesicles (33, 47), i.e., the hypovirulence-associated dsRNAs are not encapsidated in a discrete proteinaceous virus particle.

Numerous surveys of hypovirulent *C. parasitica* field isolates conducted over the next decade revealed considerable variation in the number, size, and concentrations of hypovirulence-associated dsRNAs, the degree of hypovirulence, and the spectrum of hypovirulence-associated traits. While these correlative and descriptive studies were extremely valuable, it became clear to investigators in the field that molecular genetic approaches were required if additional progress was to be made in understanding and applying hypovirulence.

HYPOVIRUS MOLECULAR GENETICS

Hypovirus Genome Organization and Taxonomy

Direct analysis of individual dsRNA species isolated from hypovirulent *C. parasitica* strains originating from Europe (strain EP713) (55) and North America (strain GH2) (101) demonstrated similar terminal structures. A 25- to 50-residue long 3′poly(A) tract base-paired to a 5′poly(U) tract was found at one terminus of each of the five major large EP713 dsRNA species (~8–12.7 kbp) and several minor small species (~0.3–0.5 kbp), while the other end was found to contain a consensus 28 nucleotide 3′ terminal sequence. Moreover, all dsRNA species isolated from strain EP713 hybridized to a common cDNA probe. Tartaglia et al. (101) reported that the three dsRNA species isolated from strain GH2, designated L (~9 Kbp), M (~3.5 kbp), and S (~1 kbp), also contained a homopolymeric terminus consisting of a 3′poly(A) tail and complementary 5′poly(U) and that the M species, but not the S species, hybridized with the L species. Identification of the 3′poly(A):5′poly(U) terminal structure, unprecedented for any dsRNA genetic element, prompted Tartaglia et al. (101) to suggest that the hypovirulence-associated dsRNAs resembled replicative forms of single-stranded RNA virus genomes. Subsequent cloning and sequence analysis of the small and intermediate-sized dsRNA species present in both hypovirulent strains confirmed that GH2 S dsRNA has the properties of a satellite RNA (50), whereas the GH2 M species (50) and other faster migrating EP713 dsRNA species (94) were clearly derived from the largest species in each strain as a result of internal deletion events. The generation of defective RNAs appears to be a common feature of hypoviruses, which contributed significantly to the confusion associated with the early molecular analysis of hypovirulence-associated dsRNAs.

The direct analyses of hypovirus dsRNAs provided valuable structural landmarks for several subsequent cDNA cloning efforts (23, 25, 87), culminating in the 1991 report of the complete nucleotide sequence, genome organization, and partial expression strategy for the largest dsRNA present in hypovirulent strain EP713 (93). The virus-like genome organization of this dsRNA (Figure 2), coupled with emerging evidence of similar genome organizations for several other hypovirulence-associated dsRNA under investigation, led the International Committee on Taxonomy of Viruses to establish a new family, Hypoviridae (51). The family has a single genus, *Hypovirus*, that contains numbered species based on differences in genome structure and sequence relatedness. Designations used in hypovirus nomenclature include CHV for *Cryphonectria parasitica* hypovirus, a number indicating species relatedness and, following a hyphen, the fungal host from which the virus was isolated, e.g., the designation for the prototypic hypovirus found in hypovirulent strain EP713 is CHV1-EP713.

Complete nucleotide sequences have now been reported for four members of the Hypoviridae distributed among three species that differ in organization, as shown in Figure 3 (see color insert). The genome organization for hypovirus strains belonging to the type species CHV1, CHV1-EP713, and CHV1-Euro7, consists of two large open reading frames designated ORF A and ORF B. ORF A encodes a polyprotein, p69, that is autocatalytically cleaved to produce p29 and p40 by a papain-like cysteine protease domain located within the p29 portion of the polyprotein (25) (Figures 2, 3). Expression of ORF B also involves an autocatalytic event in which a 48-kDa protein, p48, containing a second papain-like protease catalytic site, is released from the growing polypeptide chain (95). Processing events

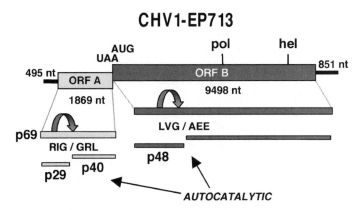

Figure 2   Cartoon depicting the genomic organization and expression strategy of prototypic hypovirus CHV1-EP713. The coding strand consists of 12,712 nucleotides, excluding the poly(A) tail, and contains two major coding domains, ORF A and ORF B. The junction between ORF A and ORF B consists of the pentanucleotide 5'-UAAUG-3' where the UAA portion serves as the ORF A termination codon and the AUG portion represents the 5'-proximal initiation codon for ORF B. Details of the known processing events for the polyproteins encoded by the two ORFs are discussed in the text. The positions of polymerase (pol) and helicase (hel) motifs related to conserved domains within plant potyvirus polyproteins are also indicated (66). This figure has been adapted from Shapira et al. (93).

involved in the expression of the conserved polymerase and helicase domains present in ORF B remain to be elucidated.

Identification of the papain-like proteases p29 and p48 also provided insight into the possible origin of hypoviruses. Choi et al. (23, 25) noted that p29 resembled the papain-like protease HC-Pro encoded by plant potyviruses in terms of the conserved amino acid sequences around the essential catalytic cysteine (Cys-162) and histidine (His-215) residues, the nature of the cleavage dipeptide (Gly-248/ Gly-249), and the distances between the two sites. A similar pattern was observed for p48, leading to the suggestion that p29 and p48 are products of a gene duplication event (95). These observations prompted Koonin et al. (66) to conduct a detailed sequence alignment analysis that identified five distinct domains within the CHV1-EP713 coding regions that showed significant sequence similarity with conserved motifs previously described for potyvirus-encoded proteins. In addition to the two proteases, these included a cysteine-rich domain at the N terminus of p29 and HC-Pro, an RNA helicase, and an RNA-dependent RNA polymerase (RDRP). This information, coupled with results of an extensive phylogenetic analysis of the RDRP domain, suggested a common ancestry for hypoviruses and plant potyviruses. The role of p29 as a symptom determinant and additional similarities with HC-Pro are addressed in a subsequent section.

The junction between ORFs A and B consists of the pentanucleotide 5'-UAAUG-3' where the UAA portion serves as the termination codon for ORF A and the

AUG portion forms the 5'-proximal translation initiation codon for ORF B (93) (Figure 2). Horvath et al. (56) reported a similar dicistronic organization for segment 7 mRNA of influenza B virus and showed that the pentanucleotide facilitated ribosome termination and reinitiation with an efficiency level of 25%. Interestingly, recent studies by Suzuki et al. (99), to be described later in this review, indicate that CHV1-EP713 can tolerate considerable modifications around the pentanucleotide-containing junction and still maintain replication competence.

The genome of hypovirus CHV2-NB58 (52), type strain of species CHV2, also consists of a two ORF configuration with a UAAUG junction (Figure 3). However, while the CHV2-NB58 ORF A shows sequence similarity to the N-terminal cysteine-rich portion of p29 and to the highly basic p40 region of CHV1-EP713 ORF A, it lacks the p29 papain-like catalytic or cleavage sites (Figure 3). In vitro translation studies confirmed that CHV2-NB58 ORF A directs the translation of a 50-kDa protein product that fails to undergo any autocatalytic processing under conditions in which CHV1-EP713-encoded p69 is readily processed (52). The organization of CHV2-NB58 ORF B is very similar to that of CHV1-EP713, including the presence of a homologue of the N-terminal p48 protease, p52, and helicase and RDRP motifs of high sequence conservation (74% and 78%, respectively) in the same relative genomic positions (52).

Completion of the sequence analysis of CHV3-GH2, the type strain of species CHV3, revealed a genome that is considerably smaller than those of the CHV1 and CHV2 species, 9.8 kbp versus 12.5 to 12.7 kbp (97), and a single rather than a two ORF organization (Figure 3). Alignment analysis at the amino acid level showed that CHV1 and CHV2 strains are more closely related to each other than CHV3-GH2 is to either of the other species (97). In fact, the CHV3-GH2 polymerase region is more closely related to that of the potyvirus barley yellow mosaic virus than it is to the polymerase domains of CHV1 and CHV2. However, the N-terminal portion of the CHV3-GH2 coding domain contains a papain-like protease, p32, that is more closely related to p29 of CHV1-EP713 than to either the p48 protease of CHV1-EP713 or p52 of CHV2-NB58. The reader is referred to Smart et al. (97) for a discussion of the implications that these differences in genome organization and sequence similarities present for understanding the evolutionary relationships of the different hypovirus species.

The coding strand of each hypovirus species contains a long untranslated leader sequence (UTL) with multiple small ORFs (miniORFs). The 5'-UTLs of CHV1-EP713 and CHV2-NB58 show striking similarities in length, 495 nt vs 487 nt, number of miniORFs, 7 versus 9, and nucleotide sequence, 65% nucleotides in common with 23 of the first 24 being identical (54, 87). This contrasts with the CHV3-GH2 5'-UTL, which is 369 nt in length, containing 6 miniORFs and no detectable sequence similarity with either CHV1 or CHV2 (97). The 5'-UTLs of CHV1-EP713 and CHV2-NB58 have also been shown to depress in vitro translation (54, 87). Although a role for the hypovirus 5'-UTLs in regulating in vivo translation has not been demonstrated, multiple miniORFs and depression of in vitro translation are properties often associated with internal ribosome entry sites (IRES) on some viral mRNAs (76, 110).

Although the four hypovirus strains described above all cause a very significant reduction of *C. parasitica* virulence, they each cause a quite different constellation of host phenotypic changes. For example, CHV1-EP713 severely reduces pigmentation, sporulation, and laccase production in its infected host, whereas these processes are only minimally reduced for CHV3-GH2-infected colonies (97). The development of infectious cDNA clones of several hypoviruses, as described in the next section, has provided the opportunity to exploit this diversity to begin identifying virus-encoded domains responsible for altered fungal phenotype. This comparative virology approach has also provided the means for engineering hypoviruses to fine-tune the interaction between *C. parasitica* and its plant host, with a view to enhancing biological control potential.

Development of a Hypovirus Infectious cDNA Clone

Two technical advances, completion of the CHV1-EP713 nucleotide sequence (93) and the development of a robust DNA-mediated transformation system for *C. parasitica* (27), set the stage for construction of an infectious hypovirus cDNA clone (22). The delivery strategy involved transformation of *C. parasitica* spheroplasts with a plasmid that contained: (*a*) the full-length cDNA copy of CHV1-EP713 RNA fused between the *C. parasitica* glyceraldehyde-3-phosphate dehydrogenase (GPD) promoter and terminator; and (*b*) a hygromycin B phosphotransferase gene as a selectable marker. Characterization of resulting hygromycin-resistant transformants confirmed chromosomal integration of the viral cDNA, phenotypic changes identical to CHV1-EP713–infected strains, and the presence of cDNA-derived hypovirus dsRNA in the cytoplasm. Thus, these "transgenic" hypovirulent *C. parasitica* strains were essentially identical to natural hypovirulent strains except that they contained a chromosomally integrated full-length hypovirus cDNA. Chen et al. (14) subsequently demonstrated that the integrated viral cDNA was stably maintained through repeated rounds of asexual sporulation (conidiation) and was faithfully transmitted to ascospore (sexual spores) progeny. Mendelian inheritance of the viral cDNA followed by resurrection of the cDNA-derived viral RNA provides a novel mode of hypovirus transmission not observed in nature that is predicted to result in enhanced biological potential. Efforts to test these predictions under actual field conditions are discussed in detail in a later section.

In preparation for field testing of the newly constructed transgenic hypovirulent strains, Chen et al. (16) showed that resurrection of functional hypovirus RNA from the integrated viral cDNA involved the precise trimming of nonviral vector nucleotides from a large nuclear transcript. They also made the surprising observation that the cDNA-derived viral dsRNA contained a 73-base pair deletion within a dispensable portion of the 5'-UTL, an apparent result of a pre-mRNA splicing event. Although further inspection of the CHV1-EP713 sequence revealed a total of five potential splice sites, only the 5'-proximal site was found to undergo splicing in vivo. However, this spliced viral dsRNA was eventually replaced by the unspliced version over time in transgenic strains recovered from the field or cultured in the laboratory (4). This result suggests that a portion of the cDNA-derived viral RNA that is continually being produced in the nucleus of the transgenic strains escapes all splicing events during transport to the cytoplasm and is able to out-compete the deleted version during replication over time.

The demonstration that cDNA-derived viral dsRNA was generated by processing of a large nuclear transcript suggested the possibility that hypovirus infection could be initiated more simply by introducing a synthetic copy of hypovirus coding strand RNA directly into fungal spheroplasts. This prediction was confirmed by Chen et al. (15) with the development of a transfection protocol that involved electroporation of in vitro synthesized transcripts of the viral coding strand into virus-free *C. parasitica* spheroplasts, plating of surviving spheroplasts in a hyperosmotic regeneration medium, and efficient recovery of infected hyphae without reliance on any selectable marker. The robust nature of this simple protocol rested in the fact the hyphal structures that were regenerated from the transfected spheroplasts fuse at a very high frequency resulting in an extensive cytoplasmic network. Consequently, replicating hypovirus RNA generated from even a single transfection event is able to effectively migrate throughout the regenerating colony, and infected mycelia can be readily obtained by transferring a portion of the colony to a new plate.

This protocol has been used to effectively extend hypovirus infection to fungal species taxonomically related to *C. parasitica* (13, 15). This includes *Cryphonectria cubensis*, a serious pathogen of *Eucalyptus* in plantation settings (106). Results of recent transfection studies with highly virulent South African strains of *C. cubensis* suggest possible opportunities for the application of hypoviruses to control *Cryphonectria* canker in that country. Hypovirus infections resulted in a dramatic reduction in virulence on *Eucalyptus*, and virus was readily transmitted by hyphal anastomosis to field strains representing a broad range of vegetative compatibility groups.

In spite of the fact that fungi in all major taxa harbor viruses (12), some of which are associated with interesting phenotypic traits and potential utility, hypoviruses remain the only viral agents within the entire Kingdom Fungi for which a reverse genetics system has been developed. Consequently, they are also the only fungal virus for which Koch's postulate has been rigorously completed. Fortunately, progress in the development of tools for engineering hypoviruses has been complemented by advancements in the genetic manipulation of *C. parasitica*. Refinements in transformation plasmid vector design has resulted in enhanced expression of endogenous and foreign genes (24, 107, 114) and made targeted gene disruption in this haploid organism an effective and routine procedure (40, 41, 58, 64, 114). Gene mapping (59) and in vivo structure/function analyses (42) have benefited from the development of efficient genetic complementation vectors. As a result, the hypovirus/*C. parasitica* system now provides the rare capability of efficiently manipulating the genomes of both a eukaryotic virus and its host (81, 82). The following section is included to provide the reader with an appreciation of the unique opportunities that this experimental system provides for uncovering fundamental processes underlying fungal pathogenesis. That discussion also sets the stage for an account of how the development of the hypovirus reverse genetics system has facilitated the mapping of hypovirus-encoded symptom determinants and identification of essential and dispensable elements of hypovirus replication.

HYPOVIRUS-MEDIATED ALTERATION OF HOST PHENOTYPE

Phenotypic Consequences of Hypovirus Infection

Phenotypic alterations, in addition to hypovirulence, that distinguish infected colonies from isogenic ds-RNA-free virulent strains can include altered colony morphology (as evidenced by a reduced growth rate and a lobed appearance to the general circular shape of colonies), female infertility, reduced asexual sporulation (conidiation), and reduced pigmentation (Figure 4; see color insert) (reviewed in 3, 75, 80, 81). The pleiotropic nature of these phenotypic changes suggested the possibility that hypoviruses perturb one or several regulatory pathways (81). Moreover, preliminary studies by Hillman et al. (53) demonstrated a potential convergence in pathways resulting in these phenotypes, with the observation that high light intensity could ameliorate the severity of several hypovirus-associated characteristics. The cloning and analysis of various signal transduction components, as described in this section, has furthered understanding of the relationship of these pathways to fungal virulence.

The first molecular evidence of an alteration of fungal gene expression by hypovirus infection was demonstrated by the differential accumulation of a number of fungal proteins and polyA+ RNAs in isogenic virus-free and virus-infected strains (85, 86). Later, specific gene products were shown to be reduced in the presence of hypovirus at either the protein or mRNA level, including CPG-1 [a G-protein α-subunit (20)], LAC-1 [laccase (67, 89)], CBH-1 [cellobiohydrolase (108)], Vir1 and Vir2 [two sex pheromones (63)], and CRP [cryparin, a cell wall hydrophobin (63)]. The best studied locus of this group is laccase, whose level is reduced by virtue of lowered transcription in CHV-infected mycelium (67). Further studies in the same report implicate the IP3-calcium signaling pathway. This represented the first direct evidence of specific modulation of a defined signal transduction pathway by hypovirus infection.

Colonization of plant tissue by a pathogenic fungus must involve the continuous subversion of plant defense mechanisms and the degradation of tissue to provide the necessary nutrients for continued growth, in addition to the normal environmental responses. Therefore, the circumstances experienced by an invading mycelium of *C. parasitica* on a chestnut stem are likely very dynamic and require continuous monitoring of and response to the specific triggers that will result in adjusted growth or development. Considering the changes observed in a number of cellular components in response to hypovirus and the hypovirulent fungal phenotype that ensues, further explorations of signal transduction mechanisms were initiated to examine those genes that are important for chestnut blight virulence in particular and filamentous fungal pathogenesis in general.

Role of G-protein Signal Transduction in Fungal Virulence

Heterotrimeric G-proteins are a large and growing family of proteins that play an essential role in response to environmental stimuli in all eukaryotic cells (43). As such, they represented ideal candidates for further investigation. In 1995, Choi et al. cloned the first two such genes from *C. parasitica*, *cpg-1* and *cpg-2* (20). Both of these are members of the large and diverse Gα subunit family. CPG-1 is 98% identical to the Giα from *Neurospora crassa*, GNA-1 (102). CPG-2, meanwhile, is only 49% identical to CPG-1 and appears to be a member of a separate subfamily. Intriguingly, western blot analysis with specific CPG-1 polyclonal antibodies demonstrated that CPG-1 protein levels were significantly reduced in hypovirulent strains when compared to an isogenic virus-free culture. A reduction of CPG-1 protein level in the uninfected background was achieved by co-suppression (using a sense copy of *cpg-1*) and resulted in reduced mycelial growth and attenuation of fungal virulence (20). Deletion of either *cpg-1* or *cpg-2* resulted in a reduction in growth and altered colony morphology, but the changes for the $\Delta cpg-1$ strain were much more severe, with this mutant proving avirulent on chestnut while $\Delta cpg-2$ was only slightly less invasive than wild type (41). Additional characteristics of the $\Delta cpg-1$ strain included reduced pigmentation, loss of conidiation, female infertility, and a reduction in laccase production, thus correlating aspects of the hypoviral phenotype with the absence of the CPG-1 protein (41). As previously mentioned, the IP3-calcium signaling pathway is involved in the control of laccase accumulation (67), suggesting that CPG-1 controls a variety of fungal phenotypes by interacting with multiple signal transduction pathways. However, since a reduction in, or loss of, CPG-1 does not completely mimic hypovirus infection, there are likely to be multiple fungal targets of viral proteins that result in the overall CHV-associated fungal phenotype.

Since the $\Delta cpg-1$ phenotype was completely rescued by complementation with the wild-type sequence (42), the effects of various mutations that alter the signaling properties of CPG-1 could be investigated. Sequence consensus suggested the location of myristoylation and palmitoylation sites that determine the membrane association of Gα-proteins. Mutation of these sites caused a redistribution of CPG-1 such that a greater proportion was in the cytosol. These changes also caused differences in growth rate, sporulation, pigmentation, and virulence, but to different degrees depending on the mutation (42). This differential ability of the mutants to propagate signals that regulate different traits represented further evidence that CPG-1 is involved in the regulation of multiple pathways.

Further mutagenesis has recently been undertaken to examine the effect of the activation state of CPG-1 on phenotype. This was achieved by mutating Q-204 to L to create a constitutively activated Gα that is unable to reassociate with the $\beta\gamma$ subunit because of an inability to hydrolyze GTP and therefore continuously

TABLE 1   A comparison of phenotypes observed in different filamentous fungi with deleted or activated Giα alleles. Percentages are relative to those values obtained with a wild-type Giα[a]

Mutation	Trait	Phenotype		
		C. parasitica (CPG-1)	A. nidulans (FadA)	N. crassa (GNA-1)
ΔGiα	Radial growth	45%	70%	53%
	Biomass on solid medium cellophane overlay	12%	63%	18%
	Conidiation	None	Slightly decreased	100%
	Pigmentation	Decreased	Not determined	Increased
Constitutively activated Giα	Growth	Decreased	Increased	Increased
	Aerial mycelium	None	Present	Present
	Conidiation	None	Strongly reduced	Slightly reduced
	Osmotic stress sensitivity	Increased	Not determined	Unchanged
	Heat stress sensitivity	Increased	Not determined	Increased
	Oxidative stress sensitivity	Reduced	Not determined	Increased

[a]Prepared from (7), (41), (57), (58), (68), (91), (111), (112) and unpublished observations of G Segers & DL Nuss.

stimulates downstream pathways (28). This mutant was avirulent and exhibited a severe phenotype of decreased growth and absence of aerial hyphae and asexual spores (G. Segers & D.L. Nuss, unpublished observations). However, this differed considerably from the observed phenotypes of similar changes in *A. nidulans* and *N. crassa* (Table 1). In these closely related fungi, analogous mutations or deletions of the Giα subunit often caused completely opposite effects on developmental processes such as growth, pigmentation, and conidiation. These differences illustrate the difficulty of relying on homology to infer specific function, yet are perhaps not surprising. *A. nidulans* and *N. crassa* differ in the developmental regulation of conidiation, for instance (1, 98), and, of course, *C. parasitica* is a plant pathogen whereas the other two are not. Therefore while the basic signal transduction components used may be homologous, the uses to which they are put may vary depending on the lifestyle requirements of the organism in question.

As described above, the sequence homology between CPG-1 and other Gα subunits clearly places this protein as a member of the Giα family (20). Members of this family are generally believed to function as negative regulators of adenylate cyclase. Chen et al. (17) provided confirmation of a similar role for CPG-1, having observed elevated cAMP levels in the strain co-suppressed for *cpg-1* and for strains infected with CHV1-713 (that show reduced CPG-1 levels). Moreover, differential display analysis showed that a large subset of those transcripts whose level was increased or decreased in the presence of hypovirus behaved similarly when virus-free strains were treated with phosphodiesterase inhibitors to elevate cAMP concentrations, or when CPG-1 levels were decreased by co-suppression (17). Thus, the current working model suggests that a major contributor to fungal phenotype in the presence of hypovirus is the perturbation in cAMP concentrations caused by the reduction in the amount of CPG-1 protein leading to an alteration in the pattern of gene transcription.

Although the second Gα subunit, CPG-2, appeared to be less important for fungal virulence and virus-induced phenotypes, as described above, recent evidence indicates that it may fall within an emerging fungal Gα subfamily. Similar sequences from *Podospora anserina* [MOD-D (74)], *Magnaporthe grisea* [MAGA (71)], *Ustilago maydis* [Gpa3 (88)], *U. hordei* [Fil1 (70)], and *Cryptococcus neoformans* [Gpa1 (103)] all grouped together when combined with other Gα subunits in a phylogenetic analysis (11, 74). Examples from this group are demonstrating functions that correlate with the behavior of Gsα subunits that act as positive regulators of adenylate cyclase. For instance, concentrations of cAMP were shown to be reduced in the Δ*cpg-2* strain (41) and in a Δ*gna-3* mutant of *N. crassa* (62). Similarly, the phenotype of a *P. anserina mod-D1* and *mod-D3* mutants can be partially rescued by exogenous cAMP. Also, the adenylate cyclase gene from this organism, *PaAC*, was identified as a suppressor of the *mod-D1* mutant phenotype (74). Therefore, it is possible that CPG-2 may represent the positive regulatory pathway of adenylate cyclase activity and that the interplay between CPG-1– and CPG-2–mediated signals helps to maintain the levels of cAMP and, hence, gene transcription required by the fungus for proper growth and development.

A third G-protein subunit, CPGB-1, was reported by Kasahara & Nuss (58). This protein was found to be approximately 66% identical to human, *Drosophila*, and *Dictyostelium* β-subunits and represented the first such family member identified in a filamentous fungus. Other examples have followed, with SFAD from *A. nidulans* (91) and GPB-1 from *C. neoformans* (109). Disruption of the *cpgb-1* had a profound effect on colony morphology highlighted by an almost complete absence of pigmentation and increased biomass production, contrasting starkly with the previously mentioned observations of Δ*cpg-1*. Older colonies did develop some pigmentation, but aerial hyphae collapsed in the center and appeared to autolyse. Virulence on chestnut tissue was greatly reduced, with cankers only attaining about 10% of the diameter of the wild-type strain (58). An additional mutation was also identified that conferred an almost indistinguishable phenotype from the Δ*cpgb-1* strain. Subsequent analysis revealed this to be a lesion in a second gene, *bdm-1* (for beta-disruption mimic; 59). The predicted protein shows some identity (23%) to mammalian phosducins, a family of proteins that are believed to negatively regulate signal transduction through interactions with the βγ subunits (69, 113). However, since Δ*bdm-1* was shown to phenocopy the Gβ disruptant and levels of CPGB-1 are not measurably altered in the absence of BDM-1, BDM-1 most likely functions as a positive regulator of Gβ function. Similarities between the phenotypes of the two fungal disruptants extended to the reduction in accumulation of CPG-1 (59), although the phenotypes bear little resemblance to that of the Δ*cpg-1* or co-suppressed strains. Intriguingly, preliminary two-hybrid analysis has demonstrated an interaction between CPGB-1 and BDM-1, but the strength of the association appears to be weak (A.L. Dawe, S. Kasahara & D.L. Nuss, unpublished observations). There is also the possibility that these two proteins could perform some functions independently of one another. Overexpression of BDM-1 in the Δ*cpgb-1* strain causes some amelioration of the Δβ phenotype (an increase in pigmentation and reduced autolysis), although the reverse scenario causes no apparent changes (A.L. Dawe & D.L. Nuss, unpublished observations). Further analysis of this interaction is under way and will doubtless yield more clues as to the interrelationships between the various signaling components.

Understanding the involvement of G-protein–linked signal transduction during fungal virulence has been greatly facilitated by the *C. parasitica* system. Employing the modern tools of molecular biology to both hypovirus and fungus will likely open new avenues of investigation as recent genes identified from this organism such as a homolog of *A. nidulans* FLBA (an RGS protein; G. Segers & D.L. Nuss, unpublished observations), adenylate cyclase (S. Kasahara, T. Parsley & D.L. Nuss, unpublished observation), and the catalytic and regulatory subunits of protein kinase A (L.M. Geletka & D.L. Nuss, unpublished observations), are fully explored.

FUNCTIONAL DISSECTION OF THE HYPOVIRUS GENOME

Mapping Hypovirus-Encoded Symptom Determinants

The discovery of hypovirulence-associated dsRNAs by Day et al. (32) immediately raised a number of very basic questions. Principal among these was whether the phenotypic changes exhibited by a hypovirulent *C. parasitica* strain are the result of a general response of the host to the physical presence of the replicating dsRNA or whether the changes are dependent upon functions encoded by specific dsRNA sequences. An answer to this and several related questions was provided by Choi & Nuss (21) even prior to the development of the CHV1-EP713 infectious cDNA clone. These investigators found that transformation of virus-free strain EP155 with a cDNA copy of ORF A generated from CHV1-EP713 dsRNA under the transcriptional control of the *C. parasitica gpd-1* promoter resulted in a subset of the phenotypic traits that were exhibited by CHV1-EP713-infected colonies, e.g., reduced orange pigment production, reduced conidiation, and reduced laccase production. Thus, it was possible to show that a hypovirus-associated dsRNA coding domain could cause several hypovirulence-associated traits in the absence of dsRNA replication. Moreover, these transformants were not reduced in growth rate on synthetic medium or in virulence, providing the first indication that it might be possible to uncouple hypovirulence from hypovirulence-associated traits such as reduced conidiation.

Using an extension of the transformation analysis, Craven et al. (29) were able to map the region responsible for the ORF A-mediated phenotypic changes to the p29 coding domain. The availability of the infectious CHV1-EP713 cDNA clone allowed these investigators to perform the converse experiment by testing the consequence of p29 deletion on the ability of the virus to alter host phenotype. Fortunately, p29 was found to be dispensable for viral replication; deletion mutant Δp29, lacking 88% of the p29 coding region, was replication competent. Relative to *C. parasitica* isolates infected with wild-type CHV1-EP713, fungal colonies infected with this mutant exhibited a near-complete restoration of orange pigment production, a moderate increase in conidiation, and a slight increase in laccase production. It should be emphasized that the level of virus-mediated hypovirulence was not altered by p29 deletion. It was concluded from this study that, although not essential for virus replication or hypovirulence, p29 does contribute to virus-mediated reductions in fungal pigmentation, asexual sporulation, and laccase accumulation.

Suzuki et al. (99) recently mapped the polypeptide domain required for p29-mediated symptom expression to a region extending from Phe-25 to Gln-73 by progressively repairing the Δp29 viral mutant in a gain-of-function assay. These investigators noted with interest that this region of p29 corresponded to the cysteine-rich domain that Koonin et al. (66) had identified as having a moderate level of sequence similarity with the potyvirus HC-Pro protease that included four conserved cysteine residues, Cys-38, Cys-48, Cys-70, and Cys-72. Substitution of a glycine for Cys-72 resulted in loss of p29-mediated symptom expression whereas substitution for Cys-70 caused a very severe phenotype that included significantly reduced mycelial growth and profoundly altered colony morphology. Substitutions for the other two cysteine residues had no effect on virus-mediated symptom expression.

HC-Pro has received considerable attention recently because of reports that it suppresses a posttranscriptional gene silencing mechanism used by plants to defend against virus infection (60). It is becoming more likely that many of the functions tentatively assigned to HC-Pro, e.g., genome activation (61,65), long-distance movement (30), transactivation of heterologous virus replication (96), may be manifestations of this newly identified property. In light of the proposed evolutionary relationship between HC-Pro and p29 (66), it will be instructive to look for parallels in the effect of the two proteins on specific cellular signaling pathways in the respective hosts.

Insights into Essential and Dispensable Elements of Hypovirus Replication

The observation by Craven et al. (29) that 88% of p29 was dispensable for virus replication suggested the possibility that the deleted portion could be replaced with foreign gene sequences, i.e., that hypoviruses could be developed as gene expression vectors. This was an appealing idea given the proposed and actual use of modified animal viruses for delivery of foreign genes and of plant viruses for large-scale in planta production of protein products. Moreover, the fact that

CHV1-EP713 RNA is not encapsidated suggested that constraints on the size of heterologous inserts would be minimal.

Suzuki et al. (100) explored four basic strategies to construct 20 hypovirus expression vectors. Genes of non-hypovirus origin used in the study included two reporter genes, the green fluorescent protein (GFP) from *Aequorea victoria* and the *E. coli* hygromycin B phosphotransferase gene, and two non-reporter genes, the foot and mouth virus 2A protease gene (2A) and the Mat-2 mating-type pheromone gene (PH) from *C. parasitica*. Vectors in Group I were constructed by inserting the reporter genes in frame into previously characterized p29 deletion mutants of the CHV1-EP713 infectious cDNA clone. Group II vectors all contained modifications or gene insertions at the precise N terminus of ORF A. Construction of Group III vectors involved insertion of foreign sequences in the 3'-terminal portion of the p40 coding domain near or adjacent to the pentanucleotide UAAUG that separates ORF A from ORF B. The final pair of vectors, Group IV, contained radical modifications in which nearly all of ORF A, including the entire p40 coding domain, was replaced with foreign sequences, and the pentanucleotide separating ORFs A and B was mutated to remove the ORF A stop codon, thereby creating an in-frame single ORF genome organization.

Transient expression was observed for a subset of vectors that contained the GFP gene, while none of the vector transcripts containing the hygromycin gene were infectious. Although long-term expression was not observed for any of the vectors, analysis of the RNAs recovered from the transfected fungal colonies provided unexpected new insights into essential and disposable elements of hypovirus replication (Figure 5; see color insert). Comparison of the 11 replication-competent vectors revealed one obvious distinguishing feature: Each contained at least 66 nucleotides, specifying codons 1 through 24, of the 5'-portion of the p29 coding domain. Further deletion of this region to codon 12 or complete removal of the N-terminal coding domain in the context of a replication-competent foreign gene-free Δp29 mutant infectious cDNA clone resulted in loss of replication competency. These authors concluded that the 5'-terminal portion of the p29 coding domain, extending perhaps as far as codon 24, is required for CHV1-EP713 replication. They also raised the interesting possibility that, given the IRES-like properties associated with the 5'-UTL, the 5'-UTL and the N-terminal region of p29 might function coordinately to facilitate an IRES-guided translation mechanism.

Two surprising observations resulted from testing the Group IV vectors: Both the p40 coding domain and the UAAUG pentanucleotide were found to be dispensable for CHV1-EP713 replication. The deduced p40 amino acid sequence has a predicted pKa of 11.96, a feature consistent with a possible role in viral RNA binding or replication, and is conserved among hypovirus species CHV1 and CHV2. Thus, the observation that p40 was not required for viral replication was unexpected and prompted Suzuki et al. (manuscript in preparation) to further analyze the consequences of p40 deletion on virus-mediated alterations of fungal phenotype. Mutant CHV1-EP713 viral transcripts lacking the p40 coding domain, Δp40, retained the ability to confer hypovirulence, but caused much less reduction in asexual sporulation and pigmentation than wild-type CHV1-EP713. A gain of function analysis, similar to that performed for p29 (99), mapped the p40 suppressive activity to the N-terminal domain extending from Thr-288 to Asn-313 (N. Suzuki & D.L. Nuss, unpublished observation). However, accumulation of viral RNA in Δp40-infected colonies was found to be reduced by more than 50%, and restoration of suppressive activity correlated with increased accumulation of viral RNA. These preliminary results suggested that the reductions in symptom expression accompanying p40 deletion were due to reduced viral RNA accumulation and that p40 provides an accessory function in viral RNA amplification (Figure 5).

The more surprising observation was that the two Group IV vectors with a single ORF configuration were infectious and that the single ORF progeny generated from vector Δp69bEGFP by deletion of the entire GFP sequence, thereby fusing Pro-24 of p29 directly in frame with Met-1 of p48, maintained replication competence. Recent studies showing that other single ORF CHV1-EP713-derived mutant transcripts can replicate have confirmed that the UAAUG pentanucleotide is dispensable for viral replication (N. Suzuki & D.L. Nuss, unpublished observation). Efforts to develop hypoviruses as gene vectors also provided unexpected observations concerning the region immediately upstream of the pentanucleotide. Hillman et al. (52) had noted the presence of an A/U-rich string of nucleotides immediately preceding the pentanucleotide for members of CHV1 and CHV2 species: 5′-AAAAUAAAAUUUUAAUG-3′ for CHV1-EP713 and 5′-AAAAUUUUAAUUAAUUAAUG-3′ for CHV2-NB58. Replacement of this A/U-rich sequence with the pheromone sequence was tolerated while introduction of the 2A sequence, which has the potential for forming a 10-residue hairpin structure, adjacent to the pentanucleotide, abolished replication competence. Several vector progeny were also found to have undergone the deletion of a uracil residue immediately adjacent to the UAAUG pentanucleotide resulting in a frameshift mutation and a two-codon extension of ORF A.

The fact that only transient expression of foreign genes was achieved by Suzuki et al. (100) does not exclude the possibility that hypoviruses can be developed for stable expression of foreign genes. Insertions have been restricted to ORF A thus far. Insertion into ORF B or the use of defective RNA platforms represent unexplored alternative strategies. It is anticipated that such efforts will lead to additional unexpected revelations about hypovirus molecular biology.

Exploiting Hypovirus Diversity

Hypovirulent *C. parasitica* field isolates recovered from populations in Asia (84), North America (26, 34–36, 83), and Europe (2) exhibited considerable variability in virulence levels and the magnitude and constellation of hypovirulence-associated traits. Correspondingly, hypoviruses associated with many of these same populations exhibited a significant degree of nucleotide sequence diversity (2, 26). The potential of this natural genetic diversity for practical and fundamental application has only recently been tapped through the construction of an infectious cDNA clone of a second hypovirus CHV1-Euro7 (19).

The impetus for cloning CHV1-Euro7 derived primarily from the properties of the *C. parasitica* field isolate, Euro7, from which this hypovirus was isolated. As indicated in Figure 6 (see color insert), virus-free *C. parasitica* strains aggressively expand after colonization of chestnut tissue forming necrotic cankers covered with orange spore-containing stromal pustules. Such cankers expand until the stem is girdled. In stark contrast, CHV1-EP713-infected *C. parasitica* strain EP713 is severely reduced in its ability to expand on chestnut tissue and forms small superficial cankers that produce few to no pustules. Canker formation by strain Euro7, isolated by William MacDonald (West Virginia University) in 1978 near Florence, Italy, was quite different. Colonization is followed by aggressive canker expansion that slows or ceases as the canker reaches a size three- to fourfold larger than that attained by EP713 cankers (Figure 6). These cankers are distinguished by ridged margins that form concomitantly with reduced canker expansion and a surface covered with a significant level of spore-forming pustules. Consistent with the differences in the ability to colonize and produce spores on chestnut tissue, Euro7 was reported (31) to be more readily disseminated after introduction into several North American forest ecosystems than has generally been observed for highly debilitated strains such as EP713, properties that are discussed in more detail in the next section.

Direct comparisons of Euro7 and EP713 under defined laboratory growth conditions also revealed a number of interesting phenotypic differences (19). As shown in Figure 4, colonies formed by EP713 grew slightly slower than the virus-free counterpart and were characterized by hyphae that penetrate into the medium, by irregular margins and by the general absence of asexual spores. In contrast, Euro7 colonies expanded faster than the corresponding virus-free isolate, produced abundant aerial hyphae and regular margins, and produced asexual spores at a level intermediate between the corresponding virus-free isolate and EP713.

Although Euro7 was phenotypically quite different from EP713, the viruses harbored by the two fungal isolates, CHV1-Euro7 and CHV1-EP713, were found to be highly conserved at the nucleotide and amino acid levels, 87–93% and 90–98%, respectively (Figure 3). The construction of an infectious CHV1-Euro7 cDNA allowed Chen & Nuss (19) to ask whether the phenotypic differences observed for isolates Euro7 and EP713 were a function of the relative contributions of the two viral genomes or attributable to additional contributions by the genomes of the two fungal hosts. This was accomplished by transfecting transcripts of CHV1-Euro7 and CHV1-EP713 independently into virus-free isolates derived from Euro7 and EP713. The results clearly showed that the contributions of the viral genomes are predominant, with minor, but measurable contributions from the host genomes. Based on the differences in phenotypic changes caused by the two viruses, Chen & Nuss (19) suggested that, by analogy with plant viruses, CHV1-Euro7 and CHV1-EP713 could be viewed as mild and severe hypovirus strains, respectively.

Chen & Nuss (19) also demonstrated the feasibility of constructing viable chimeras of the CHV1-Euro7 and CHV1-EP713 cDNA clones. By interchanging ORFs A and B, it was possible to show that the determinants responsible for the differences in canker and colony morphology conferred by the two viruses reside predominantly within ORF B. Chen et al. (18) showed that it was possible to extend this experimental approach for mapping of viral symptom determinants by swapping portions of ORF B to generate a series of stable chimeras that caused a spectrum of defined canker and colony morphologies. For example, differences in the mild and severe colony morphologies mapped predominantly to a region extending from a position just downstream of the p48 coding domain (map position 3575) to map position 9879, with clear indications of multiple discrete determinants (Figure 5). More specifically, the region extending from position 3575 to 5310 conferred a EP713-like colony morphology when inserted into a CHV1-Euro7 genetic background. The CHV1-EP713 p48 coding region was found to be a dominant determinant contributing to suppression of pustule formation on the canker face. The construction of these chimeras demonstrated the feasibility of engineering hypoviruses to fine-tune the interaction between a pathogenic fungus and its host. For example, fungal isolates infected with a chimera that contained nucleotides 2363 through 5310 from CHV1-Euro7 in a CHV1-EP713 background formed a small canker similar to that formed by EP713, but retained the capacity to produce asexual spores at levels approaching that of Euro7. Using fungal hosts transformed with pathway-specific promoter-reporter constructs, it has also been possible to use the severe/mild hypovirus chimeras to begin correlating virus-induced changes in specific cellular signaling pathways with virus-induced alterations in fungus-host interactions and specific phenotypic traits (T. Parsley, L.M. Geletka, B. Chen & D.L. Nuss, unpublished observations).

ENGINEERING HYPOVIRUSES FOR ENHANCED BIOLOGICAL CONTROL

Contrasting Hypovirulence in North America and Europe

A comprehensive review of the chestnut blight epidemic in Europe through 1994 by Heiniger & Rigling (49) provides the following general conclusions:

- Hypovirulence has greatly reduced the severity of chestnut blight in Europe.
- Superficial and healing cankers are widespread in chestnut stands throughout Europe where the blight is well established.
- European *C. parasitica* populations are characterized by a generally high incidence of hypovirus infection and a uniformly low level of vegetative compatibility (vc) diversity.

The results of several recent surveys support and extend this general view of hypovirulence and its influence on the European chestnut blight epidemic.

Studying the population structure and disease development in two 6-year-old chestnut coppices in southern Switzerland over a 4-year period, Bissegger et al. (10) found that 59% and 40% of isolates recovered from the respective sites were infected with hypoviruses. A total of 21 vc types were identified in the two plots. However, 63% of the total isolates fell within three vc types. Moreover, hypovirus transmission was observed between isolates within five of the six most common vc types in laboratory pairings, consistent with previous conclusions that vegetative incompatibility is not a significant barrier to hypovirus transmission in southern Switzerland. The authors concluded that the reduction in disease severity observed during the 4-year study was attributable to the ability of hypoviruses to infect large portions of the fungal population, presumably as a result of frequent virus transmission within and between vc groups. A survey of six regions of southern France by Robin et al. (90) found between 27% and 98% of cankers to be healing, with an average of 49%. A much wider range of incidence of hypovirus infections, 2% to 67%, was observed, with an overall average of 25%. An average of six vc types were found for each subpopulation. In 15 of the 25 subpopulations tested, the dominant vc type represented in excess of 50% of the isolates. Allemann et al. (2) recently examined the genetic variability of hypoviruses isolated from nine European countries between 1975 and 1997. This study found that a single CHV1-subtype related to CHV1-EP747 and the mild strain CHV1-Euro7 (the Italian subtype) was widespread and the most dominant subtype in the European hypovirus population. Hypoviruses related to the French-derived severe species CHV1-EP713 were not well represented.

The success of the intensive biological control program instituted by the French Ministry of Agriculture has resulted in the commercialization of hypovirulence for control of chestnut blight in chestnut orchards. Tubes containing pastes consisting of multiple hypovirulent C. parasitica strains representing several different vc groups are available from the Italian company f.lli Delfino S.p.A., Division Applicazioni Biologiche Milano. Although labor intensive, the protocol of introducing plugs of hypovirulent strain in closely spaced holes around the margin of newly formed cankers on an annual basis provides quite effective biological control (38, 49, 90). Interestingly, Robin et al. (90) found no increased incidence of hypovirus infection in regions treated for biological control, and RFLP analysis of a limited number of the hypovirus isolates indicated the dominance of mild Italian strains rather than the more severe French hypovirus subtypes used in the biological control treatments.

Reports of natural dissemination of hypoviruses in Europe and the successful use of hypovirulent strains to control chestnut blight in orchards sustain efforts to develop durable hypovirulence-mediated biological control in North America. Evidence of natural hypovirulence in North America is limited to the state of Michigan, well outside of the natural range of the American chestnut, and was discovered some 80 years after the beginning of the North American chestnut blight epidemic (38). In addition to hypovirus-mediated hypovirulence, a new form of virus-independent hypovirulence involving mitochondrial dysfunction was also found to be operating in Michigan chestnut stands (78). Interestingly, hypoviruses associated with the Michigan hypovirulent strains are also distinctive, typified by the type member of the recently recognized third hypovirus species CHV3-GH2 (97). Hypoviruses related to the CHV3 type species have only been encountered in and around Michigan (83, 84, 97) and differ from CHV1 and CHV2 species in that they do not significantly reduce host pigmentation or sporulation.

Repeated efforts to establish sustained hypovirulence-mediated biological control in North American forest ecosystems by the artificial introduction of hypovirulent strains have been unsuccessful to date (3, 38, 46, 75). The contrasting efficacy of hypovirulence-mediated biological control in Europe and North America has been attributed to differences in the blight susceptibility of the European and American chestnut species, differences in the population structures of the pathogens in the two geographic areas, and differences in the properties of the indigenous hypoviruses. Heiniger & Rigling (49) have stated that *C. sativa*, the European chestnut, is less susceptible than the American chestnut *C. dentata* and that a slower rate of canker expansion on the former trees may provide a wider window of opportunity for canker conversion by compatible hypovirulent strains. Unfortunately, direct comparative analyses of the susceptibility of *C. dentata* and *C. sativa* have not been rigorously performed to confirm this possibility. The most widely held theory suggests that the higher level of vc diversity observed for North American *C. parasitica* populations relative to that found in European populations (6, 10, 72, 73) severely limits hypovirus dissemination. Fungal strains with identical alleles at all or most of the six genetic loci that control vc readily anastomose, allowing transmission of cytoplasmically replicating hypoviruses. As the number of dissimilar alleles increases, the ability of the fungal strains to fuse also decreases. Thus, a high level of vc diversity in a *C. parasitica* population would be predicted to hamper hypovirus transmission. Consistent with this prediction, Liu & Milgroom (73) reported a negative correlation between hypovirus transmission and the number of vc genes that differed among isolates of a natural *C. parasitica* population. A third consideration is differences in the properties of the hypoviruses associated with effective biological control in Europe and those naturally occurring or introduced in North America. MacDonald & Fulbright (75) have noted that the hypovirulent strains used in most attempts at biological control of chestnut blight in North America have contained highly debilitating severe hypovirus strains. These authors have also articulated the generally held view that successful hypovirulence-mediated biological control is likely to require a continual source of hypovirulent inoculum, i.e., there must be a balance between the effect of hypovirus infection on reduced virulence and reduced ecological fitness. This view is supported by recent reports of high levels of persistence and dissemination of hypovirulent strains containing the mild hypovirus strain CHV1-Euro7 after release in a mature North American chestnut stand outside of the natural chestnut range (31) and the dominance of the mild Italian hypovirus subtype in healing chestnut stands in Europe (2). As discussed in the next section, transgenic hypovirulent *C. parasitica* strains constructed with infectious hypovirus cDNAs provide potential for overcoming two of these three considerations.

Field Testing of Transgenic Hypovirulent *C. parasitica* Strains

Hypovirus RNA is not transmitted during mating to ascospores. This is unfortunate because the ascospore progeny represent a spectrum of vc groups as a result of allelic rearrangement at the vc genetic loci (14). As mentioned in a preceding section, the chromosomally integrated viral cDNA present in transgenic hypovirulent strains is transmitted into ascospore progeny where the viral RNA is resurrected and replicates in the cytoplasm. Moreover, every asexual spore produced by a transgenic strain contains a nuclear copy of the viral cDNA and cDNA-derived viral RNA. This contrasts with natural hypovirulent strains that produce virus-free conidia at rates ranging from 10% to 90% (13, 36, 92), thereby providing a continual source of virulent inoculum. These combined transmission properties for transgenic hypovirulent strains were predicted to circumvent barriers to virus transmission and enhance dissemination of the hypovirulence phenotype (14). This contrasts, however, with predictions of a recently developed computer model that suggested that while more effective than natural hypovirulent strains in some circumstances, transgenic hypovirulent strains were only marginally better at establishing a significant level of hypovirus infection in populations with a high vc diversity (72).

Studies designed to test the actual field performance of transgenic hypovirulent strains were initiated in July 1994 under conditions specified in United States Department of Agriculture (USDA) Biotechnology Permit # 94-010-01. This involved a single season, limited environmental release of a transgenic hypovirulent *C. parasitica* strain containing the cDNA of severe hypovirus CHV1-EP713 into a Connecticut forest site. The results of that study confirmed hypovirus transmission to ascospore progeny under actual field conditions and provided the first indication of cytoplasmic spread of cDNA-derived viral RNA to nontransgenic endogenous *C. parasitica* strains (4). However, the transgenic CHV1-EP713 strains failed to persist in the release site for more than two years after introduction. These results provided a firm base for extended field trials beginning in the summer 1997 under USDA Biotechnology Permit # 96-275-01.

The protocol approved for the second field release involved a population replacement strategy. Three indigenous virulent *C. parasitica* isolates recovered from the test site, representing three different vc groups and both mating types, were transformed with cDNA of the severe hypovirus CHV1-EP713. The transgenic strains were applied as a sprayed spore mixture to the test plot for three summers beginning in 1998. While this study is still in progress, there is now good evidence for transmission of cDNA-derived hypovirus RNA independent of the input transgenic strains and for transmission of viral cDNA through mating to ascospore progeny within the treatment plot. However, no evidence has yet been obtained

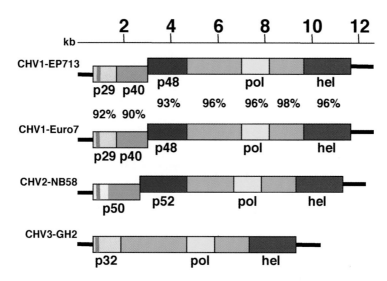

Figure 3 Organizations for the four fully sequenced hypovirus genomes representing the three species within the genus hypovirus. The percent amino acid identity for different coding regions of the two sequenced members of species CHV1, CHV1-EP713 and CHV1-Euro7, is indicated between representations of the two viral genomes. Coding regions homologous to p29, p40 p48, pol and hel are color coded. The magenta regions represents a short cysteine-rich domain found near the N terminus of CHV1-p29, CHV2-p50 and CHV3-p32. Note that the genome CHV3-GH2 contains a single ORF rather than the two ORF configuration found for members of the CHV1 and CHV2 species. The reader is referred to the following GenBank accession numbers and publications for detailed sequence information for the four hypovirus genomes: CHV1-EP713: M57938 (93), CHV1-Euro7: AF082191 (19), CHV2-NB58: L29010 (52), CHV3-GH2: AF188515 (97). Elements of this figure were modified from Chen & Nuss (19) and Smart et al (97).

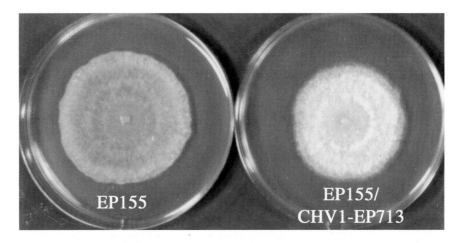

Figure 4 Effect of hypovirus infection on *C. parasitica* pigmentation and colony morphology. Virus-free virulent strain EP155 is shown on the left. Isogenic CHV1-EP713-infected strain EP713 is shown on the right. The colonies were grown in parallel for 5 days on potato dextrose agar (PDA) on the laboratory bench with a light intensity of approximately 2000 lx and a temperature of 22 to 25°.

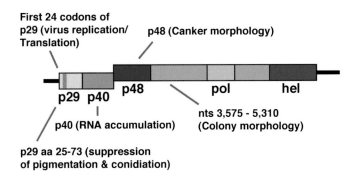

Figure 5 Emerging map of CHV1-EP713 symptom determinant domains and essential/dispensible replication elements as described in the text.

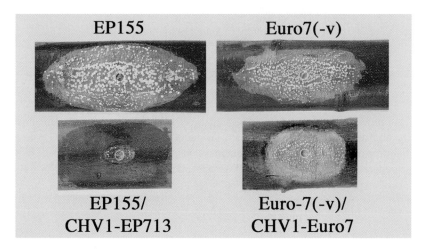

Figure 6 Comparison of canker morphologies formed by *C. parasitica* strains infected with severe hypovirus CHV1-EP713 or mild hypovirus CHV1-Euro7. (*Top row*) Typical cankers caused by virus-free strains EP155 and Euro7(-v). (*Bottom row*) Cankers caused by strain EP155 transfected with CHV1-EP713 and strain Euro7(-v) transfected with CHV1-Euro7. Cankers were photographed 30 days postinoculation after wetting with ethanol to enhance color contrast of cankers and surrounding area. Stromal pustules (stromata that contain asexual spore-forming bodies termed pycnidia) are prominent features on the surface of cankers caused by the virus-free strains and the CHV1-Euro7-transfected strain. Note the ridged margins of the canker formed by the CHV1-Euro7 transfectant, suggestive of callus formation. The reader is referred to Chen & Nuss (19) for additional details.

for spread of transgenic strains or derived viral RNA to the control plot or outside of the treatment plot (C.J. Balbalian, R.E. Bierman, L.M. Geletka, C.M. Root & D.L. Nuss, unpublished observations).

All of the CHV1-EP713 transgenic isolates used in both field releases form very small cankers on chestnut tissue and are severely reduced in the ability to conidiate, i.e., their ecological fitness is significantly compromised. As noted above, CHV1-Euro7-infected *C. parasitica* strains differ from strains infected with CHV1-EP713 in precisely those properties that are expected to have a direct impact on ecological fitness: colonization and spore production on bark tissue. Thus, the full-length infectious CHV1-Euro7 cDNA clone provides the means for constructing second-generation transgenic hypovirulent strains that combine properties of enhanced colonization and spore production with a novel mode of virus transmission to ascospore progeny. The most effective use of hypovirus-mediated biological control in North America may be as part of an integrated program to facilitate reforestation with less-susceptible chestnut trees derived from the ongoing chestnut blight resistance breeding program (48).

ACKNOWLEDGMENTS

Portions of this review were adapted in part from Nuss (81, 82). We thank L.M. Geletka for critical reading of the manuscript and G. Segers for the comparative analysis of Giα alleles in Table 1. Support from the NIH (GM 55981) and the USDA (NRICGP95-37312-1638) is gratefully acknowledged.

Visit the Annual Reviews home page at www.AnnualReviews.org

LITERATURE CITED

1. Adams TH, Wieser JK, Yu JH. 1998. Asexual sporulation in *Aspergillus nidulans*. *Microbiol. Mol. Biol. Rev.* 62:35–54

2. Allemann C, Hoegger P, Heiniger U, Rigling D. 1999. Genetic variation of *Cryphonectria* hypoviruses (CHV1) in Europe, assessed using restriction fragment length polymorphism (RFLP) markers. *Mol. Ecol.* 8:843–54

3. Anagnostakis SL. 1982. Biological control of chestnut blight. *Science* 215:466–71

4. Anagnostakis SL, Chen B, Geletka LM, Nuss DL. 1998. Hypovirus transmission to ascospore progeny by field-released transgenic hypovirulent strains of *Cry-*phonectria parasitica. *Phytopathology* 88:598–604

5. Anagnostakis SL, Day PR. 1979. Hypovirulence conversion in *Endothia parasitica*. *Phytopathology* 69:1226–29

6. Anagnostakis SL, Kranz J. 1987. Population dynamics of *Cryphonectria parasitica* in a mixed-hardwood forest in Connecticut. *Phytopathology* 77:751–54

7. Baasiri RA, Lu X, Rowley PS, Turner GE, Borkovich KA. 1997. Overlapping functions for two G protein α subunits in *Neurospora crassa*. *Genetics* 147:137–45

8. Biraghi A. 1946. Il cancro del castagno causato da *Endothia parasitica*. *Ital. Agric.* 7:1–9

9. Biraghi A. 1953. Possible active resistance to *Endothia parasitica* in *Castanea sativa*. *Rep. Congr. Int. Union For. Res. Org.*, 11th, pp. 149–57, Rome

10. Bissegger M, Rigling D, Heiniger U. 1997. Population structure and disease development of *Cryphonectria parasitica* in European chestnut forests in the presence of natural hypovirulence. *Phytopathology* 87:51–59

11. Bolker M. 1998. Sex and crime: heterotrimeric G proteins in fungal mating and pathogenesis. *Fungal Genet. Biol.* 25:143–50

12. Buck KW. 1986. Fungal virology—an overview. In *Fungal Virology*, ed. KW Buck, pp. 2–84. Boca Raton: CRC Press

13. Chen B, Chen C-H, Bowman BH, Nuss DL. 1996. Phenotypic changes associated with wild-type and mutant hypovirus RNA transfection of plant pathogenic fungi phylogenetically related to *Cryphonectria parasitica*. *Phytopathology* 86:301–10

14. Chen B, Choi GH, Nuss DL. 1993. Mitotic stability and nuclear inheritance of integrated viral cDNA in engineered hypovirulent strains of the chestnut blight fungus. *EMBO J.* 12:2991–98

15. Chen B, Choi GH, Nuss DL. 1994. Attenuation of fungal virulence by synthetic infectious hypovirus transcripts. *Science* 264:1762–64

16. Chen B, Craven MG, Choi GH, Nuss DL. 1994. cDNA-derived hypovirus RNA in transformed chestnut blight fungus is spliced and trimmed of vector nucleotides. *Virology* 202:441–48

17. Chen B, Gao S, Choi GH, Nuss DL. 1996. Extensive alteration of fungal gene transcript accumulation and alteration of G-protein regulated cAMP levels by virulence-attenuating hypovirus. *Proc. Natl. Acad. Sci. USA* 93:7996–8000

18. Chen B, Geletka LM, Nuss DL. 2000. Using chimeric hypoviruses to fine-tune the interaction between a pathogenic fungus and its plant host. *J. Virol.* 74:7562–67

19. Chen B, Nuss DL. 1999. Infectious cDNA clone of hypovirus CHV1-Euro7: a comparative virology approach to investigate virus-mediated hypovirulence of the chestnut blight fungus *Cryphonectria parasitica*. *J. Virol.* 73:985–92

20. Choi GH, Chen B, Nuss DL. 1995. Virus-mediated or transgenic suppression of a G-protein α subunit and attenuation of fungal virulence. *Proc. Natl. Acad. Sci. USA* 92:305–9

21. Choi GH, Nuss DL. 1992. A viral gene confers hypovirulence-associated traits to the chestnut blight fungus. *EMBO J.* 11:473–77

22. Choi GH, Nuss DL. 1992. Hypovirulence of chestnut blight fungus conferred by an infectious viral cDNA. *Science* 257:800–3

23. Choi GH, Pawlyk DM, Nuss DL. 1991. The autocatalytic protease p29 encoded by a hypovirulence-associated virus of the chestnut blight fungus resembles the potyvirus-encoded protease HC-Pro. *Virology*, 183:747–52

24. Choi GH, Pawlyk DM, Rae B, Shapira R, Nuss DL. 1993. Molecular analysis and overexpression of the gene encoding endothiapepsin, an aspartic protease from *Cryphonectria parasitica*. *Gene* 125:135–41

25. Choi GH, Shapira R, Nuss DL. 1991. Co-translational autoproteolysis involved in gene expression from a double-stranded RNA genetic element associated with hypovirulence of the chestnut blight fungus. *Proc. Natl. Acad. Sci. USA* 88:1167–71

26. Chung P, Bedker PJ, Hillman BI. 1994. Diversity of *Cryphonectria parasitica* hypovirulence-associated double-stranded RNAs within a chestnut population in New Jersey. *Phytopathology* 84:984–90

27. Churchill ACL, Ciufetti LM, Hansen DR, Van Etten HD, Van Alfen NK. 1990. Transformation of the fungal pathogen *Cryphonectria parasitica* with a variety

of heterologous plasmids. *Curr. Genet.* 17:25–31

28. Coleman DE, Berghuis AM, Lee E, Linder ME, Gilman AG, Sprang SR. 1994. Structures of active conformations of Giα-1 and the mechanism of GTP hydrolysis. *Science* 265:1405–12

29. Craven MG, Pawlyk DM, Choi GH, Nuss DL. 1993. Papain-like protease p29 as a symptom determinant encoded by a hypovirulence-associated virus of the chestnut blight fungus. *J. Virology* 67:6513–21

30. Cronin S, Verchot J, Haldeman-Cahil R, Schaad MC, Carrington JC. 1995. Long-distance movement factor: a transport function of potyvirus helper component proteinase. *Plant Cell* 7:549–59

31. Cummings-Carlson J, Fulbright DW, MacDonald WL, Milgroom MG. 1998. West Salem: a research update. *J. Am. Chestnut Found.* 12:24–26

32. Day PR, Dodds JA, Elliston JE, Jaynes RA, Anagnostakis SL. 1977. Double-stranded RNA in *Endothia parasitica*. *Phytopathology* 67:1393–96

33. Dodds JA. 1980. Association of type 1 viral-like dsRNA with club-shaped particles in hypovirulent strains of *Endothia parasitica*. *Virology* 107:1–12

34. Elliston JE. 1978. Pathogenicity and sporulation in normal and diseased strains of *Endothia parasitica* in American chestnut. In *Proc. Am. Chestnut Symp.*, ed. WL MacDonald, FC Cech, J Luchok, C Smith, pp. 95–100. Morgantown: West Virginia Univ. Press

35. Elliston JE. 1985. Characterization of dsRNA-free and ds-RNA-containing strains of *Endothia parasitica* in relation to hypovirulence. *Phytopathology* 75:151–58

36. Enebak SA, MacDonald WL, Hillman BI. 1994. Effect of dsRNA associated with isolates of *Cryphonectria parasitica* from the central Appalachians and their relatedness to other dsRNAs from North America and Europe. *Phytopathology* 84:528–34

37. Fahima T, Kazmierczak P, Hansen DR, Pfeiffer P, Van Alfen NK. 1993. Membrane-associated replication of the unencapsidated double-stranded RNA of the fungus *Cryphonectria parasitica*. *Virology* 195:81–89

38. Fulbright DW. 1999. Hypovirulence to control fungal pathogenesis. In *Handbook of Biological Control*, ed. TW Fisher, TS Bellows, pp. 691–98. San Diego: Academic

39. Fulbright DW, Weidlich WH, Haufler KZ, Thomas CS, Paul CP. 1983. Chestnut blight and recovering American chestnut trees in Michigan. *Can. J. Bot.* 61:3164–71

40. Gao S, Choi CH, Shain L, Nuss DL. 1996. Cloning and targeted disruption of *epng-1*, encoding the major in vitro extracellular endopolygalacturonase of the chestnut blight fungus, *Cryphonectria parasitica*. *Appl. Environ. Microbiol.* 62:1984–90

41. Gao S, Nuss DL. 1996. Distinct roles for two G protein α subunits in fungal virulence, morphology and reproduction revealed by targeted gene disruption. *Proc. Natl. Acad. Sci. USA* 93:14122–27

42. Gao S, Nuss DL. 1998. Mutagenesis of putative acylation sites alters function, localization and accumulation of a Giα subunit of the chestnut blight fungus *Cryphonectria parasitica*. *Mol. Plant-Microbe Interact.* 11:1130–35

43. Gilman AG. 1987. G proteins: transduction of receptor-generated signals. *Annu. Rev. Biochem.* 56:615–49

44. Grente J. 1965. Les forme hypovirulentes d'Endothia parasitica et les espoirs de lutte contre le chancre du chataignier. *C. R. Seances Acad. Agric. Fr.* 51:1033–37

45. Grente J, Sauret S. 1969. L'hypovirulence exclusive, est-elle controlée par des determinants cytoplasmiques? *C. R. Acad. Sci. Paris Ser. D* 268:3173–76

46. Griffin GJ. 1986. Chestnut blight and its control. *Hortic. Rev.* 8:291–336

47. Hansen DR, Van Alfen NK, Gillies K,

Powell WA. 1985. Naked dsRNA associated with hypovirulence of *Endothia parasitica* is packaged in fungal vesicles. *J. Gen. Virol.* 66:2605–14

48. Hebard FV. 1994. Inheritance of juvenile leaf and stem morphological traits in crosses of Chinese and American chestnut. *J. Hered.* 85:440–46

49. Heiniger U, Rigling D. 1994. Biological control of chestnut blight in Europe. *Annu. Rev. Phytopathol.* 32:581–99

50. Hillman BI, Foglia R, Yuan W. 2000. Satellite, and defective RNAs of *Cryphonectria* hypovirus 3, a virus species in the family *Hypoviridae* with a single open reading frame. *Virology* 276:181–89

51. Hillman BI, Fulbright DW, Nuss DL, Van Alfen NK. 1995. *Hypoviridae.* In *Virus Taxonomy*, ed. FA Murphy, pp. 261–64. New York: Springer Verlag

52. Hillman BI, Halpern BT, Brown MP. 1994. A viral dsRNA element of the chestnut blight fungus with a distinct genetic organization. *Virology* 201:241–50

53. Hillman BI, Shapira R, Nuss DL. 1990. Hypovirulence-associated suppression of host functions in *Cryphonectria parasitica* can be partially relieved by high light intensity. *Phytopathology* 80:950–56

54. Hillman BI, Tian Y, Bedker PJ, Brown MP. 1992. A North American hypovirulent isolate of the chestnut blight fungus with European isolate-related dsRNA. *J. Gen. Virol.* 73:681–86

55. Hiremath S, L'hostis BL, Ghabrial SA, Rhoads RE. 1986. Terminal structure of hypovirulence-associated dsRNA in the chestnut blight fungus *Endothia parasitica. Nucleic Acids Res.* 14:9877–96

56. Horvath CM, Williams MA, Lamb RA. 1990. Eukaryotic coupled translation of tandem cistrons: identification of the influenza B virus BM2 polypeptide. *EMBO J.* 9:2639–47

57. Ivey FD, Hodge PN, Turner GE, Borkovich KA. 1996. The Gαi homologue *gna-1* controls multiple differentiation pathways in *Neurospora crassa. Mol. Biol. Cell.* 7:1283–97

58. Kasahara S, Nuss DL. 1997. Targeted disruption of a fungal G-protein β subunit gene results in increased vegetative growth but reduced virulence. *Mol. Plant Microbe Interact.* 8:984–93

59. Kasahara S, Wang P, Nuss DL. 2000. Identification of *bdm-1*, a gene involved in G-protein β-subunit function and α-subunit accumulation. *Proc. Natl. Acad. Sci. USA.* 97:412–17

60. Kasschau KD, Carrington JC. 1998. A counter-defensive strategy of plant viruses: suppression of post-transcriptional gene silencing. *Cell* 95:461–70

61. Kasschau KD, Cronin S, Carrington JC. 1997. Genome amplification and long-distance movement functions associated with the central domain of tobacco etch potyvirus helper component-proteinase. *Virology* 228:251–62

62. Kays AM, Rowley PS, Baasiri RA, Borkovich KA. 2000. Regulation of conidiation and adenylyl cyclase levels by the Gα protein GNA-3 in *Neurospora crassa. Mol. Cell Biol.* 20:7693–705

63. Kazmierczak P, Pfeiffer P, Zhang L, Van Alfen NK. 1996. Transcriptional repression of specific host genes by the mycovirus *Cryphonectria* hypovirus 1. *J. Virol.* 70:1137–42

64. Kim DH, Rigling D, Zhang L, Van Alfen NK. 1995. A new extracellular laccase of *Cryphonectria parasitica* is revealed by deletion of *Lac1. Mol. Plant-Microbe Interact.* 8:259–66

65. Klein PG, Klein RR, Rodriguez-Cerezo E, Hunt AG, Shaw JG. 1994. Mutational analysis of tobacco vein mottling virus genome. *Virology* 204:759–69

66. Koonin EV, Choi GH, Nuss DL, Shapira R, Carrington JC. 1991. Evidence for common ancestry of a chestnut blight hypovirulence-associated double-stranded RNA and a group of positive-strand RNA plant viruses. *Proc. Natl. Acad. Sci. USA* 88:10647–51

67. Larson TG, Choi GH, Nuss DL. 1992. Regulatory pathways governing modulation of fungal gene expression by a virulence-attenuating mycovirus. *EMBO J.* 11:4539–48

68. Lee BN, Adams TH. 1994. Overexpression of *flbA*, an early regulator of *Aspergillus* asexual sporulation, leads to activation of *brlA* and premature initiation of development. *Mol. Microbiol.* 14:323–34

69. Lee RH, Ting TD, Lieberman BS, Tobias DE, Lolley RN, Ho YK. 1992. Regulation of retinal cGMP cascade by phosducin in bovine rod photoreceptor cells. Interaction of phosducin and transducin. *J. Biol. Chem.* 267:25104–12

70. Lichter A, Mills D. 1997. Fil1, a G-protein α-subunit that acts upstream of cAMP and is essential for dimorphic switching in haploid cells of *Ustilago hordei. Mol. Gen. Genet.* 256:426–35

71. Liu S, Dean RA. 1997. G protein α subunit genes control growth, development and pathogenicity of *Magnaporthe grisea. Mol. Plant-Microbe Interact.* 9:1075–86

72. Liu Y-C, Durrett R, Milgroom MG. 1999. A spatially-structured stochastic model to simulate heterogenous transmission of viruses in fungal populations. *Ecol. Model.* 127:291–308

73. Liu Y-C, Milgroom MG. 1996. Correlation between hypovirus transmission and the number of vegetative incompatibility (vic) genes different among isolates from natural populations of *Cryphonectria parasitica. Phytopathology* 86:79–86

74. Loubradou G, Beguereet J, Turcq B. 1999. MOD-D, a Gα subunit of the fungus *Podospora anserina*, is involved in both regulation of development and vegetative incompatibility. *Genetics* 152:519–28

75. MacDonald WL, Fulbright DW. 1991. Biological control of chestnut blight: use and limitation of transmissible hypovirulence. *Plant Dis.* 75:656–61

76. McBratney S, Chen CY, Sarnow P. 1993. Internal initiation of translation. *Curr. Opin. Cell Biol.* 5:961–65

77. Merkel HW. 1906. A deadly fungus on the American chestnut. *NY Zool. Soc. Annu. Rep.* 10:97–103

78. Monteiro-Vitorello CB, Bell JA, Fulbright DW, Bertrand H. 1995. A cytoplasmically-transmissible hypovirulence phenotype associated with mitochondrial DNA mutations in the chestnut blight fungus *Cryphonectria parasitica. Proc. Natl. Acad. Sci. USA* 92:5935–39

79. Newhouse JR, Hoch HC, MacDonald WL. 1983. The ultrastructure of *Endothia parasitica*: comparison of a virulent isolate with a hypovirulent isolate. *Can. J. Bot.* 61:389–99

80. Nuss DL. 1992. Biological control of chestnut blight: an example of virus mediated attenuation of fungal pathogenesis. *Microbiol. Rev.* 56:561–76

81. Nuss DL. 1996. Using hypoviruses to probe and perturb signal transduction processes underlying fungal pathogenesis. *Plant Cell* 8:1845–53

82. Nuss DL. 2000. Hypovirulence and chestnut blight: from the field to the laboratory and back. In *Fungal Pathology*, ed. JW Kronstad, pp. 149–70. The Netherlands: Kluwer. 404 pp.

83. Peever TL, Liu Y-C, Milgroom MG. 1997. Diversity of hypoviruses and other double-stranded RNAs in *Cryphonectria parasitica* in North America. *Phytopathology* 87:1026–33

84. Peever TL, Liu Y-C, Milgroom MG. 1998. Incidence and diversity of hypoviruses and other double-stranded RNAs occurring in the chestnut blight fungus *Cryphonectria parasitica*, in China and Japan. *Phytopathology* 88:811–17

85. Powell WA, Van Alfen NK. 1987. Differential accumulation of poly(A)+ RNA between virulent and double-stranded RNA-induced hypovirulent strains of *Cryphonectria (Endothia) parasitica. Mol. Cell. Biol.* 7:3688–93

86. Powell WA, Van Alfen NK. 1987. Two

nonhomologous viruses of *Cryphonectria* (*Endothia*) *parasitica* reduce accumulation of specific virulence-associated polypeptides. *J. Bacteriol.* 169:5324–26

87. Rae B, Hillman BI, Tartaglia J, Nuss DL. 1989. Characterization of double-stranded RNA genetic elements associated with biological control of chestnut blight: organization of terminal domains and identification of gene products. *EMBO J.* 8:657–63

88. Regenfelder E, Spellig T, Hartmann A, Lauenstein S, Bolker M, Kahmann R. 1997. G proteins in *Ustilago maydis*: transmission of multiple signals? *EMBO J.* 16:1934–42

89. Rigling D, Van Alfen NK. 1991. Regulation of laccase biosynthesis in the plant pathogenic fungus *Cryphonectria parasitica* by double-stranded RNA. *J. Bacteriol.* 173:8000–3

90. Robin C, Anziani C, Cortesi P. 2000. Relationship between biological control, incidence of hypovirulence, and diversity of vegetative compatibility types of *Cryphonectria parasitica* in France. *Phytopathology* 90:730–37

91. Rosen S, Yu JH, Adams TH. 1999. The *Aspergillus nidulans sfaD* gene encodes a G-protein beta subunit that is required for normal growth and repression of sporulation. *EMBO J.* 18:5592–600

92. Shain L, Miller JB. 1992. Movement of cytoplasmic hypovirulence agents in chestnut blight cankers. *Can. J. Bot.* 70:557–61

93. Shapira R, Choi GH, Nuss DL. 1991. Virus-like genetic organization and expression strategy for double-stranded RNA genetic element associated with biological control of chestnut blight. *EMBO J.* 10:731–39

94. Shapira R, Choi GH, Hillman BI, Nuss DL. 1991. The contribution of defective RNAs to complexity of viral-encoded double-stranded RNA populations present in hypovirulent strains of the chestnut blight fungus, *Cryphonectria parasitica*. *EMBO J.* 10:741–46

95. Shapira R, Nuss DL. 1991. Gene expression by a hypovirulence-associated virus of the chestnut blight fungus involves two papain-like protease activities. *J. Biol. Chem.* 266:19419–25

96. Shi XM, Miller H, Verchot J, Carrington JC, Vance VB. 1996. Mutational analysis of the potyviral sequence that mediates potato virus X/potyviral synergistic disease. *Virology* 231:35–42

97. Smart CD, Yuan W, Foglia R, Nuss DL, Fulbright DW, Hillman BI. 2000. *Cryphonectria* hypovirus 3, a virus species in the family Hypoviridae with a single open reading frame. *Virology* 265:66–73

98. Springer ML. 1993. Genetic control of fungal differentiation: the three sporulation pathways of *Neurospora crassa*. *BioEssays* 15:365–74

99. Suzuki N, Chen B, Nuss DL. 1999. Mapping of a hypovirus p29 protease symptom determinant domain with sequence similarity to potyvirus HC-Pro protease. *J. Virol.* 73:9478–84

100. Suzuki N, Geletka LM, Nuss DL. 2000. Essential and dispensable virus-encoded replication elements revealed by efforts to develop hypoviruses as gene expression vectors. *J. Virol.* 74:7568–77

101. Tartaglia J, Paul CP, Fulbright DW, Nuss DL. 1986. Structural properties of double-stranded RNAs associated with biological control of chestnut blight fungus. *Proc. Natl. Acad. Sci. USA* 83:9109–13

102. Turner GE, Borkovich KA. 1993. Identification of a G protein α subunit from *Neurospora crassa* that is a member of the G_i family. *J. Biol. Chem.* 268:14805–11

103. Tolkacheva T, McNamara P, Piekarz E, Courchesne W. 1994. Cloning of a *Cryptococcus neoformans* gene, GPA1, encoding a G-protein alpha-subunit homolog. *Infect. Immunol.* 62:2849–56

104. Van Alfen NK. 1986. Hypovirulence of *Endothia* (*Cryphonectria*) *parasitica* and

Rhizoctonia solani. See Ref. 12, pp. 143–62

105. Van Alfen NK, Jaynes RA, Anagnostakis SL, Day PR. 1975. Chestnut blight: biological control by transmissible hypovirulence in *Endothia parasitica. Science* 189:890–91

106. Van Heerden SW, Geletka LM, Preisig O, Nuss DL, Wingfield BD, Wingfield MJ. 2001. Characterization of South African *Cryphonectria cubensis* isolates infected with a *Cryphonectria parasitica* hypovirus. *Phytopathology.* In press

107. Wang P, Larson TG, Chen C-H, Pawlyk DM, Clark JA, Nuss DL. 1998. Cloning and characterization of a general amino acid control transcriptional activator from the chestnut blight fungus *Cryphonectria parasitica. Fungal Genet. Biol.* 23:81–94

108. Wang P, Nuss DL. 1995. Induction of a *Cryphonectria parasitica* cellobiohydrolase I genesis suppressed by hypovirus infection and regulation by a G-protein-linked signaling pathway involved in fungal pathogenesis. *Proc. Natl. Acad. Sci. USA* 92:11529–33

109. Wang P, Perfect JR, Heitman J. 2000. The G-protein beta subunit GPB1 is required for mating and haploid fruiting in *Cryptococcus neoformans. Mol. Cell. Biol.* 20:352–62

110. Wimmer E, Heller CUJ, Cao X. 1993. Genetics of poliovirus. *Annu. Rev. Genet.* 27:353–36

111. Yang Q, Borkovich KA. 1999. Mutational activation of a Gαi causes uncontrolled proliferation of aerial hyphae and increased sensitivity to heat and oxidative stress in *Neurospora crassa. Genetics* 147:137–45

112. Yu JH, Wieser J, Adams TH. 1996. The *Aspergillus* FlbA RGS domain protein antagonizes G protein signaling to block proliferation and allow development. *EMBO J.* 15:5184–90

113. Yoshida T, Willardson BM, Wilkins JF, Jensen GJ, Thornton BD, Bitensky MW. 1994. The phosphorylation state of phosducin determines its ability to block transducin subunit interactions and inhibit transducin binding to activated rhodopsin. *J. Biol. Chem.* 269:24050–57

114. Zhang L, Churchill ACL, Kazmierczak P, Kim D, Van Alfen NK. 1993. Hypovirulence-associated traits induced by a mycovirus of *Cryphonectria parasitica* are mimicked by targeted inactivation of a host gene. *Mol. Cell. Biol.* 13:7782–92

Annu. Rev. Genet. 2001. 35:31–52

GENETICS AND THE FITNESS OF HYBRIDS

John M. Burke[1] and Michael L. Arnold[2]

[1]Department of Biology, Indiana University, Bloomington, Indiana 47405;
e-mail: jmburke@indiana.edu
[2]Department of Genetics, University of Georgia, Athens, Georgia 30602;
e-mail: arnold@dogwood.botany.uga.edu

Key Words   adaptation, epistasis, hybridization, speciation, transgressive segregation

■ **Abstract**   Over the years, the evolutionary importance of natural hybridization has been a contentious issue. At one extreme is the relatively common view of hybridization as an evolutionarily unimportant process. A less common perspective, but one that has gained support over the past decade, is that of hybridization as a relatively widespread and potentially creative evolutionary process. Indeed, studies documenting the production of hybrid genotypes exhibiting a wide range of fitnesses have become increasingly common. In this review, we examine the genetic basis of such variation in hybrid fitness. In particular, we assess the genetic architecture of hybrid inferiority (both sterility and inviability). We then extend our discussion to the genetic basis of increased fitness in certain hybrid genotypes. The available evidence argues that hybrid inferiority is the result of widespread negative epistasis in a hybrid genetic background. In contrast, increased hybrid fitness can be most readily explained through the segregation of additive genetic factors, with epistasis playing a more limited role.

CONTENTS

0066-4197/01/1215-0031$14.00

INTRODUCTION

> The total weight of evidence contradicts the assumption that hybridization plays a major evolutionary role among higher animals.... Successful hybridization is indeed a rare phenomenon among animals.
>
> *Mayr (75, p. 133)*

> To be sure, the occasional production of an interspecific hybrid occurs frequently in plants. However, most of these hybrids seem to be sterile, or do not backcross with the parent species for other reasons.
>
> *Mayr (76, p. 233)*

The preceding quotes illustrate the historically common view of natural hybridization as an evolutionarily unimportant process. This view is largely based on the observation that crosses between divergent lineages often give rise to progeny with decreased levels of viability and/or fertility (e.g., 14, 75, 76, 105, 106). Despite the apparent rarity of "successful" hybrids in nature, a large body of literature on natural hybridization has accumulated over the years. The majority of this work has focused on either inferring evolutionary relationships based on the ability (or inability) to hybridize (e.g., 28, 51, 56, 69, 111), or deciphering the mechanisms that limit gene flow (e.g., 11, 12, 58, 59, 61, 67, 88, 96). When hybridization has been directly implicated in the evolutionary process, it has traditionally been for the role it may play in finalizing speciation (e.g., 40, 75; reviewed in 60). Hybrid zones, therefore, have generally been viewed as transient, with selection on mating preferences ultimately giving rise to "good" species or, if stable, as little more than an impediment to continuing divergence.

A less common viewpoint is that of hybridization as a relatively widespread and potentially creative evolutionary process (e.g., 3, 4, 44, 52, 70, 107). Indeed, introgression (the transfer of genetic material from one species into another via hybridization) has been documented in a wide variety of both plant and animal taxa (5, 93), and there is evidence that it may serve as a source of adaptive genetic variation (e.g., 52, 53, 70, 107). In addition, there are a number of well-documented cases of homoploid hybrid speciation, suggesting that natural hybridization may play an important role in evolutionary diversification (reviewed in 89).

The creative potential of natural hybridization depends critically on the production of recombinant genotypes that can outperform their parents in at least some habitats (6). Broadly speaking, hybrid fitness can be influenced by either endogenous or exogenous selection. Endogenous selection refers to that which acts against certain hybrid genotypes regardless of the environment in which they occur. This inherent loss of fitness is assumed to result from either meiotic irregularities or physiological/developmental abnormalities in individuals of mixed ancestry (hybrid incompatibility). Exogenous selection, on the other hand, refers to environment-specific fitness differences. According to this view, the distribution of genotypes across a hybrid zone is assumed to be governed by adaptation to different habitats. Consequently, although many hybrid genotypes will fail to find suitable habitat, hybridization could lead to the production of recombinant genotypes that outperform their parents in certain habitats (2, 6, 78). Because the fitness of hybrids relative to their parents has been discussed extensively elsewhere (e.g., 6, 7, 13, 54, 101), we summarize it only briefly here. In general, the pattern that has emerged is one in which many, if not most, hybrids perform poorly as compared to their parents. Although hybrids tend to perform poorly on average, some fraction of hybrid genotypes are often found to outperform their parental counterparts under certain conditions (e.g., 18, 19, 47, 64, 91, 107). Indeed, recent theoretical work confirms that a small fraction of hybrid genotypes will likely outperform their parents, even in parental habitats (13). Therefore, we no longer need to ask whether or not hybrids will exhibit a wide range of fitnesses. They do. Of greater interest are the mechanisms underlying such fitness variation. As pointed out elsewhere, the outcome of hybridization depends not only on the distribution of hybrid fitness, but also on the underlying genetics (13).

In this review, we examine the genetic basis of hybrid fitness differences. Although chromosomal factors influence hybrid fitness, generally through their negative effects on hybrid fertility (e.g., 66, 99, 110), our focus is on the role of genic factors in determining the fitness of hybrids. As such, we begin with a summary of what is known about the genetic basis of hybrid inferiority. Although hybrids may suffer reduced fitness due to exogenous factors such as a lack of suitable habitat, the best genetic data on hybrid inferiority come from analyses of endogenous selection. We therefore focus on the genetics of hybrid incompatibility. We then review what is known about the genetic architecture of increased hybrid fitness. We are particularly interested in examining the extent to which loci governing hybrid fitness interact, both with other nuclear loci, as well as with some component of the cytoplasm. We close with a discussion of the likelihood that hybridization will lead to the establishment and spread of novel hybrid genotypes.

HYBRID INFERIORITY

The role of epistasis in adaptive evolution has been a controversial issue ever since Sewall Wright and R.A. Fisher first formalized their views in the early 1930s. According to Wright (113, 114), natural selection retains favorably interacting gene combinations. Therefore, as a result of the highly integrated nature of the genome, selection may lead to the production of what Dobzhansky (43) has termed "coadapted" gene complexes. In contrast, Fisher (48) argued that natural selection acts primarily on single genes, rather than on gene complexes. In Fisher's view, therefore, selection favors alleles that elevate fitness, on average, across all possible genetic backgrounds within a lineage. Such alleles have been termed "good mixers" (75). Regardless of the role of epistasis within lineages, however, negative epistasis in a hybrid genetic background, or hybrid incompatibility, is fully consistent with both the Wrightian and Fisherian worldviews. This is because allelic fixation occurs in any one lineage without regard to the compatibility (or lack thereof) of new alleles with those in any other lineage. Hybridization then produces a vast array of recombinant genotypes that have never before been subjected to selection. On average, these genotypes will be less well adapted than their parents, giving rise to some level of selection against hybrids.

Hybrid breakdown, or the reduction in fitness of segregating hybrid progeny that often results from intercrossing genetically divergent populations or taxa, has long been taken as evidence of unfavorable interactions between the genomes of the parental individuals (e.g., 39, 42, 43, 75, 80). The most widely accepted genetic model for the occurrence of such incompatibilities was first described by Bateson (15, as cited in 83), and later by Dobzhansky (39) and Muller (79, 80). In short, the Bateson-Dobzhansky-Muller (BDM) model assumes that an ancestral population consisting solely of individuals of the genotype *aa/bb* is broken into two parts that are temporarily isolated from each other. In one subpopulation, a new allele (*A*) is then assumed to arise at the first locus. Meanwhile, a new allele (*B*) is assumed to arise in the other subpopulation. Because individuals of the genotype *aa/bb*, *Aa/bb*, and *AA/bb* can interbreed freely, the *A* allele can then spread to fixation in the first subpopulation; likewise, individuals of the genotype *aa/bb*, *aa/Bb*, and *aa/BB* can interbreed freely, and the *B* allele spreads to fixation in the second subpopulation. However, although *A* is compatible with *b*, and *B* is compatible with *a*, the interaction of *A* with *B* is assumed to produce some sort of developmental or physiological breakdown, such that hybridization between the two subpopulations leads to the production of offspring with decreased levels of viability and/or fertility. Although this model focuses on negative interactions between differentiated regions of the nuclear genome, similar interactions between one or more regions of the nuclear genome and some component of the cytoplasm (e.g., the chloroplast or mitochondrial genome) could also play an important role in hybrid incompatibility. Unfortunately, the BDM model does not provide any mechanistic explanation as to how mutations that are neutral (or beneficial) within a given lineage will produce strongly disadvantageous incompatibilities when combined in a hybrid background.

More recently, Werth & Windham (109) proposed a model for the generation of incompatibilities at the polyploid level. This model states that allopatric populations of a single tetraploid species may experience silencing of the same gene, but in different parental genomes (reciprocal silencing; see Figure 1). Because every individual in this model carries two full genomic complements, these silencing events occur within either lineage with no detrimental effect on fitness. If these populations were to come back into reproductive contact, however, 25% of all gametes produced by first-generation hybrids would carry only nonfunctional copies of such a gene. If one or more of these genes were required for the function of gametes, there would be a marked decrease in hybrid fertility. Conversely, if there were no gametic problems, 6.25% of all F_2 hybrid individuals would carry no functional copies of such a gene. If one or more of these genes were required for survival, the F_2 generation would experience a reduction in viability. Lynch & Force (72) have since extended this model (and dubbed it duplication, degeneration, and complementation,

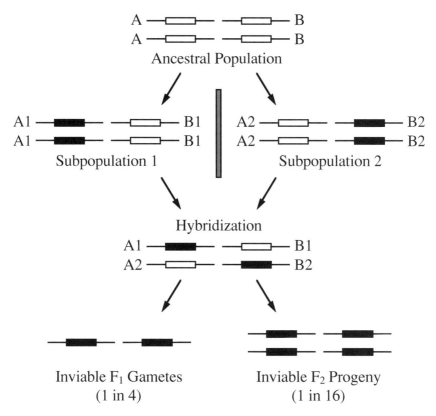

Figure 1 Hypothetical example of reciprocal silencing of gene duplicates. Open boxes correspond to functional gene copies, whereas closed boxes denote silenced copies. Subpopulation 1 loses function at the A locus and retains it at the B locus, while the opposite occurs in subpopulation 2. One fourth of the F_1 gametes and one sixteenth of the F_2 zygotes resulting from crosses between subpopulations will carry only nonfunctional copies of the gene (Adapted from Figure 1 in Reference 72).

or DDC) to the silencing or functional divergence of any type of gene duplicate, not just those resulting from polyploidy. Unlike the BDM model, the DDC provides a simple mechanism by which hybrid-incompatibility can arise. Although this model is attractive in principle, data confirming or refuting it are lacking.

Despite our lack of insight into the genetic mechanisms underlying hybrid inferiority, a variety of empirical studies have confirmed the role of relatively widespread negative epistasis (reviewed in 116). What follows are recent examples from the animal and plant literature that have helped illuminate the nature of such genetic interactions. In some cases, these studies have also shed light on the within-lineage processes that may have given rise to the observed incompatibilities between lineages.

Drosophila simulans and *D. mauritiana*

The best direct evidence on the role of gene interactions in hybrid sterility and inviability comes from *Drosophila* (see 29, 115, 116 for reviews). In particular, recent studies of hybrid incompatibility within the *D. simulans* clade have provided detailed insight into the complexity of such incompatibilities (25, 36, 57, 84–86, 104). In general terms, these studies have involved the introgression of small regions of *D. mauritiana* or *D. sechellia* chromosomes into the *D. simulans* genetic background. The results of this work indicate that hybrid sterility results from a large number of genetic interactions. Indeed, Palopoli & Wu (84) estimate that there are at least 40 loci that influence hybrid male sterility on the X chromosome alone, and Hollocher & Wu (57) found that the density of autosomal factors contributing to hybrid sterility is comparable to the density of X chromosome factors (but see 104). Moreover, many of these interactions involve more than two loci (25, 57, 86, 104). These sorts of higher-order interactions are, in fact, predicted by theory (82). Interestingly, the general pattern that has emerged is one in which conspecific genes often interact strongly (and negatively) when placed together in a hybrid genetic background (25, 57, 86). This body of work is also noteworthy in that it has led to the cloning and characterization of a major gene (*Odysseus*) involved in the production of male sterility in crosses between *D. simulans* and *D. mauritiana* (103). Although the function of this gene is still unknown, the authors found that it contains a homeobox with high sequence similarity to known genes from mice, rats, and nematodes. They also found that *Odysseus* has undergone extremely rapid sequence divergence over the past half million years, suggesting that positive selection has played a role in the evolution of hybrid incompatibility within the *D. simulans* clade.

Helianthus annuus and *H. petiolaris*

Analyses of both synthetic hybrid lineages and natural hybrid zones between the annual sunflower species *Helianthus annuus* and *H. petiolaris* have provided some of the best evidence on the role of gene interactions in hybrid incompatibility between plant species (49, 91). In general, decreased hybrid fitness between these species results from reduced hybrid fertility. Although the genomes of these two species differ by three inversions and at least seven translocations (92), suggesting that hybrid sterility may result from chromosomal variation, genetic map-based analyses of experimental backcross hybrids indicate that most *H. petiolaris* markers in colinear (i.e., non-rearranged) genomic regions introgress into the *H. annuus* background at frequencies significantly lower than expected (91). This result suggests that loci in these regions of the *H. petiolaris* genome interact unfavorably with some component of the *H. annuus* genome. The majority of multilocus interactions (measured as linkage disequilibrium) were positive, indicating that *H. petiolaris* alleles interact favorably when placed together on the *H. annuus* genetic background. This finding is consistent with predictions of the

BDM model, in that selection should act in favor of those individuals that retain combinations of complementary genes from a given species (i.e., parental types). In addition, consistent with findings in *Drosophila* (see above), there were also a number of cases in which conspecific (i.e., *H. petiolaris*) genes interacted negatively when placed together in a hybrid background, suggesting the occurrence of deleterious higher-order interactions. In related work, Rieseberg and colleagues (49, 94) analyzed patterns of introgression in wild hybrid populations between these species. This work is especially noteworthy in that it represents the first application of a large number of mapped molecular markers to the analysis of natural hybrid zones. Overall, their findings mirrored the experimental hybrid lineages. Considering only the seven colinear linkage groups, there were at least eight (possibly ten) *H. petiolaris* chromosomal regions that introgressed at frequencies lower than expected (94). Analyses of hybrid fertility revealed a possible mechanism for this pattern: The majority of these underrepresented blocks were significantly associated with reduced pollen fertility. This result confirms fertility selection against certain hybrid genotypes as the most likely cause of reduced introgression.

Oryza sativa ssp. *japonica* and *O. s.* ssp. *indica*

Both F_1 sterility and later-generation hybrid breakdown have been documented in intersubspecific crosses of rice (*Oryza sativa* L.; 81, 100). Interestingly, although F_1 sterility and hybrid breakdown often coincide in rice, hybrid breakdown sometimes occurs in the advanced-generation progeny of compatible (fully fertile) F_1 hybrids. In an attempt to elucidate the genetic basis of these phenomena, Li et al. (71) applied quantitative trait locus (QTL) mapping techniques to a cross between *O. s.* ssp. *japonica* and *O. s.* ssp. *indica*. In terms of F_1 sterility, the authors found evidence for the existence of "supergenes," or groups of tightly linked, favorably interacting genes within which recombination causes decreased fitness (35). Although the occurrence of cryptic structural rearrangements cannot be ruled out, F_1 hybrid sterility in rice is not generally associated with cytologically detectable abnormalities (27). This result suggests that F_1 sterility may, at least in part, be a genic phenomenon in rice. Regardless of the role of cryptic structural rearrangements, such incompatibilities arise as a result of recombination within differentiated regions. In order to account for the occasional occurrence of compatible F_1 hybrids, therefore, Li et al. (71) posited the existence of genes that influence recombination rates in rice. Indeed, Ikehashi & Araki (62) and Sano (95) both found evidence for genetic control of recombination rates in rice. These results suggest that hybrid fitness may be directly influenced by genes that regulate recombination. Interestingly, the phenotypic effect of the putative supergenes was strongly dependent on cytoplasmic background, providing evidence for the role of cytonuclear interactions in hybrid sterility. Consistent with theoretical predictions (82), Li et al. (71) also found evidence for widespread negative epistasis between the *indica* and *japonica* genomes. This result led the authors to conclude that "hybrid breakdown may involve large numbers of genes and complex higher-order interactions, as reported in *Drosophila*" (p. 1146; see above).

Tigriopus californicus

Hybrid breakdown in the marine copepod *Tigriopus californicus* has been the subject of numerous studies since the mid-1980s (20–24, 45). In short, Burton and colleagues have found that crosses between divergent, isolated populations of this species give rise to F_1 hybrids whose performance is virtually indistinguishable from their parents'. When these individuals are crossed to the F_2 generation, however, the resulting progeny often exhibit hybrid breakdown in terms of both development time and response to osmotic stress. Of particular interest is the observation that crosses between certain populations result in some sort of nearly lethal epistatic selection involving the region of the genome marked by the Me^F allozyme (21). Indeed, Me^F homozygotes are extremely rare in the F_2 generation, with an estimated viability of less than 8% of the interpopulational heterozygote. This result is reminiscent of the "synthetic lethal" systems discussed by Dobzhansky (41), in that otherwise harmless loci seem to interact to produce a lethal (or nearly so, in this case) phenotype. Another intriguing finding of this work is the observation of decreasing levels of cytochrome *c* oxidase (COX) activity as the mitochondrial genome of one population is introgressed into the nuclear background of another (45). Because COX is composed of subunits encoded by both nuclear and mitochondrial genes, its activity may reflect the coordinated function of the two genomes. Although their results varied among crosses, Edmands & Burton (45) found strong support for the hypothesis of deleterious nuclear-mitochondrial interactions in certain crosses. In view of the critical role of COX in the electron transport chain (it catalyzes the final step), cytonuclear coadaptation of this enzyme provides a plausible mechanism by which some degree of hybrid incompatibility may arise.

Iris fulva and I. brevicaulis

Work on natural and experimental Louisiana iris hybrid populations indicates that both nuclear and cytonuclear interactions influence hybrid viability (19, 33, 34). In their analysis of the genetic structure of a natural hybrid zone between *Iris fulva* and *I. brevicaulis*, Cruzan & Arnold (33) documented the occurrence of differential seed abortion among hybrid genotypes. More specifically, intermediate hybrid seeds experienced markedly higher rates of abortion relative to parental-like seeds. This result suggests that intermediate hybrid genotypes are selected against owing to incompatibilities between the *I. fulva* and *I. brevicaulis* genomes. In a related study, Burke et al. (19) investigated the role of both nuclear and cytonuclear interactions in determining the frequencies of F_2 genotypes produced in crosses between the same two species. Consistent with the findings of Cruzan & Arnold (33), there was an overall deficit of intermediate hybrid genotypes in the F_2 generation (Figure 2). Analyses of single and multilocus segregation patterns also revealed a variety of nuclear and cytonuclear interactions. Of particular interest was the complete absence of individuals homozygous for the *I. brevicaulis* allele

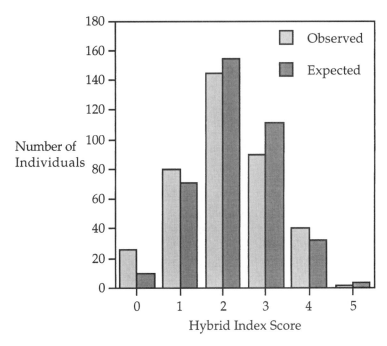

Figure 2   Observed and expected genotypic distributions of F_2 seedlings resulting from a cross between *Iris fulva* and *I. brevicaulis*. Hybrid index scores are based on genotypes at a series of species-specific nuclear markers. The observed distribution is significantly different from the expected distribution ($\chi^2 = 31.75$, df $= 4$, $P < 0.001$; Redrawn from Figure 1*C* in Reference 19).

at a marker known as L180. As in the case of *Tigriopus* (see above), this result suggests that the region of the *I. brevicaulis* genome marked by L180 interacts with one or more different regions of the *I. fulva* genome to produce synthetic lethality (41). The potential negative effect of cytonuclear interactions on hybrid viability was illustrated by a significant excess of *I. fulva* alleles (and a deficit of *I. brevicaulis* alleles) in one particular region of the genome on the *I. fulva* cytoplasmic background, but not on the *I. brevicaulis* cytoplasmic background. The nonreciprocal nature of this deviation suggests that hybridization has disrupted favorable interactions between this region of the *I. fulva* nuclear genome and some component of the *I. fulva* cytoplasm, leading to selection against the *I. brevicaulis* genotype in this region. Once again, this result indicates the possibility of hybrid incompatibility due to cytonuclear coadaptation.

Gossypium hirsutum and *G. barbadense*

Jiang et al. (64) investigated the role of multilocus interactions in restricting introgression between two polyploid species of cotton, *Gossypium hirsutum* and *G. barbadense*. After three generations of backcrossing with *G. hirsutum*, the authors found large and widespread deficiencies of *G. barbadense* chromatin. In fact, there were no *G. barbadense* alleles at nearly 30% of the loci under study, and seven independent *G. barbadense* chromosomal regions were entirely absent. Because the genomes of these two species appear to be colinear, this result led the authors to conclude that unfavorable genic interactions in certain hybrid genotypes protect these regions of the *G. hirsutum* genome from introgression. However, the observed absence of certain *G. barbadense* chromosomal blocks could also be explained in completely nonepistatic terms. In other words, *G. hirsutum* may simply harbor better alleles in these regions, leading to the selective loss of *G. barbadense* alleles, regardless of genetic background. This being said, it seems unlikely that *G. hirsutum* alleles would outperform *G. barbadense* alleles in all seven regions, making epistasis the most plausible explanation. In addition to these seven "protected" chromosomal regions, the authors detected significantly more interactions among unlinked pairs of loci than expected by chance. Because the occurrence and/or magnitude of these interactions varied across backcross families, there may have been additional, higher-order interactions that went undetected. In view of this widespread epistasis, it is especially interesting to note that a disproportionate number of the negative interactions detected by Jiang et al. (64) occurred between subgenomes. Although this result is superficially consistent with the DDC, there is no evidence to suggest that these negative interactions tend to occur between homoeologous loci (P. Chee & A.H. Paterson, personal communication).

HYBRID SUPERIORITY

In general, natural hybridization can contribute to adaptation and/or speciation in one of two ways. First, introgression may lead to the transfer of adaptations from one taxon into another, perhaps allowing for range expansion of the introgressed form (70; but see 16, 50). Alternatively, hybridization may lead to the founding of new evolutionary lineages (see 6, 89 for references). As alluded to above, the creative potential of natural hybridization depends not only on the production of relatively fit hybrid genotypes, but also on the genetic architecture of such hybrid superiority. The importance of genetic architecture lies in the likelihood of establishment and spread of favorable hybrid genotypes. Indeed, as Barton (13) has pointed out, natural selection can pick out rare, relatively fit hybrid individuals, but only if their offspring are also fit.

At a genetic level, increased hybrid fitness can arise in several ways. First-generation hybrids often exhibit heterosis, or hybrid vigor, especially if their parents are inbred. Depending on the genetic basis of such heterosis, however, the increase in fitness may be short-lived, breaking down with the passing of generations. In later generations, relatively fit hybrids may result from either the production of novel, favorably interacting (epistatic) gene combinations, or through the combining of advantageous alleles across noninteracting (additive) loci. In any case, exogenous selection is believed to play a central role in the establishment of relatively fit hybrids. The main reason for this is that, in the absence of niche differentiation, new hybrid genotypes are likely to be overwhelmed by competition and/or gene flow from the parental populations (17, 90).

Unfortunately, although a number of studies have documented the production of relatively fit hybrids (see 6, 8 for examples), there are few data on the genetic basis of increased hybrid fitness. There are two reasons for this. First, as pointed out previously, the initial production of relatively fit hybrid individuals is expected to be a rare occurrence. Similar to the study of beneficial mutations, which are also exceedingly rare, the genetic analysis of increased hybrid fitness has therefore been difficult. Second, many of the studies dealing with hybrid fitness have relied on statistical comparisons of performance across classes (e.g., parental, F_1, F_2, backcross, etc.) rather than on detailed analyses of specific hybrid genotypes (7, but see HYBRID INFERIORITY above). Recently, however, experimental crosses as well as analyses of natural hybrid zones have begun to provide data on the genetic architecture of increased hybrid fitness.

Evidence for the role of epistasis in the production of relatively fit hybrids, although limited, comes from several sources. As was the case for hybrid incompatibility, the best data on the genetics of increased hybrid fitness in plants come from the work of Rieseberg and colleagues on the annual sunflower species *H. annuus* and *H. petiolaris* (49, 91). In addition to providing evidence for the role of epistasis in hybrid sterility, genetic map-based analyses of synthetic hybrid lineages between these species have revealed favorable heterospecific gene interactions (91). This result led the authors to conclude that "a small percentage of alien genes do appear to interact favorably in hybrids" (p. 744). Consistent with this finding, analyses of natural hybrid zones between these species also uncovered evidence of favorable heterospecific gene interactions in the form of significant, negative disequilibrium between certain pairs and triplets of unlinked markers (49). In addition, Burke et al. (19) documented the occurrence of favorable heterospecific cytonuclear interactions in crosses between *Iris fulva* and *I. brevicaulis*. This work suggests that increased hybrid fitness can arise not only as a result of interactions among nuclear loci, as in the case of sunflower, but also as a result of interactions between the nuclear genome and some component of the cytoplasm. Finally, in addition to the restricted introgression described above (see HYBRID INFERIORITY), Jiang et al. (64) detected several instances of higher than expected rates of introgression when *Gossypium barbadense* is backcrossed against *G. hirsutum*. This discovery led the authors to conclude that "genomic interactions do not always favor host chromatin" (p. 798). Rather, it appears that favorable heterospecific interactions may encourage the introgression of certain chromosomal blocks from one taxon into another. However, introgression analyses such as this should generally be interpreted with caution. In the absence of detectable associations between the chromosomal region of interest and one or more regions of the recipient genome, nonepistatic explanations are fully consistent with the data.

Because of the emphasis placed on niche divergence by many students of hybridization, the potential for the production of relatively fit hybrids is often tied to the production of novel or extreme hybrid phenotypes (e.g., 17, 55, 70, 102). As

TABLE 1 Hypothetical example of transgressive segregation due to the complementary action of genes with additive effects (from Reference 90)

QTL	Phenotypic values			
	Species A	Species B	Transgressive F_2	Transgressive F_2
1	+1	−1	+1 (A)	−1 (B)
2	+1	−1	+1 (A)	−1 (B)
3	+1	−1	+1 (A)	−1 (B)
4	−1	+1	+1 (B)	−1 (A)
5	−1	+1	+1 (B)	−1 (A)
Total	+1	−1	+5	−5

it turns out, such transgressive segregation appears to occur frequently in crosses between divergent lineages in both plants and animals (reviewed in 90). Moreover, QTL mapping studies have consistently implicated the additive effects of complementary genes, rather than epistasis, as the mechanism by which transgressive phenotypes arise (see Table 1) (e.g., 38, 65, 74, 108). In fact, QTL alleles often have effects that are opposite in direction to that expected on the basis of overall trait values (90). In other words, alleles reducing a trait are sometimes found in species with a high trait value, whereas alleles increasing a trait are sometimes found in species with low trait values. Unfortunately, most studies to date have gone only as far as documenting the production of extreme phenotypic variants. Therefore, the connection between transgressive segregation and the production of relatively fit hybrid genotypes remains tenuous. A number of studies have documented transgressive segregation for traits such as fecundity, as well as tolerance to various biotic and abiotic stresses (see 90 for references). Given the right environmental conditions, these sorts of traits clearly have adaptive significance. It therefore seems likely that transgressive segregation, presumably due to the additive effects of alleles across loci, has the potential to contribute to the success of certain hybrid lineages.

Although exogenous selection is generally assumed to play a crucial role in the production and establishment of relatively fit hybrid genotypes, it is also possible that hybridization could give rise to genotypes that are intrinsically more fit than their parents. One mechanism by which this may occur is through the purging of mutational load (46). Due to the constraints of finite population size, mildly deleterious alleles can become fixed within lineages, leading to the gradual erosion of fitness (inbreeding depression; e.g., 68, 77). Hybridization between lineages could, therefore, lead to the production of heterotic F_1 hybrids due to the masking of deleterious recessive alleles. In later generations, one possible outcome of such heterosis would be the introgression of favorable alleles from one parental population into the other (63). Alternatively, if the hybrid offspring are isolated in some way from their parents, the joint effects of recombination and natural selection may decrease the frequency of deleterious alleles, ultimately giving rise to a true-breeding hybrid lineage with increased fitness relative to its parents. Although the so-called "dominance hypothesis" of inbreeding depression has received considerable support (e.g., 26, 31, 32, 68), the application of this idea to the potential adaptive consequences of natural hybridization has received little attention. Such purging of mutational load via hybridization has, however, been suggested to play a role in the evolution of invasiveness in plants (46). Furthermore, common garden experiments in *Helianthus* have documented hybrid lineages that exhibit higher fecundity than their parents (LH Rieseberg, unpublished data).

EVOLUTIONARY IMPLICATIONS

The main conclusion that can be drawn from genetic analyses of hybrid incompatibility is that postzygotic reproductive isolation generally results from widespread, negative epistasis in a hybrid genetic background. Such analyses cannot, however, distinguish between interactions that were initially involved in reproductive isolation and those that arose later. The main reason for this is that hybrid sterility and inviability are predicted to evolve nonlinearly (i.e., faster) with respect to time (82). This "snowballing" effect could, therefore, lead to an overestimate of the number of genes required for speciation. What we do know is that the complex and widespread nature of these incompatibilities makes them relatively effective barriers to genetic exchange between taxa. Indeed, as the number of loci that contribute to reduced hybrid fitness increases, the likelihood of producing relatively fit hybrid genotypes decreases, and the proportion of the genome protected from gene flow increases. However, the efficacy of low hybrid fitness as a barrier to gene flow varies across the genome, and even strong postzygotic barriers can fail to preclude successful hybridization. For example, in spite of the extremely strong sterility barriers between members of the *Drosophila simulans* clade (see HYBRID INFERIORITY above), a phylogenetic analysis of mtDNA sequences placed certain interspecific mtDNA types together (Figure 3) (98). One clade in this phylogeny contained mitochondrial haplotypes of all three species, whereas the other clade contained haplotypes of both *D. simulans* and *D. mauritiana*. This result led the authors to conclude that introgressive hybridization had occurred between these well-isolated species, a finding that was later supported by the results of experimental crosses (9).

When considering the evolutionary importance of natural hybridization, we are thus faced with an odd dichotomy of data. On the one hand, there is a substantial body of evidence documenting that, in the majority of cases, hybrid matings give rise to progeny with decreased levels of fertility and/or viability (e.g., 14, 40, 75, 105, 106, 112). In fact, along with various prezygotic barriers to hybridization, this sort of reproductive isolation is the very reason why we have distinct species. On

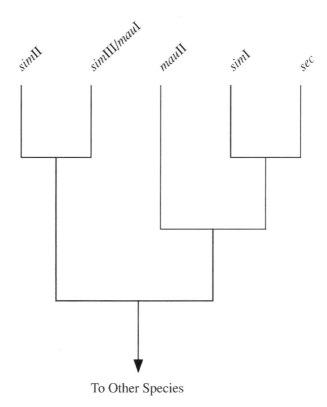

Figure 3   Mitochondrial DNA phylogeny of *Drosophila simulans* (*sim*I, *sim*II, *sim*III), *D. mauritiana* (*mau*I, *mau*II), and *D. sechellia* (*sec*) (Redrawn from Figure 2*B* in Reference 98).

the other hand, there are numerous examples of potentially adaptive introgression (5, 93), as well as the production of new hybrid species (reviewed in 89). This dichotomy underscores a fundamental aspect of the evolutionary process, namely the overwhelming importance of rare events. Indeed, it is widely recognized that evolution proceeds as a direct consequence of extremely rare events—the occurrence of beneficial mutations, long-distance migration, founder events, etc.,—yet, with reference to the production of relatively fit hybrid genotypes, it has been argued that the rarity of such individuals precludes an important role for hybridization in evolution (75, 76). Barton (13) has gone on to argue that, in abundant species, mutation is not limiting and favored variants are therefore more likely to arise via mutation than hybridization. What this view neglects to recognize is that such mutations may arise anywhere within the range of a species, but may only be favored in certain locales. Although the same might be said of novel variants resulting from hybridization, hybrid matings often occur near the edge of a species range or in marginal habitats—just the sort of places where new variants may be most likely to survive and thrive (2, 4, 78). In contrast to mutation, which will generally occur at the same low rate across the range of a species, hybridization thus provides a mechanism by which genetic variation can be generated in areas where the resulting variants are most likely to invade and utilize novel habitats.

But what about the likelihood that favored hybrid variants will persist and spread? As pointed out previously, this depends on the underlying genetic basis of increased hybrid fitness, as well as the circumstances under which these rare individuals occur (13). In contrast to hybrid incompatibility, where widespread epistasis seems to be the rule, the genetic basis of increased hybrid fitness is less clear. Although there is evidence that favorable epistatic interactions may be involved in the production of relatively fit hybrids, explaining the persistence and spread of these genotypes is somewhat problematic. Indeed, when considering the impact of epistatic factors on the creative potential of natural hybridization, the argument closely mirrors the Wright/Fisher debate over the importance of epistasis in adaptation (48, 113, 114). Alleles at unlinked loci will be rapidly dissociated by recombination. Thus, in the absence of an extremely strong selective advantage, epistatic selection will be effective only during the first few generations following the onset of hybridization (97). In short, favorably interacting gene complexes resulting from hybridization will face the same difficulties that adaptive gene combinations face in Wright's shifting balance theory of evolution (30). The spread of novel epistatic gene combinations out of a hybrid zone should, therefore, occur only under extremely rare circumstances. Before rejecting the importance of favorably interacting gene complexes in influencing the outcome of hybridization, however, it is important to consider the possibility that the conditions appropriate for the establishment and/or spread of favorably interacting hybrid gene combinations can (and do) occur, albeit rarely. For example, hybrid founder events may play a critical role in the establishment of new hybrid lineages, and episodes of high migration or strong selection could aid in the spread of favorable gene combinations from one population to another. These rare instances may have a tremendous impact on the trajectory of existing or new evolutionary lineages.

The evolutionary importance of nonepistatic factors, on the other hand, is less difficult to explain. Alleles that are favored regardless of genetic background will readily introgress across hybrid zones, even if they are initially associated with loci involved in decreased hybrid fitness (87). Because such alleles can rapidly spread to fixation, at least in certain habitats, adaptive trait introgression may be difficult to detect. Therefore, the numerous examples of introgression documented in the hybridization literature may be just the tip of the iceberg. Perhaps more important, however, is the role that additive factors may play in niche divergence. Indeed, the most likely way in which hybrids will initially become established and achieve isolation from their parents is via adaptation to a novel habitat. In fact, most stabilized introgressants and hybrid species are ecologically isolated from their parents (1, 6, 89). Moreover, these lineages often occur in habitats that are extreme, rather than intermediate, with respect to the requirements of their parents. So how does this niche divergence arise? Most evidence suggests that the production of extreme (transgressive) phenotypes occurs through the additive effects of alleles segregating at complementary genes (see HYBRID SUPERIORITY above). Adaptation of hybrids to a novel habitat may, therefore, simply result from: (*a*) the generation of adaptive (and additive) genetic variation via hybridization, and (*b*) selection favoring extreme phenotypes following hybridization. Alternatively, hybridization could lead to the production of individuals with unique character combinations (55). Some fraction of these recombinant types may be especially well suited to an available, unique habitat and will therefore be able to increase numerically, at least locally. Because these ecologically isolated recombinants will be more or less released from the competitive and swamping effects of close contact with their parents, they will be free to evolve independently and, perhaps, develop some degree of reproductive isolation.

CONCLUSIONS AND FUTURE DIRECTIONS

The prominent role of epistasis in postzygotic reproductive isolation was first proposed well over half a century ago (15, 39, 79, 80). Since that time, numerous investigators have confirmed that negative epistasis in a hybrid genetic background is, indeed, largely responsible for the common observation of hybrid inferiority. Unfortunately, although these studies have provided insight into the widespread nature and complexity of such genetic interactions, the underlying mechanisms are still largely unknown. To more fully understand the nature of these barriers, we must move beyond the abstract notion of locus A interacting negatively with locus B (see HYBRID INFERIORITY above). One promising area of inquiry would be to investigate the role of gene duplicates in producing hybrid incompatibility. Perhaps the easiest approach would be to map hybrid sterility in a cross between polyploid species. If complementary sterility loci map to the same position on homoeologous chromosomes, the results would provide evidence that the silencing or functional divergence of gene duplicates (in this case derived through polyploidy) plays a role in the evolution of species incompatibilities.

Another promising line of research is the identification of the loci that interact to produce incompatibilities. Great progress has been made on this front by Wu and colleagues, who have identified and characterized a locus involved in hybrid male sterility (*Odysseus*) in *Drosophila* (85, 86, 103). This is, however, but a single example. Before any generalizations can be made about the types of loci likely to be involved in hybrid incompatibility, we need additional examples from a variety of study systems. Moreover, although the introgression approach utilized in the identification of *Odysseus* has proven to be a powerful technique, interacting loci derived from the other species cannot be identified. The logical next step in such studies, therefore, might be to use introgressed individuals carrying a heterospecific incompatibility factor as a tool to identify other genomic regions that interact with the factor of interest. In the case of hybrid male sterility, introgressed females of species B carrying a male sterility factor from species A could be crossed against a panel of advanced-generation (and fertile) male hybrids segregating for the species A genome on a species B background. By scoring male fertility in the resulting progeny, it may be possible to localize interacting factors. These candidate regions could then be targeted for introgression analyses in much the same way that *Odysseus* was identified in *Drosophila*.

In terms of the genetic basis of increased hybrid fitness, much of the evidence (empirical as well as theoretical) points to the importance of additive factors in the establishment and spread of favorable hybrid genotypes. However, the QTL mapping approaches generally employed in studies that have documented transgressive segregation have low power for detecting epistasis (73). Thus, epistasis may play a larger role in transgressive segregation than previously believed, but may have gone largely undetected. In order to investigate this possibility more rigorously, modified mapping approaches may be necessary. Perhaps the most useful approach would be to use recombinant inbred lines (RILs; 10) or recombinant congenic (RC) strains (37), which control for the effects of genetic background. These strategies, therefore, enhance the ability to identify interactions between a given marker and the background on which it occurs. Whatever the cause of transgressive segregation, the connection between phenotype and fitness needs to be made before we can be certain that this phenomenon plays an important role in the production of relatively fit hybrid lineages. One approach might be to compare QTL combinations found in naturally occurring hybrid lineages with those required for the production of the most extreme phenotypes (90). Correspondence would provide convincing, albeit indirect, evidence that transgressive segregation played a primary role in the evolution of such lineages. Alternatively, transplant experiments could be used to test the fitness effects of various phenotypes (or QTL combinations) in the wild. This approach would provide direct evidence on the potential fitness effects of transgressive segregation.

We have now entered a new phase of research on natural hybridization. Instead of focusing on this process as an impediment to "normal" divergence, or as an evolutionary epiphenomenon, numerous investigators are approaching their studies assuming that natural hybridization can be evolutionarily important in its own right. Indeed, the future is bright for studies of natural hybridization. The tools required for detailed genetic analyses are widely available, and there is now a freedom to investigate previously discounted questions and hypotheses.

ACKNOWLEDGMENTS

We thank Kevin Livingstone, Loren Rieseberg, and members of the Wade lab for numerous discussions that helped clarify our thinking on various aspects of hybrid fitness. This work was supported by grants from the USDA Plant Genome Program (00-35300-9244 to JMB) and the National Science Foundation (DEB-0074159 to MLA).

Visit the Annual Reviews home page at www.AnnualReviews.org

LITERATURE CITED

1. Abbott RJ. 1992. Plant invasions, interspecific hybridization and the evolution of new plant taxa. *Trends Ecol. Evol.* 7:401–5
2. Anderson E. 1948. Hybridization of the habitat. *Evolution* 2:1–9
3. Anderson E. 1949. *Introgressive Hybridization.* New York: Wiley
4. Anderson E, Stebbins GL Jr. 1954. Hybridization as an evolutionary stimulus. *Evolution* 8:378–88
5. Arnold ML. 1992. Natural hybridization as an evolutionary process. *Annu. Rev. Ecol. Syst.* 23:237–61
6. Arnold ML. 1997. *Natural Hybridization and Evolution.* Oxford: Oxford Univ. Press
7. Arnold ML, Hodges SA. 1995. Are natural hybrids fit or unfit relative to their parents? *Trends Ecol. Evol.* 10:67–71
8. Arnold ML, Kentner EK, Johnston JA, Cornman S, Bouck AC. 2001. Natural hybridization and fitness. *Taxon* 50:93–104
9. Aubert J, Solignac M. 1990. Experimental evidence for mitochondrial DNA introgression between *Drosophila* species. *Evolution* 44:1272–82
10. Bailey DW. 1971. Recombinant-inbred strains. An aid to identify linkage and function of histocompatibility and other genes. *Transplantation* 11:325–27
11. Baker MC, Baker AEM. 1990. Reproductive behavior of female buntings: isolating mechanisms in a hybridizing pair of species. *Evolution* 44:332–38
12. Ball RW, Jameson DL. 1966. Premating isolating mechanisms in sympatric and allopatric *Hyla regilla* and *Hyla californiae.* *Evolution* 20:533–51
13. Barton NH. 2001. The role of hybridisation in evolution. *Mol. Ecol.* 10:551–68
14. Barton NH, Hewitt GM. 1985. Analysis

of hybrid zones. *Annu. Rev. Ecol. Syst.* 16:113–48
15. Bateson W. 1909. Heredity and variation in modern lights. In *Darwin and Modern Science,* ed. AC Seward, pp. 85–101. Cambridge: Cambridge Univ. Press
16. Birch LC, Vogt WG. 1970. Plasticity of taxonomic characters of the Queensland fruit flies *Dacus tryoni* and *Dacus neohumeralis* (Tephritidae). *Evolution* 24:320–43
17. Buerkle CA, Morris RJ, Asmussen MA, Rieseberg LH. 2000. The likelihood of homoploid hybrid speciation. *Heredity* 84:441–51
18. Burke JM, Carney SE, Arnold ML. 1998. Hybrid fitness in the Louisiana irises: analysis of parental and F_1 performance. *Evolution* 52:37–43
19. Burke JM, Voss TJ, Arnold ML. 1998. Genetic interactions and natural selection in Louisiana iris hybrids. *Evolution* 52:1304–10
20. Burton RS. 1986. Evolutionary consequences of restricted gene flow in the intertidal copepod *Tigriopus californicus.* *Bull. Mar. Sci.* 39:526–35
21. Burton RS. 1987. Differentiation and integration of the genome in populations of the marine copepod *Tigriopus californicus.* *Evolution* 41:504–13
22. Burton RS. 1990. Hybrid breakdown in developmental time in the copepod *Tigriopus californicus.* *Evolution* 44:1814–22
23. Burton RS. 1990. Hybrid breakdown in physiological response: a mechanistic approach. *Evolution* 44:1806–13
24. Burton RS, Rawson PD, Edmands S. 1999. Genetic architecture of physiological phenotypes: empirical evidence for coadapted gene complexes. *Am. Zool.* 39:451–62
25. Cabot EL, Davis AW, Johnson NA, Wu

C-I. 1994. Genetics of reproductive isolation in the *Drosophila simulans* clade: complex epistasis underlying hybrid male sterility. *Genetics* 137:175–89

26. Charlesworth D, Charlesworth C. 1987. Inbreeding depression and its evolutionary consequences. *Annu. Rev. Ecol. Syst.* 18:237–68

27. Chu YE, Morishima H, Oka HI. 1969. Reproductive barriers distributed in cultivated rice species and their wild relatives. *Jpn. J. Genet.* 44:207–33

28. Clausen J, Keck DD, Hiesey WH. 1939. The concept of species based on experiment. *Am. J. Bot.* 26:103–6

29. Coyne JA. 1992. Genetics and speciation. *Nature* 355:511–15

30. Coyne JA, Barton NH, Turelli M. 1997. A critique of Sewall Wright's shifting balance theory of evolution. *Evolution* 51:643–71

31. Crow JF. 1948. Alternative hypotheses of hybrid vigor. *Genetics* 33:477–87

32. Crow JF. 1952. Dominance and overdominance. In *Heterosis*, ed. JE Gowen, pp. 282–97. Ames, IA: Iowa State College Press

33. Cruzan MB, Arnold ML. 1994. Assortative mating and natural selection in an *Iris* hybrid zone. *Evolution* 48:1946–58

34. Cruzan MB, Arnold ML. 1999. Consequences of cytonuclear epistasis and assortative mating for the genetic structure of hybrid populations. *Heredity* 82:36–45

35. Darlington CD, Mather K. 1949. *The Elements of Genetics*. London: Allen & Unwin

36. Davis AW, Noonburg EG, Wu C-I. 1994. Evidence for complex genic interactions between conspecific chromosomes underlying hybrid female sterility in the *Drosophila simulans* clade. *Genetics* 137:191–99

37. Démant P, Hart AAM. 1986. Recombinant congenic strains—a new tool for analyzing genetic traits determined by more than one gene. *Immunogenetics* 24:416–22

38. deVicente MC, Tanksley SD. 1993. QTL analysis of transgressive segregation in an interspecific tomato cross. *Genetics* 134:585–96

39. Dobzhansky T. 1936. Studies on hybrid sterility. II. Localization of sterility factors in *Drosophila pseudoobscura* hybrids. *Genetics* 21:113–35

40. Dobzhansky T. 1940. Speciation as a stage in evolutionary divergence. *Am. Nat.* 74:312–21

41. Dobzhansky T. 1946. Genetics of natural populations. XIII. Recombination and variability in populations of *Drosophila pseudoobscura*. *Genetics* 31:269–90

42. Dobzhansky T. 1950. Genetics of natural populations. XIX. Origin of heterosis through natural selection in populations of *Drosophila pseudoobscura*. *Genetics* 35:288–302

43. Dobzhansky T. 1970. *Genetics of the Evolutionary Process*. New York: Columbia Univ. Press

44. Dowling TE, DeMarais BD. 1993. Evolutionary significance of introgressive hybridization in cyprinid fishes. *Nature* 362:444–46

45. Edmands S, Burton RS. 1999. Cytochrome *c* oxidase activity in interpopulation hybrids of a marine copepod: a test for nuclear-nuclear or nuclear-cytoplasmic coadaptation. *Evolution* 53:1972–78

46. Ellstrand NC, Schierenbeck KA. 2000. Hybridization as a stimulus for the evolution of invasiveness in plants? *Proc. Natl. Acad. Sci. USA* 97:7043–50

47. Emms SK, Arnold ML. 1996. The effect of habitat on parental and hybrid fitness: transplant experiments with Louisiana irises. *Evolution* 51:1112–19

48. Fisher RA. 1930. *The Genetical Theory of Natural Selection*. Oxford: Clarendon Press

49. Gardner K, Buerkle A, Whitton J, Rieseberg LH. 2000. Inferring epistasis in wild sunflower hybrid zones. In *Epistasis and*

the Evolutionary Process, ed. JB Wolf, ED Brodie III, MJ Wade, pp. 264–79. Oxford: Oxford Univ. Press

50. Gibbs GW. 1968. The frequency of interbreeding between two sibling species of *Dacus* (Diptera) in wild populations. *Evolution* 22:667–83

51. Gillett GW. 1966. Hybridization and its taxonomic implications in the *Scaevola gaudichaudiana* complex of the Hawaiian islands. *Evolution* 20:506–16

52. Grant BR, Grant PR. 1996. High survival of Darwin's finch hybrids: effects of beak morphology and diets. *Ecology* 77:500–9

53. Grant PR, Grant BR. 1994. Phenotypic and genetic effects of hybridization in Darwin's finches. *Evolution* 48:297–316

54. Grant V. 1963. *The Origin of Adaptations*. New York: Columbia Univ. Press

55. Grant V. 1981. *Plant Speciation*. New York: Columbia Univ. Press

56. Heiser CB Jr, Smith DM, Clevenger SB, Martin WC Jr. 1969. The North American sunflowers (*Helianthus*). *Mem. Torrey Bot. Club* 22:1–213

57. Hollocher H, Wu C-I. 1996. The genetics of reproductive isolation in the *Drosophila simulans* clade: X *vs.* autosomal effects and male *vs.* female effects. *Genetics* 143:1243–55

58. Hopper SD, Burbidge AH. 1978. Assortative pollination by red wattlebirds in a hybrid population of *Anigozanthos Labill.* (Haemodoraceae). *Aust. J. Bot.* 26:335–50

59. Howard DJ. 1986. A zone of overlap and hybridization between two ground cricket species. *Evolution* 40:34–43

60. Howard DJ. 1993. Reinforcement: origin, dynamics and fate of an evolutionary hypothesis. In *Hybrid Zones and the Evolutionary Process*, ed. RG Harrison, pp. 46–69. New York: Oxford Univ. Press

61. Howard DJ, Waring GL. 1991. Topographic diversity, zone width, and the strength of reproductive isolation in a zone of overlap and hybridization. *Evolution* 45:1120–35

62. Ikehashi H, Araki H. 1986. Genetics of F_1 sterility in the remote crosses of rice. *Proc. Int. Rice Genet. Symp.* pp. 119–30

63. Ingvarsson PK, Whitlock MC. 2000. Heterosis increases the effective migration rate. *Proc. R. Soc. London Ser. B* 267:1321–26

64. Jiang C-X, Chee PW, Draye X, Morrell PL, Smith CW, Paterson AH. 2000. Multilocus interactions restrict gene introgression in interspecific populations of polyploid *Gossypium* (cotton). *Evolution* 54:798–814

65. Kim S-C, Rieseberg LH. 1999. Genetic architecture of species differences in annual sunflowers: implications for adaptive trait introgression. *Genetics* 153:965–77

66. King M. 1993. *Species Evolution: The Role of Chromosome Change*. Cambridge: Cambridge Univ. Press

67. Lamb T, Avise JC. 1986. Directional introgression of mitochondrial DNA in a hybrid population of tree frogs: the influence of mating behavior. *Proc. Natl. Acad. Sci. USA* 83:2526–30

68. Lande R. 1995. Mutation and conservation. *Conserv. Biol.* 9:782–91

69. Lenz LW. 1958. A revision of the Pacific Coast irises. *Aliso* 4:1–72

70. Lewontin RC, Birch LC. 1966. Hybridization as a source of variation for adaptation to new environments. *Evolution* 20:315–36

71. Li Z, Pinson SRM, Paterson AH, Park WD, Stansel JW. 1997. Genetics of hybrid sterility and hybrid breakdown in an intersubspecific rice (*Oryza sativa* L.) population. *Genetics* 145:1139–48

72. Lynch M, Force AG. 2000. The origin of interspecific genomic incompatibility via gene duplication. *Am. Nat.* 156:590–605

73. Lynch M, Walsh B. 1998. *Genetics and Analysis of Quantitative Traits*. Sunderland, MA: Sinauer

74. Mansur LM, Lark KG, Kross H, Oliveira A. 1993. Interval mapping of quantitative trait loci for reproductive, morphological,

and seed traits of soybean (*Glycine max* L.). *Theor. Appl. Genet.* 86:907–13

75. Mayr E. 1963. *Animal Species and Evolution.* Cambridge, MA: Belknap

76. Mayr E. 1992. A local flora and the biological species concept. *Am. J. Bot.* 79:222–38

77. Mills LS, Smouse P. 1994. Demographic consequences of inbreeding in remnant populations. *Am. Nat.* 144:412–31

78. Moore WS. 1977. An evaluation of narrow hybrid zones in vertebrates. *Q. Rev. Biol.* 52:263–77

79. Muller HJ. 1942. Isolating mechanisms, evolution and temperature. *Biol. Symp.* 6:71–125

80. Muller HJ, Pontecorvo G. 1940. Recombinants between *Drosophila* species the F_1 hybrids of which are sterile. *Nature* 146:199–200

81. Oka HI. 1988. *Origin of Cultivated Rice.* New York: Elsevier

82. Orr HA. 1995. The population genetics of speciation: the evolution of hybrid incompatibilities. *Genetics* 139:1805–13

83. Orr HA. 1996. Dobzhansky, Bateson, and the genetics of speciation. *Genetics* 144:1331–35

84. Palopoli MF, Wu C-I. 1994. Genetics of hybrid male sterility between *Drosophila* sibling species: a complex web of epistasis is revealed in interspecific studies. *Genetics* 138:329–41

85. Perez DE, Wu C-I. 1993. Genetics of reproductive isolation in the *Drosophila simulans* clade: DNA marker-assisted mapping and characterization of a hybrid-male sterility gene, *Odysseus* (*Ods*). *Genetics* 133:261–75

86. Perez DE, Wu C-I. 1995. Further characterization of the *Odysseus* locus of hybrid sterility in *Drosophila*: one gene is not enough. *Genetics* 140:201–6

87. Pialek J, Barton NH. 1997. The spread of an advantageous allele across a barrier: the effects of random drift and selection against heterozygotes. *Genetics* 145:493–504

88. Rand DM, Harrison RG. 1989. Ecological genetics of a mosaic hybrid zone: mitochondrial, nuclear, and reproductive differentiation of crickets by soil type. *Evolution* 43:432–49

89. Rieseberg LH. 1997. Hybrid origins of plant species. *Annu. Rev. Ecol. Syst.* 28:359–89

90. Rieseberg LH, Archer MA, Wayne RK. 1999. Transgressive segregation, adaptation, and speciation. *Heredity* 83:363–72

91. Rieseberg LH, Sinervo B, Linder CR, Ungerer MC, Arias DM. 1996. Role of gene interactions in hybrid speciation: evidence from ancient and experimental hybrids. *Science* 272:741–45

92. Rieseberg LH, Van Fossen C, Desrochers AM. 1995. Hybrid speciation accompanied by genomic reorganization in wild sunflowers. *Nature* 375:313–16

93. Rieseberg LH, Wendel JF. 1993. Introgression and its consequences in plants. In *Hybrid Zones and the Evolutionary Process*, ed. RG Harrison, pp. 70–109. New York: Oxford Univ. Press

94. Rieseberg LH, Whitton J, Gardner K. 1999. Hybrid zones and the genetic architecture of a barrier to gene flow between two sunflower species. *Genetics* 152:713–27

95. Sano Y. 1990. The genetic nature of gametic eliminator in rice. *Genetics* 125:183–91

96. Shaw DD, Wilkinson P. 1980. Chromosome differentiation, hybrid breakdown and the maintenance of a narrow hybrid zone in *Caledia*. *Chromosoma* 80:1–31

97. Slatkin M. 1995. Epistatic selection opposed by immigration in multiple locus genetic systems. *J. Evol. Biol.* 8:623–34

98. Solignac M, Monnerot M. 1986. Race formation, speciation, and introgression within *Drosophila simulans*, *D. mauritiana*, and *D. sechellia* inferred from mitochondrial DNA analyses. *Evolution* 40:531–39

99. Spirito F. 1998. The role of chromosomal change in speciation. In *Endless Forms:*

Species and Speciation, ed. DJ Howard, SH Berlocher, pp. 320–29. Oxford: Oxford Univ. Press

100. Stebbins GL Jr. 1958. The inviability, weakness, and sterility of interspecific hybrids. *Adv. Genet.* 9:147–215

101. Stebbins GL Jr. 1959. The role of hybridization in evolution. *Proc. Am. Philos. Soc.* 103:231–51

102. Templeton AR. 1981. Mechanisms of speciation—a population genetic approach. *Annu. Rev. Ecol. Syst.* 12:23–48

103. Ting C-T, Tsaur S-C, Wu M-L, Wu C-I. 1998. A rapidly evolving homeobox at the site of a hybrid sterility gene. *Science* 282:1501–4

104. True JR, Weir BS, Laurie CC. 1996. A genome-wide survey of hybrid incompatibility factors by the introgression of marked segments of *Drosophila mauritiana* chromosomes into *Drosophila simulans. Genetics* 142:819–37

105. Wagner WH Jr. 1969. The role and taxonomic treatment of hybrids. *BioScience* 19:785–95

106. Wagner WH Jr. 1970. Biosystematics and evolutionary noise. *Taxon* 19:146–51

107. Wang H, McArthur ED, Sanderson SC, Graham JH, Freeman DC. 1997. Narrow hybrid zone between two subspecies of big sagebrush (*Artemisia tridentata*: Asteraceae). IV. Reciprocal transplant experiments. *Evolution* 51:95–102

108. Weller JI, Soller M, Brody T. 1988. Linkage analysis of quantitative traits in an interspecific cross of tomato (*Lycopersicon esculentum X Lycopersicon pimpinellifolium*) by means of genetic markers. *Genetics* 118:329–39

109. Werth CR, Windham MD. 1991. A model for divergent, allopatric speciation of polyploid pteridophytes resulting from silencing of duplicate-gene expression. *Am. Nat.* 137:515–26

110. White MJD. 1978. *Modes of Speciation.* San Francisco: Freeman

111. Wiegand KM. 1935. A taxonomist's experience with hybrids in the wild. *Science* 81:161–66

112. Wilson EO. 1965. The challenge from related species. In *The Genetics of Colonizing Species*, ed. HG Baker, GL Stebbins Jr, pp. 7–24. Orlando, FL: Academic

113. Wright S. 1931. Evolution in Mendelian populations. *Genetics* 16:97–159

114. Wright S. 1932. The roles of mutation, inbreeding, crossbreeding and selection in evolution. *Proc. Int. Congr. Genet., 6th*, 1:356–66

115. Wu C-I, Davis AW. 1993. Evolution of postmating reproductive isolation: the composite nature of Haldane's rule and its genetic bases. *Am. Nat.* 142:187–212

116. Wu C-I, Palopoli MF. 1994. Genetics of postmating reproductive isolation in animals. *Annu. Rev. Genet.* 27:283–308

Annu. Rev. Genet. 2001. 35:53–82

RECOMBINATIONAL DNA REPAIR OF DAMAGED REPLICATION FORKS IN *ESCHERICHIA COLI*: Questions

Michael M. Cox

Department of Biochemistry, University of Wisconsin-Madison, Madison, Wisconsin 53706-1544; e-mail: cox@biochem.wisc.edu

Key Words   replication, recombination, RecA protein, PriA protein, repair

■ **Abstract**   It has recently become clear that the recombinational repair of stalled replication forks is the primary function of homologous recombination systems in bacteria. In spite of the rapid progress in many related lines of inquiry that have converged to support this view, much remains to be done. This review focuses on several key gaps in understanding. Insufficient data currently exists on: (*a*) the levels and types of DNA damage present as a function of growth conditions, (*b*) which types of damage and other barriers actually halt replication, (*c*) the structures of the stalled/collapsed replication forks, (*d*) the number of recombinational repair paths available and their mechanistic details, (*e*) the enzymology of some of the key reactions required for repair, (*f*) the role of certain recombination proteins that have not yet been studied, and (*g*) the molecular origin of certain in vivo observations associated with recombinational DNA repair during the SOS response. The current status of each of these topics is reviewed.

CONTENTS

0066-4197/01/1215-0053$14.00

INTRODUCTION

Homologous genetic recombination is a substantial part of the molecular basis of the science of genetics. However, it has been clear for decades that homologous genetic recombination systems did not evolve to generate genetic diversity in populations or to assist geneticists in their efforts to map genes. A primary role in DNA repair has been the major alternative hypothesis (11, 12, 22, 26, 29, 36, 44, 67, 88, 100, 122). This view has now been refined. The major function of homologous genetic recombination in bacteria, and a major function in virtually all cells, is the nonmutagenic recombinational DNA repair of stalled or collapsed replication forks. This hypothesis is built on a recent convergence of many independent lines of research (28–32, 64, 67–69, 97, 98).

The DNA metabolism of every cell is replete with connections between replication and recombination. Replication is part of the recombination that accompanies bacterial conjugation or transduction (149, 150). In bacteriophage T4, the two processes are tightly linked, with recombination essential to the process of replication initiation after the first few replication cycles (82, 108, 109). In eukaryotes, replication accompanies the repair of programmed double-strand breaks in meiosis and the miscellaneous double-strand breaks that may occur as a result of exposure to ionizing radiation (46). In some cases, break (recombination)-induced replication can replicate major parts of a chromosome (15, 85, 106, 166) or permit telomere maintenance in eukaryotic cells lacking telomerase (15, 71).

The repair of replication forks is more than just another process to fill out this list. In bacteria, replication forks appear to require recombinational DNA repair often under normal growth conditions. Best current estimates indicate that a replication fork undergoes such repair in nearly every cell in every generation, making this the most important bacterial application of homologous genetic recombination on a frequency-of-use basis (29, 31, 32). In mammals, perhaps ten forks undergo recombinational DNA repair in every mitotic division (46). The need for nonmutagenic repair when replication forks encounter template damage provides an excellent rationale for the evolution of recombination systems. From this standpoint, the repair of replication forks might be considered the original or determinative function of homologous genetic recombination. The systems that have evolved in bacteria (and in eukaryotes, where many enzymes probably remain to be discovered) are elaborate and often redundant. Subsets of the recombination enzymes provide the potential for an adaptable response to whatever DNA structure is found at a stalled fork. Recombination during conjugation or transduction must then be considered a byproduct of the presence of this repair system.

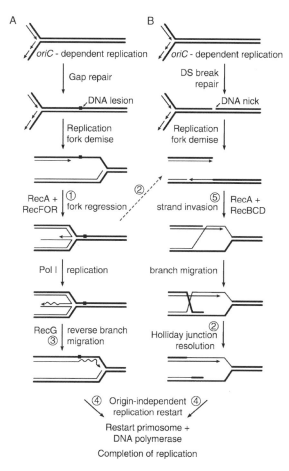

Figure 1   Pathways for the recombinational DNA repair of stalled or collapsed replication forks. The path in column *A* describes some of the processes proposed for the repair of forks that encounter an unrepaired DNA lesion. Column *B* illustrates some of the steps in the repair of double-strand breaks resulting from the encounter of a fork with a template strand break. Circled numbers correspond to five key processes described in the text.

There are five major types of reactions that underlie most of the proposed pathways for fork repair (Figure 1). If a fork encounters one of the spontaneous DNA lesions that occur in every cell (3000–5000 lesions per bacterial cell per generation under normal aerobic growth conditions), replication may halt and leave the lesion in a single-strand gap (Figure 1*A*). Four of the basic reactions can be seen in the possible pathways for repair of such a gap. First, fork regression involves the re-pairing of the DNA template strands such that the fork is moved backwards. The nascent DNA strands are eventually paired to generate a Holliday structure with one abbreviated branch. Such structures have been observed for nearly 30 years and have recently been labeled "chicken feet" (121). Second, the Holliday structure can be cleaved by the RuvC or a related enzyme. This leads to a double-strand break. Third, as an alternative to cleavage, the structure can be subjected to enzymatic branch migration. In principle, this can move the branch back toward the original point where replication stalled, or further regress the fork. Fourth, all recombinational repair processes must include replication restart.

Many of the same processes can be seen in the likely pathways for repairing a double-strand break (Figure 1*B*). In addition, there is one new process, the invasion of the 3′ end of a single-stranded DNA into a homologous duplex DNA to generate a branched intermediate that can be processed to a form compatible with replication restart (#5 in Figure 1*B*). This must be considered one of the central reactions in recombination, playing a role in the repair of free DNA ends in all organisms in many contexts. This aspect of replication fork repair gives rise to key steps in conjugation and transduction in bacteria, and meiotic recombination in eukaryotes. Once invasion has occurred, the now paired 3′ end can be utilized as a primer for replication.

The new focus on replication fork repair is a significant paradigm shift within the broader study of DNA metabolism, and an illuminating one. Replication fork repair can take its place with excision repair, mismatch repair, base excision repair, and direct repair [such as the reaction catalyzed by DNA photolyase (135)] as a major cellular DNA repair pathway. When placed in this context, fork repair is readily seen as the least understood of the major cellular DNA repair processes.

Genetic recombination has often been viewed as a way to generate genetic diversity (thus altering genomes), but the application of recombination systems to replication fork repair provides another perspective. The recombinational repair of replication forks functions to maintain genome integrity. Most DNA repair processes (such as excision repair) are enabled by the double-stranded character of DNA. When a lesion occurs in one strand, it can be cut out. The other strand can be used as a template for replication to replace the damaged strand with correct genomic information. Replication fork encounters with strand breaks or lesions tend to generate structures in which both DNA strands are damaged. When a replication fork encounters a DNA lesion, the lesion is left in a single-strand gap. The recombination steps in Figure 1*A* do not directly repair the DNA lesion, but instead create the situation (an undamaged complementary strand) needed to effect later repair. When a replication fork encounters a template strand break, the recombination steps (Figure 1*B*) can be more readily seen as a true repair process designed to reconstruct the fork and permit a restart of replication. In either case, the recombination process is needed to permit a continuation of replication without introducing alterations in the DNA.

Many different enzymes participate in recombinational DNA repair. When DNA degradation processes and the action of a variety of enzymes with ATPase activities are factored in, the repair of one stalled replication fork consumes a considerable amount of chemical energy in the form of expended dNTPs and rNTPs. Such energetic expenditures are a trademark of DNA repair systems such as mismatch repair (102) and the direct repair of O^6-alkylguanine lesions (118), presumably reflecting the generally low tolerance of biological systems for genome damage.

Our current understanding of nonmutagenic replication fork repair has been detailed in a number of recent reviews (30–32, 64, 68, 69, 97). This review attempts to explore some of the more impressive gaps in understanding, focusing entirely on the fork repair process in bacteria.

SPONTANEOUS DNA DAMAGE

Replication fork repair is probably most often predicated by a collision of a fork with DNA damage. The precise path taken by the repair process may depend to a large extent on what type of damage is encountered and whether it is on the leading or lagging strand template. Surprisingly little information is available about DNA damage frequencies in bacterial cells under different sets of growth conditions. Estimates of the spontaneous frequency of events such as depurination and cytosine deamination are readily available (43, 74). However, the majority of DNA lesions that occur spontaneously in a cell growing aerobically are oxidative lesions, arising from the action of hydroxyl radicals (57, 153). Indeed, many normally inviable cells lacking certain combinations of replication and recombination functions are able to grow anaerobically (52, 75, 104).

Oxidative damage includes a wide range of lesions, not all of which have been characterized. Under normal aerobic growth conditions, the best numbers available suggest that an *Escherichia coli* culture suffers 3000–5000 DNA lesions per cell per generation. Spent culture media from an *E. coli* culture harvested at an $A_{600} = 1.0$ yields sufficient 8-oxo-7,8-dihydro-2′-deoxyguanosine (oxo8dG) to account for several hundred of the corresponding lesions per cell per generation (116). Other work indicates that oxo8dG represents about 5% of the oxidative lesions in a typical spectrum of oxidative damage (126), providing the final estimate cited above. More detailed estimates that quantify the occurrence of a broader range of lesions under a variety of growth conditions would be very useful.

The potential barriers to replication forks do not begin and end with DNA lesions themselves, but include bound proteins, unusual DNA structures, and natural replication pause sites in the DNA (51, 97, 130). Recent studies suggest that RNA polymerase complexes themselves stalled at the sites of DNA lesions may be important barriers to replication in cells that have been irradiated with UV light (92). In general, a complete understanding of replication fork repair pathways will not be possible without a more comprehensive understanding of the situations in which replication forks are halted.

HOW OFTEN IS THE FORK HALTED?

Some controversy has persisted in the recombinational DNA repair literature concerning the capacity of replication forks to bypass DNA lesions. Bypass could take at least two forms. In some instances, the polymerase may insert a nucleotide opposite the lesion, potentially creating a mutation at that position. One can consider lesions as being of two types. Certain base analogues may cause mispairing and thus replication errors, but not block replication. These can include some naturally occurring lesions such as O^6-methylguanine. Lesions causing significant distortion in the DNA will instead block the progress of the replication fork. If the polymerase halts at the site of a lesion, replication might be restarted upstream by the same or a reconstituted DNA polymerase, leaving the lesion in a single-strand gap. In principle, this might occur more often on the lagging than the leading strand (94, 95) (Figure 2).

Much of the work to date on the effects of DNA lesions on replication has focused on UV-irradiated cells. A transient halt or inhibition of DNA replication can be readily observed in irradiated bacterial cells (131). The inhibition is amplified if the cells to be irradiated lack the uvrA gene function and thus cannot make use of DNA excision repair. A similar phenomenon is seen in mammalian cells (16, 72). Sedimentation of chromosomal DNA following UV irradiation provided some early evidence that gaps appeared in the DNA at the sites of UV lesions (16, 72, 73, 131, 132). This in turn suggested that DNA synthesis continued upstream of the lesion following a short lag, and models for the repair of the resulting gaps were constructed based on that assumption (174). Later work demonstrated that RecA protein was directly required in some capacity to restart replication after UV irradiation (58, 179). A need for recombination functions to get significant

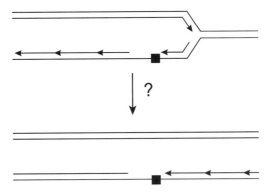

Figure 2   A potential path for bypass of DNA lesions on the lagging strand template of the replication fork. If the replication complex stays intact, it might be possible to abandon the synthesis of DNA at the lesion and revert to the synthesis of newly primed Okazaki fragments.

replication in the irradiated cells was also evident in the original experiments of Howard-Flanders and colleagues (131). This suggests that replication might proceed only after recombinational DNA repair was completed. It is well established that a variety of lesions, and particularly cyclobutane pyrimidine dimers, effectively halt DNA replication in vivo and in vitro (7, 9, 56, 70, 103, 123, 133, 155). Some replicational bypass of pyrimidine dimers in vitro by bacterial DNA polymerase III has been observed (77), but it is unclear if this would occur in a fully coupled replication fork or in vivo.

How and when replication restart can occur is still controversial, in spite of much research. Additional work is needed to determine how often (if ever) and under what circumstances replication can continue downstream of a lesion without being preceded by recombinational DNA repair. The pathway for recombinational repair of DNA gaps proposed by Howard-Flanders and coworkers required a nuclease activity to create the DNA ends needed for productive recombinational exchanges (174). The required nuclease has never been identified, although a candidate activity has been detected in bacterial extracts (19).

THE REPLICATION FORK TRAIN WRECK

Certain of my colleagues are fond of using photos of old train wrecks to accompany their descriptions of what occurs when a replication fork encounters DNA damage in one of the template strands. Unfortunately, these photos provide too apt a summary of the current understanding of the structure of forks following these encounters. There are multiple questions here, including the DNA structures present and the status of the replication enzymes.

There are only three instances in the current literature in which a stalled replication fork has been carefully characterized (Figure 3). Cordeiro-Stone and colleagues examined the fate of SV40 replication forks when they encountered pyrimidine dimers on the leading strand in vitro (25). Most of the stalled and deproteinized forks examined by electron microscopy exhibited the structure shown in Figure 3A. The leading strand synthesis had been halted at the site of the lesion, but the lagging strand had continued, perhaps uncoupled. The resulting fork typically had a leading strand gap on the order of 1000–2000 bases in length. This particular structure has been used as a starting point in a number of proposed models for fork repair, and is used in Figure 1A. The general structure of this stalled fork is compatible with stalled eukaryotic replication forks characterized by more indirect methods. Several additional studies of replication in cell-free extracts of mammalian cells produced encounters with a pyrimidine dimer or other bulky lesion, leading to the uncoupling of leading and lagging strand DNA synthesis and selective replication of the lagging strand (154, 157, 158). Another type of fork structure was found by Sogo and colleagues, who examined the structure of forks stalled at natural replication fork barriers near the rRNA gene clusters in vivo in yeast (45). In this case, the stalled fork also featured a lagging strand that had been completed ahead

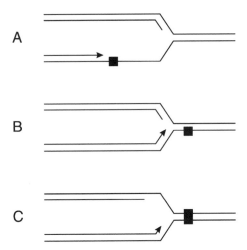

Figure 3 Characterized structures of stalled replication forks. *A*. The major structure found after an SV40 replication fork encounters a cyclobutane pyrimidine dimer in the leading strand (25). The single-strand gaps found at the forks were typically on the order of 1000 nucleotides or more. *B*. Structure of replication forks stalled at the rRNA replication fork barrier in vivo in yeast (45). Here, the newly synthesized lagging strand typically extends just a few nucleotides beyond the leading strand. *C*. Structure of a bacterial replication fork stalled at a replication termination site embedded in a plasmid substrate in vitro (49). The nascent leading strand extended 50–70 nucleotides beyond the lagging strand in these structures.

of the leading strand progress, but in this case the difference was only a few base pairs (Figure 3*B*). In the final instance, Hill & Marians examined the structure of forks stalled in vitro at a terminator site (ter), embedded in a plasmid template and bound with the terminator protein Tus (49). Here, the leading strand was ahead of the lagging strand, leaving a gap of 50–70 nucleotides in the lagging strand branch (Figure 3*C*).

If this handful of studies is any indication, the structures present at stalled forks are likely to be varied. The particular structures present are also likely to depend upon what sort of lesion or other barrier is encountered and the template strand (leading or lagging) it is found on. Since the stalled fork structure represents the starting point for any repair process, a better understanding of the diversity of structures that occur is a prerequisite to a full appreciation of the diversity in repair mechanisms.

In addition to the DNA structures present, it is also necessary to determine the disposition of replication enzymes after an encounter with an unreplicable barrier. It has not been determined if the DNA polymerase complexes disassemble, and which barriers trigger what degree of disassembly. A transition presumably occurs here where the replication complex gives way to recombination enzymes. The mechanistic details of this transition will depend on what initially happens to the replication complex.

It is conceivable that certain proteins facilitate the replication to recombination transition. The RecF and RecR proteins might be good candidates for such transition mediator functions. In vitro, these proteins can bind to dsDNA as a complex and can block the extension of assembling RecA filaments (171, 172). This has suggested a role in modulating RecA filament formation. However, random binding to dsDNA in vivo would be unlikely to permit useful interaction with RecA filaments, and localization to stalled replication forks would require interaction with other proteins, perhaps the replication complex itself (172). Both proteins are co-transcribed with DNA polymerase subunits (13, 41, 119), and additional roles in recombinational DNA repair may remain to be elucidated.

RECOMBINATION PATHWAYS

The pathways outlined in Figure 1 are oversimplified, and a number of additional pathways or pathway variations have been proposed. Many of the variants begin with fork regression (process 1 in Figure 1A) and differ primarily in the starting point for regression and/or fate of the resulting "chicken foot."

Cells with mutations in certain replication functions are often dependent on recombination enzymes for survival and exhibit a hyperrec phenotype. This is one of the general observations linking recombination to replication fork repair (98). The replication defects that make bacterial growth recombination dependent are generally those that result in more frequent stalling of the replication fork.

Enzymes with defects in the accessory replicative helicase Rep make the cells dependent on a functional RecBCD enzyme, but not on RecA protein (140). A similar phenomenon is seen in cells with a mutation in the holD gene, which encodes the ψ subunit in the clamp-loading complex of DNA polymerase III (40). The properties of these mutant cells have given rise to a proposed pathway for fork repair in the absence of RecA protein. The tail produced by fork regression is degraded by RecBCD to generate a structure suitable for replication restart (Figure 4).

The chicken foot is a classical Holliday intermediate that can be processed by the RuvABC proteins. RuvA and RuvB form a complex that binds to a Holliday intermediate and promotes branch migration (173). The RuvC protein is a Holliday intermediate resolvase (173). The chicken foot intermediate can thus be cleaved to produce a double-strand break as shown in Figure 5. In *rep recBC* mutant cells, double-strand breaks accumulate (40, 130, 140). The appearance of these breaks depends on the presence of the RuvABC proteins, providing some of the evidence that the stalling of a replication fork can give rise to a Holliday intermediate (chicken foot) (40, 130, 140). This provides a pathway for the generation of double-strand breaks that is distinct from the double-strand break generation that occurs when replication forks encounter template strand breaks as demonstrated in vivo

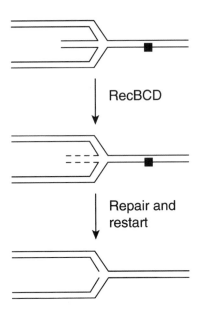

Figure 4 Pathway for the nucleolytic processing of a chicken foot structure by the RecBCD enzyme. The nuclease activity of RecBCD is proposed to eliminate the short arm of the structure, generating a structure that can be used to restart replication. If this pathway is to be successful, the lesion would have to be repaired by some process like excision repair while the fork was regressed and prior to replication restart.

by Kuzminov (69a) (Figure 1*B*). Regardless how they are generated, the repair of double-strand breaks requires the action of the RecBCD enzyme and the RecA protein. How replication fork repair events are distributed between pathways that do or do not involve double-strand breaks (Figure 1) is not known, and probably varies depending on factors such as growth conditions and the associated spectrum of spontaneous DNA damage.

While the list of potential repair pathways may already seem complicated, it is likely to grow. A determination of which repair pathways are most important may be difficult to achieve. A focus on one or a few pathways in individual research reports often reflects the particulars of the investigation. In bacterial cells growing aerobically, most of the replication forks undergo repair, with the potential for use of a wide range of repair pathways. Most in vivo studies directed at the elucidation of these repair paths make use of some strategy to amplify the signal by increasing the frequency of fork stalling. This may involve the use of replication mutants, or DNA-damaging treatments such as UV irradiation, or the use of growth conditions that lead to an increased number of replication barriers involving bound proteins. If the number of fork stalling events in the cell increases substantially, the cells may become completely dependent on the particular path that is used to bypass whatever barriers are presented to the replication forks, and the enzymes needed for

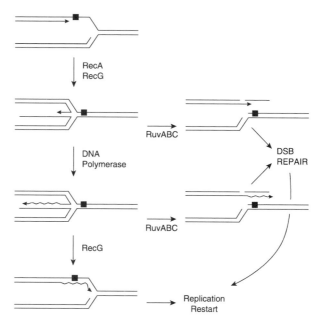

Figure 5   Some alternative paths for the processing of stalled replication forks. The chicken foot is a Holliday structure that can be processed and cleaved by the RuvABC enzymes. Cleavage at any stage funnels the process into the double-strand break repair pathway.

that path. Thus, individual in vivo research efforts have elucidated the molecular course of certain repair pathways, but generally shed little light on the question of which pathways are most important to the cell under normal growth conditions.

FIVE KEY REACTIONS IN THE RECOMBINATIONAL DNA REPAIR OF REPLICATION FORKS

Replication Fork Regression

Fork regression [also called fork reversal (98)] involves the re-pairing of the recently replicated template strands at a stalled fork, such that the fork is moved backwards (Figure 6). The newly synthesized strands are displaced and then paired as the fork is regressed to create a 4-branched structure, the chicken foot. Evidence for replication fork regression has been available in the literature for nearly three decades. The first proposal of a fork regression process as a means of repairing a replication fork was made by Higgins and coworkers (48). Many other reports of regressed fork structures were made in the years following the work of Higgins (53, 111, 156, 170, 181), although the significance of the structures was not always

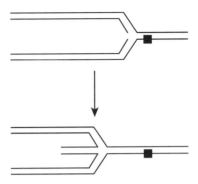

Figure 6   Replication fork regression to generate a chicken foot.

apparent. In bacteria, it is now clear that there are at least three paths to fork regression, and each of them may play a role in the repair of stalled replication forks. The simplest path is nonenzymatic. The positive supercoiling that builds up ahead of a replication fork will lead to spontaneous fork regression to produce a chicken foot structure (121). Such spontaneous fork regression during sample preparation may be the origin of some of the early reports of regressed forks. The spontaneous process may also play a direct role in repair. If a stalled fork undergoes regression, even limited to a few dozens or hundreds of base pairs, it can in principle provide loading sites for enzymes that bind to and process Holliday intermediates.

A second pathway for fork regression is provided by the activity of the RecG helicase. This enzyme is an ATPase and promotes branch migration at DNA junctions with 3 and 4 branches (79, 91, 92, 175). When a stalled fork has few or no single-strand gaps, this may be the principal enzymatic path to fork regression. Thus, RecG may be particularly important when replication forks are stalled by replication pause sites [such as the rDNA fork barriers in yeast (Figure 3*B*)] or bound proteins. Lloyd and coworkers have provided evidence that replication can be halted at bound RNA polymerases themselves stalled at DNA lesions, and that RecG protein plays a key role in the resulting fork repair process (92). More recent work establishes a role for RecG in the regression of forks stalled in vitro (93).

The final pathway for fork regression is provided by the RecA protein. If the stalled fork contains a substantial single-strand gap, such as a fork with the structure in Figure 3*A*, RecA protein can form a filament in the gap and promote fork regression. This reaction has been demonstrated in vitro and is quite efficient (127). RecA is a DNA-dependent ATPase, and RecA-promoted regression exhibits a requirement for ATP hydrolysis. When coupled to ATP hydrolysis, RecA protein–promoted DNA strand exchange reactions are unidirectional, with branch movement proceeding 5′ to 3′ relative to the bound single-stranded DNA in the gap

(55). If a RecA filament loads onto a leading strand gap such as that in Figure 3*A*, this direction of branch movement will result in fork regression. There is also in vivo evidence for a role of RecA in fork regression in some situations. The primary helicase at replication forks is the DnaB protein, and cells with a *dnaB*ts mutation exhibit high levels of fork stalling. The generation of recombination intermediates that can be processed by RuvABC (presumably chicken feet generated by fork regression) in these strains depends upon RecA function (141). However, it is as yet unclear whether the in vitro experiments demonstrating RecA-mediated fork regression (127) provide an appropriate model for the repair of stalled forks in *dnaB*ts cells. The DnaB helicase moves along the lagging strand template at the replication fork, and a defective helicase may cause fork stalling that leaves gaps opposite the lagging strand rather than the leading strand template (141). A RecA filament formed in a lagging strand gap would be expected to promote a DNA strand exchange in the direction opposite to that required for fork regression.

The in vitro efforts to date do not address the considerable topological complexities of the fork regression process. In the cell, fork regression is likely to require the action of helicases, topoisomerases, and other enzymes beyond those currently being investigated.

Holliday Intermediate Cleavage

The four-branched Holliday structure is a signature recombination intermediate, associated with a wide range of recombination processes. Enzymes that specifically cleave such structures to generate viable recombinant products have been found in many types of organisms (4). In *E. coli*, there are at least two such enzymes, RuvC and RusA (10, 18). The reaction generally involves symmetric cleavage of two opposing strands of the intermediate so that two branches remain with each product (Figure 7) (10). Thus, a given Holliday intermediate can be resolved in two ways, depending on which pair of strands is cleaved.

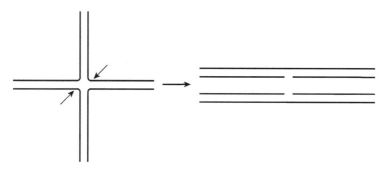

Figure 7   Cleavage of Holliday intermediates by resolvases like RuvC enzyme.

The resolution of Holliday intermediates can have interesting genomic consequences for the cell. If a Holliday intermediate is formed behind the replication fork during fork repair, cleavage of that intermediate can lead to the formation of chromosomal dimers (Figure 8). Such dimers occur in about 15% of all bacterial cells under normal growth conditions (152). Conversion of chromosomal dimers to monomeric chromosomes is the task of a specialized site-specific recombinase called the XerCD enzyme, acting at specific sites near the DNA replication termini (144). If XerCD is mutagenically inactivated, the unprocessed chromosomal dimers block cell division. Interestingly, the RuvABC proteins appear to bias the resolution of Holliday intermediates such that the chromosomal dimers formed in

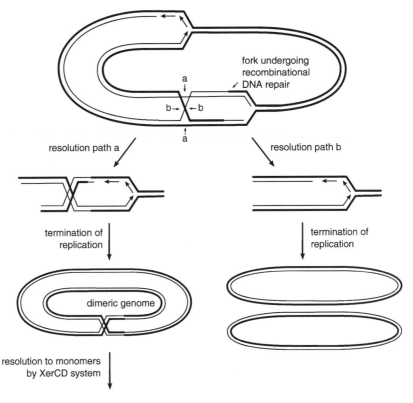

Figure 8   The generation of chromosome dimers by recombination and replication. In principle, cleavage of a Holliday intermediate placed behind the replication fork could occur in two ways, labeled *a* and *b*. Cleavage by pathway *a* generates a crossover, and the subsequent replication leads to the formation of a contiguous chromosomal dimer that must be converted to monomers by a specialized site-specific recombination system (XerCD).

15% of the cells represent less than half of the Holliday intermediate resolution events (8, 99). The mechanism by which that bias is preserved has been established (33, 99, 163).

Holliday Junction Branch Migration

This process is closely related to fork regression, but it deserves some additional discussion. The migration of DNA branches has been studied for several decades and has been a recognized part of recombination processes for a similar period of time. Spontaneous branch migration in isolate DNA proceeds in a random walk (112, 113, 159). With respect to replication fork repair, there is a need instead for directed branch migration. The fork regression process is one branch migration that can only proceed in one direction, and it can create a Holliday intermediate. Migration of the Holliday intermediate in the direction opposite to regression (or fork reversal; terminology gets tricky here as one begins to talk about the reverse of a reversal) can reset the replication fork (Figure 9). If the chicken foot structure is not cleaved, then this branch migration offers a conservative path to restoration of the replication fork.

There are at least three enzymes that might catalyze this process, RuvAB, RecG, and PriA. The RuvAB complex binds to a Holliday intermediate and promotes branch migration, with the direction of branch movement depending on the way RuvAB loads onto the DNA (117). In some reaction contexts, this loading appears to preferentially occur so as to reverse a DNA strand exchange reaction promoted by RecA protein (54). The RecG helicase has also been characterized as a branch migration activity (79, 175, 176), with a tendency to reverse RecA-mediated DNA strand exchange reactions in some reaction systems (175). RecA protein and either RuvAB or RecG could thus participate in a regression/reversal cycle, with intermediate replication and repair steps as needed to restore the fork. In some instances, RecG may act alone to promote both regression and its reversal (92). The PriA protein also has a helicase activity and may play a role in branch movement. There is some in vivo and in vitro evidence that PriA and RecG proteins may process branched structures in different ways (1, 90). One of the more substantial and interesting problems in fork repair will come in determining how these proteins interact and how their order and mode of action is determined to

Figure 9   Resetting the replication fork by reverse branch migration in the chicken foot. The short branch of the chicken foot is shown with thick lines so that the fate of these strands can be seen in the product of the branch migration reaction.

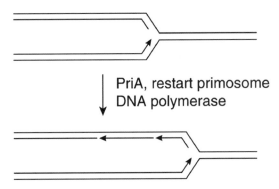

Figure 10 The origin-independent replication restart process makes use of branched recombination products.

bring about the end result of fork restoration. Additional proteins may well play a role in these molecular decisions.

Replication Restart Pathways

Every repaired replication fork must go through a restart process to resume replication (Figure 10). This process is by necessity origin-independent and distinct from the replication initiation that occurs at *oriC*. Origin-independent replication restart requires the activity of a seven-protein complex. This was originally known as the ϕX174-dependent primosome, reflecting the particulars of its discovery (147, 177) [see (31) for review], but has more recently been renamed the restart primosome (137). The assembly of this complex on certain types of recombination intermediates has been reconstituted in vitro (76, 87, 137). The entire process also must involve DNA polymerase III, and the cryptic DNA polymerase II may also play a role (124).

The restart primosome consists of seven proteins, with historical names given in parentheses: PriA (protein n′, factor Y), PriB (protein n), PriC (protein n″), DnaT (protein i), DnaC, the DnaB helicase, and the DnaG primase. The PriA protein plays a key role in the assembly of the complex on a recombination intermediate (76). As might be expected if nearly all cells must go through fork repair and replication restart, cells lacking priA function are nearly inviable (86, 138). An unstated assumption that all fork restoration must go through a PriA-mediated restart process has served to link estimates of the frequency of cellular restart to the phenotype of priA mutant cells (29, 30, 32). Recently, evidence has appeared for the existence of priA-independent paths for replication restart (136). Thus, the frequency of replication fork stalling and recombinational DNA repair may be higher than the estimates drawn from the *priA* phenotype might suggest, perhaps involving several fork events per cell per generation.

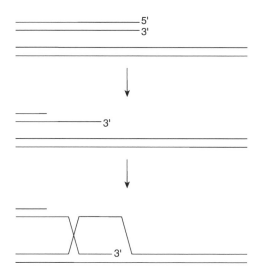

Figure 11 The DNA strand invasion reaction. In *E. coli*, the processing of the double-strand end in the first step is promoted by the RecBCD enzyme. The second step (invasion) is promoted by RecA protein.

3′ End Invasion

In some respects, this is the best-understood process within the broader topic of replication fork repair. The invasion of 3′ ends into a homologous duplex DNA (Figure 11) has been a key component of published recombination models since the early constructs of Meselson & Weigle (96). The process is central to the late stages of bacteriophage T4 replication (82, 108, 109), and it has been studied in some detail in vitro (66, 105). This same reaction was among the first reactions investigated in the early study of purified RecA protein (89, 145). When RecA protein is bound to linear single-stranded DNAs, there is a pronounced tendency for any subsequent invasion of a complementary duplex DNA to take place on the 3′ end of the single strand (37, 63). This reaction may be a necessary step in the repair path any time a replication fork branch is severed by a fork encounter with a strand break (Figure 1*B*) or a chicken foot repair intermediate is cleaved.

The inherent bias for 3′ ends in RecA-mediated strand invasion reactions comes about as a function of the directional bias observed in RecA filament formation and disassembly reactions. Once one or a few RecA monomers form a nucleus on a ssDNA molecule, filaments are extended 5′ to 3′ from that point (125, 142). Little addition of RecA monomers on the "wrong" 5′-proximal end can be detected (5). RecA filaments also disassemble in the 5′ to 3′ direction, such that monomers are added to and subtracted from filaments at opposite ends (5, 142). Both the assembly and disassembly processes help to assure that the 3′ ends of linear single-stranded DNAs are more likely to be bound by RecA protein, and thus to react in strand invasion, than 5′ ends. The presence of SSB protein also helps ensure that RecA protein will not be present at the 5′ ends (5, 63, 142). Interestingly, the 3′ end bias is largely eliminated if the RecO and RecR proteins are included in the reaction mixture (J. Bork & M. Cox, unpublished observations). Possible functions for a RecAOR-mediated 5′ end invasion have not been explored.

The repair of double-strand breaks in *E. coli* generally represents a collaboration between the RecBCD enzyme and the RecA protein (11). *E. coli* mutant cells deficient in the activities necessary to mature Okazaki fragments on the lagging strand during replication (e.g., DNA ligase and DNA polymerase I) become completely dependent on the function of both the RecBCD enzyme and RecA protein (75, 101, 107, 169, 178) as a result of the many strand breaks present. The pathway leading to strand invasion has been reconstituted in vitro by Kowalczykowski and coworkers (3, 20). The RecBCD complex has both nuclease and helicase activities. RecBCD binds at a free duplex DNA end, simultaneously degrading and unwinding it. When it encounters an 8 nucleotide sequence called Chi, its DNA degradation properties are altered so that degradation of the 3′-ending strand is greatly reduced (2). The result is a processed end with a long single-strand extension with a 3′ end. The RecBCD protein plays a direct role in loading the RecA protein onto this single-strand extension, leading to strand invasion (3, 20).

Some interesting problems arise when an attempt is made to couple strand invasion to a restart of DNA synthesis. After RecA protein promotes DNA strand invasion, the continued presence of a RecA filament blocks access by DNA polymerases to the introduced 3′ end (K. Marians, personal communication). Extension of this 3′ end by any polymerase requires the prior disassembly of the RecA filament that created it. A complete reconstitution of DNA strand invasion with establishment of a complete replication fork should introduce many more interesting complexities.

THE ROLES OF MANY RECOMBINATION FUNCTIONS ARE IMPERFECTLY UNDERSTOOD

The recombinational DNA repair of stalled replication forks involves many proteins, and in many cases the activities of the proteins and their precise role in repair are not yet understood. I focus here only on a selection of proteins not mentioned in the preceding discussion. In the following section, gene map locations reflect the current published *E. coli* genome sequence (14).

The RadA protein (also called Sms; M_r 49,477, 99.7 min on the *E. coli* chromosome map) is encoded by a gene that exhibits homology to both the RecA protein and the lon protease (128). Mutations in the gene for RadA confer sensitivity to UV- and X-ray irradiation when the cells are grown in rich media (35, 151). Little is known about the activity of this protein, except that it appears to have a DNA-dependent ATPase activity (S. Lovett, personal communication). Defects in the RadC protein (M_r 25,573, 82.1 min) also confer sensitivity to UV and X rays in rich media. RadC function is required to observe elevated levels of tandem repeat recombination induced by replication fork defects (139). There is thus a clear link between RadC and replication fork repair. Nothing is known about the molecular role of RadC protein.

The recX gene (also oraA) is located just downstream of the gene encoding the RecA protein and encodes a polypeptide with M_r 19,425, 60.8 min. The function of the RecX protein is not known, and the gene has been little studied. In other bacteria, a very closely related gene is found just downstream of the recA gene, and it is generally co-transcribed with recA (114, 115, 164). In *Streptomyces lividans*, the RecX protein function is required in cells in which RecA protein is overproduced, suggesting it moderates some toxic effect of RecA overexpression (164). The protein thus may interact with and perhaps regulate the RecA protein and/or the filaments it forms on DNA.

Somewhat more is known about some of the other auxiliary functions of recombinational DNA repair (65, 128). The RecF [M_r 40,518; 83.6 min (50)], RecO [M_r 27,393, 58.2 min (62)], and RecR [M_r 21,965, 10.6 min (83, 84)] proteins historically have helped define a recombination pathway distinct from that defined by the RecBCD enzyme (21, 22). The phenotypes of all three of these genes are quite similar, and there has been a general association between the proteins and the repair of DNA gaps (22). Additional in vivo data have suggested a direct interaction between these proteins and RecA protein. In particular, certain mutants of RecA protein (RecA441 and RecA803) suppress deficiencies in all three of the RecFOR functions (168). in vitro, the three proteins do not form a heterotrimer but instead form two pairwise RecOR and RecFR complexes. The activities of these proteins studied to date in vitro involve the modulation of RecA filament assembly and disassembly. The RecOR complex promotes RecA filament nucleation on ssDNA substrates that are already bound with the single-strand DNA binding protein (SSB) (160). The same complex also inhibits a net end-dependent disassembly of RecA filaments (142). The RecFR complex binds randomly to dsDNA and blocks the extension of growing RecA filaments (172). Together, these activities could suffice to constrain RecA filaments to DNA gaps where repair was to take place (172). However, RecOR is not needed for RecA function in fork regression either in vivo (141) or in vitro (127). As already noted, an interaction of RecFR with some component of the replication complex may be needed for this complex to function at the replication fork. Clearly, more remains to be discovered about the functions of these proteins.

Three other proteins have been loosely associated with the RecF pathway of recombination, although they may function in other contexts as well (22, 65). These are the RecJ [M_r 63,396, 65.4 min (80)], RecN [M_r 61,377, 59.3 min (78)], and RecQ proteins [M_r 68,441, 86.3 min (110)]. These three appear to be functionally quite distinct.

The RecJ protein is a 5′ to 3′ single-strand DNA exonuclease (81). The protein could play a variety of roles in the processing of displaced 5′ ends during recombination processes. RecJ can also play a role in DNA mismatch repair (24, 165).

Although initially associated with the RecF pathway, the properties of cells deficient in RecN function suggested a role in double-strand break repair (120). There are no reports of in vitro activities of RecN protein. This is primarily because the protein has proven to be insoluble and quite hard to work with. Partially purified preparations of RecN protein exhibit a weak ATPase activity and some nuclease activity, (T. Arenson & M. Cox, unpublished results), but the purity of the preparations is insufficient to ascribe these activities to RecN with a high degree of confidence. The RecN protein exhibits a sequence relationship to the eukaryotic Smc proteins (134). This family of proteins generally has various roles in chromosome maintenance (23), such as chromosome condensation. These proteins have a predicted tertiary structure consisting of two domains connected by a hinge. Two parts of the ATPase active site are distributed between the two domains, such that the domains presumably must come together for ATP hydrolysis to occur (23, 134). Interestingly, the RecN protein is one of the most prominent proteins induced during the SOS response, expressed at levels similar to those of RecA protein (39, 68, 129, 146, 148, 167). Clearly, new approaches are needed to further investigate this interesting protein.

The RecQ protein is a DNA helicase (161, 162). Defects in certain mammalian homologues of RecQ helicase are associated with the human Werner's (180) and Bloom's (38) syndromes. In eukaryotes, the RecQ family of helicases appear to play roles in genome maintenance (17, 42, 143). The molecular role of RecQ helicase in the repair of replication forks has not yet been elucidated. Kowalczykowski and colleagues have demonstrated that RecQ will unwind covalently closed DNA circles in vitro and stimulate the strand passage activity of topoisomerase III. The end result is the catenation of the DNA circles (47). Given the topological constraints of many of the processes outlined in Figure 1 when they occur on the *E. coli* chromosome, such activities could play roles in several different processes in the repair of replication forks.

A major challenge in the way of a complete understanding of replication fork repair is the integration of these many protein functions, and probably more that remain undiscovered, into the various repair pathways.

THE POTENTIAL FOR ORIGIN-SPECIFIC RECOMBINATION-INDUCED REPLICATION DURING THE SOS RESPONSE

Many of the original clues that helped to elucidate the role of recombination in the repair of replication forks came from the study of replication restart during the SOS response (31), particularly the work of Kogoma (59). SOS is a complex physiological response, and the replication restart that occurs in it has both nonmutagenic and mutagenic components induced in separate stages (34, 68). The nonmutagenic responses that can be associated with the nonmutagenic restoration of replication forks are more important (58, 179).

Kogoma and coworkers found that DNA replication could be initiated without protein synthesis under conditions that induced the SOS response (60, 61), a phenomenon that was labeled stable DNA replication (SDR). Later, SDR associated with the SOS response was designated induced stable DNA replication or iSDR. This aspect of replication restart was also shown to be recombination dependent (6). The study of these phenomena provided much information about the proteins involved in the nonmutagenic repair of replication forks (31, 59), but some interesting mysteries remain. In particular, some of the many distinct recombination-dependent replication processes studied by Kogoma and coworkers were initiated largely at specific origins (59). These workers suggested that specific nucleases might be induced during the SOS response that would introduce site-specific double-strand breaks in the chromosome to initiate recombination. Such nucleases would be of considerable interest, and the origin-dependence of stable DNA replication during SOS deserves additional investigation.

ACKNOWLEDGMENTS

The author acknowledges with pleasure the many conversations with colleagues that helped shape this review. In particular, the author thanks Susan Lovett, Ken Marians, Benedicte Michel, David Sherratt, and Andrei Kuzminov for communication of results prior to publication and Ken Marians and Andrei Kuzminov for helpful discussions and comments. Work in the author's laboratory is supported by grants from the National Institutes of Health (GM32335 and GM52725).

Visit the Annual Reviews home page at www.AnnualReviews.org

LITERATURE CITED

1. Al Deib A, Mahdi AA, Lloyd RG. 1996. Modulation of recombination and DNA repair by the RecG and PriA helicases of *Escherichia coli* K-12. *J. Bacteriol.* 178:6782–89

2. Anderson DG, Kowalczykowski SC. 1997. The recombination hot spot chi is a regulatory element that switches the polarity of DNA degradation by the RecBCD enzyme. *Genes Dev.* 11:571–81

3. Anderson DG, Kowalczykowski SC. 1997. The translocating RecBCD enzyme stimulates recombination by directing RecA protein onto ssDNA in a chi-regulated manner. *Cell* 90:77–86

4. Aravind L, Makarova KS, Koonin EV. 2000. Survey and summary: Holliday junction resolvases and related nucleases: identification of new families, phyletic distribution and evolutionary trajectories. *Nucleic Acids Res.* 28:3417–32

5. Arenson TA, Tsodikov OV, Cox MM. 1999. Quantitative analysis of the kinetics of end-dependent disassembly of RecA filaments from ssDNA. *J. Mol. Biol.* 288:391–401

6. Asai T, Bates DB, Kogoma T. 1994. DNA replication triggered by double-stranded breaks in *E. coli*: dependence on homologous recombination functions. *Cell* 78:1051–61

7. Banerjee SK, Christensen RB, Lawrence CW, LeClerc JE. 1988. Frequency and spectrum of mutations produced by a

single cis-syn thymine-thymine cyclobutane dimer in a single-stranded vector. *Proc. Natl. Acad. Sci. USA* 85:8141–45

8. Barre FX, Soballe B, Michel B, Aroyo M, Robertson M, Sherratt DJ. 2001. Circles: the replication-recombination-chromosome segregation connection. *Proc. Natl. Acad. Sci. USA* 98:8189–95

9. Belguise-Valladier P, Maki H, Sekiguchi M, Fuchs RP. 1994. Effect of single DNA lesions on in vitro replication with DNA polymerase III holoenzyme. Comparison with other polymerases. *J. Mol. Biol.* 236:151–64

10. Bennett RJ, West SC. 1995. RuvC protein resolves Holliday junctions via cleavage of the continuous (noncrossover) strands. *Proc. Natl. Acad. Sci. USA* 92:5635–39

11. Bernstein H, Byerly HC, Hopf FA, Michod RE. 1985. Genetic damage, mutation, and the evolution of sex. *Science* 229:1277–81

12. Bernstein H, Hopf FA, Michod RE. 1988. Is meiotic recombination an adaptation for repairing DNA, producing genetic variation, or both? In *The Evolution of Sex: An Examination of Current Ideas*, ed. RE Michod, BR Levin, pp. 139–60. Sunderland, MA: Sinauer

13. Blanar MA, Sandler SJ, Armengod ME, Ream LW, Clark AJ. 1984. Molecular analysis of the recF gene of *Escherichia coli. Proc. Natl. Acad. Sci. USA* 81:4622–26

14. Blattner FR, Plunkett GR, Bloch CA, Perna NT, Burland V, et al. 1997. The complete genome sequence of *Escherichia coli* K-12. *Science* 277:1453–74

15. Bosco G, Haber JE. 1998. Chromosome break-induced DNA replication leads to nonreciprocal translocations and telomere capture. *Genetics* 150:1037–47

16. Buhl SN, Stillman RM, Setlow RB, Regan JD. 1972. DNA chain elongation and joining in normal human and xeroderma pigmentosum cells after ultraviolet irradiation. *Biophys. J.* 12:1183–91

17. Chakraverty RK, Hickson ID. 1999. Defending genome integrity during DNA replication: a proposed role for RecQ family helicases. *BioEssays* 21:286–94

18. Chan SN, Vincent SD, Lloyd RG. 1998. Recognition and manipulation of branched DNA by the RusA Holliday junction resolvase of *Escherichia coli. Nucleic Acids Res.* 26:1560–66

19. Chiu SK, Low KB, Yuan A, Radding CM. 1997. Resolution of an early RecA-recombination intermediate by a junction-specific endonuclease. *Proc. Natl. Acad. Sci. USA* 94:6079–83

20. Churchill JJ, Anderson DG, Kowalczykowski SC. 1999. The RecBC enzyme loads RecA protein onto ssDNA asymmetrically and independently of chi, resulting in constitutive recombination activation. *Genes Dev.* 13:901–11

21. Clark AJ. 1974. Progress toward a metabolic interpretation of genetic recombination of *Escherichia coli* and bacteriophage lambda. *Genetics* 78:259–71

22. Clark AJ, Sandler SJ. 1994. Homologous genetic recombination: the pieces begin to fall into place. *Crit. Rev. Microbiol.* 20:125–42

23. Cobbe N, Heck MM. 2000. Review: SMCs in the world of chromosome biology—from prokaryotes to higher eukaryotes. *J. Struct. Biol.* 129:123–43

24. Cooper DL, Lahue RS, Modrich P. 1993. Methyl-directed mismatch repair is bidirectional. *J. Biol. Chem.* 268:11823–29

25. Cordeiro-Stone M, Makhov AM, Zaritskaya LS, Griffith JD. 1999. Analysis of DNA replication forks encountering a pyrimidine dimer in the template to the leading strand. *J. Mol. Biol.* 289:1207–18

26. Cox MM. 1993. Relating biochemistry to biology: how the recombinational repair function of the RecA system is manifested in its molecular properties. *BioEssays* 15:617–23

27. Deleted in proof

28. Cox MM. 1997. Recombinational crossroads—eukaryotic enzymes and the limits

of bacterial precedents. *Proc. Natl. Acad. Sci. USA* 94:11764–66

29. Cox MM. 1998. A broadening view of recombinational DNA repair in bacteria. *Genes Cells* 3:65–78

30. Cox MM. 1999. Recombinational DNA repair in bacteria and the RecA protein. *Prog. Nucleic Acids Res. Mol. Biol.* 63:310–66

31. Cox MM. 2001. Historical overview: searching for replication help in all the rec places. *Proc. Natl. Acad. Sci. USA* 98:8173–80

32. Cox MM, Goodman MF, Kreuzer KN, Sherratt DJ, Sandler SJ, Marians KJ. 2000. The importance of repairing stalled replication forks. *Nature* 404:37–41

33. Cromie GA, Leach DRF. 2000. Control of crossing over. *Mol. Cell* 6:815–26

34. Defais M, Devoret R. 2000. *SOS Response: Encyclopedia of Life Sciences* (online). Hampshire, UK: Macmillan

35. Diver WP, Sargentini NJ, Smith KC. 1982. A mutation (radA100) in *Escherichia coli* that selectively sensitizes cells grown in rich medium to x- or u.v.-radiation, or methyl methanesulphonate. *Int. J. Rad. Biol. Rel. Stud. Phys. Chem. Med.* 42:339–46

36. Dougherty EC. 1955. Comparative evolution and the origin of sexuality. *Syst. Zool.* 4:145–69, 90

37. Dutreix M, Rao BJ, Radding CM. 1991. The effects on strand exchange of 5' versus 3' ends of single-stranded DNA in RecA nucleoprotein filaments. *J. Mol. Biol.* 219:645–54

38. Ellis NA, Groden J, Te T-Z, Straughen J, Lennon DJ, et al. 1995. The Bloom's syndrome gene product is homologous to RecQ helicases. *Cell* 83:655–66

39. Finch PW, Chambers P, Emmerson PT. 1985. Identification of the *Escherichia coli* recN gene product as a major SOS protein. *J. Bacteriol.* 164:653–58

40. Flores MJ, Bierne H, Ehrlich SD, Michel B. 2001. Impairment of lagging strand synthesis triggers the formation of a RuvABC substrate at replication forks. *EMBO J.* 20:619–29

41. Flower AM, McHenry CS. 1991. Transcriptional organization of the *Escherichia coli* dnaX gene. *J. Mol. Biol.* 220:649–58

42. Frei C, Gasser SM. 2000. RecQ-like helicases: the DNA replication checkpoint connection. *J. Cell Sci.* 113:2641–46

43. Friedberg EC, Walker GC, Siede W. 1995. *DNA Repair and Mutagenesis.* Washington, DC: ASM Press. 698 pp.

44. Galitski T, Roth JR. 1997. Pathways for homologous recombination between chromosomal direct repeats in *Salmonella typhimurium. Genetics* 146:751–67

45. Gruber M, Wellinger RE, Sogo JM. 2000. Architecture of the replication fork stalled at the 3' end of yeast ribosomal genes. *Mol. Cell. Biol.* 20:5777–87

46. Haber JE. 1999. DNA recombination: the replication connection. *Trends Biochem. Sci.* 24:271–75

47. Harmon FG, DiGate RJ, Kowalczykowski SC. 1999. RecQ helicase and topoisomerase III comprise a novel DNA strand passage function: a conserved mechanism for control of DNA recombination. *Mol. Cell* 3:611–20

48. Higgins NP, Kato K, Strauss B. 1976. A model for replication repair in mammalian cells. *J. Mol. Biol.* 101:417–25

49. Hill TM, Marians KJ. 1990. *Escherichia coli* Tus protein acts to arrest the progression of DNA replication forks in vitro. *Proc. Natl. Acad. Sci. USA* 87:2481–85

50. Horii Z, Clark AJ. 1973. Genetic analysis of the RecF pathway to genetic recombination in *Escherichia coli* K12: isolation and characterization of mutants. *J. Mol. Biol.* 80:327–44

51. Hyrien O. 2000. Mechanisms and consequences of replication fork arrest. *Biochimie* 82:5–17

52. Imlay JA, Linn S. 1986. Bimodal pattern of killing of DNA-repair-defective or anoxicallygrown *Escherichia coli* by hydrogen peroxide. *J. Bacteriol.* 166:519–27

53. Inman RB. 1984. Methodology for the study of the effect of drugs on development and DNA replication in *Drosophila melanogaster* embryonic tissue. *Biochim. Biophys. Acta* 783:205–15

54. Iype LE, Inman RB, Cox MM. 1995. Blocked RecA protein-mediated DNA strand exchange reactions are reversed by the RuvA and RuvB proteins. *J. Biol. Chem.* 270:19473–80

55. Jain SK, Cox MM, Inman RB. 1994. On the role of ATP hydrolysis in RecA protein-mediated DNA strand exchange III. Unidirectional branch migration and extensive hybrid DNA formation. *J. Biol. Chem.* 269:20653–61

56. Kaufmann WK, Cleaver JE. 1981. Mechanisms of inhibition of DNA replication by ultraviolet light in normal human and xeroderma pigmentosum fibroblasts. *J. Mol. Biol.* 149:171–87

57. Keyer K, Gort AS, Imlay JA. 1995. Superoxide and the production of oxidative DNA damage. *J. Bacteriol.* 177:6782–90

58. Khidhir MA, Casaregola S, Holland IB. 1985. Mechanism of transient inhibition of DNA synthesis in ultraviolet-irradiated *E. coli*: Inhibition is independent of *recA* whilst recovery requires RecA protein itself and an additional, inducible SOS function. *Mol. Gen. Genet.* 199:133–40

59. Kogoma T. 1997. Stable DNA replication: interplay between DNA replication, homologous recombination, and transcription. *Microbiol. Mol. Biol. Rev.* 61:212–38

60. Kogoma T, Lark KG. 1970. DNA replication in *Escherichia coli*: replication in absence of protein synthesis after replication inhibition. *J. Mol. Biol.* 52:143–64

61. Kogoma T, Lark KG. 1975. Characterization of the replication of *Escherichia coli* DNA in the absence of protein synthesis: stable DNA replication. *J. Mol. Biol.* 94:243–56

62. Kolodner R, Fishel RA, Howard M. 1985. Genetic recombination of bacterial plasmid DNA: effect of RecF pathway mutations on plasmid recombination in *Escherichia coli. J. Bacteriol.* 163:1060–66

63. Konforti BB, Davis RW. 1990. The preference for a 3′ homologous end is intrinsic to RecA-promoted strand exchange. *J. Biol. Chem.* 265:6916–20

64. Kowalczykowski SC. 2000. Initiation of genetic recombination and recombination-dependent replication. *Trends Biochem. Sci.* 25:156–65

65. Kowalczykowski SC, Dixon DA, Eggleston AK, Lauder SD, Rehrauer WM. 1994. Biochemistry of homologous recombination in *Escherichia coli. Microbiol. Rev.* 58:401–65

66. Kreuzer KN. 2000. Recombination-dependent DNA replication in phage T4. *Trends Biochem. Sci.* 25:165–73

67. Kuzminov A. 1996. *Recombinational Repair of DNA Damage.* Georgetown, TX: RG Landes. 210 pp.

68. Kuzminov A. 1999. Recombinational repair of DNA damage in *Escherichia coli* and bacteriophage lambda. *Microbiol. Mol. Biol. Rev.* 63:751–813

69. Kuzminov A. 2001. DNA replication meets genetic exchange: generation of chromosomal damage and its repair by homologous recombination. *Proc. Natl. Acad. Sci. USA* 98:8461–68

69a. Kuzminov A. 2001. Single-strand interruptions in replicating chromosomes cause double-strand breaks. *Proc. Natl. Acad. Sci. USA* 98:8241–46

70. Lawrence CW, Borden A, Banerjee SK, LeClerc JE. 1990. Mutation frequency and spectrum resulting from a single abasic site in a single-stranded vector. *Nucleic Acids Res.* 18:2153–57

71. Le S, Moore JK, Haber JE, Greider CW. 1999. RAD50 and RAD51 define two pathways that collaborate to maintain telomeres in the absence of telomerase. *Genetics* 152:143–52

72. Lehmann AR. 1972. Postreplication repair of DNA in ultraviolet-irradiated mammalian cells. *J. Mol. Biol.* 66:319–37

73. Lin PF, Howard-Flanders P. 1976. Genetic exchanges caused by ultraviolet photoproducts in phage lambda DNA molecules: the role of DNA replication. *Mol. Gen. Genet.* 146:107–15

74. Lindahl T. 1993. Instability and decay of the primary structure of DNA. *Nature* 362:709–15

75. Linn S, Imlay JA. 1987. Toxicity, mutagenesis and stress responses induced in *Escherichia coli* by hydrogen peroxide. *J. Cell Sci. Suppl.* 6:289–301

76. Liu J, Marians KJ. 1999. PriA-directed assembly of a primosome on D loop DNA. *J. Biol. Chem.* 274:25033–41

77. Livneh Z. 1986. Replication of UV-irradiated single-stranded DNA by DNA polymerase III holoenzyme of *Escherichia coli*: evidence for bypass of pyrimidine photodimers. *Proc. Natl. Acad. Sci. USA* 83:4599–603

78. Lloyd RG, Picksley SM, Prescott C. 1983. Inducible expression of a gene specific to the RecF pathway for recombination in *Escherichia coli* K12. *Mol. Gen. Genet.* 190:162–67

79. Lloyd RG, Sharples GJ. 1993. Dissociation of synthetic Holliday junctions by *E. coli* RecG protein. *EMBO J.* 12:17–22

80. Lovett ST, Clark AJ. 1984. Genetic analysis of the recJ gene of *Escherichia coli* K-12. *J. Bacteriol.* 157:190–96

81. Lovett ST, Kolodner RD. 1989. Identification and purification of a single-stranded-DNA-specific exonuclease encoded by the recJ gene of *Escherichia coli*. *Proc. Natl. Acad. Sci. USA* 86:2627–31

82. Luder A, Mosig G. 1982. Two alternative mechanisms for initiation of DNA replication forks in bacteriophage T4: priming by RNA polymerase and by recombination. *Proc. Natl. Acad. Sci. USA* 79:1101–5

83. Mahdi AA, Lloyd RG. 1989. Identification of the recR locus of *Escherichia coli* K-12 and analysis of its role in recombination and DNA repair. *Mol. Gen. Genet.* 216:503–10

84. Mahdi AA, Lloyd RG. 1989. The recR locus of *Escherichia coli* K-12: molecular cloning, DNA sequencing and identification of the gene product. *Nucleic Acids Res.* 17:6781–94

85. Malkova A, Ivanov EL, Haber JE. 1996. Double-strand break repair in the absence of RAD51 in yeast: a possible role for break-induced DNA replication. *Proc. Natl. Acad. Sci. USA* 93:7131–36

86. Marians KJ. 1999. PriA: at the crossroads of DNA replication and recombination. *Prog. Nucleic Acids Res. Mol. Biol.* 63:39–67

87. Marians KJ. 2000. PriA-directed replication fork restart in *Escherichia coli*. *Trends Biochem. Sci.* 25:185–89

88. Maynard Smith J. 1978. *The Evolution of Sex.* Cambridge, UK: Cambridge Univ. Press

89. McEntee K, Weinstock GM, Lehman IR. 1980. RecA protein-catalyzed strand assimilation: stimulation by *Escherichia coli* single-stranded DNA-binding protein. *Proc. Natl. Acad. Sci. USA* 77:857–61

90. McGlynn P, Al DA, Liu J, Marians KJ, Lloyd RG. 1997. The DNA replication protein PriA and the recombination protein RecG bind D-loops. *J. Mol. Biol.* 270:212–21

91. McGlynn P, Lloyd RG. 1999. RecG helicase activity at three- and four-strand DNA structures. *Nucleic Acid Res.* 27:3049–56

92. McGlynn P, Lloyd RG. 2000. Modulation of RNA polymerase by (p)ppGpp reveals a RecG-dependent mechanism for replication fork progression. *Cell* 101:35–45

93. McGlynn P, Lloyd RG, Marians KJ. 2001. Formation of Holliday junctions by regression of nascent DNA in intermediates containing stalled replication forks: RecG stimulates regression even when the DNA is negatively supercoiled. *Proc. Natl. Acad. Sci. USA* 98:8235–40

94. Meneghini R. 1976. Gaps in DNA synthesized by ultraviolet light-irradiated WI38

human cells. *Biochim. Biophys. Acta* 425:419–27

95. Meneghini R, Hanawalt P. 1976. T4-endonuclease V-sensitive sites in DNA from ultraviolet-irradiated human cells. *Biochim. Biophys. Acta* 425:428–37

96. Meselson M, Weigle JJ. 1961. Chromosome breakage accompanying genetic recombination in bacteriophage. *Proc. Natl. Acad. Sci. USA* 47:857–68

97. Michel B. 2000. Replication fork arrest and DNA recombination. *Trends Biochem. Sci.* 25:173–78

98. Michel B, Flores M-J, Viguera E, Grompone G, Seigneur M, Bidnenko V. 2001. Rescue of arrested replication forks by homologous recombination. *Proc. Natl. Acad. Sci. USA* 98:8181–88

99. Michel B, Recchia GD, Penel-Colin M, Ehrlich SD, Sherratt DJ. 2000. Resolution of Holliday junctions by RuvABC prevents dimer formation in rep mutants and UV-irradiated cells. *Mol. Microbiol.* 37:180–91

100. Michod RE. 1995. *Eros and Evolution: A Natural Philosophy of Sex.* Menlo Park, CA: Addison Wesley

101. Miguel AG, Tyrrell RM. 1986. Repair of near-ultraviolet (365 nm)-induced strand breaks in *Escherichia coli* DNA. The role of the polA and recA gene products. *Biophys. J.* 49:485–91

102. Modrich P. 1989. Methyl-directed DNA mismatch correction. *J. Biol. Chem.* 264:6597–600

103. Moore PD, Bose KK, Rabkin SD, Strauss BS. 1981. Sites of termination of in vitro DNA synthesis on ultraviolet- and N-acetylaminofluorene-treated phi X174 templates by prokaryotic and eukaryotic DNA polymerases. *Proc. Natl. Acad. Sci. USA* 78:110–14

104. Morimyo M. 1982. Anaerobic incubation enhances the colony formation of a polA recB strain of *Escherichia coli* K-12. *J. Bacteriol.* 152:208–14

105. Morrical S, Hempstead K, Morrical M, Chou KM, Ando R, Grigorieva O. 1994. Mechanisms of assembly of the enzyme-ssDNA complexes required for recombination-dependent DNA synthesis and repair in bacteriophage T4. *Ann. NY Acad. Sci.* 726:349–50

106. Morrow DM, Connelly C, Hieter P. 1997. "Break copy" duplication: a model for chromosome fragment formation in *Saccharomyces cerevisiae. Genetics* 147:371–82

107. Morse LS, Pauling C. 1975. Induction of error-prone repair as a consequence of DNA ligase deficiency in *Escherichia coli. Proc. Natl. Acad. Sci. USA* 72:4645–49

108. Mosig G. 1987. The essential role of recombination in phage T4 growth. *Annu. Rev. Genet.* 21:347–71

109. Mosig G. 1998. Recombination and recombination-dependent DNA replication in bacteriophage T4. *Annu. Rev. Genet.* 32:379–413

110. Nakayama H, Nakayama K, Nakayama R, Irino N, Nakayama Y, Hanawalt PC. 1984. Isolation and genetic characterization of a thymineless death-resistant mutant of *Escherichia coli* K12: identification of a new mutation (recQ1) that blocks the RecF recombination pathway. *Mol. Gen. Genet.* 195:474–80

111. Nilsen T, Baglioni C. 1979. Unusual base-pairing of newly synthesized DNA in HeLa cells. *J. Mol. Biol.* 133:319–38

112. Panyutin IG, Hsieh P. 1993. Formation of a single base mismatch impedes spontaneous DNA branch migration. *J. Mol. Biol.* 230:413–24

113. Panyutin IG, Hsieh P. 1994. The kinetics of spontaneous DNA branch migration. *Proc. Natl. Acad. Sci. USA* 91:2021–25

114. Papavinasasundaram KG, Colston MJ, Davis EO. 1998. Construction and complementation of a recA deletion mutant of *Mycobacterium smegmatis* reveals that the intein in *Mycobacterium tuberculosis*

recA does not affect RecA function. *Mol. Microbiol.* 30:525–34

115. Papavinasasundaram KG, Movahedzadeh F, Keer JT, Stoker NG, Colston MJ, Davis EO. 1997. Mycobacterial recA is cotranscribed with a potential regulatory gene called recX. *Mol. Microbiol.* 24:141–53

116. Park EM, Shigenaga MK, Degan P, Korn TS, Kitzler JW, et al. 1992. Assay of excised oxidative DNA lesions: isolation of 8-oxoguanine and its nucleoside derivatives from biological fluids with a monoclonal antibody column. *Proc. Natl. Acad. Sci. USA* 89:3375–79

117. Parsons CA, Stasiak A, Bennett RJ, West SC. 1995. Structure of a multisubunit complex that promotes DNA branch migration. *Nature* 374:375–78

118. Pegg AE, Byers TL. 1992. Repair of DNA containing O6–alkylguanine. *FASEB J.* 6:2302–10

119. Perez-Roge RL, Garcia-Sogo M, Navarro-Avino J, Lopez-Acedo C, Macian F, Armengod ME. 1991. Positive and negative regulatory elements in the dnaA-dnaN-recF operon of *Escherichia coli*. *Biochimie* 73:329–34

120. Picksley SM, Attfield PV, Lloyd RG. 1984. Repair of DNA double-strand breaks in *Escherichia coli* K12 requires a functional recN product. *Mol. Gen. Genet.* 195:267–74

121. Postow L, Ullsperger C, Keller RW, Bustamante C, Vologodskii AV, Cozzarelli NR. 2001. Positive torsional strain causes the formation of a four-way junction at replication forks. *J. Biol. Chem.* 267:2790–96

122. Potter H, Dressler D. 1988. Genetic recombination: molecular biology, biochemistry, and evolution. In *The Recombination of Genetic Material*, ed. KB Low, pp. 217–82. San Diego, CA: Academic

123. Rajagopalan M, Lu C, Woodgate R, O'Donnell M, Goodman MF, Echols H. 1992. Activity of the purified mutagenesis proteins UmuC, UmuD′, and RecA in replicative bypass of an abasic DNA lesion by DNA polymerase III. *Proc. Natl. Acad. Sci. USA* 89:10777–81

124. Rangarajan S, Woodgate R, Goodman MF. 1999. A phenotype for enigmatic DNA polymerase II: a pivotal role for pol II in replication restart in UV-irradiated *Escherichia coli*. *Proc. Natl. Acad. Sci. USA* 96:9224–29

125. Register JC III, Griffith J. 1985. The direction of RecA protein assembly onto single strand DNA is the same as the direction of strand assimilation during strand exchange. *J. Biol. Chem.* 260:12308–12

126. Richter C, Park JW, Ames BN. 1988. Normal oxidative damage to mitochondrial and nuclear DNA is extensive. *Proc. Natl. Acad. Sci. USA* 85:6465–67

127. Robu ME, Inman RB, Cox MM. 2001. RecA protein promotes the regression of stalled replication forks in vitro. *Proc. Natl. Acad. Sci. USA* 98:8211–18

128. Roca AI, Cox MM. 1997. RecA protein: structure, function, and role in recombinational DNA repair. *Prog. Nucleic Acid Res. Mol. Biol.* 56:129–223

129. Rostas K, Morton SJ, Picksley SM, Lloyd RG. 1987. Nucleotide sequence and LexA regulation of the *Escherichia coli* recN gene. *Nucleic Acids Res.* 15:5041–49

130. Rothstein R, Michel B, Gangloff S. 2000. Replication fork pausing and recombination or "gimme a break". *Genes Dev.* 14:1–10

131. Rupp WD, Howard-Flanders P. 1968. Discontinuities in the DNA synthesized in an excision-defective strain of *Escherichia coli* following ultraviolet irradiation. *J. Mol. Biol.* 31:291–304

132. Rupp WD, Wilde CEd, Reno DL, Howard-Flanders P. 1971. Exchanges between DNA strands in ultraviolet-irradiated *Escherichia coli*. *J. Mol. Biol.* 61:25–44

133. Sagher D, Strauss B. 1983. Insertion of nucleotides opposite apurinic/apyrimidinic sites in deoxyribonucleic acid during in vitro synthesis: uniqueness

of adenine nucleotides. *Biochemistry* 22:4518–26

134. Saitoh N, Goldberg IG, Wood ER, Earnshaw WC. 1994. ScII: an abundant chromosome scaffold protein is a member of a family of putative ATPases with an unusual predicted tertiary structure. *J. Cell Biol.* 127:303–18

135. Sancar A, Sancar GB. 1988. DNA repair enzymes. *Annu. Rev. Biochem.* 57:29–67

136. Sandler SJ. 2000. Multiple genetic pathways for restarting DNA replication forks in *Escherichia coli* K-12. *Genetics* 155:487–97

137. Sandler SJ, Marians KJ. 2000. Role of PriA in replication fork reactivation in *Escherichia coli. J. Bacteriol.* 182:9–13

138. Sandler SJ, Samra HS, Clark AJ. 1996. Differential suppression of priA2::kan phenotypes in *Escherichia coli* K-12 by mutations in priA, lexA, and dnaC. *Genetics* 143:5–13

139. Saveson CJ, Lovett ST. 1999. Tandem repeat recombination induced by replication fork defects in *Escherichia coli* requires a novel factor, RadC. *Genetics* 152:5–13

140. Seigneur M, Bidnenko V, Ehrlich SD, Michel B. 1998. RuvAB acts at arrested replication forks. *Cell* 95:419–30

141. Seigneur M, Ehrlich SD, Michel B. 2000. RuvABC-dependent double-strand breaks in dnaBts mutants require RecA. *Mol. Microbiol.* 38:565–74

142. Shan Q, Bork JM, Webb BL, Inman RB, Cox MM. 1997. RecA protein filaments: end-dependent dissociation from ssDNA and stabilization by RecO and RecR proteins. *J. Mol. Biol.* 265:519–40

143. Shen JC, Loeb LA. 2000. The Werner syndrome gene: the molecular basis of RecQ helicase-deficiency diseases. *Trends Genet.* 16:213–20

144. Sherratt DJ, Arciszewska LK, Blakely G, Colloms S, Grant K, et al. 1995. Site-specific recombination and circular chromosome segregation. *Philos. Trans. R. Soc. London Ser. B: Biol. Sci.* 347:37–42

145. Shibata T, Das Gupta C, Cunningham RP, Radding CM. 1980. Homologous pairing in genetic recombination: formation of D loops by combined action of RecA protein and a helix-destabilizing protein. *Proc. Natl. Acad. Sci. USA* 77:2606–10

146. Shinagawa H. 1996. SOS response as an adaptive response to DNA damage in prokaryotes. *Exs* 77:221–35

147. Shlomai J, Polder L, Arai K, Kornberg A. 1981. Replication of phi X174 dna with purified enzymes. I. Conversion of viral DNA to a supercoiled, biologically active duplex. *J. Biol. Chem.* 256:5233–38

148. Smith BT, Walker GC. 1998. Mutagenesis and more: umuDC and the *Escherichia coli* SOS response. *Genetics* 148:1599–610

149. Smith GR. 1991. Conjugational recombination in *E. coli*: myths and mechanisms. *Cell* 64:19–27

150. Smith GR. 1998. DNA double-strand break repair and recombination in *E. coli*. In *DNA Damage and Repair, Vol 1: DNA Repair in Prokaryotes and Lower Eukaryotes*, ed. JA Nickoloff, MF Hoekstra, pp. 135–62. Totowa, NJ: Humana

151. Song Y, Sargentini NJ. 1996. *Escherichia coli* DNA repair genes radA and sms are the same gene. *J. Bacteriol.* 178:5045–48

152. Steiner WW, Kuempel PL. 1998. Sister chromatid exchange frequencies in *Escherichia coli* analyzed by recombination at the dif resolvase site. *J. Bacteriol.* 180:6269–75

153. Storz G, Imlay JA. 1999. Oxidative stress. *Curr. Opin. Microbiol.* 2:188–94

154. Svoboda DL, Vos JMH. 1995. Differential replication of a single, UV-induced lesion in the leading or lagging strand by a human cell extract—fork uncoupling or gap formation. *Proc. Natl. Acad. Sci. USA* 92:11975–79

155. Tang M, Bruck I, Eritja R, Turner J, Frank EG, et al. 1998. Biochemical basis of SOS-induced mutagenesis in *Escherichia coli*: reconstitution of in vitro lesion bypass dependent on the UmuD'2C

mutagenic complex and RecA protein. *Proc. Natl. Acad. Sci. USA* 95:9755–60

156. Tatsumi K, Strauss B. 1978. Production of DNA bifilarly substituted with bromodeoxyuridine in the first round of synthesis: branch migration during isolation of cellular DNA. *Nucleic Acids Res.* 5:331–47

157. Thomas DC, Veaute X, Fuchs RP, Kunkel TA. 1995. Frequency and fidelity of translesion synthesis of site-specific N-2-acetylaminofluorene adducts during DNA replication in a human cell extract. *J. Biol. Chem.* 270:21226–33

158. Thomas DC, Veaute X, Kunkel TA, Fuchs RP. 1994. Mutagenic replication in human cell extracts of DNA containing site-specific N-2-acetylaminofluorene adducts. *Proc. Natl. Acad. Sci. USA* 91:7752–56

159. Thompson BJ, Camien MN, Warner RC. 1976. Kinetics of branch migration in double-stranded DNA. *Proc. Natl. Acad. Sci. USA* 73:2299–303

160. Umezu K, Kolodner RD. 1994. Protein interactions in genetic recombination in *Escherichia coli*. Interactions involving RecO and RecR overcome the inhibition of RecA by single-stranded DNA-binding protein. *J. Biol. Chem.* 269:30005–13

161. Umezu K, Nakayama H. 1993. RecQ DNA helicase of *Escherichia coli*. Characterization of the helix-unwinding activity with emphasis on the effect of single-stranded DNA-binding protein. *J. Mol. Biol.* 230:1145–50

162. Umezu K, Nakayama K, Nakayama H. 1990. *Escherichia coli* RecQ protein is a DNA helicase. *Proc. Natl. Acad. Sci. USA* 87:5363–67

163. van Gool AJ, Hajibagheri NM, Stasiak A, West SC. 1999. Assembly of the *Escherichia coli* RuvABC resolvasome directs the orientation of Holliday junction resolution. *Genes Dev.* 13:1861–70

164. Vierling S, Weber T, Wohlleben W, Muth G. 2000. Transcriptional and mutational analyses of the *Streptomyces lividans* recX gene and its interference with

RecA activity. *J. Bacteriol.* 182:4005–11

165. Viswanathan M, Lovett ST. 1998. Single-strand DNA-specific exonucleases in *Escherichia coli*. Roles in repair and mutation avoidance. *Genetics* 149:7–16

166. Voelkel-Meiman K, Roeder GS. 1990. Gene conversion tracts stimulated by HOT1-promoted transcription are long and continuous. *Genetics* 126:851–67

167. Walker GC, Smith BT, Sutton MD. 2000. The SOS response to DNA damage. In *Bacterial Stress Responses*, ed. G Storz, R HenggeAronis, pp. 131–44. Washington, DC: Am. Soc. Microbiol.

168. Wang TC, Chang HY, Hung JL. 1993. Cosuppression of recF, recR and recO mutations by mutant recA alleles in *Escherichia coli* cells. *Mutat. Res.* 294:157–66

169. Wang TC, Smith KC. 1986. Inviability of dam recA and dam recB cells of *Escherichia coli* is correlated with their inability to repair DNA double-strand breaks produced by mismatch repair. *J. Bacteriol.* 165:1023–25

170. Wanka F, Brouns RM, Aelen JM, Eygensteyn A, Eygensteyn J. 1977. The origin of nascent single-stranded DNA extracted from mammalian cells. *Nucleic Acids Res.* 4:2083–97

171. Webb BL, Cox MM, Inman RB. 1995. An interaction between the *Escherichia coli* RecF and RecR proteins dependent on ATP and double-stranded DNA. *J. Biol. Chem.* 270:31397–404

172. Webb BL, Cox MM, Inman RB. 1997. Recombinational DNA repair—the RecF and RecR proteins limit the extension of RecA filaments beyond single-strand DNA gaps. *Cell* 91:347–56

173. West SC. 1997. Processing of recombination intermediates by the RuvABC proteins. *Annu. Rev. Genet.* 31:213–44

174. West SC, Cassuto E, Howard-Flanders P. 1981. Heteroduplex formation by RecA protein: polarity of strand exchanges. *Proc. Natl. Acad. Sci. USA* 78:6149–53

175. Whitby MC, Ryder L, Lloyd RG. 1993. Reverse branch migration of Holliday junctions by RecG protein: a new mechanism for resolution of intermediates in recombination and DNA repair. *Cell* 75:341–50

176. Whitby MC, Vincent SD, Lloyd RG. 1994. Branch migration of Holliday junctions: identification of RecG protein as a junction specific DNA helicase. *EMBO J.* 13:5220–28

177. Wickner S, Hurwitz J. 1975. Association of phiX174 DNA-dependent ATPase activity with an *Escherichia coli* protein, replication factor Y, required for in vitro synthesis of phiX174 DNA. *Proc. Natl. Acad. Sci. USA* 72:3342–46

178. Witkin EM, Roegner MV. 1992. Overproduction of DnaE protein (alpha subunit of DNA polymerase III) restores viability in a conditionally inviable *Escherichia coli* strain deficient in DNA polymerase I. *J. Bacteriol.* 174:4166–68

179. Witkin EM, Roegner MV, Sweasy JB, McCall JO. 1987. Recovery from ultraviolet light-induced inhibition of DNA synthesis requires *umuDC* gene products in *recA718* mutant strains but not in *recA*$^+$ strains of *Escherichia coli*. *Proc. Natl. Acad. Sci. USA* 84:6805–9

180. Yu C-E, Oshima J, Fu Y-H, Wijsman E, Hisama F, et al. 1996. Positional cloning of the Werner's syndrome gene. *Science* 272:258–62

181. Zannis-Hadjopoulos M, Persico M, Martin RG. 1981. The remarkable instability of replication loops provides a general method for the isolation of origins of DNA replication. *Cell* 27:155–63

Annu. Rev. Genet. 2001. 35:83–101

SIR FRANCIS GALTON AND THE BIRTH OF EUGENICS

Nicholas W. Gillham

DCMB Group, Department of Biology, Box 91000, Duke University, Durham, North Carolina 27708-1000; e-mail: gillham@duke.edu

Key Words Galton, eugenics, pedigrees, biometrics, correlation

■ **Abstract** The eugenics movement was initiated by Sir Francis Galton, a Victorian scientist. Galton's career can be divided into two parts. During the first, Galton was engaged in African exploration, travel writing, geography, and meteorology. The second part began after he read the *Origin of Species* by his cousin Charles Darwin. The book convinced Galton that humanity could be improved through selective breeding. During this part of his career he was interested in the factors that determine what he called human "talent and character" and its hereditary basis. Consequently, he delved into anthropometrics and psychology and played a major role in the development of fingerprinting. He also founded the field of biometrics, inventing such familiar statistical procedures as correlation and regression analysis. He constructed his own theory of inheritance in which nature and not nurture played the leading role. He actively began to promote eugenics and soon gained important converts.

CONTENTS

INTRODUCTION

On January 11, 1999, *Time* magazine ran a series entitled "The Future of Medicine" devoted to the effects of the genetic revolution on the human race. In an article entitled "Cursed by Eugenics" (33), Paul Gray wrote that the "rise and fall of the theory known as eugenics is in every respect a cautionary tale. The early

0066-4197/01/1215-0083$14.00

eugenicists were usually well-meaning progressive types. They had imbibed their Darwin and decided that the process of natural selection would improve if it were guided by human intelligence. They did not know they were shaping a rationale for atrocities." With the rapid advances in modern human genetics the specter of eugenics is with us once more, although dressed in somewhat different garb. Insurance companies are interested in data relating to genetic maladies with regard to risk assessment. Although most governments currently limit or deny the use of genetic-test information by insurers, a British government committee recently decided to grant insurers access to the results of the genetic test for Huntington's disease (43). Thus a precedent is now in place for access that could eventually widen to more genetic diseases. Meanwhile, employers worry about susceptibility genes. Sperm banks exist, as does the possibility of elective abortion of genetically defective children and, in some countries, embryos of an unwanted sex.

In view of current concerns about the emergence of a new eugenics, it seems appropriate to present a profile of Sir Francis Galton (1822–1911) (Figure 1) who coined the word and initiated the original eugenics movement. Galton was a man of diverse interests. To those who follow the history of Africa, he is remembered as an explorer and geographer. He was also a well-known travel writer who authored *The Art of Travel* (12), an immensely popular guide book for amateur and professional alike who ventured into the bush over a century ago, that was just reissued by Phoenix Press. Meteorologists recognize Galton as the man who discovered the anticyclone. Those who delve into the history of statistics will find Galton's name associated with regression, correlation, and the founding of biometrics. Psychologists, especially those whose research is on mental imagery, acknowledge their debt to Galton. Forensic experts recognize Galton's central role in giving fingerprinting scientific legitimacy. And last, but certainly not least, Francis Galton's name will always be associated with the dawn of human genetics and, of course, eugenics.

Because of his vast array of interests Galton may appear a dilettante, but this is not the case. His research and published work revolve around two central themes. During the first part of his career Galton was absorbed in exploration, geography, and travel writing. Meteorology was a natural extension of this interest as the explorer is often at the mercy of the vagaries of the weather. The second part of Galton's career opened when he read *On The Origin of Species* (8) and concluded that it might be possible to improve the human race through selective breeding. In order to study the heritability of fitness, what Galton referred to as "talent and character," he made use of pedigrees, twin studies, and anthropometric measurements. Although he believed that favorable physical characteristics signaled above-average mental capacity, he had no way to measure the latter, the IQ test being in the distant future. Hence, he tried to probe personal characteristics through studies of mental imagery, composite photographs of men with similar backgrounds, whether in crime or in military service, and by developing rigorous criteria for fingerprint identification. To analyze the masses of anthropometric data he also

Figure 1   Francis Galton photographed at age 80. (Reprinted from K Pearson. 1930. *The Life, Labours and Letters of Francis Galton*, 3A, with permission from Cambridge University Press.)

succeeded in accumulating, Galton invented new statistical tools including regression and correlation analysis.

Although this catalog of pursuits may seem astonishingly broad to a modern scientist, the diversity of Galton's manifold investigations was not atypical for a Victorian scientist. For example, William Whewell (1794–1866), the Master of Trinity College, Cambridge, while Galton was a student, there had written tracts about philosophy, mathematics, mechanics, theology, and moral philosophy. He published a book on his theory of Gothic design, taught mineralogy, and wrote a text on the classification of minerals. Galton, like Charles Darwin, was independently wealthy and could spend full time on whatever attracted him, but most scientists were not so lucky. T.H. Huxley, who had no fortune to depend on, had to work hard to support himself and his family as a scientist and teacher. Many Victorians studied science in their spare time, depending on other occupations to put bread on the table. For instance, Galton's friend, the mathematician William Spottiswoode, was Printer to the Queen, and Charles Booth, whose 17-volume work *Life and Labour of the People of London* (7) is a classic in early sociology, founded and chaired a successful steamship company with his brother. If one were to choose two words to characterize Francis Galton, they would be optimism and quantification. He was by nature enthusiastic about his work and where it would lead. He also believed in the power of numerical data whether they related to longitudes, latitudes, or altitudes; measurements of arm, leg length, etc., which led him to the concept of correlation; or counting the frequency of fidgits of his friends and colleagues at some august meeting of the British Association of the Advancement of Science or the Royal Institution.

The focus of this review is on the second half of Galton's career, following his reading of the *Origin of Species* when the ideas that led him eventually to the notion of eugenics began to crystallize. However, it is first necessary to make a few points concerning Francis Galton's background and education, which are critical to understanding his overall philosophy.

BACKGROUND AND EDUCATION

Francis Galton was born on February 16, 1822, the youngest of nine children, seven of whom survived infancy. His mother, Violetta Darwin Galton, was the daughter of Erasmus Darwin by his second marriage to Elizabeth Collier Sacheveral-Pole, whereas Charles Darwin, 13 years older than Galton, was a grandson of Erasmus Darwin by his first marriage to Mary (Polly) Howard. His father, Samuel Tertius Galton, was the scion of an old and wealthy Quaker family that converted to Anglicanism, an important point as dissenters were not admissable to Oxford or Cambridge.

The relationship between Charles Darwin and his younger cousin proved crucial at several points in Galton's life, most notably when Galton started to think seriously about improving humanity through selective breeding. Like Darwin,

Galton initially pursued medical studies and seems to have been better adapted to them than Darwin ever was. After apprenticing at the General Hospital in Birmingham, Galton moved to King's College in London to continue his work. Darwin was living nearby at Macaw Cottage on Upper Gower Street, his *Beagle Journal* just having been published. Seemingly, Darwin played a crucial role not only in convincing Galton to apply to Cambridge, his own alma mater, to study mathematics, but also in helping to make the arguments necessary to convince Galton's father that Cambridge would prove a useful hiatus prior to completing his medical studies. In fact, Galton never returned to his medical studies after Cambridge.

Cambridge represented a formative experience in Galton's life. He decided to sit for the tripos (honors examination) in mathematics, and to study for this examination, as well as to prepare in classics, he hired coaches (private tutors). Since Cambridge dons were required to be celibate, many of the brightest young scholars in and around the university became coaches instead. For a time Galton studied with the greatest of the early Victorian mathematics coaches, William Hopkins, who was known as "the Senior Wrangler maker." The Senior Wrangler was the student with the highest marks on the mathematics tripos, followed by the 1st, 2nd, etc., Wranglers. Regular physical exercise, particularly long-distance walking, was seen as a necessary adjunct to hard study as the Cambridge undergraduate equated physical fitness with mental agility. This perceived relationship later took on a particular significance for Galton as he used physical ability as a kind of surrogate indicator of mental capacity. Galton also made many life-long friends at Cambridge. However, while studying for the tripos he had the equivalent of a nervous breakdown (not at all uncommon among students cramming for the tripos), dropped out of the honors program, and graduated with an ordinary (poll for hoi polloi) degree. Galton had a high regard for Cambridge and its students, especially the mathematically gifted, all of his life.

HEREDITARY TALENT AND CHARACTER

Galton was approaching middle age when Darwin published the *Origin of Species*. He "devoured its contents and assimilated them as fast as they were devoured, a fact which may be ascribed to an hereditary bent of mind that both its illustrious author and myself have inherited from our common grandfather, Dr. Erasmus Darwin" (30). Consequently, Galton was encouraged to investigate questions that had long interested him that "clustered round the central topics of Heredity and the possible improvement of the Human Race" (30).

From the outset Galton seems to have been convinced that nature, and not nurture, determined hereditary ability, but how was he to show it? He hit upon a fairly simple device, the pedigree, which would remain an analytical mainstay for the rest of his life. He reasoned that if ability was determined by heredity, a famous man's closest male relatives were the most likely to exhibit exceptional abilities, with this characteristic diluting out with hereditary distance. Women were omitted in his analysis because, from his Victorian viewpoint, notable achievement was principally a male prerogative. Except in literature, this was certainly true in Great Britain and elsewhere at the time as opportunities for female advancement beyond the home were virtually absent. Galton's opening statement of his thesis appeared in 1865 in "Hereditary Talent and Character" (13), a two-part article in *MacMillan's Magazine*, one of a stable of high-quality and popular Victorian magazines. To collect pedigree data he examined works like *The Million of Facts* by Sir Thomas Phillips. From this he culled a select biography of 605 notable persons who lived between the years 1453 and 1853. He was pleased to find there were 102 notable relationships for a frequency of 1 in 6. Although he was strongly prejudiced in nature's favor, Galton acknowledged that nurture might also play a role since the son of a great man "will be placed in a more favourable position for advancement, than if he had been the son of an ordinary person" (14). Nevertheless, this was a problem that never bothered Galton much. As a kind of control, Galton tried to estimate the frequency of men of ability in the population as a whole by rough determination of the number of students educated in Europe during the four preceding centuries. He calculated that only 1 in 3000 of these "randomly" selected persons achieved notability, concluding that "everywhere is the enormous power of hereditary influence forced upon our attention" (15). One of the most remarkable ideas elaborated in "Hereditary Talent and Character," for which no scientific justification was presented, was that the embryos of the next generation sprang forth from the embryos of the preceding generation. This anticipated by almost 20 years August Weismann's experimentally supported theory of the continuity of the germ line.

The *MacMillan's* papers were precursors for Galton's book *Hereditary Genius* (16). There he used the same method of gathering pedigree data on a much grander scale, but he also applied the normal distribution as an evaluative technique for the first time. He had been introduced to "the Gaussian Law of Probable Error" by William Spottiswoode. When Spottiswoode explained the normal distribution to his friend, Galton was delighted by the "far-reaching application of that extraordinarily beautiful law which I fully apprehended" (32). Galton now familiarized himself with the work of Adolph Quetelet, the Astronomer Royal of Belgium, who gained his scientific reputation not so much for astronomy, but as a statistician and population biologist. Quetelet used published data on chest measurements from 5738 Scottish soldiers to calculate the proportions of soldiers in each size class demonstrating that they were normally distributed. To get some idea of whether mental ability was normally distributed, Galton obtained examination marks for admission to the Royal Military College at Sandhurst for 1868. Inspection of the data revealed a clear fit to the normal distribution at the upper tail of the curve and in its center, but for dunces getting low scores there were no numbers as they had either eschewed competition or been "plucked." This seems to be the first time that the bell curve was applied to some indicator of mental ability.

Galton thought highly of the method of pedigree analysis he had first used in "Hereditary Talent and Character." In his autobiography, published in 1908 when he was 86, Galton wrote that "on re-reading these articles . . . considering the novel conditions under which they were composed . . . I am surprised at their justness and comprehensiveness" (31). Karl Pearson, in his enormous four-volume biography of Galton, agreed. It "is really the epitome of the great bulk of Galton's work for the rest of his life; in fact all his labours on heredity, anthropometry, psychology and statistical method seem to take their roots in the ideas of this paper. It might almost have been written as a résumé of his labours after they were completed, rather than as prologue to the yet to be accomplished" (35).

THE SECOND-BEST THEORY OF HEREDITY

Galton was putting the finishing touches on *Hereditary Genius* when he began to leaf through Darwin's new work, *Variation in Animals and Plants under Domestication* (9). Volume 1 was replete with examples of the results of artificial selection yielding new animal breeds and cultivated plants. Galton was familiar with these from the *Origin of Species*, but volume 2 was a different story. The first three chapters dealt with inheritance, the fourth the laws of variation, and the last presented Darwin's "Provisional Hypothesis of Pangenesis." It was the pangenesis chapter that particularly excited Galton. Pangenesis was the next logical step in Darwin's theory of evolution for he needed to explain how the variations arose upon which natural selection acted. Most contemporary theories of heredity were of the "paint pot" or blending type. The difficulty for Darwin was this: If a variant is likened to a few drops of black paint added to a bucket of white paint and stirred, it will vanish when mixed (crossed) into the bucket. Hence, Darwin assumed that the variants upon which natural selection acted must be particulate.

Darwin named his hypothetical particles "gemmules," proposing that they were gathered from all parts of the organism "to constitute the sexual elements, and their development in the next generation forms a new being." To account for reversion, the occasional appearance in a pure-breeding strain of an ancestral character, Darwin assumed the existence of dormant elements that might be expressed in future generations. This intrigued Galton and would become embedded in his own theory of heredity. Darwin imagined two mechanisms by which variations might arise. First, when the reproductive organs were "injuriously affected by changed conditions," gemmules from different parts of the body might fail to aggregate properly so some were in excess whereas others were in deficit, resulting in modification and variation. Darwin's second mechanism assumed that gemmules could be modified "by direct action of changed conditions." This caused the affected part of the body to "throw off modified gemmules, which are transmitted to the offspring." Although Darwin hypothesized that exposure to modified environmental conditions over several generations was required for an alteration to become heritable, this was acquired characteristics pure and simple, and Galton did not like this. Given his penchant for quantification, Galton embraced Darwin's notion that the hereditary factors were particulate and scrambled to add a chapter on pangenesis to *Hereditary Genius*.

Galton was anxious to test his cousin's hypothesis, and he did so with rabbits. He interpreted Darwin's hypothesis as meaning that gemmules might be transmitted in the bloodstream. Galton chose a strain of pure-breeding rabbits called silver-grays as recipients for blood from donors having different characteristics such as color. One or both parents were transfused and then crossed to see whether their progeny had inherited any of the characteristics of the blood donor. The experiments were carried out with considerable rigor, and Darwin must have been quite hopeful of a positive outcome, as judged by his involvement in the project and his extensive correspondence with Galton concerning the results. They did not work, of course, and Galton wrote a paper, seemingly with no input from Darwin, that he published in the *Proceedings of the Royal Society* (17). In his introduction, he concluded in unmistakable terms that the pangenesis hypothesis was wrong. "I have now made experiments of transfusion and cross circulation on a large scale in rabbits, and have arrived at definite results, negativing, in my opinion, beyond all doubt, the truth of the doctrine of Pangenesis." Although he temporized in his discussion by narrowing his conclusion to disproving pangenesis "pure and simple as I have interpreted it," the damage was done. Darwin got annoyed and let his cousin know it, writing that he intended to go public in *Nature*. In his letter to *Nature* (10), Darwin pointedly remarked that he had "not said one word about the blood, or about any fluid proper to any circulating system." After all, pangenesis applied to protozoa and plants too, but later he was more charitable and admitted that Galton's experiments seemed plausible as Darwin "saw not the difficulty of believing in the presence of gemmules in the blood." Afterwards Galton publically apologized in *Nature* (18) for his brash conclusion, saying he had been led astray by certain aspects of Darwin's phrasing in presenting his hypothesis. His letter ended "Vive Pangenesis!"

Meanwhile, Galton was constructing his own theory of heredity (19). He approved of Darwin's distinction between species of gemmules and also recognized two classes of elements, which he called latent and patent. The latent elements were the equivalent of Darwin's latent gemmules and could explain reversion. As diagrammed by Galton, both kinds of elements derived from a common group of structureless elements in the fertilized ovum (Figure 2). A subset, the patent elements, were selected for development into embryonic elements. These subsequently differentiated into adult elements while the residue of latent elements followed a parallel course, first as latent embryonic elements and then as latent adult elements. The two parallel pathways converged when a subset of patent and latent elements were selected to yield the structureless elements of the offspring. A key assumption in Galton's hypothesis was that the patent elements in the adult could be supplemented from the latent pool, but the reverse process, while it might occur in the embryo, did not occur in the adult. By making information transfer unidirectional, Galton was attempting to rule out

Figure 2 The development of patent and latent elements from structureless elements according to Galton's scheme of inheritance in F Galton. 1872. On blood-relationship. *Proc. R. Soc.* 20:394–402 as modified and reprinted from K Pearson. 1924. *The Life, Labours and Letters of Francis Galton*, 2:172, with permission from Cambridge University Press.

acquired characteristics, for if patent elements could be environmentally modified and differentiated back into latent elements, acquired characteristics would be a reality.

In succeeding papers Galton further refined his hypothesis. Shorn of confusing terminology and analogies, Galton's message boiled down to this (see 21). Within the fertilized ovum is the sum total of gemmules (now referred to as germs by Galton) or stirp, which derives from the Latin, stirpes, a root. Galton then stated the four postulates on which his hypothesis was based. First "each of the enormous number of quasi-independent units the body is made up of, has a separate origin, or germ." The modern equivalent would be that each gene specifies a protein. Second, the germs in the stirp are much greater in number and variety than the structural units derived from them. Only a few germs actually developed, and these were sterile. This is like saying the protein specified by a gene lacks the hereditary information required for its own transmission. Also embedded in this postulate is a vague glimmer of the notion of dominance and recessiveness. Galton needed this postulate to account for variation and reversion. Third, the undifferentiated germs propagated themselves in the latent state and contributed to the stirp of the offspring. That is, genes, and not their products, were transmitted from one generation to the next. Fourth, the final structure, organization, and appearance of the adult organism depended "on the mutual affinities and repulsions of separate germs" within the stirp during development. The modern analogy might be to the operation of positive and negative control loops during development.

Galton also recognized that there must be some mechanism for halving the size of the stirp contributed by each parent to the fertilized ovum to prevent its doubling in size at each generation. But, unaware of meiosis, Galton argued that "one half" of the "possible heritage" in the progeny "must have been suppressed." So Galton had enunciated a form of the germ-line theory normally credited to the German biologist, August Weismann. Weismann acknowledged this in a letter to Galton dated February 23, 1889. "It was Mr Herdman of Liverpool who—some years ago—directed my attention to this paper of yours. . . . I regret not to have known it before, as you have exposed in your paper an idea which is in one essential point nearly allied to the main idea contained in my theory of the continuity of the germ plasm" (38).

ANTHROPOMETRICS AND *NATURAL INHERITANCE*

Having constructed a theory of inheritance that suited him and eliminated the possibility of acquired characteristics, Galton was anxious to obtain quantitative anthropometric measurements that he could analyze statistically. He also wanted to obtain these measurements in actual pedigrees so that he could test their heritability. While he was publishing articles and making plans to obtain the data he desired, Galton turned to a model system once again, as he had when he used rabbits to test Darwin's pangenesis hypothesis. However, this time he picked sweet peas on the advice of Darwin and the botanist Joseph Hooker. He cited three reasons. Sweet peas had little tendency to cross-fertilize; they were hardy and prolific; and seed weight did not vary with humidity. His first experimental crop, planted at Kew in the spring of 1874, failed. To avoid this outcome a second time, Galton dispersed his seeds widely the next time to friends and acquaintances all over Great Britain. "Each set (of seeds) contained seven little packets, and in each packet were ten seeds, precisely the same weight" (22). The packets were lettered K, L, M, N, O, P, and Q, with K containing the heaviest seeds, L the next heaviest, and so forth down to packet Q, which had the lightest seeds. At the end of the season the plants were uprooted, tied together, labeled, and sent to Galton.

Having weighed and measured some 490 seeds, Galton summarized his results. What he discovered was something very interesting. Seed sizes were distributed normally among the progeny as they had been among the parents, but he also found that the mean size of progeny seeds obtained, for example, from parental seeds selected for large size, had reverted (or regressed) toward the original mean seen for the entire population. He also discovered that if he plotted the average diameters of the parental seeds on the X axis against the average diameter of the progeny seeds on the Y axis, he could connect the points with a straight line (Figure 3). This was the first regression line ever plotted, and Galton referred to its slope as the "coefficient of reversion." However, on finding that this coefficient was not a hereditary property, but the result of his own statistical manipulation, he changed the name to coefficient of regression. Discovery of regression to the mean was crucially important to Galton's thinking about evolution by natural selection for he could not see how this process could succeed, since regression to the mean would thwart the incremental effects of natural selection.

The International Health Exhibition, which opened on Thursday, May 8, 1884, in the Gardens of the Royal Horticultural Institution in South Kensington, was a major event with hordes of visitors scrutinizing various exhibits and sampling food and drink. Galton seized upon this venue to establish an Anthropometric Laboratory. Visitors to the laboratory were given a card with various entries on it for the measurements to be taken over which a thin transfer paper had been stretched with a piece of carbon paper in between. After passing through the laboratory, the visitor was given the original and the laboratory kept the copy. During a half-hour in the laboratory the visitor was measured and took various tests in succession using devices designed to detect "Keenness of Sight and of Hearing; Colour Sense;

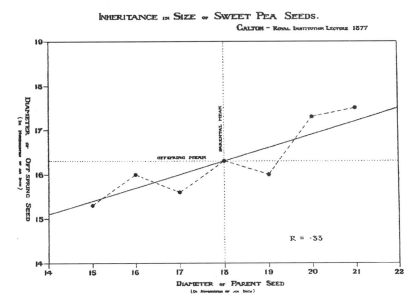

Figure 3   The first regression line. (Reprinted from K Pearson. 1930. *The Life, Labours and Letters of Francis Galton*, 3A:4, with permission from Cambridge University Press.)

Judgment of Eye; Breathing Power; Reaction Time; Strength of Pull and of Squeeze; Force of Blow; Reaction Time; Span of Arms; Height, both standing and sitting; and Weight" (29). By the time the International Health Exhibition closed in 1885, Galton had compiled data on 9337 individuals, each measured in 17 different ways. Subsequently, at Galton's request, the Anthropometric Laboratory was moved to a room in the Science Galleries of the South Kensington Museum. In summarizing the achievements of the Anthropometric Laboratory in 1924, Karl Pearson noted that Galton had succeeded in collecting an immense amount of data, "which only forty years later is being adequately reduced" (36).

As Galton had been assiduous in attempting to collect partial pedigree data, he was able to demonstrate that regression to the mean for a character such as height was operative in people as well as sweet peas. One day, while plotting forearm length against height, he discovered that they were correlated. He extended these observations to other physical parameters such as head width versus head breadth, head length versus height, etc., and in 1888 published his results in a classic paper entitled "Co-relations and their Measurements, chiefly from Anthropometric Data" (24). He also determined the first set of correlation coefficients, using the now familiar symbol *r*. Most of his correlation coefficients were pleasingly high, between 0.7 and 0.9.

Galton's paper on correlation and *Natural Inheritance* (25), a book he published the same year, are probably his two most influential scientific works. As

Galton wrote early in Chapter 1 of *Natural Inheritance*, "I have long been engaged upon certain problems that lie at the base of the science of heredity, and . . . have published technical memoirs concerning them. . . . This volume contains the more important of the results, set forth in an orderly way, with more completeness than has hitherto been possible, together with a large amount of new matter" (26). The book followed a logical progression from heredity, to a description of statistical methods (frequency distributions and normal variation), to Galton's anthropometric data and their statistical analysis.

GALTON'S DISCIPLES

It is surprising that *Natural Inheritance* is a largely forgotten book for in many ways it provides the foundation of biometrics. The book (and Galton's correlation paper) served as inspirations for the great statistician Karl Pearson and his zoologist friend and colleague at University College, London, Raphael Weldon. They were particularly attracted to the statistical and quantitative possibilities opened up in *Natural Inheritance*. Weldon applied the normal distribution to various metrics he derived from the marine shrimps and crabs he studied. He also calculated masses of correlation coefficients. Pearson, a much better mathematician than Galton, began the rigorous development of statistical theory. Like Galton, they were interested in continuously varying characters. In 1897, Galton elaborated a new theory of heredity that concerned itself not with the mechanistic aspects of the problem, but with the contribution of each ancestor to the total heritage of an individual (28). In its simplest form, the Ancestral Law is easy to comprehend. Galton contemplated a continuous series with parents contributing one half (0.5) the heritage of their offspring; the four grandparents one quarter $(0.5)^2$; the eight great-grandparents one eighth $(0.5)^8$, etc. The whole series $(0.5) + (0.5)^2 + (0.5)^3$ . . . sums to one. Pearson found Galton's paper intriguing, dressed it up in fancy mathematics, and sent it to Galton as his New Year's greeting in 1898. Galton was suitably impressed, and the Ancestral Law, supported by Weldon and Pearson, became the great rival of Mendelism at the outset of the twentieth century.

Unlike Pearson and Weldon, the young British biologist William Bateson took a very different message from his reading of *Natural Inheritance*. Like Weldon, Bateson attended Cambridge, but by his own admission "mathematics were my difficulty. Being destined for Cambridge I was specially coached, but failed. Coached once more I passed, having wasted, not one, but several hundred hours on that study" (1). Weldon became Bateson's closest friend at Cambridge, and they both came under the influence of the brilliant morphologist Francis Balfour. Balfour encouraged Bateson to do research on *Balanoglossus*, a marine organism perceived to be allied to the vertebrates that was abundant in Chesapeake Bay. Weldon helped Bateson make contact with W.K. Brooks at Johns Hopkins University. Brooks did most of his research at the Chesapeake Zoological Laboratory, a movable marine station on the shores of the bay. In long evenings spent with Brooks, there was much conversation about discontinuous variation for Brooks had proposed a new theory of heredity, designed to replace pangenesis, that permitted discontinuous evolution by jumps.

Bateson became hooked on discontinuous variation, which meant that he was intellectually primed to accept Mendel's principles upon their rediscovery in 1900. What Bateson discovered in reading *Natural Inheritance* was that Galton also believed that evolution must proceed discontinuously. The problem for Galton was regression to the mean. He did not see how natural selection could proceed in small steps as it would be thwarted by regression to the mean. Hence, he proposed that evolution must occur discontinuously via "sports" (essentially mutations).

THE MENDELIANS TRUMP THE BIOMETRICIANS

Galton's *Natural Inheritance* had succeeded in inspiring two diametrically opposed views of the importance of continuous versus discontinuous variation. Pearson and Weldon, on one hand, believed that continuous variation was the stuff of evolution and heredity, whereas Bateson was convinced that discontinuous variation fulfilled this role. Both Bateson and Galton thought that evolution had to proceed by jumps or saltations. Meanwhile, Bateson, intently searching for examples of discontinuous variation, "ransacked museums, libraries, and private collections; he attended every kind of 'show' mixing freely with gardeners, shepherds and drovers learning all they had to teach him" (2). In 1894, he published his great monograph *Materials for the Study of Variation* (4). Bateson's conundrum was that members of a species were similar to, but distinct from those belonging to another. Although transitional forms between related species were sometimes recognizable, in most cases none was detected so "the forms of living things do . . . most certainly form a discontinuous rather than a continuous series" (5). He also argued that because such "Discontinuity is not in the environment; may it not be in the living thing itself" (6). Often he cited Galton's view that evolution must proceed discontinuously.

Despite its intimidating size and density, 886 examples of discontinuous variation, Galton was so enthusiastic about *Materials* that he published an article in *Mind* (27), entitled "Discontinuity in Evolution," where he summarized his own supportive views on the subject. In contrast, Weldon's review in *Nature* (45) was largely negative. To be sure, Bateson had done a nice job of assembling a great many facts, but his interpretation of them was flawed, reflecting a lack of familiarity with the "history" of his chosen subject. He also disputed Bateson's contention that discontinuity was not environmental, saying that he referred only to the continuity of the physical environment. It seems reasonable to suppose that Bateson was stung by his erstwhile friend's review and that the falling out between the two men can probably be traced to this moment. It was exacerbated further the following year by a public controversy in *Nature* concerning the nature of variation in the plant *Cineraria* [for an excellent blow-by-blow account of the *Cineraria* conflict see (40)]. This was touched off by the distinguished botanist Sir William

Thiselton-Dyer, the Director of Kew Gardens. Dyer's initial letter to *Nature* was followed by a rebuttal from Bateson, with Weldon soon weighing in to defend Dyer's viewpoint. The crux of the matter was whether the colorfully flowered horticultural varieties of *Cineraria* had arisen through small variations (Dyer and Weldon) or through discontinuous alterations (Bateson). Disputatious letters popped up regularly in *Nature* for several months thereafter.

Meanwhile, the Council of the Royal Society had proposed formation of a "Committee for conducting Statistical Inquiries into the Measurable Characteristics of Plants and Animals." The committee was appointed on January 18, 1894, with Galton as chairman, Weldon as secretary, and several distinguished scientists as members. The first report from the committee was read at the February 28, 1895, meeting of the Royal Society and summarized quantitative research that Weldon had been doing on the frontal breadth of the carapace in shore crabs. Bateson detected flaws in the work and wrote four critical letters to Galton, offering to have them printed and sent to the committee members. Galton passed the letters on to Weldon who became incensed. Angry communiques whizzed back and forth, and Galton, caught in the middle, tried to apply balm. He was immensely supportive of Weldon's groundbreaking quantitative studies on variation, but he was also impressed by Bateson's arguments for discontinuous variation. Galton next imposed a solution that pleased no one. He wrote to Weldon (and probably did some arm-twisting), suggesting that Bateson be brought onto the committee. Bateson was duly invited to come on board, but refused. However, Galton soldiered on, presumably with serious foot-dragging by Pearson and Weldon, and convinced Bateson that the committee would change its direction. In late January 1897, Bateson, along with several other nonquantitative types, were brought onto the committee whose name was changed to the Evolution Committee of the Royal Society.

On May 8, 1900, Bateson read a paper before the Royal Horticultural Society in which he discussed Galton's Ancestral Law and its application, but he also presented a revolutionary new concept of heredity that he had learned about from Professor Hugo de Vries in Holland. As Bateson's wife wrote, her husband's "delight and pleasure on his first introduction to Mendel's work were greater than I can describe. . . . He was fortified with renewed faith in the largeness of his research; he found in it new interest, new possibilities, and drew from it new inspiration" (3). Bateson communciated his enthusiasm to Galton on August 8, 1900, suggesting he look up Mendel's paper "in case you miss it. Mendel's work seems to me one of the most remarkable investigations yet made on heredity and it is extraordinary that it should have been forgotten" (11). But the elderly Galton either did not take advantage of this suggestion or failed to appreciate the significance of Mendel's hypothesis.

Relations between Bateson and Weldon were already bad, and now they worsened with Pearson as well. The precipitating factor was a gargantuan memoir submitted by Pearson to the Royal Society on a new theory for which he coined the term "homotyposis" (34). The theory itself is not important in the context of this review, but Bateson's reaction is. Pearson read an abstract describing his theory before the Royal Society on November 15, 1900. Bateson was in the audience and, in Pearson's words in a letter to Galton, Bateson "came to the R.S. at the reading and said there was nothing in the paper" (37). The editors then sent Pearson's manuscript to Bateson to review who rejected it while identifying himself as a referee. Worse, they sent his negative comments to the other reviewers, ensuring they would be biased against the manuscript. Pearson complained about his treatment to Bateson and to the Royal Society, but he decided that a new journal would have to be founded if he was going to get his work published. Thus the distinguished journal *Biometrika* was born as a Bateson avoidance mechanism, with Pearson, Weldon, and the American, Charles Davenport, as editors. Galton was listed on the masthead as a person whose advice the editors would seek.

There is simply not space to recount properly what now took place, but in a few sentences it was this [for an account of the various controversies between the Ancestrians and the Mendelians, especially the key inheritance experiments in mice carried out by Weldon's student Darbishire, see (41)]. Weldon singly, and with his student Darbishire, published a series of papers in *Biometrika* whose purpose was to undermine Mendelism and support Galton's law of Ancestral Inheritance. Bateson wrote papers refuting Weldon's papers. A single example will give the flavor of dispute. Bateson had used the presence or absence of hairs on the leaves of Campion (*Lychnis*) as an example of a simple pair of Mendelian alternatives. Weldon counted the number of hairs per centimeter in leaves at a similar stage of development and observed wide variations in hair numbers. Weldon argued that the arbitrary "adoption of such a category as 'hairy' conceals the facts of variation within the races studied" (46). Hence, Bateson's statements were "utterly inadequate, either as a description of their own experiments, or as a demonstration of Mendel's or of any other laws" (46). So it was the battle over continuous versus discontinuous variation all over again. But Bateson knew that the presence or absence of hairs represented Mendelian alternatives, a fact that Weldon refused to acknowledge. In fact, Weldon was slowly losing the battle as Bateson systematically extended the frontiers of Mendelism. For all intents and purposes, the Ancestrians' last stand was at the British Association for the Advancement of Science meeting that commenced on August 18, 1904. The pertinent action took place in Section D, Zoology, of which Bateson was the president. Bateson, Weldon, and others spoke, but the Ancestrians were ever more on the defensive. Bateson's comments on what Weldon had to say give the flavor of the Mendelians' view of the Ancestrians. "Without doubt ... disputants in the past had maintained the flatness of the earth before applauding crowds, much as Professor Weldon has to-day upheld the view of the Ancestrians. The paths of the heavenly bodies had been harmonised with the theory of the flat earth, as some of the facts of heredity had been with the law of Ancestry; but as the theory of gravitation had brought together great ranges of facts into one coherent whole, so had Mendelian theory begun to co-ordinate the facts of heredity, till then utterly incoherent and apparently contradictory" (44).

Weldon died unexpectedly in his mid-forties in 1906, and Pearson retreated behind the walls of statistics, *Biometrika*, and the Francis Galton Laboratory for the Study of National Eugenics at University College, London, where he was installed as Director. The Mendelians had won the day and the significance of this was that at the First International Congress of Eugenics held in London in 1912, the year after Galton's death, the black and white pedigrees of Mendel, often badly misapplied to presumptive human maladies, reigned supreme.

EUGENICS

Francis Galton's scientific career did not proceed in the linear fashion outlined here, but was instead a jumble of interconnected interests and approaches. Yet to make sense out of it, one must tease out specific sequences of events and leave out others entirely. All of the various threads outlined here ultimately were directed at a single goal: the improvement of humanity through selective breeding. The word eugenics was coined in an 1883 book by Galton entitled *Inquiries into Human Faculty and its Development* (23). There he pulled together the results of his twin studies, his thoughts on anthropometrics and statistics, as well as psychometrics (another term we owe to Galton), psychology, race, and population. His purpose in writing the book was "to touch on various topics more or less connected with that of the cultivation of race, or, as we might call it with 'eugenic' questions." In a footnote Galton defined his new word. Eugenics, he wrote, deals with "questions bearing on what is termed in Greek, *eugenes*, namely, good in stock, hereditarily endowed with noble qualities." Nine years earlier, in an inconsequential little book called *English Men of Science: Their Nature and Nurture* (20), Galton had coined another famous phrase, writing that the "phrase 'nature and nurture' is a convenient jingle of words, for it separates under two distinct heads the innumerable elements of which personality is composed. Nature is all that a man brings with himself into the world; nurture is every influence from without that affects him after birth."

The opening decade of the twentieth century found the educated classes in England primed to welcome eugenics (42). There were two main reasons for this. First, there was an overriding concern about biological degeneration in the country. Statistical data suggested that the birth rate in England was declining, with the process being far more pronounced in the upper and middle classes than among the lower classes. Second, the battle between the Church of England and the Darwinians was mostly over, and evolution by natural selection was now widely accepted. Hence, it seemed logical to many that the quality of the British population as a whole could be improved by reducing the reproductive rate of those perceived as less fit while increasing the propagation of those of good stock. This notion soon became popular not only in England, but in much of Europe and the United States. It required only that men like Galton and Pearson popularize eugenics and justify their conclusions with apparently sound scientific arguments. During this last decade of Galton's life, he gave lectures promoting eugenics and in 1904, made financial arrangements for the establishment of the Eugenics Records Office at University College, London. In 1907, this was converted into the Eugenics Laboratory, with Pearson as its Director. Upon his death, Galton arranged to endow a Galton professorship. The first occupant of this chair was Pearson. He has been followed by a parade of distinguished geneticists over the years. Because of the pejorative nature of the term eugenics today, the distinguished laboratory Galton created is simply known by his name.

In the last year of his life Galton embarked on a novel, never published, called *Kantsaywhere*, which described his view of a eugenic utopia [See Pearson (39) for detailed description of *Kantsaywhere* with much of the remaining text interpolated]. Inhabitants of Kantsaywhere were required to take an examination that vetted them genetically. Failures had inferior genetic material and were segregated in labor colonies where conditions were not onerous, but celibacy was enforced. Those passing the examination with a "second-class certificate" could propagate "with reservations." Those who did well took the honors examination at the Eugenics College of Kantsaywhere and were granted "diplomas for heritable gifts physical and mental." These elite individuals were encouraged to intermarry. In an eerie throwback to Galton's unpublished novel, the 1997 movie *Gattaca* sketches a modern version of a eugenic paradise. Sex and procreation are totally decoupled in Gattaca. In vitro fertilization is used instead, and the genetic profile of each embryo is vetted, with only those lacking genetic defects implanted. The protagonist is a product of the old-fashioned genetic lottery by which we bear children today. He has numerous potential genetic defects and is consigned to sweep floors. The rest of the story is not important for this article, but the analogy is. Kantsaywhere and Gattaca are very similar places. The ethical problems posed in each case are the same. The only difference is 100 years of technological advancement.

ACKNOWLEDGMENTS

I am indebted to Ms Gillian Furlong and her staff, the Manuscripts Room, University College, London, for access to the papers and correspondence of Sir Francis Galton. I also wish to thank Dr. June Rathbone of the Galton Laboratory, University College, London, for allowing me to examine Galton's personal book collection. Dr. Andrew Tatham, Keeper, the Royal Geographical Society, was most helpful in providing manuscripts and correspondence related to Galton's service with the Society. I am also indebted to the Duke University Research Council for providing me with a travel grant that partially offset the expenses of my research trips to London.

NOTE FOR FURTHER READING

As this is not the usual sort of *Annual Review* article, but rather a biographical sketch, I have only cited the pertinent literature in cases where specific papers are mentioned or quotations from papers are given. Those wishing a more comprehensive biographical treatment of Francis Galton are directed to my forthcoming (2001) biography, *A Life of Sir Francis Galton: From African Exploration to the Birth of Eugenics*. New York: Oxford University Press. Existing biographies of Francis Galton have been published by DW Forrest. 1974. *Francis Galton: The Life and Work of a Victorian Genius*. New York: Taplinger, 340 pp. and K Pearson. 1914–30. *The Life, Letters and Labours of Francis Galton*, I (1914), II (1924), IIIA and IIIB (1930). Cambridge: Cambridge University Press. A series of essays covering different aspects of Galton's career is to be found in M Keynes, ed. 1993. *Sir Francis Galton, FRS: The Legacy of His Ideas*. London: Macmillan. 237 pp. Complete bibliographies of all of Galton's papers and books are given in the volumes of Forrest and Keynes.

Visit the Annual Reviews home page at www.AnnualReviews.org

LITERATURE CITED

1. Bateson B. 1928. *William Bateson, F.R.S.: Naturalist*, p. 10. Cambridge: Cambridge Univ. Press
2. Bateson B. 1928. See Ref. 1, p. 28
3. Bateson B. 1928. See Ref. 1, p. 73
4. Bateson W. 1894. *Materials for the Study of Variation Treated with Especial Regard to Discontinuity in the Origin of Species*. London: Macmillan. 598 pp.
5. Bateson W. 1894. See Ref. 2, p. 2
6. Bateson W. 1894. See Ref. 2, p. 17
7. Booth C. 1902–1903. *Life and Labour in London*. London: Macmillan. 17 vols.
8. Darwin CR. 1859. *On the Origin of Species by Means of Natural Selection, or the Preservation of Favored Races in the Struggle for Life*. London: J. Murray. 502 pp.
9. Darwin CR. 1875. *Variation in Animals and Plants under Domestication*. London: J. Murray. 2nd ed. Reprinted in *The Works of Charles Darwin*, 1988, ed. PH Barrett, RB Freeman, p. 321. London: William Pickering
10. Darwin CR. 1871. Pangenesis. *Nature* 3: 502–3
11. Galton Archive, Manuscripts Room, Univ. College, London. List No. 198
12. Galton F. 1855. *The Art of Travel; or Shifts and Contrivances Available in Wild Countries*. London: J. Murray. 360 pp.
13. Galton F. 1865. Hereditary talent and character. *Macmillan's Mag.* 12:157–66, 318–27
14. Galton F. 1865. See Ref. 13, p. 161
15. Galton F. 1865. See Ref. 13, p. 163
16. Galton F. 1869. *Hereditary Genius: An Inquiry into its Laws and Consequences*. London: Macmillan. 390 pp.
17. Galton F. 1871. Experiments in pangenesis, by breeding from rabbits of a pure variety, into whose circulation blood taken from other varieties had previously largely been transfused. *Proc. R. Soc.* 19:393–410
18. Galton F. 1871. Pangenesis. *Nature* 4:5–6
19. Galton F. 1872. On blood-relationship. *Proc. R. Soc.* 20:392–402
20. Galton F. 1874. *English Men of Science: Their Nature and Nurture*. London: Macmillan. 206 pp.
21. Galton F. 1875. A theory of heredity. *Contemp. Rev.* 27:80–95
22. Galton F. 1877. Typical laws of heredity. *Proc. R. Inst.* 8:282–301
23. Galton F. 1883. *Inquiries into Human Faculty and its Development*. London: Macmillan. 261 pp.
24. Galton F. 1888. Co-relations and their measurements, chiefly from anthropometric data. *Proc. R. Soc.* 45:135–45

25. Galton F. 1889. *Natural Inheritance*. London: Macmillan. 259 pp.
26. Galton F. 1889. See Ref. 25, p. 1
27. Galton F. 1894. Discontinuity in evolution. *Mind* (n.s.) 3:362–72
28. Galton F. 1897. The average contribution of each several ancestor to the total heritage of the offspring. *Proc. R. Soc.* 61:401–13
29. Galton F. 1909. *Memories of My Life*, pp. 245–47. London: Methuen. 3rd ed.
30. Galton F. 1909. See Ref. 29, pp. 287–88
31. Galton F. 1909. See Ref. 29, p. 289
32. Galton F. 1909. See Ref. 29, p. 304
33. Gray P. 1999. Cursed by eugenics. *Time* 153 (no. 1):84–85
34. Pearson K. 1901. Mathematical contributions to the theory of evolution.–IX. On the principle of homotyposis and its relation to heredity, to the variability of the individual, and to that of the race. Part I. Homotyposis in the vegetable kingdom. *Philos. Trans. R. Soc. Ser. A* 197:285–379
35. Pearson K. 1924. *The Life, Letters and Labours of Francis Galton*, II: 86–87. Cambridge: Cambridge Univ. Press
36. Pearson K. 1924. See Ref. 35, II:375
37. Pearson K. 1930. See Ref. 35, IIIA:241
38. Pearson K. 1930. See Ref. 35, IIIA:340–41
39. Pearson K. 1930. See Ref. 35, IIIA:413–25
40. Provine WB. 1971. *The Origins of Theoretical Population Genetics*, pp. 45–48. Chicago: Univ. Chicago Press
41. Provine WB. 1971. See Ref. 40, pp. 70–89
42. Soloway RA. 1995. *Demography and Degeneration*. Chapel Hill: Univ. NC Press. 443 pp.
43. *The Economist*. 2000. 357:23, 93–94
44. *The Times*. 1904. Aug. 20
45. Weldon WFR. 1894. The study of animal variation. *Nature* 50:25–26
46. Weldon WFR. 1903. "On the ambiguities of Mendel's characters." *Biometrika* 2:44–55

Annu. Rev. Genet. 2001. 35:103–23

BUILDING A MULTICELLULAR ORGANISM

Dale Kaiser

Departments of Biochemistry and of Developmental Biology, Stanford University School of Medicine, Stanford, California 94305; e-mail: luttman@cmgm.stanford.edu

Key Words cell-cell interaction, signal transduction, development, differentiation, dispersion

■ **Abstract** Multicellular organisms appear to have arisen from unicells numerous times. Multicellular cyanobacteria arose early in the history of life on Earth. Multicellular forms have since arisen independently in each of the kingdoms and several times in some phyla. If the step from unicellular to multicellular life was taken early and frequently, the selective advantage of multicellularity may be large. By comparing the properties of a multicellular organism with those of its putative unicellular ancestor, it may be possible to identify the selective force(s). The independent instances of multicellularity reviewed indicate that advantages in feeding and in dispersion are common. The capacity for signaling between cells accompanies the evolution of multicellularity with cell differentiation.

CONTENTS

INTRODUCTION

When we look at life around us, multicellular organisms—plants, animals, and colonial microorganisms—meet our eyes. It is generally believed that single cells were first to evolve, but the oldest fossils of ancient life, 3500 million years

0066-4197/01/1215-0103$14.00

(Myr) of age, show multicellular cyanobacterial filaments (44). The 2-methyl-bacteriohopanepolyols that characterize the cyanobacteria are abundant in organic rich sediments as old as 2500 Myr (48). The transition to multicellularity is unusual among the major transitions in evolution (33) in that it has occurred numerous times. Novel multicellular forms are still being discovered (37). By a multicellular organism, we understand one in which the activities of the individual cells are coordinated and the cells themselves are either in contact or close enough to interact strongly. According to this definition, a bacterial colony is not a multicellular organism, even though it may show patterned growth, because it apparently lacks overall coordination of function.

The multicellular condition has independently arisen in each of the organic kingdoms, and in some phyla several times in the course of their evolution. For example, the sponges are believed to have arisen from the choanoflagellates separately from all other animals; the seed plants, the fungi, the brown algae, and the red algae all gained their multicellularity in separate events (53). The same is true among microorganisms: After the cyanobacteria, the cellular slime molds, the myxomycetes, and the myxobacteria independently adopted multicellularity (5). These transitions can be viewed as repetitions of an experiment of nature that tests the advantages of multicellular life. The repetition challenges biologists to identify their similarities. An aim of this review is to compare, for a sample of three independent steps to multicellularity, the growth and fitness of the multicellular organism with those of a possible unicellular ancestor.

Life's unicellular origin implies that the fundamental genetic and epigenetic systems of all animals and plants were originally devised for those single cells (5). If so, how were those systems enhanced to manage a multicellular life cycle? A second aim of this review is to look for enhancements by comparing the regulatory circuits of single and multicellular forms. Were any new modes of epigenetic control needed? Practically complete parts lists are now available for a multicellular cyanobacterium (*Nostoc*), for *Caenorhabditis elegans*, for *Drosophila melanogaster*, for *Homo sapiens*, and soon for mice and zebrafish. Already, specific enhancements have been proposed for the signaling in metazoa (34, 38, 47). Scrutiny of the completed multicellular genomes may reveal traces of their origins. The challenge is to read those traces.

CYANOBACTERIA

Filamentous cyanobacteria are the Earth's oldest known multicellular organisms (44). Fossilized remains discovered in the 3465-Myr Warrawoona sedimentary rocks of northwestern Australia record organisms that may have lived when the Earth was only 1 billion years old (Earth's estimated creation at 4700 mya). The fossils show unsheathed filaments with terminal and medial cell shapes that resemble modern species of Oscillatoria, such as *Oscillatoria grunowiana*, *O. chalybea*, and *O. antillarum* (44). Whether the fossilized organisms carried out oxygenic photosynthesis with chlorophyll a and photosystems I and II, like their modern counterparts, or were anaerobic has not been settled (45). However, the 2-methyl-bacteriohopanepolyols that characterize modern aerobic cyanobacteria are abundant in organic rich sediments as old as 2500 Myr (48). Moreover, the necessary CO_2, H_2O, and light would have been available when the fossils were alive, and the isotopic carbon ratios contained within the organic matter of the fossil (kerogen) are compatible with oxygenic photosynthesis by these organisms (16). Assuming that life began with single cells, the antiquity of these multicellular filaments, as well as their morphological diversity, imply that multicellularity was advantageous on the Archaen Earth.

In addition to photosynthesis, many extant species of cyanobacteria fix atmospheric dinitrogen. However, because nitrogenase, the fixation catalyst, is very oxygen-sensitive, cyanobacteria arrange to separate nitrogen fixation from photosynthesis either temporally or spatially. In some genera, like *Nostoc*, aerobic nitrogen fixation is confined to differentiated cells called heterocysts. The ability to form heterocysts probably evolved over 2 billion years ago, but following the first filamentous forms (13). O_2 sensitivity of nitrogenase may have been a factor in their selection, but the atmospheric O_2 concentration at the time of their appearance is a much discussed matter (45). The data, despite their uncertainties, suggest that a major selective force in the evolution of the multicellular cyanobacteria was access to a more efficient and a more universal source of environmental nutrition. They became autotrophs capable of living on substances of high abundance at the surface of the earth: CO_2, H_2O, N_2, light, and inorganic ions.

Figure 1*A* shows a long chain of *Nostoc* vegetative cells in which several heterocysts are visible as larger darker cells. Figure 1*B* is an electron micrograph of a section through three vegetative cells and a thick-walled heterocyst at the right. Glutamine carries fixed nitrogen from the heterocyst to neighboring vegetative cells in the filament, whereas the photosynthetic vegetative cells supply organic carbon and the reducing power (NADPH) necessary for N_2-fixation in the heterocysts (1, 49). Heterocysts lack the oxygen-evolving photosystem II activity. They also have a laminated glycolipid layer outside the cell envelope that reduces the diffusion of gases (52). Heterocysts are thus able to maintain a relatively anoxic microenvironment in a filament that is oxygen rich. Heterocysts also lack ribulose bisphosphate carboxylase and so do not fix CO_2. A *Nostoc* filament thus lives as a nitrogen-fixing photoautotroph by virtue of its multicellular condition.

Recently, the Joint Genome Institute released a draft of the 9.76-Mb *Nostoc punctiforme* genome: 7432 ORFs were detected (18). The ensemble of the putative ORF functions sketches a rough metabolic and regulatory picture of *Nostoc*. For comparison, the full sequence of the unicellular, phototrophic cyanobacterium, *Synechocystis*, of 3.6 Mb has also recently been released (11).

Synechocystis, shown in Figure 2, can also use NH_4^+ and NO_3^- as nitrogen sources, but cannot fix N_2. A genomic relationship to the multicellular *Nostoc* is revealed by the fact that 80% of the *Synechocystis* genes are significantly similar,

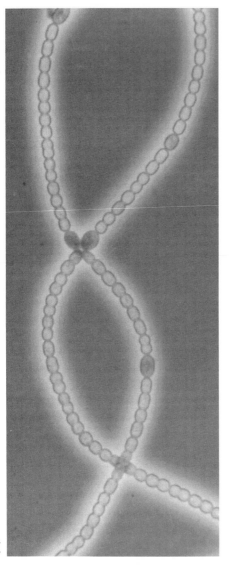

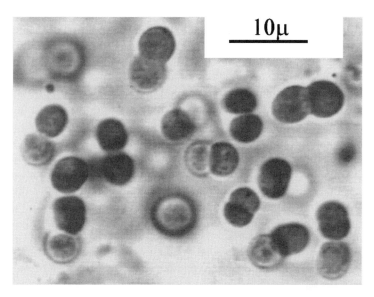

Figure 2 Culture of *Synechocystis* sp. Several cells are dividing. Photo courtesy of Cyanosite. Web address <www.cyanosite.bio.purdue.edu/images/images.html>

with BLAST expectation values less than e^{-5}, to one and often to several *Nostoc* genes that are also assigned the same function (11). Both 16S rRNA sequences of *Synechocystis* are 89% identical in BLAST alignments with the 16S rRNA sequences of *Nostoc*, and so it is possible but uncertain whether the latter arose from the former. Nevertheless, comparison of the parts list can suggest which functions are required to manage multicellularity. It is generally believed that core metabolic, transcriptional, translational, and replicative functions evolved for unicells (5). A filamentous form may then have arisen by failure of two daughter cells to separate after division. If the filament had some advantage over the unicell, it required preservation, and various regulatory devices would have evolved to manage what had become a true multicellular organism.

What might those devices be? There are 7400 protein genes in the sequenced strain *Nostoc* PCC 73102, and 3200 protein genes in *Synechocystis* PCC 6803, leaving 4200 genes, some of which might encode new devices. Functional analysis of those 4200 genes awaits a final annotation, but differences between the *Nostoc*

←――――――――――――――――――――――――――――――――

Figure 1 *A. Nostoc punctiforme*, light photomicrograph. *B.* is an electron micrograph of a section through three vegetative cells and a thick-walled heterocyst at the right. Note the connection between the heterocyst and the adjacent vegetative cell. Both photographs courtesy Dr John C. Meeks, University of California, Davis.

TABLE 1   Signaling proteins

Function	*Synechocystis* 6803	*Nostoc* 73102
Total	3200	7400
Histidine kinase	20	146
Response regulator, receiver	48	168
Ser/Thr protein kinase	8	51
HTH regulator	1	100
Sigma-70 forms		13

draft and the final annotation of *Synechocystis* are already apparent in the numbers and proportions of genes that encode components of signal transduction pathways. These are components of signal production and signal reception circuits. There are histidine protein kinases, often associated with sensory function, reception domains of response regulators, serine/threonine protein kinases, helix-turn-helix proteins, and alternative sigma factors. Table 1 offers a comparison of the numbers and proportions of these classes of proteins in the two genomes. Differences between the two are apparent in the numbers and in the proportions of all genes.

Evidence for added regulatory devices to manage the multicellular state also follows from genetic studies on the differentiation of heterocysts. *Anabaena* (also known as *Nostoc* PCC 7120) forms chains of photosynthesizing vegetative cells, punctuated with an occasional heterocyst. The sequenced multicellular *Nostoc* PCC 73102 shares 73% of its genes with *Anabaena*. If *Anabaena* is grown in a medium that provides ample fixed nitrogen as NH_4^+, nitrogen fixation is not required and no heterocysts are formed, only vegetative cells. But when cells that have been grown with ample NH_4^+ are washed, then resuspended in medium free of fixed nitrogen, they develop heterocysts during the next 24 h. Since the vegetative generation time under these conditions is also about 24 h, heterocyst differentiation can be thought of as growing a new cell with a different wall and a somewhat different set of enzymes. The RNA hybridization data suggest as many as 1000 protein differences between vegetative cells and heterocysts (7).

When nitrogen fixation is needed, about 1 cell in 10 becomes a heterocyst. N_2 fixation and respiration in the heterocyst require a supply of reductant and of carbon from the adjacent, photosynthesizing, vegetative cells. Reductant and carbon are provided in the form of maltose, sucrose, or other disaccharides. In return, the heterocyst releases fixed nitrogen in the form of glutamine to its vegetative neighbors.

The intercellular exchange of metabolites illustrates the metabolic interdependence of vegetative cells and heterocysts. Wilcox (54) noted that these cells are also developmentally interdependent in that the heterocysts differentiate at fairly regular spatial intervals, which is evident in Figure 1*A*. Chain-breaking experiments show that many more cells have the potential to become heterocysts than normally do so (54). Because all the cells in a chain have received the same environmental cue—the paucity of ammonium ion, some kind of cell interaction has been sought to explain the regular spatial pattern of heterocysts. If all cells in a chain of vegetative cells grow then divide, the chain elongates. To maintain a fixed ratio of heterocysts to vegetative cells, new heterocysts would need to differentiate in proportion to the new vegetative cells that form. Moreover, a regulatory system that appropriately selects particular vegetative cells to become heterocysts would seem to be necessary so that fixed nitrogen (glutamine) will be available to all the vegetative cells. Since heterocysts and vegetative cells differ in their levels of hundreds of proteins, the process needs coordination. It is hard to escape the inference that a cell interaction triggers one daughter cell to become a proheterocyst and its sister to remain a vegetative cell.

Each new heterocyst forms very near the center of a segment of vegetative cells after their number has doubled. This location maintains a stable ratio and spatial distribution of the two cell types, presumably optimized for exchange of fixed carbon and fixed nitrogen between vegetative cells and heterocysts. But what are the cell interactions and how do they generate the pattern? A promise of answers to this question comes with the discovery of several genes that alter the heterocyst pattern. Major candidates are *hetR*, *patA*, and *patS*. Mutations in *hetR* eliminate heterocysts, whereas multiple copies of *hetR* stimulate the formation of heterocysts in the presence of combined nitrogen, and the formation of clusters of heterocysts in its absence (6). *patA* mutants are unable to develop interstitial heterocysts; they form terminal heterocysts only (30). A *patA* mutation suppresses the multiple heterocyst phenotype of *hetR* mutants. Overexpression of *patS* blocks heterocyst differentiation, while a *patS* null mutant has an increased frequency of heterocysts—clustered abnormally along the chain of cells (56). Since *hetR* and *patA* are increasingly expressed after nitrogen step-down in vegetative cells, this set of properties suggests that *hetR* and *patA* combine to induce heterocyst differentiation, while *patS* encodes an inhibitor of that process (2, 55).

Moreover, *patS* is expressed in proheterocysts but not in vegetative cells. The *patS* gene can encode a peptide of 17 amino acids, and a synthetic peptide corresponding to its C-terminal pentapeptide has the capacity to inhibit heterocyst development. Yoon & Golden (56) propose that heterocysts synthesize and secrete a *patS* peptide that prevents neighboring vegetative cells from becoming heterocysts.

MYXOBACTERIA

Some organisms become multicellular by aggregation rather than by growth and cell division. Myxobacterial cells aggregate to build sometimes quite elaborate fruiting bodies within which they sporulate. The life cycle of these prokaryotes is surprisingly similar to development of the (eukaryotic) cellular slime molds, such as *Dictyostelium* (4, 20). All myxobacteria form fruiting bodies and no unicellular species have been found. Perhaps the multicellular myxobacteria replaced their unicellular progenitors from a common ecologic niche.

Myxobacteria lie on the boundary between uni- and multicellular organisms. Although they grow and divide as proper Gram-negative bacteria, they constitute a primitive multicellular organism whose cells feed socially as multicellular swarm units. They eat particulate organic matter in the soil using a variety of secreted hydrolytic enzymes: lysozymes, proteases, and cellulases. They feed like packs of microbial wolves. Their cooperation is demonstrated when they are provided with a nutrient polymer like casein. Growth on casein is faster at higher cell density (42). When fed casein digested with the proteases in a culture supernatant, then low-density cultures grow at the high-density rate.

They also build multicellular fruiting bodies, five species of which are shown in Figure 3 (see color insert). When they have exhausted their food, 100,000 cells build these compact and symmetrical structures, often with stalks, branches, and multiple cysts evident in the Figure.

According to the 16S ribosomal RNA sequence of these organisms and to other chemotypic markers, the various species in this photo are phylogenetically related. Molecular genetic studies show that both *Myxococcus* and *Stigmatella* have circular genomes of about 9.5 Mb (8, 35). Immediately the question arises: Why do myxobacteria have more DNA than *Bacillus subtilis*, for example, which efficiently makes notoriously good spores (46)? Each myxobacterial cyst is a package of spores. As soon as food becomes available, the spores germinate and organize a feeding swarm that is thus able to grow. A selective advantage for their multicellular state is their cooperative feeding. Myxobacteria feed on particulate organic matter in the soil, which they digest with a battery of secreted hydrolytic enzymes, proteases, lysozymes, nucleases, and in some cases cellulase (43).

The selective advantage of multicellular fruiting bodies is probably twofold. One advantage is that immediately after the spores germinate from the same fruiting body, they can cooperate for efficient feeding. A second advantage relates to the macroscopic dimension of a fruiting body, which effectively gives legs to the spores. The fruiting body package of 0.2 mm, or more, is large enough to adhere to an animal that happens to brush by it in the soil and then to be carried away by that animal. Fruiting bodies have been seen on the backs of mites (40). A passing insect or worm in the soil is likely to be searching for food. The animal carries, then deposits, that fruiting body on its food, providing food for the myxobacteria as well. While sporulation is a great strategy for survival when food runs out, multicellular sporulation adds the possibility of transport to a new place where starving cells can refresh themselves and grow.

A spore-transport/cooperative-feeding advantage is suggested by the fact that only the cells inside the rounded masses of a *Stigmatella* or *Chondromyces* fruiting body, or the spherical mound of *Myxococcus xanthus* or *M. stipitatis* differentiate into spores (19). The point is that the masses are more likely to be plucked for purely mechanical reasons (see Figure 3). Spores outside the masses or adhering directly to the substrate are less likely to be picked up by a passing animal. Though plausible, this suggestion needs to be tested.

A- and C-Signaling

How spores become localized is explained by cell-to-cell signaling. Sporulation is the final step in a developmental program that starts from a disorganized biofilm of growing cells (Figure 4, *top*). In 4 h, irregular aggregates of $\sim$1000 cells appear. Then, many cells stream into these foci from all directions so that by 24 h a hemispherical mound of 10^5 cells has been created. When such a mound is cracked open it is seen to be densely packed with spores. At least two extracellular signals are necessary for making spores. Signaling to coordinate sporulation might require an even larger set of genes than the structural changes that differentiate a spore. For this purpose, myxobacteria may have many genes to make and to export the signals, as well as to receive and to interpret them accurately.

Isolated and purified from medium conditioned by developing cells, A-factor is a set of 6 particular amino acids: trp, pro, phe, tyr, leu, and ile. A-factor, like the homoserine-lactones in enteric bacteria, is a quorum sensor. Each cell produces a fixed amount of A-factor. However, only cells whose nutritional state indicates severe starvation release A-factor. Moreover, there is a response threshold for the A-signal. Therefore, a certain minimum number of cells must agree that the population should commit itself to fruiting body development even before aggregation starts. Note that the signaling concentration of A-factor amino acids is about tenfold lower than the concentration necessary to support growth (29).

Purified C-factor, by contrast to A-factor, is an approximately 20-kDa protein. It has a hydrophobic N-terminal that keeps it associated with the cell surface. The properties of C-factor (water insoluble and cell surface bound) are appropriate to the disposition and density of cells at the time of C-signaling. In the beginning when the cells are at relatively low density, A-factor amino acids must diffuse from cell to cell. Later, when the cells have aggregated to a 1000-fold higher density within a nascent fruiting body, C-signaling starts. C-factor is bound to the cell surface, and C-signaling requires end-end cell contact for signal transmission (22).

Mutants defective in A-signaling (*asg* mutants) fail to assemble fruiting bodies and fail to sporulate. Mutants defective in C-signaling (*csg* mutants) are blocked with only the early asymmetric aggregates having formed. They fail to assemble hemispherical mounds and they fail to sporulate.

Response to Signals

Starvation, the A-signal, and the C-signal order the time of gene expression for the program of fruiting body development. This diagram (Figure 5), invented by Kroos (21), shows how developmentally regulated genes are induced at appropriately different points in time as a consequence of their signal dependence. The signals act directly or indirectly as transcription factors. Starvation, i.e., the lack of any

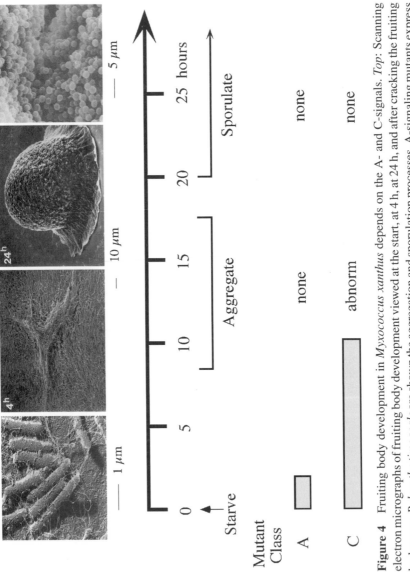

Figure 4 Fruiting body development in *Myxococcus xanthus* depends on the A- and C-signals. *Top:* Scanning electron micrographs of fruiting body development viewed at the start, at 4 h, at 24 h, and after cracking the fruiting body open. *Below the time scale* are shown the aggregation and sporulation processes. A-signaling mutants express a few early developmentally regulated genes, but neither aggregate nor sporulate. C-signaling mutants express more developmentally regulated genes, aggregate at the 4-h stage, but do not sporulate.

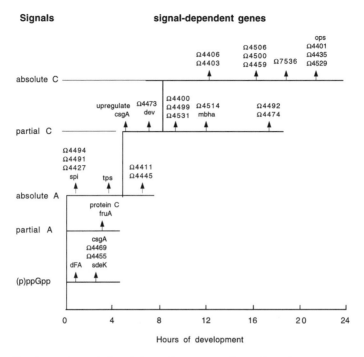

Figure 5 Expression of signal dependent genes. Genes have lower case names or Ω numbers. The arrow beneath a gene indicates the time at which it begins to be expressed. The signals are listed at the left in the order, from bottom to top, in which they are produced during development. The horizontal lines at the level of each signal divide the genes into those above the line, which are expressed with each signal and those below the line, which are independent of the signal. Absolute means that there is no expression in the absence of the signal; partial means that there is some expression in the absence of the signal but much more when the signal is present (26, 28).

one or more of the amino-acylated tRNAs induces a stringent response and the synthesis of (p)ppGpp by *Myxococcus* (14). This highly phosphorylated guanosine nucleotide apparently initiates the developmental program. Some genes, such as *sdeK*, require only (p)ppGpp for expression and they are expressed early, before 2 h of development. The next set requires A-factor, which is released around 2 h of development in addition to (p)ppGpp (29). The various developmental gene products are indicated by name in the diagram of Figure 5, and promoters that fire at the time shown on the horizontal axis by an omega 4-digit number. Genes that are indicated as "partial A" on the vertical axis begin to be expressed before A-factor is produced, but their expression increases twofold or more when A-factor is released at 2 h (28). Others, denoted "absolute A," are totally dependent on A-factor: no signal, no expression.

Consequently, A-factor-deficient mutants arrest development having expressed only those genes below the "absolute A" line shown on Figure 5. No gene above that line is expressed. As mentioned above, A-signal helps to evaluate starvation. The proteins charted in this diagram, and other proteins (since the screen for such genes is not yet saturated) are synthesized during assembly of the fruiting body and differentiation of spores. To be able to synthesize these developmental proteins, aggregation and sporulation must be initiated while the cells still retain some capacity for protein synthesis. In other words, a starving population must have anticipated the future absence of nutrient when it started to build a fruiting body. Individual cells register their vote in favor of building a fruiting body by releasing the set of A-factor amino acids, once they have made a grim evaluation of the nutrient available.

The C-signal is a morphogen that organizes the fruiting body and limits sporulation to its interior. C-factor-deficient mutants are blocked below the "absolute C" line in Figure 5, expressing the early A-signal-dependent, but C-signal-independent genes. Some essential sporulation genes are absolutely C-signal dependent. The spatial confinement of sporulation to the interior of the fruiting body is shown with the aid of a transcriptional fusion between the green fluorescent protein (GFP) and a sporulation promoter. The fusion strain was induced to develop fruiting bodies. A fruiting body in which sporulation has begun is shown photographed in visible light in the left panel of Figure 6 (see color insert).

Localized Gene Expression

The fruiting body is dense because the cells within are close packed. The fruiting body is surrounded by cells at much lower density. The surrounding rod-shaped cells are organized into many raft-like clusters. Even as many of the fruiting body cells are differentiating into spherical spores, none of the peripheral cells are doing so; all retain their rod shape. The very same microscopic field under UV illumination to excite the GFP is shown in the right panel of Figure 6. Only the fruiting body fluoresces; none of the peripheral cells fluoresce. All the cells in this experiment, whether inside the fruiting body or out, carry the same GFP transcriptional fusion. Why might an apparatus that localizes sporulation have been selected in evolution? The fruiting body/sporulation process is induced by starvation; by the time (24 h) that sporulation commences, the amino acid reserves are in short supply, and yet new proteins must be made. The very limited capacity of the cells for protein synthesis at this point requires it to be used where it counts most. Perhaps the rigors of starvation explain why sporulation is spatially restricted: Spores at the top of the fruiting body are in the best position to be picked up and carried to a new place.

Natural selection for social behavior has been explored in long-term population experiments. Twelve parallel suspension cultures were propagated continuously for many generations under conditions that required neither cooperative feeding nor multicellular fruiting body development. Each culture was repeatedly sampled on the way to 1000 generations. A majority of the cultures contained mutants that had lost fruiting body development and sporulation efficiency, suggesting that these social behaviors are detrimental to fitness under such asocial growth conditions (50). Maynard Smith (33) pointed out that when the multicellular stage of myxobacteria is formed anew in each cycle by aggregation, there is a danger that selfish mutant cells will disrupt multicellular organization. Indeed, Velicer et al. showed that some developmental mutants are cheaters (51). When such mutants were mixed in low proportion with wild-type (WT) cells, and the mixture induced to develop fruiting bodies, the mutants represented a disproportionately large number of the spores. Cheaters had never been reported to accumulate in laboratory cultures that arise from the spores of fruiting bodies. However, these anecdotes need to be replaced by systematic observations. Cheaters may be infrequent in natural populations because Velicer and colleagues also observed that fruiting bodies with many cheaters contained fewer spores than pure WT cultures (51).

Cooperative feeding is not always linked to the formation of fruiting bodies. Many gliding bacteria in the soil, such as the *Cytophagas* and the *Flexibacters*, feed with extracellular enzymes but do not build fruiting bodies. Although some, such as *Sporocytophaga*, do sporulate, they do so as single cells. Myxobacteria are found in all climate zones, vegetation belts, and altitudes (41). They compete effectively with the cellular slime molds, which inhabit the same ecological niche in soil. Well-cultivated and aerated soils often contain 10^6 myxobacteria per gram. Two forces appear to have combined in selecting the myxobacterial grade of multicellularity: more efficient feeding and more efficient dispersal.

To what extent does an infrastructure required for social behavior explain why *M. xanthus* has a genome of 9.5 megabase pairs, larger than almost all other bacteria? The gene density, average gene size, and absence of repeat sequences are comparable to *Escherichia coli* K12 in the 1% of the *M. xanthus* genome whose sequence is published. They are *E. coli*-like overall, according to the Monsanto/Cereon sequencing group who have completed a 4X sequencing of the *M. xanthus* genome. Annotation is under way that will give a draft inventory of gene functions. That inventory will test whether cell-cell signaling plays a significant role in determining its genome size.

VOLVOX

This photosynthetic green alga (Figure 7; see color insert) lives in sunlit, standing waters. Many plants compete for light and nutrients in such places, including unicellular and various grades of multicellullar green algae. The family *Volvocaceae* contains about 40 multicellular species, all closely related to one of the unicellular members of that family, like *Chlamydomonas reinhardtii*, which is also in Figure 7. The relationships of their cytology, physiology, and phylogenetics support the hypothesis that *Volvox* evolved from *Chlamydomonas*. The richness of the cell biological and physiological data for the photosynthetic green algae allows comprehensive comparisons to be made between *Chlamydomonas*, *Volvox*, and their relatives.

Phylogeny

Fossils of unicellular algae have been reported from the Precambrian (9). Although the identity of these species has yet to be established, the fossils imply that green algae had arisen by that time. Based on sequence comparisons of the complete small ribosomal RNA gene (39), *Volvox carteri* is as closely related to *Chlamydomonas reinhardtii* as the two grasses, corn (*Zea mays*) and rice (*Oryza sativa*), are to each other. Kirk (23) estimates that corn and rice shared their last common ancestor about 50 mya, providing an estimate for the divergence between *Chlamydomonas* and *Volvox*. Extensive comparisons of the two internal transcribed spacer regions of the nuclear rDNA confirm the notion that *C. reinhardtii*, among a series of other green unicellular members of the *Volvocaceae*, is the most similar to *V. carteri* (10, 32).

Historically, the genus *Volvox* has been defined with morphologic, not phylogenetic, criteria. Later ribosomal RNA comparisons as well as the sequences of the genes for the ATP synthase beta-subunit and the large subunit of the CO_2-fixing enzyme, ribulose 1,5-bisphosphate carboxylase (36), indicate that the morphologic genus *Volvox* is probably polyphyletic (10, 36). Nevertheless, according to these and other molecular comparisons, the various *Volvox* clades are all related to *C. reinhardtii*. Thus, *Chlamydomonas*, or its progenitor, may have gained a morphologically similar grade of multicellularity several times in the past 50 million years.

A variety of multicellular states are represented by *Volvox*, and by *Gonium*, *Pandorina*, *Eudorina*, and *Pleodorina*, which differ in the number of *Chlamydomonas*-like cells they contain. This variety may exist because, being photosynthetic, sunlight provides the bulk of their energy, and each cell in the community absorbs light; their evolution need not have been constrained by the need for inventing a communal feeding apparatus, which was needed for the sponges (5). The wide range of multicellular forms represented by *Volvox*, *Gonium*, *Pandorina*, *Eudorina*, and *Pleodorina* implies either that the multicellular transition occurs easily, or that the multicellular state has a substantial selective advantage. Ease is suggested by the tendency of sister *Chlamydomonas* cells to cohere to each other after division, surrounding themselves with a mucilaginous polysaccharide, and spontaneously generating a multicellular array of nonmotile cells. This tendency is common enough to be named a palmelloid stage. Similarly, *Gonium* forms disks of 4, 8, 16, or 32 *Chlamydomonas*-like cells; the flagella of each cell beat independently; each cell divides independently and is capable of regeneration. However, the evidence to be discussed suggests that such arrays are not functionally integrated in the way that *Volvox* is. This crucial difference motivates the distinction between colonial and multicellular organisms made for this review.

Modern *Chlamydomonas* lives with modern *Volvox*. Mixed populations of green algae are found in small pools, puddles, and temperate lakes. These waters tend to be unmixed and turbid. They are typically nutrient rich, providing sources of fixed nitrogen, phosphorus, and sulfur (23). Cohabitation with *Chlamydomonas* suggests that any multicellular advantage in growth has not caused *Volvox* to replace the unicells with which they compete. This may result from the facts that illumination and nutrient availability tend to change with the season, and puddles tend to dry up or freeze. Cyclic changes in temperature and light may have prevented replacement or even the establishment of an equilibrium population. In lakes and ponds, *Volvox*, *Pandorina*, and *Eudorina* fluorish only briefly each year (23). With such seasonal changes, motility and phototaxis are essential for photosynthetic organisms.

Advantages of Multicellularity

Certain advantages accrue simply from a larger size. The larger *Gonium, Eudorina*, and *Volvox* colonies escape from predation by filter-feeding rotifers and small crustaceans (23). Daughter colonies of *Volvox*, which would be small enough to be eaten by these animals, are kept internally, protected inside their mother colony. Another important advantage is that the larger colonies can absorb and store essential nutrients more efficiently (23). Inorganic phosphate is often a limiting nutrient for algae (3). Large multicellular algae have an advantage in phosphate uptake, storing any excess as polyphosphate in the extracellular matrix that separates the cells (25). Other nutrients may also be retained in the matrix, such as minerals, ions, and water that would help protect against desiccation. Many algae are dispersed by waterbirds (23); the larger colonies may have a better chance than unicells to be carried.

Multicellularity Exacted A Price

Volvox obtained the advantages of multicellularity after having paid a price. Insofar as that price measures the balancing advantages, that price is of interest. The vast majority of the 50,000 cells in a *Volvox* colony are specialized somatic cells that are sterile. If they are removed from the colony, they are incapable of regeneration (27). Only the germ cells of *Volvox* have offspring. By contrast, individual cells of the multicellular *Eudorina* are capable of regeneration (27). Thus the loss of reproductive potential is not inherent in this colonial form; *Eudorina* is, however, smaller and has many fewer cells than *Volvox*.

A second element is the cost of reorganizing basic cytoskeletal structures required for *Volvox* to survive. To keep the intensity of light they receive within a range favorable for photosynthesis, all swimming photosynthetic organisms strive to maintain a euphotic position in the water column. The favored intensity also must not be so high as to produce photo damage. Phototaxis, both positive and negative, is used to maintain that optimal position in the water column. However, the problem for photosynthetic flagellates is that cell division and motility are incompatible. The flagellum is constructed from microtubules and the centriole. The mitotic spindle is a mutually exclusive arrangement of the microtubules and the centriole (23a). Both *Chlamydomonas* and *Volvox* need to be motile to efficiently utilize light even as they grow and divide. Ordinarily, flagella are lost during the mitotic cycle. Having a buoyant density greater than one, nonmotile organisms fall out of their euphotic zone.

The flagellates have variously resolved the competition for their microtubular cytoskeleton. Whereas almost all cells grow twofold before dividing once, *Chlamydomonas* and its relatives, in response to their special need for motility, regularly grow up to 32-fold, then rapidly divide (up to) five times in succession (31), apparently shortening the period of their nonmotility. Under illumination conditions resembling nature, daylight hours are devoted to growth and *Chlamydomonas reinhardtii* tends to confine division events to the night, when maintaining height is not critical. Some multicellular species use the *Chlamydomonas* solution, but not those, like Volvox, with more than 32 cells (23).

Colonies larger than 32 cells apparently need a different way to sort out the conflict between motility and cell division. The *Volvox* solution is to differentiate a subset of cells in the anterior end of the growing organism that do not divide, continue beating their flagella, and thereby provide the colony with a continuous source of photo-responsive motility. Meanwhile the rest of the cells in an immature colony divide and produce progeny. After growth is complete, the mature colony is always prepared to swim because its somatic cells no longer divide (24). The specialized germ cells are the only ones that retain the need and capacity to divide, but they are small in number. In addition to creating cells that can no longer reproduce, another cost is incurred for a cytoskeletal rearrangement that permits the cells in the mature colony to coordinate the rowing movements of all their individual flagellae.

The Cytoskeleton of All Cells in a *Volvox* Colony Must be Coordinated

Flagellar motility is necessary for *Volvox* to maintain a proper height in the water column in order to carry out photosynthesis, and this necessity appears to have played an important role in its evolution from the motile, photosynthetic *Chlamydomonas* (24). For the mature somatic cells to move the multicellular colony as a unit, they must each undergo a remarkable structural differentiation. A spherical *Volvox* colony consists of 500–60,000 individual cells, depending on the species, embedded in a common matrix. The cells are connected to each other by fine cytoplasmic bridges that remain from incomplete cell separations at cytokinesis. Those strands break down in *V. carteri* after embryogenesis, but by then the pole of each cell has become correctly oriented relative to the head of the spheroidal *Volvox* colony. Those cytoplasmic bridges had fixed the orientation of the axis of each individual cell relative to the anterior-posterior (A-P) axis of the colony producing the organized arrangement diagrammed in Figure 8.

To execute phototaxis, the colony depends on the oriented beating of all 100,000 flagella (2 flagella per cell times 50,000 cells). Both flagella on each cell must be properly oriented with respect to that cell's latitude, longitude, and the A-P axis of the spheroidal colony (Figure 8). The eyespot of each cell, which detects the incoming light, is also systematically oriented. The pair of flagella in *Chlamydomonas*

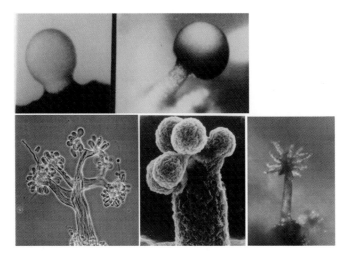

Figure 3 Fruiting bodies (*top row left to right*), *Myxococcus fulvus*, *M.stipitatis*; (*bottom row left to right*), *Chondromyces crocatus*, *Stigmatella aurantiaca*, and *C. apiculatus*. Photographs courtesy Dr. Hans Reichenbach and Dr. Martin Dworkin, University of Minnesota.

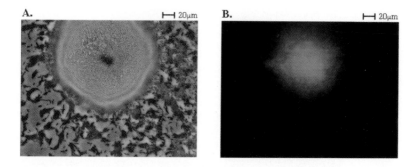

Figure 6 Expression of sporulation genes is localized to the fruiting body. *Panel A*: a single fruiting body photographed from above with visible light. *Panel B*: the same field photographed with ultraviolet light to excite the fluorescence of GFP. Photograph by Bryan Julien (19).

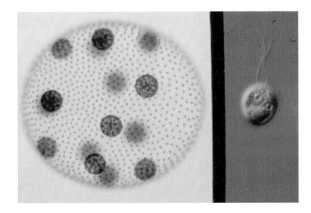

Figure 7 (*Left*) A spheroidal colony of *Volvox carteri*. (*Right*) A single *Chlamydomonas reinhardtii* cell at higher magnification showing its pair of flagella. Photograph courtesy of Dr. David Kirk and Dr. Ursula Goodenough, Washington University, St Louis, MO.

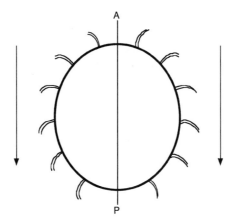

Figure 8   Orientation of cells within a *V. carteri* spheroid indicated by their pairs of flagella. A plane containing the anterior/posterior axis is illustrated. Somatic cells are oriented so that all flagella beat with their effective strokes directed from the anterior pole (*A*) toward the posterior pole of the spheroid (*P*), as indicated by the two arrows at the left and right. The posterior pole contains the reproductive cells, or gonidia, inside the spheroid. Modified after Hoops (15).

and in the immature somatic cells of *Volvox* are related to each other by 180° rotational symmetry about the cell's own axis. By means of this symmetry, a *Chlamydomonas* cell swims with a kind of breaststroke as its two flagella beat in opposite directions but in the same plane (15). By contrast, the two flagella of each mature somatic cell in a *Volvox* colony are oriented to beat in parallel planes, and in the same direction. That direction must vary systematically from cell to cell in such a way that the whole colony can progress in the direction of its (fixed) A-P axis (Figure 8) (15). As a consequence of this organized beating of flagella, the colony rotates slowly about its A-P axis as it moves anteriorly, justifying its name as the "fierce roller."

Because the direction of the effective stroke of each flagellum is fixed by the arrangement of its microtubules, maturation of *Volvox* somatic cells involves a one-time rotation of their constituent flagellar axonemes. Rotation is complete before the developing embryo breaks free of its parental spheroid. The two flagellar axonemes rotate in opposite directions, transforming the 180° rotational symmetry of *Chlamydomonas* into the requisite parallel orientation of mature *Volvox*. When the two flagella of each cell beat in the same direction, they can contribute appropriately to the movement of the entire colony. Also, the photosensitive eyespot moves in each cell from a position nearer one of the flagellar bases to a position equidistant between them (15).

Mutational studies of *Volvox* by Huskey (17) showed that the individual cellular units must indeed be properly oriented in the colony to enable phototactic behavior. This regular orientation is demonstrably lacking in the *eye* mutants of Huskey (17).

The flagella of an *eye* mutant beat, but as the organisms are unable to move coordinately in any direction, they lose the capacity for phototaxis. It thus appears that the rearrangements of centriole, microtubules, and eyespot are essential and developmentally programmed. Flagella also reorient in *Pleodorina* (12). Assessing the net cost of generating the proper cytoskeleton in an adult *Volvox* awaits an understanding of the underlying molecular events. These events will be clarified by the *Chlamydomonas* genome program, now under way. In any case, the rearrangement of the cytoskeleton is programmed and coordinated with the arrest in cytokinesis. The cost of this set of adaptations is likely to have been significant, and by inference, the advantage of multicellularity must also have been significantly large to offset that cost.

VOLVOX CONCLUSIONS

Despite the costs of the specializations required to make a *Volvox carterii* colony a functional unit, multicellularity arose several times in the genus *Volvox*. Repeat occurrences suggest therefore that selection strongly favors the multicellular state. *Chlamydomonas* must have a cytoskeleton with strict polarity, yet one that permits systematic axoneme rotation. More efficient resource acquisition in eutrophic environments and protection of the offspring and parents from predation could explain most of the major evolutionary trends in *Volvox* and its relatives: increased size, increased cell number, increased matrix volume, and differentiation of germ and soma. The transient abundance of nutrients in quiet ponds each spring followed by late-season scarcity would provide a selective advantage for multicells that could execute phototaxis, photosynthesize efficiently, store precious phosphate against later shortages, and survive an abundance of predators. These advantages combined must have been strong enough to drive the complex cytological and regulatory changes required.

GENERAL CONCLUSIONS

Multicellular cyanobacteria like *Nostoc* enjoy a clear nutritional advantage. The myxobacterial grade of multicellularity has both a feeding advantage and an advantage in dispersion. *Volvox* benefits in feeding by means of its efficient phototaxis, in dispersion, and in protection from predation. Advantages in feeding and in dispersion seem to be common threads in this sample of three organisms.

ACKNOWLEDGMENTS

The general viewpoint on multicellularity adopted for this article owes much to the writings of John Tyler Bonner. Published and unpublished work on myxobacteria from the authors laboratory was generously supported by the National Institute of General Medical Sciences, under grant NIH GM23441. Preliminary sequence data for *N. punctiforme* was obtained from The DOE Joint Genome Institute (JGI).

Visit the Annual Reviews home page at www.AnnualReviews.org

LITERATURE CITED

1. Adams DG. 2000. Cyanobacterial phylogeny and development: questions and challenges. See Ref. 5a, pp. 51–81
2. Adams DG. 2000. Heterocyst formation in cyanobacteria. *Curr. Opin. Microbiol.* 3:618–24
3. Bell G. 1985. The origin and early evolution of germ cells as illustrated by the *Vovocales*. In *The Origin and Evolution of Sex*, ed. HO Halvorson, A Monroy, pp. 221–56. New York: Liss
4. Bonner JT. 1963. *Morphogenesis, an Essay on Development*, pp. 165–73. New York: Atheneum. 296 pp.
5. Bonner JT. 1974. *On Development—The Biology of Form*, pp. 80–109. Cambridge, MA: Harvard Univ. Press
5a. Brun YV, Shimkets LJ, eds. 2000. *Prokaryotic Development*. Washington, DC: ASM Press
6. Buikema WJ, Haselkorn R. 1991. Characterization of a gene controlling heterocyst differentiation in the cyanobacterium *Anabaena* 7120. *Genes Dev.* 5:321–30
7. Buikema WJ, Haselkorn R. 1993. Molecular genetics of cyanobacterial development. *Annu. Rev. Plant Physiol. Plant Mol. Biol.* 44:33–52
8. Chen H, Kuspa A, Keseler IM, Shimkets LJ. 1991. Physical map of the *Myxococcus xanthus* chromosome. *J. Bacteriol.* 173:2109–15
9. Cloud PE, Licari GR, Wright LA, Troxel BW. 1969. Proterozoic eucaryotes from Eastern California. *Proc. Natl. Acad. Sci. USA* 62:623–30
10. Coleman AW. 1999. Phylogenetic analysis of "Volvocacae" for comparative genetic studies. *Proc. Natl. Acad. Sci. USA* 96:13892–97
11. CyanoBase. Sequence of *Synechocystis*. *http://www.kazusa.or.jp/cyano/*
11a. Dworkin M, Kaiser D, eds. 1993. *Myxobacteria II*. Washington, DC: ASM. 404 pp.
12. Gerisch G. 1959. Die Zellendifferenzierung bei *Pleodorina californica* Shaw und die Organization der Phytomonadineenkolnien. *Arch. Protistenkd.* 104:292–358
13. Giovannoni SJ, Turner S, Olsen GJ, Barns S, Lane DJ, Pace NR. 1988. Evolutionary relationships among cyanobacteria and green chloroplasts. *J. Bacteriol.* 170:3584–92
14. Harris BZ, Kaiser D, Singer M. 1998. The guanosine nucleotide (p)ppGpp initiates development and A-factor production in *Myxococcus xanthus*. *Genes Dev.* 12:1022–35
15. Hoops HJ. 1993. Flagellar, cellular and organismal polarity in *Volvox carteri*. *J. Cell Sci.* 104:105–17
16. House CH, Schopf W, McKeegan KD, Coath CD, Harrison M, Stetter KO. 2000. Carbon isotopic composition of individual precambrian microfossils. *Geology* 28:707–10
17. Huskey RJ. 1979. Mutants affecting vegetative cell orientation in *Volvox carteri*. *Dev. Biol.* 72:236–43
18. Joint Genome Institute. 2001. Sequence of *Nostoc punctiforme*. http://genome.ornl.gov/microbial/npun/
19. Julien B, Kaiser AD, Garza A. 2000. Spatial control of cell differentiation in *Myxococcus xanthus Proc. Natl. Acad Sci. USA* 97:9098–103

20. Kaiser D. 1986. Control of multicellular development: *Dictyostelium* and *Myxococcus. Annu. Rev. Genet.* 20:539–66

21. Kaiser D, Kroos L. 1993. Intercellular signaling. See Ref. 11a, pp. 257–83

22. Kim SK, Kaiser D. 1990. Cell alignment required in differentiation of *Myxococcus xanthus. Science* 249:926–28

23. Kirk DL. 1998. *Volvox*, pp. 30–60. New York: Cambridge Univ. Press. 381 pp.

23a. Kirk DL. 1998. *Volvox*, See Ref. 23, pp. 102–8

24. Koufopanou V. 1994. The evolution of soma in the *Vovocales. Am. Nat.* 143:907–31

25. Koufopanou V, Bell G. 1993. Soma and germ: an experimental approach using *Volvox. Proc. R. Soc. London Ser. B. Biol. Sci.* 254:107–13

26. Kroos L, Kaiser D. 1987. Expression of many developmentally regulated genes in *Myxococcus* depends on a sequence of cell interactions. *Genes Dev.* 1:840–54

27. Kuhn A. 1971. *Lectures on Developmental Physiology*, pp. 112–18. New York: Springer-Verlag. 535 pp.

28. Kuspa A, Kroos L, Kaiser D. 1986. Intercellular signaling is required for developmental gene expression in *Myxococcus xanthus. Dev. Biol.* 117:267–76

29. Kuspa A, Plamann L, Kaiser D. 1992. Identification of heat-stable A-factor from *Myxococcus xanthus. J. Bacteriol.* 174:3319–26

30. Liang J, Scappino L, Haselkorn R. 1992. The *patA* gene product, which contains a region similar to CheY of *Escherichia coli*, controls heterocyst pattern formation in the cyanobacterium *Anabaena* 7120. *Proc. Natl. Acad. Sci. USA* 89:5655–59

31. Lien T, Knutsen G. 1979. Synchronous growth of *Chlamydomonas reinhardtii* (Chlorophyceae): a review of optimal conditions. *J. Phycol.* 15:191–200

32. Mai JC, Coleman AW. 1997. The internal transcribed spacer 2 exhibits a common secondary structure in green algae and flowering plants. *J. Mol. Evol.* 44:258–71

33. Maynard Smith J, Szathmary E. 1995. *Major Transitions in Evolution.* New York: Oxford. 346 pp.

34. Müller CI, Blumbach B, Krasko A, Schröder HC. 2001. Receptor protein-tyrosine phosphatases: origin of domains (catalytic domain, Ig-related domain, fibronectin type III module) based on the sequence of the sponge *Geodia cydonium. Gene* 262:221–30

35. Neumann B, Pospiech A, Schairer HU. 1993. A physical and genetic map of the *Stigmatella aurantiaca* DW4/3.1 chromosome. *Mol. Microbiol.* 10:1087–99

36. Nozaki H, Ohta N, Takano H, Watanabe MM. 1999. Reexamination of phylogenetic relationships within the colonial *volvocales* (chlorophyta): an analysis of *atoB* and *rbcL* gene sequences. *J. Phycol.* 35:104–12

37. Olive LS. 1978. Sorocarp development by a newly discovered ciliate. *Science* 202:530–32

38. Ono K, Suga H, Iwabe N, Kuma K, Miyata T. 1999. Multiple protein tyrosine phosphatases in sponges and explosive gene duplication in the early evolution of animals before the Parazoan-Eumetazoan split. *J. Mol. Evol.* 48:654–62

39. Rausch H, Larsen N, Schmitt R. 1989. Phylogenetic relationships of the green alga *Volvox carteri* deduced from small-subunit ribosomal RNA comparisons. *J. Mol. Evol.* 29:255–65

40. Reichenbach H. 1984. Myxobacteria: a most peculiar group of social prokaryotes. See Ref. 41a, pp. 1–50

41. Reichenbach H. 1993. Biology of the myxobacteria: ecology and taxonomy. See Ref. 11a, pp. 13–62

41a. Rosenberg E, ed. 1984. *Myxobacteria.* New York: Springer-Verlag

42. Rosenberg E, Keller K, Dworkin M. 1977. Cell-density dependent growth of *Myxococcus xanthus* on casein. *J. Bacteriol.* 129:770–77

43. Rosenberg E, Varon M. 1984. Antibiotics and lytic enzymes. See Ref. 41a, pp. 109–25

44. Schopf JW. 1993. Microfossils of the early archean apex chert: new evidence of the antiquity of life. *Science* 260:640–46

45. Schopf JW. 2000. The paleobiologic record of cyanobacterial evolution. See Ref. 5a, pp. 105–29

46. Sonenshein AL. 1999. Endospore-forming bacteria: an overview. See Ref. 5a, pp. 133–50

47. Suga H, Koyanagi M, Hoshiyama D, Ono K, Iwabi N, et al. 1999. Extensive gene duplication in the early evolution of animals before the Parazoan-Eumetazoan split demonstrated by G proteins and protein tyrosine kinases from sponge and hydra. *J. Mol. Evol.* 48:646–53

48. Summons RE, Jahnke LL, Hope JM, Logan GA. 1999. 2-Methylhopanoids as biomarkers for cyanobacterial photosynthesis. *Nature* 400:554–57

49. Thiel T, Pratte B. 2001. Effect on heterocyst differentiation of nitrogen fixation in vegetative cells of the cyanobacterium *Anabaena variabilis* ATCC 29413. *J. Bacteriol.* 183:280–86

50. Velicer G, Kroos L, Lenski RE. 1998. Loss of social behaviors by *Myxococcus xanthus* during evolution in an unstructured habitat. *Proc. Natl. Acad. Sci. USA* 95:12376–80

51. Velicer GJ, Kroos L, Lenski RE. 2000. Developmental cheating in the social bacterium *Myxococcus xanthus*. *Nature* 404:598–601

52. Walsby AE. 1985. The permeability of heterocysts to the gases nitrogen and oxygen. *Proc. R. Soc. London Ser. B. Biol. Sci.* 226:345–66

53. Whittaker RH. 1969. New concepts of kingdoms of organisms. *Science* 163:150–60

54. Wilcox M, Mitchison GJ, Smith RJ. 1973. Pattern formation in the blue-green alga, Anabaena. *J. Cell. Sci.* 12:707–23

55. Wolk CP. 2000. Heterocyst formation in Anabaena. See Ref. 5a, pp. 83–104

56. Yoon HS, Golden JW. 1998. Heterocyst pattern formation controlled by a diffusible peptide. *Science* 282:935–38

Annu. Rev. Genet. 2001. 35:125–48

THE INHERITANCE OF GENES IN MITOCHONDRIA AND CHLOROPLASTS: Laws, Mechanisms, and Models

C. William Birky, Jr.

Department of Ecology and Evolutionary Biology, and Graduate Interdisciplinary Program in Genetics, The University of Arizona, Tucson, Arizona 85721;
e-mail: birky@u.arizona.edu

Key Words   organelles, transmission genetics, vegetative segregation, uniparental inheritance, recombination

■ **Abstract**   The inheritance of mitochondrial and chloroplast genes differs from that of nuclear genes in showing vegetative segregation, uniparental inheritance, intracellular selection, and reduced recombination. Vegetative segregation and some cases of uniparental inheritance are due to stochastic replication and partitioning of organelle genomes. The rate and pattern of vegetative segregation depend partly on the numbers of genomes and of organelles per cell, but more importantly on the extent to which genomes are shared between organelles, their distribution in the cell, the variance in number of replications per molecule, and the variance in numerical and genotypic partitioning of organelles and genomes. Most of these parameters are unknown for most organisms, but a simple binomial probability model using the effective number of genomes is a useful substitute. Studies using new cytological, molecular, and genetic methods are shedding some light on the processes involved in segregation, and also on the mechanisms of intracellular selection and uniparental inheritance in mammals. But significant issues remain unresolved, notably about the extent of paternal transmission and mitochondrial fusion in mammals.

CONTENTS

0066-4197/01/1215-0125$14.00

INTRODUCTION

The literature on the inheritance of genes in mitochondria and chloroplasts (here-after, organelle genes) has changed, grown, and advanced tremendously since I first reviewed it for the *Annual Review of Genetics* (13). That review focused on the new discoveries made with *Saccharomyces cerevisiae* (hereafter, yeast or budding yeast) and on the models developed to account for these data. Some of these models have long since been discarded and others have been modified in light of new information from yeast and an increasing number of other organisms. Fortunately, this paper can build on more recent reviews that, together, cover various aspects of the subject in greater depth than is possible here (15–18, 32). The focus is on the search for molecular and cellular mechanisms responsible for the patterns of organelle gene inheritance. The aim is to explain the unique features of organelle inheritance, just as the chromosome theory of heredity and its extensions explain the most basic features of the inheritance of nuclear genes.

Some of the most exciting advances since the previous reviews have been made in understanding the molecular and cellular mechanisms of organelle division and distribution between daughter cells (partitioning) in yeast, animals, and plants; genetic studies of segregation and within-generation selection of mitochondrial genes in mammals and *Drosophila*; and the controversial subject of mitochondrial bottlenecks in mammals. Other exciting discoveries dealt with the mechanisms of uniparental inheritance in *Chlamydomonas* and mammals, and a controversy over whether there is a low level of biparental inheritance and recombination in humans. The growing excitement about mitochondrial genetics in humans and mammals has been driven in large part by their application to human diseases caused by mitochondrial mutations, and by the widespread use of mitochondrial genes to study the population genetics and evolution of humans and other animals. These subjects have also been reviewed in the past three years, but mainly as separate subjects and not in the context of organelle heredity in general.

The review begins with a brief reminder of the basic rules of inheritance of nuclear genes (Mendelian genetics) and the cellular mechanisms behind them. These are contrasted with the non-Mendelian rules of inheritance of organelle genes. This is followed by discussions of the molecular and cellular mechanisms of non-Mendelian inheritance: vegetative segregation, uniparental inheritance, and limited recombination. Besides satisfying our curiosity about the mechanisms of heredity, an understanding of these phenomena is essential for plant and animal breeding, including genetic engineering of organelle genomes; for diagnosing inherited mitochondrial diseases and counseling patients; and to enable the use of organelle genomes to study population genetics and evolution, including human evolution.

MENDELIAN VERSUS NON-MENDELIAN GENETICS

Mendel's Five Laws and the Mechanisms of Mendelian Inheritance

Mendel's model of heredity is still the accepted description of most of the important features of heredity for nuclear genes. As biological models go, Mendel's has great generality, although we now know of many exceptions (e.g., linkage). Textbooks commonly speak of Mendel's first and second laws, but the model that he used to explain his data actually had five components all of which are explicit or implicit in his paper (82). In 1909, Erwin Baur (5) showed that chloroplast genes in *Pelargonium* (geraniums) violate four of Mendel's five laws:

1. During asexual reproduction, alleles of nuclear genes do not segregate: Heterozygous cells produce heterozygous daughters. We now know that this is because nuclear genomes are stringent genomes (16) in which (*a*) all chromosomes are replicated once and only once in interphase; and (*b*) mitosis ensures that both daughter cells get one copy of each chromosome. In contrast, alleles of organelle genes in heteroplasmic cells segregate during mitotic as well as meiotic divisions to produce homoplasmic cells. This vegetative segregation occurs because organelles are relaxed genomes (15, 16) in which some copies of the organelle genome can replicate more often than others by chance or in response to selective pressures or intrinsic advantages in replication, and alleles can segregate by chance during cytokinesis.

2. Alleles of a nuclear gene always segregate during meiosis, with half of the gametes receiving one allele and half the other ("Mendel's first law"). Alleles of organelle genes may or may not segregate during meiosis; the mechanisms are the same as for vegetative segregation.

3. Inheritance of nuclear genes is biparental. Organelle genes, in contrast, are often inherited from only one parent (uniparental inheritance).

4. Alleles of different nuclear genes segregate independently ("Mendel's second law"), as a result of the independent segregation of chromosomes at meiosis and of recombination between genes on the same chromosome. In

contrast, organelle genes are nearly always on a single chromosome and recombination is often severely limited by uniparental inheritance or failure of organelles to fuse and exchange genomes.

5. Fertilization is random with respect to the genotype of the gametes. This is the only part of Mendel's model that applies to organelle as well as nuclear genes.

VEGETATIVE SEGREGATION

Vegetative segregation is the most general characteristic of the inheritance of organelle genes, occurring in both mitochondria and chloroplasts in all individuals or clones of all eukaryotes. I emphasize four systems that have been the focus of much of the research: chloroplast genes in plants and the alga *Chlamydomonas*, and mitochondrial genes in yeast and mammals. However, the conclusions about mechanisms of organelle heredity apply to many, and probably all, eukaryotes.

Plant Chloroplasts: Vegetative Segregation Due to Stochastic Partitioning of Organelles

AN ASIDE ON TERMINOLOGY   Organelle genomes are physically divided up between daughter cells at every cell division, but alleles do not necessarily segregate at every division. To avoid confusion, I use partitioning for the physical separation of genomes or alleles, and reserve segregation for those cases in which different alleles end up in different cells.

STOCHASTIC PARTITIONING OF ORGANELLES   The first, and simplest, formal model of organelle inheritance was applied to *Epilobium* and other plants with two or more chloroplasts per cell (62). This model assumes that the plastid is the unit of mutation and inheritance (Figure 1; see color insert). (We speak of plastids because that term includes the proplastids in the embryo plants as well as fully differentiated chloroplasts.) Each plastid is assumed to divide once per cell cycle. At cytokinesis, the two daughter cells are assumed to receive equal numbers of chloroplasts but a strictly random sample of the chloroplast genotypes. This physical model of partitioning corresponds to the mathematical model of sampling without replacement and the hypergeometric probability distribution (62, 95). Although it was developed for plant plastids, the physical model and the hypergeometric distribution have been applied to mitochondria, to mitochondrial and chloroplast genomes, and to other organisms. It has been known for three decades that no part of this simple physical model is strictly correct (for reviews see 11, 13, 15, 20). However, it is a reasonable approximation for many plants, correctly predicting that vegetative segregation will be complete within about one plant generation, given the number of proplastids seen in plant embryo cells (62). Although plastids do fuse (59), this is so rare in plants that it can be ignored. Of course, the plastid cannot always be the unit of inheritance because each plastid has tens to hundreds of genomes, and a new mutation only affects one. However, a genome with a new mutation increases in frequency in one or more plastids until they become homoplasmic for the mutant allele, after which the plastid is the unit of inheritance. We also know that daughter cells do not always receive equal numbers of organelles, and consequently, some plastids may have to divide more often than others to restore equality, but partitioning is numerically equal in most cases (14). It is to be expected that mechanisms have evolved to ensure that a cell has a better-than-random chance of receiving half of the parent cell's organelles (or organelle genomes), because this reduces the chances of its receiving no organelles.

ORGANELLES CAN BE DIVIDED INTO APPROXIMATELY EQUAL PARTS   Plastids must divide into two equal parts if they are to be considered units of inheritance that are identical in all respects except genotype; recent studies are beginning to show how this happens. In many organisms, division of mitochondria or chloroplasts is preceded by the appearance of a filamentous plastid division ring, or mitochondrial division ring, around the division furrow at the middle of the organelle (48). Recent studies detected two different molecular systems for the division of organelles in different organisms. One is FtsZ, which is used to divide bacterial and archaeal cells (73) as well as mitochondria and chloroplasts in plants and at least some eukaryotic protists (7, 45, 57; reviewed in 6, 70). In bacteria and archaea, the FtsZ protein polymerizes as a ring of filaments similar to tubulin around cells at their midpoint and constricts during cell division. In plants and algae, there are probably two FtsZ genes, one each for the division rings on the inner and outer plastid membranes (70, 71).

In bacteria, several proteins are necessary for positioning the ring made by FtsZ in the middle of the cell (reviewed in 25). Homologues of two of these are encoded by plastid genes in the green alga *Chlorella* (94). A nuclear gene encodes one in *Arabidopsis*, where it is required for correct positioning of the division ring at the midpoint of the plastid (25).

ANOTHER ASIDE ON TERMINOLOGY: DETERMINISTIC, STOCHASTIC, AND RANDOM MODELS   The study of organelle genetics has been plagued by confusion about the roles of chance and determinism in genetics. Scientists tend to favor deterministic models (hypotheses, laws; e.g., first and fourth Mendelian laws) because they make unambiguous predictions: A specific event is invariably succeeded by a specific outcome. But much of the world is unpredictable and must be described by stochastic models, which give only the probability of a specific outcome. Such models are also commonly called random, but that term is also used to describe a specific kind of stochastic process in which all outcomes have the same probability. I use strictly random to describe this specific class of models. The third and fifth Mendelian laws are strictly random. We now know that the segregation of different genes is not always strictly random because of linkage, but it is still stochastic.

Chlamydomonas Chloroplasts: Stochastic Replication and Partitioning of Genomes

VEGETATIVE SEGREGATION IS RAPID IN *CHLAMYDOMONAS* CHLOROPLASTS *Chlamydomonas reinhardtii* has been used extensively for chloroplast genetics since the pioneering studies of Sager (74, 75) and Gillham (31, 33). In contrast to the plants discussed above, *Chlamydomonas* cells have one chloroplast, which divides into two equal parts just before the cell divides; consequently, vegetative segregation cannot be explained by the partitioning of chloroplasts. Most data come from crosses of antibiotic-resistant by sensitive clones. Vegetative segregation can be studied in vegetative zygotes, which divide by mitosis instead of meiosis, or in the meiotic and early mitotic divisions of the small percentage of zygospores that show biparental inheritance. In either case, segregation is complete within a few cell generations. This is much too fast to be accounted for by random partitioning of the approximately 50–100 genomes.

GENOME PARTITIONING IS PROBABLY STOCHASTIC BUT NOT STRICTLY RANDOM One possible explanation for rapid segregation is that when the two gamete chloroplasts fuse in the zygote, the plastid genomes from the parents tend to remain in different parts of the chloroplast and consequently tend to segregate together rather than strictly randomly (92). The chloroplast genomes are grouped in about 5–15 nucleoids, and it is possible that the 10 or more genomes in each nucleoid tend to be replication products of one genome. In other words, genome partitioning is stochastic but not strictly random; like molecules tend to segregate together because they are joined in nucleoids and/or the nucleoids from the gametes are not completely mixed in the zygote.

GENOME REPLICATION IS STOCHASTIC Different *Chlamydomonas* zygotes from the same mating give rise to clones with very different frequencies of alleles from the two parents. Some zygote clones are uniparental, with organelle genomes from only one parent or the other. Frequency distributions of gene frequencies in a large number of zygote clones bear a striking resemblance to the gene frequency distributions of Mendelian populations undergoing random genetic drift (21). When the mitotic division of vegetative zygotes (93), or the meiotic divisions of zygospores (76), was delayed for a time by starvation, the variance in gene frequencies increased and more uniparental zygote clones were produced. These data suggested that plastid genomes continue to replicate during starvation and that replication is stochastic, with some genomes replicating more often than others by chance. The result is that gene frequencies within cells undergo stochastic changes, which I called intracellular random drift by analogy to random drift of nuclear gene frequencies in populations of organisms (13). Stochastic replication by itself will not completely eliminate an allele from a cell or clone, but may reduce it to a frequency too low to detect. Alternatively, there may be some degradation of organelle DNA molecules, which will then be replaced by additional replications of other molecules (turnover). Note that the stochastic replication of genomes, and the stochastic partitioning of genomes into daughter organelles when an organelle divides, can also explain how a mutant genome becomes homoplasmic in plant plastids. Figure 2 (see color insert) illustrates vegetative segregation due to a combination of stochastic replication and partitioning of organelle genomes.

Yeast Mitochondria

Much has been learned about organelle heredity from the study of another model genetic system, mitochondrial genes in budding or baker's yeast (*Saccharomyces cerevisiae*). The best markers are mutant genes conferring antibiotic resistance; respiration-deficient mutants (*petites*) are also used but their inheritance is strongly affected by selection. When heteroplasmic zygotes are produced by mating yeast strains that differ in one or more mitochondrial alleles, the majority of diploid progeny are homoplasmic after no more than 20 cell generations. Strictly random partitioning could only explain this rate of segregation if there were no more than 2 to 5 segregating units (19). This is much smaller than the number of mtDNA molecules in diploid cells [approximately 100] and slightly smaller than the number of nucleoids. Mitochondria from the two parents cannot be the segregating units because they fuse in the zygote. Consequently, vegetative segregation in yeast must be explained by some combination of the same factors that were invoked above for chloroplast genes in *Chlamydomonas*: (*a*) partitioning of genes that is stochastic but not strictly random, with similar molecules tending to remain together; (*b*) stochastic replication; or (*c*) turnover. There is experimental evidence only for the first two processes, but it is likely that all three are involved.

MITOCHONDRIAL FUSION AND FISSION   A yeast cell may contain a single large mitochondrial network, or a network plus a few small separate mitochondria, or many small discrete mitochondria, depending on its physiological state. Yeast mitochondrial genomes undergo multiple pairings with recombination in zygotes, showing that genomes from the two parents can interact extensively. Considerable progress has been made in understanding mitochondrial fusion and fission in yeast. Fission is accomplished by the dynamin system in yeast and animals (reviewed in 23). The dynamin Dnmp1p localizes to mitochondria at division sites and tips and is required for normal mitochondrial morphology. Mitochondrial fusion requires the *fzo1* (*fuzzy onion*) gene, a homologue of the *fuzzy onion* gene that is required for mitochondrial fusion in *Drosophila*. In yeast, normal mitochondrial morphology requires a balance between the activities of Dnm1p and Fzo1 (78).

BUD POSITION EFFECTS: NONRANDOM PARTITIONING   Early models of mitochondrial gene inheritance in yeast assumed that fusion was so frequent that a cell is effectively a single population of freely interacting genomes. That this could not be strictly true was demonstrated by pedigree studies of zygotes (19, 22, 83), which showed that (*a*) when the first bud comes from one end of the zygote, the majority of its mitochondrial genes come from the parent which formed that end of the zygote; and (*b*) buds that arise from the neck of the zygote receive markers from both parents, as well as a higher frequency of recombinant genotypes. This indicates that the mixing of mitochondrial genomes from the two parents is incomplete when the first bud is formed; later buds usually include markers from both parents, indicating more complete mixing. This interpretation was verified by showing that labeled mtDNA from one parent failed to enter the opposite side of the zygote until some time after the first bud was formed, although it did enter first center buds (68). The mitochondrial membranes from the two parents fused quickly, so delayed mixing of mtDNA was not due to delayed mitochondrial fusion; evidently, the movement of mtDNA across the zygote involves a different mechanism from the movement of mitochondria. Mitochondrial proteins also move more quickly through the mitochondrial network than does mtDNA (3, 68, 69).

MITOCHONDRIAL MOVEMENT FROM MOTHER TO BUD   Because *Saccharomyces* cells bud rather than undergoing binary fission, a mechanism is required to move mitochondria and their genes from the mother into the growing bud. The experimental studies of this process have been reviewed (23). Mitochondria are actively transported from the mother cell into the bud, where they are immobilized at the tip of the bud until cytokinesis is complete. Mitochondria probably move along actin filaments by a motor that depends on actin polymerization (80), and movement also requires intermediate filaments encoded by the MDM gene (58). It is not surprising that a mechanism evolved which ensures that buds receive at least some mitochondria, which are required for survival, and mitochondrial genomes, which are required for respiratory competence.

STOCHASTIC REPLICATION   As was the case for *Chlamydomonas* chloroplast genes, yeast cells can become homoplasmic for mitochondrial genes without dividing, owing to random genetic drift of gene frequencies within the cell (reviewed in 16). This was demonstrated using delayed division experiments with both budding and fission yeast (91), analogous to those in *Chlamydomonas*. Birky and colleagues (20) reported that many first central buds are uniparental, producing clones with mitochondrial genes from only one parent; however, when wild-type cells were mated with ρ^o mutants that have mitochondria but no mtDNA, all first central buds receive mtDNA. They suggested that all first central buds probably receive mtDNA from both parents but that stochastic replication (possibly combined with turnover) eliminates genes from one parent or the other. Stochastic replication is almost certainly a major contributor to the production of homoplasmic cells during asexual reproduction in yeast, i.e., to vegetative segregation.

NUCLEOID STRUCTURE AFFECTS MITOCHONDRIAL GENE INHERITANCE   It was suggested that the segregating units in yeast mitochondria might be nucleoids (19), and recent studies suggest that nucleoid structure does affect the inheritance of mitochondrial genes. The mtDNA molecules in a nucleoid appear to be held together by Holliday structures (46, 51, 53, 97), perhaps because mtDNA replication is initiated by recombination (8, 56, 77) as it is in T-even phage (66). Mutations that affect the resolution of the Holliday structures also modify the inheritance of neutral ρ^- genomes in $\rho^- \times \rho^+$ crosses (54, 98).

Mammalian Mitochondria

Vegetative segregation is difficult to study in humans and other mammals with uniparental inheritance, because the only sources of heteroplasmic cells are new mutations. Early studies of mitochondrial genetics in mammals bypassed uniparental inheritance by fusing cultured animal cells or enucleated cytoplasts of different mitochondrial genotypes and following the proportions of the two genotypes over time. The interpretation of these studies is complicated by the use of human-rodent and other interspecific hybrids that may have been affected by incompatibility of nuclear and mitochondrial genes, or by the use of antibiotic resistance mutants that were subject to selection. The discovery of mitochondrial mutations segregating in a herd of dairy cattle showed that a mitochondrial mutation can be fixed in a few generations (1, 47). More recently, vegetative segregation has been studied in heteroplasmic mice created by cytoplast fusion. The offspring of such mice can have dramatically different levels of heteroplasmy; this is not due to selection because the mean allele frequency among the progeny is the same as that in the mother (24). Single-cell PCR has been used to demonstrate that the increase in the variance of allele frequencies takes place in maturing oocytes and in dividing germline cells (41). Colonic crypts were used for another elegant demonstration of drift in heteroplasmic mice (42). Each crypt is derived from a single founder cell that produces stem cells, which in turn continually divide to replace crypt cells. The proportion of donor mitochondrial genomes was determined in individual crypts from heteroplasmic embryos aged 4 and 15 months. The mean gene frequency was about 4% at both times, indicating the absence of selection, but the variance of donor mtDNA proportion among crypts increased greatly. The frequency distributions of genotype frequencies at the two times strongly resemble frequency distributions of allele frequencies in populations of organisms undergoing random genetic drift.

There is no way of telling to what extent vegetative segregation in animals is due to stochastic partitioning as opposed to stochastic replication. There is some interesting evidence that the choice of mtDNA molecules for replication is not strictly random but is biased in favor of molecules near the nucleus (26, 61, 63).

The studies on cattle and on mouse models show that vegetative segregation requires one or a few organismal generations to complete. If there are about 20 cell generations per organismal generation, segregation in animals is substantially slower than in yeast, *Chlamydomonas*, and plants. Paradoxically, a clone of animal cells carrying wild-type and respiration-deficient genomes produced almost no homoplasmic cells (39, 50). Unrealistic models were proposed to explain this as a case of no, or very slow, vegetative segregation. However, the apparently stable heteroplasmy may actually represent an equilibrium between intracellular selection tending to increase the frequency of respiration-deficient genomes (see Paradoxical Intracellular Selection . . . below) and intercellular selection against cells with very high frequencies of respiration-deficient genomes.

Do Mammalian Mitochondria Fuse?

A continuing controversy about mammalian mitochondrial genetics is the question of the extent to which mitochondria fuse and share genomes and other components. Fusion is difficult to prove using the static pictures from electron microscopy, and light microscopy cannot distinguish between permanent fusion and transient contacts of these tiny organelles. A number of authors have created cells heteroplasmic for two different mitochondrial genotypes and looked for complementation, which would indicate sharing of genes or gene products. Complementation was found in some but not all cases. I pointed out that in the experiments where no complementation was observed, there was no independent evidence that the two mutants could complement each other (18). This potential problem was highlighted by a recent paper (27). Human cells with two different mutations, one in tRNA(Lys) and one in ND4, were mixed and treated to promote fusion. Fused cells were selected using nuclear drug-resistant mutant genes in glucose medium, which does not select for respiratory competence. Cells with complementation of the mitochondrial mutations were selected in galactose-containing medium in which respiratory competence is required for growth. The frequency of cell fusion was much greater than the frequency of complementation, leading the authors to conclude that no more than 1.5% of the fusion products showed complementation. But the cells showing complementation grew slowly, suggesting that complementation might not be complete enough to be detected in many of the cells in which it occurred. This potential problem was avoided in another experiment by creating cybrids that were heteroplasmic for a genome with a deletion of several tRNAs and a genome with a point mutation in another tRNA (87). The cybrids did not have sufficient respiratory competence to grow in selective medium, but complementation was demonstrated in medium that did not select for respiratory competence, by finding fusion peptides that could only be transcribed from the deletion genome and translated with the help of normal tRNAs from both genomes. It has been suggested that when complementation was found, it might be due to transient fusion of mitochondria induced by the PEG used to fuse the cells (27). But this does not explain why complementation was found only in medium that did not select for respiratory competence (87). I conclude that the evidence for complementation continues to be more convincing than the evidence against it. Unfortunately, these studies do not indicate the frequency of mitochondrial fusion. Mitochondrial DNA from wild-type cytoplasts spread throughout mitochondria from ρ^0 cells within 6 h after fusing the cells (35). However, we do not know if such rapid fusion and sharing of genomes would be seen between ρ^+ mitochondria.

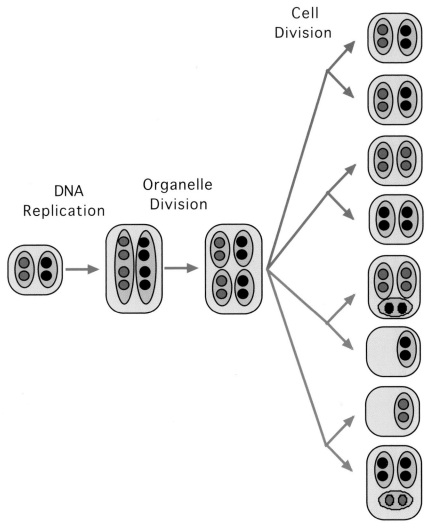

Figure 1 Vegetative segregation: simple plant model. A heteroplasmic cell has two organelles, one with two wild-type genomes (*black circles*) and one with two mutant genomes (*red circles*). DNA replication is stringent (each genome replicates once). When the organelles divide, the genomes are partitioned equally. When the cell divides the organelles are partitioned numerically equally (*red and green arrows*) or unequally (*blue arrows*). Organelle partitioning is genetically stochastic (relaxed) and can produce cells that are homoplasmic. If partitioning were deterministic with sister organelles always going to different cells (stringent partitioning; *red arrows*), there would be no vegetative segregation.

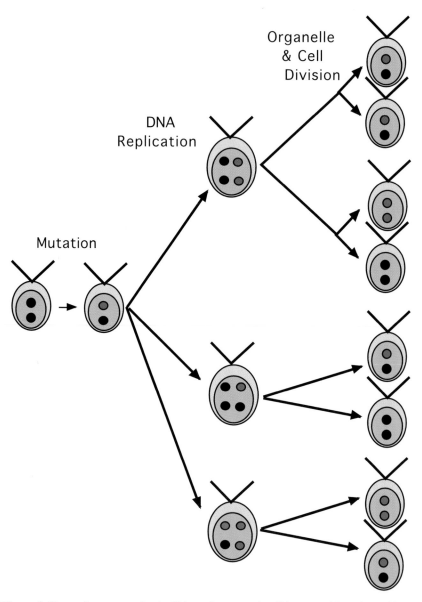

Figure 2 Vegetative segregation in Chlamydomonas. A cell has one chloroplast with two wild-type genomes; mutation produces a heteroplasmic cell. DNA replication is stochastic (relaxed). Genome partitioning is numerically equal but genetically stochastic. Relaxed replication increases the probability that a daughter cell will be homoplasmic because it produces cells with 3:1 or 1:3 ratios of wild-type:mutant and these always produce a homoplasmic daughter.

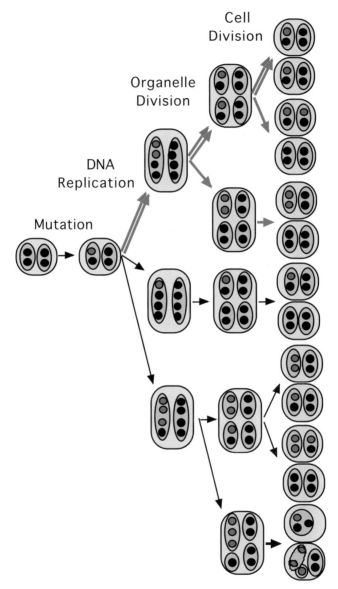

Figure 3 Vegetative segregation with discrete organelles: a more sophisticated model. After a mutation produces a heteroplasmic organelle, relaxed replication and partitioning of genomes produces homoplasmic organelles; thereafter the organelle is the unit of segregation. Red and green double arrows show the results of stringent replication and partitioning; green arrows collectively show the simple plant model; the black arrows add the additional cases that are possible if genome replication is relaxed and if organelle partitioning is relaxed with respect to number.

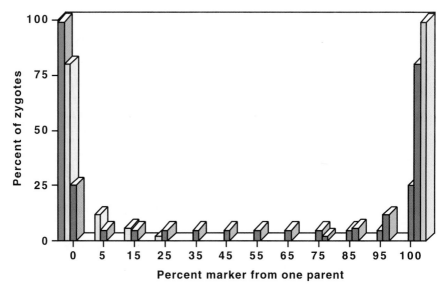

Figure 4 Uniparental inheritance is a quantitative phenomenon. Frequency distributions of the frequency of alleles from one parent (e.g. the paternal parent) in the zygotes produced by matings of different organisms. Blue graph shows strict uniparental maternal inheritance; yellow graph shows mixture of maternal and biparental zygotes; red graph shows maternal, biparental, and paternal zygotes; purple graph is paternal plus biparental zygotes; and grey graph illustrates strictly paternal inheritance.

The Simplest Model of Vegetative Segregation

It is apparent from the discussion of vegetative segregation that stochastic replication and partitioning of genomes during the division of organelles and cells, and stochastic partitioning of organelles at cell division work together to produce homoplasmic daughter cells from heteroplasmic mothers (Figure 3; see color insert). Stochastic replication of organelles may also play a role but this has not been clearly demonstrated. Two other stochastic processes, gene conversion and turnover, can assist in making homoplasmic cells heteroplasmic even in the absence of cell division. Evidently, the rate of vegetative segregation is determined by many factors: the number of organelles and the extent to which they share organelles; the number of genomes per organelle and per cell; the variance in number of times a genome replicates; the variance in numbers of genomes that are partitioned into daughter organelles and daughter cells; and the degree of mixing of organelles of different genotypes in organelles and in cells. It is extremely unlikely that we will ever know all of these parameters exactly for any organism. This is especially true for animals and plants, where the parameters probably vary among different cell types and at different stages of development. Even if we did know all of the relevant parameters, an exact mathematical model would be impossible to solve and computer simulations would be tedious at best. Fortunately, we can borrow a simple mathematical model that requires only two measurable parameters from Mendelian population genetics. In this model, the cell has an effective number n_e of organelle genomes. Stochastic replication and partitioning are modeled by giving each daughter cell a strictly random sample with replacement from the mother cell. The allele frequencies in the daughter cells follow the binomial distribution. Starting with a cell in which the allele frequency is p_0, binomial sampling is continued for a number c of cell generations, at which time the variance in the frequency of the allele among the cells is given by

$$V_c = p_0(1 - p_0)[1 - (1 - 1/n_e)^c] \qquad 1.$$

The literature reflects a great deal of misunderstanding about the parameter n_e. This is an effective number, which is an unspecified function of the real number of genomes in a cell. For example, if the increase in variance were due entirely to random partitioning, n_e could be replaced by the hypergeometric distribution and the real number n of genomes per cell, as in the simplified plant model described earlier. The important point is that n_e is not the number of genomes or of any other biological entity in any cell, and in fact it is unlikely ever to correspond closely to the number of anything. Its utility lies in the fact that it can be estimated from V_c, p_0, and c using Equation 1, after which it can be used to predict the rate of vegetative segregation or to compare the rate of segregation in different systems. It should be measured using neutral alleles so that it reflects drift alone, not intracellular or intercellular selection. After that it can be used as a null model; if other alleles segregate more rapidly, one can suspect that the alleles are subject to selection or have some effect on other factors that can affect segregation such as the number of genomes or their distribution in the cell.

Bottlenecks: Are they Real, and Do We Need Them?

The literature on mammalian mitochondrial genetics is full of references to bottlenecks, most of which reflect a misunderstanding of the effective number of genomes in Equation 1. Many authors have used this equation or something analogous to estimate n_e from the variance in gene frequencies in a clone of cells or the offspring of a single female. Then they compare this to the real number of genomes, nucleoids, or organelles per cell to see if any real number matches n_e; this is then taken to be the effective segregating unit. But n_e does not represent any real physical entity. For example, if n_e is smaller than the real number of genomes per cell (which it always is), one possibility is that genome replication is strictly random but partitioning is not because genomes are not well mixed in the cell. We do not need bottlenecks to explain why n_e is smaller than the number of genomes. But this does not mean that bottlenecks are not real. A review of published electron micrographs led to the conclusion that there are fewer than 10 mitochondria per primordial germ cell (40); these eventually give rise to primary oocytes with many hundreds of mitochondria.

INTRACELLULAR SELECTION

Intracellular Selection Based on Phenotype

Birky (4, 10) showed that mitochondrial genomes carrying antibiotic-resistance point mutations replicate and replace wild-type genomes when cells are exposed to antibiotic. In the absence of the antibiotic, wild-type genomes out-replicate those with resistance markers. Selection within an organism has also been demonstrated in mice that are heteroplasmic for mitochondrial genomes from different strains (e.g., 42, 88). Unfortunately, studies on whole animals cannot distinguish between intracellular selection and intercellular selection, i.e., selective growth of cells that acquired more copies of the favored genotype. Clear evidence of intercellular selection has been seen in plants that are heteroplasmic for green and white plastids.

Intracellular Selection Based on Genome Structure

A superficially similar phenomenon was seen in the earliest experiments on yeast mitochondrial genetics, in which wild-type genomes are out-replicated by highly suppressive *petite* genomes. A highly suppressive *petite* genome consists of a small segment of the wild-type genome, repeated to produce a molecule of approximately normal size. It is now almost certain that these genomes out-replicate wild-type genomes because they have more copies of a replication origin (52). Molecules with more replication origins may also have an advantage in other systems such as cultured human cells (89). To distinguish cases such as this from intracellular selection based on the phenotypes of the molecules, I suggest that the two phenomena be called phenotypic intracellular selection and structural intracellular selection.

DO SHORT GENOMES HAVE A REPLICATIVE ADVANTAGE?   It is often assumed that shorter genomes will generally have a replicative advantage over longer ones in the same cell because they can complete replication and re-initiate more quickly. The experimental evidence is mixed: in heteroplasmic *Drosophila*, selection can favor either shorter (72) or longer genomes (44, 81). However, it is not clear whether selection was intracellular or intercellular in these cases, or whether it is based on some phenotypic effect of the deletion. Even without experimental evidence, it is not necessarily expected that smaller molecules can complete replication faster that larger ones. In human cells from a KSS patient, heteroplasmic for wild-type mtDNA and genomes with a deletion, the same number of wild-type genomes and deletion genomes were synthesized during a 5-h period (65). This suggests that the original amplification of the deletion mutant genome in the patient (or the patient's mother) was not due to the more rapid completion of replication of the smaller genome. The authors suggested that the overall rate of replication might be limited by the rate of initiation rather than the rate of completion of mtDNA synthesis. Perhaps the structure of the replicating genome is modified so as to prevent additional initiations until the first replication is completed.

HOW IS MTDNA REPLICATION CONTROLLED?   The outcome of intracellular selection in heteroplasmic cells could be affected not only by the rate of replication initiation or the time to completion, but also by the mechanism used by cells to stop replication when the appropriate number of genomes has been reached. Recall that genomes are selected stochastically for replication until the number of genomes is doubled (or reaches some other predetermined value). It is likely that either the number of genomes or the mass of organelle DNA is counted, directly or indirectly. In yeast, studies on *petite* genomes suggest that mass, or the total number of base pairs, is titrated (30, 37, 67). Human cell lines containing wild-type mtDNA or mtDNA mutants with complete or no impairment of respiration, very different sizes, and different numbers of replication origins all had the same total mass of mtDNA per cell (90). The authors suggested that mtDNA may be replicated until tightly regulated dNTP pools are depleted.

If cells limit mtDNA replication by titrating total DNA mass, then there is an alternative explanation for the replicative advantage of smaller genomes. Consider a simple model in which mtDNA molecules are selected randomly in a series of rounds of replication until the total mass of genomes is increased to a certain value. If long genomes are chosen for replication more often than short ones, by chance, the final mass will be reached in a smaller number of replication events.

If short genomes are chosen more often, they will have to undergo more rounds of replication to reach the same mass. After two or more doublings, the result will be an increase in the proportion of short genomes. The behavior of this model needs to be confirmed mathematically or by simulations, but it is easily shown to be correct for a simple case of cells that are heteroplasmic for two genomes, one half as long as the other (C.W. Birky, Jr., unpublished).

Paradoxical Intracellular Selection Based on Respiration

Many patients with respiration-deficient mitochondria acquired the mutant genomes as a new mutation early in embryogenesis or in their maternal germline (e.g., 49). This mutation must have been amplified in the cells to the point where it causes clinical symptoms. Many of these are deletions, and one could suppose that this gives them an advantage, as discussed above. However, respiration-deficient point mutations can also have a selective advantage (96). In a review of data from a large number of human pedigrees in which one or another of six common pathogenic point mutations were segregating, at least three point mutations showed significant selection in favor of the mutant allele (24). Mutant and wild-type genomes have the same mass in this case, but the replicative advantage of this mutant can be explained if it is respiration that is titrated. The model for this case is formally the same as the one for mass differences: It takes more mutant molecules than wild type for a cell to achieve any specific level of respiration, so mutant molecules replicate more often. This explanation may also apply to respiration-deficient mutations such as *kalilo* in *Neurospora* (9), which increase in frequency as a mycelium ages, until they kill it.

UNIPARENTAL INHERITANCE

Patterns of Uniparental Inheritance

MATERNAL INHERITANCE IS NOT A GENERAL FEATURE OR LAW OF ORGANELLE HEREDITY   Baur's work on plants showed that maternal inheritance was not a general law of organelle heredity, since some plants produce a mixture of maternal, paternal, and biparental progeny. More recently, uniparental inheritance has been seen in organisms that have no differentiation of maternal or paternal sexes. Moreover, in plants and algae, mitochondrial and chloroplast genes may be inherited preferentially from different sexes or mating types. The most general statement we can make about uniparental inheritance is that in most organisms, some or all progeny inherit organelle genes from only one parent.

UNIPARENTAL INHERITANCE IS OFTEN A QUANTITATIVE PHENOMENON   When there is some degree of biparental inheritance, one can estimate the frequency of one allele in a large number of offspring from a single mating or from a group of matings of the same genotypes. A frequency distribution of the numbers of progeny with different allele frequencies usually shows a continuous distribution, which may be unimodal, bimodal, or occasionally trimodal but is without sharp discontinuities (Figure 4; see color insert).

LARGE SAMPLE SIZES ARE NEEDED TO DEMONSTRATE STRICT UNIPARENTAL INHERITANCE   This is a consequence of the variation among the progeny of a mating in the degree of biparental transmission. For example, assume that an animal transmits paternal mitochondrial genes such that 1% of all of the mitochondrial genes in the progeny of a cross come from the father. If there are few or no biparental zygotes, only about 1% of the progeny will have the paternal genotype and a sample of about 300 progeny with no paternal genotypes would be required to demonstrate that there was less than 1% paternal inheritance (64). A more likely scenario would be that all of the paternal genes were in biparental progeny, but in that case it is possible that there are only a few such individuals, or that most of the biparental progeny have very low frequencies of paternal genes. Furthermore, most of the minority markers may be localized in one or a few tissues. The best approach is to use sensitive molecular methods to detect the marker in the pooled tissue of many progeny; if paternal transmission is found, one should then screen individual progeny to determine how the paternal markers are distributed. Selection for streptomycin-resistance genes was used to detect paternal transmission in 1/1500 progeny of a cross in *Nicotiana* (60), but this method is rarely practical.

Mechanisms of Uniparental Inheritance

THERE ARE MANY DIFFERENT MECHANISMS OF UNIPARENTAL INHERITANCE   The striking variation among different organisms in the extent and pattern of uniparental inheritance is mirrored in a remarkable diversity of mechanisms. The transmission of organelle genes from one parent to the offspring can be blocked at any stage of sexual reproduction [see (12, 17) for evidence and examples]:

1. Gametogenesis: organelles may be segregated from the gamete during premeiotic or meiotic divisions; organelles or organelle DNA may be degraded in the gamete.
2. Fertilization: organelle DNA is shed from gametes before fertilization, or does not enter the egg.
3. Postfertilization: selective silencing (degradation) of organelles or organelle DNA in the zygote; stochastic or directed segregation of organelles into extraembryonic tissues during early cleavages; or loss of alleles from one parent due to stochastic replication and/or turnover of organelle genomes.

SELECTIVE SILENCING OF PATERNAL MITOCHONDRIAL GENES IN MAMMALS   In mussels, mtDNA occurs in two separate lineages that are inherited differently; this is not reviewed here because the mechanism remains a mystery. Apart from this exception, mtDNA is always inherited maternally in crosses in all the animals that have been studied to date. However, in most cases too few offspring have been examined to detect low levels of paternal inheritance. This is especially true in studies of human pedigrees in which fewer than 2500 offspring have been examined in all of the available pedigree data (28). This seems like a lot, but because the sperm contains about 1/1000 times as many mtDNA molecules as the oocyte, one expects to find fewer than 2.5 uniparental paternal individuals in this sample, or somewhat more biparental individuals. When different species of mice were crossed and the hybrids were repeatedly backcrossed to the male parent to amplify small paternal contributions, paternal mtDNA was detectable by PCR (34). Another group used PCR to detect small paternal contributions in both inter- and intraspecific crosses (43). In intraspecific matings, paternal mtDNA was detected in the majority of embryos at the early pronucleus stage in intraspecific crosses but disappeared in all of 48 embryos by late pronucleus through blastocyst stages. But in interspecific crosses, paternal mtDNA was detected in some embryos at every stage, including 24 of 45 neonates. Subsequent studies of an interspecific mouse cross by this group (79) used a PCR method sufficiently sensitive to detect a few molecules of paternal mtDNA in a background of 10^8 molecules of maternal mtDNA. They again detected paternal mtDNA in embryos but when the hybrid mice were reared to maturity, paternal mtDNA could be found in only one or a few tissues in each individual; this is expected if the paternal and maternal genomes segregated during development. Few hybrid animals had paternal mtDNA in the ovary and none had it in their unfertilized eggs, nor was any detected in backcross progeny. The investigators conclude that some mechanism recognizes and specifically destroys paternal mtDNA in eggs of intraspecific crosses, but partially fails in interspecific hybrid eggs so that a small number of paternal genomes escape degradation.

In fact, electron microscopy indicates that the entire sperm midpiece degenerates in the mammalian egg (e.g., 36, 84, 86). Recent studies have provided strong evidence that sperm are marked for degradation by ubiquitination. Ubiquitin is a protein that binds to other proteins and marks them for degradation by the 26S proteasome. It also marks for engulfment and lysis by lysosomes or vacuoles. Sutovsky and collaborators (85) demonstrated that ubiquitin, detected by fluorescence-labeled anti-ubiquitin antibodies, is bound to sperm mitochondria during spermatogenesis and in the oocyte of cows and rhesus monkeys. Ubiquitin was not detected in sperm on the surface of the egg or soon after entering the egg; the authors proposed that ubiquitin is masked in these sperm, but they could not rule out the possibility that the ubiquitin on the sperm is lost somewhere between spermatogenesis and fertilization, then is re-established in the egg. The ubiquitinated sperm subsequently disappear, typically between the third and fourth cell division. The authors proposed that ubiquitin marks the mitochondria for subsequent degradation by proteasomes and lysosomes. Ubiquitination is evidently required for this degradation, because degradation is prevented by injecting anti-ubiquitin antibodies into the fertilized egg, or by treating the egg with ammonium chloride, which is "lysosomotropic" (84). Ubiquitination of sperm was not observed when cow eggs were fertilized with sperm of the wild gaur, and the ubiquitin-labeled sperm could be detected in eight-cell embryos (85). The results of this interspecific cross parallel the transmission of paternal sperm in crosses between *Mus musculus* and *M. spretus* (43).

RECOMBINATION

Between-Lineage Recombination of Organelle Genes is Limited

The majority of animals and plants, and many or most fungi and eukaryotic protists, reproduce sexually at least occasionally. During sexual reproduction, nuclear genes are inherited biparentally and genes from different parents recombine due to crossing-over, gene conversion, and independent segregation of chromosomes. But organelle genes from different lineages rarely or never recombine in most of these same organisms. This is because organelle genomes are usually inherited uniparentally; and if they are inherited biparentally, the organelles from the two parents fail to fuse and share genomes. In many angiosperms, for example, mitochondrial and chloroplast genes from different individuals do not recombine in crosses where they are inherited uniparentally. If cells from two parents are fused, recombinant genotypes are readily detected for mitochondrial genes, but recombinant chloroplast genomes are rare and can only be detected after stringent selection. When recombinants for two different antibiotic resistance genes are selected, other markers on the selected genomes show extensive recombination (59). This shows that plant chloroplasts do have the enzymatic machinery required for recombination but the chloroplasts rarely fuse.

Although plant organelle genomes from different lineages rarely have an opportunity to recombine, intramolecular and intermolecular recombination within a lineage can still occur. These forms of recombination can be very important in rearranging genomes. Intramolecular and intermolecular recombination of cpDNA maintains the sequence identity of the inverted repeats, inverts the order of genes in the single-copy regions, and produces dimeric genomes; while inter- and intramolecular recombination in mtDNA produces subgenomic circles (reviewed in 32). Repeated rounds of random pairing and gene conversion could cause intracellular random drift (12), but so far it has not been possible to determine how important this is, relative to stochastic replication. One might imagine that when organelle genes are inherited biparentally, they would still show less recombination on average than nuclear genes because they are all on one chromosome. But this is not necessarily so: In *Saccharomyces cerevisiae*, the genomes in a zygote and its early buds undergo repeated rounds of pairing and recombination, resulting in recombination frequencies of about 1% per 100 bp (97).

Do Hominid Mitochondrial Genes Have a Low Level of Biparental Inheritance and Recombination?

A potentially more powerful approach to detect paternal inheritance is to analyze population and evolutionary genetic data for evidence of recombination between different mtDNA lineages. In principle, this approach could detect paternal inheritance and recombination because it analyzes the pooled results of very large numbers of matings taking place over long time periods. But the results have been controversial. Eyre-Walker et al. (29) argued that the substantial homoplasy seen in mtDNA trees was more likely due to recombination between different lineages than to multiple mutations at a site, as was previously assumed. However, their data sets contained significant errors, and when these were corrected the argument that homoplasy is due to recombination was weakened (55). Awadalla et al. (2) then provided more compelling evidence for recombination, showing that the amount of linkage disequilibrium between pairs of mitochondrial sites decreased as the distance between the sites increased. Although the logic of this test is correct, the results were challenged on methodological grounds, and some additional data sets were found not to show a significant negative correlation between distance and disequilibrium (see discussion at http://www.sciencemag.org/cgi/content/full/288/5473/1931a). The subject was nicely reviewed by Eyre-Walker (28), who noted that only 9 of 14 human data sets show the negative correlation and none was significant. On the other hand, a chimpanzee data set did show a significant negative correlation, and the majority of the data seem to point in that direction. If this is confirmed by future studies, the assumption that mtDNA can be used as a clonal maternal lineage in studies of human evolution will have to be reconsidered. Another important consequence of paternal leakage would be that mitochondria would not be strictly asexual and would be less susceptible to the accumulation of detrimental mutations (Muller's ratchet) than has been assumed.

Do human mitochondria contain the enzymes necessary for recombination? Tang et al. (89) found that a human cell line homoplasmic for dimeric mtDNA molecules gave rise to monomeric wild-type and deletion genomes, as expected if there were intramolecular recombination of the dimers. They review biochemical evidence that mammalian mitochondria contain at least some of the enzymes required for intramolecular recombination. However, when they cultured cells that were heteroplasmic for the wild-type and deletion genomes, they found none of the dimers that would result from intermolecular recombination. They suggest that this may be because the two genomes were initially in separate cells that were fused to make heteroplasmic cybrids; consequently, they may have remained physically isolated from each other, in different organelles or different regions of an organelle. On the other hand, they argue that the dimeric mutants were almost certainly originally formed by a combination of intramolecular and intermolecular recombination between two molecules in the same cell. The data of Tang et al. (89) suggest that the conclusion that mammalian mitochondria do not recombine

(18) is incorrect, at least for genomes in the same organelle. These authors (89) also point out that triplicated mitochondrial genomes found by Holt et al. (38) in cell cultures initially homoplasmic for duplicated genomes probably arose by intermolecular recombination.

WHAT NEEDS TO BE DONE

Most of the genetic phenomena unique to organelles have been known for over a decade, and the possible cellular and molecular mechanisms have been identified. Finally, the time is ripe to apply a combination of genetic, molecular, and cytological methods to determining the relative importance of these mechanisms in specific cases. New genetic methods such as directed mutagenesis and transfection and new selection methods now enable us to obtain mutants that are defective in a variety of processes affecting the transmission of organelles and organelle DNA. New molecular and cytological tools such as green fluorescence protein, fluorescence in situ hybridization, and confocal microscopy enable us to measure parameters such as organelle and genome number and visualize the effects of mutants. Now we need to apply these methods. For example:

1. It is not yet clear to what extent vegetative segregation in plant plastids is due to strictly random partitioning of plastids. For this, the new molecular and cytological techniques must be used to determine the numbers of organelles in eggs, embryos, and meristem cells where most segregation occurs. Plastids need to be counted in pairs of daughter cells to determine how often partitioning really is numerically equal.

2. *Saccharomyces cerevisiae* is the only organism in which extensive data on the inheritance of mitochondrial genes can be matched with detailed cytological pictures of the movement of mitochondrial membranes and mtDNA and molecular genetic analysis of the role of specific proteins. The combined use of mutants and molecular genetic methods for which yeast is famous, coupled with new high-resolution cytology, could lead to a detailed picture of the mechanisms underlying vegetative segregation and uniparental inheritance in yeast. Surprisingly, this has not been done. There have been almost no studies of the effect of mutants defective in fusion or fission, or mitochondrial and mtDNA movement in zygotes, on the inheritance of mitochondrial genes. Much of what has been done used *petite* mutants whose inheritance is too strongly affected by intracellular and intercellular selection to give a clear picture of mtDNA inheritance.

3. Before we can understand intracellular selection, which is so important in human mitochondrial gene diseases, we must understand how cells control organelle DNA replication. We cannot expect a single model to suffice. The available data suggest that replication control mechanisms differ not only between organisms, but also between cell types in the same organism. It also

appears that different kinds of mutants may have a replicative advantage for different reasons; perhaps one cell type can measure organelle genome replication in more than one way.

Visit the Annual Reviews home page at www.AnnualReviews.org

LITERATURE CITED

1. Ashley MV, Laipis PJ, Hauswirth WW. 1989. Rapid segregation of heteroplasmic bovine mitochondria. *Nucleic Acids Res.* 17:7325–31
2. Awadalla P, Eyre-Walker A, Maynard Smith J. 1999. Linkage disequilibrium and recombination in hominid mitochondrial DNA. *Science* 286:2524–25
3. Azpiroz R, Butow RA. 1993. Patterns of mitochondrial sorting in yeast zygotes. *Mol. Biol. Cell* 4:21–36
4. Backer JS, Birky CWJ. 1985. The origins of mutant cells: mechanisms by which *Saccharomyces cerevisiae* produces cells homoplasmic for new mitochondrial mutants. *Curr. Genet.* 9:627–40
5. Baur E. 1909. Das Wesen und die Erblichkeitsverhältnisse der "Varietates albomarginatae hort" von *Pelargonium zonale.* *Z. Vererbungsl.* 1:330–51
6. Beech PL, Gilson PR. 2000. FtsZ and organelle division in protists. *Protist* 151:11–16
7. Beech PL, Nheu T, Schultz T, Herbert S, Lithgow T, et al. 2000. Mitochondrial FtsZ in a chromophyte alga. *Science* 287:1276–79
8. Bendich AJ. 1996. Structural analysis of mitochondrial DNA molecules from fungi and plants using moving pictures and pulsed-field gel electrophoresis. *J. Mol. Biol.* 255:564–88
9. Bertrand H, Griffiths AJF, Court DA, Cheng CK. 1986. An extrachromosomal plasmid is the etiological precursor of kalDNA insertion sequences in the mitochondrial chromoosome of senescent Neurospora. *Cell* 47:829–37
10. Birky CW Jr. 1973. On the origin of mito-

chondrial mutants: evidence for intracellular selection of mitochondria in the origin of antibiotic-resistant cells in yeast. *Genetics* 74:421–32
11. Birky CW Jr. 1975. Mitochondrial genetics in fungi and ciliates. In *Genetics and Biogenesis of Mitochondria and Chloroplasts*, ed. CW Birky Jr, PS Perlman, TJ Byers, pp. 182–224. Columbus, Ohio: Ohio State Univ. Press
12. Birky CW Jr. 1976. The inheritance of genes in mitochondria and chloroplasts. *BioScience* 26:26–33
13. Birky CW Jr. 1978. Transmission genetics of mitochondria and chloroplasts. *Annu. Rev. Genet.* 12:471–512
14. Birky CW Jr. 1983. The partitioning of cytoplasmic organelles at cell division. *Int. Rev. Cytol. Suppl.* 15:49–89
15. Birky CW Jr. 1983. Relaxed cellular controls and organelle heredity. *Science* 222:468–75
16. Birky CW Jr. 1994. Relaxed and stringent genomes: Why cytoplasmic genes don't obey Mendel's laws. *J. Hered.* 85:355–65
17. Birky CW Jr. 1996. Uniparental inheritance of mitochondrial and chloroplast genes: Mechanisms and evolution. *Proc. Natl. Acad. Sci. USA* 92:11331–38
18. Birky CW Jr. 1998. Inheritance of mitochondrial mutations. In *Mitochondrial DNA Mutations in Aging, Disease and Cancer*, ed. KK Singh, pp. 85–99. Austin, TX: Landes Biosciences
19. Birky CW Jr, Strausberg RL, Perlman PS, Forster JL. 1978. Vegetative segregation of mitochondria in yeast: estimating parameters using a random model. *Mol. Gen. Genet.* 158:251–61

20. Birky CW Jr, Acton AR, Dietrich R, Carver M. 1982. Mitochondrial transmission genetics: replication, recombination, and segregation of mitochondrial DNA and its inheritance in crosses. In *Mitochondrial Genes*, ed. P Slonimski, P Borst, G Attardi, pp. 333–48. Cold Spring Harbor, NY: Cold Spring Harbor Lab.

21. Birky CW Jr, VanWinkle-Swift KP, Sears BB, Boynton JE, Gillham NW. 1981. Frequency distributions for chloroplast genes in *Chlamydomonas* zygote clones: evidence for random drift. *Plasmid* 6:173–92

22. Callen DF. 1974. Recombination and segregation of mitochondrial genes in *Saccharomyces cerevisiae*. *Mol. Gen. Genet.* 143:49–63

23. Catlett NL, Weisman LS. 2000. Divide and multiply: organelle partitioning in yeast. *Curr. Opin. Cell Biol.* 12:509–16

24. Chinnery PF, Thorburn DR, Samuels DC, White SL, Dahl H-HM, et al. 2000. The inheritance of mitochondrial DNA heteroplasmy: random drift, selection or both? *Trends Genet.* 16:500–5

25. Colletti KS, Tattersall EA, Pyke KA, Froelich JE, Stokes KD, Osteryoung KW. 2000. A homologue of the bacterial cell division site-determining factor MinD mediates placement of the chloroplast division apparatus. *Curr. Biol.* 10:510–16

26. Davis AF, Clayton DA. 1996. In situ localization of mitochondrial DNA replication in intact mammalian cells. *J. Cell Biol.* 135:883–93

27. Enriquez JA, Cabezas-Herrera J, Bayona-Bafaluy MP, Attardi G. 2000. Very rare complementation between mitochondria carrying different mitochondrial DNA mutations points to intrinsic genetic autonomy of the organelles in cultured human cells. *J. Biol. Chem.* 275:11207–15

28. Eyre-Walker A. 2000. Do mitochondria recombine in humans? *Philos. Trans. R. Soc. London Ser. B* 355:1573–80

29. Eyre-Walker A, Smith NH, Maynard Smith J. 1999. How clonal are human mitochondria? *Proc. R. Soc. London Ser. B* 266:477–83

30. Fukuhara H. 1969. Relative proportions of mitochondrial and nuclear DNA in yeast under various conditions of growth. *Eur. J. Biochem.* 11:135–39

31. Gillham NW. 1963. Transmission and segregation of a nonchromosomal factor controlling streptomycin resistance in diploid *Chlamydomonas*. *Nature* 200:294

32. Gillham NW. 1994. *Organelle Genes and Genomes*. New York: Oxford Univ. Press

33. Gillham NW, Levine RP. 1962. Studies on the origin of streptomycin resistant mutants in *Chlamydomonas reinhardi*. *Genetics* 47:1463–74

34. Gyllensten U, Wharton D, Josefsson A, Wilson AC. 1991. Paternal inheritance of mitochondrial DNA in mice. *Nature* 352:255–57

35. Hayashi J-I, Takemitsu M, Goto Y-i, Nonaka I. 1994. Human mitochondria and mitochondrial genome function as a single dynamic cellular unit. *J. Cell Biol.* 125:43–50

36. Hiraoka J-i, Hirao Y-h. 1988. Fate of sperm tail components after incorporation into the hamster egg. *Gamete Res.* 19:369–80

37. Hollenberg CP, Borst P, van Bruggen EF. 1972. Mitochondrial DNA from cytoplasmic petite mutants of yeast. *Biochim. Biophys. Acta* 277:35–43

38. Holt IJ, Dunbar DR, Jacobs HT. 1997. Behavior of a population of partially duplicated mitochondrial DNA molecules in cell culture: segregation, maintenance and recombination dependent on nuclear background. *Hum. Mol. Genet.* 6:1251–60

39. Jacobs HT, Lehtinen SK, Spelbrink JN. 2000. No sex please, we're mitochondria: a hypothesis on the somatic unit of inheritance of mammalian mtDNA. *BioEssays* 22:564–72

40. Jansen RP. 2000. Germline passage of mitochondria: quantitative considerations and possible embryological sequelae. *Hum. Reprod.* 2 (Suppl.):112–28

41. Jenuth JP, Peterson AC, Fu K, Shoubridge EA. 1996. Random genetic drift in the

female germline explains the rapid segregation of mammalian mitochondrial DNA. *Nat. Genet.* 14:146–51

42. Jenuth JP, Peterson AC, Shoubridge EA. 1997. Tissue-specific selection for different mtDNA genotypes in heteroplasmic mice. *Nat. Genet.* 16:93–95

43. Kaneda H, Hayashi J-I, Takahama S, Taya C, Fischer Lindahl K, Yonekawa H. 1995. Elimination of paternal mitochondrial DNA in intraspecific crosses during early mouse embryogenesis. *Proc. Natl. Acad. Sci. USA* 92:4542–46

44. Kann LM, Rosenblum EB, Rand DM. 1998. Aging, mating, and the evolution of mtDNA heteroplasmy in *Drosophila melanogaster*. *Proc. Natl. Acad. Sci. USA* 95:2372–77

45. Kiessling J, Kruse S, Rensing SA, Harter K, Decker EL, Reski R. 2000. Visualization of a cytoskeleton-like FtsZ network in chloroplasts. *J. Cell Biol.* 151:945–50

46. Kleff S, Kemper B, Sternglanz R. 1992. Identification and characterization of yeast mutants and the gene for a cruciform cutting endonuclease. *EMBO J.* 11:699–704

47. Koehler CM, Lindberg GL, Brown DR, Beitz DC, Freeman AE, et al. 1991. Replacement of bovine mitochondrial DNA by a sequence variant within one generation. *Genetics* 129:247–55

48. Kuroiwa T, Uchida H. 1996. Organelle division and cytoplasmic inheritance. *BioScience* 46:827–35

49. Larsson N-G, Holme E, Kristiansson B, Oldfors A, Tulinius M. 1990. Progressive increase of the mutated mitochondrial DNA fraction in Kearns-Sayre syndrome. *Pediatr. Res.* 28:131–36

50. Lehtinen SK, Hance N, El Meziane A, Juhola MK, Juhola KMI, et al. 2000. Genotypic stability, segregation and selection in heteroplasmic human cell lines containing np 3243 mutant mtDNA. *Genetics* 154:363–80

51. Lockshon D, Zweifel SG, Freeman-Cook LL, Lorimer HE, Brewer BJ, Fangman WL. 1995. A role for recombination junctions in the segregation of mitochondrial DNA in yeast. *Cell* 81:947–55

52. MacAlpine DM, Kolesar J, Okamoto K, Perlman PS, Butow RA. 2001. Replication and preferential inheritance of hypersuppresive petite mitochondrial DNA. *EMBO J.* In press

53. MacAlpine DM, Perlman PS, Butow RA. 1998. The high mobility group protein Abf2p influences the level of yeast mitochondrial DNA recombination intermediates in vivo. *Proc. Natl. Acad. Sci. USA* 95:6739–43

54. MacAlpine DM, Perlman PS, Butow RA. 2000. The numbers of individual mitochondrial DNA molecules and mitochondrial DNA nucleoids in yeast are co-regulated by the general amino acid control pathway. *EMBO J.* 19:767–75

55. Macaulay V, Richards M, Sykes B. 1999. Mitochondrial DNA recombination—no need to panic. *Proc. R. Soc. London Ser. B* 266:2037–39

56. Maleszka R, Skelly PJ, Clark-Walker GD. 1991. Rolling circle replication of DNA in yeast mitochondria. *EMBO J.* 10:3923–29

57. Martin W. 2000. A powerhouse divided. *Science* 287:1219

58. McConnell SJ, Yaffe MP. 1993. Intermediate filament formation by a yeast protein essential for organelle inheritance. *Science* 260:687–89

59. Medgyesy P, Fejes E, Maliga P. 1985. Interspecific chloroplast recombination in a *Nicotiana* somatic hybrid. *Proc. Natl. Acad. Sci. USA* 82:6960–64

60. Medgyesy P, Pay A, Marton L. 1986. Transmission of paternal chloroplasts in *Nicotiana*. *J. Gen. Genet.* 204:195–98

61. Meirelles FV, Smith LC. 1998. Mitochondrial genotype segregation during preimplantation development in mouse heteroplasmic embryos. *Genetics* 148:877–83

62. Michaelis P. 1971. The investigation of plasmone segregation by the pattern-analysis. *Nucleus* 10:1–14

63. Mignotte F, Tourte M, Mounolou J-C. 1987. Segregation of mitochondria in the

cytoplasm of *Xenopus laevis* vitellogenic oocytes. *Biol. Cell.* 60:97–102

64. Milligan B. 1992. Is organelle DNA strictly maternally inherited? Power analysis of a binomial distribution. *Am. J. Bot.* 79:1325–28

65. Moraes CT, Schon EA. 1995. Replication of a heteroplasmic population of normal and partially-deleted human mitochondrial genomes. *Prog. Cell Res.* 5:209–15

66. Mosig G. 1998. Recombination and recombination-dependent DNA replication in bacteriophage T4. *Annu. Rev. Genet.* 32: 379–413

67. Nagley P, Linnane AW. 1972. Biogenesis of mitochondria. XXI. Studies on the nature of the mitochondrial genomes in yeast: the degenerative effects of ethidium bromide on mitochondrial genetic information in a respiratory competent strain. *J. Mol. Biol.* 66:181–93

68. Nunnari J, Marshall WF, Straight A, Murray A, Sedat JW, Walter P. 1997. Mitochondrial transmission during mating in *Saccharomyces cerevisiae* is determined by mitochondrial fusion and fission and the intramitochondrial segregation of mitochondrial DNA. *Mol. Biol. Cell* 8:1233–42

69. Okamoto K, Perlman PS, Butow RA. 1998. The sorting of mitochondrial DNA and mitochondrial proteins in zygotes: preferential transmission of mitochondrial DNA to the medial bud. *J. Cell Biol.* 142:613–23

70. Osteryoung KW. 2000. Organelle fission. Crossing the evolutionary divide. *Plant Physiol.* 123:1213–16

71. Osteryoung KW, Stokes KD, Rutherford SM, Percival AL, Lee WY. 1998. Chloroplast division in higher plants requires members of two functionally divergent gene families with homology to bacterial *ftsZ. Plant Cell* 10:1991–2004

72. Petit N, Touraille M, Lécher P, Alziari S. 1998. Developmental changes in heteroplasmy level and mitochondrial gene expression in a *Drosophila subobscura* mi-tochondrial deletion mutant. *Curr. Genet.* 33:330–39

73. Rothfield L, Justice S, García-Lara J. 1999. Bacterial cell division. *Annu. Rev. Genet.* 33:423–48

74. Sager R. 1954. Mendelian and non-Mendelian inheritance of streptomycin resistance in *Chlamydomonas reinhardtii. Proc. Natl. Acad. Sci. USA* 40:356–63

75. Sager R, Ramanis Z. 1968. The pattern of segregation of cytoplasmic genes in *Chlamydomonas. Proc. Natl. Acad. Sci. USA* 61:324–31

76. Sears BB. 1980. Changes in chloroplast genome composition and recombination during the maturation of zygospores of *Chlamydomonas reinhardtii. Curr. Genet.* 2:1–8

77. Sena EP, Revet B, Moustacchi E. 1986. In vivo homologous recombination intermediates of yeast mitochondrial DNA analyzed by electron microscopy. *Mol. Gen. Genet.* 202:421–28

78. Sesaki H, Jensen RE. 1999. Division versus fusion: Dnm1p and Fzo1 antagonistically regulate mitochondrial shape. *J. Cell Biol.* 147:699–706

79. Shitara H, Hayashi J-I, Takahama S, Kaneda H, Yonekawa H. 1998. Maternal inheritance of mouse mtDNA in interspecific hybrids: segregation of the leaked paternal mtDNA followed by the prevention of subsequent paternal leakage. *Genetics* 148:851–57

80. Simon VR, Karmoln SL, Pon LA. 1997. Mitochondrial inheritance: cell cycle and actin cable dependence of polarized mitochondrial movements in *Saccharomyces cerevisiae. Cell Motil. Cytoskelet.* 37:199–210

81. Solignac M, Génermont J, Monnerot M, Mounolou J-C. 1987. Drosophila mitochondrial genetics: evolution of heteroplasmy through germ line cell divisions. *Genetics* 117:687–96

82. Stern C, Sherwood ER, eds. 1966. *The Origin of Genetics.* San Francisco: Freeman

83. Strausberg RL, Perlman PS. 1978. The

effect of zygotic bud position on the transmission of mitochondrial genes in *Saccharomyces cerevisiae. Mol. Gen. Genet.* 163:131–44

84. Sutovsky P, Moreno RD, Ramalho-Santos J, Dominko T, Simerly C, Schatten G. 2000. Ubiquitinated sperm mitochondria, selective proteolysis, and the regulation of mitochondrial inheritance in mammalian embryos. *Biol. Reprod.* 63:582–90

85. Sutovsky P, Moreno RD, Ramalho-Santos J, Dominko T, Simerly C, Schatten G. 1999. Ubiquitin tag for sperm mitochondria. *Nature* 402:371–72

86. Szollosi D. 1965. The fate of sperm middle-piece mitochondria in the rat egg. *J. Exp. Zool.* 159:367–78

87. Takai D, Isobe K, Hayashi J. 1999. Trans-complementation between different types of respiration-deficient mitochondria with different pathogenic mutant mitochondrial DNAs. *J. Biol. Chem.* 274:11199–202

88. Takeda K, Takahashi S, Onishi A, Hanada H, Imai H. 2000. Replicative advantage and tissue-specific segregation of RR mitochondrial DNA between C57BL/6 and RR heteroplasmic mice. *Genetics* 155:777–83

89. Tang Y, Manfredi G, Hirano M, Schon EA. 2000. Maintenance of human rearranged mitochondrial DNAs in long-term cultured transmitochondrial cell lines. *Mol. Biol. Cell* 11:2349–58

90. Tang Y, Schon EA, Wilichowski E, Vazquez-Memije ME, Davidson E, King MP. 2000. Rearrangements of human mitochondrial DNA (mtDNA): new insights into the regulation of mtDNA copy number and gene expression. *Mol. Biol. Cell* 11:1471–85

91. Thrailkill KM, Birky CW Jr, Lückemann

G, Wolf K. 1980. Intracellular population genetics: Evidence for random drift of mitochondrial gene frequencies in *Saccharomyces cerevisiae* and *Schizosaccharomyces pombe. Genetics* 96:237–62

92. Van Winkle-Swift K. 1980. A model for the rapid vegetative segregation of multiple chloroplast genomes in *Chlamydomonas*: assumptions and predictions of the model. *Curr. Genet.* 1:113–25

93. Van Winkle-Swift KP. 1978. Uniparental inheritance is promoted by delayed division of the zygote in *Chlamydomonas. Nature* 275:749–51

94. Wakasugi T, Nagai T, Kapoor M, Sugita M, Ito M, et al. 1997. Complete nucleotide sequence of the chloroplast genome from the green alga *Chlorella vulgaris*: the existence of genes possibly involved in chloroplast division. *Proc. Natl. Acad. Sci. USA* 94:5967–72

95. Wright S. 1968. *Evolution and the Genetics of Populations.* Chicago: Univ. Chicago Press. 969 pp.

96. Yoneda M, Chomyn A, Marinuzzi A, Hurko O, Attardi G. 1992. Marked replicative advantage of human mtDNA carrying a point mutation that causes the MELAS encephalomyopathy. *Proc. Natl. Acad. Sci. USA* 89:11164–68

97. Zelenaya-Troitskaya O, Newman SM, Okamoto K, Perlman PS, Butow RA. 1998. Functions of the high mobility group protein, Abf2p, in mitochondrial DNA segregation, recombination and copy number in *Saccharomyces cerevisiae. Genetics* 148:1763–76

98. Zweifel SG, Fangman WL. 1991. A nuclear mutation reversing a biased transmission of yeast mitochondrial DNA. *Genetics* 128:241–49

Annu. Rev. Genet. 2001. 35:149–91

THE ACTION OF MOLECULAR CHAPERONES IN THE EARLY SECRETORY PATHWAY

Sheara W. Fewell[1], Kevin J. Travers[2], Jonathan S. Weissman[2], and Jeffrey L. Brodsky[1]

[1]*Department of Biological Sciences, University of Pittsburgh, Pittsburgh, Pennsylvania 15260; e-mail: jbrodsky@pitt.edu*
[2]*Department of Cellular and Molecular Pharmacology and the Howard Hughes Medical Institute, University of California, San Francisco, California 94143*

Key Words Hsp70, Hsp40, calnexin, oxidative protein folding, UPR, ERAD

■ **Abstract** The endoplasmic reticulum (ER) serves as a way-station during the biogenesis of nearly all secreted proteins, and associated with or housed within the ER are factors required to catalyze their import into the ER and facilitate their folding. To ensure that only properly folded proteins are secreted and to temper the effects of cellular stress, the ER can target aberrant proteins for degradation and/or adapt to the accumulation of misfolded proteins. Molecular chaperones play critical roles in each of these phenomena.

CONTENTS

INTRODUCTION

In addition to a number of growth factors and other serum proteins, the liver secretes about 1% of its total weight in albumin each day. Although the secretory capacity of other cells may be lower, every eukaryotic cell exports a variety of proteins either to meet its nutritional needs (proteases and glycanases), for cell-cell communication (growth hormones and pheromones), or for defense (toxins and antibodies). The first compartment in which this heterogeneous assortment of factors matures is the endoplasmic reticulum (ER). In addition, nearly all proteins that ultimately reside in this and other compartments of the secretory pathway or in the plasma membrane are first translocated (imported) into the ER. To fulfill its role as the primary gateway to the secretory pathway, the ER houses enzymes that process polypeptides and catalyze protein folding. Also associated with the ER is a group of molecular chaperones, defined here as proteins that facilitate the folding of nascent polypeptides. In this review, we discuss the mechanism of action of ER-associated chaperones, emphasizing studies undertaken in the yeast *Saccharomyces cerevisiae*. When appropriate, specific experiments are discussed using other organisms.

MECHANISM OF ACTION OF THE MAJOR CLASSES OF CHAPERONES INVOLVED IN PROTEIN SECRETION

The stress-inducible 70-kDa heat shock proteins (Hsp70s) and the constitutively expressed heat shock cognate proteins (Hsc70s) belong to a family of molecular chaperones that bind and release polypeptides in an ATP-dependent cycle (32). Early partial proteolysis experiments suggested that Hsc70 could be divided into three regions: a 44-kDa amino-terminal ATPase domain and a 15-kDa peptide-binding domain, followed by a poorly conserved, 10-kDa carboxy-terminus (40, 257). Structural studies on the peptide-binding domain indicated that peptides are trapped in a channel that is gated by a flexible helical lid (166, 286). When ATP is bound, the helical lid pivots to expose this channel and Hsc70 exhibits weak affinity for substrates. Transient interactions with peptide can stimulate ATP hydrolysis and thus trigger a conformational change in the peptide-binding domain that increases the stability of the peptide-Hsc70 complex by closing the helical lid to trap the bound substrate. The exchange of ADP for ATP is then critical for release of the substrate (151, 219). This cycle of substrate binding and release, combined with the preference of Hsc70s for hydrophobic stretches of amino acids (18, 70, 211) that may become solvent-exposed in unfolded proteins, enable Hsc70s to interact transiently with polypeptides as they progress through the secretory pathway.

Hsc70s are inherently weak ATPases (0.03–0.27 min^{-1}), but their activity is greatly enhanced by members of the GrpE and DnaJ (Hsp40) families. Identified first in bacteria, GrpE stimulates the exchange of ATP for ADP on DnaK (138), thus facilitating release of substrate and permitting the commencement of a new binding cycle (241). However, a GrpE-like exchange factor does not appear to be critical for the early secretory pathway. In contrast, DnaJ chaperones, which stimulate the ATPase activity of Hsc70 and thus promote stable substrate binding (151, 219), are critical at many steps in the eukaryotic secretory pathway. In some cases, DnaJ chaperones may even deliver specific substrates to Hsc70s (126, 136, 160, 240, 267) and anchor Hsc70s at the ER or mitochondrial membrane (24, 206).

The DnaJ (Hsp40) family of chaperones is defined by the presence of an ~70 amino acid sequence called the J domain that mediates Hsc70 interaction. Structural studies of the J domains from bacterial DnaJ (106, 193, 242), human Hdj1p (203) and the SV40 and polyomavirus large T antigens (14, 131) indicate that this domain folds into four α-helices, two of which are packed tightly against each other to form a finger-like projection. An invariant HPD motif is positioned in a loop at the tip of this finger between the second and third α-helices and is critical for interactions with the Hsc70 ATPase domain (59, 68, 247, 254). Additional regions of DnaJ chaperones may also contact a site near or at the substrate-binding domain of Hsc70s (80, 237, 238). With the exception of the J domain, DnaJ family members share little else in common. There are three broad subtypes, defined by their similarity to the bacterial DnaJ protein (41). Type 1 DnaJ homologs are the most similar to DnaJ. In addition to the J domain, they contain a 30–40 amino acid glycine/phenylalanine-rich region that facilitates Hsc70 interaction (128, 240, 255), and a cysteine-rich region that binds zinc and forms a peptide-binding pocket (5, 240). Type II DnaJ family members lack the cysteine-rich region and Type III proteins lack both the cysteine-rich and glycine/phenylalanine-rich regions. *Escherichia coli* DnaJ was recently observed to bind preferentially to hydrophobic, 8-amino acid motifs (212), providing further evidence that this chaperone can deliver polypeptide substrates to Hsc70s.

The Hsp90 chaperones are abundant, essential proteins that have been most intensely studied for their role in kinase and steroid hormone receptor maturation (reviewed in 35). Although mammalian cells contain both cytoplasmic and lumenal Hsp90s, yeast lack a lumenal Hsp90 homologue. Hsp90 function depends upon two peptide-binding domains at the amino and carboxy termini, both of which can mediate dimerization. The amino terminus also contains a weak ATPase activity (202, 234) that is essential for Hsp90 function (183, 188). Peptide binding (and dimerization) at the amino terminus is regulated by ATP (234). Whereas both peptide binding domains of Hsp90 can prevent aggregation of denatured proteins in vitro, recent evidence indicates that full-length Hsp90 is required to enhance the refolding of denatured substrate in vitro (124). Still controversial, however, is whether Hsp90s are generally involved in protein folding. Support for their involvement in protein folding includes the observation that Hsp90 can maintain aggregation-prone substrates in a refolding-competent state in vitro (21, 74) and that it associates with unassembled immunoglobulin chains in the mammalian ER (158, 217). In addition, Hsp90 can enhance the Hsp70- and Hsp40-mediated folding of luciferase in vitro and in vivo (87, 125, 223). However, the folding of most proteins is unaffected in yeast lacking functional Hsp90 (173).

The peptidyl prolyl isomerases (PPIases) comprise another family of chaperones, with members found in virtually all cellular compartments where protein folding occurs. The PPIases catalyze the *cis/trans* isomerization of the peptide bond immediately preceding proline residues, which is kinetically unfavorable when uncatalyzed (135). These chaperones can be divided into two structurally distinct classes based on homology: the immunophilin family, whose founding member is the target of the immunosuppressant cyclosporin, and the FK-binding protein (FKBP) family, whose founding member is the target of the compound FK506 (280). PPIases have multiple functions, some with little to do with protein folding, which has led to controversy regarding the requirement for PPIases in folding reactions. However, other observations support their significance for protein folding in the ER. First, PPIases are induced by the unfolded protein response (UPR; see below) (39, 245). Second, the physiological significance of a cyclophilin family member from *Drosophila melanogaster* known as *ninaA* is well established (3, 45, 233). As *ninaA* is specifically required in the folding of a subset of rhodopsins, it is discussed in the final section of this review.

MOLECULAR CHAPERONES AND PROTEIN TRANSLATION

The influence of chaperones on the maturation of both secretory and nonsecretory pathway-targeted proteins begins during translation. In yeast, chaperones of the Hsp70 (Ssb1/2p; 174) and Hsp40 families (Sis1p, zuotin, Ydj1p; 28, 275, 284) have been implicated in translation by their association with ribosomes and/or by their requirement for translation initiation, efficient protein synthesis, or the translation of heterologous proteins. Although the Ssb chaperones crosslink directly to the nascent polypeptide-ribosome complex and might prevent protein misfolding (194), the mechanisms by which these and possibly other chaperones facilitate translation remain largely unknown. Studies examining early events during protein translation in mammalian and in heterologous systems have uncovered ribosome-nascent polypeptide chain interactions with the Hsp70 and Hsp40 chaperones, as well as with the TriC/CCT chaperonin complex (12, 63, 76, 196). Recently, a ribosome-associated Hsp70-Hsp40 complex formed by Ssz1p/Pdr13p and zuotin was shown to facilitate the translocation of a ribosome-bound mitochondrial precursor protein into the mitochondria in vitro (81). However, a ribosome-associated Hsp70-Hsp40 complex that similarly facilitates preprotein translocation into the ER has not been identified.

PROTEIN TRANSLOCATION INTO THE YEAST ER IS FACILITATED BY MOLECULAR CHAPERONES IN THE CYTOPLASM AND ER LUMEN

Protein translocation (import) into the ER proceeds either cotranslationally or post-translationally. During cotranslational translocation, translation is attenuated after the emergence of the signal peptide through the action of the signal recognition particle (SRP). Upon interaction with the SRP receptor (SR, also known as "Docking Protein"), GTP-dependent release of SRP permits the re-initiation of translation, and the signal sequence of the preprotein is presented to the translocation machinery at the ER membrane (reviewed in 123). During posttranslational translocation, the signal sequence-containing secreted preprotein is synthesized in its entirety before interaction with the translocation machinery (reviewed in 205). One of many advantages provided by the examination of translocation in yeast is that the pathways operate in parallel, and a survey by Ng et al. (176) concluded that signal sequences with greater hydrophobic cores co-opt preferentially the cotranslational pathway, or utilize both pathways.

When genes encoding components of SRP and SR are deleted, yeast grow ~fourfold more slowly than wild-type cells and accumulate several cotranslationally targeted preproteins in the cytoplasm (97, 185). A recent microarray analysis from Mutka & Walter (170) indicates that the ability of yeast to survive the loss of SRP arises from the induction of ER-associated chaperones that may prevent preproteins from aggregating, and from an attenuation of protein synthesis, perhaps permitting the translocation machinery to better couple translation and translocation.

Because the translocation of preproteins utilizing the posttranslational pathway is uncoupled from translation, and because the diameter of the translocation channel in the ER membrane precludes the translocation of folded proteins (see below), the preprotein must translocate in a nonnative conformation. Thus, cytoplasmic molecular chaperones are required for the translocation of preproteins into the ER posttranslationally even when SRP function is proficient.

The first evidence that chaperones facilitate posttranslational protein translocation emerged from both biochemical and genetic studies. Chirico et al. (43) utilized an in vitro assay in which the dependence on yeast cytosol of the translocation of a wheat germ–synthesized yeast mating factor prepheromone, pre-pro α factor (ppαF), into yeast ER-derived microsomes was observed (261). The cytoplasmic Hsc70s Ssa1p and Ssa2p were then purified based on their ability to substitute for cytosol (43). The cytosol dependence could also be replaced by denaturation of the substrate in urea prior to the translocation assay, suggesting that the chaperones maintained ppαF in an unfolded conformation. Deshaies and colleagues (56) simultaneously found that the depletion of Ssa1p in yeast harboring knockout alleles of the *SSA1*, *SSA2*, and *SSA4* genes (which encodes a third, related Hsc70) led to the accumulation of ppαF and a mitochondrial-targeted preprotein in the cytoplasm. Support for the requirement of cytoplasmic chaperones in protein translocation also emerged from examining translocation in other systems. First, although the posttranslational pathway is rarely utilized in higher eukaryotes, short preproteins that cannot interact with SRP during translation required Hsc70 for translocation into mammalian microsomes (288). Second, Gross and co-workers (270) found that *dnaJ* and *dnaK* mutants failed to secrete bacterial preproteins that utilize the Sec-independent secretory pathway, and that overexpression of DnaJ and DnaK facilitated the secretion of Sec-dependent preproteins in *secB* mutants. Thus, as in yeast, cytoplasmic chaperones apparently compensate for defects in an alternate secretory pathway.

To determine whether the action of Ssa1/2p on posttranslational translocation requires an Hsp40 co-chaperone, Caplan et al. (36) examined yeast containing a temperature-sensitive allele of *YDJ1*, an ER-associated Type I Hsp40, and found that posttranslationally translocated preproteins accumulated in the cytoplasm at the nonpermissive temperature. Both genetic (10) and biochemical (53, 54) data indicate that Ssa1p and Ydj1p interact, an interaction that is required to support translocation (10, 36, 153). These combined results led to a model in which release of the Ssa1p-bound preprotein is catalyzed by Ydj1p at the ER membrane, resulting in the delivery of the preprotein in an unfolded conformation to the translocation machinery.

However, a recent study suggests that chaperones may free preproteins spontaneously before interacting with the translocation apparatus at the ER (196). Consistent with these data, a fusion protein heterologously expressed in yeast was folded in the cytoplasm prior to its translocation (192), and based on studies of luciferase folding in yeast lysates, Bush & Meyer (33) proposed that Ssa1/2p catalyze the folding of preproteins in the cytoplasm before ER targeting. In each case, however, the preprotein must be unfolded again before it can insert into the translocation channel. Whether this process is re-engineered by Ssa1p/Ydj1p or by other chaperones is not clear.

After a nascent secretory preprotein is targeted to the cytosolic face of the ER membrane, its signal sequence is engaged by the Sec61p translocation complex (169, 197, 214), composed of the Sec61p, Sss1p, and Sbh1p proteins in yeast (99). High-resolution electron micrographs of both the yeast and mammalian pore complexes suggest that the Sec61p complex assembles into a tetramer, forming a central pore that is likely to be the channel through which preproteins are translocated (96). Measurements of truncated preproteins containing fluorescently tagged amino acids at various positions within the polypeptide indicate that the mammalian pore may be as large as 80 Å (90), a value somewhat higher than that obtained by the EM studies (11, 96). Regardless, the pore is of sufficient diameter to transport polypeptides in an α-helical conformation, or containing some secondary structure, but not the sizes adopted by most fully folded proteins. Because the interior of the pore is an aqueous channel (52, 83), transmembrane domains of integral membrane proteins must laterally diffuse into the lipid bilayer during their translocation (61, 103, 148, 167).

Elegant biochemical studies using proteoliposomes lacking lumenal components indicate that the driving force for cotranslational translocation into the ER can be provided by the ribosome (149). Thus, the nascent polypeptide is "pushed" into the lumen. In contrast, lumenal components most likely "pull" polypeptides into the ER during posttranslational translocation. In yeast, numerous studies have led to the conclusion that the Hsc70 and Hsp40 homologues, BiP (Kar2p) and Sec63p, respectively, are responsible for driving polypeptides into the ER (reviewed in 24). Yeast BiP is $\sim$50% identical to *E. coli* DnaK (182, 209), whereas

Sec63p (a Type III Hsp40) is a polytopic membrane protein in which only one ~70 amino acid lumenal segment is homologous to DnaJ (68, 213). Strains containing temperature-sensitive mutations in *KAR2* and *SEC63* accumulate preproteins at the nonpermissive temperature (210, 252), and microsomes or reconstituted proteoliposomes prepared from *kar2* and *sec63* mutant strains are defective for translocation in vitro (27, 29, 210, 214). A role for BiP during translocation into washed or reconstituted ER-derived mammalian vesicles has also been suggested (60, 178).

Two prevailing models to account for the action of the BiP-Sec63p complex have been proposed, based primarily on studies of the mechanics of posttranslational translocation into isolated mitochondria (reviewed in 24). In the Brownian ratchet model, the translocation pore is passive and the interaction of the preprotein with BiP in the ER lumen prevents retro-translocation (228). Polypeptide oscillations within the pore provided by thermal energy lead to the emergence of longer segments of the preprotein in the ER, allowing for a higher stoichiometry of BiP binding and the prevention of retro-translocation. Support for this model for translocation into the yeast ER comes from a study by Matlack et al. (149) in which reconstituted proteoliposomes lacking lumenal components but containing antibodies against ppαF could support the translocation of ppαF through the Sec61 complex and into vesicles. This impressive proof-of-principle and a recent mathematical modeling study (139) indicate that a ratchet is sufficient for posttranslational translocation in a highly defined in vitro system.

The second model depicts the BiP-Sec63p complex as a motor in which successive rounds of ATP binding and hydrolysis are coupled to the interaction with, and pulling and release of, a preprotein in the lumen of the ER (84). Consistent with this model, mutations in BiP that prevent an ATP-dependent conformational change and interaction with Sec63p, but not peptide binding, inhibit translocation both in vivo and in vitro in a dominant manner (152). The motor model may also be supported by the observations that mitochondrial Hsp70 cannot support ppαF translocation into ER-derived reconstituted vesicles even though it interacts with ppαF (25) and that mutations in the BiP-binding domain of Sec63p abrogate BiP interaction and posttranslational translocation (29, 47, 146). Thus it seems that BiP must interact with the substrate and be anchored to the inner face of the ER membrane to exert force and pull preproteins into the ER. In contrast, if BiP is simply a molecular "glue," its interaction with Sec63p may not be essential for import. However, an examination of BiP-peptide interactions in the presence or absence of Sec63p by surface plasmon resonance studies led to the intriguing conclusion that Sec63p may expand BiP's peptide-binding repertoire (160). Thus, BiP is bound near the translocation channel via its interaction with Sec63p in order to receive the preprotein, and Sec63p, in turn, signals BiP to bind to a wider spectrum of chemically diverse peptide segments. This result resolves the problem of how a single chaperone can bind either with a high enough affinity or in sufficient amounts to effect import, given that Hsp70s bind preferably to highly hydrophobic sequences (18, 70, 211) that may not be abundant in all secreted proteins.

The original identification of Hsp70 and Hsp40 in bacteria as DnaK and DnaJ—factors required for the ATP-dependent activation of the DnaB helicase at the replication origin during λ phage DNA replication (289)—suggests that the BiP-Sec63p complex may also be thought of as an energy-dependent, regulatory machine. Early support for this model came from the observation by Sanders et al. (214) that the interaction of a preprotein with Sec61p at the cytoplasmic face of the ER was compromised in a *kar2* mutant at the nonpermissive temperature. This result was confirmed using a more refined system in which the transfer of a signal sequence-containing preprotein from the integral membrane signal sequence-binding components of the translocation machinery (encoded by the *SEC62*, *SEC71*, and *SEC72* genes) to the translocation pore could be assessed in a solubilized system (147). These later studies suggested that BiP activates the Sec62p-Sec71p-Sec72p complex upon its ATP-dependent interaction with Sec63p, although data supporting an alternate theory have been presented (197). Nevertheless, the retention of BiP via Sec63p adjacent to the translocation channel may couple the recognition of a preprotein at the ER membrane, the entry of a preprotein into the channel, and its subsequent translocation into the ER.

Alternatively, the chaperones may be required to gate the pore. Using an in vitro assay in which lumenal access of a preprotein confined within the Sec61 complex could be assessed in mammalian microsomes, Johnson and colleagues showed that BiP seals the pore until the translocating polypeptide reaches a length of ~50 amino acids. Because the ability of BiP to seal the pore was ATP-dependent, and because of the recent identification of Sec63p homologues in the mammalian ER (159, 231, 249), it will be interesting to determine whether BiP-dependent gating in the mammalian ER is through a direct interaction with Sec61p, or through a Sec63p homologue.

One expectation from these models is that the chaperones should be required for both cotranslational and posttranslational translocation: Signal sequence recognition at the translocation machinery and/or gating of the channel are required regardless of how the preprotein is delivered to the ER membrane. Consistent with this hypothesis, microsomes derived from thermosensitive *kar2* and *sec63* strains were defective for the import of both a co- and posttranslationally translocated substrate (26). More recently, Stirling and colleagues selected for yeast specifically defective for the import of a cotranslationally translocated substrate and recovered mutations primarily in *SEC61* and *SEC63* (278). The authors also characterized further a group of *kar2* mutants required for co- and posttranslational translocation in vitro (26, 214) and confirmed that BiP function is necessary for both pathways in vivo. The biochemical characterization of the corresponding Sec63 and Kar2 mutant proteins should help elucidate how the lumenal chaperones might regulate and facilitate co- and posttranslational preprotein import into the ER.

As discussed in the Introduction, the action of Hsp70-Hsp40 pairs can be regulated further by other co-chaperones. Mutations in a lumenal, ~100-kDa protein with some homology to Hsp70 (known variably as Lhs1p/Cer1p/Ssi1p) are synthetically lethal with a translocation-defective allele of *KAR2*, and strains lacking the *LHS1/CER1/SSI1* gene are unable to posttranslationally translocate some preproteins into the ER (7, 51, 89). In addition, a genetic selection to uncover factors associated with the translocation machinery in *Yarrowia lipolytica* led to the identification of Sls1p, and *Y. lipolytica* strains deleted for *SLS1* are defective for the translocation of a secreted alkaline protease (19). Both the *Y. lipolytica* Sls1 protein and *S. cerevisiae* homologue of Sls1p (known as Per100p, Sil1p, or scSls1p) interact with the ATPase domain of BiP and regulate its activity (127, 245, 250), possibly by acting as a nucleotide exchange factor for BiP (127). While deletion of *SIL1* alone does not compromise translocation (250), suggesting that Sil1p may modulate another aspect of BiP function in *S. cerevisiae*, it exacerbates the translocation defect when combined with a translocation-defective *kar2* mutant (127). These recent results point to additional chaperone modulators required for protein translocation.

THE ENDOPLASMIC RETICULUM ENSURES PROPER FOLDING AND MATURATION

As soon as nascent chains enter the ER they face an environment dramatically different from that within the cytoplasm, but more similar to that present outside the cell. As the primary regulator of cellular Ca^{++} levels (235), the ER possesses a dramatically higher concentration of free Ca^{++} than found in the cytosol (1 mM in the ER compared with 100 nM in the cytosol; 39). Also, the ER is significantly more oxidizing than the cytoplasm (with a redox potential of -230 mV versus -150 mV; 114), which in turn means that disulfide bond formation is favored within the ER, whereas disulfide bonds are virtually absent in the cytoplasm (208). In fact, a hallmark of secretory proteins is the presence of disulfide bonds that in many cases are absolutely required for folding and/or activity (208). Additionally, many secretory proteins are N-glycosylated in the ER, a modification frequently required for proper folding and/or activity (104).

The ER serves as both a protein-folding compartment and a gatekeeper, guaranteeing the structural integrity of each protein before it is presented extracellularly (93, 113). As such, the ER is highly enriched in factors that promote efficient protein folding and prevent improperly folded proteins from progressing through the ER. Members of virtually all classes of chaperones, except the Hsp60/GroEL family, are found within the ER (235). The central role of the most abundant Hsc70 in the lumen, BiP, in protein folding has been well-documented (81a, 262), and yeast with reduced levels of lumenal Hsp70 activity exhibit protein folding defects (228a) and synthetic interactions with mutated alleles of genes encoding the lumenal Hsp40 chaperones and components of the oligosaccharyl transferase (179, 218, 242a). BiP most likely aids folding by preventing off-pathway intermediates from forming. Both BiP and another Hsp70 homologue in yeast, Lhs1p (Hsp170), also play an active role in the re-folding of heat-damaged secreted proteins in the lumen (120, 215).

Protein folding in the ER presents challenges not faced by the folding of proteins in other cellular compartments. For example, the chemical steps involved in disulfide bond formation and rearrangement are intrinsically slow compared to conformational rearrangements. Moreover, partial native structure can dramatically inhibit access to buried cysteines, further slowing disulfide rearrangement (264). Similarly, transmembrane domains must be inserted with correct topology into the ER membrane if a protein is to adopt its native structure. This problem is solved in part through the cotranslational insertion of transmembrane domains. However, as illustrated by the recently described maturation of aquaporin-1 (144), transmembrane domains are not always correctly oriented until after synthesis of nearly the entire polypeptide. How this reversal of topology is accomplished remains obscure. Given these particular demands on folding in the ER, there are, not surprisingly, several chaperone systems unique to the ER: general chaperones, such as calnexin and its soluble homolog calreticulin and the machinery responsible for the introduction of disulfide bonds into proteins, as well as chaperones dedicated to assisting the maturation of a single or a limited number of proteins.

OXIDATIVE PROTEIN FOLDING IN THE ER

Since the classical protein folding studies performed by Anfinsen, it has been clear that oxidation can proceed spontaneously in an aerobic environment. However, the observation that protein folding occurs far more rapidly in vivo than in vitro suggests that protein oxidation must be catalyzed within living cells (1). Indeed, it was this observation that led to the initial identification of protein disulfide isomerase (PDI) over 30 years ago as a chaperone that catalyzed the rearrangement of disulfide bonds (85). Oxidative protein folding in vitro requires only a source of oxidizing equivalents and an enzymatic activity to rearrange disulfide bonds (71). These two activities can be supplied by an appropriate redox buffer, usually a mixture of reduced and oxidized glutathione (GSH and GSSG, respectively), and PDI. The ratio of GSH to GSSG in vivo varies between the cytoplasm and the secretory pathway, with the cytoplasm having a ratio of GSH:GSSG of $\sim$100:1 and the secretory pathway having a ratio of $\sim$3:1 (114).

Since glutathione ratios differ between the secretory pathway and the cytoplasm, the presence of this buffer alone was considered sufficient to maintain an environment favorable for protein oxidation. However, every disulfide bond formed in a nascent chain introduces reducing equivalents into the ER that must be disposed of to keep the environment in an oxidized state. A number of explanatory models have been proposed, including import of oxidizing equivalents from the cytoplasm, secretion of reducing equivalents, and different enzymatic activities (71, 114, 287), none entirely satisfactory.

Work from two labs using *S. cerevisiae* has led to the identification of factors that are required for proper oxidative protein folding (72, 201). Central to these studies is the fact that the redox balance of living cells can be manipulated through treatment with the membrane-permeable reducing agent dithiothreitol (DTT) (22, 119). Whereas wild-type cells tolerate limited quantities of DTT, cells with defective oxidation machinery exhibit an increased sensitivity to this drug. Screening for mutant yeast cells either with an increased sensitivity to DTT (201) or with specific defects in secretion (72) resulted in the identification of a novel factor, termed *ER* Oxidoreductase 1 (*ERO1*). This gene is specifically required for oxidative protein folding, as mutations in *ERO1* result in a folding defect for three substrates that contain disulfide bonds, while proteins that do not depend on disulfide formation for their folding are secreted with normal kinetics (72, 201). Furthermore, *ERO1* activity determines the overall redox balance within cells: Whereas *ero1* mutation results in DTT sensitivity, overexpression of *ERO1* leads to increased resistance to DTT. The central role played by *ERO1* is underscored by the fact that homologs of the Ero1p protein can be found in all eukaryotic organisms examined (34, 72, 187, 201).

Much evidence indicates that a direct interaction between Ero1p and PDI is responsible for the introduction of disulfide bonds in ER lumenal proteins (Figure 1). First, whereas PDI is normally found in an oxidized state in vivo, mutations in *ERO1* result in the steady-state accumulation of PDI in its reduced form (73). Second, mutations in the active-site cysteines of PDI result in the isolation of mixed-disulfide complexes between Ero1p and PDI (73, 248). This complex is likely to be an intermediate formed in wild-type cells during the oxidation of PDI by Ero1p. Further support for the conservation of this oxidation machinery is supplied by the finding that mammalian homologs of Ero1p can also be isolated in mixed-disulfide cross-links with PDI (13). Even more compelling is the recent development of an in vitro oxidative folding system in which PDI is a required component for Ero1p-dependent oxidation of RNase A (248).

What about the role of glutathione in oxidation? Recent genetic and biochemical data counter the long-held belief that glutathione in the ER served as the source of oxidizing potential. Kaiser and co-workers removed glutathione from yeast using a strain lacking a gene required for its biosynthesis, *GSH1*, and found that oxidative protein folding was proficient (52a). Surprisingly, this mutation suppressed the folding defect of *ero1-1* mutant strains, suggesting that the presence of GSH places a burden on the oxidation system. Removal of this burden through the deletion of *GSH1* allows *ero1-1* mutant cells to generate enough oxidizing equivalents to grow again. In an in vitro system, the addition of oxidized glutathione was dispensable for oxidation of RNase A (248), indicating that Ero1p itself can generate oxidizing equivalents. Furthermore, the addition of reduced glutathione at concentrations up to 2 mM had no effect on the ability of Ero1p to oxidize substrates. Also, reduced glutathione was not oxidized by Ero1p in vitro unless PDI was present, suggesting that Ero1p does not oxidize glutathione and that the presence of GSSG within the secretory pathway is in fact a by-product of the reduction of secretory proteins and/or PDI. This view is consistent with the genetic experiments suggesting that the role of glutathione in the secretory pathway is actually to prevent overoxidation (52a).

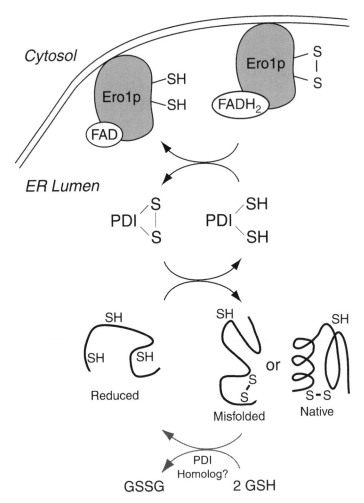

Figure 1   Disulfide bonds are introduced into folding proteins through the actions of Ero1p and protein disulfide isomerase. Oxidizing equivalents are generated on Ero1p using FAD as a co-factor. These oxidizing equivalents are passed on to PDI, but not other PDI homologs, which is then able to oxidize substrate proteins. Ero1p cannot oxidize substrates directly, nor does it directly oxidize GSH. Instead, GSH oxidation appears to be a consequence of reduction of pre-existing disulfide bonds. See text for details.

An important outstanding question is how oxidizing equivalents are introduced into the ER. As all redox reactions involve the movement of electrons from one molecule to another, there must be a "sink" for the excess electrons generated within the ER. An oxidation system analogous to that in the ER of eukaryotic cells is found in the periplasm of *E. coli* (208). In this system, electrons are disposed of through the respiratory chain (134), ultimately passing through either cytochrome bo oxidase or cytochrome bd oxidase to oxygen (2a). However, in contrast to the bacterial oxidation system, when components of the respiratory chain in *S. cerevisiae* are inactivated the oxidation machinery is unaffected (248). Instead, oxidative protein folding depends exquisitely on cellular levels of FAD. The significance of FAD was confirmed in vitro, as oxidation of RNase A by Ero1p required the addition of FAD.

Although FAD is sufficient for oxidation in vitro, its precise role in vivo is not clear. FAD may act as a co-factor for Ero1p. In this scenario, another protein is likely to bind Ero1p and catalyze a reaction leading to the regeneration of FAD from the $FADH_2$ that is produced during the oxidation of PDI. Alternatively, $FADH_2$ may be released, allowing Ero1p to bind a different, oxidized molecule of FAD. In either case, some mechanism must regenerate FAD within the cell. $FADH_2$ reacts very rapidly with free oxygen to generate FAD and H_2O, but oxidative protein folding occurs even under anaerobic conditions in a manner that is dependent on functional Ero1p (248). This suggests that another oxidizing source must be present. This has also been seen in the oxidizing environment of the bacterial periplasm, where fumarate acts as the terminal electron acceptor under anaerobic conditions (2a).

PROTEIN QUALITY CONTROL: RETENTION

The primary step in ER quality control is retention of misfolded or misassembled proteins, and the most common mechanism for retaining misfolded proteins is through association with other proteins that are themselves normally retained (65). As the exposure of epitopes rendering a protein susceptible to recognition by molecular chaperones is a feature likely to be common to all misfolded or misassembled proteins, chaperones are particularly good candidates for retention molecules. In fact, the calnexin/calreticulin system (282), BiP (92, 113), and PDI (207) have been implicated in aspects of quality control (for review, see 65).

The processing of the core oligosaccharyl glycan in the ER plays a major role in protein quality control and exemplifies one mechanism by which misfolded proteins may be retained in the ER. Initially, a branched chain of sugars of the composition $Glc_3Man_9GlcNAc_2$ is added to asparagines within the sequence motif NX(S/T). However, proteins that fold rapidly in the ER are less prone to this modification (109), suggesting that a steric block arising from secondary structures may occlude the consensus sequence. Glycoproteins with the attached $Glc_3Man_9GlcNAc_2$ moiety undergo a rapid trimming, in which the three external glucose residues are removed sequentially through the actions of glucosidase I and II (104). A single glucose residue can be added back by UDP-glucose:glycoprotein glucosyltransferase (UGGT; 69, 191), but UGGT only recognizes misfolded proteins. The monoglycosylated glycan is recognized by the lectins calnexin and its soluble homolog calreticulin, which retain the glycoprotein in the ER and facilitate its folding (104). Removal of the terminal glucose triggers dissociation of the calnexin/calreticulin-glycoprotein complex and the correctly folded glycoprotein can then exit the ER (102); however, misfolded proteins are recognized by UGGT and re-glycosylated, leading to re-association with the lectins and permitting another chance at folding (91). Thus, unlike modification of glycosyl groups in the Golgi, which is used to create diversity, modification of the basic glycosylation structure in the ER appears to be used primarily by the quality control machinery to distinguish folded from misfolded proteins.

Recent studies using a GFP-tagged secretory protein and fluorescence photo-bleaching experiments in mammalian cells provide evidence for another mechanism of retention: lack of a positive secretion signal. The glycoprotein of vesicular stomatitis virus (VSVG) is normally secreted; however, a temperature-sensitive variant of this protein (tsO45) leads to misfolding and ER retention at 40°C. By tagging this VSVG mutant with GFP and examining the kinetics of recovery after photo-bleaching, Nehls et al. observed that the retained protein is highly mobile, with a mobility equivalent to that seen with the mutant protein examined at the secretion-permissive temperature of 32°C (175). Furthermore, repetitive photo-bleaching of cells expressing the mutant VSVG at 40°C showed a gradual loss of total fluorescence, consistent with a model in which mutant VSVG cannot be exported from the ER but remains free to diffuse within it. When cells are depleted of ATP or express an ATPase-defective BiP mutant, the VSVG mutant becomes immobile. Consistent with previous suggestions from studies of the influenza hemagglutinin (HA) protein (112), these results indicate that the ER may form a dense matrix that inhibits the movement of normally mobile proteins. With VSVG, the aggregates were held together by disulfide bonds, as mobility could be restored through treatment of the cells with DTT (175).

In their studies of the subcellular localization of VSVG by indirect immunofluorescence, Hammond & Helenius provide another suggestion for how the temperature-sensitive VSVG might be retained in the ER (92). At the nonpermissive temperature, VSVG was found throughout the ER as well as in ER-Golgi intermediate compartments. The only marker that co-localized with the VSVG protein was BiP. When BiP localization was examined in non-transfected cells or in cells expressing VSVG at the permissive temperature, BiP was seen only in the ER. This suggests that misfolded VSVG can pull BiP out of the ER, but that VSVG-BiP complexes are returned to the ER, thus keeping VSVG mutant protein from being secreted.

Given the broad spectrum of chaperone systems in the ER, how is it determined which chaperone system is selected by translocating nascent chains? Insight into this question has come from recent work by Molinari & Helenius, who observed that the E1 nascent chain of Semliki forest virus associated with BiP, while the p62 viral nascent chains bound to calnexin (161). If interaction with calnexin was blocked, both proteins immunoprecipitated with BiP, suggesting that the lack of interaction between p62 and BiP was not due to a lack of a suitable BiP binding site. This hypothesis was confirmed by examining interactions between BiP and HA, which contains numerous glycosylation sites near its N terminus. Consistent with glycosylation being the determining factor, the native HA protein did not bind BiP; however, when the most N-terminal glycosylation sites were mutated, strong interactions with BiP could be detected. Thus, the presence of N-terminal glycosylation sites seems to direct nascent chains to the calnexin system, while the absence of such sites directs nascent chains to BiP.

One of the first recognized needs for a quality control system is in the retention of misassembled protein complexes; polypeptides that normally associate with other polypeptides cannot progress through the secretory pathway until they are bound to their partners (113). A molecular basis for this observation is suggested by recent studies that analyzed the trafficking of subunits of a mammalian ATP-sensitive potassium channel (281). This channel is composed of four regulatory subunits (known as SUR1/2A/2B) and four potassium ion channel subunits (known as Kir6.1/2), and proper surface expression of the channel requires the co-expression of all eight subunits. Sequence analysis combined with mutagenesis identified a sequence, RXR, whose presence blocks the trafficking of channel subunits to the cell surface and retains them in the ER (281). The similarity to KKXX motifs, whose presence at the C terminus of transmembrane proteins of the secretory pathway leads to ER retention (117), suggests that a similar mechanism may prevent the trafficking of proteins containing either motif.

However, unlike KKXX motifs, the use of an RXR motif as a retrieval mechanism has been documented only in proteins whose final location is not the ER (281). Importantly, the RXR motif functions only when present in unassembled or partially unassembled complexes (281), suggesting that the RXR motif is buried in the assembled complexes. Thus, the masking of RXR-containing sequences in the folded or mature state of a protein appears to be the factor that allows the RXR motif to act in a quality control mechanism, rather than a constitutive retention mechanism. These results do not differentiate between a retention mechanism and a retrieval mechanism. A more general role for RXR motifs is suggested by the observation of several such sequences in the cytosolic loops of CFTR, the transmembrane conductance-regulator that is responsible for all inherited cases of cystic fibrosis. When these sequences are altered in a ΔF508 CFTR mutant, which is ordinarily retained in the ER, the protein can be secreted (38).

PROTEIN QUALITY CONTROL: ER ASSOCIATED PROTEIN DEGRADATION (ERAD)

Because unfolded proteins cannot always achieve their native state in the ER, eukaryotic cells have evolved a constitutively active quality control system to rid the ER of misfolded proteins. This process, termed ER-associated protein degradation (ERAD; 154), involves three key steps: (*a*) recognition of the aberrant polypeptide; (b) export of soluble proteins to the cytoplasm back through the translocation pore ("retrotranslocation"); and (c) degradation by the proteasome. The conformational diversity of proteins entering the ER and the requirement for ERAD to selectively degrade only misfolded or regulated proteins suggests that this process is complex. However, biochemical and genetic studies together are elucidating the cellular players and mechanisms in this elaborate safety net.

The prevailing evidence suggests that many of the same molecular chaperones involved in folding proteins in the ER are also involved in the removal of ERAD substrates. Aberrant secretory proteins may have exposed structural motifs (221, 230, 271) or hydrophobic patches that could prolong chaperone interactions and trigger their destruction. Consistent with this model, the chaperones BiP (30, 198), calnexin (154), and protein disulfide isomerase (PDI; 82) are required for the degradation of some ERAD substrates in yeast. In addition, two lumenal Hsp40 homologs in yeast, Scj1p and Jem1p, interact with BiP (179, 218) and help prevent the aggregation of misfolded proteins prior to their retrotranslocation (180). Biochemical studies indicate that mammalian calnexin prevents the aggregation of unfolded (and unglycosylated) proteins in solution (115), an activity that also promotes protein folding in vitro in conjunction with yeast BiP and the Sec63p J domain (D. Williams, personal communication). Aggregation of ERAD substrates prior to their export would preclude their transit through the translocation channel.

Is there a signal that distinguishes between slowly folding proteins and those that are terminally misfolded? Recent studies indicate that competition between enzymes that attach or remove sugar moieties may function as a timer for the folding of individual glycoproteins in the mammalian ER (65, 140). As discussed above, the trimming of glucose residues on the branched $Glc_3Man_9GlcNAc_2$ oligosaccharide triggers dissociation of the calnexin/calreticulin-glycoprotein complex so that the correctly folded glycoprotein can exit the ER (102). After prolonged retention of a misfolded protein in the ER, the trimming of mannose residues may divert the protein from the calnexin-catalyzed folding pathway into the degradation pathway, which may or may not be dependent on further interactions with calnexin (2, 42, 44, 57, 140, 244, 274). Consistent with this model, degradation of a yeast ERAD substrate, a mutated form of the vacuolar-targeted carboxypeptidase Y (CPY*), depends upon glycosylation and requires the mannosidase I-generated $Man_7GlcNAc_2$ moiety (118, 133), but there is limited evidence for a calreticulin or calnexin binding cycle in *S. cerevisiae*. Instead, factors like the recently identified α-mannosidase-like protein, Mnl1p, may identify glycoproteins containing $Man_7GlcNAc_2$ linkages as ERAD substrates (172). In contrast, *Schizosaccharomyces pombe* calnexin is essential and more homologous to calnexin in higher eukaryotes (121, 189), and glucosyltransferase activity is required for *S. pombe* viability under stress conditions (66). If the calnexin/calreticulin cycle is a general feature of ERAD, an as yet undiscovered lectin must target glycoproteins containing trimmed mannoses for ERAD. However, not all ERAD substrates are glycosylated, suggesting the existence of multiple mechanisms for identifying terminally misfolded proteins.

The export of soluble ERAD substrates occurs by retrotranslocation (or dislocation) through the Sec61p translocation pore. The strongest evidence supporting this hypothesis is the stabilization of yeast ERAD substrates in vitro (195) and in vivo (198, 285) in *sec61* mutant microsomes or cells, respectively. Despite using the same channel and requiring BiP, the isolation of ERAD-specific mutations in *KAR2* (BiP; 30) and *SEC61* (273, 285) suggests that translocation and retrotranslocation are mechanistically distinct. As mentioned above, Sec63p is an Hsp40 homolog that cooperates with BiP in the import of proteins into the ER, but it appears to play a less prominent role in ERAD (180, 195, 198). Conversely, Scj1p and Jem1p are required for ERAD but not for translocation (180). In addition, because signal sequences are cleaved concomitant with translocation, there must be a different mechanism for targeting ERAD substrates to the lumenal face of the Sec61 pore. Several studies suggest that BiP may deliver misfolded proteins to the Sec61 channel (30, 132, 221, 230) and perhaps gate the pore to regulate opposing traffic (90a, 200). Römisch and coworkers also propose a role for PDI in targeting one ERAD substrate to BiP at the translocation pore (82), and Norgaard et al. (181a) report that expression of any one of four other PDI homologues restores ERAD activity in yeast lacking PDI.

The mechanism(s) governing export and degradation of transmembrane proteins from the ER may be distinct from that controlling the ERAD of soluble proteins. Membrane proteins, like soluble proteins, might exit the ER through the Sec61 channel because they can be co-immunoprecipitated with a component of the mammalian ER translocation channel (9, 58, 268) or are stabilized in *sec61* mutant yeast (199, 285). However, some transmembrane proteins may be directly extracted by the proteasome (150, 255a) or attacked by other proteases (67, 78, 143, 165, 168, 269, 272). The proteasome might also "shave" the cytoplasmic portions of integral ER membrane proteins, as Jentsch and colleagues have recently reported that the proteasome may be able to clip polypeptide "loops" (111). The resulting transmembrane domains might be unstable and spontaneously dissociate to the cytoplasm or could be cleaved further (263). In addition, BiP is not required to degrade four known transmembrane ERAD substrates, whereas the cytosolic hsc70 Ssa1p is necessary to degrade the integral membrane proteins Ste6p*, CFTR and Vph1p (105, 199, 283; S. Michaelis, personal communication) (see Table 1). In contrast, Ssa1p is dispensable and BiP is required for the ERAD of three soluble proteins, PαF, carboxypeptidase Y (CPY*), and mammalian α-1 protease (A1PiZ) in yeast (Table 1). We have suggested that Ssa1p may be required to prevent aggregation of the large cytoplasmic domains in these transmembrane proteins (283). Consistent with this hypothesis, the degradation of Sec61-2p, which contains significantly fewer amino acids in the cytoplasm than the other transmembrane ERAD substrates described above, is only modestly affected in *ssa1* mutant cells (180).

TABLE 1   ERAD requirements for substrates in yeast[a]

Soluble			Integral membrane		
Substrate	**Required**	**Dispensable**	**Substrate**	**Required**	**Dispensable**
PαF	BiP	Ssa1p	Ste6p*	Ssa1p	BiP
	PDI	Ubiquitination		Ubc6/7p	
	Sec61p	Sec63p			
	Cne1p	Scj1p	CFTR	Ssa1p	BiP
	Scj1p/Jem1p	Cer1p/Lhs1p/		Ubc6/7p	Cne1p
		Ssi1p			
		Hsp90	Vph1p	Ssa1p	BiP
		Ssh1p			
		Eug1p	Hmg2p	Ubc7p	Der1p
CPY*	BiP	Ssa1p		Hrd1p/Der3p	Ubc6p
	Png1p	Cne1p			
	Der1p	PDI	Sec61p	Der3p/Hrd1p	Scj1p/Jem1p
	Hsp90			Ubc6/7p	
	Der3p/Hrd1p			Cue1p	
	Hrd3p				
	Sec61p		Pdr5p*	Hrd3p	BiP
	Cue1p			Ubc6/7p	Der1p
	Pmr1p			Der3p/Hrd1p	
	Sec63p			Sec61p	
	Mns1p				
	Scj1p/Jem1p				
	Ubc6/7p				
A1PiZ	BiP	Ssa1p			

[a]Updated from Brodsky & McCracken (28a).

Multiple studies indicated that ERAD substrates are degraded in the cytoplasm by the proteasome (16, 94, 107, 122, 184, 204, 232, 265, 268). This complex proteolytic machine consists of a catalytic 20S cylindrical core particle and two copies of the 19S (PA700) regulatory particle that "caps" the 20S subunit (6). Ubiquitination is necessary for proteasomal processing of most (17, 95, 107, 122, 142, 260, 285), but not all ERAD substrates (156, 265, 279). Two ubiquitin-conjugating enzymes, Ubc6p and Ubc7p (16, 17, 107, 232), and a ubiquitin ligase, Hrd1p/Der3p (8, 55, 79), reside at the yeast ER membrane and are required to degrade many ERAD substrates (see Table 1). In addition to targeting substrates to the proteasome, ubiquitination is also required for the retrotranslocation of some proteins (16, 20, 58). Likewise, Mayer and colleagues (150) and Plemper et al. (199) were unable to detect cytosolic, ubiquitinated forms of ERAD substrates in yeast proteasome mutants. The proteasome may provide the energy, via its six resident ATPases, to directly extract ERAD proteins concomitantly with their ubiquitination (150, 255a).

Before interacting with the proteasome, ERAD substrates are de-glycosylated and de-ubiquitinated and must be unfolded to fit through a small aperture at the tip of the catalytic core of the proteasome (239, 268; reviewed in 253). The 19S subunit of the proteasome is a molecular chaperone, capable of binding and preventing the aggregation of unfolded proteins (23, 236). In addition, other cytosolic molecular chaperones including Ssa1p (see above) and Hsp90 (77, 88, 116, 222) may maintain ERAD substrates in an aggregation-free state for attack by the proteasome or help deliver substrates to the proteasome, as these chaperones have been found associated with the yeast 19S subunit (251). In mammalian cells, a putative E3 ubiquitin ligase, CHIP, could mediate the delivery of misfolded proteins from these cytosolic chaperones to the proteasome (4, 46, 157). Likewise, the mammalian nucleotide exchange factor BAG-1 may help target hsc70-bound substrates to the proteasome (108, 145).

The elucidation of the ERAD pathway provided a model for how several toxins are able to transit from the ER to the cytoplasm (101). These toxins, which include ricin, pertussis toxin, Shiga toxin, *Pseudomonas* exotoxin A, cholera toxin, and yeast killer toxins, enter host cells through the endocytic pathway and ultimately reside in the ER by virtue of harboring ER retrieval sequences at their C termini. Once in the ER, they are exported to the cytoplasm via the Sec61 channel (64, 220, 229, 266). Consistent with the ERAD machinery being required for toxin action, yeast containing mutations in the genes encoding BiP and calnexin exhibit increased resistance to the K28 killer toxin (64), and prior to export, PDI is required for toxin unfolding (246). Inhibition of proteasome activity sensitizes both yeast and mammalian cells to toxins (229, 266), suggesting that a fraction of the retro-translocated toxin is recognized as an ERAD substrate. However, most of the toxin may escape degradation because the proteins are lysine-poor (101), thus minimizing their probability of being ubiquitinated.

Additional components of the ERAD machinery have been identified in three independent yeast genetic screens. Stabilization of hydroxymethylglutaryl-coenzyme A reductase (HMG-R) in mutant yeast strains (94) led to the discovery of Hrd1p (also known as Der3p, see below) and Hrd3p, which form a stoichiometric complex spanning the ER membrane (79) and preferentially ubiquitinate misfolded proteins (8). Also identified in this screen was Hrd2p, a component of the 19S regulatory subunit of the proteasome. A screen by Wolf and coworkers for mutants in which CPY* is stabilized uncovered three *DER* genes, (see above; 133). Der1p is an integral ER membrane protein of unknown function (133) and *DER2* and *DER3* encode for Ubc7p and Hrd1p, respectively, factors involved in ERAD (20, 133). Finally, mutants that accumulate a heterologously expressed variant of the mammalian ERAD substrate, Alpha-1 protease inhibitor (AlPiZ; 204), have identified seven complementation groups that may represent novel genes involved in ERAD (155). Combining the continued analysis of these and other genes required for ERAD with powerful biochemical tools will ensure a finer dissection of the ERAD pathway.

THE UNFOLDED PROTEIN RESPONSE (UPR)

A primary mechanism by which eukaryotic cells counteract the accumulation of misfolded proteins within the lumen of the ER is the unfolded protein response (UPR). This response was initially recognized in mammalian cells by the induction of a specific set of proteins in response to glucose starvation, which results in protein misfolding through the under-glycosylation of nascent polypeptides (39). The proteins induced through this treatment were designated GRPs as a consequence of their glucose regulation (e.g., GRP78 was the original name given to BiP), and consisted largely of molecular chaperones. Other treatments were soon discovered that increased the transcription of the same set of genes, including tunicamycin (an inhibitor of N-linked glycosylation), DTT, and calcium-ionophores. However, other general stress conditions, including heat shock, do not induce the expression of the same set of genes. This stereotyped response to ER-specific folding stressors is shared among all eukaryotic cells.

Rapid progress in detailing the mechanism of UPR activation became possible with the discovery of this response in *S. cerevisiae*. A promoter element, termed the UPRE, was found upstream of UPR targets in *S. cerevisiae* (164) and was subsequently used to begin genetically defining the signaling pathway between the ER and nucleus that is responsible for activation of UPR target gene expression (48, 163). The first screens identified molecules at the extreme ends of the signaling pathway (Figure 2). The signal originates in the lumen of the endoplasmic reticulum with the activation of the transmembrane serine/threonine kinase Ire1p (48, 163). When unfolded proteins begin to accumulate in the ER, the Ire1p kinase dimerizes and is autophosphorylated in *trans* (224). At the other end of the signaling pathway lies Hac1p, a member of the bZIP family of transcription factors (49, 162). Both factors are absolutely required for UPR induction, as deletion of either gene results in a strain unable to increase the expression of known UPR targets in response to ER folding stress.

The discovery of the pathway linking Ire1p and Hac1p awaited the convergence of a number of different observations. First, *HAC1* mRNA migrates differently when isolated from UPR-induced or noninduced cells (49, 162). Second, Hac1p can only be detected in cells under conditions that induce the UPR (49, 162). Finally, another genetic screen implicated *RLG1*, a tRNA ligase, in induction of the UPR (226). When combined with the observation that Ire1p contains a domain with homology to nucleases (48, 163), a model emerged in which Ire1p becomes an active nuclease when unfolded proteins accumulate within the ER. Ire1p then cleaves the transcribed *HAC1* message (termed *HAC1^u*) at specific locations near the 3′ end, removing a nonconventional intron (130, 227). The alternative splicing of the *HAC1* mRNA is completed through the action of Rlg1p, which ligates the alternative exon to the *HAC1* message, forming a new message designated *HAC1^i* (226, 227). Only the protein encoded by the alternatively spliced message accumulates in cells. This reaction has since been reconstituted in vitro using only Ire1p, *HAC1* mRNA, and Rlg1p (86).

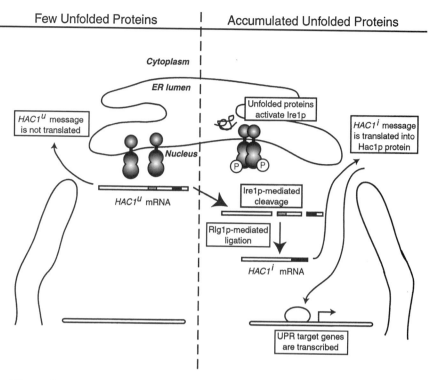

| Few Unfolded Proteins | Accumulated Unfolded Proteins |

Figure 2   A schematic model of the unfolded protein response pathway as defined in the yeast *S. cerevisiae*.Upon the accumulation of unfolded proteins within the ER, Ire1p becomes activated through dimerization. This initiates an alternative splicing event, ultimately producing the *HAC1^i* mRNA. Only the *HAC1^i* message is efficiently translated. The Hac1p protein enters the nucleus, binds to promoter elements upstream of the UPR target genes, and activates their transcription. Although Ire1p is depicted here with its endonuclease domain localized to the nucleus, this has not been demonstrated experimentally. See text for details (modified from 223a).

One model to explain how cells might sense unfolded proteins predicts that BiP binds to the Ire1p lumenal domain during normal growth, preventing the dimerization of Ire1p molecules. As unfolded proteins accumulate in the ER, increasing amounts of BiP are recruited from Ire1p. Eventually, Ire1p dimerizes, initiating the UPR signaling pathway. This model was recently tested by Ron and co-workers (15), who detected a physical association between Ire1p and BiP in extracts from a rat pancreas–derived cell line under normal growth conditions; under conditions of UPR induction, a physical interaction between Ire1p and BiP was absent. A similar mechanism is also likely to exist in the yeast ER (186).

The pathway leading to UPR activation in mammalian cells is more complex than that in *S. cerevisiae* and has been less clearly defined. A number of groups have identified Ire1p homologs in higher eukaryotes, including Ire1α (identified in humans) (243), Ire1β (identified in mouse cells) (258), and PERK (98, 225). Whereas both Ire1α and Ire1β show homology to Ire1p throughout their entire lengths, PERK is homologous to Ire1p only in its ER lumenal domain and has a kinase domain more like that of eIF2α than that of Ire1p. Consistent with these data, Ire1α, Ire1β, and PERK respond to the same inducers, but diverge in the downstream signaling events that they mediate. Whereas Ire1α and Ire1β induce the expression of BiP and CHOP (another UPR target), PERK responds to the accumulation of unfolded proteins by phosphorylating eIF2α, leading to a decrease in translation (98).

In contrast to the situation in *S. cerevisiae*, activation of the mammalian UPR appears to involve a proteolysis step at the level of Ire1 activation. Upon stimulation of the UPR, both Ire1α and Ire1β are cleaved from the membrane, and the newly released, soluble form redistributes to the nucleus (181). This redistribution seems to depend on the activity of presinilin-1 (PS1), as cells lacking PS1 activity are unable to produce the soluble form of Ire1. In addition, in at least some cell lines, lack of PS1 decreases the level of UPR induction as measured by BiP expression (129, 181). However, although two groups have observed a role for PS1, a third report finds no effect of PS1 on UPR activation (216). As the conditions used in these experiments are not identical, the full significance of PS1 in UPR activation will await future experiments.

Proteolysis has also been implicated in the activation of at least one transcription factor responsible for the ER stress response in metazoan cells. ATF6, a Type-II transmembrane protein, is cleaved into two fragments in response to treatments that lead to the accumulation of misfolded proteins, and the released cytosolic domain translocates into the nucleus and induces the transcription of several chaperones (31, 100). Goldstein and co-workers subsequently demonstrated that S1P and S2P, the proteases responsible for cleavage of the sterol-starvation transcription factors of the SREBP family, are necessary for cleavage of ATF6 and for a normal ER stress response (276). However, unlike the SREBP targets of S1P and S2P, sterols do not affect activation of gene expression through ATF6.

At this point, the relationship between the ATF6 and IRE1α/IRE1β pathways is unclear. Data from Kaufman and co-workers suggest that ATF6 activation lies downstream of Ire1α activation and that the response to ER stress begins with activation of Ire1α (259). However, ATF6 does not appear to be alternatively spliced under conditions of ER stress (277). As both Ire1α and Ire1β show homology to the nuclease domain of *S. cerevisiae IRE1* (243, 258), another transcription factor in mammalian cells, yet to be identified, may be activated in the same fashion as *HAC1* in *S. cerevisiae*. Indeed, both Ire1α and Ire1β can cleave yeast *HAC1* mRNA in vitro (181, 243).

INTERACTION BETWEEN THE UPR AND ERAD

Although the initial characterization of the UPR suggested that its targets would be limited to chaperones and factors required to maintain ER homeostasis, such as lipid biosynthesis (50), a growing body of evidence now suggests that the UPR regulates many aspects of secretory pathway function. By taking advantage of the genetic requirements of UPR activation in *S. cerevisiae* and oligonucleotide microarray technology, a list has been compiled detailing the breadth of the transcriptional output of the UPR (37, 245). From this analysis, nearly 400 genes were identified as UPR targets, of which 208 were of known or inferred function. Of these 208 genes, approximately half play roles in the secretory pathway. Thus, genes encoding chaperones that exist entirely outside the secretory pathway in yeast, such as Hsp104, Hsp60, and Hsp90, were not identified as UPR targets. Of chaperone families with members found in every cellular compartment, only those genes encoding ER-localized chaperones were identified as UPR targets. For example, of the 15 DnaJ homologs encoded in the yeast genome, only the three homologs encoding ER-localized DnaJ homologs are induced by the UPR. The functional categorization of UPR targets is depicted in Figure 3.

The UPR activates the expression of genes encoding proteins acting throughout the secretory pathway and spanning virtually all activities. How these UPR-induced targets improve the state of folding within the ER is unclear. The various activities induced by the UPR may well act in concert to reduce the lumenal concentration of misfolded protein, by either directly refolding proteins or removing them from the ER. This "fix or clear" model suggests that all activities required for folding are up-regulated, such that chaperones bind to misfolded species, prevent aggregation, and promote folding, while glycosylation enzymes assist in the folding of proteins that require carbohydrate modification to attain their proper conformation. Consistent with this suggestion, mutations that compromise either addition of GPI anchors or protein glycosylation are lethal in the absence of UPR function (177). Moreover, UPR induction in mammalian cells accelerates synthesis of the dolichol-oligosaccharides employed in asparagine-linked glycosylation (62).

In addition to up-regulating factors that directly promote folding, UPR activation may also induce factors to clear misfolded proteins from the ER. The induction of specific COPII or coatomer components might facilitate the packaging of cargo proteins into anterograde vesicles, or simply increase the overall capacity of anterograde transport. This increase in anterograde transport might catalyze the passage of misfolded species to the vacuole for degradation (110, 137), consistent with the observation that several genes involved in vacuolar targeting are also UPR targets, or the retrieval of ERAD substrates from the Golgi that must be returned to the ER for degradation (D. Ng, personal communication). Similarly, induction of lipid synthesis may lead to an increase in the volume of the ER, diluting the concentration of unfolded proteins. Finally, the induction of ERAD components directly enhances the clearance of misfolded proteins from the ER.

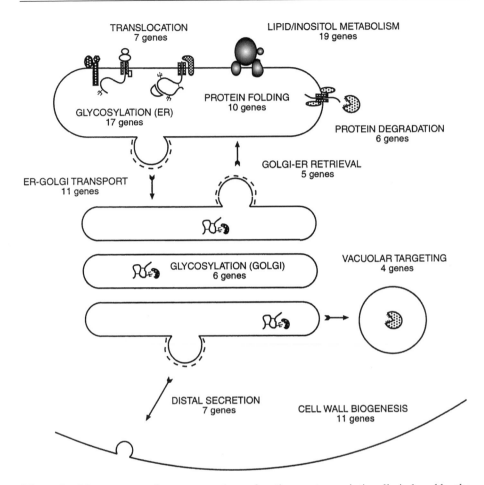

Figure 3   Many aspects of secretory pathway function are transcriptionally induced by the UPR. A schematic diagram of the secretory pathway is shown. The number of genes whose function is either known or can be inferred from homology to characterized genes is indicated underneath each functional category (reproduced from 245).

The link between the UPR and ERAD suggests the existence of a previously unrecognized connection between two pathways that deal with the consequences of misfolded proteins. Mechanistic studies from a number of research groups have now confirmed the physiological significance of these findings. First, efficient ERAD requires an intact UPR. In particular, deletion of *IRE1* decreased the ERAD of CPY* (245) and MHC class I heavy chain (H-2K[b]) in yeast (37). Second, loss of ERAD function leads to chronic UPR induction. Mutants defective for CPY* degradation show a small but significant induction of the UPR (75, 133, 245). Alleles of *SEC61* with specific defects in ERAD, as well as deletions of several other ERAD components, also caused constitutive UPR induction (285). Thus, the chronic accumulation of misfolded proteins in the ER appears to be a general consequence of loss of ERAD. Third, simultaneous loss of ERAD and UPR function greatly decreases cell viability. For example, Ng and co-workers conducted a screen to identify genomic mutations that are synthetically lethal with loss of the UPR (177). A large panel of mutants were isolated and were then further classified based on functional analysis. Analysis of the rate of CPY* degradation indicated that one third of the identified mutants have defects in ERAD. The functional significance of this genetic interaction is emphasized by the synthetic lethality between mutations of genes in the UPR pathway and in a number of components required for ERAD [*SON1*, *UBC1*, *UBC7*, *HRD1*, *HRD3*, and *DER1* (75, 177, 245)] that act at multiple steps in the ERAD pathway.

In sum, these findings suggest that protein folding in the ER is inefficient, and the removal of misfolded proteins is an essential process performed together by the UPR and ERAD machineries. In the absence of the UPR, ERAD deals with the consequences of protein misfolding by retro-translocating these species to the cytoplasm where they are degraded. Conversely, in the absence of ERAD, the UPR deals with the consequences of protein misfolding by activating the expression of factors involved in protein folding, anterograde vesicular transport, or an alternate site of degradation, such as the vacuole. Thus, the UPR and ERAD systems provide partially overlapping functions in the same essential process: the removal of misfolded proteins from the ER.

SUBSTRATE-SPECIFIC CHAPERONES

Several genes identified in yeast are required for the biogenesis of specific secreted proteins. Although the genes do not encode classical chaperones, and in many cases their specific functions are unknown, they apparently evolved to facilitate the folding or quality control of selected secreted substrates. This class of protein was first recognized genetically through the identification of a mutant strain of yeast that showed defects in amino acid uptake. The gene implicated in this study, identified as *SHR3*, was found to be ER-localized, and resulted in retention of amino acid permeases in the ER (141). Several substrate-specific chaperones have subsequently been identified in *D. melanogaster*, *S. cerevisiae*, *C. elegans*, and mammalian cells, all of which are ER-localized and required for proper secretion of only one or a subset of proteins (reviewed in 65).

In most cases, the level at which the proteins act is not clear; they may be required for folding or secretion, or they may act as a specific quality control mechanism. For example, Naik & Jones (171) screened for mutants defective for the processing of the vacuolar-targeted proteinase B (Prb1p) and isolated a gene encoding an ER-localized, integral membrane protein named Pbn1p. Two-hybrid analysis indicated that Pbn1p interacts with the Prb1p pro-peptide. *PBN1* is essential, unlike the gene encoding its substrate (*PRB1*), suggesting that Pbn1p may play a role in the biogenesis of essential factors and/or that it is required for ER homeostasis. Interestingly, Prb1p becomes an ERAD substrate when Pbn1p function is ablated.

In contrast, in at least two other cases, homology between the substrate-specific chaperone and a class of general chaperones implies that the required activity is indeed related to protein folding. One example of a substrate-specific chaperone with homology to a specific chaperone class is a PDI homolog (*EPS1*) whose activity was identified in studies of the yeast plasma membrane ATPase, encoded by the *PMA1* gene. A dominant mutation was described that prevented the ER export of both the mutant and wild-type forms of the Pma1 protein from the ER, which became ERAD substrates. Wang & Chang (256) screened for suppressors of the dominant phenotype and uncovered the nonessential gene, *EPS1*. Wild-type and mutant Pma1p are stabilized in cells deleted for *EPS1* because they are no longer retained in the ER, suggesting that Eps1p is a quality control gatekeeper in the ER, preventing the secretion of misfolded Pma1p.

The second example involves the PPIase homolog *ninaA* that was identified in studies of rhodopsin folding in *D. melanogaster* (45, 233). *ninaA* mutant flies show a greatly reduced level of rhodopsin in the outer photoreceptor cells. Upon closer examination, it was observed that only the Rh1 and Rh2 rhodopsins were affected by the lack of *ninaA* activity, while the Rh3 rhodopsin was not (233). Cloning of *ninaA* revealed that it was a transmembrane protein and highly homologous to vertebrate cyclophilin. Like other substrate-specific chaperones, mutation of *ninaA* results in the accumulation of its substrates in the ER, although rhodopsin was also found in vesicles distributed throughout the cytoplasm (45).

Although little is known about the substrate-specific chaperones, they are clearly a growing class of proteins with important roles in the maturation of a wide variety of secretory proteins. These proteins provide a unique insight into the particular demands on protein folding in the secretory pathway. More generally, an understanding of substrate-specific chaperones might reveal previously overlooked aspects of protein folding in the ER.

SUMMARY

By virtue of their endogenous biochemical properties and their promiscuity, chaperones have adapted as critical factors in the eukaryotic secretory pathway. Not only do molecular chaperones act as central players in each of the processes discussed in this review, but an individual chaperone may play critical roles in several of these processes, sometimes simultaneously. This has complicated genetic analyses of chaperone action in the cell. However, the isolation of mutants that are specifically defective for a single process, and/or the use of strong, conditionally acting mutants, has permitted a better molecular dissection of chaperone action during protein secretion. Equally powerful has been the use of in vitro assays in which the functions of wild-type and mutant chaperones can be ascertained in defined systems. The continued development of biochemical assays that measure unique aspects of secretory pathway function, combined with the isolation and construction of new mutants, should permit researchers to define further how chaperones can exert their pleiotropic effects. In addition, a relatively recent, but rapidly progressing field is the solution of chaperone structures using biophysical techniques. Such undertakings will further catalyze biochemical and genetic experiments.

ACKNOWLEDGMENTS

The authors thank Drs. Mehdi Kabani, Davis Ng, Chris Patil, David Williams, Sabine Rospert, Elizabeth Jones, and Ardythe McCracken for helpful suggestions and for providing unpublished data. Work in the authors' laboratories was funded by grants from the National Science Foundation and the American Cancer Society (JLB), and the National Institutes of Health and the Howard Hughes Medical Research Foundation (JSW). SWF acknowledges a National Research Service Award from the National Institutes of Health.

Visit the Annual Reviews home page at www.AnnualReviews.org

LITERATURE CITED

1. Anfinsen CB. 1973. Principles that govern the folding of protein chains. *Science* 181:223–30

2. Ayalon-Soffer M, Shenkman M, Lederkremer GZ. 1999. Differential role of mannose and glucose trimming in the ER degradation of asialoglycoprotein receptor subunits. *J. Cell Sci.* 112:3309–18

2a. Bader M, Muse W, Ballo DP, Gassner C, Bardwell JC. 1999. Oxidative protein folding is driven by the electron transport system. *Cell* 98:217–27

3. Baker EK, Colley NJ, Zuker CS. 1994. The cyclophilin homolog NinaA functions as a chaperone, forming a stable complex in vivo with its protein target rhodopsin. *EMBO J.* 13:4886–95

4. Ballinger CA, Connell P, Wu Y, Hu Z, Thompson LJ, et al. 1999. Identification of CHIP, a novel tetratricopeptide repeat-containing protein that interacts with heat shock proteins and negatively regulates chaperone function. *Mol. Cell. Biol.* 19:4535–45

5. Banecki B, Liberek K, Wall D, Wawrzynow A, Georgopoulos C, et al. 1996. Structure-function analysis of the zinc finger region of the DnaJ molecular chaperone. *J. Biol. Chem.* 271:14840–48

6. Baumeister W, Walz J, Zühl F, Seemüller E. 1998. The proteasome: paradigm of a self-compartmentalizing protease. *Cell* 92:367–80

7. Baxter BK, James P, Evans T, Craig EA. 1996. SSI1 encodes a novel hsp70 of the *Saccharomyces cerevisiae* endoplasmic reticulum. *Mol. Cell. Biol.* 16:6444–56

8. Bays NW, Gardner RG, Seelig LP, Joazeiro CA, Hampton RY. 2001. Hrd1p/Der3p is a membrane-anchored ubiquitin ligase required for ER-associated degradation. *Nat. Cell Biol.* 3:24–29

9. Bebök Z, Mazzochi C, King SA, Hong JS, Sorscher EJ. 1998. The mechanism underlying cystic fibrosis transmembrane conductance regulator transport from the endoplasmic reticulum to the proteasome includes Sec61β and a

cytosolic, deglycosylated intermediary. *J. Biol. Chem.* 273:29873–78

10. Becker J, Walter W, Yan W, Craig EA. 1996. Functional interaction of cytosolic hsp70 and a DnaJ-related protein, Ydj1p, in protein translocation in vivo. *Mol. Cell. Biol.* 16:4378–86

11. Beckmann R, Bubeck D, Grassucci R, Penczek P, Verschoor A, et al. 1997. Alignment of conduits for the nascent polypeptide chain in the ribosome-Sec61 complex. *Science* 278:2123–26

12. Beckmann RP, Mizzen LA, Welch WJ. 1990. Interaction of Hsp70 with newly synthesized proteins; implications for protein folding and assembly. *Science* 248:850–54

13. Benham AM, Cabibbo A, Fassio A, Bulleid N, Sitia R, Braakman I. 2000. The CXXCXXC motif determines the folding, structure and stability of human Ero1-Lalpha, *EMBO J.* 19:4493–502

14. Berjanskii MV, Riley MI, Xie A, Semenchenko V, Folk WR, Van Doren SR. 2000. NMR structure of the N-terminal J domain of murine polyomavirus. *J. Biol. Chem.* 275:36094–103

15. Bertolotti A, Zhang Y, Hendershot LM, Harding HP, Ron D. 2000. Dynamic interaction of BiP and ER stress transducers in the unfolded-protein response. *Nat. Cell Biol.* 2:326–32

16. Biederer T, Volkwein C, Sommer T. 1996. Degradation of subunits of the Sec61p complex, an integral component of the ER membrane, by the ubiquitin-proteasome pathway. *EMBO J.* 15:2069–76

17. Biederer T, Volkwein C, Sommer T. 1997. Role of Cue1 in ubiquitination and degradation at the ER surface. *Science* 278:1806–9

18. Blond-Elguindi S, Cwirla SE, Dower WJ, Lipshutz RJ, Sprang SR, et al. 1993. Affinity panning of a library of peptides displayed on bacteriophages reveals the binding specificity of BiP. *Cell* 75:717–28

19. Boisrame A, Kabani M, Beckerich JM, Hartmann E, Gaillardin C. 1998. Interaction of Kar2p and Sls1p is required for efficient co-translational translocation of secreted proteins in the yeast *Yarrowia lipolytica*. *J. Biol. Chem.* 27:30903–8

20. Bordallo J, Plemper RK, Finger A, Wolf DH. 1998. Der3p/Hrd1p is required for endoplasmic reticulum associated degradation of mis-folded lumenal and integral membrane proteins. *Mol. Biol. Cell* 9:209–22

21. Bose S, Weikl T, Bugl H, Buchner J. 1996. Chaperone function of Hsp90-associated proteins. *Science* 274:1715–17

22. Braakman I, Helenius J, Helenius A. 1992. Manipulating disulfide bond formation and protein folding in the endoplasmic reticulum. *EMBO J.* 11:1717–22

23. Braun BC, Glickman M, Kraft R, Dahlman B, Kloetzel P-M, et al. 1999. The base of the proteasome regulatory particle exhibits chaperone-like activity. *Nat. Cell Biol.* 1:221–26

24. Brodsky JL. 1996. Post-translational protein translocation: Not all Hsp70s are created equal. *Trends Biochem. Sci.* 21:121–26

25. Brodsky JL, Bauerle M, Horst M, McClellan AJ. 1998. Mitochondrial Hsp70 cannot replace BiP in driving protein translocation into the yeast endoplasmic reticulum. *FEBS Lett.* 435:183–86

26. Brodsky JL, Goeckeler J, Schekman R. 1995. Sec63p and BiP are required for both co- and post-translational protein translocation into yeast microsomes. *Proc. Natl. Acad. Sci. USA* 92:9643–46

27. Brodsky JL, Hamamoto S, Feldheim D, Schekman R. 1993. Reconstitution of protein translocation from solubilized yeast membranes reveals topologically distinct roles for BiP and cytosolic hsc70. *J. Cell Biol.* 120:95–102

28. Brodsky JL, Lawrence JG, Caplan AJ. 1998. Mutations in the cytosolic DnaJ

homologue, YDJ1, delay and compromise the efficient translation of heterologous proteins in yeast. *Biochemistry* 37:18045–55

28a. Brodsky JL, McCracken AA. 1999. ER protein quality control and proteasome-mediated protein degradation. *Semin. Cell Dev. Biol.* 10:507–13

29. Brodsky JL, Schekman R. 1993. A Sec63p-BiP complex from yeast is required for protein translocation in a reconstituted proteoliposome. *J. Cell Biol.* 123:1355–63

30. Brodsky JL, Werner ED, Dubas ME, Goeckler JL, Kruse KB, McCracken AA. 1999. The requirement for molecular chaperones during endoplasmic reticulum-associated protein degradation demonstrates that protein export and import are mechanistically distinct. *J. Biol. Chem.* 274:3453–60

31. Brown MS, Ye J, Rawson RB, Goldstein JL. 2000. Regulated intramembrane proteolysis: a control mechanism conserved from bacteria to humans. *Cell* 100:391–98

32. Bukau B, Horwich AL. 1998. The Hsp70 and Hsp60 chaperone machines. *Cell* 92:351–66

33. Bush GL, Meyer DI. 1996. The refolding activity of the yeast heat shock proteins Ssa1p and Ssa2p defines their role in protein translocation. *J. Cell Biol.* 135:1229–37

34. Cabibbo A, Pagani M, Fabbri M, Rocchi M, Farmery MR, et al. 2000. ERO1-L, a human protein that favors disulfide bond formation in the endoplasmic reticulum. *J. Biol. Chem.* 275:4827–33

35. Caplan AJ. 1999. Hsp90's secrets unfold: new insights from structural and functional studies. *Trends Cell Biol.* 9:262–68

36. Caplan AJ, Cyr DM, Douglas MG. 1992. YDJ1 facilitates polypeptide translocation across different intracellular membranes by a conserved mechanism. *Cell* 71:1143–55

37. Casagrande R, Stern P, Diehn M, Shamu C, Osario M, et al. 2000. Degradation of proteins from the ER of *S. cerevisiae* requires an intact unfolded protein response pathway. *Mol. Cell* 5:729–35

38. Chang XB, Cui L, Hou YX, Jensen TJ, Aleksandrov AA, et al. 1999. Removal of multiple arginine-framed trafficking signals overcomes misprocessing of delta F508 CFTR present in most patients with cystic fibrosis. *Mol. Cell* 4:137–42

39. Chapman R, Sidrauski C, Walter P. 1998. Intracellular signaling from the endoplasmic reticulum to the nucleus. *Annu. Rev. Cell Dev. Biol.* 14:459–85

40. Chappell TG, Konforti BB, Schmid SL, Rothman JE. 1987. The ATPase core of a clathrin uncoating protein. *J. Biol. Chem.* 262:746–51

41. Cheetham ME, Caplan AJ. 1998. Structure, function and evolution of DnaJ: conservation and adaptation of chaperone function. *Cell Stress Chaperones* 3:28–36

42. Chillaron J, Adan C, Haas IG. 2000. Mannosidase action, independent of glucose trimming, is essential for proteasome-mediated degradation of unassembled glycosylated Ig light chains. *Biol. Chem.* 381:1155–64

43. Chirico WJ, Waters MG, Blobel G. 1988. 70K heat shock related proteins stimulate protein translocation into microsomes. *Nature* 332:805–10

44. Chung DH, Ohashi K, Watanabe M, Miyasaka N, Hirosawa S. 2000. Mannose trimming targets mutant $\alpha2$-plasmin inhibitor for degradation by the proteasome. *J. Biol. Chem.* 275:4981–87

45. Colley NJ, Baker EK, Stamnes MA, Zuker CS. 1991. The cyclophilin homolog ninaA is required in the secretory pathway. *Cell* 67:255–63

46. Connell P, Ballinger CA, Jiang J, Wu Y, Thompson LJ, et al. 2001. The co-chaperone CHIP regulates protein triage decisions mediated by heat-shock proteins. *Nat. Cell Biol.* 3:93–96

47. Corsi A, Schekman R. 1997. The lumenal domain of Sec63p stimulates the ATPase activity of BiP and mediates BiP recruitment to the translocon in *Saccharomyces cerevisiae*. *J. Cell Biol.* 137:1483–93

48. Cox JS, Shamu CE, Walter P. 1993. Transcriptional induction of genes encoding endoplasmic reticulum resident proteins requires a transmembrane protein kinase. *Cell* 73:1197–206

49. Cox JS, Walter P. 1996. A novel mechanism for regulating activity of a transcription factor that controls the unfolded protein response. *Cell* 87:391–404

50. Cox JS, Chapman RE, Walter P. 1997. The unfolded protein response coordinates the production of endoplasmic reticulum protein and endoplasmic reticulum membrane. *Mol. Biol. Cell* 8:1805–14

51. Craven RA, Egerton M, Stirling CJ. 1996. A novel hsp70 of the yeast ER lumen is required for the efficient translocation of a number of protein precursors. *EMBO J.* 15:2640–50

52. Crowley KS, Reinhart GD, Johnson AE. 1993. The signal sequence moves through a ribosomal tunnel into a noncytoplasmic aqueous environment at the ER membrane early in translocation. *Cell* 73:1101–16

52a. Cuozzo JW, Kaiser CA. 1999. Competition between glutathione and protein thiols for disulphide-bond formation. *Nat. Cell Biol.* 1:130–35

53. Cyr DM. 1995. Cooperation of the molecular chaperone Ydj1p with specific Hsp70 homologs to suppress protein aggregation. *FEBS Lett.* 359:129–32

54. Cyr DM, Lu X, Douglas MJ. 1992. Regulation of eucaryotic hsp70 function by a dnaJ homolog. *J. Biol. Chem.* 267:20927–31

55. Deak PM, Wolf DH. 2001. Membrane topology and function of Der3/Hrd1p as an ubiquitin-ligase (E3) involved in

endoplasmic reticulum degradation. *J. Biol. Chem.* 276:10663–69

56. Deshaies RJ, Koch BD, Werner-Washburne M, Craig EA, Schekman R. 1998. A subfamily of stress proteins facilitates translocation of secretory and mitochondrial precursor proteins. *Nature* 332:800–5

57. de Virgilio M, Kitzmüller C, Schwaiger E, Klein M, Kreibich G, Ivessa NE. 1999. Degradation of a short-lived glycoprotein from the lumen of the endoplasmic recticulum: the role for N-linked glycans and the unfolded protein response. *Mol. Biol. Cell* 10:4059–73

58. de Virgilio M, Weninger H, Ivessa NE. 1998. Ubiquitination is required for the retro-translocation of short-lived luminal endoplasmic reticulum glycoprotein to the cytosol for degradation by the proteasome. *J. Biol. Chem.* 273:9734–43

59. Dey B, Caplan AJ, Boschelli F. 1996. The Ydj1p molecular chaperone facilitates formation of active p60 v-src in yeast. *Mol. Biol. Cell* 7:91–100

60. Dierks T, Volkmer J, Schlenstedt G, Jung C, Sandholzer U, et al. 1996. A microsomal ATP-binding protein involved in efficient protein transport into the mammalian endoplasmic reticulum. *EMBO J.* 15:6931–42

61. Do H, Falcone D, Lin J, Andrews DW, Johnson AE. 1996. The cotranslational integration of membrane proteins into the phospholipid bilayer is a multistep process. *Cell* 85:369–78

62. Doerrler WT, Lehrman MA. 1999. Regulation of the dolichol pathway in human fibroblasts by the endoplasmic reticulum unfolded protein response. *Proc. Natl. Acad. Sci. USA* 96:13050–55

63. Eggers DK, Welch WJ, Hansen WJ. 1997. Complexes between nascent polypeptides and their molecular chaperones in the cytosol of mammalian cells. *Mol. Biol. Cell* 8:1559–73

64. Eisfeld K, Riffer F, Mentges J, Schmitt MJ. 2000. Endocytotic uptake and

retrograde transport of a virally encoded killer toxin in yeast. *Mol. Microbiol.* 37:926–40

65. Ellgaard L, Molinari M, Helenius A. 1999. Setting the standards: quality control in the secretory pathway. *Science* 286:1882–88

66. Fanchiotti S, Fernandez F, D'Alessio C, Parodi AJ. 1998. The UDP-Glc: Glycosyltransferase is essential for *Schizosaccharomyces pombe* viability under conditions of extreme endoplasmic reticulum stress. *J. Cell Biol.* 143:625–35

67. Fayadat L, Siffroi-Fernandez S, Lanet J, Franc JL. 2000. Degradation of human thyroperoxidase in endoplasmic reticulum involves two different pathways depending on the folding state of the protein. *J. Biol. Chem.* 275:15948–54

68. Feldheim D, Rothblatt J, Schekman R. 1992. Topology and functional domains of Sec63p, an endoplasmic reticulum membrane protein required for secretory protein translocation. *Mol. Cell. Biol.* 12:3288–96

69. Fernandez F, D'Alessio C, Fanchiotti S, Parodi AJ. 1998. A misfolded protein conformation is not a sufficient condition for in vivo glucosylation by the UDP-Glc: glycoprotein glucosyltransferase. *EMBO J.* 17:5877–86

70. Flynn GC, Pohl J, Flocco MT, Rothman JE. 1991. Peptide binding specificity of the molecular chaperone BiP. *Nature* 353:726–30

71. Frand AR, Cuozzo JW, Kaiser CA. 2000. Pathways for protein disulphide bond formation. *Trends Cell Biol.* 10:203–10

72. Frand AR, Kaiser CA. 1998. The ERO1 gene of yeast is required for oxidation of protein dithiols in the endoplasmic reticulum. *Mol. Cell* 1:161–70

73. Frand AR, Kaiser CA. 1999. Ero1p oxidizes protein disulfide isomerase in a pathway for disulfide bond formation in the endoplasmic reticulum. *Mol. Cell* 4:469–77

74. Freeman BC, Morimoto RI. 1996. The human cytosolic molecular chaperones hsp90, hsp70 (hsc70) and hdj-1 have distinct roles in recognition of a non-native protein and protein refolding. *EMBO J.* 5:2969–79

75. Friedlander R, Jarosch E, Urban J, Volkwein C, Sommer T. 2000. A regulatory link between ER-associated protein degradation and the unfolded-protein response. *Nat. Cell Biol.* 2:379–84

76. Frydman J, Nimmesgern E, Ohtsuka K, Hartl FU. 1994. Folding of nascent polypeptide chains in a high molecular mass assembly with molecular chaperones. *Nature* 370:111–17

77. Fuller WV, Cuthbert AW. 2000. Posttranslational disruption of the ΔF508 cystic fibrosis transmembrane conductance regulator (CFTR)-molecular chaperone complex with geldanamycin stabilizes ΔF508 CFTR in the rabbit reticulocyte lysate. *J. Biol. Chem.* 275: 37462–68

78. Gardner AM, Aviel S, Argon Y. 1993. Rapid degradation of an unassembled immunoglobulin light chain is mediated by a serine protease and occurs in a pre-Golgi compartment. *J. Biol. Chem.* 268:25940–47

79. Gardner RG, Swarbrick GM, Bays NW, Cronin SR, Wilhovsky S, et al. 2000. Endoplasmic reticulum degradation requires lumen to cytosol signaling: transmembrane control of Hrd1 by Hrd3p. *J. Cell Biol.* 151:69–82

80. Gässler CS, Buchberger A, Laufen T, Mayer MP, Schröder, et al. 1998. Mutations in the DnaK chaperone affecting interaction with the DnaJ cochaperone. *Proc. Natl. Acad. Sci. USA* 95:15229–34

81. Gautschi M, Lilie H, Fünfschilling U, Mun A, Ross S, et al. 2001. RAC, a stable ribosome-associated complex in yeast formed by the DnaK-DnaJ homologs Ssz1p and zuotin. *Proc. Natl. Acad. Sci. USA* 98:3762–67

81a. Gething MJ, Sambrook J. 1992. Protein folding in the cell. *Nature* 355:33–45

82. Gillece P, Luz JM, Lennarz WJ, de La Cruz FJ, Römisch K. 1999. Export of a cysteine-free misfolded secretory protein from the endoplasmic reticulum for degradation requires interaction with protein disulfide isomerase. *J. Cell Biol.* 147:1443–56

83. Gilmore R, Blobel G. 1985. Translocation of secretory proteins across the microsomal membrane occurs through an environment accessible to aqueous perturbants. *Cell* 42:497–95

84. Glick BS. 1995. Can hsp70 proteins act as force-generating motors? *Cell* 80:11–14

85. Goldberger RF, Epstein CJ, Anfinsen CB. 1963. Purification and properties of a microsomal enzyme system catalyzing the reactivation of reduced ribonuclease and lysozyme. *J. Biol. Chem.* 238:1406–10

86. Gonzalez TN, Sidrauski C, Dörfler S, Walter P. 1999. Mechanism of non-spliceosomal mRNA splicing in the unfolded protein response pathway. *EMBO J.* 18:3119–32

87. Grenert JP, Johnson BD, Toft DO. 1999. The importance of ATP binding and hydrolysis by hsp90 in formation and function of protein heterocomplexes. *J. Biol. Chem.* 274:17525–33

88. Gusarova V, Caplan AJ, Brodsky JL, Fisher EA. 2001. Apoprotein B degradation is promoted by the molecular chaperones hsp90 and hsp70. *J. Biol. Chem.* 276:24891–900

89. Hamilton TG, Flynn GC. 1996. Cer1p, a novel hsp70-related protein required for posttranslational endoplasmic reticulum translocation in yeast. *J. Biol. Chem.* 271:30610–13

90. Hamman BD, Chen JC, Johnson EE, Johnson AE. 1997. The aqueous pore through the translocon has a diameter of 40–60 Å during cotranslational protein translocation at the ER membrane. *Cell* 89:535–44

90a. Hamman BD, Hendershot LM, Johnson AE. 1998. BiP maintains the permeability barrier of the ER membrane by sealing the lumenal end of the translocation pore before and early in translocation. *Cell* 92:747–58

91. Hammond C, Braakman I, Helenius A. 1994. Role of N-linked oligosaccharide recognition, glucose trimming, and calnexin in glycoprotein folding and quality control. *Proc. Natl. Acad. Sci. USA* 91:913–17

92. Hammond C, Helenius A. 1994. Quality control in the secretory pathway: retention of a misfolded viral membrane glycoprotein involves cycling between the ER, intermediate compartment, and Golgi apparatus. *J. Cell Biol.* 126:41–52

93. Hammond C, Helenius A. 1995. Quality control in the secretory pathway. *Curr. Opin. Cell Biol.* 7:523–29

94. Hampton RY, Gardner RG, Rine J. 1996. Role of the 26S proteasome and HRD genes in the degradation of 3-hydroxy-3-methylglutaryl-CoA reductase, an integral endoplasmic reticulum membrane protein. *Mol. Biol. Cell* 7:2029–44

95. Hampton RY, Bhakta H. 1997. Ubiquitin-mediated regulation of 3-hydroxy-3-methylglutaryl-CoA reductase. *Proc. Natl. Acad. Sci. USA* 94:12944–48

96. Hanein D, Matlack KE, Jungnickel B, Plath K, Kalies KU, et al. 1996. Oligomeric rings of the Sec61p complex induced by ligands required for protein translocation. *Cell* 87:721–32

97. Hann BC, Walter P. 1991. The signal recognition particle in yeast. *Cell* 67:131–43

98. Harding HP, Zhang Y, Ron D. 1999. Protein translation and folding are coupled by an endoplasmic-reticulum-resident kinase. *Nature* 397:271–74

99. Hartmann E, Sommer T, Prehn S, Görlich D, Jentsch S, Rapoport TA. 1994.

Evolutionary conservation of components of the protein translocation complex. *Nature* 367:654–57

100. Haze K, Yoshida H, Yanagi H, Yura T, Mori K. 1999. Mammalian transcription factor ATF6 is synthesized as a transmembrane protein and activated by proteolysis in response to endoplasmic reticulum stress. *Mol. Biol. Cell* 10:3787–99

101. Hazes B, Read RJ. 1997. Accumulating evidence suggests AB-toxins subvert the endoplasmic reticulum-associated protein degradation pathway to enter target cells. *Biochemistry* 36:111051–54

102. Hebert DN, Foellmer B, Helenius A. 1995. Glucose trimming and reglucosylation determine glycoprotein association with calnexin in the endoplasmic reticulum. *Cell* 81:425–33

103. Heinrich SU, Mothes W, Brunner J, Rapoport TA. 2000. The Sec61p complex mediates the integration of a membrane protein by allowing lipid partitioning of the transmembrane domain. *Cell* 102:233–44

104. Helenius A, Aebi M. 2001. Intracellular functions of N-linked glycans. *Science* 291:2364–69

105. Hill K, Cooper AA. 2000. Degradation of unassembled Vph1p reveals novel aspects of the yeast ER quality control system. *EMBO J.* 19:550–61

106. Hill RB, Flanagan JM, Prestegard JH. 1995. 1H and 15N magnetic resonance assignments, secondary structure, and tertiary fold of *Escherichia coli* DnaJ (1–78). *Biochemistry* 34:5587–96

107. Hiller MM, Finger A, Schweiger M, Wolf DH. 1996. ER degradation of a misfolded luminal protein by the cytosolic ubiquitin-proteasome pathway. *Science* 273:1725–28

108. Höhfeld J, Jentsch S. 1997. GrpE-like regulation of the hsc70 chaperone by the anti-apoptotic protein BAG-1. *EMBO J.* 16:6209–16

109. Holst B, Bruun AW, Kielland-Brandt MC, Winther JR. 1996. Competition

between folding and glycosylation in the endoplasmic reticulum. *EMBO J.* 15:3538–46

110. Hong E, Davidson AR, Kaiser CA. 1996. A pathway for targeting soluble misfolded proteins to the yeast vacuole. *J. Cell Biol.* 135:623–33

111. Hoppe T, Matuschewski K, Rape M, Schlenker S, Ulrich HD, Jentsch S. 2000. Activation of a membrane-bound transcription factor by regulated ubiquitin/proteasome-dependent processing. *Cell* 102:577–86

112. Hurtley SM, Bole DG, Hoover-Litty H, Helenius A, Copeland CS. 1989. Interactions of misfolded influenza virus hemagglutinin with binding protein (BiP). *J. Cell Biol.* 108:2117–26

113. Hurtley SM, Helenius A. 1989. Protein oligomerization in the endoplasmic reticulum. *Annu. Rev. Cell Biol.* 5:277–77

114. Hwang C, Sinskey AJ, Lodish HF. 1992. Oxidized redox state of glutathione in the endoplasmic reticulum. *Science* 257:1496–92

115. Ihara Y, Cohen-Doyle MF, Saito Y, Williams DB. 1999. Calnexin discriminates between protein conformational states and functions as a molecular chaperone in vitro. *Mol. Cell* 4:331–41

116. Imamura T, Haruta T, Yasumitsu T, Usui I, Iwata M, et al. 1998. Involvement of heat shock protein 90 in the degradation of mutant insulin receptors by the proteasome. *J. Biol. Chem.* 273:11183–88

117. Jackson MR, Nilsson T, Peterson PA. 1993. Retrieval of transmembrane proteins to the endoplasmic reticulum. *J. Cell Biol.* 121:317–33

118. Jakob CA, Burda P, Roth J, Aebi M. 1998. Degradation of mis-folded endoplasmic reticulum glycoproteins in *Saccharomyces cerevisiae* is determined by a specific oligosaccharide structure. *J. Cell Biol.* 142:1223–33

119. Jämsä E, Simonen M, Makarow M.

1994. Selective retention of secretory proteins in the yeast endoplasmic reticulum by treatment of cells with a reducing agent. *Yeast* 10:355–70

120. Jämsä E, Vakula N, Arffman A, Kilpelainen I, Makarow M. 1995. In vivo reactivation of heat-denatured protein in the endoplasmic reticulum of yeast. *EMBO J.* 14:6028–33

121. Jannatipour M, Rokeach LA. 1995. The *Schizosaccharomyces pombe* homologue of the chaperone calnexin is essential for viability. *J. Biol. Chem.* 270:4845–53

122. Jensen TJ, Loo MA, Pind S, Williams DB, Goldberg AL, Riordan JR. 1995. Multiple proteolytic systems, including the proteasome, contribute to CFTR processing. *Cell* 83:129–35

123. Johnson AE, van Waes MA. 1999. The translocon: a dynamic gateway at the ER membrane. *Annu. Rev. Cell Dev. Biol.* 15:799–842

124. Johnson BD, Chadli A, Felts SJ, Bouhouche I, Catelli MG, Toft DO. 2000. Hsp90 chaperone activity requires the full-length protein and interaction among its multiple domains. *J. Biol. Chem.* 275:32499–507

125. Johnson BD, Schumacher RJ, Ross ED, Toft DO. 1998. Hop modulates Hsp70/Hsp90 interactions in protein folding. *J. Biol. Chem.* 273:3679–86

126. Johnson JL, Craig EA. 2001. An essential role for the substrate-binding region of Hsp40s in *Saccharomyces cerevisiae*. *J. Cell Biol.* 152:851–56

127. Kabani M, Beckerich JM, Gaillardin C. 2000. Sls1p stimulates the Sec63p-mediated activation of Kar2p in a conformation dependent manner in the yeast endoplasmic reticulum. *Mol. Cell. Biol.* 20:6923–34

128. Karzai AW, McMacken R. 1996. A bipartite signaling mechanism involved in DnaJ-mediated activation of the *Escherichia coli* DnaK protein. *J. Biol. Chem.* 271:11236–46

129. Katayama T, Imaizumi K, Sato N, Miyoshi K, Kudo T, et al. 1999. Presenilin-1 mutations downregulate the signalling pathway of the unfolded-protein response. *Nat. Cell Biol.* 1:479–85

130. Kawahara T, Yanagi H, Yura T, Mori K. 1997. Endoplasmic reticulum stress-induced mRNA splicing permits synthesis of transcription factor Hac1p/Ern4p that activates the unfolded protein response. *Mol. Biol. Cell* 8:1845–62

131. Kim H-Y, Ahn B-Y, Cho Y. 2001. Structural basis for the inactivation of retinoblastoma tumor suppressor by SV40 large T antigen. *EMBO J.* 20:295–304

132. Knittler M, Dirks RS, Haas IG. 1995. Molecular chaperones involved in protein degradation in the endoplasmic reticulum: quantitative interaction of the heat shock cognate protein BiP with partially folded immunoglobulin light chains that are degraded in the endoplasmic reticulum. *Proc. Natl. Acad. Sci. USA* 92:764–68

133. Knop M, Finger A, Braun T, Hellmuth K, Wolf DH. 1996. Der1, a novel protein specifically required for endoplasmic reticulum degradation in yeast. *EMBO J.* 15:753–63

134. Kobayashi T, Kishigami S, Sone M, Inokuchi H, Mogi T, Ito K. 1997. Respiratory chain is required to maintain oxidized states of the DsbA-DsbB disulfide bond formation system in aerobically growing *Escherichia coli* cells. *Proc. Natl. Acad. Sci. USA* 94:11857–62

135. Lang K, Schmid FX, Fischer G. 1987. Catalysis of protein folding by prolyl isomerase. *Nature* 329:268–70

136. Langer T, Lu C, Echols H, Flanagan J, Hayer MK, Hartl FU. 1992. Successive action of dnaK, dnaJ and GroEL along the pathway of chaperone-mediated folding. *Nature* 356:683–89

137. Li Y, Kane T, Tipper C, Spatrick P, Jenness DD. 1999. Yeast mutants affecting

possible quality control of plasma membrane proteins. *Mol. Cell. Biol.* 19:3588–99

138. Liberek K, Marszalek J, Ang D, Georgopoulos C, Zylicz M. 1991. *Escherichia coli* dnaJ and grpE heat shock proteins jointly stimulate ATPase activity of dnaK. *Proc. Natl. Acad. Sci. USA* 88:2874–78

139. Liebermeister W, Rapoport TA, Heinrich R. 2001. Ratcheting in post-translation protein translocation: a mathematical model. *J. Mol. Biol.* 305:643–56

140. Liu Y, Choudhury P, Cabral CM, Sifers RN. 1999. Oligosaccharide modification in the early secretory pathway directs the selection of a mis-folded glycoprotein for degradation by the proteasome. *J. Biol. Chem.* 274:5861–67

141. Ljungdahl PO, Gimeno CJ, Styles CA, Fink GR. 1992. SHR3: a novel component of the secretory pathway specifically required for localization of amino acid permeases in yeast. *Cell* 71:463–78

142. Loayza D, Tam A, Schmidt WK, Michaelis S. 1998. Ste6p mutants defective in exit from the endoplasmic reticulum (ER) reveal aspects of an ER quality control pathway in *Saccharomyces cerevisiae*. *Mol. Biol. Cell* 9:2767–84

143. Loo TW, Clarke DM. 1998. Quality control by proteases in the endoplasmic reticulum. Removal of a protease-sensitive site enhances expression of human P-glycoprotein. *J. Biol. Chem.* 273:32373–76

144. Lu Y, Turnbull IR, Bragin A, Carveth K, Verkman AS, Skach WR. 2000. Reorientation of aquaporin-1 topology during maturation in the endoplasmic reticulum. *Mol. Biol. Cell* 11:2973–85

145. Luders J, Demand J, Höhfeld J. 2000. The ubiquitin-related BAG-1 provides a link between the molecular chaperones Hsc70/Hsp70 and the proteasome. *J. Biol. Chem.* 275:4613–17

146. Lyman SK, Schekman R. 1995. Interaction between BiP and Sec63p is required for the completion of protein translocation into the ER of *S. cerevisiae*. *J. Cell Biol.* 131:1163–71

147. Lyman SK, Schekman R. 1997. Binding of secretory precursor polypeptides to a translocon subcomplex is regulated by BiP. *Cell* 88:85–96

148. Martoglio B, Hofmann MW, Brunner J, Dobberstein B. 1995. The protein conducting channel in the membrane of the endoplasmic reticulum is open laterally toward the lipid bilayer. *Cell* 81:207–14

149. Matlack KE, Misselwitz B, Plath K, Rapoport TA. 1999. BiP acts as a molecular ratchet during posttranslational transport of prepro-alpha factor across the ER membrane. *Cell* 97:553–56

150. Mayer TU, Braun T, Jentsch S. 1998. Role of the proteasome in membrane extraction of a short-lived ER transmembrane protein. *EMBO J.* 17:3251–57

151. McCarty JS, Buchberger A, Reinstein J, Bukau B. 1995. The role of ATP in the functional cycle of the DnaK chaperone system. *J. Mol. Biol.* 249:126–37

152. McClellan AJ, Endres J, Vogel JP, Palazzi D, Rose MD, Brodsky JL. 1998. Specific molecular chaperone interactions and an ATP-dependent conformational change are required during post-translational protein translocation into the yeast ER. *Mol. Biol. Cell* 9:3533–45

153. McClellan AJ, Brodsky JL. 2000. Mutation of the ATP binding pocket of SSA1 indicates that a functional interaction between Ssa1p and Ydj1p is required for post-translational translocation into the yeast endoplasmic reticulum. *Genetics* 156:501–12

154. McCracken AA, Brodsky JL. 1996. Assembly of ER-associated protein degradation in vitro: dependence on cytosol, calnexin, and ATP. *J. Cell Biol.* 132:291–98

155. McCracken AA, Karpichev IV, Ernaga JE, Werner ED, Dillin AG, Courchesne W. 1996. Yeast mutants deficient in

ER-associated degradation of the Z variant of alpha-1-protease inhibitor. *Genetics* 144:1355–62

156. McGee TP, Cheng HH, Kumagai H, Omura S, Simoni RD. 1996. Degradation of 3-hydroxymethyl-3-glutaryl-CoA reductase in endoplasmic reticulum membranes is accelerated as a result of increased susceptibility to proteolysis. *J. Biol. Chem.* 271:25630–38

157. Meacham GC, Patterson C, Zhang W, Younger M, Cyr DM. 2001. The hsc70 co-chaperone CHIP targets immature CFTR for proteasomal degradation. *Nat. Cell Biol.* 3:100–5

158. Melnick J, Dul JL, Argon Y. 1994. Sequential interaction of the chaperones BiP and GRP94 with immunoglobulin chains in the endoplasmic reticulum. *Nature* 370:373–75

159. Meyer HA, Grau H, Kraft R, Kostka S, Prehn S, et al. 2000. Mammalian Sec61 is associated with Sec62 and Sec63. *J. Biol. Chem.* 275:14550–57

160. Misselwitz B, Staeck O, Rapoport TA. 1998. J proteins catalytically activate Hsp70 molecules to trap a wide range of peptide sequences. *Mol. Cell* 2:593–603

161. Molinari M, Helenius A. 2000. Chaperone selection during glycoprotein translocation into the endoplasmic reticulum. *Science* 288:331–33

162. Mori K, Kawahara T, Yoshida H, Yanagi H, Yura T. 1996. Signalling from endoplasmic reticulum to nucleus: transcription factor with a basic-leucine zipper motif is required for the unfolded protein-response pathway. *Genes Cells* 1:803–17

163. Mori K, Ma W, Gething MJ, Sambrook J. 1993. A transmembrane protein with a cdc2+/CDC28-related kinase activity is required for signaling from the ER to the nucleus. *Cell* 74:743–56

164. Mori K, Sant A, Kohno K, Normington K, Gething MJ, Sambrook JF. 1992. A 22 bp cis-acting element is necessary and sufficient for the induction of the yeast KAR2 (BiP) gene by unfolded proteins. *EMBO J.* 11:2583–93

165. Moriyama T, Sather SK, McGee TP, Simoni RD. 1998. Degradation of HMG-CoA reductase in vitro. Cleavage in the membrane domain by a membrane-bound cysteine protease. *J. Biol. Chem.* 273:22037–43

166. Morshauser RC, Wang H, Flynn GC, Zuiderweg ERP. 1995. The peptide-binding domain of the chaperone protein hsc70 has an unusual secondary structure topology. *Biochemistry* 34:6261–66

167. Mothes W, Heinrich SU, Graf R, Nilsson I, von Heijne G, et al. 1997. Molecular mechanism of membrane protein integration into the endoplasmic reticulum. *Cell* 89:523–33

168. Mullins C, Lu Y, Campbell A, Fang H, Green N. 1995. A mutation affecting signal peptidase inhibits degradation of an abnormal membrane protein in *Saccharomyces cerevisiae*. *J. Biol. Chem.* 270:17139–47

169. Müsch A, Wiedmann M, Rapoport TA. 1992. Yeast Sec proteins interact with polypeptides traversing the endoplasmic reticulum membrane. *Cell* 69:343–52

170. Mutka SC, Walter P. 2001. Multifaceted physiological response allows yeast to adapt to the loss of the signal recognition particle-dependent protein-targeting pathway. *Mol. Biol. Cell* 12: 577–88

171. Naik RR, Jones EW. 1998. The *PBN1* gene of *Saccharomyces cerevisiae*: an essential gene that is required for the post-translational processing of the protease B precursor. *Genetics* 149:1277–92

172. Nakatsukasa K, Nishikawa S, Hosokawa N, Nagata K, Endo T. 2001. Mnl1p, an α-mannosidase-like protein in yeast *Saccharomyces cerevisiae*, is required for ER associated degradation of glycoproteins. *J. Biol. Chem.* 276:8635–38

173. Nathan DF, Vos MH, Lindquist S. 1997. In vivo functions of the *Saccharomyces*

cerevisiae Hsp90 chaperone. *Proc. Natl. Acad. Sci. USA* 94:12949–56

174. Nelson RJ, Ziegelhoffer T, Nicolet C, Werner-Washburne M, Craig EA. 1992. The translation machinery and 70 kd heat shock protein cooperate in protein synthesis. *Cell* 71:97–105

175. Nehls S, Snapp EL, Cole NB, Zaal KJ, Kenworthy AK, et al. 2000. Dynamics and retention of misfolded proteins in native ER membranes. *Nat. Cell Biol.* 2:288–95

176. Ng DTW, Brown JD, Walter P. 1996. Signal sequences specify the targeting route to the endoplasmic reticulum membrane. *J. Cell Biol.* 134:269–78

177. Ng DTW, Spear ED, Walter P. 2000. The unfolded protein response regulates multiple aspects of secretory and membrane protein biogenesis and endoplasmic reticulum quality control. *J. Cell Biol.* 150:77–88

178. Nicchitta CV, Blobel G. 1993. Lumenal proteins of the mammalian endoplasmic reticulum are required to complete protein translocation. *Cell* 73:989–98

179. Nishikawa S, Endo T. 1997. The yeast Jem1p is a DnaJ-like protein of the endoplasmic reticulum membrane required for nuclear fusion. *J. Biol. Chem.* 272:12889–92

180. Nishikawa S, Fewell SW, Kato Y, Brodsky JL, Endo T. 2001. Molecular chaperones in the yeast ER maintain the solubility of proteins for retrotranslocation and degradation. *J. Cell Biol.* 153:1061–69

181. Niwa M, Sidrauski C, Kaufman RJ, Walter P. 1999. A role for presenilin-1 in nuclear accumulation of Ire1 fragments and induction of the mammalian unfolded protein response. *Cell* 99:691–702

181a. Norgaard P, Westphal V, Christine T, Alsøe L, Holst B, Winther JR. 2001. Functional differences in yeast protein disulfide isomerases. *J. Cell Biol.* 152:553–62

182. Normington K, Kohno K, Kozutsumi Y, Gething JM, Sambrook J. 1989. *S. cerevisiae* encodes an essential protein in sequence and function to mammalian BiP. *Cell* 57:1223–36

183. Obermann WM, Sondermann H, Russo AA, Pavletich NP, Hartl FU. 1998. In vivo function of Hsp90 is dependent on ATP binding and ATP hydrolysis. *J. Cell Biol.* 143:901–10

184. Oda K, Ikehara Y, Omura S. 1996. Lactacystin, an inhibitor of the proteasome, blocks the degradation of a mutant precursor of glycosylphosphatidylinositol-linked protein in a pre-Golgi compartment. *Biochem. Biophys. Res. Commun.* 219:800–5

185. Ogg SC, Poritz MA, Walter P. 1992. Signal recognition particle receptor is important for cell growth and protein secretion in *Saccharomyces cerevisiae*. *Mol. Biol. Cell* 3:895–911

186. Okamura K, Kimata Y, Higashio H, Tsuru A, Kohno K. 2000. Dissociation of Kar2p/BiP from an ER sensory molecule, Ire1p, triggers the unfolded protein response in yeast. *Biochem. Biophys. Res. Commun.* 279:445–50

187. Pagani M, Fabbri M, Benedetti C, Fassio A, Pilati S, et al. 2000. Endoplasmic reticulum oxidoreductin 1-lbeta (ERO1-Lbeta), a human gene induced in the course of the unfolded protein response. *J. Biol. Chem.* 275:23685–92

188. Panaretou B, Prodromou C, Roe SM, O'Brien R, Ladbury JE, et al. 1998. ATP binding and hydrolysis are essential to the function of the Hsp90 molecular chaperone in vivo. *EMBO J.* 17:4829–36

189. Parlati F, Dignard D, Bergeron JJM, Thomas DY. 1995. The calnexin homologue cnx1+ in *Schizosaccharomyces pombe*, is an essential gene which can be complemented by its soluble ER domain. *EMBO J.* 14:3064–72

190. Parodi AJ. 2000. Protein glycosylation and its role in protein folding. *Annu. Rev. Biochem.* 69:69–93

191. Parodi AJ, Lederkremer GZ, Mendelzon DH. 1983. Protein glycosylation in *Trypanosoma cruzi*. The mechanism of glycosylation and structure of protein-bound oligosaccharides. *J. Biol. Chem.* 258:5589–95

192. Paunola E, Suntio T, Jämsä E, Makarow M. 1998. Folding of active β-lactamase in the yeast cytoplasm before translocation into the endoplasmic reticulum. *Mol. Biol. Cell* 9:817–27

193. Pellecchia M, Szyperski T, Wall D, Georgopoulos C, Wüthrich K. 1996. NMR structure of the J-domain and the Gly-Phe-rich region of the *Escherichia coli* DnaJ chaperone. *J. Mol. Biol.* 260:235–50

194. Pfund C, Lopez-Hoyo N, Ziegelhoffer T, Schilke BA, Lopez-Buesa P, et al. 1998. The molecular chaperone Ssb from *Saccharomyces cerevisiae* is a component of the ribosome-nascent chain complex. *EMBO J.* 17:3981–89

195. Pilon M, Schekman R, Römisch K. 1997. Sec61p mediates export of a misfolded secretory protein from the endoplasmic reticulum to the cytosol for degradation. *EMBO J.* 16:4540–48

196. Plath K, Rapoport TA. 2000. Spontaneous release of cytosolic proteins from posttranslational substrates before their transport into the endoplasmic reticulum. *J. Cell Biol.* 151:167–78

197. Plath K, Mothes W, Wilkinson BM, Stirling CJ, Rapoport TA. 1998. Signal sequence recognition in posttranslational protein transport across the yeast ER membrane. *Cell* 94:795–807

198. Plemper RK, Bohmler S, Bordallo J, Sommer T, Wolf DH. 1997. Mutant analysis links the translocon and BIP to retrograde protein transport for ER degradation. *Nature* 388:891–95

199. Plemper RK, Egner R, Kuchler K, Wolf DH. 1998. Endoplasmic reticulum degradation of a mutated ATP-binding cassette transporter Pdr5 proceeds in a concerted action of Sec61 and the proteasome. *J. Biol. Chem.* 273:32848–56

200. Plemper RK, Wolf DH. 1999. Retrograde protein translocation: ERADication of secretory proteins in health and disease. *Trends Biochem. Sci.* 24:266–70

201. Pollard MG, Travers KJ, Weissman JS. 1998. Ero1p: a novel and ubiquitous protein with an essential role in oxidative protein folding in the endoplasmic reticulum. *Mol. Cell* 1:171–82

202. Prodromou C, Roe SM, O'Brien R, Ladbury JE, Piper PW, Pearl LH. 1997. Identification and structural characterization of the ATP/ADP-binding site in the Hsp90 molecular chaperone. *Cell* 90:65–75

203. Qian YQ, Patel D, Hartl FU, McColl DJ. 1996. Nuclear magnetic resonance solution structure of the human hsp40 (HDJ-1) J-domain. *J. Mol. Biol.* 260:224–35

204. Qu D, Teckman JH, Omura S, Perlmutter DH. 1996. Degradation of a mutant secretory protein, α1-antitrypsin Z, in the endoplasmic reticulum requires proteasome activity. *J. Biol. Chem.* 271:22791–95

205. Rapoport TA, Matlack KES, Plath K, Misselwitz B, Staeck O. 1999. Posttranslational protein translocation across the membrane of the endoplasmic reticulum. *Biol. Chem.* 380:1143–50

206. Rassow J, von Ahsen O, Bömer U, Pfanner N. 1997. Molecular chaperones: towards a characterization of the heat-shock protein 70 family. *Trends Cell Biol.* 7:129–33

207. Reddy PS, Corley RB. 1998. Assembly, sorting, and exit of oligomeric proteins from the endoplasmic reticulum. *BioEssays* 20:546–54

208. Rietsch A, Beckwith J. 1998. The genetics of disulfide bond metabolism. *Annu. Rev. Genet.* 32:163–84

209. Rose MD, Misra LM, Vogel JP. 1989. *KAR2*, a karyogamy gene, is the yeast

homolog of the mammalian BiP/GRP78 gene. *Cell* 57:1211–21

210. Rothblatt JA, Deshaies RJ, Sanders SL, Daum G, Schekman R. 1989. Multiple genes are required for proper insertion of secretory proteins into the endoplasmic reticulum in yeast. *J. Cell Biol.* 109:2641–52

211. Rüdiger S, Germeroth L, Schneider-Mergener J, Bukau B. 1997. Substrate specificity of the DnaK chaperone determined by screening cellulose-bound peptide libraries. *EMBO J.* 16:1501–7

212. Rüdiger S, Schneider-Mergener J, Bukau B. 2001. Its substrate specificity characterizes the DnaJ co-chaperone as a scanning factor for the DnaK chaperone. *EMBO J.* 20:1042–50

213. Sadler I, Chiang A, Kurihara T, Rothblatt J, Way J, Silver P. 1989. A yeast gene important for protein assembly into the endoplasmic reticulum and the nucleus has homology to DnaJ, an *Escherichia coli* heat shock protein. *J. Cell Biol.* 109:2665–75

214. Sanders SL, Whitfield KM, Vogel KP, Rose MD, Schekman RW. 1992. Sec61p and BiP directly facilitate polypeptide translocation into the ER. *Cell* 69:353–65

215. Saris N, Holkeri H, Craven RA, Stirling CJ, Makarow M. 1997. The Hsp70 homologue Lhs1p is involved in a novel function of the yeast endoplasmic reticulum, refolding and stabilization of heat-denatured protein aggregates. *J. Cell Biol.* 137:813–24

216. Sato N, Urano F, Yoon Leem J, Kim SH, Li M, et al. 2000. Upregulation of BiP and CHOP by the unfolded-protein response is independent of presenilin expression. *Nat. Cell Biol.* 2: 863–70

217. Schiebel T, Buchner J. 1997. The Hsp90 family—an overview. In *Guidebook to Molecular Chaperones and Protein Folding Catalysts*, ed. MJ Gething, pp. 147–51. London: Oxford Univ. Press

218. Schlenstedt G, Harris S, Risse B, Lill R, Silver PA. 1995. A yeast DnaJ homologue, Scj1p, can function in the endoplasmic reticulum with Bip/Kar2p via a conserved domain that specifies interactions with hsp70s. *J. Cell Biol.* 129:979–88

219. Schmid D, Baici Gehring H, Christen P. 1994. Kinetics of molecular chaperone action. *Science* 263:971–73

220. Schmitz A, Herrgen H, Winkler A, Herzog V. 2000. Cholera toxin is exported from microsome by the Sec61p complex. *J. Cell Biol.* 148:1203–12

221. Schmitz A, Maintz M, Kehle T, Herzog V. 1995. in vivo iodination of a misfolded proinsulin reveals co-localized signals for Bip binding and for degradation in the ER. *EMBO J.* 14:1091–98

222. Schneider C, Sepp-Lorenzino L, Nimmesgern E, Ouerfelli O, Danishefsky S, et al. 1996. Pharmacologic shifting of a balance between protein protein refolding and degradation mediated by Hsp90. *Proc. Natl. Acad. Sci. USA* 93:14536–41

223. Schumacher RJ, Hansen WJ, Freeman BC, Alnemri E, Litwack G, Toft DO. 1996. Cooperative action of Hsp70, Hsp90, and DnaJ proteins in protein renaturation. *Biochemistry* 35:14889–98

223a. Shamu CE. 1998. Splicing: HACking into the unfolded-protein response. *Curr. Biol.* 8:R121–23

224. Shamu CE, Walter P. 1996. Oligomerization and phosphorylation of the Ire1p kinase during intracellular signaling from the endoplasmic reticulum to the nucleus. *EMBO J.* 15:3028–39

225. Shi Y, Vattem KM, Sood R, An J, Liang J, Stramm L, Wek RC. 1998. Identification and characterization of pancreatic eukaryotic initiation factor 2 alpha-subunit kinase, PEK, involved in translational control. *Mol. Cell. Biol.* 18:7499–509

226. Sidrauski C, Cox JS, Walter P. 1996.

tRNA ligase is required for regulated mRNA splicing in the unfolded protein response. *Cell* 87:405–13

227. Sidrauski C, Walter P. 1997. The transmembrane kinase Ire1p is a site-specific endonuclease that initiates mRNA splicing in the unfolded protein response. *Cell* 90:1031–39

228. Simon SM, Peskin CS, Oster GF. 1992. What drives the translocation of proteins? *Proc. Natl. Acad. Sci. USA* 89: 3770–74

228a. Simons JF, Ferro-Novick S, Rose MD, Helenius A. 1995. BiP/Kar2p serves as a molecular chaperone during carboxypeptidase Y folding in yeast. *J. Cell Biol.* 130:41–49

229. Simpson JC, Roberts LM, Römisch K, Davey J, Wolf DH, Lord MJ. 1999. Ricin A chain utilizes the endoplasmic reticulum-associated protein degradation pathway to enter the cytosol of yeast. *FEBS Lett.* 459:80–84

230. Skowronek MH, Hendershot LM, Haas IG. 1998. The variable domain of nonassembled Ig light chains determines both their half-life and binding to the chaperone BiP. *Proc. Natl. Acad. Sci. USA* 95:1574–78

231. Skowronek MH, Rotter M, Haas IG. 1999. Molecular characterization of a novel mammalian DnaJ-like Sec63p homolog. *Biol. Chem.* 380:1133–38

232. Sommer T, Jentsch S. 1993. A protein translocation defect linked to ubiquitin conjugation at the endoplasmic reticulum. *Nature* 365:176–79

233. Stamnes MA, Shieh BH, Chuman L, Harris GL, Zuker CS. 1991. The cyclophilin homolog ninaA is a tissue-specific integral membrane protein required for the proper synthesis of a subset of Drosophila rhodopsins. *Cell* 65:219–27

234. Stebbins CE, Russo AA, Schneider C, Rosen N, Hartl FU, Pavletich NP. 1997. Crystal structure of an Hsp90-geldanamycin complex: targeting of a protein chaperone by an antitumor agent. *Cell* 89:239–50

235. Stevens FJ, Argon Y. 1999. Protein folding in the ER. *Semin. Cell Dev. Biol.* 10:443–54

236. Strickland E, Hakala K, Thomas PJ, DeMartino GN. 2000. Recognition of misfolding proteins by PA700, the regulatory subcomplex of the 26S proteasome. *J. Biol. Chem.* 275:5565–72

237. Suh W-C, Burkholder WF, Lu CZ, Zhao X, Gottesman ME, Gross CA. 1998. Interaction of the Hsp70 molecular chaperone, DnaK, with its cochaperone DnaJ. *Proc. Natl. Acad. Sci. USA* 95:15223–28

238. Suh W-C, Lu CZ, Gross CA. 1999. Structural features required for the interaction of the Hsp70 molecular chaperone DnaK with its cochaperone DnaJ. *J. Biol. Chem.* 274:30534–39

239. Suzuki T, Park H, Hollingsworth NM, Sternglanz R, Lennarz WJ. 2000. *PNG1*, a yeast gene encoding a highly conserved peptide:N-glycanase. *J. Cell Biol.* 149:1039–51

240. Szabo A, Korszun R, Hartl FU, Flanagan J. 1996. A zinc finger-like domain of the molecular chaperone DnaJ is involved in binding to denatured protein substrates. *EMBO J.* 15:408–17

241. Szabo A, Langer T, Schröder H, Flanagan J, Bukau B, Hartl FU. 1994. The ATP hydrolysis dependent cycle of the *Escherichia coli* Hsp70 system-DnaK, DnaJ, and GrpE. *Proc. Natl. Acad. Sci. USA* 91:10345–49

242. Szyperski T, Pellecchi M, Wall D, Georopolous C, Wütrich K. 1994. NMR structure determination of the *Escherichia coli* DnaJ molecular chaperone: secondary structure and backbone fold of the N-terminal region (residues 2–108) containing the highly conserved J domain. *Proc. Natl. Acad. Sci. USA* 91:11343–47

242a. te Heesen S, Aebi M. 1994. The genetic interaction of kar2 and wbp1 mutations. Distinct functions of binding protein BiP and N-linked glycosylation in the processing pathway of secreted proteins in *Saccharomyces cerevisiae*. *Eur. J. Biochem.* 222:631–37

243. Tirasophon W, Welihinda AA, Kaufman R. 1998. A stress response pathway from the endoplasmic reticulum to the nucleus requires a novel bifunctional protein kinase/endoribonuclease (Ire1p) in mammalian cells. *Genes Dev.* 12:1812–24

244. Tokunaga F, Brostrom C, Koide T, Arvan P. 2000. Endoplasmic reticulum (ER)-associated degradation of misfolded glycoprotein is suppressed upon inhibition of ER mannosidase I. *J. Biol. Chem.* 275:40757–64

245. Travers KJ, Patil CK, Wodicka L, Lockhart DJ, Weissman JS, Walter P. 2000. Functional and genomic analyses reveal an essential coordination between the unfolded protein response and ER-associated degradation. *Cell* 101:249–58

246. Tsai B, Rodighiero C, Lencer WI, Rapoport TA. 2001. Protein disulfide isomerase acts as a redox-dependent chaperone to unfold cholera toxin. *Cell* 104:937–48

247. Tsai J, Douglas MG. 1996. A conserved HPD sequence of the J domain is necessary for YDJ1 stimulation of Hsp70 ATPase activity at a site distinct from substrate binding. *J. Biol. Chem.* 271: 9347–54

248. Tu BP, Ho-Schleyer SC, Travers KJ, Weissman JS. 2000. Biochemical basis of oxidative protein folding in the endoplasmic reticulum. *Science* 290:1571–74

249. Tyedmers J, Lerner M, Bies C, Dudek J, Skowronek MH, et al. 2000. Homologs of the yeast Sec complex subunits Sec62p and Sec63p are abundant proteins in dog pancreas microsomes. *Proc. Natl. Acad. Sci. USA* 97:7214–19

250. Tyson JR, Stirling CJ. 2000. LHS1 and SIL1 provide lumenal function that is essential for protein translocation into the endoplasmic reticulum. *EMBO J.* 30:794–97

251. Verma R, Chen S, Feldman R, Schieltz D, Yates J, et al. 2000. Proteasomal proteomics: identifcation of nucleotide-sensitive proteasome-interacting proteins by mass spectrometric analysis of affinity-purified proteasomes. *Mol. Biol. Cell* 11:3425–39

252. Vogel JP, Misra LM, Rose MD. 1990. Loss of BiP/GRP78 function blocks translocation of secretory proteins in yeast. *J. Cell Biol.* 110:1885–95

253. Voges D, Zwickl P, Baumeister W. 1999. The 26S proteasome: a molecular machine designed for controlled proteolysis. *Annu. Rev. Biochem.* 68:1015–68

254. Wall D, Zylicz M, Georgopoulos C. 1994. The NH2-terminal 108 amino acids of the *Escherichia coli* DnaJ protein stimulate the ATPase activity of DnaK and are sufficient for λ replication. *J. Biol. Chem.* 269:5446–51

255. Wall D, Zylic M, Georgopoulos C. 1995. The conserved G/F motif of the DnaJ chaperone is necessary for the activation of the substrate binding properties of theDnaK chaperone. *J. Biol. Chem.* 270:2139–44

255a. Walter J, Urban J, Volkwein C, Sommer T. 2001. Sec61p-independent degradation of the tail-anchored ER membrane protein Ubc6p. *EMBO J.* 20:3124–31

256. Wang Q, Chang A. 1999. Eps1, a novel PDI-related protein involved in ER quality control in yeast. *EMBO J.* 18:5972–5982

257. Wang TF, Chang JH, Wang C. 1993. Identification of the peptide binding domain of hsc70. 18-kilodalton fragment located immediately after ATPase domain is sufficient for high affinity binding. *J. Biol. Chem.* 268:26049–51

258. Wang XZ, Harding HP, Zhang Y, Jolicoeur EM, Kuroda M, Ron D. 1998. Cloning of mammalian Ire1 reveals diversity in the ER stress responses. *EMBO J.* 17:5708–17

259. Wang Y, Shen J, Arenzana N, Tirasophon W, Kaufman RJ, Prywes R.

2000. Activation of ATF6 and an ATF6 DNA binding site by the endoplasmic reticulum stress response. *J. Biol. Chem.* 275:27013–20

260. Ward CL, Omura S, Kopito RR. 1995. Degradation of CFTR by the ubiquitin proteasome pathway. *Cell* 83:121–27

261. Waters MG, Blobel G. 1986. Secretory protein translocation in a yeast cell-free system can occur post-translationally and requires ATP hydrolysis. *J. Cell Biol.* 102:1543–50

262. Wei J, Hendershot LM. 1996. Protein folding and assembly in the endoplasmic reticulum. *EXS* 77:41–55

263. Weihofen A, Lemberg MK, Ploegh HL, Bogyo M, Martoglio B. 2000. Release of signal peptide fragments into the cytosol requires cleavage in the transmembrane region by a protease activity that is specifically blocked by a novel cysteine protease inhibitor. *J. Biol. Chem.* 275:30951–56

264. Weissman JS, Kim PS. 1992. Kinetic role of nonnative species in the folding of bovine pancreatic trypsin inhibitor. *Proc. Natl. Acad. Sci. USA* 89:9900–4

265. Werner ED, Brodsky JL, McCracken AA. 1996. Proteasome-dependent ER-associated protein degradation: an unconventional route to a familiar fate. *Proc. Natl. Acad. Sci. USA* 93:13797–801

266. Wesche J, Rapak A, Olsnes S. 1999. Dependence of ricin toxicity on translocation of the toxin A-chain from the endoplasmic reticulum to the cytosol. *J. Biol. Chem.* 274:34443–49

267. Wickner S, Hoskins J, McKenney K. 1991. Function of DnaJ and DnaK as chaperones in origin-specific DNA binding by RepA. *Nature* 350:165–67

268. Wiertz EJ, Tortorella D, Bogyo M, Yu J, Mothes W, et al. 1996. Sec61p-mediated transfer of a membrane protein from the endoplasmic reticulum to the proteasome for destruction. *Nature* 384:432–38

269. Wikstrom L, Lodish HF. 1992. Endoplasmic reticulum degradation of a subunit of the asialoglycoprotein receptor in vitro. Vesicular transport from endoplasmic reticulum is unnecessary. *J. Biol. Chem.* 267:5–8

270. Wild J, Altman E, Yura T, Gross CA. 1992. DnaK and DnaJ heat shock proteins participate in protein export in *Escherichia coli*. *Genes Dev.* 6:1165–72

271. Wileman T, Carson GR, Shih FF, Concino MF, Terhorst C. 1990. The transmembrane anchor of the T-cell antigen receptor beta chain contains a structural determinant of pre-Golgi proteolysis. *Cell Regul.* 1:907–19

272. Wileman T, Kane LP, Terhorst C. 1991. Degradation of T-cell receptor chains in the endoplasmic reticulum is inhibited by inhibitors of cysteine proteases. *Cell Regul.* 2:753–65

273. Wilkinson BM, Tyson JR, Reid PJ, Stirling CJ. 2000. Distinct domains within yeast Sec61p involved in post-translational translocation and protein dislocation. *J. Biol. Chem.* 275:521–29

274. Wilson CM, Farmery MR, Bulleid NJ. 2000. Pivotal role of calnexin and mannose trimming in regulating the endoplasmic reticulum-associated degradation of major histocompatibility complex class I heavy chain. *J. Biol. Chem.* 275:21224–32

275. Yan W, Schilke B, Pfund C, Walter W, Kim S, Craig EA. 1998. Zuotin, a ribosome-associated DnaJ molecular chaperone. *EMBO J.* 16:4809–17

276. Ye J, Rawson RB, Komuro R, Chen X, Dave UP, et al. 2000. ER stress induces cleavage of membrane-bound ATF6 by the same proteases that process SREBPs. *Mol. Cell* 6:1355–64

277. Yoshida H, Haze K, Yanagi H, Yura T, Mori K. 1998. Identification of the cis-acting endoplasmic reticulum stress response element responsible for transcriptional induction of mammalian glucose-regulated proteins. Involvement

of basic leucine zipper transcription factors. *J. Biol. Chem.* 273:33741–49

278. Young BP, Craven RA, Reid PJ, Willer M, Stirling CJ. 2001. Sec63p and Kar2p are required for the translocation of SRP-dependent precursors into the yeast endoplasmic reticulum in vivo. *EMBO J.* 20:262–71

279. Yu H, Kaung G, Kobayashi S, Kopito R. 1997. Cytosolic degradation of T-cell receptor α chains by the proteasome. *J. Biol. Chem.* 272:20800–4

280. Zapun A, Jakob CA, Thomas DY, Bergeron JJ. 1999. Protein folding in a specialized compartment: the endoplasmic reticulum. *Struct. Fold. Des.* 7:R173–82

281. Zerangue N, Schwappach B, Jan YN, Jan LY. 1999. A new ER trafficking signal regulates the subunit stoichiometry of plasma membrane K(ATP) channels. *Neuron* 22:537–48

282. Zhang JX, Braakman I, Matlack KE, Helenius A. 1997. Quality control in the secretory pathway: the role of calreticulin, calnexin and BiP in the retention of glycoproteins with C-terminal truncations. *Mol. Biol. Cell* 8:1943–54

283. Zhang Y, Nijbroek G, Sullivan ML, Mc-Cracken AA, Watkins SC, et al. 2001. The Hsp70 molecular chaperone facilitates the ER associated protein degradation of the cystic fibrosis transmembrane conductance regulator in yeast. *Mol. Cell. Biol.* 12:1303–14

284. Zhong T, Arndt KT. 1993. The yeast SIS1 protein, a DnaJ homolog, is required for the initiation of translation. *Cell* 73:1175–86

285. Zhou M, Schekman R. 1999. The engagement of Sec61p in the ER dislocation process. *Mol. Cell* 4:925–34

286. Zhu X, Zhao X, Burkholder WF, Gragerov A, Ogata CM, et al. 1996. Structural analysis of substrate binding by the molecular chaperone DnaK. *Science* 272:1606–14

287. Ziegler DM, Poulsen LL. 1977. Protein disulfide bond synthesis: a possible intracellular mechanism. *Trends Biochem. Sci.* 2:79–81

288. Zimmermann RM, Sagstetter M, Lewis MJ, Pelham HR. 1988. Seventy-kilodalton heat shock proteins and an additional component from reticulocyte lysate stimulate import of M13 procoat protein into microsomes. *EMBO J.* 7:2875–80

289. Zylicz M, Ang D, Liberek K, Georgopoulos C. 1989. Initiation of λ DNA replication with purified host and bacteriophage encoded proteins: the role of the dnaK, dnaJ, and grpE heat shock proteins. *EMBO J.* 8:1601–8

Annu. Rev. Genet. 2001. 35:193–208

CHROMATIN INSULATORS AND BOUNDARIES: Effects on Transcription and Nuclear Organization

Tatiana I. Gerasimova and Victor G. Corces

Department of Biology, The Johns Hopkins University, 3400 North Charles Street, Baltimore, Maryland 21218; e-mail: tgerasimova@jhu.edu; Corces@jhu.edu

Key Words DNA, chromatin, insulators, transcription, nucleus

■ **Abstract** Chromatin boundaries and insulators are transcriptional regulatory elements that modulate interactions between enhancers and promoters and protect genes from silencing effects by the adjacent chromatin. Originally discovered in *Drosophila*, insulators have now been found in a variety of organisms, ranging from yeast to humans. They have been found interspersed with regulatory sequences in complex genes and at the boundaries between active and inactive chromatin. Insulators might modulate transcription by organizing the chromatin fiber within the nucleus through the establishment of higher-order domains of chromatin structure.

CONTENTS

INTRODUCTION

Insulators or chromatin boundaries are DNA sequences defined operationally by two characteristics: They interfere with enhancer-promoter interactions when present between them, and they buffer transgenes from chromosomal position effects (diagrammed in Figures 1 and 2, see color insert) (30). These two properties must be manifestations of the normal role these sequences play in the control of gene expression. The former property suggests insulators might be one more regulatory sequence, in the same class as enhancers and promoters, at the service of genes to ensure their proper temporal and spatial transcription. The latter attribute

0066-4197/01/1215-0193$14.00

suggests that insulators might play a role in the organization of the chromatin fiber into functional domains, such that genes present in one domain are not affected by regulatory sequences present in a different one.

A function for insulators in the organization of the chromatin within the eukaryotic nucleus would fill a long-standing void in our understanding of nuclear biology. Results from cytological and molecular studies have long suggested the existence of a structural organization of the DNA within the nucleus. For example, the reproducible banding pattern of insect polytene chromosomes is suggestive of an underlying structural organization, perhaps imposed by the DNA sequence on the higher-order organization of chromatin. This specific structural layout might have a functional significance based on the correlation between transcriptional activation and decondensation of particular polytene bands (72). Similarly, the finding of active genes in the loops of lampbrush chromosomes was taken early as an indication of a direct relationship between activation of gene expression and location within a specific structural chromosomal domain (10). More recently, biochemical studies have identified DNA sequences possibly involved in the structural organization of the DNA within the nucleus. When histones and other chromosomal proteins are extracted from nuclei of interphase cells, loops of DNA containing negative unrestrained supercoils can be observed. The bases of these loops are attached to a matrix or scaffold through sequences termed MARs (matrix attachment regions) or SARs (scaffold attachment regions) (49). MARs or SARs are A/T-rich DNA sequences, often containing topoisomerase II cleavage sites, that mediate the anchoring of the chromatin fiber to the chromosome scaffold or nuclear matrix and that might delimit the boundaries of discrete and topologically independent higher-order domains. Although some of these sequences play a role in the expression of particular genes, the question of whether they are merely structural components or whether they play a functional role is still unanswered. At least some insulator elements seem to have properties that bridge those of MARs/SARs and of standard transcriptional regulatory elements, opening the possibility that the function of both types of sequences is related. Here we examine in detail the structural and functional properties of known insulators in a variety of organisms. We then review models that attempt to bring together all the characteristics of insulators and offer suggestions on their possible role in the cell.

SPECIFIC EXAMPLES OF INSULATOR ELEMENTS

The two defining properties of insulators, i.e., their ability to interfere with promoter-enhancer interactions and their capacity to buffer transgenes from silencing effects of the adjacent chromatin, have been used as experimental assays for their identification and characterization. As interest in these sequences has grown in the past few years, they have been characterized in a variety of organisms, ranging from yeast to humans (4, 15). Rather than reviewing all known insulators, we concentrate on those that have been studied in more detail and whose analysis might offer insights into the function of these sequences.

Insulator Elements in *Drosophila*

A variety of chromatin boundaries or insulator elements have been described in *Drosophila*, including the *Mcp*, *Fab-6*, *Fab-7*, and *Fab-8* elements present in the bithorax complex (1, 36, 53, 77), the scs and scs′ elements flanking the 87A7 heat shock gene locus (47, 48), the *gypsy* insulator present in the *gypsy* retrotransposon (34, 42), and an insulator present in the *even-skipped* promoter that contains a binding site for the GAGA protein (60). These different insulators have some common characteristics that might suggest shared mechanisms of action, while at the same time they display idiosyncratic properties suggestive of particular roles in chromatin organization and regulation of gene expression.

THE SCS AND SCS′ ELEMENTS OF THE HSP70 HEAT SHOCK LOCUS   The first experimental evidence of a specific DNA sequence having insulator activity was obtained with the identification of the scs (specialized chromatin structure) and scs′ elements of *Drosophila*. These sequences were identified at the borders of the 87A7 heat shock puff, suggesting that they might demarcate the extent of chromatin that decondenses after induction of transcription by temperature elevation (73). The scs and scs′ sequences contain two strong nuclease-hypersensitive sites surrounding a nuclease-resistant core, which is flanked by additional weaker nuclease cleavage sites present at intervals corresponding to the length of a nucleosome. A similar pattern of strong hypersensitive sites at the location of the proposed boundary elements is observed at the sites of the chicken β-globin 5′ boundary (14) and the insulator present in the *gypsy* retrovirus (11), and suggests that this chromatin organization might play a role in boundary function. The scs and scs' insulators differ in their DNA structure and require different proteins to mediate their function. In scs, sequences associated with DNase I hypersensitive sites are essential for complete blocking activity of enhancer function, whereas the central nuclease-resistant A/T-rich region is dispensable for this effect. Deletion of sequences associated with some hypersensitive sites leads to a reduction in enhancer blocking, whereas multimerization of subfragments with partial activity restores full boundary function (74). Further insights into the specific sequences required for boundary function have come from the identification of SBP (scs binding protein) as the product of the *zeste-white* 5 (*zw5*) gene (27). SBP binds to a 24-bp sequence of scs in vitro, and multiple copies of this sequence have insulator activity as determined by their ability to block enhancer-promoter interactions in vivo. Mutations in the sequence that disrupt SBP binding also disrupt insulator function. In addition, mutations in the *zw5* gene decrease the enhancer-blocking activity of these sequences. The ZW5 protein contains Zn finger motifs and is essential for cell viability. Null mutations in the gene are recessive lethal, but hypomorphic alleles display a variety of pleiotropic effects on wing, bristle, and eye development consistent with a role for this protein in chromatin organization (27).

Sequences responsible for the boundary function of the scs' element have also been characterized in detail. A series of CGATA repeats that interact with the BEAF-32 proteins are responsible for the insulator activity of the scs' sequences (16, 75). Mutations in this sequence that interfere with binding of the BEAF-32 protein also abolish insulator activity, whereas multimers containing several copies of the sequence display boundary function. The latter results are similar to those obtained with the *Drosophila* scs sequences and the *gypsy* insulator, and suggest that the effect of boundary elements on transcription might require the binding of a critical number of proteins that somehow cause chromatin alterations as a consequence of their interaction with DNA.

Two related 32-kDa proteins termed BEAF-32A and BEAF-32B (for boundary element associated factor of 32 kDa) have been purified from nuclei of a *Drosophila* cell line and found to interact with scs' sequences (40, 75). These proteins bind with high affinity to a site containing three copies of the CGATA motif that flanks the two hypersensitive regions in the scs' sequence. The DNA binding activity resides in the amino-terminal region, which is different in the two proteins; the carboxy terminus is shared and it is involved in heterocomplex formation. The sequence containing BEAF-32 binding sites acts as a typical boundary element and blocks the activity of both heat shock and ecdysone responsive enhancers in stably transfected cells (75). Immunolocalization of BEAF-32 using antibodies shows the presence of this protein in specific subnuclear regions and its exclusion from the nucleolus. BEAF-32 is present in the interband regions that separate the highly reproducible and characteristic polytene bands of *Drosophila* third instar larval chromosomes. Interbands contain lower amounts of DNA than bands, and are presumed to be regions of partial unfolding of the 30-nm chromatin fiber. As expected, BEAF-32 is present at the scs'-containing border of the 87A7 chromomere, and is also found at one of the edges of many developmental puffs typically seen in polytene chromosomes at this stage of larval development (75). This observation suggests that BEAF-32 might have general structural and functional roles in defining many boundary elements throughout the *Drosophila* genome.

Recent results suggest that the ZW5 and BEAF-32 proteins can interact with each other in vitro, supporting the possibility that the scs and scs' insulators influence transcription by creating higher-order domains of chromatin organization (see below) (7). A second protein capable of interacting with endogenous scs' insulators has been recently identified (39). This protein is the transcription factor DREF; it binds to a sequence overlapping that recognized by BEAF, suggesting that the two proteins might compete for DNA binding in vivo. DREF participates in the regulation of genes encoding proteins required for DNA replication and cell proliferation. Displacement of BEAF by binding of DREF might occur during the time of rapid proliferation, and competition between the two proteins for binding to insulator sites would open the possibility for regulation of boundary function.

INSULATOR ELEMENTS OF THE BITHORAX COMPLEX   The *Ultrabithorax* (*Ubx*), *Abdominal-A* (*Abd-A*) and *Abdominal-B* (*Abd-B*) genes of the bithorax complex are expressed in a parasegmental-specific pattern by a complex set of regulatory sequences arranged over 300 kb of DNA in a linear fashion, corresponding to the order of expression along the anterior-posterior axis. These parasegment-specific regulatory sequences appear to be separated by boundaries initially identified owing to the dominant gain-of-function phenotypes observed in "boundary deletion mutants" that result in the fusion of two adjacent parasegment-specific regulatory elements into one single functional unit (54). The best studied of these boundaries is the *Fab-7* element located between the *iab-6* and *iab-7* regulatory sequences that control expression of the *Abd-B* gene in parasegments PS11 and PS12 (36, 53, 77). Deletion of the boundary in the chromosomal DNA results in cross-talk between the *iab-6* and *iab-7* regulatory regions, causing homeotic phenotypes in the adult fly. These results indicate that the *Fab-7* region contains an insulator element that is involved in the normal regulation of the *Abd-B* gene. The location of the insulator has been narrowed down to a 1.2-kb DNA that contains one weak and two strong DNase I hypersensitive sites (36, 53).

In the bithorax complex, the role of the insulators that separate different parasegment-specific regulatory sequences is to avoid interactions between these sequences and to maintain proper segmental expression of the genes. This organization nevertheless poses the problem of how these regulatory elements can overcome the effect of the insulators to activate transcription of the *Abd-B* gene when appropriate. A solution to this problem might lie in a recently described sequence named the PTS (promoter-targeting sequence). This sequence, found within the *Fab-8* element, which also contains an insulator, allows distal enhancers to overcome the blocking effects of the *Fab-8* insulator (1, 78).

AN INSULATOR ELEMENT IN THE GYPSY RETROVIRUS   Another insulator element found in *Drosophila* is present in the *gypsy* retrovirus. This insulator is 350 bp in length and is located in the 5′ transcribed, untranslated region of *gypsy*, upstream from the start of the *gag* open reading frame (28). Insertion of *gypsy* into noncoding regions of genes causes a tissue-specific mutant phenotype due to the inability of specific enhancers to interact with the promoter (34, 42, 44). The *gypsy* insulator does not inactivate the adjacent enhancer as this can still activate transcription of a gene located on the other side (8, 69). The *gypsy* insulator can also buffer the expression of a transgene from position effects from adjacent sequences in the genome (66), and it protects the replication origin of the *Drosophila* chorion genes from similar position effects (52). This insulator contains 12 copies of a 26-bp sequence that is the binding site for the Zn finger Su(Hw) protein. The strength of the insulator depends on the number of copies of the 26-bp basic motif: One copy causes a very small effect on enhancer activation of transcription, whereas additional copies result in a stronger effect, with an apparent linear relationship between number of copies and enhancer blocking (68, 70). As in other boundary elements, the *gypsy* insulator also contains a series of three strong DNase I hypersensitive sites indicative of a special chromatin organization (11). This *Drosophila* insulator has recently been shown to function in *S. cerevisiae* (19).

The *gypsy* insulator is perhaps the best-studied system with respect to the characterization of protein components that interact with insulator DNA. One of these components, the Suppressor of Hairy-wing [Su(Hw)] protein, contains 12 zinc fingers involved in DNA binding and an α-helical region homologous to the second helix-coiled coil region of basic HLH-zip proteins that is absolutely required for insulator function (38). This domain mediates interactions between Su(Hw) and a second component of the *gypsy* insulator, Modifier of mdg4 [Mod(mdg4)], which contains a BTB domain (20, 32). The BTB domain is required for dimerization of the Mod(mdg4) protein and these dimers can then interact with the leucine zipper region of Su(Hw) through the carboxy-terminal region of Mod(mdg4) (35).

The Chicken β-Globin Locus and Other Vertebrate Boundary Elements

The first insulator element discovered in vertebrates is located at the 5' end of the chicken β-globin locus and was initially characterized through its ability to interfere with activation of transcription of a reporter gene by the LCR (Locus Control Region) (14). Like the *Drosophila* insulators described above, the chicken β-globin element contains a strong DNase I hypersensitive site (64). In the genome, this element marks a boundary between the open, DNase I-sensitive and acetylated chromatin of the β-globin locus and the more condensed, DNase I-resistant and hypoacetylated chromatin located outside of the locus (41). The insulator activity was originally mapped to a 250-bp DNA fragment (13), and subsequent experiments identified a single binding site for the protein CTCF that was sufficient to confer enhancer-blocking activity (3). The CTCF protein contains 11 zinc fingers and has been previously reported to act as a repressor or activator of transcription (25). A second boundary or insulator element, also marked by hypersensitivity to DNase I, is present at the 3' end of the chicken β-globin gene, and this element also contains CTCF binding sites (67). The fact that the β-globin locus is flanked by insulator elements supports a role for these sequences in the establishment of an open functional domain that allows the transcriptional activation of the globin genes (62).

The chicken 5' β-globin boundary element has also been tested for its ability to protect against position effects (24, 43, 61). A reporter gene expressing a cell surface marker was introduced by stable transformation into a pre-erythroid chicken cell line under conditions in which expression of the reporter was variable from line to line. When two copies of the complete 1.2-kb β-globin boundary element surround the reporter gene, expression is quite uniform among different transformed lines. This behavior is similar to the protection against heterochromatic position effects by insulators observed in *Drosophila*. It has previously been shown that these types of position effects are associated with loss of histone acetylation (12), and the presence of the chicken β-globin insulator protects against deacetylation of histones H3 and H4 (61). This suggests that the boundary elements either promote

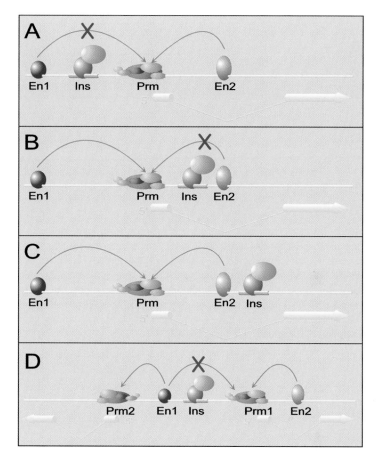

Figure 1 Polar effect of an insulator on enhancer-promoter interactions. The DNA and a hypothetical gene with two exons and one intron are shown in *yellow*. En1 and En2 represent two different enhancers and their associated transcription factors bound to nucleosomal DNA. Prm is the promoter of the gene where the different components of the transcription complex are present. Ins is an insulator element with its associated proteins. Solid arrows indicate a positive activation of transcription by the enhancer element; an X on the arrows indicates the inability of the enhancer to activate transcription. (*A*) An insulator (Ins), with two associated proteins, located in the 5′ region of the gene inhibits its transcriptional activation by an upstream enhancer (En1) without affecting the function of a second enhancer (En2) located in the intron of the gene. (*B*) When the insulator is located in the intron, expression from the downstream enhancer (En2) is blocked, whereas the upstream enhancer (En1) is active. (*C*) When the insulator is located in the intron but distal to the En2 enhancer, both enhancers are active and transcription of the gene is normal. This property distinguishes an insulator from a typical silencer. (*D*) If a second gene is located upstream of the En1 enhancer, although this enhancer cannot act on the Prm1 promoter, it is still functional and able to activate transcription from the upstream Prm2 promoter.

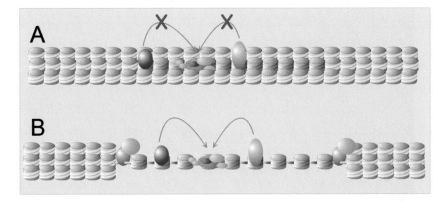

Figure 2 Insulator elements buffer gene expression from repressive effects of adjacent chromatin. Symbols are as in Figure 1. (*A*) A transgene (represented by the *blue* DNA) integrated in the chromosome in a region of condensed chromatin is not properly expressed; the repressive chromatin structure of the surrounding region presumably spreads into transgene sequences, inhibiting enhancer-promoter interactions. (*B*) If the transgene is flanked by insulator elements (in *brown*), these sequences inhibit the spreading of the repressive chromatin, allowing an open chromatin conformation and normal transcription of the gene.

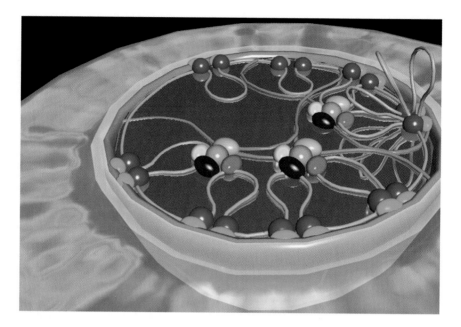

Figure 3 Schematic model explaining the role of trxG and PcG proteins in the function of the *gypsy* insulator. The diagram represents a section though a cell (*blue*) with a nucleus (*dark gray*) surrounded by the nuclear membrane (*light blue*) and the nuclear lamina, which is also located on the inside of the nucleus (*red*). The chromatin fiber is represented as a *gold* line and proteins are represented as ovals colored in *dark blue* [Su(Hw)] and *green* [Mod(mdg4)]; members of the trxG and PcG are represented as *dark purple, red, pink, yellow* and *light green* ovals.

acetylation of the protected region or prevent the action of histone deacetylases. Interestingly, the boundary properties of the β-globin insulator are not associated with sequences that bind CTCF, and other DNA sequence elements within the boundary are required instead (63).

Binding sites for CTCF similar to those present in the chicken β-globin insulator have recently been found to be responsible for the parent-of-origin–specific expression of the *Igf2* and *H19* genes in mice. The presence of an insulator between these two genes had been proposed earlier as an explanation for the inability of enhancers present 3′ to the *H19* gene to activate expression of the *Igf2* maternally transmitted allele. Methylation of DNA sequences located between the two genes, where the putative insulator resides, in the paternally transmitted allele would lead to inactivation of the insulator and activation of *Igf2* in the paternal chromosome (51). The region targeted for methylation has now been shown to contain a series of CTCF binding sites that possess strong insulating activity (2, 37, 45, 46, 71). Mutations in these sites that prevent binding of CTCF abolish the enhancer-blocking activity. More importantly, methylation of these sites abolishes CTCF binding and insulator activity. Therefore, the imprinting phenomenon at the *H19/Igf2* locus closely correlates with the activity of the CTCF insulator, and this activity can be modulated by methylation. The ability to control the activity of an insulator by methylation opens the possibility that other mechanisms might exist in the cell to control the function of these sequences at different times of the cell cycle or during cell differentiation.

Several other elements with enhancer-blocking activity have been identified in vertebrates. Two different human MARs from the apolipoprotein B and alpha1-antitrypsin loci can work as insulators in *Drosophila* by insulating a transgene from position effects. Both elements reduced variability in transgene expression without enhancing levels of the *white* reporter gene expression (59). An insulator has also been described in the human T-cell receptor α/δ locus; this sequence, designated BEAD-1, prevents a δ-specific enhancer from acting on the α genes early in T cell development (76). A binding site for CTCF has been detected within BEAD-1, and deletion of this site abolishes enhancer-blocking effects (3). Similarly, a site within the *Xenopus* ribosomal RNA gene repeat that has limited enhancer-blocking activity when assayed in *Xenopus* oocytes (23) has been identified as a CTCF binding site (3).

Yeast Boundary Elements

Yeast insulator elements have been found at the telomeres and the mating-type loci, where they appear to separate active from silenced chromatin. Genes inserted at yeast telomeric regions or the *HM* mating-type loci are subject to silencing in a manner similar to position effect variegation in *Drosophila*. Surprisingly, the yeast *TEF1* and *TEF2* genes, when present at the *HM* loci, are resistant to this silencing. This resistance can be attributed to the presence of the upstream activation site for ribosomal protein genes (UASrpg) (5). This sequence behaves as a boundary or insulator element, since it blocks the spread of the repressive chromatin structure associated with *HM* silencing when interposed between the *HML α* genes and the E silencer. The insulator activity has been mapped to a 149-bp fragment containing three tandemly repeated binding sites for the Rap1 protein (5).

Insulator elements are normally present flanking the *HM* loci, where they delimit the region subject to silencing by the *HMR* locus (18). Deletion of these elements causes spreading of the silenced chromatin. In addition, when these elements were inserted between a silencer and a promoter, they blocked the repressive effect of the silencer on the promoter. These elements contain a *Ty1* LTR, although the presence of LTR sequences is not sufficient to confer full insulator activity, and additional sequences from a tRNA gene are required for full boundary function. The function of this insulator requires Smc proteins, which constitute structural components required for chromosome condensation, as mutations in the *SMC1* and *SMC3* genes, but not *RAP1*, affect its activity. The structure of this insulator has been characterized in detail recently (19), and the insulator activity has been mapped to transcriptional regulatory sequences of the tRNA gene, where they normally play an important role in the regulation of the expression of the adjacent *GIT1* gene. Mutations in promoter elements of the tRNA gene, or in genes that affect the assembly of the RNA polymerase III transcription complex, affect insulator function. These results suggest that the transcriptional potential of the tRNA gene is essential for its insulator activity. Interestingly, mutations in genes encoding histone acetyltransferases, such as *GCN5* and *SAS2*, reduce insulator activity, whereas tethering Gal4-Sas2 or Gal4-Gcn5 fusion proteins to specific sites results in the formation of a robust insulator (19). These results have important implications for understanding the mechanisms of insulator function.

Additional sequences with the functional hallmarks of insulators have been found at the yeast telomeres. These sequences, called STARS (for subtelomeric anti-silencing regions), can buffer against silencing effects of both telomeric and HML sequences. In addition, when placed flanking a reporter gene, STARs can buffer its expression from surrounding silencing elements. STARS contain binding sites for Tbf1p and Reb1p, and the insulator activity can be reproduced by fragments containing multiple copies of the binding sites for these proteins (26).

MECHANISMS OF INSULATOR FUNCTION

Insulator elements are defined by their ability to interfere with enhancer-promoter interactions and to buffer transgenes against chromosomal position effects. Given these broad standards, it would not be surprising if a variety of sequences with very different roles in normal nuclear function can still fulfill the operational requirements required to be considered a boundary or insulator. For example, the boundaries characterized in yeast play a role in halting the spread of a silenced chromatin, whereas some insulators identified in *Drosophila* and vertebrates might be involved in the establishment of functional domains of gene expression. Genes in higher eukaryotes have proved to be more complex than genes in lower eukaryotes, and this complexity requires that sequences responsible for transcriptional regulation be flexible in the way they operate. Enhancers have thus been designed to control transcription in a distance- and orientation-independent manner and, although this property allows an enhancer great flexibility in where it is positioned with respect to the gene it controls, it also entails the possibility of promiscuous interactions with neighboring genes. Insulators might keep intergenic interactions from taking place by forming boundaries that establish functional domains of gene expression. Some of this function might already be included in the promoter itself; in fact, it was found early on that enhancers could not efficiently transcribe a gene when a second gene was interposed in between (17). It might then be unwise to try to unify all the observed phenomena pertaining to insulators into a single coherent model that explains their role in transcription and, possibly, nuclear organization. This conclusion is also supported by findings suggesting that, at least in some insulators, the ability to interfere with enhancer-promoter interactions can be separated from that of buffering from position effects (63).

Two different types of models have been proposed to explain insulator function; these two models reflect the two types of activity found in insulators and perhaps also reflect the possibility that two different types of sequences are being classified as insulators when, in fact, they play different roles in nuclear function. The two models differ more in the conceptual implications for the normal role of insulators than in the actual mechanisms of how they work. The "promoter decoy" model proposes that insulators act as barriers against a signal that is propagated on the DNA from the enhancer to the promoter (33). According to this model, insulators can imitate the promoter, perhaps by interacting with some or all protein components of the transcription complex, and trick the enhancer into interacting with the insulator instead of the promoter. Although there is no evidence suggesting that this is the case for many insulators, it could certainly be true for yeast insulators found to contain promoter elements. This model seems incompatible with models of enhancer action that do not require tracking of a signal from the enhancer to the promoter. For example, experiments in *Xenopus* oocytes have shown that an enhancer can activate a promoter when the two are on separate but interlinked closed circular plasmids (22). Also difficult to explain with this type of model is the finding that surrounding the enhancer or the promoter in interlocked plasmids with insulators is sufficient to block enhancer action (23). One could argue that the insulator could also trap an enhancer as it loops out the intervening sequences to interact with the promoter. Nevertheless, this argument is not supported by the fact that the strength of at least some insulators is not affected by their position relative to the enhancer and the promoter (44).

An alternative view suggests that insulators exert their effects on transcription through changes in higher-order chromatin structure. This model is supported by the observation that insulators are usually associated with strong DNase I hypersensitive sites and tend to separate chromatin domains with different degrees of condensation (62, 73). A role of at least the *gypsy* insulator in chromatin organization is also supported by the properties of one of its protein components, Mod(mdg4).

The *mod(mdg4)* gene is involved in two different phenomena related to changes in chromatin structure; mutations in this gene act as classical *enhancers of position effect variegation* [*E(var)*] and have the properties characteristic of *trithorax-Group* (*trx-G*) genes (20, 32). Additional evidence supporting this type of model comes from analysis of the subnuclear distribution of *gypsy* insulator proteins. Results from immunofluorescence experiments, using antibodies against Su(Hw) and Mod(mdg4), indicate that these proteins are present at hundreds of sites in polytene chromosomes from salivary glands (31). Given the large number of sites and their regular distribution along the chromosome arms, one would expect to observe a diffuse homogeneous scattering of insulator sites in the nuclei of interphase diploid cells. Surprisingly, this is not the case; instead, *gypsy* insulator proteins accumulate at a small number of nuclear locations. This has led to the suggestion that each of the locations where Su(Hw) and Mod(mdg4) proteins accumulate in the nucleus is made up of several individual sites that come together, perhaps through interactions among protein components of the insulator. Interactions among individual insulator sites would thus lead to a specific arrangement of the chromatin fiber within the nucleus (Figure 3, see color insert). This role for the *gypsy* insulator in nuclear organization is supported by the finding that mutations in the *su(Hw)* gene result in an increase in the frequency at which double-strand breaks are repaired, suggesting that the genome-wide homology search of broken DNA ends for homologous template sequences is affected when the *gypsy* insulator is not functional (50).

Interestingly, the locations where individual insulator sites appear to aggregate in the nucleus are not random; approximately 75% of them are present immediately adjacent to the nuclear lamina (29). This finding suggests that the formation of *gypsy* insulator aggregates may require a substrate for attachment, and that physical attachment might play a role in the mechanism by which this insulator affects enhancer-promoter interactions. The nuclear lamina itself might serve as a substrate for attachment, perhaps through interactions between lamin and protein components of the insulator. The preferential aggregation of insulator sites at the nuclear periphery and the possibility that this targeting might take place through interactions with the nuclear lamina led to the idea that the *gypsy* insulator might be equivalent to MARs/SARs (31). This hypothesis is directly supported by the finding of MAR activity within the DNA sequences containing the *gypsy* insulator (58). This attachment might impose a topological or physical constraint on the DNA that interferes with the transmission of a signal from an enhancer located in one domain to a promoter located in an adjacent one. According to this model, the primary role of the insulator is to organize the chromatin fiber within the nucleus, and its effect on enhancer-promoter interactions is only a secondary consequence of this organization. An important question arising from these results is whether the organization imposed by the *gypsy* insulator is static, and has a mostly structural role, or whether the organization is dynamic and has direct functional significance. In the latter case, modulation of insulator activity could mediate global changes in nuclear organization and gene expression.

A series of recently published experiments underscore the complexity of the mechanisms involved in insulator function. When a direct tandem repeat of insulators was used instead of a single copy, not only was the insulating effect not reinforced, it was indeed abolished and the enhancer was able to activate transcription (9, 57). Control experiments involving other enhancers demonstrated that the loss of insulator activity is independent of the enhancer studied. Also, the distance between insulators did not affect the results. It is difficult to reconcile these observations with a transcriptional insulator model. For example, a reasonable prediction from the decoy model would be a reinforcement of the trapping of the enhancer by a dual insulator configuration. Similarly, if insulators were entry points for chromatin-modifying enzymatic complexes, the doubling of the insulator should lead to a significant increase in its efficiency. The results seem to support models that suggest a role for insulators in establishing higher-order chromatin domains. If an enhancer-promoter pair has to reside within the same domain to be able to interact, two tandemly repeated insulators may have a tendency to preferentially interact with each other to the exclusion of other insulators because of their physical proximity, thus canceling each other (55).

OTHER FACTORS INVOLVED IN INSULATOR FUNCTION

The study of the properties of chromatin boundaries or insulators should lead to a better understanding of the mechanisms by which enhancers activate transcription in eukaryotes and of the role of complex levels of chromatin organization in the control of gene expression. Transcriptional activation in eukaryotic organisms involves changes in chromatin structure that are probably a prerequisite for the ensuing interactions of enhancer-bound transcription factors with the transcription complex present at the promoter. These changes in chromatin structure might involve alterations of higher-order levels of organization as well as changes in nucleosome structure/organization in the primary chromatin fiber involving histone acetylases/deacetylases or other chromatin remodeling complexes (6). Much of our knowledge on these issues comes from studies carried out in yeast, where upstream activating sequences are located relatively close to the promoter. But in most eukaryotes, including *Drosophila*, enhancer elements are located tens or even hundreds of kilobases away from the promoters of genes. How do eukaryotic enhancers activate transcription over such long distances? Since insulators regulate this interaction, studies on the mechanisms of insulator function should shed light into how enhancers activate transcription over long distances, and some of these studies are already giving important insights (21). Studies of the effects of the *gypsy* insulator on the regulation of the *cut* gene by the wing margin enhancer have led to the identification of Chip, a protein that appears to regulate enhancer-promoter interactions (56). Chip is a homolog of the mouse Nli/Ldb1/Clim-2 family and can also interact with nuclear LIM domain proteins. Chip is widely distributed on *Drosophila* polytene chromosomes and it is required for the expression of many genes, although it does not participate directly in transcriptional activation. These results have led to the suggestion that Chip facilitates enhancer-promoter interactions by stabilizing the formation of chromatin structures that bring enhancers located far upstream in close proximity with the promoter (21). The analysis of Chip and Nipped-B (65), both of which affect insulator function, will shed light on the mechanisms of long-range interactions between enhancers and promoters.

A connection between insulators and other proteins involved in the establishment of particular chromatin structures was made by the observation that the Mod(mdg4) protein of the *gypsy* insulator has properties of E(var) and trx-G proteins. Interestingly, the function of the *gypsy* insulator is affected by mutations in *trx-G* and *Pc-G* genes. This genetic interaction correlates with changes in the ability of *gypsy* insulator sites to form aggregates in the nuclei of interphase diploid cells; in the background of mutations in *trx-G* and *Pc-G* genes, these aggregates fail to form and the insulator sites appear to be distributed throughout the nucleus. These observations have been interpreted in the context of a model in which trx-G and Pc-G proteins participate and help insulator proteins in the establishment and maintenance of higher-order chromatin domains (Figure 3, see color insert) (31).

Other factors that are more directly involved in regulating insulator activity must be present in the nucleus. If insulators play a role in establishing higher-order domains of chromatin organization, their activity might be modulated during both cell division and cell differentiation. There must then be proteins that are either constitutive insulator components or are functionally linked to alter the properties of insulators by modifying their protein components. Such proteins have not yet been identified and their existence would lend support to the idea that insulators play important roles in global aspects of gene regulation.

Visit the Annual Reviews home page at www.AnnualReviews.org

LITERATURE CITED

1. Barges S, Mihaly J, Galloni M, Hagstrom K, Muller M, et al. 2000. The Fab-8 boundary defines the distal limit of the bithorax complex iab-7 domain and insulates iab-7 from initiation elements and a PRE in the adjacent iab-8 domain. *Development* 127:779–90

2. Bell AC, Felsenfeld G. 2000. Methylation of a CTCF-dependent boundary controls imprinted expression of the Igf2 gene [see comments]. *Nature* 405:482–85

3. Bell AC, West AG, Felsenfeld G. 1999. The protein CTCF is required for the enhancer blocking activity of vertebrate insulators. *Cell* 98:387–96

4. Bell AC, West AG, Felsenfeld G. 2001. Insulators and boundaries: versatile regulatory elements in the eukaryotic genome. *Science* 291:447–50

5. Bi X, Broach JR. 1999. UASrpg can function as a heterochromatin boundary element in yeast. *Genes Dev.* 13:1089–101

6. Blackwood EM, Kadonaga JT. 1998. Going the distance: a current view of enhancer action. *Science* 281:61–63

7. Blanton J, Gaszner M, Schedl P. 2001. Interaction between two boundary proteins, zeste-white 5 and BEAF. In *Annu. Drosophila Res. Conf., 42nd,* p. a20. Washington, DC

8. Cai H, Levine M. 1995. Modulation of enhancer-promoter interactions by insulators in the Drosophila embryo. *Nature* 376:533–36

9. Cai HN, Shen P. 2001. Effects of cis arrangement of chromatin insulators on ennhancer-blocking activity. *Science* 291: 493–95

10. Callan HG. 1986. Lampbrush chromosomes. *Mol. Biol. Biochem. Biophys.* 36:1–252

11. Chen S, Corces VG. 2001. The gypsy insulator affects chromatin structure in a directional manner. Submitted

12. Chen WY, Bailey EC, McCune SL, Dong JY, Townes TM. 1997. Reactivation of silenced, virally transduced genes by inhibitors of histone deacetylase. *Proc. Natl. Acad. Sci. USA* 94:5798–803

13. Chung JH, Bell AC, Felsenfeld G. 1997. Characterization of the chicken beta-globin insulator. *Proc. Natl. Acad. Sci. USA* 94: 575–80

14. Chung JH, Whiteley M, Felsenfeld G. 1993. A 5′ element of the chicken beta-globin domain serves as an insulator in human erythroid cells and protects against position effect in Drosophila. *Cell* 74:505–14

15. Corces VG, Felsenfeld G. 2000. Chromatin boundaries. In *Chromatin Structure and Gene Expression*, ed. SCR Elgin, JL Workman, pp. 278–99. Frontiers Mol. Biol. Oxford: Oxford Univ. Press

16. Cuvier O, Hart CM, Laemmli UK. 1998. Identification of a class of chromatin boundary elements. *Mol. Cell. Biol.* 18: 7478–86

17. de Villiers J, Olson L, Banerji J, Schaffner W. 1983. Analysis of the transcriptional enhancer effect. *Cold Spring Harbor Symp. Quant. Biol.* 47:911–19

18. Donze D, Adams CR, Rine J, Kamakaka RT. 1999. The boundaries of the silenced HMR domain in *Saccharomyces cerevisiae*. *Genes Dev.* 13:698–708

19. Donze D, Kamakaka RT. 2001. RNA polymerase III and RNA polymerase II promoter complexes are heterochromatin barriers in *Saccharomyces cerevisiae*. *EMBO J.* 20:520–31

20. Dorn R, Krauss V, Reuter G, Saumweber H. 1993. The enhancer of position-effect variegation of Drosophila, E(var)3-93D, codes for a chromatin protein containing a conserved domain common to several transcriptional regulators. *Proc. Natl. Acad. Sci. USA* 90:11376–80

21. Dorsett D. 1999. Distant liaisons: long-range enhancer-promoter interactions in Drosophila. *Curr. Opin. Genet. Dev.* 9:505–14

22. Dunaway M, Droge P. 1989. Transactivation of the Xenopus rRNA gene promoter by its enhancer. *Nature* 341:657–59

23. Dunaway M, Hwang JY, Xiong M, Yuen HL. 1997. The activity of the scs and scs′ insulator elements is not dependent on chromosomal context. *Mol. Cell. Biol.* 17:182–89

24. Emery DW, Yannaki E, Tubb J, Stamatoyannopoulos G. 2000. A chromatin insulator protects retrovirus vectors from chromosomal position effects. *Proc. Natl. Acad. Sci. USA* 97:9150–55

25. Filippova GN, Fagerlie S, Klenova EM, Myers C, Dehner Y, et al. 1996. An exceptionally conserved transcriptional repressor, CTCF, employs different combinations of zinc fingers to bind diverged promoter sequences of avian and mammalian c-myc oncogenes. *Mol. Cell. Biol.* 16:2802–13

26. Fourel G, Revardel E, Koering CE, Gilson E. 1999. Cohabitation of insulators and silencing elements in yeast subtelomeric regions. *EMBO J.* 18:2522–37

27. Gaszner M, Vazquez J, Schedl P. 1999. The Zw5 protein, a component of the scs chromatin domain boundary, is able to block enhancer-promoter interaction. *Genes Dev.* 13:2098–107

28. Gdula DA, Gerasimova TI, Corces VG. 1996. Genetic and molecular analysis of the gypsy chromatin insulator of Drosophila. *Proc. Natl. Acad. Sci. USA* 93:9378–83

29. Gerasimova TI, Byrd K, Corces VG. 2000. A chromatin insulator determines the

nuclear localization of DNA. *Mol. Cell.* 6:1025–35

30. Gerasimova TI, Corces VG. 1996. Boundary and insulator elements in chromosomes. *Curr. Opin. Genet. Dev.* 6:185–92

31. Gerasimova TI, Corces VG. 1998. Polycomb and trithorax group proteins mediate the function of a chromatin insulator. *Cell* 92:511–21

32. Gerasimova TI, Gdula DA, Gerasimov DV, Simonova O, Corces VG. 1995. A Drosophila protein that imparts directionality on a chromatin insulator is an enhancer of position-effect variegation. *Cell* 82:587–97

33. Geyer PK. 1997. The role of insulator elements in defining domains of gene expression. *Curr. Opin. Genet. Dev.* 7:242–48

34. Geyer PK, Corces VG. 1992. DNA position-specific repression of transcription by a Drosophila zinc finger protein. *Genes Dev.* 6:1865–73

35. Ghosh D, Gerasimova TI, Corces VG. 2001. Interactions between the Su(Hw) and Mod(mdg4) proteins required for gypsy insulator function. *EMBO J.* 20:2518–27

36. Hagstrom K, Muller M, Schedl P. 1996. Fab-7 functions as a chromatin domain boundary to ensure proper segment specification by the Drosophila bithorax complex. *Genes Dev.* 10:3202–15

37. Hark AT, Schoenherr CJ, Katz DJ, Ingram RS, Levorse JM, Tilghman SM. 2000. CTCF mediates methylation-sensitive enhancer-blocking activity at the H19/Igf2 locus. *Nature* 405:486–89

38. Harrison DA, Gdula DA, Coyne RS, Corces VG. 1993. A leucine zipper domain of the suppressor of Hairy-wing protein mediates its repressive effect on enhancer function. *Genes Dev.* 7:1966–78

39. Hart CM, Cuvier O, Laemmli UK. 1999. Evidence for an antagonistic relationship between the boundary element-associated factor BEAF and the transcription factor DREF. *Chromosoma* 108:375–83

40. Hart CM, Zhao K, Laemmli UK. 1997. The scs' boundary element: characterization of boundary element-associated factors. *Mol. Cell. Biol.* 17:999–1009

41. Hebbes TR, Clayton AL, Thorne AW, Crane-Robinson C. 1994. Core histone hyperacetylation co-maps with generalized DNase I sensitivity in the chicken beta-globin chromosomal domain. *EMBO J.* 13:1823–30

42. Holdridge C, Dorsett D. 1991. Repression of hsp70 heat shock gene transcription by the suppressor of hairy-wing protein of *Drosophila melanogaster. Mol. Cell. Biol.* 11:1894–900

43. Inoue T, Yamaza H, Sakai Y, Mizuno S, Ohno M, et al. 1999. Position-independent human beta-globin gene expression mediated by a recombinant adeno-associated virus vector carrying the chicken beta-globin insulator. *J. Hum. Genet.* 44:152–62

44. Jack J, Dorsett D, Delotto Y, Liu S. 1991. Expression of the cut locus in the Drosophila wing margin is required for cell type specification and is regulated by a distant enhancer. *Development* 113:735–47

45. Kaffer CR, Srivastava M, Park KY, Ives E, Hsieh S, et al. 2000. A transcriptional insulator at the imprinted H19/Igf2 locus. *Genes Dev.* 14:1908–19

46. Kanduri C, Pant V, Loukinov D, Pugacheva E, Qi CF, et al. 2000. Functional association of CTCF with the insulator upstream of the H19 gene is parent of origin-specific and methylation-sensitive. *Curr. Biol.* 10:853–56

47. Kellum R, Schedl P. 1991. A position-effect assay for boundaries of higher order chromosomal domains. *Cell* 64:941–50

48. Kellum R, Schedl P. 1992. A group of scs elements function as domain boundaries in an enhancer-blocking assay. *Mol. Cell. Biol.* 12:2424–31

49. Laemmli UK, Kas E, Poljak L, Adachi Y. 1992. Scaffold-associated regions: *cis*-acting determinants of chromatin structural loops and functional domains. *Curr. Opin. Genet. Dev.* 2:275–85

50. Lankenau DH, Peluso MV, Lankenau S.

2000. The Su(Hw) chromatin insulator protein alters double-strand break repair frequencies in the Drosophila germ line. *Chromosoma* 109:148–60

51. Leighton PA, Ingram RS, Eggenschwiler J, Efstratiadis A, Tilghman SM. 1995. Disruption of imprinting caused by deletion of the H19 gene region in mice. *Nature* 375:34–39

52. Lu L, Tower J. 1997. A transcriptional insulator element, the su(Hw) binding site, protects a chromosomal DNA replication origin from position effects. *Mol. Cell. Biol.* 17:2202–6

53. Mihaly J, Hogga I, Barges S, Galloni M, Mishra RK, et al. 1998. Chromatin domain boundaries in the Bithorax complex. *Cell Mol. Life Sci.* 54:60–70

54. Mishra R, Karch F. 1999. Boundaries that demarcate structural and functional domains of chromatin. *J. Biosci.* 24:377–89

55. Mongelard F, Corces VG. 2001. Two insulators are not better than one. *Nat. Struct. Biol.* 8:192–94

56. Morcillo P, Rosen C, Baylies MK, Dorsett D. 1997. Chip, a widely expressed chromosomal protein required for segmentation and activity of a remote wing margin enhancer in Drosophila. *Genes Dev.* 11:2729–40

57. Muravyova E, Golovnin A, Gracheva E, Parshikov A, Belenkaya T, et al. 2001. Loss of insulator activity by paired Su(Hw) chromatin insulators. *Science* 291:495–98

58. Nabirochkin S, Ossokina M, Heidmann T. 1998. A nuclear matrix/scaffold attachment region co-localizes with the gypsy retrotransposon insulator sequence. *J. Biol. Chem.* 273:2473–79

59. Namciu SJ, Blochlinger KB, Fournier RE. 1998. Human matrix attachment regions insulate transgene expression from chromosomal position effects in *Drosophila melanogaster*. *Mol. Cell. Biol.* 18:2382–91

60. Ohtsuki S, Levine M. 1998. GAGA mediates the enhancer blocking activity of the eve promoter in the Drosophila embryo. *Genes Dev.* 12:3325–30

61. Pikaart MJ, Recillas-Targa F, Felsenfeld G. 1998. Loss of transcriptional activity of a transgene is accompanied by DNA methylation and histone deacetylation and is prevented by insulators. *Genes Dev.* 12:2852–62

62. Prioleau MN, Nony P, Simpson M, Felsenfeld G. 1999. An insulator element and condensed chromatin region separate the chicken beta-globin locus from an independently regulated erythroid-specific folate receptor gene. *EMBO J.* 18:4035–48

63. Recillas-Targa F, Bell AC, Felsenfeld G. 1999. Positional enhancer-blocking activity of the chicken beta-globin insulator in transiently transfected cells. *Proc. Natl. Acad. Sci. USA* 96:14354–59

64. Reitman M, Felsenfeld G. 1990. Developmental regulation of topoisomerase II sites and DNase I-hypersensitive sites in the chicken beta-globin locus. *Mol. Cell. Biol.* 10:2774–86

65. Rollins RA, Morcillo P, Dorsett D. 1999. Nipped-B, a Drosophila homologue of chromosomal adherins, participates in activation by remote enhancers in the cut and Ultrabithorax genes. *Genetics* 152:577–93

66. Roseman RR, Pirrotta V, Geyer PK. 1993. The su(Hw) protein insulates expression of the *Drosophila melanogaster* white gene from chromosomal position-effects. *EMBO J.* 12:435–42

67. Saitoh N, Bell AC, Recillas-Targa F, West AG, Simpson M, et al. 2000. Structural and functional conservation at the boundaries of the chicken beta-globin domain. *EMBO J.* 19:2315–22

68. Scott KC, Taubman AD, Geyer PK. 1999. Enhancer blocking by the Drosophila gypsy insulator depends upon insulator anatomy and enhancer strength. *Genetics* 153:787–98

69. Scott KS, Geyer PK. 1995. Effects of the su(Hw) insulator protein on the expression of the divergently transcribed Drosophila yolk protein genes. *EMBO J.* 14:6258–67

70. Spana C, Corces VG. 1990. DNA bending

is a determinant of binding specificity for a Drosophila zinc finger protein. *Genes Dev.* 4:1505–15

71. Szabo P, Tang SH, Rentsendorj A, Pfeifer GP, Mann JR. 2000. Maternal-specific footprints at putative CTCF sites in the H19 imprinting control region give evidence for insulator function. *Curr. Biol.* 10:607–10

72. Tissieres A, Mitchell HK, Tracy UM. 1974. Protein synthesis in salivary glands of *Drosophila melanogaster*: relation to chromosome puffs. *J. Mol. Biol.* 84:389–98

73. Udvardy A, Maine E, Schedl P. 1985. The 87A7 chromomere. Identification of novel chromatin structures flanking the heat shock locus that may define the boundaries of higher order domains. *J. Mol. Biol.* 185:341–58

74. Vazquez J, Schedl P. 1994. Sequences required for enhancer blocking activity of scs are located within two nuclease-hypersensitive regions. *EMBO J.* 13:5984–93

75. Zhao K, Hart CM, Laemmli UK. 1995. Visualization of chromosomal domains with boundary element-associated factor BEAF-32. *Cell* 81:879–89

76. Zhong XP, Krangel MS. 1997. An enhancer-blocking element between alpha and delta gene segments within the human T cell receptor alpha/delta locus. *Proc. Natl. Acad. Sci. USA* 94:5219–24

77. Zhou J, Barolo S, Szymanski P, Levine M. 1996. The Fab-7 element of the bithorax complex attenuates enhancer-promoter interactions in the Drosophila embryo. *Genes Dev.* 10:3195–201

78. Zhou J, Levine M. 1999. A novel *cis*-regulatory element, the PTS, mediates an anti-insulator activity in the Drosophila embryo. *Cell* 99:567–75

Annu. Rev. Genet. 2001. 35:209–41

ANIMAL MODELS OF TUMOR-SUPPRESSOR GENES

Razqallah Hakem and Tak W. Mak

Amgen Institute, Ontario Cancer Institute and the University of Toronto, Toronto, Ontario, Canada M5G 2C1; e-mail: tmak@oci.utoronto.ca

Key Words   cancer, tumor suppressor, animal models, gene-targeting

■ **Abstract**   The development of cancer requires multiple genetic alterations perturbing distinct cellular pathways. In human cancers, these alterations often arise owing to mutations in tumor-suppressor genes whose normal function is to either inhibit the proliferation, apoptosis, or differentiation of cells, or maintain their genomic integrity. Mouse models for tumor suppressors frequently provide definitive evidence for the antitumorigenic functions of these genes. In addition, animal models permit the identification of previously unsuspected roles of these genes in development and differentiation. The availability of null and tissue-specific mouse mutants for tumor-suppressor genes has greatly facilitated our understanding of the mechanisms leading to cancer. In this review, we describe mouse models for tumor-suppressor genes.

CONTENTS

0066-4197/01/1215-0209$14.00   **209**

INTRODUCTION

The design of appropriate therapies for cancer patients and indeed, cancer prevention, will best be achieved through a thorough understanding of the mechanisms leading to tumor development. Cancer is a complex disease with both genetic and environmental components. Rather than a single mutation, multiple genetic alterations (hits) accumulated over the lifetime of an individual are required for complete development of a tumor. Cancers are classified as either familial (inherited) or sporadic, the latter comprising the vast majority of cases. In familial cancers, every cell of an individual has a germline mutation that predisposes that individual to tumorigenesis. When the function of remaining wild-type allele is lost in a somatic cell, additional mutations can accumulate that lead to tumor formation. In sporadic cancers, all tumorigenic mutations are somatic and are present only in the patient's neoplastic cells.

In humans, mutations leading to gain of functions of proto-oncogenes or loss of functions of tumor-suppressor genes (TSGs) predispose to cancer. TSGs can be divided into two groups: "gatekeepers" and "caretakers" (77). Gatekeepers regulate the growth of tumors by inhibiting their proliferation or promoting their apoptosis. Caretakers control cellular processes that repair genetic alterations and maintain genomic integrity. Mutations of caretakers can result in an increase in the overall mutation rate in a given cell, an apparent prerequisite for tumorigenesis. The accumulation of mutations in a dysregulated cell favors the subsequent clonal selection of variant progeny with aggressive growth properties leading to malignancy.

Familial cancers often result from the initial germline mutation of one allele of a TSG followed later by somatic mutation or loss of the second allele, a process known as loss of heterozygosity (LOH). Knudson's two-hit cancer model (reviewed in 78) is based on this repeated observation. Although LOH was originally a prerequisite for the identification of canonical TSGs, the remaining wild-type allele can also be transcriptionally silenced by hypermethylation of its CpG regions. Functional loss then leads to tumorigenesis.

Animal models have been instrumental in the study of genes involved in human cancer initiation and progression. Spontaneous as well as carcinogen-induced malignancies have been studied in dogs, rats, and mice, and TSGs identified in Drosophila have led to the discovery of mammalian orthologues. The mouse remains the animal model of choice for several reasons. First, mice and humans have roughly the same number of genes, and intracellular signaling pathways are highly conserved between the two species. Second, the success of genetic engineering in mice has allowed the study of gene functions in vivo. Mice can be genetically manipulated to overexpress (by transgenesis), or not express (by gene targeting), a specific cancer gene. These animals offer unique opportunities to uncover cellular pathways controlled by specific TSGs and to dissect mechanisms underlying malignancies that may closely resemble the human situation. In addition, the interbreeding of an increasing number of mice with specific mutations in both TSGs and oncogenes allows the assessment of how these muations cooperate to produce cancer. In this review, we focus on mouse models for TSGs that are involved in apoptosis and the cell cycle, DNA damage repair, or cell signaling and differentiation.

TUMOR SUPPRESSORS INVOLVED IN APOPTOSIS AND THE CELL CYCLE

Retinoblastoma Gene

Retinoblastoma is an eye tumor that occurs in children. Mutations of the gatekeeper TSG *Rb* have been found in both inherited and sporadic cases. *Rb* mutations can also predispose individuals to osteosarcomas and prostate and breast cancers. The *Rb* gene is located on chromosome 13q14.2 and encodes a 928-amino acid (aa) nuclear protein of 105 kDa (reviewed in 54). Phosphorylation of the Rb protein, which is critical for regulation of its function, is mediated by cyclin/cyclin dependent kinase (CDK) complexes (Figure 1). in vivo phosphorylation of Rb is thus cell cycle dependent, with the hypophosphorylated, active form being present in G0/G1, and the hyperphosphorylated, inactive form dominant in late G1.

Inactivation of Rb by germline mutation of one allele and LOH of the second allele is often found in Rb-associated cancers. However, functional inactivation of the Rb protein in the absence of an *Rb* mutation can also lead to tumorigenesis. Tumors exist in which an amplification of the 11q13 region results in cyclin D1 overexpression, which in turn leads to activation of cyclinD-CDK4/6 activity and hyperphosphorylation and inactivation of Rb. Functional inactivation of Rb can also be caused by a lack of p16[INK4a] function (see below). Finally, viral oncoproteins such as adenovirus E1A and SV40 large tumor antigen can bind to the Rb protein and inactivate it.

The function of Rb is to repress the transcription of genes required for DNA replication and cell division. At least two different mechanisms may be involved. Binding sites for the transcription factor E2F are present in the promoters of many genes whose expression is essential and limiting for entry into S phase. Rb binds members of the E2F family, forming a complex that inhibits transcriptional activation. The Rb-E2F complex can also actively repress transcription of genes further downstream. In addition to cell cycle regulation, Rb plays a role in apoptosis. Increased apoptosis is observed in gene-targeted Rb-deficient ($Rb^{-/-}$) embryos,

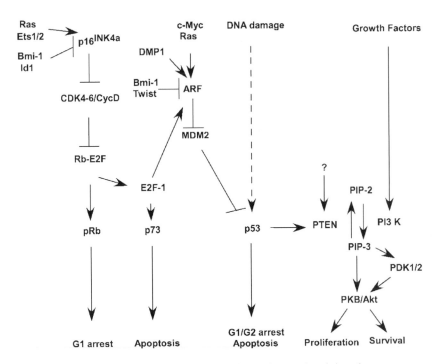

Figure 1 Schematic representation of cellular pathways involving the tumor suppressors Rb, p16^{INK4a}, p19^{ARF}, p53, and Pten. Lines ending with arrowheads or bars indicate activating or inhibitory effects, respectively. The genes and their activities are defined in the text.

and murine embryonic fibroblasts (MEFs) from $Rb^{-/-}$ mice show activation of E2F-responsive genes and apoptosis (2). A possible link between the loss of Rb function and the activation of apoptosis is E2F-1. Overproduction of E2F-1, although capable of driving S phase entry, also leads to p53-dependent apoptosis (1). E2F-1 also activates p73 transcription (64, 90, 150), activating p53-responsive target genes and apoptosis (150).

Rb MOUSE MODELS   Transgenic mice overexpressing human wild-type Rb in the germline exhibit dwarfism as early as day 15 of embryonic development (E15) (9), indicating that regulation of Rb expression is required for normal development.

Gene-targeted $Rb^{-/-}$ embryos die in utero between E13.5–E15.5 (23, 65, 86). Increased apoptosis in the nervous system is seen as early as E11.5 and is particularly evident in the hindbrain, spinal cord, and trigeminal and dorsal root ganglia. Ectopic mitoses are also observed, especially in the hindbrain. In addition to defective neurogenesis, $Rb^{-/-}$ embryos exhibit defective hematopoiesis, manifested as an increased number of immature nucleated erythrocytes. Significantly, expression of a human Rb transgene in $Rb^{-/-}$ mice fully rescues their developmental defects (9). Interestingly, apoptosis in lens fiber cells deficient for Rb is dependent on p53 since $Rb^{-/-}p53^{-/-}$ embryos show a complete suppression of apoptosis (106).

Analyses of viable chimeric mice derived from $Rb^{-/-}$ embryonic stem (ES) cells revealed that the Rb-deficient cells contribute widely to adult tissues (97, 173). The chimeric erythroid and central nervous system (CNS) compartments appeared normal, but the developing retina was defective and displayed ectopic mitoses. However, unlike human Rb patients, the Rb chimeras developed pituitary gland tumors rather than retinoblastomas. A similar phenotype was observed in $Rb^{+/-}$ mice that, at age 8–10 months, developed tumors in the brain and pituitary gland. These tumors exhibited LOH of the remaining wild-type Rb allele, demonstrating that Rb is a TSG in mice as well as in humans.

Although $Rb^{+/-}$ mice are cancer-prone, they do not accurately recapitulate the tumor spectrum observed in human Rb patients. This discrepancy was recently resolved with the identification of other members of the Rb family that can compensate for Rb loss in the murine eye. Chimeras possessing cells deficient for both Rb and its family member p107 developed retinoblastomas during the early postnatal months (124). Only $Rb^{-/-}p107^{-/-}$ chimeras, and not $Rb^{+/-}p107^{-/-}$ chimeras or germline $Rb^{+/-}p107^{-/-}$ mice, developed retinoblastomas, suggesting that the low number of target cells in the murine retina precludes the acquisition of the required number of mutations to inactivate the remaining Rb allele.

Mice homozygous for loss of Rb family members $p107$ or $p130$ are viable, fertile, and healthy (25, 86). However, $p107^{-/-}p130^{-/-}$ mice experience neonatal lethality (25), and most $Rb^{+/-}p107^{-/-}$ mice are growth-retarded, and increased mortality of these mice is observed within the first three weeks of birth (86). Although $Rb^{+/-}p107^{-/-}$ pups that survive to adulthood do not show increased cancer predisposition compared to $Rb^{+/-}$ mice, they develop multiple dysplastic lesions of the retina (86). Thus, unlike Rb, p107 and p130 are not required for embryonic development, and $p107^{+/-}$, $p130^{+/-}$, $p107^{-/-}$, and $p130^{-/-}$ mice do not show increased incidence of tumor development. ES cells with a simultaneous deficiency of Rb, p107, and p130 (triple knockout, TKO) have normal growth characteristics but impaired differentiation capacity (28, 127). TKO MEFs have a shorter cell cycle compared to controls and can spontaneously immortalize. TKO MEFs are also resistant to G1 arrest following DNA damage, contact inhibition or serum starvation.

The effects of $E2F-1$ mutation on Rb mutant phenotypes in mice have been examined. $Rb^{-/-}E2F-1^{-/-}$ embryos die in utero at approximately E17 with anemia and defective skeletal muscle and lung development (163). Significant tissue-specific suppression of apoptosis, S phase entry, and p53 activation was observed in the double mutant cells. The fact that mutation of E2F-1 did not fully rescue the Rb developmental defects indicates that these abnormalities are not entirely E2F-1 dependent. Conditional mutant mice for Rb (101) have been generated using the Cre-*LoxP* system and should prove useful in defining stage- and tissue-specific roles of Rb.

P53

P53 is the most commonly mutated TSG in human cancers (reviewed in 168). The gene encoding this gatekeeper is located on chromosome 17p13.1. Inherited *p53* mutations are associated with Li-Fraumeni syndrome, which is characterized by an increased risk for breast and lung carcinomas, soft tissue sarcomas, brain tumors, osteosarcoma, and leukemia. Genetic alterations leading to inactivation of p53 in human tumors are primarily base substitutions resulting in missense mutations. In some cervical cancers, viral proteins such as human papillomavirus (HPV) E6 can functionally inactivate p53.

The *p53* transcript of 2.8 kb encodes a 393-aa nuclear phosphoprotein of 53 kDa. The p53 protein exists as a tetramer that exerts its tumor-suppressive effect by binding to DNA at specific sites. P53 activity is modulated by protein stability, regulated largely through interactions with the E3 ligase MDM2. Binding of MDM2 to p53 leads to p53 degradation and loss of activity, a process that can be inhibited by the gatekeeper protein ARF (see below) (Figures 1, 2). P53 activity is also regulated by posttranslational modifications mediated by phosphorylation, acetylation, sumolation, and glycosylation.

Two major pathways lead to p53 activation. The first is triggered by mitogenic signals mediated through deregulated Myc, Ras, or E2F-1. These signals

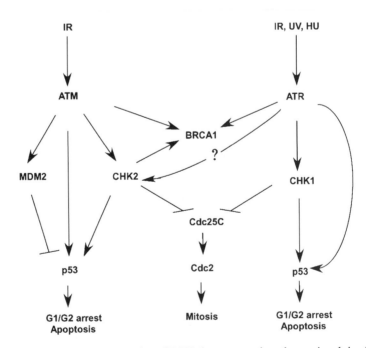

Figure 2   Schematic representation of DNA damage repair pathways involving TSGs. Lines ending with arrows or bars indicate activating or inhibitory effects, respectively. The genes and their activities are defined in the text.

induce ARF expression, which promotes p53 stabilization and activation. The second pathway is triggered in response to DNA damage and involves the regulatory kinases ATM, CHK2, and ATR (see below). When activated, p53 functions as a transcription factor. By controlling the transcription of its myriad target genes, p53 regulates multiple important cellular pathways such as the cell cycle, apoptosis, and differentiation. Several p53 transcriptional targets have been identified, including *p21, bax, PIG8, MDM2, GADD45, PIDD, Noxa, p53R2, DR5,* and *CD95.* Conversely, p53 activation can repress the expression of bcl-2 and PCNA.

p53 controls the G1 checkpoint of the cell cycle by inducing transcription of the cyclin/CDK inhibitor *p21.* Cells with damaged DNA are prevented from entering S phase and replicating the defective chromosomes. P53 controls the G2 checkpoint by regulating expression of p21 and the protein 14-3-3σ (16). Arrest of the cell cycle at these checkpoints allows the repair of the damaged DNA. Should repair be impossible, p53 induces the apoptosis of the defective cell. P53-dependent apoptosis involves the transcriptional activation of pro-apoptotic genes including *bax, PIDD, DR5, Noxa,* and *CD95.* P53 also influences the expression of Pten, a TSG involved in the control of the PKB/Akt survival pathway (V. Stambolic & T. Mak, unpublished).

P53 MOUSE MODELS   P53 is the best-studied TSG in mice. Several lines of transgenic mice overexpressing mutated or wild-type p53 have been generated. Mice bearing a human *p53* gene with an Arg193Pro or Ala135Val substitution exhibited a high incidence of lung, bone, and lymphoid tumors (84). When infected with the polycythemia-inducing strain of Friend virus, these animals progressed to the late stage of erythroleukemia more rapidly than controls (83). In contrast, heterozygous mice expressing mutant mouse p53 with an Arg172His substitution (corresponding to the Arg175His mutation "hot-spot" in human tumors) developed osteosarcomas, carcinomas, and lymphomas with high metastatic potential (93). Mice with a *p53* allele altered at Leu25 and Trp26, residues essential for transcriptional transactivation and MDM2 binding, synthesized a p53 protein that was stable but did not accumulate after DNA damage (18, 70). These mice were cancer-prone and their cells showed defects in cell cycle regulation and apoptosis.

Several strains of mice bearing null mutations of *p53* have also been created (35, 66, 121, 164). Although initial studies of *p53*$^{-/-}$ mice concluded that p53 had no role in development, subsequent work has revealed that a subset of p53-deficient embryos dies in utero. This embryonic death is associated with exencephaly marked by defects in neural tube closure and overgrowth of neural tissue in the midbrain region (3, 128). The finding that *Mdm2*$^{-/-}$ mutants, which die before E6.5, can be fully rescued in a *p53* null background suggests a requirement for tight regulation of p53 activity during development.

As expected, *p53*$^{+/-}$ mice are predisposed to tumorigenesis. These animals remain cancer-free for the first 9 months but half develop osteosarcomas, soft tissue sarcomas, and lymphomas by 18 months of age. The majority of tumors from $p53^{+/-}$ mice show LOH of the wild-type $p53$ allele. In $p53^{-/-}$ mice, the onset of tumor development is earlier than in heterozygotes; more than 75% of $p53^{-/-}$ mice develop tumors before 6 months of age. While thymic T cell lymphomas are the major tumor type, $p53^{-/-}$ mice also develop B cell lymphomas, sarcomas, and testicular teratomas. The enhanced susceptibility to cancer of $p53^{-/-}$ mice results in their death before 10 months of age. p53-deficient mice are also extremely susceptible to tumorigenesis induced by ionizing radiation (IR) or carcinogens (56, 75). The generation of p53 conditional mutant mice (101) is expected to shed further light on p53 tumor-suppressor function in vivo.

Patients with Li-Fraumeni syndrome have a 50% chance of tumor incidence by age 30 (100), which is comparable to the tumor incidence in middle-aged (18 months) $p53^{+/-}$ mice. The spectrum of tumors that arises in Li-Fraumeni patients is generally similar to that in $p53^{+/-}$ mice. However, human patients also develop breast and brain tumors, which are rarely observed in $p53^{+/-}$ mice. These differences were puzzling until a $p53^{-/-}$ mouse was crossed into the BALB/c background. Spontaneous mammary tumors developed in half of the heterozygous females of this strain (80), and the tumors showed the loss of the remaining wild-type $p53$ allele. Furthermore, when mammary glands from $p53^{-/-}$ BALB/c mice were transplanted into wild-type BALB/c hosts, 75% of the transplanted mice developed mammary tumors. Thus, differences in genetic background can have profound effects on the types of tumors associated with $p53$ mutation in mice.

In addition to their use as an animal model for p53-associated cancers, $p53^{-/-}$ mice have also been helpful in determining p53 functions in vivo. With respect to apoptosis, p53 is required for thymocyte apoptosis induced by double-strand DNA breaks (DSBs) in response to stimuli such as ionizing radiation and Adriamycin, but dispensable for the apoptosis that mediates negative selection of immature thymocytes, as well as thymocyte apoptosis induced in response to UV irradiation, dexamethasone, or anti-CD95 (24, 94). With respect to cell cycle control, cells from $p53^{-/-}$ mice show enhanced proliferation in culture compared to controls (57, 164). With respect to maintenance of genomic integrity, $p53^{-/-}$ MEFs cultured in vitro show a high degree of aneuploidy, a hallmark for genomic instability (57). p53 has also been implicated as a component of a spindle checkpoint that ensures the maintenance of diploidy. $p53^{-/-}$ fibroblasts exposed to spindle inhibitors performed multiple rounds of DNA synthesis without completing chromosome segregation, resulting in tetraploid and octaploid cells (27). Increased genomic instability is also seen in mice deficient for Gadd45α, a transcriptional target of p53 (61).

Tumor suppressors and oncogenes are presumed to cooperate in the induction of tumorigenesis. A $p53^{-/-}$ background dramatically increases tumor frequency and reduces latency in c-Myc transgenic mice (11). A cooperative effect on transformation between $p53$ and Rb mutations has also been reported. Mice carrying mutations of both $p53$ and Rb have reduced viability and exhibit pathologies not observed with either single mutant, including pinealoblastomas, islet cell tumors, bronchial epithelial hyperplasia, and retinal dysplasia (58).

P73 AND P63 MOUSE MODELS   P73 and p63 are members of the p53 family of proteins and thus attractive candidates for TSGs. P73 and p63 share significant amino acid identity with p53 in the transactivation domain, the DNA-binding domain, and the oligomerization domain. Like p53, p73 and p63 can recognize canonical p53 DNA-binding sites and, when overproduced, can activate p53-responsive target genes and induce apoptosis. P73 is localized to chromosome 1p36.3, a region of frequent aberrations in human cancer. However, there is no convincing clinical evidence that p73 and p63 play a significant role in human tumorigenesis.

E2F-1 directly activates transcription of *p73* (Figure 1), leading to activation of p53-responsive target genes and apoptosis (64, 90, 150). Furthermore, disruption of p73 function inhibits E2F-1-induced apoptosis in p53-deficient cells, suggesting a role for p73 in p53-independent apoptosis. Similarly, T cell receptor-activation-induced cell death, which is p53 independent, is inhibited in the absence of p73 (90). Interestingly, a truncated isoform of p73 that is expressed in developing neurons appears to have an anti-apoptotic function (120). However, other evidence weighs against a tumor-suppressive role for p73. Knockout mice null for p73 do not have an increased incidence of cancer (177). Several p73 isoforms and two different p73 promoters have been described. Mice deficient for specific isoforms remain to be reported.

Mice deficient for p63 show defects in limb and epidermal morphogenesis and die within hours of birth (104, 176). $p63^{+/-}$ mice are healthy and show no increase in tumorigenesis. Thus, although p53, p63, and p73 share structural homologies, studies of animal models for the mutation of these genes indicate that these proteins have divergent functions, and so far only p53 has a tumor-suppressor function.

INK4a Locus

The INK4a locus on chromosome 9p21 is frequently disrupted in human cancers (126). Germline mutations of this locus predispose an individual to familial melanomas, whereas somatic mutations increase the chance of sporadic malignancies of the pancreas and brain. In mice, the INK4a locus includes two independent but overlapping genes that encode the gatekeeper proteins $p16^{INK4a}$ and $p19^{ARF}$ (ARF). Each gene has its own promoter that precedes three coding exons. The first exons for $p16^{INK4a}$ (E1α) and $p19^{ARF}$ (E1β) are specific to each gene. Exons 2 and 3 are shared, although they are read in different frames and produce two different proteins (137). Most mutations of the INK4a locus were originally thought to inactivate $p16^{INK4a}$. However, the identification of ARF, and the finding that *ARF* and $p16^{INK4a}$ share two exons, suggest that some mutations in the INK4a locus may affect only ARF or only $p16^{INK4a}$, whereas others may affect both proteins.

The $p16^{INK4a}$ protein is a cyclin-dependent kinase inhibitor (CKI) that specifically binds to and inhibits CDK4 and CDK6, proteins that promote the G1/S transition (133). Inhibition of CDKs leads to maintenance of Rb in its active, hypophosphorylated form (Figure 1). Thus, $p16^{INK4a}$ performs its tumor-suppressor function through the functional inactivation of Rb. Interestingly, Rb represses p16^{INK4a} expression, and upregulation of p16^{INK4a} expression is observed in Rb-deficient cells (88, 159). Expression of p16^{INK4a} can also be repressed by the polycomb family member Bmi-1 (69), and by the helix-loop-helix protein Id1 (110). p16^{INK4a} expression is enhanced by the transcription factors Ets1 and Ets2, two downstream targets of Ras-Raf-MEK signaling (110).

The second tumor suppressor encoded by the murine INK4a locus is p19^{ARF} (the human homolog is called p14^{ARF}). p19^{ARF} has been functionally linked to p53 because it physically interacts with MDM2 and promotes its rapid degradation (Figure 1), leading to p53 stabilization and activation (73, 119, 152, 181). Interestingly, p53 may control ARF expression since cells express high levels of ARF in the absence of functional p53. The main activators of ARF expression are oncoproteins such as myc, adenovirus E1A, Ras, and v-abl, consistent with a role for ARF in sensing hyperproliferative signals. Bmi-1 (69), TBX2 (68), and Twist (98) repress ARF expression, whereas DMP1 increases it (63). ARF expression can also be induced by E2F-1 (7, 182), providing a functional link between the p16^{INK4a}/cyclin D/CDK4–6/Rb and the ARF/MDM2/p53 tumor-suppression pathways.

In addition to p16^{INK4a} and p19^{ARF}, the INK4 family of CKIs includes p15^{INK4b} and p18^{INK4c}. These proteins share ~40% identity, contain four tandem ankyrin motifs, and specifically inhibit CDK4 and CDK6.

INK4a MOUSE MODELS   A mouse strain (INK4a/ARF$^{ex2–3}$) in which both p16^{INK4a} and p19^{ARF} are deficient owing to deletion of their common exons 2 and 3 has been created (134). Mice heterozygous for the *INK4a/ARF$^{ex2–3}$* mutation show a moderate increase in fibrosarcomas, lymphomas, squamous cell carcinomas, and angiosarcomas. Forty percent of these tumors exhibit LOH at the INK4a locus (111). Deletion of exon 1β specific to ARF is also seen in some of these tumors, suggesting that in these cases it is a deficiency of ARF and not p16^{INK4a} that leads to tumor development. Susceptibility to tumorigenesis induced by the carcinogen dimethylbenz(a)anthrancene (DMBA) alone or in combination with UVB irradiation is only slightly increased in *INK4a/ARF$^{ex2–3}$* heterozygous mice compared to controls.

Mice homozygous for the *INK4a/ARF$^{ex2–3}$* mutation are viable, suggesting that this locus is not essential for embryonic development or survival. In fact, overexpression of the INK4a locus can be detrimental to mouse development. Increased expression of p16^{INK4a} and ARF occurs in Bmi-deficient mice, which are underdeveloped and have a cerebellum and lymphoid organs of reduced size. Interestingly, these defects are either completely or partially rescued in a homozygous *INK4a/ARF$^{ex2–3}$* background (69).

As expected, *INK4a/ARF$^{ex2–3–/–}$* mice show an increased susceptibility to cancer, with most developing sarcomas and lymphomas by age 7 months. Earlier onset of these malignancies (at about 13 weeks) is observed in *INK4a/ARF$^{ex2–3–/–}$* mice treated with DMBA or UVB. Surprisingly, in contrast to humans, mutations of the INK4a locus do not predispose mice to melanoma development.

However, $INK4a/ARF^{ex2-3-/-}$ mice that overexpress an activated H-ras^{G12V} gene in melanocytes develop cutaneous melanomas with high penetrance by age 5.5 months (20). This finding suggests that a loss of function of the INK4a locus coupled with a gain of function of ras can result in mice predisposed to melanomas.

Although studies of $INK4a/ARF^{ex2-3-/-}$ mice indicate that the INK4a locus is important for tumor suppression, the specific roles of p16^{INK4a} and p19^{ARF} cannot be determined in these animals. Mice deficient for p19^{ARF}, but competent for p16^{INK4a}, have been generated by deleting the ARF-specific exon 1β (72, 74). Mice heterozygous for this mutation develop lymphomas, sarcomas, and hemangiomas after a long latency. The tumors show loss of the remaining ARF allele and/or lack of its expression, confirming ARF's role in tumor suppression. Homozygous $ARF^{-/-}$ mice are viable and fertile but most develop sarcomas ($\sim$40%) and T cell lymphomas ($\sim$30%) at around 6 months of age and often die by 12 months. Interestingly, $ARF^{-/-}$ mice differ from $INK4a/ARF^{ex2-3-/-}$ mice in several phenotypes. The latency period for tumor formation is shorter in untreated and DMBA-treated $INK4a/ARF^{ex2-3-/-}$ mice ($\sim$32 and $\sim$12 weeks, respectively) than in untreated and DMBA-treated $ARF^{-/-}$ mice ($\sim$38 and $\sim$24 weeks, respectively). In addition, $ARF^{-/-}$ mice have carcinomas and tumors of the nervous system that do not appear in $INK4a/ARF^{ex2-3-}$ mice. A mouse deficient only for p16^{INK4a} remains to be reported.

p15^{INK4b} AND p18^{INK4c} MOUSE MODELS   There is no strong evidence from human disease to support a role for p15^{INK4b} and p18^{INK4c} in tumor suppression. Although the p15^{INK4b} locus is often deleted in human tumors, its deletion is concomitant with that of the neighboring INK4a/ARF locus. Gene-targeted mice hemizygous for $p18^{INK4c}$ or $p15^{INK4b}$ do not exhibit increased incidence of cancer (40, 82). Like other INK4 family members, p18^{INK4c} and p15^{INK4b} are not required for embryonic development. $p18^{INK4c-/-}$ mice develop pituitary adenomas and testicular tumors that lead to the death of about 40% of these animals before 18 months of age. In contrast, the cancer susceptibility of $p15^{INK4b-/-}$ mice is only slightly increased over controls; angiosarcomas are observed in fewer than 10% of older $p15^{INK4b-/-}$ mice. $p18^{INK4c}$ $p15^{INK4b}$ double mutants develop tumors representative of each single mutant (82). However, mice lacking both p18^{INK4c} and another CKI, p27, develop pituitary tumors and die by 3.5 months of age (40). Mice with nullizygous or heterozygous mutations of $p27$ are predisposed to tumors in multiple tissues when challenged with γ-irradiation or a chemical carcinogen (37).

PTEN

The gatekeeper PTEN (phosphatase and tensin homologue deleted from chromosome 10, also known as MMAC1 and TEP1) is one of the most commonly mutated TSG in human cancers (reviewed in 147). The $PTEN$ gene is located on chromosome 10q23, a genomic region that suffers LOH in many human tumors. Germline mutations of *PTEN* are believed to cause three related autosomal-dominant hamartoma disorders: Cowden disease (CD), Lhermitte-Duclos disease (LDD), and Bannayan-Zonana syndrome (BZS). These disorders share pathological features such as multiple benign tumors and an increased susceptibility to breast, thyroid, and brain cancers. Somatic mutation or deletion of *PTEN* is commonly found in human sporadic cancers, including cases of glioblastoma, endometrial carcinoma, prostate carcinoma, and melanoma.

The PTEN protein contains 403 aa and shares homology with dual specificity phosphatases. However, phosphatidylinositol (3,4,5)-triphosphate (PIP-3) is the primary PTEN substrate, and tumor-suppressor activity of PTEN depends only on its lipid phosphatase activity. PIP-3 levels are very low in cells, but rapidly increase when phosphatidylinositol 3′ kinase (PI3′K) is activated by growth factor stimulation (Figure 1). Accumulation of PIP-3 at the plasma membrane allows the recruitment of proteins containing a pleckstrin homology domain that can bind PIP-3, such as the protein kinases PDK1 and PKB/Akt. Upon membrane recruitment, PDK1 phosphorylates PKB/Akt and activates it. Activated PKB/Akt acts as a survival factor by transducing signals that regulate multiple biological processes, including survival and proliferation. PKB/Akt phosphorylation targets include Bad, the Forkhead transcription factors FKHR, FKHRL1, and AFX, and glycogen synthase kinase-3β (GSK-3β). Thus, a major PTEN function is to maintain a low threshold of cellular PIP-3, antagonizing PI3′K signaling and activation of PKB/Akt. Dysregulation of this function is thought to lead to tumorigenesis.

PTEN MOUSE MODELS   Mice heterozygous for gene-targeted mutations of *Pten* are both highly susceptible to tumors and develop a polyclonal autoimmune disorder similar to that observed in CD95-deficient mice (33, 34, 117, 148, 156, 157). The predominant malignancies in young (<6 months) $Pten^{+/-}$ mice are thymic and peripheral lymphomas, with more varied types of tumors developing in older mice (32, 149). LOH of the remaining wild-type *Pten* allele has been demonstrated in many of these tumors.

Tumors in $Pten^{+/-}$ mice often exhibit hallmark features of PTEN-associated human hamartoma syndromes. The majority of $Pten^{+/-}$ females over 6 months of age develop mammary and endometrial neoplasias (149). Increased frequencies of gastrointestinal tract hamartomas, prostate hyperplasia, and adrenal gland neoplasia, as well as hyperplasia and dysplasia of the skin and colon, occur in aged $Pten^{+/-}$ mice. PKB/Akt is hyperphosphorylated in tumor cells of $Pten^{+/-}$ mice (156), supporting the hypothesis that Pten suppresses tumorigenesis by modulating the PKB/Akt survival pathway.

Null mutation of *Pten* results in early embryonic lethality between E6.5–E9.5. $Pten^{-/-}$ embryos show poorly organized ectodermal and mesodermal layers and an overgrowth in the cephalic and caudal regions (156). $Pten^{-/-}$ ES cells have decreased levels of p27 (154), and $Pten^{-/-}$ embryos show increased BrdU incorporation (148). Apoptosis induced in response to various apoptotic stimuli is decreased in $Pten^{-/-}$ MEFs, which have elevated concentrations of PIP-3 and hyperactivated PKB/Akt (148). Mice with a T cell–specific loss of Pten show increased thymocyte numbers, hyperproliferation of T cells, and spontaneous activation of CD4$^+$ T cells leading to loss of immune tolerance (157). These animals develop T cell lymphomas by 17 weeks of age. Targeted disruption of $Pten$ in mouse brain causes seizures and ataxia associated with neuronal cell size defects resembling those in human LDD (S. Backman & T. Mak, unpublished).

LATS

The Drosophila $LATS$ (large tumor suppressor) gene was identified in a mosaic screen for hyperproliferative mutants. A mammalian homologue, LATS1, is a serine threonine kinase capable of modulating CDC2 activity (160). Human LATS1 can suppress tumor growth and rescue all developmental defects in LATS-deficient flies, but further studies are required to demonstrate a tumor-suppressor function of LATS1 in humans.

Lats1 MOUSE MODEL   Whereas mice heterozygous for gene-targeted disruption of $Lats1$ are healthy, $Lats1^{-/-}$ mice are viable but underdeveloped and exhibit soft tissue sarcomas and ovarian stromal cell tumors (146). Susceptibility to carcinogen-induced cancers is also drastically enhanced in $Lats1^{-/-}$ mice.

Apaf1 and Caspase 8

Studies of human melanomas and neuroblastomas have revealed the silencing in these tumors of the $Apaf1$ (145) or $caspase\ 8$ (161) genes, respectively. These genes are major players in mammalian apoptotic pathways, but more evidence is required to demonstrate TSG functions for these genes in humans.

Apaf-1/CASPASE MOUSE MODELS   Mutation of $Apaf-1$ (15, 179) or $caspase\ 8$ (166) in mice leads to embryonic lethality precluding the analysis of cancer development in the homozygous mutant mice. Furthermore, heterozygous mice mutant for $Apaf1$ and $caspase\ 8$ do not exhibit increased incidence of tumor development.

TUMOR SUPPRESSORS INVOLVED IN DNA DAMAGE REPAIR

Proteins involved in DNA damage repair are key candidates for TSGs. DNA damage results from external insults such as chemical agents, as well as from internal problems such as cellular oxidative stress, chromosomal breaks, and errors during DNA replication and V(D)J recombination (reviewed in 131). Mammalian cells have developed a network of proteins to sense and repair damaged DNA or induce apoptosis of the cell if the damage is beyond repair. Defective repair of damaged

DNA is thought to allow cells to acquire mutations conferring a growth advantage, the first step toward malignancy. Tumor-suppressive functions have been confirmed for several caretaker genes involved in DNA damage repair.

ATM

Mutations of the *ATM* (ataxia telangiectasia mutated) gene are responsible for the rare human genetic disorder ataxia-telangiectasia (AT) (reviewed in 139). AT patients are characterized by immunodeficiency, progressive cerebellar ataxia, radiosensitivity, defects in cell cycle checkpoints, and predisposition to leukemia and lymphoma. AT cells exhibit chromosomal instability, telomere shortening, and defects in cellular responses to ionizing radiation. *ATM* mutation accompanied by LOH of the wild-type allele has been shown in cases of sporadic T prolymphocytic leukemia (151, 169) and in mantle cell lymphoma (130).

The human *ATM* gene encodes a 3056-aa protein containing a PI3′K-like domain. In response to DSBs, ATM phosphorylates, among other sites, Ser 15 in p53 as well as Thr 68 in CHK2 (Figure 2). Activated CHK2 in turn phosphorylates Ser 20 of p53, inhibiting p53/MDM2 interaction and stabilizing p53. DSBs also trigger ATM to phosphorylate MDM2, preventing p53 nuclear transport and degradation (76). Phosphorylation of CHK2 by ATM induces this enzyme to phosphorylate and inactivate Cdc25C, a dual specificity phosphatase that maintains G2 progress by dephosphorylating and activating the CDK Cdc2. Phosphorylated Cdc25C (Ser 216) binds to the 14-3-3 protein to form a complex that is exported to the cytoplasm, arresting the cell cycle at the G2/M checkpoint.

Activated ATM and CHK2 also phosphorylate BRCA1 (see below). Both BRCA1 and ATM are found in the BASC (BRCA1-associated genome surveillance complex), a large protein complex containing several DNA damage repair proteins.

ATM MOUSE MODELS $Atm^{+/-}$ mice generated by gene targeting (5, 36, 175) are healthy but exhibit increased sensitivity to sublethal doses of IR, as manifested by decreased survival and premature graying of the hair (6). This finding is reminiscent of the radiosensitivity of AT patients.

$Atm^{-/-}$ mice are viable but display growth retardation, neurologic dysfunction, male and female infertility, a small thymus and defective T lymphocyte maturation, and increased sensitivity to γ-irradiation. $Atm^{-/-}$ mice are highly predisposed to cancer, with the majority dying of widely metastasized malignant thymic lymphomas by 4–5 months of age.

Cytogenetic analysis of $Atm^{-/-}$ thymic lymphomas has consistently identified chromosomal abnormalities involving the TCRα/δ locus, suggesting aberrant repair of the DNA DSBs that occur during V(D)J recombination. Two recombinases, RAG-1 and RAG-2, are required for V(D)J recombination, and deficiency for either results in a complete lack of V(D)J recombination. Tumors do not develop in <9-month-old $Atm^{-/-}Rag1^{-/-}$ mice, and the survival of these animals is increased compared to $Atm^{-/-}$ mice (89). $Atm^{-/-}Rag2^{-/-}$ mice do develop thymomas, although at a lower frequency and with a longer latency than $Atm^{-/-}$ mice (116). Non-thymic malignancies also occur in $Atm^{-/-}Rag2^{-/-}$ mice.

The possible synergy between Atm mutations and those affecting other TSGs has been explored. $Atm^{-/-} p53^{-/-}$ mice exhibit accelerated tumor formation compared to $p53^{-/-}$ or $Atm^{-/-}$ mice, suggesting collaboration between the two mutations (170). Apoptosis of thymocytes in response to IR is suppressed in $Atm^{-/-} p53^{-/-}$ mice but not in $Atm^{-/-} p21^{-/-}$ mice, suggesting that IR-induced apoptosis of $Atm^{-/-}$ thymocytes is p53 dependent (4).

CHK2

CHK2 (CDS1) is a 65-kDa checkpoint protein that halts the cell cycle when activated by phosphorylation (reviewed in 139). Phosphorylation of CHK2 is triggered by DNA damage induced by IR or hydroxyurea (HU) (13). Functional ATM protein is required for CHK2 phosphorylation following IR, but not HU, treatment. Heterozygous germline *CHK2* mutations have been found in some Li-Fraumeni syndrome patients without *p53* mutations, suggesting that inactivation of CHK2 can predispose to human cancers (8). However, loss of the wild-type allele of *CHK2* in tumors from these patients was not investigated, and definitive evidence that CHK2 is a typical TSG is lacking.

CHK2 MOUSE MODEL   Studies of gene-targeted $Chk2^{-/-}$ ES cells have shown that Chk2 is required for the maintenance, but not for the establishment, of G2 arrest in response to DNA damage (60). Somatic chimeras generated by Rag1 blastocyst complementation with $Chk2^{-/-}$ ES cells were used to study the function of Chk2 in the T cell lineage. Unlike Atm, but like p53, a lack of Chk2 does not perturb T cell lineage development. Furthermore, Chk2 is required to trigger p53 activation in response to DSBs, and $Chk2^{-/-}$ thymocytes, like $p53^{-/-}$ thymocytes, are resistant to apoptosis induced by γ-irradiation or chemotherapeutic drugs. To date, no increase in spontaneous tumor development has been observed in young $Chk2$ homozygous mutant mice (A. Hirao & T. Mak, unpublished).

ATR

The *ATR* (ATM and Rad3-related) gene encodes a kinase that plays an important role in the DNA damage response (reviewed in 139). In contrast to ATM, which responds exclusively to DSBs, ATR responds to DSBs, UV damage, and replication arrest. Activated ATR controls the G1 checkpoint through its phosphorylation of p53 (Ser 15), and the G2 checkpoint by phosphorylation/activation of CHK1 (Ser 345) (Figure 2). Activated CHK1, like CHK2, inactivates Cdc25C by phosphorylation, triggering the G2/M checkpoint. Although no cancer has yet been associated with an ATR mutation in humans, the involvement of this molecule in p53 activation, its control of G1 and G2 checkpoints, as well as its phosphorylation of BRCA1 (162), predict a role for ATR as a TSG.

Atr MOUSE MODELS   Homozygous mutations of *Atr* in mice result in embryonic lethality before E8.5, accompanied by a loss of genomic integrity (14, 29). Increased tumor incidence is observed in aged $Atr^{+/-}$ mice (29). However, LOH of the *Atr* wild-type allele has not been demonstrated in these tumors.

BRCA1

In humans, inheritance of one mutated allele of *BRCA1* significantly increases the risk of breast or ovarian cancer (reviewed in 122). Mutations of the *BRCA1* gene, located on chromosome 17q21, are a predisposing factor in approximately 50% of families with hereditary breast cancer and in over 80% of families with hereditary breast-ovarian cancer. The majority of tumors isolated from predisposed individuals demonstrate LOH for the wild-type *BRCA1* allele.

BRCA1 expression is increased in the S phase of the cell cycle, and the BRCA1 protein is phosphorylated in a cell cycle-dependent manner and in response to DNA damage (reviewed in 132). Several lines of evidence support a role for BRCA1 in DNA damage repair. BRCA1 is phosphorylated following activation of ATM, CHK2, and ATR-dependent DNA damage signaling pathways (Figure 2). Furthermore, BRCA1 interacts or forms a complex with multiple proteins involved in DNA damage repair, including RAD51, RAD50–Mre11–p95, MSH2, MSH6, MLH1, ATM, CHK2, BRCA2, and BLM (reviewed in 31, 32). BRCA1 also interacts with proteins that are involved in transcription, such as the RNA polymerase II holoenzyme complex, RNA helicase A, CtIP, and CBP/p300.

BRCA1 MOUSE MODELS   Unlike humans with an inherited *BRCA1* mutation, mice hemizygous for a *Brca1* mutation do not show increased incidence of tumors. Homozygous *Brca1* mutations inevitably lead to postimplantation embryonic lethality (47, 52, 92, 95). However, phenotypic differences exist among different $Brca1^{-/-}$ strains, such as a range of the onset of lethality from E6.5 to E13. Such differences could be due to different types of targeted *Brca1* mutations and/or to the frequent alternative splicing of this gene during embryogenesis.

$Brca1^{-/-}$ embryos exhibit defective cellular proliferation and activation of p53-dependent pathways (52). Mutation of *p53* or its transcription target *p21* partially delays the lethality of the $Brca1^{-/-}$ embryos (50, 95). Hypomorphic *Brca1* mutants with a partial loss of Brca1 function show spina bifida and anencephaly accompanied by increased apoptosis in the neuroepithelium and die at E10–E13 (47). Hypomorphic *Brca1* mutants ES cells and MEFs are hypersensitive to IR (46, 136) and have a defect in transcription-coupled DNA repair. Human *BRCA1* transgenes can rescue the embryonic lethality of *Brca1* mutant mice (17, 81).

Mice with a mammary gland–specific partial loss of Brca1 function show increased mammary cell apoptosis and abnormal ductal development (174) and develop mammary gland tumors after 10 to 13 months of age. These tumors exhibit genetic instability characterized by aneuploidy and chromosomal rearrangements. Loss of p53 function accelerates the formation of mammary tumors in female mice with the mammary gland–specific *Brca1* mutation, paralleling the frequent loss of p53 in tumors from human BRCA1 patients. Mice with a T cell lineage–specific null mutation of *Brca1* exhibit depletion of T lineage cells, abnormal p53 activation, and decreased cell cycle progression and apoptosis (99). The genomic instability in Brca1-deficient T cells likely triggers p53 activation, leading to increased spontaneous and γ-irradiation-induced apoptosis. Interestingly, $Brca1^{-/-}$ thymocyte development is completely rescued in a $p53^{-/-}$ or *Bcl2* transgenic background. Predisposition to thymomas is increased in T cell–specific $Brca1^{-/-}$ mice and the frequency of tumors is drastically increased by p53 mutation (P. McPherson & R. Hakem, unpublished).

BRCA2

The human *BRCA2* gene is located on chromosome 13q12-q13, and encodes a major transcript of 11 kb that is translated into a highly charged 3418 aa protein (reviewed in 51, 132). In women, mutations of *BRCA2* are responsible for 32% of hereditary breast cancers. *BRCA2* mutations are also related to increased breast cancer in men. Like BRCA1, BRCA2 is thought to act as a caretaker involved in DNA damage repair and the maintenance of genomic integrity. BRCA1 and BRCA2 co-immunoprecipitate and co-localize with RAD51 in subnuclear foci and on the axial elements of developing synaptonemal complexes. Furthermore, BRCA2, like BRCA1 and RAD51, relocates to PCNA-positive replication sites following exposure of S phase cells to HU or UV (19).

BRCA2 MOUSE MODELS $Brca2^{-/-}$ mice die in utero between E7.5 and E9.5 (95, 135, 155). The onset of abnormalities is seen as early as E5.5 and the mutant embryos remain underdeveloped until death. Defective cell proliferation occurs in vivo and in vitro. Some hypomorphic *Brca2* mutants survive to adulthood (26, 41, 115). These animals are smaller than their control littermates, show abnormal tissue differentiation, lack germ cells, and are infertile. *Brca2* hypomorphs develop lethal thymic lymphomas by 12–14 weeks of age. Mammary gland cancers were not observed, possibly because the animals suffer an early death. It is not yet clear whether *Brca2* mutations in mice will accurately reproduce human pathologies.

Brca2-deficient cells have defects in their ability to repair DNA damage induced by genotoxic agents such as γ-, X- and UV-irradiation, and MEFs of Brca2 hypomorphic mutants exhibit genomic instability (41, 115, 165, 180). Unlike *Brca1*, T cell–specific mutation of *Brca2* does not affect T cell development (A. Cheung & T. Mak, unpublished). However, $Brca2^{-/-}$ T cells exhibit chromosomal aberrations, and an increase in T cell lymphomas is observed in T cell–specific $Brca2^{-/-}p53^{-/-}$ mice.

Mismatch Repair and Blm Genes

Mismatch repair genes and the Blm gene are not usually considered TSGs. However, based on the gatekeeper/caretaker model, these genes qualify as caretakers.

Mutation of mismatch repair genes predisposes humans to hereditary non-polyposis colon cancer (HNPCC).

MISMATCH REPAIR AND *Blm* GENE MOUSE MODELS   Mice mutated in certain mismatch repair genes, such as *Msh2*, *Msh6*, *PMS2*, *Mlh1* (reviewed in 55), or the DNA helicase *Blm* (mutated in human Bloom syndrome) (96), are predisposed to cancer.

TUMOR SUPPRESSORS INVOLVED IN CELL SIGNALING AND DIFFERENTIATION

Several human tumor suppressors inhibit cell growth by regulating intracellular signaling and/or differentiation.

Neurofibromatosis Type 1

Neurofibromatosis type 1 (NF1) is a common autosomal-dominant disease associated with benign peripheral nerve sheath tumors termed neurofibromas (reviewed in 21, 114). NF1 patients are also predisposed to astrocytomas, glioblastomas, optic gliomas, pheochromocytomas, and myeloid leukemia. Other symptoms of NF1 patients include intellectual deficits, bone deformations, and hyperpigmentation defects of the skin. The disease results from a germline mutation of the *NF1* gene, which is localized on chromosome 17q11.2. LOH of the wild-type *NF1* allele has been found in some tumors in NF1 patients, consistent with a function for *NF1* as a TSG. Neurofibromin, the protein encoded by the *NF1* gene, is highly homologous to mammalian p120Ras GTPase-activating protein (GAP). GAPs are known to regulate the GTP status of the proto-oncogene p21-Ras by converting p21-Ras GTP (active form) to p21-Ras GDP (inactive form). A lack of neurofibromin results in inappropriate activation of the Ras signaling pathway, which in turn leads to aberrant cell growth.

NF1 MOUSE MODELS $Nf1^{-/-}$ embryos die between E11–E13.5 with heart defects and enlarged superior ganglia (12, 67). One-year-old $Nf1^{+/-}$ mice have an increased predisposition to pheochromocytoma and myeloid leukemia, but not to the human hallmark neurofibromas. A study of $Nf1$ chimeric mice showed that loss of both $Nf1$ alleles is required for the formation of neurofibromas (22).

There are close parallels between *NF1* patients and $Nf1^{-/-}$ mice. Certain astrocytomas of human NF1 patients have mutations of p53. $Nf1^{+/-}$ $p53^{+/-}$ mice have a reduced lifespan compared to $Nf1$ or $p53$ hemizygous controls and develop sarcomas characteristic of neuronal crest derivatives, astrocytomas, and glioblastomas (22, 167). These tumors show LOH at both the $p53$ and $Nf1$ loci. Children with *NF1* mutations are predisposed to juvenile myelomonocytic leukemia. Some $Nf1^{+/-}$ mice develop a similar myeloproliferative disorder, and adoptive transfer of Nf1-deficient fetal liver cells consistently induced myeloproliferative disease in recipient mice (10). Finally, loss of NF1 in humans causes learning defects, and $Nf1^{+/-}$ mice have a memory deficit and a learning defect (141).

Neurofibromatosis Type 2

Neurofibromatosis 2 (NF2) is a severe inherited disorder that is rare in frequency compared to NF1 (reviewed in 48). The major clinical features of NF2 are vestibular schwannoma, meningioma, and spinal tumors. The $NF2$ gene, located on chromosome 22q12, encodes the merlin protein, a member of the 4.1 family of cytoskeleton-associated proteins. These proteins link membrane proteins to the cytoskeleton and are involved in dynamic cytoskeletal reorganization.

NF2 MOUSE MODELS $Nf2^{-/-}$ embryos do not initiate gastrulation and die between E6.5–7.0 (103). $Nf2^{+/-}$ mice develop a variety of highly metastatic tumors at age 10–30 months, in contrast to the narrow spectrum of benign tumors observed in human NF2 patients (102). Tumors in $Nf2^{+/-}$ mice include osteosarcoma, lymphoma, lung adenocarcinoma, hepatocellular carcinoma, and fibrosarcoma. Cells of these tumors show LOH of the wild-type $Nf2$ allele. Although $Nf2^{+/-}$ mice are cancer-prone, they do not develop the schwannomas, meningiomas, or ependymomas seen in human NF2 patients. However, mice with a Schwann cell-specific homozygous mutation of $Nf2$ exhibit schwannomas, Schwann cell hyperplasia, cataracts, and osseous metaplasia (43). The absence of certain types of tumors in $Nf2^{+/-}$ mice is likely due to inefficient LOH rather than to a species difference in $Nf2$ gene function.

Another mouse model for NF2 was generated by expressing a mutated merlin protein in Schwann cells (42). These transgenic mice develop Schwann cell-derived tumors and Schwann cell hyperplasia.

Adenomatous Polyposis Coli

Several inherited predispositions to colorectal cancer have been described, including familial adenomatous polyposis coli (APC) associated with mutation of the APC gene. The majority of sporadic and hereditary colorectal tumors show loss of APC function. The human APC gene is localized on chromosome 5q21 and encodes a 2843-aa protein that contains several domains and motifs (reviewed in 118). Seven 20-aa repeats phosphorylated by GSK-3 are critical for APC binding to β-catenin (Figure 3). APC also contains Ser-Ala-Met-Pro (SAMP) motifs associated with downregulation of intracellular β-catenin levels.

APC plays a major regulatory role in the Wnt signaling pathway. The binding of Wnt to the Frizzled receptor activates Dishevelled, which inhibits GSK-3 β phosphorylation of β-catenin and prevents its proteosomal degradation. Free β-catenin interacts with several transcription factors that activate transcription of cell cycle genes such as *cyclin D* and *c-myc*. However, APC forms a complex with axin, which recruits β-catenin and facilitates its phosphorylation by GSK-3β. Phosphorylated

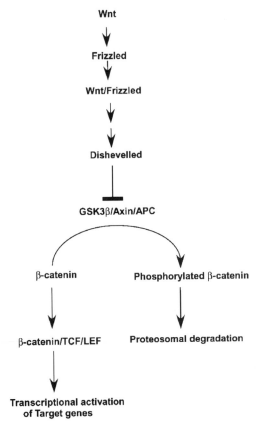

Figure 3   Schematic representation of the Wnt signaling pathway. Lines ending with arrowheads or bars indicate activating or inhibitory effects, respectively. The genes and their activities are defined in the text.

β-catenin is ubiquitinated and degraded. Therefore, by destabilizing β-catenin, APC controls the Wnt signaling pathway and suppresses cell growth. Tumor cells mutated for *APC* contain elevated levels of β-catenin that can be downregulated by expression of exogenous wild-type APC (109).

Apc MOUSE MODELS   Several mouse models for *Apc* mutation have been established. Multiple intestinal neoplasia (Min) mice were derived from a C57BL/6J male treated with ethylnitrosourea to induce random germline mutagenesis (107, 153). Gene-targeting of *Apc* in ES cells results in the production of truncated Apc proteins unable to perform Apc functions (38, 113). Homozygosity for these *Apc* mutations leads to embryonic lethality before E8 (38, 108, 113). *Apc*$^{+/-}$ mice develop multiple intestinal neoplasia (38, 107, 113, 144, 153). A hypomorphic *Apc* mutation (*Apc*1638T) results in expression of a stable 182-kDa truncated protein containing the N-terminal 1638 aa of Apc (143). Homozygous *Apc*1638T mice are viable but growth retarded and do not show increased incidence of tumors. Thus, only the N-terminal half of the Apc protein is required for its functions in embryogenesis and tumor suppression. Conditional mutant mice with a colorectal-specific mutation of *Apc* develop adenomas within 4 weeks of Apc loss (138).

Genetic backgrounds capable of modifying tumorigenesis in *Apc*-deficient mice have been investigated. Smad4, a candidate human tumor suppressor in colorectal cancers, is a key player in TGF-β signaling. *Smad4*$^{-/-}$ mutant mice die before E8.5 (142, 158, 178). *Apc*$^{+/-}$*Smad4*$^{+/-}$ mice develop a greater number of intestinal polyps than *Apc*$^{+/-}$ mice (158). Similarly, mutations of *Msh2* (123), *Mlh1* (140), *Tcf1* (125), and *p53* (53) have a synergistic effect on tumor development in *Apc*$^{+/-}$ mice. Interestingly, null mutation of *cyclooxygenase 2* dramatically reduces the number and size of the intestinal polyps in *Apc*$^{+/-}$ mice (112).

The Wilms Tumor-Suppressor Protein

Wilms tumor is a pediatric kidney cancer associated with germline mutations of the Wilms TSG (*WT1*) localized on chromosome 11p13 (reviewed in 91). WT1 inactivation has also been linked to defective renal development. WT1 activates transcription of *p21* but suppresses expression of the paired box transcription factor PAX-2. Analysis of WT1 transcriptional targets using oligonucleotide microarrays identified *p21*, *HSP70* and *amphiregulin* (87). *Amphiregulin*, which encodes a growth and differentiation factor belonging to the epidermal growth factor (EGF) family, is the principal Wt1 transcriptional target, suggesting a role for Wt1 in the EGF pathway.

Wt1 MOUSE MODELS   Gene-targeted *Wt1*$^{-/-}$ mice of the C57BL/6 (B6) background die between E13.5–E15.5 (79). The kidneys, gonads, and adrenal glands of *Wt1*$^{-/-}$ embryos fail to develop and the heart is malformed (79, 105). Interestingly, *Wt1*$^{-/-}$ animals of the MF1, BALB/c, or C3H backgrounds are born but die immediately after birth with a failure of spleen development (59). These observations suggest that a modifier gene(s) can influence the Wt1 phenotype. Although *Wt1* mutation in mice reproduces some of the developmental defects observed in WT1 patients, *Wt1*$^{+/-}$ mice do not show increased incidence of tumors. This species discrepancy has not been yet resolved and requires the generation of mice with *Wt1* mutations targeted to the kidneys.

von Hippel-Lindau

von Hippel-Lindau (VHL) is a dominantly inherited cancer syndrome associated with the formation of hypervascularized neoplasms of the kidney, retina, pancreas, adrenal gland, and CNS (reviewed in 71). *VHL* mutations are also observed in sporadic renal cancers. The mechanism of action of VHL remains poorly understood. VHL forms a multimeric complex with cullin-2, rbx1, and the Elongins B and C. VHL also negatively regulates hypoxia response genes through ubiquitination of the α-subunit of the heterodimeric transcription factor HIF (hypoxia-inducible factor).

Vhl MOUSE MODELS   Gene-targeted $Vhl^{-/-}$ embryos die between E10.5–E12.5 with a lack of placental vasculogenesis (44). In this particular study, the $Vhl^{+/-}$ littermate mice were normal and did not show increased incidence of tumors. However, $Vhl^{+/-}$ mice of a different genetic background and bearing a different targeted Vhl mutation showed increased cancer susceptibility (49). After 12 months, the majority of these animals developed vascular lesions and cavernous hemangiomas in the liver. Mice with a liver-specific mutation of Vhl appear sick and die at 6–12 weeks. Vascular histopathologies such as hepatocellular steatosis and foci of increased vascularization within the hepatic parenchyma were observed, but not tumors. HIF-2α and VEGF expression was increased in $Vhl^{-/-}$ hepatocytes. It remains unresolved why VHL patients develop a spectrum of tumors whereas only liver tumors are seen in some $Vhl^{+/-}$ mice.

Patched

Patched (Ptc) was first identified as a key regulator of Hedgehog signaling during Drosophila development. The Ptc protein opposes Hedgehog-induced gene transcription and sequesters Hedgehog protein (reviewed in 30). Ptc thus negatively regulates the expression of several genes, including the TGF-β family member *Decapentaplegic* and the Wnt family member *wingless*. A human *Ptc* homologue encoding a receptor for the hedgehog protein is responsible for naevoid basal cell carcinoma syndrome (NBCCS/Gorlin Syndrome). NBCCS is an autosomal-dominant condition in which patients suffer from multiple basal cell carcinomas and a wide spectrum of developmental abnormalities. These patients are also predisposed to medulloblastoma and rhabdomyosarcoma. LOH inactivating the wild-type *Ptc* allele has been found in some NBCCS tumors, indicating that *Ptc* is a TSG.

Ptc MOUSE MODELS   Targeted disruption of *Ptc* in mice leads to altered neural tube closure, abnormal cardiac development, and death by E10.5 (45). About 14% of $Ptc^{+/-}$ mice develop spontaneous cerebellar medulloblastomas by 10 months of age (45, 172). LOH of *Ptc* in these tumors could not be detected, and full-length *Ptc* transcripts were present, suggesting that haploinsufficiency of Ptc can promote oncogenesis in the murine CNS. Mutation of *Ptc* cooperates with p53 deficiency since an increased incidence (95%) of medulloblastoma is seen in $Ptc^{+/-}p53^{-/-}$ mice (171).

FHIT

The *FHIT* (fragile histidine triad) gene is located at the FRA3B fragile site of chromosome 3p14.2 (reviewed in 62). This gene, which encodes a nucleoside hydrolase, is altered by deletion or translocation in various cancers, including cervical, lung, pancreatic, and gastric tumors. Although FHIT protein is absent or decreased in many of these tumors, it is still debated whether *FHIT* is a typical TSG. The mechanism of growth suppression by FHIT is not yet fully understood.

Fhit MOUSE MODEL   Gene-targeted $Fhit^{+/-}$ mice exhibit increased susceptibility to carcinogen-induced visceral and skin tumors (39). The remaining wild-type *Fhit* allele is inactivated in these tumors as demonstrated by the absence of Fhit protein expression. $Fhit^{-/-}$ mice are viable, but their characterization and cancer predisposition have not yet been reported.

CONCLUSION

Mouse models for TSGs, particularly models generated by gene targeting, offer the advantage of studying a genetically modified animal bearing one or only a few mutations. The use of these animals has revealed previously unsuspected developmental roles for TSGs such as Rb, Pten, Brca1, and Brca2, and has advanced our overall understanding of tumor-suppressor functions. Mouse models for TSGs have also facilitated the identification and characterization of cellular pathways controlled by these genes. This knowledge will be invaluable in choosing a pathway or molecule to target for therapy of a specific cancer.

How well do these mouse models represent the human situation? Most TSG mouse models exhibit some degree of predisposition to cancer development. A closer relationship between the murine and human situations may be precluded by several factors. Tumorigenesis associated with homozygous, but not heterozygous, mutation of a mouse TSG strongly suggests that LOH and inactivation of a wild-type allele are rare and limiting events. In addition, the mouse lifespan may be too short to allow the inactivation of the wild-type allele followed by the accumulation of other genetic hits necessary for tumor development. The frequent thymic lymphomas in the mice model for TSG might preclude these mice from developing other tumors owing to their early death associated with lymphoma development. Finally, different genetic backgrounds give rise to different cancer spectra associated with a given TSG mutation. It should not, therefore, be surprising that phenotypic differences exist between human diseases and their mouse models.

Even though the mouse models covered in this review do not always reproduce all the human pathologies associated with mutation of a specific TSG, they are still extremely useful. The engineering of mutant mice with combinations of mutated TSGs and oncogenes can provide powerful systems for identifying synergistic effects of specific mutations on tumorigenesis. Where the embryonic lethality of a mutation hampers the study of gene function in vivo, conditional mutants can be generated in which deletion or mutation of a TSG is induced at a time or in a tissue of choice. In the future, mouse models in which mutation of both alleles of a TSG is induced in a very limited number of somatic cells will replicate sporadic human cancers.

Mouse models have already had a major effect on cancer research. The sequencing of the human genome will no doubt lead to the identification of still more TSGs and oncogenes, and thus even more sophisticated mouse models. The study of such models should ultimately be of great benefit to human cancer patients.

ACKNOWLEDGMENTS

We thank our colleagues for reviewing the manuscript, and M. Saunders for scientific editing. We also apologize to those whose work was not cited directly owing to space limitations.

Visit the Annual Reviews home page at www.AnnualReviews.org

LITERATURE CITED

1. Adams PD, Kaelin WG. 1996. The cellular effects of E2F overexpression. *Curr. Top. Microbiol. Immunol.* 208:79–93
2. Almasan A, Yin Y, Kelly RE, Lee EY, Bradley A, et al. 1995. Deficiency of retinoblastoma protein leads to inappropriate S-phase entry, activation of E2F-responsive genes, and apoptosis. *Proc. Natl. Acad. Sci. USA* 92:5436–40
3. Armstrong JF, Kaufman MH, Harrison DJ, Clarke AR. 1995. High-frequency developmental abnormalities in p53-deficient mice. *Curr. Biol.* 5:931–36
4. Barlow C, Brown KD, Deng CX, Tagle DA, Wynshaw-Boris A. 1997. Atm selectively regulates distinct p53-dependent cell-cycle checkpoint and apoptotic pathways. *Nat. Genet.* 17:453–56
5. Barlow C, Hirotsune S, Paylor R, Liyanage M, Eckhaus M, et al. 1996. Atm-deficient mice: a paradigm of ataxia telangiectasia. *Cell* 86:159–71
6. Barlow C, Liyanage M, Moens PB, Deng CX, Ried T, Wynshaw-Boris A. 1997. Partial rescue of the prophase I defects of Atm-deficient mice by p53 and p21 null alleles. *Nat. Genet.* 17:462–66
7. Bates S, Phillips AC, Clark PA, Stott F, Peters G, et al. 1998. p14ARF links the tumour suppressors RB and p53. *Nature* 395:124–25
8. Bell DW, Varley JM, Szydlo TE, Kang DH, Wahrer DC, et al. 1999. Heterozygous germ line hCHK2 mutations in Li-Fraumeni syndrome. *Science* 286:2528–31
9. Bignon YJ, Chen Y, Chang CY, Riley DJ, Windle JJ, et al. 1993. Expression of a retinoblastoma transgene results in dwarf mice. *Genes Dev.* 7:1654–62
10. Birnbaum RA, O'Marcaigh A, Wardak Z, Zhang YY, Dranoff G, et al. 2000. Nf1 and Gmcsf interact in myeloid leukemogenesis. *Mol. Cell* 5:189–95
11. Blyth K, Terry A, O'Hara M, Baxter EW, Campbell M, et al. 1995. Synergy between a human c-myc transgene and p53 null genotype in murine thymic lymphomas: contrasting effects of homozygous and heterozygous p53 loss. *Oncogene* 10:1717–23
12. Brannan CI, Perkins AS, Vogel KS, Ratner N, Nordlund ML, et al. 1994. Targeted disruption of the neurofibromatosis type-1 gene leads to developmental abnormalities in heart and various neural crest-derived tissues. *Genes Dev.* 8:1019–29
13. Brown AL, Lee CH, Schwarz JK, Mitiku N, Piwnica-Worms H, et al. 1999. A human Cds1-related kinase that functions downstream of ATM protein in the cellular response to DNA damage. *Proc. Natl. Acad. Sci. USA* 96:3745–50
14. Brown EJ, Baltimore D. 2000. ATR disruption leads to chromosomal fragmentation and early embryonic lethality. *Genes Dev.* 14:397–402

15. Cecconi F, Alvarez-Bolado G, Meyer BI, Roth KA, Gruss P. 1998. Apaf1 (CED-4 homolog) regulates programmed cell death in mammalian development. *Cell* 94:727–37

16. Chan TA, Hwang PM, Hermeking H, Kinzler KW, Vogelstein B. 2000. Cooperative effects of genes controlling the G(2)/M checkpoint. *Genes Dev.* 14:1584–88

17. Chandler J, Hohenstein P, Swing D, Tessarollo L, Sharan S, et al. 2001. Human BRCA1 gene rescues the embryonic lethality of Brca1 mutant mice. *Genesis* 29:72–77

18. Chao C, Saito S, Kang J, Anderson CW, Appella E, Xu Y. 2000. p53 transcriptional activity is essential for p53-dependent apoptosis following DNA damage. *EMBO J.* 19:4967–75

19. Chen J, Silver DP, Walpita D, Cantor SB, Gazdar AF, et al. 1998. Stable interaction between the products of the BRCA1 and BRCA2 tumor suppressor genes in mitotic and meiotic cells. *Mol. Cell* 2:317–28

20. Chin L, Pomerantz J, Polsky D, Jacobson M, Cohen C, et al. 1997. Cooperative effects of INK4a and ras in melanoma susceptibility in vivo. *Genes Dev.* 11:2822–34

21. Cichowski K, Jacks T. 2001. NF1 tumor suppressor gene function: narrowing the GAP. *Cell* 104:593–604

22. Cichowski K, Shih TS, Schmitt E, Santiago S, Reilly K, et al. 1999. Mouse models of tumor development in neurofibromatosis type 1. *Science* 286:2172–76

23. Clarke AR, Maandag ER, van Roon M, van der Lugt NM, van der Valk M, et al. 1992. Requirement for a functional Rb-1 gene in murine development. *Nature* 359:328–30

24. Clarke AR, Purdie CA, Harrison DJ, Morris RG, Bird CC, et al. 1993. Thymocyte apoptosis induced by p53-dependent and independent pathways. *Nature* 362:849–52

25. Cobrinik D, Lee MH, Hannon G, Mulligan G, Bronson RT, et al. 1996. Shared role of the pRB-related p130 and p107 proteins in limb development. *Genes Dev.* 10:1633–44

26. Connor F, Bertwistle D, Mee PJ, Ross GM, Swift S, et al. 1997. Tumorigenesis and a DNA repair defect in mice with a truncating Brca2 mutation. *Nat. Genet.* 17:423–30

27. Cross SM, Sanchez CA, Morgan CA, Schimke MK, Ramel S, et al. 1995. A p53-dependent mouse spindle checkpoint. *Science* 267:1353–56

28. Dannenberg JH, van Rossum A, Schuijff L, te Riele H. 2000. Ablation of the retinoblastoma gene family deregulates G(1) control causing immortalization and increased cell turnover under growth-restricting conditions. *Genes Dev.* 14:3051–64

29. de Klein A, Muijtjens M, van Os R, Verhoeven Y, Smit B, et al. 2000. Targeted disruption of the cell-cycle checkpoint gene ATR leads to early embryonic lethality in mice. *Curr. Biol.* 10:479–82

30. Dean M. 1997. Towards a unified model of tumor suppression: lessons learned from the human patched gene. *Biochim. Biophys. Acta* 1332:M43–52

31. Deng CX, Brodie SG. 2000. Roles of BRCA1 and its interacting proteins. *BioEssays* 22:728–37

32. Di Cristofano A, De Acetis M, Koff A, Cordon-Cardo C, Pandolfi PP. 2001. Pten and p27KIP1 cooperate in prostate cancer tumor suppression in the mouse. *Nat. Genet.* 27:222–24

33. Di Cristofano A, Kotsi P, Peng YF, Cordon-Cardo C, Elkon KB, Pandolfi PP. 1999. Impaired Fas response and autoimmunity in Pten+/− mice. *Science* 285:2122–25

34. Di Cristofano A, Pesce B, Cordon-Cardo C, Pandolfi PP. 1998. Pten is essential for embryonic development and tumour suppression. *Nat. Genet.* 19:348–55

35. Donehower LA, Harvey M, Slagle BL, McArthur MJ, Montgomery CA, et al.

1992. Mice deficient for p53 are developmentally normal but susceptible to spontaneous tumours. *Nature* 356:215–21

36. Elson A, Wang Y, Daugherty CJ, Morton CC, Zhou F, Campos-Torres J, Leder P. 1996. Pleiotropic defects in ataxia-telangiectasia protein-deficient mice. *Proc. Natl. Acad. Sci. USA* 93: 13084–89

37. Fero ML, Randel E, Gurley KE, Roberts JM, Kemp CJ. 1998. The murine gene p27Kip1 is haplo-insufficient for tumour suppression. *Nature* 396:177–80

38. Fodde R, Edelmann W, Yang K, van Leeuwen C, Carlson C, et al. 1994. A targeted chain-termination mutation in the mouse Apc gene results in multiple intestinal tumors. *Proc. Natl. Acad. Sci. USA* 91:8969–73

39. Fong LY, Fidanza V, Zanesi N, Lock LF, Siracusa LD, et al. 2000. Muir-Torre-like syndrome in Fhit-deficient mice. *Proc. Natl. Acad. Sci. USA* 97:4742–47

40. Franklin DS, Godfrey VL, Lee H, Kovalev GI, Schoonhoven R, et al. 1998. CDK inhibitors p18(INK4c) and p27(Kip1) mediate two separate pathways to collaboratively suppress pituitary tumorigenesis. *Genes Dev.* 12:2899–911

41. Friedman LS, Thistlethwaite FC, Patel KJ, Yu VP, Lee H, et al. 1998. Thymic lymphomas in mice with a truncating mutation in Brca2. *Cancer Res.* 58:1338–43

42. Giovannini M, Robanus-Maandag E, Niwa-Kawakita M, van der Valk M, Woodruff JM, et al. 1999. Schwann cell hyperplasia and tumors in transgenic mice expressing a naturally occurring mutant NF2 protein. *Genes Dev.* 13:978–86

43. Giovannini M, Robanus-Maandag E, van der Valk M, Niwa-Kawakita M, Abramowski V, et al. 2000. Conditional biallelic Nf2 mutation in the mouse promotes manifestations of human neurofibromatosis type 2. *Genes Dev.* 14:1617–30

44. Gnarra JR, Ward JM, Porter FD, Wagner JR, Devor DE, et al. 1997. Defective placental vasculogenesis causes embryonic lethality in VHL-deficient mice. *Proc. Natl. Acad. Sci. USA* 94:9102–7

45. Goodrich LV, Milenkovic L, Higgins KM, Scott MP. 1997. Altered neural cell fates and medulloblastoma in mouse patched mutants. *Science* 277:1109–13

46. Gowen LC, Avrutskaya AV, Latour AM, Koller BH, Leadon SA. 1998. BRCA1 required for transcription-coupled repair of oxidative DNA damage. *Science* 281:1009–12

47. Gowen LC, Johnson BL, Latour AM, Sulik KK, Koller BH. 1996. Brca1 deficiency results in early embryonic lethality characterized by neuroepithelial abnormalities. *Nat. Genet.* 12:191–94

48. Gusella JF, Ramesh V, MacCollin M, Jacoby LB. 1999. Merlin: the neurofibromatosis 2 tumor suppressor. *Biochim. Biophys. Acta* 1423:M29–36

49. Haase VH, Glickman JN, Socolovsky M, Jaenisch R. 2001. Vascular tumors in livers with targeted inactivation of the von Hippel-Lindau tumor suppressor. *Proc. Natl. Acad. Sci. USA* 98:1583–88

50. Hakem R, de la Pompa JL, Elia A, Potter J, Mak TW. 1997. Partial rescue of Brca1 (5-6) early embryonic lethality by p53 or p21 null mutation. *Nat. Genet.* 16:298–302

51. Hakem R, de la Pompa JL, Mak TW. 1998. Developmental studies of Brca1 and Brca2 knock-out mice. *J. Mammary Gland Biol. Neoplasia* 3:431–45

52. Hakem R, de la Pompa JL, Sirard C, Mo R, Woo M, et al. 1996. The tumor suppressor gene Brca1 is required for embryonic cellular proliferation in the mouse. *Cell* 85:1009–23

53. Halberg RB, Katzung DS, Hoff PD, Moser AR, Cole CE, et al. 2000. Tumorigenesis in the multiple intestinal neoplasia mouse: redundancy of negative regulators and specificity of modifiers. *Proc. Natl. Acad. Sci. USA* 97:3461–66

54. Harbour JW, Dean DC. 2000. The Rb/E2F pathway: expanding roles and emerging paradigms. *Genes Dev.* 14:2393–409

55. Harfe BD, Jinks-Robertson S. 2000. DNA mismatch repair and genetic instability. *Annu. Rev. Genet.* 34:359–99

56. Harvey M, McArthur MJ, Montgomery CA, Butel JS, Bradley A, Donehower LA. 1993. Spontaneous and carcinogen-induced tumorigenesis in p53-deficient mice. *Nat. Genet.* 5:225–29

57. Harvey M, Sands AT, Weiss RS, Hegi ME, Wiseman RW, et al. 1993. In vitro growth characteristics of embryo fibroblasts isolated from p53-deficient mice. *Oncogene* 8:2457–67

58. Harvey M, Vogel H, Lee EY, Bradley A, Donehower LA. 1995. Mice deficient in both p53 and Rb develop tumors primarily of endocrine origin. *Cancer Res.* 55:1146–51

59. Herzer U, Crocoll A, Barton D, Howells N, Englert C. 1999. The Wilms tumor suppressor gene wt1 is required for development of the spleen. *Curr. Biol.* 9:837–40

60. Hirao A, Kong YY, Matsuoka S, Wakeham A, Ruland J, et al. 2000. DNA damage-induced activation of p53 by the checkpoint kinase Chk2. *Science* 287:1824–27

61. Hollander MC, Sheikh MS, Bulavin DV, Lundgren K, Augeri-Henmueller L, et al. 1999. Genomic instability in Gadd45a-deficient mice. *Nat. Genet.* 23:176–84

62. Huebner K, Garrison PN, Barnes LD, Croce CM. 1998. The role of the FHIT/FRA3B locus in cancer. *Annu. Rev. Genet.* 32:7–31

63. Inoue K, Roussel MF, Sherr CJ. 1999. Induction of ARF tumor suppressor gene expression and cell cycle arrest by transcription factor DMP1. *Proc. Natl. Acad. Sci. USA* 96:3993–98

64. Irwin M, Marin MC, Phillips AC, Seelan RS, Smith DI, et al. 2000. Role for the p53 homologue p73 in E2F-1-induced apoptosis. *Nature* 407:645–48

65. Jacks T, Fazeli A, Schmitt EM, Bronson RT, Goodell MA, Weinberg RA. 1992. Effects of an Rb mutation in the mouse. *Nature* 359:295–300

66. Jacks T, Remington L, Williams BO, Schmitt EM, Halachmi S, et al. 1994. Tumor spectrum analysis in p53-mutant mice. *Curr. Biol.* 4:1–7

67. Jacks T, Shih TS, Schmitt EM, Bronson RT, Bernards A, Weinberg RA. 1994. Tumour predisposition in mice heterozygous for a targeted mutation in Nf1. *Nat. Genet.* 7:353–61

68. Jacobs JJ, Keblusek P, Robanus-Maandag E, Kristel P, Lingbeek M, et al. 2000. Senescence bypass screen identifies TBX2 which represses Cdkn2a (p19(ARF)) and is amplified in a subset of human breast cancers. *Nat. Genet.* 26:291–99

69. Jacobs JJ, Kieboom K, Marino S, DePinho RA, van Lohuizen M. 1999. The oncogene and Polycomb-group gene bmi-1 regulates cell proliferation and senescence through the ink4a locus. *Nature* 397:164–68

70. Jimenez GS, Nister M, Stommel JM, Beeche M, Barcarse EA, et al. 2000. A transactivation-deficient mouse model provides insights into Trp53 regulation and function. *Nat. Genet.* 26:37–43

71. Kaelin WG, Maherr R. 1998. The VHL tumour-suppressor gene paradigm. *Trends Genet.* 14:423–26

72. Kamijo T, Bodner S, van de Kamp E, Randle DH, Sherr CJ. 1999. Tumor spectrum in ARF-deficient mice. *Cancer Res.* 59:2217–22

73. Kamijo T, Weber JD, Zambetti G, Zindy F, Roussel MF, Sherr CJ. 1998. Functional and physical interactions of the ARF tumor suppressor with p53 and Mdm2. *Proc. Natl. Acad. Sci. USA* 95:8292–97

74. Kamijo T, Zindy F, Roussel MF, Quelle DE, Downing JR, et al. 1997. Tumor suppression at the mouse INK4a locus mediated by the alternative reading frame product p19ARF. *Cell* 91:649–59

75. Kemp CJ, Wheldon T, Balmain A. 1994. p53-deficient mice are extremely susceptible to radiation-induced tumorigenesis. *Nat. Genet.* 8:66–69

76. Khosravi R, Maya R, Gottlieb T, Oren M, Shiloh Y, Shkedy D. 1999. Rapid ATM-dependent phosphorylation of MDM2 precedes p53 accumulation in response to DNA damage. *Proc. Natl. Acad. Sci. USA* 96:14973–77

77. Kinzler KW, Vogelstein B. 1997. Cancer-susceptibility genes. Gatekeepers and caretakers. *Nature* 386:761–63

78. Knudson AG. 2000. Chasing the cancer demon. *Annu. Rev. Genet.* 34:1–19

79. Kreidberg JA, Sariola H, Loring JM, Maeda M, Pelletier J, et al. 1993. WT-1 is required for early kidney development. *Cell* 74:679–91

80. Kuperwasser C, Hurlbut GD, Kittrell FS, Dickinson ES, Laucirica R, et al. 2000. Development of spontaneous mammary tumors in BALB/c p53 heterozygous mice. A model for Li-Fraumeni syndrome. *Am. J. Pathol.* 157:2151–59

81. Lane TF, Lin C, Brown MA, Solomon E, Leder P. 2000. Gene replacement with the human BRCA1 locus: tissue specific expression and rescue of embryonic lethality in mice. *Oncogene* 19:4085–90

82. Latres E, Malumbres M, Sotillo R, Martin J, Ortega S, et al. 2000. Limited overlapping roles of P15(INK4b) and P18(INK4c) cell cycle inhibitors in proliferation and tumorigenesis. *EMBO J.* 19:3496–506

83. Lavigueur A, Bernstein A. 1991. p53 transgenic mice: accelerated erythroleukemia induction by Friend virus. *Oncogene* 6:2197–201

84. Lavigueur A, Maltby V, Mock D, Rossant J, Pawson T, Bernstein A. 1989. High incidence of lung, bone, and lymphoid tumors in transgenic mice overexpressing mutant alleles of the p53 oncogene. *Mol. Cell. Biol.* 9:3982–91

85. Lee EY, Chang CY, Hu N, Wang YC, Lai CC, et al. 1992. Mice deficient for Rb are nonviable and show defects in neurogenesis and haematopoiesis. *Nature* 359:288–94

86. Lee MH, Williams BO, Mulligan G, Mukai S, Bronson RT, et al. 1996. Targeted disruption of p107: functional overlap between p107 and Rb. *Genes Dev.* 10:1621–32

87. Lee SB, Huang K, Palmer R, Truong VB, Herzlinger D, et al. 1999. The Wilms tumor suppressor WT1 encodes a transcriptional activator of amphiregulin. *Cell* 98:663–73

88. Li Y, Nichols MA, Shay JW, Xiong Y. 1994. Transcriptional repression of the D-type cyclin-dependent kinase inhibitor p16 by the retinoblastoma susceptibility gene product pRb. *Cancer Res.* 54:6078–82

89. Liao MJ, Van Dyke T. 1999. Critical role for Atm in suppressing V(D)J recombination-driven thymic lymphoma. *Genes Dev.* 13:1246–50

90. Lissy NA, Davis PK, Irwin M, Kaelin WG, Dowdy SF. 2000. A common E2F-1 and p73 pathway mediates cell death induced by TCR activation. *Nature* 407:642–45

91. Little M, Holmes G, Walsh P. 1999. WT1: What has the last decade told us? *BioEssays* 21:191–202

92. Liu CY, Flesken-Nikitin A, Li S, Zeng Y, Lee WH. 1996. Inactivation of the mouse Brca1 gene leads to failure in the morphogenesis of the egg cylinder in early postimplantation development. *Genes Dev.* 10:1835–43

93. Liu G, McDonnell TJ, Montes de Oca Luna R, Kapoor Mims B, et al. 2000. High metastatic potential in mice inheriting a targeted p53 missense mutation. *Proc. Natl. Acad. Sci. USA* 97:4174–79

94. Lowe SW, Schmitt EM, Smith SW, Osborne BA, Jacks T. 1993. p53 is required for radiation-induced apoptosis in mouse thymocytes. *Nature* 362:847–49

95. Ludwig T, Chapman DL, Papaioannou VE, Efstratiadis A. 1997. Targeted

mutations of breast cancer susceptibility gene homologs in mice: lethal phenotypes of Brca1, Brca2, Brca1/Brca2, Brca1/p53, and Brca2/p53 nullizygous embryos. *Genes Dev.* 11:1226–41

96. Luo G, Santoro IM, McDaniel LD, Nishijima I, Mills M, et al. 2000. Cancer predisposition caused by elevated mitotic recombination in Bloom mice. *Nat. Genet.* 26:424–29

97. Maandag EC, van der Valk M, Vlaar M, Feltkamp C, O'Brien J, et al. 1994. Developmental rescue of an embryonic-lethal mutation in the retinoblastoma gene in chimeric mice. *EMBO J.* 13:4260–68

98. Maestro R, Dei Tos AP, Hamamori Y, Krasnokutsky S, Sartorelli V, et al. 1999. Twist is a potential oncogene that inhibits apoptosis. *Genes Dev.* 13:2207–17

99. Mak TW, Hakem A, McPherson JP, Shehabeldin A, Zablocki E, et al. 2000. Brca1 is required for T cell lineage development but not TCR loci rearrangement. *Nat. Immunol.* 1:77–82

100. Malkin D. 1993. p53 and the Li-Fraumeni syndrome. *Cancer Genet. Cytogenet.* 66: 83–92

101. Marino S, Vooijs M, van Der Gulden H, Jonkers J, Berns A. 2000. Induction of medulloblastomas in p53–null mutant mice by somatic inactivation of Rb in the external granular layer cells of the cerebellum. *Genes Dev.* 14:994–1004

102. McClatchey AI, Saotome I, Mercer K, Crowley D, Gusella JF, et al. 1998. Mice heterozygous for a mutation at the Nf2 tumor suppressor locus develop a range of highly metastatic tumors. *Genes Dev.* 12:1121–33

103. McClatchey AI, Saotome I, Ramesh V, Gusella JF, Jacks T. 1997. The Nf2 tumor suppressor gene product is essential for extraembryonic development immediately prior to gastrulation. *Genes Dev.* 11:1253–65

104. Mills AA, Zheng B, Wang XJ, Vogel H, Roop DR, Bradley A. 1999. p63 is a p53 homologue required for limb and epidermal morphogenesis. *Nature* 398:708–13

105. Moore AW, McInnes L, Kreidberg J, Hastie ND, Schedl A. 1999. YAC complementation shows a requirement for Wt1 in the development of epicardium, adrenal gland and throughout nephrogenesis. *Development* 126:1845–57

106. Morgenbesser SD, Williams BO, Jacks T, DePinho RA. 1994. p53-dependent apoptosis produced by Rb-deficiency in the developing mouse lens. *Nature* 371:72–74

107. Moser AR, Pitot HC, Dove WF. 1990. A dominant mutation that predisposes to multiple intestinal neoplasia in the mouse. *Science* 247:322–24

108. Moser AR, Shoemaker AR, Connelly CS, Clipson L, Gould KA, et al. 1995. Homozygosity for the Min allele of Apc results in disruption of mouse development prior to gastrulation. *Dev. Dyn.* 203:422–33

109. Munemitsu S, Albert I, Souza B, Rubinfeld B, Polakis P. 1995. Regulation of intracellular beta-catenin levels by the adenomatous polyposis coli (APC) tumor-suppressor protein. *Proc. Natl. Acad. Sci. USA* 92:3046–50

110. Ohtani N, Zebedee Z, Huot TJ, Stinson JA, Sugimoto M, et al. 2001. Opposing effects of Ets and Id proteins on p16INK4a expression during cellular senescence. *Nature* 409:1067–70

111. Orlow I, Rabbani F, Chin L, Pomerantz J, Ligeois N, et al. 1999. Involvement of the Ink4a gene (p16 and p19arf) in murine tumorigenesis. *Int. J. Oncol.* 15:17–24

112. Oshima M, Dinchuk JE, Kargman SL, Oshima H, Hancock B, et al. 1996. Suppression of intestinal polyposis in Apc delta716 knockout mice by inhibition of cyclooxygenase 2 (COX-2). *Cell* 87:803–9

113. Oshima M, Oshima H, Kitagawa K, Kobayashi M, Itakura C, et al. 1995. Loss of Apc heterozygosity and abnormal

tissue building in nascent intestinal polyps in mice carrying a truncated Apc gene. *Proc. Natl. Acad. Sci. USA* 92:4482–86

114. Parada LF. 2000. Neurofibromatosis type 1. *Biochim. Biophys. Acta* 1471:M13–19

115. Patel KJ, Yu VP, Lee H, Corcoran A, Thistlethwaite FC, et al. 1998. Involvement of Brca2 in DNA repair. *Mol. Cell* 1:347–57

116. Petiniot LK, Weaver Z, Barlow C, Shen R, Eckhaus M, et al. 2000. Recombinase-activating gene (RAG) 2–mediated V(D)J recombination is not essential for tumorigenesis in Atm-deficient mice. *Proc. Natl. Acad. Sci. USA* 97:6664–69

117. Podsypanina K, Ellenson LH, Nemes A, Gu J, Tamura M, et al. 1999. Mutation of Pten/Mmac1 in mice causes neoplasia in multiple organ systems. *Proc. Natl. Acad. Sci. USA.* 96:1563–68

118. Polakis P. 2000. Wnt signaling and cancer. *Genes Dev.* 14:1837–51

119. Pomerantz J, Schreiber-Agus N, Liegeois NJ, Silverman A, Alland L, et al. 1998. The Ink4a tumor suppressor gene product, p19Arf, interacts with MDM2 and neutralizes MDM2's inhibition of p53. *Cell* 92:713–23

120. Pozniak CD, Radinovic S, Yang A, McKeon F, Kaplan DR, Miller FD. 2000. An anti-apoptotic role for the p53 family member, p73, during developmental neuron death. *Science* 289:304–6

121. Purdie CA, Harrison DJ, Peter A, Dobbie L, White S, et al. 1994. Tumour incidence, spectrum and ploidy in mice with a large deletion in the p53 gene. *Oncogene* 9:603–9

122. Rahman N, Stratton MR. 1998. The genetics of breast cancer susceptibility. *Annu. Rev. Genet.* 32:95–121

123. Reitmair AH, Cai JC, Bjerknes M, Redston M, Cheng H, et al. 1996. MSH2 deficiency contributes to accelerated APC-mediated intestinal tumorigenesis. *Cancer Res.* 56:2922–26

124. Robanus-Maandag E, Dekker M, van der

Valk M, Carrozza ML, Jeanny JC, et al. 1998. p107 is a suppressor of retinoblastoma development in pRb-deficient mice. *Genes Dev.* 12:1599–609

125. Roose J, Huls G, van Beest M, Moerer P, van der Horn K, et al. 1999. Synergy between tumor suppressor APC and the beta-catenin-Tcf4 target Tcf1. *Science* 285:1923–26

126. Ruas M, Peters G. 1998. The p16INK4a/CDKN2A tumor suppressor and its relatives. *Biochim. Biophys. Acta* 1378:F115–77

127. Sage J, Mulligan GJ, Attardi LD, Miller A, Chen S, et al. 2000. Targeted disruption of the three Rb-related genes leads to loss of G(1) control and immortalization. *Genes Dev.* 14:3037–50

128. Sah VP, Attardi LD, Mulligan GJ, Williams BO, Bronson RT, Jacks T. 1995. A subset of p53-deficient embryos exhibit exencephaly. *Nat. Genet.* 10:175–80

129. Deleted in proof

130. Schaffner C, Idler I, Stilgenbauer S, Dohner H, Lichter P. 2000. Mantle cell lymphoma is characterized by inactivation of the ATM gene. *Proc. Natl. Acad. Sci. USA* 97:2773–78

131. Schar P. 2001. Spontaneous DNA damage, genome instability, and cancer—when DNA replication escapes control. *Cell* 104:329–32

132. Scully R, Livingston DM. 2000. In search of the tumour-suppressor functions of BRCA1 and BRCA2. *Nature* 408:429–32

133. Serrano M, Hannon GJ, Beach D. 1993. A new regulatory motif in cell-cycle control causing specific inhibition of cyclin D/CDK4. *Nature* 366:704–7

134. Serrano M, Lee H, Chin L, Cordon-Cardo C, Beach D, DePinho RA. 1996. Role of the INK4a locus in tumor suppression and cell mortality. *Cell* 85:27–37

135. Sharan SK, Morimatsu M, Albrecht U, Lim DS, Regel E, et al. 1997. Embryonic lethality and radiation hypersensitivity mediated by Rad51 in mice lacking Brca2. *Nature* 386:804–10

136. Shen SX, Weaver Z, Xu X, Li C, Weinstein M, et al. 1998. A targeted disruption of the murine Brca1 gene causes gamma-irradiation hypersensitivity and genetic instability. *Oncogene* 17:3115–24

137. Sherr CJ. 2000. The Pezcoller lecture: cancer cell cycles revisited. *Cancer Res.* 60:3689–95

138. Shibata H, Toyama K, Shioya H, Ito M, Hirota M, et al. 1997. Rapid colorectal adenoma formation initiated by conditional targeting of the Apc gene. *Science* 278:120–23

139. Shiloh Y. 2001. ATM and ATR: networking cellular responses to DNA damage. *Curr. Opin. Genet. Dev.* 11:71–77

140. Shoemaker AR, Haigis KM, Baker SM, Dudley S, Liskay RM, Dove WF. 2000. Mlh1 deficiency enhances several phenotypes of Apc(Min)/+ mice. *Oncogene* 19:2774–79

141. Silva AJ, Frankland PW, Marowitz Z, Friedman E, Lazlo G, et al. 1997. A mouse model for the learning and memory deficits associated with neurofibromatosis type I. *Nat. Genet.* 15:281–84

142. Sirard C, de la Pompa JL, Elia A, Itie A, Mirtsos C, et al. 1998. The tumor suppressor gene Smad4/Dpc4 is required for gastrulation and later for anterior development of the mouse embryo. *Genes Dev.* 12:107–19

143. Smits R, Kielman MF, Breukel C, Zurcher C, Neufeld K, et al. 1999. Apc1638T: a mouse model delineating critical domains of the adenomatous polyposis coli protein involved in tumorigenesis and development. *Genes Dev.* 13:1309–21

144. Smits R, van der Houven van Oordt W, Luz A, Zurcher C, et al. 1998. Apc1638N: a mouse model for familial adenomatous polyposis-associated desmoid tumors and cutaneous cysts. *Gastroenterology* 114:275–83

145. Soengas MS, Capodieci P, Polsky D, Mora J, Esteller M, et al. 2001. Inactivation of the apoptosis effector Apaf-1 in malignant melanoma. *Nature* 409:207–11

146. St John MA, Tao W, Fei X, Fukumoto R, Carcangiu ML, et al. 1999. Mice deficient of Lats1 develop soft-tissue sarcomas, ovarian tumours and pituitary dysfunction. *Nat. Genet.* 21:182–86

147. Stambolic V, Mak TW, Woodgett JR. 1999. Modulation of cellular apoptotic potential: contributions to oncogenesis. *Oncogene* 18:6094–103

148. Stambolic V, Suzuki A, de la Pompa JL, Brothers GM, Mirtsos C, et al. 1998. Negative regulation of PKB/Akt-dependent cell survival by the tumor suppressor PTEN. *Cell* 95:29–39

149. Stambolic V, Tsao MS, Macpherson D, Suzuki A, Chapman WB, et al. 2000. High incidence of breast and endometrial neoplasia resembling human Cowden syndrome in pten+/− mice. *Cancer Res.* 60:3605–11

150. Stiewe T, Putzer BM. 2000. Role of the p53-homologue p73 in E2F1-induced apoptosis. *Nat. Genet.* 26:464–69

151. Stilgenbauer S, Schaffner C, Litterst A, Liebisch P, Gilad S, et al. 1997. Biallelic mutations in the ATM gene in T-prolymphocytic leukemia. *Nat. Med.* 3:1155–59

152. Stott FJ, Bates S, James MC, McConnell BB, Starborg M, et al. 1998. The alternative product from the human CDKN2A locus, p14(ARF), participates in a regulatory feedback loop with p53 and MDM2. *EMBO J.* 17:5001–14

153. Su LK, Kinzler KW, Vogelstein B, Preisinger AC, Moser AR, et al. 1992. Multiple intestinal neoplasia caused by a mutation in the murine homolog of the APC gene. *Science* 256:668–70

154. Sun H, Lesche R, Li DM, Liliental J, Zhang H, et al. 1999. PTEN modulates cell cycle progression and cell survival by regulating phosphatidylinositol 3,4,5,-trisphosphate and Akt/protein kinase B signaling pathway. *Proc. Natl. Acad. Sci. USA* 96:6199–204

155. Suzuki A, de la Pompa JL, Hakem R, Elia A, Yoshida R, et al. 1997. Brca2 is

required for embryonic cellular proliferation in the mouse. *Genes Dev.* 11:1242–52

156. Suzuki A, de la Pompa JL, Stambolic V, Elia AJ, Sasaki T, et al. 1998. High cancer susceptibility and embryonic lethality associated with mutation of the PTEN tumor suppressor gene in mice. *Curr. Biol.* 8:1169–78

157. Suzuki A, Yamaguchi MT, Ohteki T, Sasaki T, Kaisho T, et al. 2001. T cell specific loss of Pten leads to defects in central and peripheral tolerance. *Immunity* 14:523–34

158. Takaku K, Oshima M, Miyoshi H, Matsui M, Seldin MF, Taketo MM. 1998. Intestinal tumorigenesis in compound mutant mice of both Dpc4 (Smad4) and Apc genes. *Cell* 92:645–56

159. Tam SW, Shay JW, Pagano M. 1994. Differential expression and cell cycle regulation of the cyclin-dependent kinase 4 inhibitor p16Ink4. *Cancer Res.* 54:5816–20

160. Tao W, Zhang S, Turenchalk GS, Stewart RA, St John MA, et al. 1999. Human homologue of the *Drosophila melanogaster* lats tumour suppressor modulates CDC2 activity. *Nat. Genet.* 21:177–81

161. Teitz T, Wei T, Valentine MB, Vanin EF, Grenet J, et al. 2000. Caspase 8 is deleted or silenced preferentially in childhood neuroblastomas with amplification of MYCN. *Nat. Med.* 6:529–35

162. Tibbetts RS, Cortez D, Brumbaugh KM, Scully R, Livingston D, et al. 2000. Functional interactions between BRCA1 and the checkpoint kinase ATR during genotoxic stress. *Genes Dev.* 14:2989–3002

163. Tsai KY, Hu Y, Macleod KF, Crowley D, Yamasaki L, Jacks T. 1998. Mutation of E2f-1 suppresses apoptosis and inappropriate S phase entry and extends survival of Rb-deficient mouse embryos. *Mol. Cell* 2:293–304

164. Tsukada T, Tomooka Y, Takai S, Ueda Y, Nishikawa S, et al. 1993. Enhanced proliferative potential in culture of cells from p53-deficient mice. *Oncogene* 8:3313–22

165. Tutt A, Gabriel A, Bertwistle D, Connor F, Paterson H, et al. 1999. Absence of Brca2 causes genome instability by chromosome breakage and loss associated with centrosome amplification. *Curr. Biol.* 9:1107–10

166. Varfolomeev EE, Schuchmann M, Luria V, Chiannilkulchai N, Beckmann JS, et al. 1998. Targeted disruption of the mouse Caspase 8 gene ablates cell death induction by the TNF receptors, Fas/Apo1, and DR3 and is lethal prenatally. *Immunity* 9:267–76

167. Vogel KS, Klesse LJ, Velasco-Miguel S, Meyers K, Rushing EJ, et al. 1999. Mouse tumor model for neurofibromatosis type 1. *Science* 286:2176–79

168. Vogelstein B, Lane D, Leviner J. 2000. Surfing the p53 network. *Nature* 408:307–10

169. Vorechovsky I, Luo L, Dyer MJ, Catovsky D, Amlot PL, et al. 1997. Clustering of missense mutations in the ataxia-telangiectasia gene in a sporadic T-cell leukaemia. *Nat. Genet.* 17:96–99

170. Westphal CH, Rowan S, Schmaltz C, Elson A, Fisher DE, Leder P. 1997. atm and p53 cooperate in apoptosis and suppression of tumorigenesis, but not in resistance to acute radiation toxicity. *Nat. Genet.* 16:397–401

171. Wetmore C, Eberhart DE, Curran T. 2001. Loss of p53 but not ARF accelerates medulloblastoma in mice heterozygous for patched. *Cancer Res.* 61:513–16

172. Wetmore C, Eberhart DE, Curran T. 2000. The normal patched allele is expressed in medulloblastomas from mice with heterozygous germ-line mutation of patched. *Cancer Res.* 60:2239–46

173. Williams BO, Schmitt EM, Remington L, Bronson RT, Albert DM, et al. 1994. Extensive contribution of Rb-deficient cells to adult chimeric mice with limited histopathological consequences. *EMBO J.* 13:4251–59

174. Xu X, Wagner KU, Larson D, Weaver

Z, Li C, et al. 1999. Conditional mutation of Brca1 in mammary epithelial cells results in blunted ductal morphogenesis and tumour formation. *Nat. Genet.* 22:37–43

175. Xu Y, Ashley T, Brainerd EE, Bronson RT, Meyn MS, Baltimore D. 1996. Targeted disruption of ATM leads to growth retardation, chromosomal fragmentation during meiosis, immune defects, and thymic lymphoma. *Genes Dev.* 10:2411–22

176. Yang A, Schweitzer R, Sun D, Kaghad M, Walker N, et al. 1999. p63 is essential for regenerative proliferation in limb, craniofacial and epithelial development. *Nature* 398:714–18

177. Yang A, Walker N, Bronson R, Kaghad M, Oosterwegel M, et al. 2000. p73-deficient mice have neurological, pheromonal and inflammatory defects but lack spontaneous tumours. *Nature* 404:99–103

178. Yang X, Li C, Xu X, Deng C. 1998. The tumor suppressor SMAD4/DPC4 is essential for epiblast proliferation and mesoderm induction in mice. *Proc. Natl. Acad. Sci. USA* 95:3667–72

179. Yoshida H, Kong YY, Yoshida R, Elia AJ, Hakem A, et al. 1998. Apaf1 is required for mitochondrial pathways of apoptosis and brain development. *Cell* 94:739–50

180. Yu VP, Koehler M, Steinlein C, Schmid M, Hanakahi LA, et al. 2000. Gross chromosomal rearrangements and genetic exchange between nonhomologous chromosomes following BRCA2 inactivation. *Genes Dev.* 14:1400–6

181. Zhang Y, Xiong Y, Yarbrough WG. 1998. ARF promotes MDM2 degradation and stabilizes p53: ARF-INK4a locus deletion impairs both the Rb and p53 tumor suppression pathways. *Cell* 92:725–34

182. Zindy F, Eischen CM, Randle DH, Kamijo T, Cleveland JL, et al. 1998. Myc signaling via the ARF tumor suppressor regulates p53-dependent apoptosis and immortalization. *Genes Dev.* 12:2424–33

Annu. Rev. Genet. 2001. 35:243–74

HOMOLOGOUS RECOMBINATION NEAR AND FAR FROM DNA BREAKS: Alternative Roles and Contrasting Views

Gerald R. Smith

Fred Hutchinson Cancer Research Center, 1100 Fairview Avenue North, Seattle, Washington 98109; e-mail: gsmith@fhcrc.org

Key Words   recombination hotspots, meiosis, RecBCD enzyme, Chi, heteroduplex DNA, DNA replication, enzyme induction

■ **Abstract**   Double-strand breaks and other lesions in DNA can stimulate homologous genetic recombination in two quite different ways: by promoting recombination near the break (roughly within a kb) or far from the break. Recent emphasis on the repair aspect of recombination has focused attention on DNA interactions and recombination near breaks. Here I review evidence for recombination far from DNA breaks in bacteria and fungi and discuss mechanisms by which this can occur. These mechanisms include entry of a traveling entity ("recombination machine") at a break, formation of long heteroduplex DNA, priming of DNA replication by a broken end, and induction of recombination potential in *trans*. Special emphasis is placed on contrasting views of how the RecBCD enzyme of *Escherichia coli* promotes recombination far (tens of kb) from a double-strand break. The occurrence of recombination far from DNA breaks and of correlated recombination events far apart suggests that "action at a distance" during recombination is a widespread feature among diverse organisms.

CONTENTS

0066-4197/01/1215-0243$14.00

INTRODUCTION

It has long been recognized that homologous genetic recombination has at least two roles—reassortment of alleles and repair of broken DNA (e.g., see 16, 54, 82). A currently popular view holds that the repair of double-strand (ds) breaks in DNA is the primary role of homologous recombination and that the generation of diversity is a byproduct of repair (e.g., see 25). This emphasis on repair of ds breaks has focused attention on recombination occurring near the ends of broken DNA, typically within ~1 kb. Although recombination does frequently occur within such a limited region, recombination can also occur far (>30 kb) from DNA ends or the presumptive initiating lesion.

This review discusses two examples of recombination initiated by ds breaks and extensively studied at the molecular level. In the first example, meiotic recombination in the budding yeast *Saccharomyces cerevisiae*, the ends of the broken DNA are processed and invade an intact homologous duplex: resolution of the joint molecule and therefore the recombinational exchanges often occur within ~1 kb of the initial ds break. In the second example, recombination by the RecBCD pathway of the bacterium *Escherichia coli*, the ends of the broken DNA serve as entry sites for an enzyme that can travel tens of kb before initiating joint molecule formation; recombination therefore typically occurs far from the initial ds break.

There are additional ways in which recombination can occur far from a ds break or another initiating lesion. (*a*) The hybrid DNA of a joint molecule formed near a break can be extended by branch migration; resolution of the joint molecule and repair of base mismatches within it can produce recombinational exchanges separated from the initial break by tens of kb. (*b*) The end of a broken DNA molecule can invade an intact duplex and prime DNA synthesis, which may proceed to the end of the chromosome. In this case the recombinational exchange point is at the site of the initial ds break, but all markers in the replicated region, potentially tens or hundreds of kb, undergo gene conversion, or nonreciprocal recombination. (*c*) A ds break can activate or induce the synthesis of recombination-promoting enzymes, potentially acting throughout the genome. Distant recombination events are expected to be statistically correlated in such "hot cells." After discussing the two well-studied cases of recombination near and far from ds breaks, I review additional data that appear to reflect each of these mechanisms.

RECOMBINATION NEAR DOUBLE-STRAND BREAKS IN *S. CEREVISIAE* MEIOSIS

Meiotic recombination in *S. cerevisiae*, extensively studied at the genetic and molecular levels, provides a clear example of recombination occurring near ds breaks, although there appear to be exceptions noted later. Most meiotic recombination in *S. cerevisiae* can be accounted for by ds breaks that occur at hotspots of gene conversion (reviewed in 65). In brief, recombination at these hotspots occurs as follows (Figure 1) [for more extensive discussion and references, see (79, 83)].

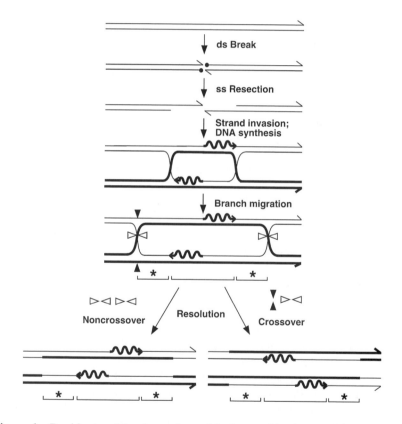

Figure 1 Double-strand break repair model of recombination, after Szostak et al. (112). For explanation, see the text. Alternative resolutions at the last step are not shown. Arrowheads indicate 3′ ends; solid dots, Spo11 protein linked to 5′ ends at the DNA break (56); wavy arrows, newly synthesized DNA; thin lines, DNA from one parent; thick lines, DNA from the other parent; open and closed triangles, points of resolution by strand cutting, swapping, and ligation. Brackets with and without an asterisk indicate regions of symmetric and asymmetric hybrid DNA, respectively, which can give rise to aberrant 4:4, 5:3, or 6:2 segregations.

After the induction of meiosis the chromatin structure surrounding a hotspot (e.g., that at *ARG4* or *HIS4*) is altered, as assayed by the accessibility of micrococcal nuclease to the DNA in isolated chromatin. This alteration appears to occur during premeiotic replication and may allow access to the DNA by recombination-promoting proteins induced during meiosis (14). One of these proteins, Spo11, in conjunction with several others, makes a ds break at one of several potential sites spread over a region of a few hundred bp defining the hotspot. Spo11 remains covalently bound to the 5′ end of the DNA on each side of the break (56). Resection of the 5′ end and the bound protein requires Mre11 and Rad50, although they may not be the catalysts. Typically, ~600 to 800 nucleotides are resected in wild-type cells, but resection continues farther (~2 kb) in mutants lacking Rad51 or Dmc1, proteins acting at the next stage (13, 109).

The Rad51 and Dmc1 proteins, like the *E. coli* RecA protein, bind to single-stranded (ss) DNA and, in the presence of a ss DNA binding protein (SSB), promote strand exchange between the protein · ss DNA filament and homologous duplex DNA (110). The purified proteins promote formation of D-loops (displacement loops), in which the ss DNA of the filament replaces its identical (or nearly identical) strand in the duplex. Strand exchange may continue beyond the region of resection and allow formation of symmetric hybrid DNA and the cross-stranded structure called a Holliday junction (Figure 1). DNA forms consistent with D-loops and Holliday junctions have been isolated from meiotic cells (55a). In addition, there are molecules with double Holliday junctions, formed by the invasion of both ends at the initial ds break into the same duplex (10, 92). The distance between these two Holliday junctions is determined by a combination of the length of resection and the length of symmetric strand exchange, also called branch migration. This distance is reported to range from ~0.1 to 1 kb (10).

Double Holliday junction molecules are presumably resolved into unbranched recombinant molecules, but the mechanism of this resolution is unclear. There may be multiple mechanisms, but two are widely considered. (A third is proposed later in this review; see Figure 6.) In the first mechanism, two strands of identical polarity in each Holliday junction are cut, and the ss ends are exchanged and ligated (100, 112). If the two junctions are resolved in the same "plane" (horizontal or vertical as diagrammed in Figure 1), DNA flanking the junctions remains in parental (non-crossover) configuration, whereas if the two junctions are resolved in opposite planes (one horizontal and one vertical), the flanking DNA is recombined, producing a crossover. In the second mechanism, the double Holliday junction is resolved without further covalent exchanges between the DNA molecules (51, 77). Conceptually, the two entwined duplexes are simply pulled apart, made possible by the unlinking action of a topoisomerase. Flanking DNA remains in parental configuration (non-crossover). Genetic markers flanking the Holliday junctions therefore undergo reciprocal recombination (crossover) or not, depending upon the mode of resolution of the junctions.

Genetic markers within the hybrid DNA region can undergo more complex recombination reactions, depending upon the fate of base mismatches within the hybrid DNA (hDNA) spanning the two Holliday junctions. Each allelic difference between the two parents in this interval will produce a mismatch either on one duplex—in the region of ss resection and asymmetric strand exchange, or on both duplexes—in the region of symmetric strand exchange (Figure 1). In the absence of mismatch correction each duplex with a mismatch will segregate two genetic types, designated $+$ and $-$, after one postmeiotic replication. Counting the eight duplexes present at this stage, an allelic difference in the asymmetric region will segregate $5^+:3^-$ or $3^+:5^-$. A difference in the symmetric region will generate two postmeiotic segregations (PMS), designated an aberrant $4^+:4^-$ (Ab $4^+:4^-$) segregation. (The absence of any recombination event produces normal $4^+:4^-$ segregation, i.e., without PMS.)

Gene conversion arises from correction of mismatches within the hDNA. Depending upon the direction of this correction, an incipient $5^+:3^-$ (or $3^+:5^-$) can be converted to a $6^+:2^-$ (or $2^+:6^-$) or restored to a normal $4^+:4^-$. Similarly, an incipient Ab $4^+:4^-$ with its two mismatches will produce a $5^+:3^-$ or $3^+:5^-$ if there is one correction, or a $6^+:2^-$, $2^+:6^-$, or normal $4^+:4^-$ if there are two corrections.

By definition, markers near a hotspot convert at higher than average frequency. Double-strand DNA breaks occur near or coincident with the hotspot in several well-studied cases (65, 79). The finding of double Holliday junctions encompassing the hotspot and the frequent association of crossovers with conversion at these hotspots strongly support the ds break repair model of Szostak et al. (112), an elaboration of an earlier ds break repair model of Resnick (82). Szostak et al. (112) postulated that both strands on each side of the ds break are degraded and that resynthesis of the lost strands using an intact homolog as a template is a major source of gene conversion. At the hotspots investigated, the 3′ ends on each side of the gap appear to be separated by 0 to 2 bp (27, 66, 129, 130), indicating that conversion rarely, if ever, arises from ds gapping. Furthermore, the loss of mismatch correction enzymes increases the frequency of PMS at the expense of 6:2 segregations (128). These results strongly support the model of Holliday (53) in which gene conversion results from correction of mismatches in hDNA.

Within the context of the model of Szostak et al. (112) (Figure 1) the length of hDNA extending from the initial ds break determines the distance over which markers convert at high frequency (the spread or gradient of the hotspot) and the distance over which markers are corrected in the same event (co-convert). Many genetic studies in *S. cerevisiae* confirm this prediction: meiotic gene conversion generally occurs within ~1 kb of the initiating ds break (e.g., 28, 90). Also as predicted, crossovers are statistically associated with conversion at hotspots, but the positions of the crossovers have not, to my knowledge, been precisely determined. The paucity of Ab $4^+:4^-$ tetrads in *S. cerevisiae* suggests, within the context of the model in Figure 1, that Holliday junctions do not frequently migrate from their point of formation at the end of the ss resection. The observed distance between double Holliday junctions, whose resolution presumably gives rise to the associated crossovers, also suggests that crossovers generally occur within ~1 kb of the initiating ds break. Thus, the genetic and physical evidence is consistent with

S. cerevisiae meiotic recombination frequently occurring near a ds DNA break, but some exceptions are discussed later.

RECOMBINATION FAR FROM DOUBLE-STRAND BREAKS BY THE RecBCD PATHWAY OF *E. COLI*

Like meiotic recombination in *S. cerevisiae*, recombination by the major pathway in *E. coli* is also initiated by ds breaks and occurs at high frequency at special sites (hotspots). But in sharp contrast, although the activity of these hotspots is dependent on the ds breaks, the hotspots are separable from the ds breaks: recombination at a hotspot can occur >30 kb from the activating ds break. This pathway (RecBCD) with its hotspot Chi provides an especially well-studied example of recombination far from a ds break.

Genetic and Biochemical Observations

Studies primarily by F. W. Stahl and his colleagues established the following picture of the RecBCD pathway recombining phage λ in *E. coli* cells (Figure 2) (reviewed in 76, 95, 98, 121). These studies used λ Red⁻ Gam⁻ mutant phage to block recombination by the λ-encoded Red pathway and to avoid inhibition, by the λ-encoded Gam protein, of the *E. coli* RecBCD enzyme, an essential component of the RecBCD pathway. After infection, such phages replicate their DNA in the circular (θ) mode. Conversion of the circles into linear forms for packaging into mature phage particles occurs by a ds cut at a special site *cos* by the λ-encoded Ter protein complex, which remains bound to the genetically defined left end of the DNA. The unbound right end is available to RecBCD enzyme, which binds tightly to ds ends and subsequently travels unidirectionally along the DNA until it meets the asymmetric hotspot sequence Chi (5′ GCTGGTGG 3′) from the right (as written here; present in the "top" strand of λ as conventionally written). Assisted by other proteins including RecA and SSB proteins, RecBCD enzyme then promotes recombination at Chi and, with decreasing probability, to its left.

The remoteness of recombination from the ds break at *cos* stems from two sources: the distance from *cos* over which RecBCD enzyme travels before encountering an active Chi site and the distance from Chi over which the recombination events occur. RecBCD enzyme can travel tens of kb before encountering Chi. The activity of a Chi site (χ^+A) 30 to 35 kb from the activating ds break at *cos* is nearly equivalent to that of a Chi site closer to *cos* (χ^+D at 3.5 kb or χ^+C at 10.0 kb) (105). Recombination stimulated by Chi is detectable as far as 5 to 10 kb to the left of Chi (e.g., 22, 101). Thus, recombination can occur at least 35 kb from the ds break that provoked the event.

Recombination of the *E. coli* chromosome, which occurs by the RecBCD pathway in wild-type cells, may occur even farther from a ds break. In Hfr conjugation there can be up to ∼100 kb of Chi-free F-factor DNA at the leading end of the

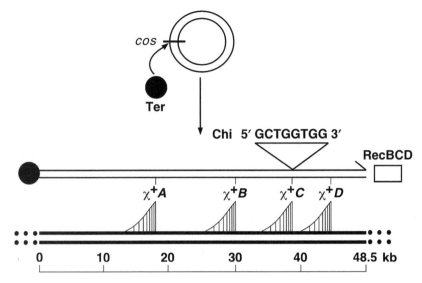

Figure 2 Overview of Chi-stimulated RecBCD pathway recombination far from a DNA break at the *cos* site in phage λ. The λ Ter protein complex (*filled circle*) cuts circular λ DNA at *cos* and remains bound to the left end, allowing entry of RecBCD enzyme (*open box*) at the right end of the 48.5 kb linear λ DNA (*thin lines*). Upon encountering the Chi sequence 5′ GCTGGTGG 3′ at any one of the sites χ^+A, χ^+B, χ^+C, or χ^+D, the traveling RecBCD enzyme, with other proteins, promotes recombination to the left of Chi with decreasing probability (*shaded graphs*). The lower parental DNA (*thick lines*) may be either linear or circular, as indicated by the dotted lines at its ends.

transferred DNA. Presumably, RecBCD enzyme traverses this distance from the origin of transfer (*oriT*) before encountering Chi-containing DNA homologous to the recipient chromosome (96).

The evidence for the preceding picture rests on both genetic studies of recombination in λ-infected *E. coli* cells and biochemical studies of purified RecBCD enzyme, RecA and SSB proteins, and DNA. The salient genetic observations are that RecBCD pathway recombination is stimulated at and to the left of Chi (e.g., 22, 101, 105), the activity of Chi depends upon its orientation with respect to *cos* (37, 57), and a *cos* site activates Chi only if it can be cut (58, 59). These observations indicate that some entity travels from a ds break at *cos* to Chi before stimulating recombination. Since Chi is specific to the RecBCD pathway (42, 104) and its only known unique component is RecBCD enzyme (24), this enzyme was implicated as the traveling entity. Certain mutations in the *recB*, *recC*, and *recD* genes, encoding the three subunits of RecBCD enzyme, reduce or abolish Chi activity without abolishing recombination (18, 67, 91). These observations suggested a direct interaction between RecBCD enzyme and Chi.

The salient biochemical observations are the following. RecBCD enzyme binds tightly and specifically to ds DNA ends [$K_D \approx 0.1$ nM or $\sim 1/10$ the concentration of one DNA end per cell (115, 117)]. Starting from a ds end, it unwinds DNA rapidly (300–500 bp/sec) and for long distances (>20–30 kb) (85, 114). Upon encountering Chi in the proper orientation (that predicted by the genetic studies), it generates, by a mechanism discussed below, ss DNA extending to the left of Chi and with a 3′ OH end at or near Chi (31, 80, 113). Mutations in Chi or in *recBCD* coordinately reduce Chi activity in cells and generation of this end by purified components (1, 8, 20, 21, 33, 80). After encountering Chi, RecBCD enzyme continues to travel along the DNA, elongating the ss DNA with Chi at its 3′ end (31, 80). RecBCD enzyme loads RecA protein onto this ss DNA, which forms a joint molecule with homologous supercoiled DNA (7, 84). Although it is not yet clear how the RecBCD-, RecA-promoted joint molecule is converted to a recombinant molecule, the genetic and biochemical observations are consistent with the joint molecule giving rise to the observed recombinants to the left of Chi, far from the ds break at *cos*. [The DNA strand polarity of ss "patches" in λ recombinants measured genetically has been reported to be opposite to that predicted by cutting of the "top" strand of λ discussed below (50, 86, 87), but the polarity of ss "splices" and of hDNA measured physically has been reported to be that predicted (52, 93). The "patches" may reflect double splices (93), or the determination of their polarity may have been influenced by mismatch correction.] The mechanism by which Chi regulates the formation of joint molecules is discussed next.

Multiple Effects of the Chi-RecBCD Enzyme Interaction

Biochemical studies have shown that Chi regulates at least three activities of RecBCD enzyme: (*a*) the generation of 3′-ended ss DNA extending to the left of Chi, (*b*) the loading of RecA protein onto this ss DNA, and (*c*) the ability to act at a subsequently encountered Chi site. After describing the biochemical observations, I discuss related genetic observations bearing on the question of Chi's role in *E. coli* recombination.

The mechanism by which RecBCD enzyme makes a DNA end at or near Chi depends upon the reaction conditions, most notably the ratio of the ATP and Mg^{++} concentrations. With (ATP) > (Mg^{++}) RecBCD enzyme unwinds the DNA from the ds end to Chi, nicks the "top" strand a few nucleotides to the 3′ ("upstream") side of Chi, and continues unwinding DNA to the end (Figure 3, top five panels) (80, 113, 116). The products of this reaction, two ss DNA fragments and one full-length ss DNA, are observed with or without RecA and SSB proteins. In contrast, with (Mg^{++}) > (ATP) RecBCD enzyme degrades the upper strand up to Chi, at which degradation of this strand is attenuated, the "bottom" strand is cut, and degradation of the bottom strand is augmented; continued travel and degradation produces two ss DNA fragments, the top strand to the left of Chi and the bottom strand to the right of Chi (Figure 4, top three panels) (6, 31, 118). These products are seen only in the presence of RecA and SSB proteins; in their absence all of

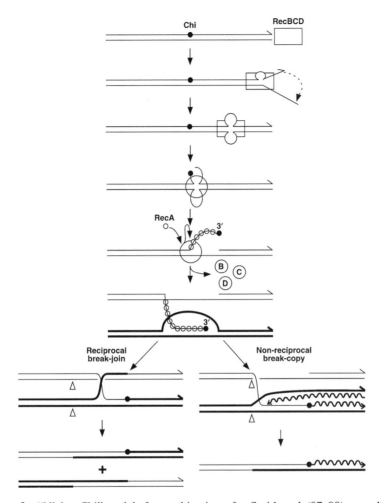

Figure 3 "Nick at Chi" model of recombination, after Smith et al. (97, 99) as modified by Smith (96). For explanation, see the text. The open box indicates RecBCD enzyme; solid dot, Chi; large open circle, RecBCD enzyme after its change at Chi; small open circles, RecA protein; circled B, C, and D, the disassembled subunits of RecBCD enzyme after its dissociation from DNA. Other symbols are as in Figure 1. Alternative resolutions at the bottom are not shown.

the DNA is degraded to oligonucleotides. Both reaction conditions produce the ss DNA extending to the left of Chi (the 3′ "Chi tail") that forms a joint molecule, a likely precursor to recombinants. Thus, both reactions are compatible with the genetic observations cited above.

After encountering Chi, RecBCD enzyme loads RecA protein specifically onto the 3′ Chi tail. It was first noted that RecBCD enzyme efficiently forms joint

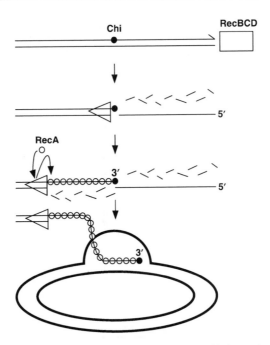

Figure 4 Reaction of RecBCD enzyme and formation of joint molecules under conditions of $(Mg^{++}) > (ATP)$, after Anderson & Kowalczykowski (7). Broken lines indicate DNA strands degraded by RecBCD enzyme. The open triangle indicates RecBCD enzyme after its change at Chi. Other symbols are as in Figures 1 and 3.

molecules between homologous supercoiled DNA and linear, Chi-containing DNA only if RecA and SSB proteins are present during RecBCD enzyme's unwinding of the linear DNA; addition of RecA protein after the initial unwinding reaction is less effective (7). A direct assay for loading of RecA protein onto ss DNA by RecBCD enzyme—protection of the ss DNA from digestion by exonuclease I—showed that efficient loading requires Chi, occurs exclusively on the 3′ Chi tail, and requires the presence of RecA protein during the unwinding phase of the reaction. RecBCD enzyme may load each of the RecA monomers to form a continuous RecA · ssDNA filament, or it may load only a minority, which then nucleate the spontaneous $5′ \rightarrow 3′$ polymerization of RecA protein (81). Mutations in *recBCD* coordinately block loading of RecA protein and recombination (3, 5). Thus, self-assembly of RecA protein onto ss DNA, which can occur with purified components (e.g., 81), is not adequate for RecBCD pathway recombination in *E. coli* cells.

The nature of the Chi-dependent change in RecBCD enzyme that permits it to load RecA protein is unclear. The *recB1080* (D1080A) mutation, which abolishes the nuclease activities (132), blocks RecA loading (3, 5), suggesting that some aspect of the Chi-nuclease interaction must precede RecA loading. In accord with a model hypothesizing the release of RecD at Chi, discussed below, RecBC enzyme loads RecA onto the DNA strand with a 3' end at the entry point, independent of Chi [i.e., "constitutively" (23)], and genetic elimination of RecD from the RecB (D1080A) mutant enzyme renders it proficient for RecA loading and recombination (3). Other evidence discussed below does not, however, support the release of RecD at Chi. Perhaps a conformational change in RecB or RecD at Chi exposes a surface on RecBCD enzyme that allows RecA loading (3, 23, 131).

After encountering Chi, RecBCD enzyme not only gains RecA loading activity, it also gains or loses other activities, either on the same DNA molecule (in *cis*) or on a separate DNA molecule (in *trans*). As noted previously, in reactions with $(Mg^{++}) > (ATP)$ top strand degradation after Chi is attenuated and bottom strand degradation is augmented (6). This alteration of RecBCD enzyme's nuclease activity in *cis* is not apparent in *trans*, presumably because upon leaving the DNA substrate the enzyme quickly reverts to its former state under this reaction condition [$(Mg^{++}) > (ATP)$] (30). With $(ATP) > (Mg^{++})$ Chi-dependent changes both in *cis* and in *trans* are apparent: after nicking a DNA substrate at a Chi site, RecBCD enzyme does not detectably nick at a subsequently encountered Chi site in *cis*, although it does unwind the DNA beyond the first Chi and cleaves a distal hairpin end (116). Subsequently added DNA is neither unwound nor hydrolyzed in *trans* (116, 119).

The Chi-dependent inactivation of RecBCD enzyme in *trans* persists for at least 2 h and results from disassembly of the three enzyme subunits (119). Examination of the Chi-inactivated enzyme by centrifugation or native gel electrophoresis revealed that all three subunits were separate (119). The disassembly of the enzyme appears not to occur at Chi, however: unwinding and hairpin cleavage activities are maintained after Chi, but the isolated subunits have low or undetectable unwinding activity, and subunit combinations other than the holoenzyme have low or undetectable nuclease activity (19, 71). Experiments using polystyrene beads attached to RecBCD enzyme and monitored by light microscopy during unwinding also failed to detect release at Chi of the one subunit tested (RecD) (32). The Chi-dependent disassembly may occur upon dissociation of the enzyme from the DNA substrate, perhaps at the end of the substrates used.

In summary, Chi induces complex changes in RecBCD enzyme, some of which occur at Chi and some subsequently. The nature of these changes depends upon the reaction conditions. These alternative results and other observations have given rise to contrasting views of how Chi and RecBCD enzyme promote recombination in *E. coli* cells. In the following section I evaluate the genetic and biochemical support for these views.

Contrasting Views of the Role of Chi in *E. coli* Recombination

Two views of how RecBCD enzyme and Chi promote recombination can be called the "nick at Chi" model and the "stop degradation at Chi" model.

In the "nick at Chi" model, RecBCD enzyme enters a ds DNA end, unwinds the DNA up to Chi, nicks one strand at Chi, and continues unwinding, thereby producing a recombinogenic 3′ ss DNA Chi tail (Figure 3) (97, 99). This model parallels the action of RecBCD enzyme under conditions with (ATP) > (Mg++). Based on the findings of Anderson & Kowalczykowski (7), an important addition to this model is the loading of RecA protein specifically onto the Chi tail. Consequently, the Chi tail, unlike the 3′ end at which RecBCD enzyme entered the DNA, is recombinogenic; recombination is more frequent at Chi than at *cos* (e.g., 101). Another addition is the ability of RecBCD enzyme to cut at only one Chi site and its disassembly upon leaving the DNA (116, 119). Consequently, one RecBCD enzyme promotes one recombinational exchange on each end of linear DNA, the minimum required to effect recombination or ds break repair of *E. coli*'s circular chromosome (96).

In the "stop degradation at Chi" model, RecBCD enzyme is a destructive nuclease until it reaches Chi, at which it is changed into a "recombinase" (Figure 5) (61, 76, 107, 120). This change is hypothesized to be the release of the RecD subunit, and the Chi-altered enzyme is hypothesized to be equivalent to RecBC enzyme, i.e., the holoenzyme lacking RecD. In this model, RecA and SSB proteins promote at Chi a nonreciprocal exchange by either a break-join or a break-copy reaction and to the left of Chi a reciprocal exchange with a third homolog by an unspecified mechanism.

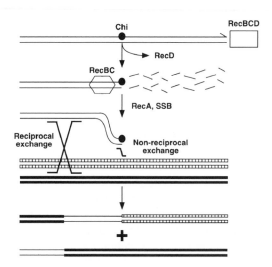

Figure 5 "Stop degradation at Chi" model of recombination, after Thaler et al. (120) as modified by Stahl et al. (107), Kuzminov et al. (61), and Myers & Stahl (76). RecBCD enzyme (*open box*) degrades DNA up to Chi (*solid dot*), at which point RecD is ejected and degradation ceases. RecBC enzyme (*open hexagon*) continues to travel along the DNA. RecA and SSB proteins promote a non-reciprocal exchange at Chi with a second parental DNA (*hatched lines*) and a reciprocal exchange to the left of Chi with a third parental DNA (*thick lines*).

The hypothesis that RecD is ejected at Chi (120) was based in part on the observation that *recD* null mutants are nuclease-deficient but recombination-proficient, although they lack Chi activity (2, 18). In other words, RecBC enzyme is viewed as constitutively Chi-activated. In some ways this model parallels the action of RecBCD enzyme under conditions with $(Mg^{++}) > (ATP)$ (Figure 4): RecBCD enzyme degrades the top DNA strand up to Chi, attenuates that degradation, and loads RecA protein onto the Chi tail. The Chi-independent (constitutive) loading of RecA protein by RecBC enzyme supports the RecD-ejection aspect of this model. Retention of nuclease activity on the bottom strand and on a distal hairpin and retention of the RecD subunit after Chi, however, are not so easily compatible with this model (4, 32, 118).

A major difference in these two models is the fate of the DNA between *cos* and Chi (to the right of Chi). In the first model this DNA survives, but in the second it is degraded. The following genetic evidence bears on the distinction between these models.

Reciprocality of recombination, the generation of two complementary recombinants in one event, requires that the DNA bearing the markers in question not be destroyed. Thus, the "nick at Chi" model readily accommodates reciprocality, whereas the "stop degradation at Chi" model does not. The evidence for reciprocality of Chi-stimulated recombination, all from λ crosses, has oscillated over the years: some evidence has supported reciprocality (58, 89, 102, 106), other evidence nonreciprocality (63, 105), and other evidence both (33a, 103, 107). Determining reciprocality in phage infections is complicated by the inability to count all of the input and output DNA molecules, as can be done in meiotic tetrad analysis, and by the possibility of phage DNA undergoing more than one recombination event in an infected cell. The latter appears to contribute frequently to Chi-stimulated recombination, since triparental recombinants are detected and the apparent reciprocality depends on the multiplicity of infection (103, 107). When observed, nonreciprocality could be accounted for either by degradation up to Chi (Figure 4) or by a break-copy event in which the 3′ OH end at Chi primes replication templated by a homolog (Figure 3, right) (96). It seems difficult to make a firm conclusion about the degradation of DNA between *cos* and Chi based on these observations.

There is evidence that Chi "protects" intracellular DNA, much as it protects the top DNA strand to its left in reactions with $(Mg^{++}) > (ATP)$, but this intracellular protection does not appear to be the simple inactivation of the RecBCD nuclease activity, as hypothesized by the "stop degradation at Chi" model. The first evidence reported was the Chi-dependent accumulation of high-molecular-weight DNA (HMW) forms of a plasmid in *E. coli* (26). In these experiments an adventitious origin of replication was activated by induction of a replication protein at high temperature. Presumably, activation of this origin initiates unidirectional rolling circle replication and entry of RecBCD enzyme into the linear tail. If the plasmid contains a properly oriented Chi site, HMW accumulates; otherwise, only circular plasmid DNA is observed. This is expected if Chi inactivates RecBCD nuclease. The accumulation of HMW, however, requires RecA protein (26), indicating that

Chi does not simply inactivate RecBCD nuclease. RecA protein could protect the DNA and allow HMW accumulation in either of two ways. (*a*) RecA protein loaded onto the DNA could protect it from degradation by other intracellular nucleases, but RecA protein only modestly protects dsDNA against RecBCD enzyme or λ exonuclease (127). (*b*) RecA protein could promote recombination of the Chi tail with a circular plasmid form; the resultant "dumbbell," a linear molecule with a circle at each end, would be resistant to intracellular exonucleases, including RecBCD enzyme.

Kuzminov and colleagues (61, 62) reported more direct evidence for protection of intracellular linear DNA by Chi. In these experiments a plasmid containing *cos* was linearized after induction of Ter at high temperature. A few percent of the initial DNA survives but only if it contains a properly oriented Chi site. As in the preceding experiments, survival of this DNA requires RecA and SSB proteins. Thus, the RecBCD nuclease does not appear to be frequently inactivated at Chi, for in that case Chi-containing DNA should survive at high level and whether or not the RecA and SSB proteins are present. Furthermore, the partial length linear molecules predicted by the "stop degradation at Chi" model were not observed. The protection of DNA "upstream" of Chi, even at low level, suggested a *trans* effect of Chi; indeed, Chi on one linearized plasmid partially protects a second, Chi-free linearized plasmid (61). The protection observed in these and the preceding experiments may reflect only the *trans* effect of Chi or recombination (as noted above) or both.

Additional experiments have shown intracellular *trans* effects of Chi, presumably by an alteration of RecBCD enzyme. (*a*) λ crosses were conducted in *E. coli* cells bearing a *cos* plasmid previously linearized by Ter; if the plasmid contains Chi, the λ crosses manifest reduced Chi activity (75). In replication-blocked crosses in these cells recombination is focused at the right end of λ, as it is in *recD* mutants, suggesting that Chi on the linearized plasmid converted RecBCD enzyme into RecBC enzyme. However, all known homologous recombination independent of RecBCD enzyme, e.g., that by the RecF pathway, also focuses recombination at the right end of nonreplicating λ (121). (*b*) *E. coli* cells were treated with bleomycin, which presumably breaks the chromosome and allows RecBCD enzyme access to the many Chi sites on it; in subsequent λ crosses Chi activity is strongly reduced for 1 to 2 h after bleomycin-treatment (60). In addition, phage T4 gene 2 mutant phage can multiply in these cells, just as these phage can in *recB*, *recC*, or *recD* null mutants, reflecting the lack of RecBCD nuclease activity (2, 18, 60, 78).

These results suggest that RecBCD enzyme is altered by Chi such that its action on another DNA molecule (in *trans*) is changed. The alteration was inferred to be the loss of the RecD subunit (60, 75). The long-term inactivation of RecBCD enzyme, however, is reminiscent of the Chi-dependent disassembly of all three subunits with (ATP) > (Mg^{++}) (119). In this case the retention of Chi-independent recombination proficiency may reflect DNA break-dependent induction of, for example, the RecF pathway, which is not stimulated by Chi (104). The restoration of RecBCD enzyme activity after ∼1 to 2 cell generations may reflect *de novo* synthesis of RecBCD enzyme.

In summary, recombination is stimulated far from the DNA break at which RecBCD enzyme enters the λ or *E. coli* chromosome. Precisely how RecBCD enzyme processes the broken DNA into recombinants within *E. coli* remains to be determined. The observations cited here support some aspects, but not others, of two models (Figures 3, 5). Further studies of *recBCD* mutants and of recombination intermediates from *E. coli* cells may help elucidate this complex process.

ADDITIONAL EVIDENCE FOR RECOMBINATION FAR FROM DNA BREAKS IN PHAGE AND FUNGI

In this section I discuss additional examples of recombination that occur far from DNA breaks or that can be interpreted as doing so. In some cases no DNA break has been demonstrated, but the initiating lesion may be a DNA break; in these examples, the recombination event spans a long distance on the chromosomes, often >10 kb.

Long Hybrid DNA and Meiotic Gene Conversion Tracts

Within the context of models such as those in Figures 1 and 3, hybrid DNA (hDNA), that with one strand from each parent, is a precursor to crossing over and gene conversion. The length of hDNA from an initiating DNA break determines the distance from the break over which recombination can occur. Thus, long hDNA can give rise to recombination far from a DNA break.

In phage λ, hDNA can extend nearly the length of the chromosome, 48.5 kb (40). Long hDNA can be detected physically when DNA replication is blocked and the λ Red recombination pathway and the *E. coli* RecA protein are active. From mixed infections by density-labeled phage, DNA in progeny phage containing closely linked markers (O^+, P^+) from each parent has a wide range of densities, indicating that either parent can contribute anywhere from ∼10% to 90% of the nucleotides (126). Much of the DNA has hybrid sections that can be nearly the length of the chromosome (40). Long hDNA can also be detected genetically when replication is permitted and the progeny from a six-factor cross are plated on *mutL* bacteria deficient in mismatch repair. Of phages with hDNA spanning a central marker (*cI*), 50% have hDNA >10 kb long, and 25% > 17 kb long (55). Thus, long hDNA is common in λ recombination.

Although the event initiating hDNA formation was not investigated in these studies, other studies indicate that the λ Red exonuclease initiates ss DNA digestion from a ds break at *cos*; the resultant 3′ ss DNA end invades an intact duplex to form hDNA (121). This hDNA may be extended by branch migration the length of the λ chromosome. Correction of mismatches within the hDNA can produce a recombinational exchange tens of kb from *cos*, the site of the presumed initiating DNA break.

Perhaps surprisingly, the hDNA in λ crosses can encompass deletion heterologies of 0.7 or 1.3 kb (64); larger deletions were not tested. hDNA encompassing a 0.6-kb deletion heterology is also formed by the Chi-stimulated RecBCD pathway (52). Although deletion heterologies of this size block spontaneous branch migration, purified RecA protein can promote formation of hDNA across large deletion heterologies at low frequency (12). Intracellular activities such as RuvAB (125) may aid the formation of long hDNA containing large deletion heterologies. In fungi, large deletions or insertions undergo gene conversion at frequencies comparable to those of single bp mutations (38, 88, 133). Furthermore, frequent PMS of 1.5-kb insertions and occasional PMS of 5.6-kb insertions are observed in *S. cerevisiae* strains deficient in Rad1 or Rad10, proteins involved in nucleotide excision repair (55b). Thus, hDNA formation in phage and fungi does not appear to be dramatically impeded by deletion heterologies of ~1 kb or more.

Long hDNA initiated by DNA breaks provides a plausible mechanism for long co-conversion tracts produced during *S. cerevisiae* meiosis. A study employing 12 restriction site mutations scattered over a 16-kb interval of chromosome III showed that meiotic gene conversion tracts were usually continuous and often <4 kb long, but of 64 tracts analyzed 15% were at least 4 to 7 kb long and 5% were >12 kb long (111). Another study employed unsequenced mutations in four genes spanning 14 kb of chromosome VI (29). Many conversion tracts included only one marker, but of those that included an internal marker (*sup6*) 7.8% included three markers and 2.3% included all four. The minimum lengths of these tracts were, respectively, 3 to 9 kb and 8 to 14 kb, the uncertainty being due to the unknown positions of the markers within the genes. Furthermore, 6% of the tetrads manifesting PMS of *sup6* manifested PMS of three markers, indicating hDNA at least 6 to 8 kb long. In both studies the absence of markers outside the clusters analyzed precludes putting an upper limit on the length of hDNA, but these studies suggest that hDNA can occasionally be >10 kb long in *S. cerevisiae* meiosis.

Meiotic Recombination Far from Prominent DNA Breaks

Although there is substantial evidence that *S. cerevisiae* meiotic recombination frequently occurs near ds DNA breaks (see above), there is suggestive evidence that some occurs far from breaks in *S. cerevisiae* and in the fission yeast *Schizosaccharomyces pombe*.

At the *S. cerevisiae HIS4* locus there is a prominent DNA break site that accounts for some but not all meiotic recombination at this locus (28, 36, 79). The frequency of gene conversion and the amount of meiotic DNA breakage are reduced by mutations eliminating transcription factors or their binding sites and increased by mutations introducing telomeric DNA sequences. There is a good linear relation between the conversion frequency and the amount of breakage, but this line indicates 10% conversion at zero breakage (36). In this strain background the "wild-type" conversion frequency is ~30%, and that of the most active hotspot is ~70%. Thus, about one third of the "wild-type" conversion appears independent of the prominent DNA breaks at *HIS4*. This recombination may result from prominent breaks at distant sites or from low-level breaks throughout the *HIS4* region.

There also appears to be a discrepancy between the amount of meiotic DNA breakage and the amount of recombination in another region of chromosome III of *S. cerevisiae* (9). Across this 340-kb chromosome there are two "hot" regions of ~70 kb and ~100 kb in which many sites of prominent DNA breakage occur. At each site within these regions the frequency of observed breaks ranges from 0.2% to 8.8%; the cumulative frequencies of breaks in these regions are 32% and 44%. Between these two regions is a "cold" region of ~90 kb in which there are only three sites of observed breakage; these occur within a 3-kb interval and have a cumulative frequency of breaks of only 1.7%. Although the amount of DNA breakage in the "hot" and "cold" regions differs by a factor of ~20, the amount of recombination differs by a factor of only 2 to 3: ~35 cM and ~55 cM in the two "hot" regions and 15 to 20 cM in the "cold" region (9, 15). The available data thus suggest that recombination frequently occurs in the "cold" region tens of kb from the observed breaks. Recombination in this region could stem from widely distributed low-level, undetected breaks or from the distant, observed breaks.

A similar discrepancy between meiotic recombination and DNA breaks appears in *S. pombe*. Meiotic DNA breaks occur in *S. pombe* and are dependent on the products of eight *rec* genes, which are also required for meiotic recombination (17; R. Schreckhise, J. Young, & G. Smith, unpublished data). Thus, these breaks appear to be required for meiotic recombination, but the wide spacing of the breaks does not seem to coincide with the more uniform distribution of recombination. A 501-kb region of chromosome I has been examined extensively (J. Young, G. Hovel-Miner, C. Rubio & G. Smith, unpublished data). There are six sites of prominent DNA breakage spaced ~30 to 100 kb apart; at each site ~2% to ~12% of the DNA is broken, or ~25% cumulative. Recombination between eight single-bp transition markers in this region has been studied; the most distant markers, *lys3-37* and *pro1-1*, are separated by 476 kb physically and 45 cM genetically. The ratio of ~25% breaks: 45 cM is similar to that observed in the "hot" region of *S. cerevisiae* chromosome III mentioned above (9). The intensity of recombination is 0.9 cM/10 kb for the entire *lys3-37–pro1-1* interval. For 12 subintervals this value ranges from 0.7 to 1.6 cM/10 kb, whether the subinterval studied contains no prominent DNA break site or one to five such sites. Recombination thus appears to be more uniformly distributed than the prominent break sites, implying that recombination must occur far from these prominent sites. As for the 90-kb "cold" region of *S. cerevisiae* chromosome III, there may be widely distributed, low-level breaks responsible for some *S. pombe* meiotic recombination. The available evidence suggests, however, that recombination frequently occurs tens of kb from DNA breaks.

How might recombination occur so far from DNA breaks? One possibility is that a DNA break relieves torsion within a domain of the chromosome and that recombination can occur, by an unspecified mechanism, anywhere within the relaxed domain. Another possibility is the entry of a "recombination machine," exemplified by RecBCD enzyme (Figures 2–5), at a DNA break and its travel to a distant point before promoting recombination.

A third possibility for recombination far from a ds DNA break is outlined in Figure 6: a double Holliday junction is formed at the site of a ds DNA break, as in the model in Figure 1 (112). The junctions migrate in tandem to one or the other side for a substantial distance before being resolved with or without a crossover. Heteroduplex DNA at the site of the initial break is confined to one chromatid and is not flanked by an adjacent crossover. This outcome is seen in ~10% of events at the *S. cerevisiae ARG4* locus and has been attributed to the "topoisomerase pullout" resolution of double Holliday junctions mentioned earlier (41). The scheme in Figure 6 provides an alternative that has the same genetic consequences close to the initial break but, in addition, can account for crossovers far from the DNA break.

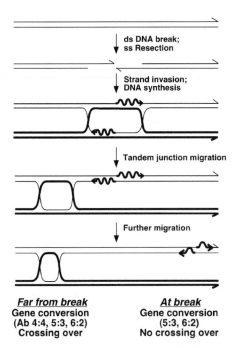

Figure 6 Proposal for recombination far from a DNA break due to long-distance branch migration of a double Holliday junction, established from a ds break as in Figure 1. Symbols are as in Figure 1. At the site of the initial break, PMS (5:3 segregation) or gene conversion (6:2) can occur, but crossing over cannot (*bottom right*). At the site of resolution of the double Holliday junction, PMS (aberrant 4:4 or 5:3) or gene conversion (6:2) with or without crossing over can occur (*bottom left*; see Figure 1).

Long Mitotic Conversion Tracts: Priming of DNA Replication by a Broken DNA End

In *S. cerevisiae*, a DNA ds break introduced by the HO endonuclease normally promotes localized mitotic gene conversion of the adjacent *MAT* locus, i.e., mating-type switching (49). But in abnormal circumstances the broken DNA can be processed in alternative ways. For example, if the cells lack Rad51 protein, the conversion tracts can extend to the end of the chromosome, a distance of 120 kb (68). Diploid cells heterozygous for four markers spanning 300 kb of chromosome III were transiently induced for *HO* expression; viable cells, recovered at high frequency, were tested for homozygosity of the markers. In ~60% of the cells there was homozygosity for *MAT* and the two centromere-distal markers used; a marker on the other arm of the chromosome remained heterozygous. In the remaining cells the two arms of the broken chromosome were lost. Slightly modifying a previous model for break-copy recombination of phage λ (94) and T4 (39, 74), the authors proposed that the broken end of the centromere-containing arm frequently invades the homolog and primes DNA replication to the telomere (Figure 7). For this to give homozygosis of the *MAT* locus, immediately centromere-proximal to the break, the broken end must be degraded at least 1 kb (to remove the *MAT* locus) and be processed into a 3′ ss DNA end (to produce a primer for replication).

A subsequent study (70) indicated that degradation frequently proceeds >13 kb from the HO-induced ds break. A marker, *URA3*, was inserted at sites 3, 13, 48, or 78 kb from the break toward the centromere. At the first two sites *URA3* was rarely,

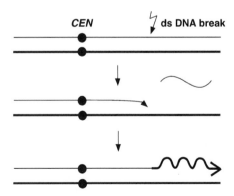

Figure 7   Break-induced replication can give rise to long gene conversion tracts. A ds DNA break (*jagged arrow*) generates an invasive 3′ end (arrowhead), which primes DNA synthesis. The replication fork (not shown) progresses to the end of the invaded chromosome (see Figure 3, bottom right). The broken fragment (*sinuous line*) distal to the centromere (*CEN, solid dot*) is lost. The thin line indicates duplex DNA from one homolog, and thick from the other. The wavy line indicates newly replicated duplex DNA.

if ever, retained, but at the latter two sites it was retained in ~50% or ~80% of the cells that repaired the broken chromosome. Located between these two pairs of sites, at 44 kb from the break, is a site important for the retention of a centromere-proximal *URA3* marker. This site is close to an origin of replication *ARS310* but is functionally separable from it. Thus, the inferred gene conversion tracts can extend ~50 to 100 kb to each side of the break. This event also occurs in Rad⁺ cells, but at lower frequency (~2%). [See below for previous, similar observations in Rad⁺ cells and the same proposed mechanism (123).]

These marker arrangements were also tested for meiotic gene conversion prompted by the DNA break at the HO cut site (69). A *rad50* mutation blocked formation of the normal meiotic breaks, and expression of *HO* by the meiosis-specific promoter of *SPO13* led to a meiotic DNA break at the HO cut site. A *spo13* deletion mutation allowed recovery of viable spores in dyads. Among these dyads at least 29% had converted one or both markers (*URA3* and *THR4*) flanking *MAT* and the HO cut site. Thus, these events, like the mitotic events, can be accounted for by degradation >10 kb from the DNA break, followed by replication to the end of the chromosome, a distance of >100 kb.

Distant, Correlated Mitotic Recombination Events

Analysis of spontaneous mitotic recombination in diploid cells is not as simple as that of meiotic recombination, since the presumed initiating lesion is not known, recovery of all the participating chromosomes can be difficult, and recombination can occur in the pre- or postreplication phase of the cell division cycle or both. Nevertheless, the occurrence of widely separated mitotic recombination events suggests that recombination can occur far from an initiating lesion.

The *HOT1* mitotic recombination hotspot of *S. cerevisiae* promotes the formation of long, continuous conversion tracts (123), as seen by the following evidence. *HOT1* is part of the rDNA gene cluster and includes the enhancer-promoter complex (122) and overlaps a replication pause site (124). Presumably, ss or ds DNA breaks at one or both of these sites trigger mitotic recombination at high frequency. One study employed five heterozygous markers on chromosome III and selected cells homozygous for a marker (Ura⁻) 48 kb centromere-distal to a 570-bp transplaced *HOT1* fragment (123). Of these cells, >90% were homozygous for all markers between the *URA3* marker and *HOT1*, and more than half of these were also homozygous for a marker 27 kb centromere-proximal to *HOT1*. Only rarely, <5% of the time, were these events loss of the entire *URA3* chromosome. *HOT1* in *cis* to *URA3* stimulated the formation of Ura⁻ cells much more than *HOT1* in *trans*, indicating that the events were rarely, if ever, due to crossing over. The simplest interpretation is that the homozygosity results from conversion tracts including *HOT1* and extending >48 kb to one side and more than half the time >27 kb to the other side, or ≥75 kb total. The inferred conversion tract covering *HOT1* need not have an end at *HOT1*; the ends can be >27 kb from *HOT1*. The authors proposed that the centromere-proximal end of the chromosome broken at *HOT1* is degraded by an exonuclease (sometimes a substantial distance) and that this end primes DNA replication (templated by the homolog) to the centromere-distal end of the chromosome, thereby giving rise to long conversion tracts covering *HOT1*.

In addition, there is a substantial class of *HOT1*-stimulated events that are homozygous Ura⁻ but remain heterozygous for two other markers between *URA3* and *HOT1*; these appear to have conversion tracts both of whose ends are >20 kb from *HOT1* but not covering *HOT1*. They evidently did not arise by the proposed break-copy mechanism; they may have arisen by an entity moving from *HOT1* and promoting recombination at a distance, perhaps by a mechanism such as that in Figure 3 or 6.

In the following examples of spontaneous mitotic recombination, no initiating lesion is known, but events covering a long distance or widely separated from each other can be interpreted within the framework of the preceding examples.

A long series of studies examined mitotic recombination in diploid cells carrying up to six markers spanning ~425 kb on one arm of chromosome VII of *S. cerevisiae* (Figure 8) (11, 34, 43–47). Heteroalleles at *LEU1*, *TRP5*, or *MET13* allowed the selection of prototrophic intragenic recombinants ("convertants"). Heterozygosity at *ADE5* allowed the detection of red-white sectored colonies, since the cells were homozygous for the unlinked *ade2* mutation. (*ADE5 ade2* colonies are red, and *ade5 ade2* colonies are white.) Sectored colonies arise from a cell in which recombination happened immediately before or shortly after plating; examination of the progeny within this colony allows inferences about all of the participating chromosomes and the reciprocal or nonreciprocal nature of the recombination event(s).

As noted below, the recombination event(s) often span long distances (>100 kb). These were initially interpreted as a reflection of long hDNA (43, 44), and some events may indeed reflect this. But three other interpretations are: (*a*) break-copy (Figures 3, 7), (*b*) a moving entity (Figures 3, 6), and (*c*) multiple, independent events that occur in a subpopulation of cells especially proficient for recombination ("hot cells"). All four factors may contribute, perhaps some concurrently.

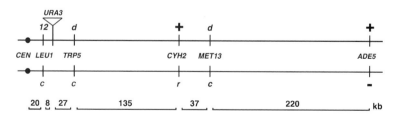

Figure 8   Multiply marked chromosome VII of *S. cerevisiae* used to demonstrate widely separated mitotic recombination events. The *LEU1*, *TRP5*, *CYH2*, *MET13*, and *ADE5* genes were marked on each homolog with alleles *c, d, r, 12,* ⁺ and ⁻ as indicated. The *URA3* insertion was present in some strains but not others. The distances in kb between loci are indicated. A single line indicates duplex DNA, and a filled circle the centromere.

An early observation was the high coincidence of doubly prototrophic cells reflecting recombination at widely separated loci (44). For example, a doubly heteroallelic strain (Figure 8) produces Leu$^+$ cells at a rate of 3.4×10^{-6} per cell division and Trp$^+$ at 2.6×10^{-6}. Leu$^+$ Trp$^+$ are produced at 1.1×10^{-8}, or 1200 times higher than the rate expected from independent events. Similarly, Leu$^+$ Met$^+$ are produced at a rate 200 times higher than expected. Thus, conversions at loci 35 kb or 207 kb apart are statistically correlated. These could arise from hDNA spanning these long distances, followed by mismatch correction. They are not easily accounted for by the break-copy model (Figure 7): a break near *leu1*, for example, followed by hDNA formation and mismatch correction might produce Leu$^+$, but the subsequent copying would produce a cell homozygous for either *trp5-c* or *trp5-d*. Two separate events in "hot cells" is an alternative discussed in the next section.

To test whether the events at *LEU1* and *TRP5* were connected, *URA3* was inserted between them on one homolog (45, 46). Among doubly selected Leu$^+$ Trp$^+$ cells 39% became homozygous Ura$^-$, and 12% became homozygous *URA3/URA3*. If the *URA3* insertion was flanked by a direct repeat of 6.1 kb, thereby allowing intrachromosomal events such as a deletion, 70% became homozygous Ura$^-$, and 2% homozygous *URA3/URA3*. (The *URA3* markers are lost in <0.2% of unselected cells, indicating that their conversion is strongly associated with events at *LEU1* or *TRP5*.) These high frequencies of conversion of the marker between the selected Leu$^+$ Trp$^+$ events suggest that frequently a single event spans the 35-kb region. If these events result from hDNA spanning the 35 kb, then hDNA must frequently encompass 5.5- or 11.6-kb insertion heterologies, since these insertions do not significantly alter the rates of Leu$^+$, Trp$^+$, or coincident Leu$^+$ Trp$^+$ recombinant formation (45, 46). Among singly selected Leu$^+$ cells, *URA3* converted in 9%, or in 30% if *URA3* was flanked by direct repeats. Thus, singly selected events appear frequently to extend >8 kb.

Many of the mitotic events studied appear to result from a conversion, perhaps near a spontaneous DNA break, and a nearby reciprocal exchange ("crossover"). This event is readily seen as a selected Trp$^+$ (or Leu$^+$) colony that is red-white sectored. The conversion at *TRP5* (or at *LEU1*) gives the selected prototrophy, and, with appropriate segregation of centromeres at mitosis, the associated crossover gives homozygosity for *CYH2*, *MET13*, and *ade5* on the white side of the sector. Sectoring is highly correlated with conversion: for example, 1.1% of selected Trp$^+$, Leu$^+$, or Met$^+$ colonies are sectored, whereas the frequency of sectors among unselected colonies is ~0.03% (11, 34). Analysis of cells in the red sector and in the white sector confirms, in ~50% of the cases, the recovery of markers expected for a conversion and a nearby crossover (11, 43).

In the other cases (~50%), more complex events appear. In one study, ~25% of the Trp$^+$ red-white sectored colonies had apparently converted all markers centromere-distal to *TRP5* but the centromere-proximal marker was rarely altered (11); these can be accounted for by the break-copy mechanism (Figures 3, 7), with a break near *TRP5*. But ~7% (16/227) had recombinational exchange points between *CYH2* and *MET13* or between *MET13* and *ADE5*; these points are at least 135 kb or 170 kb, respectively, from *TRP5*. In another study (43), ~10% (7/71) were of the latter type. Thus, widely separated recombination events occur fairly frequently; these events, like the coincident Leu$^+$ Trp$^+$ events, are not accounted for by the break-copy mechanism.

More complex events also appear: in ~14% of the Trp$^+$ sectored colonies there are three to six genotypes ("mosaics"). Among these, ~15% have exchanges in the *CYH2-MET13* or *MET13-ADE5* interval or both; i.e., >135 kb from the selected Trp$^+$ event (11, 47). Similar values are obtained among Leu$^+$ or Met$^+$ selected colonies. Some of these complex outcomes could result from a single event, such as long hDNA resolved by a crossover and one or more mismatch corrections. Arguing for a single event spanning a long distance is the "polarization" of the mosaic genotypes: markers centromere-distal to a selected event are more frequently recombinant in the mosaics than are centromere-proximal markers (11). A similar polarization occurs among the non-mosaic sectored colonies (43).

Other mosaic colonies, such as those containing cells with five or six genotypes, must involve multiple events. Furthermore, some statistically correlated events involve two nonhomologous or three homologous chromosomes (11, 47, 73). These events are most easily accounted for by the "hot cell" notion, discussed next.

Induction of Recombination Potential in *trans*: "Hot Cells"

DNA damaging agents stimulate recombination apparently in two ways: the DNA lesions may be sites of enhanced recombination (in *cis*), and they may induce recombination potential on undamaged DNA (in *trans*). A similar induction of DNA repair capacity is well established as exemplified by the DNA damage-inducible SOS repair regulon of *E. coli*. In the absence of exogenous DNA damaging agents, a subpopulation of cells may be induced for spontaneous recombinational potential. In these "hot cells" recombination occurs at higher frequency than in most cells in this population. Two separate events in these cells may appear as a single event covering a long distance.

Evidence for the induction of recombination potential is seen in X-irradiated *S. cerevisiae* cells (35). If a haploid *ade6-21,45* double mutant is irradiated and then mated with an unirradiated *ade6-21/ade6-45* heteroallelic diploid, the frequency of Ade$^+$ recombinants increases as a function of X-ray dose. Most of the Ade$^+$ recombinants appear to arise from a single recombination event between the unirradiated chromosomes of the diploid, since their frequency is about 3000 times higher than expected from the frequency of two recombination events between the doubly mutant chromosome and each singly mutant chromosome. Furthermore, radiation of the haploid increases recombination even when nuclear fusion is largely prevented by a *kar1* mutation. UV-irradiation of the haploid also increases recombination, but this effect is diminished when the UV-induced lesions are removed by exposure to visible light (photoreactivation) before mating. These observations imply that a DNA lesion induces a diffusible agent that promotes recombination, presumably anywhere in the genome.

There is also evidence for an induced state of hyper-recombination ("hyper-Rec") in a subpopulation of untreated yeast cells. With appropriate markers one can select cells with recombination event #1, or with #2, or with both. If the events occur independently, the frequency of the double event should equal the product of the frequencies of the individual events ($f_{1,2} = f_1 \cdot f_2$). If the events occur preferentially in a subpopulation of cells, they will not occur independently, and $f_{1,2} = C \cdot f_1 \cdot f_2$, where C, the coefficient of coincidence, is >1. If the events occur exclusively and independently within a distinct subpopulation, the fraction of the total cells in this subpopulation is $1/\sqrt{C}$.

Several studies of spontaneous mitotic recombination show that C is >1 in both *S. cerevisiae* and *S. pombe*. For example, one study (73) used strains similar to those discussed previously (Figure 8) but also heteroallelic at loci on other chromosomes. The rates of recombination to produce prototrophy (e.g., *TRP5*) and to produce white colonies (*ade5/ade5*) were measured separately and simultaneously. For *TRP5* and white (both on chromosome VII), C = 340. For *LYS2*, *TYR1*, or *URA3* (on chromosomes II, II, or V, respectively) and white, C = 1200, 800, or 720. Thus, the coincidence of events on separate chromosomes is as great as those on the same chromosome, which implies that this coincidence must be due to separate recombination events, presumably in a subset of "hot" cells. From the assumptions in the preceding paragraph, this subset is a few percent of the total cells. These observations call into question the interpretation that the prototrophic red-white sectored colonies discussed previously arise from one event covering a long distance. The polarization of the events and the high frequency of a second event among cells with a selected event (up to 70%), however, argue that some single events cover a long distance.

In *S. pombe* also there is coincidence between spontaneous mitotic recombination events on the same or separate chromosomes; this holds for pairs of intragenic or intergenic events or one of each (48, 72). The measured coefficients of coincidence are similar for the three types of double events, ranging from $\sim$10 to 100.

The fraction of hot cells in the total population appears to be increased in mutants defective in DNA metabolism since C is decreased in such mutants. For example, in an *S. cerevisiae cdc9* mutant deficient in DNA ligase, C is reduced to 7–36 for the events discussed above (73). This suggests that the *cdc9* mutation increases the hot cell population from $\sim$3% to $\sim$20% of the total population. Similarly, in an *S. pombe rad2* mutant deficient in the "flap" endonuclease important for completing DNA replication, C for two intragenic events on different chromosomes is reduced to 4 from 29 in *rad2*[+] cells; for two intergenic events on different chromosomes C is reduced to $\sim$10 from $\sim$100 (48). Presumably, the defects in DNA metabolism increase the level of unrepaired DNA damage, which induces recombination potential throughout the genome in a substantial fraction of the population. This effect is similar to that of X rays discussed at the beginning of this section.

CONCLUSION AND PERSPECTIVE

In numerous organisms, including phage, bacteria, and fungi, recombination can occur either near a DNA break or far from it. The relative frequencies of these two types of recombination are still unclear, as too few studies have focused on recombination remote from DNA breaks. In no case is the molecular mechanism completely understood; in some cases the genetic and biochemical observations are not in complete accord. Further investigations, including reconstruction of a complete pathway of recombination in cell-free preparations, are needed to answer these questions.

There are multiple mechanisms by which recombination can occur far from a DNA break. More than one of these mechanisms may play a role in a given event. For example, a recombination machine may enter a ds DNA break, travel to the left, and establish a replication fork toward the right (e.g., Figure 3); this event may also form long hDNA and induce recombination-promoting enzymes. Sorting out the molecular mechanism underlying such events may require multiple genetic and physical approaches.

There is circumstantial evidence for a cellular state conferring high recombination potential ("hyperRec" state). Additional studies are needed to confirm and elucidate this state. There is also evidence for a hyper-mutable state ("hyperMut"), for example, in starved *E. coli* cells (39a). Perhaps in some circumstances the hyperRec and hyperMut states are coincident, creating a hyper-unstable state. A suggestion of such a state comes from the high rate of mutation near recombination events stimulated by a ds DNA break (108). It is conceivable that the hyperRec or hyperMut states, or both, persist for several cell generations after the occurrence of DNA damage. Such a hyper-unstable state, especially if persistent, may contribute to the development of cancer.

ACKNOWLEDGMENTS

I am grateful to Françis Fabre, John Golin, Jim Haber, Phil Hastings, Galadriel Hovel-Miner, Neil Hunter, Mike Lichten, Tom Petes, Claudia Rubio, Randy Schreckhise, Lorraine Symington, and Jennifer Young for helpful discussions and unpublished information; to Sue Amundsen, Jim Haber, Mike Lichten, Walt Steiner, Andrew Taylor, and Jeff Virgin for helpful comments on the manuscript; and Karen Brighton for preparing it. Research in my laboratory is supported by grants GM31693 and GM32194 from the National Institutes of Health.

Visit the Annual Reviews home page at www.AnnualReviews.org

LITERATURE CITED

1. Amundsen SK, Neiman AM, Thibodeaux SM, Smith GR. 1990. Genetic dissection of the biochemical activities of RecBCD enzyme. *Genetics* 126:25–40
2. Amundsen SK, Taylor AF, Chaudhury AM, Smith GR. 1986. *recD*: the gene for an essential third subunit of exonuclease V. *Proc. Natl. Acad. Sci. USA* 83:5558–62
3. Amundsen SK, Taylor AF, Smith GR. 2000. The RecD subunit of the *Escherichia coli* RecBCD enzyme inhibits RecA loading, homologous recombination and DNA repair. *Proc. Natl. Acad. Sci. USA* 97:7399–404
4. Anderson DG, Churchill JJ, Kowalczykowski SC. 1997. Chi-activated RecBCD enzyme possesses $5' \rightarrow 3'$ nucleolytic activity, but RecBC enzyme does not:

evidence suggesting that the alteration induced by Chi is not simply ejection of the RecD subunit. *Genes Cells* 2:117–28

5. Anderson DG, Churchill JJ, Kowalczykowski SC. 1999. A single mutation, RecB$_{D1080A}$, eliminates RecA protein loading but not Chi recognition by RecBCD enzyme. *J. Biol. Chem.* 274: 27139–44

6. Anderson DG, Kowalczykowski SC. 1997. The recombination hot spot χ is a regulatory element that switches the polarity of DNA degradation by the RecBCD enzyme. *Genes Dev.* 11:571–81

7. Anderson DG, Kowalczykowski SC. 1997. The translocating RecBCD enzyme stimulates recombination by directing RecA protein onto ssDNA in a χ-regulated manner. *Cell* 90:77–86

8. Arnold DA, Bianco PR, Kowalczykowski SC. 1998. The reduced levels of χ recognition exhibited by the RecBC^{1004}D enzyme reflect its recombination defect in vivo. *J. Biol. Chem.* 273:16476–86

9. Baudat F, Nicolas A. 1997. Clustering of meiotic double-strand breaks on yeast chromosome III. *Proc. Natl. Acad. Sci. USA* 94:5213–18

10. Bell LR, Byers B. 1982. Homologous association of chromosomal DNA during yeast meiosis. *Cold Spring Harbor Symp. Quant. Biol.* 47:829–40

11. Bethke BD, Golin J. 1994. Long-tract mitotic gene conversion in yeast: evidence for a triparental contribution during spontaneous recombination. *Genetics* 137:439–53

12. Bianchi ME, Radding CM. 1983. Insertions, deletions and mismatches in heteroduplex DNA made by RecA protein. *Cell* 35:511–20

13. Bishop DK, Park D, Xu L, Kleckner N. 1992. *DMC1*: A meiosis-specific homolog of E. coli *recA* required for recombination, synaptonemal complex formation, and cell cycle progression. *Cell* 69: 439–56

14. Borde V, Goldman ASH, Lichten M.

2000. Direct coupling between meiotic DNA replication and recombination initiation. *Science* 290:806–9

15. Borde V, Wu T-C, Lichten M. 1999. Use of a recombination reporter insert to define meiotic recombination domains on chromosome III of *Saccharomyces cerevisiae*. *Mol. Cell. Biol.* 19:4832–42

16. Campbell A. 1984. Types of recombination: common problems and common strategies. *Cold Spring Harbor Symp. Quant. Biol.* 49:839–44

17. Cervantes MD, Farah JA, Smith GR. 2000. Meiotic DNA breaks associated with recombination in *S. pombe*. *Mol. Cell* 5:883–88

18. Chaudhury AM, Smith GR. 1984. A new class of *Escherichia coli recBC* mutants: implications for the role of RecBC enzyme in homologous recombination. *Proc. Natl. Acad. Sci. USA* 81:7850–54

19. Chen H-W, Randle DE, Gabbidon M, Julin DA. 1998. Functions of the ATP hydrolysis subunits (RecB and RecD) in the nuclease reactions catalyzed by the RecBCD enzyme from *Escherichia coli*. *J. Mol. Biol.* 278:89–104

20. Cheng KC, Smith GR. 1984. Recombinational hotspot activity of Chi-like sequences. *J. Mol. Biol.* 180:371–77

21. Cheng KC, Smith GR. 1987. Cutting of Chi-like sequences by the RecBCD enzyme of *Escherichia coli*. *J. Mol. Biol.* 194:747–50

22. Cheng KC, Smith GR. 1989. Distribution of Chi-stimulated recombinational exchanges and heteroduplex endpoints in phage lambda. *Genetics* 123:5–17

23. Churchill JJ, Anderson DG, Kowalczykowski SC. 1999. The RecBC enzyme loads RecA protein onto ssDNA asymmetrically and independently of χ, resulting in constitutive recombination activation. *Genes Dev.* 13:901–11

24. Clark AJ. 1973. Recombination deficient mutants of *E. coli* and other bacteria. *Annu. Rev. Genet.* 7:67–86

25. Cox MM, Goodman MF, Kreuzer KN,

Sherratt DJ, Sandler SJ, et al. 2000. The importance of repairing stalled replication forks. *Nature* 404:37–41

26. Dabert P, Ehrlich SD, Gruss A. 1992. χ sequence protects against RecBCD degradation of DNA in vivo. *Proc. Natl. Acad. Sci. USA* 89:12073–77

27. de Massy B, Rocco V, Nicolas A. 1995. The nucleotide mapping of DNA double-strand breaks at the *CYS3* initiation site of meiotic recombination in *Saccharomyces cerevisiae. EMBO J.* 14:4589–98

28. Detloff P, White MA, Petes TD. 1992. Analysis of a gene conversion gradient at the *HIS4* locus in *Saccharomyces cerevisiae. Genetics* 132:113–23

29. DiCaprio L, Hastings PJ. 1976. Gene conversion and intragenic recombination at the *SUP6* locus and the surrounding region in *Saccharomyces cerevisiae. Genetics* 84:697–721

30. Dixon DA, Churchill JJ, Kowalczykowski SC. 1994. Reversible inactivation of the *Escherichia coli* RecBCD enzyme by the recombination hotspot χ in vitro: Evidence for functional inactivation or loss of the RecD subunit. *Proc. Natl. Acad. Sci. USA* 91:2980–84

31. Dixon DA, Kowalczykowski SC. 1993. The recombination hotspot χ is a regulatory sequence that acts by attenuating the nuclease activity of the E. coli RecBCD enzyme. *Cell* 73:87–96

32. Dohoney KM, Gelles J. 2001. χ-sequence recognition and DNA translocation by single RecBCD helicase/nuclease molecules. *Nature* 409:370–74

33. Eggleston AK, Kowalczykowski SC. 1993. Biochemical characterization of a mutant recBCD enzyme, the recB^{109}CD enzyme, which lacks χ-specific, but not non-specific, nuclease activity. *J. Mol. Biol.* 231:605–20

33a. Ennis DG, Amundsen SK, Smith GR. 1987. Genetic functions promoting homologous recombination in *Escherichia coli*: a study of inversions in phage λ. *Genetics* 115:11–24

34. Esposito MS. 1978. Evidence that spontaneous mitotic recombination occurs at the two-strand stage. *Proc. Natl. Acad. Sci. USA* 75:4436–40

35. Fabre F, Roman H. 1977. Genetic evidence for inducibility of recombination competence in yeast. *Proc. Natl. Acad. Sci. USA* 74:1667–71

36. Fan Q, Xu F, Petes TD. 1995. Meiosis-specific double-strand DNA breaks at the *HIS4* recombination hot spot in the yeast *Saccharomyces cerevisiae*: control in *cis* and *trans. Mol. Cell. Biol.* 15:1679–88

37. Faulds D, Dower N, Stahl MM, Stahl FW. 1979. Orientation-dependent recombination hotspot activity in bacteriophage λ. *J. Mol. Biol.* 131:681–95

38. Fink GR, Styles CA. 1974. Gene conversion of deletions in the *HIS4* region of yeast. *Genetics* 77:231–44

39. Formosa T, Alberts BM. 1986. DNA synthesis dependent on genetic recombination: characterization of a reaction catalyzed by purified bacteriophage T4 proteins. *Cell* 47:793–806

39a. Foster PL. 1999. Mechanisms of stationary phase mutation: a decade of adaptive mutation. *Annu. Rev. Genet.* 33:57–88

40. Fox MS, Dudney CS, Sodergren EJ. 1978. Heteroduplex regions in unduplicated bacteriophage λ recombinants. *Cold Spring Harbor Symp. Quant. Biol.* 43:999–1007

41. Gilbertson LA, Stahl FW. 1996. A test of the double-strand break repair model for meiotic recombination in *Saccharomyces cerevisiae. Genetics* 144:27–41

42. Gillen JR, Clark AJ. 1974. The RecE pathway of bacterial recombination. See Ref. 47a, pp. 123–26

43. Golin JE, Esposito MS. 1981. Mitotic recombination: mismatch correction and replication resolution of Holliday structures formed at the two strand stage in *Saccharomyces. Mol. Gen. Genet.* 183:252–63

44. Golin JE, Esposito MS. 1984. Coincident

gene conversion during mitosis in *Saccharomyces*. *Genetics* 107:355–65

45. Golin JE, Falco SC. 1988. The behavior of insertions near a site of mitotic gene conversion in yeast. *Genetics* 119:535–40

46. Golin JE, Falco SC, Margolskee JP. 1986. Coincident gene conversion events in yeast that involve a large insertion. *Genetics* 114:1081–94

47. Golin JE, Tampe H. 1988. Coincident recombination during mitosis in *Saccharomyces*: distance-dependent and -independent components. *Genetics* 119:541–47

47a. Grell RF, ed. 1974. *Mechanisms in Recombination.* New York: Plenum

48. Grossenbacher-Grunder A-M. 1985. Spontaneous mitotic recombination in *Schizosaccharomyces pombe. Curr. Genet.* 10:95–101

49. Haber JE. 1998. Mating-type gene switching in *Saccharomyces cerevisiae. Annu. Rev. Genet.* 32:561–99

50. Hagemann AT, Rosenberg SM. 1991. Chain bias in Chi-stimulated heteroduplex patches in the λ *ren* gene is determined by the orientation of λ *cos. Genetics* 129:611–21

51. Hastings PJ. 1988. Recombination in the eukaryotic nucleus. *BioEssays* 9:61–64

52. Holbeck SL, Smith GR. 1992. Chi enhances heteroduplex DNA levels during recombination. *Genetics* 132:879–91

53. Holliday R. 1964. A mechanism for gene conversion in fungi. *Genet. Res.* 5:282–304

54. Howard-Flanders P, Theriot L. 1966. Mutants of *Escherichia coli* K12 defective in DNA repair and genetic recombination. *Genetics* 53:1137–50

55. Huisman O, Fox MS. 1986. A genetic analysis of primary products of bacteriophage lambda recombination. *Genetics* 112:409–20

55a. Hunter M, Kleckner N. 2001. The single-end invasion: an asymmetric intermediate at the double-strand break to double-Holliday junction transition of meiotic recombination. *Cell* 106:59–70

55b. Kearney HM, Kirkpatrick DT, Gerton JL, Petes TD. 2001. Meiotic recombination involving heterozygous large insertions in *S. cerevisiae*: formation and repair of large, unpaired DNA loops. *Genetics.* In press

56. Keeney S, Giroux CN, Kleckner N. 1997. Meiosis-specific DNA double-strand breaks are catalyzed by Spo11, a member of a widely conserved protein family. *Cell* 88:375–84

57. Kobayashi I, Murialdo H, Crasemann JM, Stahl MM, Stahl FW. 1982. Orientation of cohesive end site *cos* determines the active orientation of χ sequence in stimulating recA•recBC mediated recombination in phage λ lytic infections. *Proc. Natl. Acad. Sci. USA* 79:5981–85

58. Kobayashi I, Stahl MM, Fairfield FR, Stahl FW. 1984. Coupling with packaging explains apparent nonreciprocality of Chi-stimulated recombination of bacteriophage lambda by RecA and RecBC functions. *Genetics* 108:773–94

59. Kobayashi I, Stahl MM, Stahl FW. 1984. The mechanism of the Chi-*cos* interaction in RecA-RecBC-mediated recombination in phage λ. *Cold Spring Harbor Symp. Quant. Biol.* 49:497–506

60. Köppen A, Krobitsch S, Thoms B, Wackernagel W. 1995. Interaction with the recombination hot spot χ in vivo converts the RecBCD enzyme of *Escherichia coli* into a χ-independent recombinase by inactivation of the RecD subunit. *Proc. Natl. Acad. Sci. USA* 92:6249–53

61. Kuzminov A, Schabtach E, Stahl FW. 1994. χ sites in combination with RecA protein increase the survival of linear DNA in *Escherichia coli* by inactivating exoV activity of RecBCD nuclease. *EMBO J.* 13:2764–76

62. Kuzminov A, Stahl FW. 1997. Stability of linear DNA in *recA* mutant *Escherichia coli* cells reflects ongoing chromosomal DNA degradation. *J. Bacteriol.* 179:880–88

63. Lam ST, Stahl MM, McMilin KD,

Stahl FW. 1974. Rec-mediated recombinational hotspot activity in bacteriophage lambda. II. A mutation which causes hotspot activity. *Genetics* 77:425–33

64. Lichten M, Fox MS. 1984. Evidence for inclusion of regions of nonhomology in heteroduplex products of bacteriophage λ recombination. *Proc. Natl. Acad. Sci. USA* 81:7180–84

65. Lichten M, Goldman ASH. 1995. Meiotic recombination hotspots. *Annu. Rev. Genet.* 29:423–44

66. Liu J, Wu T-c, Lichten M. 1995. The location and structure of double-strand DNA breaks induced during yeast meiosis: evidence for a covalently linked DNA-protein intermediate. *EMBO J.* 14:4599–608

67. Lundblad V, Taylor AF, Smith GR, Kleckner N. 1984. Unusual alleles of *recB* and *recC* stimulate excision of inverted repeat transposons Tn*10* and Tn*5*. *Proc. Natl. Acad. Sci. USA* 81:824–28

68. Malkova A, Ivanov EL, Haber JE. 1996. Double-strand break repair in the absence of *RAD51* in yeast: a possible role for break-induced DNA replication. *Proc. Natl. Acad. Sci. USA* 93:7131–36

69. Malkova A, Ross L, Dawson D, Hoekstra MF, Haber JE. 1996. Meiotic recombination initiated by a double-strand break in *rad50Δ* yeast cells otherwise unable to initiate meiotic recombination. *Genetics* 143:741–54

70. Malkova A, Signon L, Schaefer CB, Naylor ML, Theis JF, et al. 2001. *RAD51*-independent break-induced replication to repair a broken chromosome depends on a distant enhancer site. *Genes Dev.* 15:1055–60

71. Masterson C, Boehmer PE, McDonald F, Chaudhuri S, Hickson ID, et al. 1992. Reconstitution of the activities of the RecBCD holoenzyme of *Escherichia coli* from the purified subunits. *J. Biol. Chem.* 267:13564–72

72. Minet M, Grossenbacher-Grunder A-M, Thuriaux P. 1980. The origin of a centromere effect on mitotic recombination. *Curr. Genet.* 2:53–60

73. Montelone BA, Prakash S, Prakash L. 1981. Spontaneous mitotic recombination in *mms8-1*, an allele of the *CDC9* gene of *Saccharomyces cerevisiae. J. Bacteriol.* 147:517–25

74. Mosig G. 1983. Relationship of T4 DNA replication and recombination. In *Bacteriophage T4*, ed. CK Mathews, EM Kutter, G Mosig, PB Berget, pp. 120–30. Washington, DC: Am. Soc. Microbiol.

75. Myers RS, Kuzminov A, Stahl FW. 1995. The recombination hot spot χ activates RecBCD recombination by converting *Escherichia coli* to a *recD* mutant phenocopy. *Proc. Natl. Acad. Sci. USA* 92:6244–48

76. Myers RS, Stahl FW. 1994. χ and the RecBCD enzyme of *Escherichia coli*. *Annu. Rev. Genet.* 28:49–70

77. Nasmyth KA. 1982. Molecular genetics of yeast mating type. *Annu. Rev. Genet.* 16:439–500

78. Oliver DB, Goldberg EB. 1977. Protection of parental T4 DNA from a restriction exonuclease by the product of gene *2*. *J. Mol. Biol.* 116:877–81

79. Petes TD. 2001. Meiotic recombination hot spots and cold spots. *Nat. Rev. Genet.* 2:360–70

80. Ponticelli AS, Schultz DW, Taylor AF, Smith GR. 1985. Chi-dependent DNA strand cleavage by RecBC enzyme. *Cell* 41:145–51

81. Register JC III, Griffith J. 1985. The direction of RecA protein assembly onto single strand DNA is the same as the direction of strand assimilation during strand exchange. *J. Biol. Chem.* 260:12308–12

82. Resnick MA. 1976. The repair of double-strand breaks in DNA: a model involving recombination. *J. Theor. Biol.* 59:97–106

83. Roeder GS. 1997. Meiotic chromosomes: it takes two to tango. *Genes Dev.* 11:2600–21

84. Roman LJ, Dixon DA, Kowalczykowski SC. 1991. RecBCD-dependent joint

molecule formation promoted by the *Escherichia coli* RecA and SSB proteins. *Proc. Natl. Acad. Sci. USA* 88:3367–71

85. Roman LJ, Eggleston AK, Kowalczykowski SC. 1992. Processivity of the DNA helicase activity of *Escherichia coli* recBCD enzyme. *J. Biol. Chem.* 267:4207–14

86. Rosenberg SM. 1987. Chi-stimulated patches are heteroduplex, with recombinant information on the phage λ *r* chain. *Cell* 48:855–65

87. Rosenberg SM. 1988. Chain-bias of *Escherichia coli* Rec-mediated λ patch recombinants is independent of the orientation of λ cos. *Genetics* 120:7–21

88. Rossignol J-L, Nicolas A, Hamza H, Langin T. 1984. Origins of gene conversion and reciprocal exchange in *Ascobolus*. *Cold Spring Harbor Symp. Quant. Biol.* 49:13–21

89. Sarthy PV, Meselson M. 1976. Single burst study of rec- and red-mediated recombination in bacteriophage lambda. *Proc. Natl. Acad. Sci. USA* 73:4613–17

90. Schultes NP, Szostak JW. 1990. Decreasing gradients of gene conversion on both sides of the initiation site for meiotic recombination at the *ARG4* locus in yeast. *Genetics* 126:813–22

91. Schultz DW, Taylor AF, Smith GR. 1983. *Escherichia coli* RecBC pseudorevertants lacking Chi recombinational hotspot activity. *J. Bacteriol.* 155:664–80

92. Schwacha A, Kleckner N. 1995. Identification of double Holliday junctions as intermediates in meiotic recombination. *Cell* 83:783–91

93. Siddiqi I, Stahl MM, Stahl FW. 1991. Heteroduplex chain polarity in recombination of phage λ by the Red, RecBCD, RecBC(D⁻) and RecF pathways. *Genetics* 128:7–22

94. Skalka A. 1974. A replicator's view of recombination (and repair). See Ref. 47a, pp. 421–32

95. Smith GR. 1987. Mechanism and control of homologous recombination in *Escherichia coli. Annu. Rev. Genet.* 21:179–201

96. Smith GR. 1991. Conjugational recombination in *E. coli*: myths and mechanisms. *Cell* 64:19–27

97. Smith GR, Amundsen SK, Chaudhury AM, Cheng KC, Ponticelli AS, et al. 1984. Roles of RecBC enzyme and Chi sites in homologous recombination. *Cold Spring Harbor Symp. Quant. Biol.* 49:485–95

98. Smith GR, Amundsen SK, Dabert P, Taylor AF. 1995. The initiation and control of homologous recombination in *Escherichia coli. Philos. Trans. R. Soc. London* 347:13–20

99. Smith GR, Schultz DW, Taylor AF, Triman K. 1981. Chi sites, RecBC enzyme, and generalized recombination. *Stadler Genet. Symp.* 13:25–37

100. Sobell HM. 1972. Molecular mechanism for genetic recombination. *Proc. Natl. Acad. Sci. USA* 69:2483–87

101. Stahl FW, Crasemann JM, Stahl MM. 1975. Rec-mediated recombinational hot spot activity in bacteriophage lambda. III. Chi mutations are site-mutations stimulating Rec-mediated recombination. *J. Mol. Biol.* 94:203–12

102. Stahl FW, Lieb M, Stahl MM. 1984. Does Chi give or take? *Genetics* 108:795–808

103. Stahl FW, Shurvinton CE, Thomason LC, Hill S, Stahl MM. 1995. On the clustered exchanges of the RecBCD pathway operating on phage λ. *Genetics* 139:1107–21

104. Stahl FW, Stahl MM. 1977. Recombination pathway specificity of Chi. *Genetics* 86:715–25

105. Stahl FW, Stahl MM, Malone RE, Crasemann JM. 1980. Directionality and nonreciprocality of Chi-stimulated recombination in phage λ. *Genetics* 94:235–48

106. Stahl FW, Stahl MM, Young L, Kobayashi I. 1982. Chi-stimulated recombination between phage λ and the plasmid λ dv. *Genetics* 102:599–613

107. Stahl FW, Thomason LC, Siddiqi I,

Stahl MM. 1990. Further tests of a recombination model in which Chi removes the RecD subunit from the RecBCD enzyme of *Escherichia coli. Genetics* 126:519–33

108. Strathern JN, Shafer BK, McGill CB. 1995. DNA synthesis errors associated with double-strand-break repair. *Genetics* 140:965–72

109. Sun H, Treco D, Szostak JW. 1991. Extensive 3′-overhanging, single-stranded DNA associated with the meiosis-specific double-strand breaks at the *ARG4* recombination initiation site. *Cell* 64:1155–61

110. Sung P, Trujillo KM, Van Komen S. 2000. Recombination factors of *Saccharomyces cerevisiae. Mutat. Res.* 451:257–75

111. Symington LS, Petes TD. 1988. Expansions and contractions of the genetic map relative to the physical map of yeast chromosome III. *Mol. Cell. Biol.* 8:595–604

112. Szostak JW, Orr-Weaver TL, Rothstein RJ, Stahl FW. 1983. The double-strand-break repair model for recombination. *Cell* 33:25–35

113. Taylor AF, Schultz DW, Ponticelli AS, Smith GR. 1985. RecBC enzyme nicking at Chi sites during DNA unwinding: location and orientation dependence of the cutting. *Cell* 41:153–63

114. Taylor AF, Smith GR. 1980. Unwinding and rewinding of DNA by the RecBC enzyme. *Cell* 22:447–57

115. Taylor AF, Smith GR. 1985. Substrate specificity of the DNA unwinding activity of the RecBC enzyme of *Escherichia coli. J. Mol. Biol.* 185:431–43

116. Taylor AF, Smith GR. 1992. RecBCD enzyme is altered upon cutting DNA at a Chi recombination hotspot. *Proc. Natl. Acad. Sci. USA* 89:5226–30

117. Taylor AF, Smith GR. 1995. Monomeric RecBCD enzyme binds and unwinds DNA. *J. Biol. Chem.* 270:24451–58

118. Taylor AF, Smith GR. 1995. Strand specificity of nicking of DNA at Chi sites

by RecBCD enzyme: modulation by ATP and magnesium levels. *J. Biol. Chem.* 270:24459–67

119. Taylor AF, Smith GR. 1999. Regulation of homologous recombination: Chi inactivates RecBCD enzyme by disassembly of the three subunits. *Genes Dev.* 13:890–900

120. Thaler DS, Sampson E, Siddiqi I, Rosenberg SM, Stahl FW, et al. 1988. A hypothesis: Chi-activation of RecBCD enzyme involves removal of the RecD subunit. In *Mechanisms and Consequences of DNA Damage Processing*, ed. E Friedberg, P Hanawalt, pp. 413–22. New York: Liss

121. Thaler DS, Stahl FW. 1988. DNA double-chain breaks in recombination of phage λ and of yeast. *Annu. Rev. Genet.* 22:169–97

122. Voelkel-Meiman K, Keil RL, Roeder GS. 1987. Recombination-stimulating sequences in yeast ribosomal DNA correspond to sequences regulating transcription by RNA polymerase I. *Cell* 48:1071–79

123. Voelkel-Meiman K, Roeder GS. 1990. Gene conversion tracts stimulated by *HOT1*-promoted transcription are long and continuous. *Genetics* 126:851–67

124. Ward TR, Hoang ML, Prusty R, Lau CK, Keil RL, et al. 2000. Ribosomal DNA replication fork barrier and *HOT1* recombination hot spot: shared sequences but independent activities. *Mol. Cell. Biol.* 20:4948–57

125. West SC. 1997. Processing of recombination intermediates by the RuvABC proteins. *Annu. Rev. Genet.* 31:213–44

126. White RL, Fox MS. 1974. On the molecular basis of high negative interference. *Proc. Natl. Acad. Sci. USA* 71:1544–48

127. Williams JG, Shibata T, Radding CM. 1981. *Escherichia coli* RecA protein protects single-stranded DNA or gapped duplex DNA from degradation by RecBC DNase. *J. Biol. Chem.* 256:7573–82

128. Williamson MS, Game JC, Fogel S. 1985. Meiotic gene conversion mutants

in *Saccharomyces cerevisiae*. I. Isolation and characterization of *pms1-1* and *pms1-2*. *Genetics* 110:609–46

129. Wu T-C, Lichten M. 1995. Factors that affect the location and frequency of meiosis-induced double-strand breaks in *Saccharomyces cerevisiae*. *Genetics* 140:55–66

130. Xu F, Petes TD. 1996. Fine-structure mapping of meiosis-specific double-strand DNA breaks at a recombination hotspot associated with an insertion of telomeric sequences upstream of the *HIS4* locus in yeast. *Genetics* 143:1115–25

131. Yu M, Souaya J, Julin DA. 1998. The 30-kDa C-terminal domain of the RecB protein is critical for the nuclease activity, but not the helicase activity, of the RecBCD enzyme from *Escherichia coli*. *Proc. Natl. Acad. Sci. USA* 95:981–86

132. Yu M, Souaya J, Julin DA. 1998. Identification of the nuclease active site in the multifunctional RecBCD enzyme by creation of a chimeric enzyme. *J. Mol. Biol.* 283:797–808

133. Zahn-Zabal M, Lehmann E, Kohli J. 1995. Hot spots of recombination in fission yeast: inactivation of the *M26* hot spot by deletion of the *ade6* promoter and the novel hotspot *ura4-aim*. *Genetics* 140:469–78

Annu. Rev. Genet. 2001. 35:275–302

MECHANISMS OF RETROVIRAL RECOMBINATION

Matteo Negroni[1] and Henri Buc[2]

[1]*Unité de Regulation Enzymatique des Activités Cellulaires, FRE 2364–CNRS, and* [2]*URA 1960–CNRS; Institut Pasteur, 25-28 rue du Dr. Roux, 75724 Paris cedex 15, France; e-mail: matteo@pasteur.fr*

Key Words strand transfer, reverse transcription, genetic variability, retroviruses, RNA

■ **Abstract** Recombination is a major source of genetic variability in retroviruses. Each viral particle contains two single-stranded genomic RNAs. Recombination mostly results from a switch in template between these two RNAs during reverse transcription. Here we emphasize the main mechanisms underlying recombination that are emerging from recent advances in biochemical and cell culture techniques. Increasing evidence supporting the involvement of RNA secondary structures now complements the predominant role classically attributed to enzyme pausing during reverse transcription. Finally, the implications of recombination on the dynamics of emergence of genomic aberrations in retroviruses are discussed.

CONTENTS

0066-4197/01/1215-0275$14.00 **275**

INTRODUCTION

The infectious cycle of retroviruses is characterized by the alternate use of DNA and RNA as genetic material (8, 148). During their extracellular life, retroviruses store genetic information as a linear positive RNA strand. After entering the cell, the genomic RNA is converted into a double-stranded DNA structure that is then integrated into the host's genome. The integrated form, called the "provirus," is replicated as part of the cellular genome and is used to synthesize the viral mRNAs as well as the new genomic RNAs (29).

A salient characteristic of retroviral replication is the high degree of genetic recombination observed after co-infection of a cell by two genetically distinct viruses, a feature first observed in avian tumor viruses (151) and later in other retroviruses as well (26, 160, 166). The impact of recombination on retroviral evolution has become increasingly a matter of investigation. Recently, the growing number of human immunodeficiency virus (HIV) genomes sequenced has documented the importance of recombination in the dynamics of viral populations. At least 10% of the infectious strains of this virus have been generated by recombination among different viral subtypes (91, 107, 130). In mice, recombination among endogenous retroviruses was shown to be responsible for generating pathogenic strains (134). Most recombination events involve homologous regions of the genomes (homologous recombination). Homologous recombination is the most frequent source of genetic rearrangements in retroviruses; the frequency of its appearance is at least as high as the cumulative occurrence of all other types of genomic rearrangements (Table 1). Recombination can also involve regions of homology as short as a few nucleotides (25, 176), leading to "non-homologous recombination." Although this process is 100- to 1000-fold less efficient than homologous recombination in vivo (175), it allows more extensive genomic rearrangements. Nonhomologous recombination has been invoked as being operative

TABLE 1 In vivo estimates of the frequency of occurrence of different types of genomic modifications in retroviruses

Virus	Recombination frequency[a,b]	Cumulative frequency[a] of point mutations and other genomic aberrations
SNV[c]	4×10^{-5} (66)	1.2×10^{-5} (105, 106)
MLV	4.7×10^{-5} (3)	
MoMLV		3×10^{-5} (103)
HIV	2.4×10^{-4} (73)	3×10^{-5} (88, 90)
BLV		5×10^{-6} (89)

[a]Values per nt and per infectious cycle.

[b]Values for markers less than, or equal to 1 kb apart.

[c]Abbreviations: SNV, spleen necrosis virus; MLV, murine leukemia virus; MoMLV, Moloney murine leukemia virus; HIV, human immunodeficiency virus; BLV, bovine leukemia virus.

at several turning points of retroviral evolution: the capture of oncogenes by trans-duction of cellular sequences (53, 141, 142), the generation of gene duplications, and the acquisition of the *vpx* gene in the simian immunodeficiency virus from sooty mangabeys (SIV_{SM}) and HIV type 2 (HIV-2) (131).

When does recombination occur? Retroviruses are pseudo-diploid, since two copies of genomic RNA are present in each virion. The hypothesis that the forma-tion of heterozygous virions is necessary for genetic recombination to occur was first formulated in 1973 (155). Definitive evidence that copackaging of two genet-ically distinct genomic RNAs is required in the same viral particle was provided by Hu & Temin, using an experimental system where retroviral infection was lim-ited to a single cycle (66, 67). These experiments demonstrated that recombinant genotypes were produced during reverse transcription. In principle, recombina-tion could occur during synthesis of the first or second DNA strand, called (−) and (+)DNA strands, respectively. Although results in support of both hypotheses have been obtained (14, 65), mounting data indicate that retroviral recombination is due mostly to template switching during synthesis of the (−)DNA strand, when the template is an RNA molecule (3, 73, 171, 174). Recombination during (+)DNA synthesis has been extensively reviewed by Boone & Skalka (14). Recombination occurring on RNA templates results from the transfer, during reverse transcrip-tion, of the nascent DNA from one molecule of genomic RNA to the other (strand transfer). The various mechanisms proposed to account for these kinds of event, and globally known as "copy choice" (152), are reviewed here. These models are derived from studies made by infecting cells in culture (ex vivo observations) or by using purified proteins and nucleic acids (in vitro observations). Before discussing these mechanistic models we briefly describe the main actors and scenarios that occur during the process of reverse transcription.

REVERSE TRANSCRIPTION

Retroviral particles have an outer membrane composed of lipids and envelope gly-coproteins, and an internal protein core. This core contains the two copies of the genomic RNA, the viral DNA polymerase (reverse transcriptase, RT) and several components required for their reverse transcription and integration of the resulting products into the host's genome (153). [For a detailed description of the process of reverse transcription, see (145)]. Reverse transcription begins near the 5′ end of the genomic RNA using as primer a specific host-encoded tRNA (reviewed in 86), the 3′ terminal 18 nucleotides of which are annealed to a complementary region of the genome, called the primer binding site (PBS) (Figure 1a). DNA synthesis soon reaches the 5′ end of the genomic RNA (Figure 1a). To copy the internal regions of the genome, the nascent DNA must be transferred to the 3′ end of the genomic RNA, on the terminal repeated sequence R (Figure 1a). This process, known as "(−)DNA strong stop strand transfer," is detailed later. During reverse transcription, the RT-encoded RNase H activity degrades the genomic RNA once it is copied (145), leaving intact one or more (depending on the virus) specific RNA

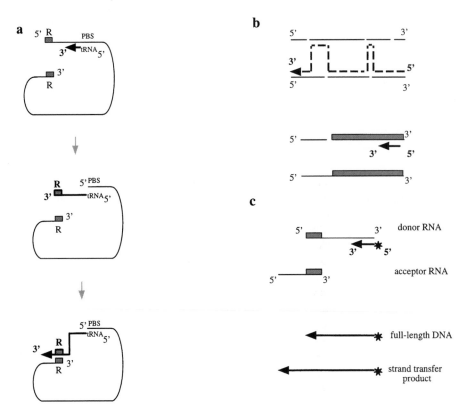

Figure 1 (−)DNA strong stop strand transfer and forced copy choice. Thin lines and plain case text, RNA; thick line and bold case text, DNA. Panel *A*: Strong stop strand transfer. DNA synthesis begins at the tRNA/PBS complex (*top panel*) and is arrested at the 5′ end of the template. The template RNA is degraded by the RNase H activity (*middle panel*), and the nascent DNA is transferred (*bottom panel*) onto the 3′ copy of the repeated region R (*hatched box*). The size of R varies among different retroviruses, from 15 to 247 nucleotides. Panel *B*: Forced copy choice model. Top drawing, thin lines: genomic RNAs. When a break is encountered, DNA synthesis is transferred to the other genomic RNA, as shown by the thick dotted line indicating the path followed by reverse transcription. The extent of the region of homology for each transfer varies depending on the positions of the breaks on the genomic RNAs. This is detailed in the lower drawing, where the region of homology is drawn as a hatched box, in analogy with the representation of the region R in panel *A*. Panel *C*: Top drawing: experimental system generally used to study strong stop or forced copy choice in vitro. Arrow with a star, labeled primer for DNA synthesis; hatched box, region of homology between donor and acceptor templates. Strand transfer is monitored by the different size of the product of reverse transcription (*bottom drawing*).

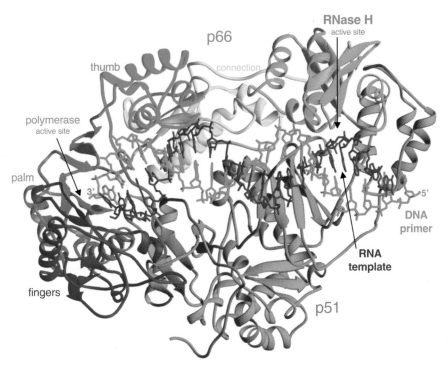

Figure 2 A sketch of the interaction of HIV reverse transcriptase with its primer template. (Figure kindly provided by S. Sarafianos & E. Arnold)

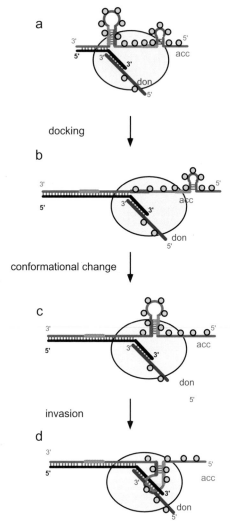

Figure 4 Pause independent model for strand transfer proposed in Reference 97. Symbols: *gray ellipse*: RT; *red and blue lines*: acceptor and donor RNA templates, respectively; *black lines*: nascent DNA; acc: acceptor template; don: donor template. The light and dark boxes on the acceptor template represent stretches of inverted repeats. In the example given the hairpin on the left in (*a*) is disrupted. The process begins with the docking of the acceptor template onto the nascent DNA, allowing a more extensive hybridization of the two partners (*b*). The presence of NC (*yellow circles*) on the acceptor template allows new folding with the formation of a stable hairpin, containing the sequence drawn in green in the loop portion (*c*). Strand exchange proceeds via an invasion mechanism, with the acceptor template displacing the donor RNA from the DNA primer (*d*). The displacement begins with a sequence competent for the invasion, likely in a single stranded form, arbitrarily chosen here as the loop portion of the hairpin.

oligomer(s), rich in purine bases and called the PPT (polypurine tract) (reviewed in 145). These RNA oligonucleotides are then used as primers to synthesize the second, or (+)DNA, strand. Completion of synthesis of (+)DNA also requires a strand transfer event—(+)DNA strong stop strand transfer—that involves the annealed tRNA/PBS sequences. The full-length double-stranded DNA is then competent for integration into the host's genome (16).

The Reverse Transcription Complex

Completion of retroviral DNA synthesis can be observed, albeit inefficiently, in detergent-permeabilized virions supplemented with deoxyribonucleotides triphosphate (dNTPs), indicating that all the components required for reverse transcription, apart from the dNTPs, are present in the virions (52, 125, 173). Reverse transcription occurs almost exclusively in the cytoplasm of the infected cell and, initially, within the confined environment provided by the viral core. For Moloney murine leukemia virus (MoMLV), although the virions seem to travel across the cytoplasm as basically intact nonenveloped viral cores (121), a gradual dissociation of the core as reverse transcription progresses was suggested by a decrease in the size of this complex (46). For HIV type 1 (HIV-1), by contrast, the core is more fragile and is disrupted soon after entry in the cell (45, 75, 93). However, complementation of RT⁻ mutants by expressing a wild-type RT in *trans* is extremely inefficient, indicating that DNA synthesis is essentially carried out by the enzyme already present in the complex (5). Most available data on the components of the reverse transcription complex come from studies on HIV-1, where mature cores contain the reverse transcriptase, the integrase, the nucleocapsid protein (NC), Vpr, Vif, and the protease (156, and references therein). The physical association with the reverse transcription complex of the mature integrase (165), NC (92), and Vpr (114) has been demonstrated. Participation of the same viral proteins has also been indirectly inferred from perturbations of reverse transcription processes observed in mutant virions such as Nef (129), Vif (17), the integrase (150), and, again, NC (see below).

Most studies on retroviral recombination in reconstituted in vitro systems have been performed either on naked RNA templates or in the presence of the NC protein alone. The NC is the most abundant component of the complex inasmuch as the genomic RNAs in the viral particle are coated by several hundred NC molecules (32, 40, 92, 154). This ribonucleoprotein complex constitutes the template for reverse transcription. The possible role of other components of the reverse transcription complex in recombination is under investigation in several laboratories.

The Reverse Transcriptase

Reverse transcriptases are directly involved in recombination generated by template switching. RTs possess RNA- and DNA-dependent DNA polymerase activities, as well as a distinct RNase H activity (21). All RTs share common structural motifs with other DNA polymerases (reviewed in 15). Our knowledge on the structure-function relationships of RT is mostly due to the wealth of information accumulated from crystal structures of the HIV-1 RT bound either to DNA duplexes

(69, 71) or to RNA-DNA hybrids (126). This enzyme is a heterodimer composed of two polypeptide chains, p66 and p51 (reviewed in 6). Only the p66 subunit possesses enzymatic activity. Because of p66's resemblance to a right hand, three of its subdomains have been called "finger," "palm," and "thumb." A fourth domain ("connection") links the thumb to the RNase H domain. In all crystal structures, the primer-template duplex extends from the polymerase catalytic site, located in the palm, through a deep cleft defined by the thumb and the fingers, to the RNase H site (Figure 2; see color insert). In the structure of HIV-1 RT complexed with the RNA-DNA duplex, the RNase H site is located 18 nucleotides behind the polymerase catalytic site (126). The main contacts between the enzyme and the nucleic acids lie at the two catalytic sites, ahead of the polymerase active site [where the overhanging template kinks away from the orientation of the primer-template, and slides up the finger subdomain (69)], and on the nascent primer strand. Here a loop between the palm and the thumb, called the primer-grip (144), contacts the primer strand while the thumb inserts a series of essential amino-acid side chains within the minor groove (69, 71) (Figure 2). Polymerization by RT is frequently interrupted. This phenomenon, called "pausing," leads occasionally to the dissociation of the enzyme from the primer-template. The number of nucleotides incorporated before dissociation defines the processivity of the enzyme, which is low for RTs, compared with cellular "replicative" DNA polymerases (145). Pausing of reverse transcription is sequence specific (60). On RNA, pausing has been correlated with the presence of stable secondary structures on the template RNA, either on the RNA template ahead the RT, or on the nascent DNA behind it (60, 138, 139). All the nucleic acid-binding motifs reviewed above contribute to the processivity of the enzyme (9, 79, 84), an important parameter for strand transfer.

Another structural feature of possible relevance for template switching is that RT/primer-template binary complexes can bind an additional single-stranded RNA molecule (19, 109). Indeed, the surface potential in the enzyme groove allows the insertion of another nucleic acid (19). The possible occurrence of extended interactions outside the canonical primer/template-binding site is also suggested by the crystal structure of a complex between HIV-1 RT and an RNA pseudo-knot for which it displays a high affinity (72). In both cases, the pseudo-knot and the additional single-stranded RNA were shown to interact also with the p51 subunit.

RNase H

The RNase H activity carried by reverse transcriptases is an endonuclease (80, 100) that can follow two alternative modes of action. One mode allows the same RT molecule engaged in reverse transcription to hydrolyze the RNA template concomitantly to DNA synthesis (polymerization-dependent RNase H). This cut occurs at a fixed distance from the 3'-OH of the nascent DNA, as determined by the spacing between the polymerase and RNase H catalytic sites. Contrasting results have been obtained on the temporal coupling between DNA synthesis and RNA degradation, described as being either concomitant or delayed with respect to DNA polymerization (55, 76). The RNase H activity also displays a certain site preference for cleavage (49). As a result, residual RNA oligomers occasionally remain hybridized to the nascent DNA. These residual oligonucleotides are then further hydrolyzed by the second mode of action of the RNase H activity through a series of endonucleolytic cuts performed in the absence of DNA synthesis [polymerization-independent RNase H (10, 158)]. The short RNA stretches produced will then dissociate from the DNA.

The Nucleocapsid Protein

The nucleocapsid protein (NC) is a small basic Zn-binding protein (13, 22, 137) tightly bound to the genomic RNA of almost all retroviruses (31, 33, 48, 92, 115). NC contain one or two stretches of 14 amino acids organized in a Cys-X_2-Cys-X_4-His-X_4-Cys array (61), called a Zn-finger. Each Zn-finger binds one Zn^{2+} ion and structurally constitutes a rigid domain, flanked by flexible N-terminal and C-terminal regions (94, 117). The resolution of the structure of the HIV-1 NC bound to the stem-loop SL3 of the packaging signal in HIV-1 RNA by nuclear magnetic resonance (NMR) indicates that the Zn-finger and the flanking domains both interact with the RNA (34).

NC tightly binds nucleic acids in a nonspecific manner (51, 169), although a systematic evolution of ligands by an exponential enrichment (SELEX) approach (11) and measurements of its affinity for short DNA or RNA oligonucleotides in low ionic strength buffers have disclosed a moderate preference for certain sequences (47). However, in vivo the genomic RNA is completely coated with NC molecules (32, 40, 92), and the relevance of such sequence preferences for the reverse transcription process has yet to be assessed.

NC enhances the annealing of complementary strands of nucleic acids (39, 149), but also promotes the reverse reaction, the unwinding of complex structures of nucleic acids (77) as those responsible for self priming during (−)DNA strong stop strand transfer (58). Therefore, NCs are considered to be RNA chaperones (62, 118), proteins that assist RNA folding by preventing, or resolving, misfolded structures, a point recently demonstrated by elegant in vivo and in vitro assays (27, 157a). The Zn-finger and the basic N-terminal residues are both required to enhance the annealing activity (35, 59, 82, 119, 132).

Since strand transfer occurs during reverse transcription and is strongly enhanced by NC (see below), the effect of NC on DNA synthesis must also be evaluated. This analysis is particularly difficult in vivo, since NC mutants are perturbed at several steps preceding reverse transcription, including the assembly of the viral particles (reviewed in 118). HIV-1 NC mutants were shown to synthesise lower levels of DNA in vivo (56, 172). Most reverse transcription products were shorter than full-length products and carried major aberrations that prevent their integration (56, 57). Accordingly, MoMLV NC mutant virions also demonstrated significant defects in DNA synthesis in vivo. However, in this case no defects were observed for reverse transcription in permeabilized virions (54), leading the authors to suggest that the mutations affected reverse transcription only indirectly, probably because of problems at the level of uncoating the virion after cell entry (54).

The effect of NC on reverse transcription has been intensively studied in vitro. Two main parameters have been monitored: the amount of synthesized product and the degree of pausing during reverse transcription. Conflicting results have been obtained. In some cases, a reduction in pausing, particularly at structured regions of the template, was clearly observed (74, 143, 163). In another study, varying the concentration of NC yielded either an increase or a decrease in the amount of product, and no effect on processivity (124). In other work, the presence of NC did not induce major modifications in the pattern of pausing (43, 44, 96, 116) or of the amount of DNA synthesized (43, 44, 116). In contrast, NC decreased the amount of full-length product when reverse transcription was performed on long RNAs (96). Despite these uncertainties, the NC is currently referred to as a protein enhancing reverse transcription.

STRAND TRANSFER ON FRAGMENTED TEMPLATES

Several lines of evidence suggest that the genomic RNAs are fragmented within the viral particle (see 28), an observation that led Coffin to formulate the "forced copy choice" model for retroviral recombination (28). Here, the trigger for strand transfer is an arrest of reverse transcription imposed by a break on the RNA. The only solution for continuing DNA synthesis is to transfer the nascent DNA onto the homologous sequence present on the second genomic RNA strand (Figure 1*b*). Retroviral recombination would then mostly depend on the integrity of the retroviral genomic RNA. Although the contribution of RNA breaks seems extremely plausible, a correlation between the degree of RNA breakage in the virions and the frequency of recombination has yet to be established. The most direct attempt to demonstrate such a relationship, a study on recombination in virions treated with γ radiation, produced no conclusive results (68). No significant enhancement in the frequency of recombination was observed for increasing doses of irradiation, and the modification of the genomic RNAs induced by this treatment was the cross-linking of proteins to the genomic RNA, rather than its hydrolysis (68).

Strong-Stop and Forced Copy Choice Strand Transfers

In the absence of model systems to study forced copy choice in vivo, the parameters that govern the process can be inferred by analogy with a well-known case of strand transfer occurring from the end of an RNA template, (−)DNA strong stop strand transfer (Figure 1*a*). What are the main parameters distinguishing strong stop from forced copy-choice strand transfers, and how relevant are these differences in vivo? One difference lies in the size of the region of homology involved, which is limited to the repeated sequence R for strong stop but is more extensive for forced copy choice (Figures 1*a*, *b*). Deletion analyses on R from murine leukemia viruses (MLV) and HIV-1 show that the size of R does not limit the efficiency of strong stop strand transfer in vivo (12, 30). Longer regions of homology are therefore probably not required to assure efficient strand transfer. Thus the difference in size of the regions of homology involved in these two cases could not be an important parameter, and strong stop strand transfer and forced copy choice might have comparable efficiencies.

Another point of diversity concerns the sequence involved in the transfer. While strong stop always occurs within the R sequence region, forced copy choice potentially occurs wherever a break is present in the genomic RNA. The following experiments suggest that the efficiency of strand transfer from the end of an RNA template can vary according to the sequence involved. Recently, the repeated sequence R from MoMLV was successfully replaced, ex vivo, by a nonviral sequence (25). However, as judged by the viral titer, the efficiency of strand transfer was six times lower with respect to wild-type viruses. In addition, in vitro observations show that the yield of strand transfer product generated by MoMLV RT varies after replacing R with other sequences (1). In conclusion, sequence-specific effects could affect the efficiency of forced copy choice.

Mechanisms of Terminal Strand Transfer

How does the transfer of the nascent DNA proceed? Dissection of the mechanisms of terminal strand transfer in vitro has mostly been carried out for ($-$)DNA strong stop. Classically, the experimental procedure was to reverse transcribe an RNA in the presence of another RNA whose sequence is homologous to the 5′ terminal region of the donor template (see Figure 1*c*). This approach not only reproduces strong stop strand transfer but also mimics the situation found for the forced copy choice model (Figure 1*c*). Once reverse transcription has reached the 5′ end of the RNA, the 3′ end of the nascent DNA is still annealed to an RNA fragment, which spans the distance from the polymerase to the RNase H RT active sites. The main problem is to make the nascent DNA strand available for hybridization on the complementary sequence of the acceptor template. According to the forced copy-choice model, the solution to this problem is provided by stalling the RT at the 5′ end of the donor RNA template. Since continuing DNA synthesis is not possible, stalling allows the RNase H polymerase-independent activity of the RT to degrade the residual donor RNA template, yielding short fragments that dissociate from the nascent DNA (55). The ultimate degradation of the donor RNA could occur after dissociation of the enzyme from the primer/template followed by a re-binding event, as demonstrated by an elegant series of in vivo experiments on MoMLV. Here a mutation in the RNase H domain of the RT could be complemented for strong stop strand transfer by a RNase H$^+$/polymerization-defective form of RT (146). Polymerization and hydrolysis of the RNA template can therefore be performed by distinct molecules of RT. The nascent DNA, now single-stranded, will then anneal to the complementary sequence of the acceptor template, allowing reverse transcription to continue (Figure 1a). In vitro, both the annealing of the nascent DNA onto the acceptor RNA (1, 44, 58, 63, 108) and the hydrolysis of the RNA (18, 108) are enhanced by NC.

Under conditions of strong stop strand transfer, the translocation of the nascent DNA is currently assumed to take place in the absence of bound RT. However, during this process the acceptor RNA could be cross-linked to a molecule of HIV-1 RT that is still bound to its primer/template (109). Cross-linking and appearance of the strand transfer product were kinetically correlated. These observations led to the proposal for an alternative model where the acceptor RNA, annealed to the trailing part of the nascent DNA, interacts with the RT before strand exchange is completed (109). Strand exchange might therefore occur while the RT is still bound onto the nascent DNA, as previously proposed for strand transfer occurring on homopolymeric sequences (70).

PAUSE-DRIVEN STRAND TRANSFER

In the models discussed above, the trigger for strand transfer is an obvious arrest of DNA polymerization as the enzyme reaches the end of the template. A similar situation is encountered in the presence of a strong, albeit non-definitive, obstacle to reverse transcription. The possibility that strand transfer could be promoted by pausing of DNA synthesis was originally proposed to account for the high frequency of deletions observed during transposition of the Ty element in *Saccharomyces cerevisiae* (167). For retroviruses this eventuality has been investigated in vitro on a region of the HIV-1 nef gene spanning 88 nucleotides, where four prominent pause sites of reverse transcription were identified (Figure 3) (37). Reverse transcription of the donor RNA was performed in the presence of an acceptor RNA homologous to 27 internal nucleotides of the donor template, as detailed in Figure 3a. This work demonstrated that the most prominent pause site contributed significantly to efficient strand transfer, a conclusion further supported by an experiment where abolishing this pause site led to a significant decrease of the frequency of strand transfer (Figure 3b) (162). An accurate analysis of the integrity of the RNA templates during these assays demonstrated that strand transfer occurred efficiently in the absence of detectable breaks in the RNA (37). To account for pause-driven strand transfer, it has been proposed that the nascent DNA dissociates from the donor RNA before annealing to the acceptor template (36). Alternatively, the acceptor RNA could actively displace the nascent DNA from the donor template (36). In all cases the transfer was envisaged as an "enzyme-free" process involving only the nucleic acids (36, 124).

In vitro experiments performed with HIV-1 or MoMLV RTs show that decreasing the concentration of dNTPs in the reverse transcription reaction also enhances

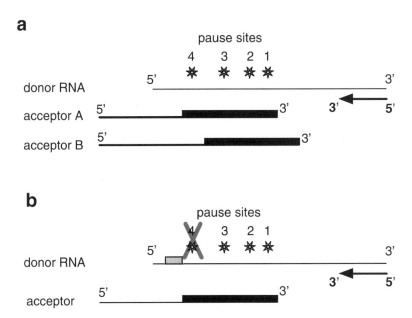

Figure 3 Pause-driven strand transfer. RNA and DNA are represented as in Figure 1; arrow: primer for reverse transcription; stars represent pause sites for reverse transcription and are numbered following the sense of reverse transcription; dashed boxes: regions of homology between donor RNA and acceptor template (27 nucleotides). Strand transfer products yield DNA molecules of a different size from those generated by copying the donor RNA. Panel *A*. Two alternative types of acceptor DNA were used (37). Using acceptor A, instead of B, increased the amount of strand transfer product by a factor 1.25 to 1.5. This suggested that the fourth site, overlapped only by acceptor A, significantly contributed to efficient strand transfer. Panel *B*. Replacing 12 nucleotides downstream of pause site 4 on the donor template by another sequence (*gray box*) eliminated pausing by RT and, concomitantly, decreased the frequency of strand transfer (162).

strand transfer (37, 96). Under these conditions the rate of incorporation of nucleotides into the growing DNA chain is reduced, thereby increasing the probability of enzyme stalling. A parallel observation was made for the occurrence of deletions after treating cultured cells with hydroxyurea. As described below, the mechanism responsible for deletions is probably similar to the one responsible for homologous recombination. Hydroxyurea treatment disrupts the pool of nucleotides, thereby reducing their concentration (50). In MoMLV, the rate of deletion of regions framed by tandem repeated sequences was approximately threefold higher in hydroxyurea-treated cells than in untreated controls (113). Physiologically this parameter could be important, since the concentration of nucleotides varies significantly according to cell type and to the phase of the cellular cycle (50, 85). These observations have been considered as an indication that pausing of DNA synthesis is a major cause of strand transfer.

What is the role played by NC in the framework of this model? On the templates used to study the mechanism of pause-driven strand transfer, NC reduced the intensity of pausing of reverse transcription and concomitantly enhanced template switching (124). If the presence of a pause during copying of the donor template were the only parameter to be considered, NC should rather lower the frequency of template switching. To resolve this paradox, it was suggested that the enhancement of strand transfer by NC is the result of two antagonistic effects: a positive effect on strand transfer exerted through the enhanced resumption of DNA synthesis on the acceptor RNA and a negative one induced by the reduced degree of pausing (124). The positive effect would take precedence over the negative.

PAUSE-INDEPENDENT STRAND TRANSFER

In the models discussed above, changing template is a choice by default, since reverse transcription cannot be continued on the original RNA template. Another possibility is that transferring DNA synthesis onto the acceptor RNA becomes favorable without requiring the existence of an obstacle to reverse transcription on the donor template. An early indication suggesting uncoupling between template switching and pausing of reverse transcription came from experiments performed on the transactivation response element (TAR) hairpin element of HIV-1. Despite the presence of a strong pause site at the base of the hairpin, most strand transfer events took place 45 to 59 nucleotides after the base of the hairpin, in the descending portion of the stem (78). The secondary structure of this RNA region was postulated to promote strand transfer by bringing the acceptor RNA and the nascent DNA into close proximity (78). Recent data on recombination in another region of the HIV-1 genome, the DIS (dimer initiation site) sequence, also implicate specific RNA structures (7). In this case the intermolecular interaction would involve donor and acceptor RNAs, forming a "kissing-loop" structure necessary for the dimerization of genomic RNAs (83, 95, 101, 102, 133). Again, strand transfer would be favored by virtue of close spatial proximity (7).

A mechanism of recombination independent of pre-existing pauses of reverse transcription emerges from another experimental approach. Copy-choice was studied in vitro by reverse transcribing two RNA templates (donor and acceptor templates) that share a region of homology 0.9 kb in size. This region was flanked by two bacterial genetic markers, one on the donor and the other on the acceptor RNAs. After bacterial transformation of the reverse transcription products, recombination was followed by noting the rearrangement of these genetic markers (99). This clonal approach and the presence of distinct mutations on the donor and on the acceptor RNA made it possible to map the region of the template involved in generating each individual recombinant molecule and to determine the frequency of recombination on each portion of these model templates. Significant fluctuations in recombination rates were observed along the template, with values ranging from 1×10^{-5} to 3.7×10^{-4} per nt (97). The regions of most frequent strand transfer, hotspots, were common to the two RTs used (from HIV-1 and from MoMLV). Neither the intensity nor the frequency of pauses during reverse transcription correlated, within a resolution of 50–160 nucleotides, with the positions of the recombination hotspot (97; M. Negroni, L. Polomack & H. Buc, unpublished data). Furthermore, although HIV-1 NC increased the frequency of recombination to varying extents, depending on the region of template under consideration, no corresponding change was observed in the pausing pattern, under the same experimental conditions (96, 97). Finally, whereas coating the acceptor template by NC was sufficient to enhance strand transfer, binding NC to the donor RNA completely failed to enhance strand transfer (97). The enhancement of strand transfer by NC was therefore not related to changes in the reverse transcription process.

A model compatible with these results is that template switching begins with docking of the acceptor RNA on the nascent DNA, a process that yields a long, and therefore stable, heteroduplex (Figure 4a, see color insert). The use of heterologous RNA chaperones and RTs indicated that the increase in strand transfer observed with the NC is mostly due to its RNA chaperone activity (97). Therefore, NC could improve docking by destabilizing intramolecular structures formed by the nascent DNA or by the acceptor template (97), and possibly also by removing residual fragments of donor RNA resistant to degradation by the RNase H. Once docking is achieved, strand exchange in the heteroduplex containing the residual donor RNA and the nascent DNA becomes an intramolecular process (Figure 4c) that is more efficient than an intermolecular reaction. Whether the exchange requires the dissociation of the enzyme from the nascent DNA is not known. If the chasing of the donor RNA takes place in the presence of bound RT (Figure 4d), this displacement could be mediated by a steric hindrance within the enzyme cleft. The displacement step might vary in efficiency according to the structure of the acceptor template, which could contract transitory interactions with the RT (19, 109). The folding of both RNA templates (donor and acceptor) is a dynamic process involving new sequences as reverse transcription progresses (Figure 4b, c), since the hybridization of the nascent DNA onto the acceptor RNA becomes more extensive. The change in the pattern of recombination hotspots observed after NC addition might be due to a change in the folding of the RNA templates, again promoted by the chaperone activity of NC (97).

This model does not exclude the possibility that strand exchange could lead to a temporary arrest of reverse transcription. However, pausing of reverse transcription would not be the cause of strand transfer, as in the previous models, but rather its consequence. Here, temporary pausing at a given position would be restricted to the limited population of molecules that recombine at that position, and recombination hotspots could not be predicted by looking at the pattern of pausing during reverse transcription of the donor RNA.

GENETIC CONSIDERATIONS

Multiple Recombinants and Negative Interference

A screening for recombination events generated during a single infectious cycle by MLV showed that the frequency of recombination does not increase linearly with the distance between the genetic markers (3), because of the high frequency of multiple crossing-over (67). This deviation is particularly evident because multiple events of template switching occur at a higher than expected frequency for independent events (64). This phenomenon, known as "negative interference" (24, 157), can be explained if both genomic RNAs are initially reverse transcribed in vivo. Transferring DNA synthesis from one genomic RNA to the other would then block concomitant reverse transcription of this second molecule, as shown by the schematic diagram in Figure 5. In vitro, restriction of reverse transcription to the donor RNAs increases the frequency of recombination, an effect particularly evident with MoMLV RT (96). Therefore, a first template switching, abolishing concomitant reverse transcription, indirectly increases the probability that the subsequent template switching events will occur.

Deletions

Strand transfer may involve two regions presenting a limited homology and located on the same molecule of genomic RNA. If so, the resulting DNA will carry a deletion. The sequences flanking the simple deletions are sometimes completely divergent, indicating that, as in recombination (25, 176), strand transfer can involve templates with little or no sequence similarity (103).

During infection by MLV, the frequency of deletions observed between two direct repeats of 700 nt increased with the length of the intervening sequence, suggesting the model depicted in Figure 6 for template switching. Mutagenesis of MLV RT also made it possible to pinpoint structural determinants of the enzyme involved in this process in vivo. Mutations of the polymerase catalytic site that reduced the rate of reverse transcription increased the frequency of deletions, a result consistent with those obtained after treating the cells with hydroxyurea (113). However, when mutations were placed in regions important for the processivity of the enzyme, no significant trend was detectable in the correlative change in deletion frequencies (112, 140). Increasing the probability of dissociation of the enzyme from the primer/template did not clearly favor the occurrence of deletions.

Fidelity of Strand Transfer

When assayed in vitro, strong stop, forced copy-choice, and pause-driven strand transfer events display a marked infidelity (38, 109, 110, 162, 164). This is due to the addition of nontemplated nucleotides to the growing DNA chain once reverse transcription is arrested (104, 109). After strand transfer, these extra nucleotides might lead to the formation of a mismatched heteroduplex at the 3′ end of the

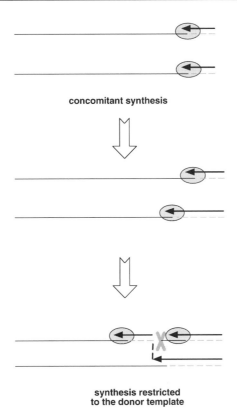

concomitant synthesis

**synthesis restricted
to the donor template**

Figure 5   Possible mechanistic basis for negative interference in retroviruses. Thin line, RNA; thick line, DNA; dotted gray line, RNA template degraded after reverse transcription; gray ellipse, RT. When both genomic RNAs are reverse transcribed (*top panel*), strand transfer can only occur from an RNA where DNA synthesis gains an advance (as for the lower RNA in the *middle panel*) over the concurrent synthesis on the other RNA. Lower panel; the transfer will block (*gray cross*) reverse transcription of the upper RNA. Transfer from the upper RNA onto the lower one becomes unlikely since the region of homology left available is small. This would restrict DNA synthesis to a single template, thereby increasing the probability of a second strand transfer.

primer DNA. Elongation of these mismatched primers would finally introduce point mutations in the resulting DNA (111, 120, 170). In contrast "pause-independent" strand transfer is a faithful process (78; M. Negroni & H. Buc, unpublished observations).

Sequence analyses of recombination junctions generated during intracellular reverse transcription of internal regions of the genome in MoMLV (25, 30, 176) or in endogenous murine retroviruses (134) indicate that the process is extremely accurate. For strong stop strand transfer the situation is less clear. During MoMLV

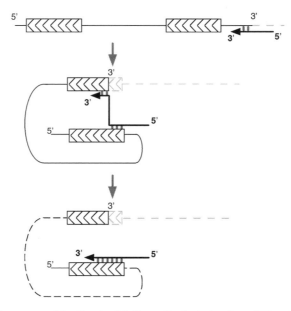

Figure 6 From recombination to deletions: the "missing loop." Two direct repeats, represented by open boxes, are present on the template RNA. Dotted gray line, RNA template degraded after reverse transcription. The crucial step is step 2, where nascent DNA pairs with the complementary RNA sequence present in the 5′ direct repeat. Alignment of the repeats requires flexibility in the intervening sequence (a feature increasing with its length), and the creation of a loop (140). The dotted black line represents the part of viral RNA that will be deleted. Redrawn from Reference 140.

infection, strong stop strand transfer was observed to be error-prone, leading in particular to the addition of nontemplated guanine residues in approximately 10% of the cases (81). However, in a recent analysis on the same retrovirus no mutations were found in 40 products of strong stop strand transfer (30). The reason for this apparent discrepancy is not yet clear.

Sequence Preferences for Strand Transfer

As mentioned above, most infectious forms of HIV are of recombinant origin (122, 123). However, assessing the occurrence of preferential sites for recombination by sequence comparison among the infectious strains of this virus is precluded, since analysis of distribution of the junctions is difficult (4) and only takes into account recombinants that have survived the selective pressure. Therefore, each identified breakpoint results from a process involving an initial recombination event and the adaptability of the resulting recombinant virus in the infected organism. Limiting infection of cultured cells to a single cycle is more appropriate (66, 67, 136, 161, 171). Using this approach, it was shown that recombination in spleen necrosis virus (SNV) is nonrandom and that the insertion of a homopolymeric sequence increased the frequency of strand transfer in a region of the RNA template located 5′ of the insert (161). For HIV-1, the available ex vivo data indicate that, at a resolution of roughly 0.8 kb, recombination rates fluctuate within a five- to eightfold range along the genome, around a value of 3×10^{-4}/nt (73).

RECOMBINATION REVISITED

We have reviewed here three principal mechanisms leading to copy choice, the most frequent type of recombination in retroviruses. These models are not mutually exclusive and might all be operative in vivo. Their relative contribution to most recombination events observed in retroviruses is still a matter of speculation, as recently discussed (98). However, in view of the intensive studies carried out over the past decade, the role of pauses during DNA synthesis as a prerequisite for recombination must be reconsidered. The concept of a pre-existing pause during reverse transcription being mandatory for strand transfer has been challenged recently by the results emerging from totally different approaches (7, 78, 97). It is well established, in vivo (113, 140) and in vitro (37, 96), that a reduction in polymerization rate increases the frequency of strand transfer. However, this finding does not necessarily imply that pausing of reverse transcription is a prerequisite for template switching. In fact, since template switching can occur only while the RT performs synthesis through the region of homology between the two model templates, the longer it takes to reverse transcribe across this region, the higher the probability of strand transfer whatever the underlying mechanism. Apart from this consideration, the requirement for pausing during reverse transcription has been documented only for one specific case: strand transfer on the portion of the nef gene of HIV-1 described above (37, 162). In all other cases, whether local analyses on specific model templates (7, 78), or random searches for a correlation between pausing and strand transfer on long RNAs (97), there is no evidence to support a pause-driven mechanism for strand transfer.

The molecular mechanism by which NC enhances strand transfer remains to be defined, and its elucidation would be of considerable interest. As depicted above, the effect of NC on reverse transcription itself is also far from clear. Its role in strand transfer could be resolved if the molecular processes that occur during reverse transcription of an RNA/NC complex were better understood. For instance, what happens to the molecules of NC coating the RNA when the RT polymerizes through the region of template to which they are bound? Does the NC leave the RNA and rebind to the trailing nascent DNA in a manner similar to the movement of the nucleosome during DNA replication in eukaryotes (135)? That the NC influences the processivity of the RT, and template switching, by modulating the structures of the trailing nascent DNA cannot be ruled out in this case. The mechanism might be similar to that operating during attenuation of transcription in the tryptophan operon in *Escherichia coli* (168).

Mechanistic considerations aside, intensive research on retroviral recombination over the past few years has also improved our understanding of how genetic recombination shapes retroviral evolution. The high frequency of occurrence of recombination requires the assumption generally used to construct phylogenetic trees be re-examined, namely that the divergence of viral strains is essentially governed by the rates of appearance and stabilization of point mutations. The distortions that recombination introduces in how these mutations spread must be taken into account (20, 127, 128). These distortions might be substantial since recombination allows the transfer, in a single cycle of replication, of a whole set of nucleotide changes in a new combination of alleles and occurs with a frequency higher, on average, than that of point mutations. In addition, recombination has a considerably lower probability of generation of nonviable genomes than have point mutations. More specifically, new observations highlight the role of genetic diversity during the progression of HIV infection (87). Viral fitness can be defined as the ability of an infectious progeny to replicate. In the absence of significant environmental changes, genetic drift is often paralleled by a decrease in the fitness of the dominant type of the viral population (2) and in the progression of the disease (159). This is consistent with the theory elaborated for RNA viruses (41, 42) whereby small samples of highly variable RNA viruses accumulate mutations that lead to the generation of subpopulations with a lower fitness (23). However, this population of retroviruses can be suddenly challenged during infection by a strong new selective pressure, as in the infection of a new individual. Most suboptimal forms will be eliminated and recombination, by reshuffling the mutations present in the population, can accelerate the generation of virions that are fit for the new environment. In this context, even mild fluctuations in the local frequency of recombination could bias the generation of specific recombinant forms.

A decade ago, Temin proposed that RTs evolved to be slow polymerases in order to optimize their ability to promote recombination (147). This is consistent with the idea that a slow polymerization rate increases recombination frequency. The mechanism of template switching is probaby more complex and involves secondary, and possibly tertiary, structures of the RNA as well as structures of nucleoprotein complexes. Structural features of RTs, other than those that lower the rate of DNA synthesis, could then have been selected to perform efficient template switching on structured RNAs.

ACKNOWLEDGMENTS

We are grateful to Rob Gorelick, Judith Levin, Roland Marquet, Robert Bambara, Mini Balakhrishnan, Abdeladim Moumen, Miria Ricchetti, and Michel Veron for critical reading of the manuscript. We also acknowledge Robert Bambara, Mini Balakhrishnan, and Stephen Goff for communicating results previous to publication, and Edward Arnold and Stefan Sarafianos for providing Figure 2.

LITERATURE CITED

1. Allain B, Rascle JB, de Rocquigny H, Roques B, Darlix JL. 1998. *Cis* elements and *trans*-acting factors required for minus strand DNA transfer during reverse transcription of the genomic RNA of murine leukemia virus. *J. Mol. Biol.* 277:225–35

2. Allen TM, O'Connor DH, Jing P, Dzuris JL, Mothe BR, et al. 2000. Tat-specific cytotoxic T lymphocytes select for SIV escape variants during resolution of primary viraemia. *Nature* 407:386–90

3. Anderson JA, Bowman EH, Hu WS. 1998. Retroviral recombination rates do not increase linearly with marker distance and are limited by the size of the recombining subpopulation. *J. Virol.* 72:1195–202

4. Anderson JP, Rodrigo AG, Learn GH, Madan A, Delahunty C, et al. 2000. Testing the hypothesis of a recombinant origin of human immunodeficiency virus type 1 subtype E. *J. Virol.* 74:10752–65

5. Ansari-Lari MA, Gibbs RA. 1996. Expression of human immunodeficiency virus type 1 reverse transcriptase in *trans* during virion release and after infection. *J. Virol.* 70:3870–75

6. Arts EJ, Le Grice SFJ. 1998. Interaction of retroviral reverse transcriptase with template-primer duplexes during reverse transcription. *Prog. Nucleic Acid Res. Mol. Biol.* 58:339–93

7. Balakrishnan M, Fay PJ, Bambara RA. 2001. The kissing hairpin sequence promotes recombination with the HIV-1 5' leader region. *J. Biol. Chem.* In press

8. Baltimore D. 1970. RNA-dependent DNA polymerase in virions of RNA tumour viruses. *Nature* 226:1209–11

9. Bebenek K, Beard WA, Darden TA, Li L, Prasad R, et al. 1997. A minor groove binding track in reverse transcriptase. *Nat. Struct. Biol.* 4:194–97

10. Ben-Artzi H, Zeelon E, Amit B, Wortzel A, Gorecki M, Panet A. 1993. RNase H activity of reverse transcriptases on substrates derived from the 5' end of retroviral genome. *J. Biol. Chem.* 268:16465–71

11. Berglund JA, Charpentier B, Rosbash M. 1997. A high affinity binding site for the HIV-1 nucleocapsid protein. *Nucleic Acids Res.* 25:1042–49

12. Berkhout B, van Wamel J, Klaver B. 1995. Requirements for DNA strand transfer during reverse transcription in mutant HIV-1 virions. *J. Mol. Biol.* 252:59–69

13. Bess JW, Powell PJ, Issaq HJ, Schumack LJ, Grimes MK, et al. 1992. Tightly bound zinc in human immunodeficiency virus type 1, human T-cell leukemia virus type I, and other retroviruses. *J. Virol.* 66:840–7

14. Boone LR, Skalka AM. 1993. Strand displacement synthesis by reverse transcriptase. See Ref. 132a, pp. 119–33

15. Brautigam CA, Steitz TA. 1998. Structural and functional insights provided by crystal structures of DNA polymerases and their substrate complexes. *Curr. Opin. Struct. Biol.* 8:54–63

16. Brown PO. 1997. Integration. See Ref. 29, pp. 161–203

17. Camaur D, Trono D. 1996. Characterization of human immunodeficiency virus type 1 Vif particle incorporation. *J. Virol.* 70:6106–11

18. Cameron CE, Ghosh M, Le Grice SF, Benkovic SJ. 1997. Mutations in HIV reverse transcriptase which alter RNase H activity and decrease strand transfer efficiency are suppressed by HIV nucleocapsid protein. *Proc. Natl. Acad. Sci. USA* 94:6700–5

19. Canard B, Sarfati R, Richardson CC. 1997. Binding of RNA template to a

complex of HIV-1 reverse transcriptase/ primer/template. *Proc. Natl. Acad. Sci. USA* 94:11279–84

20. Casci T. 2000. Not such a variable clock? *Nat. Rev. Genet.* 1:86

21. Champoux JJ. 1993. Roles of ribonuclease H in reverse transcription. See Ref. 132a, pp. 103–17

22. Chance MR, Sagi I, Wirt MD, Frisbie SM, Scheuring E, et al. 1992. Extended x-ray absorption fine structure studies of a retrovirus: equine infectious anemia virus cysteine arrays are coordinated to zinc. *Proc. Natl. Acad. Sci. USA* 89: 10041–45

23. Chao L. 1990. Fitness of RNA virus decreased by Muller's ratchet. *Nature* 348:454–55

24. Chase M, Doermann AH. 1958. High negative interference over short segments of the genetic structure of bacteriophage T4. *Genetics* 43:332–53

25. Cheslock SR, Anderson JA, Hwang CK, Pathak VK, Hu WS. 2000. Utilization of nonviral sequences for minus-strand DNA transfer and gene reconstitution during retroviral replication. *J. Virol.* 74: 9571–79

26. Clavel F, Hoggan MD, Willey RL, Strebel K, Martin MA, Repaske R. 1989. Genetic recombination of human immunodeficiency virus. *J. Virol.* 63:1455–59

27. Clodi E, Semrad K, Schroeder R. 1999. Assaying RNA chaperone activity in vivo using a novel RNA folding trap. *EMBO J.* 18:3776–82

28. Coffin JM. 1979. Structure, replication, and recombination of retrovirus genomes: some unifying hypotheses. *J. Gen. Virol.* 42:1–26

29. Coffin JM, Hughes SH, Varmus HE. 1997. *Retroviruses*, pp. 121–60. Cold Spring Harbor, NY: Cold Spring Harbor Lab. Press

30. Dang Q, Hu WS. 2001. Effects of homology length in the repeat region on minus-strand DNA transfer and retroviral replication. *J. Virol.* 75:809–20

31. Darlix JL, Lapadat-Tapolsky M, de Rocquigny H, Roques BP. 1995. First glimpses at structure-function relationships of the nucleocapsid protein of retroviruses. *J. Mol. Biol.* 254:523–37

32. Darlix JL, Spahr PF, Prats AC, Roy C, Wang PA, et al. 1982. Binding sites of viral protein P19 onto Rous sarcoma virus RNA and possible controls of viral functions. *J. Mol. Biol.* 160:147–61

33. Davis NL, Rueckert RR. 1972. Properties of a ribonucleoprotein particle isolated from Nonidet P-40-treated Rous sarcoma virus. *J. Virol.* 10:1010–20

34. De Guzman RN, Wu ZR, Stalling CC, Pappalardo L, Borer PN, Summers MF. 1998. Structure of the HIV-1 nucleocapsid protein bound to the SL3 psi-RNA recognition element. *Science* 279:384–88

35. De Rocquigny H, Gabus C, Vincent A, Fournie-Zaluski MC, Roques B, Darlix JL. 1992. Viral RNA annealing activities of human immunodeficiency virus type 1 nucleocapsid protein require only peptide domains outside the zinc fingers. *Proc. Natl. Acad. Sci. USA* 89:6472–76

36. DeStefano JJ, Bambara RA, Fay PJ. 1994. The mechanism of human immunodeficiency virus reverse transcriptase-catalyzed strand transfer from internal regions of heteropolymeric RNA templates. *J. Biol. Chem.* 269:161–68

37. DeStefano JJ, Mallaber LM, Rodriguez-Rodriguez L, Fay PJ, Bambara RA. 1992. Requirements for strand transfer between internal regions of heteropolymer templates by human immunodeficiency virus reverse transcriptase. *J. Virol.* 66:6370–78

38. DeStefano JJ, Raja A, Cristofaro JV. 2000. In vitro strand transfer from broken RNAs results in mismatch but not frameshift mutations. *Virology* 276:7–15

39. Dib-Hajj F, Khan R, Giedroc DP. 1993. Retroviral nucleocapsid proteins possess potent nucleic acid strand renaturation activity. *Protein Sci.* 2:231–43

40. Dickson C, Eisenman R, Fan H, Hunter E, Reich N. 1985. Protein biosynthesis and assembly. In *RNA Tumor Viruses*, ed. R Weiss, N Teich, HE Varmus, J Coffin, pp. 513–648. Cold Spring Harbor, NY: Cold Spring Harbor Lab. Press

41. Domingo E. 2000. Viruses at the edge of adaptation. *Virology* 270:251–53

42. Domingo E, Escarmis C, Sevilla N, Moya A, Elena SF, et al. 1996. Basic concepts in RNA virus evolution. *FASEB J.* 10:859–64

43. Driscoll MD, Golinelli MP, Hughes SH. 2001. In vitro analysis of human immunodeficiency virus type 1 minus-strand strong-stop DNA synthesis and genomic RNA processing. *J. Virol.* 75:672–86

44. Driscoll MD, Hughes SH. 2000. Human immunodeficiency virus type 1 nucleocapsid protein can prevent self-priming of minus-strand strong stop DNA by promoting the annealing of short oligonucleotides to hairpin sequences. *J. Virol.* 74:8785–92

45. Fassati A, Goff S. 2001. Characterization of intracellular reverse transcription complexes of human immunodeficiency virus type-1. *J Virol:* 75:3626–35

46. Fassati A, Goff SP. 1999. Characterization of intracellular reverse transcription complexes of Moloney murine leukemia virus. *J. Virol.* 73:8919–25

47. Fisher RJ, Rein A, Fivash M, Urbaneja MA, Casas-Finet JR, et al. 1998. Sequence-specific binding of human immunodeficiency virus type 1 nucleocapsid protein to short oligonucleotides. *J. Virol.* 72:1902–9

48. Fleissner E, Tress E. 1973. Isolation of a ribonucleoprotein structure from oncornaviruses. *J. Virol.* 12:1612–15

49. Furfine ES, Reardon JE. 1991. Reverse transcriptase. RNase H from the human immunodeficiency virus. Relationship of the DNA polymerase and RNA hydrolysis activities. *J. Biol. Chem.* 266:406–12

50. Gao WY, Cara A, Gallo RC, Lori F.

1993. Low levels of deoxynucleotides in peripheral blood lymphocytes: a strategy to inhibit human immunodeficiency virus type 1 replication. *Proc. Natl. Acad. Sci. USA* 90:8925–28

51. Gelfand CA, Wang Q, Randall S, Jentoft JE. 1993. Interactions of avian myeloblastosis virus nucleocapsid protein with nucleic acids. *J. Biol. Chem.* 268:18450–56

52. Gilboa E, Goff S, Shields A, Yoshimura F, Mitra S, Baltimore D. 1979. In vitro synthesis of a 9 kbp terminally redundant DNA carrying the infectivity of Moloney murine leukemia virus. *Cell* 16:863–74

53. Goldfarb MP, Weinberg RA. 1981. Generation of novel, biologically active Harvey sarcoma viruses via apparent illegitimate recombination. *J. Virol.* 38:136–50

54. Gonsky J, Bacharach E, Goff SP, Burns DP, Temin HM. 2001. Identification of residues of the Moloney murine leukemia virus nucleocapsid critical for viral DNA synthesis in vivo. *J. Virol.* 75:2616–26

55. Gopalakrishnan V, Peliska JA, Benkovic SJ. 1992. HIV-1 reverse transcriptase: spatial and temporal relationship between the polymerase and RNase H activities. *Proc. Natl. Acad. Sci. USA* 89:10763–67

56. Gorelick RJ, Chabot DJ, Ott DE, Gagliardi TD, Rein A, et al. 1996. Genetic analysis of the zinc finger in the Moloney murine leukemia virus nucleocapsid domain: replacement of zinc-coordinating residues with other zinc-coordinating residues yields noninfectious particles containing genomic RNA. *J. Virol.* 70:2593–97

57. Gorelick RJ, Fu W, Gagliardi TD, Bosche WJ, Rein A, et al. 1999. Characterization of the block in replication of nucleocapsid protein zinc finger mutants from moloney murine leukemia virus. *J. Virol.* 73:8185–95

58. Guo J, Henderson LE, Bess J, Kane B,

Levin JG. 1997. Human immunodeficiency virus type 1 nucleocapsid protein promotes efficient strand transfer and specific viral DNA synthesis by inhibiting TAR-dependent self-priming from minus-strand strong-stop DNA. *J. Virol.* 71:5178–88

59. Guo J, Wu T, Anderson J, Kane BF, Johnson DG, et al. 2000. Zinc finger structures in the human immunodeficiency virus type 1 nucleocapsid protein facilitate efficient minus- and plus-strand transfer. *J. Virol.* 74:8980–88

60. Harrison GP, Mayo MS, Hunter E, Lever AM. 1998. Pausing of reverse transcriptase on retroviral RNA templates is influenced by secondary structures both 5' and 3' of the catalytic site. *Nucleic Acids Res.* 26:3433–42

61. Henderson LE, Copeland TD, Sowder RC, Smythers GW, Oroszlan S. 1981. Primary structure of the low molecular weight nucleic acid-binding proteins of murine leukemia viruses. *J. Biol. Chem.* 256:8400–6

62. Herschlag D. 1995. RNA chaperones and the RNA folding problem. *J. Biol. Chem.* 270:20871–74

63. Hsu M, Rong L, de Rocquigny H, Roques BP, Wainberg MA. 2000. The effect of mutations in the HIV-1 nucleocapsid protein on strand transfer in cell-free reverse transcription reactions. *Nucleic Acids Res.* 28:1724–29

64. Hu WS, Bowman EH, Delviks KA, Pathak VK. 1997. Homologous recombination occurs in a distinct retroviral subpopulation and exhibits high negative interference. *J. Virol.* 71:6028–36

65. Hu WS, Pathak VK, Temin HM. 1993. Role of reverse transcriptase in retroviral recombination. See Ref. 132a, pp. 251–74

66. Hu WS, Temin HM. 1990. Genetic consequences of packaging two RNA genomes in one retroviral particle: pseudodiploidy and high rate of genetic re-

combination. *Proc. Natl. Acad. Sci. USA* 87:1556–60

67. Hu WS, Temin HM. 1990. Retroviral recombination and reverse transcription. *Science* 250:1227–33

68. Hu WS, Temin HM. 1992. Effect of gamma radiation on retroviral recombination. *J. Virol.* 66:4457–63

69. Huang H, Chopra R, Verdine GL, Harrison SC. 1998. Structure of a covalently trapped catalytic complex of HIV-1 reverse transcriptase: implications for drug resistance. *Science* 282:1669–75

70. Huber HE, McCoy JM, Seehra JS, Richardson CC. 1989. Human immunodeficiency virus 1 reverse transcriptase. *J. Biol. Chem.* 264:4669–78

71. Jacobo-Molina A, Ding J, Nanni RG, Clark ADJ, Lu X, et al. 1993. Crystal structure of human immunodeficiency virus type 1 reverse transcriptase complexed with double-stranded DNA at 3.0 Å resolution shows bent DNA. *Proc. Natl. Acad. Sci. USA* 90:6320–24

72. Jaeger J, Restle T, Steitz TA. 1998. The structure of HIV-1 reverse transcriptase complexed with an RNA pseudoknot inhibitor. *EMBO J.* 17:4535–42

73. Jetzt AE, Yu H, Klarmann GJ, Ron Y, Preston BD, Dougherty JP. 2000. High rate of recombination throughout the human immunodeficiency virus type 1 genome. *J. Virol.* 74:1234–40

74. Ji X, Klarmann GJ, Preston BD. 1996. Effect of human immunodeficiency virus type 1 (HIV-1) nucleocapsid protein on HIV-1 reverse transcriptase activity in vitro. *Biochemistry* 35:132–43

75. Karageorgos L, Li P, Burrell C. 1993. Characterization of HIV replication complexes early after cell-to-cell infection. *AIDS Res. Hum. Retrovir.* 9:817–23

76. Kati WM, Johnson KA, Jerva LF, Anderson KS. 1992. Mechanism and fidelity of HIV reverse transcriptase. *J. Biol. Chem.* 267:25988–97

77. Khan R, Giedroc DP. 1992. Recombinant Human Immunodeficiency Virus

Type 1 nucleocapsid (NCp7) protein un-winds tRNA. *J. Biol. Chem.* 267:6689–95

78. Kim JK, Palaniappan C, Wu W, Fay PJ, Bambara RA. 1997. Evidence for a unique mechanism of strand transfer from the transactivation response region of HIV-1. *J. Biol. Chem.* 272:16769–77

79. Klarmann GJ, Schauber CA, Preston BD. 1993. Template-directed pausing of DNA synthesis by HIV-1 reverse transcriptase during polymerization of HIV-1 sequences in vitro. *J. Biol. Chem.* 268:9793–802

80. Krug MS, Berger SL. 1989. Ribonuclease H activities associated with viral reverse transcriptases are endonucleases. *Proc. Natl. Acad. Sci. USA* 86:3539–43

81. Kulpa D, Topping R, Telesnitsky A. 1997. Determination of the site of first strand transfer during Moloney murine leukemia virus reverse transcription and identification of strand transfer-associated reverse transcriptase errors. *EMBO J.* 16:856–65

82. Lapadat-Tapolsky M, Pernelle C, Borie C, Darlix JL. 1995. Analysis of the nucleic acid annealing activities of nucleocapsid protein from HIV-1. *Nucleic Acids Res.* 23:2434–41

83. Laughrea M, Jette L. 1996. Kissing-loop model of HIV-1 genome dimerization: HIV-1 RNAs can assume alternative dimeric forms, and all sequences upstream or downstream of hairpin 248–271 are dispensable for dimer formation. *Biochemistry* 35:1589–98

84. Lavigne M, Buc H. 1999. Compression of the DNA minor groove is responsible for termination of DNA synthesis by HIV-1 reverse transcriptase. *J. Mol. Biol.* 285:977–95

85. Leeds JM, Slabaugh MB, Mathews CK. 1985. DNA precursor pools and ribonucleotide reductase activity: distribution between the nucleus and cytoplasm of mammalian cells. *Mol. Cell. Biol.* 5:3443–50

86. Mak J, Kleiman L. 1997. Primer tRNAs for reverse transcription. *J. Virol.* 71:8087–95

87. Malim HM, Emerman M. 2001. HIV-1 sequence variation: drift, shift, and attenuation. *Cell* 104:469–72

88. Mansky LM. 1996. Forward mutation rate of human immunodeficiency virus type 1 in a T lymphoid cell line. *AIDS Res. Hum. Retrovir.* 12:307–14

89. Mansky LM, Temin HM. 1994. Lower mutation rate of bovine leukemia virus relative to that of spleen necrosis virus. *J. Virol.* 68:494–99

90. Mansky LM, Temin HM. 1995. Lower in vivo mutation rate of human immunodeficiency virus type 1 than that predicted from the fidelity of purified reverse transcriptase. *J. Virol.* 69:5087–94

91. McCutchan FE. 2000. Understanding the genetic diversity of HIV-1. *AIDS* 14:S31–44

92. Meric C, Darlix JL, Spahr PF. 1984. It is Rous sarcoma virus protein P12 and not P19 that binds tightly to Rous sarcoma virus RNA. *J. Mol. Biol.* 173:531–38

93. Miller MD, Farnet CM, Bushman FD. 1997. Human immunodeficiency virus type 1 preintegration complexes: studies of organization and composition. *J. Virol.* 71:5382–90

94. Morellet N, Jullian N, De Rocquigny H, Maigret B, Darlix JL, Roques BP. 1992. Determination of the structure of the nucleocapsid protein NCp7 from the human immunodeficiency virus type 1 by 1H NMR. *EMBO J.* 11:3059–65

95. Muriaux D, Fosse P, Paoletti J. 1996. A kissing complex together with a stable dimer is involved in the HIV-1Lai RNA dimerization process in vitro. *Biochemistry* 35:5075–82

96. Negroni M, Buc H. 1999. Recombination during reverse transcription: an evaluation of the role of the nucleocapsid protein. *J. Mol. Biol.* 286:15–31

97. Negroni M, Buc H. 2000. Copy-choice recombination by reverse transcriptases:

reshuffling of genetic markers mediated by RNA chaperones. *Proc. Natl. Acad. Sci. USA* 97:6385–90

98. Negroni M, Buc H. 2001. Retroviral recombination: What drives the switch? *Nat. Rev. Mol. Cell. Biol.* 2:151–55

99. Negroni M, Ricchetti M, Nouvel P, Buc H. 1995. Homologous recombination promoted by reverse transcriptase during copying of two distinct RNA templates. *Proc. Natl. Acad. Sci. USA* 92:6971–75

100. Oyama F, Kikuchi R, Crouch RJ, Uchida T. 1989. Intrinsic properties of reverse transcriptase in reverse transcription. Associated RNase H is essentially regarded as an endonuclease. *J. Biol. Chem.* 264:18808–17

101. Paillart JC, Marquet R, Skripkin E, Ehresmann B, Ehresmann C. 1994. Mutational analysis of the bipartite dimer linkage structure of human immunodeficiency virus type 1 genomic RNA. *J. Biol. Chem.* 269:27486–93

102. Paillart JC, Skripkin E, Ehresmann B, Ehresmann C, Marquet R. 1996. A loop-loop "kissing" complex is the essential part of the dimer linkage of genomic HIV-1 RNA. *Proc. Natl. Acad. Sci. USA* 93:5572–77

103. Parthasarathi S, Varela-Echavarria A, Ron Y, Preston BD, Dougherty JP. 1995. Genetic rearrangements occurring during a single cycle of murine leukemia virus vector replication: characterization and implications. *J. Virol.* 69:7991–8000

104. Patel PH, Preston BD. 1994. Marked infidelity of human immunodeficiency virus type 1 reverse transcriptase at RNA and DNA template ends. *Proc. Natl. Acad. Sci. USA* 91:549–53

105. Pathak VK, Temin HM. 1990. Broad spectrum of in vivo forward mutations, hypermutations, and mutational hotspots in a retroviral shuttle vector after a single replication cycle: deletions and deletions with insertions. *Proc. Natl. Acad. Sci. USA* 87:6024–28

106. Pathak VK, Temin HM. 1990. Broad spectrum of in vivo forward mutations, hypermutations, and mutational hotspots in a retroviral shuttle vector after a single replication cycle: substitutions, frameshifts, and hypermutations. *Proc. Natl. Acad. Sci. USA* 87:6019–23

107. Peeters M, Sharp PM. 2000. Genetic diversity of HIV-1: the moving target. *AIDS* 14:S129–40

108. Peliska JA, Balasubramanian S, Giedroc DP, Benkovic SJ. 1994. Recombinant HIV-1 nucleocapsid protein accelerates HIV-1 reverse transcriptase catalyzed DNA strand transfer reactions and modulates RNase H activity. *Biochemistry* 33:13817–23

109. Peliska JA, Benkovic SJ. 1992. Mechanism of DNA strand transfer reactions catalyzed by HIV-1 reverse transcriptase. *Science* 258:1112–18

110. Peliska JA, Benkovic SJ. 1994. Fidelity of in vitro DNA strand transfer reactions catalyzed by HIV-1 reverse transcriptase. *Biochemistry* 33:3890–95

111. Perrino FW, Preston BD, Sandell LL, Loeb LA. 1989. Extension of mismatched 3′ termini of DNA is a major determinant of the infidelity of human immunodeficiency virus type 1 reverse transcriptase. *Proc. Natl. Acad. Sci. USA* 86:8343–47

112. Pfeiffer JK, Georgiadis MM, Telesnitsky A. 2000. Structure-based Moloney murine leukemia virus reverse transcriptase mutants with altered intracellular direct-repeat deletion frequencies. *J. Virol.* 74:9629–36

113. Pfeiffer JK, Topping RS, Shin NH, Telesnitsky A. 1999. Altering the intracellular environment increases the frequency of tandem repeat deletion during Moloney murine leukemia virus reverse transcription. *J. Virol.* 73:8441–47

114. Popov S, Rexach M, Zybarth G, Reiling N, Lee MA, et al. 1998. Viral protein R regulates nuclear import of the

HIV-1 pre-integration complex. *EMBO J.* 17:909–17

115. Prats AC, Roy C, Wang PA, Erard M, Housset V, et al. 1990. *cis* elements and *trans*-acting factors involved in dimer formation of murine leukemia virus RNA. *J. Virol.* 64:774–83

116. Raja A, DeStefano JJ. 1999. Kinetic analysis of the effect of HIV nucleocapsid protein (NCp) on internal strand transfer reactions. *Biochemistry* 38:5178–84

117. Ramboarina S, Morellet N, Fournie-Zaluski MC, Roques BP, Moreller N. 1999. Structural investigation on the requirement of CCHH zinc finger type in nucleocapsid protein of human immunodeficiency virus 1. *Biochemistry* 38:9600–7

118. Rein A, Henderson LE, Levin JG. 1998. Nucleic-acid-chaperone activity of retroviral nucleocapsid proteins: significance for viral replication. *Trends Biochem. Sci.* 23:297–301

119. Remy E, de Rocquigny H, Petitjean P, Muriaux D, Theilleux V, et al. 1998. The annealing of tRNA3Lys to human immunodeficiency virus type 1 primer binding site is critically dependent on the NCp7 zinc fingers structure. *J. Biol. Chem.* 273:4819–22

120. Ricchetti M, Buc H. 1990. Reverse transcriptases and genomic variability: the accuracy of DNA replication is enzyme specific and sequence dependent. *EMBO J.* 9:1583–93

121. Risco C, Menendez-Arias L, Copeland TD, Pinto da Silva P, Oroszlan S. 1995. Intracellular transport of the murine leukemia virus during acute infection of NIH 3T3 cells: nuclear import of nucleocapsid protein and integrase. *J. Cell Sci.* 108:3039–50

122. Robertson DL, Anderson JP, Bradac JA, Carr JK, Foley B, et al. http://hiv-web.lanl.gov/HTML/reviews/nomenclature/Nomen.html

123. Robertson DL, Gao F, Hahn BH, Sharp PM. http://hiv-web.lanl.gov/HTML/reviews/hahn.html

124. Rodriguez-Rodriguez L, Tsuchihashi Z, Fuentes GM, Bambara RA, Fay PJ. 1995. Influence of human immunodeficiency virus nucleocapsid protein on synthesis and strand transfer by the reverse transcriptase in vitro. *J. Biol. Chem.* 270:15005–11

125. Rothenberg E, Smotkin D, Baltimore D, Weinberg RA. 1977. In vitro synthesis of infectious DNA of murine leukaemia virus. *Nature* 269:122–26

126. Sarafianos SG, Das K, Tantillo C, Clark AD, Ding J, et al. 2001. Crystal structure of HIV-1 reverse transcriptase in complex with a polypurine tract RNA:DNA. *EMBO J.* 20:1449–61

127. Schierup MH, Hein J. 2000. Consequences of recombination on traditional phylogenetic analysis. *Genetics* 156:879–91

128. Schierup MH, Hein J. 2000. Recombination and the molecular clock. *Mol. Biol. Evol.* 17:1578–79

129. Schwartz O, Marechal V, Danos O, Heard JM. 1995. Human immunodeficiency virus type 1 Nef increases the efficiency of reverse transcription in the infected cell. *J. Virol.* 69:4053–59

130. Sharp PM, Bailes E, Robertson DL, Gao F, Hahn BH. 1999. Origins and evolution of AIDS viruses. *Biol. Bull.* 196:338–42

131. Sharp PM, Bailes E, Stevenson M, Emerman M, Hahn BH, et al. 1996. Gene acquisition in HIV and SIV. *Nature* 383:586–87

132. Shi Y, Berg JM. 1996. DNA unwinding induced by zinc finger protein binding. *Biochemistry* 35:3845–48

132a. Skalka AM, Goff SP, eds. 1993. *Reverse Transcriptase.* Cold Spring Harbor, NY: Cold Spring Harbor Lab. Press

133. Skripkin E, Paillart JC, Marquet R, Ehresmann B, Ehresmann C. 1994. Identification of the primary site of the human immunodeficiency virus type 1

RNA dimerization in vitro. *Proc. Natl. Acad. Sci. USA* 91:4945–49

134. Stoye JP, Moroni C, Coffin JM. 1991. Virological events leading to spontaneous AKR thymomas. *J. Virol.* 65:1273–85

135. Studitsky VM, Clark DJ, Felsenfeld G, Anderson JP, Rodrigo AG, et al. 1994. A histone octamer can step around a transcribing polymerase without leaving the template. *Cell* 76:371–82

136. Stuhlmann H, Berg P. 1992. Homologous recombination of copackaged retrovirus RNAs during reverse transcription. *J. Virol.* 66:2378–88

137. Summers MF, Henderson LE, Chance MR, Bess JW, South TL, et al. 1992. Nucleocapsid zinc fingers detected in retroviruses: EXAFS studies of intact viruses and the solution-state structure of the nucleocapsid protein from HIV-1. *Protein Sci.* 1:563–74

138. Suo Z, Johnson KA. 1997. Effect of RNA secondary structure on the kinetics of DNA synthesis catalyzed by HIV-1 reverse transcriptase. *Biochemistry* 36:12459–67

139. Suo Z, Johnson KA. 1997. RNA secondary structure switching during DNA synthesis catalyzed by HIV-1 reverse transcriptase. *Biochemistry* 36:14778–85

140. Svarovskaia ES, Delviks KA, Hwang CK, Pathak VK. 2000. Structural determinants of murine leukemia virus reverse transcriptase that affect the frequency of template switching. *J. Virol.* 74:7171–78

141. Swain A, Coffin JM. 1992. Mechanism of transduction by retroviruses. *Science* 255:841–45

142. Swanstrom R, Parker RC, Varmus HE, Bishop JM. 1983. Transduction of a cellular oncogene: the genesis of Rous sarcoma virus. *Proc. Natl. Acad. Sci. USA* 80:2519–23

143. Tanchou V, Gabus C, Rogemond V, Darlix J-L. 1995. Formation of stable and functional HIV-1 nucleoprotein complexes in vitro. *J. Mol. Biol.* 252:563–71

144. Tantillo C, Ding J, Jacobo-Molina A, Nanni RG, Boyer PL, et al. 1994. Locations of anti-AIDS drug binding sites and resistance mutations in the three-dimensional structure of HIV-1 reverse transcriptase. Implications for mechanisms of drug inhibition and resistance. *J. Mol. Biol.* 243:369–87

145. Telesnitsky A, Goff S. 1997. Reverse transcriptase and the generation of retroviral DNA. See Ref. 29, pp. 121–60

146. Telesnitsky A, Goff SP. 1993. Two defective forms of reverse transcriptase can complement to restore retroviral infectivity. *EMBO J.* 12:4433–38

147. Temin HM. 1993. Retrovirus variation and reverse transcription: abnormal strand transfers result in retrovirus genetic variation. *Proc. Natl. Acad. Sci. USA* 90:6900–3

148. Temin HM, Mizutani S. 1970. RNA-dependent DNA polymerase in virions of Rous sarcoma virus. *Nature* 226:1211–13

149. Tsuchihashi Z, Brown PO. 1994. DNA strand exchange and selective DNA annealing promoted by the human immunodeficiency virus type 1 nucleocapsid protein. *J. Virol.* 68:5863–70

150. Tsurutani N, Kubo M, Maeda Y, Ohashi T, Yamamoto N, et al. 2000. Identification of critical amino acid residues in human immunodeficiency virus type 1 is required for efficient proviral DNA formation at steps prior to integration in dividing and nondividing cells. *J. Virol.* 74:4795–806

151. Vogt PK. 1971. Genetically stable reassortment of markers during mixed infection with avian tumor viruses. *Virology* 46:947–52

152. Vogt PK. 1973. *The genome of avian RNA tumor viruses: a discussion of four models.* Presented at Meet. "Possible episomes in eukaryotes"

153. Vogt VM. 1997. Retroviral virions and genomes. See Ref. 29, pp. 27–69

154. Vogt VM, Simon MN, Tsurutani N, Kubo M, Maeda Y, et al. 1999. Mass determination of Rous sarcoma virus virions by scanning transmission electron microscopy. *J. Virol.* 73:7050–55

155. Weiss RA, Mason WS, Vogt PK. 1973. Genetic recombinants and heterozygotes derived from endogenous and exogenous avian RNA tumor viruses. *Virology* 52:535–52

156. Welker R, Hohenberg H, Tessmer U, Huckhagel C, Krausslich HG. 2000. Biochemical and structural analysis of isolated mature cores of human immunodeficiency virus type 1. *J. Virol.* 74:1168–77

157. White RL, Fox MS. 1974. On the molecular basis of high negative interference. *Proc. Natl. Acad. Sci. USA* 71:1544–48

157a. Williams MC, Rouzina I, Wenner JR, Gorelick RJ, Musier-Forsyth K, Bloomfield VA. 2001. Mechanism for nucleic acid chaperone activity of HIV-1 nucleocapsid protein revealed by single molecule stretching. *Proc. Natl. Acad. Sci USA* 98:6121–26

158. Wisniewski M, Balakrishnan M, Palaniappan C, Fay PJ, Bambara RA. 2000. Unique progressive cleavage mechanism of HIV reverse transcriptase RNase H. *Proc. Natl. Acad. Sci. USA* 97:11978–83

159. Wolinsky SM, Korber BT, Neumann AU, Daniels M, Kunstman KJ, et al. 1996. Adaptive evolution of human immunodeficiency virus-type 1 during the natural course of infection. *Science* 272:537–42

160. Wong PK, McCarter JA, Henderson LE, Bowers MA, Sowder RC, et al. 1973. Genetic studies of temperature-sensitive mutants of Moloney-murine leukemia virus. *Virology* 53:319–26

161. Wooley DP, Bircher LA, Smith RA. 1998. Retroviral recombination is nonrandom and sequence dependent. *Virology* 243:229–34

162. Wu W, Blumberg BM, Fay PJ, Bambara RA. 1995. Strand transfer mediated by human immunodeficiency virus reverse transcriptase in vitro is promoted by pausing and results in misincorporation. *J. Biol. Chem.* 270:325–32

163. Wu W, Henderson LE, Copeland TD, Gorelick RJ, Bosche WJ, et al. 1996. Human immunodeficiency virus type 1 nucleocapsid protein reduces reverse transcriptase pausing at a secondary structure near the murine leukemia virus polypurine tract. *J. Virol.* 70:7132–42

164. Wu W, Palaniappan C, Bambara RA, Fay PJ. 1996. Differences in mutagenesis during minus strand, plus strand and strand transfer (recombination) synthesis of the HIV-1 gene in vitro. *Nucleic Acids Res.* 24:1710–18

165. Wu X, Liu H, Xiao H, Conway JA, Hehl E, et al. 1999. Human immunodeficiency virus type 1 integrase protein promotes reverse transcription through specific interactions with the nucleoprotein reverse transcription complex. *J. Virol.* 73:2126–35

166. Wyke JA, Bell JG, Beamand JA, Henderson LE, Bowers MA, et al. 1975. Genetic recombination among temperature-sensitive mutants of Rous sarcoma virus. *Cold Spring Harbor Symp. Quant. Biol.* 39:897–905

167. Xu H, Boeke JD. 1987. High-frequency deletion between homologous sequences during retrotransposition of Ty elements in *Saccharomyces cerevisiae*. *Proc. Natl. Acad. Sci. USA* 84:8553–57

168. Yanofsky C, Anderson JP, Rodrigo AG, Learn GH, Madan A, et al. 1981. Attenuation in the control of expression of bacterial operons. *Nature* 289:751–58

169. You JC, McHenry CS. 1993. HIV nucleocapsid protein. *J. Biol. Chem.* 268:16519–27

170. Yu H, Goodman MF. 1992. Comparison of HIV-1 and avian myeloblastosis virus reverse transcriptase fidelity on

RNA and DNA templates. *J. Biol. Chem.* 267:10888–96

171. Yu H, Jetzt AE, Ron Y, Preston BD, Dougherty JP. 1998. The nature of human immunodeficiency virus type 1 strand transfers. *J. Biol. Chem.* 273:28384–91

172. Yu Q, Darlix JL. 1996. The zinc finger of nucleocapsid protein of Friend murine leukemia virus is critical for proviral DNA synthesis in vivo. *J. Virol.* 70:5791–98

173. Zhang H, Zhang Y, Spicer TP, Abbott LZ, Abbott M, Poiesz BJ. 1993. Reverse transcription takes place within extracellular HIV-1 virions: potential biological significance. *AIDS Res. Hum. Retrovir.* 9:1287–96

174. Zhang J, Tang LY, Li T, Ma Y, Sapp CM. 2000. Most retroviral recombinations occur during minus-strand DNA synthesis. *J. Virol.* 74:2313–22

175. Zhang J, Temin HM. 1993. Rate and mechanism of nonhomologous recombination during a single cycle of retroviral replication. *Science* 259:234–38

176. Zhang J, Temin HM. 1994. Retrovirus recombination depends on the length of sequence identity and is not error prone. *J. Virol.* 68:2409–14

Annu. Rev. Genet. 2001. 35:303–39

THE GENETIC ARCHITECTURE OF QUANTITATIVE TRAITS

Trudy F. C. Mackay

Department of Genetics, Box 7614, North Carolina State University, Raleigh, North Carolina 27695; e-mail: trudy_mackay@ncsu.edu

Key Words quantitative trait loci (QTL), quantitative trait nucleotides (QTN), QTL mapping, linkage disequilibrium mapping, single nucleotide polymorphisms (SNPs)

■ **Abstract** Phenotypic variation for quantitative traits results from the segregation of alleles at multiple quantitative trait loci (QTL) with effects that are sensitive to the genetic, sexual, and external environments. Major challenges for biology in the post-genome era are to map the molecular polymorphisms responsible for variation in medically, agriculturally, and evolutionarily important complex traits; and to determine their gene frequencies and their homozygous, heterozygous, epistatic, and pleiotropic effects in multiple environments. The ease with which QTL can be mapped to genomic intervals bounded by molecular markers belies the difficulty in matching the QTL to a genetic locus. The latter requires high-resolution recombination or linkage disequilibrium mapping to nominate putative candidate genes, followed by genetic and/or functional complementation and gene expression analyses. Complete genome sequences and improved technologies for polymorphism detection will greatly advance the genetic dissection of quantitative traits in model organisms, which will open avenues for exploration of homologous QTL in related taxa.

CONTENTS

0066-4197/01/1215-0303$14.00

INTRODUCTION

Most observable variation between individuals in physiology, behavior, morphology, disease susceptibility, and reproductive fitness is quantitative, with population variation often approximating a statistical normal distribution, as opposed to qualitative, where individual phenotypes fall into discrete categories. Since quantitative variation is so prevalent, an understanding of the genetic and environmental factors causing this variation is important in a number of biological contexts, including medicine, agriculture, evolution, and the emerging discipline of functional genomics.

Quantitative genetics theory invokes an explicit underlying genetic model to describe the genetic architecture of a quantitative trait. Assuming that two alleles segregate at each of multiple QTL affecting variation in the trait, individual genotypes are fully specified by the homozygous (a) and heterozygous (d) effects of each locus, and pair-wise and higher-order interactions (epistasis) between loci (57, 150). However, in contrast to traits controlled by one or a few loci with large effects, variation in quantitative traits is caused by segregation at multiple QTL with individually small effects that are sensitive to the environment. For complex traits, the relationship between genotype and phenotype is not a simple ratio, and QTL genotypes cannot be determined from segregation of phenotypes in controlled crosses or pedigrees.

Although individual QTL genotypes cannot be inferred from observations of the phenotype, deductions about the net effects of all loci affecting the trait can be made by partitioning the total phenotypic variance into components attributable to additive, dominance and epistatic genetic variance, variance of genotype-environment interactions, and other environmental variance. These variances are specific to the population being studied, due to the dependence of the genetic terms on allele frequencies at each of the contributing loci, and real environmental differences between populations. Thus, until recently, our knowledge of the genetic architecture of quantitative traits was confined to estimates of the heritability (the fraction of the total phenotypic variance attributable to additive genetic variance) and other variance components from correlations between relatives and response to selection, estimates of average degree of dominance from changes of mean on inbreeding, estimates of net pleiotropic effects from genetic correlations, and estimates of the total mutation rate from phenotypic divergence between inbred lines. If we are to determine the molecular genetic basis of susceptibility to common complex diseases and individual variation in drug response, more effectively select domestic crop and animal species for improved production traits, and understand the genetic basis of adaptation, we need to go beyond these statistical descriptions and to overcome the very impediment that necessitated the biometrical approach to the analysis of quantitative traits: We need to identify and determine the properties of the individual genes underlying variation in complex traits (200).

A comprehensive understanding of the genetic architecture of any quantitative trait requires knowledge of: (*a*) the numbers and identities of all genes in the developmental, physiological and/or biochemical pathway leading to the trait phenotype; (*b*) the mutation rates at these loci; (*c*) the numbers and identities of the subset of loci that are responsible for variation in the trait within populations, between populations, and between species; (*d*) the homozygous and heterozygous effects of new mutations and segregating alleles on the trait; (*e*) all two-way and higher-order epistatic interaction effects; (*f*) the pleiotropic effects on other quantitative traits, most importantly reproductive fitness; (*g*) the extent to which additive, dominance, epistatic, and pleiotropic effects vary between the sexes, and in a range of ecologically relevant environments; (*h*) the molecular polymorphism(s) that functionally define QTL alleles; (*i*) the molecular mechanism causing the differences in trait phenotype; and (*j*) QTL allele frequencies.

No quantitative trait has yet been described at this level of resolution, requiring as it does the integration of classical genetics, evolutionary and ecological genetics, molecular population genetics, molecular genetics, and developmental biology. Here, I review recent progress toward these goals, problems to be overcome, and future prospects. As it is not possible to summarize all of the vast literature on quantitative genetics in a few pages, I apologize in advance to authors whose work is not cited due to space constraints.

IDENTIFYING GENES AFFECTING QUANTITATIVE TRAITS

Mutagenesis

Annotation of completed eukaryotic genome sequences $(1, 75, 95, 225, 226, 231)$ has revealed how few loci have been characterized genetically, even in model organisms, and how many predicted genes exist with unknown functions. Mutagenesis is the classical genetic approach for determining the numbers and identities of all genes in the developmental, physiological, and/or biochemical pathway leading to the wild-type expression of a trait. Is it possible that the functions of the predicted, but unknown, genes have not been elucidated because mutational screens are typically biased towards mutations with large qualitative effects? The answer to this question depends on the nature of genetic variation for quantitative traits, the subject of considerable debate in the past $(161, 162, 196, 203)$.

Mather $(161, 162)$ proposed that "polygenes" affecting quantitative variation were a different class of genetic entity from the "oligogenes" of Mendelian inheritance. The polygene theory can be phrased in modern parlance as the hypothesis that there is a category of loci at which the mutational spectrum is constrained to alleles with subtle, quantitative effects, as might be expected for genes in a pathway exhibiting functional redundancy. Support for the existence of such a class of genes comes from a mammoth experiment (242), in which 6925 strains of *S. cerevisiae* were constructed, each containing a precise deletion of 1 of 2026 open reading frames (about one third of the yeast genome). Only 17% of the deleted open reading frames were essential for viability under standard conditions, and 40% showed quantitative growth defects.

In contrast, Robertson & Reeve (203) resisted the idea that polygenes were a different entity from oligogenes, and proposed that it is more parsimonious to consider that so-called oligogenes in fact have a distribution of pleiotropic mutational effects, from lethality to major morphological mutations to virtually indistinguishable isoalleles with quantitative effects. Quantitative variation could then arise as a pleiotropic side effect of segregating mutations with large effects on fitness (e.g., homozygous lethals) or on another trait. Experimental support for this contention comes from the repeated observation that response to artificial selection (and subsequent selection limits) in *Drosophila* is often attributable to lethal genes with quantitative heterozygous effects on the selected trait (30, 63). Alternatively, quantitative variation could result from the segregation of isoalleles with subtle phenotypic effects at the same loci at which major mutations affecting the trait occur. This possibility was later articulated as the "candidate gene" hypothesis (152, 201), but the nomination of candidate genes is biased by the (limited) pre-existing understanding of the underlying pathways leading to phenotypic expression of the trait.

One part of the answer to these long-standing questions must come from mutagenesis studies where subtle effects of induced mutations on quantitative trait phenotypes are assessed. Such studies are much more labor-intensive than traditional F_1 mutant screens for dominant mutations with large effects that can be scored unambiguously on single individuals. Mutations with quantitative effects that are sensitive to environmental variation can only be detected if multiple individuals bearing the same mutation are evaluated for the trait phenotype. This in turn requires that the presence of a mutation can be recognized independently of its effect, or otherwise assured; that stocks are established for each mutagenized gamete (or chromosome); and that mutagenesis is conducted in a highly inbred (homozygous) genetic background. The sensitivity of the screen is proportional to the number of replicate individuals measured and is determined by the patience of the experimenter. Mobilizing transposons to generate insertional mutations in an inbred background satisfies these criteria. Further, the transposon acts as a molecular tag, enabling the cloning of the affected QTL (47, 123, 251).

Given these constraints, few direct screens for mutations with quantitative phenotypic effects have been performed to date. Genetic variation for mouse body weight was significantly increased in lines harboring multiple retrovirus insertions relative to their co-isogenic control lines (103), but the genes responsible were not identified. In *Drosophila*, highly significant quantitative mutational variation was found for activities of enzymes involved in intermediary metabolism (27), sensory bristle number (148), and olfactory behavior (5) among single *P* transposable element insertions that were co-isogenic in a common inbred background. Insertion sites were not determined in the enzyme activity study, but statistical arguments suggested that the insertions were highly unlikely to be in enzyme-coding loci (27). Of the 50 insert lines with significant effects on bristle number, 9 were hypomorphic mutations at loci known to affect nervous system development, whereas the remaining 41 inserts did not map to cytogenetic regions containing loci with previously described effects on adult bristle number (148). Most of the mutational variance for olfactory behavior was attributable to *P* element inserts in 14 novel *smell-impaired* (*smi*) loci (5). Several of the *smi* loci disrupted by the *P* element insertions are predicted genes with hitherto unknown olfactory functions, as well as candidate genes with more obvious roles in olfactory behavior (6).

Screening for quantitative effects of induced mutations is a highly efficient method both for discovering new loci affecting quantitative traits and determining novel pleiotropic effects of known loci on these traits. The value of this approach has been explicitly touted for phenotype-driven mutagenesis screens in the mouse (174). It has been suggested (174) that mutagenesis can supplant mapping segregating QTL to determine the genetic basis of complex traits. This suggestion has merit if by genetic basis one is referring to identification of loci and pathways important for the normal expression of the complex trait. However, most of the applications of quantitative genetics require that we understand what subset of these loci affect variation of the trait in natural populations. In this case, QTL mapping is currently a necessary first step, since many loci at which mutations affecting the trait can occur will be functionally invariant in nature.

QTL Mapping

Since the effects of individual QTL are too small to be tracked by segregation in pedigrees, QTL are mapped by linkage disequilibrium with molecular markers that do exhibit Mendelian segregation. The principle of QTL mapping is simple, and was noted 80 years ago (208): If a QTL is linked to a marker locus, there will be a difference in mean values of the quantitative trait among individuals with different genotypes at the marker locus. More formally, consider an autosomal marker locus, M, and quantitative trait locus, Q, each with two alleles (M_1, M_2, Q_1, Q_2). The additive and dominance effects of the QTL are a and d, respectively, and the QTL is located c centimorgans (cM) away from the marker locus. If one crosses individuals with genotype $M_1M_1Q_1Q_1$ to those with genotype $M_2M_2Q_2Q_2$, and allows the F_1 progeny to breed at random, then the difference in the mean value of the quantitative trait between homozygous marker classes in the F_2 is $a(1 - 2c)$. Similarly, the difference between the average mean phenotype of the homozygous marker classes and the heterozygote is $d(1 - 2c)^2$. If the QTL and marker locus are unlinked, $c = 0.5$ and the mean value of the quantitative trait will be the same for each of the marker genotypes. The closer the QTL and marker locus, the larger the difference in trait phenotype between the marker genotypes, with the maximum difference when the marker genotypes coincide exactly with the QTL. This suggests the basic principle of a genome scan for QTL. Given a population that is genetically variable for the quantitative trait and a polymorphic marker linkage map, one can proceed to test for differences in trait means between marker genotypes for each marker in turn. The marker in a local region exhibiting the greatest difference in the mean value of the trait is thus the one closest to the QTL. In such a single marker analysis, the estimates of the QTL effects are confounded by the map distance between the marker and QTL. This problem is not great with a dense marker map, and is avoided by localizing the QTL to intervals between adjacent linked markers (120, 228).

The primary limitation for mapping QTL until relatively recently has been the dearth of marker loci. The availability of multiple visible Mendelian mutations in *Drosophila* facilitated early efforts to map QTL in this species (16, 210, 214, 228, 243); however, such markers were limiting even in *Drosophila*, and interpretation of the QTL effects was compromised by the potential direct effects of the mutations on the quantitative traits. The discovery of abundant polymorphism in natural populations at the level of electrophoretic mobility of proteins (82, 132) was the leading edge of a wave of studies uncovering genetic variation at the molecular level. We now know that there is abundant molecular polymorphism segregating in most species, at the level of variation at single nucleotides (single nucleotide polymorphisms, or SNPs), short di-, tri-, or tetra-nucleotide tandem repeats (microsatellites), longer tandem repeats (minisatellites), small insertions/deletions, and insertion sites of transposable elements. Methods of detection of molecular variation have evolved from restriction fragment length polymorphism analysis using Southern blots (121, 129), molecular cytogenetic methods to identify transposable element insertion site polymorphism (167) and direct sequencing (115), to high-throughput methods for both polymorphism discovery and genotyping (116). The recent discovery of over two million SNPs in the human genome (227, 231) will no doubt spur the further development of rapid, high-throughput, accurate and economical methods for genotyping molecular markers, including hybridization to high-density oligonucleotide arrays (241).

Linkage disequilibrium (LD), or a correlation in gene frequencies, between invisible QTL alleles and visible marker alleles is generated experimentally by crossing lines with divergent gene frequencies at the QTL and markers. LD between marker and QTL alleles also occurs in naturally outbreeding populations within families or extended pedigrees that are segregating for the QTL genotypes. Traditionally, linkage mapping of QTL relies on the LD between markers and trait values that occurs within mapping populations or families. However, LD also occurs in outbred natural populations in which QTL alleles are in drift-recombination equilibrium (88, 238), in populations resulting from an admixture between two populations with different gene frequencies for the marker and mean values of the trait (57, 83, 238), and in populations that have not reached drift-mutation equilibrium due to a recent mutation at the trait locus, or a recent founder event followed by population expansion (150, 238). Association mapping of QTL relies on marker-trait linkage disequilibria in such populations.

The precision with which a QTL can be localized relative to a marker locus is directly proportional to the number of meioses sampled, which dictates the number of opportunities for recombination between the marker and the trait locus. Linkage analyses, which are typically restricted in the number of individuals or families sampled, thus have a lower level of resolution than do association studies, where many generations of recombination have occurred in nature. QTL mapping is thus usually an iterative process, whereby an initial genome scan is performed using linkage analysis, followed by higher resolution confirmation studies of detected QTL, and culminating with recombination or association mapping to identify candidate genes. This stepwise approach, whereby one chooses populations for mapping in which the extent of LD is just right relative to the scale with which one wishes to localize the QTL, is not strictly necessary. Association studies at candidate genes can be conducted without *a priori* evidence for linkage. In the future, given dense polymorphic marker maps and the development of high-throughput genotyping technology, whole-genome scans with sufficient resolution to detect individual loci should be possible, in both linkage and association mapping (197) frameworks.

GENOME SCANS   Linkage mapping of QTL in organisms amenable to inbreeding begins by choosing parental inbred strains that are genetically variable for the trait of interest. Usually the parent strains will have different mean values for the trait, but this is not necessary, as two strains with the same mean phenotypic value can vary genetically owing to complementary patterns of positive and negative QTL allelic effects. A mapping population is then derived by back-crossing the F_1 progeny to one or both parents, mating the F_1 *inter se* to create an F_2 population, or constructing recombinant inbred lines (RIL) by breeding F_2 sublines to homozygosity. These methods are very efficient for detecting marker-trait associations, since crosses between inbred lines generate maximum LD between QTL and marker alleles, and ensure that only two QTL alleles segregate, with known linkage phase.

The choice of method depends on the biology of the organism, and the power of the different methods given the heritability of the trait of interest (39, 57, 150). Traits with low heritabilities benefit from methods that allow assessment of the phenotype on multiple individuals of the same genotype, as is possible using RIL, or by progeny testing (120, 210, 211, 228). Because RIL are a permanent genetic resource, they need be genotyped only once and can be subsequently used to map a number of traits and to examine QTL by environment interactions. The map expansion in RIL due to the increased number of recombination events during their construction affords more precision of mapping than a similar number of backcross or F_2 individuals, but this advantage is offset by the increased difficulty and time necessary to construct RIL populations. A variant of line cross analysis is to artificially select lines from the F_2 of a cross between divergent inbred strains, and to map QTL by linkage to marker loci that systematically change in gene frequency across replicate selection lines (102, 104, 182, 183, 219).

Concomitant with the increasing availability of abundant polymorphic molecular markers in recent years has been the improvement in statistical methods for mapping QTL and development of guidelines for experimental design and interpretation. Least-squares (LS) methods test for differences between marker class means using either ANOVA or regression (212). LS methods have the advantage that they can easily be extended to cope with QTL interactions and fixed effects using standard statistical packages, but the disadvantage that assumptions of homogeneity of variance may be violated. Maximum likelihood (ML) (110, 120) uses full information from the marker-trait distribution, and explicitly accounts for QTL data being mixtures of normal distributions. However, ML methods are less versatile and computationally intensive, and require specialized software packages. There is, in fact, little difference in power between LS and ML (80, 120), and ML interval mapping can be approximated using regressions (80, 160).

Many QTL mapping protocols are one-at-a-time methods, whereby one evaluates the association of a QTL with a marker or marker interval while ignoring the effects of other segregating QTL in the mapping population. The presence of multiple linked QTL biases both single marker and interval mapping analysis (112), and segregation of unlinked QTL inflates the within-marker class phenotypic variance, thus reducing the power of QTL detection. Composite interval mapping (CIM) methods (97, 247) combine ML interval mapping with multiple regression, using marker cofactors to reduce the bias in estimates of QTL map positions and effects introduced by multiple linked QTL, and to increase the power to detect QTL by decreasing the within marker-class phenotypic variation. The principle of CIM has been extended to multiple traits (98), enabling the evaluation of the main QTL effects as well as QTL by trait and QTL by environment interactions. Strictly speaking, CIM methods are not multiple QTL methods, in that the model for evaluating the effects of each interval depends on the marker cofactors included, which varies across intervals. A true multiple interval mapping (MIM) method (99) has been developed that converges to a stable model providing estimates of positions and main and interaction effects of multiple QTL. An important caveat regarding CIM and MIM methods is that estimates of QTL positions and effects are highly model dependent, and can vary given different numbers of marker cofactors and window sizes (the region to either side of the test interval within which no marker cofactors are fitted) (187, 247, 249). These factors are under the control of the investigator, who must bear in mind that the best fitting model, identifying the most QTL, is not necessarily the closest approximation to reality.

Outbred populations pose additional challenges for QTL mapping. In experimental populations derived from a cross of two inbred lines, all individuals are informative for both markers and QTL segregation, there are only two QTL alleles at each segregating locus, and the linkage phase between marker and QTL alleles is known. In outbred populations, one is restricted to obtaining information from existing families. Only parents that are heterozygous for both markers and linked QTL provide linkage information, and individuals may differ in QTL-marker linkage phase. Further, not all families will be segregating for the same QTL affecting the trait, and in the presence of genetic heterogeneity different families may show different associations. In outbred populations marker-trait associations are evaluated through the effects of the marker on the genetic variance of the trait, as the variance attributable to a marker is proportional to the additive variance of a linked QTL (150, 178). Methods that utilize pairs of relatives (typically sibs) have been developed for human populations (70, 84, 118, 150).

Two important statistical considerations transcend the experimental design and method of statistical analysis used to map QTL: power and significance threshold. In cases where power is low, not all QTL will be detected, leading to poor repeatability of results, and the effects of those that are detected can be overestimated (12). The second problem pertains to the multiple tests for marker-trait associations inherent in a genome scan. To maintain the conventional experiment-wise significance level of 0.05, one needs to set a more stringent significance threshold for each test performed based on the number of independent tests. With multiple correlated markers per chromosome, the number of independent tests in a genome scan is clearly greater than the number of chromosomes, but less than the total number of tests. Permutation (25, 48) or other resampling methods (249) are widely accepted as providing appropriate significance thresholds.

Genome scans for QTL have become a cottage industry since the first report of QTL localizations using a high-resolution molecular marker map (190). Consistent with the economic incentive to improve production traits in agriculturally important crop and animal species, QTL have been mapped for traits of agronomic importance that differ between wild relatives and domestic, selected strains of tomato (188, 190); heterotic yield traits in crosses between elite inbred maize strains (218); growth and fatness in pigs (4, 113, 234); and milk production traits in dairy cattle (72, 213, 250). Knowledge of the loci underpinning variation in susceptibility to complex human disease is essential if the promise of personalized medicine is to be fulfilled. QTL have been mapped in humans for susceptibility to type 1 (41) and type 2 (81) diabetes mellitus, schizophrenia (18), obesity (33), Alzheimer's disease (54, 172), cholesterol levels (111), and dyslexia (21). Given the evolutionary conservation of genes and pathways across species, QTL for medically important and behavioral traits have been mapped in model organisms, with the hope that homologous loci and/or pathways will be implicated in humans. Examples include body size (17, 23, 104, 169, 194, 230), acute alcohol withdrawal (19) and emotionality in mice (59), and longevity in *C. elegans* (9, 209) and *Drosophila* (130, 184, 232). The best-case scenario for understanding the genetic basis of quantitative variation is for phenotypes that can be assessed with essentially no measurement error and for which the developmental basis is well understood. Consequently, our laboratory has mapped QTL for sensory bristle number in *D. melanogaster* (78, 79, 142, 180). Steps toward understanding the genetic basis of speciation have been taken by mapping QTL that affect morphological changes accompanying species divergence in maize (45, 46), monkeyflowers (15), and *Drosophila* (128, 136, 248).

The numbers of QTL mapped in these studies are in all cases minimum estimates of the total number of loci that potentially contribute to variation in the traits (154). Increasing the sample size would enable mapping QTL with smaller effects and enable the separation of linked QTL by virtue of the larger number of recombinant events. The number of QTL detected in crosses between inbred lines is also a minimum because one can only map QTL at which different alleles are fixed in the two parent strains, which are a limited sample of the existing genetic variation. Methods to widen this genetic net include four-way (96, 246) and eight-way (170, 221) cross designs, combining multiple line cross experiments (245), and utilizing parent strains derived by divergent artificial selection from a large base population (79, 142, 180, 237).

QTL as defined by genome scans are not genetic loci, but relatively large chromosome regions containing one or more loci affecting the trait. For example, the average size of intervals containing significant QTL from seven recent studies in *Drosophila* (78, 79, 130, 142, 180, 184, 232) was 8.9 cM and 4459 kb, with a range from 0.1 to 44.7 cM and 98 to 19,284 kb. There are at least 13,600 genes in the 120 Mb of *Drosophila* eukaryotic DNA (1); an average gene is thus 8.8 kb. Therefore, an average QTL in these studies encompasses 507 genes, with a range from 11 to 2191 genes.

The large number of genes in intervals to which QTL map, limited genetic inferences that can be drawn from analysis of most mapping populations, and poor repeatability of many initial QTL mapping efforts that had low power due to small samples have engendered some pessimism about the value of this approach (174). However, understanding the genetic basis of naturally occurring variation is too important a problem to ignore the challenge of mapping QTL to the level of genetic locus, and there are success stories to serve as beacons guiding us through the fog.

HIGH-RESOLUTION MAPPING   The size of the genomic region within which a QTL can be located is determined by the scale of LD between the gene corresponding to the QTL and marker loci flanking the QTL. High-resolution mapping entails successively reducing the size of the region exhibiting LD with the QTL ultimately to a single gene, either by generating recombinant genotypes between the two strains of interest, or by screening for historical recombination events in natural populations. Given a high-density polymorphic molecular marker map spanning the interval in which the QTL is located, and a population of recombinant genotypes with breakpoints between adjacent markers, it is a simple matter in principle to define the QTL position relative to a pair of flanking markers. The challenge posed for high-resolution QTL mapping, in contrast to mapping Mendelian loci, is that individual QTL are expected to have small effects that are sensitive to the environment, and therefore the phenotype of a single individual is not a reliable indicator of the QTL genotype.

Successful high-resolution QTL mapping is thus contingent on increased recombination in the QTL interval, and accurately determining the QTL genotype (39). One way to increase the precision of mapping in line crosses is to use more advanced generations of the cross than the F_2 for the initial genome scan, since the number of recombinations increases proportionately with the number of generations (40). In organisms amenable to genetic manipulation, the effects of individual QTL can be magnified by constructing strains that are genetically identical except for a defined region surrounding the QTL. These strains are a permanent genetic resource from which multiple measurements of the trait phenotype can be obtained to increase the accuracy of determining the QTL genotype. The region surrounding the QTL can be a whole chromosome (chromosome substitution lines) or a smaller interval [variously called introgression lines, interval-specific congenic strains (38), and near-isoallelic lines (NIL)]. NIL are constructed by recurrent backcrosses to one parent, accompanied by selection for the marker associated with the QTL and against other markers associated with the genotype of the nonrecurrent strain. QTL can be mapped with high resolution by constructing multiple independent introgressions, ascertaining recombination breakpoints for each using a molecular marker map of the introgressions, and determining the phenotypic effect of each introgression genotype (189). In theory (38), effort spent in systematically generating informative recombinants across a QTL interval, using molecular markers to track recombination breakpoints, can result in localizing QTL to 1-cM intervals.

If constructing introgression lines is necessary for high-resolution mapping, why not dispense with the traditional low-resolution genome scan and develop methods that both map QTL and produce the first-generation introgression lines simultaneously? In *Drosophila*, the availability of balancer chromosomes enables the rapid construction of chromosome substitution lines. This technique has been adopted to substitute single homozygous chromosomes from a high-scoring strain into the homozygous genetic background of a low-scoring strain, and to subsequently map QTL one chromosome at a time (79, 142, 210, 237). Chromosome substitution strains have been constructed in mice (175), enabling chromosomal localization of QTL by screening just 20 lines.

Interval-specific introgression was first used to map QTL for *Drosophila* bristle number (16). More recently, backcrossing with selection for molecular markers was used to construct introgression lines for mapping QTL for tomato fruit traits (55) and morphological traits associated with divergence between two *Drosophila* species (128). Segregating variation in intervals surrounding several *Drosophila* bristle number candidate QTL (154) was demonstrated by constructing NIL for approximately 50 different naturally occurring alleles of each candidate gene (140, 147, 149). Introgression can also be accomplished by selecting on the quantitative trait, rather than markers, to identify QTL of large effect while producing NILs for further analysis (87). When combined with progeny testing of recombinant genotypes within intervals (55, 210, 214, 228, 243), linked QTL with small effects within each interval can be individually identified.

There are several examples of successful fine-scale recombination mapping of QTL. (*a*) *Idd3*, a susceptibility allele for type 1 diabetes mellitus in the mouse, maps to a 145-kb interval containing a single known gene, *Il2* (151). (*b*) *NIDDM1*, a susceptibility gene for human type 2 diabetes mellitus, was mapped by linkage analysis to a 5-cM region that, fortuitously, is a region of high recombination corresponding to 1.7 Mb, containing 7 known genes and 15 ESTs (91). (*c*) The tomato fruit weight QTL, *fw2.2*, was localized to a 1.6-cM interval containing four unique transcripts (65). (*d*) A tomato fruit-specific apoplastic invertase gene *Lin5* has been shown unambiguously to correspond to the *Brix9-2-5* QTL, which affects fruit glucose and fructose contents, since recombinants in a 484-bp region within this gene cosegregate with the QTL phenotype (67).

Drosophila geneticists have an additional tool for mapping QTL to sub-cM intervals: deficiency chromosomes. To fine map Mendelian mutations, the chromosome containing the mutant allele is crossed to the set of deficiencies whose breakpoints span the interval to which the gene has been localized, and complementation or failure to complement is recorded for the progeny containing the deficiency chromosomes. The location of the mutation is then delineated by the region of nonoverlap of deficiencies complementing the mutant phenotype with those that fail to complement the mutant. Deficiency complementation mapping has been extended to mapping QTL with small, additive effects and to control for the effects of other QTL outside the region uncovered by the deficiency (154, 187). This method has been used to resolve QTL for *Drosophila* life span that were initially detected by recombination mapping (184) into multiple linked QTL, one of which mapped to a 50-kb interval containing one known and two predicted genes (187). One caveat when interpreting quantitative failure to complement deficiencies is that, as for all complementation tests, the failure to complement could be attributable to an epistatic, rather than an allelic, interaction. Thus, QTL inferred using this method must be subsequently confirmed using a different approach.

In organisms where the genetic background cannot be manipulated and controlled crosses cannot be made (e.g., humans), the only option for high-resolution mapping is LD mapping, which utilizes historical recombinations, as discussed in the next section. The level of difficulty in detecting QTL in natural populations is greater than the already considerable challenge posed in genetic model organisms, since the existence of segregating genetic variation at other QTL affecting the trait translates operationally to a smaller marginal effect of any one QTL, and thus very large samples are needed.

From QTL to Gene

The high-resolution mapping methods discussed above will map QTL to genetic intervals containing several genes. How can one determine which genes correspond to the QTL? Co-segregation of intragenic recombinant genotypes in a candidate gene with the QTL phenotype, as demonstrated for the tomato apoplastic invertase gene and fruit sugar content QTL (67), constitutes clear genetic proof that the QTL corresponds to the candidate gene.

Functional complementation, in which the trait phenotype is rescued in transgenic organisms, is another gold standard for gene identification. The *fw2.2* tomato fruit weight QTL was shown to correspond to the *ORFX* gene by transforming the large-fruit domestic tomato with a cosmid containing the *ORFX* transcript and observing a significant reduction in fruit weight in the transformants (65). The multiplicity and size of intestinal tumors induced by the dominant *Apc^{Min}* mutation in mice are modified by the *Mom1* mutation. Intestinal tumors are suppressed in *Apc^{Min}* mutant mice that are heterozygous for a cosmid transgene overexpressing the *Mom1* candidate gene *Pla2g2a* (encoding a secretory phospholipase), strongly suggesting that *Mom1* is an allele of *Pla2g2a* (34). These studies were facilitated by the large effects of the QTL, and their dominant gene action. If the effect of the QTL is not large, it is necessary to construct multiple independent transgenic lines to account for changes in expression of the transgene due to different insertion sites.

Not all QTL will be identifiable with the standard of rigorous proof provided by intragenic recombination and functional complementation. The only option in these cases is to collect multiple corroborating pieces of evidence, no single one of which is convincing, but which together consistently point to a candidate gene. These lines of evidence include genetic complementation tests, LD mapping, quantitative differences of gene expression and/or protein function, and tests for orthologous QTL in other species.

GENETIC COMPLEMENTATION   Perhaps the simplest method to infer identity of a candidate gene and a QTL is genetic complementation to a mutant allele of the candidate gene. For example, a QTL explaining most of the difference in inflorescence morphology between maize and teosinte mapped to a region including the maize gene, *teosinte-branched1* (*tb1*), and the QTL effect was similar to the mutant phenotype of *tb1* (44, 45). A NIL containing the teosinte QTL in a maize background failed to complement the maize *tb1* mutant allele, suggesting that *tb1* is the QTL (46). Genetic complementation of QTL and mutant alleles of candidate genes need not be restricted to recessive QTL with a qualitative phenotypic effect. The logic of a quantitative complementation test is the same as quantitative deficiency complementation mapping (79, 141, 155). QTL for *Drosophila* sensory bristle number quantitatively failed to complement mutations at candidate genes affecting peripheral nervous system development (79, 141, 147, 149, 155).

Currently, quantitative complementation is only feasible for organisms in which controlled crosses can be made. An unbiased application of the method also requires the existence of (preferably null) mutant stocks at all loci in the region to which the QTL maps. The examples above were for cases where obvious candidate genes colocalized with the QTL. In many cases, QTL will map to regions where there are no obvious candidate genes (180), and, for some traits, one can implicate almost any locus as a candidate gene. For such regions, there is no option but to test all possible genetically defined loci. Ultimately, this must be done even for those regions containing obvious candidates, since some QTL may correspond to loci with undescribed and unexpected pleiotropic effects on the trait. Practical considerations dictate that the QTL interval must first be delimited to a manageable number of genes before a systematic quantitative complementation analysis of each gene in the region is attempted. Strains containing single gene knock-outs of all known and predicted genes in model organisms (42, 66, 76, 179, 216, 242) will be an exceptionally valuable resource for quantitative complementation tests in the future.

One caveat regarding genetic complementation tests, whether qualitative or quantitative, is that failure to complement is indicative of a genetic interaction, but cannot discriminate whether the interaction is allelic or epistatic. Thus, failure to complement imputes candidate genes for further study, but does not constitute proof that the QTL and the candidate gene are the same entity. Corroborating evidence includes gene expression patterns (47) and associations of molecular polymorphisms in the candidate gene with phenotypic variation in the trait (139, 147).

LINKAGE DISEQUILIBRIUM (ASSOCIATION) MAPPING   When a new mutation occurs in a population at a locus affecting a quantitative trait, all other polymorphic alleles in that population will initially be in complete LD with the mutation. Over time, however, recombination between the mutant allele and the other loci will create the missing haplotypes and restore linkage equilibrium between the mutant allele at all but closely linked loci. The length of the genomic fragment (in cM) surrounding the original mutation in which LD between the QTL and other loci still exists depends on the average amount of recombination per generation experienced by that region of the genome, the number of generations that have passed since the original mutation, and the population size (57, 83, 88, 238). For old mutations in large equilibrium populations, strong LD is only expected to extend over distances of the order of kilobases or less. Larger tracts of LD are expected in expanding populations derived from a recent founder event or in population isolates with small effective size. Nevertheless, the scale of LD in outbred populations is expected to be small enough that only very closely linked markers will be associated, such that screening for LD between QTL alleles and polymorphic molecular markers in outbred populations is a natural solution to the problem of generating informative recombinants within loci and between adjacent loci.

Association mapping can be used to systematically screen candidate loci in an interval defined by linkage mapping, or to evaluate associations at candidate loci even in the absence of linkage information. It has been suggested that the two million SNPs that are the fruit of the human genome project (227, 231) will herald a new era of detecting loci for susceptibility to complex human diseases using genome-wide LD mapping with a dense panel of SNP markers and a sample of individuals from a natural population, dispensing with traditional linkage mapping entirely (197).

The simplest designs for evaluating association between markers at a candidate gene and a quantitative trait only require a sample of individuals from the population of interest, each of whom has been genotyped for the marker loci and evaluated for the trait phenotype. In the case-control design for dichotomous traits, such as disease susceptibility, the population sample is stratified according to disease status, and LD between a marker and the trait is revealed by a significant difference in marker allele frequency between cases and controls (20). For continuously distributed traits, the population sample is stratified by marker genotype, and marker-trait LD is inferred if there is a significant difference in trait mean between marker genotype classes. Testing for trait associations of multiple markers again requires that an appropriate downward adjustment of the significance threshold be made. Permutation tests developed in the context of other single-marker genome scans are ideal in this regard (25, 48, 139, 147).

Unfortunately, several factors complicate this picture. First, marker-trait associations in natural populations are not necessarily attributable to linkage, but can be caused by admixture between populations that have different gene frequencies at the marker loci and different values of the trait (57, 83). This problem can be alleviated by experimental designs that control for population structure and jointly test for linkage and association. For example, the transmission-disequilibrium test (TDT) (215) for dichotomous traits utilizes population samples of trios of affected offspring and their parents. Alleles at polymorphic marker loci linked to a gene affecting the trait will be preferentially transmitted to affected offspring. This design has been extended to various sampling designs for quantitative traits (3), larger nuclear families (195), genotypic data with multiple alleles (177), and half-sib families (244).

Experimental designs to control for the potential confounding effects of admixture are more complicated and less attractive than simple samples from the population. Isolated populations descended from a small numbers of founders are thus thought to be advantageous for association studies, since LD should extend over larger distances, requiring a lower density of marker loci, and admixture can be assumed to be absent or negligible. For the case of rare disease susceptibility alleles, an additional advantage of isolated populations is that most disease-bearing chromosomes are likely to descend from a single ancestral chromosome in the founding population. A successful application of LD mapping in isolated populations is the positional cloning of the human sulfate transporter locus, associated with autosomal recessive diastrophic dysplasia (DTD) in Finland (85).

If the genotypes at all polymorphic sites in the region of interest are determined, one of them must correspond to the site causing the phenotypic effect (the QTN, or quantitative trait nucleotide) (138, 139). The power to detect an association between the QTN and the trait phenotype is much higher than linkage studies, even after correcting for multiple tests for association (197). Thus, a disease-susceptibility allele with a relative risk of 2 can be detected in association study with 1000 individuals. However, genotyping all polymorphic sites in 1 Mb (the size of interval to which fine-mapping can localize a QTL) for 1000 individuals is currently laborious and expensive. For a typical level of nucleotide heterozygosity of 0.001, the population genetic expectation is that there will be 5200 segregating sites per Mb in a sample of only 100 individuals (236). Consequently, association studies are typically based on a subset of markers in LD with, but probably not including, the QTN polymorphism. Analogous to single-marker analysis of segregating generations derived from line crosses, the estimate of the effect of a linked marker on the trait underestimates the effect of the polymorphism at the QTN (122, 177). Therefore, if one evaluates marker-trait associations for several linked marker loci in the region of a QTL, the basic theory predicts that the marker associated with the largest effect on the trait is closest to the casual polymorphic site.

The consequence of conducting association studies with a subset of marker loci is that the power to detect an association depends not only on the number of individuals sampled, but also on the density of polymorphic markers. Samples of at least 500 individuals are necessary to detect a QTN contributing 5% of the total phenotypic variance with 80% power (138, 146). Further, potentially important associations will be missed unless the polymorphic markers are spaced such that adjacent markers show some degree of LD. In theory, this will be the physical distance corresponding to $4Nc$, where N is the effective population size (117, 138), which can be estimated from plots of all possible pair-wise LD between markers in a genomic region against their physical distance (86, 94). Unfortunately from the perspective of developing general guidelines, the average physical distance corresponding to $4Nc$ varies by at least an order of magnitude among different gene regions, both in humans (20, 138) and in *Drosophila* (2, 165, 166). Further, the association between LD and physical distance breaks down over short physical distances, with some closely linked sites in linkage equilibrium and others in strong disequilibrium (28, 49, 165, 166, 176, 222). While some of the region-to-region and within-region variation in the relationship of LD to distance is attributable to statistical sampling error, much of the variation reflects the very real differences in recombination rates (2, 13), natural selection, genetic drift, marker mutations, and other genetic processes (e.g., gene conversion), both on a regional and local scale (89, 90, 238, 239). Optimally, one should adjust the spacing of markers in an association study relative to variation in LD across the region of interest. For most regions, multiple markers will be required per gene. In *Drosophila* regions of high polymorphism and recombination, the optimal spacing of markers could be as small as 200 base pairs (139). These considerations seriously compromise the feasibility of whole-genome LD mapping (117, 240), especially given simulation (117) and empirical results (49, 222) demonstrating that mean levels of LD and the relationship between LD of markers at intermediate frequency and physical distance are not significantly different in mixed and isolated populations.

An example of the successful application of LD mapping is the positional cloning of the human calpain-10 (*CAPN10*) gene, a putative susceptibility gene for type 2 diabetes in the interval containing the *NIDDM1* QTL (91). Associations between 21 SNPs and diabetes susceptibility in the 5-cM region to which *NIDDM1* mapped narrowed the search to three genes in a 66-kb interval: *CAPN10*, G-protein–coupled receptor 35 (*GPR35*), and a gene encoding a protein with homology to aminopeptidase B (*RNPEPL1*). Additional SNPs in this interval were detected by re-sequencing in 10 diabetic patients. Association tests for 63 SNPs indicated that 16 SNPs in *CAPN10* and *GPR35* were associated with diabetes, but that only 1 SNP in *CAPN10* was associated both with the disease and evidence for linkage, implicating *CAPN10* and not *GPR35* as the disease-susceptibility locus. This conclusion is supported by a haplotype analysis indicating that haplotypes at three polymorphic SNPs in *CAPN10* were associated with diabetes in the study population, a finding that was confirmed for two unrelated populations.

ORTHOLOGOUS QTL   Prospects for describing variation for quantitative traits in terms of contributions of individual loci are bright if the same loci affect variation for the trait in different populations and across taxa. Classical quantitative genetic analyses and QTL mapping both suggest that many of the same loci cause variation in bristle number in different geographical populations. If different loci caused variation for bristle number in different populations, one would expect that the limit to selection from any one population would be less than limits achieved in synthetic populations derived by crossing; this is not the case (143, 144). Further, the same genomic regions containing QTL for *Drosophila* sensory bristle number recur in studies that span a period of nearly 40 years, for populations of diverse origin, and using different mapping methodologies (78, 180, 210, 214, 243).

An intriguing observation of early QTL mapping studies was that QTL for the same traits mapped to similar locations in different species (188, 191). This suggests a strategy whereby genes corresponding to QTL can be identified in genetically tractable model organisms, then evidence for linkage and association for the orthologous locus sought in other species. The non-obese diabetic (NOD) mouse is a model for human type 1 diabetes. Interleukin-12 is functionally associated with disease progression in the NOD mouse, which motivated a search for association of the human homolog, *IL12B*, with susceptibility to type 1 diabetes (168). There was strong evidence for linkage in the region containing *IL12B*, and significant LD and linkage was confined to a 30-kb region in which *IL12B* is the only known gene.

Conversely, one could map a QTL with high resolution in one species and conduct functional assays for correspondence of the genes in the interval with the QTL in a model organism. A human asthma QTL was mapped to 1-Mb interval containing 6 cytokine genes and 17 partially characterized genes. Quantitative changes in asthma-related phenotypes were observed in transgenic mice expressing either of two human candidate genes, interleukin 4 (*IL4*) and interleukin 13 (*IL13*), but not in mice transgenic for the other candidate genes, implicating *IL4* and *IL13* as the loci corresponding to the asthma QTL (220).

From QTL to QTN

A description of the genetic architecture of quantitative traits is not complete until we can specify which polymorphic sites in the gene we have identified as corresponding to the QTL actually cause the difference in the trait phenotype—the quantitative trait nucleotides (QTN). To put this problem in perspective, consider the levels of polymorphism observed in the candidate genes discussed above. Direct sequence comparisons have revealed 42 nucleotide differences distinguishing the 2 parental alleles of the tomato fruit weight QTL, *ORFX/fw2.2* (65); 11 molecular variants in 484 bp differentiate the parental alleles of the tomato QTL affecting fruit sugar content, *Lin5/Brix9-2-5* (67); 88 variant sites in 71 individuals for the human lipoprotein lipase gene (176); and there are at least 108 variant sites in *CAPN10* (91). Polymorphism at the *Drosophila* bristle number QTL *scabrous* (*sca*) and *Delta* (*Dl*) was determined using low-resolution restriction map surveys of approximately 50 alleles derived from nature. There were 27 polymorphisms in the *sca* gene region (122) and 53 polymorphisms at *Dl* (139). In contrast, there are no fixed differences between the maize and teosinte alleles of *tb1*, despite considerable polymorphism at this locus within each species (235).

The same LD mapping methods used to focus on a candidate gene can be used to determine which of the polymorphic sites at the candidate gene is associated with the quantitative trait phenotype. Again, samples of 500 individuals or more are required to detect QTN accounting for 5% of the phenotypic variance or less. Ideally, one would obtain complete sequences of the candidate gene for each of these individuals. However, until re-sequencing technology becomes more rapid and cost-effective, associations within a candidate gene will be based on a subset of the polymorphic sites. The choice of sites should be motivated by the pattern of LD in the candidate gene region, and not by preconceived notions of what the causal sites might be (e.g., polymorphisms in coding regions). Three contrasting scenarios can be imagined. (*a*) If all polymorphic sites in the candidate gene are in complete LD, it is only necessary to examine associations with one of the sites. However, if an association is detected, it will not be possible to use LD mapping in that population to make inferences about which of the sites is causal. Either another population with a different evolutionary history must be sought, or alternate methods developed. (*b*) If there is no significant LD between any of the polymorphic sites in the candidate gene, all sites must be utilized in the association study, but associations detected at any one site are more likely to be causal. However, one can never preclude the existence of a true causal site in strong LD with the site associated with variation in the trait that is outside the surveyed region. (*c*) Some sites in the candidate gene are in LD, others are not. This is the usual case, for which sequence information for a sample of alleles would greatly inform the choice of markers.

To date, no study of genotype-phenotype associations at candidate genes has utilized the optimal density of markers. Indeed, many studies in humans have been conducted with a single marker, leading to difficulty in replicating associations due to variation among populations in the degree of LD between the marker and the putative causal variant (20). The few studies in which multiple polymorphic markers within candidate genes have been examined for association with phenotypic variation reveal some interesting features. All kinds of molecular variation (transposable element insertions, small insertions/deletions, and SNPs) of *Drosophila* bristle number candidate genes have been associated with quantitative variation in bristle number, and all significant associations have been for molecular polymorphisms in introns and noncoding flanking regions (122, 139, 140, 147, 157). Polymorphisms in noncoding regions of the *CAPN10* putative diabetes susceptibility locus in humans are associated with disease risk (91). It is thus possible that variation in regulatory sequences causes slight differences in message levels, timing and tissue-specificity of gene expression, and protein stability that lead to subtle quantitative differences in phenotype. The *CAPN10* story is unexpectedly complex. Two haplotypes of three SNPs in this gene have significantly increased risk of diabetes, but the at-risk genotype is homozygous for one SNP and heterozygous for the others; homozygotes for either haplotype do not have increased risk (91). This suggests that at least two interacting risk factors within a single gene are required to affect susceptibility to diabetes.

Correlation is not causality, and the ultimate proof that a molecular variant is functionally associated with differences in phenotype will be biological, not statistical. One such functional test is to construct by in vitro mutagenesis alleles that differ for each of the putative QTN, separately and in combination, and to assess their phenotypic effects by germ-line transformation into a null mutant background (24, 125–127, 217). Mimicking complex effects such as those observed at *CAPN10* could prove to be challenging.

PROPERTIES OF GENES AFFECTING QUANTITATIVE TRAITS

Distribution of Effects

The mathematical formalization of Mather's polygene hypothesis is the "infinitesimal" model, which assumes genetic variation for quantitative traits is caused by a very large number of QTL with very small and equal allelic effects (57, 150). Although convenient for theoretical modeling of quantitative variation, this model makes little sense genetically, as it predicts all loci equally (and trivially) affect all conceivable quantitative traits. Robertson (200) proposed that the distribution of allelic effects should be more nearly exponential, whereby a few loci have large effects and cause most of the variation in the traits, with increasingly larger numbers of loci with increasingly smaller effects making up the remainder. If Robertson's model is correct, it will be feasible to understand the salient features of the genetic architecture of quantitative traits by a detailed study of those relatively few loci that contribute most of the genetic variation. If the infinitesimal model is largely correct, understanding quantitative genetic variation in terms of individual QTL is a hopeless enterprise. Under this model, the effect of each locus contributing to variation in a quantitative trait is so small that it cannot, by definition, be measured.

Several lines of evidence indicate that variation for quantitative traits is not accounted for by a very large number of loci with equal and small effects. First, only a small fraction of the theoretical maximum response to artificial selection predicted by the infinitesimal model (199) is achieved experimentally (57). Second, distributions of effects of new *P* element insertional mutations on *Drosophila* quantitative traits are highly skewed and leptokurtic (148, 159), with most mutational variance attributable to a few mutations with large effects. Third, observed distributions of QTL effects are also in accord with Robertson's prediction (15, 50, 188, 191, 210, 223). This observation on its own is not a strong refutation of the infinitesimal model, however, since undetected QTL with small effects and overestimation of effects of detected QTL could be an artifact of small sample size (12) or LD between QTL, low heritability, and epistatic interactions between QTL (14). Further, observed unequal distributions of QTL effects could reflect variation in the precision of estimating the QTL positions. If this were true, QTL effects would be directly proportional to the fraction of the genome embraced by the confidence limits marking the location of the QTL. This is not generally true; in many cases, QTL with the largest effect map to the smallest genetic regions (180). Of course, high-resolution recombination mapping (47, 65, 67) and transformation of cloned QTL (65, 220) prove the existence of at least some QTL with large effects, and significant associations of molecular polymorphisms at candidate genes with moderate-to-large QTL effects is consistent with an exponential distribution of effects. Finally, simulation studies have shown that the distribution of effects of genes that are fixed during adaptation is exponential (185, 186).

Interaction Effects

QTL effects typically vary according to the genetic, sexual, and external environment. If the magnitude and/or the rank order of the differences in phenotypes associated with QTL genotypes also changes depending on the genetic or environmental context, then that QTL exhibits genotype by genotype interaction (epistasis), genotype by sex interaction (GSI), or genotype by environment interaction (GEI) (154). Extensive epistasis, GSI, and GEI have practical and theoretical consequences. First, estimates of QTL positions and effects are relevant only to the sex and environment in which the phenotypes were assessed and may not replicate across sexes and in different environments. Second, estimates of main QTL effects will be biased in the presence of epistasis. Third, genetic variation for quantitative traits can be maintained at loci exhibiting GSI and GEI (68, 74, 131).

GENOTYPE BY GENOTYPE INTERACTIONS (EPISTASIS)   Any developmental or biochemical pathway that culminates in the expression of a quantitative trait is comprised of networks of loci that interact at the genetic and molecular levels. Genetic variation at some of these loci is causally connected to phenotypic variation in the trait. To what extent does this translate to interaction between loci at the level of the phenotype, i.e., epistasis? In classical Mendelian genetics, epistasis between loci with alleles of large effect refers to the masking of genotypic effects at one locus by genotypes of another, as reflected by distorted segregation ratios in a di-hybrid cross (192). The usage of the term in quantitative genetics is much broader, and refers to any statistical interaction between genotypes at two (or more) loci (22, 31, 106, 163). Epistasis can refer to a modification of the homozygous or heterozygous effects of the interacting loci. Epistasis can be synergistic, whereby the phenotype of one locus is enhanced by genotypes at another locus; antagonistic, in which the difference between genotypes at a locus is suppressed in the presence of genotypes at the second locus; or even produce novel phenotypes.

The theory for estimating the contribution of epistatic interactions to the total phenotypic variance of a trait in outbred populations and crosses among inbred lines is well developed. However, epistasis is difficult to detect in these designs, partly because even strong epistatic interactions contribute little to the epistatic variance, and partly because this term has very high sampling variance, requiring huge sample sizes (10). Further, knowledge of significant epistatic interaction variance is not terribly enlightening regarding the nature of the individual effects. In *Drosophila*, interactions between whole chromosomes are detectable by constructing all possible chromosome substitution lines between pairs of strains that are genetically divergent for the trait. Contrary to the results from decomposition of variance components, interactions between chromosomes are usually found for traits that are considered to be components of fitness (101) and less often, but frequently, for traits that are less closely related to fitness (61, 107, 202).

The power to detect epistasis between QTL in mapping populations is low, for several reasons: (*a*) Even large mapping populations contain few individuals in the rarer two-locus genotype classes; (*b*) segregation for other QTL can interfere with detecting epistasis between the pair of loci under consideration; and (*c*) after adjusting the significance threshold for the multiple statistical tests involved in searching for epistatic interactions, only extremely strong interactions remain significant. Given these biases against detecting epistasis, many studies report largely additive QTL effects (51, 188, 248) or do not test for epistasis. On the other hand, strong interactions have been observed between QTL affecting *Drosophila* bristle number (142, 210, 214), mouse body size (206), grain yield in rice (133), susceptibility to diabetes in humans (37), and longevity in *Drosophila* (130) and *C. elegans* (209).

Observations of epistasis between QTL from experimental designs that are not optimal for detecting interactions hint that genotype-specific QTL effects will be rather common when more precise tests for interactions are performed. One such test is to introgress a mutant allele into several different wild-type genetic backgrounds. Typically, the expression of the mutant will be enhanced or suppressed, and the degree of dominance and pleiotropic effects on other traits can be modified (73, 173, 193, 207). These results indicate a hidden reservoir of genetic variation that is only revealed in the mutant background. Identification of such modifier loci, as has been done for the *Mom1* tumor suppressor gene in mice (34), will provide insights regarding the mechanistic basis of the interaction.

Specific tests for epistasis between QTL also reveal considerable interaction. One such test is to cross coisogenic NIL containing different QTL affecting a trait, or mutations with quantitative effects on the trait that have been derived in the same inbred background, in all possible pair-wise combinations. This test was used to show that QTL for yield-related traits in tomato exhibited antagonistic epistasis (56), and that *P* element insertions affecting odor-guided behavior in *Drosophila* could be organized in a network of enhancing and suppressing interactions (58). Tests for additive by additive epistasis require the construction of the four double homozygous genotypes at two biallelic loci, while estimating all classes of epistatic interactions involves synthesizing the nine two-locus genotypes. Such tests have revealed considerable epistasis between QTL affecting divergence in plant architecture between maize and teosinte (145) and between *P* element insertions on metabolic activity in *Drosophila* (26). Further, functional assays to identify QTN responsible for differences in protein concentration at the *Drosophila Adh* locus have revealed epistasis between QTN within this gene (217).

GENOTYPE BY SEX INTERACTIONS   At the level of the trait, a significant interaction of genotype and sex means that the cross-sex genetic correlation (r_{GS}) is less than unity, and there is genetic variation in sexual dimorphism for the trait. Such an interaction can be attributable to sex-linked loci that actually have the same effects in both sexes, or, more interestingly, to autosomal loci with different effects in males and females. In the latter case, the magnitude of the interaction determines the extent to which sex dimorphism can evolve under natural or artificial selection. Estimates of r_{GS} for mouse body weight (52), several morphometric traits in *Drosophila* (35, 36), and morphological traits in other species (150, 204) are all high and nearly unity. However, this is not true for all quantitative traits. There is genetic variation in sex dimorphism in natural populations of *Drosophila*, attributable to autosomal loci, for sensory bristle number (60, 149), olfactory behavior (156), and longevity (164).

Estimates of r_{GS} at the level of the trait refer to the summation of sex dimorphism effects of all loci affecting the trait, and can conceal considerable heterogeneity in sex dimorphism effects among loci that can be revealed by examining the effects of new mutations and QTL. In *Drosophila*, sex-specific effects have been observed for spontaneous (158) and P-element–induced (148, 159) mutations, QTL (77, 78, 142, 180), and molecular polymorphisms in candidate genes (122, 139, 140, 147, 157) affecting sensory bristle number. This phenomenon is not a peculiarity of bristle number, nor is it confined to *Drosophila*. In *Drosophila*, there is significant variation in sex-dimorphism among P-element–induced mutations affecting olfactory behavior (5). QTL for *Drosophila* longevity (130, 184, 232) are highly sex specific. Polymorphisms at the *Esterase-6* locus are associated with sex-specific effects on enzyme activity (71). Whereas most QTL for mouse body size affect both males and females, some affect only males or females (230). Similar trends will likely become apparent for QTL affecting human disease susceptibility; indeed, sex-specific associations have been reported for markers at candidate genes for longevity (43) and serum triglyceride levels (224).

GENOTYPE BY ENVIRONMENT INTERACTIONS   Detection of QTL by environment interactions requires that the same genotypes are evaluated in multiple environments, which is most readily accomplished if QTL genotypes can be replicated (e.g., RIL, NIL, selfed progeny of F_2 individuals, or molecular markers at a candidate gene). Early tests for GEI were based on whether the same QTL were significant in multiple environments. For example, of a total of 29 QTL for tomato fruit traits detected in 3 rearing environments, 4 QTL were detected in all 3 environments, 10 in 2 environments, and 15 in 1 environment (188). However, a significant QTL effect in all environments does not preclude the existence of GEI, and a QTL might actually have the same effect in all environments, but not reach statistical significance in some if the power of the experimental design is low. Several studies have implemented formal tests for GEI by examining the significance of marker by environment interactions using analysis of variance (32) or maximum likelihood (98). QTL genotype by environment interaction is common, but by no means ubiquitous (150). In maize, QTL effects for grain yield, ear height, and plant height were largely independent of the environment, while QTL for days to tassel, grain moisture, and ear number were highly environment dependent (32). In *Drosophila*, GEI has been observed for QTL affecting sensory bristle number (78), life span (130, 232), and fitness (69). Some *Drosophila* QTL have even more complicated effects, with significant three-way interactions of QTL, sex, and environment (78, 130, 232), and epistatic interactions that are sex specific and environment specific (130).

Pleiotropy

Pleiotropy is an important feature of the genetic architecture of any quantitative trait. Most loci involved in development participate in multiple developmental pathways. For example, genes affecting *Drosophila* bristle development are also involved in development of the central nervous system, sex determination, embryonic pattern formation, and eye and wing development (134). Genes once thought to have specific effects on *Drosophila* behavior are now known to affect a similarly diverse suite of characters (77). To the extent that these loci also harbor naturally occurring variants with quantitative effects, one might expect effects on several phenotypes. However, it is not possible to distinguish between pleiotropy and linkage as a cause of a correlated effect on two traits until one has mapped the QTN for each trait. For example, significant variation in *Drosophila* sternopleural and abdominal bristle number is associated with different molecular polymorphisms within each of two candidate genes (139, 140).

The most important property of a QTL from an evolutionary perspective is its pleiotropic relationship to fitness. Fitness effects at all loci contributing to variance in a trait determine the magnitude of genetic variation segregating for the trait and relative contributions of additive, dominance, and epistatic variance, and response of the trait to evolutionary forces of natural selection and inbreeding (198). Some fraction of the genetic variation for all quantitative traits must be that expected at equilibrium between the input of new deleterious alleles by mutation and their elimination by natural selection, which occurs when mutant alleles are rare (11). This mechanism probably pertains to a larger proportion of loci affecting life history traits than morphological traits (93). Intermediate allele frequencies are expected at loci at which variation is maintained by a balance of selective forces (57, 83), while the classic U-shaped distribution of allele frequencies (108) is expected for selectively neutral loci at drift-mutation balance.

Stabilizing selection for an intermediate optimum phenotype has different consequences, depending on the causal relationship of the QTL affecting the trait to fitness. If the effect on fitness is through the trait itself, selection favors genotypes with the least environmental sensitivity; reduces genetic variation by pushing gene frequencies at the relevant loci towards fixation (57, 233) and by building up repulsion linkages between linked QTL affecting the trait (57, 161), and promotes genetic canalization through antagonistic epistatic interactions between

QTL (203, 233). However, an intermediate optimum for fitness could also occur when effects on fitness of loci controlling the trait are not at all associated with the value of the trait, such as overdominance of alleles associated with intermediate trait values (11, 200) or deleterious pleiotropic fitness effects of alleles causing extreme phenotypes (11, 105, 114).

Although empirical estimates of magnitudes and patterns of genetic variance are largely in accord with our perceived relationships of traits to fitness (92, 171, 204, 205), determining the nature and magnitude of selection acting on quantitative traits is difficult in nature (7, 8, 53, 109, 119) and even in the laboratory. Efforts to deduce the relationship of *Drosophila* bristle numbers to fitness have reached contradictory conclusions, with experiments supporting strong (100, 135, 181), moderate (29, 124), or very weak (153) stabilizing selection. Given that it not easy to establish the relationship of a quantitative trait phenotype to fitness, inferring the fitness effects of the underlying QTL genotypes is exceedingly difficult. Further, there is likely heterogeneity in causal relationships to fitness among loci affecting a trait. For example, variation for sensory bristle numbers in *Drosophila* seems to be attributable to both intermediate- and low-frequency alleles, indicating different selective effects of the underlying QTN. The presence of intermediate-frequency alleles was deduced from the observation that the limits to artificial selection was not severely reduced in lines derived from single pairs of parents relative to selection limits achieved starting from large base populations (62, 200). Higher limits to selection in large populations and substantial variation in selection response among long-term selection lines (63) also suggest that rare alleles are present in natural populations. More directly, mutations with large effects on bristle number have been found segregating at low frequencies in nature (64). Both common (139, 140) and rare (140, 157) molecular polymorphisms at candidate bristle number QTL have been associated with variation in bristle number, even within a single gene (140).

We now stand at the threshold of addressing these questions empirically. High-resolution QTL mapping will ultimately reveal whether there are repulsion linkages between QTL, indicative of stabilizing selection. There is some evidence that these exist across whole chromosomes (229), between closely linked QTL (180, 187), and even within a single gene (217). Once QTL are identified, the genotypes necessary to assess epistatic interactions can be constructed; there is preliminary evidence that diminishing epistasis characteristic of genetic canalization can occur (56). It is probably unrealistic to presume that we could ever measure fitness effects of all loci affecting variation in any trait since selection acting on any one locus affecting a quantitative trait is likely to be quite weak (108, 109), and one needs to consider the whole range of environments that are ecologically relevant. However, there is a rich body of population genetics theory for inferring the action of historical selection from data on DNA sequence variation (83). When applied to sequences of cloned QTL, it will be possible to detect the signatures of purifying selection, selective sweeps, balancing selection, and neutrally evolving polymorphisms (235).

FUTURE PROSPECTS

A major challenge of post-genome biology is to forge links between DNA sequence and levels of transcripts (functional genomics) and proteins (proteomics). Quantitative genetics is concerned with associating variation in DNA sequences with variation at the level of organismal phenotype (phenomics?). Thus, achieving an understanding of the genetic basis of variation for quantitative traits in terms of molecular variation at the underlying loci will benefit enormously from efforts and technology development in functional genomics. Targeted disruptions of all known and predicted genes in several model organisms (42, 66, 76, 179, 216, 242) will provide new candidate genes for evaluation of mapped QTL, and will be valuable resources for quantitative complementation tests. The cost, efficiency, and accuracy of high-throughput genotyping is the current technical limitation to high-resolution QTL mapping. When the technology for rapidly and economically resequencing whole genomes is developed, one should be able to map QTL with high resolution by generating and phenotyping huge mapping populations and genotyping the extremes. Similar considerations apply to the mapping of QTN within candidate genes. Combining analyses of whole-genome variation in transcript levels using expression arrays (137) with information from QTL mapping should be useful for nominating candidate genes for further study. One can envision future tests for epistasis based on analyses of genome-wide changes in expression in response to NIL or single mutations affecting a common trait. Development of techniques for knocking in alternative QTL and QTN alleles at the endogenous sites, without altering genetic backgrounds, will provide the rigorous standard of proof for functional association of genotype and phenotype.

Conversely, quantitative genetics is itself a functional genomics tool to fill in details that other approaches will miss. Qualitative evaluation of phenotypes of knock-out mutants will succeed in identifying genes that are essential for viability, fertility, and normal development, but a quantitative analysis of knock-out phenotypes will find many additional genes with subtle phenotypes on these traits as well as aspects of metabolism, physiology, morphology, and behavior. Identification of QTN in noncoding regions of candidate genes will motivate functional studies to specify regulatory motifs. The necessary reliance on genetic model systems to work out the genetic basis of variation for complex traits is not limiting, given the extensive homology in genetic systems across taxa. Finally, observations of epistasis and sex- and environment-specific effects of genes affecting variation in complex traits signal a revival of quantitative genetics in the post-genome era: The emerging molecular details provide grist for the waiting mill of theoretical quantitative genetics.

ACKNOWLEDGMENTS

This work was supported by grants from the NIH. This is a publication of the W. M. Keck Center for Behavioral Biology.

Visit the Annual Reviews home page at www.AnnualReviews.org

LITERATURE CITED

1. Adams MD, Celniker S, Holt RA, Evans CA, Gocayne JD, et al. 2000. The genome sequence of *Drosophila melanogaster*. *Science* 287:2185–95

2. Aguadé M, Miyashita N, Langley CH. 1989. Reduced variation in the *yellow-achaete-scute* region in natural populations of *Drosophila melanogaster*. *Genetics* 122:607–15

3. Allison DB. 1997. Transmission-disequilibrium tests for quantitative traits. *Am J. Hum. Genet.* 60:676–90

4. Andersson L, Haley CS, Ellegren H, Knott SA, Johansson M, et al. 1994. Genetic mapping of quantitative trait loci for growth and fatness in pigs. *Science* 263:1771–74

5. Anholt RRH, Lyman RF, Mackay TFC. 1996. Effects of single *P* element insertions on olfactory behavior in *Drosophila melanogaster*. *Genetics* 143:293–301

6. Anholt RRH, Mackay TFC. 2001. The genetic architecture of odor-guided behavior in *Drosophila melanogaster*. *Behav. Genet.* In press

7. Arnold SJ, Wade MJ. 1984. On the measurement of natural and sexual selection: theory. *Evolution* 38:709–19

8. Arnold SJ, Wade MJ. 1984. On the measurement of natural and sexual selection: applications. *Evolution* 38:720–34

9. Ayyadevara S, Ayyadevara R, Hou S, Thaden JJ, Shmookler Reis RJ. 2001. Genetic mapping of quantitative trait loci governing longevity of *Caenorhabditis elegans* in recombinant-inbred progeny of a Bergerac-BO × RC301 interstrain cross. *Genetics* 157:655–66

10. Barker JSF. Inter-locus interactions: a review of experimental evidence. *Theor. Pop. Biol.* 16:323–46

11. Barton NH. 1990. Pleiotropic models of quantitative variation. *Genetics* 124:773–82

12. Beavis WD. 1994. The power and deceit of QTL experiments: lessons from comparative QTL studies. In *Annu. Corn Sorghum Res. Conf., 49th*, pp. 252–68. Washington, DC: Am. Seed Trade Assoc.

13. Begun D, Aquadro CH. 1992. Levels of naturally occurring DNA polymorphism correlate with recombination rates in *D. melanogaster*. *Nature* 356:519–20

14. Bost B, de Vienne D, Hospital F, Moreau L, Dilmann C. 2001. Genetic and nongenetic bases for the L-shaped distribution of quantitative trait loci effects. *Genetics* 157:1773–87

15. Bradshaw HD, Otto KG, Frewen BE, McKay JK, Schemske DW. 1998. Quantitative trait loci affecting differences in floral morphology between two species of monkeyflower (*Mimulus*). *Genetics* 149:367–82

16. Breese EL, Mather K. 1957. The organization of polygenic activity within a chromosome of *Drosophila*. I. Hair characters. *Heredity* 11:373–95

17. Brockmann G, Haley CS, Renne U, Knott SA, Schwerin M. 1998. QTL affecting body weight and fatness from a mouse line selected for extreme high growth. *Genetics* 150:369–81

18. Brzustowicz LM, Hodgkinson KA, Chow EWC, Honer WC, Bassett AS. 2000. Location of a major susceptibility locus for familial schizophrenia on chromosome 1q21-q22. *Science* 288:678–82

19. Buck KJ, Metten P, Belknap JK, Crabbe JC. 1997. Quantitative trait loci involved in genetic predisposition to acute alcohol withdrawal in mice. *J. Neurosci.* 17:3946–55

20. Cardon LR, Bell JI. 2001. Association study designs for complex diseases. *Nat. Rev. Genet.* 2:91–99

21. Cardon LR, Smith SD, Fulker DW, Kimberling WJ, Pennington BF, et al.

1994. Quantitative trait locus for reading disability on chromosome 6. *Science* 266:276–79

22. Cheverud JM, Routman EJ. 1995. Epistasis and its contribution to genetic variance components. *Genetics* 139:1455–61
23. Cheverud JM, Routman EJ, Duarte FA, van Swinderen B, Cothran K, et al. 1996. Quantitative trait loci for murine growth. *Genetics* 142:1305–19
24. Choudhary M, Laurie CC. 1991. Use of in vitro mutagenesis to analyze the molecular basis of the difference in *Adh* expression associated with the allozyme polymorphism in *Drosophila melanogaster. Genetics* 129:481–88
25. Churchill GA, Doerge RW. 1994. Empirical threshold values for quantitative trait mapping. *Genetics* 138:963–71
26. Clark AG, Wang L. 1997. Epistasis in measured genotypes: Drosophila *P* element insertions. *Genetics* 147:157–63
27. Clark AG, Wang L, Hulleberg T. 1995. *P*-element-induced variation in metabolic regulation in Drosophila. *Genetics* 139:337–48
28. Clark AG, Weiss KM, Nickerson DA, Taylor SL, Buchanan A, et al. 1998. Haplotype structure and population genetic inferences from nucleotide sequence variation in human lipoprotein lipase. *Am J. Hum. Genet.* 63:595–612
29. Clayton GA, Morris JA, Robertson A. 1957. An experimental check on quantitative genetical theory. Short-term responses to selection. *J. Genet.* 55:131–51
30. Clayton GA, Robertson A. 1957. An experimental check on quantitative genetical theory. II. The long-term effects of selection. *J. Genet.* 55:152–70
31. Cockerham CC. 1954. An extension of the concept of partitioning hereditary variance for analysis of covariance among relatives when epistasis is present. *Genetics* 39:859–82
32. Cockerham CC, Zeng Z-B. 1996. Design III with marker loci. *Genetics* 143:1437–56

33. Comuzzie AG, Hixson JE, Almasy L, Mitchell BD, Mahaney MC, et al. 1997. A major quantitative trait locus determining serum leptin levels and fat mass is located on human chromosome 2. *Nat. Genet.* 15:273–75
34. Cormier RT, Hong KH, Halberg RB, Hawkins TL, Richardson P, et al. 1997. Secretory phospholipase Pla2g2a confers resistance to intestinal tumorigenesis. *Nat. Genet.* 17:88–91
35. Cowley DE, Atchley WR, Rutledge JJ. 1986. Quantitative genetics of *Drosophila melanogaster.* I. Sexual dimorphism in genetic parameters for wing traits. *Genetics* 114:549–66
36. Cowley DE, Atchley WR. 1988. Quantitative genetics of *Drosophila melanogaster.* II. Heritabilities and genetic correlations between sexes for head and thorax traits. *Genetics* 119:421–33
37. Cox NJ, Frigge M, Nicolae DL, Concannon P, Hanis C, et al. 1999. Loci on chromosomes 2 (*NIDDM1*) and 15 interact to increase susceptibility to diabetes in Mexican Americans. *Nat. Genet.* 21:213–15
38. Darvasi A. 1997. Interval-specific congenic strains (ISCS): an experimental design for mapping a QTL into a 1-centimorgan interval. *Mamm. Genome* 8:163–67
39. Darvasi A. 1998. Experimental strategies for the genetic dissection of complex traits in animal models. *Nat. Genet.* 18:19–24
40. Darvasi A, Soller M. 1995. Advanced intercross lines, an experimental population for fine genetic mapping. *Genetics* 141:1199–207
41. Davies JL, Kawaguchi Y, Bennett ST, Copeman JB, Cordell HJ, et al. 1994. A genome-wide search for human type 1 diabetes susceptibility genes. *Nature* 371:130–36
42. De Angelis MH, Flaswinkel H, Fuchs H, Rathkolb B, Soewarto D, et al. 2000. Genome-wide, large-scale production of mutant mice by ENU mutagenesis. *Nat. Genet.* 25:444–47

43. De Benedictis G, Caratenuto L, Carrieri G, De Luca M, Falcone E, et al. 1998. Gene/longevity association studies at four autosomal loci (*REN, THO, PARP, SOD2*). *Eur. J. Hum. Genet.* 6:534–41

44. Doebley J, Stec A. 1991. Genetic analysis of the morphological differences between maize and teosinte. *Genetics* 129:285–95

45. Doebley J, Stec A. 1993. Inheritance of the morphological differences between maize and teosinte: comparison of results from two F_2 populations. *Genetics* 134:559–70

46. Doebley J, Stec A, Gustus C. 1995. *teosinte branched1* and the origin of maize: evidence for epistasis and the evolution of dominance. *Genetics* 141:333–46

47. Doebley J, Stec A, Hibbard L. 1997. The evolution of apical dominance in maize. *Nature* 386:485–88

48. Doerge RW, Churchill GA. 1996. Permutation tests for multiple loci affecting a quantitative character. *Genetics* 142:285–94

49. Eaves IA, Merriman TR, Barber RA, Nutland S, Tuomilehto-Wolf E, et al. 2000. The genetically isolated populations of Finland and Sardinia may not be a panacea for linkage disequilibrium mapping of common human disease genes. *Nat. Genet.* 25:320–23

50. Edwards MD, Helentjaris T, Wright S, Stuber CW. 1992. Molecular-marker-facilitated investigations of quantitative trait loci in maize. 4. Analysis based on genome saturation with isozyme and restriction fragment length polymorphisms. *Theor. Appl. Genet.* 83:765–74

51. Edwards MD, Stuber CW, Wendel JF. 1987. Molecular-marker-facilitated investigations of quantitative trait loci in maize. I. Numbers, genomic distributions and types of gene action. *Genetics* 116:113–25

52. Eisen EJ, Legates JE. 1966. Genotype-sex interaction and the genetic correlation between the sexes for body weight in *Mus musculus*. *Genetics* 54:611–23

53. Endler JA. 1986. *Natural Selection in the Wild*. Princeton, NJ: Princeton. 336 pp.

54. Ertekin-Taner N, Graff-Radford N, Younkin LH, Eckman C, Baker M, et al. 2000. Linkage of plasma $A\beta42$ to a quantitative locus on chromosome 10 in late-onset Alzheimer's disease pedigrees. *Science* 290:2303–4

55. Eshed Y, Zamir D, 1995. An introgression line population of *Lycopersicon pennellii* in the cultivated tomato enables the identification and fine-mapping of yield-associated QTL. *Genetics* 141:1147–62

56. Eshed Y, Zamir D. 1996. Less-than-additive interactions of quantitative trait loci in tomato. *Genetics* 143:1807–17

57. Falconer DS, Mackay TFC. 1996. *Introduction to Quantitative Genetics*. Harlow, Essex: Addison Wesley Longman. 464 pp.

58. Fedorowicz GM, Fry JD, Anholt RRH, Mackay TFC. 1998. Epistatic interactions between *smell-impaired* loci in *Drosophila melanogaster*. *Genetics* 148:1885–91

59. Flint J, Corley R, DeFries JC, Fulker DW, Gray JA, et al. 1995. A simple genetic basis for a complex psychological trait in laboratory mice. *Science* 269:1432–35

60. Frankham R. 1968. Sex and selection for a quantitative character in *Drosophila*. II. The sex dimorphism. *Aust. J. Biol. Sci.* 21:1225–37

61. Frankham R. 1969. Genetic analysis of two abdominal bristle selection lines. *Aust. J. Biol. Sci.* 22:1485–95

62. Frankham R. 1980. The founder effect and response to artificial selection in *Drosophila*. In *Selection Experiments in Laboratory and Domestic Animals*, ed. A Robertson, pp. 87–90. Slough: Commonw. Agric. Bur. 245 pp.

63. Frankham R, Jones LP, Barker JSF. 1968. The effects of population size and selection intensity for a quantitative character in Drosophila. III. Analysis of the lines. *Genet. Res.* 12:267–83

64. Frankham R, Nurthen RK. 1981. Forging

links between population and quantitative genetics. *Theor. Appl. Genet.* 59:251–63

65. Frary A, Nesbitt TC, Frary A, Grandillo S, van der Knaap E, et al. 2000. *fw2.2*: a quantitative trait locus key to the evolution of tomato fruit size. *Science* 289:85–88

66. Fraser AG, Kamath RS, Zipperien P, Martinez-Campos M, Sohrmann M, et al. 2000. Functional genomic analysis of *C. elegans* chromosome I by systematic RNA interference. *Nature* 408:325–30

67. Fridman E, Pleban T, Zamir D. 2000. A recombination hotspot delimits a wild-species quantitative trait locus for tomato sugar content to 484 bp within an invertase gene. *Proc. Natl. Acad. Sci. USA* 97:4718–23

68. Fry JD, Heinsohn SL, Mackay TFC. 1996. The contribution of new mutations to genotype-environment interaction for fitness in *Drosophila melanogaster*. *Evolution* 50:2316–27

69. Fry JD, Nuzhdin SV, Pasyukova EG, Mackay TFC. 1998. QTL mapping of genotype-environment interaction for fitness in *Drosophila melanogaster*. *Genet. Res.* 71:133–41

70. Fulker DW, Cardon LR. 1994. A sib-pair approach to interval mapping of quantitative trait loci. *Am. J. Hum. Genet.* 54:1092–103

71. Game AY, Oakeshott JG. 1990. The association between restriction site polymorphism and enzyme activity variation for Esterase 6 in *Drosophila melanogaster*. *Genetics* 126:1021–31

72. Georges M, Nielsen D, Mackinnon M, Mishra A, Okimoto R, et al. 1995. Mapping quantitative trait loci controlling milk production in dairy cattle by exploiting progeny testing. *Genetics* 139:907–20

73. Gibson G, Wemple M, van Helden S. 1999. Potential variance affecting homeotic *Ultrabithorax* and *Antennapedia* phenotypes in *Drosophila melanogaster*. *Genetics* 151:1081–91

74. Gillespie JH, Turelli M. 1989. Genotype-environment interactions and the main-tenance of polygenic variation. *Genetics* 121:129–38

75. Goffeau A, Barrell BG, Bussey H, Davis RW, Dujon B, et al. 1996. Life with 6,000 genes. *Science* 274:546–67

76. Gönczy P, Echeverri C, Oegema K, Coulson A. Jones SJM, et al. 2000. Functional genomic analysis of cell division in *C. elegans* using RNAi of genes on chromosome III. *Nature* 408:331–36

77. Greenspan R. 2001. The flexible genome. *Nat. Rev. Genet.* 2:383–87

78. Gurganus MC, Fry JD, Nuzhdin SV, Pasyukova EG, Lyman RF, et al. 1998. Genotype-environment interaction at quantitative trait loci affecting sensory bristle number in *Drosophila melanogaster*. *Genetics* 149:1883–98

79. Gurganus MC, Nuzhdin SV, Leips JW, Mackay TFC. 1999. High-resolution mapping of quantitative trait loci for sternopleural bristle number in *Drosophila melanogaster*. *Genetics* 152:1585–604

80. Haley CS, Knott SA. 1992. A simple regression model for interval mapping in line crosses. *Heredity* 69:315–324

81. Hanis CL, Boerwinkle E, Chakroborty R, Ellsworth DL, Concannon P, et al. 1996. A genome-wide search for human non-insulin-dependent (type 2) diabetes genes reveals a major susceptibility locus on chromosome 2. *Nat. Genet.* 13:161–66

82. Harris H. 1966. Enzyme polymorphisms in man. *Proc. R. Soc. London Ser. B.* 164:298–310

83. Hartl DL, Clark AG. 1997. *Principles of Population Genetics.* Sunderland, MA: Sinauer. 542 pp. 3rd ed.

84. Haseman JK, Elston RC. 1972. The investigation of linkage between a quantitative trait and a marker locus. *Behav. Genet.* 2:3–19

85. Hästbacka J, de la Chapelle A, Mahtani MM, Clines G, Reeve-Daly MP, et al. 1994. The diastrophic dysplasia gene encodes a novel sulfate transporter: positional cloning by fine-structure linkage

disequilibrium mapping. *Cell* 78:1073–87

86. Hey J, Wakeley J. 1997. A coalescent estimator of the population recombination rate. *Genetics* 145:833–46

87. Hill WG. 1998. Selection with recurrent backcrossing to develop congenic lines for quantitative trait loci analysis. *Genetics* 148:1341–52

88. Hill WG, Robertson A. 1968. Linkage disequilibrium in finite populations. *Theor. Appl. Genet.* 38:226:31

89. Hill WG, Weir BS. 1988. Variances and covariances of squared linkage disequilibria in finite populations. *Theor. Pop. Biol.* 33:54–78

90. Hill WG, Weir BS. 1994. Maximum likelihood estimation of gene location by linkage disequilibrium. *Am. J. Hum. Genet.* 54:705–14

91. Horigawa Y, Oda N, Cox NJ, Li X, Orho-Melander M, et al. 2000. Genetic variation in the gene encoding calpain-10 is associated with type 2 diabetes mellitus. *Nat. Genet.* 26:163–75

92. Houle D. 1992. Comparing evolvability and variability of quantitative traits. *Genetics* 130:195–204

93. Houle D, Morikawa B, Lynch M. 1996. Comparing mutational variabilities. *Genetics* 143:1467–83

94. Hudson R. 1987. Estimating the recombination parameter of a finite population model without selection. *Genet. Res.* 50:245–50

95. International Human Genome Sequencing Consortium. 2001. Initial sequencing and analysis of the human genome. *Nature* 409:860–921

96. Jackson AU, Fornés A, Galecki A, Miller RA, Burke DT. 1999. Multiple-trait quantitative trait loci analysis using a large mouse sibship. *Genetics* 151:785–95

97. Jansen RC, Stam P. 1994. High resolution of quantitative traits into multiple loci via interval mapping. *Genetics* 136:1447–55

98. Jiang C, Zeng Z-B. 1995. Multiple trait analysis of genetic mapping for quantitative trait loci. *Genetics* 140:1111–27

99. Kao C-H, Zeng Z-B, Teasdale R. 1999. Multiple interval mapping for quantitative trait loci. *Genetics* 152:1203–16

100. Kearsey MJ, Barnes BW. 1970. Variation for metrical characters in *Drosophila* populations. II. Natural selection. *Heredity* 25:11–21

101. Kearsey MJ, Kojima K-I. 1967. The genetic architecture of body weight and egg hatchability in *Drosophila melanogaster*. *Genetics* 56:23–37

102. Keightley PD, Bulfield G. 1993. Detection of quantitative trait loci from frequency changes of marker alleles under selection. *Genet. Res.* 62:195–203

103. Keightley PD, Evans MJ, Hill WG. 1993. Effects of multiple retrovirus insertions on quantitative traits of mice. *Genetics* 135:1099–106

104. Keightley PD, Hardge T, May L, Bulfield G. 1996. A genetic map of quantitative trait loci for body weight in the mouse. *Genetics* 142:227–35

105. Keightley PD, Hill WG. 1990. Variation maintained in quantitative traits with mutation-selection balance: pleiotropic side-effects on fitness traits. *Proc. R. Soc. London Ser. B* 242:95–100

106. Kempthorne O. 1954. The correlation between relatives in a random mating population. *Proc. R. Soc. London Ser. B* 143:103–13

107. Kidwell JF. 1969. A chromosomal analysis of egg production and abdominal chaeta number in *Drosophila melanogaster*. *Can. J. Genet. Cytol.* 11:547–57

108. Kimura M. 1983. *The Neutral Theory of Molecular Evolution*. Cambridge, UK: Cambridge. 367 pp.

109. Kingsolver JG, Hoekstra HE, Hoekstra JM, Berrigan D, Vignieri SN, et al. 2001. The strength of phenotypic selection in natural populations. *Am. Nat.* 157:245–61

110. Knapp SJ, Bridges WC, Birkes D. 1990. Mapping quantitative trait loci using

molecular marker linkage maps. *Theor. Appl. Genet.* 79:583–92

111. Knoblauch H, Muller-Myhsok B, Busjahn A, Ben Avi L, Bahring S, et al. 2000. A cholesterol-lowering gene maps to chromosome 13q. *Am. J. Hum. Genet.* 66:157–66

112. Knott SA, Haley CS. 1992. Aspects of maximum likelihood methods for the mapping of quantitative trait loci in line crosses. *Genet. Res.* 60:139–51

113. Knott SA, Marklund L, Haley CS, Andersson K, Davies W, et al. 1998. Multiple marker mapping of quantitative trait loci in a cross between outbred wild boar and Large White pigs. *Genetics* 149:1069–80

114. Kondrashov AS, Turelli AS, 1992. Deleterious mutations, apparent stabilizing selection and the maintenance of quantitative variation. *Genetics* 132:603–18

115. Kreitman M. 1983. Nucleotide polymorphism at the alcohol dehydrogenase locus of *Drosophila melanogaster*. *Nature* 304:412–17

116. Kristensen CN, Kelefiotis D, Kristensen T, Børresen-Dale A-L. 2001. High-throughput methods for detection of genetic variation. *BioTechniques* 30:318–32

117. Kruglyak L. 1999. Prospects for whole-genome linkage disequilibrium mapping of common disease genes. *Nat. Genet.* 22:139–44

118. Kruglyak L, Lander ES. 1995. Complete multi-point sib-pair analysis of qualitative and quantitative traits. *Am. J. Hum. Genet.* 57:439–54

119. Lande R, Arnold SJ. 1983. The measurement of selection on correlated characters. *Evolution* 37:1210–26

120. Lander ES, Botstein D. 1989. Mapping Mendelian factors underlying quantitative traits using RFLP linkage maps. *Genetics* 121:185–200

121. Langley CH, Montgomery E, Quattlebaum WF. 1982. Restriction map variation in the *Adh* region of *Drosophila*. *Proc. Natl. Acad. Sci. USA* 79:5631–35

122. Lai C, Lyman RF, Long AD, Langley

CH, Mackay TFC. 1994. Naturally occurring variation in bristle number and DNA polymorphisms at the *scabrous* locus of *Drosophila melanogaster*. *Science* 266:1697–702

123. Lai C, McMahon R, Young C, Mackay TFC, Langley CH. 1998. *que mao*, a *Drosophila* bristle locus, codes for geranlygeranyl pyrophosphate synthase. *Genetics* 149:1051–61

124. Latter BDH, Robertson A. 1962. The effects of inbreeding and artificial selection on reproductive fitness. *Genet. Res.* 3:110–38

125. Laurie-Ahlberg CC, Stam LF. 1987. Use of P-element-mediated transformation to identify the molecular basis of naturally occurring variants affecting *Adh* expression in *Drosophila melanogaster*. *Genetics* 115:129–40

126. Laurie CC, Bringham JT, Choudhary M. 1991. Associations between DNA sequence variation and variation in gene expression of the *Adh* gene in natural populations of *Drosophila melanogaster*. *Genetics* 129:489–99

127. Laurie CC, Stam LF. 1994. The effect of an intronic polymorphism on alcohol dehydrogenase expression in *Drosophila melanogaster*. *Genetics* 138:379–85

128. Laurie CC, True JR, Liu J, Mercer JM. 1997. An introgression analysis of quantitative trait loci that contribute to a morphological difference between *Drosophila simulans* and *D. mauritiana*. *Genetics* 145:339–48

129. Leigh Brown AJ. 1983. Variation at the 87A heat shock locus in *Drosophila melanogaster*. *Proc. Natl. Acad. Sci. USA* 80:5350–54

130. Leips J, Mackay TFC. 2000. Quantitative trait loci for life span in *Drosophila melanogaster*: interactions with genetic background and larval density. *Genetics* 1:5:1773–88

131. Levene H. 1953. Genetic equilibrium when more than one ecological niche is available. *Am. Nat.* 87:331–33

132. Lewontin RC, Hubby JL. 1966. A molecular approach to the study of genic heterozygosity in natural populations. II. Amount of variation and degree of heterozygosity in natural populations of *Drosophila pseudoobscura*. *Genetics* 54:595–609

133. Li Z, Pinson SR, Park WD, Paterson AH, Stansel JW. 1997. Epistasis for three grain yield components in rice (*Oryza sativa* L.). *Genetics* 145:453–65

134. Lindsley DL, Zimm GG. 1992. *The Genome of Drosophila melanogaster*. San Diego: Academic. 1133 pp.

135. Linney R, Barnes BW, Kearsey MJ. 1971. Variation for metrical characters in *Drosophila* populations. III. The nature of selection. *Heredity* 27:163–74

136. Liu J, Mercer JM, Stam LF, Gibson GC, Zeng Z-B, et al. 1996. Genetic analysis of a morphological shape difference in the male genitalia of *Drosophila simulans* and *D. mauritiana*. *Genetics* 142:1129–45

137. Lockhart DJ, Barlow C. 2001. Expressing what's on your mind: DNA arrays and the brain. *Nat. Rev. Neurosci.* 2:63–68

138. Long AD, Langley CH. 1999. Power of association studies to detect the contribution of candidate genetic loci to complexly inherited phenotypes. *Genome Res.* 9:720–31

139. Long AD, Lyman RF, Langley CH, Mackay TFC. 1998. Two sites in the *Delta* gene region contribute to naturally occurring variation in bristle number in *Drosophila melanogaster*. *Genetics* 149:999–1017

140. Long AD, Lyman RF, Morgan AH, Langley CH, Mackay TFC. 2000. Both naturally occurring insertions of transposable elements and intermediate frequency polymorphisms at the *achaete-scute* complex are associated with variation in bristle number in *Drosophila melanogaster*. *Genetics* 154:1255–69

141. Long AD, Mullaney SL, Mackay TFC, Langley CH. 1996. Genetic interactions between naturally occurring alleles at quantitative trait loci and mutant alleles at candidate loci affecting bristle number in *Drosophila melanogaster*. *Genetics* 114:1497–510

142. Long AD, Mullaney SL, Reid LA, Fry JD, Langley CH, et al. 1995. High resolution mapping of genetic factors affecting abdominal bristle number in *Drosophila melanogaster*. *Genetics* 139:1273–91

143. López-Fanjul C, Hill WG. 1973. Genetic differences between populations of *Drosophila melanogaster* for a quantitative trait. Laboratory populations. *Genet. Res.* 22:51–68

144. López-Fanjul C, Hill WG. 1973. Genetic differences between populations of *Drosophila melanogaster* for a quantitative trait. Wild and laboratory populations. *Genet. Res.* 22:69–78

145. Lukens LN, Doebley J. 1999. Epistatic and environmental interactions for quantitative trait loci involved in maize evolution. *Genet. Res.* 74:291–302

146. Luo ZW, Tao SH, Zeng Z-B. 2000. Inferring linkage disequilibrium between a polymorphic marker locus and a trait locus in natural populations. *Genetics* 156:457–67

147. Lyman RF, Lai C, Mackay TFC. 1999. Linkage disequilibrium mapping of molecular polymorphisms at the scabrous locus associated with naturally occurring variation in bristle number in *Drosophila melanogaster*. *Genet. Res.* 74:303–11

148. Lyman RF, Lawrence F, Nuzhdin SV, Mackay TFC. 1996. Effects of single *P* element insertions on bristle number and viability in *Drosophila melanogaster*. *Genetics* 143:277–92

149. Lyman RF, Mackay TFC. 1998. Candidate quantitative trait loci and naturally occurring phenotypic variation for bristle number in *Drosophila melanogaster*: the *Delta-Hairless* gene region. *Genetics* 149:983–98

150. Lynch M, Walsh B. 1998. *Genetics and Analysis of Quantitative Traits*. Sunderland, MA: Sinauer. 980 pp.

151. Lyons PA, Armitage N, Argentina F, Denny P, Hill NJ, et al. 2000. Congenic mapping of the type 1 diabetes locus, *Idd3*, to a 780-kb region of mouse chromosome 3: identification of a candidate segment of ancestral DNA by haplotype mapping. *Genome Res.* 10:446–53

152. Mackay TFC. 1985. Transposable element-induced response to artificial selection in *Drosophila melanogaster*. *Genetics* 111:351–74

153. Mackay TFC. 1985. A quantitative genetic analysis of fitness and its components in *Drosophila melanogaster*. *Genet. Res.* 47:59–70

154. Mackay TFC. 2001. Quantitative trait loci in *Drosophila*. *Nat. Rev. Genet.* 2:11–20

155. Mackay TFC, Fry JD. 1996. Polygenic mutation in *Drosophila melanogaster*: genetic interactions between selection lines and candidate quantitative trait loci. *Genetics* 144:671–88

156. Mackay TFC, Hackett JB, Lyman RF, Wayne ML, Anholt RRH. 1996. Quantitative genetic variation of odor-guided behavior in a natural population of *Drosophila melanogaster*. *Genetics* 144:727–35

157. Mackay TFC, Langley CH. 1990. Molecular and phenotypic variation in the *achaete-scute* region of *Drosophila melanogaster*. *Nature* 348:64–66

158. Mackay TFC, Lyman RF, Hill WG. 1995. Polygenic mutation in *Drosophila melanogaster*: non-linear divergence among unselected strains. *Genetics* 139:849–59

159. Mackay TFC, Lyman RF, Jackson MS. 1992. Effects of *P* element insertions on quantitative traits in *Drosophila melanogaster*. *Genetics* 130:315–32

160. Martínez O, Curnow RN. 1992. Estimating the locations and the sizes of the effects of quantitative trait loci using flanking markers. *Theor. Appl. Genet.* 85:480–88

161. Mather K. 1941. Variation and selection of polygenic characters. *J. Genet.* 41:159–93

162. Mather K, Harrison BJ. 1949. The manifold effect of selection. *Heredity* 3:3–52

163. Mather K, Jinks JL. 1977. *Introduction to Biometrical Genetics*. London: Chapman & Hall. 231 pp.

164. Maynard Smith J. 1958. Sex-limited inheritance of longevity in *Drosophila subobscura*. *J. Genet.* 56:227–35

165. Miyashita N. 1990. Molecular and phenotypic variation of the *Zw* locus region in *Drosophila melanogaster*. *Genetics* 125:407–19

166. Miyashita N, Langley CH. 1988. Molecular and phenotypic variation of the *white* locus region in *Drosophila melanogaster*. *Genetics* 120:199–212

167. Montgomery EA, Langley CH. 1983. Transposable elements in mendelian populations. II. Distribution of three *copia*-like elements in a natural population of *Drosophila melanogaster*. *Genetics* 104:473–83

168. Morahan G, Huang D, Ymer SI, Cancilla MR, Stephen K, et al. 2001. Linkage disequilibrium of a type 1 diabetes susceptibility locus with a regulatory *IL12B* allele. *Nat. Genet.* 27:218–21

169. Morris KH, Ishikawa A, Keightley PD. 1999. Quantitative trait loci for growth traits in C57BL/6J × DBA/2J mice. *Mam. Genome* 10:225–28

170. Mott R, Talbot CJ, Turri MG, Collins AC, Flint J. 2000. A method for fine mapping quantitative trait loci in outbred animal stocks. *Proc. Natl. Acad. Sci. USA* 97:12649–54

171. Mousseau TA, Roff DA. 1987. Natural selection and the heritability of fitness components. *Heredity* 59:181–97

172. Myers A, Holmans P, Marshall H, Kwon J, Meyer D, et al. 2000. Susceptibility locus for Alzheimer's disease on chromosome 10. *Science* 290:2304–5

173. Nadeau J. 2001. Modifier genes in mice and humans. *Nat. Rev. Genet.* 2:165–74

174. Nadeau JH, Frankel D. 2000. The roads

from phenotypic variation to gene discovery: mutagenesis versus QTL. *Nat. Genet.* 25:381–84

175. Nadeau JH, Singer JB, Matin A, Lander ES. 2000. Analysing complex genetic traits with chromosome substitution strains. *Nat. Genet.* 24:221–25

176. Nickerson DA, Taylor SL, Weiss KM, Clark AG, Hutchinson RG, et al. 1998. DNA sequence diversity in a 9.7-kb region of the human lipoprotein lipase gene. *Nat. Genet.* 19:233–40

177. Nielsen DM, Weir BS. 1999. A classical setting for associations between markers and loci affecting quantitative traits. *Genet. Res.* 74:271–77

178. Niemann-Sørensen A, Robertson A. 1961. The association between blood groups and several production characteristics in three Danish cattle breeds. *Acta Agric. Scand.* 11:163–96

179. Nolan PM, Peters J, Strivens M, Rogers D, Hagan J, et al. 2000. A systematic, genome-wide, phenotype-driven mutagenesis programme for gene function studies in the mouse. *Nat. Genet.* 25:440–43

180. Nuzhdin SV, Dilda CL, Mackay TFC. 1999. The genetic architecture of selection response: inferences from fine-scale mapping of bristle number quantitative trait loci in *Drosophila melanogaster*. *Genetics* 153:1317–31

181. Nuzhdin SV, Fry JD, Mackay TFC. 1995. Polygenic mutation in *Drosophila melanogaster*: the causal relationship of bristle number to fitness. *Genetics* 139:861–72

182. Nuzhdin SV, Keightley PD, Pasyukova EG. 1993. The use of retrotransposons as markers for mapping genes responsible for fitness differences between related *Drosophila melanogaster* strains. *Genet. Res.* 62:125–31

183. Nuzhdin SV, Keightley PD, Pasyukova EG, Morozova EA. 1998. Mapping quantitative trait loci affecting sternopleural bristle number in *Drosophila melanogaster* using changes of marker allele

frequencies in divergently selected lines. *Genet. Res.* 72:79–91

184. Nuzhdin SV, Pasyukova EG, Dilda CL, Zeng Z-B, Mackay TFC. 1997. Sex-specific quantitative trait loci affecting longevity in *Drosophila melanogaster*. *Proc. Natl. Acad. Sci. USA* 94:9734–39

185. Orr HA. 1998. The population genetics of adaptation: the distribution of factors fixed during adaptive evolution. *Evolution* 52:935–49

186. Orr HA. 1999. The evolutionary genetics of adaptation: a simulation study. *Genet. Res.* 74:207–14

187. Pasyukova EG, Vieira C, Mackay TFC. 2000. Deficiency mapping of quantitative trait loci affecting longevity in *Drosophila melanogaster*. *Genetics* 156:1129–46

188. Paterson AH, Damon S, Hewitt JD, Zamir D, Rabinowitch HD, et al. 1991. Mendelian factors underlying quantitative traits in tomato: comparison across species, generations, and environments. *Genetics* 127:181–97

189. Paterson AH, DeVerna JW, Lanini B, Tanksley SD. 1990. Fine mapping of quantitative trait loci using selected overlapping recombinant chromosomes, in an interspecies cross of tomato. *Genetics* 124:735–42

190. Paterson AH, Lander ES, Hewitt JD, Peterson S, Lincoln SE, et al. 1988. Resolution of quantitative traits into Mendelian factors by using a complete linkage map of restriction fragment length polymorphisms. *Nature* 335:721–26

191. Paterson AH, Lin Y-R, Li Z, Schertz KF, Doebley JF, et al. 1995. Convergent domestication of cereal crops by independent mutations at corresponding genetic loci. *Science* 269:1714–18

192. Phillips PC. 1998. The language of gene interaction. *Genetics* 149:1167–71

193. Polaczyk PJ, Gasperini R, Gibson G. 1998. Naturally occurring genetic variation affects Drosophila photoreceptor determination. *Dev. Genes Evol.* 207:462–70

194. Pomp D. 1997. Genetic dissection of obesity in polygenic animal models. *Behav. Genet.* 27:285–306

195. Rabinowitz D. 1997. A transmission disequilibrium test for quantitative trait loci. *Hum. Hered.* 47:342–50

196. Reeve ECR, Robertson FW. 1954. Studies in quantitative inheritance. VI. Sternite chaeta number in *Drosophila*: a metameric quantitative trait. *Z. Vererbungs.* 86:269–88

197. Risch N, Merikangas K. 1996. The future of genetic studies of complex human diseases. *Science* 273:1516–17

198. Robertson A. 1955. Selection in animals: synthesis. *Cold Spring Harbor Symp. Quant. Biol.* 20:225–29

199. Robertson A. 1960. A theory of limits in artificial selection. *Proc. R. Soc. London Ser. B* 153:234–49

200. Robertson A. 1967. The nature of quantitative genetic variation. In *Heritage From Mendel*, ed. A Brink, pp. 265–80. Madison, WI: Univ. Wisc.

201. Robertson DS. 1985. A possible technique for isolating genic DNA for quantitative traits in plants. *J. Theor. Biol.* 117:1–10

202. Robertson FW. 1954. Studies in quantitative inheritance. V. Chromosome analyses of crosses between selected and unselected lines of different body size in *Drosophila melanogaster*. *J. Genet.* 52:494–520

203. Robertson FW, Reeve ECR. 1952. Studies in quantitative inheritance. I. The effects of selection of wing and thorax length in *Drosophila melanogaster*. *J. Genet.* 50:414–48

204. Roff DA. 1997. *Evolutionary Quantitative Genetics.* New York: Chapman & Hall. 493 pp.

205. Roff DA, Mousseau TA. 1987. Quantitative genetics and fitness: lessons from *Drosophila. Heredity* 58:103–18

206. Routman EJ, Cheverud JM. 1997. Gene effects on a quantitative trait: two-locus epistatic effects measured at microsatellite markers and at estimated QTL. *Evolution* 51:1654–62

207. Rutherford LS, Lindquist S. 1998. Hsp90 as a capacitor of morphological evolution. *Nature* 396:336–42

208. Sax K. 1923. The association of size differences with seed-coat pattern and pigmentation in *Phaseolus vulgaris. Genetics* 8:552–60

209. Shook DR, Johnson TE. 1999. Quantitative trait loci affecting survival and fertility-related traits in *Caenorhabditis elegans* show genotype-environment interactions, pleiotropy and epistasis. *Genetics* 153:1233–43

210. Shrimpton AE, Robertson A. 1988. The isolation of polygenic factors controlling bristle score in *Drosophila melanogaster*. II. Distribution of third chromosome bristle effects within chromosome sections. *Genetics* 118:445–59

211. Soller M, Beckmann JS. 1990. Marker-based mapping of quantitative trait loci using replicated progenies. *Theor. Appl. Genet.* 80:205–8

212. Soller M, Brody T, Genezi A. 1976. On the power of experimental designs for the detection of linkage between marker loci and quantitative loci in crosses between inbred lines. *Theor. Appl. Genet.* 47:35–39

213. Spelman RJ, Coppieters W, Karim L, van Arendonk JA, Bovenhuis H. 1996. Quantitative trait loci analysis for five milk production traits on chromosome six in the Dutch Holstein-Friesian population. *Genetics* 144:1799–808

214. Spickett SG, Thoday JM. 1966. Regular responses to selection. 3. Interactions between located polygenes. *Genet. Res.* 7:96–121

215. Spielman RS, McGinnis RE, Ewens WJ. 1993. Transmission test for linkage disequilibrium: the insulin gene region and insulin-dependent diabetes mellitus (DDM). *Am. J. Hum. Genet.* 52:506–16

216. Spradling AC, Stern D, Beaton A, Rhem EJ, Laverty T, et al. 1999. The Berkeley

Drosophila genome gene disruption project: single *P* element insertions mutating 25% of vital *Drosophila* genes. *Genetics* 153:135–77

217. Stam LF, Laurie CC. 1996. Molecular dissection of a major gene effect on a quantitative trait: the level of alcohol dehydrogenase expression in *Drosophila melanogaster*. *Genetics* 144:1559–64

218. Stuber CW, Lincoln SE, Wolff DW, Helentjaris T, Lander ES. 1992. Identification of genetic factors contributing to heterosis in a hybrid from two elite maize inbred lines using molecular markers. *Genetics* 132:823–39

219. Stuber CW, Moll RH, Goodman MM, Schaffer HE, Weir BS. 1980. Allozyme frequency changes associated with selection for increased grain yield in maize. *Genetics* 95:225–36

220. Symula DJ, Frazer KA, Ueda Y, Denefle P, Stevens ME, et al. 1999. Functional screening of an asthma QTL in YAC transgenic mice. *Nat. Genet.* 23:241–44

221. Talbot CJ, Nicod A, Cherny SS, Fulker DW, Collins AC, et al. 1999. High resolution mapping of quantitative trait loci in outbred mice. *Nat. Genet.* 21:305–8

222. Taillon-Miller P, Bauer-Sardiña I, Saccone N, Putzel J, Laitinen T, et al. 2000. Juxtaposed regions of extensive linkage disequilibrium in human Xq25 and Xq28. *Nat. Genet.* 25:324–28

223. Tanksley SD. 1993. Mapping polygenes. *Annu. Rev. Genet.* 27:205–33

224. Templeton A. 1999. Uses of evolutionary theory in the human genome project. *Annu. Rev. Ecol. Syst.* 30:23–49

225. The Arabidopsis Genome Initiative. 2000. Analysis of the genome sequence of the flowering plant *Arabidopsis thaliana*. *Nature* 408:796–815

226. The *C. elegans* Sequencing Consortium. 1998. Genome sequence of the nematode *C. elegans*: a platform for investigating biology. *Science* 282:2012–18

227. The International SNP Map Working Group. 2001. A map of human genome sequence variation containing 1.42 million single nucleotide polymorphisms. *Nature* 409:928–33

228. Thoday JM. 1961. Location of polygenes. *Nature* 191:368–70

229. Thompson JN, Hellack JJ, Tucker RR. 1991. Evidence for balanced linkage of *X* chromosome polygenes in a natural population of *Drosophila*. *Genetics* 127:117–23

230. Vaughn TT, Pletscher S, Peripato A, King-Ellison K, Adams E, et al. 1999. Mapping quantitative trait loci for murine growth: a closer look at genetic architecture. *Genet. Res.* 74:313–22

231. Venter JC, Adams MD, Myers EW, Li PW, Mural RJ, et al. 2001. The sequence of the human genome. *Science* 291:1304–51

232. Vieira C, Pasyukova EG, Zeng S, Hackett JB, Lyman RF, et al. 2000. Genotype-environment interaction for quantitative trait loci affecting life span in *Drosophila melanogaster*. *Genetics* 154:213–27

233. Wagner G, Booth G, Bagheri-Chaichian H. 1997. A population genetic theory of canalization. *Evolution* 51:329–47

234. Walling GA, Visscher PM, Andersson L, Rothschild M, Wang L, et al. 2000. Combined analyses of data from quantitative trait loci mapping studies: chromosome 4 effects on porcine growth and fatness. *Genetics* 155:1369–78

235. Wang R-L, Stec A, Hey J, Lukens L, Doebley J. 1999. The limits of selection during maize domestication. *Nature* 398:236–39

236. Watterson GA. 1975. On the number of segregating sites in genetical models without recombination. *Theor. Popul. Biol.* 7:256–76

237. Weber K, Eisman R, Morey L, Patty A, Sparks J, et al. 1999. An analysis of polygenes affecting wing shape on chromosome 3 in *Drosophila melanogaster*. *Genetics* 153:773–86

238. Weir BS. 1996. *Genetic Data Analysis II*. Sunderland, MA: Sinauer. 445 pp.

239. Weir BS, Hill WG. 1986. Nonuniform recombination within the human β-globin gene cluster. *Am. J. Hum. Genet.* 38:776–78

240. Weiss KM, Terwilliger JD. 2000. How many diseases does it take to map a gene with SNPs? *Nat. Genet.* 26:151–57

241. Winzeler EA, Richards DR, Conway AR, Goldstein AL, Kalman S, et al. 1998. Direct allelic variation scanning of the yeast genome. *Science* 281:1194–97

242. Winzeler EA, Shoemaker DD, Astromoff A, Liang H, Anderson K, et al. 1999. Functional characterization of the *S. cerevisiae* genome by gene deletion and parallel analysis. *Science* 285:901–6

243. Wolstenholme DR, Thoday JM. 1963. Effects of disruptive selection. VII. A third chromosome polymorphism *Heredity* 10:413–31

244. Wu R, Zeng Z-B. 2001. Linkage and linkage disequilibrium mapping in natural populations. *Genetics* 157:899–909

245. Xie C, Gessler DDG, Xu S. 1998. Combining different line crosses for mapping quantitative trait loci using the identical by descent-based variance component method. *Genetics* 149:1139–46

246. Xu S. 1996. Mapping quantitative trait loci using four-way crosses. *Genet. Res.* 68:175–81

247. Zeng Z-B. 1994. Precision mapping of quantitative trait loci. *Genetics* 136:1457–68

248. Zeng Z-B, Liu J, Stam LF, Kao C-H, Mercer JM, et al. 2000. Genetic architecture of a morphological shape difference between two *Drosophila* species. *Genetics* 154:299–310

249. Zeng Z-B, Kao C-H, Basten CJ. 1999. Estimating the genetic architecture of quantitative traits. *Genet. Res.* 74:279–89

250. Zhang Q, Boichard D, Hoeschele I, Ernst C, Eggen A, et al. 1998. Mapping quantitative trait loci for milk production and health of dairy cattle in a large outbred pedigree. *Genetics* 149:1959–73

251. zur Lage P, Shrimpton AE, Mackay TFC, Leigh Brown AJ. 1997. Genetic and molecular analysis of *smooth*, a quantitative trait locus affecting bristle number in *Drosophila melanogaster*. *Genetics* 146:607–18

Annu. Rev. Genet. 2001. 35:341–64

NEW PERSPECTIVES ON NUCLEAR TRANSPORT

Arash Komeili and Erin K. O'Shea

Howard Hughes Medical Institute, University of California, Department of Biochemistry and Biophysics, San Francisco, California 94143; e-mail: akomeil@itsa.ucsf.edu; oshea@biochem.ucsf.edu

Key Words   nuclear import, nuclear export, import receptor, export receptor, RNA transport

■ **Abstract**   A central aspect of cellular function is the proper regulation of nucleocytoplasmic transport. In recent years, significant progress has been made in identifying and characterizing the essential components of the transport machinery. Despite these advances, some facets of this process are still unclear. Furthermore, recent work has uncovered novel molecules and mechanisms of nuclear transport. This review focuses on the unresolved and novel aspects of nuclear transport and explores issues in tRNA, snRNA, and mRNA export that highlight the diversity of nuclear transport mechanisms.

CONTENTS

INTRODUCTION

A distinguishing feature of eukaryotic cells is the compartmentalization of genetic information within a nucleus. The nucleus and the cytoplasm are separated by a nuclear envelope, and all macromolecular exchange between them takes place

0066-4197/01/1215-0341$14.00

through large protein channels termed the nuclear pore complex (NPC). The size and complexity of molecules that are exchanged between these two compartments ranges from ions and other small molecules to large complexes such as the ribosome. In contrast to ions and small proteins that diffuse across the NPC, transport of macromolecules is an active process. Active nucleocytoplasmic transport allows for the proper regulation and trafficking of nuclear proteins involved in transcription, DNA replication, and chromatin remodeling as well as mRNAs, tRNAs, and rRNAs that are transcribed in the nucleus but function in the cytoplasm.

In recent years, there has been significant progress in identifying the components and understanding the mechanics of nucleocytoplasmic transport. The development of an in vitro import assay using digitonin-permeabilized cells has led to the identification of the soluble factors required for nuclear import (2). This biochemical approach has been complemented by extensive genetic studies in *Saccharomyces cerevisiae* where multiple components of the nuclear pore complex and the machinery for nuclear export of proteins and mRNAs have been identified (17, 88). In addition, the large size of *Xenopus laevis* oocytes has provided a powerful cell biological tool for studying transport processes.

As a result of this work a general model for nuclear transport has been established (25). As is described later, active nuclear transport is mediated by a family of transport receptors that recognize targeting sequences on their cargoes and traverse the NPC through interactions with nuclear pore proteins. The small Ras-like GTPase Ran and its effector molecules establish the directionality of transport. Remarkably, it appears that the components of the nuclear transport machinery are well conserved in a variety of organisms.

The specifics of this model and the history behind the discoveries that have fueled this explosion in the understanding of nucleocytoplasmic transport have been discussed in many reviews (22, 25, 55, 64). Despite the progress in the study of nuclear transport, certain aspects of this process have not been elucidated. In this review, we examine the most recent findings regarding nucleocytoplasmic transport. We focus on energetics and mechanisms of translocation through the NPC, as well as on recent work that has put aspects of the classical model for nuclear transport in question. Finally, to highlight the variety of nuclear transport mechanisms, we provide a detailed view of recent progress in understanding tRNA, snRNA, and mRNA transport. Where possible, we emphasize the role of genetics in the study of nuclear transport.

CLASSICAL NUCLEAR TRANSPORT

General Model for Nuclear Transport

The soluble components of the nuclear transport machinery and the basic mechanisms of nuclear transport were identified through a combination of biochemical and genetic approaches (25, 55). The in vitro import assay led to the discovery of transport receptors that can bind to specific targeting sequences on their cargo and carry them across the NPC. Importin β, the first such transport receptor to be identified, mediates the import of proteins containing basic nuclear localization signals (NLS) (1, 24, 68). Importin β recognizes these targeting sequences using an adaptor molecule, importin α, which binds directly to the NLS (1, 26). Additionally, importin β can interact directly with a subset of cargoes and thus transport them independently of importin α (41). Other transport receptors that have been identified share significant homology to importin β in their N-terminal regions. Based on this sequence similarity, a group of 14 importin β-related transport receptors have been identified in *S. cerevisiae* (23). Interestingly, mutants in these homologues display a variety of defects in the import or export of proteins and RNAs, lending further support to their central role in nuclear transport (25).

Whereas the importin β family of receptors is responsible for recognition of transport cargoes, the GTPase Ran appears to be the key to establishing the directionality of transport in vivo (25). The RanGAP (GTPase activating protein) is localized to the cytoplasm and stimulates Ran's intrinsic GTPase activity, resulting in conversion of RanGTP to RanGDP (5, 6, 57). RCC1, the Ran guanine nucleotide exchange factor (RanGEF), is chromatin-associated and stimulates release of the bound nucleotide on Ran (7, 66). The asymmetric distribution of RanGEF and RanGAP implies the existence of a RanGTP gradient within the cell such that there are high levels of RanGTP in the nucleus and low levels of RanGTP in the cytoplasm. This asymmetry is thought to be essential in establishing the directionality of transport since export receptors of the importin β family require RanGTP to bind to their cargo whereas import receptors are dissociated from their cargoes in the presence of RanGTP (25). In addition to RanGAP and RanGEF, two other soluble factors are components of the RanGTP cycle and are required for efficient nuclear transport. The Ran binding protein RanBP1 helps RanGAP in dissociation of export complexes, and the small protein NTF2 is responsible for the active import of RanGDP into the nucleus, where it is converted to RanGTP (4, 71, 83).

The study of the soluble components of the nuclear transport machinery has led to the following model for the import and export of macromolecules (25) (Figure 1). An import receptor will bind to its cargo in the cytoplasm and traverse the NPC. In the nucleus RanGTP binds to the import receptor and releases the cargo into the nucleus. Export of macromolecules begins by the formation of a trimeric complex consisting of the receptor, cargo, and RanGTP in the nucleus. Subsequently, this complex translocates through the NPC and upon reaching the cytoplasm, the combined action of RanBP1 and RanGAP lead to GTP hydrolysis by Ran and release of the cargo in the cytoplasm.

Energetics of Transport

Nuclear transport of macromolecules can occur against a concentration gradient, highlighting the need for energy in this process. Early studies in vivo and with the in vitro import assay showed that nucleotides and GTP hydrolysis by Ran were necessary for the import of NLS cargoes by importin α/β (2, 93). Although these

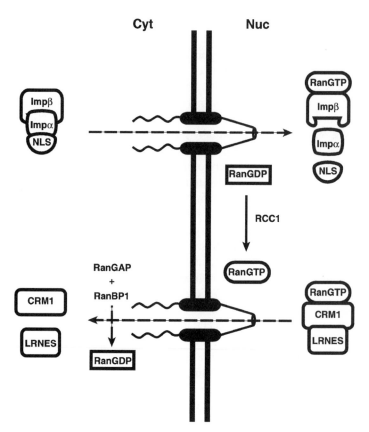

Figure 1   Classical model for nuclear transport. Importin α/β heterodimer (Imp α/ Imp β) binds to the nuclear localization signal (NLS) of the import substrate and carries it into the nucleus where RanGTP binds to Impβ and dissociates the complex. The formation of a trimeric complex between RanGTP, a nuclear export receptor (CRM1), and an export cargo (leucine rich nuclear export signal, LRNES), initiates an export cycle. In the cytoplasm the combined action of RanBP1 and RanGAP lead to the GTP hydrolysis on Ran and dissociation of the complex.

experiments show that energy is needed for nuclear transport, they do not elucidate the exact step where this energy requirement manifests itself.

A long-standing model suggested that energy and GTP hydrolysis by Ran would be used in the translocation of receptor-cargo complexes through the nuclear pore (93). However, several recent studies show that energy and GTP hydrolysis by Ran are not required for translocation of import or export complexes through the nuclear pore and might only be needed at the terminal steps of transport. One such study examined the energy requirements in transportin-mediated nuclear import (70). Transportin is an importin β homologue that acts as the import receptor for the M9 sequence of hnRNA1 (62). In an in vitro import assay Ran and energy are required for nuclear import of M9 when transportin levels are lower than M9 levels (70). However, when transportin and M9 are added in equimolar quantities, there is significant import of M9 into the nucleus independent of Ran, GTP, ATP, or an energy regenerating system (70). These results suggest that when an import receptor is present in substoichiometric amounts, RanGTP is needed for multiple rounds of transport. Thus, energy and Ran are not required for translocation through the pore but are important in the release of cargoes and recycling of receptors. This model is further supported by the observation that import of the importin β binding (IBB) domain of importin α by importin β is not dependent on nucleotides or GTP hydrolysis on Ran (60). Also, in the absence of RanGTP, importin β devoid of cargo can translocate to the nuclear side of the pore (46). Interestingly, similar energy requirements are observed for nuclear export processes. Crm1 is an importin β-like transporter that mediates the export of proteins containing a leucine-rich nuclear export signal (LRNES) (20, 85). Although RanGTP is required for the formation of a Crm1-LRNES export complex, nucleotide hydrolysis is not needed for nuclear export of this complex in an in vitro export assay (16).

Collectively, these results imply that energy and Ran are not required for translocation but are involved in the terminal steps of nuclear transport. In the nucleus, RanGTP ensures the release of import receptors from their cargoes, allowing them to return to the cytoplasm for further rounds of transport. For export, GTP hydrolysis by Ran in the cytoplasm leads to dissociation of export complexes, thereby freeing the export receptor to return to the nucleus. Although it is clear from these results that translocation of simple transporter-cargo complexes through the NPC is energy independent, larger and more complex cargoes may have different energy requirements.

Mechanisms of Translocation through the NPC

A somewhat complete picture of the soluble components and the energetics of transport has emerged in recent years. The events that initiate and terminate nuclear transport have been characterized in detail, and examination of multiple transport pathways has confirmed most aspects of this general model. However, many questions remain regarding the translocation of transport complexes through the nuclear pore complex.

This is not a trivial issue since a transport complex must travel through 200 nm of the NPC before it can be dissociated in its target compartment. The NPC is a large structure, varying in size from 60 MDa in *S. cerevisiae* to 125 MDa in vertebrates, that contains multiple copies of 30 to 50 different proteins (17, 75, 86). The size and the complexity of the NPC have made biochemical reconstitution of the translocation process very difficult. Genetic and biochemical studies with isolated nuclear pore proteins (Nups or nucleoporins) have shown that they interact directly with import and export receptors, leading to the hypothesis that these interactions form the basis for translocation through the NPC

(21, 69, 74). Thus, identifying the protein components of the NPC and their localization within this large complex is crucial to understanding the translocation process.

Recently, a major step has been taken toward elucidation of the complete protein composition of the *S. cerevisiae* NPC (73). Rout et al. identified *S. cerevisiae* proteins that were enriched in nuclear envelope preparations. The localization of each protein was determined by immunofluorescence, and those that displayed characteristic nuclear pore complex localization were categorized as members of the NPC. The results of this study indicate that each NPC in *S. cerevisiae* is composed of 16 to 32 copies of only 30 proteins, a relatively low number given the size and architectural complexity of this structure. Immunoelectron microscopy experiments were carried out to determine the distribution of these proteins within the NPC. Remarkably, all but 5 of these proteins are present on both sides of the pore. Of these 25, 4 proteins were more abundant on one side of pore and the rest were distributed equally across the nuclear envelope. The largely symmetric distribution of NPC proteins is surprising because the NPC has an asymmetric architecture consisting of a central channel, fibrils that extend into the cytoplasm, and a basket-like structure within the nucleus. It is possible that the few nucleoporins that have a biased distribution are the constituents of the nuclear basket and the cytoplasmic fibrils. Also, it is possible that some peripheral proteins that may contribute to the overall architecture of the NPC were lost during biochemical preparations of nuclear envelopes. In addition to providing a more complete catalogue of NPC proteins, this study provides a framework for examining the mechanisms of translocation through the pore.

Given the symmetric distribution of the pore proteins, it is tempting to hypothesize that directionality of transport relies more on the soluble transport components of transport rather than the actual protein composition of the NPC. In fact, the directionality of transport through the nuclear pore can be reversed (60). If a preformed Crm1-LRNES-RanQ69L export complex is added to the cytoplasmic face of permeabilized cells, the NES substrate can equilibrate between the two compartments. This "import" of an export complex relies on a viable interaction between Crm1 and the LRNES and requires functional NPCs. These results indicate that the directionality of transport may largely be determined by the Ran system and that the location of specific nucleoporins within the NPC does not play a dominant role in translocation.

These results have led to the suggestion that once an export complex has been formed in the nucleus, it moves through the NPC through successive interactions between transport receptors and nucleoporins (60, 73). The cargo-receptor complex is thought to move in both directions until it reaches the cytoplasmic side in a stochastic manner. Once the complex nears the cytoplasm, the concerted action of RanBP1 and RanGAP leads to GTP hydrolysis on Ran and dissociation of the complex into the cytoplasm. According to this model, translocation of import complexes would proceed in a similar manner until their dissociation by RanGTP in the nucleus.

However, there is evidence disputing the above model. First, although the distribution of the majority of NPC proteins is symmetric, there are some that display a biased localization (73). Second, biochemical evidence has pointed to the existence of subcomplexes within the NPC that show differential localization (75). Third, certain mutants in yeast Nups display defects in either import or export processes, but not both (58). Lastly, experiments with reconstituted nuclear pores have indicated that the direction of transport across the NPC cannot be reversed (45). These results suggest that import and export receptors may display differential interactions with nucleoporins such that movement through the NPC is vectorial. In such a model, the interactions between an export receptor and NPC components would increase in affinity as the complex approaches the cytoplasmic side of the pore. Similarly, an import complex would interact more strongly with nuclear nucleoporins than with cytoplasmic ones. Further support for this model comes from examination of receptor-nucleoporin interactions in *S. cerevisiae* (10). During transport in vivo, the export receptor Msn5 and the import receptor Pse1 show differential interactions with nucleoporins, suggesting that they use different routes for translocation through the NPC.

How can the largely symmetric composition of the NPC be reconciled with a model where translocation is driven by differential interactions between receptors and nucleoporins? Perhaps, despite their symmetric distribution, nuclear pore proteins interact asymmetrically with each other within the pore to create tracks that would be specific to either import or export complexes. This idea is supported by the observation that the yeast Nup53, Nup59, and Nup170 form a subcomplex that binds specifically to the import receptor Pse1/Kap121, but fails to bind to the closely related Kap123 or importin β/Kap95 (54). In addition to the existence of subcomplexes that bind to a subset of transport receptors, certain Nups contain domains that display specificity in their interactions with transport receptors. Nup153 in humans contains one domain that is required for importin β-dependent import and another that mediates transportin-dependent import (81). Therefore, subcomplexes and subdomains of nucleoporins might create microenvironments within the NPC that are specific for different transporters.

Despite extensive progress in defining the architecture and composition of the NPC and the abundance of models to describe translocation, our view of this process is still rather unclear. Further characterization of the subset of nucleoporin-transport receptor interactions that promote translocation through the NPC is needed to arrive at a more unified theory regarding this crucial step in nuclear transport.

NONCLASSICAL TRANSPORT

Due to the generality and the remarkable evolutionary conservation of the model described in the previous section, most studies have relied heavily on reverse genetics and homology-based techniques to identify factors involved in transport of specific macromolecules. Although most cases follow this simple model, several examples of nonclassical nuclear transport have been identified. In this section, we examine some of these cases that have introduced novel mechanisms of nuclear transport (Figure 2).

Ran-Independent Transport

The RanGTP cycle is at the center of the classical model for transport. Ran is required at three distinct points in nuclear transport (25). Without RanGTP in the nucleus, import complexes would not dissociate and export complexes would not form. Also, GTP hydrolysis by Ran is required for the dissociation of export complexes. Therefore, it has been surprising to find that several cargoes can be transported independently of RanGTP (9, 33, 90).

The most striking example of Ran-independent transport is the import of cyclin B1-Cdc2 by importin β (88). In an in vitro import assay, the import of this complex depends on importin β but does not require Ran. In contrast, the import of basic NLS-containing cargoes by importin β in the same assay is dependent on Ran. These results indicate that in contrast to classical importin β-mediated import, RanGTP is not required for the release of cyclin B1-Cdc2 from importin β in the nucleus. Accordingly, RanQ69L, a hydrolysis-deficient Ran mutant, inhibits import of basic NLS-containing cargoes by dissociating them in the cytoplasm but does not affect cyclin B1-CDC2 import. Thus, an importin β-dependent but Ran-independent mechanism is used in the import of cyclin B1-CDC2.

If RanGTP is not required for these transport processes, then alternative mechanisms must exist for the formation of export complexes and dissociation of import and export complexes. One possibility is that direct modification of cargoes in the target compartment leads to dissociation of the transport complex.

Alternative Roles of Ran

It has long been known that mutations in the Ran system lead to a variety of phenotypes such as defects in nuclear morphology and the cell cycle (37). Given the importance of nuclear transport, it has been difficult to assign a direct role for Ran in these processes. Recently, several reports have identified a direct role for the RanGTPase cycle in regulation of the mitotic spindle (31). Addition of demembranated sperm nuclei to *Xenopus* oocyte mitotic extracts induces the formation of a mitotic spindle around the DNA. Addition of high levels of RanQ69L can also induce the formation of microtubule asters that resemble those formed in the presence of sperm nuclei (29, 31, 59). Interestingly, RanGTP uses the components of the nuclear transport machinery in inducing spindle assembly (12). Mitotic *Xenopus* egg extracts that have been depleted of RanQ69L binding proteins form microtubule asters spontaneously (59). Addition of importin β to these depleted extracts inhibits their spontaneous aster formation, suggesting that RanGTP promotes spindle assembly by overcoming an inhibitory effect imposed by importin β. In support of this hypothesis, addition of an excess of SV40 NLS or IBB domain of importin α to mitotic egg extracts also induces aster

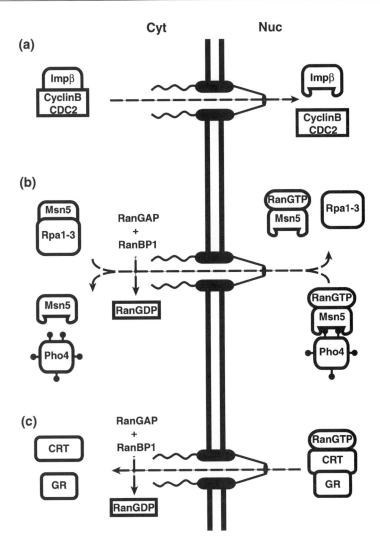

Figure 2 Nonclassical nuclear transport. (*A*) Ran-independent transport. Import of CyclinB/CDC2 by importin β does not require RanGTP. (*B*) Msn5 acts as an import and an export receptor. The importin β family member protein, Msn5, has a well-characterized role in the export of many proteins including the budding yeast transcription factor Pho4. It appears that it might also play a role in the import of the DNA damage response proteins, Rpa1, Rpa2, and Rpa3 (Rpa1-3) (*C*) Proteins not related to importin β may act as transport receptors. Calreticulin (CRT), an ER-resident protein, which is not related to importin β, is an export receptor for the glucocorticoid receptor (GR).

formation (29, 59). Thus, proteins required for an aster promoting activity (APA) are sequestered by importin α/β during interphase, resulting in a block to spindle assembly. Presumably, in interphase APA proteins and other factors required for spindle assembly are kept in separate compartments. After nuclear envelope breakdown during mitosis, RanGTP releases these proteins from importin α/β allowing them to interact with other factors to build the mitotic spindle. This mechanism ensures that the spindle forms in the proximity of the DNA where high levels of RanGTP are present as a result of the action of the chromatin-bound RanGEF.

Two potential components of APA have been identified. The nuclear mitotic apparatus protein (NuMA) is a known mitotic spindle protein and induces aster formation when added at very low concentrations to extracts that are depleted of both RanQ69L and importin β-associated proteins (59, 94). NuMA binds to importin α/β, and it is released in the presence of RanGTP. Another protein, TPX2, targets the motor protein Xklp2 to microtubules (29). Recombinant TPX2 is sufficient to induce aster formation in mitotic extracts, and this activity is inhibited by co-addition of importin α. NuMA and TPX2 are not the only components of APA. Addition of either importin α or amino acids 1-601 of importin β inhibits the spontaneous formation of asters in *Xenopus* egg extracts that have been depleted of RanGTP-binding proteins. Importin β 1-601 lacks the importin α binding domain but retains the ability to bind a group of importin α independent cargoes (41). Since TPX2 and NuMA are importin α cargoes, other APA proteins must also exist.

The common theme emerging from these studies is that RanGTP is a biochemical marker for DNA throughout the cell cycle. In addition to its role in nuclear transport during interphase and spindle formation in metaphase, RanGTP is also required for reformation of the nuclear envelope in late mitosis (32, 98). Furthermore, perturbations of RanGTP concentrations in the nucleus appear to alter chromatin structure, suggesting that Ran may play a role in DNA maintenance (37, 59). More work is needed to know if other components of the nuclear transport machinery are also involved in these Ran-dependent processes.

It is not clear if RanGTP plays a similar role in spindle assembly in *S. cerevisiae*. Yeast cells undergo a closed mitosis where the nuclear envelope does not break down. If a similar system is used by yeast, then import of APA or other spindle formation factors must be regulated across the cell cycle. The study of yeast also provides the opportunity to learn about the evolutionary origins of RanGTP function in nuclear transport and spindle assembly. It is possible that the original function of Ran and importin α/β was in mitotic spindle assembly (59). As eukaryotic cells acquired nuclear envelopes, this system may have evolved to have multiple functions, including a role in nuclear transport. Alternatively, nuclear import of APA might have been necessary for spindle formation before the evolution of nuclear envelope breakdown (94). Once cells adopted nuclear envelope breakdown then the transport machinery was used to locally release APA proteins.

Transport Receptors Dedicated to Both Import and Export

One prevailing hypothesis regarding importin β-like transporters has been that transport receptors are dedicated to either import or export of cargoes. Accordingly, import receptors have been termed importins and export receptors are called exportins.

Contrary to this hypothesis, the *S. cerevisiae* importin β homologue Msn5 has been implicated in both export and import processes (97). Msn5 was categorized as one of the 14 importin β homologues in *S. cerevisiae* (23). Msn5 is the export receptor for several *S. cerevisiae* proteins, including the transcription factors Pho4 and Mig1 and the CDK inhibitor Far1 (8, 13, 42). Affinity chromotography with Msn5 identified Rpa1-3, a nuclear complex involved in the DNA damage response, as an Msn5-associated protein complex (97). When affinity chromotography is performed in the presence of RanGTP, the interaction between Rpa1-3 and Msn5 is lost. Additionally, there is a significant mislocalization of Rpa1-3 to the cytoplasm in *msn5Δ* cells. These results suggest that Msn5 could be the import receptor for the Rpa complex. One question regarding these results is that the Rpa proteins are essential whereas Msn5 is not required for growth. In fact, in the *msn5Δ* cells, a portion of the Rpa proteins are still nuclear. Rpa proteins can interact with other import receptors, raising the possiblity that the Rpa complex can utilize multiple import pathways (97). Furthermore, these results do not show that RanGTP affects the Msn5-Rpa complex directly. To confirm this, it is important to assess the effects of RanGTP on a preformed Msn5-Rpa1-3 complex in a purified system.

These experiments raise the possibility that the same transport receptor can direct the import and export of different proteins. This idea is not unexpected since the natural cycle for a transport receptor requires it to move in both directions through the nuclear pore (25). Additionally, some transport receptors use different domains for binding to different cargoes (41). If these two sites could be affected differentially by RanGTP binding, a transport receptor might be used in both import and export. It would be interesting to investigate the generality of this phenomenon for other receptors and in other systems.

Non-Importin β Receptors

Importin β family members have been thought of as the only proteins capable of transporting cargoes across the nuclear pore. However, recent work suggests that other proteins can also use the RanGTP system to transport cargoes. Calreticulin (CRT), an ER-resident calcium binding protein, has been implicated in the nuclear export of proteins with a LRNES (35). Recombinant CRT can stimulate export of the LRNES even in the presence of leptomycin B (LMB), a specific inhibitor of Crm1-dependent nuclear export (20), showing that this pathway is distinct from the Crm1 pathway. Additionally, CRT binds to a LRNES peptide only in the presence of RanGTP with kinetics very similar to the Crm1-RanGTP-LRNES interaction (35). CRT's possible role as an export receptor is further substantiated by the discovery of CRT-specific cargoes. The export of glucocorticoid receptor (GR) is independent of Crm1, as it is not inhibited by LMB, but CRT can stimulate its export in vitro and is required for GR export in vivo (35).

Although these results indicate that CRT behaves like known importin β-like export receptors, it is not an importin β homologue (35). This is the first report of a non-importin β-like receptor being involved in RanGTP-dependent export of macromolecules. One effective method for finding the import or export receptor for proteins or RNAs in *S. cerevisiae* has been to screen a panel of strains bearing mutations in importin β homologues (42, 43). Such a reverse genetics approach is not sufficient if receptors not related to importin β are involved in nucleocytoplasmic transport. These results highlight the importance of classical nonbiased genetics approaches in studying nucleocytoplasmic transport.

One cautionary note regarding these results is that CRT has been characterized as an ER-resident protein with multiple cellular functions (47). Although biochemical fractionation studies imply that a fraction of the CRT is present in the nucleus, examination of CRT localization by immunofluorescence has produced conflicting results (47). Since the discovery of a non-importin β receptor will have major implications in the study of nucleocytoplasmic transport, it is important to resolve these outstanding issues.

RNA TRANSPORT

In this section we examine the export of tRNAs, snRNAs, and mRNAs to showcase the diversity of classical and nonclassical transport mechanisms that were discussed above (Figure 3).

tRNA Export

To function in translation, tRNAs must be transcribed, modified, and subsequently exported from the nucleus (28, 95). The export receptor, exportin-t (Xpo-t), is an importin β homologue that is responsible for tRNA export (3, 48). Similar to protein export by importin β homologues, Xpo-t binds directly to tRNAs with the help of RanGTP. En route to maturation, bases and sugar residues of tRNAs are modified, their 5' and 3' ends as well as introns are specifically removed, and a CCA sequence is added to the 3' end (95). The maturation of tRNAs appears to be required for their export since mature tRNAs are exported much more efficiently than pre-tRNAs (3, 51). This bias in export is achieved primarily through the specificity of the interaction of mature tRNAs with Xpo-t and RanGTP (3, 51). in vitro, Xpo-t can interact efficiently with mature tRNAs, whereas it fails to show significant binding to pre-tRNAs. However, differential binding to Xpo-t may not be the sole source of specificity for export of mature tRNAs (3). Intron-containing tRNAs interact with Xpo-t as well as spliced tRNAs, but they are exported with very low efficiency in vivo. Injection of excess Xpo-t stimulates the export of intron-containing tRNAs, suggesting that competition between export factors and tRNA processing enzymes within the nucleus might be another mechanism restricting the export of pre-tRNAs. It has also been proposed that 3' and 5' end

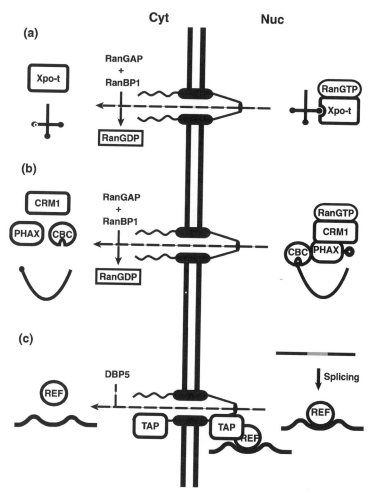

Figure 3   RNA export. (*A*) tRNA export. Importin β family member exportin-t (Xpo-t) binds tRNA molecules with the aid of RanGTP and acts as their export receptor. Hydrolysis of GTP on Ran in the cytoplasm dissociates the export complex. (*B*) snRNA export. Although CRM1 is the export receptor for snRNAs it does not bind to them directly. The cap binding complex (CBC) recognizes the methylated cap of snRNA molecules and this complex binds to CRM1 and RanGTP with the help of the phosphorylated form of the phosphorylated adapter for RNA export (PHAX). Dephosphorylation of PHAX and hydrolysis of GTP on Ran dissociate the export complex in the cytoplasm. (*C*) mRNA export. REF proteins associate with the spliced mRNA and direct the message to TAP. TAP, which is not related to importin β, is thought to export the complex to the cytoplasm in a RanGTP-independent manner. The RNA helicase DBP5 might act to release the mRNA into the cytoplasm.

processing occurs after splicing in vivo (52). Such a mechanism ensures that only spliced tRNAs are exported since mature ends are required for tRNA recognition by Xpo-t.

In *S. cerevisiae* tRNA export seems to utilize multiple pathways. The importin β homologue Los1 has been implicated in tRNA export because it is homologous to Xpo-t and displays many genetic interactions with the tRNA biogenesis pathway (23, 36, 77). For example, it was first isolated as a mutant displaying loss of suppressor activity (thus called LOS) (38). Additionally, strains lacking Los1 display accumulation of tRNAs within their nuclei (77). However, deletion of Los1 is not lethal, suggesting that it is not the sole receptor for tRNA export (38). Recently, the elongation factor eEF-1A and some tRNA synthetases have been implicated in the Los1-independent export of tRNAs (27). Deletion of these genes results in synthetic lethality when combined with Los1 deletions. Mutants defective in eEF-1A and some tRNA synthetases display nuclear accumulation of tRNAs. Thus, it is possible that components of the translation machinery are used as an alternate tRNA export pathway in yeast. However, these results do not point to a specific mechanism for how translation factors might be involved in tRNA export. Does eEF1-A target tRNAs to another importin β or does it export them directly? Furthermore, is the action of tRNA synthetases and eEF1-A indicative of a role for maturation in tRNA export or does it imply the existence of a pathway parallel to the Los1 export pathway?

snRNA Export

At the heart of the splicing machinery are specialized small nuclear RNAs (snRNAs) (56). These snRNAs associate with proteins to form ribonucleoprotein complexes (snRNPs) that catalyze splicing reactions. In metazoans, subsequent to their transcription, snRNAs are exported to the cytoplasm where they bind to snRNP proteins. The resulting snRNP complexes are then reimported into the nucleus by the import receptor Snurportin-1 so that they can participate in splicing reactions (67). The export of snRNAs from the nucleus is dependent on RanGTP since it is blocked by injection of RanGAP and RanBP1 into *Xenopus* nuclei (39). Furthermore, snRNAs use a Crm1-dependent export pathway because their export is blocked by the addition of LMB and can be competed by the addition of NES peptides (19, 20).

Unlike tRNA export, however, snRNAs do not bind directly to their export receptor. It has long been known that the cap-binding complex (CBC) proteins, CBC20 and CBC80, bind cooperatively to the m7G-cap of snRNAs and that this binding is required for efficient snRNA export (40). Other factors also appear to be required for the formation of an snRNA export complex since recombinant CBC, CRM1, and RanGTP do not form an export complex with snRNAs unless crude extract is also present (65). Recently, a 55-kDa protein, p55, has been purified as an activity responsible for formation of a complex containing CBC, snRNA, CRM1, and RanGTP (65). p55 binds cooperatively to the CBC-snRNA complex, shuttles between the nucleus and cytoplasm, and contains an NES that is recognized by Crm1. In addition to its in vitro role in formation of an export complex, p55 is also required for snRNA export in vivo since injection of antibodies against p55 blocks snRNA export without affecting tRNA or mRNA export. Additionally, injection of p55 into *Xenopus* oocyte nuclei enhances the export of snRNAs. Interestingly, p55's ability to form an snRNA-export complex is dependent on its phosphorylation state (65). Phosphorylated p55 mediates the formation of this export complex whereas dephosphorylated p55 cannot bridge the CBC-snRNA complex to Crm1 and RanGTP. Since dephosphorylation of p55 induces disassembly of the export complex, dephosphorylation of p55 could be a mechanism used to release snRNAs into the cytoplasm. In support of this idea, p55 is mainly in the phosphorylated form in the nucleus whereas it is mostly dephosphorylated in the cytoplasm. Accordingly, p55 has been named PHAX, phosphorylated adaptor for RNA export (65).

Several aspects of this system are quite intriguing. First, every step of the formation of an snRNA export complex is regulated by cooperative binding events (65). The CBC complex binds cooperatively to the cap, PHAX binds cooperatively to CBC and the snRNA molecule, and finally CRM1 and RanGTP recognize the PHAX-CBC-snRNA complex cooperatively. Presumably, these multiple layers allow for proper regulation of snRNA export. Second, the release of snRNAs by dephosphorylation of PHAX seems to be a redundant mechanism. Ran dissociates export complexes through RanGAP-stimulated GTP hydrolysis. Perhaps in snRNA export GTP hydrolysis is most important for recycling of the export receptor whereas dephosphorylation of PHAX is required for the release of the snRNA from CBC and PHAX. Third, snRNA export in *S. cerevisiae* seems to follow very different rules. Homologues of PHAX and Snurportin exist in other eukaryotes but are absent from *S. cerevisiae* (65). Thus, snRNP biogenesis in *S. cerevisiae* may be an exclusively nuclear event.

mRNA export

mRNA export appears to be a much more complicated process than protein, tRNA, or snRNA export. Studies with the Balbiani ring particles of *Chironomus tentans* showed that multiple proteins bind to and are released from mRNAs as they exit the nucleus (11). Furthermore, mRNA processing events such as capping, polyadenylation, and splicing are nuclear events and must precede mRNA export. This complexity, coupled with the lack of in vitro RNA export assays, has hampered the biochemical exploration of mRNA export. Thus, yeast genetics has been a prominent and fruitful approach in understanding mRNA export (88). Poly(A)+ mRNA can be visualized by fluorescence in situ hybridization (FISH) using labeled oligo dT probes. Poly(A)+ mRNA has a cytoplasmic localization in wild-type cells but it is localized to the nucleus in mRNA export mutants.

Genetic studies have implicated a variety of proteins in mRNA export (88). A persistent hypothesis has been the involvement of shuttling heterogeneous ribonucleoprotein particle (hnRNP) proteins in mRNA export (14, 61). These proteins have many of the characteristics expected of mRNA transport proteins. They bind RNA directly, some can shuttle between the nucleus and cytoplasm, and mutations in them lead to a block in export of poly-(A)+ mRNA. These hnRNP proteins are proposed to link mRNAs to specific transport receptors and promote their passage through the nuclear pore. A variety of importin β family members have also been implicated in mRNA export (79, 85). Several studies have linked the Crm1 (Xpo1 in yeast) pathway to mRNA export. The Xpo1 temperature-sensitive mutant, *xpo1-1*, rapidly accumulates poly-(A)+ mRNA in the nucleus at nonpermissive temperatures (85). In human cells, mRNA accumulates in the nucleus after long incubations with the Crm1 inhibitor LMB (92). However, many studies have put the role of Crm1 in mRNA transport in doubt. Neville et al. constructed LMB-sensitive versions of *S. cerevisiae* Crm1 and showed that LMB blocked mRNA export only partially and significantly later than a complete block to LRNES protein export, suggesting that Crm1 plays an indirect or redundant role in mRNA export (63). Furthermore, injection of NES proteins into *Xenopus* nuclei can compete for snRNA export but it has no effect on mRNA export (19). Another indication that importin β-like receptors are not involved in this process is that mRNA export does not require RanGTP (9). The export of snRNAs and tRNAs is rapidly blocked if nuclear pools of RanGTP are depleted by the injection of RanBP1 and RanGAP into *Xenopus* nuclei (9, 39). In contrast, mRNA transport is unaffected by these treatments. Furthermore, no detectable levels of RanGTP are found in purified mRNP complexes that are competent for export (9). The simplest interpretation of these results is that in some cases, RanGTP is not directly involved in mRNA transport. These results also imply that in contrast to protein, tRNA, and snRNA export, export of some classes of mRNAs might rely on non-importin β-like export receptors.

Recently, several lines of evidence have pointed to Mex67 as a possible export receptor for mRNA. Although not an importin β homologue, Mex67 localizes to the NPC, binds mRNA, and is required for mRNA export in yeast (80). In yeast, Mex67 dimerizes with Mtr2 and this complex can bind to several different nucleoporins (76, 87). This pathway is well conserved from yeast to humans. TAP, the human version of Mex67, is a shuttling protein that localizes to the nuclear pore complex as well as the nucleoplasm (44). TAP is involved in export of viral messages by directly binding to the constitutive transport elements (CTE) of viral mRNAs (30). Analogous to the Mex67-Mtr2 interaction, TAP interacts with p15, an NTF2 homologue. Remarkably, TAP can functionally replace yeast Mex67 (44). When expressed in yeast, TAP localizes to the nuclear pores in an Mtr2-dependent manner, and coexpression of TAP and p15 in yeast can partially suppress the growth defect of strains carrying deletions of Mex67 or Mtr2 as well as strains lacking both proteins (44). Since Mex67/TAP can interact with mRNA cargoes as well as nucleoporins, a possible model is that Mex67/TAP fulfills a role similar to importin β family members in protein export.

Although Mex67 can bind directly to some mRNAs, other factors seem to be required for Mex67-dependent nuclear transport. One such protein, Yra1, was identified in a synthetic lethal screen with a temperature-sensitive allele of Mex67 (89). Yra1 is a shuttling nuclear protein, is required for RNA export, and can directly interact with RNA and Mex67. Yra1 is a member of the REF family of proteins that are conserved from yeast to humans. Interestingly, ALY, the mouse homologue of REF/Yra1, can complement the lethality of a Yra1 deletion in yeast (89). These proteins are essential for mRNA export as injection of antibodies against REF proteins has no effect on splicing but leads to a block in mRNA export (72).

Recently, Yra1/ALY/REF has been implicated in linking mRNA processing to its export (49, 53, 99). mRNAs derived from microinjected intron-bearing transcripts are exported much more efficiently from *Xenopus* oocyte nuclei than the same mRNAs produced from cDNAs lacking introns (53). Note that the main difference between these two messages is that one has to go through the process of splicing, whereas the other never encounters the splicing machinery. When these messages are spliced in vitro, the spliced mRNA is part of a different mRNP particle than the message produced from intron-deleted cDNA. Interestingly, when the protein contents of these mRNPs are examined, REF becomes part of the mRNP complex after the completion of splicing, suggesting that REF/Yra1 might couple the maturation of mRNAs to their export (99). Additional evidence for the role of REF/Yra1 in linking splicing to mRNA export has come from the discovery of a protein complex that marks spliced mRNAs 20–24 bases upstream of exon-exon junctions in a sequence-independent manner (49). REF is part of this protein complex whereas TAP is not. These observations imply that the completion of splicing is coupled to the deposition of REF proteins on the mature mRNA. REF then directs the message to TAP at the nuclear pores and this complex moves through the NPC as a result of the interactions of TAP with nucleoporins.

Although quite elegant, it is unclear if this model applies to the export of all mRNAs. In yeast particularly, many messages do not contain introns but are still exported efficiently. Mex67 and Yra1 may participate in mRNA export differently in yeast and metazoans. However, even in metazoans some naturally occurring intronless messages are exported efficiently, suggesting that other processing events might be important for efficient mRNA export (72). Furthermore, if splicing is sufficient for making a message competent for export, messages containing multiple introns might be targeted for export after removal of only one intron. Perhaps splicing is restricted to certain locations in the cell such that messages are released only after splicing has occurred. And finally, in addition to REF, many other proteins are part of the complex marking exon-exon junctions after splicing. The role that these and other hnRNP proteins might play in mRNA export is not clear. Possibly REF and other hnRNPs such as Npl3 and Nab2 represent different export pathways devoted to specific classes of transcripts. In support of this model, mutations in the ubiquitin ligase-like protein Tom1 result in mislocalization of Nab2 but do not affect the localization or shuttling of Npl3 (15).

Since it appears that RanGTP is not involved in mRNA export, the mechanism by which mRNAs are released from export complex remains to be elucidated. Screens for mRNA export mutants led to the discovery of Dbp5, an RNA helicase that localizes to the cytoplasm and to nuclear pores (84, 91). Studies with the human homologue of Dbp5 have shown that it localizes to the cytoplasmic fibrils of the nuclear pore complex and that its ATPase activity is required for mRNA export (78). An attractive hypothesis is that Dbp5 acts at the terminal steps of RNA transport by rearranging the mRNA-protein complex and releasing the mRNA into the cytoplasm. Interestingly, overexpression of Dbp5 can suppress the temperature-sensitive phenotype of the *xpo1-1* mutant, suggesting that the proposed Crm1 function in mRNA export might be linked to the localization of Dbp5 (34).

In addition to its role in identifying specific protein components of the mRNA transport machinery, yeast genetics has been crucial in uncovering global signaling events that may regulate mRNA export. Recently, inositol signaling has been implicated in regulation of mRNA export (96). Gle1 is a nucleoporin that appears to have a specific function in mRNA export (58). A synthetic lethal screen with a *gle1* temperature-sensitive mutant identified three mutants that displayed specific defects in mRNA export with no apparent perturbation of protein import or export (96). Interestingly, all three genes have distinct roles in the maintenance of phosphoinositol levels within the cell. These genes encode phospholipase C, Plc1, and two novel inositol phosphate kinases, Ipk1 and Ipk2/Arg82. Plc1 cleaves PIP2 into diacylglycerol and IP3, which is then phosphorylated to form IP4, IP5, and IP6 (50, 82). Ipk1 and Ipk2 appear to phosphorylate these inositol phosphate precursors to form IP6 (96). In accordance with their known functions, mutations in any of the three genes lead to defects in IP6 production. Interestingly, Ipk1 is a nuclear protein with a distinct nuclear pore localization, lending further support for a direct role in mRNA export (96). Inositol phosphates may also have a role in regulating mRNA export in mammalian cells (18). Ectopic expression of SopB, a bacterial inositol phosphatase, causes a defect in mRNA export in cultured mammalian cells. Interestingly, when SopB is targeted to the nucleus via an NLS fusion, the mRNA export defect is more severe.

These results imply a role for inositol phosphate signaling in mRNA export, although the exact targets of inositol phosphates in this process are unknown. The defect in mRNA export in yeast strains unable to produce IP6 is probably not due to gross rearrangements of the NPC or the nuclear envelope since these structures appear normal in yeast strains containing mutations in *PLC1*, *IPK1*, or *IPK2* (96). However, it is possible that inositol phosphates modulate the activity of specific NPC or mRNA export proteins. This idea is supported by the synthetic lethal interactions between *plc1*, *ipk1*, and *ipk2* and *gle1* mutations. Furthermore, it is unclear if inositol signaling is required for constitutive or regulated mRNA export. For example, it is possible that the mutations in *PLC1*, *IPK1*, and *IPK2* mimic conditions that would lead to a down regulation of mRNA export in wild-type cells. Further work is needed to understand the specific role that inositol phosphates play in regulating mRNA export.

CONCLUSIONS AND FUTURE DIRECTIONS

Our knowledge of nucleocytoplasmic transport has progressed rapidly in recent years. The basic proteins involved in import and export of macromolecules are conserved in all eukaryotic systems studied thus far. However, recent results point to the existence of alternative machineries and unique modes of transport. Also, the understanding of essential aspects of transport such as translocation through the nuclear pore is still limited.

The diversity of mechanisms used in nuclear transport are clearly demonstrated by an examination of tRNA, snRNA, and mRNA nuclear export. tRNAs and snRNAs are exported in a Ran-dependent manner, whereas mRNAs use a Ran-independent pathway. Importin β-like receptors are involved in tRNA and snRNA export, but no such receptor appears to be involved in mRNA export. A simple and direct interaction with Xpo-t and RanGTP is the sole requirement for tRNA export, whereas snRNA export requires the involvement of CBC20, CBC80, and PHAX for the formation of an export complex with Crm1 and RanGTP.

The existence of novel receptors and pathways of export also highlights the central role that genetic approaches can take in the study of nuclear transport. For example, many components of mRNA export were originally identified in genetic screens (88). The future study of nuclear transport will benefit greatly from development of genetic approaches that can be used to dissect the components of a given transport pathway. The development of small molecule inhibitors of transport receptors or specific nucleoporins will be useful in understanding transport pathways in organisms that cannot readily be studied by classical genetics-based approaches. The combination of such "biochemical genetics" approaches and classical genetic approaches could have a significant impact in uncovering novel nuclear transport molecules and pathways.

ACKNOWLEDGMENTS

We would like to thank A. Belle, W. Gilbert, C. Guthrie, and K. Weis for helpful discussions and comments on the manuscript. AK is a Howard Hughes Medical Institute Predoctoral Fellow. EKO is supported by the Howard Hughes Medical Institute and by the NIH (GM59034).

Visit the Annual Reviews home page at www.AnnualReviews.org

LITERATURE CITED

1. Adam EJ, Adam SA. 1994. Identification of cytosolic factors required for nuclear location sequence-mediated binding to the nuclear envelope. *J. Cell Biol.* 125:547–55

2. Adam SA, Marr RS, Gerace L. 1990. Nuclear protein import in permeabilized mammalian cells requires soluble cytoplasmic factors. *J. Cell Biol.* 111:807–16

3. Arts GJ, Kuersten S, Romby P, Ehresmann

B, Mattaj IW. 1998 The role of exportin-t in selective nuclear export of mature tRNAs. *EMBO J.* 17:7430–41

4. Bischoff FR, Gorlich D. 1997. RanBP1 is crucial for the release of RanGTP from importin beta-related nuclear transport factors. *FEBS Lett.* 419:249–54

5. Bischoff FR, Klebe C, Kretschmer J, Wittinghofer A, Ponstingl H. 1994. RanGAP1 induces GTPase activity of nuclear Ras-related Ran. *Proc. Natl. Acad. Sci. USA* 91:2587–91

6. Bischoff FR, Krebber H, Kempf T, Hermes I, Ponstingl H 1995. Human RanGTPase-activating protein RanGAP1 is a homologue of yeast Rna1p involved in mRNA processing and transport. *Proc. Natl. Acad. Sci. USA* 92:1749–53

7. Bischoff FR, Ponstingl H. 1991. Catalysis of guanine nucleotide exchange on Ran by the mitotic regulator RCC1. *Nature* 354:80–82

8. Blondel M, Alepuz PM, Huang LS, Shaham S, Ammerer G, et al. 1999. Nuclear export of Far1p in response to pheromones requires the export receptor Msn5p/Ste21p. *Genes Dev.* 13:2284–300

9. Clouse KN, Luo MJ, Zhou Z, Reed RA. 2001. Ran-independent pathway for export of spliced mRNA. *Nat. Cell Biol.* 3:97–99

10. Damelin M, Silver PA. 2000. Mapping interactions between nuclear transport factors in living cells reveals pathways through the nuclear pore complex. *Mol. Cell* 5:133–40

11. Daneholt B. 1997. A look at messenger RNP moving through the nuclear pore. *Cell* 88:585–88

12. Dasso M. 2001. Running on Ran: nuclear transport and the mitotic spindle. *Cell* 104:321–24

13. DeVit MJ, Johnston M. 1999. The nuclear exportin Msn5 is required for nuclear export of the Mig1 glucose repressor of *Saccharomyces cerevisiae. Curr. Biol.* 9:1231–41

14. Dreyfuss G, Matunis MJ, Pinol-Roma S,

Burd CG. 1993. hnRNP proteins and the biogenesis of mRNA. *Annu. Rev. Biochem.* 62:289–321

15. Duncan K, Umen JG, Guthrie CA. 2000. Putative ubiquitin ligase required for efficient mRNA export differentially affects hnRNP transport. *Curr. Biol.* 10:687–96

16. Englmeier L, Olivo JC, Mattaj IW. 1999. Receptor-mediated substrate translocation through the nuclear pore complex without nucleotide triphosphate hydrolysis. *Curr. Biol.* 9:30–41

17. Fabre E, Hurt E. 1997. Yeast genetics to dissect the nuclear pore complex and nucleocytoplasmic trafficking. *Annu. Rev. Genet.* 31:277–313

18. Feng Y, Wente SR, Majerus PW. 2001. Overexpression of the inositol phosphatase SopB in human 293 cells stimulates cellular chloride influx and inhibits nuclear mRNA export. *Proc. Natl. Acad. Sci. USA* 9:875–79

19. Fischer U, Huber J, Boelens WC, Mattaj IW, Luhrmann R. 1995. The HIV-1 Rev activation domain is a nuclear export signal that accesses an export pathway used by specific cellular RNAs. *Cell* 82:475–83

20. Fornerod M, Ohno M, Yoshida M, Mattaj IW. 1997. CRM1 is an export receptor for leucine-rich nuclear export signals. *Cell* 90:1051–60

21. Fornerod M, van Baal S, Valentine V, Shapiro DN, Grosveld G. 1997. Chromosomal localization of genes encoding CAN/Nup214-interacting proteins—human CRM1 localizes to 2p16, whereas Nup88 localizes to 17p13 and is physically linked to SF2p32. *Genomics* 42:538–40

22. Gorlich D. 1998. Transport into and out of the cell nucleus. *EMBO J.* 17:2721–27

23. Gorlich D, Dabrowski M, Bischoff FR, Kutay U, Bork P, et al. 1997. A novel class of RanGTP binding proteins. *J. Cell Biol.* 138:65–80

24. Gorlich D, Kostka S, Kraft R, Dingwall C, Laskey RA, et al. 1995. Two different subunits of importin cooperate to recognize nuclear localization signals and bind them

to the nuclear envelope. *Curr. Biol.* 5:383–92

25. Gorlich D Kutay U. 1999. Transport between the cell nucleus and the cytoplasm. *Annu. Rev. Cell Dev. Biol.* 15:607–60

26. Gorlich D, Prehn S, Laskey RA, Hartmann E. 1994. Isolation of a protein that is essential for the first step of nuclear protein import. *Cell* 79:767–78

27. Grosshans H, Hurt E, Simos G. 2000. An aminoacylation-dependent nuclear tRNA export pathway in yeast. *Genes Dev.* 14:830–40

28. Grosshans H, Simos G, Hurt E. 2000. Review: transport of tRNA out of the nucleus-direct channeling to the ribosome? *J. Struct. Biol.* 129:288–94

29. Gruss OJ, Carazo-Salas RE, Schatz CA, Guarguaglini G, Kast J, et al. 2001. Ran induces spindle assembly by reversing the inhibitory effect of importin alpha on TPX2 activity. *Cell* 104:83–93

30. Grüter P, Tabernero C, von Kobbe C, Schmitt C, Saavedra C, et al. 1998. TAP, the human homolog of Mex67p, mediates CTE-dependent RNA export from the nucleus. *Mol. Cell* 1:649–59

31. Heald R, Weis K. 2000. Spindles get the Ran around. *Trends Cell Biol.* 10:1–4

32. Hetzer M, Bilbao-Cortes D, Walther TC, Gruss OJ, Mattaj IW. 2000. GTP hydrolysis by Ran is required for nuclear envelope assembly. *Mol. Cell* 5:1013–24

33. Hetzer M, Mattaj IW. 2000. An ATP-dependent, Ran-independent mechanism for nuclear import of the U1A and U2B″ spliceosome proteins. *J. Cell Biol.* 148:293–303

34. Hodge CA, Colot HV, Stafford P, Cole CN. 1999. Rat8p/Dbp5p is a shuttling transport factor that interacts with Rat7p/Nup159p and Gle1p and suppresses the mRNA export defect of xpo1-1 cells. *EMBO J.* 18:5778–88

35. Holaska JM, Black BE, Love DC, Hanover JA, Leszyk J, et al. 2001. Calreticulin is a receptor for nuclear export. *J. Cell Biol.* 152:127–40

36. Hopper AK, Schultz LD, Shapiro RA. 1980. Processing of intervening sequences: a new yeast mutant which fails to excise intervening sequences from precursor tRNAs. *Cell* 19:741–51

37. Hughes M, Zhang C, Avis JM, Hutchison CJ, Clarke PR. 1998. The role of the ran GTPase in nuclear assembly and DNA replication: characterisation of the effects of Ran mutants. *J. Cell Sci.* 111:3017–26

38. Hurt DJ, Wang SS, Lin YH, Hopper AK. 1987. Cloning and characterization of LOS1, a *Saccharomyces cerevisiae* gene that affects tRNA splicing. *Mol. Cell Biol.* 7:1208–16

39. Izaurralde E, Kutay U, von Kobbe C, Mattaj IW, Gorlich D. 1997. The asymmetric distribution of the constituents of the Ran system is essential for transport into and out of the nucleus. *EMBO J.* 16:6535–47

40. Izaurralde E, Lewis J, Gamberi C, Jarmolowski A, McGuigan C, et al. 1995. A cap-binding protein complex mediating U snRNA export. *Nature* 376:709–12

41. Jakel S, Gorlich D. 1998. Importin beta, transportin, RanBP5 and RanBP7 mediate nuclear import of ribosomal proteins in mammalian cells. *EMBO J.* 17:4491–502

42. Kaffman A, Rank NM, O'Neill EM, Huang LS, O'Shea EK. 1998. The receptor Msn5 exports the phosphorylated transcription factor Pho4 out of the nucleus. *Nature* 396:482–86

43. Kaffman A, Rank NM, O'Shea EK. 1998. Phosphorylation regulates association of the transcription factor Pho4 with its import receptor Pse1/Kap121. *Genes Dev.* 12:2673–83

44. Katahira J, Strasser K, Podtelejnikov A, Mann M, Jung JU, et al. 1999. The Mex67p-mediated nuclear mRNA export pathway is conserved from yeast to human. *EMBO J.* 18:2593–609

45. Keminer O, Siebrasse JP, Zerf K, Peters R. 1999. Optical recording of signal-mediated protein transport through single nuclear pore complexes. *Proc. Natl. Acad. Sci. USA* 96:11842–47

46. Kose S, Imamoto N, Tachibana T, Shimamoto T, Yoneda Y. 1997. Ran-unassisted nuclear migration of a 97-kD component of nuclear pore-targeting complex. *J. Cell Biol.* 139:841–49

47. Krause KH, Michalak M. 1997. Calreticulin. *Cell* 88:439–43

48. Kutay U, Lipowsky G, Izaurralde E, Bischoff FR, Schwarzmaier P, et al. 1998. Identification of a tRNA-specific nuclear export receptor. *Mol. Cell* 1:359–69

49. Le Hir H, Izaurralde E, Maquat LE, Moore MJ. 2000. The spliceosome deposits multiple proteins 20–24 nucleotides upstream of mRNA exon-exon junctions. *EMBO J.* 19:6860–69

50. Lee SB, Rhee SG. 1995. Significance of PIP2 hydrolysis and regulation of phospholipase C isozymes. *Curr. Opin. Cell Biol.* 7:183–89

51. Lipowsky G, Bischoff FR, Izaurralde E, Kutay U, Schafer S, et al. 1999. Coordination of tRNA nuclear export with processing of tRNA. *RNA* 5:539–49

52. Lund E, Dahlberg JE. 1998. Proofreading and aminoacylation of tRNAs before export from the nucleus. *Science* 282:2082–85

53. Luo MJ, Reed R. 1999. Splicing is required for rapid and efficient mRNA export in metazoans. *Proc. Natl. Acad. Sci. USA* 96:14937–42

54. Marelli M, Aitchison JD, Wozniak RW. 1998. Specific binding of the karyopherin Kap121p to a subunit of the nuclear pore complex containing Nup53p, Nup59p, and Nup170p. *J. Cell Biol.* 143:1813–30

55. Mattaj IW, Englmeier L. 1998. Nucleocytoplasmic transport: the soluble phase. *Annu. Rev. Biochem.* 67:265–306

56. Mattaj IW, Tollervey D, Seraphin B. 1993. Small nuclear RNAs in messenger RNA and ribosomal RNA processing. *FASEB J.* 7:47–53

57. Matunis MJ, Coutavas E, Blobel G. 1996. A novel ubiquitin-like modification modulates the partitioning of the Ran-GTPase-activating protein RanGAP1 between the cytosol and the nuclear pore complex. *J. Cell Biol.* 135:1457–70

58. Murphy R, Wente SR. 1996. An RNA-export mediator with an essential nuclear export signal. *Nature* 383:357–60

59. Nachury MV, Maresca TJ, Salmon WC, Waterman-Storer CM, Heald R, et al. 2001. Importin beta is a mitotic target of the small GTPase Ran in spindle assembly. *Cell* 104:95–106

60. Nachury MV, Weis K. 1999. The direction of transport through the nuclear pore can be inverted. *Proc. Natl. Acad. Sci. USA* 96:9622–27

61. Nakielny S, Fischer U, Michael WM, Dreyfuss G. 1997. RNA transport. *Annu. Rev. Neurosci.* 20:269–301

62. Nakielny S, Siomi MC, Siomi H, Michael WM, Pollard V, et al. 1996. Transportin: nuclear transport receptor of a novel nuclear protein import pathway. *Exp. Cell Res.* 229:261–66

63. Neville M, Rosbash M. 1999. The NES-Crm1p export pathway is not a major mRNA export route in *Saccharomyces cerevisiae*. *EMBO J.* 18:3746–56

64. Ohno M, Fornerod M, Mattaj IW. 1998. Nucleocytoplasmic transport: the last 200 nanometers. *Cell* 92:327–36

65. Ohno M, Segref A, Bachi A, Wilm M, Mattaj IW. 2000. PHAX, a mediator of U snRNA nuclear export whose activity is regulated by phosphorylation. *Cell* 101:187–98

66. Ohtsubo M, Kai R, Furuno N, Sekiguchi T, Sekiguchi M, et al. 1987. Isolation and characterization of the active cDNA of the human cell cycle gene (RCC1) involved in the regulation of onset of chromosome condensation. *Genes Dev.* 1:585–93

67. Palacios I, Hetzer M, Adam SA, Mattaj IW. 1997. Nuclear import of U snRNPs requires importin beta. *EMBO J.* 16:6783–92

68. Radu A, Blobel G, Moore MS 1995. Identification of a protein complex that is required for nuclear protein import and mediates docking of import substrate to distinct

nucleoporins. *Proc. Natl. Acad. Sci. USA* 92:1769–73

69. Radu A, Moore MS, Blobel G. 1995. The peptide repeat domain of nucleoporin Nup98 functions as a docking site in transport across the nuclear pore complex. *Cell* 81:215–22

70. Ribbeck K, Kutay U, Paraskeva E, Gorlich D. 1999. The translocation of transportin-cargo complexes through nuclear pores is independent of both Ran and energy. *Curr. Biol.* 9:47–50

71. Ribbeck K, Lipowsky G, Kent HM, Stewart M, Gorlich D. 1998. NTF2 mediates nuclear import of Ran. *EMBO J.* 17:6587–98

72. Rodrigues JP, Rode M, Gatfield D, Blencowe B, Carmo-Fonseca M, et al. 2001. REF proteins mediate the export of spliced and unspliced mRNAs from the nucleus. *Proc. Natl. Acad. Sci. USA* 98:1030–35

73. Rout MP, Aitchison JD, Suprapto A, Hjertaas K, Zhao Y, et al. 2000. The yeast nuclear pore complex: composition, architecture, and transport mechanism. *J. Cell Biol.* 148:635–51

74. Rout MP, Blobel G, Aitchison JD. 1997. A distinct nuclear import pathway used by ribosomal proteins. *Cell* 89:715–25

75. Ryan KJ, Wente SR. 2000. The nuclear pore complex: a protein machine bridging the nucleus and cytoplasm. *Curr. Opin. Cell Biol.* 12:361–71

76. Santos-Rosa H, Moreno H, Simos G, Segref A, Fahrenkrog B, et al. 1998. Nuclear mRNA export requires complex formation between Mex67p and Mtr2p at the nuclear pores. *Mol. Cell Biol* 18:6826–38

77. Sarkar S, Hopper AK. 1998. tRNA nuclear export in *Saccharomyces cerevisiae*: in situ hybridization analysis. *Mol. Biol. Cell* 9:3041–55

78. Schmitt C, von Kobbe C, Bachi A, Pante N, Rodrigues JP, et al. 1999. Dbp5, a DEAD-box protein required for mRNA export, is recruited to the cytoplasmic fibrils of nuclear pore complex via a conserved

interaction with CAN/Nup159p. *EMBO J.* 18:4332–47

79. Seedorf M, Silver PA. 1997. Importin/karyopherin protein family members required for mRNA export from the nucleus. *Proc. Natl. Acad. Sci. USA* 94:8590–95

80. Segref A, Sharma K, Doye V, Hellwig A, Huber J, et al. 1997. Mex67p, a novel factor for nuclear mRNA export, binds to both poly(A)+ RNA and nuclear pores. *EMBO J.* 16:3256–71

81. Shah S, Forbes DJ. 1998. Separate nuclear import pathways converge on the nucleoporin Nup153 and can be dissected with dominant-negative inhibitors. *Curr. Biol.* 8:1376–86

82. Shears SB. 1998. The versatility of inositol phosphates as cellular signals. *Biochim. Biophys. Acta* 1436:49–67

83. Smith A, Brownawell A, Macara IG. 1998. Nuclear import of Ran is mediated by the transport factor NTF2. *Curr. Biol.* 8:1403–6

84. Snay-Hodge CA, Colot HV, Goldstein AL, Cole CN. 1998. Dbp5p/Rat8p is a yeast nuclear pore-associated DEAD-box protein essential for RNA export. *EMBO J.* 17:2663–76

85. Stade K, Ford CS, Guthrie C, Weis K. 1997. Exportin 1 (Crm1p) is an essential nuclear export factor. *Cell* 90:1041–50

86. Stoffler D, Fahrenkrog B Aebi U. 1999. The nuclear pore complex: from molecular architecture to functional dynamics. *Curr. Opin. Cell Biol.* 11:391–401

87. Strasser K, Bassler J, Hurt E. 2000. Binding of the Mex67p/Mtr2p heterodimer to FXFG, GLFG, and FG repeat nucleoporins is essential for nuclear mRNA export. *J. Cell Biol.* 150:695–706

88. Strässer K, Hurt E. 1999. Nuclear RNA export in yeast. *FEBS Lett.* 452:77–81

89. Strässer K, Hurt E. 2000. Yra1p, a conserved nuclear RNA-binding protein, interacts directly with Mex67p and is required for mRNA export. *EMBO J.* 19:410–20

90. Takizawa CG, Weis K, Morgan DO. 1999. Ran-independent nuclear import of cyclin B1-Cdc2 by importin beta. *Proc. Natl. Acad. Sci. USA* 96:7938–43

91. Tseng SS, Weaver PL, Liu Y, Hitomi M, Tartakoff AM, et al. 1998. Dbp5p, a cytosolic RNA helicase, is required for poly(A)+ RNA export. *EMBO J.* 17:2651–62

92. Watanabe M, Fukuda M, Yoshida M, Yanagida M, Nishida E. 1999. Involvement of CRM1, a nuclear export receptor, in mRNA export in mammalian cells and fission yeast. *Genes Cells* 4:291–97

93. Weis K, Dingwall C, Lamond AI. 1996. Characterization of the nuclear protein import mechanism using Ran mutants with altered nucleotide binding specificities. *EMBO J.* 15:7120–28

94. Wiese C, Wilde A, Moore MS, Adam SA, Merdes A, et al. 2001. Role of importin-beta in coupling Ran to downstream targets in microtubule assembly. *Science* 291:653–56

95. Wolin SL, Matera AG. 1999. The trials and travels of tRNA. *Genes Dev.* 13:1–10

96. York JD, Odom AR, Murphy R, Ives EB, Wente SR. 1999. A phospholipase C-dependent inositol polyphosphate kinase pathway required for efficient messenger RNA export. *Science* 285:96–100

97. Yoshida K, Blobel G. 2001. The karyopherin kap142p/msn5p mediates nuclear import and nuclear export of different cargo proteins. *J. Cell Biol.* 152:729–40

98. Zhang C, Clarke PR. 2001. Roles of Ran-GTP and Ran-GDP in precursor vesicle recruitment and fusion during nuclear envelope assembly in a human cell-free system. *Curr. Biol.* 11:208–12

99. Zhou Z, Luo MJ, Straesser K, Katahira J, Hurt E, et al. 2000. The protein Aly links pre-messenger-RNA splicing to nuclear export in metazoans. *Nature* 407:401–5

Annu. Rev. Genet. 2001. 35:365–406

TRANSLATIONAL REGULATION AND RNA LOCALIZATION IN *DROSOPHILA* OOCYTES AND EMBRYOS

Oona Johnstone and Paul Lasko

Department of Biology, McGill University, 1205 Avenue Docteur Penfield, Montréal, Québec, Canada H3A 1B1; e-mail: Paul_Lasko@maclan.mcgill.ca

Key Words axis patterning, UTRs, cytoskeleton-dependent transport, early development

■ **Abstract** Translational control is a prevalent means of gene regulation during *Drosophila* oogenesis and embryogenesis. Multiple maternal mRNAs are localized within the oocyte, and this localization is often coupled to their translational regulation. Subsequently, translational control allows maternally deposited mRNAs to direct the early stages of embryonic development. In this review we outline some general mechanisms of translational regulation and mRNA localization that have been uncovered in various model systems. Then we focus on the posttranscriptional regulation of four maternal transcripts in *Drosophila* that are localized during oogenesis and are critical for embryonic patterning: *bicoid* (*bcd*), *nanos* (*nos*), *oskar* (*osk*), and *gurken* (*grk*). *Cis*- and *trans*-acting factors required for the localization and translational control of these mRNAs are discussed along with potential mechanisms for their regulation.

CONTENTS

INTRODUCTION

There are several possible mechanisms for localizing specific proteins within a cell. Many proteins are directly targeted as such. However, proteins can be deployed at specific locations through mechanisms that operate on the mRNA that encodes them. For example, the mRNA can be localized in an untranslatable form and translated at the site where protein activity is required. Alternatively, protein localization can be accomplished strictly at the level of translational control, through the distribution or regulation of factors involved in either translational repression or activation. Proteolysis and posttranslational modification are further mechanisms by which protein activity can be spatially regulated.

Drosophila oogenesis and embryogenesis are valuable systems for studying protein localization since spatiotemporal organization of materials deposited in the egg is critical for establishing developmental decisions in the embryo. One idea that has emerged from work in *Drosophila* is that the mechanisms listed above are not mutually exclusive. Regulation of several transcripts involves an interplay between RNA localization and translational control. Many *Drosophila* mRNAs are specifically localized with the goal of producing a localized protein. Surprisingly, however, in several cases noted thus far, when mRNA localization is abrogated, protein localization and translation remain normal, implying an additional mechanism involving translational control.

Although this review focuses on the role of translational regulation in protein localization within the *Drosophila* oocyte and embryo, mechanisms uncovered from work on other systems are also discussed, since biochemistry has lagged behind genetics and cell biology in *Drosophila* until recently. First we outline some general models for translational regulation and RNA localization and then focus on the mechanisms that regulate the localized activities of four proteins: Bicoid (Bcd), and Nanos (Nos), which localize to the embryonic anterior and posterior, respectively; Oskar (Osk), which localizes to the oocyte posterior; and Gurken (Grk), which localizes to the anterodorsal corner of the oocyte. These factors represent some of the best-understood examples of protein localization occurring through a combination of mRNA localization and translational regulation. mRNA localization

(6, 112), translational control in development (259), and axis patterning in *Drosophila* (243) have recently been reviewed.

Drosophila Oogenesis and Syncytial Embryogenesis: A Synopsis

A *Drosophila* ovary is composed of a cluster of ovarioles, each of which consists of a string of egg chambers progressing through 14 morphologically defined developmental stages in an anterior to posterior direction (reviewed in 111, 215). A stem cell at the anterior end of each ovariole divides asymmetrically to generate a cystoblast, which divides four additional times with incomplete cytokinesis. A resultant 16-cell cyst contains cystocytes connected to each other by ring canals. One of the cystocytes will become the future oocyte while the other 15 will become nurse cells whose function during oogenesis is to synthesize and transport materials required for the growth and development of the oocyte.

As cysts move through the anterior region of the ovariole, the germarium, they become enclosed in a layer of somatic follicle cells. The oocyte acquires the most posterior position in the egg chamber, thus defining the anterior-posterior axis, and accumulates specific RNAs and proteins. In stage 8 the oocyte nucleus migrates to the anterodorsal corner of the oocyte, and asymmetric localization of several RNAs and proteins within the oocyte becomes apparent. Stages 8 to 10 are marked by rapid yolk uptake by the oocyte, resulting in substantial growth. From stages 10B to 12, the nurse cells transfer their cytoplasm into the oocyte through ring canals, and the mature egg is completed in stages 13 to 14.

Upon fertilization, 13 rapid synchronous mitotic divisions occur; these consist of DNA synthesis and mitosis but not cytokinesis, resulting in approximately 8000 nuclei sharing a common maternally inherited cytoplasm. Subsequently, these nuclei migrate peripherally within the embryo and are cellularized. This transition from the syncytial stage to the cellular blastoderm stage also corresponds to the transfer of developmental control from the maternal to the zygotic genome.

The oocyte and the syncytial embryo offer unique opportunities to study RNA and protein localization mechanisms within a single cell. In the germarium and during pre-vitellogenic stages of oogenesis, selected mRNAs such as *bcd*, *nos*, and *osk* are transcribed in the nurse cells and transported into the oocyte. *grk* is transcribed in the oocyte nucleus, although early in oogenesis it may also be transcribed in the nurse cells and transported into the oocyte. Later, a complex pattern of RNA and protein localization is established within the oocyte. *bcd* is anteriorly anchored by stage 8 of oogenesis and remains dormant until fertilization. *nos* achieves posterior localization within the oocyte by around stage 12 and is not translated until egg deposition. *osk* is posteriorly localized by stage 8, and translationally activated immediately upon localization. *grk* is posteriorly localized and translated within the oocyte prior to nuclear migration. Subsequently, *grk* mRNA relocates to the anterodorsal corner and is translated in this region.

MECHANISMS OF TRANSLATIONAL REGULATION

Translational repression is usually imposed at the step of initiation and generally involves binding of *trans*-acting factors to the noncoding region of a transcript. Binding can occur in the 5′ UTR, exemplified by the well-studied case of ferritin regulation (reviewed in 185). More frequently, however, repressors bind to the 3′ UTR and influence translation in a variety of ways. In this section, we first introduce the idea of mRNA circularization and then briefly discuss some models of translational repression and subsequent derepression/activation.

The Closed-Loop Model of Translation

mRNA molecules undergoing translation are thought to be circularized via interactions between *trans*-acting factors bound to the 5′ and 3′ ends (reviewed in 92, 187). This closed-loop model explains why a poly(A) tail on a transcript can stimulate translational efficiency, as was originally observed in reticulocyte extracts (44, 148) and also in vivo by analysis of poly(A) dynamics during development in various systems (4, 267). Circularization of mRNAs has been visualized in electron micrographs and by atomic force microscopy (26, 49, 87, 104, 252, 255). Linkage between 5′ and 3′ ends of a transcript is mediated at least in part through physical interaction between eIF4G, a component of the 5′ cap-binding complex eIF4F, and poly(A)-binding protein (PABP), which is bound in multiple copies to the poly(A) tail of a transcript (230). This interaction is believed to synergistically stimulate translational activation through altering either the mRNA affinity or the structural conformation of PABP or eIF4F subunits (reviewed in 187).

The mechanistic advantage of mRNA circularization to the process of translation remains unclear (reviewed in 74, 187). One idea is that it promotes reinitiation of ribosomes such that as they terminate translation at the 3′ end they can easily reinitiate on the same mRNA molecule by transferring to the 5′ end. Circularization also confers an advantage in terms of transcript stabilization because it protects both ends of the transcript from degradation. In addition, circularization could be a means to ensure that only full-length mRNAs, with a cap and poly(A) tail, will be translated, preventing translation of truncated transcripts. Developmentally regulated *trans*-acting factors that bind to the 3′ end of specific transcripts may influence translation at the level of initiation by affecting the ability of the mRNA to circularize (79).

General Translational Repression: mRNA Masking

mRNA masking refers to keeping mRNAs concealed in mRNP particles such that they are withheld from cellular processing events such as translation and degradation (reviewed in 214). Masking is thought to operate through proteins that bind to the mRNA and alter its conformation. At the correct time or place, the masking protein is influenced by a signal that alleviates its masking effect. Y-box proteins appear to be involved in mRNA masking (reviewed in 142); in *Xenopus* oocytes, the Y-box protein FRGY2 (mRNP4) is an abundant component of masked mRNP particles (180, 229). Y-box proteins are also components of mRNPs in mammalian somatic cells (57). Mammalian Y-box proteins promote translation at low concentrations but inhibit translation at high concentrations (38, 56).

Surprisingly, in *Xenopus*, the *cis*-acting cytoplasmic polyadenylation element (CPE), located in the 3′ UTR and required for translational activation (see below), is also implicated in masking of *cyclin B1* mRNA (5, 39). In oocytes, CPE-binding protein (CPEB) exists in a complex with a protein called maskin and eIF4E (220). During oocyte maturation the interaction between maskin and eIF4E decreases, leading to the hypothesis that masking is achieved in oocytes by sequestering eIF4E in this complex. To derepress translation during oocyte maturation, eIF4E is released, allowing it to interact with eIF4G to form the cap-binding complex eIF4F, and promote initiation. This represents an example of translational repression occurring by preventing formation of the cap-binding complex.

Translational Repression by Deadenylation

Controlling the length of the poly(A) tail of specific mRNAs in the cytoplasm is a translational regulatory mechanism used during development in vertebrates and invertebrates (reviewed in 40, 178, 179). In general, extending the length of the poly(A) tail leads to translational activation whereas decreasing poly(A) tail length leads to translational repression. In *Xenopus*, no *cis*-acting elements have been identified as required for deadenylation in oocytes, but the presence of a CPE and AAUAAA hexanucleotide, both required for cytoplasmic polyadenylation (see below), prevents deadenylation, leading to the theory that all mRNAs lacking cytoplasmic polyadenylation signals are deadenylated in oocytes (62, 244). *Cis*-acting elements are required in embryos for deadenylation, and one element present in several RNAs is a 17-nt sequence called the embryonic deadenylation element, or EDEN (2, 15, 114, 115), which is sufficient for mediating deadenylation of a reporter RNA (162). Deadenylation of *Cdk2* mRNA depends on two unique elements that are distinct from EDEN (222). AU-rich elements termed AREs, usually containing the sequence AUUUA in tandem arrays, signal deadenylation in *Xenopus* embryos when present in the 3′ UTR of a chimeric mRNA (248). Two proteins are candidate *trans*-acting factors required for deadenylation: poly(A)-specific RNase (PARN) (100) and EDEN-BP (162), which binds to the EDEN sequence. An extract system has recently been developed from *Xenopus* oocytes that recapitulates ARE-mediated mRNA deadenylation (247). This work has led to the identification of a protein called embryonic poly(A)-binding protein (ePAB), which binds the ARE and is the major poly(A)-binding protein present in late oocytes and early embryos. Immunodepletion of oocyte extracts for ePAB enhances deadenylation, suggesting that ePAB has a role in stabilizing poly(A) tails and regulating deadenylation in early *Xenopus* development (247). Another example of regulation of a specific mRNA by deadenylation is the *hunchback* (*hb*) mRNA of *Drosophila*, which is deadenylated when Nos and

Pumilio (Pum) interact with 3′ UTR elements called Nos response elements (NREs) (258, 268).

Translational Repression at the 60S Subunit Joining Step

Translation initiation requires recognition of the mRNA 5′ cap by eIF4F, and the recruitment of the 43S preinitiation complex containing the small ribosomal subunit (reviewed in 74, 173). Subsequently, this complex scans the mRNA until reaching a start codon, at which point the large 60S ribosomal subunit is recruited to form an active 80S ribosome (reviewed in 164). The majority of translational regulatory events are thought to act at the cap-binding step, thereby controlling recruitment of the 43S preinitiation complex. However, recent evidence implicates the 60S subunit joining step as also subject to regulation (158). In mammalian erythroid cells, a specific mRNP complex, containing hnRNP K and hnRNP E1, represses translation of the 15-lipoxygenase (LOX) mRNA through an interaction with a specific sequence in the 3′ UTR termed the differentiation control element (DICE) (159, 160). This block in translation occurs downstream of 43S recruitment and scanning, at the 60S subunit joining step. In support of this, when a reporter mRNA bearing the DICE from the LOX mRNA is subject to translational repression, toeprinting analysis indicates that the preinitiation complex is stalled at the start codon (158). The translation block probably targets one or both of the initiation factors required for the 60S subunit joining step: eIF5 and eIF5B (22, 165, 175, 176). The *Drosophila* homologue of eIF5B is dIF2, which interacts physically and genetically with the DEAD-box RNA helicase Vasa (Vas) (19, 119), which in turn has been implicated in translational activation of germline mRNAs (69, 141, 181, 225, 238, 239), suggesting that Vas might enable translation by alleviating this type of translational repression.

Mechanisms of Derepression/Activation

When repression is conferred by a specific repressor protein binding to a transcript, derepression is likely achieved through deactivation or displacement of the repressor protein, or disruption of a translational regulatory complex containing this protein. There may also be instances of competition between binding of repressor and activator proteins to the same transcript.

Cytoplasmic Polyadenylation

Cytoplasmic polyadenylation is necessary to activate dormant maternal mRNAs during development (reviewed in 40, 178, 179). As mentioned above, cytoplasmic polyadenylation in *Xenopus* oocytes depends on the presence of a CPE, UUU-UUAU, in the 3′ UTR (61, 144). A second CPE specific for mRNAs that are activated during *Xenopus* embryogenesis consists of a minimum of 12 U residues (204–206). Characterization of oocyte CPEs led to the identification of CPEB, which contains a zinc finger domain and two RNA recognition motifs (RRMs) (80).

Immunodepletion of CPEB in extracts (80) and injection of CPEB antibody in vivo (221) both prevent polyadenylation and thus translational activation of mRNAs containing the CPE. In addition to CPEB, cleavage and polyadenylation specificity factor (CPSF), and poly(A) polymerase (PAP) are also required for cytoplasmic polyadenylation (11, 43).

In *Drosophila* development, cytoplasmic polyadenylation is required to activate translation of key embryonic axis patterning determinants such as *bcd*, *torso* (*tor*), and *Toll* (*Tl*) (190). Interestingly, the *cis*-acting sequences required for cytoplasmic polyadenylation in *Drosophila* are not the same as in vertebrates. Within the *Tl* 3′ UTR, a 192-nt segment contains sequences required for its cytoplasmic polyadenylation (192). No consensus motif that controls cytoplasmic polyadenylation has yet been identified in the different *Drosophila* targets. Lack of conservation in *cis*-acting sequences between vertebrates and invertebrates also implies that the *trans*-acting factors that bind to these sequences may not be conserved.

Nevertheless, the *Drosophila* homologue of CPEB, oo18 RNA-binding protein (Orb), is required for oogenesis (27, 108, 109). *orb* mutants demonstrate defects in oocyte determination, signaling between germline and soma, localizing mRNAs such as *Bicaudal-D* (*Bic-D*), *fs(1)K10* (*K10*), *osk*, and *grk* and in dorsal-ventral and anterior-posterior patterning of the oocyte and embryo (27, 89, 109). Recent work indicates that Orb may be involved in cytoplasmic polyadenylation and translational activation of *osk* (23). Interestingly, the Bruno (Bru) protein (253), required to repress *osk* translation, shows homology to EDEN-BP, which is thought to be required for deadenylation in *Xenopus* (162). However, in vitro translation experiments with *osk* argue against the modification of poly(A) length as being a factor in its translational regulation (20, 121). Therefore, although the CPEB homologue Orb is required for oogenesis in *Drosophila*, a definitive role for Orb in polyadenylation in this system remains to be demonstrated.

A genetic approach was undertaken in *Drosophila* to search for potential *trans*-acting factors generally required for cytoplasmic polyadenylation in vivo (122). Since this process is required for activating maternal mRNAs, mutations in some of the factors involved may cause female sterility. A screen of second chromosome female-sterile mutants for those that prevented translational activation of dormant maternal mRNAs led to the identification of *cortex* (*cort*) and *grauzone* (*grau*). Cytoplasmic polyadenylation and translational activation of *bcd* and *Tl* during embryogenesis, and of *tor* during oogenesis, are both disrupted in these mutants. Injection of a polyadenylated *bcd* transcript into *cort* mutants resulted in its translation, indicating that the defect is in the polyadenylation process or regulation of this process. Cloning of *grau* identified it as a transcription factor required to activate *cort* (25). Mutations in *grau* and *cort* exhibit aberrant chromosome segregation in meiosis I and cause an arrest in meiosis II (122, 161, 196). Thus these genes may have a role in linking the progression through meiosis with translational control during development (25). In addition, two potential *trans*-acting factors have been observed to cross-link to the 192-nt element within the *Tl* 3′ UTR, but the identity of these proteins and their relevance to cytoplasmic polyadenylation remain to be investigated (192).

The mechanism by which polyadenylation and deadenylation influence translation initiation is not well understood (reviewed in 178). Increasing the length of the poly(A) tail is thought to increase the number of molecules of poly(A)-binding protein that can bind the transcript, and it has been proposed that this stimulates translational efficiency through the interaction between poly(A)-binding protein and eIF4G. This model has been questioned because of the low abundance of PABP1 in *Xenopus* oocytes (271), but the recent discovery of ePAB (247) suggests that several poly(A)-binding proteins may function at different times during development. In *Drosophila*, PABP also appears to be quite rare in ovaries (121). An alternate possibility by which cytoplasmic polyadenylation might activate translation of some transcripts is through ribose methylation of the cap (102, 103).

MECHANISMS OF LOCALIZATION OF SPECIFIC RNAs

Here we first discuss the idea that some shuttling proteins may be involved in both exporting selected transcripts out of the nucleus and mediating their subsequent localization within the cytoplasm. Next, focusing on the *Drosophila* egg chamber, we introduce the concept of transport of mRNP complexes from the nurse cells into the oocyte in particles called sponge bodies. Third, we briefly outline the role of the cytoskeleton in active localization of mRNAs during oogenesis. Within the oocyte, localization of some mRNAs is achieved by additional mechanisms such as selective degradation or passive diffusion, which are not discussed here (reviewed in 123).

Linking Nuclear Export to mRNA Localization

Recent data from several systems indicate that some localized mRNAs associate in the nucleus and are packaged with proteins involved in their cytoplasmic localization (reviewed in 193). hnRNPs (heterogeneous nuclear ribonucleoproteins) are proteins that associate with hnRNAs in the nucleus and are involved in mRNA packaging. Some hnRNPs are restricted to the nucleus, whereas others shuttle between the nucleus and the cytoplasm (167, 246). Several hnRNPs bind to mRNAs that are localized in the cytoplasm, implicating them in mRNA transport: hnRNP A2 binds to the 21-nt *cis*-acting sequence required for localizing the myelin basic protein mRNA in oligodendrocytes (86); the *Xenopus* hnRNP I homologue, VgRBP60, binds to the vegetally localized mRNA *Vg1* (33); and the *Drosophila* Squid (Sqd) hnRNP A/B homologue binds to the localized *grk* and *fushi tarazu* (*ftz*) transcripts (106, 156; and see below). Another *trans*-acting factor that binds to *Vg1* mRNA is the *Xenopus* homologue of the zipcode-binding protein (ZBP) called Vera (42), or Vg1 RBP (81). In chick embryo fibroblasts, ZBP binds to the localized β-actin transcript (182). ZBP contains motifs required for nuclear localization and export, suggesting that, like the hnRNPs, its role in cytoplasmic localization of transcripts may begin in the nucleus. For the most part, the evidence implicating the proteins discussed above in cytoplasmic localization is indirect, with the exception of the role of Sqd in localization of the *ftz* transcript in *Drosophila* embryos, as is discussed below.

mRNP Transport from the Nurse Cells to the Oocyte in the *Drosophila* Ovary

Developing and mature germ cells in various animals are characterized by the presence of nuage particles, electron-dense structures that are associated with RNA, protein, and mitochondria (reviewed in 50, 188). In *Drosophila* oogenesis, nuage particles are present in the perinuclear region of nurse cells. These particles may be precursors to polar granules (133–135), which are morphologically similar structures found in the pole plasm, specialized cytoplasm in the posterior of the oocyte required for embryonic pole cell development. During oogenesis, nuage has been proposed to function in transport of RNAs and proteins destined for the pole plasm.

Surrounding the perinuclear nuage in nurse cells are structures referred to as sponge bodies (265). Like nuage, sponge bodies are electron-dense and contain RNA and mitochondria. Sponge bodies are distinguished from nuage by the presence of elongated bodies or vesicles and by the fact that some of their protein components differ from those of nuage particles. Sponge bodies are present in nurse cells, oocytes, and in the ring canals that link those cells, suggesting that like nuage, they also function as transport vehicles for mRNPs that move from the nurse cells into the oocyte. This hypothesis has been supported by the identification of some of the protein and RNA components of these particles. In *Xenopus*, a seemingly homologous structure to sponge bodies, called the mitochondrial cloud, migrates to the vegetal pole during development, where it is believed to form the germinal granules (83).

Exuperantia (Exu) protein is the most specific marker known for sponge bodies (265). Exu is required for the anterior accumulation of *bcd* and efficient posterior localization of *osk* within the oocyte, as well as for the apical localization of both *bcd* and *osk* mRNAs within the nurse cells (10, 218, 264). GFP-Exu is visualized in particles surrounding the nurse cell nuclei, distributed throughout the cytoplasm, and localized around the ring canals (138, 251). Time lapse laser scanning microscopy was used to track the movement of Exu-particles in living oocytes and to test the effect of microtubule or actin depolymerization on that movement (234). Within the nurse cells, random movement in the cytoplasm, perinuclear localization, and clustering at ring canals was found to require three seemingly non-overlapping populations of microtubules. The movement of these particles through ring canals and into the oocyte did not require the actin or microtubule cytoskeleton. This is in contrast to vesicle and mitochondrial transport through ring canals, which does depend on the actin cytoskeleton (13, 14). Within the oocyte the accumulation of Exu at the anterior requires microtubules (234).

Immunoprecipitation with anti-Exu antisera was used to identify and analyze Exu-containing complexes, which presumably contain other sponge body components (264). This analysis uncovered an RNase-sensitive association between Exu and least six other proteins. *osk* mRNA was also identified as a component of this complex, consistent with the requirement for Exu in *osk* localization. A direct interaction was observed between Exu and the Y-box protein Ypsilon Schachtel (Yps) (236). Yps colocalizes with Exu throughout oogenesis (264) and binds to RNA (236). The presence of Yps in these particles supports the idea that they contain masked mRNAs, because this class of proteins is involved in translational regulation (see above). Also consistent with their proposed role in transporting dormant mRNAs is the observation that sponge bodies are not seen to associate with ribosomes (265). Another protein present in sponge bodies and that forms an RNase-sensitive association with Exu and Yps is the DEAD-box protein Maternal expression at 31B (Me31B) (41; A. Nakamura & S. Kobayashi, personal communication). A *Xenopus* homologue of Me31B, Xp54, is a component of dormant mRNPs in oocytes (105). Thus sponge bodies appear to be involved in transporting translationally dormant mRNPs including localized mRNAs from the nurse cells into the oocyte.

The Role of the Cytoskeleton in RNA Localization

RNAs and proteins are often localized by associating with molecular motors, factors that move directionally along cytoskeletal tracks in an energy dependent manner (reviewed in 60, 241). Two large families of motors that use the microtubule cytoskeleton are the minus end–directed dyneins and the plus end–directed kinesins. The family of myosins are plus end-directed and make use of the actin cytoskeleton.

During *Drosophila* oogenesis, some transcripts achieve their localization by transport along the microtubule cytoskeleton. Several lines of evidence support this idea. The first is the overall polarity of microtubules during oogenesis. During stages 1 to 7, a microtubule organizing center (MTOC) is situated at the posterior of the oocyte, focusing microtubule minus ends at the posterior, while plus ends extend to the anterior and into the nurse cells (Figure 1, see color insert). This organization depends on the functions of *Bic-D* and *egalitarian* (233). Subsequently, in stages 7 to 8, there is a reversal of microtubule directionality due to relocation of the MTOC to the anterior of the oocyte (235). This event requires signaling pathways involving Notch/Delta, and protein kinase A (PKA) as well as the activity of *mago nashi* (*mago*) (107, 110, 146, 154, 186). This functional microtubule distribution has been supported by experiments showing that from stage 8 the minus end–directed kinesin-related protein, Nod-β-gal, localizes to the anterior while the plus end motor Khc-β-gal shows the opposite localization

A

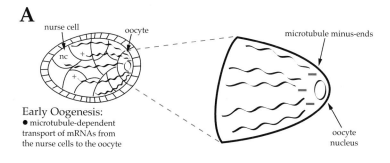

Early Oogenesis:
- microtubule-dependent transport of mRNAs from the nurse cells to the oocyte

B

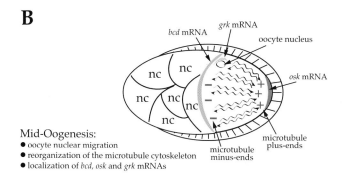

Mid-Oogenesis:
- oocyte nuclear migration
- reorganization of the microtubule cytoskeleton
- localization of *bcd, osk* and *grk* mRNAs

C

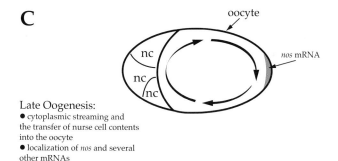

Late Oogenesis:
- cytoplasmic streaming and the transfer of nurse cell contents into the oocyte
- localization of *nos* and several other mRNAs

Figure 1 Organization of the microtubule cytoskeleton and mRNA localization. (*A*) During the early stages of oogenesis, microtubule minus-ends are concentrated at the posterior of the oocyte while plus-ends extend toward the anterior and into the nurse cells. (*B*) Subsequently a reorganization of the microtubule cytoskeleton occurs within the oocyte such that the microtubule minus-ends are anteriorly anchored and the plus-ends are located at the posterior. The oocyte nucleus migrates to the anterodorsal corner of the oocyte, and several mRNAs localize to specific positions within the oocyte. *bcd* mRNA is shown in orange, *grk* mRNA in pink, and *osk* mRNA in olive. (*C*) In the late stages of oogenesis, the nurse cells transfer their contents into the oocyte wherein rapid cytoplasmic streaming and localization of some specific factors such as *nos* mRNA, shown in blue, takes place (adapted from 60 and 109a).

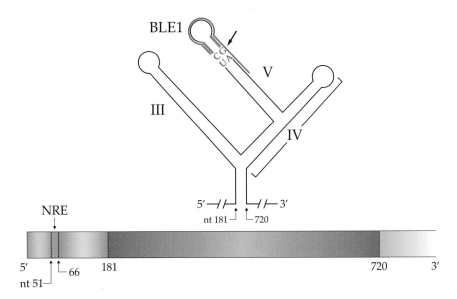

Figure 2 Schematic diagram of the *bcd* 3' UTR showing a portion of the predicted secondary structure (126, 131, 198). The location of the consensus Nos response element (NRE; 258) is indicated in dark orange. The region of stem-loop V shown in red represents the *bcd* localization element (BLE1; 129), and the region shown in green represents a stretch of nucleotides which must be double-stranded for event A localization to occur (128). At least some of the individual nucleotides shown in green are also required for this event. Substitution of the G residue, indicated by the arrow, eliminates event A localization but maintains event B and later stages of localization (127).

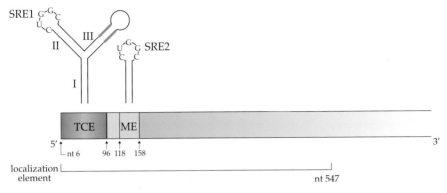

Figure 3 Schematic diagram of the *nos* 3' UTR showing the predicted secondary structure. The translational control element (TCE; 69) is shown in red, and the minimal element (ME; 9) is shown in pink. The predicted secondary structure of the TCE includes two stem-loops (35). Stem-loop II contains the Smg recognition element (SRE1; 209). The region of stem-loop III shown in green represents a UA-rich stretch of nucleotides required for translational repression (35). A second SRE (SRE2) is located within a predicted stem-loop in the ME region (209).

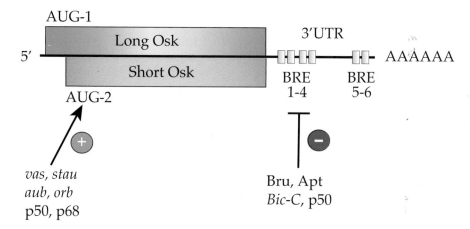

Figure 4 Schematic diagram of *osk* mRNA. Two isoforms of Osk protein, Long Osk and Short Osk, are produced from the *osk* transcript (141, 181). Genes and proteins implicated in translational regulation of *osk* are indicated. The *osk* 3' UTR contains Bru response elements (BREs), shown in yellow, which confer translational repression through the binding of Bru (96, 253).

(28, 29). Therefore, the early organization of the microtubule cytoskeleton may mediate the localization of RNAs and proteins from the nurse cells into the oocyte, whereas the later reorganization would then be required to localize a subset of these factors within the oocyte. In keeping with this idea, the requirement for Exu in localizing mRNAs into the oocyte occurs after the reorganization of the microtubule cytoskeleton, suggesting that sponge bodies may offer an alternative transport mechanism for selected transcripts entering the oocyte, after cytoskeletal reorganization (127).

The second observation to underscore the role of the microtubule cytoskeleton is that treatment with microtubule depolymerizing drugs disrupts localization of *bcd* and *osk* (28, 168, 223). Localization of these RNAs is also affected in genetic backgrounds where microtubule directionality is altered. For example, in mutants that have microtubule minus ends at both poles, *bcd* localizes to both ends, while *osk* accumulates near the center of the oocyte (28, 76, 78, 107, 183).

The third line of evidence directly links specific molecular motors to transcript localization within the oocyte. As discussed below, dynein is implicated in anterior localization of *bcd* RNA (194), and kinesin I is required for *osk* localization to the posterior (17). *osk* localization also requires the activities of several maternal effect genes that encode regulators of the actin and microtubule cytoskeletons, including *cappuccino* (*capu*), *spire* (*spir*), *chickadee* (*chic*), and *Tropomyosin II* (*TmII*). Tropomyosin binds filamentous actin and may act to stabilize microfilaments (3, 55). *capu* encodes a formin-homology-domain containing protein related to the product of the chicken *limb deformity* locus (52). Capu protein probably functions to link the microtubule cytoskeleton with the actin cytoskeleton, via an interaction with profilin, an actin-binding protein encoded by *chic* (32, 136). *spir* also encodes an actin-binding protein that interacts with Rho-family GTPases including RhoA, Rac1, and Cdc42 (254). These GTPases control the assembly and organization of the actin cytoskeleton, by responding to extracellular cues and interacting with downstream effectors (12). Recent work (202, 240) also implicates a putative kinase, PAR-1, as essential for directing *osk* to the posterior pole. Mammalian homologues of PAR-1 phosphorylate various microtubule-associated proteins (45), and in *Drosophila par-1* mutants the organization of the microtubule cytoskeleton within the oocyte is disrupted. In these oocytes, *bcd* localizes normally, but *osk* RNA is mislocalized to the center, and, consequently, abdominal patterning and pole cells are disrupted in embryos produced by *par-1* mothers. *par-1* is also required for oocyte differentiation and for microtubule organization in the early egg chamber (34, 90). In *C. elegans*, the *par-1* homologue is required maternally for the first embryonic cell division to be asymmetric (94). These results linking *osk* and *bcd* with molecular motors or molecules involved in cytoskeletal organization are consistent with the model that localization of these transcripts is mediated, at least in part, through the microtubule cytoskeleton. However, many details of this mechanism, including exactly how specific transcripts associate with motor proteins, remain to be established.

TRANSLATIONAL CONTROL AND RNA LOCALIZATION COOPERATE IN THE DEPLOYMENT OF SPECIFIC *DROSOPHILA* PROTEINS ESSENTIAL FOR EMBRYONIC AXIS DETERMINATION

Several translationally controlled RNAs have been intensively studied because of their importance in establishing the spatial axes of the embryo. Despite progress in identifying RNA-binding proteins that regulate localization and translation of these RNAs, in no case is there yet a complete understanding of the mechanism by which these specific regulators function. Localization and translational regulation of four transcripts is considered in the next sections.

bicoid

An anterior to posterior gradient of Bcd protein establishes head and thoracic development (reviewed in 219). During oogenesis the *bcd* transcript undergoes a multistep localization pathway, culminating in a strictly anterior concentration from which it will be translationally activated during embryogenesis, producing a steep concentration gradient (10, 218). *bcd* mRNA is transcribed in nurse cells and transported to the oocyte in two separable phases: event A, commencing in stages 4 to 5, and event B, commencing around stage 6. There is redundancy between events A and B in that either is thought to be able to promote later steps of localization (127).

CIS-ACTING ELEMENTS REQUIRED FOR LOCALIZATION OF BCD   *bcd* mRNA localization depends on its 3' UTR, which forms a complex secondary structure including multiple stem-loops (126, 131, 198) (Figure 2, see color insert). Stem-loops IV and V together can mediate event A completely, and are sufficient for normal localization up until embryogenesis (127). Deletion analysis identified a 50-nt segment within stem-loop V, named the *bcd* localization element 1 (BLE1), which when present in two copies can confer normal localization up to stages 9 to 10 (129). A point mutant in BLE1 eliminates event A localization but maintains event B and all later stages (127). BLE1 is hypothesized to represent a binding site for a *trans*-acting factor that may mediate event A localization, and mutational analysis has identified RNA structural elements necessary for this event (128). *Cis*-acting elements necessary for event B and later localization events have not been mapped as precisely.

TRANS-ACTING FACTORS REQUIRED FOR LOCALIZATION OF BCD   The functions of three maternal genes, *exu*, *swallow* (*swa*), and *staufen* (*stau*), are required for *bcd* mRNA concentration at the anterior of oocytes and embryos (10, 218, 223). Another protein related to Exu, Exu-like (Exl), which cross-links to the dimerized BLE1 region, may also have a role in *bcd* localization (130). As discussed above, the requirement for Exu is likely due to its association with sponge bodies, although *bcd* mRNA has not yet been identified in those structures. Microtubules are also necessary for correct *bcd* localization (168). Microtubule depolymerization produces the same phenotype as seen in *swa* mutants, whereby *bcd* is evenly distributed within the oocyte instead of anteriorly anchored (10, 218). Commencing at stage 10, Swa protein colocalizes with *bcd* mRNA in a microtubule-dependent manner at the anterior of the oocyte (84, 194). In *grk* mutants, which have microtubule minus ends at each pole, *bcd* and Swa localize to both poles, suggesting that they are transported along the microtubule cytoskeleton. Furthermore, Swa contains a coiled-coil domain through which it interacts with Ddlc-1, a light chain of cytoplasmic dynein. When this interaction is disrupted in vivo, Swa is unlocalized, leading to the hypothesis that Swa is an adaptor protein within the oocyte, connecting *bcd* mRNA to the dynein motor and enabling the transcript to be transported directionally along the microtubule cytoskeleton (194). Swa contains a region with distant homology to an RNA–binding domain (24), but has not been demonstrated to bind to *bcd* directly.

Staufen (Stau) is required for the final stage of *bcd* localization, anterior anchoring of the transcript (218). In the egg, Stau is highly concentrated at the posterior, but also shows a weak anterior accumulation, dependent on *bcd* (58, 216). RNA injection experiments demonstrated that nt 181–660 of the *bcd* 3′ UTR (consisting of stem-loops III, IV, and V) could recruit endogenous Stau into particles that migrate along microtubules (58). Dimerization or multimerization of this region appears to be required for particle formation (59). Thus Stau may bind *bcd* directly and anchor it at the anterior in a microtubule-dependent manner.

Stau contains five double-stranded RNA-binding domains (dsRBDs). dsRBD1, 3, and 4 can bind to dsRNA in vitro (145, 217), and a transgene bearing mutations within dsRBD3 that disrupt RNA binding cannot rescue mislocalization of *osk* or *bcd*, indicating that the function of Stau depends on its RNA-binding activity (171). dsRBD5 and a large insertion within dsRBD2 of Stau are both required for *bcd* localization (145). However, transgenes bearing deletions of either of these regions are still capable of rescuing head defects of *stau* mutants, indicating the presence of Bcd activity. Thus, in addition to mediating its localization, Stau is hypothesized to be involved in either prevention of degradation or translational activation of the *bcd* mRNA.

BCD IS TRANSLATIONALLY ACTIVATED THROUGH CYTOPLASMIC POLYADENYLATION
bcd is dormant throughout oogenesis. Poly(A) test (PAT) analysis has revealed that at this time *bcd* has a poly(A) tail of ∼70 nt (190). During embryogenesis, the lengthening and shortening of the *bcd* poly(A) tail correlates with its translational status, reaching ∼140 nt by 1 to 1.5 h, when *bcd* translation is active, and shortening again later. Transgenic experiments demonstrated the requirement of the poly(A) tail for *bcd* translation. The maternal-effect *bcd^{E1}* mutation results in embryos that exhibit defective anterior development due to a lack of functional Bcd, which can be rescued by injecting wild-type *bcd* mRNA into the anterior of the embryo (46, 48). Injection of a transcript lacking 537 nt of the 3′ UTR has no rescuing

ability, but the addition of 150–200 A residues to the 3' end in vitro allowed this RNA to partially rescue the morphogenetic defect, whereas addition of only 40–60 A residues did not, suggesting that an extended poly(A) tail is required for the translational activation and function of *bcd* (190). Cytoplasmic polyadenylation and translational activation of *bcd* were uncoupled by the insertion into the *bcd* 5' UTR of a 63-nt sequence complementary to a segment of the coding region (*bcd*-AS) (245). *bcd*-AS RNA was polyadenylated normally when injected into the anterior of wild-type embryos, but could not be translated in vivo, indicating that although the process of polyadenylation is necessary, it is insufficient for *bcd* translation. The structural modification caused by this mutation likely disrupted interaction between the 5' and 3' UTRs of the transcript (245).

BCD IS TRANSLATIONALLY REPRESSED BY NOS   When *nos* is ectopically expressed at the anterior, translation of *bcd* is repressed, and the RNA has a shorter poly(A) tail (67, 257). Through PAT assays and injection experiments, Nos has been shown to influence *bcd* polyadenylation (268). Injection of *bcd* into the anterior but not the posterior of wild-type embryos results in its polyadenylation, but polyadenylation and translational activation do occur upon injection into the posterior of *nos* mutants, indicating that *bcd* is not being regulated correctly. Even when *bcd* is polyadenylated in vitro and then injected into the wild-type posterior, it is not translated, suggesting that it is deadenylated by Nos, leading to translational repression, as is the case for *hb* (see below). Both the *hb* and *bcd* 3' UTRs contain NREs (258), which confer translational repression by Nos. Thus Nos appears to be capable of regulating *bcd* translation, suggesting that during embryogenesis, any *bcd* mRNA that is not restricted to the anterior can be repressed translationally, possibly via Nos-dependent deadenylation. However, the significance of this regulation in vivo is not clear since *bcd* mRNA is strictly anteriorly localized, and Nos activity is posteriorly localized.

nanos

A posterior to anterior gradient of Nos protein acts to specify posterior development. Mutants lacking Nos activity do not form an abdomen (118), and mutants with ectopic Nos activity at the anterior form a mirror image abdomen at the expense of head and thoracic development (67, 257). *nos* mRNA, like *bcd*, is concentrated by the end of oogenesis at the site at which it will be translated during embryogenesis (250). However, unlike *bcd*, which exhibits nearly complete anterior localization, and which is translated even when unlocalized (47), *nos* mRNA is found throughout the embryo; only 4% is estimated to be posteriorly localized (8). As only the localized fraction of the transcript gets translated (68), restriction of Nos activity to the posterior depends primarily on translational regulation.

During stages 1 to 7 of oogenesis, *nos* is expressed in the nurse cells and accumulates in the oocyte. Like several other oocyte-localized mRNAs, it concentrates at the anterior during stages 7 to 8 (249). As for *bcd*, the early oocyte localization of *nos* mRNA is unnecessary for its subsequent targeting (8). By stage 10, high levels of *nos* are transcribed in the nurse cells and subsequently transferred into the oocyte. Posterior concentration within the oocyte begins around stage 12, and shortly after egg deposition, translation of the posteriorly localized transcript produces a gradient of Nos protein (249).

CIS-ACTING ELEMENTS REQUIRED FOR TRANSLATIONAL REPRESSION OF *NOS*
3′ UTR elements are required both for translational repression of *nos* in the bulk cytoplasm and translational activation in the posterior (37, 68, 69, 209). Fusion of the first 184 nt of the *nos* 3′ UTR to the *tor* reading frame conferred the translational regulation profile of wild-type *nos*, consisting of translational repression outside the posterior, concentration of the transcript in the posterior, and pole plasm–dependent translational activation; thus this region was termed the translational control element (TCE) (37). However, there is redundancy between the two halves of this element such that the region encompassing nt 6–96 alone is capable of mediating strong translational repression and producing wild-type development, and so was also named the TCE, a designation used here (37, 69, 209).

Computer modeling of the TCE predicts formation of two extended stem-loops (II and III) (35) (Figure 3, see color insert), mutations which abolish translational repression. The loop portion of stem-loop II corresponds to the binding site for the translational repressor Smaug (Smg) (discussed below) (36, 37, 208, 209). At least one of the mutations in stem III that abolishes TCE-mediated translational repression does not act by affecting the ability of Smg to bind to stem-loop II, implying that stem-loop III may regulate the binding of a distinct factor also required for translational repression (35).

CIS-ACTING ELEMENTS REQUIRED FOR LOCALIZATION AND TRANSLATIONAL ACTIVA-
TION OF *NOS*   As the nurse cells are rapidly transferring their cytoplasm to the oocyte during the period when *nos* concentrates in the pole plasm, *nos* accumulation might not involve active localization but only specific anchoring (8). Furthermore, translationally repressed *nos* outside the pole plasm is unstable and degraded (7, 37, 209), thus degradation of *nos* mRNA may help restrict Nos protein to the posterior. Regardless of the mechanism by which *nos* localizes, transcripts that reach the posterior are translationally activated. A 547-nt segment of the 3′ UTR consisting of partially redundant elements is required to direct all stages of *nos* localization throughout oogenesis (66). Although this segment overlaps with the TCE, the TCE itself is not essential for localization. Mutational analysis within the TCE, including separation of stem-loops II and III from each other, has revealed that the requirements for this region in localization and translational regulation can be uncoupled. In addition, the TCE from *D. virilis* can substitute for that of *D. melanogaster* for translational repression but not localization, implying a conservation in recognition sequences for one event but not the other (35).

A 41-nt element downstream of the TCE, and highly conserved in *D. virilis*, can influence localization and translational repression, and was named the minimal element (ME) (9). Normally this region acts synergistically with the TCE to confer localization. Three mutations within the ME eliminated both its own function and that of the linked TCE with respect to localization, indicating that the ME has a long-range effect on the TCE. Such a long-range effect is not present for the translational regulatory function of the TCE. As for the TCE, the repressor and localization functions of the ME can be uncoupled.

TRANS-ACTING FACTORS REQUIRED FOR TRANSLATIONAL REPRESSION OF *NOS*   Smg was identified as a protein that binds specifically to the *nos* 3′ UTR (36, 209). Mutational analysis indicated that the Smg binding site, referred to as the Smg recognition element (SRE1), was within a predicted stem-loop (stem-loop II described above) (37, 209). A second SRE (SRE2) is located within nucleotides 97–185. Cloning of *smg* (36, 208) revealed a predicted protein lacking any previously identified RNA-binding motifs, but containing a Sterile Alpha Mating (SAM) domain, a motif implicated in protein binding (reviewed in 195).

Several lines of evidence indicate that Smg represses translation of unlocalized *nos*. Mutational analysis revealed a correlation between the ability of Smg to bind to *nos in vitro*, and translational repression in vivo (209). A wild-type *nos* transgene bearing mutations in the SREs produces ectopic *nos* activity, suggesting defective repression, whereas the addition of three stem-loops containing SREs (3xSRE) to a *nos/bcd* transgene, consisting of the *nos* open reading frame fused to the *bcd* 3′ UTR, can confer translational repression. Regulation of *nos* translation by Smg has also been demonstrated through an in vitro translation system derived from embryonic extracts (208). A reporter transcript fused to three functional SREs is translated less efficiently than one bearing point mutations in these domains. Translational efficiency is equalized when Smg is immunodepleted from embryonic extracts or when late-stage embryos, where Smg is less abundant, are used, suggesting that Smg is the repressor mediating this effect. A probable null allele for *smg* shows ectopic Nos activity, and overexpression of Smg in sensitized genetic backgrounds leads to enhanced repression of *nos in vivo* (36).

TRANS-ACTING FACTORS REQUIRED FOR LOCALIZATION AND TRANSLATIONAL ACTIVATION OF *NOS*   *nos* mRNA does not localize to the posterior if the pole plasm has not assembled there. Thus all genes required for pole plasm assembly are directly or indirectly required for *nos* localization and activation. Ectopic localization of *osk* results in ectopic Nos activity, dependent on the functions of *vas* and *tudor* (*tud*) (54, 210). When either the TCE or the entire 3′ UTR of *nos* is present, no Nos activity is ever detected in *vas* mutants, although low levels are present in *osk* mutants (69). However, when the *nos* 3′ UTR is present but the TCE is deleted, translation no longer depends on *vas*, as assessed by the production of posterior embryonic segments, implying that the role of Vas in activating *nos* translation involves overcoming repression by the TCE (37). Vas is a DEAD-box RNA helicase (119), and therefore could be directly involved in *nos* translational activation.

A 75-kDa protein (p75) that binds the ME of the *nos* 3′ UTR may mediate localization or anchoring (9). Mutations that prevent binding of p75 abolish both the localization and translational repressive properties conferred by the ME. p75 is predicted to function in localization or anchoring, and not in translational control, since it does not bind to the TCE region, and since the translational repression defect in the ME mutants could be explained by the fact that they may also disrupt formation of the Smg binding site stem-loop. Furthermore, other mutations within the ME that do not disrupt p75 binding have stronger effects on translational repression, one of which alters the SRE2 sequence.

POTENTIAL MECHANISMS FOR LOCALIZATION AND TRANSLATIONAL ACTIVATION OF *NOS*   Repression of *nos* in the bulk cytoplasm by Smg, and potentially other repressors, must be overcome in the posterior cytoplasm. Since Smg and Osk interact directly (36), a complex of pole plasm components including Osk and Vas may deactivate Smg on *nos* transcripts that enter the posterior region, leading to derepression of translation. Alternatively, mutual exclusion between the binding of translational repressors and localization factors may mediate localization of a portion of the transcript to the posterior, in keeping with data suggesting overlapping binding sites for repressors and localization factors, and the discovery of the p75 protein (8, 9, 35). Cloning and identification of p75 and/or other potential localization factors will shed more light on this function.

The different behavior of multiple TCEs in *cis* and in *trans* has been interpreted in several ways. When wild-type *nos* is overexpressed at the posterior, its translation is increased (67), and overexpressing Smg in a wild-type background has no phenotype (36). By contrast, when multiple copies of the 6–185-nt region of the *nos* 3′ UTR are added to the same transcript, localization is unaffected but posterior development is reduced, implying reduced translation (8). The presence of multiple copies of this region in *cis* might permit the simultaneous binding of translational repressors and localization factors, allowing the RNA to be both localized and repressed, and may imply that in the wild-type scenario these are mutually exclusive (8). However, multiple TCEs in *cis* may simply be more efficient at recruiting Smg than multiple copies of the same element in *trans* (36), enabling the increased Smg-bound *nos* to titrate out a derepressor in the posterior, possibly Osk, as Osk and Smg interact directly.

To investigate the mechanism of *nos* translational repression, the distribution of *nos* in polyribosomal profiles was analyzed (30). Transcripts that are being actively translated are expected to cosediment with polysomes, whereas transcripts that are blocked at initiation would be predicted to sediment with initiation complexes. Surprisingly, although only 4% of *nos* is localized and thus translated in the early embryo, over 50% of the transcript showed an EDTA- and puromycin-sensitive association with polysomes, suggesting that the translational block for the unlocalized *nos* occurs after initiation. This association is maintained in *vas* and *osk* mutants in which little or none of the *nos* transcript is translated. Substitution of the *nos* 3′ UTR had no effect on polysomal association, supporting the idea that repression does not act at this step. Substantial translational runoff following cycloheximide treatment was observed for *nos* mRNA in preblastoderm embryonic extracts, arguing against a block at either elongation or termination. These results suggest an alternative mechanism of regulation, such as premature translational termination or degradation of the nascent polypeptide after its translation (30).

Potentially relevant to this hypothesis is the *bicaudal* (*bic*) gene, which encodes the *Drosophila* beta NAC homologue (139). Some embryos from *bic* mutant females have a bicaudal phenotype that results from ectopic Nos activity, but, unlike other mutants that cause this phenotype, in *bic* mutants ectopic Nos is produced without ectopic *osk* expression. Beta NAC is a subunit of the Nascent polypeptide Associated Complex (NAC), which associates with the ribosome where it binds to emerging polypeptides (262). Thus beta NAC could play a role in regulating the translation of polypeptides on the ribosome, and the ectopic Nos in *bic* mutants may result from loss of this function. Since *bic* is a hypomorphic allele, it may particularly affect genes whose translation is most sensitive to this regulation (139).

NOS IS A TRANSLATIONAL REPRESSOR   The role of Nos in abdominal specification is achieved through translational repression of the maternal *hb* transcript, contributing to an anterior to posterior gradient of the Hb transcription factor (88, 91, 224, 231). Nos forms a translational regulatory complex with Pum, and the NRE sequences within the *hb* 3' UTR (149, 212, 258). Like *bcd*, the poly(A) profile of *hb* correlates with its translational status, and the *hb* 3' UTR is polyadenylated when injected into the posterior of *nos* or *pum* mutant embryos, but not the wild-type posterior (268). Injection of an in vitro polyadenylated *hb* 3' UTR into the wild-type posterior but not the anterior results in complete deadenylation. Thus Nos and Pum are thought to promote the deadenylation of *hb*, leading to its translational repression. This repression does not act through the cap-binding step of translation initiation, since Nos and Pum can also repress a transcript that is translated from an internal ribosomal entry site (IRES) in a cap-independent manner (256).

The complex of Nos and Pum, bound to the *hb* NRE, recruits the NHL protein Brain Tumor (Brat) (213). Specific mutations in *pum*, *nos*, or *brat* that prevent Brat recruitment abolish regulation of *hb*, and maternal effect *brat* mutants demonstrate the same posterior segmentation defects as *nos* and *pum* mutants (117, 118, 213, 256). The NHL domain is sufficient to mediate the recruitment and function of Brat for *hb* repression. This domain has not been assigned a molecular function, but it is present in several proteins implicated in RNA processing (64, 65, 207).

Nos and Pum have also been implicated in repression of *Cyclin B* translation in pole cells (1), and Pum binds to NRE sequences in the maternal *Cyclin B* 3' UTR (150, 213). This function of Nos and Pum, however, is not mediated through Brat, indicating that the ternary complex of Nos, Pum, and NRE sequences can interact with different cofactors. In *Xenopus*, a homologue of Pum (Xpum) interacts with a Nos homologue (Xcat-2) and with CPEB, and XPum binds to the 3' UTR of *cyclin B1* mRNA, suggesting conservation in the translational regulatory properties of these factors (150). Similarly, in *C. elegans* a Pum homologue (FBF) can interact with homologues of both Nos (NOS-3) and CPEB (CPB-1), and FBF binds to the 3′ UTR of at least one target transcript, *fem-3*, to mediate translational repression (101, 124, 272). In *S. cerevisiae*, two Pum-like proteins have been demonstrated to bind to 3′ UTRs of selected transcripts and mediate posttranscriptional regulation (157, 228). At least one of these functions through deadenylation, suggesting that it may operate through a mechanism similar to Pum.

oskar

Embryos lacking maternal *osk* activity do not form pole cells or develop posterior structures (116). The primary role of Osk in both these processes is demonstrated by experiments in which *osk* is ectopically expressed at the anterior through substitution of the 3′ UTR, which leads to ectopic abdominal development and pole cell formation (54). Like Nos, restriction of Osk activity to the posterior is critical, and this is achieved through a combination of mRNA localization and translational regulation. *osk* mRNA is synthesized in the nurse cells and, throughout early oogenesis, becomes concentrated in the oocyte (53, 97). Within the oocyte, a transient anterior concentration of *osk* is evident at stages 7–8, followed by posterior localization in stages 8–9 that is maintained until early embryogenesis. High levels of *osk* are synthesized in the nurse cells in stage 10 and transferred into the oocyte. Osk protein first appears in stage 8, in a restricted pattern at the posterior pole (96, 141, 181). Two isoforms of Osk, Long Osk and Short Osk, are produced from two in-frame start codons. Short Osk alone can confer wild-type development.

CIS-ACTING SEQUENCES REQUIRED FOR REGULATION OF *OSK*   As for *nos*, sequences within the *osk* 3′ UTR are necessary for its localization: Early concentration in the oocyte depends on nt 532–791; release of *osk* from the anterior is mediated by two elements in different regions of the 3′ UTR (nt 242–363 and 791–846); and posterior localization of *osk* depends on the first 242 nt (98). Translational repression of *osk* in the bulk cytoplasm also depends on sequences within the 3′ UTR and on Bru (see below), a protein that binds to three regions of the *osk* 3′ UTR (A, B, and C) (96, 181, 253). A consensus sequence of 7–9 nt is found six times within these regions, and mutating these sequences reduces binding of Bru, leading to their designation as Bru response elements (BREs). Flies bearing an *osk* transgene mutated for the BREs (*oskBRE⁻*), in an *osk* mutant background, produce Osk precociously during stages 7–8. Conversely, addition of BRE sequences to the *exu* 3′ UTR leads to a reduction in translation of this transcript, indicating that these sequences can mediate translational repression (96). Unlike *nos*, where the 3′ UTR is sufficient to confer translational activation, the *osk* 3′ UTR can mediate posterior localization of a heterologous transcript, but not translational activation (181, 200). An element between the first and second start codons (m1 and m2) is necessary for translational activation. This active region in the 5′ end is considered a derepressor element because it is only necessary for translational activation when translation is being repressed through the BREs (79).

TRANS-ACTING FACTORS REQUIRED FOR TRANSLATIONAL REPRESSION OF OSK   Four proteins and/or genes have been implicated thus far in translational repression of *osk*: Bru, Apontic (Apt), p50, and *Bicaudal-C* (*Bic-C*) (Figure 4, see color insert). Bru, an RRM-type RNA–binding protein, was identified by virtue of its specific binding to the *osk* 3′ UTR (96, 253) and is the product of the *arrest* (*aret*) gene (21, 197). Bru accumulates in the oocyte and colocalizes with *osk*. A role for Bru in *osk* regulation in vivo was demonstrated through the use of a transgene that localizes *osk* to the anterior and that contains the BREs. Ectopic *osk* expression from this transgene leads to embryonic head defects, and this phenotype is enhanced in the background of a heterozygous *aret* mutation (253). Experiments in *Drosophila* cell-free translation systems have supported the conclusion that Bru is involved in *osk* repression (20, 121). Apt interacts with Bru both in the two-hybrid system and biochemically (120). Progeny from females *trans*-heterozygous for combinations of *aret* and *apt* mutations exhibit defects in anterior patterning. This phenotype can be suppressed by reduction of *nos*, suggesting that it is caused by ectopic Osk, and thereby implying that Apt, like Bru, has a role in regulating *osk* translation.

p50 cross-links to both the 5′ and 3′ ends of the *osk* mRNA (79). Within the 3′ UTR, p50 binds to the AB region. Reduction in p50 binding enhances Bru binding, suggesting that there may be competition between p50 and Bru for binding to the same site. However, immunoprecipitation experiments showed that the two proteins are capable of binding to the same repressor element simultaneously. Transgenes carrying a mutation that specifically reduces p50 binding show precocious *osk* translation at stage 7, suggesting that p50 is also required for translational repression of *osk*.

Bic-C is a KH domain protein capable of RNA binding in vitro (189). Embryos from females heterozygous for *Bic-C* alleles demonstrate patterning defects including a bicaudal phenotype (132, 147), which would be consistent with ectopic Osk activity. In homozygous *Bic-C* mutants Osk does not localize to the posterior and *osk* is precociously translated in stages 7–8, suggesting that *Bic-C* may play a role in translational regulation of *osk* (189).

TRANS-ACTING FACTORS REQUIRED FOR LOCALIZATION AND TRANSLATIONAL ACTIVATION OF OSK   Stau is required to localize *osk* to the posterior pole of the oocyte (53, 97). Stau and *osk* colocalize within the oocyte (216) and show coincident mislocalization in mutants that disrupt oocyte microtubule polarity, suggesting that they are transported together along microtubules (76, 183). Stau is required for translational activation of *osk*, and even for *oskBRE*⁻, indicating that Stau does not operate simply as a derepressor (96). However, the expression of Short Osk from a transgene can induce endogenous *osk* translation in *stau* mutants, indicating that the requirement of Stau for *osk* translation can be bypassed (140), as is also evident when *osk* gene dosage is increased (210). Both localization and translational regulation of *osk* depend on the region of Stau containing the dsRBDs (145). The dsRBD2 insertion is required for *osk* localization, whereas dsRBD5 is required for its translational activation. Since Stau lacking dsRBD5 can activate translation of *oskBRE⁻*, but not wild-type *osk*, dsRBD5 may have a derepressor function. Components of the pole plasm are also required for translational activation of *osk*. *vas* and *tud* mutants show a reduction in levels of Short Osk. Osk also regulates its own expression (141, 181). Short Osk is present in phosphorylated and unphosphorylated forms, and its phosphorylation depends on *vas* but not *tud* (140, 141). Several *osk* alleles that disrupt Osk interaction with Vas (16) also prevent Osk phosphorylation.

Orb directly binds to *osk* mRNA, and *orb* mutants show defects in localization of *osk* mRNA to the posterior of the oocyte and a failure of *osk* translation (23, 27, 141). This could indicate a direct role for *orb* in localization or anchoring of *osk* mRNA. Alternatively, since Osk is required to maintain its own transcript at the posterior, the *orb* mutant phenotype could be due to a defect in translational activation leading to a defect in anchoring. This is supported by the observation that *osk* poly(A) tails are reduced in length in *orb* mutants (23; and see below).

Aubergine (Aub) is required for accumulation of both Osk isoforms (266). In *aub* mutants, *osk* localizes normally but fails to remain anchored at the posterior, producing low levels of protein in a broad posterior region. Ectopically expressed *osk* at the anterior, achieved through substitution of the 3′ UTR, can be robustly translated in *aub* mutants, suggesting that Aub does not control Osk stability, and that Aub is only required for translational activation of *osk* in the presence of the endogenous *osk* 3′ UTR. Aub is still required for translation of *oskBRE⁻*, suggesting a requirement in translational activation rather than derepression.

Two additional proteins, p68 and p50, both cross-link to a region of 130 nt, between m1 and m2 of *osk* (79). Large deletions that abolish derepression also reduce or eliminate the binding site of these proteins. Introduction of a transgene carrying an inversion that abrogates binding of p68 and p50 results in translational repression even when the transcript is localized. However, the presence of this derepressor element alone cannot translationally activate a transcript which contains the BREs fused to the *bcd* 3′ UTR. Similar constructs lacking BREs are translated (54, 181), suggesting that factors in the posterior pole are required in addition to the derepressor element to overcome BRE-mediated repression. The binding of p50 to both the 5′ and 3′ ends of *osk* mRNA may provide a means for circularizing the transcript and may allow *trans*-acting factors bound to the 3′ UTR to influence translation initiation at the 5′ end (79).

POTENTIAL MECHANISMS FOR TRANSLATIONAL CONTROL OF *OSK*   Cell-free translation systems using *Drosophila* ovarian extracts have recently been used to study *osk* translational regulation (20, 121). Such experiments demonstrated that reporter transcripts containing the *osk* 3′ UTR were regulated in the same way whether or not they contained a poly(A) tail, indicating that in vitro, regulation is independent of cytoplasmic polyadenylation or deadenylation. These results are contrary to in vivo studies in *orb* mutants (23) but are supported by temporal analysis of *osk* poly(A) length in vivo, which does not appear to change at the onset of translational activation (253). However, poly(A) length changes may be difficult to monitor *in vivo* because of the relatively small portion of the *osk* transcript that becomes translationally active (8). Extract systems have also been used to test whether Bru repression depends on the 5′ cap structure. The addition of free cap analogue to ovarian extracts was not found to disrupt BRE-mediated *osk* regulation of reporter transcripts, indicating that Bru inhibition also does not act through the cap-binding step of translation initiation (121).

gurken

Grk, a member of the TGF-α family of proteins, plays an integral role in two signaling processes responsible for establishing both the anterior-posterior and dorsal-ventral axes of the egg and embryo (reviewed in 155). Grk first accumulates at the posterior of the oocyte and signals to the neighboring follicle cells, specifying their fate as posterior (76, 183). This leads to the reorganization of the microtubule network within the oocyte, and the migration of the nucleus, and *grk*, to the future anterodorsal region of the oocyte. Here Grk signaling specifies adjacent follicle cells as dorsal (151, 174). As for *bcd*, *nos*, and *osk*, spatio-temporal restriction of Grk activity depends on a combination of mRNA localization and translational regulation.

grk mRNA concentrates at the posterior of the oocyte during early oogenesis and then exhibits a diffuse distribution during oocyte nuclear migration (151). Some *grk* mRNA may migrate with the nucleus, and this may occur in association with the endoplasmic reticulum (ER), since Grk is a secreted protein, and double-labeling experiments indicate that *grk* and the ER colocalize within the oocyte (191). Following this migration, *grk* exhibits an anterior ring of expression, which by stage 9 becomes restricted to the anterodorsal corner where it is maintained until at least stage 10B (151). The distribution of Grk protein is similar to that of *grk* mRNA, accumulating in the oocyte in early stages and later confined to the anterodorsal corner (153, 183).

CIS-ACTING ELEMENTS REQUIRED FOR REGULATION OF GRK   Sequences within the 5′ and 3′ UTRs, as well as within the coding region, have been implicated in directing *grk* localization. Early accumulation of the transcript within the oocyte requires elements within the *grk* promoter or 5′ UTR (237). Deletion analysis of a reporter construct bearing the *grk* 5′ UTR suggested that nt 1–35 may be important for stable *grk* localization during early to mid-oogenesis, so this region was named the *grk* localization element 1 (GLE1) (191). An element in the 5′ end of the coding region, called the anterior cortical ring (ACR) element, is required for transcript stability and/or localization during mid-oogenesis in order to achieve the anterior ring of expression (237). Sequences within the 3′ UTR are required for the final stage of *grk* localization, whereby the transcript becomes focused in the anterodorsal corner. The 3′ UTR has also been implicated in translational repression of *grk*, because a fusion construct lacking the 3′ UTR is translated across the anterior cortex at a time when a similar construct bearing the 3′ UTR shows the same mRNA distribution but is only translated in the anterodorsal corner (191). An element consisting of nt 96–155 of the *grk* 5′ UTR, called GLE2, has also been implicated in *grk* localization and translational activation; however, this element is also required for stability of *grk* mRNA after stage 8, so its other effects may be indirect.

TRANS-ACTING FACTORS REQUIRED FOR LOCALIZATION OF GRK   Several factors such as *capu*, *spir*, and *maelstrom* (*mael*) are required indirectly for *grk* mRNA localization through their role in cytoskeletal assembly (discussed above) (31, 52, 137, 151). Also, in *orb* mutants *grk* mRNA is distributed broadly across the anterior of the oocyte, rather than focused in the anterodorsal corner (27, 184). In these ovaries, Grk levels are reduced and the protein is mislocalized around the oocyte cortex (153), resulting in dorsal-ventral patterning defects (27, 184). It is not known whether Orb functions directly in *grk* localization or is involved in translational activation, as is predicted by its homology to CPEB.

Female sterile mutations in *sqd* and *K10* prevent *grk* mRNA from localizing to the anterodorsal corner. Instead, the anterior ring of *grk* mRNA persists and is actively translated, leading to a dorsalization phenotype (93, 151, 263). Sqd, or Hrp40 (93, 143), is a member of the hnRNP family of proteins (see above), some of which shuttle between the nucleus and the cytoplasm (86, 95, 113, 167, 246). Three isoforms of Sqd, termed A, B, and S, are generated through alternative splicing, and these isoforms differ in their intracellular localization and function (156). SqdA and SqdS each have distinct roles in *grk* regulation and together can fulfill this requirement for *sqd*. SqdS, which is nuclear, appears to function in *grk* mRNA localization, while SqdA, which is thought to be cytoplasmic, is implicated in *grk* translational regulation. Sqd can be cross-linked to the *grk* 3′ UTR (106, 156), and *sqd* mutants exhibit the same pattern of *grk* mRNA mislocalization as *grk* transcripts lacking the 3′ UTR (237), implying that the binding of Sqd to the *grk* 3′ UTR is required to achieve the final localization pattern of the transcript. K10 interacts directly in vitro with all three isoforms of Sqd (156). In wild-type, K10 is localized to the oocyte nucleus (170, 200) as is one of the Sqd isoforms; however, in *K10* mutants, Sqd is no longer present in the oocyte nucleus (156). Given the similar mutant phenotypes, these results suggest that K10 may function upstream of Sqd in *grk* regulation.

TRANSLATIONAL REPRESSION OF GRK   Grk is localized to the anterodorsal corner earlier than *grk* mRNA (191), implying that translational repression prevents protein synthesis from the unlocalized transcript. In support of this idea, *grk* is translated across the anterior margin in stage 8, when overexpressed, suggesting that high levels of the mRNA can titrate out a translational repressor (152). K10 has been suggested to function in translational repression of *grk* because in *K10* mutants, translation occurs from an anterior ring of *grk* mRNA, at a time when the transcript shows an anterior ring distribution in wild-type but is translationally silent (191). Since the *grk* 3′ UTR is required for its translational repression, a repressor protein is predicted to bind to this region of the transcript. Bru is a possible candidate since the *grk* 3′ UTR contains consensus BREs, and Bru can be cross-linked to *grk* mRNA (96). In addition, Bru shows a colocalization with *grk* mRNA in the anterodorsal region during stage 10, the only divergence it shows from colocalization with *osk* in the oocyte (253). In vitro, Bru interacts directly with Sqd, suggesting that Sqd may shuttle *grk* out of the nucleus and link it to Bru to ensure correct regulation (156). However, a requirement for Bru in *grk* translational regulation has not yet been reported.

TRANS-ACTING FACTORS REQUIRED FOR TRANSLATIONAL ACTIVATION OF *GRK*   Alleles of the *spindle* (*spn*) *genes*, *spnA*, *spnB*, *spnC*, *spnD*, and *spnE* (*homeless*) (232), demonstrate mislocalization of *grk* mRNA such that the anterior ring of expression persists until around stage 10 (71, 73, 77). This mislocalization produces ventralization, a phenotype opposite to that produced by a similar *grk* mRNA mislocalization in *K10* mutants, due to the fact that Grk protein levels are severely reduced at mid-oogenesis in *spn* mutants (71, 77), whereas in *K10* mutants, *grk* is actively translated from mRNA distributed across the anterior cortex (153, 201). Mutations in *okra* (*okr*) (197) show similar phenotypes to *spn* mutants but demonstrate a reduction in Grk levels in the early and middle stages of oogenesis (71). Double mutants for *K10* and alleles of *okr*, *spnB*, and *spnD* show a range of phenotypes, precluding a straightforward epistatic designation. In *okr*, *spnB*, and *spnD* mutants, levels of K10 protein in the oocyte nucleus are reduced, which may explain their *grk* mRNA mislocalization phenotype (71).

Mutations in the *spn* genes and *okr* also exhibit defects in karyosome formation (71, 77). Cloning of *okr* and *spnB* revealed homology to factors required in yeast for DNA double-strand break (DSB) repair (71). During mitosis, repair of DSBs is necessary to protect cells from damage to their DNA, whereas in meiosis, repair of DSBs is required for the processes of crossing over and recombination (166, 177). *okr* appears to be required for DSB repair in both mitosis and meiosis, whereas the function of the *spn* genes is restricted to meiosis (71).

Vas has multiple functions during oogenesis, including a role in karyosome formation resembling that of the *spindle* genes. The small fraction of oocytes that reach late stages of oogenesis in the absence of Vas activity display duplicated anterior eggshell structures at the posterior and a ventralization phenotype (225, 238, 239). In a *vas* null mutant there is a mild *grk* mRNA mislocalization phenotype but an extreme reduction in levels of Grk protein, and in a variety of *vas* allelic combinations, levels of Grk protein can be correlated with the degree of defects in axis patterning of the allele.

Vas may provide the link between the role of the *spn* genes in DSB repair and their effect on *grk* translation. In yeast, prevention of DSB repair during meiosis activates a checkpoint that leads to meiotic arrest, dependent on the gene MEC1 (125). The *Drosophila* homologue of this gene, *mei-41* (199, 203), when doubly mutant with alleles of *okr*, *spnB*, and *spnD*, can restore correct dorsoventral patterning, suggesting that these genes act through a meiotic checkpoint. *mei-41* mutations cannot suppress mutations in *vas*, potentially placing *vas* downstream of this checkpoint. In keeping with this idea, in *spnB* mutants, Vas mobility is shifted on SDS-PAGE gels, suggesting that a posttranslational modification occurs in the absence of *spnB* function. This modification is eliminated in double mutants for *spnB* and *mei-41*. These observations have led to a model whereby the detection of DSBs activates a meiotic checkpoint resulting in posttranslational modification of Vas, which reduces its ability to activate *grk* translation (72).

Mutations in *aub* and *encore* (*enc*) also cause severe reductions in Grk protein and ventralization phenotypes (82, 197, 266). In addition, both are required for karyosome formation (71, 242). Little is known about the function of *aub* in *grk* regulation. Enc colocalizes with *grk* mRNA at the posterior of the oocyte and in the anterodorsal region (242). In *enc* mutants, there is a mild *grk* mRNA mislocalization phenotype but a strong reduction in Grk levels (82). *enc* is unlikely to be involved in the meiotic checkpoint pathway, as *enc* mutations cannot be suppressed by *mei-41* mutations and do not induce modification of Vas (242).

LINKING *GRK* EXPORT, LOCALIZATION, AND REGULATION   *grk* represents one of the few transcripts whose localization has been proposed to involve vectorial export from the nucleus. Vectorial export, as opposed to transport in the cytoplasm, is envisaged as localization of a transcript in the cytoplasm by controlling the direction of its nuclear export (reviewed in 123). This model was primarily based on observations of apical localization of pair-rule transcripts in the embryonic blastoderm (63), but doubts about the validity of this general mechanism for mRNA localization have recently arisen due to experiments with the pair-rule transcript *ftz* (106). When the *ftz* transcript alone was injected into blastoderm embryos, it was not apically localized, demonstrating that the localization mechanism cannot be entirely cytoplasmic. However, preincubation of the transcript with nuclear extracts from *Drosophila* or human cells prior to injection allowed it to be apically localized in a microtubule-dependent manner. Specifically, preincubation with the *Drosophila* Sqd protein or the human proteins hnRNP A1, A2, and B was sufficient to mediate apical localization, dependent on the presence of the *ftz* 3′ UTR.

UV cross-linking demonstrated that Sqd can bind to the 3′ UTRs of both *ftz* and *grk* (156, 106). Since *grk* also requires Sqd for its localization, its localization may be cytoplasmically controlled, as for the injected *ftz* transcript. However, this has not been proven for *grk*, so it remains possible that this transcript exhibits vectorial nuclear export. The direct interaction between Sqd and Bru potentially links nuclear export, cytoplasmic localization, and translational repression of *grk* (156). Also important is the fact that Bru, Vas, Orb, and Aub are all implicated in the translational regulation of both *osk* and *grk* transcripts, indicating potential overlap in the mechanism of translational regulation of these two mRNAs. However, despite these hypotheses, both the significance of translational regulation for localizing Grk activity and the mechanism by which *grk* may be translationally regulated remain unknown.

FUTURE DIRECTIONS

Rapid progress has been made in recent years both in understanding the link between RNA localization and translational control, and in the identification of *cis*- and *trans*-acting factors required to mediate these processes for specific transcripts. Thus this field has advanced to a stage where questions about the mechanisms by which these regulatory factors influence translation can be addressed. Much remains unclear in the general models of translational regulation. Work on the *Drosophila* transcripts described here as well as on many others will contribute to refining these models.

Cytoplasmic polyadenylation is a well-established method of translational regulation during development, but the mechanism by which it influences translation initiation is not understood. In *Drosophila* it is still an open question whether cytoplasmic polyadenylation is restricted to embryogenesis or whether it is also a means of achieving translational activation during oogenesis. Also unclear is whether the differences in *cis*-acting sequences involved in cytoplasmic polyadenylation between *Drosophila* and vertebrate model systems implies differences in the underlying mechanism.

Recent work on *nos* and *osk* has demonstrated that their translational regulation does not involve cap-dependent repression (30, 121). The association of untranslated *nos* with polysomes might imply a new mechanism for translational repression (30). It is not yet known at which step *osk* or *grk* translation is regulated, but this work will be aided by the newly established in vitro translation systems from *Drosophila* ovaries and embryos (20, 70, 121). It will be of interest to discover whether the 60S ribosomal subunit joining step of translation initiation, implicated in the translation block of the mammalian LOX transcript (158), will also be important for translational control of developmentally regulated transcripts in *Drosophila*. The interaction between *Drosophila* eIF5B (dIF2) (19), one of the translation initiation factors required for this step (165), and Vas, implicated in translational activation of several germline transcripts (69, 141, 181, 225, 238, 239), could indicate that these factors act to regulate a translational block at this step of initiation.

Several of the proteins described in this review are posttranslationally modified by phosphorylation. The relevance of phosphorylation to their activity is not well understood. The kinases involved may have many targets, making it more difficult to identify their specific roles in translational control and axis patterning. PAR-1, recently characterized due to its role in *osk* mRNA localization (202, 240), is predicted to be a kinase, raising the possibility that it may phosphorylate components of the pole plasm such as Osk. It is not yet known what function of Osk is regulated by phosphorylation. Posttranslational modification of Vas, presumably by phosphorylation, may be signaled by nuclear events, and appears to be an intermediate step in *grk* translational regulation (72). It will be interesting to determine both the mechanism controlling this sequence of events and whether other functions of Vas depend on its phosphorylation. In addition, phosphorylation is involved in regulation of masking and cytoplasmic polyadenylation in several systems, but this remains to be investigated in *Drosophila*.

Many of the potential *trans*-acting factors discussed in this review cross-link to target mRNAs, defining them functionally as RNA-binding proteins. In several instances such as Smg (36, 208) and Apt (120), these proteins lack any previously characterized RNA-binding domains. Thus investigation of proteins such as these is likely to lead to new designations of motifs involved in RNA binding. A good example of this is Pum, which binds to NRE sequences in several *Drosophila* targets, and in combination with Nos mediates translational repression (reviewed in 163). The region of Pum required for RNA binding (256, 270) has been named a Puf domain and has been identified in other proteins in many different species (270, 272). This domain shows no similarity to other domains involved in RNA binding, and thus represents a new RNA-binding motif (51, 269). The polar granule component, Tud (75), contains at least 10 copies of a motif designated the tudor domain, which has subsequently been identified in several nucleic-acid binding proteins (reviewed in 169). As yet there is no direct evidence that this domain is involved in nucleic-acid binding.

Another area that warrants further investigation is how mRNP particles are loaded onto the cytoskeletal transport machinery. For both *bcd* and *osk*, Stau is thought to play a role in connecting them to the microtubule cytoskeleton. However, in the *Drosophila* nervous system, Stau appears to mediate mRNA transport via the actin cytoskeleton (18), and it is not known how this factor interacts with either cytoskeletal transport system. Mammalian Stau homologues are implicated in microtubule-dependent transport in neurons (99) and bind to tubulin via a domain that is not present in the *Drosophila* protein (261). Another protein whose requirement for mRNA localization and oocyte determination is thought to involve links with the microtubule cytoskeleton is Bic-D (172, 226, 223). *Bic-D* genetically interacts with the *Drosophila* homologue of *Lissencephaly-1* (*DLis-1*), required for localization of dynein heavy chain (Dhc) in the oocyte (227). *Bic-D*, *DLis-1*, and *Dhc* are thought to be involved in establishing a microtubule transport system early in oogenesis, necessary for growth of the oocyte. Within the oocyte, Swa appears to mediate association of *bcd* with the light chain subunit of dynein (Ddlc-1) (194). Although kinesin I is required for *osk* localization within the oocyte (17), it is not known how *osk* is linked to this motor protein. As yet no *Drosophila* transcripts have been purified in association with complexes containing molecular motors.

A great deal of attention has been given to regulatory mechanisms that involve transcriptional control of specific genes. Outside of a few specific systems, comparatively little work has been done on translational control of specific mRNAs, yet this mechanism is also fundamental to essential developmental processes. Further research seems certain to identify many other mRNAs with highly regulated translation, and other developmental processes that rely on translational control. Work in model organisms such as *Drosophila* will most likely be instrumental in extending our understanding of this field.

ACKNOWLEDGMENTS

We are grateful to Laura Nilson, Guylaine Roy, and Sylvia Styhler for critical reading of the manuscript, to Akira Nakamura and Satoru Kobayashi for sharing results prior to publication, and to Jean-Paul Acco for figure preparation. O. J. is supported by a Graduate Studentship from CIHR. P. L. is a CIHR Investigator.

Visit the Annual Reviews home page at www.AnnualReviews.org

LITERATURE CITED

1. Asaoka-Taguchi M, Yamada M, Nakamura A, Hanyu K, Kobayashi S. 1999. Maternal Pumilio acts together with Nanos in germline development in *Drosophila* embryos. *Nat. Cell Biol.* 1:431–37

2. Audic Y, Omilli F, Osborne HB. 1997. Postfertilization deadenylation of mRNAs in *Xenopus laevis* embryos is sufficient to cause their degradation at the blastula stage. *Mol. Cell. Biol.* 17:209–18

3. Ayscough KR. 1998. In vivo functions of actin-binding proteins. *Curr. Opin. Cell Biol.* 10:102–11

4. Bachvarova RF. 1992. A maternal tail of poly(A): the long and the short of it. *Cell* 69:895–97

5. Barkoff AF, Dickson KS, Gray NK, Wickens M. 2000. Translational control of *cyclin B1* mRNA during meiotic maturation: coordinated repression and cytoplasmic polyadenylation. *Dev. Biol.* 220:97–109

6. Bashirullah A, Cooperstock RL, Lipshitz HD. 1998. RNA localization in development. *Annu. Rev. Biochem.* 67:335–94

7. Bashirullah A, Halsell SR, Cooperstock RL, Kloc M, Karaiskakis A, et al. 1999. Joint action of two RNA degradation pathways controls the timing of maternal transcript elimination at the midblastula transition in *Drosophila melanogaster*. *EMBO J.* 18:2610–20

8. Bergsten SE, Gavis ER. 1999. Role for mRNA localization in translational activation but not spatial restriction of *nanos* RNA. *Development* 126:659–69

9. Bergsten SE, Huang T, Chatterjee S, Gavis ER. 2001. Recognition and long-range interactions of a minimal *nanos* RNA localization signal element. *Development* 128:427–35

10. Berleth T, Burri M, Thoma G, Bopp D, Richstein S, et al. 1988. The role of localization of *bicoid* RNA in organizing the anterior pattern of the *Drosophila* embryo. *EMBO J.* 7:1749–56

11. Bilger A, Fox CA, Wahle E, Wickens M. 1994. Nuclear polyadenylation factors recognize cytoplasmic polyadenylation elements. *Genes Dev.* 8:1106–16

12. Bishop AL, Hall A. 2000. Rho GTPases and their effector proteins. *Biochem. J.* 348:241–55

13. Bohrmann J. 1997. *Drosophila* unconventional myosin VI is involved in intra- and intercellular transport during oogenesis. *Cell. Mol. Life Sci.* 53:652–62

14. Bohrmann J, Biber K. 1994. Cytoskeleton-dependent transport of cytoplasmic particles in previtellogenic to mid-vitellogenic ovarian follicles of *Drosophila*: time-lapse analysis using video-enhanced contrast microscopy. *J. Cell Sci.* 107:849–58

15. Bouvet P, Omilli F, Arlot-Bonnemains Y, Legagneux V, Roghi C, et al. 1994. The deadenylation conferred by the 3′ untranslated region of a developmentally controlled mRNA in *Xenopus* embryos is switched to polyadenylation by deletion of a short sequence element. *Mol. Cell. Biol.* 14:1893–900

16. Breitwieser W, Markussen FH, Horstmann H, Ephrussi A. 1996. Oskar protein interaction with Vasa represents an essential step in polar granule assembly. *Genes Dev.* 10:2179–88

17. Brendza RP, Serbus LR, Duffy JB, Saxton WM. 2000. A function for kinesin I in the posterior transport of *oskar* mRNA and Staufen protein. *Science* 289:2120–22

18. Broadus J, Fuerstenberg S, Doe CQ. 1998. Staufen-dependent localization of *prospero* mRNA contributes to neuroblast daughter-cell fate. *Nature* 391:792–95

19. Carrera P, Johnstone O, Nakamura A, Casanova J, Jäckle H, Lasko P. 2000. VASA mediates translation through interaction with a *Drosophila* yIF2 homolog. *Mol. Cell* 5:181–87

20. Castagnetti S, Hentze MW, Ephrussi A, Gebauer F. 2000. Control of *oskar* mRNA translation by Bruno in a novel cell-free system from *Drosophila* ovaries. *Development* 127:1063–68

21. Castrillon DH, Gonczy P, Alexander S, Rawson R, Eberhart CG, et al. 1993. Toward a molecular genetic analysis of spermatogenesis in *Drosophila melanogaster*: characterization of male-sterile mutants generated by single P element mutagenesis. *Genetics* 135:489–505

22. Chakravarti D, Maiti T, Maitra U. 1993. Isolation and immunochemical characterization of eukaryotic translation initiation factor 5 from *Saccharomyces cerevisiae*. *J. Biol. Chem.* 268:5754–62

23. Chang JS, Tan L, Schedl P. 1999. The *Drosophila* CPEB homolog, Orb, is required for Oskar protein expression in oocytes. *Dev. Biol.* 215:91–106

24. Chao YC, Donahue KM, Pokrywka NJ, Stephenson EC. 1991. Sequence of *swallow*, a gene required for the localization of *bicoid* message in *Drosophila* eggs. *Dev. Genet.* 12:333–41

25. Chen B, Harms E, Chu T, Henrion G, Strickland S. 2000. Completion of meiosis in *Drosophila* oocytes requires transcriptional control by Grauzone, a new zinc finger protein. *Development* 127:1243–51

26. Christensen AK, Kahn LE, Bourne CM. 1987. Circular polysomes predominate on the rough endoplasmic reticulum of somatotropes and mammotropes in the rat anterior pituitary. *Am. J. Anat.* 178:1–10

27. Christerson LB, McKearin DM. 1994. *orb* is required for anteroposterior and dorsoventral patterning during *Drosophila* oogenesis. *Genes Dev.* 8:614–28

28. Clark I, Giniger E, Ruohola-Baker H, Jan LY, Jan YN. 1994. Transient posterior localization of a kinesin fusion protein reflects anteroposterior polarity of the *Drosophila* oocyte. *Curr. Biol.* 4:289–300

29. Clark IE, Jan LY, Jan YN. 1997. Reciprocal localization of Nod and kinesin fusion proteins indicates microtubule polarity in the *Drosophila* oocyte, epithelium, neuron and muscle. *Development* 124:461–70

30. Clark IE, Wyckoff D, Gavis ER. 2000. Synthesis of the posterior determinant Nanos is spatially restricted by a novel cotranslational regulatory mechanism. *Curr. Biol.* 10:1311–14

31. Clegg NJ, Frost DM, Larkin MK, Subrahmanyan L, Bryant Z, Ruohola-Baker H. 1997. *maelstrom* is required for an early step in the establishment of *Drosophila* oocyte polarity: posterior localization of *grk* mRNA. *Development* 124:4661–71

32. Cooley L, Verheyen E, Ayers K. 1992. *chickadee* encodes a profilin required for intercellular cytoplasm transport during *Drosophila* oogenesis. *Cell* 69:173–84

33. Cote CA, Gautreau D, Denegre JM,

Kress TL, Terry NA, Mowry KL. 1999. A *Xenopus* protein related to hnRNP I has a role in cytoplasmic RNA localization. *Mol. Cell* 4:431–37

34. Cox DN, Lu B, Sun TQ, Williams LT, Jan YN. 2001. *Drosophila par-1* is required for oocyte differentiation and microtubule organization. *Curr. Biol.* 11:75–87

35. Crucs S, Chatterjee S, Gavis ER. 2000. Overlapping but distinct RNA elements control repression and activation of *nanos* translation. *Mol. Cell* 5:457–67

36. Dahanukar A, Walker JA, Wharton RP. 1999. Smaug, a novel RNA-binding protein that operates a translational switch in *Drosophila. Mol. Cell* 4:209–18

37. Dahanukar A, Wharton RP. 1996. The Nanos gradient in *Drosophila* embryos is generated by translational regulation. *Genes Dev.* 10:2610–20

38. Davydova EK, Evdokimova VM, Ovchinnikov LP, Hershey JW. 1997. Overexpression in COS cells of p50, the major core protein associated with mRNA, results in translation inhibition. *Nucleic Acids Res.* 25:2911–16

39. de Moor CH, Richter JD. 1999. Cytoplasmic polyadenylation elements mediate masking and unmasking of *cyclin B1* mRNA. *EMBO J.* 18:2294–303

40. de Moor CH, Richter JD. 2001. Translational control in vertebrate development. *Int. Rev. Cytol.* 203:567–608

41. de Valoir T, Tucker MA, Belikoff EJ, Camp LA, Bolduc C, Beckingham K. 1991. A second maternally expressed *Drosophila* gene encodes a putative RNA helicase of the "DEAD box" family. *Proc. Natl. Acad. Sci. USA* 88:2113–17

42. Deshler JO, Highett MI, Abramson T, Schnapp BJ. 1998. A highly conserved RNA-binding protein for cytoplasmic mRNA localization in vertebrates. *Curr. Biol.* 8:489–96

43. Dickson KS, Bilger A, Ballantyne S, Wickens MP. 1999. The cleavage and polyadenylation specificity factor in *Xenopus laevis* oocytes is a cytoplasmic factor involved in regulated polyadenylation. *Mol. Cell. Biol.* 19:5707–17

44. Doel MT, Carey NH. 1976. The translational capacity of deadenylated ovalbumin messenger RNA. *Cell* 8:51–58

45. Drewes G, Ebneth A, Preuss U, Mandelkow EM, Mandelkow E. 1997. MARK, a novel family of protein kinases that phosphorylate microtubule-associated proteins and trigger microtubule disruption. *Cell* 89:297–308

46. Driever W, Ma J, Nüsslein-Volhard C, Ptashne M. 1989. Rescue of *bicoid* mutant *Drosophila* embryos by Bicoid fusion proteins containing heterologous activating sequences. *Nature* 342:149–54

47. Driever W, Nüsslein-Volhard C. 1988. The Bicoid protein determines position in the *Drosophila* embryo in a concentration-dependent manner. *Cell* 54:95–104

48. Driever W, Siegel V, Nüsslein-Volhard C. 1990. Autonomous determination of anterior structures in the early *Drosophila* embryo by the Bicoid morphogen. *Development* 109:811–20

49. Dubochet J, Morel C, Lebleu B, Herzberg M. 1973. Structure of globin mRNA and mRNA-protein particles. Use of dark-field electron microscopy. *Eur. J. Biochem.* 36:465–72

50. Eddy EM. 1975. Germ plasm and the differentiation of the germ cell line. *Int. Rev. Cytol.* 43:229–80

51. Edwards TA, Trincao J, Escalante CR, Wharton RP, Aggarwal AK. 2000. Crystallization and characterization of Pumilio: a novel RNA binding protein. *J. Struct. Biol.* 132:251–54

52. Emmons S, Phan H, Calley J, Chen W, James B, Manseau L. 1995. *cappuccino*, a *Drosophila* maternal effect gene required for polarity of the egg and embryo, is related to the vertebrate *limb deformity* locus. *Genes Dev.* 9:2482–94

53. Ephrussi A, Dickinson LK, Lehmann R. 1991. *oskar* organizes the germ plasm and directs localization of the posterior determinant *nanos. Cell* 66:37–50

54. Ephrussi A, Lehmann R. 1992. Induction of germ cell formation by *oskar*. *Nature* 358:387–92
55. Erdelyi M, Michon AM, Guichet A, Glotzer JB, Ephrussi A. 1995. Requirement for *Drosophila* cytoplasmic tropomyosin in *oskar* mRNA localization. *Nature* 377:524–27
56. Evdokimova VM, Kovrigina EA, Nashchekin DV, Davydova EK, Hershey JW, Ovchinnikov LP. 1998. The major core protein of messenger ribonucleoprotein particles (p50) promotes initiation of protein biosynthesis in vitro. *J. Biol. Chem.* 273:3574–81
57. Evdokimova VM, Wei CL, Sitikov AS, Simonenko PN, Lazarev OA, et al. 1995. The major protein of messenger ribonucleoprotein particles in somatic cells is a member of the Y-box binding transcription factor family. *J. Biol. Chem.* 270:3186–92
58. Ferrandon D, Elphick L, Nüsslein-Volhard C, St Johnston D. 1994. Staufen protein associates with the 3′ UTR of *bicoid* mRNA to form particles that move in a microtubule-dependent manner. *Cell* 79:1221–32
59. Ferrandon D, Koch I, Westhof E, Nüsslein-Volhard C. 1997. RNA-RNA interaction is required for the formation of specific *bicoid* mRNA 3′ UTR-STAUFEN ribonucleoprotein particles. *EMBO J.* 16:1751–58
60. Fischer JA. 2000. Molecular motors and developmental asymmetry. *Curr. Opin. Genet. Dev.* 10:489–96
61. Fox CA, Sheets MD, Wickens MP. 1989. Poly(A) addition during maturation of frog oocytes: distinct nuclear and cytoplasmic activities and regulation by the sequence UUUUUAU. *Genes Dev.* 3:2151–62
62. Fox CA, Wickens M. 1990. Poly(A) removal during oocyte maturation: a default reaction selectively prevented by specific sequences in the 3′ UTR of certain maternal mRNAs. *Genes Dev.* 4:2287–98
63. Francis-Lang H, Davis I, Ish-Horowicz D. 1996. Asymmetric localization of *Drosophila* pair-rule transcripts from displaced nuclei: evidence for directional nuclear export. *EMBO J.* 15:640–49
64. Frank DJ, Roth MB. 1998. *ncl-1* is required for the regulation of cell size and ribosomal RNA synthesis in *Caenorhabditis elegans*. *J. Cell Biol.* 140:1321–29
65. Fridell RA, Harding LS, Bogerd HP, Cullen BR. 1995. Identification of a novel human zinc finger protein that specifically interacts with the activation domain of lentiviral Tat proteins. *Virology* 209:347–57
66. Gavis ER, Curtis D, Lehmann R. 1996. Identification of *cis*-acting sequences that control *nanos* RNA localization. *Dev. Biol.* 176:36–50
67. Gavis ER, Lehmann R. 1992. Localization of *nanos* RNA controls embryonic polarity. *Cell* 71:301–13
68. Gavis ER, Lehmann R. 1994. Translational regulation of *nanos* by RNA localization. *Nature* 369:315–18
69. Gavis ER, Lunsford L, Bergsten SE, Lehmann R. 1996. A conserved 90 nucleotide element mediates translational repression of *nanos* RNA. *Development* 122:2791–800
70. Gebauer F, Corona DFV, Preiss T, Becker PB, Hentze MW. 1999. Translational control of dosage compensation in *Drosophila* by Sex-lethal: cooperative silencing via the 5′ and 3′ UTRs of *msl-2* mRNA is independent of the poly(A) tail. *EMBO J.* 18:6146–54
71. Ghabrial A, Ray RP, Schüpbach T. 1998. *okra* and *spindle-B* encode components of the RAD52 DNA repair pathway and affect meiosis and patterning in *Drosophila* oogenesis. *Genes Dev.* 12:2711–23
72. Ghabrial A, Schüpbach T. 1999. Activation of a meiotic checkpoint regulates translation of *gurken* during *Drosophila* oogenesis. *Nat. Cell Biol.* 1:354–57
73. Gillespie DE, Berg CA. 1995. *homeless* is required for RNA localization in

Drosophila oogenesis and encodes a new member of the DE-H family of RNA-dependent ATPases. *Genes Dev.* 9:2495–508

74. Gingras AC, Raught B, Sonenberg N. 1999. eIF4 initiation factors: effectors of mRNA recruitment to ribosomes and regulators of translation. *Annu. Rev. Biochem.* 68:913–63

75. Golumbeski GS, Bardsley A, Tax F, Boswell RE. 1991. *tudor*, a posterior-group gene of *Drosophila melanogaster*, encodes a novel protein and an mRNA localized during mid-oogenesis. *Genes Dev.* 5:2060–70

76. González-Reyes A, Elliott H, St Johnston D. 1995. Polarization of both major body axes in *Drosophila* by *gurken-torpedo* signalling. *Nature* 375:654–58

77. González-Reyes A, Elliott H, St Johnston D. 1997. Oocyte determination and the origin of polarity in *Drosophila*: the role of the *spindle* genes. *Development* 124:4927–37

78. González-Reyes A, St Johnston D. 1994. Role of oocyte position in establishment of anterior-posterior polarity in *Drosophila*. *Science* 266:639–42

79. Gunkel N, Yano T, Markussen FH, Olsen LC, Ephrussi A. 1998. Localization-dependent translation requires a functional interaction between the 5′ and 3′ ends of *oskar* mRNA. *Genes Dev.* 12:1652–64

80. Hake LE, Richter JD. 1994. CPEB is a specificity factor that mediates cytoplasmic polyadenylation during *Xenopus* oocyte maturation. *Cell* 79:617–27

81. Havin L, Git A, Elisha Z, Oberman F, Yaniv K, et al. 1998. RNA-binding protein conserved in both microtubule- and microfilament-based RNA localization. *Genes Dev.* 12:1593–98

82. Hawkins NC, Van Buskirk C, Grossniklaus U, Schüpbach T. 1997. Post-transcriptional regulation of *gurken* by *encore* is required for axis determination

in *Drosophila*. *Development* 124:4801–10

83. Heasman J, Quarmby J, Wylie CC. 1984. The mitochondrial cloud of *Xenopus* oocytes: the source of germinal granule material. *Dev. Biol.* 105:458–69

84. Hegde J, Stephenson EC. 1993. Distribution of Swallow protein in egg chambers and embryos of *Drosophila melanogaster*. *Development* 119:457–70

85. Hershey JWB, Mathews MB, Sonenberg N, eds. 1996. *Translational Control*. Cold Spring Harbor, NY: Cold Spring Harbor Lab. Press

86. Hoek KS, Kidd GJ, Carson JH, Smith R. 1998. hnRNP A2 selectively binds the cytoplasmic transport sequence of myelin basic protein mRNA. *Biochemistry* 37:7021–29

87. Hsu MT, Coca-Prados M. 1979. Electron microscopic evidence for the circular form of RNA in the cytoplasm of eukaryotic cells. *Nature* 280:339–40

88. Hülskamp M, Schröder C, Pfeifle C, Jäckle H, Tautz D. 1989. Posterior segmentation of the *Drosophila* embryo in the absence of a maternal posterior organizer gene. *Nature* 338:629–32

89. Huynh J, St Johnston D. 2000. The role of *BicD*, *egl*, *orb* and the microtubules in the restriction of meiosis to the *Drosophila* oocyte. *Development* 127:2785–94

90. Huynh JR, Shulman JM, Benton R, St Johnston D. 2001. PAR-1 is required for the maintenance of oocyte fate in *Drosophila*. *Development* 128:1201–9

91. Irish V, Lehmann R, Akam M. 1989. The *Drosophila* posterior-group gene *nanos* functions by repressing *hunchback* activity. *Nature* 338:646–48

92. Jacobson A. 1996. Poly(A) metabolism and translation: the closed-loop model. See Ref. 85, pp. 451–80

93. Kelley RL. 1993. Initial organization of the *Drosophila* dorsoventral axis depends on an RNA-binding protein encoded by the *squid* gene. *Genes Dev.* 7:948–60

94. Kemphues KJ, Priess JR, Morton DG, Cheng NS. 1988. Identification of genes required for cytoplasmic localization in early *C. elegans* embryos. *Cell* 52:311–20

95. Kessler MM, Henry MF, Shen E, Zhao J, Gross S, et al. 1997. Hrp1, a sequence-specific RNA-binding protein that shuttles between the nucleus and the cytoplasm, is required for mRNA 3'-end formation in yeast. *Genes Dev.* 11:2545–56

96. Kim-Ha J, Kerr K, Macdonald PM. 1995. Translational regulation of *oskar* mRNA by Bruno, an ovarian RNA-binding protein, is essential. *Cell* 81:403–12

97. Kim-Ha J, Smith JL, Macdonald PM. 1991. *oskar* mRNA is localized to the posterior pole of the *Drosophila* oocyte. *Cell* 66:23–35

98. Kim-Ha J, Webster PJ, Smith JL, Macdonald PM. 1993. Multiple RNA regulatory elements mediate distinct steps in localization of *oskar* mRNA. *Development* 119:169–78

99. Kohrmann M, Luo M, Kaether C, DesGroseillers L, Dotti CG, Kiebler MA. 1999. Microtubule-dependent recruitment of Staufen-green fluorescent protein into large RNA-containing granules and subsequent dendritic transport in living hippocampal neurons. *Mol. Biol. Cell* 10:2945–53

100. Korner CG, Wormington M, Muckenthaler M, Schneider S, Dehlin E, Wahle E. 1998. The deadenylating nuclease (DAN) is involved in poly(A) tail removal during the meiotic maturation of *Xenopus* oocytes. *EMBO J.* 17:5427–37

101. Kraemer B, Crittenden S, Gallegos M, Moulder G, Barstead R, et al. 1999. NANOS-3 and FBF proteins physically interact to control the sperm-oocyte switch in *Caenorhabditis elegans. Curr. Biol.* 9:1009–18

102. Kuge H, Brownlee GG, Gershon PD, Richter JD. 1998. Cap ribose methylation of c-*mos* mRNA stimulates translation and oocyte maturation in *Xenopus laevis. Nucleic Acids Res.* 26:3208–14

103. Kuge H, Richter JD. 1995. Cytoplasmic 3' poly(A) addition induces 5' cap ribose methylation: implications for translational control of maternal mRNA. *EMBO J.* 14:6301–10

104. Ladhoff AM, Uerlings I, Rosenthal S. 1981. Electron microscopic evidence of circular molecules in 9-S globin mRNA from rabbit reticulocytes. *Mol. Biol. Rep.* 7:101–6

105. Ladomery M, Wade E, Sommerville J. 1997. Xp54, the *Xenopus* homologue of human RNA helicase p54, is an integral component of stored mRNP particles in oocytes. *Nucleic Acids Res.* 25:965–73

106. Lall S, Francis-Lang H, Flament A, Norvell A, Schüpbach T, Ish-Horowicz D. 1999. Squid hnRNP protein promotes apical cytoplasmic transport and localization of *Drosophila* pair-rule transcripts. *Cell* 98:171–80

107. Lane ME, Kalderon D. 1994. RNA localization along the anteroposterior axis of the *Drosophila* oocyte requires PKA-mediated signal transduction to direct normal microtubule organization. *Genes Dev.* 8:2986–95

108. Lantz V, Ambrosio L, Schedl P. 1992. The *Drosophila orb* gene is predicted to encode sex-specific germline RNA-binding proteins and has localized transcripts in ovaries and early embryos. *Development* 115:75–88

109. Lantz V, Chang JS, Horabin JI, Bopp D, Schedl P. 1994. The *Drosophila* Orb RNA-binding protein is required for the formation of the egg chamber and establishment of polarity. *Genes Dev.* 8:598–613

109a. Lantz VA, Clemens SE, Miller KG. 1999. The actin cytoskeleton is required for maintenance of posterior pole plasm components in the *Drosophila* embryo. *Mech. Dev.* 85:111–22

110. Larkin MK, Holder K, Yost C, Giniger

E, Ruohola-Baker H. 1996. Expression of constitutively active Notch arrests follicle cells at a precursor stage during *Drosophila* oogenesis and disrupts the anterior-posterior axis of the oocyte. *Development* 122:3639–50

111. Lasko PF. 1994. *Molecular Genetics of Drosophila Oogenesis.* Austin: RG Landes

112. Lasko P. 1999. RNA sorting in *Drosophila* oocytes and embryos. *FASEB J.* 13:421–33

113. Lee MS, Henry M, Silver PA. 1996. A protein that shuttles between the nucleus and the cytoplasm is an important mediator of RNA export. *Genes Dev.* 10:1233–46

114. Legagneux V, Bouvet P, Omilli F, Chevalier S, Osborne HB. 1992. Identification of RNA-binding proteins specific to *Xenopus* Eg maternal mRNAs: association with the portion of Eg2 mRNA that promotes deadenylation in embryos. *Development* 116:1193–202

115. Legagneux V, Omilli F, Osborne HB. 1995. Substrate-specific regulation of RNA deadenylation in *Xenopus* embryo and activated egg extracts. *RNA* 1:1001–8

116. Lehmann R, Nüsslein-Volhard C. 1986. Abdominal segmentation, pole cell formation, and embryonic polarity require the localized activity of *oskar*, a maternal gene in *Drosophila*. *Cell* 47:141–52

117. Lehmann R, Nüsslein-Volhard C. 1987. *hunchback*, a gene required for segmentation of an anterior and posterior region of the *Drosophila* embryo. *Dev. Biol.* 119:402–17

118. Lehmann R, Nüsslein-Volhard C. 1991. The maternal gene *nanos* has a central role in posterior pattern formation of the *Drosophila* embryo. *Development* 112:679–91

119. Liang L, Diehl-Jones W, Lasko P. 1994. Localization of Vasa protein to the *Drosophila* pole plasm is independent of

its RNA-binding and helicase activities. *Development* 120:1201–11

120. Lie YS, Macdonald PM. 1999. Apontic binds the translational repressor Bruno and is implicated in regulation of *oskar* mRNA translation. *Development* 126:1129–38

121. Lie YS, Macdonald PM. 1999. Translational regulation of *oskar* mRNA occurs independent of the cap and poly(A) tail in *Drosophila* ovarian extracts. *Development* 126:4989–96

122. Lieberfarb ME, Chu T, Wreden C, Theurkauf W, Gergen JP, Strickland S. 1996. Mutations that perturb poly(A)-dependent maternal mRNA activation block the initiation of development. *Development* 122:579–88

123. Lipshitz HD, Smibert CA. 2000. Mechanisms of RNA localization and translational regulation. *Curr. Opin. Genet. Dev.* 10:476–88

124. Luitjens C, Gallegos M, Kraemer B, Kimble J, Wickens M. 2000. CPEB proteins control two key steps in spermatogenesis in *C. elegans*. *Genes Dev.* 14:2596–609

125. Lydall D, Nikolsky Y, Bishop DK, Weinert T. 1996. A meiotic recombination checkpoint controlled by mitotic checkpoint genes. *Nature* 383:840–43

126. Macdonald PM. 1990. *bicoid* mRNA localization signal: phylogenetic conservation of function and RNA secondary structure. *Development* 110:161–71

127. Macdonald PM, Kerr K. 1997. Redundant RNA recognition events in *bicoid* mRNA localization. *RNA* 3:1413–20

128. Macdonald PM, Kerr K. 1998. Mutational analysis of an RNA recognition element that mediates localization of *bicoid* mRNA. *Mol. Cell. Biol.* 18:3788–95

129. Macdonald PM, Kerr K, Smith JL, Leask A. 1993. RNA regulatory element BLE1 directs the early steps of *bicoid* mRNA localization. *Development* 118:1233–43

130. Macdonald PM, Leask A, Kerr K. 1995. Exl protein specifically binds BLE1, a *bicoid* mRNA localization element, and is required for one phase of its activity. *Proc. Natl. Acad. Sci. USA* 92:10787–91

131. Macdonald PM, Struhl G. 1988. *cis*-acting sequences responsible for anterior localization of *bicoid* mRNA in *Drosophila* embryos. *Nature* 336:595–98

132. Mahone M, Saffman EE, Lasko PF. 1995. Localized *Bicaudal-C* RNA encodes a protein containing a KH domain, the RNA binding motif of FMR1. *EMBO J.* 14:2043–55

133. Mahowald AP. 1962. Fine structure of pole cells and polar granules in *Drosophila melanogaster*. *J. Exp. Zool.* 151:201–15

134. Mahowald AP. 1971. Polar granules of *Drosophila*. IV. Cytochemical studies showing loss of RNA from polar granules during early stages of embryogenesis. *J. Exp. Zool.* 176:345–52

135. Mahowald AP. 2001. Assembly of the *Drosophila* germ plasm. *Int. Rev. Cytol.* 203:187–213

136. Manseau L, Calley J, Phan H. 1996. Profilin is required for posterior patterning of the *Drosophila* oocyte. *Development* 122:2109–16

137. Manseau LJ, Schüpbach T. 1989. *cappuccino* and *spire*: two unique maternal-effect loci required for both the anteroposterior and dorsoventral patterns of the *Drosophila* embryo. *Genes Dev.* 3:1437–52

138. Marcey D, Watkins WS, Hazelrigg T. 1991. The temporal and spatial distribution pattern of maternal Exuperantia protein: evidence for a role in establishment but not maintenance of *bicoid* mRNA localization. *EMBO J.* 10:4259–66

139. Markesich DC, Gajewski KM, Nazimiec ME, Beckingham K. 2000. *bicaudal* encodes the *Drosophila* beta NAC homolog, a component of the ribosomal translational machinery. *Development* 127:559–72

140. Markussen FH, Breitwieser W, Ephrussi A. 1997. Efficient translation and phosphorylation of Oskar require Oskar protein and the RNA helicase Vasa. *Cold Spring Harbor Symp. Quant. Biol.* 62:13–17

141. Markussen FH, Michon AM, Breitwieser W, Ephrussi A. 1995. Translational control of *oskar* generates short OSK, the isoform that induces pole plasm assembly. *Development* 121:3723–32

142. Matsumoto K, Wolffe AP. 1998. Gene regulation by Y-box proteins: coupling control of transcription and translation. *Trends Cell Biol.* 8:318–23

143. Matunis EL, Matunis MJ, Dreyfuss G. 1992. Characterization of the major hnRNP proteins from *Drosophila melanogaster*. *J. Cell Biol.* 116:257–69

144. McGrew LL, Dworkin-Rastl E, Dworkin MB, Richter JD. 1989. Poly(A) elongation during *Xenopus* oocyte maturation is required for translational recruitment and is mediated by a short sequence element. *Genes Dev.* 3:803–15

145. Micklem DR, Adams J, Grunert S, St Johnston D. 2000. Distinct roles of two conserved Staufen domains in *oskar* mRNA localization and translation. *EMBO J.* 19:1366–77

146. Micklem DR, Dasgupta R, Elliott H, Gergely F, Davidson C, et al. 1997. The *mago nashi* gene is required for the polarisation of the oocyte and the formation of perpendicular axes in *Drosophila*. *Curr. Biol.* 7:468–78

147. Mohler J, Wieschaus EF. 1986. Dominant maternal-effect mutations of *Drosophila melanogaster* causing the production of double-abdomen embryos. *Genetics* 112:803–22

148. Munroe D, Jacobson A. 1990. mRNA poly(A) tail, a 3′ enhancer of translational initiation. *Mol. Cell. Biol.* 10:3441–55

149. Murata Y, Wharton RP. 1995. Binding of Pumilio to maternal *hunchback*

mRNA is required for posterior patterning in *Drosophila* embryos. *Cell* 80:747–56

150. Nakahata S, Katsu Y, Mita K, Inoue K, Nagahama Y, Yamashita M. 2001. Biochemical identification of *Xenopus* Pumilio as a sequence-specific *cyclin B1* mRNA-binding protein that physically interacts with a Nanos homolog, Xcat-2, and a cytoplasmic polyadenylation element-binding protein. *J. Biol. Chem.* 276:20945–53

151. Neuman-Silberberg FS, Schüpbach T. 1993. The *Drosophila* dorsoventral patterning gene *gurken* produces a dorsally localized RNA and encodes a TGF alpha-like protein. *Cell* 75:165–74

152. Neuman-Silberberg FS, Schüpbach T. 1994. Dorsoventral axis formation in *Drosophila* depends on the correct dosage of the gene *gurken*. *Development* 120:2457–63

153. Neuman-Silberberg FS, Schüpbach T. 1996. The *Drosophila* TGF-alpha-like protein Gurken: expression and cellular localization during *Drosophila* oogenesis. *Mech. Dev.* 59:105–13

154. Newmark PA, Mohr SE, Gong L, Boswell RE. 1997. *mago nashi* mediates the posterior follicle cell-to-oocyte signal to organize axis formation in *Drosophila*. *Development* 124:3197–207

155. Nilson LA, Schüpbach T. 1999. EGF receptor signaling in *Drosophila* oogenesis. *Curr. Top. Dev. Biol.* 44:203–43

156. Norvell A, Kelley RL, Wehr K, Schüpbach T. 1999. Specific isoforms of Squid, a *Drosophila* hnRNP, perform distinct roles in Gurken localization during oogenesis. *Genes Dev.* 13:864–76

157. Olivas W, Parker R. 2000. The Puf3 protein is a transcript-specific regulator of mRNA degradation in yeast. *EMBO J.* 19:6602–11

158. Ostareck DH, Ostareck-Lederer A, Shatsky IN, Hentze MW. 2001. Lipoxygenase mRNA silencing in erythroid differentiation: The 3' UTR regulatory

complex controls 60S ribosomal subunit joining. *Cell* 104:281–90

159. Ostareck DH, Ostareck-Lederer A, Wilm M, Thiele BJ, Mann M, Hentze MW. 1997. mRNA silencing in erythroid differentiation: hnRNP K and hnRNP E1 regulate 15-lipoxygenase translation from the 3' end. *Cell* 89:597–606

160. Ostareck-Lederer A, Ostareck DH, Standart N, Thiele BJ. 1994. Translation of 15-lipoxygenase mRNA is inhibited by a protein that binds to a repeated sequence in the 3' untranslated region. *EMBO J.* 13:1476–81

161. Page AW, Orr-Weaver TL. 1996. The *Drosophila* genes *grauzone* and *cortex* are necessary for proper female meiosis. *J. Cell Sci.* 109:1707–15

162. Paillard L, Omilli F, Legagneux V, Bassez T, Maniey D, Osborne HB. 1998. EDEN and EDEN-BP, a *cis* element and an associated factor that mediate sequence-specific mRNA deadenylation in *Xenopus* embryos. *EMBO J.* 17:278–87

163. Parisi M, Lin H. 2000. Translational repression: a duet of Nanos and Pumilio. *Curr. Biol.* 10:R81–83

164. Pestova TV, Dever TE, Hellen CUT. 2000. Ribosomal subunit joining. See Ref. 211, pp. 425–45

165. Pestova TV, Lomakin IB, Lee JH, Choi SK, Dever TE, Hellen CU. 2000. The joining of ribosomal subunits in eukaryotes requires eIF5B. *Nature* 403:332–35

166. Petes TD, Malone RE, Symington LS. 1991. Recombination in yeast. In *The Molecular and Cellular Biology of the Yeast* Saccharomyces, ed. J Broach, JR Pringle, EW Jones, pp. 407–521. Cold Spring Harbor, NY: Cold Spring Harbor Lab. Press

167. Pinol-Roma S, Dreyfuss G. 1992. Shuttling of pre-mRNA binding proteins between nucleus and cytoplasm. *Nature* 355:730–32

168. Pokrywka NJ, Stephenson EC. 1991.

Microtubules mediate the localization of *bicoid* RNA during *Drosophila* oogenesis. *Development* 113:55–66

169. Ponting CP. 1997. Tudor domains in proteins that interact with RNA. *Trends Biochem. Sci.* 22:51–52

170. Prost E, Deryckere F, Roos C, Haenlin M, Pantesco V, Mohier E. 1988. Role of the oocyte nucleus in determination of the dorsoventral polarity of *Drosophila* as revealed by molecular analysis of the *K10* gene. *Genes Dev.* 2:891–900

171. Ramos A, Grunert S, Adams J, Micklem DR, Proctor MR, et al. 2000. RNA recognition by a Staufen double-stranded RNA-binding domain. *EMBO J.* 19:997–1009

172. Ran B, Bopp R, Suter B. 1994. Null alleles reveal novel requirements for *Bic-D* during *Drosophila* oogenesis and zygotic development. *Development* 120:1233–42

173. Raught B, Gingras AC, Sonenberg N. 2000. Regulation of ribosomal recruitment in eukaryotes. See Ref. 211, pp. 245–93

174. Ray RP, Schüpbach T. 1996. Intercellular signaling and the polarization of body axes during *Drosophila* oogenesis. *Genes Dev.* 10:1711–23

175. Raychaudhuri P, Chaudhuri A, Maitra U. 1985. Eukaryotic initiation factor 5 from calf liver is a single polypeptide chain protein of $M_r = 62,000$. *J. Biol. Chem.* 260:2132–39

176. Raychaudhuri P, Chaudhuri A, Maitra U. 1985. Formation and release of eukaryotic initiation factor 2 X GDP complex during eukaryotic ribosomal polypeptide chain initiation complex formation. *J. Biol. Chem.* 260:2140–45

177. Resnick MA. 1987. Investigating the genetic control of biochemical events in meiotic recombination. In *Meiosis*, ed. PB Moens, pp. 157–210. Orlando, FL: Academic

178. Richter JD. 1999. Cytoplasmic polyadenylation in development and be-

yond. *Microbiol. Mol. Biol. Rev.* 63:446–56

179. Richter JD. 2000. Influence of polyadenylation-induced translation on metazoan development and neuronal synaptic function. See Ref. 211, pp. 785–805

180. Richter JD, Smith LD. 1984. Reversible inhibition of translation by *Xenopus* oocyte-specific proteins. *Nature* 309:378–80

181. Rongo C, Gavis ER, Lehmann R. 1995. Localization of *oskar* RNA regulates *oskar* translation and requires Oskar protein. *Development* 121:2737–46

182. Ross AF, Oleynikov Y, Kislauskis EH, Taneja KL, Singer RH. 1997. Characterization of a beta-actin mRNA zipcode-binding protein. *Mol. Cell. Biol.* 17:2158–65

183. Roth S, Neuman-Silberberg FS, Barcelo G, Schüpbach T. 1995. *cornichon* and the EGF receptor signaling process are necessary for both anterior-posterior and dorsal-ventral pattern formation in *Drosophila*. *Cell* 81:967–78

184. Roth S, Schüpbach T. 1994. The relationship between ovarian and embryonic dorsoventral patterning in *Drosophila*. *Development* 120:2245–57

185. Rouault TA, Harford JB. 2000. Translational control of ferritin synthesis. See Ref. 211, pp. 655–70

186. Ruohola H, Bremer KA, Baker D, Swedlow JR, Jan LY, Jan YN. 1991. Role of neurogenic genes in establishment of follicle cell fate and oocyte polarity during oogenesis in *Drosophila*. *Cell* 66:433–49

187. Sachs A. 2000. Physical and functional interactions between the mRNA cap structure and the poly(A) tail. See Ref. 211, pp. 447–65

188. Saffman EE, Lasko P. 1999. Germline development in vertebrates and invertebrates. *Cell. Mol. Life Sci.* 55:1141–63

189. Saffman EE, Styhler S, Rother K, Li W, Richard S, Lasko P. 1998. Premature

translation of *oskar* in oocytes lacking the RNA-binding protein Bicaudal-C. *Mol. Cell. Biol.* 18:4855–62

190. Sallés FJ, Lieberfarb ME, Wreden C, Gergen JP, Strickland S. 1994. Coordinate initiation of *Drosophila* development by regulated polyadenylation of maternal messenger RNAs. *Science* 266:1996–99

191. Saunders C, Cohen RS. 1999. The role of oocyte transcription, the 5′ UTR, and translation repression and derepression in *Drosophila gurken* mRNA and protein localization. *Mol. Cell* 3:43–54

192. Schisa JA, Strickland S. 1998. Cytoplasmic polyadenylation of *Toll* mRNA is required for dorsal-ventral patterning in *Drosophila* embryogenesis. *Development* 125:2995–3003

193. Schnapp BJ. 1999. A glimpse of the machinery. *Curr. Biol.* 9:R725–27

194. Schnörrer F, Bohmann K, Nüsslein-Volhard C. 2000. The molecular motor dynein is involved in targeting Swallow and *bicoid* RNA to the anterior pole of *Drosophila* oocytes. *Nat. Cell Biol.* 2:185–90

195. Schultz J, Ponting CP, Hofmann K, Bork P. 1997. SAM as a protein interaction domain involved in developmental regulation. *Protein Sci.* 6:249–53

196. Schüpbach T, Wieschaus E. 1989. Female sterile mutations on the second chromosome of *Drosophila melanogaster*. I. Maternal effect mutations. *Genetics* 121:101–17

197. Schüpbach T, Wieschaus E. 1991. Female sterile mutations on the second chromosome of *Drosophila melanogaster*. II. Mutations blocking oogenesis or altering egg morphology. *Genetics* 129:1119–36

198. Seeger MA, Kaufman TC. 1990. Molecular analysis of the *bicoid* gene from *Drosophila pseudoobscura*: identification of conserved domains within coding and noncoding regions of the *bicoid* mRNA. *EMBO J.* 9:2977–87

199. Sekelsky JJ, Burtis KC, Hawley RS. 1998. Damage control: the pleiotropy of DNA repair genes in *Drosophila melanogaster*. *Genetics* 148:1587–98

200. Serano TL, Cohen RS. 1995. Gratuitous mRNA localization in the *Drosophila* oocyte. *Development* 121:3013–21

201. Serano TL, Karlin-McGinness M, Cohen RS. 1995. The role of *fs(1)K10* in the localization of the mRNA of the TGF alpha homolog *gurken* within the *Drosophila* oocyte. *Mech. Dev.* 51:183–92

202. Shulman JM, Benton R, St Johnston D. 2000. The *Drosophila* homolog of C. *elegans* PAR-1 organizes the oocyte cytoskeleton and directs *oskar* mRNA localization to the posterior pole. *Cell* 101:377–88

203. Sibon OC, Laurencon A, Hawley R, Theurkauf WE. 1999. The *Drosophila* ATM homologue Mei-41 has an essential checkpoint function at the midblastula transition. *Curr. Biol.* 9:302–12

204. Simon R, Richter JD. 1994. Further analysis of cytoplasmic polyadenylation in *Xenopus* embryos and identification of embryonic cytoplasmic polyadenylation element-binding proteins. *Mol. Cell. Biol.* 14:7867–75

205. Simon R, Tassan JP, Richter JD. 1992. Translational control by poly(A) elongation during *Xenopus* development: differential repression and enhancement by a novel cytoplasmic polyadenylation element. *Genes Dev.* 6:2580–91

206. Simon R, Wu L, Richter JD. 1996. Cytoplasmic polyadenylation of activin receptor mRNA and the control of pattern formation in *Xenopus* development. *Dev. Biol.* 179:239–50

207. Slack FJ, Basson M, Liu Z, Ambros V, Horvitz HR, Ruvkun G. 2000. The *lin-41* RBCC gene acts in the C. *elegans* heterochronic pathway between the *let-7* regulatory RNA and the LIN-29 transcription factor. *Mol. Cell* 5:659–69

208. Smibert CA, Lie YS, Shillinglaw W, Henzel WJ, Macdonald PM. 1999. Smaug, a novel and conserved protein, contributes to repression of *nanos* mRNA translation in vitro. *RNA* 5:1535–47

209. Smibert CA, Wilson JE, Kerr K, Macdonald PM. 1996. Smaug protein represses translation of unlocalized *nanos* mRNA in the *Drosophila* embryo. *Genes Dev.* 10:2600–9

210. Smith JL, Wilson JE, Macdonald PM. 1992. Overexpression of *oskar* directs ectopic activation of *nanos* and presumptive pole cell formation in *Drosophila* embryos. *Cell* 70:849–59

211. Sonenberg N, Hershey JWB, Mathews MB, eds. 2000. *Translational Control of Gene Expression.* Cold Spring Harbor, NY: Cold Spring Harbor Lab. Press

212. Sonoda J, Wharton RP. 1999. Recruitment of Nanos to *hunchback* mRNA by Pumilio. *Genes Dev.* 13:2704–12

213. Sonoda J, Wharton RP. 2001. *Drosophila* Brain Tumor is a translational repressor. *Genes Dev.* 15:762–73

214. Spirin AS. 1996. Masked and translatable messenger ribonucleotides in higher eukaryotes. See Ref. 85, pp. 319–34

215. Spradling AC. 1993. Developmental genetics of oogenesis. In *The Development of* Drosophila melanogaster, ed. M Bate, A Martinez-Arias, pp. 1–70. Cold Spring Harbor, NY: Cold Spring Harbor Lab. Press

216. St Johnston D, Beuchle D, Nüsslein-Volhard C. 1991. Staufen, a gene required to localize maternal RNAs in the *Drosophila* egg. *Cell* 66:51–63

217. St Johnston D, Brown NH, Gall JG, Jantsch M. 1992. A conserved double-stranded RNA-binding domain. *Proc. Natl. Acad. Sci. USA* 89:10979–83

218. St Johnston D, Driever W, Berleth T, Richstein S, Nüsslein-Volhard C. 1989. Multiple steps in the localization of *bicoid* RNA to the anterior pole of the *Drosophila* oocyte. *Development* 107(Suppl.):13–19

219. St Johnston D, Nüsslein-Volhard C. 1992. The origin of pattern and polarity in the *Drosophila* embryo. *Cell* 68:201–19

220. Stebbins-Boaz B, Cao Q, de Moor CH, Mendez R, Richter JD. 1999. Maskin is a CPEB-associated factor that transiently interacts with eIF-4E. *Mol. Cell* 4:1017–27

221. Stebbins-Boaz B, Hake LE, Richter JD. 1996. CPEB controls the cytoplasmic polyadenylation of *cyclin, Cdk2* and c-*mos* mRNAs and is necessary for oocyte maturation in *Xenopus. EMBO J.* 15:2582–92

222. Stebbins-Boaz B, Richter JD. 1994. Multiple sequence elements and a maternal mRNA product control *cdk2* RNA polyadenylation and translation during early *Xenopus* development. *Mol. Cell. Biol.* 14:5870–80

223. Stephenson EC, Chao YC, Fackenthal JD. 1988. Molecular analysis of the *swallow* gene of *Drosophila melanogaster. Genes Dev.* 2:1655–65

224. Struhl G. 1989. Differing strategies for organizing anterior and posterior body pattern in *Drosophila* embryos. *Nature* 338:741–44

225. Styhler S, Nakamura A, Swan A, Suter B, Lasko P. 1998. Vasa is required for GURKEN accumulation in the oocyte, and is involved in oocyte differentiation and germline cyst development. *Development* 125:1569–78

226. Suter B, Steward R. 1991. Requirement for phosphorylation and localization of the Bicaudal-D protein in *Drosophila* oocyte differentiation. *Cell* 67:917–26

227. Swan A, Nguyen T, Suter B. 1999. *Drosophila Lissencephaly-1* functions with *Bic-D* and dynein in oocyte determination and nuclear positioning. *Nat. Cell Biol.* 1:444–49

228. Tadauchi T, Matsumoto K, Herskowitz I, Irie K. 2001. Post-transcriptional

regulation through the HO 3'-UTR by Mpt5, a yeast homolog of Pumilio and FBF. *EMBO J.* 20:552–61

229. Tafuri SR, Wolffe AP. 1993. Selective recruitment of masked maternal mRNA from messenger ribonucleoprotein particles containing FRGY2 (mRNP4). *J. Biol. Chem.* 268:24255–61

230. Tarun SZ Jr, Sachs AB. 1996. Association of the yeast poly(A) tail binding protein with translation initiation factor eIF-4G. *EMBO J.* 15:7168–77

231. Tautz D. 1988. Regulation of the *Drosophila* segmentation gene *hunchback* by two maternal morphogenetic centres. *Nature* 332:281–84

232. Tearle RG, Nüsslein-Volhard C. 1987. Tübingen mutants and stock list. *Drosoph. Inf. Serv.* 66:209–69

233. Theurkauf WE, Alberts BM, Jan YN, Jongens TA. 1993. A central role for microtubules in the differentiation of *Drosophila* oocytes. *Development* 118:1169–80

234. Theurkauf WE, Hazelrigg TI. 1998. In vivo analyses of cytoplasmic transport and cytoskeletal organization during *Drosophila* oogenesis: characterization of a multi-step anterior localization pathway. *Development* 125:3655–66

235. Theurkauf WE, Smiley S, Wong ML, Alberts BM. 1992. Reorganization of the cytoskeleton during *Drosophila* oogenesis: implications for axis specification and intercellular transport. *Development* 115:923–36

236. Thieringer HA, Singh K, Trivedi H, Inouye M. 1997. Identification and developmental characterization of a novel Y-box protein from *Drosophila melanogaster. Nucleic Acids Res.* 25:4764–70

237. Thio GL, Ray RP, Barcelo G, Schüpbach T. 2000. Localization of *gurken* RNA in *Drosophila* oogenesis requires elements in the 5' and 3' regions of the transcript. *Dev. Biol.* 221:435–46

238. Tinker R, Silver D, Montell DJ. 1998. Requirement for the Vasa RNA helicase in *gurken* mRNA localization. *Dev. Biol.* 199:1–10

239. Tomancak P, Guichet A, Zavorszky P, Ephrussi A. 1998. Oocyte polarity depends on regulation of *gurken* by Vasa. *Development* 125:1723–32

240. Tomancak P, Piano F, Riechmann V, Gunsalus KC, Kemphues KJ, Ephrussi A. 2000. A *Drosophila melanogaster* homologue of *Caenorhabditis elegans par-1* acts at an early step in embryonic-axis formation. *Nat. Cell Biol.* 2:458–60

241. Vale RD, Milligan RA. 2000. The way things move: looking under the hood of molecular motor proteins. *Science* 288:88–95

242. Van Buskirk C, Hawkins NC, Schüpbach T. 2000. Encore is a member of a novel family of proteins and affects multiple processes in *Drosophila* oogenesis. *Development* 127:4753–62

243. van Eeden F, St Johnston D. 1999. The polarisation of the anterior-posterior and dorsal-ventral axes during *Drosophila* oogenesis. *Curr. Opin. Genet. Dev.* 9:396–404

244. Varnum SM, Wormington WM. 1990. Deadenylation of maternal mRNAs during *Xenopus* oocyte maturation does not require specific *cis*-sequences: a default mechanism for translational control. *Genes Dev.* 4:2278–86

245. Verrotti AC, Wreden C, Strickland S. 1999. Dissociation of mRNA cytoplasmic polyadenylation from translational activation by structural modification of the 5'-UTR. *Nucleic Acids Res.* 27:3417–23

246. Visa N, Alzhanova-Ericsson AT, Sun X, Kiseleva E, Bjorkroth B, et al. 1996. A pre-mRNA-binding protein accompanies the RNA from the gene through the nuclear pores and into polysomes. *Cell* 84:253–64

247. Voeltz GK, Ongkasuwan J, Standart N, Steitz JA. 2001. A novel embryonic

poly(A) binding protein, ePAB, regulates mRNA deadenylation in *Xenopus* egg extracts. *Genes Dev.* 15:774–88

248. Voeltz GK, Steitz JA. 1998. AUUUA sequences direct mRNA deadenylation uncoupled from decay during *Xenopus* early development. *Mol. Cell. Biol.* 18:7537–45

249. Wang C, Dickinson LK, Lehmann R. 1994. Genetics of *nanos* localization in *Drosophila. Dev. Dyn.* 199:103–15

250. Wang C, Lehmann R. 1991. Nanos is the localized posterior determinant in *Drosophila. Cell* 66:637–47

251. Wang S, Hazelrigg T. 1994. Implications for *bcd* mRNA localization from spatial distribution of Exu protein in *Drosophila* oogenesis. *Nature* 369:400–3

252. Warner JR, Rich A, Hall CE. 1962. Electron microscope studies of ribosomal clusters synthesizing hemoglobin. *Science* 138:1399–403

253. Webster PJ, Liang L, Berg CA, Lasko P, Macdonald PM. 1997. Translational repressor Bruno plays multiple roles in development and is widely conserved. *Genes Dev.* 11:2510–21

254. Wellington A, Emmons S, James B, Calley J, Grover M, et al. 1999. Spire contains actin binding domains and is related to ascidian posterior end mark-5. *Development* 126:5267–74

255. Wells SE, Hillner PE, Vale RD, Sachs AB. 1998. Circularization of mRNA by eukaryotic translation initiation factors. *Mol. Cell* 2:135–40

256. Wharton RP, Sonoda J, Lee T, Patterson M, Murata Y. 1998. The Pumilio RNA-binding domain is also a translational regulator. *Mol. Cell* 1:863–72

257. Wharton RP, Struhl G. 1989. Structure of the *Drosophila* BicaudalD protein and its role in localizing the posterior determinant *nanos. Cell* 59:881–92

258. Wharton RP, Struhl G. 1991. RNA regulatory elements mediate control of *Drosophila* body pattern by the posterior morphogen *nanos. Cell* 67:955–67

259. Wickens M, Goodwin EB, Kimble J, Strickland S, Hentze M. 2000. Translational control of developmental decisions. See Ref. 211, pp. 295–369

260. Deleted in proof

261. Wickham L, Duchaine T, Luo M, Nabi IR, DesGroseillers L. 1999. Mammalian Staufen is a double-stranded-RNA- and tubulin-binding protein which localizes to the rough endoplasmic reticulum. *Mol. Cell. Biol.* 19:2220–30

262. Wiedmann B, Sakai H, Davis TA, Wiedmann M. 1994. A protein complex required for signal-sequence-specific sorting and translocation. *Nature* 370:434–40

263. Wieschaus E, Marsh JL, Gehring W. 1978. *fs(1)K10*, a germline-dependent female sterile mutation causing abnormal chorion morphology in *Drosophila melanogaster. Roux's Arch. Dev. Biol.* 184:75–82

264. Wilhelm JE, Mansfield J, Hom-Booher N, Wang S, Turck CW, et al. 2000. Isolation of a ribonucleoprotein complex involved in mRNA localization in *Drosophila* oocytes. *J. Cell Biol.* 148:427–40

265. Wilsch-Bräuninger M, Schwarz H, Nüsslein-Volhard C. 1997. A sponge-like structure involved in the association and transport of maternal products during *Drosophila* oogenesis. *J. Cell Biol.* 139:817–29

266. Wilson JE, Connell JE, Macdonald PM. 1996. *aubergine* enhances *oskar* translation in the *Drosophila* ovary. *Development* 122:1631–39

267. Wormington M. 1993. Poly(A) and translation: development control. *Curr. Opin. Cell Biol.* 5:950–54

268. Wreden C, Verrotti AC, Schisa JA, Lieberfarb ME, Strickland S. 1997. Nanos and Pumilio establish embryonic polarity in *Drosophila* by promoting posterior deadenylation of *hunchback* mRNA. *Development* 124:3015–23

269. Zamore PD, Bartel DP, Lehmann R, Williamson JR. 1999. The PUMILIO-RNA interaction: a single RNA-binding domain monomer recognizes a bipartite target sequence. *Biochemistry* 38:596–604

270. Zamore PD, Williamson JR, Lehmann R. 1997. The Pumilio protein binds RNA through a conserved domain that defines a new class of RNA-binding proteins. *RNA* 3:1421–33

271. Zelus BD, Giebelhaus DH, Eib DW, Kenner KA, Moon RT. 1989. Expression of the poly(A)-binding protein during development of *Xenopus laevis*. *Mol. Cell. Biol.* 9:2756–60

272. Zhang B, Gallegos M, Puoti A, Durkin E, Fields S, et al. 1997. A conserved RNA-binding protein that regulates sexual fates in the *C. elegans* hermaphrodite germ line. *Nature* 390:477–84

Annu. Rev. Genet. 2001. 35:407–37

CONSERVATION AND DIVERGENCE IN MOLECULAR MECHANISMS OF AXIS FORMATION

Sabbi Lall and Nipam H. Patel

Howard Hughes Medical Institute, University of Chicago, Chicago, Illinois 60637;
e-mail: npatel@midway.uchicago.edu

Key Words axis formation, insects, *Drosophila*, evolution

■ **Abstract** Genetic screens in *Drosophila melanogaster* have helped elucidate the process of axis formation during early embryogenesis. Axis formation in the *D. melanogaster* embryo involves the use of two fundamentally different mechanisms for generating morphogenetic activity: patterning the anteroposterior axis by diffusion of a transcription factor within the syncytial embryo and specification of the dorsoventral axis through a signal transduction cascade. Identification of *Drosophila* genes involved in axis formation provides a launch-pad for comparative studies that examine the evolution of axis specification in different insects. Additionally, there is similarity between axial patterning mechanisms elucidated genetically in *Drosophila* and those demonstrated for chordates such as *Xenopus*. In this review we examine the postfertilization mechanisms underlying axis specification in *Drosophila*. Comparative data are then used to ask whether aspects of axis formation might be derived or ancestral.

CONTENTS

0066-4197/01/1215-0407$14.00

INTRODUCTION

Many organisms manifest polarity at some level; indeed, asymmetry seems essential for promoting meaningful interaction with the environment. Even the simplest unicellular organisms display temporary polarity in response to their environment; for example, in reception and response to chemotactic signals (for example, see 11). In higher animals the body plan is arranged along two axes: the anteroposterior (AP) axis, and the broadly perpendicular dorsoventral (DV) axis.

Breaking symmetry to specify these axes is one of the most basic and earliest processes in the development of higher animals. The early nature of axis formation is clear in *Drosophila melanogaster* where the process has been genetically dissected, and the cues for AP and DV axiation are established during oogenesis. Animals can utilize diverse environmental cues as the basis of axiation, suggesting that axis formation is a variable, plastic process. For instance, the physical process of axis formation can vary between closely related nematodes (46). Furthermore, *Xenopus laevis* can be induced to use gravity rather than sperm entry point as a cue for axis formation; indeed, the dorsalizing organizer forms $180°$ to sperm entry point only 70% of the time (45). How can an event that provides basic information for patterning the entire body plan be so inconstant?

In this review, we use knowledge about embryonic axis formation in the dipteran *Drosophila melanogaster* (*D. melanogaster*) to address the contention that axis formation is variable, and then to ask how this can be true of such a fundamental process. Variation in axis formation probably has an intimate relationship with alterations in embryonic morphology and changes in life-history. Hence we discuss data suggesting that variation in axis formation within the dipterans correlates phylogenetically with variation in embryonic morphology.

D. melanogaster has four maternal coordinate systems that specify the major body axes (reviewed in 165). Three specify positional information along the AP axis (the anterior, posterior, and terminal systems), whereas the fourth is involved in DV axis formation. We describe the molecular nature of the anterior and DV patterning systems, addressing which of the following might be true:

1. Does a given molecular process play a conserved role in axis formation across phyla, and could it therefore be ancient?

2. Might a particular molecular process have been co-opted into axis formation from another context prior to the origin of insects?

3. Might a given molecular process be a very recent evolutionary innovation?

Examples of all three scenarios can be drawn out of the axis formation mechanisms elucidated in *D. melanogaster*, suggesting that the process of axis formation may well have been elaborated upon multiple times. We discuss the fact that *D. melanogaster* has been shown genetically to display redundancy between two maternal coordinate systems, and suggest that this may facilitate radical changes in axis formation. Finally, we examine recent data suggesting that rapid molecular evolution of axiation can be correlated with gross morphological rearrangements within the dipterans.

Axis Formation in *Drosophila melanogaster*

Genetic screens and subsequent molecular analyses have led to a detailed understanding of the earliest embryonic events in axis formation in *D. melanogaster* (165). All four of the maternal coordinate systems are set up during oogenesis, and recognition of and elaboration upon axial maternal cues are among the earliest events after egg activation and fertilization. The anterior system involves cytoplasmic diffusion of morphogens within the syncytial *D. melanogaster* embryo. In contrast the terminal and dorsal patterning systems require signaling from the extracellular perivitelline space to the embryo. In this review, we focus on the dorsoventral patterning system, as an example of morphogenetic activity set up by extracellular signaling, and the anterior system, as an illustration of cytoplasmic diffusion to form a morphogen gradient. We begin by asking whether such mechanisms are likely to be conserved among other insect orders.

Are Models of Axis Formation Formulated in *Drosophila* Tenable in Other Insects?

D. melanogaster displays a mode of development designated long germ embryogenesis, in which the presumptive head, thoracic, and abdominal cells are present at blastoderm stage in the same proportion as in the hatching larva (reviewed in 140). *D. melanogaster* also has a relatively prolonged syncytial stage. The cues for AP axis formation are initially elaborated upon within a syncytial environment during the first 2.5 h of embryonic development. Short germ insects, on the other hand, have only specified the most anterior segments at the end of blastoderm stage, the posterior segments being generated by subsequent growth [for reviews addressing short and long germ development, see (140, 144, 177)]. Although most insects display superficial cleavage and have a syncytial phase, cellularization occurs well before gastrulation in many short and long germ embryos. For example, grasshopper nuclei cellularize almost as soon as they reach the embryonic cortex, i.e., at the onset of blastoderm stage (55).

Both short germ embryogenesis and early cellularization have profound consequences for axis formation models involving morphogen diffusion. Short germ embryogenesis suggests that patterning of posterior segments may be substantially delayed, whereas early cellularization makes it difficult to envision specification by the diffusion of a morphogen such as the *bicoid* transcription factor in *Drosophila* (see below). In contrast, systems that elaborate axes via a signaling pathway might more obviously be conserved. The insect orders discussed in this review and their approximate relationships are illustrated in Figure 1b (based on 82, 195).

Axis formation mechanisms in *D. melanogaster* are even harder to apply to animals that undergo holoblastic (complete) rather than superficial cleavage after fertilization. In embryos undergoing complete cleavage, diffusion or segregation of a cytoplasmic morphogen must occur during the first embryonic cleavages. Thus, if aspects of axial patterning in *Drosophila* are conserved across species, they will at the very least have been modified in the lineage leading to flies.

THE DORSOVENTRAL SYSTEM

DV Axiation Processes in *D. melanogaster*

Our understanding of DV patterning during early embryogenesis in *D. melanogaster* is based on the genetic dissection of the *Toll/dorsal* pathway (Figure 2) (reviewed in 100). Upon mutation, 12 genes show maternal-effect dorsoventral patterning defects in the embryo, but have normal eggshell patterning (5, 21, 152). These genes are (in putative order of action within the pathway) *windbeutel, pipe, nudel, gastrulation defective, snake, easter, spätzle, Toll, pelle, tube, dorsal,* and *cactus* (4, 22, 53, 99, 133). Normal DV development is ultimately evident by the stereotypical arrangement of denticles on the larval cuticle, and more immediately by the correct expression of molecular markers indicating specification of territories along the dorsoventral axis during embryogenesis. The strongest recessive mutations in 11 of the dorsal group genes lead to a larval cuticle that is covered in dorsal-type denticles and lacks ventral tissue such as the mesoderm. The 12th gene, *cactus*, gives the opposite phenotype: a ventralized cuticle (133, 152). Readouts of the *Toll/dorsal* pathway display dose sensitivity genetically, and depend upon differential levels of Dorsal nuclear localization along the DV axis (110, 125, 134, 136, 138, 171, 172). Such evidence suggests that fate along the DV axis depends upon the activity of the *dorsal* transcription factor.

During oogenesis, a molecular cue localized around the oocyte nucleus determines follicle cells lying on one side of the oocyte as dorsal. This leads to the development of follicle cells on the other side of the oocyte as ventral, via the expression of the *pipe* gene. *pipe* encodes a heparan sulfate 2-O-sulfotransferase and has been suggested to make a ventral extracellular modification, or perhaps modify *nudel* ventrally, as the latter behaves nonautonomously and may thus potentially mediate an extracellular signal (109, 155). These cues, set up during oogenesis, initiate a proteolytic cascade, mediated by the proteases *nudel, gastrulation defective, snake,* and *easter,* in the perivitelline space outside the fertilized embryo (Figure 2a) (see 20, 58, 79, 86, 169). Interestingly, *pipe* is not essential for initiating the proteolytic activity of Gastrulation-defective, implying that the situation is more complicated than a simple spatial cue locally activating the protease at the top of a proteolytic hierarchy (86). The proteolytic cascade results in the ventral processing

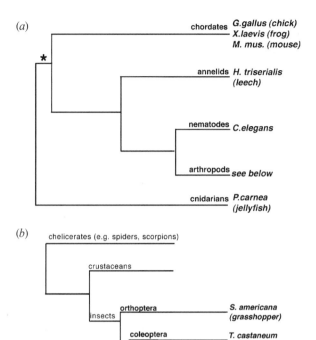

Figure 1 Phylogenetic trees illustrating the positions of the species and phyla discussed in this review. The trees provide a rough framework of (*a*) the positions of the phyla discussed (based on 1), and (*b*) the positions of the insect orders and particular dipterans (flies) used as examples in the text. The asterisk indicates the origin of bilaterians, essentially animals with a body plan arranged along two perpendicular axes. Arthropod species are: *Schistocerca americana*, *Tribolium castaneum*, *Nasonia vitripennis*, *Bombyx mori*, *Clogmia albipunctata*, *Empis livida*, *Megaselia abdita*, and *Drosophila melanogaster*.

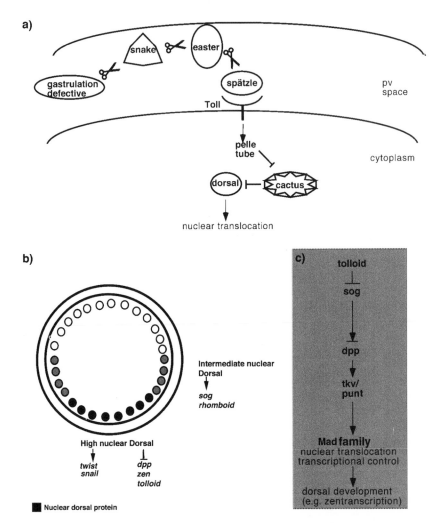

Figure 2 DV patterning in *Drosophila melanogaster*. The pathways illustrate the interactions of genes that act to pattern the DV axis of the embryo without influencing eggshell morphology (*a*). (*a*) shows maternal effect genes that act, within the perivitelline space (pv space) to control the nuclear localization of Dorsal protein along the DV axis. (*b*) Dorsal is found in a nuclear gradient. At high nuclear concentrations Dorsal represses zygotic genes that pattern the dorsal regions of the embryo while activating ventral development through *twist* and *snail* activation. At intermediate levels the Dorsal transcription factor promotes lateral fates (*b*). (*c*) Factors regulating Dpp signaling are illustrated, showing basic interactions. *sog* both antagonizes and potentiates Dpp signaling.

of Spätzle protein to a 23-kD form by Easter (99, 147). When injected into the periv-itelline space, cleaved Spätzle activates ventral development in both a site- and concentration-specific fashion (99). Genetic and biochemical evidence suggests that the cleaved and active form of Spätzle then acts as a ligand for the Toll recep-tor, which is immediately upstream of *Toll* but downstream of *easter* (Figure 2*a*) (see 22). Localized activation of the Toll receptor leads to the stimulation of an intracellular pathway involving *tube* and *pelle*, the end result of which is the phos-phorylation and degradation of the IkB orthologue *cactus* (10, 53, 126).

Cactus physically interacts with and thereby inhibits a key gene in dorsoventral axis formation, the morphogenetic transcription factor Dorsal (a *NFκB/rel* homo-logue; 170). Degradation of Cactus allows Dorsal to enter the nucleus (12, 186). Since Toll is activated ventrally, Dorsal protein enters the nucleus at highest concen-tration ventrally. Immunostaining against Dorsal protein allows direct and elegant visualization of its nucleocytoplasmic gradient, running ventral to dorsal across the embryo (134, 138, 171).

Activation of the maternal *Toll/dorsal* pathway leads to the expression of zy-gotic genes at different DV levels of the embryo (Figure 2*b*). At highest nuclear concentration (ventral), Dorsal activates *twist* and *snail*, which are required for specification of ventral fate (mesoderm) and for the inhibition of lateral fates such as neurectoderm (48, 66, 80, 87, 104, 124, 125, 180–182). Laterally, interme-diate nuclear concentrations of Dorsal activate a second group of targets including *rhomboid* and *short gastrulation* (*sog*). These genes specify lateral neurectoder-mal territories and influence the activity of Decapentaplegic (Dpp), respectively (see below; 42, 65). Dorsal also acts as a transcriptional repressor of genes such as *dpp* and *zen*, which specify dorsal fates (32, 64, 68, 125, 136). Thus, higher Dorsal nuclear activity in ventral and lateral regions restricts *dpp* expression to the dor-sal side of the embryo. *dpp* acts to pattern ectoderm, specifying different tissue territories such as amnioserosa and dorsal epidermis (39, 187).

Which aspects of these processes are conserved across species? We focus on components with known homologues in other species that can therefore be discussed in an evolutionary context. In particular, we discuss *dpp* and *twist* and examine their potential regulation by the *Toll* signaling pathway outside *D. melanogaster*.

Dpp Signaling May Play a Conserved Role in Axis Formation Across Phyla

dpp is expressed at blastoderm stage, in a longitudinal stripe restricted by Dorsal protein to the dorsal 40% of the embryo (166). Dpp protein (enhanced by the ligand Screw; 106) is responsible for the transcriptional activation of a number of tar-gets in specific dorsal territories including *zerknüllt* (*zen*) (8, 39, 68, 125, 137, 187). Dpp is a ligand for the Tkv/Punt receptor complex (17, 88, 105, 116, 135). Genera-tion of a Dpp activity gradient is also dependent on the *dpp* antagonist, *sog*. Genetic mosaic analysis suggests that *sog* is required ventrolaterally and acts non-cell autonomously. Moreover, genetic experiments addressing the interaction of *dpp* and *sog* suggest that *sog* antagonizes Dpp function dorsolaterally but intensifies Dpp activity in the extreme dorsal region of the embryo (Figure 2c) (see 7, 13, 40, 42, 92, 199). Therefore an antagonistic gradient of Sog emanating from lateral regions of the embryo may help to grade the activity of Dpp in dorsal territories.

Regulation of Dpp signaling is also mediated by the *tolloid* gene. *tolloid* embryos have a ventralized cuticle phenotype, and are missing the most dorsal structures (amnioserosa) as well as some dorsal epidermis (74, 156). *tolloid* is expressed dorsally, encodes a metalloprotease, and has been shown genetically to be upstream of *dpp* (Figure 2c). Epistasis analysis in *Drosophila* and second axis induction assays in *Xenopus* suggest that *tolloid* functions upstream of *sog* as an antagonist (92). Tolloid cleaves Sog protein, as evidenced by the fact that Sog cleavage products can be detected in the embryo upon *sog* overexpression, or when Sog and Tolloid are co-incubated in vitro (92, 196). Sog destruction by Tolloid may allow Dpp to bind to its receptor, perhaps by reducing the affinity of the Sog/Dpp interaction and thereby freeing Dpp to function independently of Sog. The isolation of interacting antimorphic mutations was used to suggest a physical interaction between the *tolloid* and *dpp* gene products. These phenotypes have not been reconciled with the current view that *tolloid* acts on *sog* (24, 40, 41, 123). Further regulation of Dpp activity is suggested by recent data concerning the *twisted gastrulation* gene, which though originally proposed to antagonize Sog might actually be an antagonist of Dpp via Sog (for more details, see 132, 196).

Components downstream of the Tkv/Punt receptor complex have also been identified. *Mad* and *Medea* mutants are enhancers of a weak *dpp* phenotype, and mediate the transcriptional response to Dpp signaling (107, 108, 123, 154). Between them, the *Mad/Medea* and *schnurri* DNA binding transcription factors mediate activation and repression of *dpp*-responsive transcriptional targets (27, 61, 77, 93, 107, 108, 123, 154, 184). *Mad* family genes seem to collaborate with a variety of transcriptional cofactors and are more generally required for Dpp signaling than *schnurri*, which functions in part by antagonizing the general repression of *dpp* transcriptional targets by *brinker* (71, 93). Some of the above components have been isolated not only from other insect species, but also from across phyla, suggesting that *dpp* may have played an ancient role in dorsoventral axis formation. We first discuss comparative insect studies which are distinct from chordate studies in focusing on Dpp expression, not activity. We then discuss data from chordates, where the regulation of Dpp signaling activity has been successfully dissected biochemically and genetically.

dpp and *zen* homologues have been isolated from more basal insects, including the beetle *Tribolium castaneum* (*Tc-dpp* and *Tc-zen*) and the grasshopper *Schistocerca americana*. Analysis of the genes has been used to determine whether they are expressed in dorsal tissue (as in *Drosophila*), and whether *dpp* expression is regulated by Dorsal protein. Expression of *Tc-dpp* and *Tc-zen* is observed in serosal cells (23, 37, 139). Serosa is an extraembryonic membrane with a putative protective function during insect embryogenesis (3). In the higher flies extraembryonic membranes are reduced and referred to as the amnioserosa, a dorsally placed tissue that, in *Drosophila*, also expresses *dpp* and *zen* (2). In more basal insects such as *Tribolium* and *Schistocerca*, the serosa is anteriorly placed in the egg. At first this may suggest that *dpp* and *zen* do not play a role in DV axis formation in basal insects but expression of *dpp* in the serosa of basal insects may still play a role in patterning dorsal tissue, given the position of serosa relative to the dorsal ectoderm. *Tc-dpp* is expressed in serosal cells surrounding the germ anlage [i.e., closest to dorsal ectoderm (72, 139)]. Positionally, this is essentially the same, relative to the dorsal ectoderm, as *dpp* and zen expression in *D. melanogaster* and may therefore constitute "dorsal" expression. Thus the serosal *dpp* domain might still be involved in patterning "dorsal" tissue in the *Tribolium* embryo. Later in development, both *Tribolium* and *Schistocerca dpp* are expressed in the dorsal ectoderm in the abdominal field of the embryo (72, 139). Thus there is potentially a more compelling argument for DV patterning by *dpp* during later development.

Evidence for repression of *Tc-dpp* by Dorsal protein can be inferred from the fact that *Tc-dpp* is not co-expressed in most cells with nuclear Tc-Dorsal (23). Although one could argue against repression of *Tc-dpp* by Tc-Dorsal as they are co-expressed in terminal cells, both *dpp* and nuclear Dorsal are also found in terminal cells of the *D. melanogaster* embryo suggesting differences in *dpp* regulation at the termini in both species. Here we should point out the limits of expression data, in that *Tribolium dpp* expression also potentially overlaps with the anterior factor *hunchback* and with the terminal system gene product *tailless* (23, 148, 193). Thus expression data could just as well be used to argue that *Tc-dpp* is regulated by the anterior or terminal systems. The growing inventory of tools available for gene expression manipulation in *Tribolium* may confirm whether the transcriptional regulation of *dpp* by Dorsal is conserved from flies to beetles.

dpp, one of the *D. melanogaster* homologues of the TGFβ superfamily, is most similar to the BMP2/4 group. Indeed, the *dpp* phenotype can be rescued using a Dpp-BMP4 fusion product (112, 113). Manipulations in *Xenopus* as well as zebrafish genetics suggest that the Dpp signaling pathway plays a conserved role in axial patterning (reviewed in 29, 56). As in *Drosophila*, regulation of Dpp protein activity may be the key to generating positional information along the DV axis in chordates. This has been shown in *Xenopus* by the injection of BMP4, which affects DV patterning in a concentration-dependent fashion: Injection of BMP4 can ventralize mesoderm (promoting fates such as blood, and having antineurogenic effects on ectodermal tissue). Similarly, injection of a dominant negative BMP receptor shows that BMP signaling is required for the development of ventral fates (47). The role of BMPs in DV patterning of zebrafish has been demonstrated genetically, where the *swirl* (BMP2) mutant has a dorsalized phenotype [broadened notochord and expanded somites (51, 78)].

Regulation of BMP signaling is also conserved in chordates in that antagonists of BMPs operate during DV axial patterning. Classical embryological experiments demonstrate that dorsal fates are promoted by organizer tissue, which upon transplantation leads to dorsalization of ventral host tissue, i.e., an ectopic DV axis (163). The *Xenopus* organizer expresses BMP antagonists including *chordin*, the functional homologue of *short gastrulation* (57, 141). *Xenopus chordin* injections induce a second axis, rescue UV ventralized embryos, and can dorsalize mesoderm by antagonizing BMPs (118, 141). Conversely, the effect of *chordin* injections alone can be mitigated by the co-injection of high concentrations of BMP4, suggesting a competitive binding interaction (189). Physical Chordin/BMP4 interaction has been confirmed by in vitro binding studies that demonstrate interaction with a binding constant sufficient to interfere with receptor binding (118). Genetic data from zebrafish confirm that the interaction of BMP/Chordin mirrors the *Drosophila* Dpp/Sog interaction in patterning the DV axis. Thus the *chordino* (*chordin*) mutant leads to ventralization, a phenotype that is suppressed by the *swirl* (BMP2) mutant (51, 78, 149).

Further levels of conservation are revealed upon examining Chordin regulation. A *Xenopus tolloid* homologue is capable of cleaving Chordin, thereby overriding Chordin inhibition of BMP4 (117). More recently, conservation of the *twisted gastrulation* gene has also been demonstrated. In contrast to initial data, *twisted gastrulation* may be a conserved inhibitor of BMP function. Some evidence indicates that it is responsible for producing a differential cleavage product of Sog/Chordin that may have increased anti-Dpp/BMP activity (19, 111, 132, 153, 196).

Downstream effectors of Dpp signaling also appear to be conserved in chordates, such that signal transduction is mediated by *Xenopus* homologues of the *Mad* transcription factor (for example, see 89). A number of zebrafish mutations leading to DV phenotypes lie in genes encoding members of the BMP signaling pathway, once again suggesting a conserved role for this signal transduction pathway in DV axis formation (for examples, see 9, 14, 54).

Thus the *dpp* ligand, its antagonist (*sog*), potentiator (*tolloid*), and downstream components (*Mad* homologues) all play a role in dorsoventral axis formation in *Drosophila*, *Xenopus*, and zebrafish, suggesting a potentially ancient role for this pathway in DV axis formation. Moreover, injection of *D. melanogaster dpp* and *sog* into *Xenopus* shows that they behave functionally as BMP4 and *chordin* (57, 144). However, not all aspects of DV axis formation are conserved from flies to chordates. Although the protein Noggin binds and antagonizes BMP4 in *Xenopus*, an orthologue has yet to be found in the genomic sequence of the fly (103, 198).

Since Dpp signaling is involved in providing axial information in both *Xenopus* and *D. melanogaster*, one might argue that it is an ancient component in dorsoventral axis formation and perhaps ancestral. However, there is evidence that the axis classically regarded as the dorsoventral axis in *Xenopus* may not actually be so. Reassessment of the position of primitive blood on the *Xenopus* fate map has been used to argue that the classical embryonic DV axis can actually be characterized as an anteroposterior axis (83). This does not necessarily argue for nonconservation, only that the BMP signaling pathway may also control some aspects of anteriorization along the "DV" axis in *Xenopus*. One could also argue that as *dpp* genetically behaves morphogenetically, it could have been recruited into axis specification in multiple lineages, given that morphogenetic activity is clearly an efficient means of generating positional information.

However, the BMP pathway is tightly coupled to the dorsoventral axis, as demonstrated by the facts that the deuterostome dorsoventral axis seems morphologically inverted when compared to protostomes, but that this inversion occurs with appropriate alterations in the expression patterns of *dpp*/BMP4 and *sog/chordin* (56). More specifically, tissue closest to the blastopore and central nervous system expresses *chordin/sog*, whereas tissue further away (be it amnioserosa or blood) expresses BMP4/*dpp*. This suggests a tight coupling of the *dpp* homologue to distinct tissue fates, wherever they lie along the DV body axis (dorsal in flies and ventral in frogs), and may reflect an ancestral role of the pathway in restricting neural fates. The intimate relationship between *dpp*/BMP expression and dorsoventral territories across phyla argues that the Dpp signaling pathway played an ancestral role in DV patterning. However, systems involved in the activation of *dpp*/BMP4 transcription have not been shown to be conserved across species in the context of axis formation, and thus may be derived (see below). This may also be true of the system for setting up ventral fates in *Drosophila*.

twist Is Involved in Mesoderm Patterning in Many Animals

twist is essential for specifying the ventral-most territory of the *Drosophila* embryo [fated as mesoderm; (Figure 2*b*)]. Genetic analysis and dissection of the *twist* promoter suggest that it is directly activated by nuclear Dorsal (182). Loss-of-function alleles of *twist* lead to loss of mesoderm, suggesting that *twist* is essential for specification of ventral tissue in flies (180). This role for *twist* seems to be conserved between flies and beetles as the *Tribolium twist* orthologue is expressed at blastoderm stage in a narrow ventral stripe overlapping the ventral embryonic domain of nuclear Dorsal (23, 160). Slightly later in embryogenesis, the *twist* expression domain retracts anteriorly and widens posteriorly in an apparently *Tc-Dorsal*-independent fashion (23). Thus in *Tribolium*, the role of twist as a specifier of ventral fate seems conserved, but its early expression pattern may not be as dependent on *dorsal* as it is in *Drosophila*.

However, unlike *dpp*, *twist* does not appear to play a role in specifying axial fate in the chordates, as evidenced by its lack of involvement in pan-mesodermal specification. Although multiple *twist* orthologues have been isolated from chordates including mouse and *Xenopus*, they are expressed in and seem to activate specific mesodermal derivatives, and may also inhibit some myogenesis (for example, see 59, 191). *twist* may therefore control submesodermal fates rather than acting as a mesoderm or ventral specification factor in the chordates. Interestingly, the jellyfish *twist* orthologue is expressed in nonmesodermal tissue (164). Jellyfish are considered to be diploblastic (i.e., have only two germ layers, lacking true mesoderm). Thus the expression of *twist* in muscle-like cells in the jellyfish might suggest that *twist* is involved in specification of a mesodermal-like layer. Alternatively, *twist* may play a role in specification of muscle-like cell fate that predates the origin of mesoderm. *twist* may have been independently recruited as an axial patterning output in the arthropod lineage, a likely hypothesis as the data from jellyfish and chordates suggest that the ancestral role of *twist* is not general mesoderm specification. As with *dpp*, regulation of *twist* by the Dorsal transcription factor may be a recent innovation, as the *Toll* signaling pathway may have been recently recruited into DV axis formation in the insects.

The *Toll/dorsal* Signaling Pathway May Have Been Co-Opted into Axis Formation from the Immune System

The role of the *Toll* signaling pathway in dorsoventral axis formation may not be ancestral. This is particularly interesting as the *D. melanogaster Toll* pathway is directly responsible for restricting transcriptional activation of *dpp* to the dorsal side of the embryo, and the latter appears to be ancestral for axis formation (see above). As with the *dpp* pathway, the components of the *Toll* signaling pathway have orthologues within the insects and across phyla. Thus *Toll* homologues have been isolated from *Tribolium castaneum*, as well as from mammals (52, 94, 130). The key downstream effector of *Toll* receptor homologues seems functionally conserved across species (*dorsal* in the insects, *NFκB* in mammals).

Expression data indicate that both *Toll* and *dorsal* play a conserved role in DV patterning of *Tribolium* (23, 94). Both *Toll* and *dorsal* are found in gradients in the ventral region of the embryo during early *Tribolium* embryogenesis. Indeed, immunostaining indicates that the *Tribolium* Dorsal protein forms a ventral nuclear localization gradient during blastoderm stage. Interestingly, the *Tribolium Toll* receptor seems at first sight to differ in some respects from its *D. melanogaster* counterpart (94). Rather than being maternally provided and ubiquitously expressed, *Tribolium Toll* is found in a ventral gradient in cells with nuclearly localized *Tc*-Dorsal. This finding was used as the basis of the proposition that the *Toll* gradient may form zygotically in response to nuclear localization of Dorsal (23, 94). Local (as opposed to ubiquitous) injections of *D. melanogaster Toll* can rescue the *Drosophila Toll* phenotype (52). Thus, although it seems important to have enough *Drosophila* Toll receptor to sequester ligand, *Toll* transcript need only be applied (and presumably expressed) relatively locally to form a normal axis. The localized expression of *Toll* in *Tribolium* does not necessarily imply a functional difference to ubiquitous *Toll* expression in *Drosophila*. Furthermore, it has been argued that there has been a shift toward maternal control of axis formation in higher insects such as *Drosophila*, which is consistent with a shift from zygotic to maternal expression of factors such as *Toll* in flies (115, 121). Essentially, the ventral increase in *Toll* levels in putative response to nuclear Dorsal may indicate extensive zygotic refinement of pattern in response to axial signaling. Thus, although the transcriptional regulation of *Tribolium Toll* appears to be different, the

Toll/dorsal pathway may play a fundamentally conserved role in axis formation from beetles to flies.

It is unknown whether these factors play a role in axis formation in other phyla. The mammalian *Toll* and *dorsal* homologues were identified for their role in the immune system. Interestingly, *Toll* and *dorsal* in *D. melanogaster* and the *IL-1R/NFκB* pathway in mammals both play a role in stimulation of the innate immune response during pathogenic aggression (84, 85, 95). Although *D. melanogaster Toll, spätzle,* and *dorsal* can stimulate dorsalization in UV-ventralized *Xenopus* embryos, there is no evidence that the chordate homologues of these genes function, or are expressed, at the correct developmental time to play a role in DV axis formation in these animals (6). Thus one can only hypothesize that the ancestral role of the *Toll/dorsal* signaling pathway was in the immune system. The pathway appears to have become involved in axis formation in the lineage leading to insects (Figure 3).

Conservation of the DV Axis Formation Cassette in Its Entirety?

Although the Dpp signaling pathway appears to play a highly conserved role in DV axis formation, this is not necessarily true of either *twist* or *dorsal*. *twist* may have played an ancestral role in some form of mesoderm development (either general

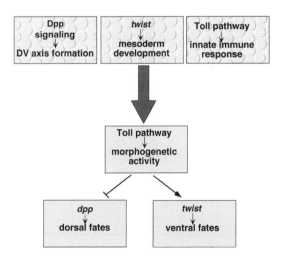

Figure 3 A simplified model for the evolution of DV patterning in *D. melanogaster*. The model considers an ancestral state (*bubbled boxes, top row*) where Dpp signaling was involved in DV patterning, *twist* in mesoderm fate specification, and *Toll* signaling in the innate immune response. In the lineage leading to flies (*big arrow*), *dpp* and *twist* fell under the regulation of *Toll* signaling during early development, and thus a single maternal axis formation system.

mesoderm specification or, more likely, specification of muscle percursors) in triploblastic lineages, and perhaps even in more primitive animals. At some point in the lineage leading to higher insects, very early *dpp* and *twist* expression may have fallen under the regulation of the *Toll* signaling pathway. The *Toll/dorsal* pathway could have been recruited from an ancestral role in the immune system for DV axis specification (as suggested by conserved usage of this signaling system in innate immunity from *Drosophila* to mammals). Assumption by *dorsal* of transcriptional regulation of factors such as *dpp* and *twist* would have led to the modern cascade of pathways specifying regions along the DV axis in *D. melanogaster* (Figure 3).

However, the above genes may already form a regulatory cassette in the chick, suggesting that regulation of *twist* and *dpp* by *dorsal* is more ancient than implied in the above scenario (18, 75). During outgrowth of the chick limb bud, *NFκB* (*dorsal*) is expressed in mesenchyme underlying the apical ectodermal ridge (AER). The AER is a morphological ridge along the developing limb bud, which is known from classical embryological experiments to be required for limb outgrowth (142, 175). Decreasing the activity of NFκB using viral overexpression of an *IκB* (*cactus*) mutant that cannot be targeted for degradation leads to defects in limb outgrowth (18, 75). Interestingly, *twist* mutants in mouse and humans lead to similar limb phenotypes, suggesting conservation of this role in limb development across vertebrate species (15, 60). *twist*, which is expressed in chick limb mesenchyme, is downregulated under these experimental conditions, implying that it is under positive regulation by *NFκB* (18, 75). Furthermore, in this experiment the expression of BMP4 is upregulated, suggesting that a *dpp* homologue is negatively controlled by *NFκB* in the context of chick limb development.

Thus the entire regulatory cassette (*dpp* and *twist* under control of a *dorsal* homologue) may actually be an ancient network that in extant animals has become critical for limb development in chordates and DV axiation in *D. melanogaster*. Whether this cassette had a function in axis formation in the chordates is not known since vestiges of *dorsal*-mediated *dpp* regulation during early embryogenesis have not been found outside the arthropods. There is as yet no indication that a *NFκB* homologue is expressed in mouse during early embryogenesis [although not all *dorsal* homologues have been tested, and the transgenic reporter approach used to analyze expression may not recapitulate the complete expression pattern of the gene (see 146)]. The fact that *D. melanogaster spätzle* and *Toll* can induce dorsalization in ventralized *Xenopus* embryos is compelling in this context (6). However, expression analysis and loss-of-function studies are required to show that the *Xenopus Toll* and putative *spätzle* orthologues are expressed at the correct time and can also generate DV phenotypes when overexpressed.

Thus far, we have discussed conservation of signal transduction systems that can easily be envisioned as being conserved in animals with very different embryogenesis from that of *D. melanogaster*. We now examine the anterior system, which is based on the diffusion of molecules in the syncytial environment of the early *Drosophila* embryo, and ask whether such an extreme mechanism for generating a morphogen gradient can be conserved.

THE ANTERIOR SYSTEM

AP Patterning in *Drosophila melanogaster*

Our understanding of the anterior system centers around the archetypal morphogen, *bicoid* (*bcd*) (Figure 4; see color insert). During oogenesis *bcd* mRNA becomes localized to the anterior of the oocyte in a process depending upon various factors (96). Diffusion of Bcd protein from its anterior location sets up a gradient that determines positional information. The homeoprotein Bcd regulates gene activity through transcriptional and translational control. An example of the former is zygotic *hunchback* (*hb*), which *bcd* transcriptionally activates in an anterior domain (33). Like Bcd, *hb* provides information for anterior patterning through gap, pair-rule, and *hox* genes (reviewed in 127, 157, 173). Bcd is also responsible for the anterior translational repression of ubiquitous maternal *caudal* mRNA leading to a posterior gradient of the homeoprotein transcription factor Caudal (34, 129). *caudal* is involved in posterior patterning through the gap genes and behaves as a homeotic gene in the posterior-most segment of the fly (90, 97, 98, 128, 151). The anterior system involves two factors, *caudal* and *hb*, that are conserved to varying extents in other species.

caudal May Play a Conserved Role in Posterior Patterning Across Phyla

caudal appears to play a role in posterior patterning throughout arthropods and across to the chordates. Within the arthropods, *caudal* has been cloned from, amongst other insects, *Tribolium castaneum* and the silkmoth *Bombyx mori* (150, 194). Both *Tribolium* and *Bombyx caudal* (RNA and protein) form posterior gradients, reminiscent of those observed in the flies, and suggestive of conserved regulation. Furthermore, heterologous expression of the *Tribolium caudal* gene in *D. melanogaster* shows that a *bcd*-dependent gradient can be generated, suggesting conserved translational regulation of *caudal* in the beetle, though a *Tribolium bicoid* homologue has not yet been identified (192). *caudal* is also expressed posteriorly in the grasshopper embryo, suggesting a conserved role in posterior patterning in a more basal insect (31). *caudal* may also play a conserved role in the wasp *Nasonia vitripennis*, where the *head only* mutation greatly resembles the phenotype of the *D. melanogaster caudal* mutant (121). Strikingly, *caudal* homologues may play conserved roles in vertebrate development, as well as in an ascidian and nematode (for example, see 36, 43, 44, 69, 73, 76, 91, 119). For example, *cdxA-C* in chick are expressed in a spatially and temporally graded fashion in posterior neural plate and midline during primitive streak stages, suggesting a conserved role in posterior patterning. Loss of *C. elegans caudal* (*pal-1*) function affects posterior blastomere and adult male tail development (36, 63). Similarly, the *caudal* homologue in an ascidian has been shown to play a functional role in tail development (76). Thus *caudal* likely played a role in posterior patterning in the ancestral insect and perhaps even in the ancestral bilaterian.

hb May Play a Conserved Role in AP Patterning Within the Insects

The status of *hunchback* (*hb*) as an ancestral AP patterning factor is less clear-cut. Orthologues have been cloned across phyla, but there is no evidence for a role in AP patterning outside the insects. As described above, maternal and zygotic *hb* are involved in AP patterning during early *D. melanogaster* development (Figure 4). Flies closely related to *D. melanogaster* (*Drosophila virilis*), as well as basal insects such as the grasshopper *Schistocerca americana*, have *hb* homologues (115, 131, 158, 159, 168, 183, 193). The fact that insect *hb* homologues are expressed maternally and zygotically in an anterior domain suggests a conserved role in AP patterning within the insects. Interestingly, the grasshopper, *S. americana*, expresses *hb* in a cellular environment in blocks of different concentration along the AP axis (115). This may reflect how positional information is conveyed in a cellular environment, as opposed to the gradients observed in the syncytial environment of the *Drosophila* blastoderm. However, outside the insects there is no clear evidence for a role in AP patterning from expression data, although *hb* orthologues are found in conserved domains in the nervous system (and are therefore probably true orthologues). For example, neither *H. triserialis* (annelid) nor *C. elegans* (nematode) *hb* is expressed in an early anterior domain that would indicate a role in AP patterning (38, 70, 143). Perhaps *hb* was co-opted relatively recently from the nervous system into its role as an anterior morphogen within the arthropods. Thus *hunchback* may have played a role in AP axis formation in the lineage leading to insects, but probably not in other phyla. *hb* has orthologues outside the insects. Expression data from the chelicerates and crustaceans should indicate whether *hb* plays a role in AP patterning in all arthropods.

bcd Function May Be a Recent Novelty in AP Patterning

The status of *bcd* in ancestral AP patterning is markedly different. Despite its importance in AP patterning in *Drosophila*, *bcd* homologues have been isolated only from higher dipterans (the cyclorrhaphan flies; 158, 168). There is evidence that *Megaselia abdita bicoid* (*Ma-bcd*) does play a functional role in axis formation: *Ma-bcd* is expressed anteriorly, and dsRNA interference experiments lead to a phenotype that resembles that of the *D. melanogaster bcd* mutant, except that the range of its action extends more posteriorly (168). These data suggest that in contrast to *Drosophila*, *bcd*'s sphere of influence may extend further posteriorly along the AP axis of *Megaselia*. Indirect evidence for the conservation of *bcd* in a more basal insect comes from *Tribolium*. As noted above, *Tc-caudal* transcript can be regulated in a *bcd*-dependent fashion when introduced into flies (192). This suggests the existence of an as yet unidentified *bicoid*-like activity in *Tribolium*. Such data would imply that *bcd* is involved in AP patterning from flies to beetles.

Molecular data, however, suggest that the *bcd* gene arose relatively recently. Phylogenetic analysis of the *Megaselia abdita bicoid* and *zen* orthologues suggests that they are products of a Hox3 duplication in an ancestor of the higher dipterans (167). *bcd* and *zen* reside together in the part of *D. melanogaster* Hox cluster

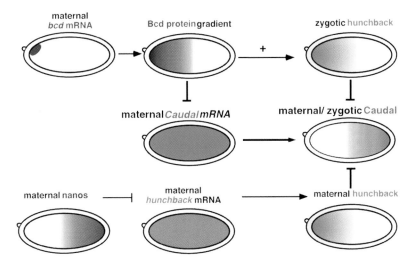

Figure 4 AP patterning in *D. melanogaster*. The diagram illustrates the interconnections between the anterior and posterior maternal co-ordinate systems. The anterior system (key gene *bicoid*) leads to the localized activation of zygotic *hb* transcription and maternal *caudal* translational control. The posterior system involves the translational repression of maternal *hunchback* transcript. The resulting Hunchback protein gradient has an effect on zygotic *caudal* transcription. Anterior is to the left, posterior to the right, and shading indicates gradient vs ubiquitous localization.

in the location where one should find a Hox3 orthologue. This adds support to the idea that the two genes are products of a Hox3 duplication. Furthermore, the region of the *Tribolium* hox cluster in which one would expect to find *bcd* has now been sequenced and no *bcd* orthologue was found (16). Although the *Tribolium* orthologue could have moved via a recombination event, the inability to isolate *bcd* from lower insects, as well as the sequence analysis of *Megaselia bcd* and *zen*, are evidence for a relatively recent origin of *bcd* in the lineage leading to higher flies. *bcd* is thus an interesting case in which a gene has arisen recently and yet become a major molecular player in axis formation (whether the ancestor of *bcd* played a role in axis formation is interesting but unresolved). The mechanism by which *bcd* might have become involved in AP axis formation is discussed below and by Schmidt-Ott (145). The Hox3 orthologue has been cloned from two chelicerates, a mite and a spider (28, 179). Hox3 in chelicerates is expressed in a discrete domain of the AP axis, but its anterior border coincides with the Hox gene *proboscopedia* (*pb*). Thus, although Hox3 in chelicerates is found in a hox-like expression domain, it is expressed more anteriorly than expected, suggesting a breakdown in colinear expression of this gene within the arthropods. Overlap with *pb* and the breakdown of colinearity are hypothesized to have allowed Hox3 to lose hox function and take on novel roles in the lineage leading to insects (28, 179).

What does this tell us about the way in which the anterior patterning system in *D. melanogaster* evolved? The expression of *hb* outside the insects suggests that its role in nervous system development is highly conserved. The ancestral patterning system in insects may have involved a posterior *caudal* gradient and an anterior *hb* gradient. The duplication of an ancestral Hox3 gene may have given rise to *bcd*, which in the modern fly plays a dual role as translational regulator of maternal *caudal*, and transcriptional activator of zygotic *hb*. In this scheme a novel gene, *bcd*, has become a major organizer of the AP axis (Figure 5) (reviewed in 30). Thus anterior patterning in *D. melanogaster* displays the same properties as the DV system: Potentially ancient (*caudal*) and more recently recruited (*hb*) regulatory factors have fallen under the global coordination of a novel gene (the anterior *bcd* gradient) (Figure 5).

EVOLUTION OF AXIS FORMATION MECHANISMS

The maternal coordinate systems in *Drosophila* appear to be a melange of molecular processes elaborated upon multiple times in evolutionary history. Modern *D. melanogaster* utilizes evolutionarily ancient factors such as *dpp* and *caudal*, which play a conserved role in axial patterning in chordates. *Drosophila* also use molecules such as *zen* and *bicoid*, which may have arisen relatively recently, in the lineage leading to higher dipterans. This suggests that axis formation in *D. melanogaster* is a plastic and rapidly evolving process at the molecular level and is constantly taking advantage of new cues and innovations. However, innovation does not necessarily mean the displacement of pathways, although the cues

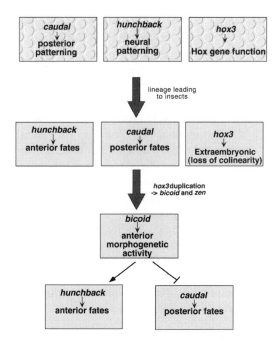

Figure 5 A potential, simplified scheme for the origin of the *D. melanogaster* anterior system. The scheme envisages an ancestral state (*bubble boxes, top row*) where *hunchback* was involved in neural patterning, *caudal* in posterior patterning, and a Hox3 gene in segmental identity along the AP axis. In the lineage leading to insects, *hunchback* and *caudal* may both have been involved in axial patterning, whereas Hox3 became expressed in a spatial pattern that is no longer colinear with other Hox genes (*middle row*). In the lineage leading to the higher dipterans, the duplication of Hox3 may have led to *bicoid*, which acts in the modern fly as a major player in AP axis formation through the regulation of *hunchback* and *caudal*.

regulating them may have changed, as exemplified by the deep-set link between Dpp signaling and DV axiation. Also, it is not simply axial cues that change, but downstream specification factors can also fall under the control of axial patterning systems. This is exemplified by *twist*, which plays a role in the development of mesoderm and its derivatives in all systems studied, but may only constitute a global mesoderm specification factor within the insects.

REDUNDANCY IN AXIS FORMATION

How can an event as important as axis formation evolve rapidly? One mechanism would be to retain old pathways in a redundant fashion. *bcd* may still be rapidly changing at the sequence level, and its role in AP axiation might have arisen relatively recently. Interestingly, the anterior system has long been known to show redundancy with the posterior maternal coordinate system. Such large-scale redundancy may be crucial to plasticity in axis formation.

Both the anterior and posterior systems lead to the same outcome, although this outcome is slightly temporally displaced: Both systems result in a Hb protein gradient with highest levels at the anterior (Figure 4) (see 33, 62, 67, 173, 176). Hb protein is detectable as a plateau until about 50% egg length, where it begins to taper off (176). As mentioned above, the Bcd gradient is crucial to generating the zygotic *hb* pattern. The earliest differential *hb* expression is, however, due to the translational repression of a uniform maternal pool of *hb* transcript (Figure 4) (see 176, 178). This translational repression is mediated by *nanos*, and its cofactor *pumilio* (62, 67, 102). Pumilio recognizes a sequence (NRE) in the 3'UTR of *hb* transcript, and forms a multiprotein complex in vitro, which includes the NRE, Nanos protein, and Pumilio itself (161, 162, 188). Since *nanos* transcript is maternally localized to the posterior of the developing oocyte and translationally repressed anteriorly, a Nanos protein gradient forms emanating from the posterior (Figure 4) (26, 185). Hence, maternal *hb* transcript is translationally repressed at the posterior, leading to an anterior gradient of Hb protein.

Thus two systems, an anterior and a posterior system, independently generate an anterior Hb gradient. Furthermore, elegant genetic data indicate that the two gradients of *hb* are redundant. Genetically, the *nanos* phenotype, in which abdominal segments are entirely lost, can be rescued by eliminating maternal *hb* transcript (62, 67). The *nanos* axis formation phenotype is therefore due to the ectopic activity of maternal *hb* in the posterior of the embryo. Furthermore, zygotic *hb* compensates entirely for loss of the maternal transcript. Therefore, the posterior maternal coordinate system can be eliminated and AP axis formation occurs normally, with only the anterior system operating.

More recent evidence suggests the converse may also be true. Elimination of *bcd* activity cannot be compensated for by the posterior system as it stands (190). Phenotypes are still observed in T2/T3 (parasegment 4), so that levels of maternal *hb* seem to be unable to compensate for the lack of zygotically activated *hb* in this domain. However, if extra copies of *hb* are supplied, and the levels of *hb* further increased by reducing levels of a transcriptional repressor of zygotic *hb* (*knirps*), the need for *bcd* in thoracic development is abrogated (190). Thus *hb* can almost entirely compensate for the anterior system during early AP axis formation (almost entirely, since the extreme anterior of the *bcd* cuticle phenotype cannot be rescued in this way).

Why does *D. melanogaster* have two virtually redundant systems carrying out such similar functions? The answer may be that one system was derived more recently than the other. Nanos response elements are found in *hb* orthologues in many insects (for example, see 158, 183, 193). Furthermore, *nanos* and *pumilio* have been cloned from multiple phyla including chordates, nematodes, and annelids (e.g., 25, 81, 101, 120, 174, 197). *nanos* may even play a role in axis specification in comparatively basal insects (S.L. & N.H.P., unpublished observation). Thus while the *bcd* system seems a recent innovation in axis formation, setting up a gradient of Hb using translational repression by Nanos seems to be a more ancestral mechanism. What we may be observing in *D. melanogaster* is an intermediate in the displacement of an ancient AP patterning system (*nanos* regulated) by a more recently innovated one (*bcd* regulated). Redundancy of the two systems could constitute a safety net that protects the integrity of AP patterning as axis formation rapidly evolves.

LINKING MORPHOLOGICAL CHANGES IN EMBRYOGENESIS TO EVOLUTION OF AXIS FORMATION

Recent data from dipterans suggest a correlation between the origin of *bcd* as an anterior patterning factor and a profound and potentially linked change in developmental morphology. Less derived insects have two extraembryonic membranes, the amnion and the serosa, that have distinct morphologies and derivations during embryogenesis (3). However, *D. melanogaster* has a fused and reduced single extraembryonic tissue called the amnioserosa (2). The link between the amnioserosa in *D. melanogaster* and the amnion and the serosa in less derived insects is clear from the expression of extraembryonic markers such as *zen*, as well as from the fate of these tissues in later embryogenesis. However, whereas *D. melanogaster* amnioserosa lies in the DV axis of the egg and is clearly specified by the DV axis formation pathway (under the control of *dpp*), serosa in basal insects originates from tissue anterior to the embryo proper. A major change may thus have occurred in specification of extraembryonic material since it has moved into a different egg axis.

Examination of the status of extraembryonic membranes within dipterans suggests that the fused and reduced amnioserosa is a character of the higher dipterans (cyclorrhaphan flies). Thus the basal cyclorrhaphan fly *Megaselia abdita* has a single extraembryonic membrane, whereas a representative species from a sister taxon, the *Empidoidea* (*E. livida*), has a separate amnion and serosa (145, 167). These data place the origin of amnioserosa and its DV axis location in the same phylogenetic position as the hypothesized duplication of a Hox3 group gene to give *bcd* and *zen* (167). This gene duplication may also have been essential for relocation of extraembryonic material from the anterior region of the egg to the DV axis under control of the *Toll* signaling system. The appearance of a novel anterior patterning factor, *bcd*, may have been key to allowing full spatial repositioning of amnioserosa, along with *zen* expression, into the DV axis of the egg (145).

Interestingly, there are also major changes in head morphology at the base of the cyclorrhaphan flies. In particular, the cyclorrhaphan flies have reduced feeding structures (the cephalopharyngeal skeleton) and larvae hatch with involuted heads (195). Although the morphology of head characters is variable amongst the cyclorrhaphan flies, displacement of an ancestral anterior patterning system by *bcd* may have been essential for (or even responsible for) these changes. Early patterning by *bcd* may have a profound effect on later head morphology, and the fact that *Megaselia* has a markedly different range of *bcd* activity may indicate that even within the cyclorrhaphan flies *bcd* has varying roles or effects (168).

Thus the molecular evolution of axis formation systems in the higher flies appears to correlate with major changes in morphology. This would suggest that rapid evolution of axis formation is potentially tied to the diversity of morphology before and perhaps even after the phylotypic stage, i.e., the stage where embryos of different species within a phylum converge on a very similar morphology.

CONCLUDING REMARKS

Recent data, such as those examining *bicoid* and *zen* in more basal insects, suggest that some factors involved in axis formation in *D. melanogaster* are derived (167). Along with data examining the morphology of axis specification in nematodes, these findings suggest that axis formation is a variable and plastic process (46). This is consistent with models that view developmental processes as an hourglass with variability at the base leading to a highly conserved phylotypic stage, followed by increasing diversity in adult body plan (see 35, 122). This analogy may understate the remarkable underlying conservation of axis formation (for example in the use of Dpp signaling in DV axis formation and *caudal* in AP patterning).

The recent data that correlate changes in morphology with molecular evolution of axis formation are particularly exciting, as they indicate the types of morphological change that rapid evolution of axis formation might allow. Early developmental processes have been investigated in the light of extreme modifications in life history, for example, in the context of the endoparasitic wasp *Copidosoma* (reviewed in 49). *Copidosoma* undergoes polyembryonic development, where early cleavage events give rise to 2000 randomly oriented embryos. Here axis formation may differ from the mechanisms elucidated in *D. melanogaster* (50). Future data from *Copidosoma* may shed light on the molecular evolution of axis formation during a derived and fundamentally different form of embryogenesis.

ACKNOWLEDGMENTS

We thank William Browne, Gregory Davis, James McClintock, and Steven Podos for reading and helpful discussion of the manuscript. NHP is an HHMI investigator.

Visit the Annual Reviews home page at www.AnnualReviews.org

LITERATURE CITED

1. Aguinaldo AM, Turbeville JM, Linford LS, Rivera MC, Garey JR, et al. 1997. Evidence for a clade of nematodes, arthropods and other moulting animals. *Nature* 387:489–93

2. Anderson DT. 1972. The development of holometabolous insects. In *Developmental Systems: Insects 1*, ed. SJ Counce,

CH Waddington, 1:166–241. New York: Academic

3. Anderson DT. 1973. *Embryology and Phylogeny in Annelids and Arthropods.* Oxford: Pergamon

4. Anderson KV, Jürgens G, Nüsslein-Volhard C. 1985. Establishment of dorsal-ventral polarity in the *Drosophila* embryo:

genetic studies on the role of the Toll gene product. *Cell* 42:779–89

5. Anderson KV, Nüsslein-Volhard C. 1986. Dorsal-group genes of *Drosophila*. In *Gametogenesis and the Early Embryo*, ed. J Gall, pp. 177–94. New York: Liss

6. Armstrong NJ, Steinbeisser H, Prothmann C, DeLotto R, Rupp RA. 1998. Conserved Spatzle/Toll signaling in dorsoventral patterning of *Xenopus* embryos. *Mech. Dev.* 71:99–105

7. Ashe HL, Levine M. 1999. Local inhibition and long-range enhancement of Dpp signal transduction by Sog. *Nature* 398:427–31

8. Ashe HL, Mannervik M, Levine M. 2000. Dpp signaling thresholds in the dorsal ectoderm of the *Drosophila* embryo. *Development* 127:3305–12

9. Bauer H, Lele Z, Rauch GJ, Geisler R, Hammerschmidt M. 2001. The type I serine/threonine kinase receptor Alk8/Lost-a-fin is required for Bmp2b/7 signal transduction during dorsoventral patterning of the zebrafish embryo. *Development* 128:849–58

10. Belvin MP, Jin Y, Anderson KV. 1995. Cactus protein degradation mediates *Drosophila* dorsal-ventral signaling. *Genes Dev.* 9:783–93

11. Berg H. 1975. How bacteria swim. *Sci. Am.* 233:36–44

12. Bergmann A, Stein D, Geisler R, Hagenmaier S, Schmid B, et al. 1996. A gradient of cytoplasmic Cactus degradation establishes the nuclear localization gradient of the Dorsal morphogen in *Drosophila*. *Mech. Dev.* 60:109–23

13. Biehs B, Francois V, Bier E. 1996. The *Drosophila short gastrulation* gene prevents Dpp from autoactivating and suppressing neurogenesis in the neuroectoderm. *Genes Dev.* 10:2922–34

14. Blader P, Rastegar S, Fischer N, Strahle U. 1997. Cleavage of the BMP-4 antagonist chordin by zebrafish tolloid. *Science* 278:1937–40

15. Bourgeois P, Bolcato-Bellemin AL, Danse JM, Bloch-Zupan A, Yoshiba K, et al. 1998. The variable expressivity and incomplete penetrance of the *twist*-null heterozygous mouse phenotype resemble those of human Saethre-Chotzen syndrome. *Hum. Mol. Genet.* 7:945–57

16. Brown S, Fellers J, Shippy T, Denell R, Stauber M, Schmidt-Ott U. 2001. A strategy for mapping *bicoid* on the phylogenetic tree. *Curr. Biol.* 11:R43–44

17. Brummel TJ, Twombly V, Marques G, Wrana JL, Newfeld SJ, et al. 1994. Characterization and relationship of Dpp receptors encoded by the *saxophone* and *thick veins* genes in *Drosophila*. *Cell* 78:251–61

18. Bushdid PB, Brantley DM, Yull FE, Blaeuer GL, Hoffman LH, et al. 1998. Inhibition of NF-κB activity results in disruption of the apical ectodermal ridge and aberrant limb morphogenesis. *Nature* 392:615–18

19. Chang C, Holtzman DA, Chau S, Chickering T, Woolf EA, et al. 2001. Twisted gastrulation can function as a BMP antagonist. *Nature* 410:483–87

20. Chasan R, Anderson KV. 1989. The role of *easter*, an apparent serine protease, in organizing the dorsal-ventral pattern of the Drosophila embryo. *Cell* 56:391–400

21. Chasan R, Anderson KV. 1993. Maternal control of dorsal-ventral polarity and pattern in the embryo. In *The Development of Drosophila melanogaster*, ed. M Bate, A Martinez Arias. New York: CSHL Press

22. Chasan R, Jin Y, Anderson KV. 1992. Activation of the easter zymogen is regulated by five other genes to define dorsal-ventral polarity in the *Drosophila* embryo. *Development* 115:607–16

23. Chen G, Handel K, Roth S. 2000. The maternal NF-kB/Dorsal gradient of *Tribolium castaneum*: dynamics of early dorsoventral patterning in a short-germ beetle. *Development* 127:5145–56

24. Childs SR, O'Connor MB. 1994. Two domains of the *tolloid* protein contribute to its unusual genetic interaction with

decapentaplegic. Dev. Biol. 162:209–20

25. Curtis D, Apfeld J, Lehmann R. 1995. nanos is an evolutionarily conserved organizer of anterior-posterior polarity. *Development* 121:1899–910

26. Dahanukar A, Wharton RP. 1996. The Nanos gradient in *Drosophila* embryos is generated by translational regulation. *Genes Dev.* 10:2610–20

27. Dai H, Hogan C, Gopalakrishnan B, Torres-Vazquez J, Nguyen M, et al. 2000. The zinc finger protein schnurri acts as a Smad partner in mediating the transcriptional response to *decapentaplegic. Dev. Biol.* 227:373–87

28. Damen WG, Tautz D. 1998. A Hox class 3 orthologue from the spider *Cupiennius salei* is expressed in a Hox-gene-like fashion. *Dev. Genes Evol.* 208:586–90

29. De Robertis EM, Larrain J, Oelgeschlager M, Wessely O. 2000. The establishment of Spemann's organizer and patterning of the vertebrate embryo. *Nat. Rev. Genet.* 1:171–81

30. Dearden P, Akam M. 1999. Developmental evolution: Axial patterning in insects. *Curr. Biol.* 9:R591–94

31. Dearden P, Akam M. 2001. Early embryo patterning in the grasshopper, *Schistocerca gregaria: wingless, dpp* and *caudal* expression. *Development.* In press

32. Doyle HJ, Kraut R, Levine M. 1989. Spatial regulation of *zerknüllt*: a dorsalventral patterning gene in *Drosophila. Genes Dev.* 3:1518–33

33. Driever W, Nüsslein-Volhard C. 1989. The bicoid protein is a positive regulator of hunchback transcription in the early Drosophila embryo. *Nature* 337:138–43

34. Dubnau J, Struhl G. 1996. RNA recognition and translational regulation by a homeodomain protein. *Nature* 379:694–99

35. Duboule D. 1994. Temporal colinearity and the phylotypic progression: a basis for the stability of a vertebrate Bauplan and the evolution of morphologies through heterochrony. *Dev. Suppl.* 135–42

36. Edgar LG, Carr S, Wang H, Wood WB. 2001. Zygotic expression of the *caudal* homolog *pal-1* is required for posterior patterning in *Caenorhabditis elegans* embryogenesis. *Dev. Biol.* 229:71–88

37. Falciani F, Hausdorf B, Schroder R, Akam M, Tautz D, et al. 1996. Class 3 Hox genes in insects and the origin of zen. *Proc. Natl. Acad. Sci. USA* 93:8479–84

38. Fay DS, Stanley HM, Han M, Wood WB. 1999. A *Caenorhabditis elegans* homologue of *hunchback* is required for late stages of development but not early embryonic patterning. *Dev. Biol.* 205:240–53

39. Ferguson EL, Anderson KV. 1992. *decapentaplegic* acts as a morphogen to organize dorsal-ventral pattern in the Drosophila embryo. *Cell* 71:451–61

40. Ferguson EL, Anderson KV. 1992. Localized enhancement and repression of the activity of the TGF-β family member, *decapentaplegic*, is necessary for dorsal-ventral pattern formation in the *Drosophila* embryo. *Development* 114:583–97

41. Finelli AL, Bossie CA, Xie T, Padgett RW. 1994. Mutational analysis of the *Drosophila tolloid* gene, a human BMP-1 homolog. *Development* 120:861–70

42. Francois V, Solloway M, O'Neill JW, Emery J, Bier E. 1994. Dorsal-ventral patterning of the *Drosophila* embryo depends on a putative negative growth factor encoded by the *short gastrulation* gene. *Genes Dev.* 8:2602–16

43. Frumkin A, Rangini Z, Ben-Yehuda A, Gruenbaum Y, Fainsod A. 1991. A chicken *caudal* homologue, *CHox-cad*, is expressed in the epiblast with posterior localization and in the early endodermal lineage. *Development* 112:207–19

44. Gamer LW, Wright CV. 1993. Murine Cdx-4 bears striking similarities to the *Drosophila caudal* gene in its homeodomain sequence and early expression pattern. *Mech. Dev.* 43:71–81

45. Gerhart J, Ubbels G, Black S, Hara K, Kirschner M. 1981. A reinvestigation of the role of the grey crescent in axis formation in *Xenopus laevis*. *Nature* 292:511–16

46. Goldstein B, Frisse LM, Thomas WK. 1998. Embryonic axis specification in nematodes: evolution of the first step in development. *Curr. Biol.* 8:157–60

47. Graff JM, Thies RS, Song JJ, Celeste AJ, Melton DA. 1994. Studies with a *Xenopus* BMP receptor suggest that ventral mesoderm-inducing signals override dorsal signals in vivo. *Cell* 79:169

48. Grau Y, Carteret C, Simpson P. 1984. Mutations and chromosomal rearrangements affecting the expression of *snail*, a gene involved in embryonic patterning in *Drosophila melanogaster*. *Genetics* 108:347–60

49. Grbic M. 2000. "Alien" wasps and evolution of development. *BioEssays* 22:920–32

50. Grbic M, Nagy LM, Carroll SB, Strand M. 1996. Polyembryonic development: insect pattern formation in a cellularized environment. *Development* 122:795–804

51. Hammerschmidt M, Serbedzija GN, McMahon AP. 1996. Genetic analysis of dorsoventral pattern formation in the zebrafish: requirement of a BMP-like ventralizing activity and its dorsal repressor. *Genes Dev.* 10:2452–61

52. Hashimoto C, Hudson KL, Anderson KV. 1988. The *Toll* gene of *Drosophila*, required for dorsal-ventral embryonic polarity, appears to encode a transmembrane protein. *Cell* 52:269–79

53. Hecht PM, Anderson KV. 1993. Genetic characterization of *tube* and *pelle*, genes required for signaling between Toll and dorsal in the specification of the dorsal-ventral pattern of the Drosophila embryo. *Genetics* 135:405–17

54. Hild M, Dick A, Rauch GJ, Meier A, Bouwmeester T, et al. 1999. The smad5 mutation *somitabun* blocks Bmp2b signaling during early dorsoventral patterning of the zebrafish embryo. *Development* 126:2149–59

55. Ho K, Dunin-Borkowski OM, Akam M. 1997. Cellularization in locust embryos occurs before blastoderm formation. *Development* 124:2761–68

56. Holley SA, Ferguson EL. 1997. Fish are like flies are like frogs: conservation of dorsal-ventral patterning mechanisms. *BioEssays* 19:281–84

57. Holley SA, Jackson PD, Sasai Y, Lu B, De Robertis EM, et al. 1995. A conserved system for dorsal-ventral patterning in insects and vertebrates involving sog and chordin. *Nature* 376:249–53

58. Hong CC, Hashimoto C. 1995. An unusual mosaic protein with a protease domain, encoded by the *nudel* gene, is involved in defining embryonic dorsoventral polarity in Drosophila. *Cell* 82:785–94

59. Hopwood ND, Pluck A, Gurdon JB. 1989. A *Xenopus* mRNA related to *Drosophila* twist is expressed in response to induction in the mesoderm and the neural crest. *Cell* 59:893–903

60. Howard TD, Paznekas WA, Green ED, Chiang LC, Ma N, et al. 1997. Mutations in TWIST, a basic helix-loop-helix transcription factor, in Saethre-Chotzen syndrome. *Nat. Genet.* 15:36–41

61. Hudson JB, Podos SD, Keith K, Simpson SL, Ferguson EL. 1998. The *Drosophila Medea* gene is required downstream of dpp and encodes a functional homolog of human Smad4. *Development* 125:1407–20

62. Hülskamp M, Schroder C, Pfeifle C, Jackle H, Tautz D. 1989. Posterior segmentation of the *Drosophila* embryo in the absence of a maternal posterior organizer gene. *Nature* 338:629–32

63. Hunter CP, Kenyon C. 1996. Spatial and temporal controls target *pal-1* blastomere-specification activity to a single blastomere lineage in *C. elegans* embryos. *Cell* 87:217–26

64. Ip YT, Kraut R, Levine M, Rushlow

CA. 1991. The dorsal morphogen is a sequence-specific DNA-binding protein that interacts with a long-range repression element in *Drosophila*. *Cell* 64:439–46

65. Ip YT, Park RE, Kosman D, Bier E, Levine M. 1992. The dorsal gradient morphogen regulates stripes of *rhomboid* expression in the presumptive neuroectoderm of the *Drosophila* embryo. *Genes Dev.* 6:1728–39

66. Ip YT, Park RE, Kosman D, Yazdanbakhsh K, Levine M. 1992. *dorsal-twist* interactions establish *snail* expression in the presumptive mesoderm of the *Drosophila* embryo. *Genes Dev.* 6:1518–30

67. Irish V, Lehmann R, Akam M. 1989. The *Drosophila* posterior-group gene *nanos* functions by repressing *hunchback* activity. *Nature* 338:646–48

68. Irish VF, Gelbart WM. 1987. The *decapentaplegic* gene is required for dorsal-ventral patterning of the *Drosophila* embryo. *Genes Dev.* 1:868–79

69. Isaacs HV, Pownall ME, Slack JM. 1998. Regulation of Hox gene expression and posterior development by the *Xenopus caudal* homologue *Xcad3*. *EMBO J.* 17:3413–27

70. Iwasa JH, Suver DW, Savage RM. 2000. The leech *hunchback* protein is expressed in the epithelium and CNS but not in the segmental precursor lineages. *Dev. Genes Evol.* 210:277–88

71. Jazwinska A, Kirov N, Wieschaus E, Roth S, Rushlow C. 1999. The *Drosophila* gene *brinker* reveals a novel mechanism of Dpp target gene regulation. *Cell* 96:563–73

72. Jockusch EL, Nulsen C, Newfeld SJ, Nagy LM. 2000. Leg development in flies versus grasshoppers: differences in *dpp* expression do not lead to differences in the expression of downstream components of the leg patterning pathway. *Development* 127:1617–26

73. Joly JS, Maury M, Joly C, Duprey P, Boulekbache H, Condamine H. 1992. Expression of a zebrafish *caudal* homeobox gene correlates with the establishment of posterior cell lineages at gastrulation. *Differentiation* 50:75–87

74. Jürgens G, Wieschaus E, Nüsslein-Volhard C, Kluding H. 1984. Mutations affecting the pattern of the larval cuticle in *Drosophila melanogaster*. *Wilhelm Roux's Arch Dev. Biol.* 193:283–95

75. Kanegae Y, Tavares AT, Izpisua Belmonte JC, Verma IM. 1998. Role of Rel/NF-κB transcription factors during the outgrowth of the vertebrate limb. *Nature* 392:611–14

76. Katsuyama Y, Sato Y, Wada S, Saiga H. 1999. Ascidian tail formation requires *caudal* function. *Dev. Biol.* 213:257–68

77. Kim J, Johnson K, Chen HJ, Carroll S, Laughon A. 1997. *Drosophila* Mad binds to DNA and directly mediates activation of *vestigial* by Decapentaplegic. *Nature* 388:304–8

78. Kishimoto Y, Lee KH, Zon L, Hammerschmidt M, Schulte-Merker S. 1997. The molecular nature of zebrafish *swirl*: BMP2 function is essential during early dorsoventral patterning. *Development* 124:4457–66

79. Konrad KD, Goralski TJ, Mahowald AP, Marsh JL. 1998. The *gastrulation defective* gene of *Drosophila melanogaster* is a member of the serine protease superfamily. *Proc. Natl. Acad. Sci. USA* 95:6819–24

80. Kosman D, Ip YT, Levine M, Arora K. 1991. Establishment of the mesoderm-neuroectoderm boundary in the *Drosophila* embryo. *Science* 254:118–22

81. Kraemer B, Crittenden S, Gallegos M, Moulder G, Barstead R, et al. 1999. NANOS-3 and FBF proteins physically interact to control the sperm-oocyte switch in *Caenorhabditis elegans*. *Curr. Biol.* 9:1009–18

82. Kristensen NP. 1991. Phylogeny of extant hexapods. In *The Insects of Australia*, ed. ID Naumann, pp. 125–40. Melbourne: Melbourne Univ. Press

83. Lane MC, Smith WC. 1999. The origins of primitive blood in *Xenopus*: implications for axial patterning. *Development* 126:423–34

84. Lemaitre B, Meister M, Govind S, Georgel P, Steward R, et al. 1995. Functional analysis and regulation of nuclear import of dorsal during the immune response in *Drosophila*. *EMBO J.* 14:536–45

85. Lemaitre B, Nicolas E, Michaut L, Reichhart JM, Hoffmann JA. 1996. The dorsoventral regulatory gene cassette spatzle/Toll/cactus controls the potent antifungal response in *Drosophila* adults. *Cell* 86:973–83

86. LeMosy EK, Tan YQ, Hashimoto C. 2001. Activation of a protease cascade involved in patterning the *Drosophila* embryo. *Proc. Natl. Acad. Sci. USA* 10:10

87. Leptin M. 1991. twist and snail as positive and negative regulators during *Drosophila* mesoderm development. *Genes Dev.* 5:1568–76

88. Letsou A, Arora K, Wrana JL, Simin K, Twombly V, et al. 1995. Drosophila Dpp signaling is mediated by the *punt* gene product: a dual ligand-binding type II receptor of the TGFβ receptor family. *Cell* 80:899–908

89. Liu F, Hata A, Baker JC, Doody J, Carcamo J, et al. 1996. A human Mad protein acting as a BMP-regulated transcriptional activator. *Nature* 381:620–23

90. Macdonald PM, Struhl G. 1986. A molecular gradient in early *Drosophila* embryos and its role in specifying the body pattern. *Nature* 324:537–45

91. Marom K, Shapira E, Fainsod A. 1997. The chicken *caudal* genes establish an anterior-posterior gradient by partially overlapping temporal and spatial patterns of expression. *Mech. Dev.* 64:41–52

92. Marqués G, Musacchio M, Shimell MJ, Wünnenberg-Stapleton K, Cho KWY, O'Connor MB. 1997. Production of a DPP activity gradient in the early *Drosophila* embryo through the opposing actions of

the SOG and TLD proteins. *Cell* 91:417–26

93. Marty T, Müller B, Basler K, Affolter M. 2000. Schnurri mediates Dpp-dependent repression of *brinker* transcription. *Nat. Cell Biol.* 2:745–49

94. Maxton-Küchenmeister J, Handel K, Schmidt-Ott U, Roth S, Jäckle H. 1999. *Toll* homologue expression in the beetle *Tribolium* suggests a different mode of dorsoventral patterning than in *Drosophila* embryos. *Mech. Dev.* 83:107–14

95. Medzhitov R, Preston-Hurlburt P, Janeway CA Jr. 1997. A human homologue of the *Drosophila* Toll protein signals activation of adaptive immunity. *Nature* 388:394–97

96. Micklem DR. 1995. mRNA localisation during development. *Dev. Biol.* 172:377–95

97. Mlodzik M, Gehring WJ. 1987. Expression of the *caudal* gene in the germ line of *Drosophila*: formation of an RNA and protein gradient during early embryogenesis. *Cell* 48:465–78

98. Moreno E, Morata G. 1999. Caudal is the Hox gene that specifies the most posterior Drosophila segment. *Nature* 400:873–77

99. Morisato D, Anderson KV. 1994. The *spätzle* gene encodes a component of the extracellular signaling pathway establishing the dorsal-ventral pattern of the *Drosophila* embryo. *Cell* 76:677–88

100. Morisato D, Anderson KV. 1995. Signaling pathways that establish the dorsal-ventral pattern of the *Drosophila* embryo. *Annu. Rev. Genet.* 29:371–99

101. Mosquera L, Forristall C, Zhou Y, King ML. 1993. A mRNA localized to the vegetal cortex of *Xenopus* oocytes encodes a protein with a nanos-like zinc finger domain. *Development* 117:377–86

102. Murata Y, Wharton RP. 1995. Binding of pumilio to maternal *hunchback* mRNA is required for posterior patterning in *Drosophila* embryos. *Cell* 80:747–56

103. Myers EW, Sutton GG, Delcher AL, Dew

IM, Fasulo DP, et al. 2000. A whole-genome assembly of *Drosophila. Science* 287:2196–204

104. Nambu JR, Franks RG, Hu S, Crews ST. 1990. The *single-minded* gene of *Drosophila* is required for the expression of genes important for the development of CNS midline cells. *Cell* 63:63–75

105. Nellen D, Affolter M, Basler K. 1994. Receptor serine/threonine kinases implicated in the control of *Drosophila* body pattern by *decapentaplegic. Cell* 78:225–37

106. Neul JL, Ferguson EL. 1998. Spatially restricted activation of the SAX receptor by SCW modulates DPP/TKV signaling in *Drosophila* dorsal-ventral patterning. *Cell* 95:483–94

107. Newfeld SJ, Chartoff EH, Graff JM, Melton DA, Gelbart WM. 1996. *Mothers against dpp* encodes a conserved cytoplasmic protein required in DPP/TGFβ responsive cells. *Development* 122:2099–108

108. Newfeld SJ, Mehra A, Singer MA, Wrana JL, Attisano L, Gelbart WM. 1997. *Mothers against dpp* participates in a DPP/TGF-β responsive serine-threonine kinase signal transduction cascade. *Development* 124:3167–76

109. Nilson LA, Schüpbach T. 1998. Localized requirements for *windbeutel* and *pipe* reveal a dorsoventral prepattern within the follicular epithelium of the *Drosophila* ovary. *Cell* 93:253–62

110. Nüsslein-Volhard C, Lohs-Schardin M, Sander K, Cremer C. 1980. A dorsoventral shift of embryonic primordia in a new maternal-effect mutant of *Drosophila. Nature* 283:474–76

111. Oelgeschlager M, Larrain J, Geissert D, De Robertis EM. 2000. The evolutionarily conserved BMP-binding protein Twisted gastrulation promotes BMP signaling. *Nature* 405:757–63

112. Padgett RW, St Johnston RD, Gelbart WM. 1987. A transcript from a *Drosophila* pattern gene predicts a protein homologous to the transforming growth factor-beta family. *Nature* 325:81–84

113. Padgett RW, Wozney JM, Gelbart WM. 1993. Human BMP sequences can confer normal dorsal-ventral patterning in the Drosophila embryo. *Proc. Natl. Acad. Sci. USA* 90:2905–9

114. Patel NH. 1994. Developmental evolution: insights from studies of insect segmentation. *Science* 266:581–90

115. Patel NH, Hayward DC, Lall S, Pirkl NR, DiPietro D, Ball EE. 2001. Grasshopper *hunchback* expression reveals conserved and novel aspects of axis formation and segmentation. *Development.* In press

116. Penton A, Chen Y, Staehling-Hampton K, Wrana JL, Attisano L, et al. 1994. Identification of two bone morphogenetic protein type I receptors in Drosophila and evidence that Brk25D is a decapentaplegic receptor. *Cell* 78:239–50

117. Piccolo S, Agius E, Lu B, Goodman S, Dale L, De Robertis EM. 1997. Cleavage of Chordin by Xolloid metalloprotease suggests a role for proteolytic processing in the regulation of Spemann organizer activity. *Cell* 91:407–16

118. Piccolo S, Sasai Y, Lu B, De Robertis EM. 1996. Dorsoventral patterning in Xenopus: inhibition of ventral signals by direct binding of chordin to BMP-4. *Cell* 86:589–98

119. Pillemer G, Epstein M, Blumberg B, Yisraeli JK, De Robertis EM, et al. 1998. Nested expression and sequential downregulation of the Xenopus *caudal* genes along the anterior-posterior axis. *Mech. Dev.* 71:193–96

120. Pilon M, Weisblat DA. 1997. A *nanos* homolog in leech. *Development* 124:1771–80

121. Pultz MA, Pitt JN, Alto NM. 1999. Extensive zygotic control of the anteroposterior axis in the wasp *Nasonia vitripennis. Development* 126:701–10

122. Raff RA. 1996. *The Shape of Life*, pp. 173–210. Chicago: Univ. Chicago Press

123. Raftery LA, Twombly V, Wharton K,

Gelbart WM. 1995. Genetic screens to identify elements of the *decapentaplegic* signaling pathway in Drosophila. *Genetics* 139:241–54

124. Rao Y, Vaessin H, Jan LY, Jan YN. 1991. Neuroectoderm in *Drosophila* embryos is dependent on the mesoderm for positioning but not for formation. *Genes Dev.* 5:1577–88

125. Ray RP, Arora K, Nüsslein-Volhard C, Gelbart WM. 1991. The control of cell fate along the dorsal-ventral axis of the *Drosophila* embryo. *Development* 113: 35–54

126. Reach M, Galindo RL, Towb P, Allen JL, Karin M, Wasserman SA. 1996. A gradient of Cactus protein degradation establishes dorsoventral polarity in the *Drosophila* embryo. *Dev. Biol.* 180:353–64

127. Rivera-Pomar R, Jackle H. 1996. From gradients to stripes in *Drosophila* embryogenesis: filling in the gaps. *Trends Genet.* 12:478–83

128. Rivera-Pomar R, Lu X, Perrimon N, Taubert H, Jackle H. 1995. Activation of posterior gap gene expression in the *Drosophila* blastoderm. *Nature* 376:253–56

129. Rivera-Pomar R, Niessing D, Schmidt-Ott U, Gehring WJ, Jackle H. 1996. RNA binding and translational suppression by bicoid. *Nature* 379:746–49

130. Rock FL, Hardiman G, Timans JC, Kastelein RA, Bazan JF. 1998. A family of human receptors structurally related to Drosophila Toll. *Proc. Natl. Acad. Sci. USA* 95:588–93

131. Rohr KB, Tautz D, Sander K. 1999. Segmentation gene expression in the moth-midge *Clogmia albipunctata* (Diptera, psychodidae) and other primitive dipterans. *Dev. Genes Evol.* 209:145–54

132. Ross JJ, Shimmi O, Vilmos P, Petryk A, Kim H, et al. 2001. Twisted gastrulation is a conserved extracellular BMP antagonist. *Nature* 410:479–83

133. Roth S, Hiromi Y, Godt D, Nüsslein-Volhard C. 1991. cactus, a maternal gene required for proper formation of the dorsoventral morphogen gradient in *Drosophila* embryos. *Development* 112:371–88

134. Roth S, Stein D, Nüsslein-Volhard C. 1989. A gradient of nuclear localization of the *dorsal* protein determines dorsoventral pattern in the *Drosophila* embryo. *Cell* 59:1189–202

135. Ruberte E, Marty T, Nellen D, Affolter M, Basler K. 1995. An absolute requirement for both the type II and type I receptors, punt and thick veins, for dpp signaling in vivo. *Cell* 80:889–97

136. Rushlow C, Frasch M, Doyle H, Levine M. 1987. Maternal regulation of *zerknüllt*: a homeobox gene controlling differentiation of dorsal tissues in *Drosophila* embryos. *Nature* 330:583–86

137. Rushlow C, Levine M. 1990. Role of the *zerknüllt* gene in dorsal-ventral pattern formation in *Drosophila*. *Adv. Genet.* 27:277–307

138. Rushlow CA, Han K, Manley JL, Levine M. 1989. The graded distribution of the dorsal morphogen is initiated by selective nuclear transport in *Drosophila*. *Cell* 59:1165–77

139. Sanchez-Salazar J, Pletcher MT, Bennett RL, Brown SJ, Dandamudi TJ, et al. 1996. The *Tribolium* decapentaplegic gene is similar in sequence, structure, and expression to the *Drosophila dpp* gene. *Dev. Genes Evol.* 206:237–46

140. Sander K. 1976. Specification of the basic body pattern in insect embryogenesis. *Adv. Insect Physiol.* 12:125–238

141. Sasai Y, Lu B, Steinbeisser H, Geissert D, Gont LK, De Robertis EM. 1994. Xenopus *chordin*: a novel dorsalizing factor activated by organizer-specific homeobox genes. *Cell* 79:779–90

142. Saunders JW. 1948. The proximo-distal sequence of origin of the parts of the chick wing and the role of the ectoderm. *J. Exp. Zool.* 108

143. Savage RM, Shankland M. 1996. Identification and characterization of a

hunchback orthologue, Lzf2, and its expression during leech embryogenesis. *Dev. Biol.* 175:205–17

144. Schmidt J, Francois V, Bier E, Kimelman D. 1995. *Drosophila short gastrulation* induces an ectopic axis in *Xenopus*: evidence for conserved mechanisms of dorsal-ventral patterning. *Development* 121:4319–28

145. Schmidt-Ott U. 2000. The amnioserosa is an apomorphic character of cyclorrhaphan flies. *Dev. Genes Evol.* 210:373–76

146. Schmidt-Ullrich R, Memet S, Lilienbaum A, Feuillard J, Raphael M, Israel A. 1996. NF-κB activity in transgenic mice: developmental regulation and tissue specificity. *Development* 122:2117–28

147. Schneider DS, Jin Y, Morisato D, Anderson KV. 1994. A processed form of the Spätzle protein defines dorsal-ventral polarity in the *Drosophila* embryo. *Development* 120:1243–50

148. Schröder R, Eckert C, Wolff C, Tautz D. 2000. Conserved and divergent aspects of terminal patterning in the beetle *Tribolium castaneum*. *Proc. Natl. Acad. Sci. USA* 97:6591–96

149. Schulte-Merker S, Lee KJ, McMahon AP, Hammerschmidt M. 1997. The zebrafish organizer requires chordino. *Nature* 387:862–63

150. Schulz C, Schroder R, Hausdorf B, Wolff C, Tautz D. 1998. A *caudal* homologue in the short germ band beetle *Tribolium* shows similarities to both the *Drosophila* and the vertebrate *caudal* expression patterns. *Dev. Genes Evol.* 208:283–89

151. Schulz C, Tautz D. 1995. Zygotic *caudal* regulation by *hunchback* and its role in abdominal segment formation of the *Drosophila* embryo. *Development* 121:1023–28

152. Schüpbach T, Wieschaus E. 1989. Female sterile mutations on the second chromosome of *Drosophila melanogaster*. I. Maternal effect mutations. *Genetics* 121:101–17

153. Scott IC, Blitz IL, Pappano WN, Maas SA, Cho KW, Greenspan DS. 2001. Homologues of Twisted gastrulation are extracellular cofactors in antagonism of BMP signalling. *Nature* 410:475–78

154. Sekelsky JJ, Newfeld SJ, Raftery LA, Chartoff EH, Gelbart WM. 1995. Genetic characterization and cloning of mothers against dpp, a gene required for decapentaplegic function in *Drosophila melanogaster*. *Genetics* 139:1347–58

155. Sen J, Goltz JS, Stevens L, Stein D. 1998. Spatially restricted expression of *pipe* in the *Drosophila* egg chamber defines embryonic dorsal-ventral polarity. *Cell* 95:471–81

156. Shimell MJ, Ferguson EL, Childs SR, O'Connor MB. 1991. The Drosophila dorsal-ventral patterning gene *tolloid* is related to human bone morphogenetic protein 1. *Cell* 67:469–81

157. Simpson-Brose M, Treisman J, Desplan C. 1994. Synergy between the hunchback and bicoid morphogens is required for anterior patterning in *Drosophila*. *Cell* 78:855–65

158. Sommer R, Tautz D. 1991. Segmentation gene expression in the housefly *Musca domestica*. *Development* 113:419–30

159. Sommer RJ, Retzlaff M, Goerlich K, Sander K, Tautz D. 1992. Evolutionary conservation pattern of zinc-finger domains of *Drosophila* segmentation genes. *Proc. Natl. Acad. Sci. USA* 89:10782–86

160. Sommer RJ, Tautz D. 1994. Expression patterns of *twist* and *snail* in *Tribolium* (Coleoptera) suggest a homologous formation of mesoderm in long and short germ band insects. *Dev. Genet.* 15:32–37

161. Sonoda J, Wharton RP. 1999. Recruitment of Nanos to *hunchback* mRNA by Pumilio. *Genes Dev.* 13:2704–12

162. Sonoda J, Wharton RP. 2001. Drosophila Brain Tumor is a translational repressor. *Genes Dev.* 15:762–73

163. Spemann H, Mangold H. 1924. Induction of embryonic primordia by implantation of organizers from a different species.

Wilhelm Roux's Arch. Dev. Biol. 100:599–638 (Transl. V Hamburger)

164. Spring J, Yanze N, Middel AM, Stierwald M, Groger H, Schmid V. 2000. The mesoderm specification factor *twist* in the life cycle of jellyfish. *Dev. Biol.* 228:363–75

165. St Johnston D, Nüsslein-Volhard C. 1992. The origin of pattern and polarity in the *Drosophila* embryo. *Cell* 68:201–19

166. St Johnston RD, Gelbart WM. 1987. Decapentaplegic transcripts are localized along the dorsal-ventral axis of the *Drosophila* embryo. *EMBO J.* 6:2785–91

167. Stauber M, Jackle H, Schmidt-Ott U. 1999. The anterior determinant bicoid of *Drosophila* is a derived Hox class 3 gene. *Proc. Natl. Acad. Sci. USA* 96:3786–89

168. Stauber M, Taubert H, Schmidt-Ott U. 2000. Function of *bicoid* and *hunchback* homologs in the basal cyclorrhaphan fly *Megaselia* (Phoridae). *Proc. Natl. Acad. Sci. USA* 97:10844–49

169. Stein D, Nüsslein-Volhard C. 1992. Multiple extracellular activities in *Drosophila* egg perivitelline fluid are required for establishment of embryonic dorsal-ventral polarity. *Cell* 68:429–40

170. Steward R. 1987. Dorsal, an embryonic polarity gene in *Drosophila*, is homologous to the vertebrate proto-oncogene, c-rel. *Science* 238:692–94

171. Steward R. 1989. Relocalization of the *dorsal* protein from the cytoplasm to the nucleus correlates with its function. *Cell* 59:1179–88

172. Steward R, Zusman SB, Huang LH, Schedl P. 1988. The dorsal protein is distributed in a gradient in early *Drosophila* embryos. *Cell* 55:487–95

173. Struhl G, Struhl K, Macdonald PM. 1989. The gradient morphogen bicoid is a concentration-dependent transcriptional activator. *Cell* 57:1259–73

174. Subramaniam K, Seydoux G. 1999. nos-1 and nos-2, two genes related to *Drosophila nanos*, regulate primordial germ cell development and survival in *Caenorhabditis elegans*. *Development* 126:4861–71

175. Summerbell D. 1974. A quantitative analysis of the effect of the excision of the AER from the chick limb bud. *J. Embryol. Exp. Morphol.* 32:651–60

176. Tautz D. 1988. Regulation of the *Drosophila* segmentation gene hunchback by two maternal morphogenetic centres. *Nature* 332:281–84

177. Tautz D, Friedrich M, Schroder R. 1994. Insect embryogenesis—what is ancestral and what is derived? *Dev. Suppl.* 193–99

178. Tautz D, Pfeifle C. 1989. A non-radioactive in situ hybridization method for the localization of specific RNAs in *Drosophila* embryos reveals translational control of the segmentation gene *hunchback*. *Chromosoma* 98:81–85

179. Telford MJ, Thomas RH. 1998. Of mites and zen: expression studies in a chelicerate arthropod confirm zen is a divergent Hox gene. *Dev. Genes Evol.* 208:591–94

180. Thisse B, El Messel M, Perrin-Schmitt F. 1987. The twist gene: isolation of a *Drosophila* zygotic gene necessary for the establishment of the dorsal-ventral pattern. *Nucleic Acids Res.* 15:3439–53

181. Thisse B, Stoetzel C, El Messal M, Perrin-Schmitt F. 1987. Genes of the Drosophila maternal dorsal group control the specific expression of the zygotic gene *twist* in presumptive mesodermal cells. *Genes Dev.* 1:709–15

182. Thisse C, Perrin-Scmitt F, Stoetzel C, Thisse B. 1991. Sequence-specific transactivation of the *Drosophila twist* gene by the *dorsal* gene product. *Cell* 65:1191–201

183. Treier M, Pfeifle C, Tautz D. 1989. Comparison of the gap segmentation gene *hunchback* between *Drosophila melanogaster* and *Drosophila virilis* reveals novel modes of evolutionary change. *EMBO J.* 8:1517–25

184. Udagawa Y, Hanai J, Tada K, Grieder NC, Momoeda M, et al. 2000. Schnurri

interacts with Mad in a Dpp-dependent manner. *Genes Cells* 5:359–69

185. Wang C, Lehmann R. 1991. Nanos is the localized posterior determinant in *Drosophila*. *Cell* 66:637–47

186. Whalen AM, Steward R. 1993. Dissociation of the Dorsal-Cactus complex and phosphorylation of the Dorsal protein correlate with the nuclear localization of Dorsal. *J. Cell Biol.* 123:523–34

187. Wharton KA, Ray RP, Gelbart WM. 1993. An activity gradient of *decapentaplegic* is necessary for the specification of dorsal pattern elements in the *Drosophila* embryo. *Development* 117:807–22

188. Wharton RP, Struhl G. 1991. RNA regulatory elements mediate control of *Drosophila* body pattern by the posterior morphogen *nanos*. *Cell* 67:955–67

189. Wilson PA, Lagna G, Suzuki A, Hemmati-Brivanlou A. 1997. Concentration-dependent patterning of the *Xenopus* ectoderm by BMP4 and its signal transducer Smad1. *Development* 124:3177–84

190. Wimmer EA, Carleton A, Harjes P, Turner T, Desplan C. 2000. Bicoid-independent formation of thoracic segments in Drosophila. *Science* 287:2476–79

191. Wolf C, Thisse C, Stoetzel C, Thisse B, Gerlinger P, Perrin-Schmitt F. 1991. The *M-twist* gene in *Mus* is expressed in subsets of mesodermal cells and is closely related to the *Xenopus X-twi* and the *Drosophila twist* genes. *Dev. Biol.* 143:363–73

192. Wolff C, Schroder R, Schulz C, Tautz D, Klingler M. 1998. Regulation of the *Tribolium* homologues of *caudal* and *hunch-*

back in *Drosophila*: evidence for maternal gradient systems in a short germ embryo. *Development* 125:3645–54

193. Wolff C, Sommer R, Schroder R, Glaser G, Tautz D. 1995. Conserved and divergent expression aspects of the *Drosophila* segmentation gene *hunchback* in the short germ band embryo of the flour beetle *Tribolium*. *Development* 121:4227–36

194. Xu X, Xu PX, Suzuki Y. 1994. A maternal homeobox gene, Bombyx *caudal*, forms both mRNA and protein concentration gradients spanning anteroposterior axis during gastrulation. *Development* 120:277–85

195. Yeates DK, Wiegmann BM. 1999. Congruence and controversy: toward a higher-level phylogeny of Diptera. *Annu. Rev. Entomol.* 44:397–428

196. Yu K, Srinivasan S, Shimmi O, Biehs B, Rashka KE, et al. 2000. Processing of the *Drosophila* Sog protein creates a novel BMP inhibitory activity. *Development* 127:2143–54

197. Zhang B, Gallegos M, Puoti A, Durkin E, Fields S, et al. 1997. A conserved RNA-binding protein that regulates sexual fates in the *C. elegans* hermaphrodite germ line. *Nature* 390:477–84

198. Zimmerman LB, De Jesus-Escobar JM, Harland RM. 1996. The Spemann organizer signal noggin binds and inactivates bone morphogenetic protein 4. *Cell* 86:599–606

199. Zusman SB, Sweeton D, Wieschaus EF. 1988. *short gastrulation*, a mutation causing delays in stage-specific cell shape changes during gastrulation in Drosophila melanogaster. *Dev. Biol.* 129:417–27

Annu. Rev. Genet. 2001. 35:439–68

REGULATION OF GENE EXPRESSION BY CELL-TO-CELL COMMUNICATION:
Acyl-Homoserine Lactone Quorum Sensing

Clay Fuqua[1], Matthew R. Parsek[2], and E. Peter Greenberg[3]

*[1]Department of Biology, Indiana University, Bloomington, Indiana 47405;
e-mail: cfuqua@bio.indiana.edu; [2]Department of Civil Engineering, Northwestern
University, Evanston, Illinois 60208; e-mail: m-parsek@nwu.edu; [3]Department of
Microbiology, University of Iowa, Iowa City, Iowa 52242;
e-mail: everett-greenberg@uiowa.edu*

Key Words quorum sensing, acylated homoserine lactones, intercellular signaling,
bacterial communication, multicellularity

■ **Abstract** Quorum sensing is an example of community behavior prevalent among
diverse bacterial species. The term "quorum sensing" describes the ability of a microor-
ganism to perceive and respond to microbial population density, usually relying on the
production and subsequent response to diffusible signal molecules. A significant number of gram-negative bacteria produce acylated homoserine lactones (acyl-HSLs) as
signal molecules that function in quorum sensing. Bacteria that produce acyl-HSLs
can respond to the local concentration of the signaling molecules, and high population
densities foster the accumulation of inducing levels of acyl-HSLs. Depending upon
the bacterial species, the physiological processes regulated by quorum sensing are extremely diverse, ranging from bioluminescence to swarming motility. Acyl-HSL quorum sensing has become a paradigm for intercellular signaling mechanisms. A flurry
of research over the past decade has led to significant understanding of many aspects
of quorum sensing including the synthesis of acyl-HSLs, the receptors that recognize
the acyl-HSL signal and transduce this information to the level of gene expression, and
the interaction of these receptors with the transcriptional machinery. Recent studies
have begun to integrate acyl-HSL quorum sensing into global regulatory networks and
establish its role in developing and maintaining the structure of bacterial communities.

CONTENTS

INTRODUCTION

A few words are necessary on the mass-action of bacteria. It is a common observation, one made by the writer at least a hundred times, that in culture-media not exactly adapted to the needs of the organism, a scanty inoculation may not give any growth—not even after a long time—whereas a copious one will lead to a growth which gradually clouds the fluid or covers the solid. The penetration of bacterial strands from cell to cell in the root nodules of Leguminosae is another example. The only explanation I can think of is that a multitude of bacteria are stronger than a few, and thus by union are able to overcome obstacles too great for the few.

Dr. Erwin F. Smith, 1905, Bacteria in Relation to Plant Disease, Vol. 1, Carnegie Institution Report.

As indicated by E. F. Smith's quote, it is apparent that a few microbiologists have considered bacterial populations as more than collections of single cells for some time. Metazoan organisms rely on communication between individual cells to coordinate many aspects of physiology and development. Likewise, prokaryotic, single-celled organisms also communicate with each other via diffusible and cell-associated signaling mechanisms. Intercellular communication in prokaryotes is often employed to monitor population size. The term "quorum-sensing" has been proposed to describe the ability of certain bacteria to monitor their own population density and modulate gene expression accordingly. There are several different examples of diffusible signaling systems that function as quorum sensors [see (27) for an exhaustive treatment of this subject]. Arguably, one of the best-characterized quorum-sensing mechanisms is exclusive to the gram-negative Proteobacteria and relies on acylated homoserine lactones (acyl-HSLs) as signal molecules (8, 38, 75). The physiological processes regulated by acyl-HSL quorum sensing in different species may vary from conjugal plasmid transfer to bioluminescence to antibiotic synthesis, to virulence. More often than not, the regulated genes involve interaction with a eukaryotic host (75).

Quorum sensing was originally discovered for the bioluminescent marine bacteria, *Vibrio harveyi* and *Vibrio fischeri*. Initially, most of the research on understanding bioluminescence regulation focused on *V. fischeri*. This bacterium colonizes the light organs of a variety of marine fishes and squids, where it occurs at very high densities (10^{10} cells/ml) and produces light (71, 93). In 1970, Nealson et al. reported that *V. fischeri* produced an extracellular factor, an autoinducer, that regulated production of the light-producing enzyme luciferase (72). The autoinducer was later shown to be 3-oxo-N-(tetrahydro-2-oxo-3-furanyl) hexanamide, more commonly known as N-3-(oxohexanoyl)homoserine lactone

TABLE 1 Examples of Acyl HSL quorum sensors[a]

Bacteria	Regulators[b]	Chain length (n)[c]	β R-group[d]	Target function
Vibrio fischeri	LuxR-LuxI	6 (1)	=O	Bioluminescence
	AinR-AinS[e]	8 (2)	−H	Bioluminescence
Pseudomonas aeruginosa	LasR-LasI	12 (4)	=O	Host interaction
	RhlR-RhlI	4 (0)	−H	Rhamnolipids
	QscR[f]	?	?	Host interaction
Agrobacterium tumefaciens	TraR-TraI	8 (4)	=O	Conjugal transfer
Pantoea stewartii	EsaR-EsaI	6 (1)	=O	Exopolysaccharide
Rhodobacter sphaeroides	CerR-CerI	14 (5)[g]	−H	Aggregation
Vibrio anguillarum	VanR-VanI	10 (3)	=O	Proteases
Yersenia enterocolitica	YenR-YenI	6 (1)	−H, =O	Target functions unknown

[a]For details and references on individual proteins and acyl HSLs, see (37).

[b]Quorum-sensing regulatory proteins, members of the LuxR-LuxI family except where indicated, transcriptional regulator–acyl HSL synthase.

[c]Corresponds to n in diagram.

[d]Moiety at β position, (R in diagram; fully reduced, H; hydroxyl, OH; carbonyl, O).

[e]AinR is homologous to two component sensor kinases; AinS is an acyl HSL synthase but is not homologous to LuxI (see text).

[f]QscR is a LuxR homologue that inhibits the activity of LasR (17).

[g]Has an unsaturated bond between positions 7–8 on the acyl chain.

[3-oxo-hexanoyl-HSL; see (28)]. This autoinduction system was thought by many to be unique to the bioluminescent marine vibrios. Identification of acyl-HSL-based quorum-sensing systems from the opportunistic human pathogen *Pseudomonas aeruginosa* and the plant pathogens *Erwinia carotovora* and *Agrobacterium tumefaciens* proved quorum sensing to be more widely distributed (7, 41, 43, 56, 79, 86, 116). In fact, there are now over 50 species recognized to produce acyl-HSLs, and these signals regulate a diverse range of bacterial processes (see Table 1 for examples). Clearly, acyl-HSL quorum sensing is a highly conserved regulatory system among the Proteobacteria.

Quorum sensing with acyl-HSLs is a relatively simple process (Figure 1). Acyl-HSLs are synthesized at a low basal level by acyl-HSL synthases (generally homologous to the *V. fischeri* LuxI-type protein). Newly synthesized acyl-HSL signals are rapidly removed from the cell by diffusion down their concentration gradient.

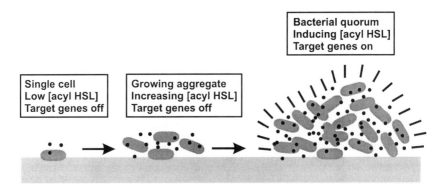

Figure 1 Population density-dependent gene regulation. A single population of microorganisms accumulating on a surface is depicted. Increasing cell number may be due to clonal growth or recruitment. Filled dots are intercellular quorum signals.

Increased population density can elevate the local acyl-HSL signal concentration. Although each individual cell still produces low levels of the acyl-HSL, in aggregate the group synthesizes higher concentrations. At a threshold level, the acyl-HSL signal interacts with a transcription factor (homologous to the LuxR protein in *V. fisheri*), and in turn the transcription factor modulates expression of quorum sensing–regulated genes. Often, the target genes include the gene encoding the LuxI homolog, creating a positive feedback circuit within the quorum sensor. Studies of acyl-HSL quorum sensors in diverse bacteria indicate that there are significant and interesting variations of this general mechanism. This review focuses on our current understanding of the molecular biology and genetics of quorum sensing with acyl-HSLs.

QUORUM SENSING IN THE REAL
WORLD: ENVIRONMENTAL CONSIDERATIONS

The environments in which quorum-sensing bacteria achieve the high population densities required to elevate acyl-HSL concentrations to inducing levels differ for every system. Symbiotic colonization of the light organs of marine animals by *V. fisheri* occurs in an environment that has evolved to foster high bacterial densities (66, 93, 95). Acyl-HSL concentrations in light organs exceed the level required to induce bioluminescence in laboratory cultures. *Euprymna scolopes*, a sepaloid squid, provides the *V. fisheri* that colonize its light organ with branched chain amino acids to support dense populations and luminescence (46). The nutrient-rich environment within the light organ contrasts with the surrounding oligotrophic seawater, where most marine bacteria, including *V. fisheri*, are found at relatively low-cell densities (a few cells/ml). Under these nutrient-limiting conditions, *V. fisheri* does not accumulate acyl-HSLs and consequently does not express the luminescence (*lux*) genes. Most of the bacteria that comprise the dense populations of *V. fischeri* reared in the light organ are expelled into seawater every morning at dawn and must re-grow within the light organ during the day.

In contrast to the structurally complex, highly evolved light organs colonized by *V. fischeri*, most bacteria that employ acyl-HSL quorum sensing have less clearly defined structures to achieve high population densities. Quorum sensing is thought to provide many plant and animal pathogens with a mechanism by which to minimize the host defense response, by delaying the production of tissue-damaging virulence factors until sufficient bacteria have been amassed (18). Pathogenesis may also involve the formation of biofilms, surface-associated communities of microorganisms, which foster large numbers of bacteria in close proximity to each other. A variety of biotic and abiotic surfaces within the host may accumulate bacterial biofilms. Biofilms or smaller, surface-associated microcolonies seem likely to provide the conditions that lead to acyl-HSL accumulation and induction of quorum-regulated genes. For example, *P. aeruginosa* is a pathogen that regulates its virulence factors via acyl-HSL quorum sensing and forms biofilms of clinical significance (83, 102). Quorum sensing is also involved in biofilm formation in *P. aeruginosa* (21).

Although there are some examples of symbiotic and pathogenic bacteria within a eukaryotic host that involve a single bacterial genus, in most environments bacterial species coexist with other microorganisms. It is virtually certain that many of these multispecies communities contain bacteria that employ acyl-HSL quorum sensing. Interspecies quorum sensing may play an important role, synergistic or competitive, in the dynamics of these microbial communities. For example, cell-free extracts of wild-type *P. aeruginosa* were reported to induce the formation of quorum sensing-regulated exoproducts in *Burkolderia cepacia*, whereas extracts from *P. aeruginosa* mutants reduced for acyl-HSL synthesis did not (67). This suggests the potential for cross-communication between these microorganisms in situ. The perception of the acyl-HSLs of one bacterial species by other bacterial species that employ quorum sensors with similar acyl-HSLs is likely and perhaps even unavoidable in certain environments (Figure 2). Bacterial species may also respond to and possibly impinge upon acyl-HSL signaling by another species in an R-protein-independent manner. Examples of this process have recently been provided by isolation of bacteria capable of degrading acyl-HSLs (25, 62). It remains unclear whether acyl-HSL degradation is strictly nutritional or represents a means of deliberate signal interference. Perhaps in an environment where two species commonly encounter each other, the ability to sense acyl-HSL production and even confuse the normal communication of the other species would provide an advantage.

MOLECULAR MECHANISMS OF QUORUM SENSING

Acyl-HSL Structures and Synthesis

STRUCTURAL DIVERSITY IN ACYL-HSLS   In 1981, the primary acyl-HSL of *Vibrio fischeri*, 3-oxo-hexanoyl-HSL, was identified (28). Over ten acyl-HSLs produced

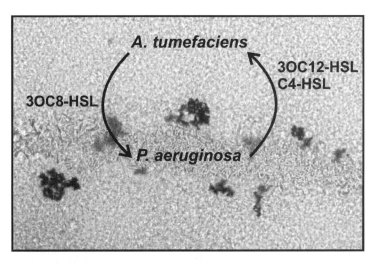

Figure 2 Dual species biofilm of quorum-sensing microbes. Bright field microscopy of *P. aeruginosa* PAO1 and *A. tumefaciens* C58 mixed species biofilm on a glass surface (640× magnification). Biofilm cultivated in a once-through format flow cell (1 mm × 4 mm × 40 mm) in defined minimal salts. *A. tumefaciens* in the biofilm was identified by green fluorescence derived from a plasmid expressing the GFP protein (*separate micrograph*). Arrows indicate signal exchange within the biofilm (C4-HSL, butyryl-HSL; 3OC8-HSL, 3-oxo-octanoyl-HSL; 3OC12-HSL, 3-oxo-dodecanoyl-HSL). The dark granules in the micrograph are likely to be precipitated iron oxides from the medium that are clustered around the *P. aeruginosa* aggregate. Microscopy was performed with a Zeiss Axio-Phot microscope and Axio-Vision software.

by different bacterial species have subsequently been identified, often through use of a combination of high performance liquid chromatography (HPLC) purification, nuclear magnetic resonance (NMR) spectroscopy, and mass spectroscopy (98). Although acyl-HSL structures all have the homoserine lactone ring moiety in common, the acyl side chain of different acyl-HSLs can vary in length, degree of substitution, and saturation (Table 1). The overall hydrophobicity of the molecule is a balance between the somewhat hydrophilic homoserine lactone ring and the hydrophobic side chain. The amphipathy of acyl-HSLs presumably allows them to navigate the phospholipid bilayer of the cell membrane as well as the aqueous intracellular and extracellular environments. All acyl-HSLs currently identified have side chains that range from 4 to 14 carbons in length, usually in increments of 2-carbon units (e.g., C4, C6, C8). The overall length of the side chain and chemical modification at the β position provide specificity to quorum-sensing systems. Bacteria that employ multiple, distinct quorum-sensing systems, such as *Pseudomonas aeruginosa*, must maintain specificity for the different acyl-HSLs. Butyryl-HSL and 3-oxo-dodecanoyl-HSL, the two primary acyl-HSLs of *P. aeruginosa*, are structurally distinct, presumably limiting the interaction between these two systems (79, 80). The tetradecanoyl-HSLs (C14 acyl side chain) produced by *Rhizobium leguminosarum* and *Rhodobacter sphaeroides* have the longest acyl-HSL side chains identified to date (48, 88, 100). However, acyl-HSLs with even longer side chains may exist, as fatty acid biosynthesis provides the acyl-HSL side chain precursors and produces fatty acyl groups greater than 14 carbons in length (see below). Acyl-HSLs appear to be dedicated signaling molecules produced solely to mediate quorum sensing.

ENZYMOLOGY OF ACYL-HSL SYNTHESIS   Genetic analyses of acyl-HSL synthesis in a variety of bacteria clearly demonstrated that the expression of LuxI-type proteins was necessary and sufficient to impart the ability to synthesize acyl-HSLs in heterologous host backgrounds (34, 69, 79, 80). Addition of *S*-adenosylmethionine (SAM) and 3-oxo-hexanoyl-CoA to crude extracts from *V. fischeri* resulted in synthesis of 3-oxo-hexanoyl-HSL, suggesting that these might be the substrates for synthesis (29). Purification and in vitro studies with affinity-tagged fusion derivatives of TraI and LuxI demonstrated that acyl-HSL synthesis required SAM and the fatty acid biosynthetic precursors provided as acylated-acyl carrier protein (ACP) conjugates (69, 99). Kinetic analysis of a highly active preparation of native RhlI has revealed that the acyl-HSL synthesis reaction is unique (77).

Although a ping-pong reaction mechanism involving an acylated enzyme intermediate was originally proposed for the acyl-HSL biosynthetic reaction, there is little evidence to support this model (69, 103). The enzymatic reaction mechanism for butyryl-HSL synthesis was studied for RhlI from *P. aeruginosa*. The reaction products, 5′-methylthioadenosine (5′-MTA) and holo-ACP, and analogs of SAM, such as *S*-adenosylhomocysteine and sinefungin, inhibited acyl-HSL synthesis. RhlI kinetics using the substrates and inhibitors revealed that this protein catalyzes butyryl-HSL synthesis via a sequential ordered reaction mechanism (Figure 3) (see 77). Binding of SAM by RhlI initiates the reaction. Butyryl-ACP is subsequently bound to the enzyme complex, followed by amide bond formation and release of holo-ACP. Lactonization of the homoserine ring occurs and the product, butyryl-HSL, is liberated. In a final step, 5′-MTA is released. Although it seems likely that other LuxI-type proteins share this reaction mechanism, it remains to be proven. SAM is a common metabolite in the cell and is used for a variety of processes, many of which are essential (49). However, enzymatic analysis suggests that RhlI, and by extension other LuxI-type proteins, utilize SAM in a novel way. Therefore, it is feasible that inhibitors of acyl-HSL synthesis designed to block interactions with SAM would not affect its other essential functions.

LuxI-Type Acyl-HSL Synthases

STRUCTURE-FUNCTION STUDIES   All LuxI-type proteins appear to be orthologues that catalyze acyl-HSL formation. These enzymes direct amide bond formation between the acyl side chain and the amino group of SAM. However, many synthesize acyl-HSLs with structurally different acyl side chains and therefore must recognize different acyl-ACP conjugates. Amino acid sequence alignments of LuxI-type proteins (194–226 amino acids long) reveal ten completely conserved amino acids,

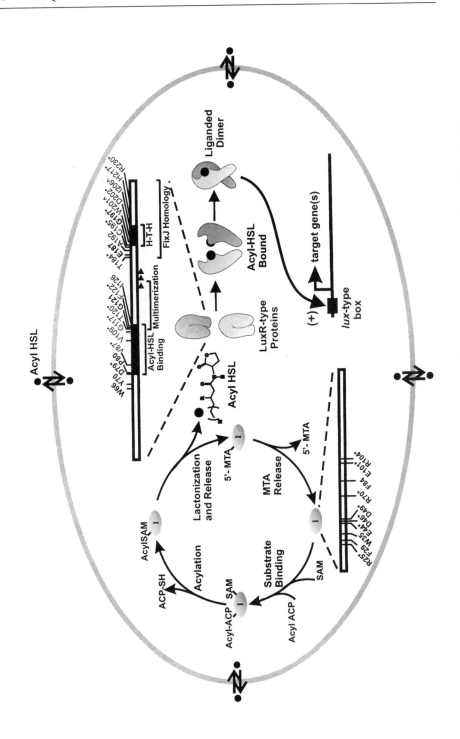

most of these within the amino-terminal halves of the proteins (37, 76). Of the ten conserved amino acid residues, seven are charged (Figure 3). The observation that the amino-terminal portions of LuxI-type proteins are the best conserved, whereas the carboxy terminus is more divergent, suggests that the C terminus may provide recognition of the different acyl chains on precursor acyl-ACPs (52).

Mutational analysis of several LuxI-type proteins has identified amino acid residues essential for function. Nonsaturation, random mutagenesis of *luxI* itself resulted in the isolation of 13 missense mutants. Eleven of these encode a protein with no detectable activity, while two of the mutant proteins display significant, but diminished activity (Figure 3) (52). The observation that 7 of the 11 mutations that resulted in nonfunctional LuxI proteins were clustered within a region spanning amino acids 25 to 70 suggested that these residues might play a role in a consistent enzymatic function of the protein, such as amide bind formation. Strikingly, several of the point mutations identified occurred at conserved residues found in all LuxI-type proteins (R25, E44, D46, D49, and R70; relative to the wild-type amino acid position on LuxI; see Figure 3). Four of these mutations (at positions R104, A133, E150, and G164) were located within a region toward the carboxy terminus of the protein (res. 104–164). The carboxy-terminal residues identified in this study might be critical for acyl side chain specificity of the acyl-HSL. Site-directed mutagenesis demonstrated that none of the cysteine residues in LuxI was critical for activity (52).

In a similar mutational analysis of RhlI, the acyl-HSL synthase in *P. aeruginosa* responsible for butyryl-HSL synthesis, eight critical residues required for acyl-HSL synthesis were identified (76). Seven of the eight residues corresponded

---

←

Figure 3   General model of acyl-HSL signal transduction in quorum sensing. A single quorum-sensing cell is shown. Tentative models for acyl-HSL synthesis cycle (*left side*) and acyl-HSL interaction with LuxR-type proteins (*right side*) are depicted. Double arrows with filled circles at the cell envelope indicate the potential two-way traffic of acyl-HSLs into and out of the cell. LuxR-type and LuxI-type proteins are expanded into open bars. Functional modules defined by mutational analyses or sequence homology are indicated by black fill and brackets. Amino acid residues conserved throughout the LuxI and LuxR families of proteins are indicated in bold (numbering relative to *V. fischeri* LuxI and LuxR). Positions where mutations reduce the activity of the proteins are labeled (*). Nonconserved residues implicated by mutational analysis are in plain text. Two residues in LuxR where mutations resulting in positive control mutants have been identified are indicated (^, see Reference 33). Three filled triangles indicate positions where deletion of the amino-terminal module results in acyl-HSL independence for LuxR and LasR. The LuxR-type protein is shown as dimerizing, although higher-order multimers may be important in other systems. Although the act of binding to the acyl-HSL and multimerization are represented as different events, these may occur simultaneously.

to mutations that also reduced LuxI activity (Figure 3). One additional mutation that reduces RhlI activity alters residue E101 (numbering relative to LuxI sequence), an amino acid residue conserved among LuxI family members. Again, the amino-terminal region of the protein was implicated as an active site for catalysis as all the *rhlI* mutations fell within a region spanning codons 24–104. No mutations in the carboxy-terminal region were isolated, and site-directed mutagenesis demonstrated that those residues at positions shown to be essential in LuxI were dispensible in RhlI (A133, E150, and G164 of LuxI). Site-directed mutagenesis also excluded the role of cysteines in enzymatic activity for RhlI and *A. tumefaciens* TraI (76; C. Fuqua, unpublished results).

Although there are as yet no published reports of the three-dimensional structure of a LuxI-type protein, these efforts are under way in several laboratories. X-ray crystallographic data should provide a wealth of information on substrate interactions. Detailed structural and mechanistic information on acyl-HSL synthesis may identify targets for rationally designed inhibitors.

THE AINS FAMILY OF ACYL-HSL SYNTHASES   The products of *ainS* (*V. fischeri*) and *luxM* (*V. harveyi*) genes, encoding 46-kDa and 25-kDa proteins, respectively, have been implicated in acyl-HSL synthesis and do not show any similarity to LuxI (9, 44). Comparison of the AinS and LuxM amino acid sequences sequences reveals 34% identity between LuxM and the C-terminal 218 residues of AinS. Null mutations in either *ainS* of *V. fischeri* or *luxM* of *V. harveyi* abolish synthesis of their respective acyl-HSLs, octanoyl-HSL and 3-hydroxybutanoyl-HSL. Although expression of *ainS* in *Escherichia coli* directs production of octanoyl-HSL, expression of *luxM* in the same background did not result in 3-hydroxybutanoyl-HSL synthesis. The *luxM* gene positively regulates bioluminescence in *V. harveyi* (9). The presumptive target genes under control of AinS and octanoyl-HSL in *V. fischeri* remain to be identified, although it has been proposed that this system modulates the LuxI-LuxR quorum sensor by competing with 3-oxohexanoyl-HSL for binding to LuxR (60). The *ainS* gene has upstream DNA sequence elements similar to the *lux* box of *V. fischeri*, a promoter element required to activate *lux* gene expression, suggesting that expression of *ainS* may be under control of the LuxI-LuxR quorum sensor (44).

The precursors for acyl-HSL synthesis by AinS, and presumably by LuxM, appear to be similar to those used by LuxI-type acyl-HSL synthases. A purified maltose binding protein-AinS fusion catalyzes acyl-HSL synthesis in vitro with SAM as a source for the homoserine lactone ring (51). Surprisingly, either octanoyl-ACP or octanoyl-CoA conjugates could serve as precursors for the acyl side chain. Unlike LuxI-type proteins, which use primarily acyl-ACP substrates, the in vitro kinetics of AinS-directed acyl-HSL synthesis suggest that both of these compounds may serve as a substrate for octanoyl-HSL synthesis in vivo. Whether this also holds true for 3-hydroxy-butanoyl-HSL synthesis by LuxM remains to be determined. Acyl-ACP conjugates are intermediates in fatty acid synthesis, and acyl-CoA compounds are typically intermediates of fatty acid β-oxidation. Perhaps the flexibility of AinS for its fatty acyl precursor would be beneficial under conditions that cause an imbalance between these substrate pools. Acyl-HSL synthesis by AinS is inhibited by S-adenosylhomocysteine and holo-ACP, compounds that also inhibit RhlI, suggesting that there may be a shared enzymatic mechanism (51).

Dynamics of Acyl-HSL Accumulation

MEMBRANE TRAFFICKING OF ACYL-HSLS   In 1985, it was demonstrated that radiolabeled 3-oxo-hexanoyl-HSL rapidly diffused into and out of bacterial cells and that the distribution of 3-oxo-hexanoyl-HSL was dictated largely by its concentration gradient across the bacterial envelope (58). Acyl-HSLs from other bacteria are also thought to cross the bacterial envelope via diffusion. The general observation that LuxR-type proteins can be activated by exogenous addition of acyl-HSLs in heterologous hosts is consistent with a transmembrane traffic mechanism that relies on passive diffusion. However, those more hydrophobic acyl-HSLs with longer chains (8–14 carbons), would be less soluble in the cytoplasm and might concentrate in membrane bilayers. In *P. aeruginosa*, the MexAB-OprD efflux system plays a role in the exit of 3-oxo-dodecanoyl-HSL from cells, but not the traffic of the shorter chain butyryl-HSL signal (35, 82). MexAB-OprD is a transporter of hydrophobic molecules from phospholipid bilayers, and therefore this observation provides indirect evidence that a substantial amount of 3-oxo-dodecanoyl-HSL partitions into the lipid bilayer. It is not known whether the release of other acyl-HSLs is assisted by efflux pumps.

ACCUMULATION OF ACYL-HSLS   An increase in population density is thought to be the primary mechanism by which acyl-HSL concentrations rise to inducing levels. The specialized environments of the light organs of symbiotic marine animals support *V. fischeri* to a high population density, fostering concomitant increases in 3-oxo-hexanoyl-HSL concentrations (94). Although the appropriate environment for quorum-sensing induction in other bacteria is less well defined, the necessary population density is probably achieved within aggregates or surface-associated biofilms. Although inducing acyl-HSL concentrations can clearly occur within a dense population, other physical and chemical factors also influence this dynamic process. Multiple pathways for induction are possible, as revealed through mathematical modeling of LuxR interactions with 3-oxo-hexanoyl-HSL (55). The dimensions and diffusion characteristics of the environment could influence induction. Similarly, the flow through a given environment will influence signal concentrations, with low flow rates fostering and high flow rates preventing acyl-HSL accumulation, respectively. Under alkaline conditions most acyl-HSLs are chemically unstable, and therefore high pH environments will accelerate signal degradation, and effectively increase the size of the quorum required to activate the associated quorum sensor. The interactions of acyl-HSLs with extracellular reactants that destabilize or sequester the signal molecules may also affect accumulation.

Different environmental and physiological conditions often regulate the expression of the acyl-HSL synthase genes and therefore directly affect the production of signal. Positive feedback of cognate LuxR-type proteins on signal production, additional diffusible signals, and other regulatory proteins are examples of this regulation. Lastly, as discussed above, the intracellular concentration of acyl-HSLs may be influenced by cellular efflux systems. It is the integration of all of these features that determines the inducing character of any given environment.

LuxR-Type Proteins

LuxR-type proteins share an end-to-end sequence identity of 18–23%. An acyl-HSL interaction region [res. 66–138] and a DNA binding motif [res. 183–229] are defined by two clusters of stronger sequence conservation (Figure 3). LuxR-type proteins are members of the larger FixJ-NarL superfamily (57). Most members of the FixJ-NarL superfamily are two component–type response regulators that differentially control DNA binding activity by phosphorylation of a conserved aspartate residue in the amino-terminal halves of these proteins. There is no significant sequence similarity between the amino-terminal halves of LuxR-type proteins and other members of the FixJ-NarL group (57). This reflects the specific function of this region in the acyl-HSL interaction of LuxR-type proteins (42). However, several mechanistic features may be shared between LuxR-type proteins and other members of the FixJ-NarL superfamily.

LuxR-type proteins facilitate responses to acyl-HSLs through a series of recognizable steps including (*a*) specific binding of cognate acyl-HSLs, (*b*) conformational changes and alterations in multimerization of the protein following binding of the signal, (*c*) binding or release of specific regulatory sequences upstream of target genes, and (*d*) often, activation of transcription (Figure 3). Although we discuss each of these as separate events below, there is clearly significant overlap and integration of processes during response to the signal.

BINDING OF ACYL-HSLS BY LUXR-TYPE RECEPTORS   Genetic and biochemical evidence has identified the LuxR protein of *V. fischeri* and its homologues as acyl-HSL receptors. Significant amounts of acyl-HSL are bound by *E. coli* cells expressing either *luxR* or *lasR* of *P. aeruginosa* [elevated expression of the GroESL chaperone complex improves the efficiency of binding; see (2, 50, 81)]. Acyl-HSLs bound by LuxR or LasR fractionate with the cells during centrifugation. Mutations in *luxR* identified by a loss of responsiveness to 3-oxo-hexanoyl-HSL also reduce or abolish its cell-association when the LuxR mutant proteins are expressed in *E. coli*. Analysis of truncated LuxR proteins suggests that the N-terminal region [res. 10–194] contains the acyl-HSL binding pocket (Figure 3) (50). Purification of LuxR-type proteins provides the best evidence for their receptor function. Purified preparations of TraR from *A. tumefaciens* and CarR from *Erwinia carotovora* stably associate with their cognate acyl-HSLs (3-oxo-octanoyl-HSL and 3-oxo-hexanoyl-HSL, respectively) in an equimolar ratio of acyl-HSL to protein (110, 119).

V. fischeri LuxR is comprised of two domains (Figure 3) (for a detailed review, see 107). Interactions with 3-oxo-hexanoyl-HSL are mediated by the amino-terminal 70% of the protein. DNA binding and transcriptional activation require sequences in the carboxy-terminal region, which includes a helix-turn-helix motif (HTH). The response to acyl-HSLs is abolished by mutations in *luxR* between residues 79–127 and 184–230 (Figure 3) (101, 104). A pair of mutations in the amino-terminal domain, V82I and H127Y, which result in loss of LuxR activity in vivo, are suppressed by the addition of higher concentrations of 3-oxo-hexanoyl-HSL (101, 104). This observation is consistent with weakened 3-oxo-hexanoyl HSL binding, suggesting a direct interaction of the ligand with these residues. A *luxR* truncation allele comprised of codons 1–195 but missing 55 codons from its carboxy terminus is sufficient for binding to 3-oxo-hexanoyl-HSL in the whole-cell binding assay described above (50). A comparison of the amino acid sequences of LuxR proteins from different bacterial genera reveals that the sequence identity within the region defined as the LuxR acyl-HSL interaction site is significantly higher than the overall sequence identity (Figure 3). Considering this evidence collectively, the amino-terminal region of other LuxR-type proteins most likely also interacts with acyl-HSLs, but this has been shown only for LuxR.

CONSEQUENCES OF ACYL-HSL INTERACTION WITH LUXR-TYPE PROTEINS   Interactions between acyl-HSLs and their binding sites are not well understood. Biologically active acyl-HSLs are released following organic extraction of either purified LuxR-type proteins bound to acyl-HSLs or whole cells that have sequestered acyl-HSLs through interactions with LuxR-type proteins (1, 50, 119). Therefore, receptor binding does not irreversibly alter the chemistry of the acyl-HSL.

Studies using acyl-HSL analogues have examined the interaction between acyl-HSLs and LuxR-type receptors (14, 30, 78, 97, 117). LuxR-type proteins recognize features of both the homoserine lactone moiety and the acyl chain. Alterations to the homoserine lactone ring dramatically reduce binding or signaling. Conversely, significant inducing activity is retained by acyl-HSL analogues such as thiolactone and lactam derivatives, which conserve the basic ring configuration. However, significant reductions in activity are observed for analogues with alterations to the size and orientation of the ring, and side chain lengths that differ greatly from the cognate acyl-HSL. Modifications at the β-position reduce, but do not abolish, activity. Direct assays of acyl-HSL binding reveal much the same pattern of recognition (78, 97). Several potential competitive inhibitors have been identified that do not induce well but retain the ability to strongly bind to LuxR-type proteins. The concentrations required for productive inhibition are often high. High-level expression of the TraR protein of *A. tumefaciens* reduced its specificity for noncognate acyl-HSLs and resulted in a general recalcitrance to competitive inhibition, as the acyl-HSL analogues themselves resulted in activation (117). However, normal expression of *traR* in *A. tumefaciens* provided far greater specificity for the cognate acyl-HSL, and in this case, acyl-HSL analogues effectively blocked activation. Whether this is the case with other LuxR family members remains to be determined.

POTENTIAL MEMBRANE INTERACTIONS OF LUXR-TYPE PROTEINS   LuxR-type proteins are cytoplasmically localized and do not possess membrane-spanning sequences. However, there is evidence that these proteins may be physically associated with the cytoplasmic membrane and that acyl-HSL interaction may alter this association. Immunoprecipitation studies indicate that LuxR is tightly membrane associated in *V. fischeri* (59). It has been proposed that LuxR contacts the interior leaflet of the cytoplasmic membrane bilayer through amphipathic interactions. Full-length LuxR or truncations retaining the amino-terminal domain are highly insoluble (50). However, removal of the amino-terminal 156 codons from *luxR* results in fully soluble truncated proteins (105). Corroboration of these findings comes from studies of *A. tumefaciens* TraR (89). In the absence of acyl-HSL, monomeric TraR cofractionates with cytoplasmic membranes, whereas in the presence of ligand, TraR appears to be largely cytoplasmic. A plausible model is that the acyl-HSL interaction sites on LuxR and TraR are associated with the membrane and bind to acyl-HSL molecules as they traverse the bilayer. Perception of hydrophobic, plant-released flavonoid Nod factors has been proposed to occur via contact of the cytoplasmic NodD regulatory protein with the ligand in the membrane bilayer (99a).

AN INTRAMOLECULAR INHIBITION MODEL FOR LUXR ACTIVATION   Interaction with acyl-HSLs radically alters the activity of LuxR-type proteins, modulating their ability to bind to DNA. Truncated derivatives of LuxR and LasR from *P. aeruginosa* that lack the acyl-HSL interaction region (LuxRΔN and LasRΔN, respectively) constitutively activate transcription (4, 15, 85). A model has been proposed in which DNA binding by LuxR and LasR, through their carboxy-terminal domains, is held in check by the amino-terminal region of each protein (Figure 3). For the full-length proteins, acyl-HSL binding presumably unmasks the carboxy-terminal domain of the protein and relieves inhibition. The carboxy-terminal regions can function as constitutive activators when expressed independently from the amino-terminal domain. Several other members of the NarL-FixJ superfamily are thought to function by an analogous mechanism. Several amino-terminal truncations of the FixJ response regulator from *Rhizobium meliloti* are constitutive transcriptional activators (20, 57). Structural studies of the *E. coli* NarL protein in the unphosphorylated state suggest that its DNA-binding domain is occluded by the amino-terminal domain (6). Presumably, phosphorylation of NarL releases this inhibition. These observations support a general model for the FixJ-NarL superfamily, including LuxR-type proteins. In this model, interaction with the ligand (acyl-HSLs or phosphoryl groups) stimulates the activity of the carboxy-terminal DNA-binding segment by relieving the inhibitory activity of the amino-terminal half of the protein.

LUXR-TYPE PROTEINS MULTIMERIZE IN RESPONSE TO ACYL-HSL BINDING   There is now ample evidence that LuxR-type proteins form multimers. Multimerization was originally implicated by the observation that deletion alleles encoding truncated

LuxR proteins lacking from 15–89 of the carboxy-terminal amino acid residues exerted a dominant negative effect on wild-type *luxR* (16). The LuxR deletion proteins retain the ability to interact with full-length LuxR, but are missing some or all of the amino acid residues required for DNA binding. Likewise, several point mutations in *luxR* that resulted in nonfunctional LuxR proteins have a dominant inhibitory effect over the wild-type protein (16). The observed dominant negative effect of the truncation and missense alleles suggested that these proteins formed inactive heterodimers with the wild-type LuxR protein. Similarly, a natural deletion allele of the *A. tumefaciens traR* gene, called *trlR*, exerts a dominant negative effect over the full-length *traR* gene. The *trlR* gene product lacks 52 amino acids from its carboxy terminus, including the HTH motif (74, 118). in vitro analysis with purified TraR and TrlR has identified heterodimers of the two proteins that are inactive for DNA binding (12, 118). The effect of dominant negative alleles suggests that LuxR and TraR form multimers. However, this information does not address the consequences of acyl-HSL interactions on multimerization, nor does it reveal a stoichiometry for the normal homomultimers.

Ligand-induced multimerization is common among regulatory proteins. Recent studies of several LuxR-type proteins suggest that acyl-HSLs promote multimerization. The active, ligand-bound form of TraR is a dimer, whereas apo-TraR is a monomer (120). In contrast, the CarR protein of *E. carotovora* exists as a dimer in the absence of ligand and is shifted to a higher-order multimer(s) in response to acyl-HSL addition (110). Despite the differences, in both cases, DNA binding is mediated by protein dimers, and multimerization is stimulated by acyl-HSL interaction. However, the truncated LuxRΔN protein is monomeric in solution, binds to DNA (see below), and is a constitutive transcriptional activator (105). This apparent discrepancy is best explained by the observation that LuxRΔN circumvents the requirement for dimerization by associating directly with RNA polymerase (RNAP). Thus far, all full-length LuxR-type proteins must dimerize as a prerequisite for DNA binding (Figure 3). Such a model is consistent with the dyad symmetry of the DNA-binding sites (*lux*-type boxes, see below) for many LuxR-type proteins (107). Ultimately, a model must be developed that incorporates the integration of ligand-binding with alterations in domain interactions and multimerization.

CIS-ACTING DNA ELEMENTS REQUIRED FOR LUXR-TYPE PROTEIN REGULATION
Members of the LuxR family generally require DNA sequences associated with their target genes to activate transcription. Originally, the DNA sequence element required for *lux* gene activation was called the *lux* operator, but it is now more often referred to as the *lux* box (24, 47). The *lux* box is a 20-base pair inverted repeat sequence located upstream of the *lux* operon transcriptional start site, and centered at position −42.5 (31). Transcriptional activation by LuxR is dependent on the position of the *lux* box relative to the promoter. Additionally, both arms of the dyad repeat are required for activation (24, 31).

There are similar inverted repeat elements ranging from 18 to 22 bp, generally called *lux*-type boxes, associated with the promoters of genes regulated by

LuxR-type proteins from several different bacteria. Surprisingly, in addition to the shared dyad configuration, the *lux*-type boxes also retain primary sequence identity with the original *lux* box (107). Often, but not always, the *lux*-type boxes are located just upstream of the −35 promoter element. However, a number of alternative configurations have also been identified. For example, there are two separate 20-bp *lux*-type boxes (one centered at −42 and the other at −102) required for maximal induction of the *lasB* gene in *P. aeruginosa* (4, 96). The mechanism by which the upstream site contributes to promoter activity is not yet known. Several 18-bp *lux*-type boxes are required to activate expression of *A. tumefaciens* target operons by TraR (39). One of the *lux*-type boxes in *A. tumefaciens* is positioned between the −35 elements of a pair of divergent *tra* promoters, occupying a 16-bp gap and overlapping each −35 element by one base pair. This centrally located *lux*-type box is required to activate both of the divergent promoters (39).

There are also several reports of promoters that are clearly regulated by LuxR-type regulators but for which there are no easily identified *lux*-type boxes. For example, the *traM* and *traR* genes of *A. tumefaciens* are activated by TraR, although more modestly than other *tra* genes (36, 39). Neither of the *traM* or *traR* genes has canonical *lux*-type box elements upstream of their promoters. Although the primary sequence similarity of *lux*-type boxes is a striking level of conservation among diverse bacteria, examples such as the *traR* and *traM* promoters demonstrate some of the different promoter architecture that can be recognized by LuxR-type proteins. The existence of these divergent sequences cautions against relying solely on DNA sequence information to identify such regulatory elements.

BINDING OF DNA BY LUXR-TYPE PROTEINS   The first in vitro evidence that LuxR-type proteins bind to DNA was provided by studies using the truncated LuxRΔN derivative. LuxRΔN is deleted for its amino-terminal 156 amino acid residues, resulting in acyl-HSL-independent transcriptional activation and enhanced solubility when expressed in *E. coli* (15, 105). These studies established that LuxRΔN bound to the *lux* box, but only in the presence of RNAP purified from *E. coli* (105). Neither protein bound the *lux* box independently. Recent experiments suggest that the dependence of LuxRΔN on RNAP is a consequence of its inability to multimerize, and it seems unlikely that full-length LuxR shares this requirement (31).

Since the original work on LuxRΔN, several other LuxR-type proteins have been examined in vitro (70, 110, 115, 119). A common theme from this work is that the acyl-HSL increases or modifies the DNA-binding capacity of the protein. Although there are differences between the reported attributes of each protein, some of these may be due to experimental design and the quality of the protein preparation analyzed.

Arguably, the *A. tumefaciens* TraR protein is the best studied LuxR-type protein in vitro. TraR was purified from *E. coli* as a complex with its cognate acyl-HSL, 3-oxooctanoyl-HSL (119). The TraR-acyl-HSL complex bound quite specifically to a single genetically defined *lux*-type box. This preparation of TraR was sufficient to activate transcription of a target promoter in vitro when combined with *A.*

tumefaciens RNAP. Subsequent removal of the acyl-HSL by detergent treatment reduced the DNA-binding activity of TraR, and this was partially restored by addition of acyl-HSL. The difficulty in removing the acyl-HSL from TraR reflects the tenacious association of this ligand with the protein. The complex is comprised of a dimer of protein with monomers in an equimolar ratio with 3-oxo-octanoyl-HSL (119, 120). TraR purified in the absence of 3-oxo-octanoyl-HSL is monomeric and does not bind to DNA. It remains to be determined whether other LuxR-type proteins share similar interactions with their target promoters.

The ExpR$_{Ech}$ protein of *Erwinia chrysanthemi* binds to a number of potential target promoters in the absence of its cognate acyl-HSL (70). Interaction with 3-oxo-hexanoyl-HSL alters the contacts of ExpR$_{Ech}$ with these promoter sequences. However, ExpR$_{Ech}$ is included in a group of LuxR-type proteins from species of *Erwinia* that, rather than functioning as transcriptional activators, appear to be acyl-HSL-responsive repressors (5, 11). LuxR-type proteins that act as repressors have significant differences with LuxR-type transcriptional activators, and mechanistic features may not be consistent between the two types of regulatory proteins (see below).

TRANSCRIPTIONAL CONTROL   Most LuxR-type regulators are transcriptional activators and as such, null mutations in genes encoding LuxR-type proteins generally result in decreases in target gene expression. in vitro, LuxRΔN and TraR are sufficient to activate target gene transcription from purified DNA templates in the presence of RNAP (106, 119). Many *lux*-type boxes are positioned just upstream of the -35 sequences of regulated promoters, suggesting that LuxR-type proteins can interact directly with RNAP. As with other regulators that function from this position, LuxR-type proteins probably have specific residues that contact RNAP and are brought into appropriate juxtaposition with the polymerase when the transcription factor binds to the *lux*-type box. LuxR and, by extension, other LuxR-type proteins may interact with the carboxy-terminal domain of the RNAP alpha subunit, or the RNAP sigma subunit, or both (31, 92). In fact, the positional dependence of the *V. fischeri lux* box itself is consistent with an ambidextrous activator that contacts both of these control sites on RNAP simultaneously (31).

A reconstructed *lux* promoter, with the *lux* box positioned between the -10 and -35 elements, is repressed by LuxR in the presence of 3-oxo-hexanoyl-HSL (32). The development of a genetic repression assay using this promoter has allowed analysis of DNA binding that is independent of transcriptional activation. Directed mutational analysis of LuxR (alanine-scanning) suggests that residues within the carboxy-terminal DNA-binding domain (W201 and I206) are required for transcriptional activation. Alanine replacements at these positions result in mutant proteins that bind to DNA as well as the wild-type protein (as assessed using the repression assay), but cannot activate transcription of the *lux* promoter (i.e., positive control or PC mutants) (33). Curiously, mutant TraR proteins with PC phenotypes, analyzed using a similar, but perhaps not as robust, artificial repression assay, have amino acid substitutions that map to the amino terminus of the protein

(63). The discrepancies between LuxR and TraR may reflect true differences between the transcriptional activation mechanisms of these proteins.

REPRESSION BY LUXR-TYPE PROTEINS   Several LuxR-type proteins from different species and subspecies of *Erwinia* act as transcriptional repressors (5, 10, 11). The most compelling evidence exists for the EsaR protein of *E. stewartii* (recently assigned to the genus *Pantoea*), which represses the expression of genes for synthesis of exopolysaccharide (EPS) (11). As with other acyl-HSL quorum sensors, the presence of the acyl-HSL signal elevates expression of the target genes. However, *esaR* null mutants constitutively express the EPS genes. EsaR may bind to DNA target sequences in the absence of the acyl-HSL, occluding the promoter from RNAP. Association with the acyl-HSL might weaken or abolish DNA binding, permitting both the recruitment of RNAP to the promoter and transcription initiation. If such a repression-derepression model is correct, the molecular consequences of ligand interaction must be substantially different for EsaR and other LuxR-type repressors than for the more common activators. There is no simple structural basis for such a different response upon acyl-HSL binding. The amino acid sequences of those LuxR-type proteins suspected to be repressors are not strikingly or consistently different from other members of the LuxR family (107).

Modulation of Quorum Sensing Through Regulatory Gene Expression

Acyl-HSL quorum sensors are almost always integrated into other regulatory circuitry. This effectively expands the range of environmental signals that influence target gene expression beyond population density. This is usually via regulation of the genes encoding the LuxR-type regulator and the LuxI-type acyl-HSL synthase.

EXPRESSION OF GENES ENCODING LUXR-TYPE PROTEINS   Control of the expression or stability of the LuxR-type protein is a common mechanism for integrating quorum sensing into other aspects of cellular physiology. For example, the expression of the *V. fischeri luxR* gene is under the influence of the cyclic AMP receptor protein (CRP) (26). Another interesting example is the strict opine-dependence of *traR* gene expression in *A. tumefaciens*. Opines are unusual compounds produced by plants infected with *A. tumefaciens*, thereby restricting the function of this quorum sensor to the plant-associated environment (40, 87). The *lasR* gene in *P. aeruginosa* is controlled by the GacA two-component response regulator, a common regulator of virulence in *P. aeruginosa*, and also by a CRP homologue, called Vfr (3, 91). The environmental conditions to which these regulators respond and their influence on the Las quorum sensor are not yet fully understood.

An interesting variation occurs in *P. aeruginosa*, where two discrete acyl-HSL quorum-sensing systems co-exist. LasR and 3-oxododecanoyl-HSL regulate expression of a second LuxR-type protein, called RhlR, responsive to butyryl-HSL

(61, 85). Although the Las and Rhl quorum sensors are in many ways discrete systems, this configuration results in a hierarchical quorum-sensing cascade. The recent identification of a third LuxR-type protein, called QscR, that limits the activity of LasR, adds further complexity to this regulatory pathway (17). Microorganisms that employ multiple acyl-HSL quorum-sensing systems are often revealed as these regulatory systems are deliberately sought after and as whole genome sequences are analyzed.

Posttranscriptional control of LuxR-type proteins will also clearly affect acyl-HSL quorum sensing. For example, the proteolytic degradation of CarR and TraR proteins is altered by interaction with their cognate acyl-HSL ligands (110, 119). TraR is degraded by several different proteases, but is stabilized by binding to 3-oxo-octanoyl-HSL (120). Changes in overall proteolytic activity within the cell may therefore affect the half-life of TraR and modulate its response to increasing levels of acyl-HSL.

CONTROL OF ACYL-HSL SYNTHASE GENE EXPRESSION   There are also many examples in which the acyl-HSL synthase genes are differentially controlled. LuxR-type proteins often activate expression of their cognate acyl-HSL synthase counterpart, establishing a positive feedback loop (38). This positive feedback is not a prerequisite for quorum sensing, and there are some LuxI-LuxR type systems in which the acyl-HSL synthase gene is not positively autoregulated (109). When present, positive regulation of the acyl-HSL gene may act to desensitize the quorum sensor and buffer the expression of the target genes from minor fluctuations in acyl-HSL concentration.

There are also examples of acyl-HSL synthase gene control in addition to positive regulation by the cognate LuxR-type protein. The GacAS two-component regulatory system (described above for *P. aeruginosa*) is required for quorum sensing control of phenazine antibiotic synthesis in the plant biocontrol agent *Pseudomonas aureofaciens* (13). In contrast to GacA regulation of *lasR* in *P. aeruginosa*, in *P. aureofaciens* GacA is required to activate the *phzI* acyl-HSL synthase gene, elevating production of hexanoyl-HSL. Acyl-HSL synthesis is regulated at several different levels in *P. aeruginosa* also. An inhibitory protein called RsaL specifically inhibits transcription of the *lasI* gene (22). The *rhlI* gene, encoding synthesis of butyryl-HSL, is negatively regulated by the stationary phase sigma factor gene *rpoS* in *P. aeruginosa* (112).

Posttranscriptional control of acyl-HSL synthases also can influence the process of quorum sensing. The amount of acyl-HSL synthesized by a given cell will be affected by turnover rates of acyl-HSL synthase mRNA transcripts and proteins. An example of the former is found in *Erwinia carotovora* subsp. *carotovora* 71. The *rsmA* gene product negatively regulates synthesis of 3-oxo-hexanoyl-HSL by destabilizing the transcript encoding the HslI acyl-HSL synthase (19). The RsmA homologue, CsrA from *E. coli*, also destabilizes specific mRNAs, and homologues of RsmA are widespread in bacteria (111). Regulated stability of acyl-HSL synthase transcripts may therefore be relatively common.

Fine Tuning of Quorum Sensing

A number of diffusible antagonists as well as other regulatory proteins can impinge upon the function of acyl-HSL quorum sensors through interactions with LuxR-type proteins. Several such factors, produced by the bacteria themselves or host organisms, have now been identified.

REGULATORY PROTEINS THAT INFLUENCE QUORUM SENSING   In almost all quorum-sensing systems that have been characterized, LuxR-LuxI type regulatory proteins are sufficient to impart population-density responsive gene regulation. However, in several cases additional regulatory proteins directly influence quorum sensing by interfering with the activity of the LuxR-type protein. Two separate mechanisms that modulate TraR-dependent transcriptional activation in A. tumefaciens provide excellent examples. Both of these mechanisms function through formation of in-active heterocomplexes with TraR. TrlR is highly similar to the amino-terminal 181-amino acid residues of TraR, but lacks the DNA-binding domain (118). TrlR forms inactive heterodimers with TraR (120). The expression of trlR is regulated by a subset of the plant tumor–released opines, thereby influencing TraR activity in response to the specific opine composition of the environment. TraR is also inhibited by the TraM protein, which is required to prevent TraR from activating target genes under noninducing conditions (36, 54, 64). As with TrlR, TraM occupies the transcription factor in inactive heteromultimers, although TraM shares no clear sequence similarity with TraR, and the stoichiometry of the proteins in the complex is not known. In further contrast to TrlR, the TraM antiactivator can inhibit the activity of TraR dimers prebound to the acyl-HSL signal (64). TraM and TrlR have only been identified in A. tumefaciens and closely related bacteria, but other examples of proteins that modulate LuxR-type transcription factors through direct interactions are likely to exist.

DIFFUSIBLE SIGNALS THAT INFLUENCE QUORUM SENSING   LuxR-type proteins weakly recognize noncognate acyl-HSLs. In fact, noncognate acyl-HSLs can function as effective competitors of the correct signal (30, 97, 117). In V. fischeri, octanoyl-HSL is produced by the AinS synthase (51). ainS mutants of V. fischeri that cannot synthesize octanoyl-HSL are induced to luminesce at lower population densities than the wild-type cells (60). This observation might be explained by a model in which, in the wild-type background, 3-oxo-hexanoyl-HSL competes against octanoyl-HSL for the binding site on LuxR. Such competition would necessitate higher concentrations of the inducing acyl-HSL to activate lux gene expression, and modulation of the octanoyl-HSL inhibitor levels would influence the population density at which the lux genes were activated. It is not clear which other quorum-sensing systems employ additional acyl-HSLs to modulate the activity of LuxR type proteins. Noncognate acyl-HSLs are produced by many different LuxI-type proteins (37, 113). These secondary acyl-HSLs may affect the activity of their associated LuxR-type protein.

Competitive interactions between different acyl-HSLs could also function as a mechanism of cell-to-cell crosstalk between different species of bacteria. High-level production of a noncognate, competing acyl-HSL by one type of bacteria could impinge upon the function of LuxR-type proteins in adjacent bacteria (Figure 2). Positive crosstalk between coresident bacteria is well documented, and therefore such competitive interactions are certainly possible in complex bacterial communities (67, 114).

Several bacteria produce cyclic dipeptides (diketopiperazines) that can act as weak inducers of LuxR-type proteins in the laboratory (53). The weakly inducing cyclic dipeptides can also act as inhibitors of LuxR-acyl-HSL interactions. However, the concentration of the cyclic dipeptides required to activate and inhibit LuxR-type proteins is notably higher than that for acyl-HSLs. It has not been determined whether these compounds are physiologically relevant signal molecules.

P. aeruginosa produces a different diffusible signal molecule, called the *P. aeruginosa* quinolone signal (2-heptyl-3-hydroxy-4-quinolone or PQS). PQS production is integrated with the Las and Rhl quorum-sensing systems (84). In wild-type *P. aeruginosa*, PQS concentrations peak during stationary phase, significantly later than the acyl-HSLs. This delayed production is probably due in part to that fact that the LasR protein regulates synthesis of the PQS signal. In turn, PQS itself increases expression of the *rhlR* gene, thereby potentiating expression of RhlR-regulated genes (68). It is not known whether this molecule acts through the documented LasR-dependent control of *rhlR* (85). It is speculated that PQS elevates RhlR-dependent gene expression during slow growth or starvation.

ACYL-HSL INTERFERENCE BY EUKARYOTIC SIGNAL MOLECULES   A family of compounds, known as halogenated furanones, is produced by the marine red alga *Delisea pulchra*. These furanones prevent the association of microorganisms as well as metazoan colonizers with surfaces and, hence, are effective antifouling agents (23, 90). Although these compounds have a broad range of activities, one mechanism of furanone activity may be to inhibit acyl-HSL quorum sensing (45). In whole-cell binding assays with *E. coli* expressing *luxR*, the acyl-HSL signal is displaced by a subset of the furanones (65). This displacement is proposed to function through competition for binding sites on LuxR and other LuxR-type proteins. The exact mechanism of action for the furanones remains an area of investigation. However, development of these natural compounds and synthetic derivatives as antifouling materials is already under way. *D. pulchra* is unlikely to be the only metazoan organism to have developed the ability to confuse bacterial invaders. Modulation of quorum sensing could enable host organisms to control their associated bacterial populations. Recently, several varieties of pea and a number of other higher plants were reported to produce inhibitors of quorum sensing (108). These compounds do not appear to be acyl-HSLs or furanones, although their chemical structures have not been determined. Given the importance of acyl-HSLs and other signaling systems to host-microorganism interactions, host organisms might well have evolved the ability to inhibit or even augment interbacterial signaling.

PERSPECTIVE

It has been nearly 10 years since we coined the term quorum sensing to describe bacterial signaling systems that control gene expression in a population density-dependent manner (42). The original use of the term referred to the acyl-HSL signaling systems described in this review. Acyl-HSLs are dedicated signals, produced for the purpose of communication, and these signals have specific receptors. There are other unrelated systems with similar basic characteristics. Peptide signals are produced by certain gram-positive bacteria, an example of which is cyclic peptide signaling in *Staphylococcus aureus* (73). These are species- and strain-specific signals that appear to have but one function, that of signaling. The past 10 years have seen an explosion in research activity and interest in quorum sensing. Many other types of systems are being studied, and other specific signaling systems have been or will be discovered. However, we raise two concerns.

(*a*) For many systems currently under study, we do not yet have information on the chemical nature of the signal or the mechanism of signal reception. Many of the less-well-described systems may turn out to be nonspecific. The signals may be common metabolites or even toxic metabolic end products. The response may be stress induced. Although these phenomena may be of great interest, the question arises as to whether they are really communication systems in the strict sense. In human terms, we use speech, gestures, and pheromones in specific communication systems. We can sense and detect many additional features of other individuals and we respond, but this is not really considered active communication.

(*b*) Upon its first usage, the term quorum sensing immediately invaded the lexicon of microbiology. Ten years hence, it seems time to raise a note of caution. Any communication system at its heart requires a quorum. If an individual is alone, there can be no communication, even if a signal is produced. Systems like the acyl-HSL quorum-sensing systems are specific communication mechanisms. They are now described as quorum sensing, and in fact we have provided several examples where the communication system allows expression of specific genes that provide functions important only when cells are in a community of sufficient population density. However, we are concerned that consideration of bacterial communication strictly in terms of population density-dependent responses may overly restrict this research area.

Although we now have a reasonable understanding of the details of acyl-HSL signaling and response, this new field is just opening up, and there are many exciting areas of investigation. Important details about the mechanism of acyl-HSL quorum-sensing await discovery. We do not yet have crystal structures for any members of the LuxR or LuxI families of proteins. Such information has become particularly important because of commercial interest in quorum sensing as a target for therapeutic intervention in persistent bacterial infections. We do not understand the determinants of specificity for either signal generation or signal reception. We now have evidence that at least some bacteria use quorum-sensing signals as cues in biofilm development. Just how signaling functions in biofilm development remains a mystery. Which quorum-sensing-controlled genes are involved in biofilm physiology is not yet known. Furthermore, we continue to discover acyl-HSL systems, and many have interesting twists on the basic theme. Aside from the details of well-established quorum-sensing systems like the acyl-HSL systems, an emerging area is discovery of novel interbacterial signaling systems. Finally, there is an accelerating interest in the concept of interspecies communication and signaling dynamics among natural bacterial populations. All the evidence suggests that in the area of bacterial cell-cell communication, the next decade will be as exciting as the last.

ACKNOWLEDGMENTS

We thank our many colleagues who have contributed to the explosion of information in the field of acyl-HSL quorum sensing and interbacterial signaling in general. It is striking that an area once comprehensively covered in a short review now requires whole books. We therefore apologize to those individuals whose contributions were not included in our necessarily brief and restricted summary of the field. CF receives support from the National Science Foundation (MCB-9974863). EPG acknowledges the generous support of the National Institutes of Health, National Science Foundation, the Cystic Fibrosis Foundation, and the Procter & Gamble Company.

Visit the Annual Reviews home page at www.AnnualReviews.org

LITERATURE CITED

1. Adar YY, Simaan M, Ulitzur S. 1992. Formation of the LuxR protein in the *Vibrio fischeri lux* system is controlled by HtpR through the GroESL proteins. *J. Bacteriol.* 174:7138–43

2. Adar YY, Ulitzur S. 1993. GroESL proteins facilitate binding of externally added inducer by LuxR protein-containing *Escherichia coli* cells. *J. Biolumin. Chemilumin.* 8:261–66

3. Albus AM, Pesci EC, Runyen-Janecky LJ, West SEH, Iglewski BH. 1997. Vfr controls quorum sensing in *Pseudomonas aeruginosa*. *J. Bacteriol.* 179:3928–35

4. Anderson RM, Zimprich CA, Rust L.

1999. A second operator is involved in *Pseudomonas aeruginosa* elastase (*lasB*) activation. *J. Bacteriol.* 181:6264–70

5. Andersson RA, Eriksson ARB, Heikinheimo R, Mae A, Pirhonen M, et al. 2000. Quorum-sensing in the plant pathogen *Erwinia carotovora* subsp. *carotovora*: the role of expR$_{Ecc}$. *Mol. Plant-Microbe Interact.* 13:384–93

6. Baikalov I, Schroder I, Kaczor-Grzeskowiak M, Grzekowiak K, Gunsalus RP, Dickerson RE. 1996. Structure of the *Escherichia coli* response regulator NarL. *Biochemistry* 35:11053–61

7. Bainton NJ, Bycroft BW, Chhabra SR,

Stead P, Geldhill L, et al. 1992. A general role for the *lux* autoinducer in bacterial cell signalling: control of antibiotic biosynthesis in *Erwinia*. *Gene* 116:87–91

8. Bassler BL. 1999. How bacteria talk to each other: regulation of gene expression by quorum sensing. *Curr. Opin. Microbiol.* 2:582–87

9. Bassler BL, Wright M, Showalter RE, Silverman MR. 1993. Intercellular signalling in *Vibrio harveyi*: sequence and function of genes regulating expression of luminescence. *Mol. Microbiol.* 9:773–86

10. Beck von Bodman S, Farrand SK. 1995. Capsular polysaccharide biosynthesis and pathogenicity in *Erwinia stewartii* require induction by an *N*-acylhomoserine lactone autoinducer. *J. Bacteriol.* 177:5000–8

11. Beck von Bodman S, Majerczak DR, Coplin DL. 1998. A negative regulator mediates quorum-sensing control of exopolysaccharide production in *Pantoea stewartii* subsp. *stewartii*. *Proc. Natl. Acad. Sci. USA* 95:7687–92

12. Chai Y, Zhu J, Winans SC. 2001. TrlR, a defective TraR-like protein of *Agrobacterium tumefaciens*, blocks TraR function in vitro by forming inactive TrlR:TraR dimers. *Mol. Microbiol.* 40:414–21

13. Chansey ST, Wood DW, Pierson LS III. 1999. Two-component transcriptional regulation of *N*-acyl-homoserine lactone production in *Pseudomonas aureofaciens*. *Appl. Environ. Microbiol.* 65:2585–91

14. Chhabra SR, Stead P, Bainton NJ, Salmond GPC, Stewart GSAB, et al. 1993. Autoregulation of carbapenem biosynthesis in *Erwinia carotovora* by analogues of *N*-(3-oxohexanoyl)-L-homoserine lactone. *J. Antibiot.* 46:441–45

15. Choi SH, Greenberg EP. 1991. The C-terminal region of the *Vibrio fischeri* LuxR protein contains an inducer-independent *lux* gene activating domain. *Proc. Natl. Acad. Sci. USA* 88:11115–19

16. Choi SH, Greenberg EP. 1992. Genetic evidence for multimerization of LuxR, the transcriptional activator of *Vibrio fischeri* luminescence. *Mol. Mar. Biol. Biotechnol.* 1:408–13

17. Chugani SA, Whiteley M, Lee KM, D'Argenio D, Manoil C, Greenberg EP. 2001. QscR, a modulator of quorum-sensing signal synthesis and virulence in *Pseudomonas aeruginosa*. *Proc. Natl. Acad. Sci. USA* 98:2752–57

18. Costerton JW, Stewart PS, Greenberg EP. 1999. Bacterial biofilms: a common cause of persistent infections. *Science* 284:1318–22

19. Cui Y, Chatterjee A, Liu Y, Dumenyo CK, Chatterjee AK. 1995. Identification of a global repressor gene *rsmA*, of *Erwinia carotovora* subsp. *carotovora* that controls extracellular enzymes, *N*-(3-oxohexanoyl)-L-homoserine lactone, and pathogenicity in soft-rotting *Erwinia* spp. *J. Bacteriol.* 177:5108–15

20. Da Re S, Bertagnoli S, Fourment J, Reyrat J-M, Kahn D. 1994. Intramolecular signal transduction within the FixJ transcriptional activator: in vitro evidence for the inhibitory effect of the phosphorylatable regulatory domain. *Nucleic Acids Res.* 9:1555–61

21. Davies DG, Parsek MR, Pearson JP, Iglewski BH, Costerton JW, Greenberg EP. 1998. The involvement of cell-to-cell signals in the development of a bacterial biofilm. *Science* 280:295–98

22. De Kievit T, Seed PC, Nezezon J, Passador L, Iglewski BH. 1999. RsaL, a novel repressor of virulence gene expression in *Pseudomonas aeruginosa*. *J. Bacteriol.* 181:2175–84

23. de Nys R, Steinberg PD, Willemsen P, Dworjanyn SA, Gabelish CL, King RJ. 1995. Broad spectrum effects of secondary metabolites from the red alga *Delisea pulchra* in antifouling assays. *Biofouling* 8:259–71

24. Devine JH, Shadel GS, Baldwin TO. 1989. Identification of the operator of the *lux* regulon from the *Vibrio fischeri* strain

ATCC7744. *Proc. Natl. Acad. Sci. USA* 86:5688–92

25. Dong Y-H, Xu J-L, Li X-Z, Zhang L-H. 2000. AiiA, an enzyme that inactivates the acylhomoserine lactone quorum-sensing signal and attenuates virulence of *Erwinia carotovora*. *Proc. Natl. Acad. Sci. USA* 97:3526–31

26. Dunlap PV, Greenberg EP. 1985. Control of *Vibrio fischeri* luminescence gene expression in *Escherichia coli* by cyclic AMP and cyclic AMP receptor protein. *J. Bacteriol.* 164:45–60

27. Dunny GM, Winans SC. 1999. *Cell-Cell Signaling in Bacteria*. Washington, DC: ASM Press

28. Eberhard A, Burlingame AL, Eberhard C, Kenyon GL, Nealson KH, Oppenheimer NJ. 1981. Structural identification of autoinducer of *Photobacterium fischeri* luciferase. *Biochemistry* 20:2444–49

29. Eberhard A, Longin T, Widrig CA, Stranick SJ. 1991. Synthesis of the *lux* gene autoinducer in *Vibrio fischeri* is positively autoregulated. *Arch. Microbiol.* 155: 294–97

30. Eberhard A, Widrig CA, McBath P, Schineller JB. 1986. Analogs of the autoinducer of bioluminescence in *Vibrio fischeri*. *Arch. Microbiol.* 146:35–40

31. Egland KA, Greenberg EP. 1999. Quorum sensing in *Vibrio fischeri*: elements of the *luxI* promoter. *Mol. Microbiol.* 31:1197–204

32. Egland KA, Greenberg EP. 2000. Conversion of the *Vibrio fischeri* transcriptional activator, LuxR, to a repressor. *J. Bacteriol.* 182:805–11

33. Egland KA, Greenberg EP. 2001. Quorum sensing in *Vibrio fischeri*: analysis of the LuxR DNA binding region by alanine-scanning mutagenesis. *J. Bacteriol.* 183:382–86

34. Engebrecht J, Nealson KH, Silverman M. 1983. Bacterial bioluminescence: isolation and genetic analysis of the functions from *Vibrio fischeri*. *Cell* 32:773–81

35. Evans K, Passador L, Srikumar R, Tsang E, Nezezon J, Poole K. 1998. Influence of the MexAB-OprM multidrug efflux system on quorum sensing in *Pseudomonas aeruginosa*. *J. Bacteriol.* 180:5443–47

36. Fuqua C, Burbea M, Winans SC. 1995. Activity of the *Agrobacterium* Ti plasmid conjugal transfer regulator TraR is inhibited by the product of the *traM* gene. *J. Bacteriol.* 177:1367–73

37. Fuqua C, Eberhard A. 1999. Signal generation in autoinduction systems: synthesis of acylated homoserine lactones by LuxI-type proteins. See Ref. 27, pp. 211–30

38. Fuqua C, Greenberg EP. 1998. Self perception in bacteria: quorum sensing with acylated homoserine lactones. *Curr. Opin. Microbiol.* 1:183–89

39. Fuqua C, Winans SC. 1996. Conserved *cis*-acting promoter elements are required for density-dependent transcription of *Agrobacterium tumefaciens* conjugal transfer genes. *J. Bacteriol.* 178:435–40

40. Fuqua C, Winans SC. 1996. Localization of the OccR-activated and TraR-activated promoters that express two ABC-type permeases and the *traR* gene of the Ti plasmid pTiR10. *Mol. Microbiol.* 120:1199–210

41. Fuqua WC, Winans SC. 1994. A LuxR-LuxI type regulatory system activates *Agrobacterium* Ti plasmid conjugal transfer in the presence of a plant tumor metabolite. *J. Bacteriol.* 176:2796–806

42. Fuqua WC, Winans SC, Greenberg EP. 1994. Quorum sensing in bacteria: the LuxR/LuxI family of cell density-responsive transcriptional regulators. *J. Bacteriol.* 176:269–75

43. Gambello MJ, Iglewski BH. 1991. Cloning and characterization of the *Pseudomonas aeruginosa lasR* gene, a transcriptional activator of elastase expression. *J. Bacteriol.* 173:3000–9

44. Gilson L, Kuo A, Dunlap PV. 1995. AinS and a new family of autoinducer synthesis proteins. *J. Bacteriol.* 177:6946–51

45. Givskov M, de Nys R, Manefield M, Gram L, Maximilien R, et al. 1996.

Eukaryotic interference with homoserine lactone-mediated prokaryotic signalling. *J. Bacteriol.* 178:6618–22

46. Graf J, Ruby EG. 1998. Host-derived amino acids support the proliferation of symbiotic bacteria. *Proc. Natl. Acad. Sci. USA* 95:1818–22

47. Gray KM, Passador L, Iglewski BH, Greenberg EP. 1994. Interchangeability and specificity of components from the quorum-sensing regulatory systems of *Vibrio fischeri* and *Pseudomonas aeruginosa. J. Bacteriol.* 176:3076–80

48. Gray KM, Pearson JP, Downie JA, Boboye BEA, Greenberg EP. 1996. Cell-to-cell signalling in the symbiotic nitrogen-fixing bacterium *Rhizobium leguminosarum*: autoinduction of a stationary phase and rhizosphere-expressed genes. *J. Bacteriol.* 178:372–76

49. Greene RC. 1996. Biosynthesis of methionine. In *Escherichia coli and Salmonella: Cellular and Molecular Biology*, ed. FC Neidhardt, 1:542–60. Washington, DC: ASM Press

50. Hanzelka BL, Greenberg EP. 1995. Evidence that the N-terminal region of the *Vibrio fischeri* LuxR protein constitutes an autoinducer-binding domain. *J. Bacteriol.* 177:815–17

51. Hanzelka BL, Parsek MR, Val DL, Dunlap PV, Cronan JE, Jr Greenberg EP. 1999. Acylhomoserine lactone synthase activity of the *Vibrio fischeri* AinS protein. *J. Bacteriol.* 181:5766–70

52. Hanzelka BL, Stevens AM, Parsek MR, Crone TJ, Greenberg EP. 1997. Mutational analysis of the *Vibrio fischeri* LuxI polypeptide: critical regions of an autoinducer synthase. *J. Bacteriol.* 179:4882–87

53. Holden MTG, Chhabra SR, de Nys R, Stead P, Bainton NJ, et al. 1999. Quorum-sensing crosstalk: isolation and chemical characterization of cyclic dipeptides from *Pseudomonas aeruginosa* and other Gram-negative bacteria. *Mol. Microbiol.* 33:1254–66

54. Hwang I, Smyth AJ, Luo Z-Q, Farrand SK. 1999. Modulating quorum sensing by antiactivation: TraM interacts with TraR to inhibit activation of Ti plasmid conjugal transfer genes. *Mol. Microbiol.* 34:282–94

55. James S, Nilsson P, James G, Kjelleberg S, Fagerstrom T. 2000. Luminescence control in the marine bacterium *Vibrio fischeri*: an analysis of the dynamics of *lux* regulation. *J. Mol. Biol.* 296:1127–37

56. Jones S, Yu B, Bainton NJ, Birdsall M, Bycroft BW, et al. 1993. The *lux* autoinducer regulates the production of exoenzyme virulence determinants in *Erwinia carotovora* and *Pseudomonas aeruginosa. EMBO J.* 12:2477–82

57. Kahn D, Ditta G. 1991. Modular structure of FixJ: homology of the transcriptional activation domain with the −35 binding domain of sigma factors. *Mol. Microbiol.* 5:987–97

58. Kaplan HB, Greenberg EP. 1985. Diffusion of autoinducer is involved in regulation of the *Vibrio fischeri* luminescence system. *J. Bacteriol.* 163:1210–14

59. Kolibachuk D, Greenberg EP. 1993. The *Vibrio fischeri* luminescence gene activator LuxR is a membrane-associated protein. *J. Bacteriol.* 175:7307–12

60. Kuo A, Callaghan SM, Dunlap PV. 1996. Modulation of luminescence operon expression by *N*-octanoyl-L-homoserine lactone in *ainS* mutants of *Vibrio fischeri. J. Bacteriol.* 178:971–76

61. Latifi A, Foglino M, Tanaka K, Williams P, Lazdunski A. 1996. A hierarchical quorum-sensing cascade in *Pseudomonas aeruginosa* links the transcriptional activators LasR and RhlR (VsmR) to expression of the stationary-phase sigma factor RpoS. *Mol. Microbiol.* 21:1137–46

62. Leadbetter JR, Greenberg EP. 2000. Metabolism of acyl-homoserine lactone quorum-sensing signals by *Variovorax paradoxus. J. Bacteriol.* 182:6921–26

63. Luo Z-Q, Farrand SK. 1999. Signal-dependent DNA binding and functional domains of the quorum-sensing activator

TraR as identified by repressor activity. *Proc. Natl. Acad. Sci. USA* 96:9009–14

64. Luo Z-Q, Qin Y, Farrand SK. 2000. The antiactivator TraM interferes with the autoinducer-dependent binding of TraR to DNA by interacting with the C-terminal region of the quorum-sensing activator. *J. Biol. Chem.* 275:7713–22

65. Manefield M, de Nys R, Kumar N, Read R, Givskov M, et al. 1999. Evidence that halogenated furanones from *Delisea pulchra* inhibit acylated homoserine lactone (AHL)-mediated gene expression by displacing the AHL signal from its receptor protein. *Microbiol.* 145:283–91

66. McFall-Ngai MJ, Ruby EG. 1991. Symbiotic recognition and subsequent morphogenesis as early events in an animal-bacterial mutualism. *Science* 254: 1491–94

67. McKenney D, Brown KE, Allison DG. 1995. Influence of *Pseudomonas aeruginosa* exoproducts on virulence factor production in *Burkholderia cepacia*: evidence of interspecies communication. *J. Bacteriol.* 177:6989–92

68. McKnight SL, Iglewski BH, Pesci EC. 2000. The *Pseudomonas* quinolone signal regulates *rhl* quorum sensing in *Pseudomonas aeruginosa. J. Bacteriol.* 182:2702–8

69. Moré MI, Finger LD, Stryker JL, Fuqua C, Eberhard A, Winans SC. 1996. Enzymatic synthesis of a quorum-sensing autoinducer through use of defined substrates. *Science* 272:1655–58

70. Nasser W, Bouillant ML, Salmond G, Reverchon S. 1998. Characterization of the *Erwinia chrysanthemi expI-expR* locus directing the synthesis of two *N*-acyl-homoserine lactone signal molecules. *Mol. Microbiol.* 29:1391–405

71. Nealson KH, Hastings JW. 1979. Bacterial bioluminescence: its control and ecological significance. *Microbiol. Rev.* 43:496–518

72. Nealson KH, Platt T, Hastings JW. 1970. Cellular control of the synthesis and ac-

tivity of the bacterial luminescent system. *J. Bacteriol.* 104:313–22

73. Novick RP. 1999. Regulation of pathogenicity in *Staphylococcus aureus* by a peptide-based density-sensing system. See Ref. 27, pp. 129–46

74. Oger P, Kim K-S, Sackett RL, Piper KR, Farrand SK. 1998. Octopine-type Ti plasmids code for a mannopine-inducible dominant-negative allele of *traR*, the quorum-sensing activator that regulates Ti plasmid conjugal transfer. *Mol. Microbiol.* 27:277–88

75. Parsek MR, Greenberg EP. 2000. Acylhomoserine lactone quorum sensing in gram-negative bacteria: a signaling mechanism involved in associations with higher organisms. *Proc. Natl. Acad. Sci. USA* 97: 8789–93

76. Parsek MR, Schaefer AL, Greenberg EP. 1997. Analysis of random and site-directed mutations in *rhlI*, a *Pseudomonas aeruginosa* gene encoding an acylhomoserine lactone synthase. *Mol. Microbiol.* 26:301–10

77. Parsek MR, Val DL, Hanzelka BL, Cronan JE, Jr., Greenberg EP. 1999. Acyl homoserine-lactone quorum-sensing signal generation. *Proc. Natl. Acad. Sci. USA* 96:4360–65

78. Passador L, Tucker KD, Guertin KR, Journet MP, Kende AS, Iglewski BH. 1996. Functional analysis of the *Pseudomonas aeruginosa* autoinducer PAI. *J. Bacteriol.* 178:5995–6000

79. Pearson JP, Gray KM, Passador L, Tucker KD, Eberhard A, et al. 1994. Structure of the autoinducer required for expression of *Pseudomonas aeruginosa* virulence genes. *Proc. Natl. Acad. Sci. USA* 91:197–201

80. Pearson JP, Passador L, Iglewski BH, Greenberg EP. 1995. A second *N*-acyl-homoserine lactone signal produced by *Pseudomonas aeruginosa. Proc. Natl. Acad. Sci. USA* 92:1490–94

81. Pearson JP, Pesci EC, Iglewski BH. 1997. Roles of *Pseudomonas aeruginosa las* and

rhl quorum-sensing systems in the control of elastase and rhamnolipid biosynthesis genes. *J. Bacteriol.* 179:5756–67

82. Pearson JP, Van Delden C, Iglewski BH. 1999. Active efflux and diffusion are involved in transport of *Pseudomonas aeruginosa* cell-to-cell signals. *J. Bacteriol.* 181:1203–10

83. Pesci EC, Iglewski BH. 1999. Quorum sensing in *Pseudomonas aeruginosa*. See Ref. 27, pp. 147–55

84. Pesci EC, Milbank JBJ, Pearson JP, McKnight S, Kende AS, et al. 1999. Quinolone signaling in the cell-to-cell communication system of *Pseudomonas aeruginosa*. *Proc. Natl. Acad. Sci. USA* 96:11229–34

85. Pesci EC, Pearson JP, Seed PC, Iglewski BH. 1997. Regulation of *las* and *rhl* quorum sensing in *Pseudomonas aeruginosa*. *J. Bacteriol.* 179:3127–32

86. Piper KR, Beck von Bodman S, Farrand SK. 1993. Conjugation factor of *Agrobacterium tumefaciens* regulates Ti plasmid transfer by autoinduction. *Nature* 362:448–50

87. Piper KR, Beck von Bodman S, Hwang I, Farrand SK. 1999. Hierarchical gene regulatory systems arising from fortuitous gene associations: controlling quorum sensing by the opine regulon in *Agrobacterium*. *Mol. Microbiol.* 32:1077–89

88. Puskas A, Greenberg EP, Kaplan S, Schaefer AL. 1997. A quorum-sensing system in the free-living photosynthetic bacterium *Rhodobacter sphaeroides*. *J. Bacteriol.* 179:7530–37

89. Qin Y, Luo Z-Q, Smyth AJ, Gao P, Beck von Bodman S, Farrand SK. 2000. Quorum-sensing signal binding results in dimerization of TraR and its release from membranes into the cytoplasm. *EMBO J.* 19:5212–21

90. Reichelt JL, Borowitzka MA. 1984. Antimicrobial activity from marine algae: results of a large scale screening programme. *Hydrobiology* 116/117:158–68

91. Reimmann C, Beyeler M, Latifi A, Win-

teler H, Foglini M, et al. 1997. The global activator GacA of *Pseudomonas aeruginosa* PAO positively controls the production of the autoinducer *N*-butyryl-homoserine lactone and the formation of the virulence factors pyocyanin, cyanide, and lipase. *Mol. Microbiol.* 24:309–19

92. Rhodius VA, Busby SJW. 1998. Positive activation of gene expression. *Curr. Opin. Microbiol.* 1:152–59

93. Ruby EG. 1996. Lessons from a cooperative bacterial-animal association: the *Vibrio fischeri-Euprymna scolopes* light organ symbiosis. *Annu. Rev. Microbiol.* 50:591–624

94. Ruby EG, Asato LM. 1993. Growth and flagellation of *Vibrio fischeri* during initiation of the sepolid squid light organ symbiosis. *Arch. Microbiol.* 159:160–67

95. Ruby EG, McFall-Ngai MJ. 1992. A squid that glows in the night: development of an animal-bacterial mutualism. *J. Bacteriol.* 174:4865–70

96. Rust L, Pesci EC, Iglewski BH. 1996. Analysis of the *Pseudomonas aeruginosa* elastase (*lasB*) regulatory region. *J. Bacteriol.* 178:1134–40

97. Schaefer AL, Hanzelka BL, Eberhard A, Greenberg EP. 1996. Quorum-sensing in *Vibrio fischeri*: probing autoinducer-LuxR interactions with autoinducer analogs. *J. Bacteriol.* 178:2897–901

98. Schaefer AL, Hanzelka BL, Parsek MR, Greenberg EP. 2001. Detection, purification and structural elucidation of acyl-homoserine lactone inducer of *Vibrio fischeri* luminescence and other related molecules. *Methods Enzymol.* 336:41–47

99. Schaefer AL, Val DL, Hanzelka BL, Cronan JE, Jr., Greenberg EP. 1996. Generation of cell-to-cell signals in quorum sensing: acyl homoserine lactone synthase activity of a purified *Vibrio fischeri* LuxI protein. *Proc. Natl. Acad. Sci. USA* 93:9505–9

99a. Schlaman HRM, Okker RJH, Lugtenberg BJJ. 1989. Subcellular localization of the *nodD* gene product in *Rhizobium legumi-nosarum. J. Bacteriol.* 174:4686–93

100. Schripsema J, de Rudder KEE, van Vliet TB, Lankhorst PP, de Vroom E, et al. 1996. Bacteriocin *small* of *Rhizobium leguminosarum* belongs to the class of *N*-acyl-L-homoserine lactone molecules, known as autoinducers and as quorum sensing co-transcription factors. *J. Bacteriol.* 178:366–71

101. Shadel GS, Young R, Baldwin TO. 1990. Use of regulated cell lysis in a lethal genetic selection in *Escherichia coli*: identification of the autoinducer-binding region of the LuxR protein from *Vibrio fischeri* ATCC 7744. *J. Bacteriol.* 172:3980–87

102. Singh PK, Schaefer AL, Parsek MR, Moninger TO, Welsh MJ, Greenberg EP. 2000. Quorum-sensing signals indicate that cystic fibrosis lungs are infected with bacterial biofilms. *Nature* 407:762–64

103. Sitnikov D, Schineller JB, Baldwin TO. 1995. Transcriptional regulation of bioluminescence genes from *Vibrio fischeri. Mol. Microbiol.* 17:801–12

104. Slock J, Kolibachuk D, Greenberg EP. 1990. Critical regions of the *Vibrio fischeri* LuxR protein defined by mutational analysis. *J. Bacteriol.* 172:3974–79

105. Stevens AM, Dolan KM, Greenberg EP. 1994. Synergistic binding of the *Vibrio fischeri* LuxR transcriptional activator domain and RNA polymerase to the *lux* promoter region. *Proc. Natl. Acad. Sci. USA* 91:12619–23

106. Stevens AM, Greenberg EP. 1997. Quorum sensing in *Vibrio fischeri*: essential elements for activation of the luciferase genes. *J. Bacteriol.* 179:557–62

107. Stevens AM, Greenberg EP. 1999. Transcriptional activation by LuxR. See Ref. 27, pp. 231–42

108. Teplitski M, Robinson JB, Bauer WD. 2000. Plants secrete substances that mimic bacterial *N*-acyl homoserine lactone signal activities and affect population density-dependent behaviors in asso-

ciated bacteria. *Mol. Plant-Microbe Interact.* 13:637–48

109. Throup JP, Camara M, Briggs GS, Winson MK, Chhabra SR, et al. 1995. Characterisation of the *yenI/yenR* locus from *Yersinia enterocolitica* mediating the synthesis of two *N*-acylhomoserine lactone signal molecules. *Mol. Microbiol.* 17:345–56

110. Welch M, Todd DE, Whitehead NA, Mc-Gowan SJ, Bycroft BW, Salmond GPC. 2000. *N*-acyl homoserine lactone binding to the CarR receptor determines quorum-sensing specificity in *Erwinia. EMBO J.* 19:631–41

111. White D, Hart ME, Romeo T. 1996. Phylogenetic distribution of the global regulatory gene *csrA* among eubacteria. *Gene* 182:221–23

112. Whiteley M, Parsek MR, Greenberg EP. 2000. Regulation of quorum sensing by RpoS in *Pseudomonas aeruginosa. J. Bacteriol.* 182:4356–60

113. Winson MK, Camara M, Latifi A, Foglino M, Chhabra SR, et al. 1995. Multiple *N*-acyl-L-homoserine lactone signal molecules regulate production of virulence determinants and secondary metabolites in *Pseudomonas aeruginosa. Proc. Natl. Acad. Sci. USA* 92:9427–31

114. Wood DW, Gong F, Daykin MM, Williams P, Pierson LS III. 1997. *N*-acylhomoserine lactone-mediated regulation of phenazine gene expression by *Pseudomonas aureofaciens* 30-84 in the wheat rhizosphere. *J. Bacteriol.* 179:7663–70

115. You Z, Fukushima J, Ishiwata T, Chang B, Kurata M, et al. 1996. Purification and characterization of LasR as a DNA-binding protein. *FEMS Microbiol. Lett.* 142:301–7

116. Zhang L, Murphy PJ, Kerr A, Tate ME. 1993. *Agrobacterium* conjugation and gene regulation by *N*-acyl-L-homoserine lactones. *Nature* 362:446–48

117. Zhu J, Beaber JW, Moré MI, Fuqua C, Eberhard A, Winans SC. 1998. Analogs

of the autoinducer 3-oxooctanoylhomo-serine lactone strongly inhibit activity of the TraR protein of *Agrobacterium tumefaciens*. *J. Bacteriol.* 180:5398–405

118. Zhu J, Winans SC. 1998. Activity of the quorum-sensing regulator TraR of *Agrobacterium tumefaciens* is inhibited by a truncated, dominant defective TraR-like protein. *Mol. Microbiol.* 27:289–97

119. Zhu J, Winans SC. 1999. Autoinducer binding by the quorum-sensing regulator TraR increases affinity for target promoters in vitro and decreases TraR turnover rates in whole cells. *Proc. Natl. Acad. Sci. USA* 96:4832–37

120. Zhu J, Winans SC. 2001. The quorum-sensing transcriptional regulator TraR requires its cognate signaling ligand for protein folding, protease resistance, and dimerization. *Proc. Natl. Acad. Sci. USA* 98:1507–12

Annu. Rev. Genet. 2001. 35:469–99

MODELS AND DATA ON PLANT-ENEMY COEVOLUTION

Joy Bergelson, Greg Dwyer, and J. J. Emerson

Department of Ecology and Evolution, University of Chicago, Chicago, Illinois 60637;
e-mail: jbergels@midway.uchicago.edu, gdwyer@midway.uchicago.edu,
jje@midway.uchicago.edu

Key Words   coevolution, arms race, herbivore, disease polymorphism, adaptive
dynamics

■ **Abstract**   Although coevolution is complicated, in that the interacting species
evolve in response to each other, such evolutionary dynamics are amenable to mathematical modeling. In this article, we briefly review models and data on coevolution
between plants and the pathogens and herbivores that attack them. We focus on "arms
races," in which trait values in the plant and its enemies escalate to more and more
extreme values. Untested key assumptions in many of the models are the relationships
between costs and benefits of resistance in the plant and the level of resistance, as well
as how costs of virulence or detoxification ability in the enemy change with levels of
these traits. A preliminary assessment of these assumptions finds only mixed support
for the models. What is needed are models that are more closely tailored to particular
plant-enemy interactions, as well as experiments that are expressly designed to test
existing models.

CONTENTS

0066-4197/01/1215-0469$14.00   **469**

INTRODUCTION

In evolutionary biology, progress has historically been aided by the construction of mathematical models that describe how allele frequencies change within populations. Most such population-genetic models have concentrated on the evolution of traits that affect how a species is adapted to its physical environment (68), rather than how it interacts with other species. Understanding the evolution of traits that affect interactions between species, however, is complicated by the possibility of coevolution, in which each species evolves in response to the other. This complication has limited both the development and the application of coevolutionary models (60). Nevertheless, recent work has led to the development of promising new modeling approaches (1). Moreover, the advent of evolutionary genomics may soon allow us to test these models more thoroughly than before.

In this article, we consider the usefulness of mathematical models for understanding coevolution between plants and their natural enemies. By "natural enemies" we mean herbivores or pathogens, as opposed to competitors such as other plants, or mutualists, such as pollinators. Much of the interest in coevolutionary dynamics of victim-exploiter interactions has to do with the possibility of so-called arms races (25), or Red-Queen dynamics (18, 92). An arms race occurs when a trait in each species evolves toward ever more extreme values in response to the evolution of a corresponding trait in the other species. The simplest such example might be evolution toward increased host resistance to a pathogen, in response to increased virulence in the pathogen. Local adaptation is said to have occured either when a natural enemy strain has higher fitness when attacking a plant strain from the same area than when attacking a plant strain from a different area, or when a plant strain has higher fitness when being attacked by a natural enemy strain from the same area than when being attacked by a natural enemy from a different area. Because local adaptation has recently been reviewed elsewhere (13, 51, 84, 93), here we focus primarily on arms races.

An important feature of recent modeling work on coevolution is the proposed solution to a perceived problem in the usual definition of an arms race (77). Specifically, limitless escalation in trait values would naturally lead to infinite values, which is of course implausible. One possible solution to this problem is to have trait values cycle, so that victim and exploiter trait values rise for several generations, then fall to their previous values in a dynamic polymorphism (3, 28). It has been argued that this kind of coevolutionary dynamic is an arms race because trait values escalate at least some of the time. Our intent in this review is to show how arms race models have been developed out of classical population-genetic models, and to indicate one way that such models may be compared with data.

We first briefly review mathematical models that describe coevolution in victim-exploiter interactions, a category that includes predator-prey, host-parasite, host-pathogen, and plant-herbivore interactions. Second, we review data sets for plants and their enemies, focusing mostly on insect herbivores, in order to examine the extent to which basic assumptions of the models accurately describe real plant-enemy interactions. Third, we assess models in terms of the accuracy of their predictions. And, finally, we highlight some interactions between plants and enemies that have thus far not been incorporated into models but that have the potential of generating interesting dynamics. Our goal is to foster a better connection between models and data in the study of plant-enemy coevolution.

MATHEMATICAL MODELS OF THE EVOLUTION OF VICTIM-EXPLOITER INTERACTIONS

For practical reasons, most empirical studies of victim-exploiter interactions focus on cases in which either the victim has a strong effect on the fitness of the exploiter or, conversely, the exploiter has a strong effect on the fitness of the victim. Most mathematical models of arms races in victim-exploiter interactions therefore focus on natural selection as the primary evolutionary force, often to the exclusion of mutation, genetic drift, or migration. In the absence of these other forces, the standard population-genetic model for selection acting on two alleles at a single locus is

$$p_{t+1} = \frac{p_t w_t}{\bar{w}_t}, \qquad\qquad 1.$$

where p_t is the frequency of the allele in generation t, w_t is the allele's fitness in generation t averaged over genotypes, and $\bar{w}_t$ is the average fitness of the entire population in generation t (68). If fitnesses are constant and unequal, then p_t will approach 1, unless the population is diploid and there is heterozygote overdominance. A basic feature of loci that affect species interactions, however, is that the fitnesses of different alleles are not constant. In particular, the fitness of an allele may depend on the frequency of the other alleles in the population. To see why this might be true, consider a haploid host in which allele A confers resistance to pathogen strain 1, and allele a confers resistance to pathogen strain 2. If allele A is at high frequency for several generations, it is likely that its fitness will decline as the density of pathogen 1 increases. High fitness of allele a may thus be associated with a high frequency of allele A.

One of the simplest ways to model this kind of interaction is to use a one-locus, two-allele model for the hosts, to assume that the pathogens are haploid, and then to assign fitnesses to each host and pathogen combination. For example, Levin (58) presents a generalization of several basic early models, according to which the host alleles A and a have frequencies p_t and $1 - p_t$, and the pathogen alleles B and b have frequencies q_t and $1 - q_t$. Pathogens of type B that attack hosts of genotype AA have fitness α, those that attack hosts of genotype Aa have fitness β, and those that attack genotype aa have fitness γ. The fitness of the alternative pathogen type b is antisymmetric, meaning that pathogens of type b that attack hosts of type AA have fitness γ, those that attack Aa have fitness β, and those that attack aa have fitness α. The probability that a particular pathogen genotype will attack a particular host genotype is assumed to depend only on the frequency of the host genotype, so that the mean fitnesses of the two pathogen types in generation t are

$$v_B = p_t^2\alpha + 2p(1 - p_t)\beta + (1 - p_t)^2\gamma, \qquad 2.$$

$$v_b = p_t^2\gamma + 2p_t(1 - p_t)\beta + (1 - p_t)^2\alpha. \qquad 3.$$

Host fitnesses follow similar rules, in that the fitness of host genotype AA when attacked by pathogen type B is $1 - \alpha$, the fitness of host genotype Aa when attacked by pathogen type B is $1 - \beta$, etc. The usefulness of this type of model is that it can be used to understand the circumstances under which polymorphism is possible. It turns out that polymorphism in the host is maintained whenever

$$\beta < \frac{\alpha + \gamma}{2}. \qquad 4.$$

In other words, the fitness of the pathogen that attacks the heterozygote host must be less than the fitness of the pathogen that attacks either homozygote. Polymorphism of both host and pathogen, however, requires the stronger condition,

$$\beta^2 < \left(\frac{\alpha + \gamma}{2}\right)^2 - \left(\frac{\alpha - \gamma}{2}\right)^2. \qquad 5.$$

Because we shortly address the issue of whether polymorphisms are dynamic, here we point out that much of this early literature focuses strictly on conditions under which polymorphism occurs and apparently does not consider whether cycling in allele frequencies might also occur.

Although in this model the emphasis is on the assignment of fitnesses to particular host-pathogen combinations, it is important to remember that victim fitness is also affected by the probability of attack, and thus by the ecology of the victim and the exploiter. Because ecological theory emphasizes the importance of density for species interactions, an important next step in the development of coevolutionary models of victim-exploiter interactions was the incorporation of density effects. The simplest way to do this is to begin with host-pathogen interactions, focusing specifically on the classic epidemic model (22):

$$\frac{dS}{dt} = -\beta SI, \qquad 6.$$

$$\frac{dI}{dt} = \beta SI - \alpha I. \qquad 7.$$

Here, S and I are the densities of the uninfected and infected host populations, β is the rate of horizontal transmission, and α is the rate of disease-induced death. Because this model describes the progress of the epidemic, one can use it to assign fitnesses to different victim and exploiter genotypes. To see how this is done, we focus on the simplest case, in which the host is haploid and there is only one pathogen strain, so that allele A confers resistance, a confers susceptibility, and p_t

is the frequency of the A allele in generation t, following Gillespie (43). If we then allow $t \to \infty$ in Equations 6 and 7, the fraction of hosts that become infected, i, can be expressed by the implicit equation

$$1 - i = e^{-R_0 i (1 - p_t)}.$$ 8.

Here, R_0 is defined as the absolute fitness of the pathogen, which can be derived from Equations 6 and 7 as

$$R_0 = \frac{\beta K}{\alpha},$$ 9.

where K is the population density of the host, which is assumed to be fixed. From this expression, in the haploid case the fitnesses of the susceptible and resistant hosts are

$$W_a = (1 - i(p_t)) + i(p_t)(1 - t),$$ 10.

$$W_A = (1 - s),$$ 11.

where $(1 - t)$ is the fitness of susceptible hosts that have survived infection, and $(1 - s)$ is the fitness of resistant hosts. As in the simpler models, one can then consider the conditions under which host polymorphism occurs; specifically, polymorphism will be maintained if the ratio of selection coefficients is less than what the fraction infected, i, would be if the entire population were susceptible, so that $s/t \leq i(p_t = 0)$. The key innovation, however, is that now polymorphism depends on host population density. For example, for high values of absolute pathogen fitness R_0, the equilibrium frequency of susceptibility p_t drops very rapidly with increasing population density K. Note that again there is no consideration of whether or not cycling will occur.

Although this model assumes that there is only one pathogen strain, it can be extended easily to allow for multiple pathogen strains by rewriting Equation 8 as

$$1 - i = e^{-R_j i f_{j,t}},$$ 12.

where R_j is the basic reproductive rate for strain j ($j = 1$ for pathogen strain 1, 2 for pathogen strain 2), and $f_{j,t}$ is the frequency of the host strain that is susceptible to pathogen strain j in generation t (64). Note that by proper definition of $f_{j,t}$, one can use this model to describe either a haploid host or a diploid host with complete dominance. As with the earlier models, this model can be used to consider conditions under which polymorphism occurs. The interesting feature of the work by May & Anderson (64), however, was that they were able to show that for some parameter values the model produces fluctuations in the frequency of the two host alleles. In fact, in the case for which one host strain is completely resistant, and thus has no pathogen, the May & Anderson model reduces to the Gillespie model, showing that the Gillespie model can also show dynamic polymorphism.

A second simplifying assumption of the Gillespie model is that the annual epidemic is begun by an extremely small density of the pathogen. An additional extension of this model is therefore to explicitly keep track of the density of overwintering spores or other infectious stages from generation to generation by replacing Equation 8 with

$$1 - i = e^{-K(R_0 i(1-p_t)+Kz_t)},$$ 13.

where z_t is the density of the infectious, possibly free-living stage of the pathogen (83). In addition to the usual equations for the frequency of host alleles, as in Equation 1, we have an additional equation for pathogen density,

$$z_{t+1} = \phi K p_t i(p_t, z_t) + \gamma z_t,$$ 14.

where γ is the probability that the infectious stage of the pathogen will survive over the long term, and ϕ is the population growth of the pathogen between epidemics. Because this model assumes that there is only one pathogen strain, like the Gillespie model but unlike the May & Anderson model, it requires a cost of resistance to permit polymorphism. Also, like the May & Anderson model, it shows cycles in the frequency of resistance. An interesting difference between the dynamic polymorphisms in the two models, however, is that in the model of Stahl et al. (83) the pathogen can build up to high levels over several generations, leading to long-period fluctuations in the frequency of resistance. In contrast, fluctuations in the May & Anderson model invariably have a short period.

The models we have discussed all assume a single locus. Because in nature the traits in question are often affected by many loci, an important next step was to allow for several loci. Perhaps the most direct way to do this is to extend the epidemic model to include multiple genotypes of hosts and pathogens. For example, Frank (38) presented the model

$$\Delta h_i = h_i \left(r_i - \sum_{k=0}^{2^N-1} r_k h_k - m \sum_{k=0}^{2^N-1} \lambda_{ik} p_k \right),$$ 15.

$$\Delta p_j = p_j \left(-s + b_j \sum_{k=0}^{2^N-1} \lambda_{kj} h_k \right).$$ 16.

Here, h_i and p_j are the densities of hosts and pathogens of strain i and j, respectively, r_i is the host-strain–specific reproductive rate, N is the number of host and pathogen loci, m is the rate of disease transmission, s is the pathogen death rate, and b_j is the pathogen birth rate, the rate at which infected hosts produce pathogens of type j. Note that in Equations 15 and 16, infection will occur unless the host has a resistance allele and the pathogen has an avirulence allele at a particular locus, following the gene-for-gene interactions typical of many plants and their pathogens (reviewed in 44). This model is effectively a generalization of the earlier models we discussed in which there were two strains of the host and two strains of the pathogen. To allow for the additional complication of multiple genotypes, Equations 15 and 16 introduce the variable λ_{ik}, which is equal to zero if the host has a resistance allele and the pathogen has a virulence allele, otherwise it is equal to one.

Although Equations 15 and 16 are more complicated than the earlier models we discussed, it turns out they are fundamentally similar to Equations 6 and 7. To see this, note first that Equations 15 and 16 assume a time step of one generation. If we instead allow for time steps of any number of generations, and we allow for only one strain of the host or the pathogen, we have,

$$\Delta h = (rh - rh^2 - mhp)\Delta t, \qquad\qquad 17.$$

$$\Delta p = (bhp - s)\Delta t. \qquad\qquad 18.$$

If we then set $r = 0$ in Equations 15 and 16, divide both sides by Δt and let $\Delta t \rightarrow 0$, we again have Equations 17 and 18. For $r \neq 0$, we have instead a discrete-generation version of the Lotka-Volterra predator-prey equations in which the prey experiences intraspecific competition for resources. An important difference, however, is that all the models we have discussed so far have assumed that host population sizes are constant. Equations 15 and 16 instead allow host population densities to fluctuate freely. An additional difference is that the Frank model allows for a full range of population genetic processes, although the earlier models can in general be extended to allow for some of these complexities [see, for example, Damgaard (23) for gene flow]. First, host genotypes that have fallen to 0.1% of their carrying capacity are set to a density of zero but are reintroduced randomly whether they are extinct or not. The overall effect is thus to allow for either mutation or immigration from outside the population. Also, a sexual version of the model assumes that there is a recombination fraction of 0.5 between adjacent loci.

Like the other models we have described, this model can show polymorphism in both victim and exploiter. For the earlier models, costs of resistance were assigned on a case-by-case basis, but for this model one instead specifies a function that describes how the cost of resistance increases with an increasing number of alleles for any host genotype. There is also a cost of virulence, in that pathogens with more virulence alleles have lower reproductive rates, which is again specified by a function rather than by case-by-case assignment. Both costs are thus quantitative functions of the number of alleles.

This model is substantially more complex than the earlier models we described, even though Frank restricted the model to eight loci. Because of this complexity, summarizing the model's behavior is more difficult. The major results first have to do with the probability that a randomly selected host genotype will be resistant to a randomly selected pathogen genotype. If host and pathogen densities are stable, this probability will be low, but if host and pathogen densities fluctuate, then this probability will fluctuate as well. Second, the average number of resistance alleles carried by a host strain increases with the slope of the relationship between the cost of virulence and the number of virulence alleles. This occurs because lower costs of virulence lead to a higher number of avirulence alleles, which in turn favors resistance alleles. The average number of virulence alleles carried by a pathogen strain instead increases with 1 minus the rate at which resistance costs increase with increasing numbers of resistance alleles. This occurs because large costs of resistance lead to a reduction in the number of resistance alleles, which in turn favors avirulence alleles. Surprisingly, the number of resistance alleles is unaffected by the costs of resistance, and the number of virulence alleles is unaffected by the costs of virulence.

This model begins to make clear some of the difficulties of constructing co-evolutionary models. To begin with, all multilocus population genetic models that attempt to assign fitnesses to each genotype face the problem that the number of possible genotypes increases rapidly with the number of loci. For example, if there are two alternative loci at each of n loci, the number of possible genotypes is 2^n. As we have discussed, however, much of the interest in coevolutionary models of victim-exploiter interactions focuses on evaluating the possibility of "arms races" (25), or "Red-Queen dynamics" (18, 92). Understanding such an evolutionary dynamic clearly requires that there be no a priori limit on the number of possible genotypes. At the same time, however, plausibility requires that arms races do not lead to infinitely large trait values or to infinite numbers of genotypes (77). Much recent work has therefore focussed on modeling approaches that can show arms races without putting arbitrary limits on the number of possible genotypes, yet without leading to infinite trait values.

The basic approach taken in most of these models is first to assume that victim and exploiter traits are quantitative rather than all- or-none, so that at the population level phenotypes can be described by continuous probability distributions. Second, various simplifying assumptions are made about the genetics of the relevant traits. For example, one recent approach known as "adaptive dynamics" assumes that both victims and exploiters are asexual and haploid, and that offspring are identical to their parents unless a mutation occurs (27, 28). The distribution of phenotypes is identical to the distribution of genotypes, and it changes only because of differential net reproduction, or mutation. Mutations are assumed to be of small effect, in the sense that the probability that the offspring trait is greatly different from the parent's trait is small. Fitnesses are described by an ecological model, typically a predator-prey model that is similar to Equations 17 and 18, except with completely overlapping generations.

Although the resulting model is complex, it turns out its dynamics can be closely approximated by a deterministic model in which victim and exploiter are monomorphic, but for which the phenotype of each species can change from generation to generation (62). To understand the full model, we can therefore focus on the simpler approximate model. To derive the equations of change for the approximate model, one begins with a predator-prey model with overlapping generations:

$$\frac{dh}{dt} = h(r - \alpha(s_h) - \beta(s_h, s_p)p), \qquad 19.$$

$$\frac{dp}{dt} = p(\gamma(s_h, s_p)h - s). \qquad 20.$$

Here, s_h and s_p are the victim and exploiter phenotypes, so the density-dependence in the victim reproductive rate $\alpha(s_h)$ is a function of victim phenotype, and the rates of victim deaths, $\beta(s_h, s_p)$, and exploiter births, $\gamma(s_h, s_p)$, are functions of both host and pathogen phenotypes. It turns out that for fixed phenotypes, allowing for continuous time ensures that host and pathogen densities will be constant, a feature that simplifies the model results and permits a focus on the evolutionary dynamics rather than the ecological dynamics. The specific functions used assume that fitness is highest at intermediate trait values. To begin with, the density-dependence term $\alpha(s_h)$ is assumed to be parabolic, so that in the absence of predation, the optimal victim trait is at an intermediate value of the victim phenotype. Next, one assumes that $\gamma(s_h, s_p) = k\beta(s_h, s_p)$, where k is a constant of proportionality, and that β (and thus γ) is a bivariate Gaussian distribution (normal curve) with respect to the host and pathogen phenotypes s_h and s_p. In the context of predator-prey interactions, s_h and s_p are often taken to be equivalent to prey and predator body sizes. Predator fitness is then maximized when the predator has the same body size as the prey. The prey, however, faces a trade-off because its birth rate is maximized at an intermediate size, through the density-dependence term $\alpha(s_h)$, but its death rate is maximized when it is the same size as the predator. These trade-offs cause the model to show an interesting array of behaviors. Specifically, if h, the constant of proportionality between victim and exploiter body sizes, is such that $0.05 < k < 0.098$, victim and exploiter phenotypes will reach a stable equilibrium, but for $0.098 < k < 0.148$ and for sufficiently large values of r, the model shows cycles in host and pathogen phenotypes. If k is increased still further, the model will again reach equilibrium. In this range of k, however, there are two alternative equilibria, and the one that is reached depends on the starting values of the host and pathogen phenotypes.

As described above, Dieckmann and coauthors (27, 28) have argued that the dynamic in this model, in which trait values cycle, represents a plausible arms race because trait values sometimes escalate without reaching infinite values. Clearly, however, the assumptions of simple genetics and of trait-matching in the predator at least superficially appear to be restrictive, which suggests that the models may not be widely applicable. It turns out, however, that neither assumption is quite as restrictive as it seems. For example, Sasaki & Godfray (80) used an adaptive-dynamics approach to modeling the evolution of the interactions between a host insect and its parasitoid. In their model, host resistance (x) and parasitoid virulence (y) are quantitative characters, and the probability that a host is able to encapsulate an attacking parasitoid, which allows the host to survive but kills the parasitoid, is described by the function

$$\eta(x - y) = \frac{1}{1 + e^{-2A(x-y)}}. \qquad 21.$$

If the resistance trait and the virulence trait are equal, so that $x = y$, the probability of encapsulation is 0.5. If resistance is greater than virulence, so that $x > y$, the host has a better chance than the parasitoid of surviving, whereas if $x < y$ the reverse is true. As with the previous models, this model also assumes that there is a cost to resistance, in that the number of hosts produced by a surviving host declines with increasing resistance x, so that the total fecundity of surviving hosts is $a(x) = G\exp(-c_h x)$. Similarly, there is a cost of virulence, in that the fraction of emerging parasitoids that survive to reproduce is $b(y) = \exp(-c_p y)$. The full model is then

$$N_{t+1}(x) = a(x)N_t(x)\left(F_t + (1 - F_t)\int_0^\infty \eta(x - y)Q_t(y)dy\right), \qquad 22.$$

$$P_{t+1}(y) = b(y)\int_0^\infty N_t(x)(1 - \eta(x - y))(1 - F_t)Q_t(y)\,dx. \qquad 23.$$

Here, $N_t(x)$ and $P_t(x)$ are the densities of hosts and parasitoids that have resistance level x or virulence level y, respectively, in generation t. $Q_t(y)$ is the fraction of parasitoids that have virulence level y, so that $Q_t(y) = P_t(y)/\bar{P}_t$, where $\bar{P}_t$ is the total parasitoid density in generation t. F_t is the fraction of hosts that become parasitized in generation t, which can be described by standard host-parasitoid population-dynamic models. Equation 22 thus says that of the fraction F_t of hosts that become parasitized, a fraction $Q_t(y)$ are parasitized by parasitoids of virulence level y, and so a fraction $\eta(x - y)$ survive parasitization.

Like the work of Dieckmann and coworkers (27, 28), this model assumes that victim and exploiter are haploid and asexual, and that mutations are of small effect. An additional similarity to the models of Dieckmann and coworkers is that this model can also show coevolutionarily cycling, in which resistance and virulence oscillate between low levels and high levels. The important difference, however, is that this arms-race dynamic occurs without the assumption that selection favors exploiters that match the trait value of the victim. It thus appears that the key ecological assumption governing the dynamics of this class of models is not the assumption of trait matching. Instead, the key assumption is that victim and exploiter traits can be expressed in the same currency, thus permitting the construction of a function such as Equation 21, that relates differences in victim and exploiter trait values to their respective fitnesses. Although it remains to be seen whether one could define a virulence or resistance metric that would permit parameterization of such a function, the approach nevertheless may be generally useful.

Whether or not one can relax the second restrictive assumption of adaptive dynamic models, that the genetics of victim and exploiter are simple, is less clear. An alternative method of simplifying the genetics that allows for diploidy is to use a quantitative genetic model (1, 2). As with adaptive dynamic models, quantitative genetic models generally assume that there is some continuous distribution of phenotypes, except that the distribution of phenotypes is normal with fixed variance (but see 29, 30). Quantitative genetic models also allow for the possibility that offspring are not identical to their parents, although the ratio of additive genetic to phenotypic variance is assumed to be constant over time (4). Because of these simplifying assumptions, these models need keep track only of changes in the mean phenotype. For such models, it turns out that the rate of change in the mean phenotype is proportional to the derivative of fitness with respect to the phenotype value evaluated at the mean phenotype of all interacting populations, according to

$$\Delta \bar{z} = \frac{V_a}{\bar{W}} \left(\frac{\partial W}{\partial z} \bigg|_{\bar{z}} \right),$$ 24.

where $\Delta \bar{z}$ is the rate of change of the mean phenotype $\bar{z}$, V_a is the additive genetic variance of the distribution of phenotypes, W is the fitness function, and $\bar{W}$ is the mean fitness. Using this result, one can construct a full coevolutionary model by again expressing victim and exploiter fitnesses in terms of a predator-prey model. For example, Abrams and coauthors (3, 4) used a continuous-generation quantitative-genetic model defined according to

$$\frac{dx}{dt} = V_x \frac{\partial W_h}{\partial x} \bigg|_{\bar{x}},$$ 25.

$$\frac{dy}{dt} = V_y \frac{\partial W_p}{\partial y} \bigg|_{\bar{y}},$$ 26.

where x and y are the prey and predator trait values, respectively, and V_x, V_y, W_x, and W_y are the respective additive genetic variances (V) and fitnesses (W) of prey and predator. The fitness function W_x is then derived from a generalized form of the right-hand side of the prey Equation 19, divided by prey density h, whereas predator fitness W_y is derived similarly from the predator Equation 20. As in the adaptive dynamic models, a key step is the expression that translates differences in the victim and exploiter traits into victim and exploiter fitnesses. Abrams & Matsuda (3) used trait matching as expressed by a bivariate Gaussian distribution, and like the adaptive dynamic models, the resulting model shows coevolutionary cycling for some range of parameter values. It turns out, however, that cycling is also achieved when one drops the bivariate Gaussian assumption and instead uses the function

$$M(x, y) = M_{max} \left(0.5 + \frac{1}{\pi} \arctan[k(x - y)] \right).$$ 27.

Here, M_{max} is the maximum rate at which predators capture prey, and k is a constant. This function approaches zero as $x - y$ becomes larger and more negative, and it approaches M_{max} as $x - y$ approaches larger positive values. Exploiter fitness thus increases as the exploiter's trait value becomes larger relative to the victim's value.

The fact that this model also shows coevolutionary cycling suggests that the genetic assumptions of the adaptive dynamics approach are not as restrictive as they seem, at least in that the approach ostensibly allows for diploidy and for environmental influences on offspring phenotypes. Abrams et al. (2) showed that the genetic assumptions of adaptive dynamics models and quantitative genetic models are mathematically almost identical. It is probably the case that neither model is terribly realistic in this regard. Indeed, the basic assumption of quantitative genetic models, that the distribution of phenotypes is unchanging from generation to generation, has been shown to be only a rough approximation under strong truncating selection (90). Perhaps what is needed is an exploration of the extent to which more genetically explicit models can show arms race dynamics, which might be accomplished through an extension of the Frank model to allow for more loci. An important point, however, is that the behavior of the Frank model can, for the most part, only be analyzed using computer simulations, whereas adaptive dynamic and quantitative genetic models can be analyzed mathematically. Although this issue might seem arcane, in fact it is very important because mathematical analyses permit a much deeper understanding of the models.

The models we have discussed so far have emphasized how natural selection will affect the possibility of polymorphism, including the possibility of cycles in trait values or allele frequencies. An additional way to extend coevolutionary models is to allow for gene flow among populations. It turns out that models that allow for metapopulation structure make interesting testable predictions about the degree to which victim and exploiter are adapted to each other. For example, Gandon et al. (41) added metapopulation structure to Equations 15 and 16, such that a varying number of populations were linked via stepping stone migration. The authors then showed that when the ratio of host migration to parasite migration is <1, and when host migration is relatively small, then parasites tend to be locally adapted, and when the ratio is >1, then the hosts are locally adapted. A potential explanation for this observation is that relatively high migration introduces genetic variation at a rate sufficient to allow rapid adaptation to a wide array of antagonist genotypes. That is, migration acts as a proxy for gene flow in these models. One twist to the prediction that elevated migration rates (or gene flow) enhance local adaptation has been proposed in a model by Nuismer et al. (69). In this model, communities selecting for antagonism and for mutualism are linked by migration. By using simple population genetics recursion relations for allele frequencies with spatially varying allele fitnesses, the authors observe a wide variety of behavior. Notably, though, they find that local trait mismatching can commonly result when selection differs markedly between communities (such as the difference between selection for antagonism and mutualism assumed in this model) and when migration is high. The extent to which varying levels of selection for antagonism among communities linked by migration can similarly generate local maladaptation remains to be explored.

ASSESSING THE ASSUMPTIONS OF VICTIM-EXPLOITER COEVOLUTION MODELS

Our brief review of mathematical models of victim-exploiter interactions is intended to show the range of modeling techniques applied to the problem of understanding coevolutionary dynamics. An important additional issue is that most of the models in question have had little connection to data. This lack of connection has occurred for two reasons. First, a primary goal of some coevolutionary modeling has been to show simply that arms race dynamics can occur without infinite trait values. In these cases, there is little motivation to tailor a model to a particular system. Second, traits involved in coevolutionary interactions are likely to be multilocus with many alternative alleles and the organisms in question are likely to be diploid, thereby violating the assumptions of many of the models that we have described. Frank's work has shown that computer simulations of modest numbers of loci can give some insight into coevolutionary problems, but it is clear that truly realistic models will be very hard to understand. Nevertheless, such models should increase in importance as genomic level characterization of resistance alleles becomes feasible. Such models may be useful in statistical analyses of genomic data, but the general goal of most modeling in ecological genetics and population genetics is understanding, and so we may ultimately be faced with a trade-off between models that are realistic and models that can be understood.

This trade-off between realism and comprehensibility does not necessarily mean that simple models cannot be usefully applied to data. That is, in spite of their simplifying assumptions, coevolutionary models may provide useful descriptions of the dynamics of real coevolutionary interactions. Indeed, in spite of the fact that his models include only eight loci, Frank (38, 39) argues quite effectively that his models nevertheless provide accurate descriptions of important features of the interactions between plants and their pathogens. In that spirit, here we show some examples of data that could be used to parameterize some of the coevolutionary models that we have discussed. Although the application of simple models to complex data sets is in general a challenging statistical topic (16), here our intent is to simply rough out some basic ways in which these models may be compared with data.

Costs of Plant Resistance

With few exceptions (but see 23, 29, 30), most models of plant-enemy coevolution assume a cost of resistance; that is, they assume that resistant individuals are less fit than are susceptible individuals in the absence of attack. As we have described, the occurrence of costs is generally essential for maintaining polymorphisms, and thus for achieving intermediate, rather than maximal, levels of resistance. Given that resistance is generally neither ubiquitous nor absolute, the ability to mimic these patterns under a wide range of parameter values is a desirable attribute of any coevolutionary model. It is therefore not surprising that there is substantial evidence that many resistance characters can be costly. In a review of published studies, Bergelson & Purrington (11) found that 56% and 29% of published studies detected a statistically significant reduction in the fitness of plants resistant to pathogens and herbivores, respectively, relative to their susceptible counterparts. Many more studies showed costs but did not reach statistical significance. For the subset of studies in which costs occurred, magnitudes were variable, ranging from no cost to a 20% reduction in the fitness of resistant plants relative to nonresistant plants.

Although empiricists have been acutely aware that most coevolutionary models assume the presence of a cost of resistance, they have not always appreciated that

TABLE 1 Summary of representative victim-exploiter coevolutionary models[a]

Type	Citation	Damage	Victim costs	Model behavior
Single-locus	58	Genotype specific	Genotype specific	Polymorphism
	43	Genotype specific	Genotype specific	Polymorphism
	64	Genotype specific	Genotype specific	Host cyclic polymorphism, short period
	83	Genotype specific	Genotype specific	Host cyclic polymorphism, pathogen density fluctuations
Multilocus	39	Gene-for-gene	Multiplicative ($\approx$exponential)	Cycles in host & pathogen density & frequencies
AD	28	$\exp(-a(x-y)^2/\sigma^2)$	$a - bx + cx^2$	Cycles in traits
	80	$1/(1 + \exp(-2a(x-y)))$	$\exp(ax)$	Cycles in traits & densities
QG	3	$a/(b + (x-y)^2)$	fitness $= \exp(-ax)$	Cycles in traits & densities
	3	$a(0.5 + \frac{1}{\pi}\arctan(x-y))$	ax^3	Cycles in traits & densities
	42	$\exp(-a(x-y)^2/\sigma^2)$	$\exp(-a(x-b)^4)$	Cycles in traits
	30	$\exp(-a(x-y)^2/\sigma^2)$	ax	Cycles in traits & densities

[a]AD, adaptive dynamic models; QG, quantitative genetic models. Among the latter two groups of models, x is the value of the victim trait, and y is the value of the enemy trait.

the models assume a particular shape to the cost function. Table 1 summarizes the shapes of the cost functions assumed in a variety of published models, including the models described above. These functions take one of two forms. First, several assume a monotonic decrease in fitness, measured in the absence of attack, as resistance increases. Within this group, some functions can be additionally distinguished according to whether they assume a positive (38, 80) or negative (3) second derivative at low costs. A positive second derivative, for example, is found with an exponential decline, whereas a negative second derivative leads to a more gradual decline. The other set of models (3, 27, 29, 30, 42, 62) assumes that fitness is greatest at an intermediate level of resistance, even though natural enemies are absent. Although not an obviously sensible assumption, we show below that several datasets actually seem to have an intermediate optimum.

To assess the appropriateness of these model assumptions, we examined how damage, assumed to be correlated with fitness in the presence of the enemy, changes with the level of resistance for published and unpublished data made available to us by personal communication (45, 63, 79, 81). We restricted our attention to natural rather than agricultural systems so that the levels of resistance would not be the result of artificial selection. The results are illustrated in Figure 1, where fitness, or some proxy for fitness, is plotted against level of resistance for nine characters, and in each a best-fit polynomial is drawn. Note, first, that the fit of these relationships is often poor. With this in mind, we find general consistency in that fitness decreases with increasing levels of resistance. When relationships are monotonically decreasing, it appears that most data are more consistent with a negative second derivative than a positive second derivative (Figure 1*B–E*). And somewhat surprisingly, we find several examples more or less consistent with an intermediate optimum (Figure 1*F–H*) (although some of the cases that we have classified as showing a monotonic decline in fact sometimes show very slight humps or are flat at low trait values). Only one curve (Figure 1*I*) appears superficially inconsistent with assumed functional forms, but this is

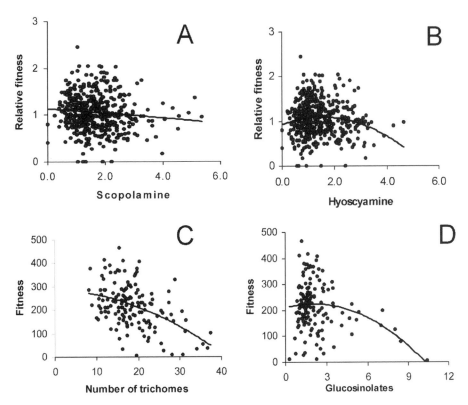

Figure 1 Plots of host plant fitness versus resistance in the absence of natural enemies. (*A, B*) *Datura stramonium* [from Shonle & Bergelson (81)]; (*C, D*) *Arabidopsis thaliana* [from Mauricio (63)]; (*E*) *Diplaucus aurantiacus* [from Han & Lincoln (45)]; (*F*) *Silene alba* [from Biere & Antonovics (12)]; (*G, H*) *Psychotria horizontalis* [from Sagers & Coley (79)]. Curves were fit in MS Excel, assuming a polynomial functional form.

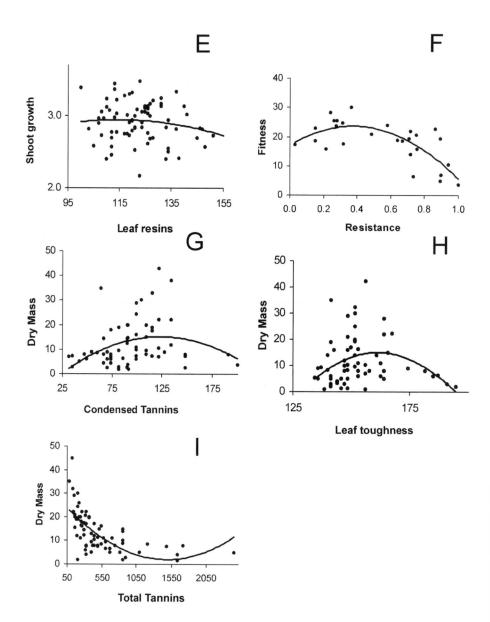

Figure 1 (*Continued*)

certainly the result of poor model fit. In short, the existing data are roughly in agreement with the assumptions of many of the models. This suggests that for the systems for which we have data, there is at least the potential for arms races to occur.

Benefits of Resistance

A second assumption made in models of coevolution is that plant defenses are beneficial to plants and that, in some cases, a particular function describes the relationship between damage and plant resistance. These functions are given in Table 1. Here, x designates a trait in the enemy that makes particular genotypes more or less tolerant of a defensive trait, and y designates the level of that defense in a particular plant. With only minor deviation, these functions, when plotted for a particular value of x, all describe a monotonically decreasing curve with a positive second derivative for the relationship between damage (dependent variable) and the level of defense (independent variable). However, in all these cases, the data were collected without reference to the herbivore's genotype. This is important because in the models, the functions relating damage to levels of resistance are invariably expressed in terms of the difference between the plant's trait and the enemy's trait. It is therefore difficult to assess how well these functions fit field data, as we could find no cases in which the damage imposed by one genotype of an herbivore species was measured for plants that vary in their levels of defense. When variation in levels of tolerance among enemy genotypes is small, however, one would expect that damage imposed by a population of enemies would follow the same functional forms as those indicated.

We obtained data from four studies that enabled us to begin to explore the shape of the relationship between damage and plant resistance (12, 63, 81; T.E. Juenger, unpublished data). We restrict attention to those studies that distinguished types of damage, at least according to general classes of herbivores (e.g., flea beetles, leaf beetles), or to studies that reported that the vast majority of damage was inflicted by one type of enemy species. We were able to find data for the damage associated with only one disease (12). As is apparent in Figure 2, we found, first, that these data are tremendously messy and provide a poor fit to virtually any function. One plot, Figure 2A, shows an increasing relationship between resistance and damage, a nonsensical pattern that presumably results from noisy data. Of the remaining plots, three patterns are apparent: a more or less flat relationship (Figure 2B–E), a monotonically decreasing curve with a negative second derivative (Figure 2F, G), and a curve with an intermediate optimum (Figure 2H). Notably, none of these plots match the functional form assumed in most models. Whether this is because the model assumptions are incorrect, because the experiments did not control for the genotype of the natural enemy, or because the data are simply messy is unclear.

Costs of Virulence

In addition to costs of resistance for the victim, several of the models described incorporate costs for the exploiter. In the context of host-pathogen interactions, these costs typically consist of trade-offs between virulence and transmissibility (40, 64). Within plants in particular, the best-studied host-pathogen systems are gene-for-gene interactions. As is well known, in gene-for-gene interactions, pathogen avirulence is conferred by a particular allele that, among other things, alerts the plant to the growth of the pathogen in the plant's tissues. Because pathogen strains that lack this allele can infect a wider range of host strains, the allele is said to confer avirulence and, hence, is known as an *avirulence* allele, or an *avr*. Clearly *avr* alleles must confer a fitness advantage under some circumstances, otherwise their frequency would decline to zero. Nevertheless, little is known about the fitness benefit of these alleles, and hence about the costs of virulence. What is known is that many *avr*'s are turned on by the so-called type III secretion system, the genes for which are homologous to genes in animal pathogens that play a key role in allowing bacterial growth and hence pathogenicity (48). The few *avr*'s that have been shown to confer fitness benefits do indeed enhance either bacterial growth, disease symptoms, or both (55, 61, 75, 88, 95). Notably, though, none of these studies has been completed under field conditions, and thus the magnitude of the benefit is unknown.

In short, to our knowledge there is generally too little data on plant-pathogen systems to allow model assumptions about costs of virulence to be compared with data. Moreover, the best-known host-pathogen systems appear to lead to qualitative resistance and, thus, would appear to be unsuitable for the quantitative models we have described. Ultimately, it may be the case that the possession of multiple *avr*'s leads to fitness that is higher under some circumstances than others, which would therefore allow for direct application of at least the Frank model, and possibly modified versions of some other models. Currently, however, not enough is known to permit such an application.

ASSESSING THE PREDICTIONS OF VICTIM-EXPLOITER COEVOLUTION MODELS

Our treatment of coevolutionary models focuses on their prediction of escalating arms races, at least under defined parameter ranges. The fact that this dynamic occurs in many different models is not surprising because, as we described, several

Figure 2   Plots of damage versus putative resistance characters. (*A–D*) From Shonle & Bergelson (81); (*E*) from T. Juenger & J. Bergelson, unpublished data; (*F–G*) from Mauricio (63); (*H*) from Biere & Antonovics (12). The enemies inflicting damage are indicated on the *Y* axis label for *A–F*. (*F, G*) Damage was measured as the number of holes in the leaves. Most of this damage was caused by two species of flea beetles, although other minor herbivores were present. (*H*) Damage was measured as the loss in fitness between damaged and undamaged plants. This damage was caused by anther smut. Curves were fit in Exel, assuming a polynomial functional form.

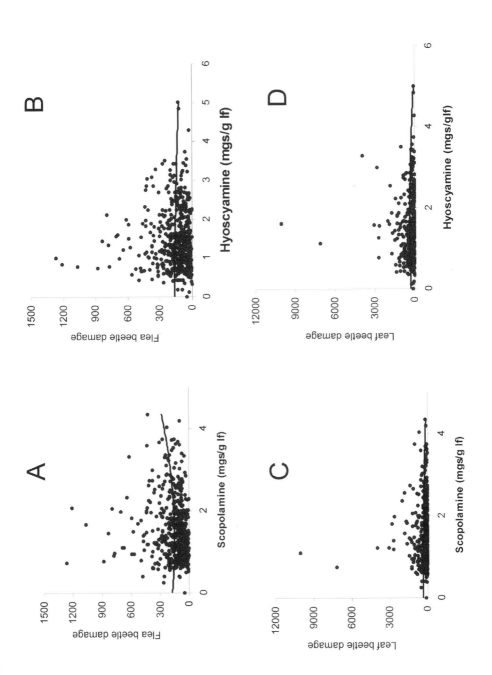

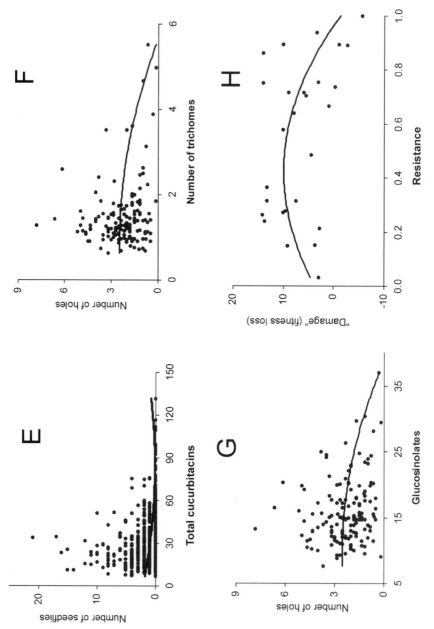

Figure 2   *(Continued)*

of these models were designed with the intent of demonstrating that arms race dynamics are plausible. Despite the widespread interest in arms races that motivates these models, it is surprising that there is little in the way of evidence that arms races actually occur in plant-enemy systems. Perhaps the most widely cited example of an arms race involves Berenbaum's hypothesis that the sequence of evolution of the coumarins, from p-coumaric acid to hydroxycoumarins, to linear furanocoumarins, to angular furanocoumarins, entails a progression in levels of defense (6). There are additional, less-celebrated examples of escalation that have also involved comparisons of plant taxa to demonstrate that derived host plant species are more toxic than are less-derived species (21, 34). These examples occur, however, on an evolutionary timescale that is not particularly relevant to extant victim-exploiter coevolutionary models that focus on the dynamics of single host species.

Within species, there is little evidence for an arms race between plants and their enemies. One suggestive example is the report that levels of sphondin are found at higher concentrations in contemporary Pastinaceae than in roughly 100-year-old herbarium specimens (7). Because sphondin, a furanocoumarin, is known to confer resistance to webworms, this pattern suggests an escalation in defense since the first reported occurrence of webworms in 1883. The other primary form of evidence involving species-level comparisons is molecular evolutionary data. There are now several examples of accelerated rates of adaptive evolution in plant resistance genes, which is suggestive of an arms race (reviewed in 10). However, as Bergelson et al. (10) discussed, it is important to note that other aspects of these data are inconsistent with a classical arms race model of a continual turnover of resistance specificity.

A much more restrictive prediction of coevolutionary models is that of local adaptation. A great number of studies find local adaptation (Table 2) (reviewed recently in 13, 51, 84, 93), although this outcome is by no means universal. For example, a meta-analysis by Van Zandt & Mopper (93) concludes that breeding mode (which affects rates of recombination) and feeding mode strongly influence the extent of local adaptation, whereas dispersal does not. In particular, these authors find that local adaptation occurs most frequently for parthenogenic and haplodiploid pathogens, and for endophagous insects that interact in a relatively pairwise manner with their host. The review by Kaltz & Shykoff (51) finds a more significant role for migration (a proxy for gene flow) but nonetheless concludes that local adaptation is an average consequence of metapopulation dynamics, migration, and variable selection pressure. Thus, the collection of studies examined does not support strongly a pivotal role of dispersal in determining local adaptation. Before rejecting the conclusion of coevolutionary models, however, it is important to note that in few systems was gene flow, or even migration, measured directly. In addition, it is not clear at what scale definitions of dispersal (or gene flow) should be applied. Thus, more comprehensive data are necessary before model predictions can be tested rigorously.

TABLE 2 Empirical studies of local adaptation[a]

Plant host species	Enemy species	Local adaptation[b]	Exploitation type	Relative dispersal rates (plant/enemy)	Relative generation time (plant/enemy)	Host breadth of enemy	Trait	Experiment type	References
Perennial herb	Anther smut fungus	E	Anther fungal infection	>1	~1	Family-species	Infectivity	Cross inoculation	17
Sand-live oak and myrtle oak	Mobile leaf miner	E	Leafmining	<1	>1	Species	Insect survival	Egg transfer	66
Annual vine	Fungal pathogen	E	Fungal infection	Not measured	~1	Genus	Infectivity	Transplant	70
Ponderosa pine	Black pineleaf scale insect	E	Folivory	>1	>>1	11 host species	Insect survival	Egg transfer	74
Sea daisy	Gall-forming midge	E	Gall formation	Not measured	>1	Not measured	Gall abundance, gall size	Reciprocal transplant	84
Beech	Beech scale	E	Exophagus needle herbivory	>1	>1	Not measured	Larval survival and adult fecundity of insects	Larvae transfer	94
Herbaceous perennial	Rust fungus	E	Fungal infection	<<1	>1	Species	Infectivity	Reciprocal cross inoculation	15, 49
White mulberry	Armored scale insect	E	Exophagus needle herbivory	>1	>1	55 families	Insect survival	Egg transfer	46, 47
Seaside daisy	Thrips	E	Folivory	>1	>1	Unclear	Insect survival	Immature insect transplanting across clones	52–54

Rockcress	Rust fungi	P	Fungal infection	<1	~1	Species	Infectivity and general herbivory	Reciprocal transplant	78
Northern red oak	General herbivores: all herbivory	P	Folivory	<1	>1	Not measured	Percent leaf damage	Transplant	82
Weedy perennial	Anther smut fungus	P	Fungal infection	>1	~1	Family-species	Infectivity	Cross inoculation	14, 26, 50
Stinging nettle	Holoparisitic plant	Varies	Xylem and phloem parasitized with haustoria	Not measured	>1	Generalist	Parasite infectivity and biomass & host biomass	Reciprocal cross infection	57
Pinyon pine	Pinyon needle scale	None	Exophagus needle herbivory	>1	>>1	Not measured	Scale survival and abundance	Egg transfer	19
Prarie cordgrass	Two rust fungi	None	Fungal infection	Not measured	>1	Generalists	Infectivity and plant survival	Reciprocal transplant	24
Scots pine	Fungal canker pathogen	None	Invasion of phloem, cambium, and xylem by mycelium	~1	>1	Genus	Selective value (in relation to other genotypes)	Reciprocal transfer of mixed inocula	33
Canyon grape	Leaf galling insect	None	Folivory and root herbivory	>1	>1	Genus	Insect survival and fecundity	Insects choose host	56
Mexican cypress	Aphid	None	Herbirvory	<1	>1	Not measured	Insect survival	Nymph aphid transfer	65

(*Continued*)

TABLE 2 (*Continued*)

Plant host species	Enemy species	Local adaptation[b]	Exploitation type	Relative dispersal rates (plant/enemy)	Relative generation time (plant/enemy)	Host breadth of enemy	Trait	Experiment type	References
Long-lived perennial grass	Root hemi-parasitic plant	None	Root parasitism	Not measured	>1	Not measured	Host and parasite performance and reproduction and infectivity	Reciprocal cross infection	67
Mayapple	Rust fungus	None	Nonsystemic rust infection	<1	>1	Species	Infectivity	Reciprocal transplant	71
Rhus glabra	Mobile herbivorous beetle	None	Herbivory	<1	>1	Genus	Insect survival and weight	Reciprocal egg transfer	86
Pinyon pine	Scale insect	None	Exophagus needle herbivory	>1	>1	Not measured	Insect survival	Egg transfer	91
Ponderosa pine and other pine	Black pineleaf scale insect	None[c]	Exophagus needle herbivory	<1	>1	11 conifer species	Insect survival	Scale transfer	5, 31
Herbaceous perennial herb	Rust pathogen	None	Rust fungus infection	<<1	>1	Species	Resistance and virulence	Reciprocal cross inoculation	T&B[d]

[a] Adapted from a compilation of local adaptation studies treated elsewhere (51, 93), as well as some located subsequent to publication of (51, 93). Personal communication supplements that data found in the references when necessary.

[b] E, enemy species; P, plant species. "None" signifies no strong pattern for either.

[c] Note: Edmunds & Alstad (31) found local adaptation, but this conclusion is reversed in Alstad (5).

[d] T&B, P.H. Thrall & J.J. Burdon, submitted for publication.

CONCLUSIONS AND SUGGESTIONS
FOR FURTHER MODELING

In the interests of brevity, our review of mathematical models of coevolutionary dynamics has only skimmed the surface of the relevant literature; for example, we have not touched at all on how arms races may favor the evolution of sexual recombination (72). Nevertheless, we hope to have shown that the relevant models have moved far beyond the classical single-locus, two-allele approach. As we have described, in some cases, the cost of particular innovations has been a superficial description of the genetic basis of the traits of interest. The conclusion that arms races are possible through coevolutionary cycling is certainly of basic interest, but given the simple genetics that are typically assumed, it remains to be seen whether these models will be of more practical use in the future. By comparing the models to data sets for particular plant-herbivore interactions, we further hope to have shown one way to relate coevolutionary models to data. Our preliminary conclusion is that the data provide only scanty support for the models, but there are two important caveats. First, almost none of the articles in question considers alternative functional forms for damage functions, so it is hard to know what it means when data do not support an assumption of a model. Second, the data we reviewed were produced by experimental designs that had little or no ties to any models. We therefore argue that what is needed is a closer tie between theoretical and experimental work.

An additional issue is that consideration of natural plant-enemy interactions makes clear that many systems do not fit into existing modeling frameworks. Here, we briefly describe four phenomena that to our knowledge have not been incorporated into coevolutionary models. First, resistance is often more complex than is assumed in typical models, particularly in terms of costs. Alternatively, many plants show tolerance, meaning they are able to maintain high reproductive output in the face of enemy attack. Trade-offs between resistance (the ability to fend off attack) and tolerance may substitute for costs of resistance (36, 85; but see 89). The coevolutionary dynamics of plants capable of both tolerance and resistance are likely to be different from the dynamics of plants that show resistance or tolerance alone. Tolerance traits thus present important opportunities for coevolutionary modelers.

Second, most models assume two-way interactions between a plant and its enemy, but in some cases a third species, which is not necessarily an enemy, can modify the two-species interaction in a complex way. For example, pollinators of *Brassica rapa* discriminate between individual plants according to levels of resistance (87). Specifically, the pollinators avoid plants from lines that have been selected for high resistance, as measured by myrosinase concentrations, because such plants have less attractive floral displays. Pollinators, however, also discriminate against plants that have been selected for low resistance but that have experienced damage by herbivores. Similarly, attacks by a fungal pathogen against the wildflower *Silene alba* interact with pollination in complex ways (12). Male flowers, but not females, show a significant cost of resistance that appears to be mediated through late onset of flowering. More generally, levels of nutrient and competition stress in the environment can strongly affect costs of resistance (8, 9), and there can be large effects of genetic background on the magnitude and demonstration of costs (reviewed in 11). These complex effects, especially those changing mating behavior, have not to our knowledge been incorporated into existing theoretical frameworks.

Third, an alternative to the usual fitness cost of resistance is an ecological trade-off among resistance characters. For example, at the *Arabidopsis thaliana Rpp8* (resistance to *Peronospora parasitica* 8) locus, one allele confers resistance to a fungal disease, whereas an alternative allele confers resistance to a virus disease (20). Other trade-offs in the ability of plants to defend against multiple enemies have been observed, although these are not allelic variants. For example, it has been demonstrated that the interaction between a resistance (*R*) gene in *A. thaliana* and its corresponding pathogen avirulence (*avr*) gene interferes with the interaction of another *avr-R* gene pair (73, 76). This interaction appears to result from competition for a common element of the signal transduction pathway, so that plants carrying both *R* genes are fully resistant to each pathogen, provided the two pathogen isolates do not attack the plant at the same time. More generally, signal cross-talk (reviewed in 35) between two pathways involved in defense against pathogens and herbivores has been demonstrated. In a variety of systems, induction of defense through one pathway reciprocally alters induction of defense through the other pathway. This improved understanding of the molecular biology of plant-pathogen interactions will, we hope, lead to new and exciting interactions between empiricists and modelers in their efforts to understand plant-enemy coevolution.

Fourth, many insect-herbivore interactions involve multiple defenses on the part of the host; indeed, the poor fit of the lines in Figures 2*A* and *B* occurs partly because of an interaction between the production of scopolamine and hyoscyamine (81). To date, however, little modeling work has been done on coevolution between herbivores and multiple plant defenses. Clearly there is much room for additional work, and Levin et al. (59) suggest some promising lines of attack.

A final point is that in fact there are many interesting coevolutionary interactions between plants and their natural enemies that fall more naturally into the kind of qualitative framework provided by classical single- or possibly two-locus models. For example, the plant *Datura wrightii* has two phenotypes, one known as sticky and the other as velvety, that differ in the relative proportions of two types of glandular trichomes. Although the interactions between this plant and its herbivores have been studied extensively from the perspectives of costs as well as damage (32, 37), as yet we are unaware of the application of coevolutionary models to predict the future evolutionary dynamics of the system. Indeed, the specifics of this system echo our underlying argument that future modeling efforts would benefit by the construction of models tailored to particular systems but based on existing general models.

ACKNOWLEDGMENTS

Joy Bergelson gratefully acknowledges the support of the National Institute of Health in the form of grant GM57994. Greg Dwyer gratefully thanks the Ecology Panel of the National Science Foundation for grant DEB-0075461. Joy Bergelson and Greg Dwyer together also thank the National Institutes of Health for grant GM62504. The following colleagues kindly shared their data and/or their expertise: May Berenbaum, Art Zangerl, Dan Hare, Cindy Sagers, David Lincoln, Tom Juenger, Irene Shonle, Rodney Mauricio, Wendy Fineblum, Arjen Biere, Bitty Roy, Oliver Kaltz, Peter Thrall, R.A. Enos, T. Koskela, Susan Mopper, Rick Karban, and Peter Price.

Visit the Annual Reviews home page at www.AnnualReviews.org

LITERATURE CITED

1. Abrams PA. 2001. Modelling the adaptive dynamics of traits involved in inter- and intraspecific interactions: an assessment of three methods. *Ecol. Lett.* 4:166–75

2. Abrams PA, Harada Y, Matsuda H. 1993. On the relationship between quantitative genetic and ESS models. *Evolution* 47:982–85

3. Abrams PA, Matsuda H. 1997. Fitness minimization and dynamic instability as a consequence of predator-prey coevolution. *Evol. Ecol.* 11:1–20

4. Abrams PA, Matsuda H, Harada Y. 1993. Evolutionarily unstable fitness maxima and stable fitness minima of continuous traits. *Evol. Ecol.* 7:465–87

5. Alstad D. 1998. Local adaptation: empirical evidence from case studies. See Ref. 66a, pp. 3–21

6. Berenbaum MR. 1983. Coumarins and caterpillars: a case for coevolution. *Evolution* 37:163–79

7. Berenbaum MR, Zangerl AR. 1998. Chemical phenotype matching between a plant and its insect herbivore. *Proc. Natl. Acad. Sci. USA* 95:13743–48

8. Bergelson J. 1994. Changes in fecundity do not predict invasiveness: a model study of transgenic plants. *Ecology* 75:249–52

9. Bergelson J. 1994. The effects of genotype and the environment on costs of resistance in lettuce. *Am. Nat.* 143:349–59

10. Bergelson J, Kreitman M, Stahl EA, Tian D. 2001. Evolutionary dynamics of plant *R*-genes. *Science* 292:2281–85

11. Bergelson J, Purrington CB. 1996. Surveying patterns in the cost of resistance in plants. *Am. Nat.* 148:536–58

12. Biere A, Antonovics J. 1996. Sex-specific costs of resistance to the fungal pathogen *Ustilago violacea* (*Microbotryum violaceum*) in *Silene alba*. *Evolution* 50:1098–110

13. Boecklen WJ, Mopper S. 1998. Local adaptation in specialist herbivores: theory and evidence. See Ref. 66a, pp. 64–90

14. Bucheli E, Gaurschi B, Shykoff JA. 1998. Isolation and characterization of microsatellite loci in the anther smut fungus *Microbotryum violaceum*. *Mol. Ecol.* 7:665–66

15. Burdon JJ, Jarosz AM. 1991. Host-pathogen interactions in natural populations of *Linum marginale* and *Melampsora lini*. 1. Patterns of resistance and racial variation in a large host population. *Evolution* 45:205–17

16. Burnham KP, Anderson DR. 1998. *Model Selection and Inference: A Practical Information Theoretic Approach.* New York: Springer-Verlag

17. Carlsson Graner U. 1997. Anther-smut disease in *Silene dioica*: variation in susceptibility among genotypes and populations, and patterns of disease within populations. *Evolution* 51:1416–26

18. Clay K, Kover PX. 1996. The Red Queen hypothesis and plant/pathogen interactions. *Annu. Rev. Phytopathol.* 34:29–50

19. Cobb NS, Whitham TG. 1998. Prevention of deme formation by the Pinyon needle scale: problems of specializing in a dynamic system. See Ref. 66a, pp. 37–63

20. Cooley MB, Pathirana S, Wu H-J, Kachroo P, Klessig DF. 2000. Members of the Arabidopsis *HRT/RPP8* family of resistance genes confer resistance to both viral and oomycete pathogens. *Plant Cell* 12:663–76

21. Cronquist A. 1977. On the taxonomic significance of secondary metabolites in angiosperms. *Plant Syst. Evol. Suppl.* 1:179–89

22. Daley DJ, Gani J. 1999. *Epidemic Modelling: An Introduction*. Cambridge, UK: Cambridge Univ. Press

23. Damgaard C. 1999. Coevolution of a plant host-pathogen gene-for-gene system in a metapopulation model without cost of resistance or cost of virulence. *J. Theor. Biol.* 201:1–12

24. Davelos AL, Alexander HM, Slade NA. 1996. Ecological genetic interactions between a clonal host plant (*Spartina pectinata*) and associated rust fungi (*Puccinia seymouriana* and *Puccinia sparganioides*). *Oecologia* 105:205–13

25. Dawkins R, Krebs JR. 1979. Arms races between and within species. *Proc. R. Soc. London Ser. B* 202:489–511

26. Delmotte F, Bucheli E, Shykoff JA. 1999. host and parasite population structure in a natural plant-pathogen system. *Heredity* 82:300–8

27. Dieckmann U, Law R. 1996. The dynamical theory of coevolution: a derivation from stochastic processes. *J. Math Biol.* 34:579–612

28. Dieckmann U, Marrow P, Law R. 1995. Evolutionary cycling in predator-prey interactions: population dynamics and the Red Queen. *J. Theor. Biol.* 176:91–102

29. Doebeli M. 1996. Quantitative genetics and population dynamics. *Evolution* 50:532–46

30. Doebeli M. 1997. Genetic variation and the persistence of predator-prey interactions in the Nicholoson-Bailey model. *J. Theor. Biol.* 188:109–20

31. Edmunds GF, Alstad DN. 1978. Coevolution in insect herbivores and conifers. *Science* 199:941–45

32. Elle E, van Dam NM, Hare JD. 1999. Cost of glandular trichomes, a "resistance" character in *Datura wrightii* Regel (Solanaceae). *Evolution* 53:22–35

33. Ennos RA, McConnell KC. 1995. Using genetic markers to investigate natural selection in fungal populations. *Can. J. Bot.* 73:S302–10

34. Feeny PP. 1977. Defensive ecology of the Cruciferae. *Ann. Miss. Bot. Gard.* 64:221–34

35. Felton GW, Korth KL. 2000. Trade-offs between pathogen and herbivore resistance. *Curr. Opin. Plant Biol.* 3:309–14

36. Fineblum WL, Rausher MD. 1995. Tradeoff between resistance and tolerance to herbivore damage in a morning glory. *Nature* 377:517–20

37. Forkner RE, Hare JD. 2000. Genetic and environmental variation in acyl glucose ester production and glandular and nonglandular trichome densities in Datura wrightii. *J. Chem. Ecol.* 26:2801–23

38. Frank SA. 1993. Coevolutionary genetics of plants and pathogens. *Evol. Ecol.* 7:45–75

39. Frank SA. 1994. Coevolutionary genetics of hosts and parasites with quantitative inheritance. *Evol. Ecol.* 8:74–94

40. Frank SA. 1996. Models of parasite virulence. *Q. Rev. Biol.* 71:37–78

41. Gandon S, Capowiez Y, Dubois Y, Michalakis Y, Olivieri I. 1996. Local adaptation and gene-for-gene coevolution

in a metapopulation model. *Proc. R. Soc. London Ser. B* 263:1003–9

42. Gavrilets S. 1997. Coevolutionary chase in exploiter-victim systems with polygenic characters. *J. Theor. Biol.* 186:527–34

43. Gillespie JH. 1975. Natural selection for resistance to epidemics. *Ecology* 56:493–95

44. Hammond-Kosack KE, Jones JDG. 1997. Plant disease resistance genes. *Annu. Rev. Plant Physiol. Plant Mol. Biol.* 48:575–607

45. Han KP, Lincoln DE. 1994. The evolution of plant carbon allocation to plant secondary metabolites: a genetic analysis of cost in *Diplaucus aurantiacus*. *Evolution* 48:1550–63

46. Hanks LM, Denno RF. 1993. Natural enemies and plant water relations influence the distribution of an armored scale insect. *Ecology* 74:1081–91

47. Hanks LM, Denno RF. 1994. Local adaptation in the armored scale insect *Pseudaulacaspis pentagona* (Homoptera: Diaspidadae). *Ecology* 75:2301–10

48. Innes RW, Bent AF, Kunkel BN, Bisgrove SR, Staskiawicz BJ. 1993. Molecular analysis of avirulence gene *avrRpt2* and identification of a putative regulatory sequence common to all known *Pseudomonas syringae* avirulence genes. *J. Bacteriol.* 175:4859–69

49. Jarosz AM, Burdon JJ. 1991. Host-pathogen interactions in natural populations of *Linum marginale* and *Melampsora lini*. 2. Local and regional variation in patterns of resistance and racial structure. *Evolution* 45:1618–27

50. Kaltz O, Gandon S, Michalakis Y, Shykoff JA. 1999. Local maladaptation in the anther-smut fungus *Microbotryum violaceum* to its host plant *Silene latifolia*: evidence from a cross-inoculation experiment. *Evolution* 53:395–407

51. Kaltz O, Shykoff JA. 1998. Local adaptation in host-parasite systems. *Heredity* 81:361–70

52. Karban R. 1989. Community organization of *Erigeron glaucus* folivores: effects of competition, predation, and host plant. *Ecology* 70:1028–39

53. Karban R. 1989. Fine-scale adaptation of herbivorous thrips to individual host plants. *Nature* 340:60–61

54. Karban R, Strauss SY. 1994. Colonization of new host-plant individuals by locally adapted thrips. *Ecography* 17:82–87

55. Kearney B, Staskawicz BJ. 1990. Widespread distribution and fitness contribution of *Xanthomonas campestris* avirulence gene *avrBs2*. *Nature* 346:385–86

56. Kimberling DN, Price PW. 1996. Variability in grape phylloxera preference and performance on canyon grape (*Vitis arizonica*). *Oecologia* 107:553–59

57. Koskela T, Salonen V, Mutikainen P. 2000. Local adaptation of a holoparasitic plant, *Cuscuta europaea*: variation among populations. *J. Evol. Biol.* 13:749–55

58. Levin SA. 1983. Some approaches to the modelling of coevolutionary interactions. In *Coevolution*, ed. M Nitecki, pp. 50–65. Chicago: Univ. Chicago Press

59. Levin SA, Segel LA, Adler FR. 1990. Diffuse coevolution in plant-herbivore communities. *Theor. Pop. Biol.* 37:171–91

60. Levin SA, Udovic JD. 1977. A mathematical model of coevolutionary populations. *Am. Nat.* 111:657–75

61. Lorang JM, Shen H, Kobashi D, Cooksey S, Keen NT. 1994. *avrA* and *avrE* in *Pseudomonas syringae* pv. *tomato PT23* play a role in virulence on tomato plants. *Mol. Plant-Microbe Interact.* 7:726–39

62. Marrow P, Law R, Cannings C. 1992. The coevolution of predator prey interactions—Esss and Red Queen dynamics. *Proc. R. Soc. London Ser. B* 250:133–41

63. Mauricio R. 1998. Costs of resistance to natural enemies in field populations of the annual plant *Arabidopsis thaliana*. *Am. Nat.* 151:20–28

64. May RM, Anderson RM. 1983. Epidemiology and genetics in the coevolution of parasites and hosts. *Proc. R. Soc. London Ser. B* 219:281–313

65. Memmott J, Day RK, Godfray HCJ. 1995. Intraspecific variation in host-plant quality: the aphid *Cinara cupressi* on the Mexican cypress, *Cupressus lusitanica*. *Ecol. Entomol.* 20:153–58

66. Mopper S, Beck M, Simberloff D, Stiling P. 1995. Local adaptation and agents of selection in a mobile insect. *Evolution* 49:810–15

66a. Mopper S, Strauss SY, eds. 1998. *Genetic Structure and Local Adaptation in Natural Insect Populations*. New York: Chapman & Hall

67. Mutikainen P, Salonen V, Puustinen S, Koskela T. 2000. Local adaptation, resistance, and virulence in a hemiparasitic plant-host plant interaction. *Evolution* 54:433–40

68. Nagylaki T. 1992. *Introduction to Theoretical Population Genetics*. Berlin: Springer-Verlag

69. Nuismer SL, Thompson JN, Gomulkiewicz R. 1999. Gene flow and geographically structured coevolution. *Proc. R. Soc. London Ser. B* 266:605–9

70. Parker MA. 1985. Local population differentiation for compatibility in an annual legume and its host-specific fungal pathogen. *Evolution* 39:713–23

71. Parker MA. 1989. Disease impact and local genetic diversity in the clonal plant *Podophyllum peltatum*. *Evolution* 43:540–47

72. Peters AD, Lively CM. 1999. The Red Queen and fluctuating epistasis: a population genetic analysis of antagonistic coevolution. *Am. Nat.* 154:393–405

73. Reuber TL, Ausubel FM. 1996. Isolation of Arabidopsis genes that differentiate between resistance responses mediated by the *RPS2* and *RPM1* disease resistance genes. *Plant Cell* 8:241–49

74. Rice WR. 1983. Parent offspring pathogen transmission: a selective agent promoting sexual reproduction. *Am. Nat.* 121:187–203

75. Ritter C, Dangl JL. 1995. The *avrRpm1* gene of *Pseudomonas syringae* pv. *maculicola* is required for virulence on Arabidopsis. *Mol. Plant-Microbe Interact.* 8:444–53

76. Ritter C, Dangl JL. 1996. Interference between two specific pathogen recognition events mediated by distinct plant disease resistance genes. *Plant Cell* 8:251–57

77. Rosenzweig ML, Brown JS. 1987. Red Queen and ESS: the coevolution of evolutionary rates. *Evol. Ecol.* 1:59–94

78. Roy BA. 1998. Differentiating the effects of origin and frequency in reciprocal transplant experiments used to test negative frequency-dependent selection hypotheses. *Oecologia* 115:73–83

79. Sagers CL, Coley PD. 1995. Benefits and costs of defense in a neotropical shrub. *Ecology* 76:1835–43

80. Sasaki A, Godfray HCJ. 1999. A model for the coevolution of resistance and virulence in coupled host-parasitoid interactions. *Proc. R. Soc. London Ser. B* 266:455–63

81. Shonle I, Bergelson J. 2000. Evolutionary ecology of the tropane alkaloids of *Datura stramonium* L. (Solanaceae). *Evolution* 54:778–88

82. Sork VL, Stowe KA, Hochwender C. 1993. Evidence for local adaptation in closely adjacent subpopulations of northern red oak (Quercus rubra L) expressed as resistance to leaf herbivores. *Am. Nat.* 142:928–36

83. Stahl EA, Dwyer G, Mauricio R, Kreitman M, Bergelson J. 1999. Dynamics of disease resistance polymorphism at the *Rpm1* locus of *Arabidopsis*. *Nature* 400:667–71

84. Stiling P, Rossi AM. 1998. Deme formation in a dispersive gall forming midge. See Ref. 66a, pp. 22–36

85. Stowe KA. 1998. Experimental evolution of resistance in *Brassica rapa*: correlated response of tolerance in lines selected for

glucosinolate content. *Evolution* 52:703–12

86. Strauss SY. 1997. Lack of evidence for local adaptation to individual plant clones or site by a mobile specialist herbivore. *Oecologia* 110:77–85

87. Strauss SY, Siemens DH, Decher MB, Mitchell-Olds T. 1999. Ecological costs of plant resistance to herbivores in the currency of pollination. *Evolution* 53:1105–13

88. Swarup S, De Feyter R, Brlansky RH, Gabriel DW. 1991. A pathogenicity locus from *Xanthomonas citri* enables strains from several pathovars of *X. campestris* to elicit cankerlike lesions on citrus. *Phytopathology* 81:802–9

89. Tiffin P, Rausher MD. 1999. Genetic constraints and selection acting on tolerance to herbivory in the common morning glory *Ipomoea purpurea*. *Am. Nat.* 154:700–16

90. Turelli M, Barton NH. 1994. Genetic and statistical analysis of strong selection on polygenic traits: what, me normal? *Genetics* 138:913–41

91. Unruh TR, Luck RF. 1987. Deme formation in scale insects: a test with the Pinyon needle scale and a review of other evidence. *Ecol. Entomol.* 12:439–49

92. van Valen L. 1973. A new evolutionary law. *Evol. Theory* 1:1–30

93. Van Zandt PA, Mopper S. 1998. A meta-analysis of adaptive deme formation in phytophagous insect populations. *Am. Nat.* 152:595–604

94. Wainhouse D, Howell RS. 1983. Intraspecific variation in Beech scale populations and in susceptibility of their host *Fagus sylvatica*. *Ecol. Entomol.* 8:351–59

95. Yang Y, De Feyter R, Gabriel DW. 1994. Host specific symptoms and increased release of *Xanthomonas citri* and *X. campestris* pv. *malvacaerum* from leaves are determined by the 102bp tandem repeats of *pthA* and *avrb6*, respectively. *Mol. Plant-Microbe Interact.* 7:345–55

Annu. Rev. Genet. 2001. 35:501–38

BIOLOGY OF MAMMALIAN L1 RETROTRANSPOSONS

Eric M. Ostertag and Haig H. Kazazian Jr

Department of Genetics, University of Pennsylvania School of Medicine, Philadelphia, Pennsylvania 19104; e-mail: ostertag@mail.med.upenn.edu, kazazian@mail.med.upenn.edu

Key Words transposable elements, reverse transcription, genome structure, human mutation, mouse mutation

■ **Abstract** L1 retrotransposons comprise 17% of the human genome. Although most L1s are inactive, some elements remain capable of retrotransposition. L1 elements have a long evolutionary history dating to the beginnings of eukaryotic existence. Although many aspects of their retrotransposition mechanism remain poorly understood, they likely integrate into genomic DNA by a process called target primed reverse transcription. L1s have shaped mammalian genomes through a number of mechanisms. First, they have greatly expanded the genome both by their own retrotransposition and by providing the machinery necessary for the retrotransposition of other mobile elements, such as Alus. Second, they have shuffled non-L1 sequence throughout the genome by a process termed transduction. Third, they have affected gene expression by a number of mechanisms. For instance, they occasionally insert into genes and cause disease both in humans and in mice. L1 elements have proven useful as phylogenetic markers and may find other practical applications in gene discovery following insertional mutagenesis in mice and in the delivery of therapeutic genes.

CONTENTS

0066-4197/01/1215-0501$14.00

INTRODUCTION

Preliminary analysis of the human genome sequence has already provided several major surprises. For one, the human genome contains less than twice as many genes as the fly, worm, and Arabidopsis genomes. Equally surprising is the observation that 45% of the human genome consists of transposable elements, a much greater percentage than the 3% to 10% observed in the genomes of the three other organisms (107). Since sequences that have been in the genome longer than 200 My have diverged to the point where they are unidentifiable, it is likely that even more than 45% of the human genome is composed of transposable elements (107, 191). Transposable elements have contributed greatly to what we now realize is a highly dynamic genome.

Although this review concentrates on a particular type of transposable element, the L1 retrotransposon, a general introduction to the topic is warranted. Mammalian transposable elements consist of DNA transposons and retrotransposons (Figure 1, see color insert). DNA transposons have structures similar to bacterial transposons. They have inverted terminal repeats and encode a transposase activity. They generally move by a "cut and paste" mechanism utilizing the transposase (143, 192). Although roughly 3% of the human genome is composed of DNA transposons, they are remnants or fossils of ancient elements, and it is unlikely that any remain transpositionally active (107). In contrast to DNA transposons, retrotransposons encode a reverse transcriptase (RT) activity and move by a "copy and paste" process involving an RNA intermediate. The original retrotransposon is maintained *in situ* where it is transcribed. The transcript is then reverse transcribed and integrated into a new genomic location (115, 217). Approximately 42% of the human genome is composed of retrotransposons (107), and, although most of these elements are inactive, some retain the ability to retrotranspose (172).

Retrotransposable elements can be classified as either autonomous or nonautonomous. Elements are considered autonomous if they encode certain activities necessary for their mobility. However, it is unlikely that they are strictly autonomous because host proteins, such as DNA repair enzymes, are probably also required for their retrotransposition. There are two classes of autonomous retrotransposons: LTR (long terminal repeat) retrotransposons and non-LTR retrotransposons. Mammalian LTR retrotransposons are structurally similar to retroviruses, but they lack a functional *env* gene. These retrotransposons include elements such as mouse intracisternal A-particles (IAPs) (103) and human endogenous retroviruses (HERVs) (13, 191, 219), both of which are unlikely to include autonomously active elements. About 8% of the human genome is composed of defective endogenous retroviruses and solitary LTRs derived from recombination between the 5′ LTR and the 3′ LTR of these elements (107). The non-LTR retrotransposon class contains LINEs (long interspersed nucleotide elements), which include inactive elements, such as L2 in humans, and active elements, such as L1 in humans and mice (107, 120, 191). Approximately 21% of the human genome is composed of autonomous non-LTR retrotransposons (107).

In addition to the autonomous retrotransposons, there are a large number of nonautonomous retrotransposons in mammalian genomes. These elements do not encode any proteins. Therefore, they require activities encoded by other autonomous retrotransposons for their mobility. The most prominent members of this class are Alu elements in humans and their B1 counterparts in mice (169). The greater than 1 million Alu elements in the human genome account for about 11% of its mass, whereas roughly 100,000 B1 elements populate the mouse genome (107, 191). Other nonautonomous retrotransposons in the human genome include processed pseudogenes and SVA elements. Thus, transposable elements and transposon-derived sequences make up about 45% of the total mass of the human genome, or 40 times the 1.1% of the genome that is composed of protein-coding sequences (107).

L1 elements are the master retrotransposons in mammalian genomes. Besides duplicating themselves, they likely have been responsible for the genomic expansion of nonautonomous retrotransposons, specifically Alu elements, processed pseudogenes, and SVA elements in the human genome. Over evolutionary time they have not only expanded greatly in number, but also have acquired other roles, some of which are quite useful to the organism, whereas others are detrimental to individual members of the species (90). Many of these roles of L1 retrotransposons are discussed in this review.

MECHANISM OF L1 RETROTRANSPOSITION

In mammals, the great majority of L1 retrotransposons are inactive, defective elements, owing to 5′ truncation, inversion, or point mutations. Of the 520,000 L1 sequences in humans, only about 3000–5000 represent full-length elements (66, 107). The discovery of a full-length mouse L1 element with intact open reading frames (112), and the creation of a consensus alignment of many human L1 sequences (177), helped to elucidate the anatomy of the full-length, 6-kb element. The consensus sequence revealed that L1 elements have a 5′ untranslated region (UTR) with internal promoter activity, two open reading frames (ORFs), a 3′ UTR that ends in an AATAAA polyadenylation signal, and a polyA tail (Figure 1). The discovery that several full-length, retrotranspositionally active elements had the predicted ORFs validated the consensus sequence (147).

In some cases, careful scrutiny of L1 structure has offered insight into the mechanism of L1 retrotransposition. In other cases, an understanding of similar retrotransposons in other organisms has suggested hypotheses for the L1 mechanism. The development of limited functional assays has strengthened some of these hypotheses (41, 52, 127a). Furthermore, the development of a cell culture–based retrotransposition assay was instrumental in demonstrating L1 functions necessary for retrotransposition (147) (Figure 2, see color insert). However, many aspects of the retrotransposition mechanism remain unknown. The general steps of retrotransposition include transcription, RNA processing, mRNA export, translation, posttranscriptional modifications and RNP formation, return to the nucleus, and reverse transcription and integration (Figure 3, see color insert). Here we summarize what is known and what is theorized regarding the mechanism of L1 retrotransposition.

Transcription

The 5′ UTR of human L1 contains internal promoter activity independent of upstream sequences (199), but the machinery responsible for transcribing L1 in vivo remains undetermined. A reporter gene driven by an L1 5′ UTR fragment apparently can be transcribed by RNA polymerase III (Pol III) (105). However, inconsistent with Pol III–mediated transcription, the L1 transcript is much larger than a typical Pol III transcript and encodes proteins. Moreover, the presence of internal Pol III termination sequences in L1 argues against Pol III–mediated transcription. With few exceptions, Pol III in higher eukaryotes terminates within clusters of four or more consecutive T residues in the noncoding DNA strand (16, 58a, 101). Pol III transcripts rarely contain four consecutive internal U residues and almost never contain five consecutive U residues. However, the noncoding strand of the consensus sequence of active human L1 elements contains a stretch of six T residues and a stretch of seven T residues. Furthermore, the active L1 consensus element contains a functional AATAAA polyadenylation (polyA) signal (146). The AATAAA polyA signal is required for RNA polymerase II (Pol II) termination (160), and proper cleavage and polyadenylation after the AATAAA polyA signal requires Pol II (77, 128), suggesting that Pol II transcribes L1. Lastly, the 5′ UTR of both mouse and human L1 elements can be replaced functionally with a heterologous Pol II promoter in the cultured cell assay for retrotransposition (147, 151).

Pol II and its associated transcription factors are recruited to a core promoter element, typically the TATA motif located 25–30 base pairs upstream of the transcription initiation site. Early steps in transcription include binding of TATA Binding Protein (TBP), a component of TFIID, to the TATA motif and subsequent binding of TFIIB to TFIID. Other transcription factors and Pol II bind this complex to form a complete transcription complex (232). However, some Pol II–dependent promoters lack TATA sequences, instead containing initiator (Inr) elements capable of independently directing transcription initiation (215). Pol II transcription from an internal promoter was first demonstrated for *jockey*, a retrotransposon from *Drosophila melanogaster* (142). It has since been shown that *jockey*, as well as the related Drosophila retrotransposons, I factor, F, and Doc, contain conserved downstream promoter elements (DPEs) that direct transcription initiation at an Inr-like sequence by directly recruiting TFIID (24, 141). A similar arrangement may allow Pol II transcription from an internal promoter in human L1 elements.

Several groups have demonstrated protein binding to a downstream core promoter element essential for human L1 transcription (127, 140). This protein has been identified as the Yin Yang-1 (YY-1) transcription factor (12, 105), and the conserved protein binding sequence, base pairs +13 to +21 of the L1 5′ UTR coding strand, is a perfect match to the YY-1 core binding sequence (83, 188). YY-1 can act as a transcriptional activator, repressor, or initiator (179, 186). Interestingly, YY-1 can initiate transcription in vitro in the absence of TFIID (179, 206), binding TFIIB directly and directing Pol II to the transcription initiation site (207). The human L1 YY-1 binding site may be analogous to the Drosophila DPE, and YY-1 binding may either recruit TFIID or bind TFIIB directly during the formation of a Pol II preinitiation complex. The TATA-less human DNA polymerase β gene promoter has an Inr with an overlapping YY-1 binding site. Careful mutational analysis of this promoter demonstrates that YY-1 binding is not required for transcription, but rather may serve to position the transcription complex or regulate promoter activity (216). Such a mechanism cannot be ruled out for L1 transcription, but limited mutational analysis of the L1 promoter suggests that YY-1 binding is very important for efficient transcription (12, 140). Additional promoter studies including further mutational analysis should help to elucidate the mechanism of L1 transcription.

Unlike humans, rats and mice have several distinct L1 subfamilies that are defined by differences in their 5′ structures. Most murine L1 subfamilies have 5′ UTRs notably different from the human 5′ UTR in that they contain a variable number of tandemly repeated units of 205 to 210 bp, called monomers, followed by a short non-monomeric region (51, 55, 112, 163). The V subfamily has no identifiable monomers, and, because its members diverge significantly in sequence from each other, it is the oldest subfamily (4, 87). The F subfamily has a large number of inactive members, and it is the oldest L1 subfamily with monomers evident at its 5′ end (163, 174, 220). A consensus F monomer has been resurrected and shown to have promoter activity (4). The A subfamily (112, 182) contains about 50,000 truncated and 6000 to 8000 full-length members per diploid mouse genome (173). These elements have monomers (112) and share about 95%–97% sequence similarity (174). The recently discovered T_F subfamily (34, 151) contains about 3000 full-length members per diploid genome (151, 173). These elements were called T_F because most are transposable and their monomers are 70% identical in sequence to the F monomers (151). T_F L1s share greater than 99% sequence similarity (most are greater than 99.6% identical to each other) (34). The most recently discovered L1 subfamily is the G_F subfamily. These L1s are 93–99+% identical to each other and contain monomers that, like T_F, are about 70% identical to F monomers. Although both G_F and T_F monomers differ from F monomers by about

30%, they also differ from each other by 33%. There are roughly 1500 full-length G_F elements in the diploid mouse genome (59).

Experiments using cultured cells transiently transfected with various regions of the mouse T_F 5′ UTR fused to a reporter gene have revealed that the promoter activity lies within the monomers and that promoter strength is proportional to the number of monomers. L1s of the T_F subfamily contain a conserved YY-1 binding site within each monomer. In the genome, many of these elements begin within or near the YY-1 site, supporting a possible role for YY-1 in positioning of the transcription complex (35). However, G_F L1s contain a YY-1 binding site that differs from consensus by one nucleotide and, in the genome, these elements tend not to begin near this site (59). L1s of the A subfamily do not contain a YY-1 binding site, suggesting that the mechanism of transcription may vary among L1 elements. The mechanism by which monomers are created and maintained is also an interesting, albeit unsolved, puzzle.

Evidence to date suggests that L1 expression is germ line specific. Full-length, sense-stranded L1 transcripts have been detected in prepuberal spermatocytes, but are rare in normal somatic tissues (18). Recently, a mouse model of retrotransposition was created using a tagged human L1 element under the control of its endogenous promoter. Strand-specific RT-PCR designed to detect the full-length tagged transcript demonstrated expression in the male and female germ line, but not in multiple somatic tissues. Studies in the transgenic mouse model indicate that retrotransposition of the tagged human L1 element under the control of its endogenous promoter occurs in late-meiotic and post-meiotic male germ cells (161). The transcription factors that determine the germ line specificity of L1 transcription have not been defined. However, indirect evidence suggests that the SOX family of transcription factors may be involved (201). Many groups have also suggested that transcription of L1 elements may be controlled by methylation at CpG dinucleotides in the L1 5′ untranslated region (72, 156, 202, 221, 230).

If L1 were a genetic parasite, as hypothesized, then one would expect a germ line–specific pattern of L1 expression. L1 retrotransposition in the germ line is likely to lead to an expansion in the number of L1 elements in the genome, whereas L1 retrotransposition events in somatic tissues cannot be passed on to future generations and are likely to be detrimental to the host. For example, unchecked retrotransposition in somatic tissues could result in an insertion into a tumor suppressor gene that could ultimately promote oncogenesis. Interestingly, in a cultured cell assay, retrotransposition is detected in a large variety of transformed cells but not in primary cell lines or ES cells (J.L. Goodier, E.T. Prak & H.H. Kazazian Jr, unpublished data). The initial mutations that cause a cell to become precancerous might occasionally activate L1 transcription. The resultant somatic retrotransposition events could increase the likelihood of accumulating additional mutations that would ultimately produce cancer. In fact, there is at least one example of an authentic retrotransposition event contributing to cancer; an L1 element inserted into the APC gene of tumor cells in a patient who had developed colon cancer (139).

RNA Processing and Nuclear Export

RNAs transcribed by Pol II are typically modified by cleavage and addition of a polyA tail, by addition of a 7-methylguanosine cap, and by splicing of introns (204). As mentioned previously, L1 elements contain a functional AATAAA polyadenylation signal and likely use the cleavage and polyadenylation machinery typical of Pol II transcripts. However, there are two unusual features of the L1 polyadenylation signal. First, the AATAAA polyadenylation signal of L1 elements is immediately followed by the presumed polyA tail. This observation can be interpreted in one of two ways: Either cleavage and polyadenylation of L1 elements occur immediately after the AATAAA signal, as opposed to the usual 10–25 nucleotides downstream (26), or they occur at the typical number of bases downstream of the AATAAA signal and the A residues found between the polyA signal and the polyA tail are encoded. If the latter is true, it begs the question why L1 elements have evolved to contain a stretch of A residues after the polyA signal. One possible answer is that the A-rich region may be positively selected by the L1 integration process (see discussion of TPRT below). Analysis of mouse L1 elements in the genome database suggests that the A-rich region is encoded because many elements have the sequence AATGG A(n) following the polyA signal (J.L. Goodier, personal communication). The second unusual feature of the L1 polyadenylation process is that L1s are usually lacking important sequences downstream of the polyadenylation site. These conserved, GU-rich sequences are 20–60 nucleotides 3' of the polyadenylation signal and promote efficient cleavage and polyadenylation (129, 130). The sequence 20–60 nucleotides downstream of retrotransposed L1 elements depends upon the insertion site and is highly variable (often this sequence is the polyA tail). One would predict that subsequent retrotransposition events would therefore polyadenylate inefficiently after the AATAAA signal. In fact, this appears to be the case. L1 elements frequently bypass their own polyA signal and use a downstream signal (146). This process results in the retrotransposition of genomic sequence 3' of the L1 element and is called L1-mediated transduction (see section on L1-mediated transduction below).

Additional modifications of L1 RNA are not known. For example, it is not known whether L1 transcripts are modified by the addition of a 7-methylguanosine cap. L1 transcripts do not contain introns and therefore do not require splicing. Although nearly all mammalian mRNAs contain introns, there are notable exceptions, such as members of the human G-protein–coupled receptor genes and type I interferon genes (57, 167). Interestingly, in a cultured cell assay, tagged L1 transcripts containing an intron are spliced appropriately and are able to retrotranspose (147).

Recent experiments are beginning to elucidate the mechanism of mRNA export from the nucleus in vertebrates. Cells have evolved a mechanism to prevent export of unspliced mRNAs, presumably because the export of unspliced RNAs would be inefficient and could result in protein products that are deleterious to the cell. Unspliced RNAs contain splice sites that are bound by splicing factors called commitment factors and retained in the nucleus (33). Splicing and nuclear export are therefore coupled. It has been suggested that L1 mRNA might contain *cis*-acting elements required for its nuclear export (159). This is a reasonable speculation based upon the observations that some mRNAs expressed from transfected cDNAs lacking intron sequences are not exported efficiently, and that several viruses have evolved *cis*-acting elements to facilitate export of unspliced RNA. However, mRNAs expressed from some intronless cDNAs are exported well. In addition, unlike retroviruses (32), L1 does not face the problem of exporting an unspliced RNA containing splice sites, an RNA species that is normally retained in the nucleus by commitment factors. One would not expect L1 RNA to be retained because it normally does not contain splice sites, and therefore, may not need *cis*-acting elements for nuclear export.

Translation

L1 mRNAs are atypical of mammalian mRNAs because they are bicistronic. In humans, the two ORFs are in frame and separated by a 63-bp noncoding spacer region that contains stop codons in all three reading frames. In mice, the two ORFs are also nonoverlapping but in different reading frames, whereas in rats, the ORFs are overlapping. The mechanism of translation remains largely a mystery. Experiments suggest that the ORF1 protein is translated by ribosomal initiation at the 5' UTR followed by ribosomal scanning (131). Whether initiation is cap-dependent is unknown. Indirect evidence suggests that the human ORF2 is not translated by termination and reinitiation (131). Frameshifting is unnecessary in humans and has been ruled out in rats (84). In addition, there is no evidence of an ORF1/ORF2 fusion protein in either species. Therefore, translation of ORF2 may occur by some form of internal ribosomal entry such as by the use of an internal ribosomal entry site (IRES) or by ribosomal shunting.

ORF1 protein is apparently translated much more efficiently than ORF2 protein. ORF1 protein has been detected in the cytoplasm of a number of human testicular germ cell tumors and in breast carcinoma and medulloblastoma (9, 19–21). ORF1 has also been detected in mouse embryonal carcinoma cell lines, male and female mouse germ cells, Leydig cells of embryonic mouse testis, theca cells of adult mouse ovary, and a large variety of transformed mouse and human cell lines (18, 124, 205). However, ORF2 has escaped detection despite efforts using several antibodies that detect either Baculovirus-produced or bacterial-produced ORF2 protein.

ORF1 encodes an approximately 40-kDa protein (p40) with RNA binding activity (80, 98, 126). The exact size of the ORF1 protein varies among species and occasionally within species. For example, in the mouse, the ORF1 protein has a length polymorphism region (LPR) that causes a length difference of up to 28 amino acids (3, 59, 132). The amino acid sequence of the COOH-terminal half of the ORF1 protein is relatively well conserved across species, whereas the NH_2-terminal half of the protein varies considerably. However, the ability of the NH_2-terminal half of the protein to form an α-helical structure is conserved. Human p40 contains a leucine zipper motif, rabbit ORF1 protein contains a coiled-coil domain, and rat and mouse proteins each have unique α-helical structures (39, 79, 82). It was hypothesized that the conserved COOH-terminal end is involved in RNA binding and the conserved NH-terminal α-helical structure is involved in protein-protein interaction (see section on RNP formation below). A series of elegant experiments using mouse ORF1 protein strongly support this hypothesis (126). In addition to these functions, mouse ORF1 protein has nucleic acid chaperone activity in vitro (125) (see section on TPRT below).

ORF2 encodes an approximately 150-kDa protein with three conserved domains, an NH_2-terminal endonuclease (EN) domain (52), a central reverse transcriptase (RT) domain (127a), and a COOH-terminal zinc knuckle-like domain (50). The L1 EN domain cleaves one strand of double-stranded DNA at a large number of genomic sites characterized by the loose consensus sequence, AA|TTTT (28, 52, 88). Cleavage site preference may be affected by the local chromatin structure (30). The EN domain is evolutionarily related in a subset of non-LTR retrotransposons and shares critical amino acids at positions corresponding to the catalytic sites of Exonuclease III, an endonuclease of *Escherichia coli* (52, 65, 120, 122, 144).

The L1 RT domain is related in all non-LTR retrotransposons (120). Non-LTR retrotransposon RT shares sequence similarities with more distantly related RTs from LTR retrotransposons and retroviruses, yet functions in a very different way (225). Retroviral and LTR retrotransposon RTs function in the cytoplasm within particles, use a tRNA primer, and carry out reverse transcription through a complex process requiring a number of steps (217). On the other hand, non-LTR retrotransposon RTs are thought to function in the nucleus, use genomic DNA as a primer, and carry out reverse transcription through the relatively simple process of target primed reverse transcription (TPRT) (115).

A third conserved domain of ORF2 is a COOH-terminal, cysteine-rich domain (50). This domain is conserved in all known mammalian L1 elements. Comparison with *Swimmer 1* and Zorro, related L1-like non-LTR retrotransposons from teleost fish and *Candida albicans*, respectively, suggests that it may be a conserved CCHC zinc knuckle structure (45, 61). CCHC zinc knuckles are present in all retroviral nucleocapsid proteins, except spumaviruses (14, 95, 198), and are found in other proteins that bind single-stranded RNA (8, 11). These observations suggest a possible role for ORF2 in protein-nucleic acid interaction, specifically the interaction of ORF2 protein with L1 RNA during the formation of retrotransposition intermediates. Interestingly, in addition to a role in nucleic acid binding, mutational analysis of various retroviral CCHC zinc knuckles suggests they are important for reverse transcription, perhaps required for unfolding structured RNA (62, 63, 68). Such a role cannot be ruled out for the L1 ORF2 protein.

Posttranslational Modifications and Ribonucleoprotein Formation

Posttranslational modifications or protein processing of ORF1 and ORF2 proteins are currently unknown. It is believed that the ORF1 protein, ORF2 protein, and L1 RNA associate to form ribonucleoprotein (RNP) particles that are intermediates in retrotransposition. L1 RNA has been found associated with ORF1 protein in RNP particles in human and mouse teratocarcinoma cells (79, 123). However, ORF2 protein has not yet been detected in these particles. Both mouse and human ORF1 have been demonstrated to form higher-order homomultimers. The protein-protein interaction is likely mediated by the leucine zipper in humans and may be stabilized by interchain disulfide bonds (79). ORF1 proteins from mouse and other mammals lack the leucine zipper, but likely use their α-helices for protein-protein interaction (126).

Entry into the Nucleus

As a consequence of the mechanism of reverse transcription and integration (see section on TPRT below), ORF2 protein and L1 RNA must both gain access to genomic DNA. Proteins larger than approximately 60 kDa are too large to enter the nucleus by passive diffusion through the nuclear pore (64). The ORF2 protein alone is predicted to be about 150 kDa. Therefore, access to genomic DNA must either occur by energy-dependent, active transport through a nuclear pore, or by entry during nuclear membrane breakdown at mitosis or meiosis. Although several mechanisms are known for the active nuclear import of proteins, the classical pathway is mediated by proteins called importins (also called karyopherins), which bind to specific amino acid sequences called nuclear localization signals (64, 153). Experiments suggest that the ORF1 protein does not contain any functional nuclear localization signals (E.M. Ostertag & H.H. Kazazian, Jr, unpublished data) and, as mentioned previously, ORF1 protein has not been detected in the nucleus by immunostaining techniques. ORF2 protein may contain one or more nuclear localization signals, but evidence for their function is lacking because of the difficulty in detecting full-length ORF2 protein. If the ORF2 protein does encode a functional nuclear localization signal, it would be an interesting case of an RNA gaining access to the nucleus by encoding its own nuclear import protein. Another possibility is that the ORF1 or ORF2 proteins bind an additional protein that itself contains a nuclear localization signal or that the L1 RNA is required for nuclear import. Also, retrotransposition may depend upon nuclear breakdown. If retrotransposition takes place only in dividing cells, this would be a second factor, along with reduced transcription, in greatly reducing insertions in differentiated, rarely dividing tissues.

Target Primed Reverse Transcription (TPRT)

L1 elements are likely reverse transcribed and integrated into the genome by a coupled reverse transcription/integration process called target primed reverse transcription (TPRT). TPRT was originally demonstrated for the R2 element, a site-specific, non-LTR retrotransposon found in arthropods (115). R2 retrotransposons have a single ORF that encodes a protein with Type II restriction endonuclease (228) and reverse transcriptase activity. Elegant in vitro experiments using a bacterially produced R2 protein demonstrated that the endonuclease domain of the protein cleaves the noncoding strand of its target site, a sequence in the 28S rRNA gene. The reverse transcriptase domain of the R2 protein then uses the free 3′-OH at the DNA nick as a primer and the R2 RNA as a template for the reverse transcription reaction. Reverse transcription of the RNA is followed by cleavage of the coding strand and integration. TPRT produces a perfect duplication of the original target site, which flanks the newly inserted element (115) (Figure 4, see color insert).

There are several reasons to believe that L1 elements use TPRT as their mechanism of reverse transcription and integration, although it has not yet been demonstrated definitively. First, recent in vitro experiments using Baculovirus-produced, full-length L1 ORF2 protein produce limited TPRT reactions (29). Second, L1 elements in the genome are often flanked by perfect 7- to 20-bp target site duplications, a typical consequence of the TPRT reaction. Lastly, the nucleotides at the predicted cleavage site are often T-rich, which are complementary to the polyA tail at the 3′ end of an L1 element, suggesting that they could indeed be used as a primer for reverse transcription of the L1 RNA.

The vast majority of L1 insertions in vivo are highly truncated at the 5′ end such that the average insertion length is only about 1 kb, or one sixth that of a full-length element (107). L1 truncation has long been explained by an inability of the L1 reverse transcriptase to copy the entire L1 RNA before disassociating from the RNA. Truncation may also be due to the action of a cellular RNAse H competing with L1 reverse transcriptase. Digestion of the RNA before the completion of reverse transcription followed by integration would result in an insertion truncated at the 5′ end.

Roughly 25% of recent L1 insertions also contain an inversion of a few hundred to fifteen hundred nucleotides of L1 sequence (161a). The inversion always involves the 5′ terminal end of the L1 element and is 5′ truncated itself. In other words, if L1 sequence is 5′-A-B-C-D-E-3′, then an inversion-containing insertion may be 5′-C-B-D-E-3′. The point of inversion may contain a deletion, a duplication, or neither. We suggest that inversion is a consequence of the L1 TPRT mechanism and describe a proposed model here.

If cleavage of the second DNA strand occurs before reverse transcription has been completed, an additional 3′ hydroxyl would be available for the priming of reverse transcription. This potential primer could invade the L1 RNA internally and prime reverse transcription at a site distinct from the reverse transcription occurring at the 3′ end of the L1 RNA. The L1 RNA template would therefore be primed by two different primers at two separate locations, a possibility that we call twin priming. Resolution of the RNA/cDNA structure that has undergone twin priming would produce a typical L1 inversion with a 5′ truncation (Figure 5, see color insert). Depending on the extent of reverse transcription from the primer at the 3′ end of the L1 RNA, the point of inversion would contain a deletion, a duplication, or neither. This model predicts all of the typical L1 inversion structures that are found in the genome database and does not predict structures that are not found (such as internal inversions).

If this model is correct, then the bases at the 3′ end of the internal primer should complement the bases on the L1 RNA template just proximal to the point of inversion (the bases at the orange arrow in Figure 5). This prediction is strongly supported by analysis of recent L1 insertions found in the genome database (161a). Additionally, cleavage of the second DNA strand must occur before reverse transcription has been completed. During in vitro experiments on the R2 TPRT process, the cleavage of the second DNA strand occurs after reverse transcription (115). The fact that R2 elements do not undergo L1-like inversions supports the possibility that the R2 and L1 TPRT mechanisms differ in this regard (T.H. Eickbush, personal communication).

The roughly 200-bp 3′UTR of human L1 appears to lack sequences that are important for reverse transcription, even though the 3′ 250 bps of R2 are critical for the reverse transcriptase activity of that element (114). Nearly all of this sequence can be deleted from human L1 elements with little effect on retrotransposition in HeLa cells (147). In addition, there are now many examples of retrotransposition of sequences flanking the 3′ ends of L1 elements in which these flanking sequences bear no resemblance to the L1 3′ UTR (60) (see section on L1-mediated transduction below). The L1 TPRT model predicts that the L1 polyA tail interacts with the RT domain of ORF2 protein during the initiation of reverse transcription, but the evidence on this point is indirect. The necessary pairing of the A-rich sequence at the 3′ end of an L1 element with the T-rich primer created at the integration site might explain the presence of A residues immediately after the L1 polyadenylation signal (see section on L1 RNA processing above). The A-rich regions might be positively selected over time if they are occasionally used during the priming of reverse transcription. The human SVA element, a nonautonomous retrotransposon thought to use the same L1 TPRT mechanism, also contains an A-rich region immediately following the presumed polyA signal (E.M. Ostertag & H.H. Kazazian Jr, unpublished data) (see section on genomic expansion sponsored by L1 retrotransposons below).

It has recently been demonstrated that mouse p40 (ORF1 protein) contains nucleic acid chaperone activity in vitro. Specifically, p40 promoted annealing of complementary DNA strands and aided strand exchange to form the most stable hybrids, facilitating the melting of imperfect duplexes and stabilizing perfect duplexes (125). The authors suggested that p40 protein might play a role in strand transfer during L1 reverse transcription. However, p40 protein, although readily detected in the cytoplasm, has yet to be detected within the nucleus.

IMPACT OF L1 RETROTRANSPOSONS ON THE MAMMALIAN GENOME

L1 retrotransposons have affected the genome in numerous ways, some beneficial and others detrimental (Figure 6, see color insert). However, the total effect of these actions has been major structural remodeling of the genome, occasionally altering gene expression. Here we present what is known about L1 and the genome, along with what has been proposed and remains unproven.

Genomic Expansion Sponsored by L1 Retrotransposons

There is considerable evidence that L1-encoded proteins preferentially act upon the L1 element that encoded them (*cis* preference). Strong indirect evidence comes from the following facts. Only about 1% of full-length L1s are active (172). If the retrotransposition machinery from the few active L1 elements could retrotranspose RNA from the many defective elements *in trans*, then one might expect most precursors of recent insertions to be inactive elements. However, this is not the case. Indeed, the precursors of three *de novo* human L1 insertions into the factor VIII, dystrophin, and CYBB genes have been isolated, and they are all active elements (42, 81, 135). Moreover, two full-length disease-causing insertions in humans, $L1_{\beta\text{-thal}}$ and $L1_{RP}$, and two full-length disease-causing insertions in mice, $L1_{spa}$ and $L1_{orl}$, are all active L1 elements (93, 151). Direct evidence of *cis* preference comes from the work of two groups using the retrotransposition assay (49, 214). Wei et al. have found that mutations in either ORF1 or ORF2 can be *trans* complemented by an active L1 at less than 1% of control levels. How the nascent proteins interact with L1 RNA or each other is a mystery. Perhaps the ribonucleoprotein particle forms as the proteins come off the ribosome, thereby limiting their availability to other L1 RNAs. In any case, *cis* preference in L1 retrotransposition is important in greatly limiting genomic expansion by defective L1s. On the other hand, preferential propagation of active L1s increases the likelihood that L1 elements will remain active in mammalian genomes (107, 214). An exception to the *cis* preference rule is the expansion of Alu elements, SVA elements, and processed pseudogenes, which almost certainly results from low-level *trans* complementation by L1 endonuclease and reverse transcriptase activities.

The 1.1 million Alu elements in the human genome contain roughly 300 bps and are composed of two similar 150-bp segments, the 3' half of which ends in a polyA tail (169) (Figure 1). A subset of human-specific Alus from four closely related subfamilies, Alu Y, Ya5, Ya8, and Yb8, are retrotranspositionally competent (38). Most remaining Alu subfamilies are either not transcribed or weakly transcribed (180). Alu elements are concentrated in GC-rich regions of the human genome, but young Alus have a more uniform distribution across the genome (107). This latter fact has suggested that there might be positive selection for Alu sequences in regions of substantial gene expression, i.e., GC-rich DNA. Schmid has suggested that an increase in Alu transcription promotes general translation of proteins under conditions of cellular stress (175). His group has shown that Alu-mediated inhibition of PKR (double-stranded RNA-regulated protein kinase) activation results in an increase in translation (27). Under this hypothesis, Alus could be under positive selection in the readily transcribed, open chromatin near genes. This could explain the large number of "old" Alus in gene-rich GC-rich regions.

Considerable circumstantial evidence suggests that L1 elements provide key enzymatic activities for Alu insertion. (*a*) Alu sequences end in a polyA tail, which is thought to be required for L1-mediated TPRT. (*b*) Both types of elements are usually flanked by a target site duplication of 7–20 bp, which is probably created by TPRT. (*c*) The insertion sites of Alu elements have the same general consensus sequence as the consensus sequence for L1 endonuclease (88). It was previously difficult to explain why Alu elements are concentrated in GC-rich DNA, while L1s are concentrated in AT-rich DNA, if the L1 machinery was responsible for the retrotransposition of both elements (100). The observation that recently inserted Alus and L1s are both concentrated in similar genomic sites, i.e., in AT-rich DNA, and the finding that the GC-rich distribution of Alu elements likely represents post-insertion selection, eliminates this argument against the role of L1 machinery in Alu retrotransposition (107).

Processed pseudogenes are DNA copies of RNA polymerase II-derived mRNAs that have been inserted into the genome at locations that resemble L1 target sites. These pseudogenes lack intronic RNA, usually have polyA tails, and are flanked by 7–20-bp target site duplications (209). Their sequences make up roughly 0.5% of the genome (44, 107). Two groups have presented evidence that cotransfection with an active L1 element in addition to a cDNA sequence tagged with a retrotransposition marker cassette can lead to insertion of the tagged cDNA at a low frequency (49, 214). The inserted sequences characterized by one of these groups appeared quite similar to those of endogenous processed pseudogenes (214). In addition, mutations in the L1 ORF1 or ORF2 proteins eliminated processed pseudogene formation, providing strong evidence for the role of both L1 proteins in this process (49, 214).

SVA elements are nonautonomous retrotransposons present in 2000 to 5000 copies in the human genome (158, 183; E.M. Ostertag & H.H. Kazazian Jr, unpublished data). At their 5′ ends, full-length SVA elements have up to 40 hexameric (CCCTCT) repeats. This region is followed by (*a*) a region containing antisense Alu sequence, (*b*) a VNTR region containing multiple copies of a 35–50-bp repeat, (*c*) a SINE-R sequence with similarity to the 3′ end of an endogenous retrovirus, and (*d*) a polyadenylation signal and polyA tail (183) (Figure 1). Since full-length SVA elements are all >89% similar in sequence to each other (E.M. Ostertag & H.H. Kazazian Jr, unpublished data) and are present only in hominoid primates (92), these elements are quite young by evolutionary standards, probably <15 My old. Many characteristics of SVA insertions are reminiscent of L1 insertions. Some insertions are 5′ truncated, they end in a polyA tail directly following a polyA signal, and they are flanked by target site duplications that are similar in length and sequence to L1 TSDs (E.M. Ostertag & H.H. Kazazian Jr, unpublished data).

Alus, processed pseudogenes, and SVA elements share features that make them candidates for retrotransposition by the L1 machinery. They are all likely transcribed in germ line cells where L1 elements are expressed and are able to retrotranspose. Furthermore, they all end in a polyA tail and are flanked by L1-like TSDs, suggesting that they are all inserted into the genome by TPRT. Of the three elements, only Alus and SVAs (see section on human disease below) have resulted in *de novo* insertions, indicating that they are currently retrotranspositionally active. Interestingly, Alus and SVAs both have Alu sequence components. Perhaps the Alu sequences are important in the *trans*-complementation by L1, placing the element RNA in close proximity to the L1 machinery either on the ribosome or within a ribonucleoprotein particle (15).

L1 Retrotransposons Can Cause Human Disease

L1 retrotransposons can cause human disease by a number of mechanisms, including promoting unequal homologous recombination, direct L1 insertion into genes, and providing the machinery for insertion of other retrotransposons into genes (Figure 6). Homologous recombination due to mispairing of repeated sequences has been rarely observed for L1 elements and more commonly seen for Alu elements (37). L1 mispairing and unequal crossing over has caused three recent deletions producing disease (25, 178; R Gatti, personal communication). It also produced an ancient duplication having important evolutionary consequences—the duplication of γ-globin genes that occurred in New World monkeys (54). Over 40 instances of Alu mispairing and crossing over leading to deletion have been reported in various disease states (37).

Several explanations have been proposed as to why homologous recombination is observed more frequently among Alu elements than it is among L1 elements (37). First, Alus contain sequences that are recombinogenic in other contexts (170); however, their role in Alu/Alu recombination is debatable (184). Second, the average genomic distance between two L1s is greater than that between two Alus (107), making L1/L1 events both less likely to occur and more likely to result in lethal mutations. Third, L1s tend to reside in more AT-rich DNA, while Alus reside in GC-rich DNA (107). Since AT-rich DNA is relatively gene poor, L1/L1 homologous recombination events may occur more frequently than suspected, but rarely result in deletion of gene sequences.

Recent insertions of retrotransposons are associated with 35 isolated cases of a variety of disease states in human beings (Table 1). Of these, 13 are L1 insertions, 19 are Alu insertions, 2 are SVA insertions, and 1 is an insertion of a sequence transduced by an SVA element. Another very recent L1 insertion is a normal variant in a family segregating hemophilia A (JH-25 in Table 1). There have not been recent insertions of processed pseudogenes, although some of these sequences are polymorphic as to presence within human populations and are less than 100,000 years old (6).

Although 16 of the 35 total insertions (44%) have occurred into the X chromosome, most of the excess of X chromosome insertions is due to L1 elements. Eleven of 14 recent L1 insertions (79%) are into X chromosomal genes, whereas 6 of 19 Alu insertions (32%) and 1 of 2 SVA insertions are into the X chromosome. Some L1 insertions causing X-linked disease are *de novo* events, i.e., mothers of affected males are noncarriers. A detection bias clearly exists for genes whose disruption by a single hit causes disease, i.e., X-linked and autosomal dominant disorders.

TABLE 1

Human insertions
1. Non-LTR retrotransposons

L1 insertions

Inserted element	Disrupted gene	Insertion size	3′ Transduction (yes or no)	Insertion site	Orientation of insertion	Reference
JH-27	Factor VIII	3.8 kb	No	Exon	Sense	(91)
JH-28	Factor VIII	2.2 kb	No	Exon	Sense and rearranged	(91)
JH-25	Factor VIII	681 nts	No	Intron	Sense	(222)
APC	APC	538 nts	Yes	Exon	—	(139)
Dystrophin	Dystrophin	608 nts	No	Exon	Sense	(154)
Dystrophin	Dystrophin	878 nts	No	Exon	Sense	(E Bakker & G van Omenn, personal communication)
JH-1001	Dystrophin	2.0 kb	Yes	Exon	Sense and rearranged	(81)
L1$_{\beta\text{-thal}}$	β-Globin	6.0 kb	No	Intron	Antisense	(40)
L1$_{XLCDM}$	Dystrophin	524 nts	No	Exon	Antisense	(231)
L1$_{RP}$	RP2	6.0 kb	No	Intron	Antisense	(176)
L1$_{CYB}$	CYBB	1.7 kb	Yes	Exon	Sense and rearranged	(134, 135)
L1$_{CYB}$	CYBB	940 nts	No	Intron	Sense	(133)
L1$_{FCMD}$	Fukutin	1.1 kb	No	Intron	Sense	(99)
L1$_{FIX}$	FIX	520 bp (460 bp of L1)	No	Exon	Sense — No target site duplication—7 nts deletion	(226)

2. Nonautonomous retrotransposons

Alu insertions

Gene	Disorder	Alu subfamily	Insertion site	Orientation	De novo (yes or no)	Reference
NF1	Neurofibromatosis	Ya5	Intron	Antisense	Yes	(212)
BCHE	Acholinesterasemia	Yb8	Exon	Sense	No	(149)
F9	Hemophilia B	Ya5	Exon	Sense	Yes	(211)
CASR	Familial hypocalciuric hypercalcemia	Ya4	Exon	Antisense	No	(85)
BRCA2	Breast cancer	Y	Exon	Sense	—	(138)
APC	Hereditary desmoid disease	Yb8	Exon	Sense	No	(69)
BTK	X-linked agammaglobulinemia	Y	Exon	Antisense	Yes	(109)
IL2RG	X-linked severe combined immunodeficiency	Ya5	Intron	Antisense	No	(109)
EYA1	Branchio-oto-renal syndrome	Ya5	Exon	Antisense	Yes	(1)
FGFR2	Apert syndrome	Ya5	Intron	Antisense	Yes	(157)
FGFR2	Apert syndrome	Yb8	Exon	Antisense	Yes	(157)
ADD1	Huntington disease	—	Intron	Sense	No	(58)
GK	Glycerol kinase deficiency	Ya5	Intron	Antisense	—	(233)
C1NH	C1 inhibitor deficiency	Y	Intron	Sense	No	(197)
PBGD	Acute intermittent porphyria	Ya5	Exon	Antisense	No	(150)
MIVI-2	Associated with leukemia	Ya5	?	?	Yes (somatic?)	(46)
FIX	Hemophilia B	Ya3a1	Exon	Antisense	No	(226)
FIX	Hemophilia B	—	Exon	Sense	No	(224)
FVIII	Hemophilia A	Yb8	Exon	Antisense	No	(224)

(Continued)

TABLE 1   (*Continued*)

SVA-related insertions

Disrupted gene	Insertion type	Reference
FcMB	Full-length SVA	(96)
BTK	5′ Truncated SVA (SINE R)	(168)
α-spectrin	SVA-mediated transduction	(71)

Mouse insertions

1. LTR-retrotransposons

Intracisternal A-particle (IAP) insertions

Locus	Allele or disorder name	Reference
Agouti	A^{iy}	(43)
Agouti	A^{vy}	(43)
Agouti	A^{vapy}	(137)
Agouti	A^{rivy}	(7)
Mgca-mahogany	Mg	(67)
Mgca-mahogany	mg^{L}	(67)
Pale ear	Hermansky-Pudlak	(56)
Vibrator	Vibrator	(70)
LamB3	Epidermolysis bullosa	(106)
Tyrosinase	Somatic mosaicism	(223)
Eyal	BOR syndrome	(86)
Fused	Fused	(210)
Fused	Knobby	(210)

DNA Transposons

ITR **Transposase** **ITR**

DR Tc1-mariner (1.4 kb) **DR**

Retrotransposons
-Autonomous
a) LTR

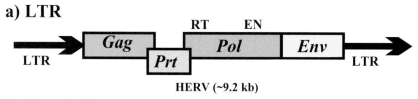

RT   EN

Gag **Pol** **Env**

LTR **Prt** LTR

HERV (~9.2 kb)

b) Non-LTR

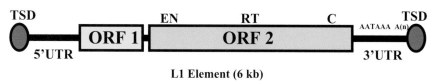

TSD EN   RT   C TSD

ORF 1 **ORF 2** AATAAA A(n)

5'UTR 3'UTR

L1 Element (6 kb)

-Non-Autonomous

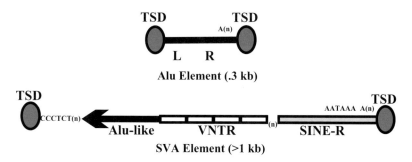

TSD   TSD
A(n)

L   R

Alu Element (.3 kb)

TSD TSD
CCCTCT(n) AATAAA A(n)

Alu-like **VNTR** (n) **SINE-R**

SVA Element (>1 kb)

See text page C-2

Figure 1 (page C-1) Structure of mammalian transposable elements. Mammalian transposable elements consist of DNA transposons and retrotransposons. DNA transposons are flanked by inverted terminal repeats (ITRs) and have a single open reading frame (ORF) that encodes a transposase. They are also flanked by short direct repeats (DRs) created during the integration process. An example of a DNA transposon is the Tc1-mariner transposon. Retrotransposons can be divided into autonomous and nonautonomous elements based upon whether they have ORFs (*colored rectangles*) that encode proteins required for their retrotransposition. Autonomous retrotransposons are classified as (*a*) long terminal repeat (LTR) or (*b*) non-LTR. An example of an LTR retrotransposon is the human endogenous retrovirus (HERV). The LTR retrotransposons are flanked by LTRs and have partially overlapping ORFs for their group-specific antigen (*gag*), protease (*prt*), polymerase (*pol*), and envelope (*env*) genes. Also shown are the reverse transcriptase (RT) and endonuclease (EN) domains of the polymerase protein. An L1 element is an example of a non-LTR retrotransposon. L1s consist of a 5′ untranslated region (5′UTR), two ORFs separated by a short intergenic region, a 3′UTR, a polyA signal (AATAAA), and a polyA tail (A(n)). L1 elements are often flanked by 7–20-bp target site duplications (TSD)s. Shown are the RT and EN domains of the ORF2 protein, as well as a conserved cysteine-rich motif (C). The Alu element and the SVA element are examples of nonautonomous retrotransposons. Alu elements contain two similar sequences, the left monomer (L) and the right monomer (R) and end in a polyA tail. SVA elements consist of CCCTCT hexameric repeats, an antisense Alu-like region, a VNTR region, a region (SINE-R) with homology to the end of a HERV, a polyA signal and a polyA tail. Alus and SVAs are flanked by L1-like TSDs. The approximate size in kilobases (kb) of a full-length element of each example is indicated in parentheses.

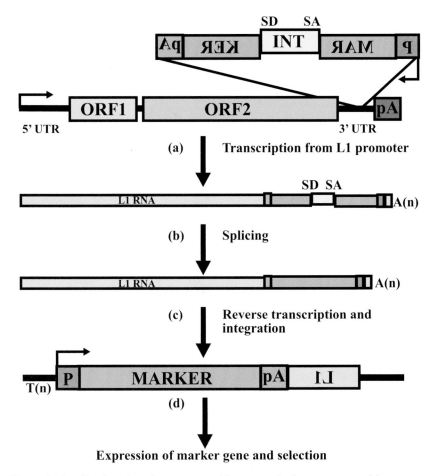

Figure 2 A cell culture-based retrotransposition assay. In the retrotransposition assay, a full-length L1 element is tagged with a retrotransposition cassette. A retrotransposition cassette consists of a marker gene (*light green rectangle*) interrupted by an intron (*light yellow rectangle*) in the opposite transcriptional orientation. The splice donor (SD) and splice acceptor (SA) sites are indicated. Transcripts directed from the marker's promoter (*light gray rectangle and bent arrow*) cannot remove the intron by splicing and will not produce functional protein. The entire cassette is cloned into the 3′ untranslated region (3′UTR) of an L1 element in the orientation opposite that of the L1 promoter (*5′ UTR and bent arrow*). The marker can only become activated (*a*) when a full-length L1 element is transcribed (L1 RNA is represented in *pink*, the color of marker components have been maintained), (*b*) the intron is removed by splicing, and (*c*) the RNA is reverse transcribed and integrated into the genome. (*d*) Transcripts directed from the marker's promoter after a retrotransposition event produce functional protein. The tagged-L1 construct is transiently transfected into cultured cells. Typical markers include the neomycin phosphotransferase (*neo*) gene or the Enhanced Green Fluorescent Protein (EGFP) gene. Positive cells are selected by growth in G418 containing media or by analysis for fluorescence, respectively (147, 162).

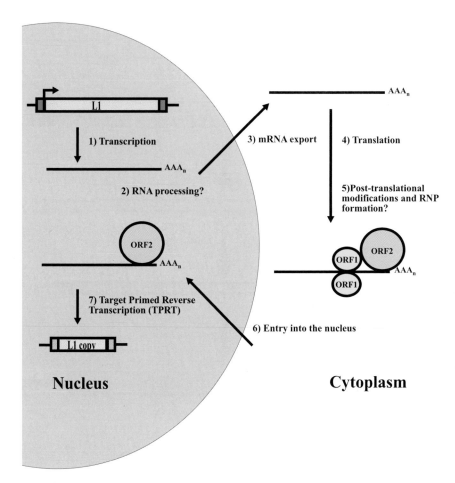

Figure 3 The steps in L1 retrotransposition. A full-length active L1 element is transcribed from its internal promoter (*bent arrow*) to produce a bicistronic mRNA. It is currently unknown if the RNA undergoes processing or how the RNA is exported from the nucleus. Once in the cytoplasm, the ORF1 and ORF2 proteins are translated and specifically function on the RNA that transcribed them (*cis* preference). At least one L1 RNA molecule, one ORF2 molecule, and one or more ORF1 molecules may assemble into a ribonucleoprotein (RNP) complex that is an intermediate in retrotransposition. Both the ORF2 protein and associated L1 RNA must gain access to the nucleus, where the L1 RNA is reverse transcribed and integrated into a new genomic location by a process called target primed reverse transcription (TPRT). Many L1 elements undergo 5′ truncation or 5′ inversion and truncation during the TPRT process, resulting in an inactive DNA copy of the original element. The TPRT process creates 7–20-bp target site duplications that flank the L1 element (*blue and pink rectangles*).

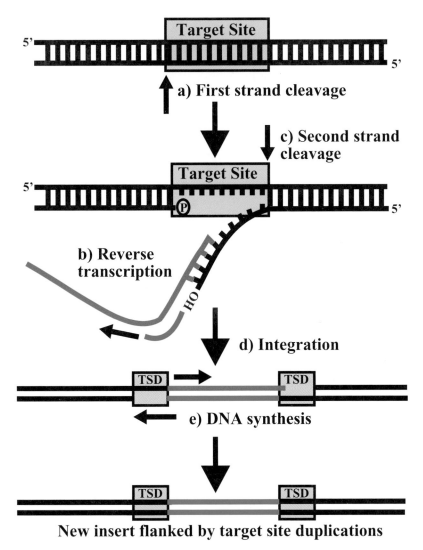

Figure 4 Target primed reverse transcription (TPRT). The non-LTR retrotransposons are thought to integrate by TPRT. This TPRT model is based upon the mechanism worked out *in vivo* for the R2 retrotransposon (115). (*a*) During TPRT, the retrotransposon's endonuclease cleaves one strand of genomic DNA at its target site (*blue rectangle*), producing a 3′ hydroxyl (OH) at the nick. (*b*) The retrotransposon RNA (*red line*) inserts at the nick and the retrotransposon's reverse transcriptase uses the free 3′OH to prime reverse transcription. Reverse transcription proceeds, producing a cDNA of the retrotransposon RNA (*green line*). (*c*) The endonuclease cleaves the second DNA strand of the target site to produce a staggered break. (*d*) The cDNA inserts into the break by an unknown mechanism. (*e*) Removal of RNA and completion of DNA synthesis produces a complete insertion flanked by target site duplications (TSDs).

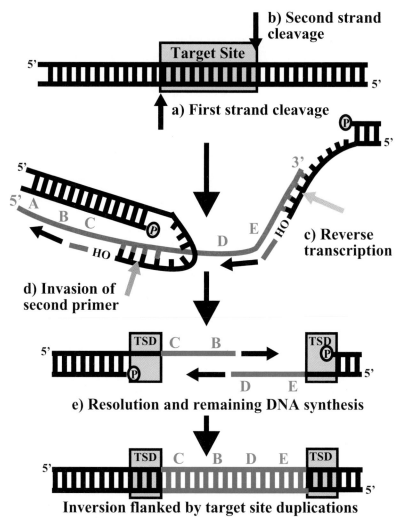

Figure 5 Twin priming. This model demonstrates our proposed mechanism by which L1 5′ inversions are created during the TPRT process. (*a*) The L1 endonuclease cleaves one strand of genomic DNA at its target site (*blue rectangle*), creating a 3′ hydroxyl (OH) at the nick. (*b*) The endonuclease performs second strand cleavage before reverse transcription has been completed, creating a second 3′ hydroxyl and staggered break. (*c*) The L1 RNA inserts into the break and the L1 reverse transcriptase uses the first 3′OH to initiate reverse transcription (*pink arrow*). (*d*) The second 3′OH invades the RNA internally and is used to prime reverse transcription at a second site (*orange arrow*). (*e*) Resolution of the RNA/cDNA structure and completion of DNA synthesis produces an insertion with a 5′ inversion. The entire insertion is flanked by target site duplications (TSDs). The L1 RNA sequence is represented by 5′-A-B-C-D-E-3′. After the inversion, the insertion sequence is 5′-C-B-D-E-3′.

a) Retrotransposition (*cis*)

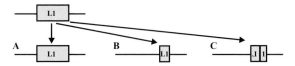

b) Retrotransposition (*trans*)

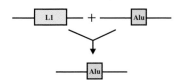

c) Insertional mutagenesis

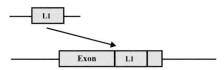

d) Unequal Homologous Recombination

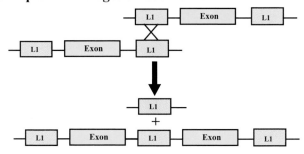

e) L1-mediated transduction

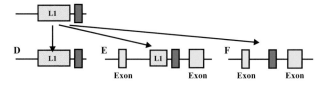

f) Effects on gene expression

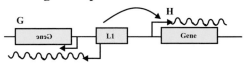

See text page C-8

Figure 6 (page C-7) Impact of L1 retrotransposons on the mammalian genome. L1 elements have had a variety of effects on the human genome. (*a*) L1 elements expand the genome by their retrotransposition. L1 elements replicate by a "copy and paste" mechanism. The L1 proteins work preferentially on the RNA that transcribed them (*cis* preference). Therefore, only full-length elements with two open reading frames are active. The active L1 is first transcribed *in situ*, and then the RNA is reverse transcribed and integrated into a new location. L1 elements can integrate as full-length copies (*A*), however, they frequently 5′ truncate (*B*), or 5′ invert and truncate (*C*), producing an inactive copy of the original element. (*b*) Some nonautonomous retrotransposons, such as Alu, are an exception to the *cis* preference rule and are retrotransposed *in trans* by the L1 machinery, further contributing to genomic expansion. (*c*) L1 elements occasionally insert into gene sequences, thereby causing genetic disorders. (*d*) L1 elements can also cause disease by creating deletions and duplications after unequal homologous recombination. (*e*) L1 elements often bypass their own polyA signal and use a downstream signal. This results in the transduction of 3′ sequences, potentially gene exon sequences (*green rectangle*), upon their retrotransposition (*D*). Retrotransposition of a 3′ exon into another gene could result in exon shuffling (*E*). L1 elements could even shuffle exons without leaving evidence of themselves if they severely 5′ truncate during retrotransposition of a 3′ exon (*F*). (*f*) The expression of genes can be affected by the presence of an L1 element. Some L1s have antisense Pol II promoters, which can affect the expression of nearby genes (*G*). Other L1s have acquired an enhancer function and can regulate expression of local genes (*H*).

Early transposon (Etm) insertions

Locus	Allele or disorder name	Reference
ob	ob^{2j}	(145)
T	Mesoderm formation	(75)
Muscle chloride channel	Myotonic	(196)
Fas	lpr	(2)
MIP	Cataract	(187)
Adenylyl cyclase (Adcy1) Type 1	Barrelless	(108)
Fidgetin	Fidget	(31)
Nude (whn)	nu-BC	(78)
Tyrosinase (Tyr)	c-3BC	(78)
Gli 3	Polydactyly Nagoya	(203)
Stargazin (neuronal Ca^{2+}–channel γ subunit)	Stargazer	(110)

MaLR insertions

Locus	Allele or disorder name	Reference
Cholesterol homeostasis gene	Niemann Pick C (NPC1)	(113)

2. Non-LTR retrotransposons

L1 insertions

Gene	Disorder or mouse name	Insertion length	Subfamily	Reference
Glycine receptor β-subunit	Spastic	Full-length (7.5 kb)	T$_F$	(94, 148)
Reelin	Orleans reeler	Full-length	T$_F$	(200)
S$_{cn}$8a	Med	<100 bp	Unknown	(97)
Beige ($\beta\gamma$)	Chediak-Higashi	1.1 kb	T$_F$	(165)
Mitf [mi-bw]	Black-eyed white	Full-length	T$_F$	(227)

The recent L1 insertions have certain characteristics that mirror those of older L1s residing in the human genome. All but one of these L1 insertions is flanked by typical target site duplications. Twelve of the 14 insertions (86%) contain truncated L1 elements ranging from 500 bp to 3.8 kb in length, while two are full-length with intact ORFs. Three of the 14 insertions are associated with transduction of 3′ flanking sequences (81, 134, 139) (see section on L1-mediated transduction below).

Of the 13 disease-producing L1 insertions, 9 are into gene exons and presumably introduce nonsense codons into the coding sequence or produce skipping of the disrupted exon. The four disease-producing L1 insertions into introns cause exon skipping, decreased transcription, or decreased stability of the primary transcript. For example, an L1 insertion into an intron of the fukutin gene in Fukuyama-type muscular dystrophy patients resulted in alternative splicing (99), as did an L1 insertion into an intron of the CYBB gene in a patient with chronic granulomatous disease (133). Full-length insertions into a β-globin gene intron in a patient with β-thalassemia and into the *RP2* gene in a retinitis pigmentosa patient caused low or absent mRNA levels without aberrant splicing (40, 176).

While 13 of the L1 insertions occurred either in the germ line or very early in development, an L1 insertion into an exon of the adenomatous polyposis coli (APC) gene in a colon cancer (mentioned previously) was a somatic event in dedifferentiated cells. The insertion clearly occurred in the cancer tissue since it was not present in normal colonic tissue of the patient (139).

A striking observation concerning recent human L1 insertions is that nearly all arise from a single, relatively small, subset of human L1s called the Ta subset. This subset is characterized by substitution of ACA for GAG 92–94 bp upstream of the polyA tail (172, 189). Of the 14 recent L1 insertions, 13 (93%) are Ta subset elements. The Ta subset has recently been subdivided into Ta-0 and Ta-1 subgroups based on nucleotide sequence (17). The Ta-1 subset is younger and presently accounts for the majority of Ta elements; about 70% of the Ta-1 insertions are polymorphic as to presence in the human population, whereas only about 30% of the Ta-0 insertions are polymorphic (17). The fourteenth recent insertion is a pre-Ta element containing ACG instead of ACA at the diagnostic trinucleotide (JH-28 in Table 1) (91). Blot hybridization estimates placed the number of full-length Ta elements in the diploid human genome at about 200 copies (172). Sassaman et al. isolated a number of full-length Ta elements from a genomic library and found that 50% had two intact ORFs and roughly one half of these, or about one quarter of the total, were retrotranspositionally active in the cell culture assay. This led to an estimate of 30 to 60 active L1s in the human genome (172). In an update of the recent analysis of the working draft sequence covering roughly 90% of the human genome (the euchromatic portion), there were 57 full-length Ta elements and 22 full-length pre-Ta elements with intact ORFs (107; R.M. Badge & J.V. Moran, personal communication). If we assume that 50% of the 57 Ta elements and 10% of the pre-Ta elements are retrotranspositionally active, then there are 30 active L1s in roughly 90% of the haploid genome, or about 65 in the full diploid genome, an estimate that is similar to that of Sassaman et al. (172).

We can use the total number of recent insertions along with the number of non-recurrent mutations in the Human Gene Mutation Database (102) (http://archive. uwcm.ac.uk/uwcm/mg/hgmd0.html) to estimate the fraction of human mutations that are retrotransposition events. There are reasons why this estimate may be too low, e.g., the inability of investigators to detect all insertions greater than 1 kb in length by PCR, or too high, e.g., the failure of the database to count recurrent mutations. However, the estimate calculated in this way is roughly 1/600, with 1/1100 as the estimate for Alu insertions and 1/1500 as the estimate for L1 insertions (89).

The frequency of retrotransposition events per individual has also been estimated in a number of ways, using mutation rates in specific genes and overall mutation rates in germ cells. All estimates range between 1 retrotransposon insertion in every 4 individuals and 1 insertion in every 100 individuals (36, 89, 226). For the L1 retrotransposon, the estimate translates into 1 insertion in every 10 to 250 individuals.

L1 Retrotransposition Can Cause Mouse Disease

Although it is likely that in the mouse unequal homologous recombination events between either non-LTR retrotransposons or LTR-retrotransposons exist, none has yet been reported. On the other hand, retrotransposon insertions account for a substantial proportion of disease-producing mutations in the mouse. Mice have many more active retrotransposons of different types than humans do, and have a correspondingly much larger fraction of spontaneous mutations due to insertion of retrotransposons.

In contrast to humans, mice are burdened by insertions of LTR retrotransposons whose origins derive from endogenous retroviruses. These are intracisternal A particles (IAPs), early transposons (Etns), and mammalian LTR-retrotransposons (MaLRs) (Table 1). The estimated 1000–2000 IAPs in the mouse genome are defective retroviral-like elements with *gag-*, *pol-*, and *env*-like similarity regions in their sequence (103). Most IAPs, but not all, lack intact ORFs and nearly all have major deletions in their *env* genes. Although low-level retrotransposition of a defective IAP element has been demonstrated in cultured cells (74), no autonomous retrotransposable IAP has been isolated and characterized to date. There have been at least 13 instances of IAP insertions in spontaneous mouse disease, all of which involve defective IAPs. Presumably IAP RNA is reverse transcribed within cytoplasmic particles by a retroviral reverse transcriptase, but the source of the enzymes and other machinery necessary for IAP mobilization is unknown.

The origin of Etns has recently been elucidated (118). These elements contain two LTRs with roughly 500 bp of sequence unrelated to retroviral sequence between the LTRs (194). Mager & Freeman have shown that the LTRs and a portion of the intervening sequence are derived from murine Type D retroviruses. At one point, Type D retroviruses underwent deletion of most of their internal sequences and later acquired new genomic sequences. Presumably, present-day Etn elements use the reverse transcriptase of their ancestor, the Type D retrovirus, to carry out their own retrotransposition (118). Etn insertions account for at least 11 instances of mouse disease.

MaLRs are the largest family of mouse retroviral-like elements. They have an unusual mosaic structure consisting of an origin region repeat (ORR1) and a mouse transposon element (MT). They have an open reading frame of 1.3–1.6 kb internal to LTRs. The ORF does not encode reverse transcriptase or any other apparent protein (190). A single insertion of a partially deleted MaLR has been found (113). Although there are less than 5000 IAPs, Etns, and MaLRs in toto in the mouse genome, insertions of these elements account for roughly 7% to 8% of all mutations in the mouse.

L1 insertions make up another 2% to 3% of mutations in the mouse. L1 retro-transposition events are responsible for spontaneous disease in five mouse lines, the spastic mouse, the Orleans reeler mouse, the black-eyed white mouse, the beige mouse, and the med mouse (Table 1). Three of these five disorders are due to insertion of full-length L1s, $L1_{spa}$, $L1_{orl}$, and $L1_{bw}$. Of the five L1 insertions, four are derived from T_F elements, a young and expanding subfamily, and the fifth is too short to classify (59). Many T_F elements have retrotransposed so recently that they are highly polymorphic as to presence in various mouse subspecies and lab strains. *Mus spretus* and *Mus musculus* appear to have similar numbers of T_F elements in their genomes, but the data suggest that many, if not all, of these elements are present at different locations in the genomes of these two subspecies (34).

Although at least 4 of 5 recent L1 insertions in the mouse belong to the T_F subfamily, the A and G_F subfamilies also contain a number of active elements. The number of active mouse L1 elements has been estimated by the cell culture assay to be 1800 T_F (34), 900 A, and 400 G_F (59), for a total of 3100. This number is 50 to 60 times the estimate of active human L1s, and is close to the excess of the proportion of L1 insertions causing mouse disease compared to that producing human disease (2.5% vs. 0.07%, or 35-fold).

L1-Mediated Transduction

Because cleavage of L1 transcripts at the polyA site is often inconsistent, sequences flanking L1 3′ ends may be carried along in a retrotransposition event (146). These "stowaway" sequences are called transduced sequences. Holmes et al. first recognized an L1 transduction event, a 3′ transduction of over 500 bp included with an L1 insertion into the human dystrophin gene (81). Since then two other 3′ transductions have been recognized in association with recent L1 disease-producing insertions in humans (134, 139). Moran et al. demonstrated 3′ transduction experimentally in the cell culture assay. They placed the retrotransposition marker cassette 3′ of the L1 polyA signal, and showed that L1s are able to retrotranspose sequences from their 3′ flanks to new genomic sites. Further, L1s could retrotranspose a promoterless marker cassette into a transcribed gene, leading to formation of new fusion proteins (146). The results indicated that exons downstream of active L1s could be shuffled into new sites, thereby creating new genes. When human genome databases were analyzed, it turned out that roughly 20% of L1 insertions contain 3′ transduced sequences (60, 166). In humans, these transductions range from 30 bp to 1 kb and may account for 25 Mbp or about 0.8% of the haploid genome. In mice, 3′ transductions have been observed in about 10% of L1 insertion events, ranging from 500 bp to 3 kb (60). Because there is usually substantial truncation of L1 sequence associated with L1 retrotranspositions, 3′ transductions could easily produce insertions completely lacking in L1 sequence (Figure 6). In addition, inversions within the transduced sequence create the possibility of a wide variety of inserted sequences originating from a single 3′ flank of a retrotransposon.

Since retrotransposons can be transcribed from upstream promoters, sequences flanking their 5′ ends can also be observed following retrotransposition. A few 5′ transductions have now been found associated with insertions of L1s in the human genome (107).

Effects of Retrotransposon Insertions on Gene Expression

L1 sequences alter expression of some human genes (Figure 6). Three examples are a proposed enhancer activity for the apolipoprotein Lp(a) gene (229), an age-regulatory activity for the factor IX gene (104), and a locus control region activity for the growth hormone gene cluster (185). However, all of the L1s involved have considerable sequence differences from the L1 consensus sequence. Thus, L1s in general do not appear to contain sequence involved in gene regulation.

Although most L1 elements lack enhancer activity, pol II promoter activity has recently been discovered on the antisense strand of L1 DNA between nucleotides 400 to 600 within the 5′ UTR (195) (Figure 6). Fifteen cDNAs were isolated from a human teratocarcinoma cell line that contained L1 5′ UTRs spliced to the sequences of known genes or non-protein coding sequences. Four selected chimeric transcripts were found in total RNA of other cell lines. The author suggests that many human L1s contain an antisense promoter that is capable of interfering with the normal expression of neighboring genes, and that this type of transcriptional control may be quite common. Dispersion of L1 elements may have provided many opportunities for pol II transcription at new genomic locations. In an analogous situation, the mouse B2 SINE element contains pol II promoter activity, which is responsible for the transcription of at least one gene, *Lama 3* (53).

It has recently been proposed that low-level transcription of L1 elements in a subset of somatic cells may affect the expression of neighboring genes. Individual variation in the proportion of cells with retrotransposon transcription, in the location of the retrotransposons, and in the level of transcription could, in theory, lead to individual variation in susceptibility to oncogenesis or complex diseases (218). Although no experimental evidence supports this hypothesis, it would be interesting to study the effect that polymorphic L1 elements may have on neighboring gene expression.

A Proposed Role for L1 Retrotransposons in X Chromosome Inactivation

In humans, the density of L1 sequences on the X chromosome is twice that of the average density of L1 sequences on autosomes (26% of total sequence versus 13%) (10). In every somatic cell of the mammalian female, one of the two X chromosomes is mostly inactivated (73). Lyon has proposed that L1 elements serve as "booster stations," helping to propagate the signal transmitted by *Xist* RNA (116, 117). *Xist* RNA is thought to play an important role in X inactivation because it is expressed from (22), and interacts specifically with (164), the inactive X chromosome. The evidence for the "Lyon repeat hypothesis," and an alternative version in which X chromosome heterochromatization spreads from one L1 to another through physical interaction(121), is circumstantial. (*a*) There is significant clustering of L1s around the X inactivation center (10). (*b*) L1s are in short supply in regions of the X chromosome that escape inactivation (10). (*c*) At sites of X-autosome translocations in mice, there is a positive correlation between the number of L1 elements on the autosome and the extent of heterochromatization of autosomal genes (116).

Non-LTR Retrotransposon RT and Cellular Telomerase

The relationship of the RT of non-LTR retrotransposons to the catalytic subunit of telomerase is quite striking (111, 136). In a mechanism similar to the TPRT reaction of non-LTR retrotransposons, telomerase adds deoxyribonucleotides to the ends of chromosomes using the 3'OH end of a DNA strand as primer and a telomerase-associated RNA as template (48). Telomerase has a number of sequence domains similar to those of retrotransposon RTs, along with an additional domain not found in the RTs (111). Phylogenetic analysis suggests a close relationship between these enzymes (120), but there is controversy as to whether the catalytic subunit of telomerase is derived from retrotransposon RT or vice versa (48, 152). We favor the idea that eukaryotic cells recruited retrotransposon RT to acquire telomerase activity. The evolutionary age of non-LTR retrotransposons goes along with the very early eukaryotic origins of telomerase. A phylogenetic tree of eukaryotic RTs rooted by prokaryotic mobile elements also suggests that telomerase RT was derived from retrotransposon RT (120). There are other interesting examples of human proteins that have either evolved from transposable element proteins or have incorporated transposable element protein domains during their evolution (107, 191), notably the RAG proteins, which are responsible for V(D)J recombination (5, 76).

A Proposed Role for L1s in DNA Repair

In the late 1980s, Edgell and colleagues proposed that double-strand break (DSB) repair is a major role for L1 insertions (47). Such a role for L1s would maintain the integrity of the genome and have important evolutionary consequences. One would expect that L1 insertions into genomic double-stranded breaks would not be flanked by the perfect target site duplications that are created by the action of the L1 endonuclease during the TPRT process. Since many L1 elements in the genome have perfect target site duplications and because mutations in the active site residues of the L1 endonuclease greatly reduce retrotransposition capability in the cell culture assay, it is not likely that L1 elements play a major role in DSB repair in humans. However, a minor endonuclease-independent pathway may exist for L1 insertions. Indeed, examples both in humans and in mice may represent DSB repair mediated by endonuclease-independent retrotransposition of L1 elements (23, 119, 208).

APPLICATIONS OF L1 RETROTRANSPOSONS

1. Phylogenetic markers. The presence of L1 retrotransposons in the genome for at least several hundred million years, their continuous and recent retrotransposition activity, and their stable integration are properties of L1 elements that make them excellent phylogenetic markers. Old L1 insertions can be used to perform phylogenetic analysis between species (155) and recent L1 insertions, which are polymorphic as to presence or absence in human populations, can be used to study recent human population dynamics (171, 181). Alu elements share many of the same desirable properties for use as phylogenetic markers and have been used successfully to study human diversity (213).

2. Random mutagenesis system. Weak target site preference means that L1 elements retrotranspose relatively randomly throughout the genome, and there is no bias against L1s inserting into gene sequences (146). The ability to disrupt genes randomly and stably makes L1 elements potentially very attractive for use in a random mutagenesis system in mouse. The recent development of an enhanced green fluorescent protein (EGFP)-based retrotransposition cassette that can detect single-cell retrotransposition events in vivo opens the door for such a system (162). Additionally, recent experiments demonstrate that L1 elements are able to retrotranspose in the mouse germ line at a frequency of greater than 1 in 100 sperm (161; E.M. Ostertag, unpublished data). Incorporation of gene-trapping technology may create a powerful and simple system for making mouse mutants without the requirement for embryonic stem cell-based strategies.

3. Gene delivery vector. The ability to stably integrate into the genome, the ability to carry 3′ sequences via L1-mediated transduction, and the lack of proteins that are not endogenous to the genome (and therefore potentially immunogenic) are the L1 properties that have created interest in using L1 elements as gene delivery vehicles. In fact, an L1 element packaged in a gutted adenoviral vector has been used to deliver marker genes to transformed cells in culture (193).

ACKNOWLEDGMENTS

We thank B.L. Brouha, J.L. Goodier, E.T. Luning Prak, and M.C. Seleme for helpful review of this manuscript and A.B. Downend for help in its preparation. E.M. Ostertag is supported by a Howard Hughes Predoctoral Fellowship and H.H. Kazazian Jr is supported by grants from the NIH.

Visit the Annual Reviews home page at www.AnnualReviews.org

LITERATURE CITED

1. Abdelhak S, Kalatzis V, Heilig R, Compain S, Samson D, et al. 1997. Clustering of mutations responsible for branchio-oto-renal (BOR) syndrome in the eyes absent homologous region (eyaHR) of EYA1. *Hum. Mol. Genet.* 6:2247–55
2. Adachi M, Watanabe-Fukunaga R, Nagata S. 1993. Aberrant transcription caused by the insertion of an early transposable element in an intron of the Fas antigen gene of lpr mice. *Proc. Natl. Acad. Sci. USA* 90:1756–60
3. Adey NB, Schichman SA, Graham DK, Peterson SN, Edgell MH, Hutchison CA. 1994. Rodent L1 evolution has been driven by a single dominant lineage that has repeatedly acquired new transcriptional regulatory sequences. *Mol. Biol. Evol.* 11:778–89
4. Adey NB, Tollefsbol TO, Sparks AB, Edgell MH, Hutchison CA. 1994. Molecular resurrection of an extinct ancestral promoter for mouse L1. *Proc. Natl. Acad. Sci. USA* 91:1569–73
5. Agrawal A, Eastman QM, Schatz DG. 1998. Transposition mediated by RAG1 and RAG2 and its implications for the evolution of the immune system. *Nature* 394:744–51
6. Anagnou NP, Antonarakis SE, O'Brien SJ, Modi WS, Nienhuis AW. 1988. Chromosomal localization and racial distribution of the polymorphic human dihydrofolate reductase pseudogene (DHFRP1). *Am. J. Hum. Genet.* 42:345–52
7. Argeson AC, Nelson KK, Siracusa LD. 1996. Molecular basis of the pleiotropic phenotype of mice carrying the hypervariable yellow (Ahvy) mutation at the agouti locus. *Genetics* 142:557–67
8. Arning S, Gruter P, Bilbe G, Kramer A. 1996. Mammalian splicing factor SF1 is encoded by variant cDNAs and binds to RNA. *RNA* 2:794–810
9. Asch HL, Eliacin E, Fanning TG, Connolly JL, Bratthauer G, Asch BB. 1996. Comparative expression of the LINE-1 p40 protein in human breast carcinomas and normal breast tissues. *Oncol. Res.* 8:239–47
10. Bailey JA, Carrel L, Chakravarti A, Eichler EE. 2000. Molecular evidence for a relationship between LINE-1 elements and X chromosome inactivation: the Lyon repeat hypothesis. *Proc. Natl. Acad. Sci. USA* 97:6634–39
11. Barabino SM, Hubner W, Jenny A, Minvielle-Sebastia L, Keller W. 1997. The 30-kD subunit of mammalian cleavage and polyadenylation specificity factor and its yeast homolog are RNA-binding zinc finger proteins. *Genes Dev.* 11:1703–16
12. Becker KG, Swergold GD, Ozato K, Thayer RE. 1993. Binding of the ubiquitous nuclear transcription factor YY1 to a cis regulatory sequence in the human LINE-1 transposable element. *Hum. Mol. Genet.* 2:1697–702
13. Benit L, Lallemand JB, Casella JF, Philippe H, Heidmann T. 1999. ERV-L elements: a family of endogenous

retrovirus-like elements active through-out the evolution of mammals. *J. Virol.* 73:3301–8

14. Berg JM. 1990. Zinc fingers and other metal-binding domains. Elements for interactions between macromolecules. *J. Biol. Chem.* 265:6513–16

15. Boeke JD. 1997. LINEs and Alus—the polyA connection. [letter; comment]. *Nat. Genet.* 16:6–7

16. Bogenhagen DF, Brown DD. 1981. Nucleotide sequences in Xenopus 5S DNA required for transcription termination. *Cell* 24:261–70

17. Boissinot S, Chevret P, Furano AV. 2000. L1 (LINE-1) retrotransposon evolution and amplification in recent human history. *Mol. Biol. Evol.* 17:915–28

18. Branciforte D, Martin SL. 1994. Developmental and cell type specificity of LINE-1 expression in mouse testis: implications for transposition. *Mol. Cell. Biol.* 14:2584–92

19. Bratthauer GL, Cardiff RD, Fanning TG. 1994. Expression of LINE-1 retrotransposons in human breast cancer. *Cancer* 73:2333–36

20. Bratthauer GL, Fanning TG. 1992. Active LINE-1 retrotransposons in human testicular cancer. *Oncogene* 7:507–10

21. Bratthauer GL, Fanning TG. 1993. LINE-1 retrotransposon expression in pediatric germ cell tumors. *Cancer* 71:2383–86

22. Brown CJ, Lafreniere RG, Powers VE, Sebastio G, Ballabio A, et al. 1991. Localization of the X inactivation centre on the human X chromosome in Xq13. *Nature* 349:82–84

23. Browning VL, Chaudhry SS, Planchart A, Dixon MJ, Schimenti JC. 2001. Mutations of the mouse Twist and sy (Fibrillin 2) genes induced by chemical mutagenesis of ES cells. *Genomics* 73:291–98

24. Burke TW, Kadonaga JT. 1996. Drosophila TFIID binds to a conserved downstream basal promoter element that is present in many TATA-box-deficient promoters. *Genes Dev.* 10:711–24

25. Burwinkel B, Kilimann MW. 1998. Unequal homologous recombination between LINE-1 elements as a mutational mechanism in human genetic disease. *J. Mol. Biol.* 277:513–17

26. Chen F, MacDonald CC, Wilusz J. 1995. Cleavage site determinants in the mammalian polyadenylation signal. *Nucleic Acids Res.* 23:2614–20

27. Chu WM, Ballard R, Carpick BW, Williams BR, Schmid CW. 1998. Potential Alu function: regulation of the activity of double-stranded RNA-activated kinase PKR. *Mol. Cell. Biol.* 18:58–68

28. Cost GJ, Boeke JD. 1998. Targeting of human retrotransposon integration is directed by the specificity of the L1 endonuclease for regions of unusual DNA structure. *Biochemistry* 37:18081–93

29. Cost GJ, Feng Q, Boeke JD. 2000. *Initiation of L1 transposition* in vitro. Presented at Keystone Symp. Transposition and Other Genome Rearrangements, Santa Fe, NM. (Abstr. 119)

30. Cost GJ, Golding A, Schlissel MS, Boeke JD. 2001. Target DNA chromatinization modulates nicking by L1 endonuclease. *Nucleic Acids Res.* 29:573–77

31. Cox GA, Mahaffey CL, Nystuen A, Letts VA, Frankel WN. 2000. The mouse fidgetin gene defines a new role for AAA family proteins in mammalian development. *Nat. Genet.* 26:198–202

32. Cullen BR. 1998. Retroviruses as model systems for the study of nuclear RNA export pathways. *Virology* 249:203–10

33. Cullen BR. 2000. Nuclear RNA export pathways. *Mol. Cell. Biol.* 20:4181–87

34. DeBerardinis RJ, Goodier JL, Ostertag EM, Kazazian HH Jr. 1998. Rapid amplification of a retrotransposon subfamily is evolving the mouse genome. *Nat. Genet.* 20:288–90

35. DeBerardinis RJ, Kazazian HH Jr. 1999. Analysis of the promoter from an expanding mouse retrotransposon subfamily. *Genomics* 56:317–23

36. Deininger PL, Batzer MA. 1995. SINE master genes and population biology. In *The Impact of Short, Interspersed Elements (SINEs) on the Host Genome*, ed. R Maraia, pp. 43–60. Georgetown, TX: Landes

37. Deininger PL, Batzer MA. 1999. Alu repeats and human disease. *Mol. Genet. Metab.* 67:183–93

38. Deininger PL, Batzer MA, Hutchison CA, Edgell MH. 1992. Master genes in mammalian repetitive DNA amplification. *Trends Genet.* 8:307–11

39. Demers GW, Brech K, Hardison RC. 1986. Long interspersed L1 repeats in rabbit DNA are homologous to L1 repeats of rodents and primates in an open-reading-frame region. *Mol. Biol. Evol.* 3:179–90

40. Divoky V, Indrak K, Mrug M, Brabec V, Huisman THJ, Prchal JT. 1996. A novel mechanism of beta thalassemia: the insertion of L1 retrotransposable element into beta globin IVS II. *Blood* 88:148a (Abstr. 580)

41. Dombroski BA, Feng Q, Mathias SL, Sassaman DM, Scott AF, et al. 1994. An in vivo assay for the reverse transcriptase of human retrotransposon L1 in *Saccharomyces cerevisiae. Mol. Cell. Biol.* 14:4485–92

42. Dombroski BA, Mathias SL, Nanthakumar E, Scott AF, Kazazian HH. 1991. Isolation of an active human transposable element. *Science* 254:1805–8

43. Duhl DM, Vrieling H, Miller KA, Wolff GL, Barsh GS. 1994. Neomorphic agouti mutations in obese yellow mice. *Nat. Genet.* 8:59–65

44. Dunham I, Shimizu N, Roe BA, Chissoe S, Hunt AR, et al. 1999. The DNA sequence of human chromosome 22. *Nature* 402:489–95. Erratum. 2000. *Nature* 404(6780):904

45. Duvernell DD, Turner BJ. 1998. Swimmer 1, a new low-copy-number LINE family in teleost genomes with sequence similarity to mammalian L1. *Mol. Biol. Evol.* 15:1791–93

46. Economou-Pachnis A, Tsichlis PN. 1985. Insertion of an Alu SINE in the human homologue of the Mlvi-2 locus. *Nucleic Acids Res.* 13:8379–87

47. Edgell MH, Hardies SC, Loeb DD, Shehee WR, Padgett RW, et al. 1987. The L1 family in mice. *Prog. Clin. Biol. Res.* 251:107–29

48. Eickbush TH. 1997. Telomerase and retrotransposons: Which came first? [letter; comment]. *Science* 277:911–12

49. Esnault C, Maestre J, Heidmann T. 2000. Human LINE retrotransposons generate processed pseudogenes. *Nat. Genet.* 24:363–67

50. Fanning T, Singer M. 1987. The line-1 DNA-sequences in 4 mammalian orders predict proteins that conserve homologies to retrovirus proteins. *Nucleic Acids Res.* 15:2251–60

51. Fanning TG. 1983. Size and structure of the highly repetitive BAM HI element in mice. *Nucleic Acids Res.* 11:5073–91

52. Feng Q, Moran JV, Kazazian HH, Boeke JD. 1996. Human L1 retrotransposon encodes a conserved endonuclease required for retrotransposition. *Cell* 87:905–16

53. Ferrigno O, Virolle T, Djabari Z, Ortonne JP, White RJ, Aberdam D. 2001. Transposable B2SINE elements can provide mobile RNA polymerase II promoters. *Nat. Genet.* 28:77–81

54. Fitch DH, Bailey WJ, Tagle DA, Goodman M, Sieu L, Slightom JL. 1991. Duplication of the gamma-globin gene mediated by L1 long interspersed repetitive elements in an early ancestor of simian primates. *Proc. Natl. Acad. Sci. USA* 88:7396–400

55. Furano AV, Robb SM, Robb FT. 1988. The structure of the regulatory region of the rat L1 (L1Rn, long interspersed repeated) DNA family of transposable elements. *Nucleic Acids Res.* 16:9215–31

56. Gardner JM, Wildenberg SC, Keiper NM, Novak EK, Rusiniak ME, et al. 1997. The mouse pale ear (ep) mutation

is the homologue of human Hermansky-Pudlak syndrome. *Proc. Natl. Acad. Sci. USA* 94:9238–43

57. Gentles AJ, Karlin S. 1999. Why are human G-protein-coupled receptors predominantly intronless? *Trends Genet.* 15: 47–49

58. Goldberg YP, Rommens JM, Andrew SE, Hutchinson GB, Lin B, et al. 1993. Identification of an Alu retrotransposition event in close proximity to a strong candidate gene for Huntington's disease. *Nature* 362:370–73

58a. Goodier JL, Maraia RJ. 1998. Terminator-specific recycling of a B1-Alu transcription complex by RNA polymerase III is mediated by the RNA terminus-binding protein La. *J. Biol. Chem.* 273:26110–16

59. Goodier JL, Ostertag EM, Du K, Kazazian HH Jr. 2001. Characterization of a novel active L1 retrotransposon subfamily in the mouse. *Genome Res.* In press

60. Goodier JL, Ostertag EM, Kazazian HH Jr. 2000. Transduction of 3′-flanking sequences is common in L1 retrotransposition. *Hum. Mol. Genet.* 9:653–57

61. Goodwin TJD, Ormandy JE, Poulter RTM. 2001. L1-like non-LTR retrotransposons in the yeast *Candida albicans. Curr. Genet.* 39:83–91

62. Gorelick RJ, Fu W, Gagliardi TD, Bosche WJ, Rein A, et al. 1999. Characterization of the block in replication of nucleocapsid protein zinc finger mutants from moloney murine leukemia virus. *J. Virol.* 73:8185–95

63. Gorelick RJ, Gagliardi TD, Bosche WJ, Wiltrout TA, Coren LV, et al. 1999. Strict conservation of the retroviral nucleocapsid protein zinc finger is strongly influenced by its role in viral infection processes: characterization of HIV-1 particles containing mutant nucleocapsid zinc-coordinating sequences. *Virology* 256:92–104

64. Gorlich D, Kutay U. 1999. Transport between the cell nucleus and the cytoplasm. *Annu. Rev. Cell Dev. Biol.* 15:607–60

65. Gorman MA, Morera S, Rothwell DG, de La Fortelle E, Mol CD, et al. 1997. The crystal structure of the human DNA repair endonuclease HAP1 suggests the recognition of extra-helical deoxyribose at DNA abasic sites. *EMBO J.* 16:6548–58

66. Grimaldi G, Skowronski J, Singer MF. 1984. Defining the beginning and end of KpnI family segments. *EMBO J.* 3:1753–59

67. Gunn TM, Miller KA, He L, Hyman RW, Davis RW, et al. 1999. The mouse mahogany locus encodes a transmembrane form of human attractin. *Nature* 398:152–56

68. Guo J, Wu T, Anderson J, Kane BF, Johnson DG, et al. 2000. Zinc finger structures in the human immunodeficiency virus type 1 nucleocapsid protein facilitate efficient minus- and plus-strand transfer. *J. Virol.* 74:8980–88

69. Halling KC, Lazzaro CR, Honchel R, Bufill JA, Powell SM, et al. 1999. Hereditary desmoid disease in a family with a germline Alu I repeat mutation of the APC gene. *Hum. Hered.* 49:97–102

70. Hamilton BA, Smith DJ, Mueller KL, Kerrebrock AW, Bronson RT, et al. 1997. The vibrator mutation causes neurodegeneration via reduced expression of PITP alpha: positional complementation cloning and extragenic suppression. *Neuron* 18:711–22

71. Hassoun H, Coetzer TL, Vassiliadis JN, Sahr KE, Maalouf GJ, et al. 1994. A novel mobile element inserted in the alpha spectrin gene: spectrin dayton. A truncated alpha spectrin associated with hereditary elliptocytosis. *J. Clin. Invest.* 94:643–48

72. Hata K, Sakaki Y. 1997. Identification of critical CpG sites for repression of L1 transcription by DNA methylation. *Gene* 189:227–34

73. Heard E, Clerc P, Avner P. 1997. X-chromosome inactivation in mammals. *Annu. Rev. Genet.* 31:571–610

74. Heidmann O, Heidmann T. 1991. Retrotransposition of a mouse IAP sequence

tagged with an indicator gene. *Cell* 64:159–70

75. Herrmann BG, Labeit S, Poustka A, King TR, Lehrach H. 1990. Cloning of the T gene required in mesoderm formation in the mouse. *Nature* 343:617–22

76. Hiom K, Melek M, Gellert M. 1998. DNA transposition by the RAG1 and RAG2 proteins: a possible source of oncogenic translocations. *Cell* 94:463–70

77. Hirose Y, Manley JL. 1998. RNA polymerase II is an essential mRNA polyadenylation factor. *Nature* 395:93–96

78. Hofmann M, Harris M, Juriloff D, Boehm T. 1998. Spontaneous mutations in SELH/Bc mice due to insertions of early transposons: molecular characterization of null alleles at the nude and albino loci. *Genomics* 52:107–9

79. Hohjoh H, Singer MF. 1996. Cytoplasmic ribonucleoprotein complexes containing human LINE-1 protein and RNA. *EMBO J.* 15:630–39

80. Hohjoh H, Singer MF. 1997. Sequence-specific single-strand RNA binding protein encoded by the human LINE-1 retrotransposon. *EMBO J.* 16:6034–43

81. Holmes SE, Dombroski BA, Krebs CM, Boehm CD, Kazazian HH. 1994. A new retrotransposable human L1 element from the LRE2 locus on chromosome 1q produces a chimaeric insertion. *Nat. Genet.* 7:143–48

82. Holmes SE, Singer MF, Swergold GD. 1992. Studies on p40, the leucine zipper motif-containing protein encoded by the first open reading frame of an active human LINE-1 transposable element. *J. Biol. Chem.* 267:19765–68

83. Hyde-DeRuyscher RP, Jennings E, Shenk T. 1995. DNA binding sites for the transcriptional activator/repressor YY1. *Nucleic Acids Res.* 23:4457–65

84. Ilves H, Kahre O, Speek M. 1992. Translation of the rat LINE bicistronic RNAs in vitro involves ribosomal reinitiation instead of frameshifting. *Mol. Cell. Biol.* 12:4242–48

85. Janicic N, Pausova Z, Cole DE, Hendy GN. 1995. Insertion of an Alu sequence in the Ca(2+)-sensing receptor gene in familial hypocalciuric hypercalcemia and neonatal severe hyperparathyroidism. *Am. J. Hum. Genet.* 56:880–86

86. Johnson KR, Cook SA, Erway LC, Matthews AN, Sanford LP, et al. 1999. Inner ear and kidney anomalies caused by IAP insertion in an intron of the Eya1 gene in a mouse model of BOR syndrome. *Hum. Mol. Genet.* 8:645–53

87. Jubier-Maurin V, Cuny G, Laurent AM, Paquereau L, Roizes G. 1992. A new 5′ sequence associated with mouse L1 elements is representative of a major class of L1 termini. *Mol. Biol. Evol.* 9:41–55

88. Jurka J. 1997. Sequence patterns indicate an enzymatic involvement in integration of mammalian retroposons. *Proc. Natl. Acad. Sci. USA* 94:1872–77

89. Kazazian HH Jr. 1999. An estimated frequency of endogenous insertional mutations in humans. *Nat. Genet.* 22:130

90. Kazazian HH Jr. 2000. L1 retrotransposons shape the mammalian genome. *Science* 289:1152–53

91. Kazazian HH Jr, Wong C, Youssoufian H, Scott AF, Phillips DG, Antonarakis SE. 1988. Haemophilia A resulting from de novo insertion of L1 sequences represents a novel mechanism for mutation in man. *Nature* 332:164–66

92. Kim HS, Takenaka O, Crow TJ. 1999. Cloning and nucleotide sequence of retroposons specific to hominoid primates derived from an endogenous retrovirus (HERV-K). *AIDS Res. Hum. Retroviruses* 15:595–601

93. Kimberland ML, Divoky V, Prchal J, Schwahn U, Berger W, Kazazian HH Jr. 1999. Full-length human L1 insertions retain the capacity for high frequency retrotransposition in cultured cells. *Hum. Mol. Genet.* 8:1557–60

94. Kingsmore SF, Giros B, Suh D, Bieniarz M, Caron MG, Seldin MF. 1994. Glycine

receptor beta-subunit gene mutation in spastic mouse associated with LINE-1 element insertion. *Nat. Genet.* 7:136–41

95. Klein DJ, Johnson PE, Zollars ES, De Guzman RN, Summers MF. 2000. The NMR structure of the nucleocapsid protein from the mouse mammary tumor virus reveals unusual folding of the C-terminal zinc knuckle. *Biochemistry* 39:1604–12

96. Kobayashi K, Nakahori Y, Miyake M, Matsumura K, Kondo-Iida E, et al. 1998. An ancient retrotransposal insertion causes Fukuyama-type congenital muscular dystrophy. *Nature* 394:388–92

97. Kohrman DC, Harris JB, Meisler MH. 1996. Mutation detection in the med and medJ alleles of the sodium channel Scn8a. Unusual splicing due to a minor class AT-AC intron. *J. Biol. Chem.* 271:17576–81

98. Kolosha VO, Martin SL. 1997. In vitro properties of the first ORF protein from mouse LINE-1 support its role in ribonucleoprotein particle formation during retrotransposition. *Proc. Natl. Acad. Sci. USA* 94:10155–60

99. Kondo-Iida E, Kobayashi K, Watanabe M, Sasaki J, Kumagai T, et al. 1999. Novel mutations and genotype-phenotype relationships in 107 families with Fukuyama-type congenital muscular dystrophy (FCMD). *Hum. Mol. Genet.* 8:2303–9

100. Korenberg JR, Rykowski MC. 1988. Human genome organization: Alu, lines, and the molecular structure of metaphase chromosome bands. *Cell* 53:391–400

101. Korn LJ, Brown DD. 1978. Nucleotide sequence of *Xenopus borealis* oocyte 5S DNA: comparison of sequences that flank several related eucaryotic genes. *Cell* 15:1145–56

102. Krawczak M, Cooper DN. 1997. The human gene mutation database. *Trends Genet.* 13:121–22

103. Kuff EL, Lueders KK. 1988. The intracisternal A-particle gene family: structure and functional aspects. *Adv. Cancer Res.* 51:183–276

104. Kurachi S, Deyashiki Y, Takeshita J, Kurachi K. 1999. Genetic mechanisms of age regulation of human blood coagulation factor IX. *Science* 285:739–43

105. Kurose K, Hata K, Hattori M, Sakaki Y. 1995. RNA polymerase III dependence of the human L1 promoter and possible participation of the RNA polymerase II factor YY1 in the RNA polymerase III transcription system. *Nucleic Acids Res.* 23:3704–9

106. Kuster JE, Guarnieri MH, Ault JG, Flaherty L, Swiatek PJ. 1997. IAP insertion in the murine LamB3 gene results in junctional epidermolysis bullosa. *Mamm. Genome* 8:673–81

107. Lander ES, Linton LM, Birren B, Nusbaum C, Zody MC, et al. 2001. Initial sequencing and analysis of the human genome. *Nature* 409:860–921

108. Leong WL, Dobson MJ, Logsdon JM Jr, Abdel-Majid RM, Schalkwyk LC, et al. 2000. ETn insertion in the mouse Adcy1 gene: transcriptional and phylogenetic analyses. *Mamm. Genome* 11:97–103

109. Lester T, McMahon C, Van Regemorter N, Jones A, Genet S. 1997. *X-linked immunodeficiency caused by insertion of Alu repeat sequences.* Presented at Br. Hum. Genet. Conf., York, England

110. Letts VA, Felix R, Biddlecome GH, Arikkath J, Mahaffey CL, et al. 1998. The mouse stargazer gene encodes a neuronal Ca^{2+}-channel gamma subunit. *Nat. Genet.* 19:340–47

111. Lingner J, Hughes TR, Shevchenko A, Mann M, Lundblad V, Cech TR. 1997. Reverse transcriptase motifs in the catalytic subunit of telomerase. *Science* 276:561–67

112. Loeb DD, Padgett RW, Hardies SC, Shehee WR, Comer MB, et al. 1986. The sequence of a large L1Md element reveals a tandemly repeated 5′ end and several features found in retrotransposons. *Mol. Cell. Biol.* 6:168–82

113. Loftus SK, Morris JA, Carstea ED, Gu JZ, Cummings C, et al. 1997. Murine model of Niemann-Pick C disease: mutation in a cholesterol homeostasis gene. *Science* 277:232–35

114. Luan DD, Eickbush TH. 1995. RNA template requirements for target DNA-primed reverse transcription by the R2 retrotransposable element. *Mol. Cell. Biol.* 15:3882–91

115. Luan DD, Korman MH, Jakubczak JL, Eickbush TH. 1993. Reverse transcription of R2Bm RNA is primed by a nick at the chromosomal target site: a mechanism for non-LTR retrotransposition. *Cell* 72:595–605

116. Lyon MF. 1998. X-chromosome inactivation: a repeat hypothesis. *Cytogenet. Cell Genet.* 80:133–37

117. Lyon MF. 2000. LINE-1 elements and X chromosome inactivation: a function for "junk" DNA? [letter; comment]. *Proc. Natl. Acad. Sci. USA* 97:6248–49

118. Mager DL, Freeman JD. 2000. Novel mouse type D endogenous proviruses and ETn elements share long terminal repeat and internal sequences. *J. Virol.* 74:7221–29

119. Mager DL, Henthorn PS, Smithies O. 1985. A Chinese G gamma + (A gamma delta beta)zero thalassemia deletion: comparison to other deletions in the human beta-globin gene cluster and sequence analysis of the breakpoints. *Nucleic Acids Res.* 13:6559–75

120. Malik HS, Burke WD, Eickbush TH. 1999. The age and evolution of non-LTR retrotransposable elements. *Mol. Biol. Evol.* 16:793–805

121. Marahrens Y. 1999. X-inactivation by chromosomal pairing events. *Genes Dev.* 13:2624–32

122. Martin F, Maranon C, Olivares M, Alonso C, Lopez MC. 1995. Characterization of a non-long terminal repeat retrotransposon cDNA (L1Tc) from *Trypanosoma cruzi*: homology of the first ORF with the ape family of DNA repair enzymes. *J. Mol. Biol.* 247:49–59

123. Martin SL. 1991. Ribonucleoprotein particles with LINE-1 RNA in mouse embryonal carcinoma cells. *Mol. Cell. Biol.* 11:4804–7

124. Martin SL, Branciforte D. 1993. Synchronous expression of LINE-1 RNA and protein in mouse embryonal carcinoma cells. *Mol. Cell. Biol.* 13:5383–92

125. Martin SL, Bushman FD. 2001. Nucleic acid chaperone activity of the ORF1 protein from the mouse LINE-1 retrotransposon. *Mol. Cell. Biol.* 21:467–75

126. Martin SL, Li J, Weisz JA. 2000. Deletion analysis defines distinct functional domains for protein-protein and nucleic acid interactions in the ORF1 protein of mouse LINE-1. *J. Mol. Biol.* 304:11–20

127. Mathias SL, Scott AF. 1993. Promoter binding proteins of an active human L1 retrotransposon. *Biochem. Biophys. Res. Commun.* 191:625–32

127a. Mathias SL, Scott AF, Kazazian HH Jr, Boeke JD, Gabriel A. 1991. Reverse transcriptase encoded by a human transposable element. *Science* 254:1800–10

128. McCracken S, Fong N, Yankulov K, Ballantyne S, Pan G, et al. 1997. The C-terminal domain of RNA polymerase II couples mRNA processing to transcription. *Nature* 385:357–61

129. McDevitt MA, Imperiale MJ, Ali H, Nevins JR. 1984. Requirement of a downstream sequence for generation of a poly(A) addition site. *Cell* 37:993–99

130. McLauchlan J, Gaffney D, Whitton JL, Clements JB. 1985. The consensus sequence YGTGTTYY located downstream from the AATAAA signal is required for efficient formation of mRNA 3′ termini. *Nucleic Acids Res.* 13:1347–68

131. McMillan JP, Singer MF. 1993. Translation of the human LINE-1 element, L1Hs. *Proc. Natl. Acad. Sci. USA* 90:11533–37

132. Mears ML, Hutchison CA. 2001. The evolution of modern lineages of mouse L1 elements. *J. Mol. Evol.* 52:51–62

133. Meischl C, Boer M, Ahlin A, Roos D. 2000. A new exon created by intronic insertion of a rearranged LINE-1 element as the cause of chronic granulomatous disease. *Eur. J. Hum. Genet.* 8:697–703

134. Meischl C, De Boer M, Roos D. 1998. Chronic granulomatous disease caused by line-1 retrotransposons. *Eur. J. Haematol.* 60:349–50

135. Meischl C, de Boer M, Ostertag EM, Zhang Y, Neijens HJ, et al. 2001. Characterization of an active L1 precursor to an insertion in the CYBB gene. In preparation

136. Meyerson M, Counter CM, Eaton EN, Ellisen LW, Steiner P, et al. 1997. hEST2, the putative human telomerase catalytic subunit gene, is up-regulated in tumor cells and during immortalization. *Cell* 90:785–95

137. Michaud EJ, van Vugt MJ, Bultman SJ, Sweet HO, Davisson MT, Woychik RP. 1994. Differential expression of a new dominant agouti allele (Aiapy) is correlated with methylation state and is influenced by parental lineage. *Genes Dev.* 8:1463–72

138. Miki Y, Katagiri T, Kasumi F, Yoshimoto T, Nakamura Y. 1996. Mutation analysis in the BRCA2 gene in primary breast cancers. *Nat. Genet.* 13:245–47

139. Miki Y, Nishisho I, Horii A, Miyoshi Y, Utsunomiya J, et al. 1992. Disruption of the APC gene by a retrotransposal insertion of L1 sequence in a colon cancer. *Cancer Res.* 52:643–45

140. Minakami R, Kurose K, Etoh K, Furuhata Y, Hattori M, Sakaki Y. 1992. Identification of an internal *cis*-element essential for the human L1 transcription and a nuclear factor(s) binding to the element. *Nucleic Acids Res.* 20:3139–45

141. Minchiotti G, Contursi C, Di Nocera PP. 1997. Multiple downstream promoter modules regulate the transcription of the *Drosophila melanogaster* I, Doc and F elements. *J. Mol. Biol.* 267:37–46

142. Mizrokhi LJ, Georgieva SG, Ilyin YV. 1988. jockey, a mobile Drosophila element similar to mammalian LINEs, is transcribed from the internal promoter by RNA polymerase II. *Cell* 54:685–91

143. Mizuuchi K. 1992. Transpositional recombination: mechanistic insights from studies of mu and other elements. *Annu. Rev. Biochem.* 61:1011–51

144. Mol CD, Kuo CF, Thayer MM, Cunningham RP, Tainer JA. 1995. Structure and function of the multifunctional DNA-repair enzyme exonuclease III. *Nature* 374:381–86

145. Moon BC, Friedman JM. 1997. The molecular basis of the obese mutation in ob2J mice. *Genomics* 42:152–56

146. Moran JV, DeBerardinis RJ, Kazazian HH Jr. 1999. Exon shuffling by L1 retrotransposition. *Science* 283:1530–34

147. Moran JV, Holmes SE, Naas TP, DeBerardinis RJ, Boeke JD, Kazazian HH. 1996. High frequency retrotransposition in cultured mammalian cells. *Cell* 87:917–27

148. Mulhardt C, Fischer M, Gass P, Simon-Chazottes D, Guenet JL, et al. 1994. The spastic mouse: aberrant splicing of glycine receptor beta subunit mRNA caused by intronic insertion of L1 element. *Neuron* 13:1003–15

149. Muratani K, Hada T, Yamamoto Y, Kaneko T, Shigeto Y, et al. 1991. Inactivation of the cholinesterase gene by Alu insertion: possible mechanism for human gene transposition. *Proc. Natl. Acad. Sci. USA* 88:11315–19

150. Mustajoki S, Ahola H, Mustajoki P, Kauppinen R. 1999. Insertion of Alu element responsible for acute intermittent porphyria. *Hum. Mutat.* 13:431–38

151. Naas TP, DeBerardinis RJ, Moran JV, Ostertag EM, Kingsmore SF, et al. 1998. An actively retrotransposing, novel subfamily of mouse L1 elements. *EMBO J.* 17:590–97

152. Nakamura TM, Cech TR. 1998. Reversing time: origin of telomerase. *Cell* 92:587–90

153. Nakielny S, Dreyfuss G. 1999. Transport of proteins and RNAs in and out of the nucleus. *Cell* 99:677–90

154. Narita N, Nishio H, Kitoh Y, Ishikawa Y, Minami R, et al. 1993. Insertion of a 5′ truncated L1 element into the 3′ end of exon 44 of the dystrophin gene resulted in skipping of the exon during splicing in a case of Duchenne muscular dystrophy. *J. Clin. Invest.* 91:1862–67

155. Nikaido M, Rooney AP, Okada N. 1999. Phylogenetic relationships among cetartiodactyls based on insertions of short and long interpersed elements: hippopotamuses are the closest extant relatives of whales. *Proc. Natl. Acad. Sci. USA* 96:10261–66

156. Nur I, Pascale E, Furano AV. 1988. The left end of rat L1 (L1Rn, long interspersed repeated) DNA which is a CpG island can function as a promoter. *Nucleic Acids Res.* 16:9233–51

157. Oldridge M, Zackai EH, McDonald-McGinn DM, Iseki S, Morriss-Kay GM, et al. 1999. De novo alu-element insertions in FGFR2 identify a distinct pathological basis for Apert syndrome. *Am. J. Hum. Genet.* 64:446–61

158. Ono M, Kawakami M, Takezawa T. 1987. A novel human nonviral retroposon derived from an endogenous retrovirus. *Nucleic Acids Res.* 15:8725–37

159. Ooi SL, Pope I, Hope TJ, Boeke JD. 2000. *LINE-1 RNA nuclear export signals.* Presented at Keystone Symp. Transposition and Other Genome Rearrangements, Santa Fe, NM (Abstr. 226)

160. Osheim YN, Proudfoot NJ, Beyer AL. 1999. EM visualization of transcription by RNA polymerase II: downstream termination requires a poly(A) signal but not transcript cleavage. *Mol. Cell* 3:379–87

161. Ostertag EM, DeBerardinis RJ, Kim K-S, Gerton G, Kazazian HH Jr. 2000. Human L1 retrotransposition in germ cells of transgenic mice. *Am. J. Hum. Genet.* 67:A102 (Abstr.)

161a. Ostertag EM, Kazazian HH Jr. 2001. Twin priming, a proposed mechanism for the creation of inversions in L1 retrotransposition. *Genome Res.* In press

162. Ostertag EM, Prak ET, DeBerardinis RJ, Moran JV, Kazazian HH Jr. 2000. Determination of L1 retrotransposition kinetics in cultured cells. *Nucleic Acids Res.* 28:1418–23

163. Padgett RW, Hutchison CA 3rd, Edgell MH. 1988. The F-type 5′ motif of mouse L1 elements: a major class of L1 termini similar to the A-type in organization but unrelated in sequence. *Nucleic Acids Res.* 16:739–49

164. Penny GD, Kay GF, Sheardown SA, Rastan S, Brockdorff N. 1996. Requirement for Xist in X chromosome inactivation. *Nature* 379:131–37

165. Perou CM, Pryor RJ, Naas TP, Kaplan J. 1997. The bg allele mutation is due to a LINE1 element retrotransposition. *Genomics* 42:366–68

166. Pickeral OK, Makalowski W, Boguski MS, Boeke JD. 2000. Frequent human genomic DNA transduction driven by LINE-1 retrotransposition. *Genome Res.* 10:411–15

167. Roberts RM, Liu L, Guo Q, Leaman D, Bixby J. 1998. The evolution of the type I interferons. *J. Interferon Cytokine Res.* 18:805–16. Erratum. *J. Interferon Cytokine Res.* 1999. 19(4):427

168. Rohrer J, Minegishi Y, Richter D, Eguiguren J, Conley ME. 1999. Unusual mutations in Btk: an insertion, a duplication, an inversion, and four large deletions. *Clin. Immunol.* 90:28–37

169. Rowold DJ, Herrera RJ. 2000. Alu elements and the human genome. *Genetica* 108:57–72

170. Rudiger NS, Gregersen N, Kielland-Brandt MC. 1995. One short well conserved region of Alu-sequences is involved in human gene rearrangements

and has homology with prokaryotic chi. *Nucleic Acids Res.* 23:256–60

171. Santos FR, Pandya A, Kayser M, Mitchell RJ, Liu A, et al. 2000. A polymorphic L1 retroposon insertion in the centromere of the human Y chromosome. *Hum. Mol. Genet.* 9:421–30

172. Sassaman DM, Dombroski BA, Moran JV, Kimberland ML, Naas TP, et al. 1997. Many human L1 elements are capable of retrotransposition. *Nat. Genet.* 16:37–43

173. Saxton JA, Martin SL. 1998. Recombination between subtypes creates a mosaic lineage of LINE-1 that is expressed and actively retrotransposing in the mouse genome. *J. Mol. Biol.* 280:611–22

174. Schichman SA, Severynse DM, Edgell MH, Hutchison CA 3rd. 1992. Strand-specific LINE-1 transcription in mouse F9 cells originates from the youngest phylogenetic subgroup of LINE-1 elements. *J. Mol. Biol.* 224:559–74

175. Schmid CW. 1998. Does SINE evolution preclude Alu function? *Nucleic Acids Res.* 26:4541–50

176. Schwahn U, Lenzner S, Dong J, Feil S, Hinzmann B, et al. 1998. Positional cloning of the gene for X-linked retinitis pigmentosa 2. *Nat. Genet.* 19:327–32

177. Scott AF, Schmeckpeper BJ, Abdelrazik M, Comey CT, O'Hara B, et al. 1987. Origin of the human L1 elements: proposed progenitor genes deduced from a consensus DNA sequence. *Genomics* 1:113–25

178. Segal Y, Peissel B, Renieri A, de Marchi M, Ballabio A, et al. 1999. LINE-1 elements at the sites of molecular rearrangements in Alport syndrome-diffuse leiomyomatosis. *Am. J. Hum. Genet.* 64:62–69

179. Seto E, Shi Y, Shenk T. 1991. YY1 is an initiator sequence-binding protein that directs and activates transcription in vitro. *Nature* 354:241–45

180. Shaikh TH, Roy AM, Kim J, Batzer MA, Deininger PL. 1997. cDNAs derived from primary and small cytoplasmic Alu (scAlu) transcripts. *J. Mol. Biol.* 271:222–34

181. Sheen FM, Sherry ST, Risch GM, Robichaux M, Nasidze I, et al. 2000. Reading between the LINEs: human genomic variation induced by LINE-1 retrotransposition. *Genome Res.* 10:1496–508

182. Shehee WR, Chao SF, Loeb DD, Comer MB, Hutchison CA 3rd, Edgell MH. 1987. Determination of a functional ancestral sequence and definition of the 5′ end of A-type mouse L1 elements. *J. Mol. Biol.* 196:757–67

183. Shen L, Wu LC, Sanlioglu S, Chen R, Mendoza AR, et al. 1994. Structure and genetics of the partially duplicated gene RP located immediately upstream of the complement C4A and the C4B genes in the HLA class III region. Molecular cloning, exon-intron structure, composite retroposon, and breakpoint of gene duplication. *J. Biol. Chem.* 269:8466–76

184. Shen MR, Deininger PL. 1992. An in vivo assay for measuring the recombination potential between DNA sequences in mammalian cells. *Anal. Biochem.* 205:83–89

185. Shewchuk BM, Cooke NE, Liebhaber SA. 2001. The human growth hormone locus control region mediates long-distance transcriptional activation independent of nuclear matrix attachment regions. *Nucleic Acids Res.* 29:3356–61

186. Shi Y, Lee JS, Galvin KM. 1997. Everything you have ever wanted to know about Yin Yang 1. *Biochim. Biophys. Acta* 1332:F49–66

187. Shiels A, Bassnett S. 1996. Mutations in the founder of the MIP gene family underlie cataract development in the mouse. *Nat. Genet.* 12:212–15

188. Shrivastava A, Calame K. 1994. An analysis of genes regulated by the multifunctional transcriptional regulator Yin Yang-1. *Nucleic Acids Res.* 22:5151–55

189. Skowronski J, Singer MF. 1986. The

abundant LINE-1 family of repeated DNA sequences in mammals: genes and pseudogenes. *Cold Spring Harbor Symp. Quant. Biol.* 51:457–64

190. Smit AF. 1993. Identification of a new, abundant superfamily of mammalian LTR-transposons. *Nucleic Acids Res.* 21:1863–72

191. Smit AF. 1999. Interspersed repeats and other mementos of transposable elements in mammalian genomes. *Curr. Opin. Genet. Dev.* 9:657–63

192. Smit AF, Riggs AD. 1996. Tiggers and DNA transposon fossils in the human genome. *Proc. Natl. Acad. Sci. USA* 93:1443–48

193. Soifer H, Higo C, Kazazian HH Jr., Moran JV, Mitani K, Kasahara N. 2001. Stable integration of transgenes delivered by a retrotransposon-adenovirus hybrid vector. *Hum. Gene Ther.* 11:1417–28

194. Sonigo P, Wain-Hobson S, Bouguel-eret L, Tiollais P, Jacob F, Brulet P. 1987. Nucleotide sequence and evolution of ETn elements. *Proc. Natl. Acad. Sci. USA* 84:3768–71

195. Speek M. 2001. Antisense promoter of human L1 retrotransposon drives transcription of adjacent cellular genes. *Mol. Cell. Biol.* 21:1973–85

196. Steinmeyer K, Klocke R, Ortland C, Gronemeier M, Jockusch H, et al. 1991. Inactivation of muscle chloride channel by transposon insertion in myotonic mice. *Nature* 354:304–8

197. Stoppa-Lyonnet D, Carter PE, Meo T, Tosi M. 1990. Clusters of intragenic Alu repeats predispose the human C1 inhibitor locus to deleterious rearrangements. *Proc. Natl. Acad. Sci. USA* 87:1551–55

198. Summers MF, South TL, Kim B, Hare DR. 1990. High-resolution structure of an HIV zinc fingerlike domain via a new NMR-based distance geometry approach. *Biochemistry* 29:329–40

199. Swergold GD. 1990. Identification, char-acterization, and cell specificity of a human LINE-1 promoter. *Mol. Cell. Biol.* 10:6718–29

199a. Sukarova E, Dimovski AJ, Tchacarova P, Petkov GI, Efremov GD. 2001. An Alu insert as the cause of a severe form of Hemophilia A. *Acta Haematol.* In press

200. Takahara T, Ohsumi T, Kuromitsu J, Shibata K, Sasaki N, et al. 1996. Dysfunction of the Orleans reeler gene arising from exon skipping due to transposition of a full-length copy of an active L1 sequence into the skipped exon. *Hum. Mol. Genet.* 5:989–93

201. Tchenio T, Casella JF, Heidmann T. 2000. Members of the SRY family regulate the human LINE retrotransposons. *Nucleic Acids Res.* 28:411–15

202. Thayer RE, Singer MF, Fanning TG. 1993. Undermethylation of specific LINE-1 sequences in human cells producing a LINE-1-encoded protein. *Gene* 133:273–77

203. Thien H, Ruther U. 1999. The mouse mutation Pdn (Polydactyly Nagoya) is caused by the integration of a retrotransposon into the Gli3 gene. *Mamm. Genome* 10:205–9

204. Tollervey D, Caceres JF. 2000. RNA processing marches on. *Cell* 103:703–9

205. Trelogan SA, Martin SL. 1995. Tightly regulated, developmentally specific expression of the first open reading frame from LINE-1 during mouse embryogenesis. *Proc. Natl. Acad. Sci. USA* 92:1520–24

206. Usheva A, Shenk T. 1994. TATA-binding protein-independent initiation: YY1, TFIIB, and RNA polymerase II direct basal transcription on supercoiled template DNA. *Cell* 76:1115–21

207. Usheva A, Shenk T. 1996. YY1 transcriptional initiator: protein interactions and association with a DNA site containing unpaired strands. *Proc. Natl. Acad. Sci. USA* 93:13571–6

208. Van de Water N, Williams R, Ockelford P, Browett P. 1998. A 20.7 kb deletion

within the factor VIII gene associated with LINE-1 element insertion. *Thrombosis Haemostasis* 79:938–42

209. Vanin EF. 1985. Processed pseudogenes: characteristics and evolution. *Annu. Rev. Genet.* 19:253–72

210. Vasicek TJ, Zeng L, Guan XJ, Zhang T, Costantini F, Tilghman SM. 1997. Two dominant mutations in the mouse fused gene are the result of transposon insertions. *Genetics* 147:777–86

211. Vidaud D, Vidaud M, Bahnak BR, Siguret V, Gispert Sanchez S, et al. 1993. Haemophilia B due to a de novo insertion of a human-specific Alu subfamily member within the coding region of the factor IX gene. *Eur. J. Hum. Genet.* 1:30–36

212. Wallace MR, Andersen LB, Saulino AM, Gregory PE, Glover TW, Collins FS. 1991. A de novo Alu insertion results in neurofibromatosis type 1. *Nature* 353:864–66

213. Watkins WS, Ricker CE, Bamshad MJ, Carroll ML, Nguyen SV, et al. 2001. Patterns of ancestral human diversity: an analysis of Alu-insertion and restriction-site polymorphisms. *Am. J. Hum. Genet.* 68:738–52

214. Wei W, Gilbert N, Ooi SL, Lawler JF, Ostertag EM, et al. 2001. Human L1 retrotransposition: cis preference versus *trans* complementation. *Mol. Cell. Biol.* 21:1429–39

215. Weis L, Reinberg D. 1992. Transcription by RNA polymerase II: initiator-directed formation of transcription-competent complexes. *FASEB J.* 6:3300–9

216. Weis L, Reinberg D. 1997. Accurate positioning of RNA polymerase II on a natural TATA-less promoter is independent of TATA-binding-protein-associated factors and initiator-binding proteins. *Mol. Cell. Biol.* 17:2973–84

217. Whitcomb JM, Hughes SH. 1992. Retroviral reverse transcription and integration: progress and problems. *Annu. Rev. Cell Biol.* 8:275–306

218. Whitelaw E, Martin DIK. 2001. Retrotransposons as epigenetic mediators of phenotypic variation in mammals. *Nat. Genet.* 27:361–65

219. Wilkinson DA, Mager DL, Leong JC. 1994. Endogenous human retroviruses. In *The Retroviridae*, ed. JA Levy, pp. 465–535. New York: Plenum

220. Wincker P, Jubier-Maurin V, Roizes G. 1987. Unrelated sequences at the 5′ end of mouse LINE-1 repeated elements define two distinct subfamilies. *Nucleic Acids Res.* 15:8593–606

221. Woodcock DM, Williamson MR, Doherty JP. 1996. A sensitive RNase protection assay to detect transcripts from potentially functional human endogenous L1 retrotransposons. *Biochem. Biophys. Res. Commun.* 222:460–65

222. Woods-Samuels P, Wong C, Mathias SL, Scott AF, Kazazian HH, Antonarakis SE. 1989. Characterization of a nondeleterious L1 insertion in an intron of the human factor VIII gene and further evidence of open reading frames in functional L1 elements. *Genomics* 4:290–96

223. Wu M, Rinchik EM, Wilkinson E, Johnson DK. 1997. Inherited somatic mosaicism caused by an intracisternal A particle insertion in the mouse tyrosinase gene. *Proc. Natl. Acad. Sci. USA* 94:890–94

224. Wulff K, Gazda H, Schroder W, Robicka-Milewska R, Herrmann FH. 2000. Identification of a novel large F9 gene mutation—an insertion of an Alu repeated DNA element in exon e of the factor 9 gene. *Hum. Mutat.* 15:299

225. Xiong Y, Eickbush TH. 1990. Origin and evolution of retroelements based upon their reverse transcriptase sequences. *EMBO J.* 9:3353–62

226. Xuemin L, Scaringe WA, Hill KA, Roberts S, Mengos A, et al. 2001. Frequency of recent retrotransposition events in the human factor IX gene. *Hum. Mutat.* 17:511–19

227. Yajima I, Sato S, Kimura T, Yasumoto K, Shibahara S, et al. 1999. An L1 element intronic insertion in the black-eyed white (Mitf[mi-bw]) gene: the loss of a single Mitf isoform responsible for the pigmentary defect and inner ear deafness. *Hum. Mol. Genet.* 8:1431–41

228. Yang J, Malik HS, Eickbush TH. 1999. Identification of the endonuclease domain encoded by R2 and other site-specific, non-long terminal repeat retrotransposable elements. *Proc. Natl. Acad. Sci. USA* 96:7847–52

229. Yang Z, Boffelli D, Boonmark N, Schwartz K, Lawn R. 1998. Apolipoprotein(a) gene enhancer resides within a LINE element. *J. Biol. Chem.* 273:891–97

230. Yoder JA, Walsh CP, Bestor TH. 1997. Cytosine methylation and the ecology of intragenomic parasites. *Trends Genet.* 13:335–40

231. Yoshida K, Nakamura A, Yazaki M, Ikeda S, Takeda S. 1998. Insertional mutation by transposable element, L1, in the DMD gene results in X-linked dilated cardiomyopathy. *Hum. Mol. Genet.* 7:1129–32

232. Zawel L, Reinberg D. 1993. Initiation of transcription by RNA polymerase II: a multi-step process. *Prog. Nucleic Acid Res. Mol. Biol.* 44:67–108

233. Zhang Y, Dipple KM, Vilain E, Huang BL, Finlayson G, et al. 2000. AluY insertion (IVS4–52ins316alu) in the glycerol kinase gene from an individual with benign glycerol kinase deficiency. *Hum. Mutat.* 15:316–23

Annu. Rev. Genet. 2001. 35:539–66

Does Nonneutral Evolution Shape Observed Patterns of DNA Variation in Animal Mitochondrial Genomes?

Anne S. Gerber[1], Ronald Loggins[1], Sudhir Kumar[2], and Thomas E. Dowling[2]

[1]Department of Biology, University of North Dakota, Grand Forks, North Dakota 58202-9019 and [2]Department of Biology, Arizona State University, Tempe, Arizona, 85287-1501; e-mail: thomas.dowling@asu.edu

Key Words selection, neutrality, mtDNA, Tajima's D, MK test

■ **Abstract** Early studies of animal mitochondrial DNA (mtDNA) assumed that nucleotide sequence variation was neutral. Recent analyses of sequences from a variety of taxa have brought the validity of this assumption into question. Here we review analytical methods used to test for neutrality and evidence for nonneutral evolution of animal mtDNA. Evaluations of mitochondrial haplotypes in different nuclear backgrounds identified differences in performance, typically favoring coevolved mitochondrial and nuclear genomes. Experimental manipulations also indicated that certain haplotypes have an advantage over others; however, biotic and historical effects and cyto-nuclear interactions make it difficult to assess the relative importance of nonneutral factors. Statistical analyses of sequences have been used to argue for nonneutrality of mtDNA; however, rejection of neutral patterns in the published literature is common but not predominant. Patterns of replacement and synonymous substitutions within and between species identified a trend toward an excess of replacement mutations within species. This pattern has been viewed as support for the existence of mildly deleterious mutations within species; however, other alternative explanations that can produce similar patterns cannot be eliminated.

CONTENTS

0066-4197/01/1215-0539$14.00

INTRODUCTION

Evolutionary research has been greatly aided by technical advances in molecular biology, allowing for more direct characterization of genetic variation. One marker system in particular, mitochondrial DNA (mtDNA), has revolutionized evolutionary studies of animals (reviewed in 5, 6). Since its initial application in the 1970s, this molecule has been widely used to examine evolutionary genetic questions such as population structure, patterns of hybridization, and evolutionary relationships of a variety of animal taxa. Although the advent of the polymerase chain reaction (PCR) created the potential for examination of a much large number of genes encoded in the nucleus, mtDNA has remained the molecule most often used in evolutionary studies.

Several attributes of mtDNA were responsible for its rise to prominence and continued popularity as a tool for evolutionary studies (reviewed in 5, 6). These characteristics were initially derived from the study of a relatively small number of vertebrate taxa, particularly mammals (reviewed in 17). Small size, duplex, circular conformation, and high copy number allowed for isolation of a sufficient quantity of mtDNA for use in direct characterization with restriction endonucleases (e.g., 9, 19). Gene content and arrangement were thought to be conserved across most animal lineages with few or no intervening noncoding sequences (reviewed in 17). Studies of transmission indicated that mtDNA was inherited only through maternal parents without recombination (e.g., 44, 65). Despite conservation of form, rates of nucleotide substitution were much higher than for nuclear genes (4, 18, 126). These properties made mtDNA an ideal marker system, and scientists utilized it to examine a more diverse group of organisms.

Later studies identified the lack of generality of many of these features in non-mammalian taxa (reviewed in 5, 6, 80). These deviations were so significant that their impact on the utility of mtDNA as a marker for evolutionary studies required additional assessment. As more information of major invertebrate and vertebrate taxa accumulated, considerable variation in gene arrangement was found (e.g., 15, 104, 109). While rearrangements among basal groups are common, order is sufficiently stable within major groups (but see for example, 15, 64, 67) to allow for production of universal primers for PCR amplification of specific fragments and genes (e.g., 58, 91). Several instances of paternal transmission have been identified (e.g., 43, 45, 59), and evidence for recombination has also been reported (e.g., 10, 32); however, levels of paternal leakage and recombination appear in most instances to be so low that they can be safely ignored (6, 63). Rates of mtDNA evolution (e.g., 72, 93, 114) and base composition (e.g., 62) were found to be significantly different among lineages, making local calibrations essential for proper use of a molecular clock.

Even considering these caveats, the general attributes of mtDNA still make it one of the premier marker systems for analysis of population genetic (e.g., measurement of gene flow and population subdivision, estimation of female effective population size) and phylogenetic (e.g., reconstruction of relationships, estimation of divergence times) questions. Most of these applications have taken place under the assumption (explicit or otherwise) that sequence variation in mtDNA is selectively neutral, but the basis and evidence supporting this assumption have not been evaluated.

Considerations of the Neutral Theory

The Neutral Theory has long been debated by evolutionary biologists, especially the role played by mildly deleterious mutations (55, 60, 90, 101). Here we present no new information relevant to the resolution of this dispute; however, we provide the necessary context within which we interpret information.

The neutral theory of molecular evolution (53, 55) attempts to explain the amount and pattern of DNA and protein polymorphism observed in natural populations. It proclaims that "a large proportion of molecular variation within populations is neutral or nearly neutral" (86, p. 3). Since the "polymorphism [within species] is just a transient phase of molecular evolution [among species]" (55), this proclamation applies to molecular sequence diversity among species as well as to within-population molecular variation. Therefore, the neutral theory applies to patterns of observed variation within and among taxa.

Considerable confusion exists in the literature about the evolutionary processes responsible for neutral evolution and their relationship with the neutral theory. To resolve this confusion, let us examine the evolution of protein-coding genes. Consider the subset of third codon positions containing fourfold degenerate sites. At these positions, the fate of almost all mutations will be governed by chance [neutral mutation-random drift (55)] because DNA sequence mutations do not manifest themselves at the protein sequence level and have no effect on fitness. In fact, fourfold degenerate sites are often considered to be completely neutral in nature. Therefore, observed variation at these sites is explained by the neutral theory. Now consider second codon positions. All second codon positions are 0-fold degenerate (i.e., all mutations produce an amino acid change); however, only a small fraction of mutations persist at such sites as purifying selection will quickly eliminate all other mutations. Because of the action of purifying selection, observed variation at 0-fold degenerate sites will be predominantly neutral or nearly so and governed mainly by neutral processes. When we contrast patterns of variation between these two categories, the overall fraction of mutations that are selectively neutral at 0-fold degenerate sites is much smaller than that at fourfold degenerate sites under the neutral theory. Likewise, the average nucleotide sequence diversity at 0-fold sites (π_0) would be much less than fourfold sites (π_4). The observed pattern of sequence variation ($\pi_0 < \pi_4$) is explained by the neutral theory, but the process generating this pattern is purifying selection that weeds out deleterious mutations. Kimura (54) specifically developed this line of reasoning in the framework of the Neutral Theory: "In my opinion, various observations suggest that as the functional constraint diminishes the rate of evolution converges to that of synonymous substitutions. If this is valid, such a convergence (or plateauing) of molecular evolutionary rates will turn out to be strong supporting evidence for the neutral theory."

The popular use of the phrase neutral evolution confounds these two aspects as it is often taken to mean that the fate of all mutations is governed by random chance alone in all codon positions. This perspective (strict-neutrality) is accurate only when considering the evolution of pseudogenes or nonfunctional genes. Therefore, the neutral theory framework (55) explains the patterns of sequence variation at different types of sites (in codons or parts of the gene) and clearly allows for the existence of purifying selection in the process. There has also been considerable debate over the incorporation of slightly deleterious mutations under the Neutral Theory (60, 90). In this paper, we interpret neutrality (or neutral evolution) in the broadest sense (90). Evolutionary processes such as positive selection or adaptive evolution produce nonneutral patterns of molecular variation and are responsible for nonneutral evolution, whereas pattern shifts due to drift, mutation, and/or purifying selection are consistent with the neutral theory.

Given these general considerations, the issue is not whether selection acts upon mtDNA, as it clearly does [the rate of synonymous substitution is many times larger than the rate of nonsynonymous (replacement) substitution]. The important questions are: Which processes act and how do they affect patterns of variation (e.g., number and types of substitutions, base composition, etc.) observed in natural populations? Is variation in mtDNA nucleotide sequences driven by positive selection or does it reflect the interaction of mutation, drift, and shifting selective constraints?

Early Justification for Neutral Evolution of Animal Mitochondrial DNA

Support for the assumption of neutral evolution of mtDNA was at first inferential, based on contrasts of evolutionary rates of nuclear and mitochondrial gene evolution. In one of the earliest reports of rapid evolution of mtDNA, Brown et al. (18) hypothesized that the accelerated rate of sequence change (relative to nuclear genes) could arise from low functional constraints on mitochondrial gene products. They also provided other alternatives that implicated a higher rate of spontaneous mutation of mtDNA or a slowdown in the mutation rate of nuclear DNA in the observed discrepancy. Cann & Wilson (22) and Cann et al. (21) suggested that the probability of fixation of mildly deleterious mutations might be higher in mtDNA owing to hitchhiking through linkage to adaptive variants. However, Cann et al. (21) concluded, ". . . the high rate of evolutionary change in mammalian mtDNA is probably the result of the blending of two forces. One could be increased mutation pressure and/or lack of recombination, both of which would enhance the frequency of change at all positions. The other could be relaxed translational constraints, which would have a major effect on the components of the translational machinery and a minor effect on the mitochondrial genes coding for proteins." Therefore, increased rates of mtDNA evolution were hypothesized to result from a higher mutation rate, relaxation of purifying selection, or a combination of these two factors.

Avise (4) evaluated the available information on causes for the rapid rate of mtDNA evolution and the issue of relaxed selective constraints. He concluded, ". . . there is currently no compelling reason to suppose that most of the mtDNA variants routinely assayed cannot be interpreted as neutral markers of the female lineages in which they occur. This has been the working assumption in most population surveys of mtDNA." Avise et al. (7) continued along this line of reasoning. Since the majority of substitutions in mtDNA were found to be synonymous or insertion/deletions in noncoding regions, they stated ". . . most of the particular mtDNA genotypic variants segregating in populations probably have, by themselves, absolutely no differential effect on organismal fitness." Avise et al. (7) added the caveat that ". . . mechanistically neutral mtDNA variants may, through linkage to selected mtDNA mutations, have evolutionary histories that are at times influenced or even dominated by effects of natural selection."

Given that the rapid rate of evolution was the original basis for assumed neutrality of mtDNA, variation in rates of evolution become pertinent to the discussion of neutrality. Several studies have demonstrated slowdowns in the rate of mtDNA evolution for various lineages (e.g., 8, 72). Templeton (116) demonstrated how evolutionary rates could vary in mtDNA because of the impact of different modes of cladogenesis, even in relatively closely related taxa such as Hawaiian *Drosophila*. In a comparison of rates of mitochondrial and nuclear DNA evolution between sea urchins and primates, Vawter & Brown (119) argued that the relatively rapid rate of vertebrate mtDNA evolution was due mainly to fluctuations in rates of nuclear DNA evolution. Findings such as these indicate that the assumption of rate constancy of mtDNA evolution is not valid.

Recently, many authors have tested for selection on mtDNA, and results of these studies have generated concern over the use and interpretation of mtDNA as a neutral marker (reviewed in 13, 83, 96, 100). Given these developments, it is important that evidence for selection on mtDNA be evaluated to allow for continued informed use of this molecular marker. Here we review statistics that have been used to test neutrality of mtDNA and hypotheses examined by these tests. We also review recent studies that have examined the role of selection on mtDNA evolution and evaluate their ability to discriminate among competing hypotheses. The extensive literature makes it difficult to incorporate all publications on this subject. Therefore, we focus our efforts on statistics and experimental approaches most commonly used to examine the role of selection on mtDNA evolution.

STATISTICAL TESTS FOR DETECTING DEPARTURES FROM NEUTRALITY

A number of tests have been developed over the past 25 years to examine whether observed patterns of DNA or protein polymorphism are consistent with predictions of the Neutral Theory (for a recent review, see 61). These measures are generally based upon the relationship of different estimates of diversity within and between populations, including the number of segregating sites, heterozygosity, gene diversity, and the ratio of synonymous and replacement substitution rates (see 55, 66, 86, 87). Here we present a brief summary of some of these tests, and discuss the null hypotheses tested when they are applied for the mtDNA analysis.

Single-Locus Tests of Protein and DNA Polymorphism Within a Species

Watterson (121, see also 29) developed a test that contrasts observed heterozygosity for a given locus with the theoretical distribution of single-locus heterozygosity given the observed number of alleles. If the fit of the observed and theoretical distributions is adequate, the Neutral Theory cannot be rejected. Such tests are extremely conservative (86). Of course, rejection of the null hypothesis of neutrality in these tests could also occur if one or more of the underlying assumptions (e.g., constant population size) are violated (111).

Tajima's (110) D statistic is widely used to examine the distribution of DNA polymorphisms within species. It is based on the examination of the relationship between the estimates of the population parameter $\theta(=N_e\mu$, where N_e is the effective population size and μ is the mutation rate), which can be calculated from the number of segregating sites (θ_S) and the nucleotide diversity (θ_π) in a sample of sequences from a panmictic population. The difference between these two estimates ($\Delta\theta = \theta_\pi - \theta_S$) and its variance provides a test for neutrality. Both estimates of θ are expected to be the same when a population achieves a neutral mutation-drift equilibrium ($\Delta\theta = 0$). Significant positive values of $\Delta\theta$ are consistent with balancing selection that may leave its signature in the form of increased allele frequencies. The existence of many deleterious mutations in the population will produce an excess of rare alleles, which will lead to a negative value for $\Delta\theta$. Interpretation of observed departures from neutrality are only valid if the assumption of neutral mutation-drift balance is satisfied. It is now well appreciated that the sign of $\Delta\theta$ depends on the population history if a population has recently undergone a bottleneck (111). Therefore, Tajima's test measures skew of allele frequency spectra (61). Fu & Li (38) developed a test similar to Tajima's D in which polymorphisms were separated into recent and ancestral categories based on a phylogenetic tree; however, it has been used less frequently for mtDNA.

Comparing Variation Within and Between Species for Two Loci

This approach takes advantage of the Neutral Theory prediction that "polymorphism [within species] is just a transient phase of molecular evolution [among species]" (55). The best example is the Hudson-Kreitman-Aguade (HKA) test (46), which compares within-species polymorphism and between-species divergence at two (or more) loci simultaneously. If the two loci are evolving at different rates, then the locus with a higher rate is expected to exhibit higher levels of polymorphism under the Neutral Theory. Kreitman (61) noted that the requirement of free recombination between loci makes this test simply a test of neutral substitution rate differences between loci when genes from mtDNA or Y-chromosome are compared, because they evolve as a linked unit. Therefore, its utility is limited to comparisons between mitochondrial (or Y-chromosome) and nuclear loci.

Comparing Synonymous and Replacement Substitutions Within or Between Species

In protein coding genes, we can estimate the rates of synonymous (r_S) and nonsynonymous (r_N) substitutions. Because synonymous substitutions are usually free from selection, r_S is often equated with the neutral substitution rate (76). For neutral evolution in general, $r_N < r_S$ because a large fraction of replacement mutations is likely to be eliminated by purifying selection (55). Therefore, a gene showing $r_N > r_S$, a pattern of molecular variation opposite from that expected under neutrality, is considered to be caused by positive Darwinian selection (47, 55, 87). Since r_N and r_S are both computed from the same set of sequences, they share the same evolutionary history and time of divergence. Therefore, a comparison of r_N and r_S can be conducted by comparing d_N and d_S, where d_S is the number of synonymous substitutions per synonymous site and d_N is the number of replacement substitutions per replacement site. These tests can be applied in the same way for within-species polymorphism as well as between-species comparisons.

Specifically, we test the null hypothesis of $d_N = d_S$ in a one-tailed test, as the alternative hypothesis being tested is $d_N > d_S$. A common practice for conducting this test is to compute the difference $\Delta d \, (= d_N - d_S)$ and construct a normal deviate test by using either analytical or bootstrap variance of Δd (reviewed in 87). Illustrated examples and relative usefulness of different methods for computing d_N and d_S are given in Nei & Kumar (87). If the sequences compared are closely related (e.g., individuals from the same population), the number of synonymous and replacement differences observed might be small and the large-sample assumption made in the normal deviate test is not satisfied. This may make the normal deviate test liberal in rejecting the null hypothesis (131). Under these circumstances, Fisher's Exact Test is more appropriate, typically performed as a 2×2 contingency table in which the nucleotide sites are partitioned into two pairs of site categories, synonymous and replacement sites with and without substitutions (87, 131).

Comparing Synonymous and Replacement Substitutions Within Species (Polymorphisms) and Between Species (Fixed Differences)

Tests in this category are built on the same principle as that for the HKA test and others: Neutral evolution within species is expected to show the same pattern as that observed between species. When substitutions are partitioned into synonymous and replacement types, then estimates of d_N/d_S within a population should be similar to that observed between species if the given gene is evolving in the same manner between species and within populations. This is exactly what the McDonald-Kreitman [MK test (75)] test examines. In this case, a 2×2 contingency table is constructed that contains the numbers of synonymous and replacement sites showing fixed differences between two species and polymorphisms within a population. Rejection of the null hypothesis of the same evolutionary pattern suggests shifts in patterns consistent with either neutral or nonneutral evolution. This can be examined by comparing the ratio of replacement to synonymous substitutions (R) computed from the elements of the 2×2 table for within-population (R_P) with that obtained from between-species analysis (R_F). The null hypothesis tested in the MK test is simply that $R_P = R_F$. The ratio R_P/R_F is the Pattern-Shift Index (PSI) that provides an estimate of the extent of evolutionary pattern shift observed within a population relative to that observed between species. PSI is exactly the same as the neutrality index of Rand & Kann (99).

There are several potential interpretations of MK test results. If a gene is evolving under positive selection among species as well as within populations, then $d_N > d_S$ ($R_F > 1$ and $R_P > 1$) for both among-species and within-population analyses. In this case, the nonneutral nature of sequence variation is obvious irrespective of results of the MK test.

For mtDNA, the value of d_S is rarely smaller than d_N, and, therefore, the application of the MK test usually leads to test of the shifts in neutral evolutionary patterns over time. In some instances, one may find $R_F > R_P$, which indicates an excess of replacement between species as compared to the population polymorphism data. If R_P is assumed to be the ratio expected under neutrality, one may invoke positive selection as the cause behind the increased number of replacement substitutions between species. However, if we assume that R_F is the neutral expectation, then an R_P lower than R_F potentially indicates depression of the replacement substitution rate or an increase in synonymous substitution rates within populations.

The other possible outcome, $R_P > R_F$, may indicate depression of the replacement substitution rate or an increase in synonymous substitution rates between species if we assume that R_P is the neutral expectation. The most popular interpretation of the MK test results assumes that R_F reflects the neutral ratio. Therefore, when $R_P > R_F$, there is an excess of replacement variants within species that is often equated with the overabundance of mildly deleterious mutations within populations. In this case, existence of a few newly arisen, positively selected replacement mutations may also lead to an overabundance of replacement substitutions within populations. This does not necessarily mean that the evolutionary patterns within species are nonneutral under Kimura's (55) framework, as the majority of replacement mutations may be still neutral in nature (2).

PAST STUDIES AND AVAILABLE EVIDENCE

Even as mtDNA was becoming an important marker for use in evolutionary genetic studies, neutrality of the molecule was called into question (e.g., 129). Several early studies focused on selective differences among length variants (e.g., 98, 126), possibly due to faster replication of smaller molecules. Because such variation appears not to affect long-term fitness (e.g., 3, 23, 132), selection on length variation is not considered further. Instead we focus on how selection may shape the distribution of observed nucleotide sequence variation within and among species.

Studies of neutrality of nucleotide sequence variation fall into three broad categories: (*a*) observed differences in performance among haplotypes, (*b*) quantification of shifts in haplotype frequency in experimental populations, and (*c*) comparison of observed patterns of nucleotide sequence variation with those expected under the neutral model. These categories are simply designated for convenience of discussion, as some overlap exists among them.

Observed Differences in Performance Among Haplotypes

Several lines of evidence have been used to examine and/or infer differential performance among mtDNA haplotypes. Some of the strongest evidence for purifying selection comes from studies of disease, with well-known disorders in humans and mice resulting from mutations in mtDNA (120). However, some of these diseases appear late in life and may have limited repercussions on the reproductive contribution of affected individuals (e.g., 77). Other diseases appear to have different fitness effects in males and females (e.g., 103), potentially leading to persistence of deleterious mutations (36).

In some instances, selection between haplotypes is inferred from patterns of variation, with limited support. Malhotra & Thorpe (69) found parallel patterns of morphological and mtDNA variation and ecological gradients in lizards from two Caribbean islands. Based on high levels ($>12\%$) of sequence divergence at cytochrome *b* (cyt*b*) and lack of plausible vicariant events, the authors concluded that observed patterns resulted from natural selection. In a study of hybridization between arctic char (*Salvelinus alpinus*) and brook char (*S. fontinalis*), Glémet et al. (40) found that introgressive hybridization had resulted in complete replacement of brook char mtDNA by that of arctic char in one river subdrainage in Québec. Physiological studies indicated nonequivalence of thermal sensitivity of cytochrome oxidase and pyruvate oxidation by red muscle mitochondria from introgressed and nonintrogressed brook char; however, tests of individual fish failed to identify a significant advantage for either haplotype (cited in 40). Duvernell & Aspinwall (28) found that haplotypic variation in the cyprinid fish *Luxilus chrysocephalus* was consistent with geography and concluded that observed patterns of variation were determined by selection. Dowling et al. (27) noted that distribution of haplotypes is affected by introgression and glacial history, making it difficult to eliminate a role for historical factors in explaining patterns of haplotypic variation.

The direct impact of mitochondrial haplotypes on various performance attributes has also been examined. Schizas et al. (105) quantified differential survivorship to pesticide exposure among three haplotype lineages (designated I, II, III) of the marine copepod, *Microarthridion littorale*. When a random mixture of adults containing these three lineages was exposed to pesticides, lineage I exhibited a significant increase in frequency over control lines, whereas lineages II and III showed nonsignificant reductions in frequency relative to the controls. These results were consistent with frequencies of these haplotypes at clean and toxic sites, suggesting a relationship between haplotype and persistence in natural populations. Studies of the association between mtDNA haplotypes and size-associated parameters (e.g., body weight, growth rate) in rainbow trout yielded mixed results (25, 33). Some haplotypes exhibited significantly enhanced growth; however, there was considerable variation among strains, leading the authors to hypothesize that time of spawning may also play a role in growth dynamics.

These studies demonstrate the significance of mtDNA gene products for organismal fitness; however, it can be difficult to eliminate the impacts of cyto-nuclear interactions in producing the observed patterns of variation. The significance of cyto-nuclear interaction has been demonstrated through examination of polymerase function in cell lines or manipulated eggs (reviewed in 80). These generally indicated improved performance when mitochondrial and nuclear gene products were derived from the same or similar species, indicating coevolution of nuclear and mitochondrial gene products. King & Attardi (56) used human cell lines depleted of mitochondria to demonstrate that source of exogenous mitochondria influenced cytochrome oxidase (COX) activity. Additional studies of human cell lines indicated that it is possible to replace human mtDNA with that of apes; however, human mtDNA always comes to predominate in lines initiated through fusion of ape cytoplasts with human cells, even when the human mtDNA has large deletions (79). The impact of cyto-nuclear interactions on performance has also been demonstrated in other organisms. Burton et al. (20) examined the interaction between mitochondrial cytochrome oxidase genes (COI, COII) and nuclear cytochrome *c* sequences in the copepod *Tigriopus californicus*. Crosses between natural populations from different regions resulted in decreased COX activity by the F_2 generation, whereas intrapopulation crosses undergoing the same treatment yielded an increase in COX activity.

Nevertheless, certain haplotypes may perform better regardless of nuclear background. Takeda et al. (113) used heteroplasmic mice created through reciprocal cytoplast transfer of two lines (C57BL/6 and RR) to examine temporal shifts and tissue-specific segregation of mtDNA haplotypes with different control regions. While equivalent amounts of the two forms were detected in early stage embryos, RR became abundant in most tissues regardless of nuclear background, possibly due to a replicative advantage of RR during development and differentiation.

Experimental Manipulation and Frequency Variation

Early experimental tests of fitness and neutrality of mtDNA were first conducted on lab populations of *Drosophila*. Although labor-intensive, this approach allows for direct examination of fitness effects due to mtDNA variation. In one of the earliest of these studies, MacRae & Anderson (68) tracked frequencies of two mtDNA haplotypes (*bogota* from Colombia, South America, and AH from Apple Hill, California) over 32 generations in one line of *D. pseudoobscura*. Initially present at 30%, the *bogota* haplotype increased to ~80% by generation 4, where it remained for 18 generations. At generation 22, addition of more AH haplotypes reduced the frequency of *bogota* to below 60%; however, the population returned to the pre-perturbation frequency within one generation. Attempts to replicate these results were unsuccessful, although mtDNA frequencies were observed to change significantly during the course of 10 generations in other experiments reported in the same paper. MacRae & Anderson concluded that nonneutral forces controlled mtDNA haplotype frequencies and suggested that sporadic selection can favor one haplotype over the other, possibly as a result of cyto-nuclear interactions. Selection was also invoked in other tests of *D. subobscura* (34) because haplotypes went to fixation faster in large populations. Haplotypes within their own nuclear background were positively selected, but in a mixed nuclear background, one particular haplotype predominated in all instances.

In early tests implicating selection, other explanations were forwarded to explain deviations from neutral behavior in mtDNA frequencies. Singh & Hale (107) note that studies of *D. pseudoobscura* employed haplotypes derived from different subspecies and that observed variation was consistent with known variation in reproductive compatibility. Similar results and conclusions were presented in studies of *D. simulans* populations established to detect selection on mtDNA (89), except that partial reproductive isolation due to maternally transmitted *Rickettsia* was considered to be sufficient to account for apparent departures from neutrality. Jenkins et al. (49) examined the potential impacts of assortative mating and maternally transmitted cytoplasmic incompatibility factors in *D. pseudoobscura* and concluded that neither of these factors could explain patterns of frequency change observed in the earlier experiments of MacRae & Anderson (68).

Several sets of experiments have been conducted that were specifically designed to account for the effects described above. Nigro (88) examined frequency variation in transplasmic lines of *Drosophila simulans* treated for bacterial infection and found that the *si*II haplotype was positively selected when placed in the nuclear background of its own strain but negatively selected when in the *si*III nuclear background. DeStordeur (26) used microinjection to study partitioning of mitochondrial haplotypes in heteroplasmic *D. simulans* and *D. mauritiania* and found results consistent with those of Nigro (88). In a study involving *D. pseudoobscura* and *D. persimilis*, Hutter & Rand (48) found asymmetric cyto-nuclear fitness associations. García-Martínez et al. (39) performed experiments on experimental populations of *D. subobscura*. One haplotype went to fixation in all four cages, yet tests of neutrality found no departure from neutral expectations (see below). Results such as these have generally been seen as evidence for selection on mtDNA haplotypes as well as cyto-nuclear interactions.

Not all experiments identified fitness differences among haplotypes. Kilpatrick & Rand (52) established population cages with two divergent, reproductively compatible strains of *D. melanogaster* in replicate populations. Mitochondrial frequencies were examined on each of the two nuclear backgrounds, as well as a hybrid nuclear background. Haplotype frequencies did not change on the three nuclear backgrounds, and perturbation to test for departures from drift revealed no evidence for purifying or positive selection of one mtDNA haplotype over the other, regardless of the nuclear environment. This result was not restricted to *Drosophila*, as Khambhampati et al. (51) also found no evidence for selection among mtDNA haplotypes in similar experiments on caged populations of the mosquito *Aedes albopictus*.

Nucleotide Sequence Variation and Neutral Theory Expectations

As characterization of DNA sequences became easier, mtDNA neutrality could be tested by statistical analyses of sequence data. Because of the general availability of PCR primers, much of the early data comes from humans and associated commensal taxa (e.g., *Drosophila* and mice). Development of universal PCR primers allowed for sequencing of noncommensal species, providing additional data for analyses of patterns of evolution.

DROSOPHILA   Ballard & Kreitman (12) examined evolution of *cytb* in *Drosophila melanogaster*, *D. simulans*, and *D. yakuba*. MK tests identified a significant excess of replacement polymorphisms within all species pooled ($R_P > R_F$), whereas the HKA test [using the nuclear genes alcohol dehydrogenase (*Adh*) and *period*] only identified departures from neutrality in *D. simulans*. The Watterson test identified an excess of rare haplotypes in *D. melanogaster* and was suggestive of such excess in *D. simulans*. The authors hypothesize recent selective sweeps in both species to explain these results.

Rand et al. (97) examined sequence variation in subunit five of the NADH dehydrogenase gene (ND5) of *D. melanogaster* and *D. simulans* from diverse localities worldwide. Three tests were employed in this study: MK, HKA (using the nuclear genes *Adh*, *Pgd*, and *period*), and Tajima's D. Although this paper is often reported as supporting the case for nonneutrality of mtDNA, neutrality was not rejected using the MK test except for one region of the gene in both species. HKA test results also were not significant using DNA sequences, whereas Tajima's D failed to reject neutrality for all tests except for the carboxy-terminal end of the ND5 gene in *D. simulans*. Rand & Kann (99) also examined sequence variation in the ND3 and ND5 genes among many lines of *D. melanogaster* and *D. simulans* sampled from around the world. As in their earlier studies, applications of the MK test failed to detect significant departures from neutrality in these data, and Tajima's D was consistent with neutrality except for replacement sites in the ND5 gene. Although these tests were not significant, the authors noted a consistent excess of amino acid polymorphism over that which they expected from neutrality ($R_P > R_F$) and attributed this pattern to selection. García-Martínez et al. (39) examined 984 bases of the ND5 gene from 45 haplotypes of *D. subobscura* sampled from a diversity of localities. Tajima's D, Fu & Li's D and F (38), and the MK test did not identify significant departures from the null expectation, inconsistent with fitness tests on observed changes in haplotype frequences.

PRIMATES   In the earliest tests of selection on human mtDNA, examination of allelic variation detected by RFLPs (125) found 71% of the diversity values from comparisons of allele frequency distributions to expected distributions fell within the range expected under the Neutral Theory. The greatest deviations from neutrality were found in protein-coding loci. Explanations for departures from neutrality were recent range expansion and historical purifying selection. Excoffier (30) found that all African human populations examined conformed to Neutral Theory expectations, but Oriental and Caucasoid populations did not. Diversity-reducing factors in Oriental and Caucasoid population dynamics, including population expansion, were proposed to account for these nonneutral patterns. Rogers & Harpending (102) found that the distribution of pairwise nucleotide differences for human mtDNA does not conform to the neutral model, due to either a rapidly expanding population size or a bottleneck caused by a selective sweep.

Nachman et al. (85) analyzed mtDNA sequence variation in humans, chimpanzees, and gorilla in several ways. First they sequenced the ND3 gene from humans, chimpanzees, and one gorilla. Using the MK test they could not reject neutrality within humans. They did reject neutrality within the chimpanzee data but concluded that this was likely due to inclusion of more than one subspecies in the sample. Using RFLPs from the whole genome and other published mtDNA sequences, they identified departures from neutrality within humans using chimpanzee and gorilla as comparative taxa. Tajima's D was also computed for the human RFLP-derived data, and significant departures from neutrality were found among the non-African human samples. Templeton (117) examined sequences from COII in hominoid primates. Contingency tests were used to detect departures from neutrality in mtDNA sequences from humans, chimpanzees, pygmy chimpanzees, gorillas, and an orangutan. Unlike the MK test, these contingency tests (first proposed in 116) allowed for examination of the ratio of replacement to silent substitutions between species among a number of categories, including "older" to "younger" haplotypes as represented by interior and tip haplotypes, respectively, from phylogenetic networks. Replacement mutations in the cytosolic regions of the COII gene appeared to be deleterious, whereas replacement mutations from the transmembrane region and all silent mutations behaved according to neutral expectations.

A similar nested contingency method was employed by Wise et al. (128) to examine variation in the ND2 gene of humans and chimpanzees. Contingency tests based on parsimony networks of haplotypes from humans and chimpanzees identified a significant excess of replacement polymorphisms within the transmembrane regions in humans but not in chimpanzees. Tajima's D and the Fu & Li tests also rejected neutrality in the human data, but not in the chimpanzee data.

MICE   Nachman et al. (84) examined sequence data from the ND3 gene from *Mus domesticus*, *M. musculus*, and *M. spretus*. Using the MK test, a significant excess of replacement mutations was found in each species. However, frequencies of protein variants were not significantly different from neutral expectations when tested by the methods of Watterson and Tajima's D, leading the authors to conclude, "it is possible that some of the amino acid substitutions in our sample are strictly neutral."

NONCOMMENSAL SPECIES   Initial results from studies discussed above have been used to argue for rejection for neutrality (e.g., 12, 85, 97, 128). The tendency toward an excess of replacement polymorphism within species ($R_P > R_F$) has been generally interpreted as evidence for the persistence of mildly deleterious mutations. Because selection against these variants is weak, replacement polymorphisms are observed within species but do not persist long enough to become fixed differences between species.

Concerns were raised by the possibility that relaxation of selection in *Homo sapiens* (112) could be responsible for the apparent excess of replacement polymorphisms in humans and their commensals, specifically *Drosophila* and mice. To determine whether the findings for humans and their commensals represent a general phenomenon, sequences of numerous noncommensal species have been reviewed and analyzed using the MK test (83, 100, 101, 123). In these analyses, the MK test was generally performed on species whose DNA was collected for phylogeographic analyses and included mammals, birds, reptiles, amphibians, fishes, and invertebrates (summarized in Table 1).

Additional comparisons in noncommensal species have subsequently been reported (also included in Table 1). Some recent reports include the data for MK tests (e.g., 14, 118), whereas others provide only discussion of the outcome

TABLE 1 Summary of MK test results

Species for polymorphism[a]	N	Species for divergence	N	Gene	bp	Polymorphic		Fixed		P[d]	PSI	Reference
						R[b]	S[c]	R[b]	S[c]			
Invertebrates												
Alpheus lottini	21	A. formusus	7	COI	564	2	131	0	27	0.55	NC[e]	99
Habronattus pugillis	81	H. geronimoi	1	ND1	440	7	71	3	15	0.39	0.5	73
Haemonchus contortus	37	H. placei	31	ND4	459	27	94	16	15	0.00	0.3	14
Heliconius erato	52	H.telesiphe	NR[f]	COI, COIII	819	16	70	2	24	0.24	2.7	83
Heterorhabditis marelatus	4	H. bacteriophera	1	ND4	474	1	14	13	45	0.27	0.3	14
Lutzomyia longipalpis (within clades)	34	L. longipalpis (among clades)	34	ND4	618	7	35	2	56	0.03	5.6	118
Mytilus edulis F lineage	9	M. edulis M lineage	10	COIII	396	21	53	12	47	0.23	1.6	95
All European Mytilus (F lineage)	26	American M. trossulus (F lineage)	4	COIII	396	21	48	0	27	0.00	NC[e]	95
Mytilus galloprovincialis (Atl.) (F lineage)	6	M. galloprovincialis (Atl.) (M lineage)	5	COIII	396	23	42	11	46	0.04	2.3	95
Mytilus galloprovincialis (Med.) (F lineage)	9	M. galloprovincialis (Med.) (M lineage)	10	COIII	396	45	79	8	31	0.05	2.2	95
Mytilus M lineage (European)	24	M. trossulus M lineage (American)	3	COIII	396	29	77	4	18	0.30	1.7	95
Teladorsagia circumcincta	39	T. boreoarcticus	8	ND4	390	14	65	11	18	0.04	0.4	14
Fishes												
Gadus morhua	236	Melanogrammus aeglefinus	NR[f]	cytb	275	0	16	5	25	0.15	0.0	83
Gadus morhua	41	G. ogac	1	cytb	300	3	22	0	10	0.25	NC[e]	100
Gila cypha	18	G. elegans	16	ND2	758	1	5	16	47	0.54	0.6	g
Gila elegans	18	G. cypha	16	ND2	758	1	5	16	47	0.54	0.6	g
Gila robusta	68	G. cypha	18	ND2	370	5	4	4	22	0.03	6.9	g

(Continued)

TABLE 1 (*Continued*)

Species for polymorphism[a]	N	Species for divergence	N	Gene	bp	Polymorphic		Fixed		P[d]	PSI	Reference
						R[b]	S[c]	R[b]	S[c]			
Gila robusta	68	*G. elegans*	16	ND2	370	4	4	6	16	0.23	2.7	g
Luxilus chrysocephalus	27	*L. cornutus*	18	ND2	1047	36	128	8	41	0.26	1.4	h
Luxilus cornutus	18	*L. chrysocephalus*	27	ND2	1047	16	50	10	37	0.45	1.2	h
Ptychocheilus oregonensis	28	*P. grandis*	20	cytb	639	1	9	1	42	0.35	4.7	i
Ptychocheilus oregonensis	28	*P. umpquae*	16	cytb	639	1	9	0	10	0.50	NC[e]	i
Reptiles and Amphibians												
Ambystoma jeffersonianum	6	*A. laterale*	NR[f]	cytb	307	4	3	0	27	0.00	NC[e]	83
Ambystoma laterale	11	*A. jeffersonianum*	5	cytb	238	3	4	0	22	0.00	NC[e]	100
Bufo marinus (E + W Andes)	27	*B. marinus* (between E and W Andes)	27	ND3	336	1	19	2	18	>0.1	0.5	108
Dendrobates pumilio	12	*D. speciosus*	NR[f]	cytb	292	0	6	4	6	0.23	0.0	83
Emoia impar Group II	8	*E. impar* Group I	NR[f]	cytb	779	3	4	12	81	0.07	5.1	83
Ensatina eschscholtzii	24	*Plethodon elongatus*	NR[f]	cytb	684	38	171	21	27	0.00	0.3	83
Phyllobates lugubris	8	*Dendrobates pumilio*	NR[f]	cytb	292	11	59	0	19	0.06	NC[e]	83
Tarentola delalandii	30	*T. boettgeri*	1	cytb	369	NR[f]	NR[f]	NR[f]	NR[f]	>0.05	NR[f]	42
Birds												
Aerodramus maximus	5	NR[e]	NR[f]	cytb	NR[f]	NR[f]	NR[f]	NR[f]	NR[f]	>0.05	0.0	37
Brachyramphus marmoratus	14	*B. brevirostris*	5	cytb	1041	3	24	0	50	0.02	NC[e]	100
Brachyramphus marmoratus	43	NR[e]	NR[f]	cytb	NR[f]	NR[f]	NR[f]	NR[f]	NR[f]	>0.05	1.4	37
Collocalia esculenta	4	NR[e]	NR[f]	cytb	NR[f]	NR[f]	NR[f]	NR[f]	NR[f]	>0.05	0.5	37
Fringella coelebs ssp.	15	*F. teydea*	1	cytb, atp6, nd5	1283	13	63	6	39	0.60	1.3	71
Grus antigone	9	*G. rubicunda*	NR[f]	cytb	1143	7	10	1	30	0.00	21.0	83
Grus antigone	9	*G. canadense*	4	cytb	1140	10	25	2	49	0.00	9.8	100
Melospiza melodia	11	*Passerella iliaca*	NR[f]	cytb	431	5	2	10	26	0.04	6.5	83

Melospiza melodia	6	NR[e]	NR[f]	*cytb*	NR[f]	NR[f]	NR[f]	NR[f]	NR[f]	>0.05	6.5	37
Passerella iliaca	19	*Melospiza melodia*	NR[f]	*cytb*	431	5	10	10	26	0.74	1.3	83
Passerella iliaca	11	NR[e]	NR[f]	*cytb*	NR[f]	NR[f]	NR[f]	NR[f]	NR[f]	>0.05	1.3	37
Pomatostomus temporalis	35	*P. isidori*	NR[f]	*cytb*	282	0	17	8	18	0.01	0.0	83
Mammals												
Holochilus brasiliensis	82	*H. vulpinus*	21	ND3	348	8	16	7	49	<0.05	3.5	50
Isothrix bistriata	10	*I. pagurus*	NR[f]	*cytb*	798	16	108	6	38	0.99	0.9	83
Isothrix bistriata	10	*I. pagurus*	1	*cytb*	798	15	103	4	33	0.76	1.2	100
Mesomys hispidus	29	*M. stimulax*	NR[f]	*cytb*	798	36	123	0	20	0.01	NC[e]	83
Mesomys hispidus	29	*M. stimulax*	2	*cytb*	798	30	126	0	14	0.08	NC[e]	100
Mesomys hispidus clade 1	10	*M. hispidus* clade 2	19	*cytb*	798	30	112	0	9	0.13	NC[e]	100
Microtis longicaudus	72	*M. pennsylvanicus* + *M. montanus*	NR[f]	*cytb*	1143	NR[f]	NR[f]	NR[f]	NR[f]	>0.05	NR[f]	24
Pan troglodytes	19	*Homo sapiens*	21	ND2	1041	7	32	10	82	0.40	1.8	128
Sciurus aberti	20	*S. niger*	1	*cytb*	1140	12	38	18	146	0.03	2.6	100
Ursus arctos	166	*Helarctos malayanus*	NR[f]	*cytb*	1140	11	44	15	68	0.83	1.1	83
Ursus arctos	28	*U. americana*	1	*cytb*	1137	11	44	15	81	0.50	1.4	100

[a] In some instances, within-species values were calculated from both groups pooled.

[b] Replacement sites.

[c] Synonymous sites.

[d] Level of significance determined by G-test, Fisher's Exact Test, or Monte Carlo.

[e] Undefined because of 0 in the denominator.

[f] Not reported.

[g] AS Gerber, CA Tibbets, TE Dowling, in press.

[h] TE Dowling, unpublished data.

[i] DG Buth, CA Tibbets, TE Dowling, manuscript in preparation.

(e.g., 24, 37, 70). An additional 28 MK test comparisons (26 significant) were reported by Peek et al. (92) on COI sequences from deep-sea clams. These results are not included in Table 1 as they represent all pairwise comparisons among small samples of eight species (from three different genera). Some comparisons in Table 1 contain partial replicates across studies; however, they were never reported as identical [e.g., *Isothrix bistriata* and *Mesomys hispidus* (83, 100)]. Of the 53 comparisons in Table 1, the MK test was significant in only 18 instances and marginally significant ($0.05 < P < 0.10$) in 3 others. In most cases with defined values, the pattern shift index was greater than 1, but was not significant (24 of 38 nonredundant comparisons, one-tailed binomial test, $P = 0.072$). Potential bias in these samples (e.g., nonindependence of comparisons) makes this result difficult to interpret. We tentatively conclude that there is a trend toward an excess of replacement mutations within species relative to replacement substitutions between species, with further analyses required to rigorously test this hypothesis.

In addition to the MK test, Tajima's and Fu & Li's approaches have been applied in recent studies (42, 92, 118). These tests generally failed to detect deviations from neutrality, with rare rejections reported in the study of deep-sea clams [one of eight tests (92] and geckos [combined clades but not independent population samples (42)].

CONCLUSIONS

Opinion on the subject of neutrality of mtDNA appears to have shifted over the past decade, with the general assumption in the literature that mtDNA has been proven to evolve nonneutrally. Studies that present evidence for nonneutral behavior in mtDNA generate attention because so many evolutionary biologists have used mtDNA loci for population genetic and phylogeographic studies under the assumption (explicit or implicit) that this marker is neutral and evolves in a clock-like fashion. The conclusion of nonneutral evolution, largely based upon results from ratio comparisons, has clearly been convincing to many.

There can be no doubt that some forms of selection act on mitochondrial DNA. Evidence from performance experiments suggests that certain haplotypes have a selective advantage; however, it is difficult to eliminate a role for historical and ecological factors as well as cyto-nuclear interactions. In fact, most evidence is consistent with coevolution of mitochondrial and nuclear gene products. Experimental tests of differential fitness among haplotypes (mostly in insects) have suggested a role for selection; however, it is difficult to exclude the influence of mating preference, cytoplasmic effects imposed by *Wolbachia*, and, most importantly, cyto-nuclear interactions.

Statistical analyses of sequences have been the most influential and have been most often cited in support of nonneutral evolution of mtDNA. Early studies focused on humans and their commensals; however, relaxation of selection pressure was proposed as an explanation for apparent departures from neutrality. This explanation was further scrutinized through analyses of DNA sequences from a variety of noncommensal organisms reflecting a diversity of animal taxa. Sequences used in these analyses were typically drawn from GenBank, often having been collected for analysis of population structure and phylogenetic relationships. Test results have been mixed, sometimes yielding patterns consistent with nonneutral evolution. One regular outcome of these comparisons has been an excess of replacement mutations within species relative to the number of replacement substitutions between species, leading to $R_P > R_F$ in contrast to patterns seen in nuclear genes (123).

Although published sequences have been an important resource for evolutionary biologists, tests of neutrality rely upon assumptions that may not be met by these data. If this happens, rejection of the null hypothesis may not mean that selection is responsible but may reflect violation of one or more of these assumptions (reviewed in 122). For example, significantly negative values for Tajima's D have been interpreted as evidence for selection; however, the same pattern can result from recent changes in population size or structure (61, 111). Other biotic factors can produce patterns that mimic selection, as exemplified by the impact of *Wolbachia* (11). Many of these tests assume that intraspecific samples are drawn from a panmictic population; however, this assumption is rarely met as samples were often drawn from several populations sampled for phylogeographic analyses. In addition, sample sizes may be too small to effectively test the null hypotheses. Simonsen et al. (106) suggested that the power of Tajima's D to detect nonneutrality is weakened in sample sizes less than 50 individuals (but see 37).

The MK test has become widely used because of its simple, elegant design, and the conclusion of nonneutrality of mtDNA is largely based on analyses using this test. However, a number of potential problems with the test have been identified. Maynard Smith (74) noted that differences in the mutation rate of transitions and transversions and codon biases can produce patterns that mimic selection (in his case positive selection) and suggested that ". . . the potential dangers of using his method [MK test] uncritically should be recognized." Graur & Li (41) argued that rules for designating sites as fixed or polymorphic may underestimate the amount of between-species variation. This has led to the development of several methods for counting sites, the choice of which may influence the result (66, 95, 100). The number of individuals used may also affect the results of the MK test. In addition to typical difficulties associated with small samples sizes (see replicate *Brachyramphus* and *Melospiza* comparisons in Table 1), use of too few individuals in one or both species may result in an overestimate of fixed sites (41).

Another potential difficulty is created by divergence among species compared (100, 122, 130) because multiple substitutions at the same site are not accounted for in the MK test (41, 124). To avoid the effects of saturation, the MK test is expected to be most useful when used between closely related taxa (75, 82). This problem is expected to be more severe for animal mtDNA as it often evolves much faster than nuclear genes, particularly for vertebrates. To illustrate the potential influence of divergence on the MK test, ND2 sequences from fishes of the North

TABLE 2 Summary statistics for MK tests of *Luxilus*

Species for polymorphism	N	Species for divergence	N	bp	Divergence[a]	Polymorphic		Fixed		p^d	PSI
						R^b	S^c	R^b	S^c		
L. chrysocephalus	27	*L.* sp.	2	1047	0.064	36	128	5	31	0.391	1.74
L. chrysocephalus	27	*L. cornutus*	1	1047	0.110	36	128	9	47	0.337	1.47
L. chrysocephalus	27	*L. cerasinus*	1	1047	0.154	36	128	13	78	0.270	1.69
L. chrysocephalus	27	*L. cardinalis*	1	1047	0.164	36	128	9	91	0.005	2.84

[a]Jukes-Cantor estimate of sequence divergence.
[b]Replacement sites.
[c]Silent sites.
[d]Level of significance determined by Fisher's Exact Test.

American minnow genus *Luxilus* (T.E. Dowling, unpublished data) were aligned and tested (Table 2). As in many of the previous examples, samples were collected for phylogeographic and phylogenetic analyses and do not represent samples from panmictic populations. This specific analysis includes geographic samples of *L. chrysocephalus*, a closely related undescribed species from the Central Highlands region in Arkansas (*L.* sp.), *L. cornutus*, *L. cerasinus*, and *L. cardinalis*. In these comparisons, the taxon being examined for polymorphism (*L. chrysocephalus*) was held constant, while several different outgroup species were applied sequentially to examine the impact of levels of divergence on outcome of the analysis. When less divergent taxa (e.g., *L.* sp., *L. cornutus*, and *L. cerasinus*) were used to identify fixed differences, the MK test failed to reject the null hypothesis (Table 2); however, use of the most divergent taxon as the between-species reference identified a significant excess of replacement substitutions within *L. chrysocephalus*. Since within-species polymorphism has been held constant among these comparisons, this result must reflect a shift in the ratio of replacement and synonymous substitutions among species, and therefore a change in the perceived null hypothesis.

In most instances, the number of fixed replacement substitutions tends to be smaller than the number of polymorphic replacement mutations, leading to $R_F < R_P$. This has generally been interpreted as support for the existence of mildly deleterious mutations within species; however, this conclusion requires that R_F reflect the ancestral condition. Alternatively, the observed patterns could be due to a downward bias in R_F away from the ancestral condition. This bias could result from relatively rapid accumulation of fixed synonymous sites that have arisen within populations during divergence. Because most replacement mutations are expected to be deleterious, this should lead to a relative increase in the number of fixed synonymous sites and deflate R_F erroneously. Comparison of distantly related taxa will add to the bias toward finding an excess of changes within species (122), leading to incorrect inference about shifts in patterns that do not reflect the impact of selection. Under these circumstances, R_F will not reflect the ancestral condition, making this comparison inappropriate.

This difficulty is further exacerbated by shifts in base composition and codon usage (1, 31). MtDNA is notorious for shifts in GC content (reviewed in 104). In insect mtDNA, there is a strong bias toward A + T, and the GC content at the third position varies among species (e.g., 115, 127). Vertebrates also show considerable variation in GC content (e.g., 57, 71, 104). The strength of compositional asymmetry at synonymous positions and between replication strands supports the conclusion that bias is driven by mutation pressure (35, 78), possibly due to asymmetric replication of the molecule (reviewed in 104). Under this scenario, shifts in ratios examined may not be indicative of selection but may instead identify shifts in mutational patterns or base compositional biases (31, 74).

Problems in interpreting MK test results and PSI reflect challenges in determining whether there are significant differences in ratios obtained from within-population and between-species comparisons. As noted by Nei & Kumar (87), "the ratio of two quantities is disturbed more easily by different factors than a difference." In general, caution is essential when drawing inferences from statistically significant results from any statistical test that compares ratios of synonymous and replacement substitutions from two phases of molecular evolution (see also 16, 81, 87).

The above example and discussion clearly demonstrate the difficulties that high levels of divergence and variation in base composition pose for the MK test. As in the *Luxilus* example, the MK test may be more likely to reject neutrality when used between divergent taxa. Many of the previously reported significant tests (e.g., human-chimpanzee, *Drosophila* species) involve divergent taxa (by necessity), making these results difficult to interpret.

Where Do We Go From Here?

In the mid-1980s, it was widely assumed that mtDNA was evolving neutrally, leading many investigators to use this locus for population-level and systematic studies. It now appears that we have exchanged one set of assumptions (neutrality of mtDNA loci) for another (nonneutrality of mtDNA). Positive selection almost surely plays some role in the evolution of mtDNA; however, if positive selection were infrequent the evidence of such events would disappear rapidly. Therefore, it would appear that existing patterns of polymorphism in mtDNA reflect some combination of mutation, drift, and selection perhaps on mildly deleterious mutations.

The available data and statistical approaches make it difficult to assess the relative significance of selection on the evolution of mitochondrial DNA. To appropriately address this question, samples need to be collected in such a way as not to violate assumptions of statistical methods used. At this time, evidence for nonneutral evolution of mitochondrial DNA (i.e., process) has been inferred from the ratios of replacement and synonymous variants within and between taxa (i.e., pattern). In order to truly make the connection between pattern and process, it is essential that two major issues be addressed: (*a*) the influence of biases in the data on outcome of statistical analyses and (*b*) the elimination of alternative explanations consistent with the anticipated pattern shifts. Until this is achieved it will not be possible to conclude that mitochondrial DNA evolves in a nonneutral manner.

Regardless of one's interpretation of neutrality and patterns of mtDNA evolution, we must understand the impact selection will have on phylogenetic and population genetic studies. Unfortunately, the influence of selection on questions such as these has not been specifically considered for mtDNA. Positive selection will clearly be disruptive to evolutionary studies, but the rapid rate of mtDNA evolution will likely reduce the time period of significant impact (if we assume that such bouts are relatively infrequent). Purifying selection (weak as well as strong) on a single site would have limited effect on the length and shape of a phylogenetic tree; however, recurrent mutation of multiple, strongly deleterious alleles will reduce the length of the tree but not affect the topology (reviewed in 94). Even though the tree is consistent with neutral expectations, estimates of S and π are influenced by weak purifying selection on single sites. Therefore, phylogenetic reconstruction is not likely to be impacted by patterns of selection generally observed for mtDNA, but estimates of divergence time and effective population size could be affected, depending upon the number of sites selected and the intensity of selection.

This still leaves us with the question: Is it appropriate to use mtDNA sequences for evolutionary studies? The answer is a qualified yes. Previous cautionary notes sounded by Ballard & Kreitman (13), Rand and coworkers (96, 100), and Nachman (83) (to name but a few) have provided a valuable service to the scientific community by forcing closer examination of the assumptions made when testing hypotheses. Clearly, further studies need to be conducted to address the impact of selection on population genetic and phylogenetic studies. Regardless, we always need to maintain a state of vigilance over assumptions behind analytical methods used to characterize evolutionary patterns and processes of molecular markers such as mitochondrial DNA.

ACKNOWLEDGMENTS

We thank Dan Garrigan, Phil Hedrick, Mark Miller, Masatoshi Nei, and Susan Masta for insightful discussion and/or critical evaluation of this manuscript. Research support has been provided by the National Science Foundation (SK, TED), National Institutes of Health (SK), US Bureau of Reclamation (TED), Burroughs-Wellcome Fund (SK), and the University of North Dakota Faculty Research Fund (ASG).

Visit the Annual Reviews home page at www.AnnualReviews.org

LITERATURE CITED

1. Akashi H. 1995. Inferring weak selection for patterns of polymorphism and divergence at "silent" sites in *Drosophila* DNA. *Genetics* 139:1067–76

2. Akashi H. 1999. Within- and between-species DNA sequence variation and the 'footprint' of natural selection. *Gene* 238:39–51

3. Allegrucci G, Cesaroni D, Venanzetti F, Cataudella S, Sbordoni V. 1998. Length variation in mtDNA control region in hatchery stocks of European sea bass subjected to acclimation experiments. *Genet. Sel. Evol.* 30:275–88

4. Avise JC. 1986. Mitochondrial DNA and the evolutionary genetics of higher animals. *Philos. Trans. R. Soc. London Ser. B* 312:325–42

5. Avise JC. 1994. *Molecular Markers, Natural History, and Evolution.* New York: Chapman & Hall

6. Avise JC. 2000. *Phylogeography: The History and Formation of Species.* Cambridge: Harvard

7. Avise JC, Arnold J, Ball RM, Bermingham E, Lamb T, et al. 1987. Intraspecific phylogeography: the mitochondrial DNA bridge between population genetics and systematics. *Annu. Rev. Ecol. Syst.* 18:489–522

8. Avise JC, Bowen BW, Lamb TA, Meylan AB, Bermingham E. 1992. Mitochondrial DNA evolution at a turtle's pace: evidence for low genetic variability and reduced microevolutionary rate in the Testudines. *Mol. Biol. Evol.* 9:457–73

9. Avise JC, Lansman RA, Shade RO. 1979. The use of restriction endonucleases to measure mitochondrial DNA sequence relatedness in natural populations. I. Population structure and evolution in the genus *Peromyscus. Genetics* 92:279–95

10. Awadalla P, Eyre-Walker A, Maynard-Smith J. 1999. Linkage disequilibrium and recombination in hominid mitochondrial DNA. *Science* 286:2524–25

11. Ballard JWO. 2000. Comparative genomics of mitochondrial DNA in *Drosophila simulans. J. Mol. Evol.* 51:64–75

12. Ballard JWO, Kreitman M. 1994. Unraveling selection in the mitochondrial genome of *Drosophila. Genetics* 138:757–72

13. Ballard JWO, Kreitman M. 1995. Is mitochondrial DNA a strictly neutral marker? *TREE* 10:485–88

14. Blouin MS. 2000. Neutrality tests on mtDNA: unusual results from nematodes. *J. Hered.* 91:156–58

15. Boore JL. 1999. Animal mitochondrial genomes. *Nucleic Acids Res.* 27:1767–80

16. Brookfield JF, Sharp PM. 1994. Neutralism and selectionism face up to DNA data. *Trends Genet.* 10:109–11

17. Brown WM. 1983. Evolution of animal mitochondrial DNA. In *Evolution of Genes and Proteins*, ed. M Nei, RK Koehn, pp. 62–88. Sunderland: Sinauer

18. Brown WM, George M, Wilson AC. 1979. Rapid evolution of animal mitochondrial DNA. *Proc. Natl. Acad. Sci. USA* 76:1967–71

19. Brown WM, Vinograd J. 1974. Restriction endonuclease cleavage maps of animal mitochondrial DNAs. *Proc. Natl. Acad. Sci. USA* 11:4671–721

20. Burton RS, Rawson PD, Edmands S. 1999. Genetic architecture of physiological phenotypes: empirical evidence for coadapted gene complexes. *Am. Zool.* 39:451–62

21. Cann RL, Brown WM, Wilson AC. 1984. Polymorphic sites and the mechanism of evolution in human mitochondrial DNA. *Genetics* 106:479–99

22. Cann RL, Wilson AC. 1983. Length mutations in human mitochondrial DNA. *Genetics* 104:699–711

23. Clark AG, Lyckegaard EMS. 1988. Natural selection with nuclear and cytoplasmic transmission. III. Joint analysis of segregation and mtDNA in *Drosophila melanogaster. Genetics* 118:471–81

24. Conroy CJ, Cook JA. 2000. Phylogeography of a post-glacial colonizer: *Microtus longicaudus* (Rodentia: Muridae). *Mol. Ecol.* 9:165–75

25. Danzmann RG, Ferguson MM. 1995. Heterogeneity in the body size of cultured Ontario rainbow trout with different mitochondrial DNA haplotypes. *Aquaculture* 137:231–44

26. DeStordeur E. 1997. Nonrandom

partition of mitochondria in heteroplasmic *Drosophila. Heredity* 79:615–23

27. Dowling TE, Broughton RE, DeMarais BD. 1997. Significant role for historical effects in the evolution of reproductive isolation: evidence from patterns of introgression between the cyprinid fishes, *Luxilus cornutus* and *Luxilus chrysocephalus. Evolution* 51:1574–83

28. Duvernell DD, Aspinwall N. 1995. Introgression of *Luxilus cornutus* mtDNA into allopatric populations of *Luxilus chrysocephalus* (Teleostei: Cyprinidae) in Missouri and Arkansas. *Mol. Ecol.* 4:173–81

29. Ewens WJ. 1972. The sampling theory of selectively neutral alleles. *Theor. Popul. Biol.* 3:87–112

30. Excoffier L. 1990. Evolution of human mitochondrial DNA: evidence for departure from a pure neutral model of populations at equilibrium. *J. Mol. Evol.* 30:125–39

31. Eyre-Walker A. 1997. Differentiating between selection and mutation bias. *Genetics* 147:1983–87

32. Eyre-Walker A. 2000. Do mitochondria recombine in humans? *Philos. Trans. R. Soc. London Ser. B* 355:1573–80

33. Ferguson MM, Danzmann RG. 1999. Inter-strain differences in the association between mitochondrial DNA haplotype and growth in cultured Ontario rainbow trout (*Oncorhynchus mykiss*). *Aquaculture* 178:245–52

34. Fos M, Domínguez MA, Latorre A, Moya A. 1990. Mitochondrial DNA evolution in experimental populations of *Drosophila subobscura. Proc. Natl. Acad. Sci. USA* 87:4198–201

35. Frank AC, Lobry JR. 1999. Asymmetric substitution patterns: a review of possible underlying mutational or selective mechanisms. *Gene* 238:65–77

36. Frank SA, Hurst LD. 1996. Mitochondria and male disease. *Nature* 383:224

37. Fry AJ. 1999. Mildly deleterious mutations in avian mitochondrial DNA: evidence from neutrality tests. *Evolution* 53:1617–20

38. Fu Y-X, Li W-H. 1993. Statistical tests of neutrality of mutations. *Genetics* 133:693–709

39. García-Martínez J, Castro JA, Ramón M, Latorre A, Moya A. 1998. Mitochondrial DNA haplotype frequencies in natural and experimental populations of *Drosophila subobscura. Genetics* 149:1377–82

40. Glémet H, Blier P, Bernatchez L. 1998. Geographical extent of arctic char (*Salvelinus alpinus*) mtDNA introgression in brook char populations (*S. fontinalis*) from eastern Québec, Canada. *Mol. Ecol.* 7:1655–62

41. Graur D, Li W-H. 1991. Neutral mutation hypothesis test. *Nature* 354:115–16

42. Gubitz T, Thorpe RS, Malhotra A. 2000. Phylogeography and natural selection in the Tenerife gecko *Tarentola delalandii*: testing historical and adaptive hypotheses. *Mol. Ecol.* 9:1213–21

43. Gyllensten UB, Wharton D, Joseffson A, Wilson AC. 1991. Paternal inheritance of mitochondrial DNA in mice. *Nature* 352:255–57

44. Gyllensten UB, Wharton D, Wilson AC. 1985. Maternal inheritance of mitochondrial DNA during backcrossing of two species of mice. *J. Hered.* 76:321–24

45. Hoeh WR, Blakley KH, Brown WM. 1991. Heteroplasmy suggests limited biparental inheritance of *Mytilus* mitochondrial DNA. *Science* 251:1488–90

46. Hudson RR, Kreitman M, Aguade M. 1987. A test of neutral molecular evolution based on nucleotide data. *Genetics* 116:153–59

47. Hughes AL. 1999. *Adaptive Evolution of Genes and Genomes.* Oxford: Oxford Univ. Press

48. Hutter CM, Rand DM. 1995. Competition between mitochondrial haplotypes in distinct nuclear genetic environments: *Drosophila pseudoobscura* vs. *D. persimilis. Genetics* 140:537–48

49. Jenkins TM, Babcock CS, Geiser DM,

Anderson WW. 1996. Cytoplasmic incompatability and mating preference in Colombian *Drosophila pseudoobscura*. *Genetics* 142:189–94

50. Kennedy R, Nachman MW. 1998. Deleterious mutations at the mitochondrial ND3 gene in South American marsh rats (*Holochilus*). *Genetics* 150:359–68

51. Khambhampati S, Rai KS, Verleye DM. 1992. Fequencies of mitochondrial DNA haplotypes in laboratory cage populations of the mosquito, *Aedes albopictus*. *Genetics* 132:205–9

52. Kilpatrick ST, Rand DM. 1995. Conditional hitchhiking of mitochondrial DNA: frequency shifts of *Drosophila melanogaster* mtDNA variants depend on nuclear genetic background. *Genetics* 141:1113–24

53. Kimura M. 1968. Evolutionary rate at the molecular level. *Nature* 217:625–26

54. Kimura M. 1977. Preponderance of synonymous changes as evidence for the neutral theory of molecular evolution. *Nature* 267:275–76

55. Kimura M. 1983. *The Neutral Theory of Molecular Evolution*. Cambridge: Cambridge Univ. Press

56. King MP, Attardi G. 1989. Human cells lacking mtDNA: repopulation with exogenous mitochondria by complementation. *Science* 246:500–3

57. Kocher TD, Conroy JA, McKaye KR, Stauffer JR, Lockwood SF. 1995. Evolution of NADH dehydrogenase subunit 2 in east African cichlid fish. *Mol. Phylogenet. Evol.* 4:420–32

58. Kocher TD, Thomas WK, Meyer A, Edwards SV, Pääbo S, et al. 1989. Dynamics of mitochondrial DNA evolution in animals: amplification and sequencing with conserved primers. *Proc. Natl. Acad. Sci. USA* 86:6196–200

59. Kondo R, Matsuura ET, Ishima H, Takahata N, Chigusa SI. 1990. Incomplete maternal transmission of mitochondrial DNA in *Drosophila*. *Genetics* 126:657–63

60. Kreitman M. 1996. The neutral theory is dead. Long live the neutral theory. *BioEssays* 18:678–83

61. Kreitman M. 2000. Methods to detect selection in populations with applications to the human. *Annu. Rev. Genom. Hum. Genet.* 1:539–59

62. Kumar S. 1996. Patterns of nucleotide substitution in mitochondrial protein coding genes of vertebrates. *Genetics* 143:537–48

63. Kumar S, Hedrick P, Dowling T, Stoneking M. 2000. Questioning evidence for recombination in human mitochondrial DNA. *Science* 288:1931a

64. Kumazawa Y, Nishida M. 1995. Variations in mitochondrial tRNA gene organization of reptiles as phylogenetic markers. *Mol. Biol. Evol.* 12:759–72

65. Lansman RA, Avise JC, Heuttel MD. 1983. Critical experiment to test the possibility of "paternal leakage" of mitochondrial DNA. *Proc. Natl. Acad. Sci. USA* 80:1969–71

66. Li W-H. 1997. *Molecular Evolution*. Sunderland: Sinauer

67. Macey JR, Larson A, Ananajeva NB, Fang Z, Papenfuss TJ. 1997. Two novel gene orders and the role of light-strand replication in rearrangement of the vertebrate mitochondrial genome. *Mol. Biol. Evol.* 14:91–104

68. MacRae AF, Anderson WW. 1988. Evidence for non-neutrality of mitochondrial DNA haplotypes in *Drosophila pseudoobscura*. *Genetics* 120:485–94

69. Malhotra A, Thorpe RS. 1994. Parallels between island lizards suggests selection on mitochondrial DNA and morphology. *Proc. R. Soc. London Ser. B* 257:37–42

70. Malhotra A, Thorpe RS. 2000. A phylogeny of the *Trimeresurus* group of pit vipers: new evidence from a mitochondrial gene tree. *Mol. Phylogenet. Evol.* 16:199–211

71. Marshall HD, Baker AJ. 1998. Rates and patterns of mitochondrial DNA sequence evolution in Fringilline finches (*Fringilla*

spp.) and the greenfinch (*Carduelis chloris*). *Mol. Biol. Evol.* 15:638–46

72. Martin AP, Naylor GJP, Palumbi SR. 1992. Rates of mitochondrial DNA evolution in sharks are slow compared with mammals. *Nature* 357:153–55

73. Masta S. 2000. Phylogeography of the jumping spider *Habronattus pugillis* (Araneae: Salticidae): recent vicariance of sky island populations? *Evolution* 54:1699–711

74. Maynard Smith J. 1994. Estimating selection by comparing synonymous and substitutional changes. *J. Mol. Evol.* 39:123–28

75. McDonald JH, Kreitman M. 1991. Adaptive protein evolution at the *Adh* locus in *Drosophila*. *Nature* 351:652–54

76. Miyata T, Yasunaga T, Nishida T. 1980. Nucleotide sequence divergence and functional constraint in mRNA evolution. *Proc. Natl. Acad. Sci. USA* 7:7328–32

77. Moilanen JS, Majamaa K. 2001. Relative fitness of carriers of the mitochondrial DNA mutation 3243 A > G. *Eur. J. Hum. Genet.* 9:59–62

78. Mooers AO, Holmes EC. 2000. The evolution of base composition and phylogenetic inference. *TREE* 15:365–69

79. Moraes CT, Kenyon L, Hao H. 1999. Mechanisms of human mitochondrial DNA maintenance: the determining role of primary sequence and length over function. *Mol. Biol. Cell* 10:3345–56

80. Moritz C, Dowling TE, Brown WM. 1987. Evolution of animal mitochondrial DNA: relevance for population biology and systematics. *Annu. Rev. Ecol. Syst.* 18:269–92

81. Moriyama EN, Powell R. 1996. Intraspecific nuclear DNA variation in *Drosophila*. *Mol. Biol. Evol.* 13:261–77

82. Moriyama EN, Powell JR. 1997. Synonymous substitution rates in *Drosophila*: mitochondrial versus nuclear genes. *J. Mol. Evol.* 45:378–91

83. Nachman MW. 1998. Deleterious mutations in animal mitochondrial DNA. *Genetica* 102/103:61–69

84. Nachman MW, Boyer SN, Aquadro CF. 1994. Nonneutral evolution at the mitochondrial NADH dehydrogenase subunit 3 gene in mice. *Proc. Natl. Acad. Sci. USA* 91:6364–68

85. Nachman MW, Brown WM, Stoneking M, Aquadro CF. 1996. Nonneutral mitochondrial DNA variation in humans and chimpanzees. *Genetics* 42:953–63

86. Nei M. 1987. *Molecular Evolutionary Genetics*. New York: Columbia Univ. Press

87. Nei M, Kumar S. 2000. *Molecular Evolution and Phylogenetics*. Oxford: Oxford Univ. Press

88. Nigro L. 1994. Nuclear background affects frequency dynamics of mitochondrial DNA variants in *Drosophila simulans*. *Heredity* 72:582–86

89. Nigro L, Prout T. 1990. Is there selection on RFLP differences in mitochondrial DNA? *Genetics* 125:551–55

90. Ohta T. 1996. The current significance and standing of neutral and nearly neutral theories. *BioEssays* 18:673–77

91. Palumbi SR. 1996. Nucleic acids II: The polymerase chain reaction. In *Molecular Systematics*, ed. DM Hillis, C Moritz, BK Mable, pp. 205–47. Sunderland: Sinauer. 2nd. ed.

92. Peek S, Gaut BS, Feldman RA, Barry JP, Kichevar RE, et al. 2000. Neutral and nonneutral mitochondrial genetic variation in deep-sea clams from the family Vesicomyidae. *J. Mol. Evol.* 50:141–53

93. Powell JR, Caccone A, Amato GD, Yoon C. 1986. Rates of nucleotide substitution in *Drosophila* mitochondrial DNA and nuclear DNA are similar. *Proc. Natl. Acad. Sci. USA* 83:9090–93

94. Przeworski M, Charlesworth B, Wall JD. 1999. Genealogies and weak purifying selection. *Mol. Biol. Evol.* 16:246–52

95. Quesada H, Warren M, Skibinski DOF. 1998. Nonneutral evolution and differential mutation rate of gender-associated

mitochondrial DNA lineages in the marine mussel *Mytilus*. *Genetics* 149:1511–26

96. Rand DM. 1996. Neutrality tests of molecular markers and the connection between DNA polymorphism, demography and conservation biology. *Conserv. Biol.* 10:665–71

97. Rand DM, Dorfsman M, Kann LM. 1994. Neutral and non-neutral evolution of *Drosophila* mitochondrial DNA. *Genetics* 138:741–56

98. Rand DM, Harrison RG. 1986. Mitochondrial DNA transmission genetics in crickets. *Genetics* 114:955–70

99. Rand DM, Kann LM. 1996. Excess amino acid polymorphism in mitochondrial DNA: contrasts among genes from *Drosophila*, mice and humans. *Mol. Biol. Evol.* 13:735–48

100. Rand DM, Kann LM. 1998. Mutation and selection at silent and replacement sites in the evolution of animal mitochondrial DNA. *Genetica* 102/103:393–407

101. Rand DM, Weinrich DM, Cezairliyan BO. 2000. Neutrality tests of conservative-radical amino acid changes in nuclear- and mitochondrially-encoded proteins. *Gene* 291:115–25

102. Rogers AR, Harpending H. 1992. Population growth makes waves in the distribution of pairwise differences. *Mol. Biol. Evol.* 9:552–69

103. Ruiz-Pesini E, Lapena A-C, Díez-Sánchez C, Pérez-Martos A, Montoya J, et al. 2000. Human mtDNA haplogroups associated with high or reduced spermatazoa motility. *Am. J. Hum. Genet.* 67:682–96

104. Saccone C, DeGiorgi C, Gissi C, Pesole G, Reyes A. 1999. Evolutionary genomics in metazoa: the mitochondrial DNA as a model system. *Gene* 238:195–209

105. Schizas NV, Chandler GT, Coull BC, Klosterhaus SL, Quattro M. 2001. Differential survival of three mitochondrial lineages of a marine benthic copepod exposed to a pesticide mixture. *Environ. Sci. Technol.* 35:535–38

106. Simonsen KL, Churchill GA, Aquadro CF. 1995. Properties of statistical tests of neutrality for DNA polymorphism data. *Genetics* 141:413–29

107. Singh RS, Hale LR. 1990. Are mitochondrial DNA variants selectively non-neutral? *Genetics* 124:995–97

108. Slade RW, Moritz C. 1998. Phylogeography of *Bufo marinus* from its natural and introduced ranges. *Proc. R. Soc. London Ser. B* 265:769–77

109. Staton JL, Daehler LL, Brown WM. 1997. Mitochondrial gene arrangement of the horshoe crab *Limulus polyphemus* L.: conservation of major features among arthropod classes. *Mol. Biol. Evol.* 14:867–74

110. Tajima F. 1989. Statistical method for testing the neutral mutation hypothesis by DNA polymorphism. *Genetics* 123:585–95

111. Tajima F. 1989. The effect of change in population size on DNA polymorphism. *Genetics* 123:597–601

112. Takahata N. 1993. Relaxed selection in human populations during the Pleistocene. *Jpn. J. Genet.* 68:539–47

113. Takeda K, Takahashi S, Onishi A, Hanada H, Imai H. 2000. Replicative advantage and tissue-specific segregation of RR mitochondrial DNA between C57BL/6 and RR heteroplasmic mice. *Genetics* 155:777–83

114. Takezaki N, Gojobori T. 1999. Correct and incorrect vertebrate phylogenies obtained by the entire mitochondrial DNA sequences. *Mol. Biol. Evol.* 16:590–601

115. Tamura K. 1992. The rate and pattern of nucleotide substitution in *Drosophila* mitochondrial-DNA. *Mol. Biol. Evol.* 9:814–25

116. Templeton AR. 1987. Genetic systems and evolutionary rates. In *Rates of Evolution*, ed. KSW Campbell, MF Day, pp. 218–34. Allen & Unwin: London

117. Templeton AR. 1996. Contingency tests of neutrality using intra/interspecific gene trees: the rejection of neutrality for

the evolution of the mitochondrial cytochrome oxidase II gene in the hominoid primates. *Genetics* 144:1263–70

118. Uribe Soto SI, Lehmann T, Rowton ED, Velez BID, Porter CH. 2001. Speciation and population structure in the morphospecies *Lutzomyia longipalpis* (Lutz & Neiva) as derived from the mitochondrial ND4 gene. *Mol. Phylogenet. Evol.* 18:84–93

119. Vawter L, Brown WM. 1986. Nuclear and mitochondrial DNA comparisons reveal extreme rate variation in the molecular clock. *Science* 234:194–96

120. Wallace DC. 1999. Mitochondrial diseases in man and mouse. *Science* 283:1482–88

121. Watterson GA. 1977. Heterosis or neutrality? *Genetics* 85:789–814

122. Wayne ML, Simonsen KL. 1998. Statistical tests of neutrality in the age of weak selection. *TREE* 13:236–40

123. Weinreich DM, Rand DM. 2000. Contrasting patterns of nonneutral evolution in proteins encoded in nuclear and mitochondrial genomes. *Genetics* 156:385–99

124. Whittam T, Nei M. 1991. Neutral mutation hypothesis test. *Nature* 354:115–16

125. Whittam TS, Clark AG, Stoneking M, Cann RL, Wilson AC. 1986. Allelic variation in human mitochondrial genes based on patterns of restriction site polymorphism. *Proc. Natl. Acad. Sci. USA* 83:9611–15

126. Wilson AC, Cann L, Carr SM, George M, Gyllensten UB, et al. 1985. Mitochondrial DNA and two perspectives on evolutionary genetics. *Biol. J. Linn. Soc.* 26:375–400

127. Wirth T, LeGuellec R, Veuille M. 1999. Directional substitution and evolution of nucleotide content in the cytochrome oxidase II gene in earwigs (dermapteran insects). *Mol. Biol. Evol.* 16:1645–53

128. Wise CA, Sraml M, Easteal S. 1998. Departure from neutrality at the mitochondrial NADH dehydrogenase subunit 2 gene in humans, but not in chimpanzees. *Genetics* 148:409–21

129. Wolstenholme DR, Clary DO. 1985. Sequence evolution of *Drosophila* mitochondrial DNA. *Genetics* 109:725–44

130. Yang Z, Bielawski JP. 2000. Statistical methods for detecting molecular adaptation. *TREE* 15:496–503

131. Zhang J, Kumar S, Nei M. 1997. Small-sample tests of episodic adaptive evolution: a case study of primate lysozymes. *Mol. Biol. Evol.* 14:1335–38

132. Zouros E, Pogson GH, Cook DI, Dadswell MJ. 1992. Apparent selective neutrality of mitochondrial DNA size variation: a test in the deep-sea scallop *Placopten magellanicus*. *Evolution* 46:1466–76

Annu. Rev. Genet. 2001. 35:567–88

IDENTIFICATION OF EPILEPSY GENES IN HUMAN AND MOUSE*

Miriam H. Meisler,[1] Jennifer Kearney,[1] Ruth Ottman,[2] and Andrew Escayg[1]

[1]*Department of Human Genetics, School of Medicine, University of Michigan, Ann Arbor, Michigan 48109-0618; e-mail: meislerm@umich.edu;* [2]*Sergievsky Center, College of Physicians and Surgeons, Columbia University, New York, NY 10032; e-mail: ro6@columbia.edu*

Key Words ion channel, seizure, sodium channel, mutation detection

■ **Abstract** The development of molecular markers and genomic resources has facilitated the isolation of genes responsible for rare monogenic epilepsies in human and mouse. Many of the identified genes encode ion channels or other components of neuronal signaling. The electrophysiological properties of mutant alleles indicate that neuronal hyperexcitability is one cellular mechanism underlying seizures. Genetic heterogeneity and allelic variability are hallmarks of human epilepsy. For example, mutations in three different sodium channel genes can produce the same syndrome, GEFS+, while individuals with the same allele can experience different types of seizures. Haploinsufficiency for the sodium channel SCN1A has been demonstrated by the severe infantile epilepsy and cognitive deficits in heterozygotes for de novo null mutations. Large-scale patient screening is in progress to determine whether less severe alleles of the genes responsible for monogenic epilepsy may contribute to the common types of epilepsy in the human population. The development of pharmaceuticals directed towards specific epilepsy genotypes can be anticipated, and the introduction of patient mutations into the mouse genome will provide models for testing these targeted therapies.

CONTENTS

---

*This chapter is dedicated to Roslyn Klaif in appreciation of her courage and inspiration.

INTRODUCTION

Epilepsy is one of the most common neurological disorders, affecting approximately 3% of individuals at some time in their lives, and is a significant medical burden to patients and to society (36). A strong genetic influence, long suspected, has been confirmed during the past few years by the mapping and isolation of more than 40 genes responsible for monogenic epilepsy in human families and mouse models.

Human epilepsy is a heterogeneous disorder defined by recurrent unprovoked seizures, the clinical manifestation of abnormal synchronized neuronal discharges in the brain. The primary seizure types are generalized seizures, which involve the entire brain from the outset, and partial (focal) seizures, which begin in a localized brain region (19). Classification of epilepsy syndromes combines information on seizure type, age at onset, etiology, clinical course, and electroencephalographic (EEG) findings (20). Idiopathic epilepsy lacks antecedent disease or injury to the central nervous system and is of presumed genetic origin. The current classifications are not well correlated with genetic causes, since the same mutations can produce different syndromes in different individuals, and a single syndrome can be generated by mutations in more than one gene.

All of the genes thus far identified as causing idiopathic epilepsy are molecular components of neuronal signaling. The functional effects of the mutant alleles provide direct evidence for neuronal hyperexcitability as one cellular mechanism underlying seizures. A major challenge for the future is to determine whether these monogenic epilepsy genes also contribute to the common epilepsies that do not have clear patterns of inheritance, and if so, to determine the identity and frequency of the responsible alleles. Identification of the genetic basis for inherited epilepsies provides new therapeutic targets for this frequently debilitating disorder.

In this chapter, we describe the recent progress in identification of idiopathic epilepsy genes in human and mouse, the functional effects of mutated alleles, and the preliminary efforts to evaluate the role of these genes in common epilepsies. Additional information can be found in several excellent reviews (3, 10, 13, 30, 33, 41, 45, 60, 64, 67, 70).

METHODS FOR ISOLATION OF MONOGENIC EPILEPSY GENES

Progress in molecular neurobiology during the past two decades identified many functional candidate genes for epilepsy based on their role in generation and transmission of electrical signals in neurons. Chromosomal map positions provided the key connection between candidate genes and human disorders. During the 1980s and 1990s, cDNA clones were isolated and mapped to specific human chromosome positions using somatic cell genetics, fluorescent in situ hybridization (FISH), and PCR analysis of radiation hybrid panels. During the same period, the development of polymorphic molecular markers for human linkage analysis made it possible to map the loci for clinical epilepsy syndromes found in large family pedigrees. The coincidence of chromosomal positions of candidate genes and disease loci led to the identification of several monogenic human epilepsy genes. The identified human epilepsy genes are listed in Table 1.

Mouse epilepsy genes have been isolated using experimental crosses with thousands of informative meioses that define small nonrecombinant regions of 0.5 to 1 Mb. Isolation of large-insert clones spanning the nonrecombinant region and identification of genes in the nonrecombinant region have been streamlined by new genomic resources. The availability of ordered BAC clone contigs spanning the mouse genome has eliminated the need to screen clone libraries, and the gene content of most regions can now be obtained electronically from the assembled sequence of the corresponding human chromosome region. With these methods, the time required to map and clone an epilepsy mutation has been greatly reduced, and several spontaneous mouse mutations associated with well-characterized seizures have been cloned (Table 2). A publicly funded initiative to generate additional seizure models in the mouse by chemical mutagenesis will provide increased opportunities for applying these methods to identification of epilepsy genes.

IDENTIFICATION OF MUTATIONS IN ION CHANNEL GENES

The propagation of the electrical impulse in neurons is initiated by the transient opening of voltage-gated sodium channels and influx of sodium ions along a concentration gradient. The impulse is terminated by the transient opening of

TABLE 1   Identified genes responsible for human monogenic idiopathic epilepsy

Year	Gene	Chromosome	MIM	Mode	Types of mutant alleles	Clinical syndrome
2001	GABRG2 GABA_A receptor	5q31	604233	AD	Missense	GEFS+3
2001	SCN2A sodium channel alpha subunit	2q24	604233	AD	Missense	GEFS+
2000	SCN1A sodium channel alpha subunit	2q24	604233	AD AD	Missense null, missense	GEFS+2 SMEI
2000	CHRNB2 acetylcholine recepter beta subunit	1p21	605375	AD	Missense	ADNFLE3
1998	SCN1B sodium channel beta 1 subunit	19q13	604233	AD	Missense	GEFS+1
1998	KCNQ2 potassium channel	20q13	602235	AD	Missense, null	BFNC1 (EBN1)
1998	KCNQ3 potassium channel	8q24	121201	AD	Missense	BFNC2 (EBN2)
1995	CHRNA4 acetylcholine receptor alpha	20q13	600513	AD	Missense	ADNFLE1

AD, autosomal dominant; GEFS+, Generalized epilepsy with febrile seizures plus; SMEI, severe myoclonic epilepsy of infancy; ADNFLE, autosomal dominant nocturnal frontal lobe epilepsy; BFNC, benign familial neonatal convulsions.

voltage-gated potassium channels that permit the efflux of potassium and restoration of the resting potential of the cell. Voltage-gated calcium channels in the axon terminal convert the electrical signal to a chemical signal via influx of calcium ions, leading to release of synaptic vesicles containing neurotransmitters. This release activates ligand-gated receptors in the postsynaptic membrane and initiates an electrical impulse in the downstream neuron. The shared domain structure of the voltage-gated potassium, sodium, and calcium channels demonstrates their evolutionary origin from a common ancestral protein (38). Predictions that mutations in these channels and receptors could produce disregulated neuronal firing have been confirmed by the identification of disease-causing mutations in human and mouse.

TABLE 2 Epilepsy genes identified in spontaneous mouse mutants. For details see the Mouse Genome Database (MGD) website at www.informatics.jax.org. The chromosomal locations of the human orthologs are indicated

Category	Gene	Mouse chr	Protein	Mutant	Mutation	Mode[a]	Seizure type	Human chromosome
Channels receptors	Cacna1a	8	Voltage-gated calcium channel α subunit	tottering leaner rolling-Nagoya	Missense truncation	AR	Spike wave, focal motor	19p13
	Cacnb4	2	Voltage-gated calcium channel β4 subunit	lethargic	Null	AR	Spike wave	2q22
	Cacna2d2	9	Voltage-gated calcium channel α2δ2 subunit	ducky torpid	Null	AR	Spike wave	3p21
	Cacng2	15	Voltage-dependent calcium channel γ2 subunit OR receptor transporter	stargazer waggler	Null	AR	Spike wave	22
	Kcnj6	16	G-protein gated inwardly-rectifying K$^+$ channel (GIRK2)	weaver	Missense	AR	Tonic-clonic	21q22
	Itpr1	6	Inositol 1,4,5-triphosphate receptor	opisthotonos	In-frame deletion	AR	Tonic-clonic	3p26
pH Homeostasis	Slc9a1	4	Na$^+$/H$^+$ exchanger	slow wave epilepsy	Null	AR	Spike wave tonic-clonic	1p36
Intracellular transport	Myo5a	9	Myosin Va	dilute-neurological	Null	AR		15q21
	Ap3d	10	Adaptor-related protein complex AP-3, delta	mocha	Null	AR		19p13
Myelination	Pmp22 Plp	11 X	Peripheral myelin protein Myelin proteolipid protein	trembler jimpy	Several Several	AD XR	Tonic-clonic	17p12 Xq21
Membrane protein	Massl	13	Monogenic audiogenic seizure susceptibility 1	Frings	Truncation	AR	Audiogenic	7

[a]AR, autosomal recessive; AD, autosomal dominant; XR, X-linked recessive.

Voltage-Gated Sodium Channels

The voltage-gated sodium channels contain a large pore-forming transmembrane α subunit of 260 kDa that is capable of generating a sodium current in response to membrane depolarization (15). The α subunit can associate with three auxiliary β subunits of 35 kDa that influence the rate of channel inactivation and intracellular localization (15). Four of the ten α subunit genes in the mammalian genome are expressed at high levels in the central nervous system: *SCN1A*, *SCN2A*, *SCN3A*, and *SCN8A* (47).

GENERALIZED EPILEPSY WITH FEBRILE SEIZURES PLUS   The first evidence of sodium channel mutations in epilepsy was obtained in 1998 by analysis of a large Australian family with 378 members, including 42 with a history of epilepsy (76). The phenotype in affected family members was highly variable and included febrile (fever-induced) seizures persisting beyond the usual termination age of six years, generalized epilepsy involving absence seizures, myoclonic, atonic or tonic-clonic seizures, and partial epilepsy. This syndrome was designated Generalized Epilepsy with Febrile Seizures Plus (GEFSP; MIM 604233) (63). The locus *GEFS+1* was mapped to chromosome 19q13, where the sodium channel β1 subunit gene *SCN1B* had been previously mapped. Exon sequencing identified a missense mutation, C121W, that cosegregated with the disease and was not observed in 96 controls (76). In functional assays, the mutant protein failed to accelerate the recovery from inactivation of the associated α subunit (76) (Table 3). Coexpression of mutant and wild-type β subunits with the α subunit produced an intermediate rate of inactivation (48), indicating that the inactive mutant subunit can compete for binding to the α subunit in heterozygotes and accounting for the dominant inheritance of the disorder. In heterozygotes, the association of inactive β subunits with α subunits is predicted to produce a population of channels that would inactivate slowly and generate "persistent current." A neuron with persistent sodium current will require a smaller depolarization to initiate firing, and thus may be considered to be in a hyperexcitable state.

A second locus, *GEFS+2*, was mapped in 1999 to a 20-cM interval of chromosome 2q24 that contained the α subunit genes *SCN1A*, *SCN2A*, and *SCN3A* by analysis of two large families (6, 50). Screening affected individuals from both families using conformation sensitive gel electrophoresis of amplified exons identified two missense mutations in the *SCN1A* gene, *R1648H* and *T875M* (27). Both mutations changed evolutionarily invariant residues located in the voltage-sensing S4 segments of the protein. Introduction of these mutations into the SCN1A channel and examination of the kinetic properties in *Xenopus* oocytes demonstrated that the *R1648H* mutation accelerated the recovery from inactivation and decreased the use dependence of channel activity (66). This accelerated recovery could lead to rapid firing patterns and neuronal hyperexcitability. Similar effects were observed when the corresponding mutation was introduced into *SCN4A* (2). The second mutation, *T875M*, increased the likelihood of inactivation by the slow inactivation

TABLE 3 Electrophysiological effects of dominantly inherited epilepsy mutations in voltage-gated sodium channels

Gene	Syndrome	Mutant allele (domain)	Properties of isolated mutant channel	Predicted cellular effects	Types of seizures in family members	Reference for electrophysiology
SCN2A	GEFS+	R187W (D2S6) trans-membrane	Reduced rate of inactivation	Increased persistent current leading to lower threshold for firing of action potentials	Febrile, brief afebrile generalized tonic and tonic/clonic	(72)
SCN2A (mouse)	Temporal lobe epilepsy	GAL/QQQ (D2S4/5) cytoplasmic linker	Reduced rate of inactivation	Increased persistent current leading to lower threshold for firing of action potentials	Focal seizures originating in hippocampus	(42)
SCN1B	GEFS+	C121W extra-cellular (beta)	Reduced rate of inactivation of associated alpha subunits	Increased persistent current leading to lower threshold for firing of action potentials	Febrile, absense, myoclonic-astatic	(48, 74)
SCN1A	GEFS+	R1648H (D4S4) trans-membrane	Rapid recovery from inactivation	Propensity for repetitive firing	Febrile, absence, myoclonic, generalized tonic/clonic	(66)
SCN1A	GEFS+	T875M (D2S4) trans-membrane	Enhanced slow inactivation	Reduced channel activity	Febrile, absense, generalized tonic/clonic atonic, clonic	(66)
SCN1A	SMEI	Null	Complete loss of activity	Reduced channel activity to 50% in affected heterozygotes	Febrile, generalized tonic/clonic, absence, myoclonic, partial	(18)

mode, which would reduce the proportion of channel protein available for opening. The effect of this "functional hypomorph" may be similar to the null mutations described below. Four additional missense mutations in *SCN1A* have been identified in families with GEFS+, but their functional effects have not been described (26, 75).

SCN1A is physically located in a 1 Mb cluster with *SCN2A* and *SCN3A* (26). The three genes share 85% amino acid sequence identity and are coexpressed in neurons, but they differ in subcellular distribution and levels of expression in different types of neurons. A mutation in *SCN2A* was recently identified in a Japanese family with GEFS+ (72). The mutation, *R187W*, resulted in delayed channel inactivation, which could increase sodium influx and neuronal excitability. The mutation was not observed in 224 alleles from unaffected individuals. Overlapping clinical syndromes thus result from certain mutations in three sodium channel genes, *SCN1A*, *SCN2A*, and *SCN1B*.

SEVERE MYOCLONIC EPILEPSY OF INFANCY   Children with Severe Myoclonic Epilepsy of Infancy (SMEI) experience febrile seizures that progress to frequent severe afebrile seizures, delayed psychomotor development, ataxia, and myoclonic episodes. The elevated incidence of epilepsy in relatives of children with SMEI suggested a genetic predisposition in some cases (9, 64a). Because of the association of SCN1A with febrile seizures in GEFS+, Claes and colleagues screened Belgian children with SMEI for mutations in *SCN1A* (18). Seven de novo mutations were identified in affected children that were not present in either parent. Six of these mutations are frameshift or nonsense mutations resulting in null alleles with complete loss of function. These observations demonstrate for the first time that quantitative deficiency in a sodium channel can cause disease. The haploinsufficiency of human *SCN1A* contrasts with the recessive inheritance of null alleles of other sodium channels in the mouse (47). This work demonstrates that de novo mutations may be responsible for sporadic cases of epilepsy. In the future, it will be worthwhile including transcriptional regulatory regions of the sodium channel genes in mutational screening; this will require that the transcription start sites of these genes be identified.

TEMPORAL LOBE EPILEPSY IN THE Q54 MOUSE   Specific disease mechanisms can be tested in mouse models by introducing a gain-of-function mutation by microinjection of a transgene construct. The mutation *GAL879-881QQQ* in sodium channel SCN2A is located in the S4-S5 linker of transmembrane domain 2 and results in delayed inactivation and increased persistent current in *Xenopus* oocytes. Transgenic mice carrying this mutation exhibit a progressive seizure disorder that begins between 1 and 2 months of age and has several features of human temporal lobe epilepsy (42). Continuous EEG and video monitoring detected focal seizure activity originating in the hippocampus. During seizures the mice exhibit behavioral arrest and stereotyped repetitive behaviors. There is progressive cell loss and gliosis in the CA1-CA3 and hilus, reminiscent of the hippocampal sclerosis

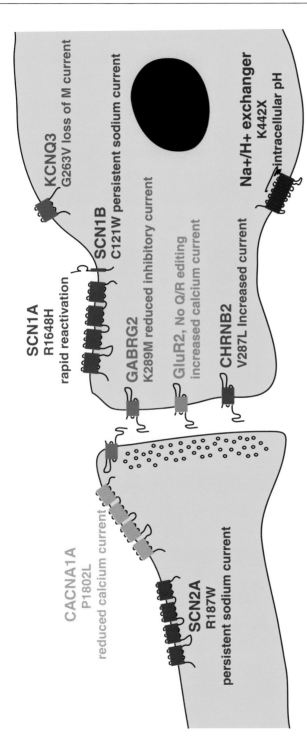

Figure 1 Mechanisms of ion channel mutations in idiopathic epilepsy. The functional effects of representative mutations in neuronal channels and their contributions to neuronal hyperexcitability are indicated.

seen in patients with temporal lobe epilepsy (40). Recordings of hippocampal CA1 neurons from presymptomatic mice detected a 50% increase in the amount of persistent sodium current between action potentials, which may increase the resting membrane potential of the cells and lead to hyperexcitability. The lifespan of these mice is greatly reduced. The Q54 mice can be genotyped presymptomatically, making them a valuable model for early interventions, and for determining whether cell loss and gliosis precede seizure activity. The Q54 mouse provides another example of seizures resulting from delayed inactivation of a voltage-gated sodium channel.

The sodium channel mutations demonstrate three types of genetic heterogeneity in epilepsy: (*a*) mutation in a single gene can generate different syndromes (*SCN1A* in GEFS+ and SMEI); (*b*) a single mutation can generate different types of seizures within a family (affected individuals in GEFS+ families); (*c*) mutations in different genes can produce the same syndrome (*SCN1A*, *SCN2A*, and *SCN1B* in GEFS+).

The electrophysiological effects of six sodium channel mutations have been measured in in vitro assays (Table 3). The abnormal properties of the isolated channels predict at least three different mechanisms of abnormal firing at the cellular level. The most common observation was delayed channel inactivation resulting in persistent sodium current. At the cellular level, persistent current may lead to seizures by reducing the threshold for firing of successive action potentials. This is also the most common defect associated with disease mutations of the skeletal and cardiac muscle sodium channels (3). An unusual mechanism was observed for the R1648H form of SCN1A, which recovers from inactivation more rapidly than wild-type channels and may increase the frequency of firing of action potentials. Surprisingly, alleles of *SCN1A* associated with decreased activity, T875M and the null mutations in SMEI, also predispose to seizures (Table 3). This effect of low activity may be unique to *SCN1A* due to aspects of intracellular localization, regional expression pattern, or a unique role in inhibitory neurons. It would be interesting to determine whether mice with reduced levels of the channels SCN2A and SCN8A are susceptible to seizures. *SCN3A* and *SCN8A*, which are expressed at high levels throughout the human brain, have not yet been screened for mutations in epilepsy families.

Voltage-Gated Potassium Channels

Two classes of voltage-gated potassium channels have been associated with seizures, the Kv channels and the KCNQ channels. The Kv channel KCN1A is involved in the recovery phase of the action potential. Like the delayed inactivation mutants of *SCN1A* and *SCN2A*, loss of function of this channel would result in prolonged sodium currents. Targeted inactivation of KCN1A in the mouse resulted in development of spontaneous tonic-clonic seizures that occur with high frequency beginning at 3 weeks of age and continue throughout adult life (65). Mutations of human *KCN1A*, which have as their primary effect an episodic ataxia (EA1), also predispose to seizures (78). Targeted inactivation of the Kv3.2 channel in the mouse resulted in increased seizure susceptibility in homozygotes (43).

The KCNQ2 and KCNQ3 proteins interact to generate the M-type current, a slowly activating and deactivating potassium conductance that contributes to sub-threshold electroexcitability of neurons and their responsiveness to synaptic inputs. The effect of the M current is to reduce neuronal excitability. Loss-of-function mutations for the potassium channels KCNQ2 and KCNQ3 have been identified in families with inherited benign neonatal convulsions, EBN1 (MIM#121200) and EBN2 (MIM#121201) (Table 1).

Voltage-Gated Calcium Channels

The voltage-gated calcium channel is composed of a large α subunit and three accessory subunits, β, γ, and $\alpha 2\delta$. Combinations of the products of multiple genes for each subunit generate a large number of molecular isoforms of the channel. Complete deficiency of the alpha subunit results in severe ataxia and late onset neurodegeneration (29a). Seven spontaneous mouse mutations in calcium channel subunits produce spike-wave epilepsy and ataxia, suggesting that human orthologs may be involved in absence epilepsy. Three mutations in the α subunit gene *Cacna1a* were identified by positional cloning. The *tottering* allele substitutes leucine for a highly conserved proline in the S5-S6 linker of domain II (29). The *Nagoya* mutation is an arginine-to-glycine substitution in the voltage-sensing S4 segment of domain III (49). Leaner mice have a splice site mutation in the coding region for the C-terminal domain of Cacna1a, which results in truncation of the open reading frame and expression of aberrant C-terminal sequences (29). A significant reduction in calcium current density was recorded from Purkinje cells of *tottering* mice (28). A splice site mutation in the $\beta 4$ subunit gene *Cacnb4* that produces a truncated, nonfunctional protein was identified in the mouse mutant *lethargic* (12). *lethargic* mice exhibit absence seizures, lethargy, and ataxia, and go through a crisis during the third week of life. Both *lethargic* and *tottering* mice exhibit decreased glutamatergic synaptic transmission in thalamic neurons, suggesting that the α and $\beta 4$ subunits are required for neurotransmitter release specific to glutaminergic synapses (14). A survey of human epilepsies identified two potential disease mutations in the ortholog *CACNB4* (25). Spontaneous mutations of the $\alpha 2\delta$ subunit arose in the mouse mutants *ducky* (4a) and *torpid*, which exhibit abnormal gait and seizures as well as dysgenesis of hindbrain and spinal cord (4, 5). Positional cloning of the mouse mutant *stargazer* identified a new protein with 25% amino acid sequence identity to the muscle-specific γ subunit of the voltage-gated calcium channel (44). Coexpression of the mutant subunit with the wild-type α subunit resulted in minor alterations in channel properties (16, 41a, 44). Recent work indicates that another function of the stargazin protein is synaptic localization of AMPA receptors, which are missing from cerebellar granule cells of *stargazer* mice (16).

Ligand-Gated GABA$_A$ Receptors

GABA$_A$ receptors are ligand-gated chloride channels composed of a pentameric assembly of homologous subunits (46). Since GABA$_A$ receptors mediate synaptic inhibition, mutations with reduced activity could produce neuronal hyperexcitability. The most abundant receptor isoform in brain is the $\alpha1\beta2\gamma2$ receptor. No mutations were detected in these subunit genes by direct mutation screening (53) or association analysis (61) in families with idiopathic generalized epilepsy. However, in 2001 the third locus for Generalized Epilepsy with Febrile Seizures[+], *GEFS+3*, was mapped to chromosome 5q34 in two large pedigrees from France and Australia (7, 74). The $\alpha1$ and $\gamma2$ genes are both located in this chromosome region, and exon sequencing identified $\gamma2$ mutations in both families. In the French family, the mutation *K289M* in the extracellular domain was present in all 13 affected members, 2 obligate carriers, and 1 asymptomatic individual (7). Affected individuals in the Australian family carry the mutation *R43Q*, located in a high-affinity benzodiazepine-binding domain (74).

The functional effects of the $\gamma2$ mutations were tested by electrophysiological analysis in *Xenopus* oocytes. Channels formed by coinjection of $\alpha1$, $\beta2$, $\gamma2$, and $\gamma2$ subunit mRNAs generate large inward currents over a range of GABA concentrations. The effect of the K289M mutation was to reduce maximum current amplitude by 90% at all GABA concentrations (7). This is consistent with the prediction that loss of GABA$_A$ receptor activity would lead to hyperexcitability and seizure susceptibility. The R43Q subunit exhibited normal activity but failed to be activated by diazepam (74). Oocytes injected with equal amounts of wild-type and mutant $\gamma2$ together with $\alpha1$ and $\beta2$ subunits displayed intermediate diazepam potentiation. The pathological effect of this mutation suggested to the authors that the GABA$_A$ receptor may be regulated in vivo by an unidentified endogenous diazepam-like molecule. The discovery of the $\gamma2$ mutations is likely to stimulate renewed attention to the other GABA$_A$ receptor subunits as candidates for epileptic disorders.

Ligand-Gated Acetylcholine Receptors

Like the GABA$_A$ receptor, the neuronal nicotinic acetylcholine receptor is a pentameric assembly of homologous subunits that mediates rapid synaptic transmission. This receptor has a nonselective cation pore. The major isoform in the brain is composed of $\alpha4$ and $\beta2$ subunits encoded by the *CHRNA4* and *CHRNB2* genes. Mutations in both of these subunits have been associated with the disorder autosomal dominant nocturnal frontal lobe epilepsy (ADNFLE, MIM 600513), characterized by brief seizures during light sleep that originate in the frontal lobe. The locus ENFL1 was mapped to chromosome 20q13 in an Australian pedigree (55). The $\alpha4$ subunit mutation *S248F* was identified in this family and was the first known human epilepsy mutation (69). A second $\alpha4$ mutation was later found in a Norwegian family (68).

The ADNFLE locus *ENFL3* was mapped to chromosome 1p21 in a three-generation pedigree from southern Italy (32). The acetylcholine $\beta 2$ subunit gene had previously been mapped to this location, and molecular analysis identified the amino acid substitution V287L (23). The mutation was present in eight affected individuals and four unaffected family members, demonstrating incomplete penetrance. The mutated residue is located in the second transmembrane domain that forms the channel pore. The electrophysiological effects of V287L were examined by patch clamp analysis of transfected HEK cells (23). Application of nicotine to wild-type $\alpha 4\beta 2$ channels results in rapid activation of an inward current that is followed by desensitization. The rate of desensitization was reduced in the V287L channel, and an intermediate rate of desensitization was observed when equal amounts of wild-type and mutant $\beta 2$ were injected. Slowed desensitization could lead to prolonged currents and increased neuronal excitability in response to cholinergic stimulation. Another mutation at the same amino acid residue, V287M, was identified in a Scottish family (54). V287M increased channel sensitivity to acetylcholine by approximately tenfold, with no change in the desensitization properties of the channel measured in *Xenopus* oocytes (54). This property is also consistent with increased neuronal excitability.

ION CHANNEL MUTATIONS AND NEURONAL HYPEREXCITABILITY

Many of the functional defects in the mutated voltage-gated and ligand-gated channels described above are predicted to increase the intrinsic excitability of neurons. Increased excitability could lead to increased neuronal firing and to episodes of synchronized firing by large numbers of neurons that constitute a seizure. The characteristics of these mutant channels strongly support this hypothesis regarding the origin of seizures. Examples of mutations predicted to predispose to neuronal hyperexcitability are shown in Figure 1 (see color insert).

SEIZURES IN MICE WITH TARGETED INACTIVATION OF ENDOGENOUS GENES

Homologous recombination in embryonic stem cells has been used for targeted inactivation (knock-out) of several thousand mouse genes during the past decade. Approximately two dozen of these lines, or 1% of the total, have been described as exhibiting spontaneous seizures (57) (for update, see Table 4 in the Supplemental Material link in the online version of this chapter or at http://www.annualreviews. org/). These genes may be considered candidate genes for human disorders mapped to the corresponding chromosome regions. If the inactivated genes are representative of the genome, several hundred genes might be targets for epilepsy mutations.

RNA Editing of Glutamate Receptor Subunits

The contribution of RNA editing of ligand-gated glutamate receptors to seizure susceptibility has been revealed by targeted mutations in the *GluR2*, *GluR6*, and *Adar2* genes of the mouse (58). *GluR2* and *GluR6* are related subunits of the neuronal AMPA and kainate glutamate receptors, respectively. The *GluR2* and *GluR6* transcripts are edited at the Q/R site, resulting in substitution of an arginine residue for glutamine that reduces calcium permeability and changes current/voltage relationships of the channel. Mice with a noneditable allele of the *GluR2* gene exhibit severe epilepsy and early lethality (11), and mice with a noneditable site in the *GluR6* gene demonstrated increased sensitivity to kainate-induced seizures (73). Severe epilepsy was also observed in mice lacking the enzyme that edits the Q/R site, ADAR2 (37). The seizures and lethality of the *Adar2* null mice could be rescued by a pre-edited allele of the *GluR2* gene, indicating that this transcript is the physiologically most important substrate for ADAR2 (37). Although glutamate receptor mutations have not been identified in human or mouse epilepsy, these experiments demonstrate that their coding sequences and intronic elements, as well as the editing enzymes, are potential targets for epileptogenic mutations. The voltage-gated sodium and calcium channels of Drosophila undergo RNA editing at multiple sites (58), but efforts to detect editing of the corresponding sites of the mammalian channels have been unsuccessful.

The Sodium/Hydrogen Transporter SLC9A

Targeted inactivation of the sodium/hydrogen transporter, a ubiquitously expressed transmembrane protein that functions in regulation of intracellular pH by export of H^+ ions in exchange for extracellular sodium ions, confirmed the earlier studies of a spontaneous mouse mutant, slow wave epilepsy (*swe*) (22) (Table 3). Homozygous *swe* mice exhibit spontaneous generalized tonic-clonic seizures as well as 3/sec spike-wave discharges accompanied by behavioral arrest that resemble human absence seizures. Survival and seizure intensity in *swe* homozygotes are influenced by genetic background. The targeted allele of *Slc9A* results in a similarly severe phenotype (8), demonstrating the sensitivity of neurons to intracellular pH.

IDENTIFYING GENES RESPONSIBLE FOR HUMAN EPILEPSIES LACKING CLEAR MODES OF INHERITANCE

The genes identified so far in human epilepsies have been found in families with clear Mendelian forms of inheritance and sufficient numbers of meioses to support positional cloning. However, Mendelian syndromes in large families comprise only a small proportion of all epilepsy. In most forms of epilepsy the genetic influences are complex and may involve the combined effects of multiple genes and environmental factors, each with a small effect on susceptibility. Identification of the genes involved in epilepsies with complex inheritance presents major challenges.

Linkage Analysis Using Collections of Small Families

Phenotype definition for linkage analysis is one of the most difficult problems in study design when multiple families are combined for analysis. Observed coseg-regation of a spectrum of clinical features in a Mendelian inheritance pattern can provide the basis for defining a syndrome, as was done for GEFS+ (64), auto-somal dominant partial epilepsy with auditory features (52), and familial partial epilepsy with variable foci (77). Alternatively, families can be selected for anal-ysis based on the correspondence of the symptoms of affected family members with the defined clinical epilepsy syndromes (20). This approach raises difficult questions about how to define the phenotype. For example, in studies of the fami-lies of probands with juvenile myoclonic epilepsy, affected family members also have a range of idiopathic generalized epilepsy syndromes such as pyknolepsy, juvenile absence, and awakening grand mal. In the absence of clear information about which syndromes should be assumed to result from the susceptibility gene, many studies have used several alternative phenotype definitions (e.g., juvenile myoclonic epilepsy only, all idiopathic generalized epilepsies, or idiopathic gen-eralized epilepsies plus EEG abnormalities without clinical seizures). Similarly, in the absence of a clear genetic model, LOD scores are sometimes estimated un-der multiple different models of penetrance and mode of inheritance. The use of multiple phenotype definitions and mode-of-inheritance assumptions inflates the type 1 error rates, and hence adjustment must be made to correct for this (39). Despite these problems, consistent evidence has been obtained for linkage of a susceptibility gene to the HLA region of chromosome 6p in families ascertained through subjects with juvenile myoclonic epilepsy.

In a linkage study of 130 families with idiopathic generalized epilepsy, Sander et al. provided evidence for a novel susceptibility locus on 3q26 (62). Suggestive LOD scores were also obtained for regions of chromosomes 2 and 14. Evidence for a susceptibility gene on chromosome 18 was obtained from a genome scan of 91 families with idiopathic generalized epilepsy (24). This study concluded that genetic classification cuts across syndrome classifications and that several interacting genes influence risk for idiopathic epilepsies (24). These complexities may explain why the gene on chromosome 6p has yet to be identified.

Association Studies Using Candidate Genes

Mild mutations in genes already identified as causing autosomal dominant forms of epilepsy, when inherited together with other predisposing mutations, may produce a state of neuronal excitability sufficient to generate seizures. To test the role of identified monogenic epilepsy genes, several studies have screened patients and controls for coding variants. These studies can potentially detect common variants that are present at higher frequencies in patients than in controls, as well as rare variants found only in patients, either of which could underlie the common epilepsies (56).

Based on the identification of *SCN1A* mutations in GEFS+2, we screened 226 additional patients for mutations in *SCN1A* (26). The sample included probands from 165 families containing multiple affected individuals (83 childhood or juvenile absence epilepsy, 72 juvenile myoclonic epilepsy, 4 generalized tonic clonic seizures, and 6 febrile seizures) and 61 sporadic patients with generalized epilepsy. One substitution affecting an evolutionary conserved residue (W1204R) was identified in a GEFS+ family with febrile seizures, JME, and other generalized seizures. Seven other coding variants were detected, but three were discordant with disease, three were observed in sporadic cases and could not be followed up, and one was a common polymorphism, T1067A, with the same allele frequencies of 0.66 and 0.33 in patients and in controls. In a similar study, Wallace and colleagues tested 53 probands with phenotypes consistent with GEFS+, including 36 familial and 17 isolated cases (75). Six mutations (three in *SCN1A* and three in *SCN1B*) were identified in the familial samples. No mutations were found in the sporadic samples. They estimate that *SCN1A* and *SCN1B* account for 17% of familial GEFS+.

Based on the calcium channel β_4 subunit gene mutation in the lethargic mouse, we screened 90 human families with dominantly inherited idiopathic generalized epilepsy for mutations in *CACNB4*, including 19 with childhood absence epilepsy, 22 with juvenile absence epilepsy, and 49 with juvenile myoclonic epilepsy. The premature terminating mutation *R482X* that eliminates 38 amino acids at the C-terminal end was identified in one patient with juvenile myoclonic epilepsy. The amino acid substitution *C104F* was identified in one family with generalized epilepsy and praxis-induced seizures, and also in another family with episodic ataxia (25). Neither mutation was observed in 510 control chromosomes. Electrophysiological analysis in *Xenopus* oocytes revealed that the R482X mutation increased the rate of inactivation of the coexpressed α1A subunit. This is predicted to reduce the net inward flow of calcium into neurons during the rapid alteration in membrane potential associated with an action potential. The *C104F* mutation did not significantly alter channel kinetics. It is difficult to prove disease causality for rare variants identified in small families that do not alter evolutionarily conserved residues or have dramatic effects on protein function. We consider *R482X* and *C104F* to be "potential disease mutations," and have proceeded to test them in a mouse model by generating transgenic mice expressing the *R482X* and *C104F* cDNAs in neurons. If these mutations are responsible for the dominant disorders in the original families, we would predict the development of seizures or seizure-susceptibility in the transgenic mice. These studies are still in progress.

KCNQ2 and *KCNQ3* are mutated in BNFC types 1 and 2. However, screening of a large collection of patients with common forms of idiopathic generalized epilepsy failed to identify additional mutations (35, 71). Analysis of noncoding polymorphisms in the voltage-gated calcium channel *CACNA1A* provided evidence for possible association with generalized idiopathic epilepsy (17).

Another approach was taken to test the acetylcholine receptor subunits as candidate genes in a group of small families with benign epilepsy of childhood with centrotemporal spikes (51). A partial genome scan was carried out using markers in the chromosome regions containing the receptor subunit genes. Analysis using an "affecteds only" model gave evidence for linkage to chromosome 15q14 in an estimated 70% of families. It will be interesting to see whether this finding is confirmed in other studies.

FUTURE PROSPECTS

Additional Monogenic Epilepsy Genes in Human

At least 15 additional loci have been mapped in human (see Table 5 in the Supplemental Material link in the online version of this chapter or at http://www. annualreviews.org), and responsible genes will likely be identified in the near future. The assembled, annotated human genome sequence provides a nearly complete list of candidate genes for the nonrecombinant intervals identified by linkage analysis in large families, and testing these genes for mutations can be carried out on a large scale by manual methods such as CSGE or SSCP gels or with automated methods such as dHPLC. In view of the severe clinical phenotypes of SMEI patients with reduced expression of SCN1A, it would be worthwhile to screen for variation in transcriptional regulatory regions. This will require experimental identification of the transcription start sites for the candidate genes, information not yet available for most of the epilepsy genes in Tables 1 and 2. Indirect evidence for hypomorphic alleles can be obtained by determining the ratio of allelic transcripts in RNA from individuals who are heterozygotes for SNPs in transcribed sequences. Underrepresentation of one allele could be caused by a mutation in regulatory sequences or reduced mRNA stability due to a transcribed variant. The mouse calcium channel mutations *lethargic* (12) and *spike wave epilepsy* (22) both reduce mRNA stability in addition to changing the protein sequence.

New Monogenic Epilepsies from Large-Scale Chemical Mutagenesis in the Mouse

A large-scale effort is in progress to generate novel mouse mutants with neurological disorders by in vivo chemical mutagenesis with ethylnitroso urea (ENU). The NIH-sponsored mutagenesis center at the Jackson Laboratory is screening for mutations that decrease the threshold for seizures after electroconvulsive shock (31). Already several seizure-prone models have emerged (e.g., http://www.jax.org/resources/documents/nmf/), and all mutants will be available to interested investigators. Genetic mapping of large numbers of mutations and isolation of the affected genes has promise for identification of additional molecular pathways involved in epileptogenesis. Improvements in imaging and automated behavioral monitoring in rodents will increase the utility of these models for understanding the pathogenesis of seizures.

Large-Scale Mutation Detection in Patients with Common Forms of Epilepsy

Application of efficient large-scale genomic screening methods to large populations of patients with common types of epilepsy will permit identification of underlying genetic variation correlated with the disease. "Resequencing chips" can detect any variant in a target sequence such as a panel of candidate genes, using genomic DNA as substrate (34). Evaluating the functional significance of rare variants is a major challenge (59), as gene frequencies may not differ between patients and controls for common susceptibility alleles and functional assays are not available for many proteins. Testing candidates by generating mouse models carrying the human mutations is feasible, but the high cost and effort involved will limit the application of this approach.

Development of Individualized Pharmacogenetic Therapies

A major motivation for research on epilepsy genetics is development of better pharmacological treatment for this debilitating disorder. An example of the future possibilities of allele-specific therapy for ion channel mutations is provided by recent work on an allele of the cardiac sodium channel gene *SCN5A* in the Brugada syndrome (1). In vitro analysis of the *D1790G* mutation in SCN5A predicted that the mutant channel would be resistant to the commonly used drug lidocaine, but would be responsive to another inhibitor, flecainide. Administration of flecainide to patients carrying the mutation confirmed the prediction, enabling these patients to receive an effective therapeutic agent that might not have been tried without knowledge of their specific mutation. This is an encouraging example of the practical applications that may be expected as the mutations responsible for epilepsy in individual patients are identified.

The past five years have seen many breakthroughs, with the identification of epilepsy genes providing new insight into molecular mechanisms of the disease. Emerging technologies for high-throughput mutation screening will greatly expand the ability to detect patient mutations. Continuing progress into the genetic basis of epilepsy will provide the basis for development of urgently needed new therapies.

ACKNOWLEDGMENTS

Jane Santoro provided expert assistance with preparation of the manuscript. We gratefully acknowledge research support from the National Institutes of Health (GM24872, NS34509, and NS20656), the March of Dimes, the Epilepsy Foundation, and the Muscular Dystrophy Association of America.

Visit the Annual Reviews home page at www.AnnualReviews.org

LITERATURE CITED

1. Abriel H, Wehrens XH, Benhorin J, Kerem B, Kass RS. 2000. Molecular pharmacology of the sodium channel mutation D1790G linked to the long-QT syndrome. *Circulation* 102:921–25

2. Alekov AL, Rahman MM, Mitrovic N, Lehmann-Horn F, Lerche H. 2000. A sodium channel mutation causing epilepsy in man exhibits subtle defects in fast inactivation and activation in vitro. *J. Physiol.* 529:533–39

3. Ashcroft F. 2000. *Ion Channels and Disease*. San Diego: Academic

4. Balaguero N, Barclay J, Mione M, Canti C, Brodbeck J, et al. 2000. Reduction in voltage-dependent calcium channel function in cerebellar purkinje cells of the mouse mutant ducky, which has a null mutation for the calcium channel accessory subunit $\alpha 2\delta 2$. *Soc. Neurosci. Abstr.* 26: 365

4a. Barclay J, Balaguero N, Mione M, Ackerman SL, Letts VA, et al. 2001. Ducky mouse phenotype of epilepsy and ataxia is associated with mutations in the *Cacna2d2* gene and decreased calcium channel current in cerebellar purkinje cells. *J. Neurosci.* 21:6095–104

5. Barclay J, Rees M. 2000. Genomic organization of the mouse and human $\alpha 2\delta 2$ voltage-dependent calcium channel subunit genes. *Mamm. Genome* 11:1142–44

6. Baulac S, Gourfinkel-An I, Picard F, Rosenberg-Bourgin M, Prud'homme JF, et al. 1999. A second locus for familial generalized epilepsy with febrile seizures plus maps to chromosome 2q21-q33. *Am. J. Hum. Genet.* 65:1078–85

7. Baulac S, Huberfeld G, Gourfinkel-An I, Mitropoulou G, Beranger A, et al. 2001. First genetic evidence of GABA$_A$ receptor dysfunction in epilepsy: a mutation in the $\gamma 2$-subunit gene. *Nat. Genet.* 28:46–48

8. Bell SM, Schreiner CM, Schultheis PJ, Miller ML, Evans RL, et al. 1999. Targeted disruption of the murine Nhe1 locus induces ataxia, growth retardation, and seizures. *Am. J. Physiol.* 276:C788–95

9. Benlounis A, Nabbout R, Feingold J, Parmeggiani A, Guerrini R, et al. 2001. Genetic predisposition to severe myoclonic epilepsy in infancy. *Epilepsia* 42:204–9

10. Berkovic S, Ottman R. 2001. Molecular genetics of the idiopathic epilepsies: the next steps. *Epileptic Disord.* 2:179–81

11. Brusa R, Zimmermann F, Koh DS, Feldmeyer D, Gass P, et al. 1995. Early-onset epilepsy and postnatal lethality associated with an editing-deficient GluR-B allele in mice. *Science* 270:1677–80

12. Burgess DL, Jones JM, Meisler MH, Noebels JL. 1997. Mutation of the Ca^{2+} channel β subunit gene *Cchb4* is associated with ataxia and seizures in the lethargic (*lh*) mouse. *Cell* 88:385–92

13. Burgess DL, Noebels JL. 2000. Calcium channel defects in models of inherited generalized epilepsy. *Epilepsia* 41:1074–75

14. Caddick SJ, Wang C, Fletcher CF, Jenkins NA, Copeland NG, Hosford DA. 1999. Excitatory but not inhibitory synaptic transmission is reduced in lethargic (*Cacnb4^{lh}*) and tottering (*Cacna1a^{tg}*) mouse thalami. *J. Neurophysiol.* 81:2066–74

15. Catterall WA. 1999. Molecular properties of brain sodium channels: an important target for anticonvulsant drugs. *Adv. Neurol.* 79:441–56

16. Chen L, Chetkovich DM, Petralia RS, Sweeney NT, Kawasaki Y, et al. 2000. Stargazin regulates synaptic targeting of AMPA receptors by two distinct mechanisms. *Nature* 408:936–43

17. Chioza B, Wilkie H, Nashef L, Blower J, McCormick D, et al. 2001. Association between the α_{1a} calcium channel gene

CACNA1A and idiopathic generalized epilepsy. *Neurology* 56:1245–46

18. Claes L, Del-Favero J, Ceulemans B, Lagae L, Van Broeckhoven C, De Jonghe P. 2001. De novo mutations in the sodium-channel gene *SCN1A* cause severe myoclonic epilepsy of infancy. *Am. J. Hum. Genet.* 68:1327–32

19. Comm. Classif. Terminol. Int. League Against Epilepsy. 1981. Proposal for revised clinical and electroencephalographic classification of epileptic seizures. *Epilepsia* 22:489–501

20. Comm. Classif. Terminol. Int. League Against Epilepsy. 1989. Proposal for revised classification of epilepsies and epileptic syndromes. *Epilepsia* 30:389–99

21. Deleted in proof

22. Cox GA, Lutz CM, Yang CL, Biemesderfer D, Bronson RT, et al. 1997. Sodium/hydrogen exchanger gene defect in slow-wave epilepsy mutant mice. *Cell* 91:139–48

23. De Fusco M, Becchetti A, Patrignani A, Annesi G, Gambardella A, et al. 2000. The nicotinic receptor *β*2 subunit is mutant in nocturnal frontal lobe epilepsy. *Nat. Genet.* 26:275–76

24. Durner M, Keddache MA, Tomasini L, Shinnar S, Resor SR, et al. 2001. Genome scan of idiopathic generalized epilepsy: evidence for major susceptibility gene and modifying genes influencing the seizure type. *Ann. Neurol.* 49:328–35

25. Escayg A, De Waard M, Lee DD, Bichet D, Wolf P, et al. 2000. Coding and noncoding variation of the human calcium-channel *β*₄-subunit gene *CACNB4* in patients with idiopathic generalized epilepsy and episodic ataxia. *Am. J. Hum. Genet.* 66:1531–39

26. Escayg A, Heils A, MacDonald BT, Haug K, Sander T, Meisler MH. 2001. A novel *SCN1A* mutation associated with generalized epilepsy with febrile seizures plus-and prevalence of variants in patients with epilepsy. *Am. J. Hum. Genet.* 68:866–73

27. Escayg A, MacDonald BT, Meisler MH, Baulac S, Huberfeld G, et al. 2000. Mutations of *SCN1A*, encoding a neuronal sodium channel, in two families with GEFS+2. *Nat. Genet.* 24:343–45

28. Fletcher CF, Frankel WN. 1999. Ataxic mouse mutants and molecular mechanisms of absence epilepsy. *Hum. Mol. Genet.* 8:1907–12

29. Fletcher CF, Lutz CM, O'Sullivan TN, Shaughnessy JD Jr, Hawkes R, et al. 1996. Absence epilepsy in tottering mutant mice is associated with calcium channel defects. *Cell* 87:607–17

29a. Fletcher CF, Tottene A, Lennon VA, Wilson SM, Dubel SJ, et al. 2001. Dystonia and cerebellar atrophy in Cacna1a null mice lacking P/Q calcium channel activity. *FASEB J.* 15:1288–90

30. Frankel WN. 1999. Detecting genes in new and old mouse models for epilepsy: a prospectus through the magnifying glass. *Epilepsy Res.* 36:97–110

31. Frankel WN, Taylor L, Beyer B, Tempel BL, White HS. 2001. Electroconvulsive thresholds of inbred mouse strains. *Genomics* 74:306–12

32. Gambardella A, Annesi G, De Fusco M, Patrignani A, Aguglia U, et al. 2000. A new locus for autosomal dominant nocturnal frontal lobe epilepsy maps to chromosome 1. *Neurology* 55:1467–71

33. Gardiner RM. 2000. Impact of our understanding of the genetic aetiology of epilepsy. *J. Neurol.* 247:327–34

34. Hacia JG. 1999. Resequencing and mutational analysis using oligonucleotide microarrays. *Nat. Genet.* 21:42–47

35. Haug K, Hallmann K, Horvath S, Sander T, Kubisch C, et al. 2000. No evidence for association between the *KCNQ3* gene and susceptibility to idiopathic generalized epilepsy. *Epilepsy Res.* 42:57–62

36. Hauser WA, Annegers JF, Kurland LT. 1993. Incidence of epilepsy and unprovoked seizures in Rochester, Minnesota: 1935–1984. *Epilepsia* 34:453–68

37. Higuchi M, Maas S, Single FN, Hartner

J, Rozov A, et al. 2000. Point mutation in an AMPA receptor gene rescues lethality in mice deficient in the RNA-editing enzyme ADAR2. *Nature* 406:78–81

38. Hille B. 2001. *Ionic Channels of Excitable Membranes.* Sunderland, MA: Sinauer. 3rd ed.

39. Hodge SE, Abreu PC, Greenberg DA. 1997. Magnitude of type I error when single-locus linkage analysis is maximized over models: a simulation study. *Am. J. Hum. Genet.* 60:217–27

40. Houser CR. 1999. Neuronal loss and synaptic reorganization in temporal lobe epilepsy. *Adv. Neurol.* 79:743–61

41. Junaid MA, Pullarkat RK. 2001. Biochemistry of neuronal ceroid lipofuscinoses. *Adv. Genet.* 45:93–106

41a. Kang M-G, Chen C-C, Felix R, Letts VA, Frankel WN, et al. 2001. Biochemical and biophysical evidence for 2 subunit association with neuronal voltage-activated Ca^{2+} channels. *J. Biol. Chem.* In press

42. Kearney JA, Plummer NW, Smith MR, Kapur J, Cummins TR, et al. 2001. A gain-of-function mutation in the sodium channel gene *Scn2a* results in seizures and behavioral abnormalities. *Neuroscience* 102:307–17

43. Lau D, Vega-Saenz de Miera EC, Contreras D, Ozaita A, Harvey M, et al. 2000. Impaired fast-spiking, suppressed cortical inhibition, and increased susceptibility to seizures in mice lacking Kv3.2 K^{+} channel proteins. *J. Neurosci.* 20:9071–85

44. Letts VA, Felix R, Biddlecome GH, Arikkath J, Mahaffey CL, et al. 1998. The mouse stargazer gene encodes a neuronal Ca^{2+} channel gamma subunit. *Nat. Genet.* 19:340–47

45. McNamara JO. 1999. Emerging insights into the genesis of epilepsy (Review). *Nature* 399:A15–22

46. Mehta AK, Ticku MK. 1999. An update on $GABA_A$ receptors. *Brain Res. Rev.* 29:196–217

47. Meisler MH, Kearney J, Escayg A, MacDonald BT, Sprunger LK. 2001. Sodium channels and neurological disease: insights from Scn8a mutations in the mouse. *Neuroscientist* 7:136–45

48. Moran O, Conti F. 2001. Skeletal muscle sodium channel is affected by an epileptogenic $\beta 1$ subunit mutation. *Biochem. Biophys. Res. Commun.* 282:55–59

49. Mori M, Konno T, Ozawa T, Murata M, Imoto K, Nagayama K. 2000. Novel interaction of the voltage-dependent sodium channel (VDSC) with calmodulin: Does VDSC acquire calmodulin-mediated Ca^{2+}-sensitivity? *Biochemistry* 39:1316–23

50. Moulard B, Guipponi M, Chaigne D, Mouthon D, Buresi C, Malafosse A. 1999. Identification of a new locus for generalized epilepsy with febrile seizures plus (GEFS+) on chromosome 2q24-q33. *Am. J. Hum. Genet.* 65:1396–400

51. Neubauer BA, Fiedler B, Himmelein B, Kampfer F, Lassker U, et al. 1998. Centrotemporal spikes in families with rolandic epilepsy: linkage to chromosome 15q14. *Neurology* 51:1608–12

52. Ottman R, Risch N, Hauser WA, Pedley TA, Lee JH, et al. 1995. Localization of a gene for partial epilepsy to chromosome 10q. *Nat. Genet.* 10:56–60

53. Peters HC, Kämmer G, Volz A, Kaupmann K, Ziegler A, et al. 1998. Mapping, genomic structure, and polymorphisms of the human GABABR1 receptor gene: evaluation of its involvement in idiopathic generalized epilepsy. *Neurogenetics* 2:47–54

54. Phillips HA, Favre I, Kirkpatrick M, Zuberi SM, Goudie D, et al. 2001. *CHRNB2* is the second acetylcholine receptor subunit associated with autosomal dominant nocturnal frontal lobe epilepsy. *Am. J. Hum. Genet.* 68:225–31

55. Phillips HA, Scheffer IE, Berkovic SF, Hollway GE, Sutherland GR, Mulley JC. 1995. Localization of a gene for autosomal dominant nocturnal frontal lobe

epilepsy to chromosome 20q13.2. *Nat. Genet.* 10:117–18

56. Pritchard JK. 2001. Are rare variants responsible for susceptibility to complex diseases? *Am. J. Hum. Genet.* 69:124–37

57. Puranam RS, McNamara JO. 1999. Seizure disorders in mutant mice: relevance to human epilepsies. *Curr. Opin. Neurobiol.* 9:281–87

58. Reenan RA. 2001. RNA world meets behavior: AI pre-mRNA editing in animals. *Trends Genet.* 17:53–56

59. Risch N, Spiker D, Lotspeich L, Nouri N, Hinds D, et al. 1999. A genomic screen of autism: evidence for a multilocus etiology. *Am. J. Hum. Genet.* 65:493–507

60. Ryan SG. 1999. Ion channels and the genetic contribution to epilepsy. *J. Child Neurol.* 14:58–66

61. Sander T, Peters C, Kammer G, Samochowiec J, Zirra M, et al. 1999. Association analysis of exonic variants of the gene encoding the $GABA_B$ receptor and idiopathic generalized epilepsy. *Am. J. Med. Genet.* 88:305–10

62. Sander T, Schulz H, Saar K, Gennaro E, Concetta Riggio M, et al. 2000. Genome search for susceptibility loci of common idiopathic generalised epilepsies. *Hum. Mol. Genet.* 9:1465–72

63. Scheffer IE, Berkovic SF. 1997. Generalized epilepsy with febrile seizures plus a genetic disorder with heterogeneous clinical phenotypes. *Brain* 120:479–90

64. Scheffer IE, Berkovic SF. 2000. Genetics of the epilepsies. *Curr. Opin. Pediatr.* 12:536–42

64a. Singh R, Andermann E, Whitehouse WP, Harvey AS, Keene DL, et al. 2001. Severe myoclonic epilepsy of infancy: extended spectrum of GEFS+? *Epilepsia* 42:837–44

65. Smart SL, Lopantsev V, Zhang CL, Robbins CA, Wang H, et al. 1998. Deletion of the Kv1.1 potassium channel causes epilepsy in mice. *Neuron* 20:809–19

66. Spampanato J, Escayg A, Meisler MH,

Goldin AL. 2001. Functional effects of two voltage-gated sodium channel mutations that cause generalized epilepsy with febrile seizures plus type 2. *J. Neurosci*

67. Stafstrom CE, Tempel BL. 2000. Epilepsy genes: the link between molecular dysfunction and pathophysiology. *Ment. Retard. Dev. Disabil. Res. Rev.* 6:281–92

68. Steinlein OK, Magnusson A, Stoodt J, Bertrand S, Weiland S, et al. 1997. An insertion mutation of the CHRNA4 gene in a family with autosomal dominant nocturnal frontal lobe epilepsy. *Hum. Mol. Genet.* 6:943–47

69. Steinlein OK, Mulley JC, Propping P, Wallace RH, Phillips HA, et al. 1995. A missense mutation in the neuronal nicotinic acetylcholine receptor alpha 4 subunit is associated with autosomal dominant nocturnal frontal lobe epilepsy. *Nat. Genet.* 11:201–3

70. Steinlein OK, Noebels JL. 2000. Ion channels and epilepsy in man and mouse. *Curr. Opin. Genet. Dev.* 10:286–91

71. Steinlein OK, Stoodt J, Biervert C, Janz D, Sander T. 1999. The voltage gated potassium channel *KCNQ2* and idiopathic generalized epilepsy. *NeuroReport* 10:1163–66

72. Sugawara T, Tsurubuchi Y, Agarwala KL, Ito M, Fukuma G, et al. 2001. A missense mutation of the Na^+ channel α_{II} subunit gene $Na_{v1.2}$ in a patient with febrile and afebrile seizures causes channel dysfunction. *Proc. Natl. Acad. Sci. USA* 98:6384–89

73. Vissel B, Royle GA, Christie BR, Schiffer HH, Ghetti A, et al. 2001. The role of RNA editing of kainate receptors in synaptic plasticity and seizures. *Neuron* 29:217–27

74. Wallace RH, Marini C, Petrou S, Harkin LA, Bowser DN, et al. 2001. Mutant $GABA_A$ receptor $\gamma2$-subunit in childhood absence epilepsy and febrile seizures. *Nat. Genet.* 28:49–52

75. Wallace RH, Scheffer IE, Barnett S, Richards M, Dibbens L, et al. 2001.

Neuronal sodium-channel α1-subunit mutations in generalized epilepsy with febrile seizures plus. *Am. J. Hum. Genet.* 68:859–65

76. Wallace RH, Wang DW, Singh R, Scheffer IE, George AL Jr, et al. 1998. Febrile seizures and generalized epilepsy associated with a mutation in the Na^+-channel beta 1 subunit gene SCN1B. *Nat. Genet.* 19:366–70

77. Xiong L, Labuda M, Li DS, Hudson TJ, Desbiens R, et al. 1999. Mapping of a gene determining familial partial epilepsy with variable foci to chromosome 22q11–q12. *Am. J. Hum. Genet.* 65:1698–710

78. Zuberi SM, Eunson LH, Spauschus A, De Silva R, Tolmie J, et al. 1999. A novel mutation in the human voltage-gated potassium channel gene (Kv1.1) associates with episodic ataxia type 1 and sometimes with partial epilepsy. *Brain* 122:817–25

Annu. Rev. Genet. 2001. 35:589–646

MOLECULAR GENETICS OF HEARING LOSS

Christine Petit, Jacqueline Levilliers, and Jean-Pierre Hardelin

Unité de Génétique des Déficits Sensoriels, CNRS URA 1968, Institut Pasteur, 25 rue du Dr Roux, 75724 Paris cedex 15, France;
e-mail: cpetit@pasteur.fr

Key Words   human genetics, deafness genes, audition, cochlea, hair cells

■ **Abstract**   Hereditary isolated hearing loss is genetically highly heterogeneous. Over 100 genes are predicted to cause this disorder in humans. Sixty loci have been reported and 24 genes underlying 28 deafness forms have been identified. The present epistemic stage in the realm consists in a preliminary characterization of the encoded proteins and the associated defective biological processes. Since for several of the deafness forms we still only have fuzzy notions of their pathogenesis, we here adopt a presentation of the various deafness forms based on the site of the primary defect: hair cell defects, nonsensory cell defects, and tectorial membrane anomalies. The various deafness forms so far studied appear as monogenic disorders. They are all rare with the exception of one, caused by mutations in the gene encoding the gap junction protein connexin26, which accounts for between one third to one half of the cases of prelingual inherited deafness in Caucasian populations.

CONTENTS

0066-4197/01/1215-0589$14.00

INTRODUCTION

Information on how the auditory system, especially the cochlea, functions at the molecular level may be anticipated once the genes underlying its dysfunction are characterized. By 1994, when the research on causative deafness genes began to bear fruit, only two proteins specific to the inner ear had been characterized (164, 184, 264) and an unconventional myosin proposed as possibly being involved in the adaptation process of auditory mechanotransduction (111, 141). Classical biochemical and molecular genetic approaches to characterizing unknown molecules have not been practical because of the small number of each cell type in the cochlea (e.g., only around 10^4 hair cells). In the retina, by contrast, the large number of photoreceptor cells (120 million) has allowed many components of the phototransduction cascade to be identified by classical approaches, even before their roles in hereditary retinopathies were recognized (255). A genetic approach to the molecular basis of inner ear function therefore seemed especially promising (242).

This review outlines recent advances in explaining human hereditary deafness in molecular terms, focusing on isolated (i.e., nonsyndromic) hearing loss. Although research into these forms of deafness is relatively recent, the number of genes identified is growing rapidly.

GENERAL CONSIDERATIONS

Hearing loss can appear at any age and with any degree of severity. A severe defect that presents in early childhood has dramatic effects on speech acquisition and literacy. Later onset of severe hearing defect seriously compromises the quality of life, as the affected individual becomes increasingly isolated socially.

Hearing impairment is classified by the degree of severity of the hearing loss (mild, moderate, severe, profound) for the better-hearing ear and by the site of the defect. Conductive hearing loss refers to external and/or middle ear defects, and sensorineural hearing loss to the other defects, i.e., anywhere from the inner ear to the cortical auditory centers of the brain. Most cases of sensorineural hearing loss are due to inner ear defects.

Approximately 1/1000 individuals is affected by severe or profound deafness at birth or during early childhood, i.e., the prelingual period. Syndromic deafness, i.e., deafness associated with other defects, contributes to about 30% of the cases and may be conductive, sensorineural, or mixed. In contrast, the prelingual nonsyndromic forms are almost exclusively sensorineural. A further 1/1000 children becomes deaf before adulthood, usually in a progressive and less severe form. Finally, 0.3% and 2.3% of the population manifest a hearing loss greater than 65 dB between the ages of 30 and 50 years and between 60 and 70 years, respectively. The late-onset forms are often conductive, with otosclerosis, defined as a fixation of the stapes footplate to the oval window (see below), accounting for a large proportion. In developed countries, between 6% to 8% of the population suffers from hearing loss.

THE STRUCTURE OF THE EAR

The three compartments of the human ear (Figure 1*A*, see color insert) are made up of the external ear, which comprises the auricle and the external auditory canal, and is closed by the tympanic membrane; the middle ear, which consists of an air cavity containing a chain of three ossicles (malleus, incus, stapes); and the inner ear, a complex membranous labyrinth in a cavity of the temporal bone (bony labyrinth), which is filled with a liquid, the endolymph, and immersed in another liquid, the perilymph. The inner ear comprises six mechanosensory organs: the snail-shaped cochlea, which is the auditory sense organ, and the five vestibular end organs, which are responsible for balance. The human cochlea detects sound frequencies between 20 Hz and 20 kHz. The vestibule is composed of the saccule, the utricle, and the three semicircular canals, which respond to linear and angular accelerations. The cochlea and the vestibule both derive from the otic placode (90, 319) and share several structural and functional features. At the junction between the utricle and the saccule is the endolymphatic canal, ending with the endolymphatic sac. This outgrowth of the membranous labyrinth ensures the resorption of the endolymph.

The six sensory patches of the inner ear are composed of sensory and supporting cells, and are covered by acellular membranes. The sensory cells are termed hair cells in reference to the bundle of stiff actin-filled microvilli, improperly named stereocilia, present at their apical surface. Thirty to 300 stereocilia form the hair bundle, which is the mechanoreceptive structure of the hair cell. The stereocilia always form into staircase or organ-pipe arrays, in which each stereociliar tip is connected to the next-taller stereocilium by a tip link (Figure 2). In addition, there are many connecting links along the length of the stereocilia (lateral links). The apical parts of the lateral walls of the hair cells are sealed by tight junctions to the walls of surrounding supporting cells, forming the reticular lamina. As a result, the hair bundle and the remaining surfaces of the hair cell are immersed in liquids of different ionic compositions, the endolymph and the perilymph, respectively. The endolymph is rich in potassium (about 150 mM), almost devoid of sodium (1 mM), and poor in calcium (0.02 mM). The perilymph has a low K^+ concentration (3.5 mM), a high Na^+ concentration (140 mM), and a higher Ca^{2+} concentration (1 mM) than the endolymph.

The sensory patch of the cochlea is called the organ of Corti (Figure 1*B*, see color insert). It lies on the basilar membrane and has two types of hair cells: the single row of inner hair cells (IHC) (Figure 1*C*, see color insert), which are the actual sensory cells that account for most of the cochlear nerve influx sent to the auditory centers, and the triple row of outer hair cells (OHC), contractile cells that amplify the auditory stimulus (see below). A great variety of epithelial cells line the membranous labyrinth of the cochlea (see legend to Figure 1*B*). The stria

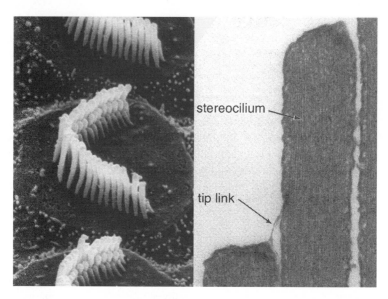

Figure 2   Stereocilia and their tip links. Scanning electron microscopy showing the hair bundle of an outer hair cell (*left panel*), and detail of the stereocilia from two adjacent rows (*right panel*). The tip link, an elastic filament that connects the apex of each stereocilium to the side of the taller adjacent one, is visible (Courtesy of R. Romand, France).

vascularis, a secretory structure in the lateral wall of the cochlea, consists of two cell barriers formed by the marginal cells and the basal cells. Each barrier consists of a continuous sheet of cells joined by tight junctional complexes. Between these barriers is the intrastrial space with a capillary bed and a discontinuous layer of intermediate cells. The tight junctions connecting the marginal, basal, and endothelial cells make the intrastrial space a separate fluid compartment. The basal cells are joined with gap junctions to intermediate cells and to fibrocytes of the adjacent connective tissue (spiral ligament) (163), indicating exchanges between these three cell types. In contrast, strial marginal cells are not coupled to each other or to other cells by gap junctions. The stria vascularis secretes K^+ into the cochlear endolymph and produces the endocochlear potential (311), i.e., about $+80$ mV differentiates the endolymphatic from the perilymphatic compartment (341).

THE FUNCTIONING OF THE COCHLEA

The external ear collects air wave pressure and transfers it to the tympanic membrane, where the vibration is picked up by the three ossicles of the middle ear and transmitted to another membrane covering a hole in the bony labyrinth (oval window), thereby producing liquid waves in the inner ear. The vibration of the basilar membrane, which peaks at a location dependent on the sound frequency (341), is transferred to the organ of Corti and thence to the hair cells. The relative motion between the hair cell apical surface and the overlaying acellular, the tectorial membrane, deflects the stereocilia bundle. The tension of the tip links increases and follows an opening of the mechanotransduction cationic channels located near the stereociliar tips (12, 206). The transducer current is carried mainly by a K^+ influx driven by an approximately 150 mV electric gradient between the endolymph and the hair cell cytoplasm. This current results in the depolarization of the hair cell. In the IHC, the depolarization induces a Ca^{2+} influx through the basal membrane, triggering fusion of the synaptic vesicles with the plasma membrane; upon neurotransmitter (probably glutamate) release, the afferent nerve fibers transmit to the brain a pattern of action potentials encoding different characteristics of the sound stimulus, including intensity, time course, and frequency (low frequencies elicit a neuronal phase-locked response, whereas fibers devoted to the high frequencies use the tonotopy to code frequency). The spectral analysis of sound frequency in mammals relies not only on the biophysical characteristics of the basilar membrane but also on the tuning response of each hair cell to a particular frequency. Hair cells responding to gradually varying frequencies are linearly positioned along the longitudinal axis of the cochlea, thus forming a tonotopic map. High frequencies are analyzed at the base of the cochlea and low frequencies at the apex. Experimental damage to OHCs has established the role of these cells in the exquisite sensitivity and frequency-resolving capacity of the mammalian hearing organ (274). The lateral membrane of these cells contracts rapidly in response to the depolarization caused by the apical transducer current (95, 102). These cyclic length changes of the OHCs are thought to amplify the acoustic stimulation by boosting the sound-induced vibration of the basilar membrane.

THE CAUSES OF HEARING LOSS

Except for embryopathies due to rubella, toxoplasmosis, or cytomegalovirus infection that can lead to polymalformations including hearing loss, syndromic forms of deafness have a genetic origin. Hundreds of syndromes associating hearing loss and miscellaneous disorders of the musculoskeletal, cardiovascular, urogenital, nervous, endocrine, digestive, or integumentary systems have been described (115), and the causative genes have been identified for ~100 [see (244) for an updated list of syndromes and corresponding genes].

Nonsyndromic forms of deafness can be caused by environmental and/or genetic factors. The hereditary nature of nonsyndromic deafness was first reported in the sixteenth century by Johannes Schenck (114, 305). In 1621, the papal physician Paolus Zacchias recommended deaf people not to marry because of the risk of having deaf children (61). Early interest in the mode of inheritance of hearing loss was stimulated by observations by the French physician Pierre Ménière in 1846 (217). In his lecture, "Upon marriage between relatives considered as the cause of congenital deaf-mutism," he was the first to recognize the autosomal recessive origin of "deaf-mutism" (218), now well established as the principal mode of transmission. In developed countries, where prelingual, nonsyndromic deafness mostly presents as sporadic cases, over two thirds have a genetic origin (71) (see below). Combined with syndromic forms, this suggests that over 80% of all cases of congenital deafness is of genetic origin in these countries.

Late-onset forms of hearing loss used to be thought to result from a combination of genetic and environmental causes. However, the increasing number of families being identified with late-onset deafness indicates that the genetic contribution has been underestimated. Recent evidence demonstrates a significant genetic basis in otosclerosis, for instance (318, 328).

All hereditary forms of hearing loss analyzed to date present as monogenic diseases, with few exceptions (5, 15, 222). However, the phenotypic heterogeneity most frequently observed for a given mutation argues for the contribution of modifier genes (100).

MOLECULAR GENETICS OF NONSYNDROMIC (ISOLATED) DEAFNESS

Nonsyndromic forms of hereditary deafness are classified by their mode of inheritance; DFN, DFNA, and DFNB refer to deafness forms inherited on the X chromosome-linked, autosomal dominant, and autosomal recessive modes of transmission, respectively. About 80% of the cases of prelingual deafness are DFNB forms, whereas most of the late-onset forms are DFNA forms.

The causative genes only began to be identified a few years ago because of difficulties in chromosomal mapping. Obstacles to mapping included (a) extreme genetic heterogeneity, (b) the absence of clinically distinctive signs for the various gene defects, and (c) the tendency of deaf people in developed countries to intermarry (see 242). Studying deaf families living in geographic isolates proved effective. Of the 60 loci now identified, 31 are for the DFNA forms, 23 for the DFNB forms, 4 for the DFN forms, and 2 linked to the mitochondrial genome. Only the 28 forms with an identified gene are discussed here. These have been classified by the primary target of the inner ear defect, i.e., hair cells, nonsensory cell types, tectorial membrane, and unknown (summarized in Table 1; see also http://www.ncbi.nlm.nih.gov/Omim/searchomim.html; http://www.uia.ac.be/dnalab/hhh).

Several mouse mutants present with an isolated inner ear defect. For all of these mutants, except *deafness* (*dn*) (153), the hearing loss is accompanied by vestibular dysfunction. Mutant behavior is characterized by circling, head tossing, and inability to swim. In humans, however, hereditary deafness is rarely associated with balance problems, probably because there is cerebral compensation for peripheral vestibular defects; the sensory substitution is based on the highly significant inputs of the visual and proprioceptive systems. Mouse mutants with only an inner ear defect involving a gene orthologous to a human deafness gene are discussed together with the corresponding human forms of deafness. Mutants for which the human orthologous gene has not yet been implicated in deafness are listed in Table 1.

Several human genes that underlie syndromic forms of deafness are also responsible for isolated forms of deafness (see below) (Table 1). No direct correlation can generally be drawn between the type of the mutation and the association with a syndromic or a nonsyndromic form of hearing loss. Moreover, phenotypic analyses coupled with mutation detection in some families affected by Pendred syndrome (209) or Usher syndrome (19, 196, 266) argue in favor of a kind of phenotypic continuum between particular forms of nonsyndromic and syndromic deafness. In these cases, modifier genes are highly likely to be a contributing factor.

Forms of Deafness Caused by a Hair Cell Defect

MYOSIN VIIA DEFECTS: DFNA11, DFNB2 (AND USHER 1B SYNDROME)   DFNB2 (MIM600060), the second reported locus responsible for an autosomal recessive form of deafness, was mapped to chromosome 11q13.5 through the study of a consanguineous family from central Tunisia (122); in the corresponding murine chromosomal region, the locus for the *shaker-1* phenotype (*sh-1*) had been assigned (36). Seven *sh-1* alleles have been described (73, 265). The locus for the Usher 1B syndrome (USH1B, MIM276903) in humans was also mapped to this region of chromosome 11 (166). Usher syndrome [reviewed in (243)] is the most frequent cause of deafness associated with blindness. Three clinical subtypes have been distinguished; the most severe is USH1, characterized by congenital deafness, vestibular dysfunction, and a loss of vision beginning in late childhood. USH1 is genetically heterogeneous, and the USH1B locus was evaluated to account for 75% of USH1 cases (32). The colocalization of USH1B and DFNB2 suggested that a single gene might be responsible for both diseases. Thereafter, DFNA11 (MIM601317), a progressive form of autosomal dominant deafness, was also mapped to the same chromosomal interval (312).

Positional cloning of *sh-1* in the mouse led to the identification of a gene, *Myo7a*, predicted to encode an unconventional myosin (myosin VIIA). Mutations in exons encoding the motor head of myosin VIIA were detected concurrently in *sh-1* mutants (110) and in USH1B patients (353). More recently, a mutation in the orthologous gene was detected in the zebrafish circler mutant *mariner* (82). The mouse and zebrafish mutants, however, do not exhibit the retinal degeneration of the USH1B syndrome. Distinct *MYO7A* mutations were subsequently shown to be responsible for DFNB2 and DFNA11 (see below).

Unconventional myosins form a large family that has been divided into 16 classes (20). These motor proteins move along the actin filaments using the energy generated by the hydrolysis of ATP. The structure of unconventional myosins

TABLE 1 Genes underlying isolated deafness

Primary defect	Gene	Protein	Human deafness	Deaf mouse mutant	Type of molecule
Hair cells	*MYO7A*	Myosin VIIA	DFNB2 ± retinopathy (Usher 1B)[b] DFNA11	*shaker-1*	Motor protein
	MYO15	Myosin XV	DFNB3	*shaker-2*	Motor protein
	MYO6	Myosin VI	DFNAi	*Snell's waltzer*	Motor protein
	USH1C	Harmonin	DFNB18 ± retinopathy (Usher 1C)[b]		PDZ domain-containing protein
	CDH23	Cadherin-23	DFNB12 ± retinopathy (Usher 1D)[b]	*waltzer*	Cell adhesion protein
	Espn[a]	Espin	?	*jerker*	Actin-bundling protein
	KCNQ4	KCNQ4	DFNA2		K$^+$ channel subunit
	Atp2b2/Pmca2[a]	Ca^{2+}-ATPase 2	?	*deafwaddler*	Calcium pump
	OTOF	Otoferlin	DFNB9		Vesicle trafficking protein
	POU4F3	POU4F3	DFNA15	*Brn3c$^{-/-}$*[c]	Transcription factor
Nonsensory cells	*CX26/GJB2*	Connexin 26	DFNB1 DFNA3 ± keratodermia[b]		Gap junction protein
	CX30/GJB6	Connexin 30	DFNA3' ± keratodermia[b]		Gap junction protein
	CX31/GJB3	Connexin 31	DFNA2' DFNBi ± peripheral neuropathy[b]		Gap junction protein
	Slc12a2/Nkcc1[a]	NKCC1	?	*sy^{ns}, Nkcc1$^{-/-}$*[c]	Na$^+$K$^+$2Cl$^-$ cotransporter
	PDS	Pendrin	DFNB4 ± thyroid goitre[b]	*Pds$^{-/-}$*[c]	Iodide/chloride transporter
	CLDN14	Claudin-14	DFNB29		Tight junction component
	COCH	Cochlin	DFNA9		Extracellular matrix component

Category	Gene	Protein	Locus / Phenotype	Mouse model	Function
	EYA4	EYA4	DFNA10		Transcriptional coactivator
	POU3F4	POU3F4	DFN3	*Brn4*$^{-/-}$[c]	Transcription factor
Tectorial membrane	*COL11A2*	Collagen XI (α2 chain)	DFNA13 ± osteochondrodysplasia[b]	*Col11a2*$^{-/-}$[c]	Extracellular matrix component
	TECTA	α-tectorin	DFNA8/12 DFNB21	*Tecta*$^{-/-}$[c]	Extracellular matrix component
	Otog[a]	Otogelin	?	*twister; Otog*$^{-/-}$[c]	Extracellular matrix component
Unknown	*TMPRSS3*	TMPRSS3	DFNB8/10		Transmembrane serine protease
	PCDH15	Protocadherin-15	DFNB23 ? ± retinopathy (Usher 1F)[b]	*Ames waltzer*	Cell adhesion protein
	HD1A1	Diaphanous-1	DFNA1		Regulator of actin cytoskeleton
	DFNA5	DFNA5	DFNA5		?
	MYH9	Myosin IIA	DFNA17 ± giant platelets[b]		Motor protein
	MTRNR1				Mitochondrial 12SrRNA
	MTTS1				Mitochondrial tRNA$^{\mathrm{ser(UCN)}}$

[a] Murine deafness genes for which the human orthologue has not yet been implicated in isolated deafness: *Espn* (371), *Atp2b2/Pmca2* (170, 308), *Slc12a2/Nkcc1* (69, 77, 94), *Otog* (297, 298), *Pcdh15* (9).

[b] Syndromic deafness.

[c] Deaf mouse mutant obtained by targeted disruption of the gene.

Figure 3   Schematic representation of myosin VIIA. The human myosin VIIA, 2215 amino acids in length (254-kDa), consists of a motor head domain of 729 amino acids, a neck region of 126 amino acids, composed of 5 isoleucine-glutamine (IQ) motifs, and a long tail of 1360 amino acids (355). The tail begins with a short coiled-coil domain (78 amino acids) (CC), which is implicated in the formation of homodimers (354). The coiled-coil domain is followed by two large repeats of about 460 amino acids, each containing a MyTH4 ("myosin tail homology 4") and a FERM ("4.1, ezrin, radixin, moesin") (53) domain, which are separated by a poorly conserved SH3 ("src homology type 3") domain (52, 213). The myosin VIIA gene consists of 48 coding exons (355).

consists of (*a*) the N-terminal motor head containing the highly conserved actin and ATP-binding sites; (*b*) a neck region, composed of a variable number of IQ (isoleucine-glutamine) motifs that are expected to bind to calmodulin; and (*c*) a tail, which differs substantially from one myosin to another. The tail sequence determines the functional specificity of each myosin because it contains various putative protein-protein interacting domains that bind to cargo molecules, regulatory factors, and components of the transduction pathways. Unconventional myosins have been implicated in the formation and the movements of cytoplasmic expansions, in the movements of vesicles and in signal transduction (219, 237, 317). The structure of the human myosin VIIA is presented in Figure 3.

Most *Myo7a* mutations in *sh-1* mouse mutants affect the stability of the protein. Interestingly, the severity of the phenotype and the histopathological anomalies in the murine mutants are well correlated with the level of protein preserved (131). Since the original report in 1995, dozens of mutations have been found in USH1B patients [reviewed in (243)]. In contrast, only three DFNB2 families have been reported so far (197, 354). In two families, the type of the mutation, namely a base substitution expected to impede normal splicing (354), and an amino acid substitution in the motor head (197), may preserve enough activity of the protein to permit normal function in the retina. In the third DFNB2 family, affected members were compound heterozygotes with a 1-bp insertion leading to a truncation in the motor head on one allele and an acceptor splice site mutation on the other (197). The absence of a retinal phenotype in this family is more difficult to explain, since the two mutations detected should lead to complete absence of the normal protein (197). This suggests an influence of the genetic background on phenotypic expression. Finally, in the single DFNA11 family reported, a 9-bp in-frame deletion located in the dimerization domain of myosin VIIA tail was detected (198). This mutation is likely to have a dominant negative effect.

Myo7a is expressed in the hair cells of the inner ear and a variety of epithelial cell types that, in most cases, harbor apical microvilli (276). In the mouse inner ear, *Myo7a* is expressed as early as E10 (276), and myosin VIIA is thereafter exclusively detected in all hair cells (78, 130). The protein is present all along the stereocilia, near the junction between hair cells and supporting cells (129), and in the synaptic region (78). In the *sh-1* mouse mutants, two types of hair cell anomalies have been detected. In the most severely affected mutants, the hair bundle is disorganized, with clumps of stereocilia projecting outside instead of forming the highly ordered U- or W-shaped structure (see Figure 2). In addition, the kinocilium has an erratic position, thus indicating a role of myosin VIIA in the polarity of the hair bundle (291). Similar anomalies of the hair bundle have been reported in the zebrafish *mariner* mutant (232). The identification of the molecules binding to the tail of myosin VIIA has provided some insight into the role of myosin VIIA in forming and stabilizing the hair bundle. In particular, myosin VIIA binds to vezatin, a ubiquitous transmembrane protein of adherens junctions that interacts with the cadherin/catenin complex (175). Myosin VIIA is thus expected to create a tension force between adherens junctions and the actin cytoskeleton, thereby helping to strengthen cell-cell adhesion. In the hair cells, vezatin is also present at the base of the stereocilia. Several lines of evidence indicate that the protein is tightly linked to a particular subset of transient lateral links (ankle links), which may play a crucial role in forming the hair bundle. Finally, inner and outer hair cells from *sh-1* mice, unlike those of the wild type, do not uptake aminoglycosides (263), thus indicating a role of myosin VIIA in endocytosis.

The finding that mutations in *MYO7A* cause Usher 1B, DFNB2, and DFNA11 was the first demonstration that a single gene underlies both syndromic and nonsyndromic forms of hearing loss. It has since been shown that *CDH23*, which encodes cadherin-23, underlies both USH1D and DFNB12, and *USH1C*, which encodes harmonin, both USH1C and DFNB18 (see below). This situation could extend to other genetic forms of Usher syndrome since the chromosomal intervals defined for USH1F (MIM602083) (8, 350) and DFNB23 (http://www.uia.ac.be/dnalab/hhh/), and for USH3A (MIM276902) (280) and DFNB15 (MIM601869) (51) overlap.

MYOSIN XV DEFECT: DFNB3   The locus for DFNB3 (MIM600316) was characterized through analysis of the population of a small village in Bali, in which 1 individual in 50 was affected by profound congenital deafness, and assigned to chromosome 17p11.2 (101). Based on synteny, the deaf mouse mutant *shaker-2* (*sh-2*) was proposed to be affected in the orthologue of the DFNB3 gene (193). Rescue of hearing and balance in *sh-2* mice was obtained by transgenesis of a bacterial artificial chromosome (BAC) covering the candidate region (253). Sequencing of this BAC revealed a gene, *Myo15*, encoding a novel unconventional myosin that defines a new class, designated myosin XV (Figure 4). A missense mutation modifying a conserved amino acid residue of the motor head was found in the *sh-2* mutant (253). Three different mutations in *MYO15* (MIM602666), one nonsense and two missense (in the tail of the protein), were detected in three DFNB3 families, i.e., the one originally described and two others from India (343).

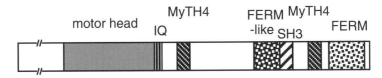

Figure 4   Schematic representation of myosin XV. The predicted myosin XV is composed of 3530 amino acids (395-kDa). The motor head contains the actin- and ATP-binding sites. Some isoforms of myosin XV contain upstream the motor head an additional sequence of 1223 amino acids. The neck contains 2 IQ motifs. The tail is 1587 amino acid long and comprises two MyTH4 domains, one FERM-like and one FERM domain, and an SH3 domain between the FERM-like domain and the second MyTH4 domain. The human myosin XV gene consists of 65 coding exons (192).

In the inner ear, expression of *Myo15/MYO15* appears to be restricted to the hair cells. The protein is detected mainly in the stereocilia and within the apical cell body, in the region of the cuticular plate (192) (see Figure 1C). This is consistent with the histopathological findings in the *sh-2* mouse mutants. Despite seemingly normal global organization of the hair bundle in these mutants, the stereocilia are particularly short. Moreover, the actin filaments of the stereocilia, unlike those from wild-type mice, extend far under the apical surface of the cell. These anomalies suggest that myosin XV is involved in the regulation of actin polymerization.

MYOSIN VI DEFECT: A DFNA FORM   The deaf mouse mutant *Snell's waltzer* (74) was identified as carrying a mutation in *Myo6*, encoding the unconventional myosin VI (14). The human orthologue, *MYO6*, maps to 6q13 (13) (MIM600970). Recently, a missense mutation in *MYO6* was detected in an Italian family affected with a dominant form of progressive hearing loss (215).

In the murine inner ear, *Myo6* is expressed only in the hair cells. The protein is concentrated in the cuticular plate (see Figure 1C) and surrounding region (129). In *Snell's waltzer* mutants, the stereocilia are shorter than in wild-type mice, and they fuse at their base (292). Myosin VI, unlike all other characterized myosins, moves toward the pointed minus end of the actin filament (356). This unique property results from the insertion of a ∼50 amino acid sequence, called converter, between the head and the neck of myosin VI (Figure 5). This discovery raises intriguing questions about the molecular mechanisms of reversal and the biological roles of this backward myosin (60, 270, 316).

HARMONIN DEFECT: DFNB18 (AND USHER 1C SYNDROME)   The DFNB18 locus (MIM602092) was mapped at 11p15.1 (147), in an interval overlapping the USH1C (MIM276904) interval (154). The USH1C gene has been isolated by (*a*) a candidate gene approach (338) and (*b*) the analysis of a chromosomal deletion (26). *USH1C* encodes a PDZ domain-containing protein, termed harmonin. The

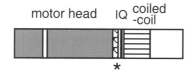

Figure 5   Schematic representation of myosin VI. The predicted myosin VI, 1263 amino acids in length (142-kDa), has a motor head containing the ATP- and actin-binding sites, a neck region containing a single IQ motif, and a short tail (424 amino acids) that begins with a coiled-coil domain and ends by a globular structure (13). Myosin VI is characterized by a hydrophobic stretch of 52 amino acids (*) located between the motor head and the IQ motif. This sequence, called converter, is responsible for the movement of myosin VI toward the minus end of the actin track (i.e., in a direction opposite to that of all other known myosins). *MYO6* is composed of 32 coding exons (7).

predicted isoforms of harmonin are represented in Figure 6. Six different mutations have been reported to date in USH1C patients (26, 338, 373). In DFNB18 patients, who present with profound congenital deafness, a missense mutation in *USH1C* has recently been detected (X.-Z. Liu, personal communication), located in an alternative exon present in inner ear transcripts but not in the retinal transcripts (338). Functional characterization of the corresponding domain of the protein should provide insight into the pathogenesis of the USH1C/DFNB18 hearing loss.

The PDZ domain-containing proteins are central organizers of high-order supra-molecular complexes located at specific emplacements of the plasma membrane.

Harmonin isoforms

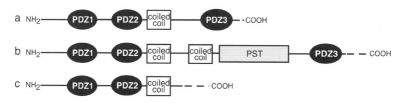

Figure 6   Schematic representation of the 3 predicted classes of harmonin isoforms. *USH1C* consists of 28 coding exons. The gene encodes a variety of alternatively spliced transcripts that result from the differential use of 8 exons. In the murine inner ear, these transcripts distribute into three subclasses, a, b, c. PDZ domains are interacting domains, 80 to 90 amino acids in length, originally detected in PSD/SAP90 (postsynaptic density protein), Dlg-A (*Drosophila* tumor suppressor) and ZO-1 (tight junction protein). PST refers to proline, serine, threonine-rich region. [Adapted from (338)].

They can form homo- or heteromeric structures and bind to transmembrane proteins, in particular ion channels or transporters, and to actin or actin-binding proteins; hence, they cluster and coordinate the activity of various plasma membrane proteins and bridge them to the cortical cytoskeleton (89, 107, 295). In the cochlea, harmonin is restricted to the hair cells, where it is present in the whole cell body and the stereocilia (338). Therefore, an interaction between myosin VIIA and harmonin is an attractive hypothesis (243).

CADHERIN-23 DEFECT: DFNB12 (AND USHER 1D SYNDROME)   The loci for a form of isolated deafness, DFNB12 (MIM601386) (50), and for a form of Usher syndrome type I, USH1D (MIM601067), were mapped to the long arm of chromosome 10 (349). The gene underlying USH1D has been isolated (*a*) based on the analysis of the *waltzer* (*v*) deaf mutants (28) proposed to be the mouse model for USH1D (40, 368) and (*b*) by a direct positional cloning (30). This gene, *CDH23*, encodes a cadherin-related protein, cadherin-23 (28, 30) (Figure 7).

Missense mutations of *CDH23* have been detected in several DFNB12 families who presented with severe-to-profound hearing loss (30). In contrast, only deletions and nonsense or frameshift mutations were found in USH1D patients. Thus, the type of mutation may well play a crucial role in the phenotypic expression.

In the murine inner ear, *Cdh23* expression is restricted to the hair cells (76). In the *waltzer* mutants, the kinocilium is mispositioned, the hair bundle becomes progressively disorganized, and the stereocilia clump together (76), i.e., a series of anomalies resembling those of *Myo7a* defective mutants. Since the myosin VIIA ligand vezatin interacts with the cadherin-catenin complex (175), an exciting possibility is that myosin VIIA and cadherin-23 belong to the same macromolecular complex required to transmit adhesion forces to the cytoskeleton (see 243).

KCNQ4 DEFECT: DFNA2   The DFNA2 locus (MIM600101) was identified in a study of three families from Indonesia, the Netherlands, and the United States, all affected by a progressive form of deafness involving preferentially the high frequencies.

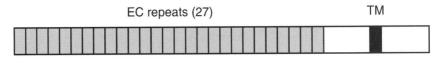

EC repeats (27)   TM

Figure 7 Schematic representation of cadherin-23. The predicted cadherin-23 has 3354 amino acids (28). Its ectodomain is composed of 27 extracellular cadherin (EC) repeats, most of which possess the characteristic Ca^{2+}-binding motifs of cadherins. These motifs are implicated in the calcium-dependent interaction between cadherins that mediates cell-cell adhesion. The single transmembrane domain is followed by a 268 amino acid cytoplasmic region that shows no homology with known proteins. *CDH23* comprises 69 coding exons (28).

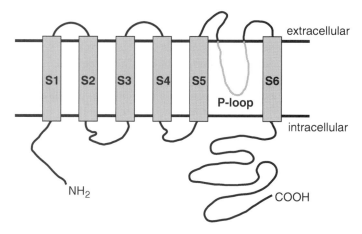

Figure 8   Schematic representation of the KCNQ4 K⁺ channel subunit. *KCNQ4* encodes a predicted 695 amino acid (77-kDa) protein (171). The structure of KCNQ subunits consists of 6 transmembrane domains with an intramembrane P-loop between domains 5 and 6, and cytoplasmic N- and C-terminal regions. The P-loop underlies the K⁺ selectivity of the pore and the 4th transmembrane domain (S4) contains the voltage sensor. The C terminus contains an "A domain," which is highly conserved between different KCNQ proteins and may be a distinctive feature of the family. This domain is followed by a short stretch thought to be involved in subunit assembly (148). *KCNQ4* consists of 14 coding exons.

This locus was mapped to chromosome 1p34 (57, 327). The discovery of a gene, *KCNQ4* (MIM603537), encoding a new member of the KCNQ potassium channel subunit family (Figure 8), which mapped to the DFNA2 chromosomal interval and was expressed in the inner ear, led to this gene being considered as a candidate for DFNA2. The finding of a missense mutation (G285S) that (*a*) affects the GYG consensus sequence required for the K⁺ selectivity of the channel pore (135), and (*b*) has a strong dominant negative effect on the potassium current observed in *Xenopus* oocytes transfected with the wild-type allele, established that *KCNQ4* underlies DFNA2 (171). Six different mutations have been reported in DFNA2 patients; all but one are missense mutations (58, 171, 329).

   KCNQ4 is the second member of the *KCNQ* family to be involved in deafness; *KCNQ1*, which is essential for K⁺ secretion into the endolymph by strial marginal cells, underlies Jervell and Lange-Nielsen syndromic deafness (MIM220400) (231). These voltage-dependent K⁺ channels are either homo- or heterotetramers, and the formation of heteromers seems to be restricted to certain combinations between the five known KCNQ proteins. In addition, KCNQ subunits can interact with KCNE proteins, i.e., small proteins with a single transmembrane domain: this interaction leads to significant changes in channel gating (17, 216, 279, 283, 314).

In the cochlea, *KCNQ4* is expressed by the OHCs, in a decreasing gradient from the basal to the apical turn. In the vestibular end organs, KCNQ4 was detected only in type I hair cells. KCNQ4 has a basal localization in OHCs, whereas it is present in almost the entire basolateral membrane of the type I hair cells as well as in the facing calyx-like postsynaptic membrane (161). Several lines of evidence suggest that KCNQ4 underlies the IK,n and IK,L currents that have been described in OHCs and type I hair cells, respectively. These currents are expected to be already active at the resting potentials of these cells and to influence their electrical properties (140, 205, 229, 259). Proof that the corresponding channels contain KCNQ4 should come from electrophysiological analysis of the knockout mouse. Intriguingly, KCNQ4 was also detected in many nuclei of the central auditory pathway (161), which raises the possibility that the DFNA2 form of deafness involves not only a defect of the peripheral sensory organ but also a central defect.

OTOFERLIN DEFECT: DFNB9   The DFNB9 locus (MIM601071) was identified in an analysis of a consanguineous Northern Lebanese family affected by a profound prelingual hearing loss, and assigned to chromosome 2p23.1 (49). The responsible gene, *OTOF* (MIM603681), was cloned by a candidate gene approach (367). The encoded protein, otoferlin (366, 367), belongs to the same protein family as dysferlin (18, 195) and myoferlin (64). These proteins exhibit homology with the *Caenorhabditis elegans* spermatogenesis factor fer-1 (4). The isolation of several *OTOF* cDNA clones has revealed alternatively spliced forms (366). A homozygous nonsense mutation was detected in four DFNB9 Lebanese families (367). Two other *OTOF* mutations have since been reported in splice sites (6, 366).

The murine *Otof* is expressed at low levels in several tissues. Strong expression is restricted to the inner ear and the brain. In the mature cochlea, the *Otof* transcript is restricted to the IHCs (367).

OTOF encodes a protein with a predicted transmembrane domain close to the C-terminal end, and six cytoplasmic C2 domains (Figure 9). C2 domains have been described in various proteins. Sequence comparison predicts that four of the six C2 domains of otoferlin bind Ca^{2+}.

The C2 domain-containing proteins interact with phospholipids and proteins (267) and fall into two functional categories, namely, the generation of the lipid second messengers involved in transduction pathways, or membrane trafficking.

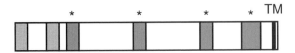

Figure 9 Schematic representation of otoferlin. The protein contains 1977 amino acids (227-kDa) (366, 367). The 6 predicted C2 domains are represented by gray boxes. Only the last four (indicated by asterisks) are expected to bind calcium. TM indicates the putative transmembrane domain. *OTOF* is composed of 48 coding exons.

This second category includes several proteins that are involved in the docking of synaptic vesicles to the plasma membrane and/or the fusion process (35, 54, 190, 241, 296, 332, 345). Based on the expression of *Otof* in the inner hair cells, and the impaired vesicle-plasma membrane fusion process in *C. elegans fer-1* mutants, it has been hypothesized that otoferlin is involved in Ca^{2+}-triggered fusion of synaptic vesicles to the plasma membrane. The sensory synapses of cochlear and vestibular hair cells, termed ribbon synapses, have specific structural and functional characteristics (239). Otoferlin may provide an entry point to understanding the molecular bases of the properties of these synapses.

POU4F3 DEFECT: DFNA15   In a large Israeli family affected by a dominant form of progressive deafness with adult onset (103), the responsible locus, DFNA15 (MIM602459), was mapped to chromosome 5q31 (325). A small deletion (8 bp) was found in the gene encoding the transcription factor POU4F3 (MIM602460) in affected individuals (325). The POU domain protein family (Figure 10) is a subclass of homeodomain proteins that exhibit cell-specific expression and control early differentiation processes in certain cell lineages (352). The deletion detected in the DFNA15 family predicts a truncated protein with a partial deletion of the POU homeodomain, including its DNA-binding site.

The role of the murine orthologue of *POU4F3*, *Brn-3.1* (or *Brn-3c*)/*Pou4f3*, in the survival of cochlear and vestibular hair cells was established by studying both the gene's expression profile and the mouse mutants with a targeted null mutation of the gene. *Pou4f3* has a restricted expression pattern; it is expressed in a few neuronal cell populations and in the inner ear postmitotic hair cells (364, 365). In *Pou4f3* null mice, no mature cochlear and vestibular hair cells can be detected at postnatal day 14, around the time mice begin to hear (81, 364). However, some immature hair cells are present (without hair bundles), that eventually undergo

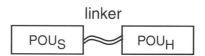

Figure 10   Schematic representation of the POU DNA-binding domain. The POU domain that characterizes POU transcription factors is a bipartite DNA-binding domain composed of a POU-specific domain (POU$_S$) and a POU-homeodomain (POU$_H$), which are about 75 amino acids and 60 amino acids in length, respectively. Both contain helix-turn-helix motifs. POU$_S$ and POU$_H$ are joined by a hypervariable linker, between 15 and 56 amino acids in length. Cooperation of the POU$_S$ and POU$_H$ domains increases the binding affinity and specificity of these transcription factors for their target DNA. The POU-domain proteins are classified into seven classes based on the sequence of the POU$_H$ domain and linker region. POU4F3 (338 amino acids) belongs to class IV, whereas POU3F4 (361 amino acids) belongs to class III (352).

apoptosis. This indicates that *POU4F3* is not involved either in the commitment of precursor cells to develop into hair cells or in their early differentiation. Rather, it seems to be required for the migration of the hair cells from the supporting cell layers to the lumenal hair cell layer, and for their maturation and survival (365). The target genes of POU4F3 are unknown.

Forms of Deafness Caused by Nonsensory Cell Defects

CONNEXIN26 DEFECTS: DFNB1, DFNA3 (AND VOHWINKEL SYNDROME)   In 1994, the first locus for an autosomal recessive form of deafness, DFNB1 (MIM220290), was reported on chromosome 13q12, from a study of two Tunisian families affected by profound prelingual deafness (123). Subsequently, a gene implicated in a family affected by a dominant form of deafness, DFNA3 (MIM601544), was localized to the same chromosomal region (48). This led to propose *CX26* (or *GJB2*) (MIM121011), the gene encoding connexin26, as the causative gene for both deafness forms. Connexin genes have a similar and remarkably simple organization, with the first exon containing most of the 5'-untranslated region and the second exon containing the complete open reading frame as well as the 3'-untranslated region (except in *CX35*). *CX26* encodes a predicted 208 amino acid (26-kDa) protein. Two distinct nonsense mutations were detected in three consanguineous Pakistani DFNB1 families, thus establishing that *CX26* is the causative gene (159). A heterozygous nucleotide change responsible for the amino acid substitution M34T was also reported in a small family affected by a dominant form of hearing impairment (159), but the causality of the M34T variant in DFNA3 was then challenged (70, 157, 286). Authentic missense mutations, W44C and C202F, were eventually detected in two families, including the family described originally, thus establishing *CX26* as the causative gene for DFNA3 (70, 225). Whether the M34T allele is involved in DFNB1 (120, 139, 207, 208, 358) is still under debate. Finally, *CX26* missense mutations have been reported in two forms of dominant syndromic deafness with skin anomalies [MIM124500 (203) and MIM148350 (134, 160, 262)].

Clinical features of DFNB1 and epidemiological data   The first evidence for the high prevalence of DFNB1 among the genetic forms of autosomal recessive deafness came soon after the locus was identified (211). Subsequent segregation analysis performed in patients from Spain and Italy established a linkage to the DFNB1 locus in 79% of the cases (108). Once *CX26* was identified as being responsible for DFNB1, mutation screening confirmed the high prevalence of *CX26* defects in the deaf populations. DFNB1 accounts for 30% to 60% of the autosomal recessive forms of isolated deafness in Europe and United States (71, 83, 105, 117, 157, 186, 200, 226, 301) and ~20% in Japan (3, 172). About 50 different *CX26* mutations have been reported [for reviews, see (158, 251, 254) and http://www.uia.ac.be/dnalab/hhh/). However, a particular mutation, 35delG, accounts for most (up to 85%) of the mutant alleles in European-Mediterranean populations (71, 72, 83, 369). This mutation has not been observed in deaf Japanese children (3, 172), nor in African American (223) and Ghanaian (127) populations, and seems to be rare in Asian populations (238). The 35delG mutation deletes a guanine (G) in a sequence of six G extending from nucleotide position 30 to 35, which results in a premature stop codon at codon position 13. Because stretches of identical nucleotides are thought to favor the slippage of the DNA polymerase, and several different haplotypes were associated with the mutation, its high prevalence was initially thought to be related to a mutational hotspot (44, 72, 223). Subsequent data on its geographical distribution argue in favor of an ancestral mutation that initially spread out around the Mediterranean Sea (109, 329b). In several European countries, the prevalence of the 35delG mutation has been estimated to 2%–4% of the population with normal hearing (11, 83, 307). Two other mutations are particularly prevalent in specific populations. The 167delT mutation is frequently detected among deaf Ashkenazi Jews (189, 223), and is associated with a specific haplotype, which indicates a single origin of the mutation. Four percent of the normal-hearing Ashkenazi Jewish population are carriers. Although less documented, the 235delC mutation seems to account for a large proportion of the *CX26* mutations in the Japanese population (3, 104, 172). It is noteworthy that for several deaf patients, a *CX26* mutation was detected on one allele only, indicating either the existence of another *CX26* mutation in the gene's unexplored region or the possible coimplication of another connexin gene, i.e., a digenic origin of the hearing loss, which could be related to the putative formation of heteromeric connexons or heterotypic channels (see below). The high frequency of the *CX26* mutations worldwide, compared with those of *CX30* and *CX31* (see below), is puzzling: Do *CX26* mutations confer a selective advantage to carrier individuals? Of most interest, the high prevalence of *CX26* mutations in autosomal recessive forms of deafness has made it possible to estimate the proportion of sporadic cases to be of genetic origin: in Southern Europe and the United States, between 35% and 40% of congenital cases of deafness presented biallelic *CX26* mutations. Thus, at least 65% to 80% of the deaf sporadic prelingual cases may be estimated to be of genetic origin.

Attempts to characterize the clinical features of DFNB1 have, to date, associated *CX26* mutations only with prelingual forms of deafness (71, 172, 301, 361), with hearing loss varying from mild to profound. The severity of hearing loss in 35delG homozygous individuals is highly variable, even among siblings (55, 71, 117). Audiometric curves are either flat or slope with a preferential loss at high frequencies (71, 361). High-resolution temporal bone computerized tomography (CT) revealed no anomalies, and the vestibular tests proved normal (71). Finally, the hearing loss has been reported to be nonprogressive in about two thirds of the cases (55, 71). These findings have important implications for genetic counseling.

Clinical features of DFNA3   Only two families unambiguously affected by DFNA3 have been described (70, 225). The phenotypes were different. In the first family, severe-to-profound prelingual hearing impairment was observed, whereas in the second family, the hearing loss was mild to moderate, postlingual, and progressed from high frequencies to mid-frequencies.

Pathophysiology   Most adjacent cells communicate via membrane channels tightly packed in aggregates at the level of gap junction plaques. In vertebrates, these channels are composed of two hemichannels, the connexons, which are hexamers of connexins. The connexons contributed by two neighboring cells, form a channel after docking (322), allowing the intercellular exchange of small diffusible molecules such as ions, metabolites, and second messengers up to a molecular mass of 1 kDa (173); this communication between adjacent cells is essential for functional synchronization, for growth, and for differentiation. Sixteen members of the connexin family, with an expected molecular mass ranging from 26 to 60 kDa and a common predicted structure, have been identified to date. These are differently expressed in tissues and during development (299). Connexins have four transmembrane domains; their N- and C-terminal regions are intracytoplasmic (Figure 11). The intracellular loop is variable in size. Whereas the current nomenclature of the connexins is based on the molecular mass deduced from their sequence, the connexins have also been classified into two groups, GJα and GJβ, characterized by long and short intracellular loops, respectively. The size of the intracytoplasmic C-terminal region, which is an interacting domain, differs for each connexin (39). Connexin26 has both a short intracellular loop and a short C-terminal region. The connexons are generally homomeric structures in which the six identical connexins are linked

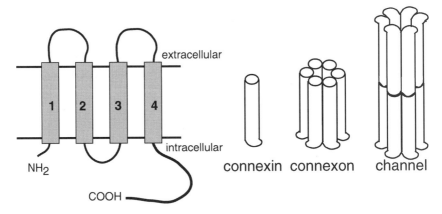

Figure 11   Schematic representation of connexins. Connexins have 4 predicted transmembrane domains. Each of the two extracellular loops contains three cysteine residues involved in intramolecular disulfide bonds. Connexins 26, 30, and 31 belong to the GJβ group (short intracellular loop) of connexins. Connexins associate (via noncovalent bonds between their extracellular loops) into homo- or heterohexamers to form connexons. To date, little is known about the functional properties of the various types of gap junction channels.

together through noncovalent bonds. However, some connexins, such as Cx26 and Cx32, can assemble to form a heteromeric connexon (182). The connexons of adjacent cells, which contribute to the formation of a given channel, may be composed of identical or different connexins, thus forming a homotypic or a heterotypic channel. The formation of heterotypic channels is limited by the compatibility between the connexins (39, 359). Connexins and their various combinations confer unique properties to the formed channels with regard to conductance, size and charge of the molecules exchanged (23, 34, 80, 113, 233, 304, 331), and regulation of their gating (by transjunctional voltage, phosphorylation, cytosolic pH, Ca^{2+}, and cyclic nucleotides) (21, 22, 75, 346). These properties have been characterized mainly in paired *Xenopus* oocyte assays and tentatively correlated with particular domains or residues of the protein (39). Connexin26 exhibits a greater permeability for positively than for negatively charged ions or molecules (43). It has no consensus sequences for phosphorylation by kinases (275). The first extracellular loop (273) and proline-87 of the second transmembrane domain (309) have been implicated in the voltage control of the gating.

Gap junctions have been analyzed in the rat cochlea by light microscopy upon immunostaining of Cx26 and by electron microscopy. Cx26 labeling was associated to all gap junction plaques of the inner ear (163). Cx26 underlies the constitution of two independent cellular networks (163). One connects the supporting cells of the organ of Corti to the adjacent epithelial cells. The other is composed of fibrocytes and mesenchymal cells, as well as basal, intermediate and endothelial cells of the stria vascularis (see Figure 1*B*). The role of these networks is unknown but they have been proposed to be the structural basis for a transcellular circulation of K^+ (163, 278, 303, 306).

The *Cx26* knockout in the mouse did not clarify the pathogenesis of the auditory defect since the mutant mice die early in embryonic life (E10), due to a deficient uptake of glucose through the placenta (106). Conditional or inducible *Cx26* knockouts in the inner ear should shed light on the role of connexin26 in cochlear development and functioning.

CONNEXIN31 DEFECTS: DFNA2′ AND A DFNB FORM   Since *CX26* was identified as the causative gene for DFNB1/DFNA3, attention has been paid to other connexin genes as possible candidates for deafness. *CX31* (*GJB3*) was found to be implicated (363). The gene encodes a 270 amino acid (31-kDa) predicted protein (MIM603324). Two mutations in *CX31* were detected in two small Chinese families affected by a dominant form of deafness (363). In both families, hearing loss starts after the second decade, affects only frequencies above 2000 Hz, and appears not to be fully penetrant. Because *CX31* maps to the same chromosomal interval as *KCNQ4*, which underlies DFNA2 (see above), the *CX31*-associated dominant form of deafness is referred to as DFNA2′. In two Chinese families affected by a recessive form of moderate or severe deafness (referred to as DFNBi in Table 1), deaf children were found to be compound heterozygotes for *CX31* mutations (199). Finally, a dominant *CX31* mutation, D66del, has recently been identified in one family with peripheral neuropathy and hearing impairment (201); auditory brain-stem responses suggested alteration in both the cochlea and the auditory nerve. Therefore *CX31*, like *CX26*, underlies both a dominant and a recessive form of iso-lated deafness, as well as a syndromic deafness. Interestingly, *CX31* also underlies an isolated skin disorder (261).

The two mutations reported in the DFNA2′ affected families were a missense mutation at the boundary between the second extracellular loop and the fourth transmembrane domain (see Figure 11), and a nonsense mutation. In the two families affected by the recessive form of deafness, the same mutations were observed, namely a 3-bp deletion, leading to the loss of an isoleucine in the third transmembrane domain, on one allele, and a valine for isoleucine substitution at the same emplacement on the other allele. Based on the parents' normal hearing in the DFNBi families, a dominant negative effect of the *CX31* mutations in DFNA2′ has been proposed (199).

Besides *CX31* expression in the inner ear (362, 363) and the keratinocytes (137, 357), the transcript has been detected in auditory and peripheral nerves (201) and in several tissues including the placenta, eye (137), kidney (321), spinal cord, and cerebral cortex (363). Whether the protein is present in all gap junctions in the inner ear has not been determined. In fact, whether *Cx31* is expressed in the adult mouse cochlea is still controversial (181, 201, 250, 362). Targeted disruption of *Cx31* in mouse causes transient placental dysmorphogenesis but does not impair hearing and skin differentiation (250). This indicates that connexin31 is essential for early placentation, but not for cochlear development and functioning in the mouse.

CONNEXIN30 DEFECT: DFNA3′   The gene encoding connexin30 (MIM604418), *CX30* (*GJB6*), is located close to *CX26* on chromosome 13q12 (156). In families with a form of deafness linked to the 13q12 region, and with no mutation detected in *CX26*, *CX30* was considered to be a good candidate because *Cx30* is expressed in the rat cochlea (181). Direct mutational screening detected a missense mutation in one family affected by a dominant form of deafness (119), hereafter referred to as DFNA3′. A second disorder, hidrotic ectodermal dysplasia (MIM129500), only rarely associated with deafness, was also shown to be caused by *CX30* missense mutations (180). Thus, like *CX26* and *CX31*, *CX30* underlies skin and inner ear diseases.

The mutation identified in the DFNA3′ family substitutes a methionine for a threonine (T5M) in the N-terminal region of connexin30 (119) (see Figure 11). Electrophysiological studies performed on paired *Xenopus* oocytes expressing the mutated *CX30* allele showed an absence of electric coupling between oocytes. The mutation also exhibited a dominant negative effect in this system (119). Se-quence comparison indicates that among the various connexins, *Cx26* and *Cx30* have the closest relationship (77% amino acid identity) (63). In addition, Cx26 and Cx30 connexons can form heterotypic channels. In the cochlea, the two connexins have the same cell distribution (181). Thus, the pathophysiological hypotheses concerning the *CX26-* and *CX30*-associated auditory defects are similar.

PENDRIN DEFECT: DFNB4 (AND PENDRED SYNDROME)   The DFNB4 (MIM 600791) locus (7q21-34) was first described in a deaf Israeli Druze family (16). When the Pendred syndrome (MIM 274600) was subsequently assigned to the same chromosomal region (59, 294), this family was clinically reexamined and diagnosed with Pendred syndrome. The autosomal recessive Pendred syndrome was initially defined as prepubertal onset thyroid goiter and congenital deafness (240). The thyroid goiter is due to intrinsic defect in iodide organification (271, 294), detectable by perchlorate discharge testing (98). However, the thyroid disease is extremely variable, even within a given family (209). Thyroid enlargement can be absent, or when present, accompanied by either normal or decreased serum levels of thyroid hormones. An increasing number of Pendred cases are also reported without abnormal perchlorate tests (257). The hearing loss is sensorineural, profound, and prelingual. Some patients also have vestibular dysfunction. CT and magnetic resonance imaging detect morphologic anomalies of the inner ear. The enlargement of the vestibular aqueduct, which houses the endolymphatic canal, is the commonest anomaly (245). An undercoiling of the cochlea is also frequently observed (245). Although the prevalence of Pendred syndrome is not known exactly, it seems to be the commonest form of syndromic deafness (98), and may account for up to 20% of all deafness cases (62). The causative gene, *PDS*, has been identified by the detection, in three affected families, of one missense mutation on a conserved amino acid residue and two single base-pair deletions predicting truncated proteins (87).

As anticipated from the intrafamilial variation of the thyroid disorder in Pendred syndrome, *PDS* also underlies the isolated deafness form DFNB4. Various missense and frameshift mutations have been reported in DFNB4 patients (42, 191, 323). The hearing loss is congenital, sometimes progressive, and affects all sound frequencies, but often predominates at high frequencies. Patients occasionally complain of vertigo. Hearing loss may fluctuate (2). Morphologic anomalies similar to those reported in Pendred syndrome, i.e., enlarged vestibular aqueduct and undercoiling of the cochlea, are frequently observed.

The predicted structure of pendrin is shown in Figure 12. Expression of pendrin in *Xenopus* oocytes and in insect cells has established that this protein is a chloride and iodide transporter (288). Furthermore, expression studies in *Xenopus* oocytes (285) and HEK-293 mammalian cells (302) have shown that pendrin can function in the chloride/formate, chloride/OH$^-$, and chloride/HCO$_3$$^-$ exchange modes. Among the various mutations reported in DFNB4 patients are some that have also been reported in patients affected by Pendred syndrome, indicating the role of modifier genes in the phenotype. However, a functional study on chloride and iodide uptake by *PDS*-transfected *Xenopus* oocytes, comparing three common Pendred syndrome allele variants with three *PDS* mutations reported only in DFNB4 individuals, showed a complete loss of transport with the Pendred alleles,

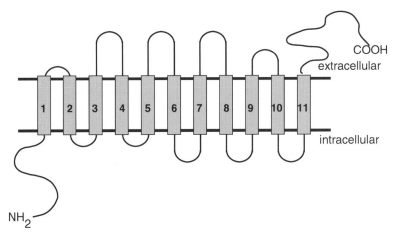

Figure 12   Schematic representation of pendrin. *PDS* comprises 21 exons. It encodes a predicted 780 amino acid (86-kDa) protein, pendrin, which possesses 11 putative transmembrane domains and a predicted extracellular C-terminal domain (87). Pendrin is a iodide/chloride transporter (288). Little is known about the structure-function relationship.

whereas a residual transport was present with the DFNB4 alleles (287). The authors suggest that this residual level of anion transport is sufficient to eliminate or postpone the onset of goiter in individuals with DFNB4.

The *PDS*/*Pds* gene has a relatively restricted pattern of expression, with the transcript detected only in the inner ear, thyroid, kidney, and placenta (24, 87). Immunolocalization studies have shown that pendrin resides within the apical membrane of thyrocytes (25, 271), suggesting that the protein functions as an apical porter of iodide in the thyroid. In the kidney, pendrin is present in the apical membrane of a subpopulation of intercalated cells in the cortical part of the collecting ducts, and functional studies in wild-type and *Pds*-knockout mice (86) have established that the protein is involved in renal bicarbonate secretion (272). In addition, pendrin may also be present in the proximal tubules (302). In the mouse inner ear, *Pds* is strongly expressed, from E13 onward, in the endolymphatic duct and sac. *Pds* is also expressed in specific nonsensory areas of the utricle, saccule, and cochlea (88). The expression of *Pds* in epithelial cells of the endolymphatic duct and sac, known to be involved in endolymph resorption, in conjunction with analyses showing that pendrin is a chloride transporter, strongly argues for a direct role of this protein in inner ear fluid ionic homeostasis. Such a role would account for the fluctuation of the hearing loss observed in certain DFNB4 patients. In the context of this hypothesis, the enlarged vestibular aqueduct, which houses the endolymphatic duct, makes sense. However, the absence of endolymphatic hydrops (distension of the membranous labyrinth) in affected individuals is suggestive of a compensatory mechanism that would develop after the morphogenesis of the inner ear. The undercoiling of the cochlea has also been proposed to result from increased endolymphatic pressure during formation of the cochlea (88). The finding that pendrin is functionally related to the renal chloride/formate exchanger (285) may provide some insight into the mechanism of morphogenetic anomalies in the inner ear in DFNB4 and Pendred syndrome. In the proximal tubule of the kidney, chloride/formate exchange provides a mechanism for NaCl and volume reabsorption (344). Therefore, a loss of chloride transport within the embryonic inner ear could lead to abnormal salt and water flux, with subsequent dilatation of the vestibular aqueduct and loss of the normal architecture of the cochlea.

Targeted disruption of *Pds* results in completely deaf mice with variable vestibular dysfunction (86). The inner ears of these mice appear to develop normally until E15, after which time severe endolymphatic dilatation occurs. Additionally, degeneration of hair cells and malformation both of the acellular membranes covering the utricular and saccular sensory epithelia (otoconial membranes) and of the overlaying biominerals (otoconia) occurs in the second postnatal week. In contrast, $Pds^{-/-}$ mice lack structural or functional evidence of thyroid disease. The $Pds^{-/-}$ mouse now provides an invaluable system for determining the role of pendrin in inner ear structure and functioning.

CLAUDIN-14 DEFECT: DFNB29   Segregation analysis of two large consanguineous Pakistani families affected by profound congenital recessive deafness defined a novel locus, DFNB29 (MIM605608), on chromosome 21q22.1 (360). The DFNB29 critical interval contained *CLDN14* (132), encoding claudin-14 (Figure 13). *CLDN14* has a single coding exon preceded by 2 noncoding exons. Two homozygous mutations have been identified (360): a single base deletion responsible for a frameshift and a missense mutation.

The claudin family consists of at least 20 members. Claudins are essential components of tight junctions. Several lines of direct experimental evidence support the role of claudins in creating the tight junction's functional barrier [reviewed in (221)]. Claudins display varied tissue distribution, consistent with the idea that differential expression might explain the variable permeability of tight junctions observed among different epithelia. The identification of claudin-16 mutations in a rare renal syndrome characterized by defective paracellular Mg^{2+} resorption (300) reinforced the idea that claudins confer specific selectivity properties to paracellular transport.

The expression of *Cldn14* in the murine inner ear has been determined by *in situ* hybridization and immunohistochemistry (360). The *Cldn14* mRNA was first detected in the early postnatal days. The transcript and the protein were restricted to the sensory epithelia. In the P4 cochlea, claudin-14 was detected in the hair cells. Between P4 and P8, the labeling decreased in the hair cells and appeared in the supporting cells of the organ of Corti.

COCHLIN DEFECT: DFNA9   The DFNA9 locus (MIM601369) was identified through analysis of a large American family affected by a progressive form of deafness

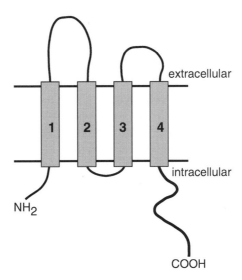

Figure 13 Schematic representation of claudin-14. The claudin family consists of at least 20 members, of approximately 22–24 kDa. These proteins have 4 transmembrane domains. They range in sequence identity from 12% to 70% and appear to group into subfamilies, which may indicate similarities in function. Claudins are essential components of tight junctions (221, 360).

(125), and mapped to chromosome 14q12-13 (204). A very abundant cochlear cDNA, *COCH* (MIM603196), isolated from a subtracted human cochlear cDNA library, was subsequently assigned to this chromosomal region (269). A total of 5 missense mutations have been identified in DFNA9 individuals (65, 96, 97, 152, 268), thereby establishing *COCH* as the causative gene. *COCH* encodes a putative extracellular matrix protein, cochlin (268). The mutations are all located in the N-terminal cysteine-rich region of the protein (Figure 14).

Figure 14 Schematic representation of cochlin. *COCH* (11 exons) encodes a predicted 550 amino acid extracellular protein. The sequence of cochlin is highly conserved between species (268). The N-terminal signal peptide (*horizontal bars*) is followed by a region of *Limulus* factor C homology (FCH) and two von Willebrand factor (vWF)-like type A domains. Each of the vWF-like domains is preceded by a region containing several cysteine residues. The presence of the vWF type A domains suggests that cochlin is an extracellular matrix component (56).

The clinical features have been reported for most of the affected families (65, 97, 125, 162, 333). The mutations are fully penetrant; age of onset varies from adolescence to mature adulthood; the hearing loss initially predominates on the high frequencies, and thereafter extends to all frequencies, with a rapid progression of 3 to 5 dBHL per year; tinnitus and occasional vestibular dysfunction, namely recurrent episodes of vertigo, and head movement-dependent oscillopsia, with vestibular areflexia or hyporeflexia, are found. Similar vestibular symptoms are observed in Ménière disease, a common affliction with endolymphatic hydrops. Although the hearing loss associated with Ménière disease is fluctuating and affects low or all sound frequencies, some defects classified as Ménière disease may be due to *COCH* mutations (65, 97, 268).

Histopathological analyses of temporal bones of DFNA9 patients revealed acidophilic deposits along the osseous spiral lamina (where the auditory nerve penetrates the membranous labyrinth), as well as in the spiral limbus and the spiral ligament (162, 194) (see Figure 1*B*). Degeneration of the organ of Corti is detected in aged patients. There is a striking correlation between the location of the acidophilic deposits and the sites of *COCH* expression (268), which suggests that cochlin might be a component of these deposits. Mutated forms of cochlin may show abnormal folding, impairing their stability or their interaction with other extracellular matrix components.

EYA4 DEFECT: DFNA10   The DFNA10 (MIM 601316) locus was mapped to chromosome 6q22.3-23.3 on the basis of linkage analysis in a large multigenerational family from the United States (235). Affected persons exhibited an inexorably progressive sensorineural hearing loss beginning in the second to fifth decade and leading ultimately to severe-profound hearing impairment. Among the known genes present in the subsequently refined DFNA10 interval (335), *EYA4* (MIM603550), a member of the *EYA* gene family homologous to the Drosophila *eyes absent*, was considered to be the best candidate because another member of this family, *EYA1*, underlies a syndromic hearing loss (1), and the murine *Eya4* is expressed in the developing inner ear (31). EYA4 belongs to a group of four transcriptional activators (EYA1-4) that interact with other proteins in a conserved regulatory hierarchy to ensure normal embryonic development. Analysis of *EYA4* revealed deleterious point mutations in two of the three reported DFNA10 families (351), both located in the eya-homologous region (Figure 15). This domain mediates the interaction of EYA proteins with *SIX* (the homologues of the Drosophila *sine oculis*) gene products, which induces the nuclear translocation of the resultant protein complex (236).

Eya genes are expressed in various tissues early in embryogenesis. Although each *Eya* gene has a unique expression pattern, there is extensive overlap, e.g., *Eya1* and *Eya4* are both expressed in the rodent otic vesicle and its derivatives (151, 351).

In contrast to the phenotype resulting from *EYA1* mutations (branchio-oto-renal syndrome), no congenital anomalies are part of the DFNA10 phenotype, even

eyaHR

Figure 15   Schematic representation of EYA4. The structure of EYA4 (640 amino acids) conforms to the basic EYA pattern, i.e., a highly conserved 271 amino acid C terminus called the eya-homologous region (eyaHR), and a more variable proline-serine-threonine-rich transactivation domain at the N terminus (31). *EYA4* contains 21 exons.

though *EYA4* is expressed early in embryogenesis (351). The late-onset hearing loss characteristic of DFNA10 might result from the loss of a cell survival role of *EYA4* in the mature cochlea. Finally, the cardiac expression of *Eya4* also makes *EYA4* a good candidate for the form of syndromic late-onset deafness associated with progressive dilated cardiomyopathy (MIM605362) that maps to the DFNA10 interval (282).

POU3F4 DEFECT: DFN3   The X chromosome-linked DFN3 locus (MIM304400) was the second locus for an isolated form of deafness to be identified, since DFN1 (MIM304700) actually corresponds to a syndromic form of deafness (149). The DFN3 hearing loss (230) is mixed, i.e., both conductive and sensorineural, with vestibular dysfunction and perilymphatic gusher on surgical ablation of the stapes (112). In males, a progressive hearing loss generally starts during youth, the sensorineural component is rapidly progressive, and results in severe-to-profound deafness. High-resolution CT reveals anomalies in stapes fixation, dilatation of the cochlea and internal auditory meatus. Heterozygous females can also present with mixed hearing loss, but with no anomalies detectable by CT scan. DFN3 was mapped to Xq21 by linkage analysis (38, 342), and the interval thereafter narrowed by deletion mapping. The DFN3 gene, *POU3F4* (MIM300039), was the first gene responsible for an isolated form of deafness to be identified (67). It consists of a single exon and encodes a POU domain transcription factor of class III. In DFN3 individuals, about half of the DNA anomalies are located within the gene and the other half, ~900 kb upstream of the gene (27, 66, 68, 99, 124). Among the mutations located within the gene, two thirds are missense mutations clustered in the POU$_H$ and POU$_S$ domains (see Figure 10). The DNA anomalies located outside the gene overlap on an 8-kb interval containing sequences that are highly conserved between species and likely to correspond to transcription regulatory elements (68).

Although *Pou3f4* is expressed in various murine tissues (128, 210), both DFN3-affected individuals and *Pou3f4* null mouse mutants only present with an ear defect (220, 246, 248). The expression of *Pou3f4* in the developing ear is restricted to the mesenchyme of the inner ear (247). The transcript appears at E10.5 when the mesenchyme starts to condense to give rise to the otic capsule, and gene expression persists in the condensed mesenchyme. Interestingly, the Pou3f4 protein remains in the nucleus of the mesenchymal cells that will form the otic capsule, but shifts to the perinuclear space in the cells from regions that will cavitate in the temporal bone to form the scala vestibuli, the scala tympani, and the internal auditory meatus (247). In adults, the gene is expressed in the fibrocytes of the spiral ligament (220).

Of the three mouse models of DFN3 described, two are the result of a targeted gene disruption (220, 248) and one of an inversion resulting from a radiation mutagenesis (246). The three mutants present with hearing losses, albeit differing in degree of severity. Curiously, these mutants, which are on different genetic backgrounds, show contrasting phenotypes and histopathological anomalies. Two have severe balance dysfunction, with vertical head bobbing that is likely due to the presence of a constriction of the superior semicircular canal, and several middle and inner ear morphogenetic anomalies (246, 248). The inner ear anomalies, such as smaller scala tympani and scala vestibuli (see Figure 1*B*) and dysplasia of the temporal bone, indicate a defect in the embryonic remodeling of the otic capsule. In addition, the spiral ligament is reduced and thinner and the Reissner's membrane is distended. In contrast, in the third mutant, no middle and inner ear anomalies were observed, and the defect only involves the fibrocytes of the spiral ligament; massive anomalies of the three types of fibrocytes and reduced endocochlear potential were found. No collapse or distension of the Reissner's membrane was reported (220). These differences among mutants are still unexplained. The target genes of this transcription factor must be identified to understand the role of POU3F4 in ear development.

Forms of Deafness Caused by a Tectorial Membrane Anomaly

COLLAGEN XI (α2 CHAIN) DEFECT: DFNA13 (AND STICKLER SYNDROME)   The DFNA13 locus has been mapped to chromosome 6p21 (MIM601868) in a study of a large family of American-European origin affected with progressive postlingual hearing loss (37). Deafness was mild to moderately severe, with a preferential mid-frequency loss (U-shaped audiogram curve). No symptomatic vestibular dysfunction was observed. However, in the two DFNA13 families analyzed, about half of the subjects with hearing impairment also had abnormal vestibular tests (174, 214). Because *COL11A2* is located at 6p21.3, and underlies two forms of syndromic deafness, namely the dominant Stickler syndrome without ocular anomalies (MIM184840) (340) and the recessive Weissenbacher-Zweymuller syndrome (MIM277610) (249), the gene was considered a candidate for DFNA13. Two distinct *COL11A2* missense mutations were detected in the two DFNA13 families (214). Collagen XI is composed of three chains, α1, α2, and α3, which interact by the repeated tripeptide GXY (where G is glycine, X and Y are often proline and hydroxyproline). *COL11A2* encodes the α2 chain. The mutations substitute either a cysteine or a glutamate for the glycine in one GXY motif, and are thus expected to impair the trimerization of the chains.

Collagen XI is a component of the tectorial membrane (see Figure 1*B*), an acellular membrane composed of several collagens (types II, V, IX, and XI), noncollagenous proteins, and proteoglycans, which is involved in deflecting the hair bundles of the cochlear hair cells upon sound stimulation. The murine *Col11a2* has been disrupted by homologous recombination. *Col11a2*$^{-/-}$ mice showed moderate-to-severe hearing impairment, whereas no hearing defect could be detected in heterozygous animals. Electron microscopy of the tectorial membrane in homozygous null mutants revealed a disorganized fibrillar structure (214).

α-TECTORIN DEFECTS: DFNA8/12 AND DFNB21   Two groups have independently identified a locus on chromosome 11q22-24 as underlying a dominant form of prelingual, nonprogressive deafness: DFNA8 (MIM601543) (167) and DFNA12 (MIM601842) (336). *TECTA* (MIM602574), which encodes α-tectorin, a specific component of the tectorial membrane (see Figure 1*B*), was located within the candidate interval for DFNA8/12. Alpha-tectorin had been isolated upon the generation of antibodies directed against components of the tectorial membrane (165, 264). Missense mutations were detected in the two DFNA8/12 families originally described (337). The locus underlying a recessive form of deafness in a consanguineous Lebanese family, DFNB21 (MIM603629), was subsequently mapped to the same chromosomal region, and a homozygous splice site mutation was detected (227).

Several cell types synthesize α-tectorin in the developing inner ear (184, 256). The *TECTA* cDNA sequence predicts that α-tectorin is synthesized as a precursor anchored to the plasma membrane via glycosylphosphatidylinositol; α-tectorin would be released from the membrane by proteolytic cleavage of the precursor. Several domains have been recognized in the protein (Figure 16).

Based on the normal auditory function of the heterozygous carriers in the DFNB21 family, the missense mutations found in DFNA8/12 patients are expected

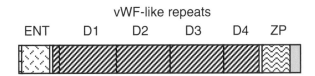

Figure 16   Schematic representation of α-tectorin. This 2155 amino acid (240-kDa) protein is a specific component of the tectorial membrane. The N-terminal signal peptide (*horizontal bars*) is followed by a region homologous to the first globular domain (G1) of entactin (ENT), 4 von Willebrand factor (vWF)-like type D domains, a zona pellucida (ZP) domain, known as an interacting domain, and a hydrophobic C-terminal region (*in gray*) (184). In the von Willebrand factor, the D domain is involved in the multimerization of the protein. Such a domain has also been found in numerous proteins that form filaments or gels (118). *TECTA* contains 23 exons (337).

to have a dominant negative effect, thereby providing genetic evidence for the idea that α-tectorin is involved in homo- or heteromeric assembly (184).

The degree of severity of the hearing loss in the five DFNA8/12 families with an identified *TECTA* mutation ranged from mild to severe, with large intrafamilial variations. In two families, the hearing loss was prelingual and stable (116, 168, 336), whereas in the three others, it was progressive (10, 15, 224), with onset ranging from prelingual to second decade. To date, no clear genotype-phenotype correlation has emerged in DFNA8/12 based on the location of the missense mutations either in the ZP domain or in the vWF type D domains.

In mice homozygous for a targeted deletion in the entactin-like G1 domain of α-tectorin, the tectorial membrane is completely detached from the sensory epithelium and lacks all noncollagenous matrix (183). In addition, this mouse model provided a unique opportunity to study how the specific and noninvasive removal of the tectorial membrane affects the mechanical behavior of the cochlea (183).

Deafness Forms of Unknown Cell Origin

TYPE II TRANSMEMBRANE SERINE PROTEASE 3 DEFECT: DFNB8/10   In 1996, two consanguineous families, one Pakistani, the other Palestinian, with autosomal recessive deafness were independently reported with linkage to chromosome 21q22.3; these families defined the DFNB8 (MIM601072) (339) and DFNB10 (MIM605316) (29) loci, respectively. *TMPRSS3*, encoding the transmembrane serine protease 3 (MIM605511) (132) was identified as the causative gene for DFNB8/10 (289). Four alternative *TMPRSS3* transcripts were detected, of which the largest, encoded by all 13 exons of the gene, was shown by RT-PCR to be expressed in a variety of tissues, including the human fetal cochlea (289). In the Palestinian family, a complex rearrangement of *TMPRSS3* was detected in exon 11, i.e., the deletion ($\sim$8 bp) and insertion of 18 monomeric ($\sim$68 bp) β-satellite repeats. In the Pakistani family, a point mutation was detected in the splice-acceptor site of intron 4 (IVS4-6 G>A), which creates an alternative splice site and results in a 4-bp insertion (frameshift) in the mRNA. This mutation may allow limited normal splicing, and thus some normal TMPRSS3 protein, which would account for the phenotypic difference between the Pakistani (childhood-onset deafness) and Palestinian (congenital deafness) families (289). Five missense mutations altering conserved amino acids of the scavenger receptor and serine protease domains (see Figure 17) have since been reported in DFNB8/10 consanguineous families affected with severe or profound congenital hearing loss (18b, 208b).

Type II transmembrane serine proteases (TTSP) represent an emerging class of cell surface proteolytic enzymes. These proteases are ideally positioned to interact with other proteins on the cell surface as well as soluble proteins and matrix components. In mammals, the TTSPs, which currently consist of 8 members (138, 289), share common structural features including (*a*) a short cytoplasmic domain, (*b*) a transmembrane domain, (*c*) a C-terminal extracellular serine protease domain, and (*d*) a variable length stem region, with a modular structure, linking

serine
TM   protease

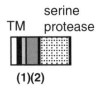

(1)(2)

Figure 17   Schematic representation of TMPRSS3. The predicted 454 amino acid type II transmembrane serine protease consists of an N-terminal cytoplasmic domain (47 amino acids) followed by a transmembrane domain (TM), one low-density lipoprotein (LDL) receptor class A domain (1) and one group A scavenger-receptor domain (2), and the C-terminal serine protease domain (289). *TMPRSS3* contains 13 exons.

the transmembrane and catalytic domains (Figure 17). In addition, these proteins lack a recognizable N-terminal signal peptide. TTSPs are synthesized as single-chain zymogens and are likely activated by cleavage within a highly conserved activation motif. Based on the predicted presence of a conserved disulfide bond linking the pro- and catalytic domains, the TTSPs are likely to remain membrane-bound following activation. However, the extracellular domains of at least some TTSPs may also be shed from the cell surface.

Most TTSPs have been identified relatively recently and have not yet been functionally characterized. The knowledge of the cell distribution of TMPRSS3 in the developing and mature inner ear is expected to provide insight into function and pathophysiology.

DIAPHANOUS-1 DEFECT: DFNA1   The first identification of a locus for an isolated form of deafness appeared in 1992 (188). In a large family from Costa Rica, a dominantly transmitted deafness, DFNA1 (MIM124900), had been traced back through eight generations (187). The hearing loss was progressive and initially affected only low frequencies. By age 40, the deafness was profound and concerned all frequencies. Intriguingly, the early low frequency hearing loss was associated with electrocochleographic findings suggestive of endolymphatic hydrops (177). Genetic linkage analysis mapped DFNA1 to chromosome 5q31 (188). The causative gene, *DIAPH1/HDIA1* (MIM602121), was identified in 1997 (202). A nucleotide substitution in a splice donor site was detected in the patients, which results in a 4 bp insertion in the transcript and a premature stop codon (202). *HDIA1* encodes a protein sharing homology with the *Drosophila* diaphanous protein.

Diaphanous-1 belongs to a family of formin-related proteins (Figure 18). These proteins are involved in cell polarization and cytokinesis (46, 85, 169). The murine diaphanous-1 has been shown to interact with the GTP-bound form of Rho and with profilin (348). The formation of this complex ensures the targeting of profilin at specific plasma membrane emplacements where profilin promotes actin polymerization (348). The human diaphanous-1 has been detected in all the tissues tested, including the cochlea (202). Preliminary results suggest that diaphanous-1 is abundant

Rho-binding poly-Pro FH2

Figure 18 Schematic representation of diaphanous-1. The predicted 1252 amino acid (139-kDa) protein belongs to a family of formin-related proteins. It is composed of a Rho GTPase-binding domain, a polyproline region, and the highly conserved formin-homology FH2 domain (202). *HDIA1* contains 26 exons.

in outer and inner hair cells, and is present at lower levels in other cochlear cell populations (370). Therefore, the primary target cells of DFNA1 could well be the hair cells themselves. It has been suggested that the *HDIA1* mutation in DFNA1 may be affecting the actin cytoskeleton in OHCs (202, 347). In addition, recent functional characterization indicates that diaphanous-1 is involved in the acetylcholine-activated, Rho-mediated signaling pathway that regulates OHC contractility (370). Generation of a mouse model of DFNA1 should help clarify the role of diaphanous-1 in this process, and reveal other functions of the protein in cochlear cells.

DFNA5   The DFNA5 locus (MIM600994) has been assigned to chromosome 7p15 through the study of a large Dutch family affected by a dominant form of progressive deafness (326). The hearing loss started between 5 and 15 years of age, and initially affected high sound frequencies only. By about 50 years of age, the hearing impairment was severe, and concerned also low frequencies (142, 143). A candidate gene was isolated in the chromosomal interval defined for DFNA5. The identification, in the affected individuals, of a complex deletion/insertion in intron 7, leading to the skipping of the downstream exon and thereby expected to result in a truncated protein, validated this gene as the causative gene (330). The gene consists of 10 exons and encodes a predicted 496-amino acid protein. Analysis of the protein sequence did not give any information on the protein function. RT-PCR analysis indicated that the gene is transcribed in the stria vascularis of the cochlea (330).

MYOSIN IIA (MYH9) DEFECT: DFNA17   The DFNA17 (MIM603622) locus (22q12-13) has been identified by studying an American family having a progressive, moderate to severe, hearing loss affecting preferentially the high frequencies (178, 179). A missense mutation (R705H) in *MYH9*, which encodes the heavy chain of a non-muscle conventional (class II) myosin, myosin IIA (Figure 19), has been detected in the single DFNA17 family reported so far (176). This mutation, which is located in the motor head, is predicted to disrupt the ATPase activity of the myosin.

Post-mortem histologic examination of the temporal bone of one patient has revealed cochleosaccular dysplasia with a degeneration of the saccular epithelium, organ of Corti and stria vascularis, and a collapsed Reissner's membrane (178). In the rat cochlea, *Myh9* is expressed in the OHCs, the Reissner's membrane and the subcentral region of the spiral ligament (176). The collapse of the Reissner's

motor head   IQ   coiled-coil

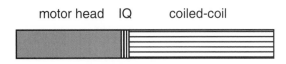

Figure 19   Schematic representation of the myosin IIA heavy chain (1962 amino acids) of the nonmuscle conventional myosin IIA. This protein consists of a motor head containing the ATP- and actin-binding sites, a neck region containing 2 IQ motifs, and a large coiled-coil tail region. *MYH9* contains 40 coding exons.

membrane indicates a perturbation of the hydro-ionic homeostasis of the cochlea, that might be related to the proposed roles of the Reissner's membrane (277, 372) and fibrocytes of the spiral ligament (303) in potassium fluxes between the endolymphatic and perilymphatic compartments. In addition, it has been proposed that some of these fibrocytes generate a tension between the basilar membrane and the spiral ligament complex (136), which maintains the mechanical cohesion of the cochlea.

Interestingly, 3 autosomal dominant syndromes, May-Hegglin anomaly (MIM155100), Fechtner (MIM153640), and Sebastian (MIM605249) syndromes, have also been shown to result from *MYH9* mutations (155, 212). These syndromes share macrothrombocytopenia and characteristic leukocyte inclusions; Fechtner syndrome, in addition includes sensorineural hearing loss, cataracts, and nephritis. At present, no genotype/phenotype correlation has been reported.

MITOCHONDRIAL GENE DEFECTS   The mitochondrial DNA (mtDNA) encodes 13 mRNAs translated into 13 proteins, which interact with approximately 60 nuclear DNA-encoded proteins to form the five enzymatic complexes required for oxidative phosphorylation. It also encodes two ribosomal RNAs (rRNA) and 22 transfer RNAs (tRNA). MtDNA-linked diseases are transmitted to both sexes through the maternal line only.

Several syndromes including a hearing loss are associated with mutations in mtDNA [see (244) for review]. The first evidence for an isolated deafness form of mtDNA origin arose in 1992, through the study of a large Arab-Israeli family affected by hearing loss. In this family, genetic analysis gave indications for the involvement of at least two mutations, a mitochondrial one and an autosomal recessive one. A homoplasmic A1555G mutation was detected in the mitochondrial 12S rRNA gene, *MTRNR1* (MIM561000) in deaf individuals (252). Most of the individuals carrying the A1555G mutation were affected by severe-to-profound hearing loss during infancy; a few had onset during adulthood, and some were normal hearing (33). This mutation was also observed in three families affected by a maternally inherited deafness induced by aminoglycoside antibiotics (93, 252). Since then, several studies have reported the presence of the A1555G mutation in deaf subjects who had been exposed to aminoglycosides [reviewed in (91, 144)].

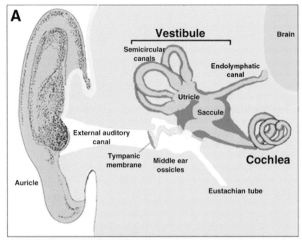

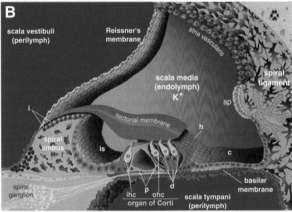

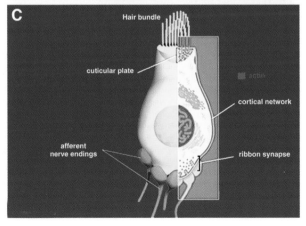

See text page C-2

Figure 1 (page C-1) (*A*) Schematic representation of the human ear. The mammalian ear is composed of three compartments: the outer ear is made up of the auricle and external auditory canal, the middle ear contains the ossicles within the tympanic cavity, and the inner ear consists of six sensory organs, namely the cochlea and the five vestibular end organs (saccule, utricle, and three semicircular canals).

(*B*) Cross-section through the cochlear duct. The membranous labyrinth of the cochlea (cochlear duct) divides the bony labyrinth in three canals, the scala vestibuli and the scala tympani, both filled with perilymph, and the scala media, filled with endolymph. The organ of Corti, which is the auditory transduction apparatus, protrudes in the scala media. This organ is made up of an array of sensory cells (in *yellow*), *i.e.* the single row of inner hair cells (ihc) and the triple row of outer hair cells (ohc); and different types of supporting cells that include pillar cells (p), cells of Deiters (d), and cells of Hensen (h). It is covered by an acellular gel, the tectorial membrane. The organ of Corti is flanked by the epithelial cells of the inner sulcus (is) on the medial side and by the cells of Claudius (c) on the lateral side. The stria vascularis, on the lateral wall of the cochlear duct, is responsible for the secretion of K^+ into the endolymph and for the generation of the endocochlear potential. Different types of fibrocytes surround the cochlear epithelium. Other abbreviations: (i) interdental cells, (sp) spiral prominence. (Adapted from a figure drawn by P. Küssel-Andermann.)

(*C*) Schematic representation of an inner hair cell. Note the highly organized hair bundle, made of several rows of stereocilia, at the apical pole of the cell. The ribbon synapse has particular structural and functional features. Three specific structures of the actin cytoskeleton are shown (in *red*), namely the filaments of the stereocilia, the cuticular plate (a dense meshwork of horizontal filaments running parallel to the apical cell surface), and the cortical network (beneath the plasma membrane).

However, three independent studies on Spanish deaf individuals reported a high incidence (up to 25%) of the A1555G mutation among families affected by sensorineural hearing loss, in whom most of the affected members had not been treated with aminoglycosides (79, 84, 281). The absence of such a high proportion of the A1555G mutation reported in other countries (185) was puzzling, especially since haplotype analysis performed on Spanish individuals suggested that this mutation had arisen from numerous independent mutational events (320). However, a recent study on the Japanese population found that 10% of profoundly deaf individuals with no history of aminoglycosides treatment also carry the A1555G mutation (324). Inhibition of mRNA translation by aminoglycosides in bacteria involves their binding to a specific G of the 16S rRNA (284). The A1555G mutation affects the analogous nucleotide in the mitochondrial 12S rRNA. Unlike normal 12S rRNA, this 12S mutated form binds to aminoglycosides with high affinity (126). The ratio of translation rates in the presence and absence of an aminoglycoside, which reflects the effect of the drug on mitochondrial protein synthesis, was reported to be significantly decreased in the lymphoblastoid cell lines derived from individuals carrying the mutation, as compared to the ratio in control cell lines (121). Finally, the recent non-parametric analysis of a large number of families with the A1555G mutation supported the role of a modifier locus on chromosome 8 (41). Another *MTRNR1* mutation, 961delT, has been discovered in a family with a hearing loss induced by aminoglycosides (45), although no direct interaction between this nucleotide and aminoglycosides has been reported. When a familial history of aminoglycoside-induced deafness is suspected, diagnosis can now be facilitated by searching for the A1555G and 961delT mutations. Screening for *MTRNR1* mutations should also help prevent a large proportion of iatrogenic deafness.

Four point mutations have been detected in *MTTS1* (MIM590080), which encodes the mitochondrial tRNA$^{ser(UCN)}$ [reviewed in (91, 144)]. An A7445G mutation was initially reported in a Scottish family affected with an isolated form of deafness (258); the same mutation was then detected in two unrelated pedigrees with deafness associated to palmoplantar keratoderma (MIM148350) (293). Second, a 7472insC mutation has been shown to underlie a syndromic form of deafness including ataxia and myoclonus (315); the same mutation was observed in a large Dutch family in which most of the patients presented with only hearing loss (334). Finally, T7510C and T7511C mutations have been found in patients suffering only from hearing loss (145, 310).

Why a defective mitochondrial tRNA^{ser} or 12S rRNA results in hearing loss is not understood. Given the damage to hair cells observed upon aminoglycoside exposure (47, 133, 313), these are presumed to be the target cells of the 12S RNA mutations. However, other inner-ear cell types that are rich in mitochondria could also be implicated. In the absence of mouse models for mitochondrial defects, this issue is difficult to resolve. Finally, some findings argue in favor of increasing mtDNA deletions and mutations in some patients affected by presbyacousis (92, 290).

CONCLUDING REMARKS

Twenty-nine genes that underlie isolated deafness in humans and/or in mouse have been identified to date. However, difficulties inherent in genetic linkage analysis, coupled with the possible involvement of environmental causes, have limited the characterization of the genes causative or predisposing to late-onset forms of deafness. Rapid progress in the isolation of the deafness genes can be expected in the near future when human and mouse genomic sequences become available. Identification of these genes holds great promise for understanding the development and functioning of the inner ear in molecular terms. The heuristic value of such an approach will rely upon our ability to integrate the molecular analysis of these genetic diseases into the conceptual framework developed over the last 40 years through remarkable advances in the biophysics, electrophysiology, and cell biology of the inner ear. In particular, this requires adaptation of biophysical and electrophysiological techniques to the two vertebrate models (mouse and zebrafish) that are amenable to genetic analysis and gene transfer. In contrast, this approach will likely contribute little to our understanding of how the central auditory pathway functions. Indeed, on the one hand, nonsyndromic hereditary deafness forms result overwhelmingly, if not entirely, from ear defects, and on the other hand, retrocochlear forms of syndromic deafness are due mainly to mitochondrial or peroxisomal defects.

Genotype/phenotype analysis of a given genetic form of deafness frequently indicates the involvement of modifier gene(s). This is particularly true for what appear to be the commonest forms, DFNB1, due to *CX26* defect, and the mitochondrial 12S RNA gene defect. Characterization of the modifier genes is of the highest priority, not only to improve the quality of genetic counseling, but also to reveal molecules or pathways that can prevent the occurrence of the hearing defect. Identification of modifier loci is much easier in mice than in humans [reviewed in (228)] especially for the rare deafness forms. Although no modifier gene for deafness has yet been isolated, several loci have been identified in mouse and humans (41, 146, 150, 234, 260). In particular, alleles reducing the penetrance of the mitochondrial A1555G mutation (41) and an unidentified mutation underlying DFNB26 (260) have been reported.

The frame of the research on hereditary deafness has now been outlined in genetic and clinical terms, moving the next substantive challenge to developing therapeutic strategies as an alternative to the prostheses currently in use. Stopping or reversing progression of hearing loss will require convergent approaches, many already under way, that include studying the pathogenic process underlying each deafness form, searching for modifier gene(s) that antagonize the effects of the mutations in the "deafness gene," finding the conditions for hair cell regeneration, and developing devices for delivery of drugs, genes, or embryonic cells specifically to the inner ear. The success of these strategies is expected to have therapeutic implications also for the nonhereditary forms of deafness.

ACKNOWLEDGMENTS

The authors thank Sébastien Chardenoux for help in drawing the protein schemes and Xue-Zhong Liu for sharing unpublished data. The studies from the Unité de Génétique des Déficits Sensoriels are supported by grants from the European Community (QLG2-CT-1999-00988), Fondation pour la Recherche Médicale (France), Fondation R. & G. Strittmatter/Retina-France, Association Française contre les Myopathies, and A. & M. Suchert and Forschung contra Blindheit-Initiative Usher Syndrome (Germany).

NOTE ADDED IN PROOF

The gene causing DFNB16 has just been identified by a candidate gene approach based on the analysis of a subtracted mouse inner ear cDNA library. The encoded protein, stereocilin, is located in the hair cell stereocilia (338a).

Visit the Annual Reviews home page at www.AnnualReviews.org

LITERATURE CITED

1. Abdelhak S, Kalatzis V, Heilig R, Compain S, Samson D, et al. 1997. A human homologue of the *Drosophila eyes absent* gene underlies Branchio-Oto-Renal (BOR) syndrome and identifies a novel gene family. *Nat. Genet.* 15:157–64

2. Abe S, Usami S, Hoover DM, Cohn E, Shinkawa H, et al. 1999. Fluctuating sensorineural hearing loss associated with enlarged vestibular aqueduct maps to 7q31, the region containing the Pendred gene. *Am. J. Med. Genet.* 82:322–28

3. Abe S, Usami S-i, Shinkawa H, Kelley PM, Kimberling WJ. 2000. Prevalent connexin 26 gene (GJB2) mutations in Japanese. *J. Med. Genet.* 37:41–43

4. Achanzar WE, Ward S. 1997. A nematode gene required for sperm vesicle fusion. *J. Cell Sci.* 110:1073–81

5. Adato A, Kalinski H, Weil D, Chaïb H, Korostishevsky M, et al. 1999. Possible interaction between USH1B and USH3 gene products as implied by apparent digenic deafness inheritance. *Am. J. Hum. Genet.* 65:261–65

6. Adato A, Raskin L, Petit C, Bonné-Tamir B. 2000. Deafness heterogeneity

in a Druze isolate from the Middle East: novel OTOF and PDS mutations, low prevalence of GJB2 35delG mutation and indication for a new DFNB locus. *Eur. J. Hum. Genet.* 8:437–42

7. Ahituv N, Sobe T, Robertson NG, Morton CC, Taggart RT, et al. 2000. Genomic structure of the human unconventional myosin VI gene. *Gene* 261:269–75

8. Ahmed ZM, Riazuddin S, Bernstein SL, Ahmed Z, Khan S, et al. 2001. Mutations of the protocadherin gene *PCDH15* cause Usher syndrome type 1F. *Am. J. Hum. Genet.* 69:25–34

9. Alagramam KN, Murcia CL, Kwon HY, Pawlowski KS, Wright CG, et al. 2001. The mouse Ames waltzer hearing-loss mutant is caused by mutation of *Pcdh15*, a novel protocadherin gene. *Nat. Genet.* 27:99–102

10. Alloisio N, Morlé L, Bozon M, Godet J, Verhoeven K, et al. 1999. Mutation in the zonadhesin-like domain of α-tectorin associated with autosomal dominant nonsyndromic hearing loss. *Eur. J. Hum. Genet.* 7:255–58

11. Antoniadi T, Rabionet R, Kroupis C,

Aperis GA, Economides J, et al. 1999. High prevalence in the Greek population of the 35delG mutation in the connexin 26 gene causing prelingual deafness. *Clin. Genet.* 55:381–82

12. Assad JA, Shepherd GM, Corey DP. 1991. Tip-link integrity and mechanical transduction in vertebrate hair cells. *Neuron* 7:985–94

13. Avraham KB, Hasson T, Sobe T, Balsara B, Testa JR, et al. 1997. Characterization of unconventional *MYO6*, the human homologue of the gene responsible for deafness in Snell's waltzer mice. *Hum. Mol. Genet.* 6:1225–31

14. Avraham KB, Hasson T, Steel KP, Kingsley DM, Russell LB, et al. 1995. The mouse Snell's waltzer deafness gene encodes an unconventional myosin required for structural integrity of inner ear hair cells. *Nat. Genet.* 11:369–75

15. Balciuniene J, Dahl N, Borg E, Samuelsson E, Koisti MJ, et al. 1998. Evidence for digenic inheritance of nonsyndromic hereditary hearing loss in a Swedish family. *Am. J. Hum. Genet.* 63:786–93

16. Baldwin CT, Weiss S, Farrer LA, De Stefano AL, Adair R, et al. 1995. Linkage of congenital, recessive deafness (DFNB4) to chromosome 7q31 and evidence for genetic heterogeneity in the Middle Eastern Druze population. *Hum. Mol. Genet.* 4:1637–42

17. Barhanin J, Lesage F, Guillemare E, Fink M, Lazdunski M, et al. 1996. K_VLQT1 and lsK (minK) proteins associate to form the I_{Ks} cardiac potassium current. *Nature* 384:78–80

18. Bashir R, Britton S, Strachan T, Keers S, Vafiadaki E, et al. 1998. A gene related to *Caenorhabditis elegans* spermatogenesis factor *fer-1* is mutated in limb-girdle muscular dystrophy type 2B. *Nat. Genet.* 20:37–42

18b. Ben-Yosef T, Wattenhofer M, Riazuddin S, Ahmed ZM, Scott HS, et al. 2001. Novel mutations of *TMPRSS3* in four DFNB8/B10 families segregating congenital autosomal recessive deafness. *J. Med. Genet.* 38:396–400

19. Ben Zina Z, Masmoudi S, Ayadi H, Chaker F, Ghorbel AM, et al. 2001. From DFNB2 to Usher syndrome: variable expressivity of the same disease. *Am. J. Med. Genet.* 101:181–83

20. Berg JS, Powell BC, Cheney RE. 2001. A millennial myosin census. *Mol. Biol. Cell* 12:780–94

21. Bevans CG, Harris AL. 1999. Direct high affinity modulation of connexin channel activity by cyclic nucleotides. *J. Biol. Chem.* 274:3720–25

22. Bevans CG, Harris AL. 1999. Regulation of connexin channels by pH. Direct action of the protonated form of taurine and other aminosulfonates. *J. Biol. Chem.* 274:3711–19

23. Bevans CG, Kordel M, Rhee SK, Harris AL. 1998. Isoform composition of connexin channels determines selectivity among second messengers and uncharged molecules. *J. Biol. Chem.* 273:2808–16

24. Bidart J-M, Lacroix L, Evain-Brion D, Caillou B, Lazar V, et al. 2000. Expression of Na^+/I^- symporter and Pendred syndrome genes in trophoblast cells. *J. Clin. Endocrinol. Metab.* 85:4367–72

25. Bidart J-M, Mian C, Lazar V, Russo D, Filetti S, et al. 2000. Expression of pendrin and the Pendred syndrome (PDS) gene in human thyroid tissues. *J. Clin. Endocrinol. Metab.* 85:2028–33

26. Bitner-Glindzicz M, Lindley KJ, Rutland P, Blaydon D, Smith VV, et al. 2000. A recessive contiguous gene deletion causing infantile hyperinsulinism, enteropathy and deafness identifies the Usher type 1C gene. *Nat. Genet.* 26:56–60

27. Bitner-Glindzicz M, Turnpenny P, Höglund P, Kääriäinen H, Sankila E-M, et al. 1995. Further mutations in *Brain 4* (POU3F4) clarify the phenotype in the X-linked deafness, DFN3. *Hum. Mol. Genet.* 4:1467–69

28. Bolz H, von Brederlow B, Ramirez A,

Bryda EC, Kutsche K, et al. 2001. Mutations of *CDH23*, encoding a new member of the cadherin gene family, causes Usher syndrome type 1D. *Nat. Genet.* 27:108–12

29. Bonne-Tamir B, Destefano AL, Briggs CE, Adair R, Franklyn B, et al. 1996. Linkage of congenital recessive deafness (gene DFNB10) to chromosome 21q22.3. *Am. J. Hum. Genet.* 58:1254–59

30. Bork JM, Peters LM, Riazuddin S, Bernstein SL, Ahmed ZM, et al. 2001. Usher syndrome 1D and nonsyndromic autosomal recessive deafness DFNB12 are caused by allelic mutations of the novel cadherin-like gene *CDH23*. *Am. J. Hum. Genet.* 68:26–37

31. Borsani G, DeGrandi A, Ballabio A, Bulfone A, Bernard L, et al. 1999. EYA4, a novel vertebrate gene related to *Drosophila eyes absent. Hum. Mol. Genet.* 8:11–23

32. Boughman JA, Vernon M, Shaver KA. 1983. Usher syndrome: definition and estimate of prevalence from two high-risk populations. *J. Chronic Dis.* 36:595–603

33. Braverman I, Jaber L, Levi H, Adelman C, Arons KS, et al. 1996. Audiovestibular findings in patients with deafness caused by a mitochondrial susceptibility mutation and precipitated by an inherited nuclear mutation or aminoglycosides. *Arch. Otolaryngol. Head Neck Surg.* 122:1001–4

34. Brissette JL, Kumar NM, Gilula NB, Hall JE, Dotto GP. 1994. Switch in gap junction protein expression is associated with selective changes in junctional permeability during keratinocyte differentiation. *Proc. Natl. Acad. Sci. USA* 91:6453–57

35. Brose N, Hofmann K, Hata Y, Südhof TC. 1995. Mammalian homologues of Caenorhabditis elegans unc-13 gene define novel family of C2–domain proteins. *J. Biol. Chem.* 270:25273–80

36. Brown KA, Sutcliffe MJ, Steel KP, Brown SDM. 1992. Close linkage of the olfactory marker protein gene to the mouse deafness mutation *shaker-1*. *Genomics* 13:189–93

37. Brown MR, Tomek MS, Van Laer L, Smith S, Kenyon JB, et al. 1997. A novel locus for autosomal dominant nonsyndromic hearing loss, DFNA13, maps to chromosome 6p. *Am. J. Hum. Genet.* 61:924–27

38. Brunner HG, van Bennekom CA, Lambermon EMM, Oei TL, Cremers CWRJ, et al. 1988. The gene for X-linked progressive mixed deafness with perilymphatic gusher during stapes surgery (DFN3) is linked to PGK. *Hum. Genet.* 80:337–40

39. Bruzzone R, White TW, Paul DL. 1996. Connections with connexins: the molecular basis of direct intercellular signaling. *Eur. J. Biochem.* 238:1–27

40. Bryda EC, Ling H, Flaherty L. 1997. A high-resolution genetic map around waltzer on mouse chromosome 10 and identification of a new allele of waltzer. *Mamm. Genome* 8:1–4

41. Bykhovskaya Y, Estivill X, Taylor K, Hang T, Hamon M, et al. 2000. Candidate locus for a nuclear modifier gene for maternally inherited deafness. *Am. J. Hum. Genet.* 66:1905–10

42. Campbell C, Cucci RA, Prasad S, Green GE, Edeal JB, et al. 2001. Pendred syndrome, DFNB4, and *PDS/SLC26A4*. Identification of eight novel mutations and possible genotype-phenotype correlations. *Hum. Mutat.* 17:403–11

43. Cao F, Eckert R, Elfgang C, Nitsche JM, Snyder SA, et al. 1998. A quantitative analysis of connexin-specific permeability differences of gap junctions expressed in HeLa transfectants and Xenopus oocytes. *J. Cell Sci.* 111:31–43

44. Carrasquillo MM, Zlotogora J, Barges S, Chakravarti A. 1997. Two different connexin 26 mutations in an inbred kindred segregating non-syndromic recessive deafness: Implications for genetic studies in isolated populations. *Hum. Mol. Genet.* 6:2163–72

45. Casano RA, Johnson DF, Bykhovskaya Y, Torricelli F, Bigozzi M, et al. 1999. Inherited susceptibility to aminoglycoside ototoxicity: genetic heterogeneity and clinical implications. *Am. J. Otolaryngol.* 20:151–56

46. Castrillon DH, Wasserman SA. 1994. Diaphanous is required for cytokinesis in Drosophila and shares domains of similarity with the products of the limb deformity gene. *Development* 120:3367–77

47. Caussé R. 1949. Action toxique vestibulaire et cochléaire de la streptomycine du point de vue expérimental. *Ann. Otol.-Laryngol. (Paris)* 66:518–38

48. Chaïb H, Lina-Granade G, Guilford P, Plauchu H, Levilliers J, et al. 1994. A gene responsible for a dominant form of neurosensory non-syndromic deafness maps to the *NSRD1* recessive deafness gene interval. *Hum. Mol. Genet.* 3:2219–22

49. Chaïb H, Place C, Salem N, Chardenoux S, Vincent C, et al. 1996. A gene responsible for a sensorineural nonsyndromic recessive deafness maps to chromosome 2p22–23. *Hum. Mol. Genet.* 5:155–58

50. Chaïb H, Place C, Salem N, Dodé C, Chardenoux S, et al. 1996. Mapping of DFNB12, a gene for a non-syndromal autosomal recessive deafness, to chromosome 10q21–22. *Hum. Mol. Genet.* 5: 1061–64

51. Chen AH, Wayne S, Bell A, Ramesh A, Srisailapathy CRS, et al. 1997. New gene for autosomal recessive non-syndromic hearing loss maps to either chromosome 3q or 19p. *Am. J. Med. Genet.* 71:467–71

52. Chen Z-Y, Hasson T, Kelley PM, Schwender BJ, Schwartz MF, et al. 1996. Molecular cloning and domain structure of human myosin-VIIa, the gene product defective in Usher syndrome 1B. *Genomics* 36:440–48

53. Chishti AH, Kim AC, Marfatia SM, Lutchman M, Hanspal M, et al. 1998. The FERM domain: a unique module involved in the linkage of cytoplasmic proteins to the membrane. *Trends Biochem. Sci.* 23:281–82

54. Clark JD, Lin LL, Kriz RW, Ramesha CS, Sultzman LA, et al. 1991. A novel arachidonic acid-selective cytosolic PLA2 contains a Ca^{2+}-dependent translocation domain with homology to PKC and GAP. *Cell* 65:1043–51

55. Cohn ES, Kelley PM, Fowler TW, Gorga MP, Lefkowitz DM, et al. 1999. Clinical studies of families with hearing loss attributable to mutations in the connexin 26 gene (GJB2/DFNB1). *Pediatrics* 103:546–50

56. Colombatti A, Bonaldo P, Doliana R. 1993. Type A modules: interacting domains found in several non-fibrillar collagens and in other extracellular matrix proteins. *Matrix* 13:297–306

57. Coucke PJ, Van Camp G, Djoyodiharjo B, Smith SD, Frants RR, et al. 1994. Linkage of autosomal dominant hearing loss to the short arm of chromosome 1 in two families. *N. Engl. J. Med.* 331:425–31

58. Coucke PJ, Van Hauwe P, Kelley PM, Kunst H, Schatteman I, et al. 1999. Mutations in the *KCNQ4* gene are responsible for autosomal dominant deafness in four DFNA2 families. *Hum. Mol. Genet.* 8:1321–28

59. Coyle B, Coffey R, Armour JAL, Gausden E, Hochberg Z, et al. 1996. Pendred syndrome (goitre and sensorineural hearing loss) maps to chromosome 7 in the region containing the nonsyndromic deafness gene *DFNB4*. *Nat. Genet.* 12:421–23

60. Cramer LP. 2000. Myosin VI: Roles for a minus end-directed actin motor in cells. *J. Cell Biol.* 150:F121–26

61. Cranefield PF, Federn W. 1970. Paulus Zacchias on mental deficiency and on deafness. *Bull. NY Acad. Med.* 46:3–21

62. Cremers WR, Bolder C, Admiraal RJ, Everett LA, Joosten FB, et al. 1998. Progressive sensorineural hearing loss and a widened vestibular aqueduct in Pendred syndrome. *Arch. Otolaryngol. Head Neck Surg.* 124:501–5

63. Dahl E, Manthey D, Chen Y, Schwarz HJ, Chang YS, et al. 1996. Molecular cloning and functional expression of mouse connexin-30, a gap junction gene highly expressed in adult brain and skin. *J. Biol. Chem.* 271:17903–10. Erratum. 1996. *J. Biol. Chem.* 271(42):26444

64. Davis DB, Delmonte AJ, Ly CT, McNally EM. 2000. Myoferlin, a candidate gene and potential modifier of muscular dystrophy. *Hum. Mol. Genet.* 9:217–26

65. de Kok YJM, Bom SJH, Brunt TM, Kemperman MH, van Beusekom E, et al. 1999. A Pro51Ser mutation in the *COCH* gene is associated with late onset autosomal dominant progressive sensorineural hearing loss with vestibular defects. *Hum. Mol. Genet.* 8:361–66

66. de Kok YJM, Cremers CW, Ropers HH, Cremers FP. 1997. The molecular basis of X-linked deafness type 3 (DFN3) in two sporadic cases: identification of a somatic mosaicism for a POU3F4 missense mutation. *Hum. Mutat.* 10:207–11

67. de Kok YJM, van der Maarel SM, Bitner-Glindzicz M, Huber I, Monaco AP, et al. 1995. Association between X-linked mixed deafness and mutations in the POU domain gene *POU3F4. Science* 267:685–88

68. de Kok YJM, Vossenaar ER, Cremers CWRJ, Dahl N, Laporte J, et al. 1996. Identification of a hot spot for microdeletions in patients with X-linked deafness type 3 (DFN3) 900 kb proximal to the DFN3 gene *POU3F4. Hum. Mol. Genet.* 5:1229–35

69. Delpire E, Lu J, England R, Dull C, Thorne T. 1999. Deafness and imbalance associated with inactivation of the secretory Na-K-2Cl co-transporter. *Nat. Genet.* 22:192–95

70. Denoyelle F, Lina-Granade G, Plauchu H, Bruzzone R, Chaïb H, et al. 1998. Connexin26 gene linked to a dominant deafness. *Nature* 393:319–20

71. Denoyelle F, Marlin S, Weil D, Moatti L, Chauvin P, et al. 1999. Clinical features of the prevalent form of childhood deafness, *DFNB1*, due to a connexin26 gene defect: implications for genetic counselling. *Lancet* 353:1298–303

72. Denoyelle F, Weil D, Maw MA, Wilcox SA, Lench NJ, et al. 1997. Prelingual deafness: high prevalence of a 30delG mutation in the connexin 26 gene. *Hum. Mol. Genet.* 6:2173–77

73. Deol MS. 1956. The anatomy and development of the mutants pirouette, shaker-1 and waltzer in the mouse. *Proc. R. Soc. London Ser. B* 145:206–13

74. Deol MS, Green MC. 1966. Snell's waltzer, a new mutation affecting behaviour and the inner ear in the mouse. *Genet. Res.* 8:339–45

75. Dhein S. 1998. Gap junction channels in the cardiovascular system: pharmacological and physiological modulation. *Trends Pharmacol. Sci.* 19:229–41

76. Di Palma F, Holme RH, Bryda EC, Belyantseva IA, Pellegrino R, et al. 2001. Mutations in *Cdh23*, encoding a new type of cadherin, cause stereocilia disorganization in waltzer, the mouse model for Usher syndrome type 1D. *Nat. Genet.* 27:103–7

77. Dixon J, Gazzard J, Chaudhry SS, Sampson N, Schulte BA, et al. 1999. Mutation of the Na-K-Cl co-transporter gene *Slc12a2* results in deafness in mice. *Hum. Mol. Genet.* 8:1579–84

78. El-Amraoui A, Sahly I, Picaud S, Sahel J, Abitbol M, et al. 1996. Human Usher IB/mouse *shaker-1*; the retinal phenotype discrepancy explained by the presence/absence of myosin VIIA in the photoreceptor cells. *Hum. Mol. Genet.* 5:1171–78

79. el-Schahawi M, Lopez de Munain A, Sarrazin AM, Shanske AL, Basirico M, et al. 1997. Two large Spanish pedigrees with nonsyndromic sensorineural deafness and the mtDNA mutation at nt 1555 in the 12s rRNA gene: evidence of heteroplasmy. *Neurology* 48:453–56

80. Elfgang C, Eckert R, Lichtenberg-Frate H, Butterweck A, Traub O, et al. 1995.

Specific permeability and selective formation of gap junction channels in connexin-transfected HeLa cells. *J. Cell Biol.* 129: 805–17

81. Erkman L, McEvilly RJ, Luo L, Ryan AK, Hooshmand F, et al. 1996. Role of transcription factors Brn-3.1 and Brn-3.2 in auditory and visual system development. *Nature* 381:603–6

82. Ernest S, Rauch G-J, Haffter P, Geisler R, Petit C, et al. 2000. *Mariner* is defective in *myosin VIIA*: a zebrafish model for human hereditary deafness. *Hum. Mol. Genet.* 9:2189–96

83. Estivill X, Fortina P, Surrey S, Rabionet R, Melchionda S, et al. 1998. Connexin-26 mutations in sporadic and inherited sensorineural deafness. *Lancet* 351:394–98

84. Estivill X, Govea N, Barcelo E, Badenas C, Romero E, et al. 1998. Familial progressive sensorineural deafness is mainly due to the mtDNA A1555G mutation and is enhanced by treatment of aminoglycosides. *Am. J. Hum. Genet.* 62:27–35

85. Evangelista M, Blundell K, Longtine MS, Chow CJ, Adames N, et al. 1997. Bni1p, a yeast formin linking Cdc42p and the actin cytoskeleton during polarized morphogenesis. *Science* 276:118–22

86. Everett LA, Belyantseva IA, Noben-Trauth K, Cantos R, Chen A, et al. 2001. Targeted disruption of mouse *Pds* provides insight about the inner-ear defects encountered in Pendred syndrome. *Hum. Mol. Genet.* 10:153–61

87. Everett LA, Glaser B, Beck JC, Idol JR, Buchs A, et al. 1997. Pendred syndrome is caused by mutations in a putative sulphate transporter gene (*PDS*). *Nat. Genet.* 17:411–22

88. Everett LA, Morsli H, Wu DK, Green ED. 1999. Expression pattern of the mouse ortholog of the Pendred's syndrome gene (Pds) suggests a key role for pendrin in the inner ear. *Proc. Natl. Acad. Sci. USA* 96:9727–32

89. Fanning AS, Anderson JM. 1999. PDZ domains: fundamental building blocks in the organization of protein complexes at the plasma membrane. *J. Clin. Invest.* 103:767–72

90. Fekete DM. 1996. Cell fate specification in the inner ear. *Curr. Opin. Neurobiol.* 6:533–41

91. Fischel-Ghodsian N. 1999. Mitochondrial deafness mutations reviewed. *Hum. Mutat.* 13:261–70

92. Fischel-Ghodsian N, Bykhovskaya Y, Taylor K, Kahen T, Cantor R, et al. 1997. Temporal bone analysis of patients with presbycusis reveals high frequency of mitochondrial mutations. *Hear. Res.* 110:147–54

93. Fischel-Ghodsian N, Prezant TR, Bu X, Oztas S. 1993. Mitochondrial ribosomal RNA gene mutation in a patient with sporadic aminoglycoside ototoxicity. *Am. J. Otolaryngol.* 14:399–403

94. Flagella M, Clarke LL, Miller ML, Erway LC, Giannella RA, et al. 1999. Mice lacking the basolateral Na-K-2Cl cotransporter have impaired epithelial chloride secretion and are profoundly deaf. *J. Biol. Chem.* 274:26946–55

95. Frank G, Hemmert W, Gummer AW. 1999. Limiting dynamics of high-frequency electromechanical transduction of outer hair cells. *Proc. Natl. Acad. Sci. USA* 96:4420–25

96. Fransen E, Verstreken M, Bom SJH, Lemaire F, Kemperman MH, et al. 2001. A common ancestor for COCH related cochleovestibular (DFNA9) patients in Belgium and The Netherlands bearing the P51S mutation. *J. Med. Genet.* 38:61–64

97. Fransen E, Verstreken M, Verhagen WIM, Wuyts FL, Huygen PLM, et al. 1999. High prevalence of symptoms of Menière's disease in three families with a mutation in the *COCH* gene. *Hum. Mol. Genet.* 8:1425–29

98. Fraser GR. 1965. Association of congenital deafness with goiter (Pendred's syndrome). A study of 207 families. *Ann. Hum. Genet.* 28:201–49

99. Friedman RA, Bykhovskaya Y, Tu G, Talbot JM, Wilson DF, et al. 1997. Molecular analysis of the POU3F4 gene in patients with clinical and radiographic evidence of X-linked mixed deafness with perilymphatic gusher. *Ann. Otol. Rhinol. Laryngol.* 106:320–25

100. Friedman T, Battey J, Kachar B, Riazuddin S, Noben-Trauth K, et al. 2000. Modifier genes of hereditary hearing loss. *Curr. Opin. Neurobiol.* 10:487–93

101. Friedman TB, Liang Y, Weber JL, Hinnant JT, Barber TD, et al. 1995. A gene for congenital, recessive deafness *DFNB3* maps to the pericentromeric region of chromosome 17. *Nat. Genet.* 9:86–91

102. Frolenkov GI, Atzori M, Kalinec F, Mammano F, Kachar B. 1998. The membrane-based mechanism of cell motility in cochlear outer hair cells. *Mol. Biol. Cell* 9:1961–68

103. Frydman M, Vreugde S, Nageris B, Weiss S, Vahava O, et al. 2000. Clinical characterization of genetic hearing loss caused by a mutation in the POU4F3 transcription factor. *Arch. Otolaryngol. Head Neck Surg.* 126:633–37

104. Fuse Y, Doi K, Hasegawa T, Sugii A, Hibino H, et al. 1999. Three novel connexin26 gene mutations in autosomal recessive non-syndromic deafness. *Neuroreport* 10:1853–57

105. Gabriel H, Kupsch P, Sudendey Jr, Winterhager E, Jahnke K, et al. 2001. Mutations in the connexin26/GJB2 gene are the most common event in non-syndromic hearing loss among the German population. *Hum. Mutat.* 17:521–22

106. Gabriel HD, Jung D, Butzler C, Temme A, Traub O, et al. 1998. Transplacental uptake of glucose is decreased in embryonic lethal connexin26–deficient mice. *J. Cell Biol.* 140:1453–61

107. Garner CC, Nash J, Huganir RL. 2000. PDZ domains in synapse assembly and signalling. *Trends Cell Biol.* 10:274–80

108. Gasparini P, Estivill X, Volpini V, Totaro A, Castellvi-Bel S, et al. 1997. Linkage of DFNB1 to non-syndromic neurosensory autosomal-recessive deafness in Mediterranean families. *Eur. J. Hum. Genet.* 5:83–88

109. Gasparini P, Rabionet R, Barbujani G, Melchionda S, Petersen M, et al. 2000. High carrier frequency of the 35delG deafness mutation in European populations. *Eur. J. Hum. Genet.* 8:19–23

110. Gibson F, Walsh J, Mburu P, Varela A, Brown KA, et al. 1995. A type VII myosin encoded by the mouse deafness gene *Shaker-1. Nature* 374:62–64

111. Gillespie PG, Wagner MC, Hudspeth AJ. 1993. Identification of a 120 kd hairbundle myosin located near stereociliary tips. *Neuron* 11:581–94

112. Glasscock ME. 1973. The stapes gusher. *Arch. Otolaryngol.* 98:82–91

113. Goldberg GS, Lampe PD, Nicholson BJ. 1999. Selective transfer of endogenous metabolites through gap junctions composed of different connexins. *Nat. Cell Biol.* 1:457–59

114. Goldstein MA. 1933. *Problems of the Deaf.* St Louis, MO: Laryngoscope Press

115. Gorlin RJ. 1995. *Hereditary Hearing Loss and Its Syndromes.* New York: Oxford Univ. Press

116. Govaerts PJ, De Ceulaer G, Daemers K, Verhoeven K, Van Camp G, et al. 1998. A new autosomal-dominant locus (DFNA 12) is responsible for a nonsyndromic, midfrequency, prelingual and nonprogressive sensorineural hearing loss. *Am. J. Otol.* 19:718–23

117. Green GE, Scott DA, McDonald JM, Woodworth GG, Sheffield VC, et al. 1999. Carrier rates in the midwestern United States for *GJB2* mutations causing inherited deafness. *JAMA* 281:2211–16

118. Greve JM, Wassarman PM. 1985. Mouse egg extracellular coat is a matrix of interconnected filaments possessing a structural repeat. *J. Mol. Biol.* 181:253–64

119. Grifa A, Wagner CA, D'Ambrosio L, Melchionda S, Bernardi F, et al. 1999. Mutations in *GJB6* cause nonsyndromic

autosomal dominant deafness at DFNA3 locus. *Nat. Genet.* 23:16–18

120. Griffith AJ, Chowdhry AA, Kurima K, Hood LJ, Keats B, et al. 2000. Autosomal recessive nonsyndromic neurosensory deafness at *DFNB1* not associated with the compound-heterozygous *GJB2* (Connexin 26) genotype M34T/167delT. *Am. J. Hum. Genet.* 67:745–49

121. Guan M-X, Fischel-Ghodsian N, Attardi C. 2000. A biochemical basis for the inherited susceptibility to aminoglycoside ototoxicity. *Hum. Mol. Genet.* 9:1787–93

122. Guilford P, Ayadi H, Blanchard S, Chaïb H, Le Paslier D, et al. 1994. A human gene responsible for neurosensory, nonsyndromic recessive deafness is a candidate homologue of the mouse *sh-1* gene. *Hum. Mol. Genet.* 3:989–93

123. Guilford P, Ben Arab S, Blanchard S, Levilliers J, Weissenbach J, et al. 1994. A non-syndromic form of neurosensory, recessive deafness maps to the pericentromeric region of chromosome 13q. *Nat. Genet.* 6:24–28

124. Hagiwara H, Tamagawa Y, Kitamura K, Kodera K. 1998. A new mutation in the POU3F4 gene in a Japanese family with X-linked mixed deafness (DFN3). *Laryngoscope* 108:1544–47

125. Halpin C, Khetarpal U, McKenna M. 1996. Autosomal dominant progressive hearing loss in a large North American family. *Am. J. Audiol.* 5:105–11

126. Hamasaki K, Rando RR. 1997. Specific binding of aminoglycosides to a human rRNA construct based on a DNA polymorphism which causes aminoglycoside-induced deafness. *Biochemistry* 36:12323–28

127. Hamelmann C, Amedofu GK, Albrecht K, Muntau B, Gelhaus A, et al. 2001. Pattern of connexin 26 (*GJB2*) mutations causing sensorineural hearing impairment in Ghana. *Hum. Mutat.* 18:84–85

128. Hara Y, Rovescalli AC, Kim Y, Nirenberg M. 1992. Structure and evolution of four POU domain genes expressed in mouse brain. *Proc. Natl. Acad. Sci. USA* 89:3280–84

129. Hasson T, Gillespie PG, Garcia JA, MacDonald RB, Zhao Y, et al. 1997. Unconventional myosins in inner-ear sensory epithelia. *J. Cell Biol.* 137:1287–307

130. Hasson T, Heintzelman MB, Santos-Sacchi J, Corey DP, Mooseker MS. 1995. Expression in cochlea and retina of myosin VIIa, the gene product defective in Usher syndrome type 1B. *Proc. Natl. Acad. Sci. USA* 92:9815–19

131. Hasson T, Walsh J, Cable J, Mooseker MS, Brown SD, et al. 1997. Effects of shaker-1 mutations on myosin-VIIa protein and mRNA expression. *Cell Motil. Cytoskel.* 37:127–38

132. Hattori M, Fujiyama A, Taylor TD, Watanabe H, Yada T, et al. 2000. The DNA sequence of human chromosome 21. The chromosome 21 mapping and sequencing consortium. *Nature* 405:311–19

133. Hawkins JEJ. 1950. Cochlear signs of streptomycin intoxication. *J. Pharmacol. Exp. Ther.* 100:38–44

134. Heathcote K, Syrris P, Carter ND, Patton MA. 2000. A connexin 26 mutation causes a syndrome of sensorineural hearing loss and palmoplantar hyperkeratosis (MIM 148350). *J. Med. Genet.* 37:50–51

135. Heginbotham L, Lu Z, Abramson T, MacKinnon R. 1994. Mutations in the K^+ channel signature sequence. *Biophys. J.* 66:1061–67

136. Henson MM, Burridge K, Fitzpatrick D, Jenkins DB, Pillsbury HC, et al. 1985. Immunocytochemical localization of contractile and contraction associated proteins in the spiral ligament of the cochlea. *Hear. Res.* 20:207–14

137. Hoh JH, John SA, Revel JP. 1991. Molecular cloning and characterization of a new member of the gap junction gene family, connexin-31. *J. Biol. Chem.* 266:6524–31

138. Hooper JD, Clements JA, Quigley JP, Antalis TM. 2001. Type II transmembrane serine proteases. Insights into an

emerging class of cell surface proteolytic enzymes. *J. Biol. Chem.* 276:857–60

139. Houseman MJ, Ellis LA, Pagnamenta A, Di W-L, Rickard S, et al. 2001. Genetic analysis of the connexin-26 M34T variant: Identification of genotype M34T/M34T segregating with mild-moderate non-syndromic sensorineural hearing loss. *J. Med. Genet.* 38:20–25

140. Housley GD, Ashmore JF. 1992. Ionic currents of outer hair cells isolated from the guinea-pig cochlea. *J. Physiol.* 448:73–98

141. Howard J, Hudspeth AJ. 1987. Mechanical relaxation of the hair bundle mediates adaptation in mechanoelectrical transduction by the bullfrog's saccular hair cell. *Proc. Natl. Acad. Sci. USA* 84:3064–68

142. Huizing EH, van Bolhuis AH, Odenthal DW. 1966. Studies on progressive hereditary perceptive deafness in a family of 335 members. I. Genetical and general audiological results. *Acta Oto-Laryngol.* 61:35–41

143. Huizing EH, van den Wijngaart WS, Verschuure J. 1983. A follow-up study in a family with dominant progressive inner ear deafness. *Acta Oto-Laryngol.* 95:620–26

144. Hutchin TP, Cortopassi GA. 2000. Mitochondrial defects and hearing loss. *Cell. Mol. Life Sci.* 57:1927–37

145. Hutchin TP, Parker MJ, Young ID, Davis AC, Pulleyn LJ, et al. 2000. A novel mutation in the mitochondrial tRNA$^{Ser(UCN)}$ gene in a family with non-syndromic sensorineural hearing impairment. *J. Med. Genet.* 37:692–94

146. Ikeda A, Zheng QY, Rosenstiel P, Maddatu T, Zuberi AR, et al. 1999. Genetic modification of hearing in tubby mice: Evidence for the existence of a major gene (*moth1*) which protects tubby mice from hearing loss. *Hum. Mol. Genet.* 8:1761–67

147. Jain PK, Lalwani AK, Li XC, Singleton TL, Smith TN, et al. 1998. A gene for recessive nonsyndromic sensorineural deafness (*DFNB18*) maps to the chromosomal region 11p14–p15.1 containing the Usher syndrome type 1C gene. *Genomics* 50:290–92

148. Jentsch TJ. 2000. Neuronal KCNQ potassium channels: physiology and role in disease. *Nat. Rev. Neurosci.* 1:21–30

149. Jin H, May M, Tranebjaerg L, Kendall E, Fontan G, et al. 1996. A novel X-linked gene, *DDP*, shows mutations in families with deafness (DFN-1), dystonia, mental deficiency and blindness. *Nat. Genet.* 14:177–80

150. Johnson KR, Zheng Qing Y, Bykhovskaya Y, Spirina O, Fischel-Ghodsian N. 2001. A nuclear-mitochondrial DNA interaction affecting hearing impairment in mice. *Nat. Genet.* 27:191–94

151. Kalatzis V, Sahly I, El-Amraoui A, Petit C. 1998. *Eya1* expression in the developing ear and kidney: towards the understanding of the pathogenesis of Branchio-Oto-Renal (BOR) syndrome. *Dev. Dyn.* 213:486–99

152. Kamarinos M, McGill J, Lynch M, Dahl H. 2001. Identification of a novel *COCH* mutation, I109N, highlights the similar clinical features observed in DFNA9 families. *Hum. Mutat.* 17:351

153. Keats BJB, Nouri N, Huang J-M, Money M, Webster DB, et al. 1995. The deafness locus (*dn*) maps to mouse chromosome 19. *Mamm. Genome* 6:8–10

154. Keats BJB, Nouri N, Pelias MZ, Deininger PL, Litt M. 1994. Tightly linked flanking microsatellite markers for the Usher syndrome type I locus on the short arm of chromosome 11. *Am. J. Hum. Genet.* 54:681–86

155. Kelley MJ, Jawien W, Ortel TL, Korczak JF. 2000. Mutation of *MYH9*, encoding non-muscle myosin heavy chain A, in May-Hegglin anomaly. *Nat. Genet.* 26:106–8

156. Kelley PM, Abe S, Askew JW, Smith SD, Usami S-i, et al. 1999. Human connexin 30 (GJB6), a candidate gene for nonsyndromic hearing loss: Molecular

cloning, tissue-specific expression, and assignment to chromosome 13q12. *Genomics* 62:172–76

157. Kelley PM, Harris DJ, Comer BC, Askew JW, Fowler T, et al. 1998. Novel mutations in the connexin 26 gene (GJB2) that cause autosomal recessive (DFNB1) hearing loss. *Am. J. Hum. Genet.* 62:792–99

158. Kelsell DP, Di W-L, Houseman MJ. 2001. Connexin mutations in skin disease and hearing loss. *Am. J. Hum. Genet.* 68:559–68

159. Kelsell DP, Dunlop J, Stevens HP, Lench NJ, Liang JN, et al. 1997. Connexin 26 mutations in hereditary non-syndromic sensorineural deafness. *Nature* 387:80–83

160. Kelsell DP, Wilgoss AL, Richard G, Stevens HP, Munro CS, et al. 2000. Connexin mutations associated with palmoplantar keratoderma and profound deafness in a single family. *Eur. J. Hum. Genet.* 8:469–72

161. Kharkovets T, Hardelin J-P, Safieddine S, Schweizer M, El-Amraoui A, et al. 2000. KCNQ4, a K⁺-channel mutated in a form of dominant deafness, is expressed in the inner ear and the central auditory pathway. *Proc. Natl. Acad. Sci.* 97:4333–38

162. Khetarpal U, Schuknecht HF, Gacek RR, Holmes LB. 1991. Autosomal dominant sensorineural hearing loss. Pedigrees, audiologic findings, and temporal bone findings in two kindreds. *Arch. Otolaryngol. Head Neck Surg.* 117:1032–42

163. Kikuchi T, Kimura RS, Paul DL, Adams JC. 1995. Gap junctions in the rat cochlea: immunohistochemical and ultrastructural analysis. *Anat. Embryol.* 191:101–18

164. Killick R, Legan PK, Malenczak C, Richardson GP. 1995. Molecular cloning of chick β-tectorin, an extracellular matrix molecule of the inner ear. *J. Cell Biol.* 129:535–47

165. Killick R, Richardson GP. 1997. Antibodies to the sulphated, high molecular mass mouse tectorin stain hair bundles

and the olfactory mucus layer. *Hear. Res.* 103:131–41

166. Kimberling WJ, Möller CG, Davenport S, Priluck IA, Beighton PH, et al. 1992. Linkage of Usher syndrome type I gene (USH1B) to the long arm of chromosome 11. *Genomics* 14:988–94

167. Kirschhofer K, Kenyon JB, Hoover DM, Franz P, Weipoltshammer K, et al. 1996. Autosomal-dominant congenital severe sensorineural hearing loss. Localisation of a disease gene to chromosome 11q by linkage in an Austrian family. *Eur. Workgroup Genet. Hear. Impair., Milan, It.,* Oct.:11–13

168. Kirschhofer K, Kenyon JB, Hoover DM, Franz P, Weipoltshammer K, et al. 1998. Autosomal-dominant, prelingual, nonprogressive sensorineural hearing loss: Localization of the gene (DFNA8) to chromosome 11q by linkage in an Austrian family. *Cytogenet. Cell Genet.* 82:126–30

169. Kohno H, Tanaka K, Mino A, Umikawa M, Imamura H, et al. 1996. Bni1p implicated in cytoskeletal control is a putative target of Rho1p small GTP binding protein in Saccharomyces cerevisiae. *EMBO J.* 15:6060–68

170. Kozel PJ, Friedman RA, Erway LC, Yamoah EN, Liu LH, et al. 1998. Balance and hearing deficits in mice with a null mutation in the gene encoding plasma membrane Ca^{2+}-ATPase isoform 2. *J. Biol. Chem.* 273:18693–96

171. Kubisch C, Schroeder BC, Friedrich T, Lütjohann B, El-Amraoui A, et al. 1999. KCNQ4, a novel potassium channel expressed in sensory outer hair cells, is mutated in dominant deafness. *Cell* 96:437–46

172. Kudo T, Ikeda K, Kure S, Matsubara Y, Oshima T, et al. 2000. Novel mutations in the connexin 26 gene (*GJB2*) responsible for childhood deafness in the Japanese population. *Am. J. Med. Genet.* 90:141–45

173. Kumar NM, Gilula NB. 1996. The gap

junction communication channel. *Cell* 84:381–8

174. Kunst H, Huybrechts C, Marres H, Huygen P, Van Camp G, et al. 2000. The phenotype of DFNA13/*COL11A2*: Nonsyndromic autosomal dominant midfrequency and high-frequency sensorineural hearing impairment. *Am. J. Otol.* 21:181–87

175. Küssel-Andermann P, El-Amraoui A, Safieddine S, Nouaille S, Perfettini I, et al. 2000. Vezatin, a novel transmembrane protein, bridges myosin VIIA to the cadherin/catenins complex. *EMBO J.* 19:6020–29

176. Lalwani AK, Goldstein JA, Kelley MJ, Luxford W, Castelein CM, et al. 2000. Human nonsyndromic hereditary deafness DFNA17 is due to a mutation in nonmuscle myosin *MYH9*. *Am. J. Hum. Genet.* 67:1121–28

177. Lalwani AK, Jackler RK, Sweetow RW, Lynch ED, Raventos H, et al. 1998. Further characterization of the DFNA1 audiovestibular phenotype. *Arch. Otolaryngol. Head Neck Surg.* 124:669–702

178. Lalwani AK, Linthicum FH, Wilcox ER, Moore JK, Walters FC, et al. 1997. A five-generation family with late-onset progressive hereditary hearing impairment due to cochleosaccular degeneration. *Audiol. Neurootol.* 2:139–54

179. Lalwani AK, Luxford WM, Mhatre AN, Attaie A, Wilcox ER, et al. 1999. A new locus for nonsyndromic hereditary hearing impairment, DFNA17, maps to chromosome 22 and represents a gene for cochleosaccular degeneration. *Am. J. Hum. Genet.* 64:318–23

180. Lamartine J, Munhoz Essenfelder G, Kibar Z, Lanneluc I, Callouet E, et al. 2000. Mutations in *GJB6* cause hidrotic ectodermal dysplasia. *Nat. Genet.* 26:142–44

181. Lautermann J, ten Cate WJ, Altenhoff P, Grummer R, Traub O, et al. 1998. Expression of the gap-junction connexins 26

and 30 in the rat cochlea. *Cell Tissue Res.* 294:415–20

182. Lee MJ, Rhee SK. 1998. Heteromeric gap junction channels in rat hepatocytes in which the expression of connexin26 is induced. *Mol. Cells* 8:295–300

183. Legan PK, Lukashkina VA, Goodyear RJ, Kossl M, Russell IJ, et al. 2000. A targeted deletion in α-tectorin reveals that the tectorial membrane is required for the gain and timing of cochlear feedback. *Neuron* 28:273–85

184. Legan PK, Rau A, Keen JN, Richardson GP. 1997. The mouse tectorins. Modular matrix proteins of the inner ear homologous to components of the sperm-egg adhesion system. *J. Biol. Chem.* 272:8791–801

185. Lehtonen MS, Uimonen S, Hassinen IE, Majamaa K. 2000. Frequency of mitochondrial DNA point mutations among patients with familial sensorineural hearing impairment. *Eur. J. Hum. Genet.* 8:315–18

186. Lench N, Housemam M, Newton V, Van Camp G, Mueller R. 1998. Connexin-26 mutations in sporadic non-syndromal sensorineural deafness. *Lancet* 351:415

187. Leon PE, Bonilla JA, Sanchez JR, Vanegas R, Villalobos M, et al. 1981. Low frequency hereditary deafness in man with childhood onset. *Am. J. Hum. Genet.* 33:209–14

188. Leon PE, Raventos H, Lynch E, Morrow J, King M-C. 1992. The gene for an inherited form of deafness maps to chromosome 5q31. *Proc. Natl. Acad. Sci. USA* 89:5181–84

189. Lerer I, Sagi M, Malamud E, Levi H, Raas-Rothschild A, et al. 2000. Contribution of connexin 26 mutations to nonsyndromic deafness in Ashkenazi patients and the variable phenotypic effect of the mutation 167delT. *Am. J. Med. Genet.* 95:53–56

190. Li C, Takei K, Geppert M, Daniell L, Stenius K, et al. 1994. Synaptic targeting of rabphilin-3A, a synaptic vesicle

Ca^{2+}/phospholipid-binding protein, depends on rab3A/3C. *Neuron* 13:885–98

191. Li XC, Everett LA, Lalwani AK, Desmukh D, Friedman TB, et al. 1998. A mutation in *PDS* causes non-syndromic recessive deafness. *Nat. Genet.* 18:215–17

192. Liang Y, Wang A, Belyantseva IA, Anderson DW, Probst FJ, et al. 1999. Characterization of the human and mouse unconventional myosin XV genes responsible for hereditary deafness *DFNB3* and Shaker-2. *Genomics* 61:243–58

193. Liang Y, Wang A, Probst FJ, Arhya IN, Barber TD, et al. 1998. Genetic mapping refines *DFNB3* to 17p11.2, suggests multiple alleles of *DFNB3*, and supports homology to the mouse model *shaker-2*. *Am. J. Hum. Genet.* 62:904–15

194. Linthicum FHJ, Foyad J, Otto SR, Galey FR, House WF. 1991. Cochlear implant histopathology. *Am. J. Otol.* 12:245–55

195. Liu J, Aoki M, Illa I, Wu C, Fardeau M, et al. 1998. Dysferlin, a novel skeletal muscle gene, is mutated in Miyoshi myopathy and limb-girdle muscular dystrophy. *Nat. Genet.* 20:31–36

196. Liu X-Z, Hope C, Walsh J, Newton V, Ke XM, et al. 1998. Mutations in the myosin VIIA gene cause a wide phenotypic spectrum, including atypical Usher syndrome. *Am. J. Hum. Genet.* 63:909–12

197. Liu X-Z, Walsh J, Mburu P, Kendrick-Jones J, Cope MJTV, et al. 1997. Mutations in the myosin VIIA gene cause non-syndromic recessive deafness. *Nat. Genet.* 16:188–90

198. Liu X-Z, Walsh J, Tamagawa Y, Kitamura K, Nishizawa M, et al. 1997. Autosomal dominant non-syndromic deafness caused by a mutation in the myosin VIIA gene. *Nat. Genet.* 17:268–69

199. Liu X-Z, Xia XJ, Xu LR, Pandya A, Liang CH, et al. 2000. Mutations in connexin31 underlie recessive as well as dominant non-syndromic hearing loss. *Hum. Mol. Genet.* 9:63–67

200. Löffler J, Nekahm D, Hirst-Stadlmann A, Günther B, Menzel H-J, et al. 2001. Sensorineural hearing loss and the incidence of *Cx26* mutations in Austria. *Eur. J. Hum. Genet.* 9:226–30

201. López-Bigas Nr, Olivé M, Rabionet R, Ben-David O, Martínez-Matos JA, et al. 2001. Connexin 31 (*GJB3*) is expressed in the peripheral and auditory nerves and causes neuropathy and hearing impairment. *Hum. Mol. Genet.* 10:947–52

202. Lynch ED, Lee MK, Morrow J, Welsh PL, Leon PE, et al. 1997. Nonsyndromic deafness DFNA1 associated with mutation of a human homolog of the *Drosophila* gene *diaphanous*. *Science* 278:1315–18

203. Maestrini E, Korge BP, Ocana-Sierra J, Calzolari E, Cambiaghi S, et al. 1999. A missense mutation in connexin 26, D66H, causes mutilating keratoderma with sensorineural deafness (Vohwinkel's syndrome) in three unrelated families. *Hum. Mol. Genet.* 8:1237–43

204. Manolis EN, Yandavi N, Nadol JBJ, Eavcy RD, McKenna M, et al. 1996. A gene for non-syndromic autosomal dominant progressive postlingual sensorineural hearing loss maps to chromosome 14q12–13. *Hum. Mol. Genet.* 5:1047–50

205. Marcotti W, Kros CJ. 1999. Developmental expression of the potassium current *I*K,n contributes to maturation of mouse outer hair cells. *J. Physiol.* 520:653–60

206. Markin VS, Hudspeth AJ. 1995. Gating-spring models of mechanoelectrical transduction by hair cells of the internal ear. *Annu. Rev. Biophys. Biomol. Struct.* 24:59–83

207. Marlin S, Garabédian E-N, Roger G, Moatti L, Matha N, et al. 2001. Connexin26 gene mutations in congenitally deaf children: Pitfalls for genetic counselling. *Arch. Otolaryngol. Head Neck Surg.* 127:927–33

208. Martin PE, Coleman SL, Casalotti SO, Forge A, Evans WH. 1999. Properties of connexin26 gap junctional proteins derived from mutations associated with non-syndromal heriditary deafness. *Hum. Mol. Genet.* 8:2369–76

208b. Masmoudi S, Antonarakis SE, Schwede T, Ghorbel AM, Grati M, et al. 2001. Novel missense mutations of TMPRSS3 in two consanguineous Tunisian families with non-syndromic autosomal recessive deafness. *Hum. Mutat.* 18:101–8

209. Masmoudi S, Charfedine I, Hmani M, Grati M, Ghorbel AM, et al. 2000. Pendred syndrome: Phenotypic variability in two families carrying the same *PDS* missense mutation. *Am. J. Med. Genet.* 90:38–44

210. Mathis JM, Simmons DM, He X, Swanson LW, Rosenfeld MG. 1992. Brain 4: a novel mammalian POU domain transcription factor exhibiting restricted brain-specific expression. *EMBO J.* 11:2551–61

211. Maw MA, Allen-Powell DR, Goodey RJ, Stewart IA, Nancarrow DJ, et al. 1995. The contribution of the DFNB1 locus to neurosensory deafness in a Caucasian population. *Am. J. Hum. Genet.* 57:629–35

212. May-Hegglin/Fechtner Syndrome C. 2000. Mutations in *MYH9* result in the May-Hegglin anomaly, and Fechtner and Sebastian syndromes. *Nat. Genet.* 26:103–5

213. Mburu P, Liu XZ, Walsh J, Saw D, Cope MJTV, et al. 1997. Mutation analysis of the mouse myosin VIIA deafness gene. *Genes Funct.* 1:191–203

214. McGuirt WT, Prasad SD, Griffith AJ, Kunst HPM, Green GE, et al. 1999. Mutations in *COL11A2* cause nonsyndromic hearing loss (DFNA13). *Nat. Genet.* 23:413–19

215. Melchionda S, Ahituv N, Bisceglia L, Sobe T, Glaser F, et al. 2001. *MYO6*, the human homologue of the gene responsible for deafness in *Snell's waltzer*

mice, is mutated in autosomal dominant nonsyndromic hearing loss. *Am. J. Hum. Genet.* 69:635–40

216. Melman YF, Domènech A, de la Luna S, McDonald TV. 2001. Structural determinants of KvLQT1 control by the KCNE family of proteins. *J. Biol. Chem.* 276:6439–44

217. Ménière P. 1846. Recherches sur l'origine de la surdi-mutité. *Gaz. Méd. Paris* 3:223

218. Ménière P. 1856. Du mariage entre parents considéré comme cause de la surdimutité congénitale. *Gaz. Méd. Paris* 3:303–6

219. Mermall V, Post PL, Mooseker MS. 1998. Unconventional myosins in cell movement, membrane traffic, and signal transduction. *Science* 279:527–33

220. Minowa O, Ikeda K, Sugitani Y, Oshima T, Nakai S, et al. 1999. Altered cochlear fibrocytes in a mouse model of DFN3 nonsyndromic deafness. *Science* 285:1408–11

221. Mitic LL, Van Itallie CM, Anderson JM. 2000. Molecular Physiology and Pathophysiology of Tight Junctions: I. Tight junction structure and function: lessons from mutant animals and proteins. *Am. J. Physiol.* 279:G250–54

222. Morell R, Spritz RA, Ho L, Pierpont J, Guo W, et al. 1997. Apparent digenic inheritance of Waardenburg syndrome type 2 (WS2) and autosomal recessive ocular albinism (AROA). *Hum. Mol. Genet.* 6:659–64

223. Morell RJ, Kim HJ, Hood LJ, Goforth L, Friderici K, et al. 1998. Mutations in the connexin 26 gene (*GJB2*) among Ashkenazi Jews with nonsyndromic recessive deafness. *N. Engl. J. Med.* 339:1500–5

224. Moreno-Pelayo MA, del Castillo I, Villamar M, Romero L, Hernandez-Calvin FJ, et al. 2001. A cysteine substitution in the zona pellucida domain of α-tectorin results in autosomal dominant, postlingual, progressive,

mid-frequency hearing loss in a Spanish family. *J. Med. Genet.* 38:e13

225. Morle L, Bozon M, Alloisio N, Latour P, Vandenberghe A, et al. 2000. A novel C202F mutation in the connexin26 gene (*GJB2*) associated with autosomal dominant isolated hearing loss. *J. Med. Genet.* 37:368–70

226. Murgia A, Orzan E, Polli R, Martella M, Vinanzi C, et al. 1999. *Cx26* deafness: mutation analysis and clinical variability. *J. Med. Genet.* 36:829–32

227. Mustapha M, Weil D, Chardenoux S, Elias S, El-Zir E, et al. 1999. An α-tectorin gene defect causes a newly identified autosomal recessive form of sensorineural pre-lingual non-syndromic deafness, DFNB21. *Hum. Mol. Genet.* 8: 409–12

228. Nadeau JH. 2001. Modifier genes in mice and humans. *Nat. Rev. Genet.* 2: 165–74

229. Nakagawa T, Kakehata S, Yamamoto T, Akaike N, Komune S, et al. 1994. Ionic properties of $I_{K,n}$ in outer hair cells of guinea pig cochlea. *Brain Res.* 661:293–97

230. Nance WE, Setleff R, McLeod A, Sweeney A, Cooper C, et al. 1971. X-linked mixed deafness with congenital fixation of the stapedial footplate and perilymphatic gusher. *Birth Defects* 7:64–69

231. Neyroud N, Tesson F, Denjoy I, Leibovici M, Donger C, et al. 1997. A novel mutation in the potassium channel gene *KVLQT1* causes the Jervell and Lange-Nielsen cardioauditory syndrome. *Nat. Genet.* 15:186–89

232. Nicolson T, Rusch A, Friedrich RW, Granato M, Ruppersberg JP, et al. 1998. Genetic analysis of vertebrate sensory hair cell mechanosensation: the zebrafish circler mutants. *Neuron* 20:271–83

233. Niessen H, Harz H, Bedner P, Kraemer K, Willecke K. 2000. Selective permeability of different connexin channels to the second messenger inositol 1,4,5-trisphosphate. *J. Cell Sci.* 113:1365–72

234. Noben-Trauth K, Zheng QY, Johnson KR, Nishina PM. 1997. *mdfw*: a deafness susceptibility locus that interacts with deaf waddler (*dfw*). *Genomics* 44:266–72

235. O'Neill ME, Marietta J, Nishimura D, Wayne S, Van Camp G, et al. 1996. A gene for autosomal dominant late-onset progressive non-syndromic hearing loss, DNFA10, maps to chromosome 6. *Hum. Mol. Genet.* 5:853–86

236. Ohto H, Kamada S, Tago K, Tominaga S-I, Ozaki H, et al. 1999. Cooperation of Six and Eya in activation of their target genes through nuclear translocation of Eya. *Mol. Cell. Biol.* 19:6815–24

237. Oliver TN, Berg JS, Cheney RE. 1999. Tails of unconventional myosins. *Cell. Mol. Life Sci.* 56:243–57

238. Park HJ, Hahn SH, Chun YM, Park K, Kim HN. 2000. Connexin26 mutations associated with nonsyndromic hearing loss. *Laryngoscope* 110:1535–38

239. Parsons TD, Lenzi D, Almers W, Roberts WM. 1994. Calcium-triggered exocytosis and endocytosis in an isolated presynaptic cell: capacitance measurements in saccular hair cells. *Neuron* 13:875–83

240. Pendred V. 1896. Deaf mutism and goitre. *Lancet* i:352

241. Perin MS, Fried VA, Mignery GA, Jahn R, Südhof TC. 1990. Phospholipid binding by a synaptic vesicle protein homologous to the regulatory region of protein kinase C. *Nature* 345:260–63

242. Petit C. 1996. Genes responsible for human hereditary deafness: *symphony of a thousand*. *Nat. Genet.* 14:385–91

243. Petit C. 2001. Usher syndrome: from genetics to pathogenesis. *Annu. Rev. Genomics Hum. Genet.* 2:271–97

244. Petit C, Levilliers J, Marlin S, Hardelin J-P. 2001. Hereditary hearing loss. In *The Metabolic and Molecular Bases of Inherited Disease*, ed. CR Scriver, AL

Beaudet, WS Sly, D Valle, pp. 6281–328. Montreal: McGraw-Hill

245. Phelps PD, Coffey RA, Trembath RC, Luxon LM, Grossman AB, et al. 1998. Radiological malformations of the ear in Pendred syndrome. *Clin. Radiol.* 53:268–73

246. Phippard D, Boyd Y, Reed V, Fisher G, Masson WK, et al. 2000. The *sex-linked fidget* mutation abolishes *Brn4/Pou3f4* gene expression in the embryonic inner ear. *Hum. Mol. Genet.* 9:79–85

247. Phippard D, Heydemann A, Lechner M, Lu L, Lee D, et al. 1998. Changes in the subcellular localization of the *Brn4* gene product precede mesenchymal remodeling of the otic capsule. *Hear. Res.* 120:77–85

248. Phippard D, Lu L, Lee D, Saunders JC, Crenshaw EB. 1999. Targeted mutagenesis of the POU-domain gene *Brn4/Pou3f4* causes developmental defects in the inner ear. *J. Neurosci.* 19:5980–89

249. Pihlajamaa T, Prockop DJ, Faber J, Winterpacht A, Zabel B, et al. 1998. Heterozygous glycine substitution in the COL11A2 gene in the original patient with the Weissenbacher-Zweymuller syndrome demonstrates its identity with heterozygous OSMED (nonocular Stickler syndrome). *Am. J. Med. Genet.* 80:115–20

250. Plum A, Winterhager E, Pesch J, Lautermann J, Hallas G, et al. 2001. Connexin31–deficiency in mice causes transient placental dysmorphogenesis but does not impair hearing and skin differentiation. *Dev. Biol.* 231:334–47

251. Prasad S, Cucci RA, Green GE, Smith RJH. 2000. Genetic testing for hereditary hearing loss: Connexin 26 (GJB2) allele variants and two novel deafness-causing mutations (R32C and 645–648delTAGA). *Hum. Mutat.* 16:502–8

252. Prezant TR, Agapian JV, Bohlman MC, Bu X, Öztas S, et al. 1993. Mitochondrial ribosomal RNA mutation associated with both antibiotic-induced and non-syndromic deafness. *Nat. Genet.* 4:289–94

253. Probst FJ, Fridell RA, Raphael Y, Saunders TL, Wang A, et al. 1998. Correction of deafness in *shaker-2* mice by an unconventional myosin in a BAC transgene. *Science* 280:1444–47

254. Rabionet R, Gasparini P, Estivill X. 2000. Molecular genetics of hearing impairment due to mutations in gap junction genes encoding beta connexins. *Hum. Mutat.* 16:190–202

255. Rattner A, Sun H, Nathans J. 1999. Molecular genetics of human retinal disease. *Annu. Rev. Genet.* 33:89–131

256. Rau A, Legan PK, Richardson GP. 1999. Tectorin mRNA expression is spatially and temporally restricted during mouse inner ear development. *J. Comp. Neurol.* 405:271–80

257. Reardon W, Coffey R, Chowdhury T, Grossman A, Jan H, et al. 1999. Prevalence, age of onset, and natural history of thyroid disease in Pendred syndrome. *J. Med. Genet.* 36:595–98

258. Reid FM, Vernham GA, Jacobs HT. 1994. A novel mitochondrial point mutation in a maternal pedigree with sensorineural deafness. *Hum. Mutat.* 3:243–47

259. Rennie KJ, Ashmore JF. 1991. Ionic currents in isolated vestibular hair cells from the guinea-pig crista ampullaris. *Hear. Res.* 51:279–91

260. Riazuddin S, Castelein CM, Ahmed ZM, Lalwani AK, Mastroianni MA, et al. 2000. Dominant modifier DFNM1 suppresses recessive deafness DFNB26. *Nat. Genet.* 26:431–44

261. Richard G, Smith LE, Bailey RA, Itin P, Hohl D, et al. 1998. Mutations in the human connexin gene *GJB3* cause erythrokeratodermia variabilis. *Nat. Genet.* 20:366–69

262. Richard G, White TW, Smith LE, Bailey RE, Compton JG, et al. 1998. Functional defects of Cx26 resulting from

a heterozygous missense mutation in a family with dominant deaf-mutism and palmoplantar keratoderma. *Hum. Genet.* 103:393–99

263. Richardson GP, Forge A, Kros CJ, Fleming J, Brown SD, et al. 1997. Myosin VIIA is required for aminoglycoside accumulation in cochlear hair cells. *J. Neurosci.* 17:9506–19

264. Richardson GP, Russell IJ, Duance VC, Bailey AJ. 1987. Polypeptide composition of the mammalian tectorial membrane. *Hear. Res.* 25:45–60

265. Rinchik EM, Carpenter DA, Selby PB. 1990. A strategy for fine-structure functional analysis of a 6– to 11–centimorgan region of mouse chromosome 7 by high-efficiency mutagenesis. *Proc. Natl. Acad. Sci. USA* 87:896–900

266. Rivolta C, Sweklo EA, Berson EL, Dryja TP. 2000. Missense mutation in the *USH2A* gene: association with recessive retinitis pigmentosa without hearing loss. *Am. J. Hum. Genet.* 6:1975–78

267. Rizo J, Südhof TC. 1998. C2–domains, structure and function of a universal Ca^{2+}-binding domain. *J. Biol. Chem.* 273:15879–82

268. Robertson NG, Lu L, Heller S, Merchant SN, Eavey RD, et al. 1998. Mutations in a novel cochlear gene cause DFNA9, a human nonsyndromic deafness with vestibular dysfunction. *Nat. Genet.* 20:299–303

269. Robertson NG, Skvorak AB, Yin Y, Weremowicz S, Johnson KR, et al. 1997. Mapping and characterization of a novel cochlear gene in human and in mouse: a positional candidate gene for a deafness disorder, DFNA9. *Genomics* 46:345–54

270. Rodriguez OC, Cheney RE. 2000. A new direction for myosin. *Trends Cell Biol.* 10:307–11

271. Royaux IE, Suzuki K, Mori A, Katoh R, Everett LA, et al. 2000. Pendrin, the protein encoded by the Pendred syndrome gene (*PDS*), is an apical porter of iodide in the thyroid and is regulated

by thyroglobulin in FRTL-5 cells. *Endocrinology* 141:839–45

272. Royaux IE, Wall SM, Karniski LP, Everett LA, Suzuki K, et al. 2001. Pendrin, encoded by the Pendred syndrome gene, resides in the apical region of renal intercalated cells and mediates bicarbonate secretion. *Proc. Natl. Acad. Sci. USA* 98:4221–26

273. Rubin JB, Verselis VK, Bennett MV, Bargiello TA. 1992. Molecular analysis of voltage dependence of heterotypic gap junctions formed by connexins 26 and 32. *Biophys. J.* 62:183–93; discussion 93–95

274. Ryan A, Dallos P. 1975. Effect of absence of cochlear outer hair cells on behavioural auditory threshold. *Nature* 253:44–46

275. Saez JC, Nairn AC, Czernik AJ, Spray DC, Hertzberg EL, et al. 1990. Phosphorylation of connexin 32, a hepatocyte gap-junction protein, by cAMP-dependent protein kinase, protein kinase C and Ca^{2+}/calmodulin-dependent protein kinase II. *Eur. J. Biochem.* 192:263–73

276. Sahly I, El-Amraoui A, Abitbol M, Petit C, Dufier J-L. 1997. Expression of myosin VIIA during mouse embryogenesis. *Anat. Embryol.* 196:159–70

277. Salt AN, Ohyama K. 1993. Accumulation of potassium in scala vestibuli perilymph of the mammalian cochlea. *Ann. Otol. Rhinol. Laryngol.* 102:64–70

278. Salt AN, Thalmann R, Marcus DC, Bohne BA. 1986. Direct measurement of longitudinal endolymph flow rate in the guinea pig cochlea. *Hear. Res.* 23:141–51

279. Sanguinetti MC, Curran ME, Zou A, Shen J, Spector PS, et al. 1996. Coassembly of K(V)LQT1 and minK (IsK) proteins to form cardiac I_{Ks} potassium channel. *Nature* 384:80–83

280. Sankila E-M, Pakarinen L, Kääriäinen H, Aittomäki K, Karjalainen S, et al. 1995. Assignment of an Usher syndrome

type III (USH3) gene to chromosome 3q. *Hum. Mol. Genet.* 4:93–98

281. Sarduy M, del Castillo I, Villamar M, Romero L, Herraiz C, et al. 1998. Genetic study of mitochondrially inherited sensorineural hearing impairment in eight large families from Spain and Cuba. In *Developments in Genetic Hearing Impairment*, ed. D Stephens, AP Read, A Martini, pp. 121–25. London: Whurr

282. Schonberger J, Levy H, Grunig E, Sangwatanaroj S, Fatkin D, et al. 2000. Dilated cardiomyopathy and sensorineural hearing loss: a heritable syndrome that maps to 6q23–24. *Circulation* 101:1812–18

283. Schroeder BC, Waldegger S, Fehr S, Bleich M, Warth R, et al. 2000. A constitutively open potassium channel formed by KCNQ1 and KCNE3. *Nature* 403:196–99

284. Schroeder R, Waldsich C, Wank H. 2000. Modulation of RNA function by aminoglycoside antibiotics. *EMBO J.* 19:1–9

285. Scott DA, Karniski LP. 2000. Human pendrin expressed in *Xenopus laevis* oocytes mediates chloride/formate exchange. *Am. J. Physiol. (Cell Physiol.)* 278:C207–11

286. Scott DA, Kraft ML, Stone EM, Sheffield VC, Smith RJH. 1998. Connexin mutations and hearing loss. *Nature* 391:32

287. Scott DA, Wang R, Kreman TM, Andrews M, McDonald JM, et al. 2000. Functional differences of the *PDS* gene product are associated with phenotypic variation in patients with Pendred syndrome and non-syndromic hearing loss (DFNB4). *Hum. Mol. Genet.* 9:1709–15

288. Scott DA, Wang R, Kreman TM, Sheffield VC, Karniski LP. 1999. The Pendred syndrome gene encodes a chloride-iodide transport protein. *Nat. Genet.* 21:440–43

289. Scott HS, Kudoh J, Wattenhofer M,

Shibuya K, Berry A, et al. 2001. Insertion of β-satellite repeats identifies a transmembrane protease causing both congenital and childhood onset autosomal recessive deafness. *Nat. Genet.* 27:59–63

290. Seidman MD, Bai U, Khan MJ, Murphy MJ, Quirk WS, et al. 1996. Association of mitochondrial DNA deletions and cochlear pathology: a molecular biologic tool. *Laryngoscope* 106:777–83

291. Self T, Mahony M, Fleming J, Walsh J, Brown SD, et al. 1998. Shaker-1 mutations reveal roles for myosin VIIA in both development and function of cochlear hair cells. *Development* 125:557–66

292. Self T, Sobe T, Copeland NG, Jenkins NA, Avraham KB, et al. 1999. Role of myosin VI in the differentiation of cochlear hair cells. *Dev. Biol.* 214:331–41

293. Sevior KB, Hatamochi A, Stewart IA, Bykhovskaya Y, Allen-Powell DR, et al. 1998. Mitochondrial A7445G mutation in two pedigrees with palmoplantar keratoderma and deafness. *Am. J. Med. Genet.* 75:179–85

294. Sheffield VC, Kraiem Z, Beck JC, Nishimura D, Stone EM, et al. 1996. Pendred syndrome maps to chromosome 7q21–34 and is caused by an intrinsic defect in thyroid iodine organification. *Nat. Genet.* 12:424–26

295. Sheng M, Pak DT. 2000. Ligand-gated ion channel interactions with cytoskeletal and signaling proteins. *Annu. Rev. Physiol.* 62:755–78

296. Shirataki H, Kaibuchi K, Sakoda T, Kishida S, Yamaguchi T, et al. 1993. Rabphilin-3A, a putative target protein for smg p25A/rab3A p25 small GTP-binding protein related to synaptotagmin. *Mol. Cell. Biol.* 13:2061–68

297. Simmler M-C, Cohen-Salmon M, El-Amraoui A, Guillaud L, Benichou J-C, et al. 2000. Targeted disruption of *Otogelin* results in deafness and severe imbalance. *Nat. Genet.* 24:139–43

298. Simmler M-C, Zwaenepoel I, Verpy E, Guillaud L, Elbaz C, et al. 2000. Twister mice are defective for otogelin, a component specific to inner ear acellular membranes. *Mamm. Genome* 11:961–66

299. Simon AM, Goodenough DA. 1998. Diverse functions of vertebrate gap junctions. *Trends Cell Biol.* 8:477–82

300. Simon DB, Lu Y, Choate KA, Velazquez H, Al-Sabban E, et al. 1999. Paracellin-1, a renal tight junction protein required for paracellular Mg^{2+} resorption. *Science* 285:103–6

301. Sobe T, Vreugde S, Shahin H, Berlin M, Davis N, et al. 2000. The prevalence and expression of inherited connexin 26 mutations associated with nonsyndromic hearing loss in the Israeli population. *Hum. Genet.* 106:50–57

302. Soleimani M, Greeley T, Petrovic S, Wang Z, Amlal H, et al. 2001. Pendrin: an apical $Cl^-/OH^-/HCO^-_3$ exchanger in the kidney cortex. *Am. J. Physiol. (Renal Physiol.)* 280:F356–64

303. Spicer SS, Schulte BA. 1996. The fine structure of spiral ligament cells relates to ion return to the stria and varies with place-frequency. *Hear. Res.* 100:80–100

304. Steinberg TH, Civitelli R, Geist ST, Robertson AJ, Hick E, et al. 1994. Connexin43 and connexin45 form gap junctions with different molecular permeabilities in osteoblastic cells. *EMBO J.* 13:744–50

305. Stephens SDG. 1985. Genetic hearing loss: A historical overview. *Adv. Audiol.* 3:3–17

306. Sterkers O. 1985. Origin and electrochemical composition of endolymph in the cochlea. In *Audiology Biochemistry*, ed. D Dresscher, pp. 473–87. Springfield, IL: Thomas

307. Storm K, Willocx S, Flothmann K, Van Camp G. 1999. Determination of the carrier frequency of the common GJB2 (connexin-26) 35delG mutation in the Belgian population using an easy and reliable screening method. *Hum. Mutat.* 14:263–66

308. Street VA, McKee-Johnson JW, Fonseca RC, Tempel BL, Noben-Trauth K. 1998. Mutations in a plasma membrane Ca^{2+}-ATPase gene cause deafness in deafwaddler mice. *Nat. Genet.* 19:390–94

309. Suchyna TM, Xu LX, Gao F, Fourtner CR, Nicholson BJ. 1993. Identification of a proline residue as a transduction element involved in voltage gating of gap junctions. *Nature* 365:847–49

310. Sue CM, Tanji K, Hadjigeorgiou G, Andreu AL, Nishino I, et al. 1999. Maternally inherited hearing loss in a large kindred with a novel T7511C mutation in the mitochondrial DNA tRNA(Ser(UCN)] gene. *Neurology* 52:1905–8

311. Takeuchi S, Ando M, Kakigi A. 2000. Mechanism generating endocochlear potential: Role played by intermediate cells in stria vascularis. *Biophys. J.* 79:2572–82

312. Tamagawa Y, Kitamura K, Ishida T, Ishikawa K, Tanaka H, et al. 1996. A gene for a dominant form of non-syndromic sensorineural deafness (*DFNA11*) maps within the region containing the *DFNB2* recessive deafness gene. *Hum. Mol. Genet.* 5:849–52

313. Theopold HM. 1977. Comparative surface studies of ototoxic effects of various aminoglycoside antibiotics on the organ of Corti in the guinea pig. A scanning electron microscopic study. *Acta Oto-Laryngol.* 84:57–64

314. Tinel N, Diochot S, Borsotto M, Lazdunski M, Barhanin J. 2000. KCNE2 confers background current characteristics to the cardiac KCNQ1 potassium channel. *EMBO J.* 19:6326–30

315. Tiranti V, Chariot P, Carella F, Toscano A, Soliveri P, et al. 1995. Maternally inherited hearing loss, ataxia and myoclonus associated with a novel point

mutation in mitochondrial tRNA$^{Ser(UCN)}$ gene. *Hum. Mol. Genet.* 4:1421–27

316. Titus MA. 2000. Cytoskeleton: Getting to the point with myosin VI. *Curr. Biol.* 10:R294–97

317. Titus MA, Gilbert SP. 1999. The diversity of molecular motors: an overview. *Cell. Mol. Life Sci.* 56:181–83

318. Tomek MS, Brown MR, Mani SR, Ramesh A, Srisailapathy CR, et al. 1998. Localization of a gene for otosclerosis to chromosome 15q25–q26. *Hum. Mol. Genet.* 7:285–90

319. Torres M, Giraldez F. 1998. The development of the vertebrate inner ear. *Mech. Dev.* 71:5–21

320. Torroni A, Cruciani F, Rengo C, Sellitto D, López-Bigas N, et al. 1999. The A1555G mutation in the 12S rRNA gene of human mtDNA: recurrent origins and founder events in families affected by sensorineural deafness. *Am. J. Hum. Genet.* 65:1349–58

321. Tucker MA, Barajas L. 1994. Rat connexins 30.3 and 31 are expressed in the kidney. *Exp. Cell Res.* 213:224–30

322. Unger VM, Kumar NM, Gilula NB, Yeager M. 1999. Three-dimensional structure of a recombinant gap junction membrane channel. *Science* 283:1176–80

323. Usami S-i, Abe S, Weston MD, Shinkawa H, Van Camp G, et al. 1999. Non-syndromic hearing loss associated with enlarged vestibular aqueduct is caused by *PDS* mutations. *Hum. Genet.* 104:188–92

324. Usami S-i, Abe S, Akita J, Namba A, Shinkawa H, et al. 2000. Prevalence of mitochondrial gene mutations among hearing impaired patients. *J. Med. Genet.* 37:38–40

325. Vahava O, Morell R, Lynch ED, Weiss S, Kagan ME, et al. 1998. Mutation in transcription factor *POU4F3* associated with inherited progressive hearing loss in humans. *Science* 279:1950–54

326. Van Camp G, Coucke P, Balemans W,

van Velzen D, van de Bilt C, et al. 1995. Localization of a gene for non-syndromic hearing loss (DFNA5) to chromosome 7p15. *Hum. Mol. Genet.* 4: 2159–63

327. Van Camp G, Coucke PJ, Kunst H, Schatteman I, Van Velzen D, et al. 1997. Linkage analysis of progressive hearing loss in five extended families maps the DFNA2 gene to a 1.25–Mb region on chromosome 1p. *Genomics* 41:70–74

328. Van Den Bogaert K, Govaerts PJ, Schatteman I, Brown MR, Caethoven G, et al. 2001. A second gene for otosclerosis, OTSC2, maps to chromosome 7q34–36. *Am. J. Hum. Genet.* 68:495–500

329. Van Hauwe P, Coucke PJ, Ensink RJ, Huygen P, Cremers CWRJ, et al. 2000. Mutations in the KCNQ4 K$^+$ channel gene, responsible for autosomal dominant hearing loss, cluster in the channel pore region. *Am. J. Med. Genet.* 93:184–87

329b. Van Laer L, Coucke P, Mueller RF, Caethoven G, Flothmann K, et al. 2001. A common founder for the 35delG *GJB2* gene mutation in connexin 26 hearing impairment. *J. Med. Genet.* 38:515–18

330. Van Laer L, Huizing EH, Verstreken M, van Zuijlen D, Wauters JG, et al. 1998. Nonsyndromic hearing impairment is associated with a mutation in *DFNA5*. *Nat. Genet.* 20:194–97

331. Veenstra RD, Wang HZ, Beblo DA, Chilton MG, Harris AL, et al. 1995. Selectivity of connexin-specific gap junctions does not correlate with channel conductance. *Circulation Res.* 77:1156–65

332. Verhage M, de Vries KJ, Roshol H, Burbach JP, Gispen WH, et al. 1997. DOC2 proteins in rat brain: complementary distribution and proposed function as vesicular adapter proteins in early stages of secretion. *Neuron* 18:453–61

333. Verhagen WIM, Bom SJH, Huygen PLM, Fransen E, Van Camp G, et al.

2000. Familial progressive vestibulocochlear dysfunction caused by a COCH mutation (DFNA9). *Arch. Neurol.* 57:1045–47

334. Verhoeven K, Ensink RJH, Tiranti V, Huygen PLM, Johnson DF, et al. 1999. Hearing impairment and neurological dysfunction associated with a mutation in the mitochondrial tRNA^Ser(UCN) gene. *Eur. J. Hum. Genet.* 7:45–51

335. Verhoeven K, Fagerheim T, Prasad S, Wayne S, De Clau F, et al. 2000. Refined localization and two additional linked families for the DFNA10 locus for nonsyndromic hearing impairment. *Hum. Genet.* 107:7–11

336. Verhoeven K, Van Camp G, Govaerts PJ, Balemans W, Schatteman I, et al. 1997. A gene for autosomal dominant nonsyndromic hearing loss (DFNA12) maps to chromosome 11q22–24. *Am. J. Hum. Genet.* 60:1168–74

337. Verhoeven K, Van Laer L, Kirschhofer K, Legan PK, Hughes DC, et al. 1998. Mutations in the human α-tectorin gene cause autosomal dominant nonsyndromic hearing impairment. *Nat. Genet.* 19:60–62

338. Verpy E, Leibovici M, Zwaenepoel I, Liu X-Z, Gal A, et al. 2000. A defect in harmonin, a PDZ domain-containing protein expressed in the inner ear sensory hair cells, underlies Usher syndrome type 1C. *Nat. Genet.* 26:51–55

338a. Verpy E, Masmoudi S, Zwaenepoel I, Leibovici M, Hutchin TP, et al. 2001. Mutations in a new gene encoding a protein of the hair bundle cause nonsyndromic deafness at the DFNB16 locus. *Nat. Genet.* 29: In press

339. Veske A, Oehlmann R, Younus F, Mohyuddin A, Mueller-Myhsok B, et al. 1996. Autosomal recessive non-syndromic deafness locus (DFNB8) maps on chromosome 21q22 in a large consanguineous kindred from Pakistan. *Hum. Mol. Genet.* 5:165–68

340. Vikkula M, Mariman ECM, Lui VCH, Zhidkova NI, Tiller GE, et al. 1995. Autosomal dominant and recessive osteochondrodysplasia associated with the COL11A2 locus. *Cell* 80:431–37

341. von Békésy G. 1960. *Experiments in Hearing.* New York: Mc Graw-Hill

342. Wallis C, Ballo R, Wallis G, Beighton P, Goldblatt J. 1988. X-linked mixed deafness with stapes fixation in a Mauritian kindred: linkage to Xq probe pDP34. *Genomics* 3:299–301

343. Wang A, Liang Y, Fridell RA, Probst FJ, Wilcox ER, et al. 1998. Association of unconventional myosin *MYO15* mutations with human nonsyndromic deafness *DFNB3*. *Science* 280:1447–51

344. Wang TG, Giebisch G, Aronson PS. 1992. Efects of formate and oxalate on volume absorption in rat proximal tubule. *Am. J. Physiol. (Renal Fluid Electrolyte Physiol.)* 263:F37–42

345. Wang Y, Okamoto M, Schmitz F, Hofmann K, Südhof TC. 1997. Rim is a putative Rab3 effector in regulating synaptic-vesicle fusion. *Nature* 388:593–98

346. Warn-Cramer BJ, Cottrell GT, Burt JM, Lau AF. 1998. Regulation of connexin-43 gap junctional intercellular communication by mitogen-activated protein kinase. *J. Biol. Chem.* 273:9188–96

347. Watanabe N, Kato T, Fujita A, Ishizaki T, Narumiya S. 1999. Cooperation between mDia1 and ROCK in Rho-induced actin reorganization. *Nat. Cell Biol.* 1:136–43

348. Watanabe N, Madaule P, Reid T, Ishizaki T, Watanabe G, et al. 1997. p140mDia, a mammalian homolog of *Drosophila* diaphanous, is a target protein for Rho small GTPase and is a ligand for profilin. *EMBO J.* 16:3044–56

349. Wayne S, Der Kaloustian VM, Schloss M, Polomeno R, Scott DA, et al. 1996. Localization of the Usher syndrome type 1D gene (Ush1D) to chromosome 10. *Hum. Mol. Genet.* 5:1689–92

350. Wayne S, Lowry RB, McLeod DR,

Knaus R, Farr C, et al. 1997. Localization of the Usher syndrome type 1F (Ush1F) to chromosome 10. *Am. J. Hum. Genet.* 61:A300

351. Wayne S, Robertson NG, DeClau F, Chen N, Verhoeven K, et al. 2001. Mutations in the transcriptional activator *EYA4* cause late-onset deafness at the DFNA10 locus. *Hum. Mol. Genet.* 10:195–200

352. Wegner M, Drolet DW, M.G R. 1993. POU-domain proteins: Structure and function of developmental regulators. *Curr. Opin. Cell Biol.* 5:488–98

353. Weil D, Blanchard S, Kaplan J, Guilford P, Gibson F, et al. 1995. Defective myosin VIIA gene responsible for Usher syndrome type 1B. *Nature* 374:60–61

354. Weil D, Küssel P, Blanchard S, Lévy G, Levi-Acobas F, et al. 1997. The autosomal recessive isolated deafness, DFNB2, and the Usher 1B syndrome are allelic defects of the myosin-VIIA gene. *Nat. Genet.* 16:191–93

355. Weil D, Lévy G, Sahly I, Levi-Acobas F, Blanchard S, et al. 1996. Human myosin VIIA responsible for the Usher 1B syndrome: a predicted membrane-associated motor protein expressed in developing sensory epithelia. *Proc. Natl. Acad. Sci. USA* 93:3232–37

356. Wells AL, Lin AW, Chen L-Q, Safer D, Cain SM, et al. 1999. Myosin VI is an actin-based motor that moves backwards. *Nature* 401:505–8

357. Wenzel K, Manthey D, Willecke K, Grzeschik KH, Traub O. 1998. Human gap junction protein connexin31: molecular cloning and expression analysis. *Biochem. Biophys. Res. Commun.* 248:910–15

358. White TW, Deans MR, Kelsell DP, Paul DL. 1998. Connexin mutations in deafness. *Nature* 394:630–31

359. White TW, Paul DL. 1999. Genetic diseases and gene knockouts reveal diverse connexin functions. *Annu. Rev. Physiol.* 61:283–310

360. Wilcox ER, Burton QL, Naz S, Riazuddin S, Smith TN, et al. 2001. Mutations in the gene encoding tight junction claudin-14 cause autosomal recessive deafness DFNB29. *Cell* 104:165–72

361. Wilcox SA, Saunders K, Osborn AH, Arnold A, Wunderlich J, et al. 2000. High frequency hearing loss correlated with mutations in the *GJB2* gene. *Hum. Genet.* 106:399–405

362. Xia A-P, Ikeda K, Katori Y, Oshima T, Kikuchi T, et al. 2000. Expression of connexin 31 in the developing mouse cochlea. *Neuroreport* 11:2449–53

363. Xia J-h, Liu C-y, Tang B-s, Pan Q, Huang L, et al. 1998. Mutations in the gene encoding gap junction protein β-3 associated with autosomal dominant hearing impairment. *Nat. Genet.* 20:370–73

364. Xiang M, Gan L, Li D, Chen ZY, Zhou L, et al. 1997. Essential role of POU-domain factor Brn-3c in auditory and vestibular hair cell development. *Proc. Natl. Acad. Sci. USA* 94:9445–50

365. Xiang M, Gao W-Q, Hasson T, Shin JJ. 1998. Requirement for Brn-3c in maturation and survival, but not in fate determination of inner hair cells. *Development* 125:3935–46

366. Yasunaga S, Grati M, Chardenoux S, Smith TN, Friedman TB, et al. 2000. *OTOF* encodes multiple long and short isoforms: genetic evidence that the long ones underlie the recessive deafness DFNB9. *Am. J. Hum. Genet.* 67:591–600

367. Yasunaga S, Grati M, Cohen-Salmon M, El-Amraoui A, Mustapha M, et al. 1999. A mutation in *OTOF*, encoding otoferlin, a FER-1 like protein, causes DFNB9, a nonsyndromic form of deafness. *Nat. Genet.* 21:363–69

368. Yonezawa S, Yoshiki A, Hanai A, Matsuzaki T, Matsushima J, et al. 1999. Chromosomal localization of a gene responsible for vestibulocochlear defects

of BUS/Idr mice: identification as an allele of waltzer. *Hear. Res.* 134:116–22

369. Zelante L, Gasparini P, Estivill X, Melchionda S, D'Agruma L, et al. 1997. Connexin26 mutations associated with the most common form of non-syndromic neurosensory autosomal recessive deafness (DFNB1) in Mediterraneans. *Hum. Mol. Genet.* 6:1605–9

370. Zhang M, Kalinec G, Urrutia R, Kalinec F. 2001. *Dia1 proteins participate in the regulation of outer hair cell motility by acetylcholine.* Presented at the Assoc. Res. Otolaryngol., St Petersburg Beach, Fl, Feb. 4–8

371. Zheng L, Sekerková G, Vranich K, Tilney LG, Mugnaini E, et al. 2000. The deaf jerker mouse has a mutation in the gene encoding the espin actin-bundling proteins of hair cell stereocilia and lacks espins. *Cell* 102:377–85

372. Zidanic M, Brownell WE. 1990. Fine structure of the intracochlear potential field. I. The silent current. *Biophys. J.* 57:1253–68

373. Zwaenepoel I, Verpy E, Blanchard S, Meins M, Apfelstedt-Sylla E, et al. 2001. Identification of three novel mutations in the USH1C gene and detection of thirty-one polymorphisms used for haplotype analysis. *Hum. Mutat.* 17:34–41

Annu. Rev. Genet. 2001. 35:647–72

GENETIC ANALYSIS OF CALMODULIN AND ITS TARGETS IN *SACCHAROMYCES CEREVISIAE*

Martha S. Cyert

Department of Biological Sciences, Stanford University, Stanford, California 94305-5020; e-mail: mcyert@stanford.edu

Key Words   calcium, signal transduction, calcineurin, myosin, spindle pole body

■ **Abstract**   Calmodulin, a small, ubiquitous Ca^{2+}-binding protein, regulates a wide variety of proteins and processes in all eukaryotes. *CMD1*, the single gene encoding calmodulin in *S. cerevisiae*, is essential, and this review discusses studies that identified many of calmodulin's physiological targets and their functions in yeast cells. Calmodulin performs essential roles in mitosis, through its regulation of Nuf1p/Spc110p, a component of the spindle pole body, and in bud growth, by binding Myo2p, an unconventional class V myosin required for polarized secretion. Surprisingly, mutant calmodulins that fail to bind Ca^{2+} can perform these essential functions. Calmodulin is also required for endocytosis in yeast and participates in Ca^{2+}-dependent, stress-activated signaling pathways through its regulation of a protein phosphatase, calcineurin, and the protein kinases, Cmk1p and Cmk2p. Thus, calmodulin performs important physiological functions in yeast cells in both its Ca^{2+}-bound and Ca^{2+}-free form.

CONTENTS

0066-4197/01/1215-0647$14.00

INTRODUCTION

Calmodulin Structure and Function

Regulated changes in the concentration of cytosolic Ca^{2+} control such diverse biological processes as muscle contraction, fertilization, secretion, cell proliferation, and apoptosis. Calmodulin, a small Ca^{2+}-binding protein, is found in all eukaryotic organisms and is highly conserved. Calmodulin serves as a major intracellular Ca^{2+} receptor and mediates many of the effects of this ion. At resting levels of Ca^{2+}, calmodulin exists in the Ca^{2+}-free, or apo-calmodulin form. In response to a Ca^{2+} signal, calmodulin binds Ca^{2+} and consequently undergoes a conformational change that allows it to bind to and activate a host of target enzymes. Since its discovery in 1970 (14), the mechanism of Ca^{2+}/calmodulin-dependent regulation of target enzymes has been characterized extensively through in vitro biochemical and structural analyses. More recently, studies of calmodulin in several genetically tractable organisms established that calmodulin is required for viability (25, 123, 142). Genetic dissection of calmodulin function in the yeast *Saccharomyces cerevisiae* has added significantly to our understanding of this regulator by identifying physiologically relevant targets of calmodulin (Table 1), and by establishing the functional significance of both Ca^{2+}-bound and Ca^{2+}-free calmodulin in vivo.

Calmodulin contains four copies of a Ca^{2+}-binding motif known as an EF-hand, each of which binds one Ca^{2+} ion. An EF-hand is made up of a 12-residue Ca^{2+}-binding loop flanked by two α-helices (60). Within the loop, Ca^{2+} is coordinated by oxygens on six different amino acid residues. Calmodulin is one member of a large class of EF-hand–containing Ca^{2+}-binding proteins (98). *S. cerevisiae* contains five calmodulin-related proteins, each of which contains four EF-hand motifs: Cdc31p, is a component of the yeast microtubule organizing center and the yeast homologue of centractin (5). Mlc1p and Mlc2p are myosin light chains, which regulate distinct yeast myosins (6, 136). Cnb1p encodes the regulatory subunit of the Ca^{2+}/calmodulin-regulated phosphatase, calcineurin (22, 62). Frq1p encodes the regulatory subunit of a phosphatidylinositol-4-OH kinase and is a homologue of frequenin, a protein found in vertebrate neurons (46).

EF-hand–containing proteins typically undergo a structural change upon binding Ca^{2+}; however, this conformational change differs substantially for each class of EF-hand protein (155). Structural analyses have established that calmodulin is a dumbbell-shaped molecule with two similar domains, each containing two EF-hand Ca^{2+}-binding motifs, connected by a short flexible linker. In the absence of Ca^{2+}, the EF-hands are in a "closed" conformation. This Ca^{2+}-free form of calmodulin is able to bind to a subset of target proteins. Ca^{2+} binding causes a change to an "open" conformation, which also results in exposure of two hydrophobic surfaces that allow calmodulin to bind to its Ca^{2+}-dependent target proteins (reviewed in 155). Binding sites for calmodulin share limited sequence homology, but are similar in structure, and are typically regions 20 amino acids long that

TABLE 1 Calmodulin targets of *S. cerevisiae*

Name	Essential	Calmodulin binding: type[a]; sequence[b]	Function	Localization	Mammalian homlogs	References
Nuf1p/Spc110p	Yes	I; 897–917	Anchors MTs to SPB	SPB	Kendrin	(34, 37, 38, 137)
Myo2p	Yes	I; IQ motifs: 790–940	Polarized growth, vacuole inheritance	Bud tip, Bud neck	Class V myosins: Dilute P190	(11, 12, 53, 65, 124, 129)
Myo4p	No	Putative	mRNA localization	Cytosolic	Class V myosins	(7, 52, 136, 144, 145)
Myo5p	No	I; IQ motifs: 725–753	Endocytosis	Actin patches	Class I myosins	(41)
Myo3p	No	Putative	Actin organization	Actin patches	Class I myosins	(42)
Arc35p	Yes	D[c]; ND	Endocytosis, spindle assembly	Actin patches	Subunit 2 of Arp2/3 complex	(126, 127, 151)
Calcineurin (Cna1p/Cna2p and Cnb1p)	No	D; Cna1p: 453–476 Cna2p: 500–523	Signaling, stress response, Ca^{2+} homeostasis, G2/M	Cytosolic	Calcineurin	(19–21, 31, 59, 75, 85, 88, 97, 134, 148)
Cmk1p, Cmk2p	No	D; Cmk1p: 313–340 Cmk2p: 323–350	Signaling, stress responses	Cytosolic	CaM Kinase II	(48, 50, 81, 93, 106, 111)
Iqg1p	Yes	I; ND	Cytokinesis	Actomyosin ring	IQGAPs	(30, 108, 130, 150)
Gad1p	No	D; ND	Oxidative stress		Glutamate decarboxylase	(16)
Dst1p/Ppr2p	No	D; ND	Transcription	Nucleus	TFIIS	(137)

ND, not determined.

[a]Calmodulin binding type: I, Ca^{2+} independent; D, Ca^{2+} dependent.

[b]Amino acid residues that bind calmodulin.

[c]Calmodulin binding is Ca^{2+} dependent in vitro, but function in endocytosis in vivo is Ca^{2+} independent (see text).

form basic amphipathic α-helices. Several types of calmodulin-binding sites can be distinguished based on the spacing of particular bulky residues and the tendency to bind apo-calmodulin (18, 125).

Identification and Characterization of Calmodulin from *S. cerevisiae*

Calmodulin was purified from *S. cerevisiae* based on the similarity of its physical properties to those of vertebrate calmodulin (25, 72, 107). Once purified, partial amino acid sequence was determined and used to synthesize oligonucleotide probes to identify the gene, *CMD1*, from a yeast genomic library (25). *S. cerevisiae* contains a single calmodulin gene that is required for viability and encodes a protein 60% identical to vertebrate calmodulins (25). The primary structure of yeast calmodulin is like its vertebrate counterpart, having four predicted helix-loop-helix EF-hand domains distributed similarly in the protein sequence. However, there are significant differences in the structure and Ca^{2+}-binding properties of yeast and vertebrate calmodulins. Vertebrate calmodulin binds four molecules of Ca^{2+} per molecule of calmodulin, whereas yeast calmodulin binds a maximum of three molecules of Ca^{2+} (72, 76, 133). The most C-terminal EF-hand in yeast calmodulin (site IV) has a deletion of one residue in the Ca^{2+}-binding loop and also contains a substitution of glutamine for a highly conserved glutamate at position 12. Thus, while this region of the protein still maintains the helix-loop-helix conformation found in other calmodulins, site IV is defective for Ca^{2+} binding (77, 133). The other EF-hands in yeast calmodulin bind Ca^{2+} with high affinity ($K_d = 2-5 \times 10^{-6}$ M), although the exchange rate of Ca^{2+} for these sites is slower than that observed for vertebrate calmodulins (133). In its Ca^{2+}-bound form, yeast calmodulin exists in a more compact form than do its vertebrate counterparts (157), owing to interactions between the N-terminal and C-terminal domains (63). Despite these differences in biochemical properties, vertebrate calmodulin is able to complement the essential function of calmodulin in yeast (24, 45, 103).

GENETIC ANALYSIS OF CALMODULIN FUNCTION

Analysis of Conditional Mutants

The identification and characterization of calmodulin in *S. cerevisiae* made possible an extensive genetic dissection of calmodulin function. Examination of conditional calmodulin mutants identified multiple distinct essential functions for this protein in vivo. Initially, two temperature-sensitive calmodulin mutants were studied: *cmd1-1*, which contains two amino acid substitutions (23) and *cmd1-101*, which contains an allele engineered in vitro to allow overexpression of a truncated calmodulin lacking its N-terminal half in the yeast genome (139). Temperature-shift experiments with both strains revealed a requirement for calmodulin primarily during nuclear division; at the nonpermissive temperature, cells with a duplicated

DNA content accumulated, and further analysis of these cells identified abnormalities in spindle morphology (23, 139). A defect in bud growth was also observed for *cmd1-1*. The consequences of depleting calmodulin in vivo, i.e., shutting off expression of *CMD1* driven by a galactose-regulated promoter, were similar to the effects observed in the temperature-sensitive calmodulin mutant strains (104). Thus, the first essential role demonstrated for calmodulin was in nuclear division.

Further genetic analysis by Ohya & Botstein established that calmodulin has several additional essential functions. A panel of conditional mutations was generated by changing conserved phenylalanine residues of calmodulin to alanine (105). The resulting collection of mutants fall into four distinct phenotypic groups, *cmd1A–D*, that also display intragenic complementation. Members of one of the groups (*cmd1C*) exhibit a defect in mitosis similar to that described for *cmd1-1* and *cmd1-101*. Other mutants reveal an essential role for calmodulin in bud emergence (*cmd1D*) and actin localization (*cmd1A*) (105). In the final group of mutants (*cmd1B*), the characteristic pattern of calmodulin localization is disrupted (105). Calmodulin localizes to sites of bud formation, bud tips, and the bud neck in vivo (10, 138). This cellular distribution overlaps in part with that of actin patches and reflects calmodulin's role in polarized growth. In *cmd1B* mutants, however, calmodulin is distributed diffusely throughout the cell.

The distinct nature of the four phenotypic groups of *cmd1* mutants and their ability to complement each other suggested that each group was compromised for activation of different essential target(s) in vivo. Further analyses have confirmed that *cmd1C* mutants are specifically compromised for regulation of Nuf1p. However, other mutant groups may be deficient for activation of more than one target (see section on Ca^{2+}-Independent Targets). Nonetheless, this panel of mutants has been a powerful tool for analyzing and characterizing the diverse functions of calmodulin in vivo.

Analysis of Ca^{2+}-Binding–Defective Mutants

The essential role of calmodulin in vivo was expected to depend on its ability to bind Ca^{2+}, because the role of this protein as an intracellular Ca^{2+} sensor was so well established. However, this notion was challenged by the finding that mutant alleles of *CMD1* that are completely defective for Ca^{2+}-binding support yeast growth (39). Ca^{2+}-binding–defective alleles were constructed by directed substitution of amino acid residues required for Ca^{2+} ion coordination, and in vitro analyses confirmed that the resulting proteins (cmd1–3p, cmd1–6p) were deficient for Ca^{2+} binding. Surprisingly, yeast cells whose only source of calmodulin are these Ca^{2+}-binding-defective–mutant proteins show minimal disruptions in growth and morphology under standard culture conditions (39). The intracellular localization of the mutant proteins is also indistinguishable from that of wild-type calmodulin (10, 92). Although Ca^{2+}-independent binding of mammalian calmodulin to several proteins had been demonstrated previously, the physiological relevance of these interactions was not fully appreciated. The studies in *S. cerevisiae* clearly established that for this organism the essential functions of calmodulin do not depend on its ability to bind Ca^{2+}, and later identification of the essential targets, Nuf1p/Spc110p and Myo2p, have confirmed their Ca^{2+}-independent interaction with calmodulin (see section on Targets of Calmodulin).

Unfortunately, these findings are often misinterpreted as indicating that there are no Ca^{2+}-dependent functions for calmodulin in *S. cerevisiae* and that this yeast is devoid of Ca^{2+}-dependent signaling pathways. However, as discussed below, the activation by calmodulin of at least two different target proteins, calcineurin, the Ca^{2+}/calmodulin-dependent phosphatase, and calmodulin-regulated kinases (Cmk1p and Cmk2p), is Ca^{2+} dependent, and the Ca^{2+}-binding–defective calmodulin mutants fail to activate these targets in vivo (20, 93). However, under standard laboratory conditions, neither the Ca^{2+}/calmodulin-dependent phosphatase nor the kinase is required for viability (21, 69, 106, 111).

TARGETS OF CALMODULIN

Ca^{2+}-Independent Targets

Nuf1p/Spc110p   Calmodulin localizes to the spindle pole body (SPB), the yeast microtubule organizing center (MTOC), throughout the cell cycle, and the essential target of calmodulin in mitosis, Nuf1/Spc110p, is a component of the SPB (38, 92, 137). The SPB is embedded in the yeast nuclear envelope and is the sole organelle responsible for nucleating nuclear and cytoplasmic microtubules. Thus, the SPB is equivalent in function to the centrosome of animal cells and there is substantial similarity among the protein components of these two organelles (reviewed in 35).

Nuf1p was identified as a target of calmodulin by genetic selections as well as by direct screening of expression libraries for calmodulin-binding proteins (38, 137). Dominant mutations of *NUF1* were selected as extragenic suppressors of the temperature-sensitive allele, *cmd1-1*. These mutations all result in truncation of the C terminus of Nuf1p (see below) and suppress the *cmd1-1* allele but not a deletion allele of *cmd1* (*cmd/Δ*) (38). *NUF1* was independently identified through a two-hybrid screen using the Ca^{2+}-binding defective *cmd1-6* as the bait (38). The calmodulin-binding site on Nuf1p, which is in the C-terminal portion of the protein, was defined through two-hybrid, biochemical, and mutational analyses (38, 137). Further studies demonstrated that a peptide derived from this region (aa. 897–917) binds calmodulin in vitro (37).

Nuf1p contains a central region predicted to form an extended coil-coil, and forms the spacer region of the SPB that lies between the "inner plaque" and the "central plaque" (57, 86). The inner plaque of the SPB lies inside the nuclear envelope and is made up of the γ-tubulin-containing microtubule nucleation complex. The central plaque, which contains the additional SPB components Spc29p and Spc42p, is embedded in the nuclear envelope and forms the core structural component of the SPB (reviewed in 35). Calmodulin binds to Nuf1p at the central plaque of the SPB (132, 141), and this interaction seems to anchor Nuf1p to the spindle pole during mitosis. Disruption of the Cmd1p-Nuf1p interaction through mutation of either Cmd1p or Nuf1p results in defective spindle formation and a loss of microtubule attachment to the spindle pole (58, 141). Thus, calmodulin forms an essential connection between microtubules and the SPB. However, *NUF1* alleles that encode a C-terminally truncated protein completely lacking the calmodulin-binding site are dominant suppressors of *cmd1-1* (38). Together with the finding that Nuf1p binds to Spc29p only in the calmodulin-bound form (28), these observations suggest that calmodulin binding to Nuf1p relieves intramolecular inhibition in Nuf1p to promote its binding to Spc29p and consequent association with the SPB central plaque. Also, calmodulin binding stabilizes Nuf1p levels in vivo (137). Genetic interactions are consistent with the physical associations observed between these gene products. Synthetic lethality is observed between *nuf1* and either *spc29* or *cmd1*, while *nuf1* conditional mutants are suppressed by *CMD1* overexpression but exacerbated by *SPC29* overexpression (28, 137, 140).

In contrast to *S. cerevisiae*, in *Schizosaccharomyces pombe* Ca^{2+} binding to calmodulin is required for its essential function, and a mutant that is compromised for Ca^{2+} binding exhibits defects in spindle function during mitosis (92, 94). This finding suggests that the same regulatory function that is Ca^{2+} independent in *S. cerevisiae* may be Ca^{2+} dependent in *S. pombe*, although the target of calmodulin at the *S. pombe* SPB has not yet been identified. Calmodulin also localizes to the MTOC of vertebrate cells, and kendrin, a component of human centrosomes that is structurally similar to Nuf1p, binds calmodulin through a site that is related to the calmodulin-binding site of Nuf1p (34).

Myo2p   The second essential, Ca^{2+}-independent target of calmodulin is Myo2p. *MYO2* is an essential gene originally identified in a screen for conditional mutants that were defective for bud emergence but continued to increase in mass at restrictive temperature (54). *MYO2* encodes the heavy chain for a non-muscle myosin that is designated as a class V myosin. Class V myosins are associated with Griscelli syndrome and deafness in humans and have been implicated in vesicle trafficking (reviewed in 124). Several other such myosins bind calmodulin (124). Myo2p is required for polarized secretion and vacuole inheritance in yeast (47).

Myo2p interacts with two different EF-hand proteins, Cmd1p and Mlc1p, both of which bind to six tandemly repeated IQ sites in its neck region. IQ sites are well established as sites of calmodulin binding and usually bind the Ca^{2+} free form of calmodulin (125). Cmd1p binds directly to the Myo2p IQ sites in vitro, and can be co-immunoprecipitated with Myo2p from extracts (11, 129). Myo2p localizes to the bud tip and bud neck of yeast cells (11, 66). Calmodulin localization to these same sites is dependent on Myo2p-binding, as polarized localization of calmodulin is disrupted in cells that contain Myo2p lacking the IQ sites (136). Surprisingly, yeast cells containing Myo2p devoid of IQ sites are viable, indicating that calmodulin binding per se is not essential for Myo2p function (136). Since studies of other myosins show that the size of the neck domain correlates with the distance the myosin head moves along actin filaments in each cycle of ATP hydrolysis, this observation also suggests that processive movement may not be required for Myo2p's essential function (124).

Genetic interactions between *CMD1* and *MYO2* are consistent with the interactions of their protein products. Conditional mutations in *cmd1*, including those in the *cmd1A* subclass defined by Ohya & Botstein, display allele-specific synthetic lethality with *myo2-66*, and fail to bind to Myo2p in vitro (11, 105, 129). In contrast, *myo2-66* phenotypes are not exacerbated in strains containing Ca^{2+}-binding–deficient calmodulin as the sole source of calmodulin (11). These genetic observations indicate that calmodulin's role in Myo2p function is Ca^{2+} independent, and are consistent with the finding that Cmd1p binds to Myo2p in the absence of Ca^{2+} (11).

Myo2p is required for the polarized growth of yeast cells and the inheritance of the vacuole by the daughter cell during mitosis (13, 47, 54). Myo2p is thought to direct polarized growth by attaching to secretory vesicles and transporting them along actin cables to the bud tip. Movement of Myo2p or secretory vesicles along actin cables has not been shown directly. However, both Myo2p and actin cables are required for polarized secretion and disruption of either leads to accumulation of vesicles in mother cells (43, 56, 120, 128). Some mutations in the C-terminal tail of Myo2p selectively disrupt either polarized secretion or vacuole inheritance, indicating that distinct regions within this domain mediate Myo2p attachment to different cargoes (12).

In addition to Myo2p, yeast contain a second class V myosin heavy chain, encoded by *MYO4*. Myo4p function is distinct from that of Myo2p: it is not essential and is involved in transport/localization of specific mRNAs (7, 52, 143–145). Like Myo2p, Myo4p contains six IQ sites, which could bind calmodulin and/or light chains. Calmodulin binding to Myo4p has not been demonstrated; however, in cells containing Myo2p that lack IQ sites, calmodulin localization is disrupted even further when *myo4* is deleted, suggesting that Cmd1p and Myo4p interact (136).

Calmodulin and endocytosis: Myo5p and Arc35p   Genetic analysis of endocytosis in yeast identified a role for calmodulin that seems to involve at least two different target proteins: the unconventional type I myosin, Myo5p, and Arc35p, a component of the Arp2/3 complex. *cmd1* mutants were identified in a search for endocytosis-defective mutants. Its role in endocytosis was shown to be Ca^{2+}-independent, as the *cmd1-3* Ca^{2+}-binding–defective mutant is competent for endocytosis (61). Further genetic analysis, using the collection of *cmd1* mutants generated by Ohya & Botstein (105), identified several calmodulin alleles that are defective for endocytosis (*cmd1-226*, *cmd1-247*, and *cmd1-228*). Intragenic complementation for the endocytosis phenotype was observed between *cmd1-247* and *cmd1-228*, suggesting that at least two distinct calmodulin targets are required for this process (41).

Myo5p, an unconventional myosin type I heavy chain, is one target for calmodulin in endocytosis. *myo5Δ* cells are defective for internalization of the plasma membrane mating factor receptor (Ste2p) at high temperature, and Myo5p contains two IQ sites that are both necessary and sufficient for its Ca^{2+}-independent interaction with calmodulin. The requirement for calmodulin in endocytosis is partially overcome in cells containing Myo5p that lacks the IQ sites (41). *cmd1-226* and *cmd1-247* both display defects in endocytosis and the mutant calmodulins fail to interact physically with Myo5p (41).

Yeast cells contain an additional type I myosin, Myo3p. Myo5p and Myo3p both localize to actin patches, and *myo3* and *myo5* are synthetically lethal. Thus Myo3p and Myo5p are redundant in executing their essential function of actin organization (42). However, *myo3* cells are not defective for endocytosis (40). Myo3p also contains two IQ sites, but has not been shown to bind calmodulin.

A second function for calmodulin in endocytosis requires an essential protein, Arc35p. Arc35p encodes the 35-kD subunit of the highly conserved Arp2/3 complex, which has been purified from several different sources and shown to stimulate actin polymerization (151). *arc35* mutants display defects in endocytosis and organization of the actin cytoskeleton (96, 152). These defects can be suppressed by overexpression of calmodulin (126, 127). However, overexpression of two mutant alleles, *cmd1-226* and *cmd1-228*, fails to suppress the *arc35-1* endocytosis defect (126, 127). These observations, together with the intragenic complementation of endocytosis defects between *cmd1-247* and *cmd1-228* (41), suggest that *cmd1-247* compromises calmodulin's interaction with Myo5p, *cmd1-228* compromises calmodulin's interaction with Arc35p, and cmd1-226p interacts with neither target. Other findings also indicate that calmodulin and Arc35p physically associate. Calmodulin localization to bud tips is disrupted at high temperature in an *arc 35-1* mutant, and interaction of Arc35p with calmodulin has been demonstrated both by two-hybrid analysis and by co-immunoprecipitation (126). Consistent with genetic findings, cmd1-226p and cmd1-228p fail to interact with Arc35p by these methods (126). However, it is not yet clear whether the association of these proteins is direct or is mediated by additional components. Another complication is that in contrast to genetic analyses that indicate a Ca^{2+}-independent role for calmodulin in endocytosis, co-immunoprecipitation of calmodulin and Arc35p does require Ca^{2+}. It is possible that Ca^{2+} is required in vitro to stabilize the calmodulin-Arc35p interaction, whereas a lower-affinity, Ca^{2+}-independent interaction is sufficient for function in vivo. Finally, although the biochemical function of Arc35p in endocytosis is unclear, it may well involve its interaction with the Arp2/3 complex as *arp2* mutants are also endocytosis defective (126). Arp2/3-mediated actin polymerization may be needed to push endocytic vesicles away from the cell surface into the cytosol (95).

Surprisingly, studies of *arc35-1* revealed that in addition to defects in the actin cytoskeleton, these mutants also exhibit a defect in metaphase spindle formation that results in cell cycle arrest (127). This cell cycle/spindle defect can also be suppressed by overexpression of *CMD1*; however, expression of mutant calmodulins that fail to suppress the *arc35* endocytosis defect (*cmd1-226 and cmd1-228*) do suppress the spindle formation defect. In contrast, overexpression of *cmd1-239* suppresses the endocytosis defect but not the cell cycle defect. *cmd1-239* falls into the *cmd1C* phenotypic class described by Ohya & Botstein that is compromised for interaction with Nuf1p/Spc110p, a component of the spindle pole body (see above). Thus, Arc35p may somehow affect Nuf1p/spindle pole body function. Alternatively, a distinct calmodulin-dependent function that affects SPB function is disrupted in both *arc35* and *cmd1-239*.

Ca^{2+}-Dependent Targets

CALCINEURIN   Calcineurin, or PP2B, is a highly conserved, Ca^{2+} calmodulin-dependent phosphoserine/phosphothreonine-specific phosphatase (reviewed in 3, 59). In mammals, calcineurin regulates many processes including NMDA signaling (64), Na$^+$/K$^+$ ATPase function (2), cardiac development and hypertrophy (26, 90, 121), learning and memory (74), T-cell activation (15, 100), and angiogenesis (44). For many of these functions, including T-cell activation, the critical target of calcineurin is the NF-AT family of transcription factors (reviewed in 17, 122). NF-AT resides in the cytosol when phosphorylated and translocates to the nucleus upon dephosphorylation by calcineurin. Thus, calcineurin regulates the activity of these transcription factors primarily through regulating their localization. Highly specific inhibitors of calcineurin, FK506 and cyclosporin A, bind to an intracellular binding protein (FKBP or cyclophilin, respectively), and form a drug-protein complex, which then binds to and inhibits calcineurin (reviewed in 68). These compounds inhibit calcineurin in a wide variety of organisms, including yeast and humans. In humans, inhibition of calcineurin by FK506 and cyclosporin A renders these compounds powerful immunosuppressants.

Calcineurin is a heterodimer composed of a catalytic A subunit and an essential regulatory or B subunit, which is an EF-hand–containing protein related to calmodulin. Under resting Ca^{2+} levels, the A and B subunits remain associated, but the enzyme is inactive due to an autoinhibitory domain at the C terminus of the A subunit. Upon elevation of [Ca^{2+}], Ca^{2+}-bound calmodulin binds to the A subunit and displaces the autoinhibitory domain, thus activating phosphatase activity (reviewed in 59).

In *S. cerevisiae*, calcineurin is encoded by three genes: *CNA1* and *CNA2/CMP2* encode functionally redundant catalytic subunits, and *CNB1* encodes the regulatory subunit (21, 22, 62, 69). Complete disruption of calcineurin activity in vivo can be achieved by mutation of both catalytic subunits (*cna1 cna2*), mutation of the regulatory subunit (*cnb1*), addition of calcineurin inhibitors (FK506, cyclosporin A, or related compounds), or expression of Ca^{2+}-binding–defective *cmd1* alleles (*cmd1-3, cmd1-6*) as the sole source of calmodulin (20–22, 62, 69, 93). Expression of a C-terminally truncated allele of calcineurin, *CNA1/2ΔC*, that removes the C-terminal autoinhibitory domain, leads to constitutive phosphatase activity in the absence of Ca^{2+}/calmodulin (84, 153, 158).

Yeast calcineurin carries out at least three different functions in yeast, regulating a stress-activated transcriptional pathway, Ca^{2+} homeostasis, and the G2 to M transition of the cell cycle. These different functions reflect the activities of distinct calcineurin substrates.

ROLE OF CALCINEURIN IN STRESS RESPONSE   Calcineurin regulates a signal transduction pathway in *S. cerevisiae* that is activated by intracellular Ca^{2+} and results in increased expression of a specific set of calcineurin-dependent genes. Under standard laboratory growth conditions, this calcineurin-dependent pathway is "off" and calcineurin is dispensable for growth. However, under specific environmental conditions, including exposure to high concentrations of ions (Ca^{2+}, OH^-, Mn^{2+}, Na^+/Li^+), mating pheromone (α-factor) and high temperature, and in mutants in which cell wall structure is compromised, calcineurin-mediated gene expression is activated (19, 75, 80, 83, 85, 134, 158). Yeast cells lacking calcineurin activity are sensitive to high pH, Mn^{2+}, Na^+/Li^+, and lose viability during prolonged exposure to mating pheromone (21, 31, 85, 93, 97, 153). Calcineurin mutants also display synthetic lethality with *fks1*, *mpk1*, *pkc1*, and several other mutations that compromise cell wall integrity (29, 36, 110). Thus, under most conditions in which calcineurin-dependent transcription is activated, calcineurin is essential for cell survival.

Ca^{2+}/calcineurin-dependent transcription is mediated by a zinc-finger transcription factor, encoded by *CRZ1/TCN1/HAL8*, that activates the expression of the structural genes for several P-type ATPases (*PMC1*, *ENA1*, *PMR1*), cell wall biosynthetic enzymes (*FKS2*) (75, 83, 134), and many other genes (H. Yoshimoto & M. Cyert, unpublished observations). Dissection of the *FKS2* promoter defined the CDRE (calcineurin dependent response element), a 24-bp DNA element that is necessary and sufficient to mediate Ca^{2+}-induced, calcineurin-dependent activation of gene expression. *CRZ1/TCN1* was identified as a multicopy suppressor that restored expression of a CDRE-*lacZ* reporter gene in a calcineurin mutant strain (134). Independently, loss-of-function alleles of *CRZ1/TCN1* were identified as mutations that eliminated Ca^{2+}-induced calcineurin-dependent gene expression of a *PMC1-lacZ* reporter gene (75). *crz1/tcn1* mutants are viable under standard laboratory growth conditions but display growth defects in the presence of high concentrations of OH^-, Mn^{2+}, or Na^+/Li^+ and lose viability during prolonged incubation with mating pheromone (α-factor) (75, 83, 134). Thus, in large part, the phenotypes of a *crz1*Δ strain are similar to, but not as severe as, those of a strain lacking calcineurin activity. For these phenotypes, the *cnb1*Δ *crz1*Δ double mutants display identical growth properties as a *cnb1*Δ mutant, suggesting that the sole mode of Crz1p regulation is calcineurin dependent. However, calcineurin mutants have additional phenotypes not shared by *crz1*Δ (see below), suggesting that there are additional roles for calcineurin distinct from Crz1p regulation. Calcineurin dephosphorylates Crz1p in vitro, establishing Crz1p as a direct substrate of the phosphatase (135). Analysis of Crz1p localization in vivo reveals that, like NF-AT, Crz1p rapidly relocalizes from the cytosol to the nucleus in a Ca^{2+}-induced, calcineurin-dependent manner (135). Recent studies have defined the mechanism of this relocalization and have established that both nuclear import and export of Crz1p depend on its phosphorylation state and are regulated by calcineurin (117; L. Boustany & M. Cyert, unpublished observations).

Activation of calcineurin in vivo requires a rise in cytosolic Ca^{2+}. The mechanisms underlying Ca^{2+} signal generation are not well characterized in *S. cerevisiae* but are best understood during the response to mating factor. When haploid **a** cells are incubated in the presence of the α-mating pheromone, Ca^{2+} uptake increases after approximately 45 minutes, resulting in an increase in cytosolic $[Ca^{2+}]$ (51, 102). Generation of this Ca^{2+} signal requires a sufficient concentration of Ca^{2+} in the extracellular medium, and in media lacking Ca^{2+}, yeast cells exposed to pheromone die (51, 102). Thus, activation of calcineurin and other Ca^{2+}-regulated processes is dependent on Ca^{2+} entry (153), which is mediated during pheromone treatment by a plasma membrane Ca^{2+} channel encoded by the *MID1* and *CCH1* genes (32, 49, 70, 109). While the mechanism of Mid1p/Cch1p activation has not been definitively demonstrated, expression of Mid1p in mammalian cells generates a novel Ca^{2+} channel that is activated by membrane stretch (55). This finding suggests that physical perturbation of the yeast cell surface may generate a Ca^{2+} signal directly through the Mid1p/Cch1p channel and activate Ca^{2+}-dependent signaling pathways. It is not yet known if Mid1p and Cch1p are required for Ca^{2+} signaling under other environmental conditions. In contrast to the well-characterized, Ca^{2+}-dependent signaling pathways in mammalian cells, it is unclear whether Ca^{2+} release from intracellular stores is required during Ca^{2+} signaling in *S. cerevisiae*. Yeast possess phospholipase C (Plc1p), which generates IP_3 from hydrolysis of PIP_2 (33, 112, 156). However, no IP_3-regulated Ca^{2+} release channel has been identified in yeast, and calcineurin/Crz1p-dependent signaling is not disrupted in *plc1* mutants (M. Cyert, unpublished observations).

Regulation of Ca^{2+} homeostasis by calcineurin   More than 95% of Ca^{2+} in yeast cells is stored in the vacuole (27), and this organelle plays a critical role in regulating Ca^{2+} homeostasis in yeast. Ca^{2+} enters the vacuole through two known routes. First, the V-type ATPase encoded by the *VMA* and *VPH1* genes acidifies the vacuole and generates an H^+ gradient across the vacuolar membrane (reviewed in 1). This proton gradient powers a Ca^{2+}/H^+ exchanger encoded by *VCX1/HUM1*, allowing rapid entry of Ca^{2+} into the vacuole (19, 101, 119). In addition to the Ca^{2+}/H^+ exchanger, yeast vacuoles contain a P-type ATPase, encoded by *PMC1*, that pumps Ca^{2+} into the vacuole against a concentration gradient (20).

Calcineurin regulates Ca^{2+} homeostasis in yeast, although its effects are not completely understood. First, calcineurin activates *PMC1* expression through the Crz1p transcription factor and thus promotes vacuolar Ca^{2+} sequestration. *crz1* mutants fail to grow on media containing high concentrations of Ca^{2+} due to this decrease in Pmc1p. However, calcineurin mutants (*cnb1*) are Ca^{2+} tolerant, i.e., they grow better than wild-type cells on Ca^{2+}-containing media, and *cnb1 crz1* double mutants display growth on Ca^{2+} that is indistinguishable from that of the *cnb1* mutant. Thus, the Ca^{2+} tolerance of *cnb1* mutants must reflect calcineurin-dependent regulation of substrate(s) other than Crz1p.

A series of observations suggest that calcineurin may regulate Vcx1p. *pmc1* mutants are Ca^{2+} sensitive, and mutations in calcineurin (*cnb1*) suppress this Ca^{2+}

sensitivity (20). However, *vcx1 pmc1* cells are more Ca^{2+} sensitive than *pmc1* mutants, and *cnb1* does not suppress *vcx1 pmc1* sensitivity. These findings suggest that *cnb1* may promote Ca^{2+} tolerance through regulation of Vcx1p, and that calcineurin decreases Vcx1p activity (19). Consistent with this idea, *cnb1* mutations cause increased vacuolar Ca^{2+} accumulation in *VCX1* cells but not in *vcx1* cells (19, 153). However, Vcx1p levels are unchanged in *cnb1* cells, and vacuolar vesicles isolated from wild-type and calcineurin-deficient cells display the same level of Ca^{2+}/H^+ exchange activity in vitro (19, 118). Therefore, there is no evidence that Vcx1p is a direct substrate of calcineurin.

Other studies indicate that calcineurin's effects on Ca^{2+} homeostasis cannot be explained solely by regulation of Vcx1p. In mutants that lack the vacuolar H^+-ATPase, no H^+ gradient is generated across the vacuolar membrane, and consequently H^+/Ca^{2+} exchange is severely compromised. Inhibiting calcineurin in these cells decreases cytosolic $[Ca^{2+}]$ (147), suggesting that calcineurin can affect Ca^{2+} homeostasis independently of Vcx1p. Also, in calcineurin mutants the rate of Ca^{2+} uptake at the plasma membrane is increased, and the activity of the plasma membrane H^+-ATPase, Pma1p, which impacts Ca^{2+} homeostasis, is decreased (153, 154). Thus, calcineurin-dependent regulation of Ca^{2+} homeostasis may be complex, and understanding its role in this process will require identification of substrates.

Calcineurin-dependent regulation of the cell cycle   Calcineurin also participates in G2/M cell cycle regulation. Mutations in *ZDS1*, which encodes a protein of unknown biochemical function, result in increased expression of *SWE1*, which encodes a kinase that phosphorylates and negatively regulates Cdc28p (the major cyclin-dependent kinase in *S. cerevisiae*) at G2/M (8). Addition of Ca^{2+} causes a delay in the onset of mitosis in *zds1*Δ cells due to high, sustained levels of *SWE1* transcription (88). This delay is relieved by mutational inactivation of *SWE1*, calcineurin (*CNB1*), or *MPK1*, the MAP kinase that acts downstream of protein kinase C (*PKC1*) (88). In *cnb1*Δ *zds1*Δ mutants, a normal pattern of *SWE1* transcription is restored; thus calcineurin is required for the Ca^{2+}-induced increase in *SWE1* transcription in *zds1*Δ mutants (88). The mechanism of this transcriptional regulation is not known and does not require Crz1p (T. Miyakawa, personal communication). Furthermore, calcineurin also seems to regulate Swe1p at a posttranslational level. Calcineurin promotes Ca^{2+}-induced degradation of Hsl1p, a kinase that inhibits Swe1p (73, 89, 146). Calcineurin co-immunoprecipitates with Hsl1p, and the phosphorylation state of Hsl1p in vivo is calcineurin dependent (89). Thus, Hsl1p may be a direct substrate of calcineurin, with dephosphorylation of Hsl1p leading to its degradation and consequently to increased Swe1p activity and slowed progression from G2 to M.

CALMODULIN-STIMULATED PROTEIN KINASES   Calmodulin regulates a number of different protein kinases in a variety of organisms, including myosin light chain kinase (MLCK), CamK I, II, and IV and CamKK. These kinases have distinct structural features and participate in different biological processes (reviewed in 81). In *S. cerevisiae*, there are only two protein kinases, the products of *CMK1* and *CMK2*, that are regulated by calmodulin (106, 111). These kinases are closely related to each other, showing 60% amino acid identity and 90% similarity. A third kinase, the product of the *CMK3/CLK1/RCK2* gene, has some sequence similarity to calmodulin-regulated kinases, but it fails to bind calmodulin, and its activity is not stimulated by calmodulin in vitro. Therefore, it should not be classified as a calmodulin-dependent kinase (82).

Cmk1p and Cmk2p most resemble the multifunctional calmodulin kinase type II from mammalian cells. In mammals this kinase has broad substrate specificity and displays a characteristic pattern of regulation. It assembles into a large oligomer (10–12 subunits) and its initial activity is dependent on Ca^{2+}/calmodulin. However, activation of the enzyme leads to its autophosphorylation on individual subunits, which results in camodulin-independent kinase activity. Mammalian CamK II has well-documented roles in learning and memory (reviewed in 81).

Cmk1p and Cmk2p display some biochemical characteristics that are similar to mammalian Cam kinase II: Although no physiological substrates for these enzymes have been documented, in the presence of Ca^{2+} and calmodulin they phosphorylate a number of different substrates in vitro and also become autophosphorylated (71, 106, 111). Cmk2p activity becomes Ca^{2+}/calmodulin independent after autophosphorylation; however, this is not observed for Cmk1p (87, 106, 111). Unlike mammalian Cam kinase II, the yeast kinases do not appear to form large oligomers, although the active form of Cmk1p may be a dimer (71).

The physiological role(s) of Cmk1p and Cmk2p are not well understood. However, like calcineurin, these enzymes seem to participate in a number of stress responses. *cmk1cmk2* double mutants lose viability during prolonged incubation with mating pheromone, and calcineurin and the calmodulin-dependent kinases act additively to promote survival of cells under these conditions (93). The specific role of calmodulin-stimulated kinases in maintaining viability during pheromone treatment is not understood. However, the expression of *CMK2* is induced in a Ca^{2+}/calcineurin/*CRZ1*-dependent manner, suggesting that conditions that lead to calcineurin activation also may stimulate calmodulin-dependent kinase activity (H. Yoshimoto & M. Cyert, unpublished observations).

Calmodulin-dependent kinases also influence two other stress responses: the ability of yeast to grow in the presence of weak organic acids and the acquisition of thermotolerance. At low pH, when exposed to weak organic acids such as sorbate or benzoate, wild-type yeast show a period of growth inhibition followed by adaptation (116). Recovery of growth is mediated largely by increased expression of *PDR12*, which encodes an ABC cassette-type transporter that catalyzes efflux of these anions (4, 116). In contrast, *cmk1* cells constitutively express resistance to organic acids, and *CMK1* seems to negatively regulate Pdr12p activity (48). The mechanism of this regulation is unclear, however, as neither the gene expression, protein levels, nor phosphorylation state of Pdr12p are altered in *cmk1*Δ mutants (48). Finally, wild-type yeast cells are also able to tolerate exposure to high temperature, and prior treatment of cells with mild stress improves their survival during subsequent heat shock. *cmk1*Δ mutants show decreased levels of this induced thermotolerance, and therefore Cmk1p somehow positively regulates this stress response (50).

Other Potential Targets

Iqg1p   *IQG1/CYK1* encodes a product with sequence similarity to mammalian IQGAP proteins that regulate the cytoskeleton and are thought to act as effectors for several small GTPases (30, 67, 108, 150). Iqg1p localizes to the bud neck during anaphase and is required for formation and contraction of the actomyosin ring during cytokinesis (30, 67, 108). Iqg1p contains IQ motifs, which are required for Iqg1p localization to the bud neck. Iqg1p recruits actin to the bud neck via an actin-binding domain and also interacts with the small GTP-binding proteins, Cdc42p and Tem1p, potentially linking these regulators to the cytoskeleton (108, 130).

Although mammalian IQGAPs bind calmodulin, it is unclear whether Iqg1p is a calmodulin target in *S. cerevisiae*. In vitro, Iqg1p binds to GST-Cmd1p in a Ca^{2+}-dependent manner; however, the IQ motifs are neither necessary nor sufficient for this interaction (130). In contrast, the IQ motifs are required for binding of Iqg1p to Mlc1p, a myosin light chain that also interacts with Myo1p and Myo2p (9, 136), and the Iqg1p-Mlc1p interaction mediates localization of Iqg1p to the actomyosin ring (9, 131). Cmd1p localization is perturbed in *iqg1* mutant cells, suggesting that these two proteins may associate in vivo (108). However, calmodulin localization may be altered in *iqg1* mutants owing to defects in actin organization rather than a disruption in calmodulin-Iqg1p binding.

VACUOLE FUSION   Several observations, mostly from in vitro biochemical studies, suggest that calmodulin plays a role in vacuole fusion in *S. cerevisiae*. The fusion of yeast vacuolar vesicles to each other, or homotypic vacuolar fusion, has been reconstituted in a cell-free system and is similar in many respects to other membrane fusion reactions (99). A priming stage is first required to activate SNAP/SNARE complexes in the two fusing membranes and is followed by a docking stage in which specific SNAP/SNARE complexes form between them. Finally, the lipid bilayers mix, resulting in formation of a single membrane/compartment (78, 79, 149). Ca^{2+} and calmodulin are required during vacuole fusion in vitro for the final stage of bilayer mixing, and some calmodulin mutants, in particular *cmd1-239*, display fragmented vacuoles in vivo at restrictive temperature (105, 115). Calmodulin also binds to vacuoles in a Ca^{2+}-dependent manner and is found together with protein phosphatase type I in a large protein complex that is required for bilayer mixing (113, 115). During vacuolar fusion, components of the vacuolar H^+-ATPase that form a proteolipid ring in each membrane interact with each other to form a dimer. Calmodulin is required for this dimerization, and Ca^{2+} is released from the vacuole during the fusion process. Thus, calmodulin is thought to act as a Ca^{2+} sensor to regulate the final stages of the fusion process (114). Although this biochemical characterization strongly supports a role for calmodulin in vacuole fusion, the in vitro findings are not completely consistent with in vivo observations. For example, the Ca^{2+}-binding–deficient calmodulin, cmd1-3p, fails to support vacuolar fusion in vitro, but shows no defects in vacuolar morphology in vivo (39, 115). These differences may reflect redundancies in vivo that do not exist in the in vitro system. In any case, a direct target for calmodulin in this process has not been established.

Gad1p   *GAD1* was identified by its ability when overexpressed to confer resistance to the oxidizing agents H_2O_2 and diamide, and *gad1* mutants are sensitive to these oxidants (16). Gad1p shows high homology to glutamate decarboxylase, an enzyme that converts glutamate to GABA. A variety of genetic evidence suggests that *GAD1* encodes a glutamate decarboxylase, although the recombinant protein purified from *Escherichia coli* had no activity in vitro (16). Gad1p binds calmodulin in vitro, suggesting that this protein may be modulated by Ca^{2+}/calmodulin. However, there is currently no evidence that indicates a role for calmodulin in the oxidative stress response.

Dst1p/Ppr2p   DST1/PPR2, which encodes the *S. cerevisiae* homologue of the TFIIS transcription factor, was identified by direct screening of a library for calmodulin-binding proteins (137) and confirmed by two-hybrid studies (M. Stark, personal communication). However, the significance of these observations remains unclear, as no role for calmodulin in regulating TFIIS function has been demonstrated.

PERSPECTIVES

Calmodulin is one of the most highly conserved proteins and is present in all eukaryotes including animals, plants, and fungi. The amino acid sequences of calmodulins from all multicellular organisms are more than 90% identical (91). In contrast, calmodulin from the yeast *S. cerevisiae* is the most divergent form of calmodulin that has been characterized and is 60% identical to vertebrate calmodulin. *S. cerevisiae* calmodulin also differs significantly from vertebrate calmodulin in its structure and its biochemical properties. This calmodulin is the only one known that binds a maximum of three rather than four molecules of Ca^{2+} (72, 76, 77, 133). The Ca^{2+}-bound form of calmodulin is also unique in exhibiting interactions between its N- and C-terminal domains (63, 157). These differences are reflected in the relatively poor ability of yeast calmodulin to activate mammalian target enzymes in vitro and in the inability of *S. cerevisae* calmodulin to complement a calmodulin mutant of *S. pombe* (72, 94, 107). Despite these differences, however, calmodulin functions are highly conserved in budding yeast. Calmodulin from vertebrates or *S. pombe* can complement the growth defect of a calmodulin mutant of *S. cerevisiae*, and all known calmodulin targets in

S. cerevisiae have a functional homologue or orthologue in multicellular eukaryotes (Table 1).

One of the most surprising findings from the studies of yeast calmodulin is the demonstration that calmodulin performs its essential functions in the budding yeast in its Ca^{2+}-free form (39). Although this finding was initially quite controversial, subsequent analyses confirmed that calmodulin association with its two essential targets, Nuf1p and Myo2p, is Ca^{2+} independent (11, 38, 137). Does Cmd1p mediate any Ca^{2+}-dependent regulation of Nuf1p or Myo2p function? In neither case can a role for Ca^{2+} be definitively ruled out. However, that Ca^{2+}-binding–defective alleles of *CMD1* support normal growth and morphology indicates that Ca^{2+}-dependent regulation of these targets by calmodulin, if it indeed exists, is not required for viability. Interestingly, in *S. pombe* calmodulin function at the SPB does require Ca^{2+} binding (92), and the ATPase activity of a mammalian class V myosin (p190) is regulated by Ca^{2+} in vitro (124). Thus, at least in some species, calmodulin's role in associating with SPB proteins or myosins is to provide some type of Ca^{2+} regulation. In *S. cerevisiae* this regulation may not occur or may only subtly affect the function of these targets.

Calmodulin participates in Ca^{2+}-dependent modulation of protein phosphorylation in yeast through activation of the calcineurin phosphatase and the Cmk1p and Cmk2p kinases (21, 22, 62, 106, 111). These enzymes are components of signaling pathways that allow yeast cells to respond to a variety of environmental stresses. Ca^{2+} signals are particularly well-suited for stress responses as cytosolic Ca^{2+} levels can be rapidly and reversibly regulated. In yeast, the rate of entry of Ca^{2+} across the plasma membrane may serve as a direct indicator of the integrity of the cell surface. A full understanding of the role that Ca^{2+} signaling plays in yeast awaits identification of the substrates of calcineurin, Cmk1p and Cmk2p, and further characterization of the mechanisms that regulate cytosolic Ca^{2+} levels.

It is highly likely that additional calmodulin targets in yeast remain to be identified. The roles of calmodulin in regulating endocytosis and vacuole fusion are only partially understood, and there may be other functions of calmodulin that have not yet been demonstrated. Ikura and colleagues have studied the interaction of calmodulin with its many different targets and have compiled a database of calmodulin-binding peptides (http://calcium.oci.utoronto.ca/). As methods for predicting calmodulin-binding domains such as these improve, and proteomic and genomic analyses evolve, the entire complement of calmodulin functions in eukaryotic cells will continue to unfold.

ACKNOWLEDGMENTS

I thank Jeremy Thorner, Michael Stark, and Trisha Davis for helpful discussion, and Michael Stark and Tokichi Miyakawa for contributing unpublished information. I gratefully acknowledge Victoria Heath and James Withee for critical reading of the manuscript. The work described in this review was supported by NIH, grant # GM-48729.

NOTE ADDED IN PROOF

Several new calmodulin binding proteins in yeast were recently discovered through proteomic analysis (159).

Visit the Annual Reviews home page at www.AnnualReviews.org

LITERATURE CITED

1. Anraku Y, Umemoto N, Hirata R, Ohya Y. 1992. Genetic and cell biological aspects of the yeast vacuolar H^+-ATPase. *J. Bioenerg. Biomem.* 24:395–405

2. Aperia A, Ibarra F, Svensson L-B, Klee C, Greengard P. 1992. Calcineurin mediates alpha-adrenergic stimulation of Na^+,K^+-ATPase activity in renal tubule cells. *Proc. Natl. Acad. Sci. USA* 89:7394–97

3. Aramburu J, Rao A, Klee CB. 2000. Calcineurin: from structure to function. *Curr. Top. Cell Regul.* 36:237–95

4. Bauer BE, Wolfger H, Kuchler K. 1999. Inventory and function of yeast ABC proteins: about sex, stress, pleiotropic drug and heavy metal resistance. *Biochim. Biophys. Acta* 1461:217–36

5. Baum P, Furlong C, Byers B. 1986. Yeast gene required for spindle pole body duplication: homology of its product with Ca^{2+}-binding proteins. *Proc. Natl. Acad. Sci. USA* 83:5512–16

6. Bi E, Caviston J, Drees B. 2000. Saccharomyces Genome Database. *http://genome-www.stanford.edu/Saccharomyces/*

7. Bobola N, Jansen RP, Shin TH, Nasmyth K. 1996. Asymmetric accumulation of Ash1p in postanaphase nuclei depends on a myosin and restricts yeast mating-type switching to mother cells. *Cell* 84:699–709

8. Booher RN, Deshaies RJ, Kirschner MW. 1993. Properties of *Saccharomyces cerevisiae* wee1 and its differential regulation of p34CDC28 in response to G1 and G2 cyclins. *EMBO J.* 12:3417–26

9. Boyne JR, Yosuf HM, Bieganowski P, Brenner C, Price C. 2000. Yeast myosin light chain, Mlc1p, interacts with both IQGAP and class II myosin to effect cytokinesis. *J. Cell Sci.* 113(Pt 24):4533–43

10. Brockerhoff SE, Davis TN. 1992. Calmodulin concentrates at regions of cell growth in *Saccharomyces cerevisiae*. *J. Cell Biol.* 118:619–29

11. Brockerhoff SE, Stevens RC, Davis TN. 1994. The unconventional myosin, Myo2p, is a calmodulin target at sites of cell growth in *Saccharomyces cerevisiae*. *J. Cell Biol.* 124:315–23

12. Catlett NL, Duex JE, Tang F, Weisman LS. 2000. Two distinct regions in a yeast myosin-V tail domain are required for the movement of different cargoes. *J. Cell Biol.* 150:513–26

13. Catlett NL, Weisman LS. 1998. The terminal tail region of a yeast myosin-V mediates its attachment to vacuole membranes and sites of polarized growth. *Proc. Natl. Acad. Sci. USA* 95:14799–804

14. Cheung WY. 1970. Cyclic $3',5'$-nucleotide phosphodiesterase: demonstration of an activator. *Biochem. Biophys. Res. Commun.* 38:533–38

15. Clipstone NA, Crabtree GR. 1992. Identification of calcineurin as a key signalling enzyme in T-lymphocyte activation. *Nature* 357:695–97

16. Coleman ST, Fang TK, Rovinsky SA, Turano FJ, Moye-Rowley WS. 2001. Expression of a glutamate decarboxylase homologue is required for normal oxidative stress tolerance in *Saccharomyces cerevisiae*. *J Biol. Chem.* 276:244–50

17. Crabtree GR. 2001. Calcium, calcineurin, and the control of transcription. *J. Biol. Chem.* 276:2313–16

18. Crivici A, Ikura M. 1995. Molecular and structural basis of target recognition by calmodulin. *Annu. Rev. Biophys. Biomol. Struct.* 24:85–116

19. Cunningham K, Fink GR. 1996. Calcineurin inhibits VCX1-dependent H$^+$/Ca2$^+$ exchange and induces Ca^{2+} ATPases in *Saccharomyces cerevisiae*. *Mol. Cell. Biol.* 16:2226–37

20. Cunningham KW, Fink GR. 1994. Calcineurin-dependent growth control in *Saccharomyces cerevisiae* mutants lacking PMC1, a homolog of plasma membrane Ca^{2+} ATPases. *J. Cell Biol.* 124: 351–63

21. Cyert MS, Kunisawa R, Kaim D, Thorner J. 1991. Yeast has homologs (*CNA1* and *CNA2* gene products) of mammalian calcineurin, a calmodulin-regulated phosphoprotein phosphatase. *Proc. Natl. Acad. Sci. USA* 88:7376–80

22. Cyert MS, Thorner J. 1992. Regulatory subunit (*CNB1* gene product) of yeast Ca^{2+}/calmodulin-dependent phosphoprotein phosphatases is required for adaptation to pheromone. *Mol. Cell. Biol.* 12: 3460–69

23. Davis TN. 1992. A temperature-sensitive calmodulin mutant loses viability during mitosis. *J. Cell Biol.* 118:607–17

24. Davis TN, Thorner J. 1989. Vertebrate and yeast calmodulin, despite significant sequence divergence, are functionally interchangeable. *Proc. Natl. Acad. Sci. USA* 86:7909–13

25. Davis TN, Urdea MS, Masiarz FR, Thorner J. 1986. Isolation of the yeast calmodulin gene: calmodulin is an essential protein. *Cell* 47:423–31

26. de la Pompa J, Timmerman L, Takimoto H, Yoshida H, Elia A, et al. 1998. Role of the NF-ATc transcription factor in morpogenesis of cardiac valves and septum. *Nature* 392:182–86

27. Eilam Y, Lavi H, Grossowicz N. 1985. Cytoplasmic Ca^{2+} transport system in the yeast *Saccharomyces cerevisiae*. *J. Gen. Microbiol.* 131:623–29

28. Elliott S, Knop M, Schlenstedt G, Schiebel E. 1999. Spc29p is a component of the Spc110p subcomplex and is essential for spindle pole body duplication. *Proc. Natl. Acad. Sci. USA* 96:6205–10

29. Eng W-K, Faucette L, McLaughlin MM, Cafferkey R, Koltin Y, et al. 1994. The yeast *FKS1* gene encodes a novel membrane protein, mutations in which confer FK506 and cyclosporin A hypersensitivity and calcineurin-dependent growth. *Gene* 151:61–71

30. Epp JA, Chant J. 1997. An IQGAP-related protein controls actin-ring formation and cytokinesis in yeast. *Curr. Biol.* 7:921–29

31. Farcasanu IC, Hirata D, Tsuchiya E, Nishiyama F, Miyakawa T. 1995. Protein phosphatase 2B of *Saccharomyces cerevisiae* is required for tolerance to manganese in blocking the entry of ions into the cell. *Eur. J. Biochem.* 232:712–17

32. Fischer M, Schnell N, Chattaway J, Davies P, Dixon G, Sanders D. 1997. The *Saccharomyces cerevisiae CCH1* gene is involved in calcium influx and mating. *FEBS Lett.* 419:259–62

33. Flick JS, Thorner J. 1993. Genetic and biochemical characterization of a phosphatidylinositol-specific phospholipase C in *Saccharomyces cerevisiae*. *Mol. Cell. Biol.* 13:5861–76

34. Flory MR, Moser MJ, Monnat RJ Jr, Davis TN. 2000. Identification of a human centrosomal calmodulin-binding protein that shares homology with pericentrin. *Proc. Natl. Acad. Sci. USA* 97:5919–23

35. Francis SE, Davis TN. 2000. The spindle pole body of *Saccharomyces cerevisiae*: architecture and assembly of the core components. *Curr. Top. Dev. Biol.* 49:105–32

36. Garrett-Engele P, Moilanen B, Cyert MS. 1995. Calcineurin, the Ca^{2+}/calmodulin-dependent protein phosphatase, is essential in yeast mutants with cell integrity defects and in mutants that lack a functional

vacuolar H$^+$-ATPase. *Mol. Cell. Biol.* 15: 4103–14

37. Geier BM, Wiech H, Schiebel E. 1996. Binding of centrins and yeast calmodulin to synthetic peptides corresponding to binding sites in the spindle pole body components Kar1p and Spc110p. *J. Biol. Chem.* 271:28366–74

38. Geiser JR, Sundberg HA, Chang BH, Muller EG, Davis TN. 1993. The essential mitotic target of calmodulin is the 110-kilodalton component of the spindle pole body in *Saccharomyces cerevisiae. Mol. Cell. Biol.* 13:7913–24

39. Geiser JR, van Tuinen D, Brockerhoff SE, Neff MM, Davis TN. 1991. Can calmodulin function without binding calcium? *Cell* 65:949–59

40. Geli MI, Riezman H. 1996. Role of type I myosins in receptor-mediated endocytosis in yeast. *Science* 272:533–35

41. Geli MI, Wesp A, Riezman H. 1998. Distinct functions of calmodulin are required for the uptake step of receptor-mediated endocytosis in yeast: the type I myosin Myo5p is one of the calmodulin targets. *EMBO J.* 17:635–47

42. Goodson HV, Anderson BL, Warrick HM, Pon LA, Spudich JA. 1996. Synthetic lethality identifies a novel myosin I gene (MYO5): myosin I proteins are required for polarization of the actin cytoskeleton. *J. Cell Biol.* 133:1277–91

43. Govindan B, Bowser R, Novick P. 1995. The role of Myo2, a yeast class V myosin, in vesicular transport. *J. Cell Biol.* 128:1055–68

44. Graef IA, Chen F, Chen L, Kuo A, Crabtree GR. 2001. Signals transduced by Ca^{2+}/calcineurin and NFATc3/c4 pattern the developing vasculature. *Cell* 105:863–75

45. Harris E, Watterson DM, Thorner J. 1994. Functional consequences in yeast of single-residue alterations in a consensus calmodulin. *J. Cell Sci.* 107:3235–49

46. Hendricks KB, Wang BQ, Schnieders EA, Thorner J. 1999. Yeast homologue of neuronal frequenin is a regulator of phosphatidylinositol-4-OH kinase. *Nat. Cell Biol.* 1:234–41

47. Hill KL, Catlett NL, Weisman LS. 1996. Actin and myosin function in directed vacuole movement during cell division in *Saccharomyces cerevisiae. J. Cell Biol.* 135:1535–49

48. Holyoak CD, Thompson S, Ortiz Calderon C, Hatzixanthis K, Bauer B, et al. 2000. Loss of Cmk1 Ca^{2+}-calmodulin-dependent protein kinase in yeast results in constitutive weak organic acid resistance, associated with a post-transcriptional activation of the Pdr12 ATP-binding cassette transporter. *Mol. Microbiol.* 37:595–605

49. Iida H, Nakamura H, Ono T, Okumura M, Anraku Y. 1994. MID1, a novel *Saccharomyces cerevisiae* gene encoding a plasma membrane protein, is required for Ca^{2+} influx and mating. *Mol. Cell. Biol.* 14:8259–71

50. Iida H, Ohya Y, Anraku Y. 1995. Calmodulin-dependent protein kinase II and calmodulin are required for induced thermotolerance in *Saccharomyces cerevisiae. Curr. Genet.* 27:190–93

51. Iida H, Yagawa Y, Anraku Y. 1990. Essential role for induced Ca^{2+} influx followed by [Ca^{2+}]$_i$ rise in maintaining viability of yeast cells late in the mating pheromone response pathway. *J. Biol. Chem.* 265:13391–99

52. Jansen RP, Dowzer C, Michaelis C, Galova M, Nasmyth K. 1996. Mother cell-specific HO expression in budding yeast depends on the unconventional myosin Myo4p and other cytoplasmic proteins. *Cell* 84:687–97

53. Johnston GC, Prendergast JA, Singer RA. 1991. The *Saccharomyces cerevisiae MYO2* gene encodes an essential myosin for vectorial transport of vesicles. *J. Cell Biol.* 113:539–51

54. Johnston GC, Prendergast JA, Singer RA. 1991. The *Saccharomyces cerevisiae*

MYO2 gene encodes an essential myosin for vectorial transport of vesicles. *J. Cell Biol.* 113:539–51

55. Kanzaki M, Nagasawa M, Kojima I, Sato C, Naruse K, et al. 1999. Molecular identification of a eukaryotic, stretch-activated nonselective cation channel. *Science* 285:882–86

56. Karpova TS, Reck-Peterson SL, Elkind NB, Mooseker MS, Novick PJ, Cooper JA. 2000. Role of actin and Myo2p in polarized secretion and growth of *Saccharomyces cerevisiae. Mol. Biol. Cell* 11:1727–37

57. Kilmartin JV, Dyos SL, Kershaw D, Finch JT. 1993. A cell cycle-regulated spacer element in the *Saccharomyces cerevisiae* spindle pole body. *J. Cell Biol.* 123:1175–84

58. Kilmartin JV, Goh PY. 1996. Spc110p: assembly properties and role in the connection of nuclear microtubules to the yeast spindle pole body. *EMBO J.* 15:4592–602

59. Klee CB, Ren H, Wang X. 1998. Regulation of the calmodulin-stimulated protein phosphatase, calcineurin. *J. Biol. Chem.* 273:13367–70

60. Kretsinger RH. 1980. Structure and evolution of calcium-modulated proteins. *Crit. Rev. Biochem.* 8:119–74

61. Kubler E, Schimmoller F, Riezman H. 1994. Calcium-independent calmodulin requirement for endocytosis in yeast. *EMBO J.* 13:5539–46

62. Kuno T, Tanaka H, Mukai J, Chang C, Hiraga K, et al. 1991. cDNA cloning of a calcineurin B homolog in *Saccharomyces cerevisiae. Biochem. Biophys. Res. Commun.* 180:1159–63

63. Lee SY, Klevit RE. 2000. The whole is not the simple sum of its parts in calmodulin from *S. cerevisiae. Biochemistry* 39:4225–30

64. Lieberman DN, Mody I. 1994. Regulation of NMDA channel function by endogenous Ca^{2+}-dependent phosphatase. *Nature* 369:235–39

65. Deleted in proof

66. Lillie SH, Brown SS. 1994. Immunofluorescence localization of the unconventional myosin, Myo2p, and the putative kinesin-related protein, Smy1p, to the same regions of polarized growth in *Saccharomyces cerevisiae. J. Cell Biol.* 125:825–42

67. Lippincott J, Li R. 1998. Sequential assembly of myosin II, an IQGAP-like protein, and filamentous actin to a ring structure involved in budding yeast cytokinesis. *J. Cell Biol.* 140:355–66

68. Liu J. 1993. FK506 and cyclosporin, molecular probes for studying intracellular signal transduction. *Immunol. Today* 14:290–95

69. Liu Y, Ishii S, Tokai M, Tsutsumi H, Ohke O, et al. 1991. The *Saccharomyces cerevisiae* genes (*CMP1* and *CMP2*) encoding calmodulin-binding proteins homologous to the catalytic subunit of mammalian protein phosphatase 2B. *Mol. Gen. Genet.* 227:52–59

70. Locke EG, Bonilla M, Liang L, Takita Y, Cunningham KW. 2000. A homolog of voltage-gated Ca^{2+} channels stimulated by depletion of secretory Ca^{2+} in yeast. *Mol. Cell. Biol.* 20:6686–94

71. Londesborough J, Nuutinen M. 1987. Ca^{2+}/calmodulin-dependent protein kinase in *Saccharomyces cerevisiae. FEBS Lett.* 219:249–53

72. Luan Y, Matsuura I, Yazawa M, Nakamura T, Yagi K. 1987. Yeast calmodulin: structural and functional differences compared with vertebrate calmodulin. *J. Biochem.* 102:1531–37

73. Ma XJ, Lu Q, Grunstein M. 1996. A search for proteins that interact genetically with histone H3 and H4 amino termini uncovers novel regulators of the Swe1 kinase in *Saccharomyces cerevisiae. Genes Dev.* 10:1327–40

74. Malleret G, Haditsch U, Genoux D, Jones MW, Bliss TV, et al. 2001. Inducible and reversible enhancement of learning, memory, and long-term potentiation by genetic inhibition of calcineurin. *Cell* 104:675–86

75. Matheos D, Kingsbury T, Ahsan U, Cunningham K. 1997. Tcn1p/Crz1p, a calcineurin-dependent transcription factor that differentially regulates gene expression in *Saccharomyces cerevisiae*. *Genes Dev.* 11:3445–58

76. Matsuura I, Ishihara K, Nakai Y, Yazawa M, Toda H, Yagi K. 1991. A site-directed mutagenesis study of yeast calmodulin. *J. Biochem.* 109:190–97

77. Matsuura I, Kimura E, Tai K, Yazawa M. 1993. Mutagenesis of the fourth calcium-binding domain of yeast calmodulin. *J. Biol. Chem.* 268:13267–73

78. Mayer A, Wickner W. 1997. Docking of yeast vacuoles is catalyzed by the Ras-like GTPase Ypt7p after symmetric priming by Sec18p (NSF). *J. Cell Biol.* 136:307–17

79. Mayer A, Wickner W, Haas A. 1996. Sec18p (NSF)-driven release of Sec17p (alpha-SNAP) can precede docking and fusion of yeast vacuoles. *Cell* 85:83–94

80. Mazur P, Morin N, Baginsky W, El-Sherbeini M, Clemas JA, et al. 1995. Differential expression and function of two homologous subunits of yeast 1,3–b-D-glucan synthase. *Mol. Cell. Biol.* 15:5671–81

81. Means AR. 2000. Regulatory cascades involving calmodulin-dependent protein kinases. *Mol. Endocrinol.* 14:4–13

82. Melcher ML, Thorner J. 1996. Identification and characterization of the CLK1 gene product, a novel CaM kinase-like protein kinase from the yeast *Saccharomyces cerevisiae*. *J. Biol. Chem.* 271:29958–68

83. Mendizabal I, Rios G, Mulet JM, Serrano R, de Larrinoa IF. 1998. Yeast putative transcription factors involved in salt tolerance. *FEBS Lett.* 425:323–28

84. Mendoza I, Quintero FJ, Bressan RA, Hasegawa PM, Pardo JM. 1996. Activated calcineurin confers high tolerance to ion stress and alters the budding pattern and cell morphology of yeast cells. *J. Biol. Chem.* 271:23061–67

85. Mendoza I, Rubio F, Rodriguez-Navarro A, Pardo JM. 1994. The protein phosphatase calcineurin is essential for NaCl tolerance of *Saccharomyces cerevisiae*. *J. Biol. Chem.* 269:8792–96

86. Mirzayan C, Copeland CS, Snyder M. 1992. The NUF1 gene encodes an essential coiled-coil related protein that is a potential component of the yeast nucleoskeleton. *J. Cell Biol.* 116:1319–32

87. Miyakawa T, Oka Y, Tsuchiya E, Fukui S. 1989. *Saccharomyces cerevisiae* protein kinase dependent on Ca^{2+} and calmodulin. *J. Bacteriol.* 171:1417–22

88. Mizunuma M, Hirata D, Miyahara K, Tsuchiya E, Miyakawa T. 1998. Role of calcineurin and Mpk1 in regulating the onset of mitosis in budding yeast. *Nature* 392:303–6

89. Mizunuma M, Hirata D, Miyaoka R, Miyakawa T. 2001. GSK-3 kinase Mck1 and calcineurin coordinately mediate Hsl1 down-regulation by Ca^{2+} in budding yeast. *EMBO J.* 20:1074–85

90. Molkentin J, Lu J, Antos C, Markham B, Richardson J, et al. 1998. A calcineurin-dependent transcriptional pathway for cardiac hypertrophy. *Cell* 93:215–28

91. Moncrief ND, Kretsinger RH, Goodman M. 1990. Evolution of EF-hand calcium-modulated proteins. I. Relationships based on amino acid sequences. *J. Mol. Evol.* 30:522–62

92. Moser MJ, Flory MR, Davis TN. 1997. Calmodulin localizes to the spindle pole body of *Schizosaccharomyces pombe* and performs an essential function in chromosome segregation. *J. Cell Sci.* 110:1805–12

93. Moser MJ, Geiser JR, Davis TN. 1996. Ca^{2+}-calmodulin promotes survival of pheromone-induced growth arrest by activation of calcineurin and Ca^{2+}-calmodulin-dependent protein kinase. *Mol. Cell. Biol.* 16:4824–31

94. Moser MJ, Lee SY, Klevit RE, Davis TN. 1995. Ca^{2+} binding to calmodulin and its role in *Schizosaccharomyces pombe* as

revealed by mutagenesis and NMR spectroscopy. *J. Biol. Chem.* 270:20643–52

95. Munn AL. 2001. Molecular requirements for the internalization step of endocytosis: insights from yeast. *Biochim. Biophys. Acta* 1535:236–57

96. Munn AL, Riezman H. 1994. Endocytosis is required for the growth of vacuolar H+-ATPase-defective yeast: identification of six new END genes. *J. Cell Biol.* 127:373–86

97. Nakamura T, Liu Y, Hirata D, Namba H, Harada S, et al. 1993. Protein phosphatase type 2B (calcineurin)-mediated, FK506-sensitive regulation of intracellular ions in yeast is an important determinant for adaptation to high salt stress conditions. *EMBO J.* 12:4063–71

98. Nakayama S, Kretsinger RH. 1994. Evolution of the EF-hand family of proteins. *Annu. Rev. Biophys. Biomol. Struct.* 23:473–507

99. Nichols BJ, Ungermann C, Pelham HR, Wickner WT, Haas A. 1997. Homotypic vacuolar fusion mediated by t- and v-SNAREs. *Nature* 387:199–202

100. O'Keefe SJ, Tamura J, Kincaid RL, Tocci MJ, O'Neill EA. 1992. FK506- and CsA-sensitive activation of the interleukin-2 promoter by calcineurin. *Nature* 357:692–94

101. Ohsumi Y, Anraku Y. 1983. Calcium transport driven by a proton motive force in vacuolar membrane vesicles of *Saccharomyces cerevisiae*. *J. Biol. Chem.* 258:5614–17

102. Ohsumi Y, Anraku Y. 1985. Specific introduction of Ca²⁺ transport activity in *MAT*a cells of *Saccharomyces cerevisiae* by a mating pheromone, a factor. *J. Biol. Chem.* 260:10482–86

103. Ohya Y, Anraku Y. 1989. Functional expression of chicken calmodulin in yeast. *Biochem. Biophys. Res. Commun.* 158:541–47

104. Ohya Y, Anraku Y. 1989. A galactose-dependent cmd1 mutant of *Saccharomyces cerevisiae*: involvement of calmodulin in nuclear division. *Curr. Genet.* 15:113–20

105. Ohya Y, Botstein D. 1994. Diverse essential functions revealed by complementing yeast calmodulin mutants. *Science* 263:963–66

106. Ohya Y, Kawasaki H, Suzuki K, Londesborough J, Anraku Y. 1991. Two yeast genes encoding calmodulin-dependent protein kinases: isolation, sequencing and bacterial expression of *CMK1* and *CMK2*. *J. Biol. Chem.* 266:12784–94

107. Ohya Y, Uno I, Ishikawa T, Anraku Y. 1987. Purification and biochemical properties of calmodulin from *Saccharomyces cerevisiae*. *Eur. J. Biochem.* 168:13–19

108. Osman MA, Cerione RA. 1998. Iqg1p, a yeast homologue of the mammalian IQGAPs, mediates Cdc42p effects on the actin cytoskeleton. *J. Cell Biol.* 142:443–55

109. Paidhungat M, Garrett S. 1997. A homolog of mammalian, voltage-gated calcium channels mediates yeast pheromone-stimulated Ca²⁺ uptake and exacerbates the cdc1(Ts) growth defect. *Mol. Cell. Biol.* 17:6339–47

110. Parent SA, Nielsen JB, Morin N, Chrebet G, Ramadan N, et al. 1993. Calcineurin-dependent growth of an FK506- and CsA-hypersensitive mutant of *Saccharomyces cerevisiae*. *J. Gen. Microbiol.* 139:2973–84

111. Pausch MH, Kaim D, Kunisawa R, Admon A, Thorner J. 1991. Multiple Ca²⁺/calmodulin-dependent protein kinase genes in a unicellular eukaryote. *EMBO J.* 10:1511–22

112. Payne WE, Fitzgerald-Hayes M. 1993. A mutation in *PLC1*, a candidate phosphoinositide-specific phospholipase C from *Saccharomyces cerevisiae*, cases aberrant mitotic chromosome segregation. *Mol. Cell. Biol.* 13:4351–64

113. Peters C, Andrews PD, Stark MJ, Cesaro-Tadic S, Glatz A, et al. 1999. Control of the terminal step of intracellular

membrane fusion by protein phosphatase 1. *Science* 285:1084–87

114. Peters C, Bayer MJ, Buhler S, Andersen JS, Mann M, Mayer A. 2001. Transcomplex formation by proteolipid channels in the terminal phase of membrane fusion. *Nature* 409:581–88

115. Peters C, Mayer A. 1998. Ca^{2+}/calmodulin signals the completion of docking and triggers a late step of vacuole fusion. *Nature* 396:575–80

116. Piper P, Mahe Y, Thompson S, Pandjaitan R, Holyoak C, et al. 1998. The Pdr12 ABC transporter is required for the development of weak organic acid resistance in yeast. *EMBO J.* 17:4257–65

117. Polizotto R, Cyert MS. 2001. Calcineurin-dependent nuclear import of the transcription factor Crz1p requires Nmd5p. *J. Cell Biol.* In press

118. Pozos TC. 1998. *HUM1, a novel yeast gene, is required for vacuolar calcium/ proton exchange in S. cerevisiae.* PhD thesis. Stanford Univ. 130 pp.

119. Pozos TC, Sekler I, Cyert MS. 1996. The product of HUM1, a novel yeast gene, is required for vacuolar Ca^{2+}/H^+ exchange and is related to mammalian Na^+/Ca^{2+} exchangers. *Mol. Cell. Biol.* 16:3730–41

120. Pruyne DW, Schott DH, Bretscher A. 1998. Tropomyosin-containing actin cables direct the Myo2p-dependent polarized delivery of secretory vesicles in budding yeast. *J. Cell Biol.* 143:1931–45

121. Ranger A, Grusby M, Hodge M, Gravallese E, de la Brousse F, et al. 1998. The transcription factor NF-ATc is essential for cardiac valve formation. *Nature* 392:186–90

122. Rao A, Luo C, Hogan PG. 1997. Transcription factors of the NFAT family: regulation and function. *Annu. Rev. Immunol.* 15:707–47

123. Rasmussen CD, Means RL, Lu KP, May GS, Means AR. 1990. Characterization and expression of the unique calmodulin gene of *Aspergillus nidulans. J. Biol. Chem.* 265:13767–75

124. Reck-Peterson SL, Provance DW Jr, Mooseker MS, Mercer JA. 2000. Class V myosins. *Biochim. Biophys. Acta* 1496:36–51

125. Rhoads AR, Friedberg F. 1997. Sequence motifs for calmodulin recognition. *FASEB J.* 11:331–40

126. Schaerer-Brodbeck C, Riezman H. 2000. Functional interactions between the p35 subunit of the Arp2/3 complex and calmodulin in yeast. *Mol. Biol. Cell* 11:1113–27

127. Schaerer-Brodbeck C, Riezman H. 2000. *Saccharomyces cerevisiae* Arc35p works through two genetically separable calmodulin functions to regulate the actin and tubulin cytoskeletons. *J. Cell Sci.* 113:521–32

128. Schott D, Ho J, Pruyne D, Bretscher A. 1999. The COOH-terminal domain of Myo2p, a yeast myosin V, has a direct role in secretory vesicle targeting. *J. Cell Biol.* 147:791–808

129. Sekiya-Kawasaki M, Botstein D, Ohya Y. 1998. Identification of functional connections between calmodulin and the yeast actin cytoskeleton. *Genetics* 150:43–58

130. Shannon KB, Li R. 1999. The multiple roles of Cyk1p in the assembly and function of the actomyosin ring in budding yeast. *Mol. Biol. Cell* 10:283–96

131. Shannon KB, Li R. 2000. A myosin light chain mediates the localization of the budding yeast IQGAP-like protein during contractile ring formation. *Curr. Biol.* 10:727–30

132. Spang A, Grein K, Schiebel E. 1996. The spacer protein Spc110p targets calmodulin to the central plaque of the yeast spindle pole body. *J. Cell Sci.* 109:2229–37

133. Starovasnik MA, Davis TN, Klevit RE. 1993. Similarities and differences between yeast and vertebrate calmodulin: an examination of the calcium-binding and structural properties of calmodulin from the yeast *Saccharomyces cerevisiae. Biochemistry* 32:3261–70

134. Stathopoulos AM, Cyert MS. 1997. Calcineurin acts through the CRZ1/TCN1-encoded transcription factor to regulate gene expression in yeast. *Genes Dev.* 11:3432–44

135. Stathopoulos-Gerontides A, Guo J, Cyert MS. 1999. Yeast calcineurin regulates nuclear localization of the Crz1p transcription factor through dephosphorylation. *Genes Dev.* 13:798–803

136. Stevens RC, Davis TN. 1998. Mlc1p is a light chain for the unconventional myosin Myo2p in *Saccharomyces cerevisiae. J. Cell Biol.* 142:711–22

137. Stirling DA, Welch KA, Stark MJ. 1994. Interaction with calmodulin is required for the function of Spc110p, an essential component of the yeast spindle pole body. *EMBO J.* 13:4329–42

138. Sun G-H, Ohya Y, Anraku Y. 1992. Yeast calmodulin localizes to sites of cell growth. *Protoplasma* 166:110–13

139. Sun GH, Hirata A, Ohya Y, Anraku Y. 1992. Mutations in yeast calmodulin cause defects in spindle pole body functions and nuclear integrity. *J. Cell Biol.* 119:1625–39

140. Sundberg HA, Davis TN. 1997. A mutational analysis identifies three functional regions of the spindle pole component Spc110p in *Saccharomyces cerevisiae. Mol. Biol. Cell* 8:2575–90

141. Sundberg HA, Goetsch L, Byers B, Davis TN. 1996. Role of calmodulin and Spc110p interaction in the proper assembly of spindle pole body compenents. *J. Cell Biol.* 133:111–24

142. Takeda T, Yamamoto M. 1987. Analysis and in vivo disruption of the gene coding for calmodulin in *Schizosaccharomyces pombe. Proc. Natl. Acad. Sci. USA* 84:3580–84

143. Takizawa PA, DeRisi JL, Wilhelm JE, Vale RD. 2000. Plasma membrane compartmentalization in yeast by messenger RNA transport and a septin diffusion barrier. *Science* 290:341–44

144. Takizawa PA, Sil A, Swedlow JR, Herskowitz I, Vale RD. 1997. Actin-dependent localization of an RNA encoding a cell-fate determinant in yeast. *Nature* 389:90–93

145. Takizawa PA, Vale RD. 2000. The myosin motor, Myo4p, binds Ash1 mRNA via the adapter protein, She3p. *Proc. Natl. Acad. Sci. USA* 97:5273–78

146. Tanaka S, Nojima H. 1996. Nik1: a Nim1-like protein kinase of *S. cerevisiae* interacts with the Cdc28 complex and regulates cell cycle progression. *Genes Cells* 1:905–21

147. Tanida I, Hasegawa A, Iida H, Ohya Y, Anraku Y. 1995. Cooperation of calcineurin and vacuolar H^+-ATPase in intracellular Ca^{2+} homeostasis of yeast cells. *J. Biol. Chem* 270:10113–19

148. Tanida I, Takita Y, Hasegawa A, Ohya Y, Anraku Y. 1996. Yeast Cls2p/Csg2p localized on the endoplasmic reticulum membrane regulates a non-exchangeable intracellular Ca^{2+} pool cooperatively with calcineurin. *FEBS Lett.* 379:38–42

149. Ungermann C, Nichols BJ, Pelham HR, Wickner W. 1998. A vacuolar v-t-SNARE complex, the predominant form in vivo and on isolated vacuoles, is disassembled and activated for docking and fusion. *J. Cell Biol.* 140:61–69

150. Weissbach L, Settleman J, Kalady MF, Snijders AJ, Murthy AE, et al. 1994. Identification of a human rasGAP-related protein containing calmodulin-binding motifs. *J. Biol. Chem.* 269:20517–21

151. Welch MD. 1999. The world according to Arp: regulation of actin nucleation by the Arp2/3 complex. *Trends Cell Biol.* 9:423–27

152. Winter D, Podtelejnikov AV, Mann M, Li R. 1997. The complex containing actin-related proteins Arp2 and Arp3 is required for the motility and integrity of yeast actin patches. *Curr. Biol.* 7:519–29

153. Withee JL, Mulholland J, Jeng R, Cyert MS. 1997. An essential role of the

pheromone-dependent Ca^{2+} signal is to activate yeast calcineurin. *Mol. Biol. Cell* 8:263–77

154. Withee JL, Sen R, Cyert MS. 1998. Ion tolerance of *Saccharomyces cerevisiae* lacking the Ca^{2+}/CaM-dependent phosphatase (calcineurin) is improved by mutations in URE2 or PMA1. *Genetics* 149:865–78

155. Yap KL, Ames JB, Swindells MB, Ikura M. 1999. Diversity of conformational states and changes within the EF-hand protein superfamily. *Proteins* 37:499–507

156. Yoko-o T, Matsui Y, Yagisawa H, Nojima H, Uno I, Toh-e A. 1993. The putative phosphoinositide-specific phospholipase C gene, *PLC1*, of the yeast *Saccharomyces cerevisiae* is important for cell growth. *Proc. Natl. Acad. Sci. USA* 90:1804–8

157. Yoshino H, Izumi Y, Sakai K, Takezawa H, Matsuura I, et al. 1996. Solution X-ray scattering data show structural differences between yeast and vertebrate calmodulin: implications for structure/function. *Biochemistry* 35:2388–93

158. Zhao C, Jung US, Garrett-Engele P, Roe T, Cyert MS, Levin DE. 1998. Temperature-induced expression of yeast FKS2 is under the dual control of protein kinase C and calcineurin. *Mol. Cell. Biol.* 18:1013–22

159. Zhu H, Bilgi M, Bangham R, Hall D, Casamayor A, et al. 2001. Global analysis of protein activities using proteome chips. *Science* 293:2101–5

Annu. Rev. Genet. 2001. 35:673–745

DISSEMINATING THE GENOME: Joining, Resolving, and Separating Sister Chromatids During Mitosis and Meiosis

Kim Nasmyth

Institute of Molecular Pathology, Dr. Bohr-Gasse 7, A-1030 Vienna, Austria;
e-mail: nasmyth@nt.imp.univie.ac.at

Key Words

■ **Abstract** The separation of sister chromatids at the metaphase to anaphase transition is one of the most dramatic of all cellular events and is a crucial aspect of all sexual and asexual reproduction. The molecular basis for this process has until recently remained obscure. New research has identified proteins that hold sisters together while they are aligned on the metaphase plate. It has also shed insight into the mechanisms that dissolve sister chromatid cohesion during both mitosis and meiosis. These findings promise to provide insights into defects in chromosome segregation that occur in cancer cells and into the pathological pathways by which aneuploidy arises during meiosis.

CONTENTS

0066-4197/01/1215-0673$14.00

THE LOGIC OF MITOSIS AND ITS IMPLICATIONS FOR THE EUKARYOTIC CELL CYCLE

In all existing living organisms, most cellular constituents, including all structural and enzymatic proteins and nucleic acids, are synthesized under instructions encoded in their genomes. The latter rarely participate directly in cell function and do so largely if not exclusively by encoding the enzymes that make the cell tick. It is hard to imagine that a distinction between "enzymatic" and "hereditary" material existed in our primordial ancestors, many of whose enzymes may have been RNAs duplicated using themselves as templates. Genomes presumably arose to ensure that the progeny of cell division inherited sufficient constituents to duplicate themselves, i.e., to solve the segregation problem. By encoding most instructions in a chemically stable form (DNA) that exists in one or only a few copies and is segregated with high fidelity to opposite poles of the cell prior to cell cleavage, our early ancestors achieved the "continuity of reproduction" that, along with mutation and selection, is an integral part of the Darwinian process. Accurate but not perfect reproduction is the raw material for evolution.

Though the mechanics of genome duplication (i.e., DNA replication) are highly conserved between bacteria and eukaryotic cells, those concerned with genome segregation have little or nothing in common. The mechanics of chromosome segregation during mitotis and meiosis in eukaryotic cells have few if any antecedents in bacteria. Bacteria clearly possess molecules that promote sister chromatid resolution [DNA gyrase and Smc-like proteins (80)], but there is no indication that they possess a cytoskeletal apparatus capable of connecting to chromosomes, let alone any mechanism for holding sister chromatids together for any significant period after their generation during DNA replication. How bacteria segregate their genomes to opposite poles of the cell without either cohesion or cytoskeletal apparatus remains mysterious. A vital clue to this mystery possibly lies in the fact that their chromosomes are replicated in a bi-directional manner from a unique origin. Instead of diverging from each other, the DNA polymerases at both forks remain in a

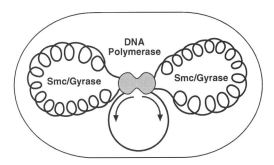

Figure 1 Bacterial chromosome segregation. The DNA polymerases associated with each end of a bi-directional replication fork remain associated and nascent chromatids that emerge are compacted through the action of DNA gyrase and Smc-like proteins.

single location in the middle of the cell (113) while nascent sequences, especially origins, rapidly move toward opposite poles (240). The implication is that bacterial chromosomes are replicated by a stationary "replisome" that uses the energy released by deoxynucleoside triphosphate hydrolysis to push nascent strands toward opposite cell poles (Figure 1).

According to this model, bacterial sister chromatids emerge from opposite sides of the replisome as loops whose distal tips contain the nascent origins. Negative supercoiling produced by DNA gyrase is thought to facilitate the packaging of each nascent chromatid into condensed nucleoid bodies via a process that is facilitated by Smc-like proteins (80). It is unclear how nascent chromatids are encouraged to move in opposite directions as they emerge from the replisome and how this is facilitated by specialized partition proteins (57, 116) that bind near origins in some but not all bacteria. What is abundantly clear, however, is that nascent origins move toward opposite poles of the cell soon after their generation and long before the bulk of the chromosome has been replicated. Thus, the bacterial equivalent of anaphase commences long before the completion of S phase and the phases of the bacterial cell cycle equivalent to the S and M phases of eukaryotic cells coexist. There is therefore no phase equivalent to G2 in the bacterial cell cycle. Though archaea possess DNA replication proteins that resemble eukaryotic ones, there is currently no evidence that their chromosomes are segregated using cohesion/cytoskeletal proteins resembling those used by eukaryotes (17).

Chromosome segregation in eukaryotes is based on a completely different principle. Both mitosis and meiosis rely on five fundamental processes (Figure 2). The first is a tubulin-based cytoskeletal apparatus (the mitotic spindle) capable of moving chromosomes around the cell by virtue of the attachment of microtubules to specialized chromosomal structures, called kinetochores (146). The second is a mechanism capable of holding together the sister chromatids produced by chromosome duplication (sister chromatid cohesion) (147). Without this, cells would not be able to ensure that their kinetochores attach to microtubules from opposite poles (bi-orientation) as opposed to the same poles (mono-orientation) (216). The third is a mechanism that detects whether sister kinetochores have indeed bi-oriented on

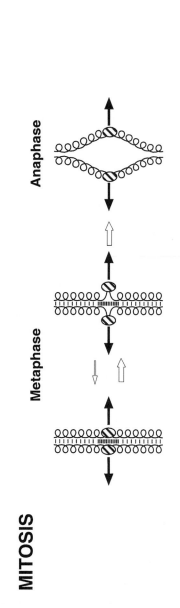

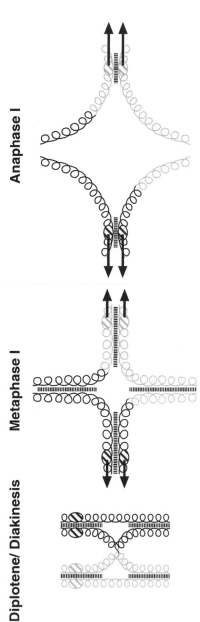

the mitotic spindle and destabilizes mono-oriented kinetochore-microtubule connections. The fourth is an apparatus that condenses chromatids (76) and partially resolves sisters from each other before the onset of chromatid separation. The fifth and last is an apparatus capable of severing once and for all the connections that hold sisters together while they are aligned on the metaphase plate, which triggers their poleward segregation during anaphase (154). Less crucial but nevertheless present in most eukaryotic cells is also a surveillance mechanism (checkpoint) (6) that blocks the destruction of sister chromatid cohesion when it detects "lagging" chromosomes, i.e., ones that have not yet bi-orientated on the mitotic spindle.

The complexities of the mitotic and meiotic processes make it easy to lose sight of their underlying logic. By some coincidence, there exists a riddle, whose origins have nothing to do with genetics, that illustrates very nicely the fundamental principle behind mitotic chromosome segregation. Two blind men enter the same department store and each orders five pairs of socks, each pair having a different color. The shop assistant is so confused by this coincidence that he/she places all ten pairs of socks (two red pairs, two blue pairs, etc.) into a single bag and sends one blind man off with all ten pairs and the other with none. By some miracle, the two blind men meet in the street as they leave the department store and discover that one has the other's socks. The question is: How do they sort out the muddle without any outside help? There is a simple solution to their problem. Socks are, of course, sold only as pairs that are joined together. As each pair of socks is removed from the bag, the two socks in each pair are pulled in opposite

---

Figure 2 Sister chromatid cohesion has a crucial role during mitosis and meiosis. Mitosis: Chromatids (*coils*) are held together by cohesin (*horizontal dashes that connect sister chromatids*), which is enriched in the vicinity of centromeres when sister kinetochores (*hatched ovals*) attach to microtubules of opposite polarity (*arrows*). During metaphase, traction exerted by microtubules on kinetochores tends to split them apart but this is resisted by cohesin concentrated in the surrounding chromatin. Anaphase is initiated by the dissolution of cohesion throughout the chromosome, which takes place due to cleavage of cohesin's Scc1 subunit by separase. Meiosis I: reciprocal crossing over between a maternal (*light gray coil*) and paternal (*black coil*) chromatid links homologous chromosomes together. These crossovers are known as chiasmata. Sister kinetochores attach to microtubules with the same polarity and as a result maternal and paternal sister kinetochore pairs are pulled toward opposite poles at the first meiotic division. During metaphase I, sister chromatid cohesion distal to the chiasmata holds homologous chromosomes together as they are pulled toward opposite poles. Chiasmata are resolved by the dissolution of cohesion along chromosome arms (due to cleavage of cohesin's Rec8 subunit by separase?), which triggers the first meiotic division. Meanwhile, cohesin in the vicinity of centromeres is protected from separase, and sister centromeres therefore remain connected until they are aligned on the meiosis II spindle.

directions by the two blind men, one of whom brandishes a pair of scissors with which he cuts the material connecting each pair. Each blind man deposits the separated socks in his own bag, which will eventually contain two red socks, two blue socks etc... once all ten pairs of socks have been separated. This then is also the logic of mitosis. The blind men pulling on each pair of socks before they are cut apart is analogous to chromosomes being aligned on the mitotic spindle, whereas their severance is analogous to the process that separates sister chromatids at the onset of anaphase. The analogies may extend even deeper, as both blind men and cells are also confronted with the problem of how they ensure that each pair of socks/chromatids are pulled in opposite directions (bi-orientation). Though we do not yet understand how cells ensure that sister kinetochores attach to spindles with opposite polarity, the analogy with the blind men and their socks would suggest that tension generated as a consequence of bi-orientation might stabilize kinetochore-microtubule connections. Indeed, there is good evidence that this is the case (157). This riddle emphasizes that mitosis is a double act that depends as much on the connections that hold sister chromatids together as it does on the cytoskeletal apparatus that actually pulls them to opposite poles of the cell during anaphase.

Sister chromatid cohesion not only makes "mitotic" chromosome segregation possible but also permits it to take place long after chromatids have been generated during S phase. The connections that hold sister chromatids together during G2 and early M phases in eukaryotic cells are, as it were, the marks by which these cells remember which chromatids are merely homologous and which are the "sister" products from the most recent round of DNA replication. Most eukaryotic cells are diploid, and homology would not be an adequate criterion for the disjunction of chromatids to opposite poles. The cohesion apparatus is sufficiently robust that in extreme cases, as in human oocytes, chromatids can still be segregated many decades after they were joined together during premeiotic DNA replication. The temporal separation of chromosome duplication and segregation in eukaryotes and hence the conventional division of their cell cycles (Figure 3) into four discrete phases (G1, S, G2, and M) is unthinkable without sister chromatid cohesion. Its absence in their bacterial cousins is presumably the reason why bacteria have little choice but to link chromosome segregation to duplication.

The temporal separation of S and M phases is in large measure responsible for much of the flexibility of eukaryotic cell cycles. It creates, for example, an opportunity to check whether the chromosome duplication process, during which DNA damage is easily generated, has been completed successfully before embarking on chromosome segregation. As one of the most powerful methods of repairing DNA damage is recombination involving an undamaged sister chromatid, it is important that cells do not separate sisters before damage has been repaired (67). Eukaryotic cells therefore possess numerous surveillance mechanisms (checkpoints) that prevent chromosome segregation if DNA is damaged or if replication is not complete. Such checkpoints either block entry into mitosis, that is, they arrest cells in G2 (211), or they prevent the onset of anaphase, that is, they arrest cells

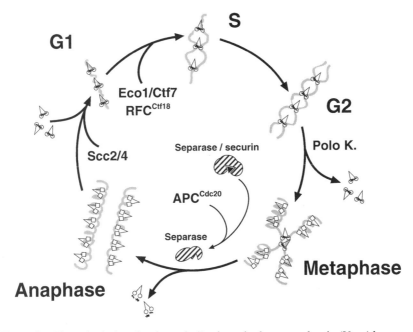

Figure 3   The cohesin/condensin cycle. During telophase, condensin (*Vs with squares at each end*) dissociates from chromatids (*gray lines*) whereas cohesin (*Vs with circles and wavy black line representing Scc1*) associates with them. The binding of cohesin to chromatin depends on a complex composed of Scc2 and Scc4. Bridges between chromatids are built with the help of Eco1/Ctf7 and RF-C(Ctf18) during passage of replication forks during S phase. Activation of mitotic protein kinases during prophase causes most cohesin to dissociate from chromosomes and triggers the binding and activity of condensin, which resolves sister chromatid arms from each other. Dissociation of cohesin depends on Polo-like kinases, whereas the association of condensin may depend on Aurora B. A small fraction of cohesin persists, largely in the vicinity of centromeres, until all chromosomes have aligned on the mitotic spindle in a bipolar fashion and congressed to the metaphase plate. This inactivates the Mad2-dependent mitotic checkpoint and triggers proteolysis of securin and B-type cyclins by the Anaphase-promoting complex (APC), which activates separase and causes cleavage of Scc1 residing at centromeres.

in metaphase (31). Neither type of checkpoint would be possible without sister chromatid cohesion.

The separation of S and M phases made possible by sister cohesion facilitates another key innovation: the packaging of eukaryotic genomes into a highly condensed and largely inactive state during the chromosome segregation process. The huge genomes of many plants and metazoa could possibly not be segregated into opposite halves of the cell at mitosis were their DNA not highly compacted.

However, the degree of compaction required is largely incompatible with transcription and presumably also DNA replication. Furthermore, the opening up of chromatin associated with transcription and replication would greatly compromise the compaction needed for segregation, except in organisms like yeast with very small genomes. By delaying chromosome segregation until well after duplication has been completed, eukaryotic cells ensure that DNA replication, transcription, and repair proceed while their chromatin is in an open or extended conformation and that mitosis only proceeds after their chromatin has been packaged into a highly condensed state. This fundamental fact has been recognized ever since chromosomes were first detected. Before the discovery of DNA replication, the eukaryotic cell cycle was divided into two phases: mitosis when chromosomes were condensed, and hence visible, and interphase when they were dispersed throughout the nucleus, and hence invisible (130). It is unclear whether the ability to compact chromosomes during mitosis was responsible for the subsequent accumulation of so much junk DNA within our genomes or whether it evolved to deal with this threat to mitosis.

Chromosome duplication and segregation are separated in eukaryotic cells functionally as well as temporarily. Origins of DNA replication and kinetochores have little or no functional connection with each other. Furthermore, kinetochores clearly function without ongoing DNA replication. This has liberated eukaryotic cells from the constraint or tyranny of needing to replicate their chromosomes using a single origin of DNA replication, as occurs in bacteria where chromosome segregation appears to be an integral part of the duplication process. The functional consequences of this emancipation have been wide ranging. Multiple origins per chromosome facilitate duplication of far larger chromosomes/genomes, which has presumably contributed to our ability to carry around a huge surfeit of junk DNA. It also permits a far more rapid execution of genome duplication than is possible in bacteria, without which embryonic cleavage divisions and hence most embryonic development in metazoa would not be possible.

The use of a cytoskeletal/cohesion process to segregate chromosomes is also responsible for another salient feature of the eukaryotic cell cycle: the invariant dependence, at least in the germline, of the reduplication of chromosomes on the segregation of sister chromatids generated during a previous round of duplication. Such linkage is not observed in bacteria, where reinitiation of chromosome duplication often commences before the previous round of duplication has been completed (41). Indeed, their ability to perform this is crucial if their doubling time is to be shorter than the time it takes to complete a single round of duplication (which is 40 min. in *Escherichia coli*). The blind men's game only works if socks are packaged as pairs. Because cohesion between sister chromatids is crucial for the attachment of sister kinetochores to microtubules pointing toward opposite poles, kinetochore reduplication prior to mitosis would create a terrible ambiguity as to which pair of chromatids should be pulled in opposite directions. Though eukaryotic cells frequently reduplicate their genomes without an intervening round of chromosome segregation (a phenomenon known as endo-duplication), they rarely if ever attempt to undergo mitosis after endo-duplication. Endo-duplication is therefore confined to somatic cells that will never contribute to the germline and have no need to regain a diploid state. The broad outlines of the mechanism by which eukaryotic cells link chromosome reduplication to the segregation process are starting to be understood (see below).

Because sister chromatid cohesion is an integral part of the chromosome segregation process and because this cohesion can only be generated during chromosome duplication, it is vital that eukaryotic cells never attempt to segregate chromosomes without their prior duplication. The dependence of M phase on S phase is therefore partly structural. Surveillance mechanisms may ensure that the attempt is rarely made in normal cells, but it could not be successful even if attempted. There is, of course, one great exception to this rule. Meiotic cells successfully undergo two rounds of chromosome segregation without any intervening round of chromosome duplication and thereby produce haploid progeny from diploid cells. They manage this extraordinary feat by using cohesion along chromosome arms for the first meiotic division and cohesion close to centromeres for the second division (26, 122), a clear case of the exception proving the rule.

THE MOLECULAR BASIS OF SISTER CHROMATID COHESION

A key question has long been whether the connections holding sister chromatids together during G2 and M phase are mediated by special proteins or simply by DNA. It has been suggested in the past, for example, that cohesion could be due either to late replication of centromeric DNA or to the intercatenation of sister DNA molecules arising from the conjunction of adjacent replication forks (152). The first of these two hypotheses was questioned by the finding that centromeres are not particularly late-replicating (33) and the second by the finding that sister chromatids from circular minichromosomes yeast are fully decatenated by Topoisomerase II despite remaining closely cohered in cells arrested in a mitotic state by agents that destabilize microtubules (106).

It is only very recently, however, that specific cohesion proteins have been identified, largely owing to the isolation of yeast mutants unable to maintain sister chromatid cohesion in cells arrested in mitosis (155). These genetic studies have now implicated six distinct classes of proteins in generating or maintaining cohesion (Table 1): (*a*), a four-subunit complex called cohesin, which possibly mediates connections between sisters (63, 118, 144, 222); (*b*) a protein called Pds5, which associates with cohesin on chromosomes (66, 160, 206, 228); (*c*) a separate complex containing at least two subunits, Scc2 and Scc4, which is necessary for cohesin's stable association with chromosomes (28); (*d*) Eco1/Ctf 7, which is neither associated with cohesin nor necessary for its association with chromosomes but is nevertheless essential for generating cohesion during DNA replication (198, 222); (*e*) a large complex related to RF-C (Replication factor C), which though not

TABLE 1   Names of cohesin and securin/separase subunits in various organisms[a]

		Cohesin subunits			Cohesin associated	Securin/separase	
S. cerevisiae	SMC1	SMC3	SCC1(MCD1) REC8	SCC3	PDS5	PDS1	ESP1
S. pombe	PSM1	PSM3	RAD21 REC8	PSC3 REC11	PDS5	CUT2	CUT1
A. nidulans		SUDA			BIMD		BIMB
S. macrospora					SPO76		
C. elegans	HIM1		COH1,2 REC8				SEP1
D. melanogaster			RAD21			PIMPLES	SSE
A. thaliana			SYN1, DIF1				
X. laevis	XSMC1	XSMC3	XRAD21	SA1, SA2	PDS5	SECURIN	SEPARASE
H. sapiens	SMC1 SMC1β	SMC3	SCC1(RAD21) REC8	SA1, SA2 STAG3	PDS5	SECURIN	SEPARASE

[a]Where appropriate, meiosis-specific variants are written below their mitotic counterparts. Though there is no need to change actual gene names, it is hoped that future authors will agree to a common nomenclature for the proteins that they encode. Thus, the *ESP1*, *cut1*, *BimB*, *SEP1*, and *SSE* genes all encode separases whereas *SMC1*, *PSM1*, and *HIM1* all encode Smc1 proteins.

essential for cohesion is necessary for its efficient generation (65, 128); and (*f*) DNA polymerase kappa, which is likewise necessary for efficient cohesion (234). Because orthologues of these proteins have been found in all fully sequenced eukaryotic genomes, it is likely that the mechanism by which sister chromatids are bound together is universal and has been inherited from the common ancestor of all eukaryotic cells. With the exception of cohesin's two Smc subunits, none of these proteins have obvious relatives in bacteria or archaea, in which sister chromatid cohesion has yet to be detected.

COHESIN: IS IT THE GLUE?

Cohesin contains four polypeptides: Smc1, Smc3, Scc1 (also known as Rad21 and Mcd1), and Scc3. Somatic cells in vertebrates express two types of Scc3 subunit, which are called SA1 and SA2 (Table 1). All four cohesin subunits form a soluble complex when not bound to chromatin (120, 206, 222) and colocalize in an interdependent manner to discrete sites on chromatin. This raises the possibility, but does not prove, that all four polypeptides always act together. Their necessity for sister chromatid cohesion was first demonstrated in the budding yeast *Saccharomyces cerevisiae*, where most if not all sister DNA sequences remain closely tethered together until metaphase. Yeast cells can be arrested at this stage of the cell cycle by inactivation of a ubiquitin protein ligase called the Anaphase-promoting complex (APC), which mediates the destruction of mitotic cyclins and anaphase inhibitory proteins called securins. APC inactivation prevents their proteolysis, and the persistence of cyclinB/Cdk1 activity prevents exit from mitosis, while securin blocks the apparatus that destroys cohesion at the metaphase to anaphase transition (see below). Mutational inactivation of any one of cohesin's subunits permits sister chromatid separation, even in the absence of APC activity. An involvement of these proteins in cohesion has since been confirmed in a variety of organisms during meiosis as well as mitosis. Thus, cohesin's depletion from extracts prepared from *Xenopus* oocytes reduces sister chromatid cohesion (118), as do mutations in *SCC1*-like genes in fission yeast (221) and *Arabidopsis thaliana* (12, 20). Furthermore, inactivation of a meiosis-specific version of Scc1 called Rec8 causes loss of sister chromatid cohesion following premeiotic DNA replication in *S. cerevisiae* (104) and *Schizosaccharomyces pombe* (148). Remarkably, inactivation of Rec8's orthologue in *Caenorhabditis elegans* by RNA interference causes the appearance of up to 24 chromatids at diakinesis/metaphase I instead of six bivalents (163).

Unlike some other proteins required for sister chromatid cohesion, cohesin subunits are required both to generate cohesion during DNA replication and to maintain cohesion between chromatids during metaphase (28, 63, 160). By turning off expression of the APC activator protein Cdc20 (see under separase regulation), it is possible to arrest either wild-type or ts *scc1* mutant yeast cells in metaphase with sister chromatid arms closely connected. Subsequently raising the temperature causes sisters to dissociate in mutant but not wild-type cells. This is consistent with the notion that cohesin may actually be part of the bridge that holds sisters together as they come under tension from the mitotic spindle.

It is, however, almost impossible to demonstrate a direct role for a protein in a given process merely by analyzing the phenotypic consequences of its inactivation. For this reason, a crucial breakthrough in this field was the observation that Scc1 is tightly associated with yeast chromatin during metaphase but suddenly disappears at the onset of anaphase (144). Subsequent analysis of this phenomenon led to the discovery that Scc1 is released from yeast chromatin due to proteolytic cleavage by a cysteine endopeptidase called separase. Furthermore, Scc1's cleavage is both necessary and sufficient to trigger anaphase (224, 226). Thus, not only genetics but also physiology points to Scc1 being the real McCoy.

If cohesin does indeed connect sisters, then it should be found at sites where sister chromatids are tightly connected. This does indeed appear to be the case, albeit only at a low level of resolution. Thus, Scc1 is concentrated between chromatids in the vicinity of centromeres in human tissue culture cells during metaphase and is clearly less abundant along chromosome arms, which are less tightly connected (83, 118, 229). Likewise, meiosis-specific forms of Scc1, Smc1, and Scc3 called Rec8, Smc1β, and STAG3, respectively, are all found between chromatids during diakinesis and metaphase I, that is, during the periods of meiosis when chiasmata and chromosome arm cohesion are vital for holding homologous chromosomes together on the metaphase I spindle (C. Heyting, personal communication; 104, 171, 176, 238). Most impressive of all, Rec8 disappears from chromosome arms at the onset of anaphase I, as cohesion is lost from these chromosomal regions but persists in the vicinity of centromeres until the onset of anaphase II, both in yeast (*S. cerevisiae* and *S. pombe*) and in mammals. Despite these impressive cytological observations linking cohesin's chromosomal distribution with cohesion itself, the direct colocalization of cohesin with bridges connecting sisters at a molecular level remains one of the holy grails in this field.

This issue has also been addressed by the identification, using chromatin immunoprecipitation, of sites along yeast chromosomes to which cohesin subunits are bound. In *S. cerevisiae*, cohesin is found at centromeres and in their vicinity and at specific loci, every 5–10 kb, along chromosome arms (22, 138, 215). Both types of site are sufficient to recruit cohesin to regions of the chromosome that normally lack cohesin, which raises the question whether they can also confer cohesion between sister chromatids. Though cohesin at centromeres and in their immediate vicinity clearly helps to promote kinetochore bi-orientation (216), possibly by providing cohesion, it is surprisingly incapable of resisting the movement of sister kinetochores toward opposite poles during metaphase (60, 72, 216). Centromeres, presumably due to their recruitment of cohesin, are nevertheless capable of conferring cohesion between sisters in the presence of drugs that destabilize microtubules (138). The implication is that the cohesin present at budding yeast centromeres is incapable of preventing the traction of sister sequences toward opposite poles once kinetochores have bi-oriented on the spindle. This could be taken to mean either that cohesin does not in fact confer cohesion between chromatids (37) or that the cohesion conferred by cohesin is insufficient to counteract the splitting force exerted in the immediate vicinity of bi-oriented kinetochores. Time lapse microscopy shows that sister centromeres but not arm sequences separate soon after formation of bipolar spindles in *S. cerevisiae* and that they only occasionally rejoin before the onset of anaphase proper (216). Such "breathing" of sister sequences in the vicinity of centromeres during metaphase is also seen in protozoa, insect cells, and even in mammalian cells and may therefore be a quite general phenomenon (139, 190, 220). In yeast, this precocious sister chromatid separation during metaphase extends for about 10 kb around centromeres and is accompanied by considerable stretching of the chromatin (72, 216), which is presumably unraveled down to nucleosomes (164). Because partial inactivation of Scc1 enlarges the interval that can be separated by bi-oriented kinetochores (T. Tanaka, personal communication), the splitting process is presumably halted by cohesin bound to flanking arm sequences. It is tempting to speculate that tension exerted along the chromsosome increases cohesion within the arms, as would be the case if one would try to separate two intertwined rubber bands. It is otherwise unclear how chromosomes prevent themselves being ripped apart once the first opening has been generated. According to this model, it is cohesion within sequences flanking kinetochores and not at kinetochores themselves that bears the brunt of the load in resisting the complete separation of sister chromatids during metaphase. It is therefore interesting that in *S. pombe*, whose centromeres are larger and more complex than those of *S. cerevisiae*, Scc1 (Rad21) is found not in the inner centromere region to which kinetochore proteins bind and at which microtubules presumably exert their action but rather in the outer centromere regions that flank these (239).

It is possibly because rather extensive regions of cohesion are necessary to oppose the spindle that insertion of individual cohesin association sites only modestly delays centromere splitting, even when present as tandem arrays (215). In conclusion, the study of cohesion association sites, though consistent with the notion that cohesin provides the connections between sister chromatids, has yet to provide a truly conclusive experiment that settles this issue once and for all.

The case for cohesin being the glue that holds sisters together is clearly a strong one: it is the only protein clearly required for cohesion that is at the right places at the right times. Furthermore, proteolytic cleavage of its scissile Scc1 subunit is both necessary and sufficient for triggering chromatid separation. How then might cohesin produce bridges between sister chromatids? What is cohesin's structure and does it alone possess activities consistent with building bridges between chromatids? Studies that address these issues are still in their infancy.

PROPERTIES OF COHESIN SUBUNITS

Cohesin's Smc1 and Smc3 subunits are both members of the SMC (structural maintenance of chromosomes) family of proteins, which are common to bacteria, archaea, and eukaryotes and have roles in chromosome condensation, sister chromatid cohesion, and DNA repair (74). All SMC proteins share five conserved domains: three globular domains separated by two long stretches of coiled-coil interrupted by a hinge region. Both X-ray crystallography (121) and electron microscopy (140) suggest that bacterial and archaeal Smc proteins are homodimers whose coiled coils are antiparallel and bring together the globular N- and C-terminal domains. These contain Walker A and B motifs, respectively, whose appropriate alignment is thought to create an active ATPase of the ABC type frequently found in membrane transporters. The Walker B motif constitutes a nucleotide-binding pocket. SMC proteins therefore form V-shaped molecules that can open or close by virtue of their flexible hinge region. Closure would bring two N- and C-terminal domain pairs together, which by bringing Walker A and B motifs into juxtaposition could modulate ATPase activity as well as create a DNA binding domain (82) (Figure 4*a, b*).

The coiled-coils of SMC proteins could either be intramolecular, in which case the N- and C-terminal domains from the same molecule would associate with each other and connections between subunits would be confined to the hinge region, or intermolecular, in which case the N-terminal domain of one Smc molecule would associate with the C-terminal domain of its partner in the complex (Figure 4*a*). According to the first geometry, the two halves of the V-shaped complex would merely be connected by homotypic interactions between the hinge regions of each Smc subunit, whereas according to the second geometry, the two halves would be connected by two continuous polypeptide chains that run in antiparallel manner from one end of the V to the other. Because it contains equal amounts of two different Smc proteins (Smc1 and Smc3), it is suspected that cohesin contains

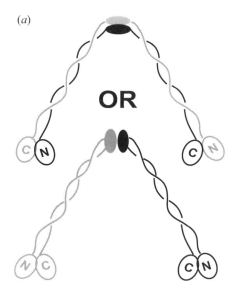

(a)

OR

Figure 4   Potential geometries of cohesin's Smc proteins. *A*, cohesin contains an Smc1/Smc3 heterodimer, whose two long stretches of coiled coil are either intermolecular (*above*) or intramolecular (*below*). In the first case, both Smc1 and Smc3 molecules stretch from one end of the V to the other end, whereas in the second case, Smc1 constitutes the left branch and Smc3 the right one. *B*, one possible mechanism by which the Smc1/Smc3 heterodimer might cooperate with Scc1 (*wavy gray lines* containing separase cleavage sites marked by an arrow) to generate bridges between sister chromatids (*thick dark lines*). *Top*: cohesin in an open configuration before it associates with chromatin. *Middle right*: cohesin clasps a DNA duplex or chromatin fiber. The Scc1 subunit, presumably with the aid of Scc3 (SA1, SA2) and Pds5, locks the chromatin's embrace by the Smc1/Smc3 heterodimer. *Bottom*: both sister chromatids are embraced by cohesin after passage of a replication fork through a "closed" cohesin complex. *Middle left*: The cohesin complex can be opened either by phosphorylation of Scc1, Scc3-SA1/SA2, or Pds5 as occurs during prometaphase or as shown by cleavage of the Scc1 subunit. The crystal structures of SMC head domains suggest that left and right Smc1/Smc3 head domains can bind ATP but that hydrolysis might only occur when both head domains are brought together as in the closed configuration. ATP binding and hydrolysis might therefore regulate the opening and closing of cohesin complexes. Closure around a chromatin fiber might also depend on the Scc2/4 complex. Condensin and DNA repair proteins like Rad50 might operate using a similar principle. For example, condensin might close around adjacent coils of the same chromatin fiber.

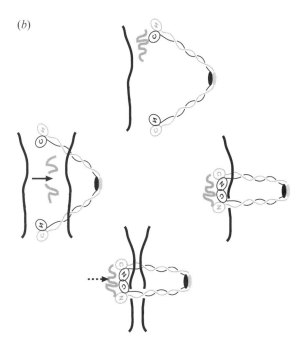

Figure 4   (*Continued*)

an Smc1/Smc3 heterodimer that would be pseudo-symmetrical if its coiled-coils were intermolecular but asymmetric if they were intramolecular. A crucial question is whether cohesion is mediated by Smc1/Smc3 heterodimers in an open or closed configuration. An open heterodimer would bridge the gap between sisters (76, 154), whereas a closed heterodimer could form a ring around them (see Figure 4*b*).

The sequence of cohesin's Scc1 subunit has thus far shed less insight into its structure. Its N- and C-terminal domains are conserved but its central domain, which contains its separase cleavage sites, is much less so and may be rather unstructured. Scc1 must nevertheless be the lynchpin of the cohesin complex because its cleavage by separase causes the sudden dissolution of cohesion at the metaphase to anaphase transition (226).

Though not strictly a subunit, because it is less stably associated with the soluble form of the complex, the Heat repeat containing protein Pds5 (156, 160) clearly has an intimate connection with cohesin. Like Scc1, it is essential for maintaining sister chromatid cohesion during mitosis in *S. cerevisiae* (66, 160) and during meiosis in Sordaria (228). Furthermore, it associates with the same chromosomal sites as cohesin subunits and is released from chromosomes at the metaphase to anaphase transition due to Scc1's proteolytic cleavage. Pds5 presumably interacts directly with cohesin because it can sometimes be coprecipitated with cohesin subunits (206). Pds5 might not be as crucial to cohesion as other cohesin subunits because it is not an essential gene in the *S. pombe*. Though *pds5* mutants are viable in *S. pombe*, they are defective in maintaining cohesion between sister chromatids during a prolonged G2 arrest (K. Tanaka, personal communication).

COHESIN IS RELATED TO CONDENSIN

Most if not all eukaryotic genomes contain at least two other SMC proteins, Smc2 and Smc4, which are more closely related to Smc1 and Smc3, respectively, than they are to any other members of this family. Remarkably, Smc2 and Smc4 also form a heterodimer that is part of a separate multisubunit complex, called condensin (75), which has an important role in the condensation and resolution of sister chromatids between prophase and metaphase (78). The implication is that unlike bacteria, whose SMC proteins form homodimers, a common ancestor of eukaryotic cells possessed an Smc heterodimer whose duplication led to the evolution of cohesin- and condensin-specific Smc heterodimers. Condensin contains three other subunits: barren/Xcap-H, X-cap-D2, and X-cap-G. It is striking that condensin's D2 and G subunits are, like Scc2 and Pds5, composed of Heat repeats. This raises the possibility that some of these proteins are descended from an ancestral complex containing not only an Smc heterodimer but also Heat repeat–containing proteins.

ACTIVITIES ASSOCIATED WITH
COHESIN AND CONDENSIN

Purified condensin and cohesin have both been associated with activities in vitro that might be relevant to their functions in vivo. While condensin is capable of imparting global positive writhe to circular DNA in an ATP-dependent manner (96), cohesin is capable of aggregating DNA molecules in a manner that facilitates intermolecular catenation in the presence of Topo II (119). Unlike the positive writhing induced by condensin, the DNA aggregation produced by cohesin in vitro is ATP-independent, and it is therefore unclear whether it requires Smc1 and Smc3 or is merely a property of its Scc1 subunit. Cohesin's ability to aggregate separate DNA molecules could clearly be relevant to its ability to hold sister chromatids together and is consistent with the notion that this complex does indeed mediate the connections between sisters. Nevertheless, further studies will be needed to demonstrate the physiological relevance of cohesin's aggregation activity, especially as it requires a very large excess of cohesin to DNA.

LOADING COHESIN ONTO CHROMOSOMES

Among the most frequent *S. cerevisiae* mutants with cohesion defects are those with mutations in the *SCC2* gene (144, 222). Mutations in its *S. pombe* orthologue *mis4* also cause cohesion defects (52), whereas mutation of the related Nipped B protein in *Drosophila. melanogaster* causes defects in long-range enhancer promoter interactions (181), and mutation of its orthologue in Ascobolus, called Rad9, causes defects in DNA repair and in meiosis (188). Scc2 is neither stoichiometrically associated with cohesin nor does it appear to associate stably with the same sites on chromatin in vivo (28). In *S. cerevisiae*, Scc2 is stably bound to a 78-kd protein called Scc4, which is also required for sister chromatid cohesion. Inactivation of Scc2 or Scc4 in *S. cerevisiae* (28) or Mis4 in *S. pombe* (52) greatly reduces the amount of cohesin associated with chromosomes, which implies that the Scc2/Scc4 complex is crucial for some as yet ill-defined aspect of cohesin's function. Possibly it catalyzes the formation of complexes between cohesin and chromatin.

Most evidence suggests that cohesin can be loaded onto chromosomes at all stages of the cell cycle apart from mitosis in organisms whose chromosomes are extensively condensed at this stage. Thus, in mammalian cells, cohesin is found stably associated with chromatin throughout interphase; it dissociates from chromosomes during prophase, reassociates during telophase, and remains on chromosomes until cells reenter mitosis (118, 206). In *S. pombe*, where mitotic chromosome condensation is rather modest, Scc1 (Rad21) is associated with chromosomes at all cell cycle stages apart from anaphase (221). In *S. cerevisiae*, where the bulk of cohesin also remains tightly associated with chromosomes until anaphase, cohesin is absent from yeast chromosomes for much of G1 (144). This is due both to a lack of *SCC1* transcription and continued proteolysis by separase during this stage of the cell cycle, and Scc1 readily binds to yeast chromosomes during G1 when ectopically expressed in separase mutants (224). Scc1 can even bind stably to yeast chromosomes when expressed in G2, though it cannot promote sister chromatid cohesion under these circumstances (225). The dependence of cohesin's association with chromosomes on a separate Scc2/Scc4 complex suggests that the structures formed between cohesin and chromatin might have a very special geometry, even when this process occurs outside S phase and does not involve the formation of sister chromatid cohesion. Cohesin is associated with chromosomes in quiescent as well as proliferating mammalian cells, which raises the possibility that it might be a key determinant of chromosome structure during G1 as well as during G2 (206).

ESTABLISHING COHESION DURING DNA REPLICATION

In yeast, whose chromosomes are not visible during mitosis by conventional cytological techniques, sister chromatid cohesion has been measured either using FISH (62) or by visualizing the location of Tet (144) or Lac (201) repressor proteins fused to GFP, which are bound to tandem operator arrays inserted in various locations within the genome. As observed by either method, most if not all sister sequences remain tightly associated, at least at the resolution of light microscopy, from their production during DNA replication until their separation at the onset of anaphase. The only exception to this rule is the precocious separation during metaphase of sequences within 5 kb of centromeres, which occurs soon after sister kinetochores bi-orient on the mitotic spindle (60, 72, 216). One of the implications of these findings is that sites of sister chromatid cohesion are rather frequent along yeast chromosomes, which fits with the observation that sites associated with cohesin are found every 5 to 10 kb. The situation is more complicated in animal cells where most sequences along chromosome arms can be resolved using FISH in G2 or early M phase cells (189), which suggests that cohesion sites might be much rarer than in yeast. Sister centromere sequences, on the other hand, remain closer to each other, at least until chromosomes align on the spindle during metaphase. There are indications, however, that some sister sequences may be very closely connected even in animal cells, for a brief period after their replication. By measuring whether cells have two or four signals, FISH has been used extensively with a view to determining replication timing (100). The problem with this approach is that "two signals" could arise either because sequences have not yet replicated or because sister sequences are so closely connected that they cannot be distinguished, as is the case in yeast. Indeed, a recent study using BrDU labeling to determine replication timing found that some but not all sequences that were "late" in producing four FISH spots were in fact early replicating (Azuara, Brown & M. Fisher, personal communication). The implication is that some but not all sequences remain closely connected with their sisters for an appreciable period after replication.

Merely looking at the association of sister sequences does not, however, address whether cohesin has established links between them or whether they are connected in a manner capable of resisting the mitotic spindle. It is perfectly conceivable, for example, that sister sequences remain close together soon after their replication due to the intertwining of sister DNA molecules (152) and that cohesin only produces proteinaceous links between chromatids after replication has been completed. An alternative approach has therefore been to address when cohesin is required during the cell cycle. Two studies, one varying the timing of *SCC1* expression during the mitotic cell cycle (225) and a second varying the timing of *REC8* expression during meiosis (239), found that neither gene can fulfil its function when expressed after DNA replication. Cohesion cannot be established between sister chromatids by Scc1 protein produced immediately after replication has been completed even though the protein is fully capable of stably associating with chromosomes under these circumstances. The simplest explanation is that cohesin can only build connections between sisters as they emerge from replication forks. One mechanism by which this could occur is shown in Figure 4*b*. However, these experiments do not exclude the possibility, albeit an improbable one, that cohesin is simply inactive when produced after S phase.

The notion that cohesive structures built by cohesin can only be produced during S phase is supported by the phenotype of *eco1/ctf7* mutants. *ECO1* is an essential gene that is crucial for establishing cohesion between chromatids during S phase (198, 222). The Eco1 protein is neither part of the cohesin complex nor does it stably colocalize with cohesin on chromosomes. *eco1/ctf7* mutants cannot establish cohesion between chromatids during S phase even though cohesin associates in normal amounts with chromosomes. Furthermore, unlike ts cohesin mutants, shifting ts *eco1* mutants to the restrictive temperature only after they have already undergone DNA replication at the permissive temperature does not destroy sister chromatid cohesion. This implies that Eco1/Ctf7 may be required to establish cohesion during DNA replication but not to maintain cohesion during G2 or M phases. The partial suppression of *ctf7* mutants by increased expression of PCNA (198) is also consistent with Eco1 acting during DNA replication. The phenotype of *eco1/ctf7* mutants implies that the mere presence of cohesin on chromatin while it is being replicated is insufficient to establish cohesion. Though there is as yet no evidence that Eco1/Ctf7 acts directly on cohesin, Eco1/Ctf7 presumably facilitates whatever special function cohesin performs soon after the passage of replication forks. Orthologues of Eco1 are found in most if not all eukaryotes, and all contain a C2H2 zinc finger as well as another conserved domain of unknown function. Eco1's homologue in *S. pombe*, called Eso1, is also an essential gene required for sister chromatid cohesion (213). Interestingly, the lethality of *eso1* mutants is suppressed by inactivation of Pds5, which is not an essential protein in *S. pombe* (K. Tanaka, personal communication). This remarkable finding implies that Eco1 cannot be a fundamental component of the cohesion system. It also suggests that Pds5 may have two roles: one that inhibits the establishment of cohesion during DNA replication and another that helps to maintain cohesion during G2. Eso1 is presumably required merely to counteract Pds5's inhibitory function during S phase.

The connection between sister chromatid cohesion and DNA replication has recently been further strengthened by the discoveries that a variant version of replication factor C (RF-C) (65, 128) and a new type of DNA polymerase encoded by the *TRF4* gene (polymerase kappa) (234) have a role in generating cohesion. RF-C is a multisubunit complex essential for loading the ring-shaped DNA polymerase clamp PCNA onto DNA and therefore has a key role in switching DNA polymerases at the replication fork. The genes for all five of its subunits, Rfc1-5, are essential. Yeast contains at least two variants of the RFC complex: one containing the checkpoint protein Rad24 and a second in which Rfc1 is replaced by an Rfc1-related protein called Ctf18 [RF-C(Ctf18)]. The latter contains two further subunits, called Ctf8 and Dcc1, not found in RFC. Deletion of *CTF8*, *CTF18*, or *DCC1* is not lethal but causes high rates of chromosome loss, an accumulation of cells in metaphase due to activation of the mitotic checkpoint, and a partial loss of sister chromatid cohesion. The loss of cohesion in these mutants could be due either to an indirect effect of interfering with DNA replication or to a direct involvement of this alternative RF-C complex in the generation of sister chromatid cohesion. If indeed RF-C(Ctf18) is directly involved, then it is curious that *CTF18* is not an essential gene. One explanation is that cohesion can be generated by two different pathways, one of which does not require RF-C(Ctf18). In this case, it is conceivable that RF-C itself or yet some other variant thereof also performs the same role, albeit inefficiently. If RF-C-like complexes are genuinely involved in generating sister chromatid cohesion, then a crucial question is whether their cohesion function involves the loading of PCNA and hence polymerase switching or the loading of some other complex such as cohesin.

A role of RF-C(Ctf18) in generating cohesion via some form of polymerase switching raises the issue of whether this switching concerns classes of DNA polymerase already known to be needed for DNA replication in eukaryotes or switching of a new class of polymerase, such as DNA polymerase kappa. Deletion of one (*TRF4*) of two genes encoding this newly discovered polymerase causes high rates of chromosome loss and cohesion defects (234), whereas inactivation of both (*TRF4* and *TRF5*) is lethal and has been reported to prevent DNA replication. However, the lack of budding as well as DNA replication in the double *trf4 trf5* mutant cells raises the possibility that the replication defect might not be genuine but instead be caused by some general cell cycle arrest.

In summary, several independent lines of evidence suggest that cohesin builds special structures during the passage of replication forks that are distinct from those found on unreplicated chromosomes. These structures are fully dependent on Eco1/Ctf7 and partially dependent on a new form of RF-C. The challenge for the future is to determine the physical form of these structures and how they are built at replication forks. It is increasingly clear that the establishment of cohesion is an integral part of the DNA replication process in eukaryotic cells. If we assume that cohesin is indeed part of the bridge linking sisters, then it will be important in the near future to find direct connections between cohesin and the functions of proteins like Eco1 and RF-C(Ctf18). It will also be important to establish the connection between cohesin and the SWI1 protein from *Arabidopsis*, which is also required for the establishment of cohesion (141).

THE SISTER CHROMATID SEPARATING PROCESS

Sister chromatids are separated from each other in two steps. The first occurs during prophase/prometaphase, when the bulk of sister sequences along chromosome arms are resolved from each other to generate parallel side by side chromatids. The partial immunity to this resolution process of sequences surrounding centromeres is responsible for the central contriction of metaphase chromosomes (207). During undisturbed mitoses, residual connections between chromatid arms hold them together along their entire length and not just at centromeres until the metaphase to anaphase transition. However, the chromatid arm resolution process continues when cells are prevented from embarking on anaphase by surveillance mechanisms that respond to spindle damage. As a result, sister chromatids lose all connection along their arms in cells blocked in metaphase by spindle poisons (155). The classic image of metaphase chromosomes in which chromatids are connected only at a central constriction is therefore largely an artifact of having treated cells with spindle poisons before spreading their chromosomes. It is nevertheless an artifact that emphasizes what are real and important differences between the processes by which centromeric and arm sequences are separated (177). Those connections between sisters that are resistant to the "prophase" resolution pathway are therefore responsible for holding sisters together while they are aligned on the mitotic spindle during metaphase. The second step in the sister separation process involves the destruction of these residual connections, which only occurs when sister chromatids disjoin at the metaphase to anaphase transition.

To those aware of the difficulties of disentangling ropes, the apparent ease with which eukaryotic cells separate their chromatids during mitosis is nothing short of miraculous. It has long been appreciated that decatenation by Topoisomerase II has an important role both in chromatid resolution during prophase (56) and in sister separation at anaphase (40, 44). To this must now be added the processes by which cohesin dissociates and condensin associates with chromosomes during prophase and cleavage of Scc1 by separase at the onset of anaphase.

SEPARATING SISTERS AT THE METAPHASE TO ANAPHASE TRANSITION

A crucial aspect of the mechanism by which sister chromatids are separated during anaphase was discovered by studying what causes the sudden disappearance of cohesin's Scc1 subunit from yeast chromosomes at the metaphase to anaphase transition. The bulk of Scc1 associated with yeast chromatin during G2 remains tightly associated with chromosomes throughout metaphase but is released at the onset of anaphase due to cleavage at two different sites by a novel cysteine protease called separase (224, 226), which is the product of the *ESP1* gene in *S. cerevisiae* (133) and *CUT1* in *S. pombe* (48). Separase cleavage sites have since been characterized in Scc1's orthologues in *S. pombe* (221) and humans (70), in its meiotic version Rec8 in *S. cerevisiae* (26), and in Slk19 (128), a protein associated with kinetochores and anaphase spindles in *S. cerevisiae* (249). All sites contain arginine at the P1 position, glutamic acid (or more rarely aspartic acid) at the P4 position, and in many cases serines or acidic residues at the P6 position (Table 2). Single mutations that replace the P1 arginine by aspartic or glutamic acid usually abolish cleavage at that site, but replacement of both P1 and P4 amino acids is needed to abolish cleavage at Rec8's second cleavage site in *S. cerevisiae*. In the yeast *S. cerevisiae*, either separase inactivation (using ts *esp1* mutants) or expression of mutant Scc1 proteins that cannot be cleaved at either site (but not mutants lacking only a single site) prevents both Scc1's disappearance from chromosomes and the separation of sister chromatids (224). It is even possible to trigger anaphase in metaphase arrested cells by induction of the foreign TEV protease in cells that express a version of Scc1 in which one of its separase sites has been replaced by that for TEV (226). Cleavage of Scc1 by separase is therefore both necessary and sufficient to trigger anaphase in yeast. Noncleavable versions of Rec8 also prevent chromosome segregation at meiosis I

TABLE 2   A list of known separase cleavage sites

Substrate	Site position	Sequence	Reference
Saccharomyces cerevisiae			
Scc1	268	DNSVEQGRRLG	224
Scc1	180	DTSLEVGRRFS	224
Rec8	431	FSSVERGRKRA	26
Rec8	453	TRSHEYGRKSF	26
Slk19	77	DRSIDYGRSSA	F. Uhlmann, personal communication
Schizosaccharomyces pombe			
Rad21	179	QLSIEAGRNAQ	221
Rad21	231	QISIEVGRDAP	221
Rec8	384	TSEVEVGRDVQ	Y. Watanabe, personal communication
Drosophila melanogaster			
Three rows	865	LQLVEPIRKQQ	C. Lehner, unpublished
Xenopus laevis			
Xrad21/Scc1	172	MDDREMMREGS	I. Waizenegger, personal communication
Separase	n.d.	DVSIEELRGSD	M. Kirschner, personal communication
Separase	n.d.	VTECEVLRRDA	M. Kirschner, personal communication
Homo sapiens			
Scc1	172	MDDREIMREGS	I. Waizenegger, personal communication
Scc1	450	PIIEEPSRLQE	I. Waizenegger, personal communication
Separase	1181	KMSFEILRGSD	M. Kirschner, personal communication
Separase	1210	SGEWELLRLDS	M. Kirschner, personal communication

(see below). Crucially, this phenotype can be suppressed by the creation of a novel cleavage site elsewhere in the protein (S.B. Buonomo, personal communication), implying that the noncleavable Rec8 is functional in all respects other than its cleavability.

These observations demonstrate what has long been suspected from biophysical studies that microtubules are already straining to pull sister chromatids to opposite poles during metaphase and that they are merely prevented from doing so by cohesion holding sisters together (179). They also suggest that Scc1 is indeed part of the bridge, if not the bridge itself, that holds sisters together and that activation of separase is the long sought after anaphase trigger. The C-terminal fragments of

Scc1 and Rec8 produced by separase in yeast contain either arginine or lysine at their N termini and are rapidly degraded by the N-end rule ubiquitin protein ligase Ubr1 (173). Indeed, their destruction is important for high-fidelity chromosome transmission, possibly because they bind to other cohesin subunits such as Smc1 and form inactive complexes.

Two key questions stem from these studies. First, is cleavage of Scc1 by separase a crucial aspect of anaphase in all eukaryotic organisms? Second, is separase solely responsible for triggering anaphase and if so does it cleave other proteins besides Scc1?

IS CLEAVAGE OF Scc1 UNIVERSAL?

The investigation of Scc1 cleavage by separase in organisms other than yeast has been greatly complicated by the fact that sometimes only a small fraction of the cell's Scc1 protein is cleaved. Furthermore, the rapid degradation of cleavage products (173) means that they can only be readily detected in cultures whose passage through mitosis is highly synchronized. In *S. pombe*, cleavage products of Scc1's orthologue Rad21 are detected at the metaphase to anaphase transition. Furthermore, expression of a Rad21 protein that cannot be cleaved at either site blocks sister chromatid separation (221). The discovery that most cohesin dissociates from chromosomes during prophase in the absence of cleavage in vertebrates raised important doubts whether cohesin could be the glue that holds sisters together during metaphase, let alone be the target for any anaphase trigger (118). However, subsequent studies, both in *D. melanogaster* (235) and in human tissue culture (229) cells, have shown that some cohesin (5% or less of the total pool) remains associated with metaphase chromosomes, in particular in the vicinity of their centromeres, but disappears from chromosomes at the metaphase to anaphase transition (229). Crucially, a similar fraction of Scc1 is cleaved at the metaphase to anaphase transition in Hela cells, which suggests that most if not all the Scc1 that remains associated with chromosomes until metaphase is cleaved at the onset of anaphase (229). Crucially, not only is human Scc1 cleaved by separase in vitro at two sites that resemble those in yeast Scc1 but also expression of mutant Scc1 protein that can be cleaved at neither site interferes with chromatid segregation at anaphase (70).

Though Scc1 cleavage clearly needs to be investigated in a wider variety of organisms, the available data are consistent with the notion that Scc1 cleavage by separase might be a universal aspect of anaphase in eukaryotic cells. If so, then all fully sequenced eukaryotic genomes should encode separase-like proteins containing toward their C termini the amino acid motifs corresponding to the protease's active site. This is indeed the case. They also encode one or more Scc1-like proteins. It is harder in this case to determine merely by sequence analysis whether these proteins contain separase cleavage sites, because amino acids in only three positions are conserved within known sites.

The yeast and human separases are large (160–180 Kd) proteins whose C-terminal domains contain their catalytic residues. These domains are conserved and found at the C termini of all separase orthologues found in GenBank. At the heart of this domain are highly conserved histidine and cysteine residues that are thought to constitute the protease's catalytic dyad. Their mutation abolishes activity (226). The pattern of amino acids immediately surrounding this dyad resembles those of the CD clan of cysteine proteases (27), which includes gingipain, a bacterial protein implicated in tooth decay (46), legumain involved in class II antigen presentation (124), and caspases whose activation triggers programmed cell death in metazoa (45). In the case of gingipain and caspase, whose crystal structures have been determined, the histidine and cysteine residues are held in juxtaposition by a pair of hydrophobic beta sheets, which are predicted to exist in equivalent positions within all separases. Like caspases, acyloxy methyl ketone derivatives of cleavage site hexapeptides act as specific inhibitors of the yeast and human enzymes, at least in vitro (226). The conserved C-terminal domain of all separase orthologues contains extensive amino acid motifs that are unique to separases. Their common ancestry with caspases must therefore predate the common ancestor of eukaryotic cells. Though fungal genomes do not encode caspases, they do all encode a caspase-like protease, called metacaspase, whose function is unknown but is far more similar to caspase than it is to separase (227). It is nevertheless remarkable that the birth and death of eukaryotic cells, two of the most irreversible events in biology, are triggered by related proteases.

DO OTHER PROTEINS NEEDED FOR SISTER SEPARATION FUNCTION VIA SEPARASE?

In the absence of readily fractionable in vitro systems for studying sister chromatid separation, the isolation and characterization of mutants in yeast and flies has been the only avenue by which new players in this process have been identified. Besides separase, genetic studies in *S. cerevisiae*, *S. pombe*, and *D. melanogaster* have identified at least two other types of protein necessary for sister chromatid separation but not for other aspects of cell cycle progression: a protein called Threerows, which has thus far only been found in flies (168, 202), and a class of proteins found in a wide variety of organisms called securins (247). If proteolytic cleavage by separase is indeed the mechanism by which sisters are separated in eukaryotic cells, then Threerows and securins should have some connection with separase. Is this the case?

Securins from *S. cerevisiae* (Pds1) (29), from *S. pombe* (Cut2) (48), from *D. melanogaster* (pimples) (253), and *vertebrates* (PTTG) (253) all bind tightly to separase. Securins are potent inhibitors of separase activity (224) and their proteolysis by the Anaphase-promoting complex shortly before the metaphase to anaphase transition is necessary for sister chromatid separation (see below) (32, 50, 112, 253, 254). They have a key role in ensuring that separase remains inactive until chromatids are fully aligned on the metaphase spindle. Besides this crucial inhibitory function, securins also have an important role in promoting separase activity. This function is important but not essential in human tissue culture cells (89) and in *S. cerevisiae* (245) but is essential for sister chromatid separation in *S. pombe* (48) and in *D. melanogaster* (202). Whether securins primarily promote separase activity in vivo by targeting the protease to correct cellular locations (29, 90, 109) or more directly by facilitating its adoption of an active conformation (89) is presently unclear. What is clear is that the sister separation defect of cells lacking securins is possibly due to a lack of separase activity.

What about Threerows? Though only found so far in *D. melanogaster*, it too is crucial for sister chromatid separation. Thus, both *Threerows* and *Pimples* mutant embryos accumulate cells in which cell cycle progression has taken place in the absence of sister chromatid separation and which therefore contain four or more (up to 32) chromatids held together at their centromeres (34, 168). Remarkably, Threerows also binds tightly to separase and may be a key regulator of its protease activity in flies (C. F. Lehner, personal communication). The separase protein in flies is about half the size of its orthologues in yeast and vertebrates, and it is possible that the missing N-terminal half of the protein is encoded by a separate polypeptide encoded by Threerows. Though genetic analyses have clearly identified a host of other proteins needed for resolving sisters once anaphase has initiated (for example, Topo II and condensin), separase, securins, and Threerows are the only proteins thus far implicated in initiating the sister separation process. Indeed, the BimB protein, which is required for the completion of mitosis but not for rereplication of DNA in *Aspergillus nidulans*, encodes a separase homologue (127). In conclusion, therefore, several independent genetic investigations of genes required for separating sister chromatids once they have been aligned on the mitotic spindle have all focused on a singe entity: separase.

DOES SEPARASE HAVE OTHER TARGETS BESIDES Scc1?

Though cleavage of Scc1 by the TEV protease is sufficient to trigger the segregation of sisters to opposite spindle poles in yeast, this by no means excludes the possibility that cleavage of other proteins might also facilitate anaphase chromosome movement. Indeed, several lines of evidence suggest that separase also targets proteins concerned with stabilizing anaphase spindles. In both *S. cerevisiae* and *S. pombe*, a sizeable fraction of separase colocalizes with mitotic spindles in a manner that depends on securins and on its own conserved C-terminal protease domain (29, 90, 109). Such an association has not, however, been seen in human cells, where most separase protein appears to be distributed throughout the cytoplasm during mitosis (J. Peters, personal communication). Nevertheless, the association of separase with spindles in yeast might be of functional significance, because the elongated spindles produced by cells induced to undergo anaphase by the TEV protease (cleaving Scc1) are much less stable than those triggered by overexpression of separase itself (226). This implies that yeast separase has at least two functions during anaphase. By cleaving Scc1, it permits microtubules attached to kinetochores to pull sister chromatids toward opposite spindle poles (known as anaphase A) and at the same time allows the poles themselves to be driven further apart by the elongation, interaction, and sliding apart of microtubules that are not associated with kinetochores (known as anaphase B). Cleavage of some protein other than Scc1 might be required for stabilizing spindle interactions in the midzone where spindles from opposite poles overlap. It might alternatively be required for a stabilization of microtubules necessary for their rapid growth during anaphase B.

Recent work has shown that yeast separase also cleaves Slk19 (204), a protein that both localizes to the spindle midzone during late anaphase and promotes the stability of late anaphase spindles (249). However, Slk19's cleavage by separase is neither essential for stabilizing anaphase spindles in otherwise wild-type cells nor sufficient for stabilizing them in cells triggered to undergo anaphase by the TEV protease. Separase must therefore stabilize anaphase spindles either by cleaving yet another protein or by a separate mechanism that does not involve proteolysis. It is still unclear whether spindle stabilization will prove to be an essential separase function in yeast, because both chromosome segregation and spindle elongation can occur in the absence of separase activity during meiosis I in cells in which crossing over been homologues has been abolished (26). In summary, separase cleaves at least two, if not more, proteins at the metaphase to anaphase transition in yeast.

SEPARASE REGULATION

Mitosis would not function if the destruction of cohesion between sister chromatids preceded their alignment on the mitotic spindle. It is therefore crucial that separase activity be very tightly controlled. As already mentioned, all known separases are bound for much of the cell cycle by a chaperone called securin, whose yeast and human orthologues have been shown to be potent inhibitors of separase's proteolytic activity. In yeast, flies, and vertebrates, securin levels rise during late G1, remain high throughout G2 and early M phase, but drop suddenly shortly before the metaphase to anaphase transition (32, 50, 112, 229, 253). The extent of securin's decline at the onset of anaphase varies amongst cells and organisms. Most if not all is destroyed in budding yeast but at most 50% in fission yeast, and possibly a similar amount in Drosophila. It is therefore conceivable that anaphase onset might only require the destruction of securin that is in the vicinity of chromosomes or mitotic spindles.

The rapid decline of securins at the onset of anaphase is due to proteolysis mediated by a multisubunit ubiquitin protein ligase called the Anaphase-promoting complex (APC) or cyclosome (248). The APC also mediates the ubiquitination and proteolysis of many cell cycle proteins, including A- and B-type cyclins, Polo-like kinases, and geminin (a regulator of DNA replication). The APC's activity depends on a pair of proteins composed of WD40 repeats called Cdc20 (Fizzy) and Cdh1 (Fizzy related), which are thought to bring substrates to the ligase complex. Cdc20 is only abundant and active during mitosis, whereas Cdh1 is only active during the subsequent G1 period. The proteolysis of securins, cyclins, and geminin shortly before anaphase onset is therefore mediated by APC-Cdc20, whereas their destruction during G1 is mediated by APC-Cdh1. APC-Cdc20 is regulated by the abundance of Cdc20 (which accumulates during G2 and M phase) (192, 241), by phosphorylation of APC subunits (108), by a surveillance mechanism called the spindle checkpoint (6), and by fluctuations in the abundance of an inhibitory protein called Emi1 (175). The spindle checkpoint works by generating an inhibitor of APC-Cdc20 called Mad2 so long as "lagging" chromosomes are present that have not yet properly attached to the spindle. This ensures that securins and cyclins are not degraded and therefore separase not activated precociously. Mutation of the spindle checkpoint is sometimes but not necessarily lethal but invariably causes high rates of chromosome loss, which might contribute to the genesis of tumors in mice (145).

According to this model, the ability to delay Scc1 cleavage (and hence loss of sister chromatid cohesion) in response to spindle damage or lagging chromosomes should be dependent on Mad2, which is needed to inhibit the APC, and on securin, whose persistence is needed to block separase activation. This is indeed the case in budding yeast (5, 87, 245). Whether this will be universally true is less clear because it has been reported that human tissue culture cells whose securin genes have been deleted by homologous recombination can still prevent loss of sister chromatid cohesion when cells are treated with poisons that cause spindle disassembly (89).

Inactivation of the APC prevents the onset of anaphase in cells from budding yeast, fission yeast, worms, flies, and mammalian tissue culture and causes them to arrest in metaphase (248). There can therefore be little question that the APC is essential for initiating anaphase in most if not all eukaryotic cells. Much but not all evidence suggests that its role in this regard is to destroy securin. For example, expression of securins that cannot be recognized and hence destroyed by APC-Cdc20 blocks sister chromatid separation in yeast (32, 50), flies (112), and vertebrate cells (253, 254). This suggests that securin's destruction by the APC is essential for the activation of separase. These findings nevertheless leave unanswered whether physiological levels of securin are sufficient to block anaphase onset and if so, whether securin is the only protein whose destruction by the APC is necessary for anaphase. In flies, a nondegradable securin (pimples) fails to block anaphase when expressed close to physiological levels but does so when expressed at twice this level (112). This could be taken to mean that securin destruction is not in fact required for anaphase in vivo and that the APC's main role is the destruction of some other anaphase inhibitor. It has been suggested, for example, that destruction of cyclin A is necessary (161, 194). An alternative explanation is that a critical concentration of securin is required to prevent separase activation in vivo and that proteolysis by the APC is indeed necessary to reduce securin to below this level shortly before the onset of anaphase. Budding yeast is the only organism where we have a reasonably definitive answer to this question. The failure of *apc* or *cdc20* mutants to enter anaphase is fully bypassed by deletion of the gene encoding yeast securin (29, 191, 246), which implies that the persistence of securin in *apc* or *cdc20* mutants is entirely responsible for their failure to enter anaphase.

In summary, most evidence is consistent with the notion that proteolysis of securin by the APC is essential for liberating sufficient separase to split sister chromatids. Moreover, inhibition of APC-Cdc20 by Mad2 and the spindle checkpoint prevents securin's destruction in the presence of lagging chromosomes or spindle damage, which in turns delays separase activation.

Ubiquitin-mediated proteolysis of securin is not, however, the sole mechanism that regulates sister separation or the Scc1 cleavage reaction. Somewhat surprisingly, deletion of genes encoding securin is not lethal either in budding yeast or in human tissue culture cells. Yeast or human cells lacking securin are defective in the process of separating sisters, probably due to their lowered separase activity, but neither sister separation nor Scc1 cleavage is precocious (4, 89). Furthermore, the lethality of securin mutants in flies and fission yeast is due to their greater dependence on securin for promoting separase activity and not due to its precocious activation. The implication is that most and possibly all eukaryotic cells possess securin-independent mechanisms that regulate Scc1 cleavage by separase. This is consistent with the finding that there is a 10–20 min delay between securin's decline and the onset of sister chromatid separation in tissue culture cells (J. Pines, personal communication) and with the discovery that disassembly of the actin cytoskeleton in *S. pombe* blocks Scc1/Rad21 cleavage without apparently inhibiting the APC (53).

One such mechanism has recently come to light in yeast, where phosphorylation of Scc1 by the Polo-like kinase Cdc5 greatly facilitates Scc1 cleavage (4). Though not essential for the cleavage of most Scc1 in wild-type cells, Cdc5 is critical for Scc1's cleavage in securin mutants in which separase activity is badly compromised. Phosphorylation by Cdc5 of serine residues six amino acids N-terminal to Scc1's cleavage sites (i.e., in the P6 position) are responsible for part but not all of Cdc5's effect. Serines at the P6 position are conserved in Scc1 orthologues from many but not all eukaryotes. An aspartic acid exists at the equivalent position of the major cleavage site in vertebrate Scc1s (70), which is consistent with the notion that residues at this position must be negatively charged, whether or not due to phosphorylation. Mitosis-specific phosphorylation of Scc1 residues other than those in the P6 position could also regulate the cleavage reaction.

In vertebrate cells, destruction of cyclins as well as securin might be necessary for sister separation. Though previous studies suggested that cyclin degradation might not be required for anaphase in *Xenopus* extracts (79), more recent studies report that expression of nondegradable versions of cyclin A in *Drosophila* (161, 194) or of cyclin B in *Xenopus* extracts (O. Stermmann & M. Kirschner, personal communication) blocks sister separation as well as exit from mitosis. The lack of chromatid disjunction caused by persistent cyclin A in *Drosophila* cannot simply be explained by an interference with APC activity because it does not affect disappearance either of cyclin B or pimples (securin). In *Xenopus* extracts, the lack of sister chromatid disjunction due to nondegradable cyclin B is largely if not entirely due to the phosphorylation of separase by Cdk1, which inhibits its ability to cleave Scc1 (O. Stermmann & M. Kirschner, personal communication). Thus, nondegradable cyclins no longer block sister separation when *Xenopus* extracts are supplemented with a mutant version of separase that cannot be phosphorylated by Cdk1. These observations suggest that activation of separase in animal cells might require destruction of both securin and cyclins by the APC. Such control of separase by Cdk1 might be responsible for the finding that human cells completely lacking securin (due to deletion of both genes) still block sister separation when the mitotic checkpoint is activated by spindle poisons (89). Direct control of separase by Cdk1 appears to be lacking in yeast because expression even of high levels of nondegradable B-type cyclin does not block anaphase (209) and because deletion of its securin gene *PDS1* permits anaphase to occur in the absence of APC-Cdc20. Inactivation of separase by Cdk1 might therefore not be a universal feature of mitotic control. It may nevertheless help ensure that sister separation never occurs when the APC is inhibited by mitotic surveillance mechanisms.

Yet another potential mechanism for regulating sister separation has been raised by the finding that human separase as well as Scc1 is cleaved around the onset of anaphase (229). The major cleavage site resembles those found in fungal Scc1s and contains a serine at the P6 position (O. Stermmann & M. Kirschner, personal communication; I. Waizenegger & J. M. Peters, personal communication). It is likely that separase cleaves itself upon securin's destruction and possible that this further promotes protease activity, as found for members of the caspase family (45). If so, phosphorylation of cleavage sites within separase itself could also regulate its activity. It is unclear whether such a mechanism also regulates separase in yeast because cleavage of the yeast enzyme has not thus far been detected.

In summary, at least four mechanisms may regulate Scc1's cleavage by separase: ubiquitin-mediated proteolysis of securin, phosphorylation of Scc1 itself, phosphorylation and inhibition of separase by Cdk1, and cleavage of separase. Separase cleavage could either activate the protease or promote its destruction due to the Ubr1 ubiquitin protein ligase (173). Indeed, both mechanisms could cooperate to generate a sudden burst of protease activity soon after securin proteolysis in vertebrate cells. Yet other mechanisms must also exist because something ensures that only Scc1 remaining on metaphase chromosomes is cleaved by separase in vivo. The bulk of Scc1, which dissociates from chromosomes during prophase, is untouched by separase. The multiplicity of mechanisms controlling separase emphasizes the importance of regulating this crucial protease.

RESOLVING SISTERS DURING PROPHASE

By the time chromosomes have aligned on the metaphase plate, the vast majority of chromatin fibers from each sister chromatid arm are packed along two distinct axes, between which there are only tenuous connections (107). This "resolution" of sister sequences coincides with and may indeed be synonymous with "mitotic" chromosome condensation, which involves an increase in the compaction of chromatin fibers from a single chromatid (76). During prophase, chromosomes emerge from the amorphous mass of chromatin fibers characteristic of interphase cells as undivided "sausages" (207). The continued disentangling of sister DNA sequences from each other during prometaphase subsequently gives rise to the paired sister chromatids that will finally be aligned on the metaphase spindle. A number of different sets of proteins have been implicated in this dramatic chromosome metamorphosis. The first are histones, in particular histone H3, whose phosphorylation is associated with chromosome condensation in some (55, 185) but not all (93) situations. The second is cohesin, which dissociates from chromosome arms during this period (118). The third is condensin, which associates with chromosomes and somehow promotes their compaction during prophase (77). The fourth is Topoisomerase II, which is required to decatenate sister chromatids (44). The fifth are the three mitosis-specific protein kinases, Cdk1, PLK, and Aurora B, which may phosphorylate and thereby regulate the activity of several of the above proteins. However, none of the roles of these different classes of proteins are understood precisely. This list would not, however, be complete without mentioning the key part played by the transcription apparatus. In most if not all cells whose chromosomes undergo a massive increase in their compaction, transcription by all three RNA polymerases is repressed as cells enter mitosis (126). This does not occur in yeast, whose chromosomes do not greatly condense during mitosis. The elimination of transcription is most probably essential for chromosome condensation, but it cannot be the trigger because mitosis-specific condensation clearly takes place in embryos undergoing cleavage divisions during which there is little or no transcription at any stage of the cell cycle.

The degree to which chromatids are resolved during prophase varies tremendously between organisms. The process is undetectable, for example, in yeast, where neither appreciable chromosome compaction nor loss of cohesion between sisters precedes the metaphase to anaphase transition (62). Indeed, it was the absence of the prophase pathway and the persistence of most if not all cohesin on yeast chromosomes during metaphase (144) that made yeast particularly suitable for studying/discovering the separase pathway.

COHESIN DISSOCIATION

Along chromosome arms, mitosis-specific condensation is invariably accompanied by a loss of sister chromatid cohesion. However, these two processes may be uncoupled in the vicinity of centromeres, where chromosomes are compacted without losing sister chromatid cohesion. Given cohesin's persistence at centromeres until metaphase, it is reasonable to suppose that the loss of cohesion along arms is caused by the dissociation during prophase of most cohesin from this region of the chromosome. Cohesin's dissociation takes place in the absence of the APC and thus presumably separase activity (206) and is not accompanied by

Scc1 cleavage (229). The finding that several cohesin subunits are phosphorylated during prophase and prometaphase (83, 120) raises the possibility that the activation of protein kinases such as Cdk1, PLK, and Aurora might trigger dissociation. Cohesin's Scc3 subunit can be phosphorylated by Cdk1 in vitro (120) but there is little or no evidence that Cdk1 is actually required for cohesin's dissociation (206). The Aurora B kinase is also a candidate because of its localization to the interchromatid zone during prometaphase (2). Recent evidence suggests, however, that PLK may be a key player, because its depletion from mitotic *Xenopus* extracts abrogates their ability to remove cohesin from chromosomes without affecting either phosphorylation of histone H3 or association of condensin (I. Sumara & J. M. Peters, personal communication), both of which may depend on Aurora B (55, 85). A role for PLK fits with the finding that this kinase also prepares yeast Scc1 for cleavage by separase. It is therefore conceivable that PLK has at least two crucial roles in chromatid separation, first in dissociating cohesin from chromosome arms during prophase and second in facilitating Scc1 cleavage at the metaphase to anaphase transition.

The persistence of cohesin at centromeres during metaphase is most likely crucial for holding sister chromatids together until separase activation triggers the onset of anaphase. This population of cohesin molecules must therefore be refractory to the process that dissociates cohesin from chromosome arms during prophase. The mechanism conferring this protection is not at all understood. Recent observations suggest that the giant filamentous protein Titin, more famous for its role in the sarcomeres of muscle, may have some role in protecting centromeric cohesin from the prophase pathway because sister chromatids separate precociously in *titin* mutants of *Drosophila* (123). In *S. pombe*, methylation of histone H3 and its binding by the HP1-like protein Swi6 is necessary for the recruitment and enrichment of cohesin to outer centromere repeats (R. Aushire, personal communication). HP1 could therefore here have a role in blocking cohesin's dissociation from pericentric heterochromatin during prophase in animal cells.

In summary then, cohesion between chromatids is destroyed in two phases in most but not all eukaryotic cells. It is thought that phosphorylation of cohesin itself or other chromosomal proteins triggers dissociation of cohesin from chromosome arms during prophase, whereas cleavage of Scc1 that persists on chromosomes causes the final loss of cohesion, which triggers chromatid segregation at the onset of anaphase. The loss of cohesion in two steps makes good biological sense. Cohesion between chromatids can only be built once during the cell cycle, presumably following passage of replication forks (225), but it must nevertheless be capable of surviving for long periods of time, as in cells with an extended G2 period. During this phase, cohesion is crucial for double-strand break repair (21, 197) and may also be necessary for maintaining chromosome structure and for regulating gene expression (43).

The amount of cohesin on G2 chromosomes is clearly greater than that required for holding sisters together as they align on the mitotic spindle. Were all of it to remain on chromosomes until metaphase, cells might be unable to disjoin chromatids as rapidly as they do at the metaphase to anaphase transition. Dissociation of the bulk of cohesin from chromosome arms during prophase enables cells to embark on the difficult task of resolving sister chromatids long before the final connections are severed at the onset of anaphase. It also permits metaphase cells to concentrate their separase on those regions where cohesin persists. By involving proteolysis, the separase pathway is ideally suited for the rapid and irreversible destruction of sister chromatid cohesion, whereas the prophase pathway is designed for the slow but sure disentanglement of the huge network of sister chromatid fibers. Yeast cells may be able to dispense with the prophase pathway because they have a genome that is small enough to be rapidly resolved into two chromatids largely, though not exclusively, by the force of the mitotic spindle alone.

CONDENSIN ASSOCIATION

The impressive acrobatics of chromosome movement mediated by the mitotic spindle tends to eclipse the extraordinary fact that much of the "work" of splitting sister chromatids is actually performed during prophase and pro-metaphase through mechanisms that do not involve microtubules. One of the major tasks facing any cell about to undergo mitosis is that of removing the intertwining of sister DNA molecules created by the conjunction of converging replication forks (208). It has long been recognized that this task is performed by Topoisomerase II (40), but it is still a mystery how this enzyme knows whether to catenate or decatenate. What then provides the directionality of Topo II action? It is clear to anyone who has seen Bayer's film documenting the effortless disengagement of circular chromatids during anaphase (13) that Topoisomerase II has little or no difficulty getting the directionality right when the two DNA strands concerned come under tension due to being pulled in opposite directions. Though never tested, this particular constellation of interlocked DNA strands is presumably the ideal substrate for Topo II. However, the vast majority of intercatenation between sister chromatids is resolved in an equally effortless process during prophase and prometaphase without the help of microtubules (44). One of the many mysteries of mitosis has therefore been the identity of the motor that drives decatenation during this stage.

With ATPases at each of its long coiled coils, condensin is an ideal candidate for this motor. Indeed, the only phenotype of condensin mutants that is consistently found in all organisms is not so much a defect in chromosome condensation but rather a failure to disengage properly sister chromatids during anaphase (19, 110, 200). It is, for example, striking that the first wave of mitotic failures in *D. melanogaster smc4* mutants are not associated with any lengthening of the chromosomal axes, as might be expected if their primary defect were chromosome compaction, but rather the accumulation of anaphase bridges arising from a failure to resolve sister chromatids (200). It is not inconceivable that the primary function of condensin is not condensation per se, which might largely be achieved by the high level coiling of chromatin fibers, but rather chromatid resolution, which clearly requires a motor to impart directionality to decatenation catalyzed by Topo II. How condensin or any other complex for that matter achieves this goal is still unclear. Such a function is not an obvious consequence of its undoubted ability to introduce positive writhe to DNA in vitro (96) without attributing other important properties to this complex. It is not unreasonable to suppose that condensin somehow achieves this goal by the same route as mitotic spindles, i.e., by bringing interlocked sister DNA molecules under tension. But, how might it perform this?

Chromatid resolution during prophase requires not only tension between sister DNA molecules to drive decatenation but also some mechanism which ensures that only sequences from the same DNA molecule are compacted or condensed together. Chromatid condensation would have little value if it failed to discriminate whether sequences belonged to one or the other chromatid. The process of chromosome condensation is therefore intimately connected with the issue of chromatid identity. What then is the process by which DNA belonging to a single molecule packs along the same axis while that belonging to its sister packs along a parallel but separate one?

There are two fundamentally different ways of thinking about chromatid identity. The first is to propose that chromosomes have an axis or core (often referred to as a scaffold), which exists in some form or another at all stages of the cell cycle, and around which all DNA from a given chromatid is organized. There is considerable evidence that mitotic and meiotic chromosomes do indeed have scaffolds (23, 125, 174), not least of which is the observation that condensin and cohesin, which are known determinants of chromosome structure, are concentrated along axial cores of chromosomes during metaphase (185) and pachytene (104, 165, 171), respectively. Nevertheless, there are potentially four different problems with the notion that a stable scaffold provides chromatid identity. The first is the lack of any serious model for how the immensely long DNA molecules from each chromosome attach to one and only one scaffold. The second is that no single protein has thus far been localized in vivo to a scaffold structure that persists throughout the cell cycle. Cohesin is clearly associated with chromosome axes during meiosis (its role in mitosis in this regard is less clear), whereas condensin is clearly associated with the two chromatid axes of metaphase chromosomes (182). However, neither complex remains associated with chromosomes throughout the cell cycle, condensin being largely absent during interphase and cohesin being largely removed during mitosis (76). Neither complex possesses the continuity required for a stable chromosome core. The third problem concerns how such cores would be replicated and the duplicates cleanly resolved from each other prior to mitosis. The fourth and possibly most serious problem is how recombination between sister DNAs (sister chromatid exchanges) would also recombine axial cores. It is clear, for example, that recombination during G2 gives rise to recombinant chromatids without any trace of discontinuity along the axes of the newly created chromatids. Thus, the axes of mitotic chromatids are determined solely by the chemical continuity of DNA, which can be created anew by recombination, and cannot trace their origin to a pre-existing core.

An alternative way of thinking about chromosome cores is that they emerge de novo from the actions of condensin during mitotic prophase (or cohesin during meiotic prophase). It is clear from cell fusion studies that the cores of mitotic chromosomes can form in the absence of DNA replication (56). Thus, the cytoplasm of mitotic cells induces single chromatids in G1 cells to form chromosomes with a single clearly defined core. The formation of cores does not therefore need some complicated structure established during DNA replication nor does it require a pair of intercatenated chromatids. They arise de novo whenever DNA or rather chromatin comes into contact with active core-forming proteins. According to this view, nucleosomal DNA and not some independent proteinaceous entity is what actually defines a chromatid's axis. The axial core or scaffold is merely a property that emerges from the activities of proteins that package chromatin and help to resolve sister chromatids from each other. If this is correct, eukaryotic chromosomes have two mysterious but interconnected properties: the ability to compact themselves (but not others) around a single axial core and the ability to resolve sister chromatids using a motor other than the mitotic spindle. It is tempting to speculate that condensin might be responsible for both of these. The challenge is to discover its mechanism.

A key question is whether condensin is itself responsible for keeping track of DNA strands (i.e., ensuring that all DNA from a single molecule ends up organized around the same axial core) or whether this crucial property emerges from the activities of other proteins. It is thought, for example, that histones alone can organize DNA into helical 30-nm fibers, in which nucleosomes are wound around a helix, with six nucleosomes per turn (219). These structures alone will tend to segregate self from nonself DNA molecules, as indeed would further coiling to produce yet higher-order fibers. Might then the simple tendency of chromosomes to coil upon themselves be the primary driving force for condensation? Might this process inevitably cause regions where chromatids are intertwined to come under the sort of tension needed to drive efficient decatenation? If so, might condensin, through its known ability to stabilize positive writhe, merely facilitate a reaction that is primarily driven by the properties of nucleosomes? Even if partially correct, at least in outline, this model explains neither why chromatids condense into cyclinders with a defined diameter nor why condensin accumulates along the axial cores of such cylinders. If condensin merely facilitates the coiling of chromatin fibers, then it must do so in a manner that both constrains the degree and form of this coiling. Nevertheless, it is quite possible that condensin has a more active role both in chromosome packaging and chromatid resolution than envisaged by the above model. Without knowing what this particular motor does when presented with a chromatin substrate, one can merely speculate about this role. One possibility is that condensin associates with the bases of small loops or coils of chromatin and enlarges these loops or coils in a processive manner, which ensures that all chromatin within the loop or coil must have been cleanly segregated from all other sequences in the genome. As this process proceeds, neighboring loops or coils would naturally converge, creating an axial core in which the bases of loops or

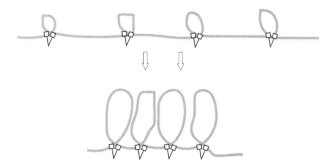

Figure 5   A model for how condensin could form axial cores and thereby help to resolve sister chromatids from each other.

coils containing condensin would alternate with a short linker (Figure 5). I give this example not so much because it is a serious candidate for the function of condensin (or cohesin for that matter) but rather because it illustrates the notion that condensin or molecules like it could have a very active role in folding and resolving chromatids. This model does help to explain many puzzling features of condensin: in particular, its crucial role in chromatid resolution, its accumulation along axial cores, and the curious finding that condensin depletion causes problems with chromatid resolution long before it has any effect on chromatid length. It also neatly explains the origin of the so-called chromosome scaffold with peripheral chromatin loops and how chromatid identity can "emerge" naturally from the actions of molecules that act processively but merely locally on the chromatin fiber. However, it is very difficult to imagine how condensin could actually perform this particular anointed task, especially as its substrate must be chromatin fibers and not naked DNA. It is conceivable that cohesin has a similar function.

In summary, then, the process of sister chromatid separation takes place in two steps in most eukaryotic cells. During the first step, some sort of processive chromatid compaction involving condensin and very probably other regulators of nucleosome packing drives the decatenation of chromatids and packages them around an axial core that contains condensin. The bulk of cohesin dissociates from chromatids as this process proceeds, and little if any remains to connect chromatids along chromosome arms by the time that chromatids are aligned on the metaphase plate. However, cohesin in the vicinity of centromeres, which is largely refractory to the process that removes it from chromatid arms, prevents resolution at centromeres and is capable of providing sufficient cohesion for the alignment of chromatids in a bipolar manner on the mitotic spindle. The second step is triggered by the activation of separase, whose cleavage of cohesin's Scc1 subunit in the vicinity of centromeres permits sisters to be pulled to opposite poles. The force supplied by microtubules now takes over from condensin in driving the decatenation process. Remarkably, the first step of chromatid resolution is almost entirely missing in yeast, where most if not all cohesin remains on chromosomes until the activation of separase, and chromatids remain tightly connected throughout their length until the metaphase to anaphase transition.

LINKING REREPLICATION WITH CHROMATID SEGREGATION

One of the most characteristic features of the eukaryotic cell cycle is the delay of chromosome reduplication until after chromatids produced by the previous round of DNA replication have been partitioned between daughter cells at mitosis. Now that we understand many of the processes required for sister separation and for DNA replication, we can also begin to understand the broad outlines of the mechanism by which these two crucial events are interlinked. The initiation of DNA replication takes place in two steps (39). The first is the loading at future origins of a hexameric DNA helicase composed of Mcm proteins, which depends on the origin recognition complex (ORC), an Mcm loading factor called Cdc6p, and a cofactor called Cdt1p. The second step is the activation of cyclin-dependent kinases along with the Dbf4-dependent Cdc7 kinase, which together trigger origin unwinding by the Mcm helicase and thereby the loading of DNA polymerase. Because the very same Cdks that trigger origin unwinding also inhibit the loading of Mcm helicase (35, 217), it is not possible for origins to reload Mcm proteins while S phase Cdks remain active, which lasts for most of S phase. Mcm helicases are likewise prevented from loading on origins during G2 and M phase by Cdks containing cyclins A and B (71). In vertebrate cells but possibly not in yeast, an additional mechanism also blocks the formation of prereplication complexes: A protein called geminin (132) accumulates during S or G2, binds to Cdt1, and blocks Mcm loading (243). Thus, preparations for a new round of DNA replication cannot begin until both cyclins and geminin are removed. Because cyclins, geminin, and securin are all destroyed by the Anaphase-promoting complex, preparations for the initiation of DNA replication cannot begin until the process of sister chromatid separation has been initiated.

SEPARATING CHROMATIDS DURING MEIOSIS

During meiotic divisions, two rounds of chromosome segregation following a single round of chromosome duplication give rise to haploid gametes from diploid germ cells. One of the most remarkable aspects of meiotic cells is their ability to undergo two rounds of chromosome segregation using only a single round of DNA replication. To do this, they must undergo the first meiotic division without fully destroying the cohesion established between sister chromatids during premeiotic DNA replication so that the residual cohesion can be utilized at the second meiotic division (149).

The first meiotic division is fundamentally different from the second one and from mitotic divisions (see Figure 2). During mitosis and meiosis II, cells attempt to pull sister kinetochores toward opposite poles of the cell but are prevented from doing so by sister chromatid cohesion until all sister kinetochore pairs have aligned on the spindle, whereupon cleavage of Scc1 or Rec8 by separase destroys this equilibrium and triggers poleward migration. During meiosis I, cells instead attempt to pull toward opposite poles homologous chromosomes, which have been joined together by recombination (158). During this process, sister kinetochores must always attach to microtubules from the same pole (58), known as syntelic or mono-orientation, which is precisely what must be avoided during mitosis. A very similar type of equilibrium is therefore established during the metaphases of meiosis I and mitosis. However, the partners being pulled in opposite directions during meiosis I are homologous chromosomes and not individual chromatids. Meanwhile, chromosome segregation at the onset of anaphase I is triggered by resolution of the chiasmata or crossovers that hold homologues together, which is invariably accompanied by loss of cohesion between sister chromatid arms (122). Another key difference between meiosis I and mitosis is that cohesion between sister chromatids in the vicinity of centromeres is always preserved at anaphase I and persists until finally destroyed at anaphase II (147). This property is not actually necessary for meiosis I but is crucial for meiosis II.

It is clear that chromosome segregation during meiosis largely depends on the same machinery used during mitosis. However, the ability of meiotic cells to reduce chromosome numbers by undergoing two rounds of chromosome segregation after only one round of DNA replication depends on several meiosis-specific innovations. Many of these involve the sister chromatid cohesion apparatus.

DO MEIOTIC AND MITOTIC CELLS USE THE SAME OR A DIFFERENT COHESION MACHINERY?

In budding yeast, all cohesin subunits apart from Scc1 are essential for meiosis I (R. K. Clyne, personal communication; 104). Scc1 declines sharply as cells enter meiosis and is replaced by Rec8, a meiosis-specific variant. Rec8 is normally never expressed in mitotic cells but it is capable of rescuing cells lacking the *SCC1* gene when expressed from the *SCC1* promoter (26, 238). Such cells undergo meiosis with high efficiency and produce largely viable spores, suggesting that Scc1 has little if any role during meiosis (F. Klein, personal communication). Rec8, in contrast, accumulates shortly before premeiotic DNA replication and, along with other cohesin subunits, is essential for maintaining sister chromatid cohesion throughout meiosis (104, 148). A similar though not identical situation prevails in *S. pombe* (238), *C. elegans* (163), mammals (C. Heyting, personal communication), and possibly also in plants (20), in which Scc1s are to a greater or lesser extent replaced by meiosis-specific variants, most of which have been called Rec8. Inactivation of Rec8 in *C. elegans* using RNA interference causes the appearance of up to 24 chromatids instead of six bivalents prior to the first meiotic division (163). Rec8 is essential for sister chromatid cohesion in the vicinity of centromeres in *S. pombe* (148), but it coexists for much of meiosis I with Scc1 (Rad21), which is relegated to chromosome arms (148). Though the replacement of Scc1s by meiosis-specific variants may be widespread in eukaryotes, it is unclear whether it is a universal phenomenon. Thus far, only a single Scc1-like protein has been detected in the (almost) complete *D. melanogaster* genome (1). Either flies use the same Scc1 subunit for mitosis and meiosis or they possess a second gene, which lurks in their unsequenced heterochromatic pericentric regions.

At least two other cohesin subunits have meiosis-specific variants. The two versions of Scc3 called SA1 and SA2, which are found in most somatic tissues, are replaced, at least in spermatocytes, by a third variant called STAG3 (171). Spermatocytes also express a meiosis-specific version of Smc1, which is called Smc1β and may be the main partner of Smc3, Rec8, and STAG3 during meiosis (C. Heyting, personal communication; 176). The *S. pombe* genome also encodes a meiosis-specific variant of Scc3 called Rec11 (115), which possibly replaces Scc3 along chromosome arms but not at centromeres (Y. Watanabe, personal communication).

Of the other proteins needed for cohesion during mitosis, Pds5's homologue in Sordaria, called Spo76, is essential for maintaining sister chromatid cohesion during diplotene/diakinesis (228), while Scc2 and its homologue in Coprinus, Rad9, are also essential for meiosis (188). The roles of other cohesion proteins such as Eco1/Ctf7and Ctf18 have not yet been investigated.

This list of meiosis-specific cohesin subunit variants (see Table 1) is presumably far from complete. It is, however, already clear that there is considerable variation between organisms in the extent to which mitotic subunits are replaced by meiosis-specific variants, which ranges from the replacement merely of Scc1 by Rec8 in yeast to that of Scc1, Smc1, and Scc3 (SA1 and SA2) by Rec8, Smc1β, and STAG3, respectively, in mammals. These replacements, or in some cases additions, presumably enable cohesin to fulfill many of its functions that are specific to meiotic cells, such as the repair of double-strand breaks using homologous chromatids instead of sisters, the creation of axial cores and synaptonemal complex during pachytene, and the persistence of cohesion at centromeres but not along arms until the second meiotic division.

COHESIN'S ROLE IN RECOMBINATION AND IN BUILDING MEIOTIC AXIAL CORES

It has long been suspected that sister chromatid cohesion has a crucial role in double-strand break repair. G2 cells, for example, are invariably far more resistant to gamma irradiation than are G1 cells (24). Furthermore, in diploid cells, sister chromatids and not homologous ones are clearly the preferred template for repair (91). However, until recently it has not been possible to test whether the greater radiation resistance of G2 cells is due to the proximity of a sister chromatid with which to repair double-strand breaks or due to other differences between these two cell cycle states, for example in the activity of repair enzymes or checkpoint proteins. The discovery that Rad21 (21), long known to be crucial for double-strand break repair, encoded a cohesin subunit homologous to Scc1 (63, 144) is consistent with the notion that the proximity of sister chromatids does indeed have a key role. Efficient double-strand break repair during G2 or M phase in budding yeast depends not only on cohesin's presence at the time of irradiation but also on its presence during the preceding S phase (197). Thus, the mere presence of cohesin on chromatin, as occurs when Scc1 is synthesized only during G2, is not sufficient for efficient repair. For cohesin to facilitate repair during G2, it must previously have participated in a process that only occurs during S phase, which is presumably the creation of sister chromatid cohesion (197). Cohesion between sisters presumably also prevents double-strand breaks from causing chromosome breakage as well as providing a ready template for repair.

Double-strand break repair has a central role during meiosis, where it is responsible for creating recombinant chromatids and thereby for joining homologues in a manner that permits them rather than sisters to be disjoined at the first meiotic division. During meiosis, the 5′ ends of double-strand breaks created by the Spo11 endonuclease after premeiotic recombination (16, 94) undergo 5′ to 3′ resection to yield 3′-OH single-strand tails, which then invade a homologous chromatid. Repair synthesis and ligation give rise to double Holliday junctions (DHJs) (180, 199). At a later stage, these structures are resolved by cleavage and ligation to yield recombinant molecules with or without exchange of flanking markers (Figure 6). These two outcomes have very different consequences for chromosome segregation. Formation of recombinant chromatids but not gene conversion results in the connection of homologous chromosomes that is so crucial for chromosome segregation at the first meiotic division.

There are several remarkable aspects about this process, which are unique to meiotic cells and are crucial for meiosis. First, the usual preference of mitotic cells to use a sister chromatid for repair is reversed in favor of homologous chromatids (187). Second, double Holliday junctions are far more frequently resolved to form crossovers during meiosis than during mitotic double-strand break repair. Third, the creation of crossovers greatly reduces the probability that neighboring double Holliday junctions will be resolved in a similar manner, a phenomenon that is called crossover interference (102, 151). Fourth, during much of the time that it takes to convert double-strand breaks into crossovers, homologous maternal and paternal chromatids are bound together along their entire lengths (synapsed) to form a structure, unique to meiotic cells, called the synaptonemal complex (SC) (252) (Figure 6).

Electron microscopic analysis of the SC after staining with silver suggests that it is composed of two axial cores, one associated with maternal and the other with paternal sister chromatids, which are connected by a central element composed of a coiled coil protein known as Zip1 in yeast (42) and Scp1 (142, 143) in mammals.

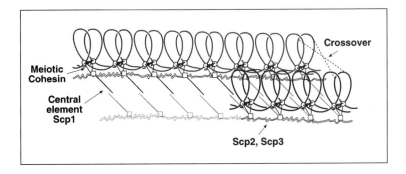

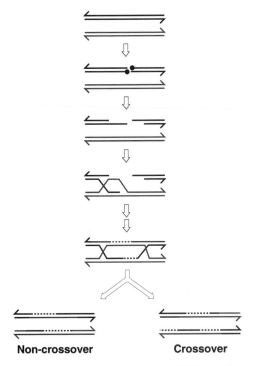

Non-crossover   **Crossover**

Figure 6   Cohesin and the synaptonemal complex (SC), in which crossing over between sister chromatids take place. Though Scp2 and Scp3 proteins run along the axial cores of the SC, meiotic cohesin composed of Smc1β, Smc3, Scc3-STAG3, and Rec8 lies at the heart of the SC's axial cores. *Below*, formation of double Holliday junctions and their resolution into crossovers, which is initiated by the Spo11 endonuclease.

Dissolution of the synaptonemal complex after the completion of recombination allows maternal and paternal sister chromatid pairs to separate except, of course, in the regions of crossovers, which are easily visible by light microscopy and are called chiasmata (252). The persistence of sister chromatid cohesion at this point ensures that crossovers now connect homologues together (122), which subsequently enables homologue pairs, and not sisters as during mitosis, to be aligned (i.e., pulled in opposite directions) by the meiosis I spindle apparatus (158).

Many if not most of these remarkable aspects of meiotic double strand repair only make sense when one considers that meiosis has two key purposes: to produce recombinant chromatids and to join homologues together via chiasmata so that they and not sisters are disjoined at the first meiotic division, which subsequently permits the formation of haploid progeny when chromatids are disjoined at the second meiotic division.

The formation of recombinant chromatids and the random assortment of centromeres from different chromosomes at the first meiotic division both contribute to the generation of gametes that differ greatly from each other and enable parents to "hedge" their genetic bets. By creating new haplotypes, some of which will lack deleterious mutation combinations, they also enable the cleansing of semideleterious mutations from diploid genomes (105). The purpose of using homologues rather than sisters to repair breaks produced by Spo11 and the resolution of double Holliday junctions as crossovers have obvious roles in halving the number of chromosomes and in producing recombinant chromatids, though the mechanism by which these goals are achieved is far from clear. The purpose of synaptonemal complexes and crossover interference is less obvious. An important clue is that some organisms, such as *S. pombe*, undergo meiosis and reciprocal recombination without forming SCs (11). Because organisms like *S. pombe* lack crossover interference, it is thought that the full synapsis of homologues might be a crucial part of the mechanism by which crossovers interfere with each other. When the number of crossovers per chromosome is low, crossover interference is crucial for ensuring that all chromosomes produce at least one crossover (note that a single crossover is sufficient to join homologues together, as long as there is sufficient cohesion between sister chromatids distal to that crossover). *S. pombe* only possesses three chromosomes along which there are very high rates of recombination, and crossover interference is unnecessary to ensure that each chromosome produces at least one chiasmata. The opposite extreme is found in *C. elegans* where there is rarely if ever more than one crossover per chromosome and crossover interference is therefore extremely high (15).

We have little or no idea how creation of a single crossover manages to inhibit the formation of others along an entire chromosome, as occurs in *C. elegans*. Signals emanating from crossovers, be they mechanical or informational, must be capable of traveling along the entire length of chromosomes and then preventing the resolution of all other double Holliday junctions (DHJs) on the same chromosome as crossovers. For this to occur, meiotic chromosomes must have a defined backbone or axial core along which these interference signals must travel. Furthermore, crossovers involving only two chromatids must also signal to DHJs elsewhere on the chromosome involving a different pair of chromatids. The synapsis of all four chromatids, as occurs in SC, presumably facilitates this remarkable process.

Because the axial core of meiotic chromosomes may have a central role in providing the correct partner for exchanges (homologues versus sisters) and in mediating crossover interference, characterization of its constituents has been an important goal. Purification of synaptonemal complex from mammals has thus far led to the identification of two meiosis-specific proteins, Scp2 and Scp3, which localize along axial cores (184). Deletion of the gene for Scp3 in mice leads also to the loss of Scp2 from chromosomes and clearly compromises the formation of cores, but it does not eliminate them entirely, suggesting that other proteins lie at the heart of these structures (C. Hoog, personal communication). Indeed, no proteins similar to Scp2 or Scp3 have yet been found in yeast, whose synaptonemal complex also contains two clearly defined axial cores.

There is a growing consensus that meiotic cohesins might be the chief architects and constituents of meiotic axial cores. They both colocalize with cores (C. Heyting, personal communication; C. Hoog, personal communication; 171) and are necessary for the formation of SC (104, 163). Thus, Rec8, Smc1, Smc3, and Scc3 all colocalize with cores during pachytene in yeast (R. K. Clyne, personal communication; 104), whereas Rec8, Smc1β, Smc3, and STAG3 do so in mammals, as does Rec8 in *C. elegans* (163). Whether condensin also participates in formation of the SC's axial cores has not been addressed. Nevertheless, there is a distinct possibility that while the cores of mitotic chromosomes are built by condensin, those of the SC are built largely if not completely by cohesin. If so, this would indicate that cohesin and condensin not only resemble each other structurally but also possess a similar capacity to organize chromatin around axial cores. There is, of course, a crucial difference between the cores of mitotic and meiosis I chromosomes: A single chromatid is organized around the former whereas a pair of sister chromatids is organized around the latter (Figure 6).

Though the cores of meiosis I chromosomes are clearly stabilized by synapsis between homologues (which is mediated by recombination, central element proteins, and by yet other more mysterious pairing mechanisms), rudimentary cores containing cohesin can clearly form in the absence of synapsis (104). In yeast, the ability to produce cores and hence SC depends on the replacement during meiosis of Scc1 by Rec8. Though Scc1 expressed from the *REC8* promoter in cells lacking an intact *REC8* gene can bind to chromatin, produce sister chromatid cohesion, and even support the monopolar attachment of sister kinetochores to meiosis I spindles during meiosis I, it cannot support the formation of SC (223). Whether the replacement of Smc1 by Smc1β and Scc3-SA1/SA2 by STAG3 also contributes to the formation of SC in mammals is not yet known.

Though essential for the formation of axial cores and SC, meiotic cohesin subunits are not, at least in yeast, required for formation of the double-strand breaks that initiate the recombination process (104). In *rec8* or *smc3* mutants, double-strand breaks occur with almost normal kinetics but are poorly repaired and fail to produce crossovers. The DNA ends produced by Spo11 in these mutants are resected more extensively than in wild-type cells, due presumably to inefficient invasion of homologous chromatids. The damage to DNA caused by this defect is thought to be detected by DNA repair surveillance mechanisms that block the first meiotic division. There is some evidence that double-strand breaks also form in the absence of Rec8 in *C. elegans*. Worms lacking Rec8 due to RNA interference accumulate chromosome fragments and this process is dependent on the Spo11 endonuclease (163). The abnormal repair of double-strand breaks in *rec8* mutants cannot simply be attributed to their defective sister chromatid cohesion because sister chromatids are not usually used to repair breaks during meiosis. Indeed, replacement of Rec8 by Scc1 fails to prevent the block to meiosis due to DNA damage, despite restoring sister chromatid cohesion (223).

In conclusion, the meiosis-specific version of cohesin, containing Rec8 instead of Scc1, is crucial for regulating the repair of double-strand breaks as well as for the formation of SC. Though the topology of meiotic chromosomes is no better understood than their mitotic counterparts, the chromatid fiber must consist of a core from which loops or coils emanate. DNA sequences within loops are presumably far more accessible than those within the core and are therefore most likely to participate in the production of double-strand breaks, though repair of these breaks might be conducted within cores. One of the key functions of cohesin during meiosis may be to organize the chromatid fiber around a cohesin-containing core, to which components like Scp2 and Scp3 might attach and reinforce the axes of meiotic chromosomes during pachytene (Figure 6). It is conceivable that cohesin would also have a similar activity during mitosis were it to remain on chromosomes. Indeed, this may be the explanation for why cohesin appears to have a role in chromosome compaction during mitosis in yeast (63), where the bulk of cohesin remains associated with chromosomes until the onset of anaphase (144).

CHIASMATA

Having regulated the production of crossovers, the SC then dissolves, which causes a dramatic change in the appearance of chromosomes. During pachytene, all four chromatids are held closely together in a single bundle by the SC. Its dissolution severs the connection between homologous chromosomes, except where they are joined by crossovers. Sister chromatids nevertheless remain associated along the entire length of chromosomes, though their proximity can vary considerably between organisms. In yeast, for example, sister chromatids remain very tightly connected (26), whereas in many animals and plants, sister chromatids start to appear as separate entities that have their own axial cores and are only connected with each other at their peripheries (242). In the ovaries of many vertebrates, meiosis is halted for long periods at this stage, which is known as diplotene. High rates of transcription, as witnessed in the famous lampbrush chromosomes of newts and salamanders, help to produce stockpiles of maternal gene products needed for embryogenesis (54).

In the absence of SC, chiasmata assume responsibility for holding homologous chromosomes together from this stage onwards. Without them, homologous chromosomes would simply drift apart and it would not be possible to align them on opposite poles of the meiosis I spindle, which is precisely what happens when the Spo11 is inactivated in yeast (26, 193) and *C. elegans* (38). In *Drosophila*, there exist alternative mechanisms, which do not use recombination, for the synapsis of homologous chromosomes both during formation of the SC and subsequent to its dissolution (134). Even though it is clearly possible to pair and disjoin homologues without using chiasmata to hold them together, the vast majority of eukaryotic cells choose to use this device, which is presumably more robust than alternative mechanisms. As a consequence, in most eukaryotic organisms recombination is obligatory for chromosome segregation at meiosis I and hence for the formation of viable gametes. This presumably helps to ensure that recombination cannot readily be abandoned when their environment might favor its elimination.

There are two theories for how chiasmata perform their crucial task of holding homologues together (149). According to the first, homologues are bound together in the vicinity of chiasmata by some as yet unidentified substance, known as the chiasma binder. According to the second, homologues are held together by chiasmata due entirely to sister chromatid cohesion that persists distal (with respect to centromeres) to the crossover (122). This second theory is not only more economical than the first but it also explains why sister chromatids remain associated with each other until the onset of anaphase I, whereupon they invariably separate from each other. If sister chromatid cohesion were responsible for the ability of chiasmata to hold homologues together from diplotene until metaphase I, then its destruction would be required to resolve chiasmata at the onset of anaphase I.

The discovery of cohesin has recently permitted this theory to be tested. If it is cohesion between sister chromatids that holds homologues together, then cohesin should be found along the interchromatid region along chromosome arms until the onset on anaphase I, whereupon its removal should trigger loss of sister chromatid cohesion and thereby the resolution of chiasmata. Remarkably, this is precisely what has recently been found. In mammals, Rec8 and STAG3 colocalize to the axis connecting sister chromatids along the entire chromosome (except in the immediate vicinity of chiasmata) during the period from diplotene until metaphase I (C. Heyting, personal communication; 171), as does Rec8 in *C. elegans* (163), and they largely disappear from this location at the first meiotic division (Figure 2). The persistence of "meiotic" cohesin along chromosome arms until the onset of anaphase I clearly contrasts with the behaviour of its mitotic cousin Scc1, which largely if not completely disappears from chromosome arms during prophase and prometaphase of mitosis. The persistence of meiotic cohesin subunits along chromosome arms even during metaphase I is particularly remarkable because meiotic chromosomes shorten and compact considerably during this phase, as they do during the equivalent period of mitosis. Whether the so-called "prophase" pathway

responsible for removing cohesin from chromosome arms during mitosis is completely or only partly inactivated during the period between the onset of diplotene
and the onset of anaphase I is unknown. It is not inconceivable that a failure to retain sufficient cohesin along chromatid arms to maintain chiasmata (i.e., to protect
it from a process analogous to the mitotic prophase pathway) could contribute to
chromosome mis-segregation (and hence to Down's syndrome) during oogenesis
in women, whose oocytes enter diplotene around birth and do not enter metaphase
I until induced to mature during menstrual cycles (see below).

ALIGNING HOMOLOGUES ON THE MEIOSIS I SPINDLE

One of the central aims of meiosis is to produce gametes with only a single complement of the chromosomes. The formation of haploid cells takes place at the second
meiotic division when cells undergo what is in fact a fairly conventional mitotic
division. What is unconventional about this division is that it takes place without a
preceding round of chromosome duplication. The unique ability of meiotic cells to
perform this extraordinary feat is due to peculiarities of their first meiotic division,
which differs radically from mitotic divisions in three key aspects: the association
through chiasmata of homologous chromosomes and not just sister chromatids,
the attachment of sister kinetochores to spindles emanating from the same pole,
and the persistence after anaphase I of sister chromatid cohesion in the vicinity of
centromeres.

Due to the formation of chiasmata and the persistence of sister chromatid cohesion along chromosome arms, homologous chromosomes and not just sister
chromatids are held together when the cell assembles its "meiotic" spindle apparatus. This creates the opportunity for spindles to pull maternal and paternal sister
kinetochore pairs in opposite directions and hence to establish an equilibrium during metaphase I in which chiasmata resist traction of homologous chromosomes to
opposite poles. One of the great mysteries of meiosis I is how cells avoid attaching
sister kinetochores to spindles of opposite polarity, as occurs during mitosis. In the
rare cases where this has been studied cytologically, sister kinetochores are seen to
be fused into a common structure until their attachment to microtubules but they
appear to split into separate structures sometime during metaphase or anaphase I
(59). There are suggestions that the unique behavior of meiosis I kinetochores is
conferred by the kinetochores themselves and not by the cytoplasm or spindles of
these cells. Meiosis I and II cells from grasshoppers can be fused to produce a single
cell with two independent spindle apparatuses. When homologous chromosomes
attached by chiasmata (that had not yet attached to the spindle) are transferred to
a meiosis II spindle, they align normally and disjoin to opposite poles at anaphase
at the same time as "native" meiosis II sister chromatid pairs disjoin from each
other on the same spindle. These and other similar experiments suggest that sister
kinetochores only acquire the ability to attach to spindles with opposing polarity
around anaphase I (159). Prior to this point, kinetochores are somehow altered so

that sisters "co-orient" (158) on the spindle. The grasshopper fusion experiments also demonstrate that the signal which triggers the resolution of chiasmata and hence the disjunction of homologues must the same as that which triggers the disjunction of sister chromatids at meiosis II.

In organisms like yeast, it has been possible to address roughly when, during the meiotic process, sister kinetochores become committed to the co-orientation (monopolar or syntelic attachment) characteristic of meiosis I. Cells undergoing meiosis can be returned to media that support vegetative growth. Surprisingly, cells only lose the ability to undergo a mitotic division after they have completed recombination; that is, it is still possible for late pachytene cells that have completed recombination to undergo mitosis instead of meiosis I if transferred to growth media (250). Thus, sister kinetochores become committed to co-orientation sometime between late pachytene and metaphase I.

The search for proteins which ensure that sister kinetochores attach to spindles with the same polarity during meiosis I has largely been undertaken on the premise that such proteins might be specific to meiotic cells. Thus far, three proteins with this property have been identified by genetic studies in yeast: Spo13 and Mam1 from *S. cerevisiae* and Rec8 from *S. pombe*. Inactivation of the *SPO13* gene, which is exclusively expressed in meiotic cells (233), causes about 50% of the chromosomes to undergo an equational rather than a reductional division at the first meiotic division (101, 195). For reasons that are still mysterious, it also prevents cells from undertaking a second meiotic division. The partial separation of sister chromatids to opposite poles at the first meiotic division in *spo13* mutants means that Spo13 is required not only for monopolar attachment but also for preventing the destruction of cohesion in the vicinity of centromeres. This implies either that these two aspects of meiosis I centromeres are intimately connected and conferred by the same set of proteins or that Spo13 has two or more separate functions. The Spo13 protein does not appear to localize exclusively to centromeres/kinetochores (A. Amon & A. Toth, personal communication) and it is therefore likely that Spo13 regulates the overall state of meiotic cells (131) or meiosis I chromosomes rather than any one particular property such as monopolar attachment. Both pairs of sister chromatids from a given homologue tend to behave in the same manner in *spo13* mutants; that is, either both or neither undergoes an equational division (86). This suggests that centromeres acquire a "reductional" state (that will ensure both monopolar attachment and cohesion protection) when all four chromatids are synapsed and can therefore communicate with each other, i.e., some time during pachytene. If so, then Spo13 must be required either to increase the probability that this state will be generated or for maintaining it. Spo13 cannot be an essential part of the monopolar attachment apparatus because this process still occurs in 50% or more of chromosomes in the complete absence of Spo13.

The Mam1 protein has recently been identified in a screen for meiosis-specific genes whose deletion causes chromosome mis-segregation (223). Unlike Spo13, Mam1 localizes to kinetochores from late pachytene until metaphase I. In its absence, the first meiotic division fails to take place despite the formation of meiotic spindles and the subsequent destruction of securin, which activates separase and removes Rec8 from chromosome arms. The second meiotic division, in contrast, takes place on schedule but does so in the presence of four spindle pole bodies (microtubule organizing centers), which leads to massive chromosome missegregation. Furthermore, the persistence of Rec8 in the vicinity of centromeres until the onset of anaphase I is largely, though possibly not entirely, unaltered in *mam1* mutants, which suggests but does not prove that Mam1 is not essential for protecting cohesion in the vicinity of centromeres. Two pieces of evidence imply that Mam1 is required very specifically to prevent sister kinetochores from attaching to spindles of opposing polarity. First, an appreciable fraction of sister kinetochores separate precociously during its first aborted division, that is, prior to securin's destruction, which is normally closely associated with the onset of anaphase I. This implies that many if not most sister kinetochores come under traction pulling them toward opposite poles during meiosis I. The failure of *mam1* mutants to segregate chromosomes at meiosis I might then be due to their failure to destroy cohesion in the vicinity of centromeres. According to this hypothesis, *mam1* mutants attempt to pull sisters to opposite poles during meiosis I but are merely prevented from doing so by the persistence of cohesion in the vicinity of centromeres. If so, loss of cohesion at centromeres at the same time as it is lost along chromosome arms (see below) should suppress the meiosis I chromosome segregation defect of *mam1* mutants and permit a fully equational division.

The discovery that Scc1 does not persist at centromeres after anaphase I when expressed instead of Rec8 (albeit in a background where recombination has been eliminated by deletion of the *SPO11* gene) and indeed cannot support sister chromatid cohesion past this point suggests that sister chromatid cohesion in the vicinity of centromeres provided by Scc1, unlike that by Rec8, is destroyed along with that along chromosome arms at the onset of anaphase I (223). Remarkably, replacement of Rec8 by Scc1 (in a *spo11* mutant background) suppresses the failure of chromosome segregation during meiosis I in *mam1* mutants and causes all sister chromatid pairs to segregate to opposite spindle poles. These data all point to the Mam1 protein having a highly specific function in regulating the orientation of sister kinetochores during meiosis I. Mam1 is not necessary for the protection of cohesion at centromeres but is essential for monopolar attachment. These two properties of meiosis I centromeres are therefore determined by separate mechanisms, even though there exist mutations, like *spo13*, that affect both. This independence might be widely conserved because inactivation of the MEI-S332 gene in *Drosophila* abolishes retention of centromere cohesion without altering monopolar attachment (117). The Mam1 protein is not well conserved, and it has therefore not yet been possible to identify homologous proteins in other organisms. This will eventually be important for studying the structural basis of monopolar attachment because yeast kinetochores are too small to be observed even at the electron microscopic level.

The last meiosis-specific protein that has been implicated in determining mono-polar attachment of sister kinetochores is the Rec8 protein in *S. pombe*. Unlike the situation in *S. cerevisiae*, deletion of *Rec8* in *S. pombe* does not abolish progress through meiosis (148). It is possible that many if not most double-strand breaks can still be repaired in Rec8's absence, due to the persistent expression of Rad21 (the homologue of Scc1), though this has never been investigated. Recombination is indeed reduced but despite this, cells proceed with both meiotic divisions. Another explanation for the very different outcome of deleting Rec8 in *S. pombe* and *S. cerevisiae* is that the former does not make SC and may therefore process its double-strand breaks in a manner that is less dependent on a Rec8-containing form of cohesin. Remarkably, the first meiotic division in *S. pombe rec8* mutants is almost entirely equational, with sister centromeres segregating to opposite poles in 90% or more of cases (238). The implication is that sister kinetochores attach to microtubules of opposing polarity in the absence of Rec8, that Rad21 (Scc1) provides sufficient cohesion between chromatids for their alignment in a bipolar fashion on the meiosis I spindle, and that all cohesion mediated by Rad21 is destroyed by separase at the onset of anaphase I.

At first glance, these data suggest that monopolar attachment in *S. pombe* might be mediated by a very different mechanism from that used by *S. cerevisiae*. Even if Mam1-like proteins exist in *S. pombe*, it is possible to eliminate completely monopolar attachment without directly inactivating such proteins. Is it possible therefore that Rec8 alone is responsible for altering the orientation of sister kine-tochores during meiosis in *S. pombe* and that Mam1-like proteins are not required? Furthermore, the monopolar attachment apparatus in *S. pombe* clearly cannot func-tion without Rec8 even when Rad21 is present in the cell, which also differs from *S. cerevisiae* where Scc1 can support monopolar attachment in Rec8's absence (223). It would not be surprising if monopolar attachment of sister kinetochores depended on cohesion between sister chromatids at centromeres, but in *S. cere-visiae* this can equally well be supplied by Scc1 as by Rec8.

Analysis of the distribution of Rad21 (Scc1) and Rec8 on *S. pombe* centromeres has shed some insight into this issue. *S. pombe* centromeres are more complex than those in *S. cerevisiae* and consist of an inner region that is associated with kinetochore proteins like the centromere-specific histone H3 variant Cenp-A and an outer region associated with HP-1-like proteins that regulate chromatin struc-ture (162, 169). Rad21 is found in the outer but not in the inner region during mitosis (221), whereas Rec8 is found in both regions during meiosis I (239). Sister chromatid cohesion along chromosome arms and in the outer centromere regions is presumably sufficient for bipolar attachment of sister kinetochores dur-ing mitosis (177) and presumably also during meiosis. The absence of Rad21 from kinetochores themselves during mitosis therefore poses no fundamental problems. Rec8's presence within the kinetochore proper (the inner region) during meiosis I in *S. pombe* is consistent with the notion that monopolar attachment depends on an intimate cohesin-dependent juxtaposition of sister kinetochores. For some reason, Rad21 is incapable of penetrating (or functioning in) this holy sanctuary either during mitosis or during meiosis and cannot therefore mediate the sister kineto-chore cohesion normally performed by Rec8, which would explain why Rec8 is essential for monopolar attachment. Rad21 is nevertheless capable of associating with chromosome arms and provides sufficient cohesion between chromatids for their disjunction to opposite poles in *rec8* mutants at the first meiotic division. It is therefore plausible that, in fact, similar principles govern the monopolar attach-ment process in both yeasts. The monopolar attachment of sister kinetochores may require not only the activity of Mam1-like proteins, which coordinate sister kine-tochores, but also cohesion between sister kinetochores without which Mam1-like proteins cannot even begin to act. Cohesion between sister kinetochores is possibly lacking during mitosis in *S. pombe* but is conferred by Rec8 during meiosis.

It is possible that monopolar attachment during meiosis I also depends on kine-tochore proteins that are not specific to meiotic cells. The Bub1 protein kinase, which is associated with kinetochores during mitosis as well as meiosis, also has some role in preventing equational segregation during meiosis I in *S. pombe*. In its absence, chromosomes segregate equationally in about 30% of cells (18). Bub1 is also needed during mitosis (and possibly also during meiosis) for delaying activa-tion of the APC when spindles are damaged or when chromosomes have failed to align on the metaphase plate (the mitotic checkpoint) (84). However, Bub1's role in promoting reductional chromosome segregation at meiosis I has little or nothing to do with its involvement in the mitotic checkpoint because mutation of other crucial components of the mitotic checkpoint such as Mad2 has little or no such effect. Bub1 clearly has multiple functions besides its role in the mitotic checkpoint and is even an essential gene in certain strains of *S. cerevisiae* (K. P. Rabitsch, personal communication). Rec8 is found at centromeres during meiosis I in *S. pombe bub1* mutants but it fails to persist at this location after anaphase I. Bub1 presumably regulates the state of Rec8 at meiosis I centromeres so that it resists destruction at the onset of anaphase I (see below). It is less clear whether this hypothetical alteration of the state of Rec8 in *bub1* mutants might be responsible for reducing, though not eliminating, monopolar attachment. It is possible that Bub1 has mul-tiple functions at the meiosis I centromere and regulates the activity of proteins involved in monopolar attachment independently of its modulation of the state of Rec8. It has also been reported that the Slk19 protein in *S. cerevisiae*, which is associated with centromeres during mitotic metaphase and with the spindle mid-zone during anaphase, is required to prevent equational chromosome segregation during meiosis I (92). Current evidence does not distinguish whether Slk19 acts primarily to protect cohesion at centromeres, to confer monopolar attachment, or is a more general regulator of the meiotic process.

Proteins like Spo12, which are neither meiosis-specific nor known to be as-sociated with centromeres, are also required to prevent equational chromosome segregation during meiosis I in *S. cerevisiae* (101, 195). Their function in regulat-ing monopolar attachment is still mysterious. Finally, little or nothing is known about the identity of proteins that confer monopolar attachment in animal or plant cells.

RESOLVING CHIASMATA

It has long been recognized that the resolution of chiasmata might be the trigger for the first meiotic division. If sister chromatid cohesion mediated by cohesin's Rec8 subunit is responsible for the ability of chiasmata to hold homologues together, then the first meiotic division could be triggered by the destruction of sister chromatid cohesion, just as occurs in mitosis. This hypothesis is consistent with the loss of sister chromatid cohesion along chromosome arms during anaphase I and with the finding in grasshoppers that homologous chromosomes disjoin at the same time as sister chromatids when transferred to the spindle of a meiosis II cell (159).

The persistence until metaphase I of Rec8 and STAG3 along the axes that lie between sister chromatid arms suggests that the process which removes cohesin from chromosome arms during prophase and prometaphase during mitosis either does not occur or occurs less efficiently during the first meiotic division. A key question is whether cleavage of Rec8 by separase or some other process removes this population of cohesin from chromosome arms at the onset of anaphase I. This issue has thus far only been investigated in *S. cerevisiae*, where several lines of evidence indicate that activation of separase might indeed be the trigger for chiasma resolution and hence anaphase I (26). Separase is required for the first meiotic division and securin is destroyed shortly before the onset of anaphase (183). Furthermore, the bulk of Rec8 is cleaved in a separase-dependent fashion at the onset of anaphase I at two sites, both of which resemble the separase cleavage sites in Scc1. Mutation of both but not just one cleavage site completely blocks meiosis I, even when only one of two copies of *REC8* are mutated. Finally, the block to meiosis I chromosome segregation imposed by separase inactivation or by nondegradable Rec8 is bypassed by deletion of the *SPO11* gene, which indicates that cleavage of Rec8 by separase is only needed for chromosome segregation during meiosis I if crossing over has previously connected homologous chromosomes. In *spo11* mutants homologous chromosomes are segregated by meiosis I spindles to the two poles at random in the complete absence of separase activity (26).

Whether cleavage of Rec8 by separase also triggers anaphase in animal cells is still unclear. On the one hand, there is clear evidence that both the APC and separase are required for meiosis I in *C. elegans*. Mutants with temperature-sensitive mutations in APC subunits arrest in metaphase of meiosis I when shifted to the restrictive temperature (51). Furthermore, inactivation of separase either by mutation or by RNA interference prevents the proper disjunction of homologues at meiosis I (M. Siomos, personal communication). On the other hand, there are indications that meiosis I in *Xenopus* oocytes might not require the APC. Injection of antibodies (166) or antisense RNA (210) directed against the APC activator protein Cdc20 (Fizzy) fails to block meiosis I, despite preventing the proteolysis of cyclin B. Furthermore, neither antibodies against Cdc27, a core APC subunit, nor high levels of the checkpoint protein Mad2, nor a nondegradable form of securin prevented the first meiotic division (166). This raises the possibility that chiasmata in vertebrates might be resolved by a mechanism that requires neither the APC nor separase. A variation on the pathway that removes cohesin during mitotic prophase is one possibility. However, this process would have to be very differently regulated in that it would have to remain inactive during diplotene, diakinesis, and metaphase I. Given the uncertainties surrounding the use of antibodies and antisense RNA, more rigorous genetic studies will be necessary to resolve whether the APC and separase really are redundant during meiosis I in vertebrates. It would be surprising though not unimaginable if chiasmata were resolved by very different mechanisms in worms and man.

In summary, it is possible though not yet certain that cleavage of Rec8 triggers the first meiotic division just as cleavage of its cousin Scc1 triggers mitotic chromosome segregation. According to this hypothesis, only two key innovations are required to convert a mitotic division into meiosis I during which homologues and not sister chromatids are segregated to opposite poles. These are the formation of chiasmata due to reciprocal recombination and the attachment of sister kinetochores to the same spindle pole. Chromosome alignment comes about because chiasmata resist the attempt of meiosis I spindles to pull homologues to opposite poles. The persistence of meiotic cohesin on chromosome arms until metaphase I, even in organisms that completely remove its mitotic counterpart from chromosome arms during prophase and prometaphase, is responsible for holding homologues together during their alignment on the meiosis I spindle. Control of chromosome arm cohesion must therefore differ between mitosis and meiosis. It is unclear whether this arises from differences between meiotic and mitotic cohesin or from other differences between mitotic and meiotic cells. Little is known about the role of condensin during the process of chiasmata resolution. One suspects that it will be found to bind to meiotic chromosomes during diplotene and diakinesis and to have a key role in resolving sister chromatid arms in preparation for their final separation at the metaphase to anaphase transition.

ACHIASMATE CHROMOSOME SEGREGATION

In many insects, there is little or no recombination during meiosis in the heterogametic sex. For example, meiosis takes place in the complete absence of recombination in *Drosophila* males (232). Despite this, homologues pair and subsequently disjoin at meiosis I. How they do this remains a mystery. Nevertheless, the very fact that they are able to pull off this feat raises questions as to why recombination is an obligate step for meiotic chromosome segregation in most other eukaryotes. There are several possible explanations. Chiasmata might just be a more effective method of holding homologues together. Alternatively, their use for chromosome segregation might provide a mechanism for ensuring that gametes are not produced in the absence of recombination, which has its own independent merits. It is also possible that by linking the production of gametes to the process of recombination, most eukaryotes make it very difficult for themselves to abandon sexual reproduction, which may have short-term advantages but be disastrous in the longer term.

It is clearly important to understand not only how homologues synapse in the absence of recombination but also how they are triggered to disjoin at the onset of anaphase I. The process of homologue pairing may have much in common with the synapsis between homologues that takes place prior to recombination in most if not all eukaryotes. Thus, both homologue pairing and formation of SC take place in mutants defective in the Spo11 endonuclease in *Drosophila* (136) and in *C. elegans* (38). Whether cohesin has a role in this process is unclear. Inactivation of Rec8 in *C. elegans* by RNA interference does not prevent initial alignment of homologues but does prevent SC formation (163).

Whereas homologues fail to remain paired in *spo11* mutants in *C. elegans* after dissolution of the SC, with disastrous consequences for chromosome segregation, they remain associated in *Drosophila* males during their alignment on the meiosis I spindle and then disjoin to opposite poles at anaphase I. Interestingly, the small fourth chromosome in *Drosophila* does not undergo recombination even in females, and its proper segregation presumably depends on the same sort of mechanism that governs segregation during male meiosis. This mechanism, known as distributive pairing, also "kicks in" when autosomes fail to form chiasmata in females. A kinesin-like protein called Nod is crucial for preventing the precocious disjunction of achiasmate homologues in *Drosophila* (3, 251). Nod is neither required during mitosis nor for preventing precocious disjunction of homologues connected by chiasmata (bivalents). Nod and its homologues in *Xenopus*, Xkid, are associated with chromosome arms and are therefore called chromokinesins (9, 49). They are thought to participate in the process by which microtubules unconnected to kinetochores manage to "blow" chromosomes toward the equator of the spindle apparatus. Xkid, for example, is essential for the proper congression of chromosomes to the metaphase plate. It would appear that in Nod's absence, achiasmate connections between homologues are insufficient to resist the tendency of kinetochore-attached microtubules to pull homologues toward opposite poles during metaphase I. Without Nod, chromosome 4 homologues disjoin either prior to or during metaphase I while all other chromosome pairs remain attached by chiasmata and align on the metaphase I spindle (135). Remarkably, Xkid is destroyed at the metaphase to anaphase transition by the APC (49). Furthermore, nondegradable versions block chromosome disjunction in *Xenopus* extracts. If Nod were destroyed likewise, the APC would not only trigger disjunction of chiasmata by activating separase but also the disjunction of achiasmate homologues by triggering destruction of Nod. This could explain how *Drosophila* oocytes manage the remarkable feat of triggering disjunction of chiasmate and achiasmate homologues at around the same time during meiosis I.

RETAINING COHESION AROUND CENTROMERES

Though monopolar attachment of sister kinetochores and the production of chiasmata that are resolved by destruction of arm cohesion can explain the disjunction of homologues and not sister chromatids to opposite poles at meiosis I, they are not sufficient to explain how meiotic cells then manage to undergo a second round of chromosome segregation without an intervening round of DNA replication. This remarkable feat depends on the retention of cohesion in the vicinity of centromeres at the onset of anaphase I while at the same time destruction of cohesion along chromatid arms triggers resolution of chiasmata. Cohesion retained at centromeres is crucial for aligning sister chromatids on the metaphase II spindle and, in all likelihood, its destruction triggers the disjunction of sister chromatids at the onset of anaphase II (148).

Though necessary for normal chromosome segregation during meiosis II, the retention of cohesion at centromeres is unnecessary for meiosis I. Cohesion between sister centromeres fails to be maintained at the onset of anaphase I in *bub1* mutants in *S. pombe* (18), in *MEI-S322* mutants in *Drosophila* (95), and in strains of *S. cerevisiae* in which Rec8 has been replaced by Scc1 (223), but homologues nevertheless segregate in a (largely) reductional manner at the first meiotic division. This indicates that, once established during metaphase I, the continued attachment of sister kinetochores to spindles from the same pole does not require cohesion to persist after the onset of anaphase I and is sufficient to draw sister centromeres toward the same spindle pole throughout anaphase I. However, loss of cohesion at centromeres does cause sister chromatids to drift apart before they can be aligned on the meiosis II spindle and as a consequence they segregate in random directions to the poles during the second meiotic division.

There are many unanswered questions about the retention of cohesion at centromeres during and after the first meiotic division. How much cohesion must be retained for successful chromatid segregation during meiosis II? By what mechanism is cohesion protected from the process that destroys cohesion along chromatid arms? How is protection propagated from centromeres and what is the signal or seed that initiates the process? How is the propagation of protection blocked by the formation of chiasmata and how far would it propagate away from centromeres in the absence of chiasmata? Finally, how does cohesion at centromeres that had been resistant to dissolution at meiosis I acquire the ability to be dissolved during meiosis II?

A clue as to how much cohesion may be sufficient for meiosis II comes from the study of mitotic cells that have been arrested in a metaphase-like state by treatment with spindle poisons and then triggered to undergo anaphase by their removal. Under these circumstances, cohesion is completely lost from chromosome arms and is only retained within centromeric heterochromatin and yet chromatid segregation takes place with reasonably high fidelity (177). There is no reason to believe that this amount of cohesion would not be equally sufficient for meiosis II.

It has never been clear until recently whether the persistence of cohesion around centromeres is due to this cohesion being of a different or similar nature to that which connects chromatid arms. However, the observation, now in a wide variety of organisms, that Rec8 persists in the vicinity of centromeres until anaphase II while disappearing from chromosome arms at the onset of anaphase I suggests (but does not yet prove) that cohesin mediates cohesion at meiotic centromeres as well as along chromosome arms. For example, Rec8 persists in the vicinity of centromeres until the onset of anaphase II not only in *S. cerevisiae* (104) and *S. pombe* (238), where this phenomenon was first described, but also in *C. elegans* (163) and in mammals (C. Heyting, personal communication). During meiosis I, centromeric cohesin must resist not only dissociation by processes analogous to the mitotic prophase pathway but also cleavage by separase. Whether the persistence of cohesin at centromeres during meiosis I is due to meiosis-specific differences in its subunit composition or due to changes in its packing (higher-order structure) or modification is not known.

In budding yeast, where it is known that cleavage of Rec8 by separase triggers resolution of chiasmata, there are strong indications that Rec8 in the vicinity of centromeres somehow escapes this fate at meiosis I but nevertheless falls victim at meiosis II. Cleavage of Scc1 during mitosis is known to be both necessary and sufficient to cause its dissociation from chromosomes during anaphase (226). If the same applies to Rec8, then its persistence in the vicinity of centromeres after meiosis I indicates that it must have escaped cleavage by separase. By the same token, Rec8's disappearance from centromeres at the onset of anaphase II indicates that it may be cleaved at this juncture (104). Furthermore, the securin that reaccumulates rapidly after anaphase I is rapidly destroyed just prior to the onset of anaphase II, which implies that separase is activated at the right time to trigger the second as well as the first meiotic division (183). Indeed, given the similarities between mitosis and the second meiotic division, it is hard to believe that chromatid separation is mediated by completely different mechanisms.

It is therefore a reasonable working hypothesis that the retention of centromeric cohesion at anaphase I is due to the resistance of Rec8 to cleavage by separase while similar if not identical molecules on chromosome arms are destroyed at the same time. This resistance must be lost after anaphase I with the result that sister chromatid disjunction can be triggered by a second round of separase activation during meiosis II. To test this model directly, it will clearly be necessary to detect intact Rec8 on meiosis II chromosomes as well as its cleavage at the onset of anaphase II, which has so far not been possible due to the lack of synchrony of meiotic cultures. Whether this model applies to centromeric cohesion in other eukaryotes depends on whether cleavage of Rec8 by separase triggers resolution of their chiasmata. The finding in grasshoppers that meiotic sister chromatids can be disjoined when placed on a meiosis I spindle (159) is certainly consistent with the hypothesis as is the finding that the Rec8 protein retained at *C. elegans* (163) and mammalian (C. Heyting, personal communication) centromeres disappears after anaphase II.

There are several potential mechanisms that could protect Rec8 in the vicinity of centromeres from separase at the first meiotic division. Rec8 might be shielded by factors or modifications that prevent access of the protease, it might be associated with a protein, which, like securin, directly inhibits the protease's activity, or it might fail to be modified in a manner necessary for its cleavage (4). Rec8, like Scc1, might need to be phosphorylated before it can serve as an efficient separase substrate and the kinase responsible for "preparing" Rec8 for cleavage might be excluded from centromeric chromatin. PLK or other mitotic kinases such as Aurora B or even Cdk1 could play a role in this process. The finding that histone H3 is phosphorylated during maize meiosis only on those chromosomal regions that will imminently lose cohesion suggests that the kinase responsible for this phosphorylation, Aurora B, might be directly involved (93). If phosphorylation of Rec8 by Aurora B were required for its cleavage by separase, then exclusion of Aurora B from centromeres at meiosis I could account for their continued cohesion until anaphase II. It is equally possible that the protection of Rec8 from separase cleavage is mediated by differences in the chromosomal distribution of other types of modification on proteins other than cohesin itself.

The finding that Scc1 can support sister chromatid cohesion and monopolar attachment during meiosis I in *S. cerevisiae* but cannot resist separase in the vicinity of centromeres provides an important clue about Rec8's resistance, at least in yeast (223). Rec8 clearly possesses special properties that are lacking in Scc1 that enable it to be protected from separase in the vicinity of centromeres. It also confirms the notion, first developed from the study of MEI-S332 in *Drosophila*, that the retention of cohesion at centromeres is conferred by a process that is independent of monopolar attachment. Because Rec8 and not Scc1 can be protected, retention of cohesion cannot be due to the general shielding of centromeric chromatin and cohesin complexes associated with it from enzymes like separase.

In the absence of any clear understanding about the biochemical basis for the retention of centromeric cohesion, the identification of proteins with a role in this process by genetics will be invaluable for providing clues as to its mechanism. One might expect that some if not most of the proteins necessary for protecting centromeric cohesion during meiosis would prove to be specific to meiotic cells. Strangely, no such protein has yet been discovered. All three proteins thus far implicated in protecting cohesion at centromeres are also expressed in mitotic cells. The first such protein to be identified was MEI-S332 from *Drosophila*, which associates with the pericentric heterochromatin adjacent to but not coincident with kinetochores from prometaphase I until the onset of anaphase II (117). In its absence, bivalents disjoin normally at anaphase I but sister chromatids soon thereafter separate and mis-segregate at meiosis II. MEI-S332 behaves in similar fashion during mitotic divisions, associating with centromeric chromatin during prometaphase and dissociating at anaphase. It is not, however, required for chromosome segregation during mitosis (150).

MEI-S332's absence from chromosomes until prometaphase (150) suggests that it is not itself part of the sister chromatid cohesion apparatus. Moreover, its presence on chromosomes during metaphase II and during mitotic metaphases implies that it does not directly protect centromeric cohesion from its imminent destruction. Though present on chromosomes during all metaphases, MEI-S332 only protects cohesion at the onset of anaphase I or during a short period thereafter. It is not immediately obvious why a protein with such a role should dissociate from chromosomes when sister chromatids separate during mitosis or meiosis II. MEI-S332's dissociation from chromatids whenever they separate raises the possibility that it might be caused by the process that triggers sister separation. If so, a conundrum arises: MEI-S332 may both regulate and be regulated by the sister separation process. It is clearly important to establish the mechanism by which MEI-S332 dissociates from chromatids as they separate as well as the mechanism by which it regulates loss of cohesion. Is its dissociation from chromosomes regulated by separase and the APC, by mitotic protein kinases, or by an entirely novel process? Does MEI-S332 regulate loss of cohesion by regulating cleavage of cohesin subunits by separase?

A model that could tie all these phenomena together is that MEI-S332 forms a complex with cohesin on centromeric heterochromatin. Disruption of this complex by cleavage of cohesin's Scc1/Rec8 subunit would explain why MEI-S332 dissociates from chromosomes whenever cohesion is destroyed. During meiosis I but not during mitosis or meiosis II, some as yet mysterious factor or set of conditions alters the MEI-S332/cohesin complex so that it, but not cohesin that has not bound MEI-S332, becomes resistant or inaccessible to separase. Thus MEI-S332 only regulates loss of cohesion during meiosis I but nevertheless dissociates from chromatin at each and every anaphase, except at anaphase I when cohesin is not cleaved in the vicinity of centromeres. Though not essential, MEI-S332 presumably does have some role in mitotic cells, possibly helping to protect cohesion under certain circumstances (150).

The Bub1 protein in *S. pombe* is also required to prevent destruction of cohesion at anaphase I (18). In *S. pombe bub1* mutants, Rec8 fails to persist in the vicinity of centromeres after anaphase I and as a result sister chromatids drift apart before anaphase II. It is unclear whether Bub1 is located at kinetochores (i.e., in the inner centromere region), or in the neighbouring heterochromatin (i.e., in the outer centromere regions). If, as in mammals, Bub1 is located within the kinetochore itself (218), it may be required for sending a signal from the kinetochore to its surrounding heterochromatin—a signal that initiates the formation (and propagation along the chromosome) of cohesion that is resistant to separase. Bub1 is also required for the surveillance mechanism (the mitotic checkpoint) that delays activation of the APC until all chromosomes have correctly aligned on the metaphase spindle. However, Bub1's role in protecting cohesion must be independent of the mitotic checkpoint because *mad2* mutants, which are equally defective in the checkpoint, are not defective in protecting centromeric cohesion. Like MEI-S332, Bub1 is not specific to meiotic cells. Some other meiosis-specific factor is presumably also required for the formation of separase-resistant cohesin at *S. pombe* centromeres during meiosis I.

The last centromeric protein to be implicated in protecting cohesion during meiosis I is Slk19 in *S. cerevisiae*. Slk19 associates with centromeres during mitosis and meiosis but a sizeable fraction of the protein relocates to the midzone of the mitotic spindle during anaphase where it has an important role in stabilizing late anaphase spindles (249). It has been reported that deletion of *SLK19* causes sister chromatids to separate during meiosis I in a high fraction of cells and causes Rec8 to disappear from centromeres precociously (92). If true, Slk19, like Spo13, must be required both for monopolar attachment and for protecting cohesion. Deletion of

SLK19 also prevents a second meiotic division. Because of this and because current studies have not carefully compared the kinetics of meiotic events in wild-type and mutant cells, it is not possible to be sure at this juncture whether the equational division observed in *slk19* mutants is due to precocious sister chromatid separation or due to an aborted reductional division. If the Slk19 protein does indeed have a role in protecting cohesion at centromeres, then a crucial question is whether this function is mechanistically related to its role in stabilizing anaphase spindles or whether these are independent functions of the same protein.

In summary, the ability of centromeric cohesion to resist the process that resolves chiasmata is conferred by a process that is to a considerable extent independent of that which causes the monopolar attachment of sister kinetochores. This is possibly not too surprising when one considers that cohesion is retained within the entire peri-centric heterochromatin whereas monopolar attachment only concerns kinetochores. Thus far, the only meiosis-specific protein to be implicated in retention of cohesion is Rec8, whose replacement by Scc1 in *S. cerevisiae* abolishes the retention of cohesion. It is likely that other meiosis-specific proteins involved in this process await discovery. No meiosis-specific Scc1-like subunit has emerged from sequencing of the *Drosophila* genome. Though it is conceivable that *Drosophila*'s Rec8 gene lurks in hitherto unsequenced heterochromatin regions, it is equally likely that flies use a single Scc1-like protein for mitosis and meiosis. Rec8, for example, is capable of complementing the complete lack of Scc1/Rad21 in both budding and fission yeast, so there is no intrinsic reason why a single gene could not suffice for meiosis and mitosis in flies. If so, then *Drosophila* at least must possess some other meiosis-specific protein that causes the protection of centromeric cohesion during meiosis I.

At least three different types of protein are implicated in protecting centromeric cohesion: (*a*) those residing at kinetochores (like Bub1 in *S. pombe*), which provide a "spatial" signal saying that the adjacent heterochromatin and not that present elsewhere on the chromosome is an appropriate substrate for protection; (*b*) those within centromeric heterochromatin itself (like MEI-S332 in flies), which might facilitate protection from separase; and (*c*) those that are specific to meiotic cells which ensure that this remarkable process only occurs during meiosis I. How the "protected" chromatin propagates along chromosomes and how this is blocked by the formation of chiasmata are genuine mysteries that await further investigation. Though crossovers must be capable of blocking propagation, they do not seem to be necessary to curb its extent to a limited region around the centromere, because the bulk of Rec8 along chromosome arms is still destroyed by separase when recombination has been eliminated by deletion of the *SPO11* gene in *S. cerevisiae* (104).

ANEUPLOIDY IN HUMANS

Mis-segregation of chromosomes during meiosis or mitosis leads to cells with altered numbers of chromosomes, which is known as aneuploidy. Aneuploidy due to mis-segregation during meiosis is usually lethal for mammalian embryos and is a leading cause of spontaneous miscarriages in humans (61). One third of all spontaneously aborted embryos are trisomic for at least one chromosome. Inheritance of an extra chromosome 21 in humans (trisomy 21) is not lethal but leads to Down's syndrome, the leading single cause of mental retardation. Meanwhile, it has long been recognized that the cells of most malignant solid tumors are highly aneuploid due presumably to frequent chromosome mis-segregation during mitosis (114). Though it has never been established whether chromosome mis-segregation and/or aneuploidy actually promotes the genesis of tumors, it could make a major contribution to the "uncovering" of recessive mutations in tumor suppressor genes. Indeed, mice heterozygous for a *mad2* mutation, which compromises control of the APC by the mitotic checkpoint, are prone to lung tumors (145), though whether this is due to their failure to regulate mitosis as opposed to other steps of the cell cycle controlled by the APC is presently unclear.

Defects in sister chromatid cohesion, resolution, and separation could all contribute to the genesis of aneuploidy. Though essential for mitosis and meiosis, partial, i.e., nonlethal defects, in cohesin or condensin subunits cause very high rates of chromosome loss in yeast (128, 144, 198, 222), and there is no reason to believe that they would not do likewise in humans. Indeed, deletion of the securin gene in a human colon carcinoma cell line with a stable karyotype is sufficient to cause the extreme karyotypic instability characteristic of many malignant colon carcinoma cell lines (89).

Defects in the sister chromatid cohesion apparatus also compromise double-strand break repair both in mitosis and meiosis (21, 104, 197) and could thereby also contribute to the genome instability of somatic tumor cells and to infertility due to defects in gametogenesis. Given that cohesin may have a fundamental role in the organization of interphase chromatin in every cell of our bodies, defects in its activity or regulation could have extremely pleiotropic and damaging consequences.

Because it gives rise to Down's syndrome, there has been extensive investigation of trisomy arising from chromosome mis-segration during meiosis. There are several potential causes: defective recombination leading to a lack of chiasmata (68), instability of chiasmata (i.e., precocious resolution), defects in sister chromatid cohesion that cause chromatids to disjoin before formation of the first meiotic spindle, a failure to retain cohesion between sister centromeres after meiosis I, defects in a back up system that facilitates disjunction of homologues even in the absence of chiasmata (135) or precocious loss of the "reductional" state needed for both monopolar attachment and retention of centromeric cohesion.

Despite extensive study of the etiology of trisomy, no single mechanism has been pinpointed. Several important conclusions have nevertheless been reached (69). First, there is considerable variation in the incidence of trisomy between chromosomes, with that of chromosome 16 by being far the most frequent among spontaneous abortions. Second, the vast majority of trisomies arise due to mis-segregation in oocytes. Third, the frequency of mis-segregation rises steeply with maternal age. Fourth, the majority of segregation errors must have occurred during the first meiotic division but those occurring at meiosis II are far from negligible especially for particular chromosomes.

Whether or not chromosomes had mis-segregated at the first or second division can to some extent be determined by scoring whether the zygote inherited centromere proximal markers from the same or different grandparents. Maternal sister centromeres normally segregate away from paternal sister centromeres at the first meiotic division. Thus, if one of the two centromeres inherited from a mis-segregating oocyte (there should have been one) has a paternal origin and the other a maternal one, there must have been an error during meiosis I. Possible causes for this "MI" mis-segregation are precocious loss of sister chromatid cohesion prior to alignment on the meiosis I spindle, an equational instead of a reductional division, a lack of chiasmata due to recombination failure, or (precocious) resolution of chiasmata before chromosome alignment on the meiosis I spindle. If, on the other hand, both centromeres derived from the misbehaving oocyte have the same origin, i.e., both are paternal or both maternal, then mis-segregation might have occurred at the second division (MII errors) and could have been caused by a failure to retain cohesion at centromeres after meiosis I or due to misalignment on the meiosis II spindle. However, it is still possible that some so-called "MII" trisomic embryos could have arisen due to abnormalities during meiosis I, if, for example, there had been a precocious loss of sister chromatid cohesion.

The fifth important finding is that many but by no means all trisomies are associated with a lack of recombination. Of embryos with MI errors, 40% of chromosome 21 trisomies were apparently achiasmatic. Furthermore, even when recombination has occurred, it tends to be more telocentric than in controls. It is conceivable that such crossovers are more likely to produce "unstable" bivalents in which sister chromatid cohesion distal to the crossover is insufficient to hold homologues together. However, a failure to recombine seems less likely to contribute to trisomy 16. The finding that achiasmatic trisomies increase in frequency with maternal age is a conundrum because recombination takes place before birth! This raises the possibility that "susceptible" chromosomes arise prenatally and are abnormally processed only much later, possibly when induced to mature. Defects in sister chromatid cohesion could contribute to this phenomenon because cohesion established prenatally during premeiotic recombination could be abnormally processed during subsequent age-dependent oocyte maturation. Furthermore, alterations in the metabolism of meiotic cohesins could contribute to recombination defects as well as sister chromatid cohesion defects.

One of the difficulties in studying human trisomy is that it is impossible to observe either of the meiotic divisions that gives rise to it. One way of overcoming this problem has been to analyze matured oocytes collected from women visiting infertility clinics (8, 244). The premise for this approach has been that some, if not a large fraction, of infertility is caused by meiotic chromosome mis-segregation. Such oocytes, which are arrested in metaphase II prior to fertilization, are far more frequently found to possess extra chromatids than they are to possess an extra pair of sister chromatids. It is unlikely, though still possible, that such chromatids arise due to a lack of recombination or due to precocious resolution of chiasmata. They are more likely to have arisen either due to an equational division of that chromosome at meiosis I (which would occur if the chromosome's reductional state were lost before meiosis I as occurs in *spo13* mutants in yeast) or to precocious loss of cohesion throughout the chromosome prior to meiosis I or to a failure to retain cohesion between sister centromeres after the first meiotic division. There is some reason to believe that the oocytes that actually give rise to trisomy might derive from this pool of oocytes with extra chromatids because their pattern of aneuploidy resembles that found in aborted fetuses. Future studies will be required to establish whether these findings also apply to oocytes from the female population at large.

SUMMARY

The chromosome movements that constitute mitosis were first properly described in 1880 by Walter Flemming (47). By noting that "the impetus causing chromosomes to split longitudinally acts simultaneously on all of them," Flemming clearly recognized the special nature of the metaphase to anaphase transition. Indeed, the discovery that chromosomes split longitudinally was an important clue that they might carry the hereditary material and that differentiation did not take place through its unequal distribution (129). For over a century, analysis of this process was confined to cytological descriptions, which clearly delineated the various phases of mitotic and meiotic chromosome morphogenesis as well as the crucial role of the spindle apparatus (36, 130, 158, 186). It is remarkable that despite major advances in our understanding of microtubules and their dynamics during the past 25 years (146), we have until very recently remained rather ignorant of the biochemical mechanisms that regulate chromosome morphology. Thus, Miyazaki and Orr-Weaver wrote as recently as 1994 that "It is critical that our understanding of sister chromatid cohesion move to a molecular level. The cell cycle signals that trigger the dissolution of sister chromatid cohesion need to be elucidated and the proteins promoting cohesion isolated" (147). This call for action was a prescient one, as the Anaphase-promoting complex was discovered merely one year later (88, 97, 203).

Though we now have a very clear picture of some of the cell cycle signals that trigger dissolution of sister chromatid cohesion, much remains to be discovered. What, for instance, is the signal that triggers dissociation of cohesin and association of condensin during prophase and by what mechanism does this occur? How is this pathway differently regulated during meiosis, which permits chiasmata to persist until the onset of anaphase I? What protects cohesin in the vicinity of centromeres from the "prophase" pathway during mitosis and from separase at the onset of anaphase I? We still have little or no idea how securin regulates separase activity, how unoccupied kinetochores inhibit the APC, or even what determines APC activation in the absence of such surveillance mechanisms. We also have little idea about the mechanism that to some limited extent confines anaphase triggers, such as active APC or separase, to a single mitotic spindle (14, 178, 231). Though many of the proteins that mediate sister chromatid cohesion (cohesin) and resolution (condensin) have been identified, we have a poor grasp, if any, as to the mechanism by which they function, without which it will not be possible to understand how cleavage of Scc1 and Rec8 triggers the destruction of cohesion and hence the onset of anaphase. The complete determination of genome sequences has tended to lull us into a false sense that we understand chromosomes. They are one of the cell's most complex organelles and one about which we remain hugely ignorant.

Even when we have answers to these questions, the mystery of how the mitotic and meiotic process evolved will remain. There is a strong tendency in evolutionary biology to view current cellular mechanisms as the result of past accidents. Though many if not most of the details might be attributable to contingency, there is an underlying logic to the mitotic and meiotic process that clearly "awaited" discovery once the first genomes had arisen (153). There is only a single solution to the blind men's riddle and nature sooner rather than later came up with it. Its evolution was therefore no accident. Far more mysterious is the pathway that facilitated its selection. The conundrum is a familiar one to evolutionary biology. Mitosis requires two fundamental processes, neither of which is much use without the other. The spindle apparatus presumably evolved first for transporting molecules around the cell and for controlling cell morphology, whereas the sister chromatid cohesion apparatus presumably evolved for double-strand break repair. Only when both of these key innovations were in place could they together be used for segregating chromosomes.

Meiosis is in a way less mysterious because the only fundamental extra innovation required was a method for systematizing crossing over between homologous chromosomes, which enables sister chromatid cohesion to hold homologous chromosomes together on the first meiotic spindle. The suggestion that sisters only separate at the second division in order to avoid "sister killers" (64) does not fully take into account the actual mechanisms by which chromatids are segregated during meiosis and how this might have evolved from mitosis. It could merely be a curious accident, from which most eukaryotic organisms profit, that sister killers are indeed minimized by the meiotic process that was most readily derived from what may have been pre-existing mitotic processes. Whether we will ever be able to answer these "why" questions will depend on the survival of missing links that confirm, for instance, that mitosis did indeed precede meiosis. Sadly, the superiority of sexual reproduction may have ensured their demise. No existing eukaryotic phylum lacks meiosis. It is, however, equally possible that the simpler process (mitosis) in fact evolved from the more complicated one (meiosis). Though this may appear counterintuitive, it is actually easier to imagine rudimentary cohesion/spindle apparatuses being used initially on a sporadic basis for sexual purposes before they were sufficiently refined for disseminating genomes in an efficient manner during cell proliferation. In the absence of missing links, we will have to remain content with addressing "how" questions, of which plenty remain.

ACKNOWLEDGMENTS

I would like to thank Hannes Tkadletz for help with drawing the figures, Alex Schleiffer for compiling the table of separase cleavage sites, Jan-Michael Peters for comments on the manuscript, Diane Turner for help preparing this manuscript, Jean-Louis Sikorav for telling me about the riddle of the blind men and their socks, and all those who kindly communicated information prior to publication.

Visit the Annual Reviews home page at www.AnnualReviews.org

LITERATURE CITED

1. Adams MD, Celniker SE, Holt RA, Evans CA, Gocayne JD, et al. 2000. The genome sequence of *Drosophila melanogaster. Science* 287:2185–95
2. Adams RR, Carmena M, Earnshaw WC. 2001. Chromosomal passengers and the (aurora) ABCs of mitosis. *Trends Cell Biol.* 11:49–54
3. Afshar K, Barton NR, Hawley RS, Goldstein LS. 1995. DNA binding and meiotic chromosomal localization of the Drosophila nod kinesin-like protein. *Cell* 81:129–38
4. Alexandru G, Uhlmann F, Mechtler K, Poupart MA, Nasmyth K. 2001. Phosphorylation of the cohesin subunit Scc1 by polo/cdc5 kinase regulates sister chromatid separation in yeast. *Cell* 105:459–72
5. Alexandru G, Zachariae W, Schleiffer A, Nasmyth K. 1999. Sister chromatid separation and chromosome re-duplication are regulated by different mechanisms in response to spindle damage. *EMBO J.* 18:2707–21
6. Amon A. 1999. The spindle checkpoint. *Curr. Opin. Genet. Dev.* 9:69–75
7. Deleted in proof
8. Angell R. 1997. First meiotic-division nondisjunction in human oocytes. *Am. J. Hum. Genet.* 61:23–32

9. Antonio C, Ferby I, Wilhelm H, Jones M, Karsenti E, et al. 2000. Xkid, a chromokinesin required for chromosome alignment on the metaphase plate. *Cell* 102:425–35
10. Deleted in proof
11. Bahler J, Wyler T, Loidl J, Kohli J. 1993. Unusual nuclear structures in meiotic prophase of fission yeast: a cytological analysis. *J. Cell Biol.* 121:241–56
12. Bai X, Peirson BN, Dong F, Xue C, Makaroff CA. 1999. Isolation and characterization of SYN1, A RAD21–like gene essential for meiosis in Arabidopsis. *Plant Cell* 11:417–30
13. Bajer A. 1958. Cine-micrographic studies on mitosis in endosperm IV. *Exp. Cell Res.* 14:245–56
14. Bajer AS, Mole-Bajer J. 1972. Spindle dynamics and chromosome movements. *Int. Rev. Cytol. Suppl.* 3:1–271
15. Barnes TM, Kohara Y, Coulson A, Hekimi S. 1995. Meiotic recombination, noncoding DNA and genomic organization in *Caenorhabditis elegans. Genetics* 141:159–79
16. Bergerat A, de Massy B, Gadelle D, Varoutas PC, Nicolas A, Forterre P. 1997. An atypical topoisomerase II from Archaea with implications for meiotic recombination. *Nature* 386:414–17

17. Bernander R. 2000. Chromosome replication, nucleoid segregation and cell division in archea. *Trends Microbiol.* 8:278–83

18. Bernard P, Maure JF, Javerzat JP. 2001. Fission yeast Bub1 is essential in setting up the meiotic pattern of chromosome segregation. *Nat. Cell Biol.* 3:522–26

19. Bhat MA, Philp AV, Glover DM, Bellen HJ. 1996. Chromatid segregation at anaphase requires the barren product, a novel chromosome-associated protein that interacts with Topoisomerase II. *Cell* 87:1103–14

20. Bhatt AM, Lister C, Page T, Fransz P, Findlay K, et al. 1999. The DIF1 gene of Arabidopsis is required for meiotic chromosome segregation and belongs to the REC8/RAD21 cohesin gene family. *Plant J.* 19:463–72

21. Birkenbihl RP, Subramani S. 1992. Cloning and characterization of rad21 an essential gene of Schizosaccharomyces pombe involved in double strand break repair. *Nucleic Acids Res.* 20:6605–11

22. Blat Y, Kleckner N. 1999. Cohesins bind to preferential sites along yeast chromosome III, with differential regulation along arms versus the centric region. *Cell* 98:249–59

23. Boy de la Tour E, Laemli UK. 1988. The metaphase scaffold is helically folded: sister chromatids have predominantly opposite helical handedness. *Cell* 55:937–44

24. Brunborg G, Williamson DH. 1978. The relevance of the nuclear division cycle to radiosensitivity in yeast. *Mol. Gen. Genet.* 162:277–86

25. Deleted in proof

26. Buonomo SB, Clyne RK, Fuchs J, Loidl J, Uhlmann F, Nasmyth K. 2000. Disjunction of homologous chromosomes in meiosis I depends on proteolytic cleavage of the meiotic cohesin Rec8 by separin. *Cell* 103:387–98

27. Chen J-M, Rawlings ND, Stevens RAE, Barrett AJ. 1998. Identification of the active site of legumain links it to caspases, clostripain and gingipains in a new clan of cysteine endopeptidases. *FEBS Lett.* 441:361–65

28. Ciosk R, Shirayama M, Shevchenko A, Tanaka T, Toth A, et al. 2000. Cohesin's binding to chromosomes depends on a separate complex consisting of Scc2 and Scc4 proteins. *Mol. Cell* 5:243–54

29. Ciosk R, Zachariae W, Michaelis C, Shevchenko A, Mann M, Nasmyth K. 1998. An Esp1/Pds1 complex regulates loss of sister chromatid cohesion at the metaphase to anaphase transition in yeast. *Cell* 93:1067–76

30. Deleted in proof

31. Cohen-Fix O, Koshland D. 1997. The anaphase inhibitor of *Saccharomyces cerevisiae* Pds1p is a target of the DNA damage checkpoint pathway. *Proc. Natl. Acad. Sci. USA* 94:14361–66

32. Cohen-Fix O, Peters J-M, Kirschner MW, Koshland D. 1996. Anaphase initiation in *Saccharomyces cerevisiae* is controlled by the APC-dependent degradation of the anaphase inhibitor Pds1p. *Genes Dev.* 10:3081–93

33. Comings DE. 1966. Centromere: absence of DNA replication during chromatid separation in human fibroblasts. *Science* 154:1463–64

34. D'Andrea RJ, Stratman R, Lehner CF, John UP, Saint R. 1993. The three rows gene of *Drosophila melanogaster* encodes a novel protein that is required for chromosomes disjunction during mitosis. *Mol. Cell Biol.* 4:1161–74

35. Dahmann C, Diffley JF, Nasmyth KA. 1995. S-phase-promoting cyclin-dependent kinases prevent re-replication by inhibiting the transition of replication origins to a pre-replicative state. *Curr. Biol.* 5:1257–69

36. Darlington CD. 1939. *The Evolution of Genetic Systems.* New York: Basic Books

37. Dej KJ, Orr-Weaver TL. 2000. Separation anxiety at the centromere. *Cell Biol.* 10:392–99

38. Dernburg AF, McDonald K, Moulder G,

Barstead R, Dresser M, Villeneuve AM. 1998. Meiotic recombination in *C. elegans* initiates by a conserved mechanism and is dispensable for homologous chromosome synapsis. *Cell* 94:387–98

39. Diffley JF. 2001. DNA replication: building the perfect switch. *Curr. Biol.* 11: R367–70

40. Dinardo S, Voelkel KA, Sternglanz RL. 1984. DNA topoisomerase mutant of *S. cerevisiae*: topoisomerase II is required for segregation of daughter molecules at the termination of DNA replication. *Proc. Natl. Acad. Sci. USA* 81:2616–20

41. Donachie WD. 1993. The cell cycle of *Escherichia coli. Annu. Rev. Microbiol.* 47:199–230

42. Dong H, Roeder GS. 2000. Organization of the yeast Zip1 protein within the central region of the synaptonemal complex. *J. Cell Biol.* 148:417–26

43. Donze D, Adams CR, Rine J, Kamakaka RT. 1999. The boundaries of the silenced HMR domain in *Saccharomyces cerevisiae. Genes Dev.* 13:698–708

44. Downes CS, Mullinger AM, Johnson RT. 1991. Inhibitors of DNA topoisomerase II prevent chromatid separation in mammalian cells but do not prevent exit from mitosis. *Proc. Natl. Acad. Sci. USA* 88: 8895–99

45. Earnshaw WC, Martins LM, Kaufmann SH. 1999. Mammalian caspases: structure, activation, substrates, and functions during apoptosis. *Annu. Rev. Biochem.* 68:383–424

46. Eichinger A, Beisel HG, Jacob U, Huber R, Medrano FJ, et al. 1999. Crystal structure of gingipain R: an Arg-specific bacterial cysteine proteinase with a caspase-like fold. *EMBO J.* 18:5453–62

47. Flemming W. 1879. Beiträge zur Kenntnisse der Zelle und ihrer Lebenserscheinungen. *Arch. Mikrosk. Anat.* 18:151–259

48. Funabiki H, Kumada K, Yanagida M. 1996. Fission yeast Cut1 and Cut2 are essential for sister chromatid separation, concentrate along the metaphase spindle and form large complexes. *EMBO J.* 15:6617–28

49. Funabiki H, Murray AW. 2000. The Xenopus chromokinesin Xkid is essential for metaphase chromosome alignment and must be degraded to allow anaphase chromosome movement. *Cell* 102:411–24

50. Funabiki H, Yamano H, Kumada K, Nagao K, Hunt T, Yanagida M. 1996. Cut2 proteolysis required for sister-chromatid seperation in fission yeast. *Nature* 381: 438–41

51. Furuta T, Tuck S, Kirchner J, Koch B, Auty R, et al. 2000. EMB-30: an APC4 homologue required for metaphase-to-anaphase transitions during meiosis and mitosis in *Caenorhabditis elegans. Mol. Biol. Cell* 11:1401–19

52. Furuya K, Takahashi K, Yanagida M. 1998. Faithful anaphase is ensured by Mis4, a sister chromatid cohesin molecule required in S phase and not destroyed in G1 phase. *Genes Dev.* 12:3408–18

53. Gachet Y, Tournier S, Millar JBA, Hyams JS. 2001. A MAP kinase-dependent actin checkpoint ensures proper spindle orientation in fission yeast. *Nature* 412: 352–55

54. Gall JG, Murphy C, Callan HG, Wu ZA. 1991. Lampbrush chromosomes. *Methods Cell Biol.* 36:149–66

55. Giet R, Glover DM. 2001. Drosophila aurora B kinase is required for histone H3 phosphorylation and condensin recruitment during chromosome condensation and to organize the central spindle during cytokinesis. *J. Cell Biol.* 152:669–82

56. Gimenez-Abian JF, Clarke DJ, Devlin J, Gimenez-Abian MI, De la Torre C, et al. 2000. Premitotic chromosome individualization in mammalian cells depends on topoisomerase II activity. *Chromosoma* 109:235–44

57. Glaser P, Sharpe ME, Raether B, Perego M, Ohlsen K, Errington J. 1997. Dynamic, mitotic-like behavior of a bacterial protein required for accurate chromosome partitioning. *Genes Dev.* 11:1160–68

58. Goldstein LSB. 1980. Mechanisms of chromosome orientation revealed by two meiotic mutants in *Drosophila melanogaster*. *Chromosoma* 78:79–111

59. Goldstein LSB. 1981. Kinetochore structure and its role in chromosome orientation during the first meiotic division in male *D. melanogaster*. *Cell* 25:591–602

60. Goshima G, Yanagida M. 2000. Establishing biorientation occurs with precocious separation of the sister kinetochores, but not the arms, in the early spindle of budding yeast. *Cell* 100:619–33

61. Griffin DK. 1996. The incidence, origin, and etiology of aneuploidy. *Int. Rev. Cytol.* 167:263–96

62. Guacci V, Hogan E, Koshland D. 1994. Chromosome condensation and sister chromatid pairing in budding yeast. *J. Cell Biol.* 125:517–30

63. Guacci V, Koshland D, Strunnikov A. 1997. A direct link between sister chromatid cohesion and chromosome condensation revealed through analysis of MCD1 in *S. cerevisiae*. *Cell* 91:47–57

64. Haig D, Grafen A. 1991. Genetic scrambling as a defence against meiotic drive. *J. Theor. Biol.* 153:531–58

65. Hanna JS, Kroll ES, Lundblad V, Spencer FA. 2001. *Sacharomyces cerevisiae* CTF18 and CTF4 are required for sister chromatid cohesion. *Mol. Cell Biol.* 21: 3144–58

66. Hartman T, Stead K, Koshland D, Guacci V. 2000. Pds5p is an essential chromosomal protein required for both sister chromatid cohesion and condensation in *Saccharomyces cerevisiae*. *J. Cell Biol.* 151:613–26

67. Hartwell LH, Weinert TA. 1989. Checkpoints: controls that ensure the order of cell cycle events. *Science* 246:629–34

68. Hassold T, Arbruzzo M, Adkins K, Griffin D, Merrill M, et al. 1996. Human aneuploidy: incidence, origin, and etiology. *Environ. Mol. Mutagen.* 28:167–75

69. Hassold T, Hunt P. 2001. To err (meiotically) is human: the genesis of human aneuploidy. *Nat. Rev. Genet.* 2:280–91

70. Hauf S, Waizenegger I, Peters JM. 2001. Cohesin cleavage by separase required for anaphase and cytokinesis in human cells. *Science.* In press

71. Hayles J, Fisher D, Woollard A, Nurse P. 1994. Temporal order of S phase and mitosis in fission yeast is determined by the state of the p34cdc2-mitotic B cyclin complex. *Cell* 78:813–22

72. He X, Asthana S, Sorger PK. 2000. Transient sister chromatid separation and elastic deformation of chromosomes during mitosis in budding yeast. *Cell* 101:763–75

73. Deleted in proof

74. Hirano T. 1998. SMC protein complexes and higher-order chromosome dynamics. *Curr. Opin. Cell Biol.* 10:317–22

75. Hirano T. 1999. SMC-mediated chromosome mechanics: a conserved scheme from bacteria to vertebrates? *Genes Dev.* 13:11–19

76. Hirano T. 2000. Chromosome cohesion, condensation, and separation. *Annu. Rev. Biochem.* 69:115–44

77. Hirano T, Kobayashi R, Hirano M. 1997. Condensins, chromosome condensation protein complexes containing XCAP-C, XCAP-E, and a Xenopus homolog of the Drosophila Barren protein. *Cell* 89:511–21

78. Hirano T, Mitchison TJ. 1994. A heterodimeric coiled-coil protein required for mitotic chromosome condensation in vitro. *Cell* 79:449–58

79. Holloway SL, Glotzer M, King RW, Murray AW. 1993. Anaphase is initiated by proteolysis rather than by the inactivation of maturation-promoting factor. *Cell* 73:1393–402

80. Holmes VF, Cozzarelli NR. 2000. Closing the ring: links between SMC proteins and chromosome partitioning, condensation, and supercoiling. *Proc. Natl. Acad. Sci. USA* 97:1322–24

81. Deleted in proof

82. Hopfer KP, Karcher A, Shin DS, Craig L, Arthur LM, et al. 2000. Structural biology of Rad50 ATPase: ATP-driven conformational control in DNA double strand break repair and the ABC-ATPase superfamily. *Cell* 101:789–800

83. Hoque MT, Ishikawa F. 2001. Human chromatid cohesin component hRad21 is phosphorylated in M phase and associated with metaphase centromeres. *J. Biol. Chem.* 276:5059–67

84. Hoyt MA, Trotis L, Roberts BT. 1991. *S. cerevisiae* genes required for cell cycle arrest in response to loss of microtubule function. *Cell* 66:507–17

85. Hsu J-Y, Sun Z-W, Li X, Reuben M, Tatchell K, et al. 2000. Mitotic phosphorylation of histone H3 is governed by Ipl1/aurora kinase and Glc7/PP1 phosphatase in budding yeast and nematodes. *Cell* 102:279–91

86. Hugerat Y, Simchen G. 1993. Mixed segregation and recombination of chromosomes and YACs during single-division meiosis in spo13 strains of *Saccharomyces cerevisiae*. *Genetics* 135:297–308

87. Hwang LH, Lau LF, Smith DL, Mistrot CA, Hardwick KG, et al. 1998. Budding yeast Cdc20: a target of the spindle checkpoint. *Science* 279:1041–44

88. Irniger S, Piatti S, Michaelis C, Nasmyth K. 1995. Genes involved in sister chromatid separation are needed for B-type cyclin proteolysis in budding yeast. *Cell* 81:269–78

89. Jallepalli PV, Waizenegger I, Bunz F, Langer S, Speicher MR, et al. 2001. Securin is required for chromosomal stability in human cells. *Cell* 105:445–57

90. Jensen S, Segal M, Clarke DJ, Reed SI. 2001. A novel role of the budding yeast separin Esp1 in anaphase spindle elongation: evidence that proper spindle association of Esp1 is regulated by Pds1. *J. Cell Biol.* 152:27–40

91. Kadyk LC, Hartwell LH. 1992. Sister chromatids are preferred over homologs as substrates for recombinational repair in *Saccharomyces cerevisiae*. *Genetics* 132:387–402

92. Kamieniecki RJ, Shanks RMQ, Dawson DS. 2000. Slk19p is necessary to prevent separation of sister chromatids in meiosis I. *Curr. Biol.* 10:1182–90

93. Kaszás É, Cande WZ. 2000. Phosphorylation of histone H3 in correlated with changes in the maintenance of sister chromatid cohesion during meiosis in maize, rather than the condensation of the chromatin. *J. Cell Sci.* 113:3217–26

94. Keeney S, Giroux CN, Kleckner N. 1997. Meiosis-specific DNA double-strand breaks are catalyzed by Spo11, a member of a widely conserved protein family. *Cell* 88:375–84

95. Kerrebrock AW, Miyazaki WY, Birnby D, Orr-Weaver TL. 1992. The Drosophila mei-S332 gene promotes sister-chromatid cohesion in meiosis following kinetochore differentiation. *Genetics* 130:827–41

96. Kimura K, Rybenkov VV, Crisona NJ, Hirano T, Cozzarelli NR. 1999. 13S condensin actively reconfigures DNA by introducing global positive Writhe: implications for chromosome condensation. *Cell* 98:239–48

97. King RW, Peters J, Tugendreich S, Rolfe M, Hieter P, Kirschner MW. 1995. A 20S complex containing CDC27 and CDC16 catalyzes the mitosis-specific conjugation of ubiquitin to cyclin B. *Cell* 81:279–88

98. Deleted in proof

99. Deleted in proof

100. Kitsberg D, Selig S, Brandeis M, Simon I, Keshet I, et al. 1993. Allele-specific replication timing of imprinted gene regions. *Nature* 364:459–63

101. Klapholz S, Esposito RE. 1980. Recombination and chromosome segregation during the single division meiosis in SPO12-1 and SPO13-1 diploids. *Genetics* 96:589–611

102. Kleckner N. 1996. Meiosis: How could

it work? *Proc. Natl. Acad. Sci. USA* 93: 8167–74

103. Deleted in proof

104. Klein F, Mahr P, Galova M, Buonomo SBC, Michaelis C, et al. 1999. A central role for cohesins in sister chromatid cohesion, formation of axial elements, and recombination during yeast meiosis. *Cell* 98:91–103

105. Kondrashov AS. 1994. The asexual ploidy cycle and the origin of sex. *Nature* 370: 213–16

106. Koshland D, Hartwell L. 1987. The structure of sister minichromosome DNA before anaphase in *Saccharomyces cerevisiae*. *Science* 238:1713–16

107. Koshland D, Strunnikov A. 1996. Mitotic chromosome condensation. *Annu. Rev. Cell Dev. Biol.* 12:305–33

108. Kramer ER, Scheuringer N, Podtelejnikov AV, Mann M, Peters J-M. 2000. Mitotic regulation of the APC activator proteins Cdc20 and Cdch1. *Mol. Biol. Cell* 11:1555–69

109. Kumada K, Nakamura T, Nagao K, Funabiki H, Nakagawa T, Yanagida M. 1998. Cut1 is loaded onto the spindle by binding to Cut2 and promotes anaphase spindle movement upon Cut2 proteolysis. *Curr. Biol.* 8:633–41

110. Lavoie BD, Tuffo KM, Oh S, Koshland D, Holm C. 2000. Mitotic chromosome condensation requires Brn1p, the yeast homologue of Barren. *Mol. Biol. Cell* 11: 1293–304

111. Deleted in proof

112. Leismann O, Herzig A, Heidmann S, Lehner CF. 2000. Degradation of Drosophila PIM regulates sister chromatid separation during mitosis. *Genes Dev.* 14: 2192–205

113. Lemon KP, Grossman AD. 2000. Movement of replicating DNA through a stationary replisome. *Mol. Cell* 6:1321–30

114. Lengauer C, Kinzler KW, Vogelstein B. 1997. Genetic instability in colorectal cancers. *Nature* 386:623–27

115. Li YF, Numata M, Wahls WP, Smith GR.

1997. Region-specific meiotic recombination in *Schizosaccharomyces pombe*: the rec11 gene. *Mol. Microbiol.* 23:869–78

116. Lin DC, Grossman AD. 1998. Identification and characterization of a bacterial chromosome partitioning site. *Cell* 92:675–85

117. Lopez JM, Karpen GH, Orr-Weaver TL. 2000. Sister-chromatid cohesion via MEI-S332 and kinetochore assembly are separable functions of the Drosophila centromere. *Curr. Biol.* 10:997–1000

118. Losada A, Hirano M, Hirano T. 1998. Identification of Xenopus SMC protein complexes required for sister chromatid cohesion. *Genes Dev.* 12:1986–97

119. Losada A, Hirano T. 2001. Intermolecular DNA interactions stimulated by the cohesin complex in vitro. Implications for sister chromatid cohesion. *Curr. Biol.* 11:268–72

120. Losada A, Yokochi T, Kobayashi R, Hirano T. 2000. Identification and characterization of SA/Scc3p subunits in the Xenopus and human cohesin complexes. *J. Cell Biol.* 150:405–16

121. Lowe J, Cordell SC, van den Ent F. 2001. Crystal structure of the SMC head domain: an ABC ATPase with 900 residues antiparallel coiled-coil inserted. *J. Mol. Biol.* 306:25–35

122. MacGuire MP. 1990. Sister chromatid cohesiveness: vital function, obscure mechanism. *Biochem. Cell Biol.* 68:1231–42

123. Machado C, Andrew DJ. 2000. D-Titin: a giant protein with dual roles in chromosomes and muscles. *J. Cell Biol.* 151:639–52

124. Manoury B, Hewitt EW, Morrice N, Dando PM, Barrett AJ, Watts C. 1998. An asparaginyl endopeptidase processes a microbial antigen for class II MHC presentation. *Nature* 396:695–99

125. Marsden MP, Laemmli UK. 1979. Metaphase chromosome structure: evidence for a radial loop model. *Cell* 17:849–58

126. Martinez-Balbas MA, Dey A, Rabindran

SK, Ozato K, Wu C. 1995. Displacement of sequence-specific transcription factors from mitotic chromatin. *Cell* 83:29–38

127. May GS, McGoldrick CA, Holt CL, Denison SH. 1992. The bimB3 mutation of *Aspergillus nidulans* uncouples DNA replication from the completion of mitosis. *J. Biol. Chem.* 267:15737–43

128. Mayer ML, Gygi SP, Aebersold R, Hieter P. 2001. Identification of RFC (Ctf18p, Ctf8p, Dcc1p): an alternative RFC complex required for sister chromatid cohesion in *S. cerevisiae. Mol. Cell* 7:959–70

129. Mayr E. 1982. *The Growth of Biological Thought*, p. 677. Cambridge: Harvard Univ. Press

130. Mazia D. 1961. *Mitosis and the Physiology of Cell Division*, pp. 77–412. New York: Academic

131. McCarroll RM, Esposito RE. 1994. SPO13 negatively regulates the progression of mitotic and meiotic nuclear division in *Saccharomyces cerevisiae. Genetics* 138:47–60

132. McGarry TJ, Kirschner MW. 1998. Geminin, an inhibitor of DNA replication, is degraded during mitosis. *Cell* 93:1043–53

133. McGrew JT, Goetsch L, Byers B, Baum P. 1992. Requirement for ESP1 in the nuclear division of *S. cerevisiae. Mol. Biol. Cell* 3:1443–54

134. McKim KS, Green-Marroquin BL, Sekelsky JJ, Chin G, Steinberg C, et al. 1998. Meiotic synapsis in the absence of recombination. *Science* 279:876–78

135. McKim KS, Hawley RS. 1995. Chromosomal control of meiotic cell division. *Science* 270:1595–601

136. McKim KS, Hayashi-Hagihara A. 1998. mei-W68 in *Drosophila melanogaster* encodes a Spo11 homolog: evidence that the mechanism for initiating meiotic recombination is conserved. *Genes Dev.* 12:2932–42

137. Deleted in proof

138. Megee PC, Mistrot C, Guacci V, Koshland D. 1999. The centromeric sister chromatid site directs Mcd1p binding to adjacent sequences. *Mol. Cell* 4:445–50

139. Melander Y. 1950. Studies on the chromosomes of Ulophysema Öresundense. *Heriditas* 36:233–55

140. Melby TE, Ciampaglio CN, Briscoe G, Erickson HP. 1998. The symmetrical structure of structural maintainance of chromosomes (SMC) and MukB proteins: Long, antiparallel coiled coils, folded at a flexible hinge. *J. Cell Biol.* 142:1595–604

141. Mercier R, Vezon D, Bullier E, Motomayor JC, Sellier A, et al. 2001. SWITCH1 (SWI1): a novel protein required for the establishment of sister chromatid cohesion and for bivalent formation at meiosis. *Genes Dev.* 15:1859–71

142. Meuwissen RL, Meerts I, Hoovers JM, Leschot NJ, Heyting C. 1997. Human synaptonemal complex protein 1 (SCP1): isolation and characterization of the cDNA and chromosomal localization of the gene. *Genomics* 39:377–84

143. Meuwissen RL, Offenberg HH, Dietrich AJ, Riesewijk A, van Iersel M, Heyting C. 1992. A coiled-coil related protein specific for synapsed regions of meiotic prophase chromosomes. *EMBO J.* 11: 5091–100

144. Michaelis C, Ciosk R, Nasmyth K. 1997. Cohesins: chromosomal proteins that prevent premature separation of sister chromatids. *Cell* 91:35–45

145. Michel LS, Liberal V, Chatterjee A, Kirchwegger R, Pasche B, et al. 2001. MAD2 haplo-insufficiency causes premature anaphase and chromosome instability in mammalian cells. *Nature* 409:355–59

146. Mitchison TJ, Salmon ED. 2001. Mitosis: a history of division. *Nat. Cell Biol.* 3:E17–21

147. Miyazaki WY, Orr-Weaver TL. 1994. Sister-chromatid cohesion in mitosis and meiosis. *Annu. Rev. Genet.* 28:167–87

148. Molnar M, Bahler J, Sipiczki M, Kohli J. 1995. The rec8 gene of *Schizosaccharomyces pombe* is involved in linear element formation, chromosome pairing and

sister-chromatid cohesion during meiosis. *Genetics* 141:61–73

149. Moore DP, Orr-Weaver TL. 1998. Chromosome segregation during meiosis: building an unambivalent bivalent. *Curr. Top. Dev. Biol.* 37:263–99

150. Moore DP, Page AW, Tang TT, Kerrebrock AW, Orr-Weaver TL. 1998. The cohesion protein MEI-S332 localizes to condensed meiotic and mitotic centromeres until sister chromatids separate. *J. Cell Biol.* 140:1003–12

151. Muller HJ. 1916. The mechanism of crossing-over. *Am. Nat.* 50:193–221

152. Murray AW, Szostak JW. 1985. Chromosome segregation in mitosis and meiosis. *Annu. Rev. Cell Biol.* 1:289–315

153. Nasmyth K. 1995. Evolution of the cell cycle. *Philos. Trans. R. Soc. London Ser. B* 349:271–81

154. Nasmyth K. 1999. Separating sister chromatids. *Trends Biochem. Sci.* 24:98–104

155. Nasmyth K, Peters JM, Uhlmann F. 2000. Splitting the chromosome: cutting the ties that bind sister chromatids. *Science* 288:1379–85

156. Neuwald AF, Hirano T. 2000. HEAT repeats associated with condensins, cohesins, and other complexes involved in chromosome-related functions. *Genome Res.* 10:1445–52

157. Nicklas RB, Ward SC. 1994. Elements of error correction in mitosis: microtubule capture, release, and tension. *J. Cell Biol.* 126:1241–53

158. Östergren G. 1951. The mechanism of co-orientation in bivalents and multivalents. *Hereditas* 37:85–156

159. Paliulis LV, Nicklas RB. 2000. The reduction of chromosome number in meiosis is determined by properties built into the chromosomes. *J. Cell Biol.* 150:1223–31

160. Panizza S, Tanaka T, Hohchwagen A, Eisenhaber F, Nasmyth K. 2000. Pds5 cooperates with cohesin in maintaining sister chromatid cohesion. *Curr. Biol.* 10:1557–64

161. Parry DH, O'Farrell PH. 2001. The schedule of destruction of three mitotic cyclins can dictate the timing of events during exit from mitosis. *Curr. Biol.* 11:671–83

162. Partridge JF, Borgstrom B, Allshire RC. 2000. Distinct protein interaction domains and protein spreading in a complex centromere. *Genes Dev.* 14:783–91

163. Pasierbek P, Jantsch M, Melcher M, Schleiffer A, Schweizer D, Loidl J. 2001. A *Caenorhabditis elegans* cohesion protein with functions in meiotic chromosome pairing and disjunction. *Genes Dev.* 15:1349–60

164. Pearson CG, Maddox PS, Salmon ED, Bloom K. 2001. Budding yeast chromosome structure and dynamics during mitosis. *J. Cell Biol.* 152:1255–66

165. Pelttari J, Hoja M, Yuan L, Liu JG, Brundell E, et al. 2001. A meiotic chromosomal core consisting of cohesin complex proteins recruits DNA recombination proteins and promotes synapsis in the absence of an axial element in mammalian meiotic cells. *Mol. Cell Biol.* 21:5667–77

166. Peter M, Castro A, Lorca T, Le Peuch C, Magnaghi-Jaulin L, et al. 2001. The APC is dispensable for first meiotic anaphase in Xenopus oocytes. *Nat. Cell Biol.* 3:83–87

167. Deleted in proof

168. Philp AV, Axton JM, Saunders RD, Glover DM. 1993. Mutations in the *Drosophila melanogaster* gene three rows permit aspects of mitosis to continue in the absence of chromatid segregation. *J. Cell Sci.* 106:87–98

169. Pidoux AL, Allshire RC. 2000. Centromeres: getting a grip of chromosomes. *Curr. Opin. Cell Biol.* 12:308–19

170. Deleted in proof

171. Prieto I, Suja JA, Pezzi N, Kremer L, Martinez C, et al. 2001. Mammalian STAG3 is a cohesin specific to sister chromatid arms during meiosis I. *Nat. Cell Biol.* In press

172. Deleted in proof

173. Rao H, Uhlmann F, Nasmyth K, Varshavsky A. 2001. Degradation of a cohesin

subunit by the N-end pathway is essential for chromosome stability. *Nature* 410: 955–59

174. Rattner JB, Lin CC. 1985. Radial loops and helical coils coexist in metaphase chromosomes. *Cell* 42:291–96

175. Reimann JD, Freed E, Hsu JY, Kramer ER, Peters JM, Jackson PK. 2001. Emi1 is a mitotic regulator that interacts with Cdc20 and inhibits the Anaphase promoting complex. *Cell* 105:645–55

176. Revenkova E, Eijpe M, Heyting C, Gross B, Jessberger R. 2001. A novel meiosis-specific isoform of mammalian SMC1. *Mol. Cell Biol.* In press

177. Rieder CL, Cole R. 1999. Chromatid cohesion during mitosis: lessons from meiosis. *J. Cell Sci.* 112:2607–13

178. Rieder CL, Khodjakov A, Paliulis LV, Fortier TM, Cole RW, Sluder G. 1997. Mitosis in vertebrate somatic cells with two spindles: implications for the metaphase/anaphase transition checkpoint and cleavage. *Proc. Natl. Acad. Sci. USA* 94: 5107–12

179. Rieder CL, Salmon ED. 1998. The vertebrate cell kinetochore and its roles during mitosis. *Trends Cell Biol.* 8:310–17

180. Roeder GS. 1997. Meiotic chromosomes: it takes two to tango. *Genes Dev.* 11:2600–21

181. Rollins RA, Morcillo P, Dorsett D. 1999. Nipped-B, a Drosophila homologue of chromosomal adherins, participates in activation by remote enhancers in the cut and Ultrabithorax genes. *Genetics* 152:577–93

182. Saitoh N, Goldberg IG, Wood ER, Earnshaw WC. 1994. ScII: an abundant chromosome scaffold protein is a member of a family of putative ATPases with an unusual predicted tertiary structure. *J. Cell Biol.* 127:303–18

183. Salah SM, Nasmyth K. 2000. Destruction of the securin Pds1p occurs at the onset of anaphase during both meiotic divisions in yeast. *Chromosoma* 109:27–34

184. Schalk JA, Dietrich AJ, Vink AC, Offen-berg HH, van Aalderen M, Heyting C. 1998. Localization of SCP2 and SCP3 protein molecules within synaptonemal complexes of the rat. *Chromosoma* 107: 540–48

185. Schmiesing JA, Gregson HC, Zhou S, Yokomori K. 2000. A human condensin complex containing hCAP-C-hCAP-E and CNAP1, a homolog of Xenopus XCAP-D2, colocalizes with phosphorylated histone H3 during the early stage of mitotic chromosome condensation. *Mol. Cell Biol.* 20:6996–7006

186. Schrader F. 1944. *Mitosis.* New York: Columbia Univ. Press. 110 pp.

187. Schwacha A, Kleckner N. 1997. Interhomolog bias during meiotic recombination: meiotic functions promote a highly differentiated interhomolog-only pathway. *Cell* 90:1123–35

188. Seitz LC, Tang K, Cummings WJ, Zolan ME. 1996. The *rad9* gene of *Coprinus cinereus* encodes a proline-rich protein required for meiotic chromosome condensation and synapsis. *Genetics* 142:1105–17

189. Selig S, Okumura K, Ward DC, Cedar H. 1992. Delineation of DNA replication time zones by fluorescence in situ hybridization. *EMBO J.* 11:1217–25

190. Shelby RD, Hahn KM, Sullivan KF. 1996. Dynamic elastic behavior of alpha-satellite DNA domains visualized in situ in living human cells. *J. Cell Biol.* 135: 545–57

191. Shirayama M, Toth A, Galova M, Nasmyth K. 1999. APC(Cdc20) promotes exit from mitosis by destroying the anaphase inhibitor Pds1 and cyclin Clb5. *Nature* 402:203–7

192. Shirayama M, Zachariae W, Ciosk R, Nasmyth K. 1998. The Polo-like kinase Cdc5p and the WD-repeat protein Cdc20p/Fizzy are regulators and substrates of the anaphase promoting complex in *Saccharomyces cerevisiae*. *EMBO J.* 17:1336–49

193. Shonn MA, McCarroll R, Murray AW.

2000. Requirement of the spindle check-point for proper chromosome segregation in budding yeast meiosis. *Science* 289:300–3

194. Sigrist S, Jacobs H, Stratmann R, Lehner CF. 1995. Exit from mitosis is regulated by Drosophila fizzy and the sequential destruction of cyclins A, B and B3. *EMBO J.* 14:4827–38

195. Simchen G, Hugerat Y. 1993. What determines whether chromosomes segregate reductionally or equationally in meiosis? *BioEssays* 15:1–8

196. Deleted in proof

197. Sjoegren C, Nasmyth K. 2001. Sister chromatid cohesion is required for post-replicative double strand break repair in *Saccharomyces cerevisiae. Curr. Biol* 11: 991–95

198. Skibbens RV, Corson LB, Koshland D, Hieter P. 1999. Ctf 7p is essential for sister chromatid cohesion and links mitotic chromosome structure to the DNA replication machinery. *Genes Dev.* 13:307–19

199. Smith KN, Nicolas A. 1998. Recombination at work for meiosis. *Curr. Opin. Genet. Dev.* 8:200–11

200. Steffensen S, Coelho PA, Cobbe N, Vass S, Costa M, et al. 2001. A role for Drosophila SMC4 in the resolution of sister chromatids in mitosis. *Curr. Biol.* 11:295–307

201. Straight AF, Belmont AS, Robinett CC, Murray AW. 1996. GFP tagging of budding yeast chromosomes reveals that protein-protein interactions can mediate sister chromatid cohesion. *Curr. Biol.* 1599–608

202. Stratmann R, Lehner CF. 1996. Separation of sister chromatids in mitosis requires the Drosophila pimples product, a protein degraded after the metaphase/anaphase transition. *Cell* 84:25–35

203. Sudakin V, Ganoth D, Dahan A, Heller H, Hershko J, et al. 1995. The cyclosome, a large complex containing cyclin-selective ubiquitin ligase activity, targets cyclins for destruction at the end of mitosis. *Mol. Biol. Cell* 6:185–98

204. Sullivan S, Lehane C, Uhlmann F. 2001. Slk19 cleavage. *Nat. Cell Biol.* In press

205. Deleted in proof

206. Sumara I, Vorlaufer E, Gieffers C, Peters BH, Peters J-M. 2000. Characterization of vertebrate cohesin complexes and their regulation in prophase. *J. Cell Biol.* 151:749–62

207. Sumner AT. 1991. Scanning electron microscopy of mammalian chromosomes from prophase to telophase. *Chromosoma* 100:410–18

208. Sundin O, Varshavsky A. 1981. Arrest of segregation leads to accumulation of highly intertwined catenated dimers: dissection of the final stages of SV40 DNA replication. *Cell* 25:659–69

209. Surana U, Amon A, Dowzer C, McGrew J, Byers B, Nasmyth K. 1993. Destruction of the CDC28/CLB mitotic kinase is not required for the metaphase to anaphase transition in budding yeast. *EMBO J.* 12:1969–78

210. Taieb FE, Gross SD, Lewellyn AL, Maller JL. 2001. Activation of the anaphase-promoting complex and degradation of cyclin B is not required for progression from meiosis I to II in Xenopus oocytes. *Curr. Biol.* 11:508–13

211. Takizawa CG, Morgan DO. 2000. Control of mitosis by changes in the subcelluluar location of cyclin-B1-Cdk1 and Cdc25C. *Curr. Opin. Cell Biol.* 12:658–65

212. Deleted in proof

213. Tanaka K, Yonekawa T, Kawasaki Y, Kai M, Furuya K, et al. 2000. Fission yeast Eso1p is required for establishing sister chromatid cohesion during S phase. *Mol. Cell. Biol.* 20:3459–69

214. Deleted in proof

215. Tanaka T, Cosma MP, Wirth K, Nasmyth K. 1999. Identification of cohesin association sites at centromeres and along chromosome arms. *Cell* 98:847–58

216. Tanaka T, Fuchs J, Loidl J, Nasmyth K. 2000. Cohesin ensures bipolar attachment

of microtubules to sister centromeres and resists their precocious separation. *Nat. Cell Biol.* 2:492–99

217. Tanaka T, Knapp D, Nasmyth K. 1997. Loading of an MCM protein onto DNA replication origins is regulated by Cdc6p and CDKs. *Cell* 90:649–60

218. Taylor SS, McKeon F. 1997. Kinetochore localization of murine Bub1 is required for normal mitotic timing and checkpoint response to spindle damage. *Cell* 89:727–35

219. Thomas JO. 1984. The higher order structure of chromatin and histone H1. *J. Cell Sci. Suppl.* 1:1–20

220. Tippet DH, Pickett-Heaps JD, Leslie RL. 1980. Cell division in two large pennate diatoms Hantzchia and Nitzchia III. A new proposal for kinetochore function during prometaphase. *J. Cell Biol.* 86:402–16

221. Tomonaga T, Nagao K, Kawasaki Y, Furuya K, Murakami A, et al. 2000. Characterization of fission yeast cohesin: essential anaphase proteolysis of Rad21 phosphorylated in the S phase. *Genes Dev.* 14:2757–70

222. Toth A, Ciosk R, Uhlmann F, Galova M, Schleifer A, Nasmyth K. 1999. Yeast cohesin complex requires a conserved protein, Eco1p (Ctf7), to establish cohesion between sister chromatids during DNA replication. *Genes Dev.* 13:320–33

223. Toth A, Rabitsch KP, Galova M, Schleifer A, Buonomo SB, Nasmyth K. 2000. Functional genomics identifies monopolin: a kinetochore protein required for segregation of homologs during meiosis I. *Cell* 103:1155–68

224. Uhlmann F, Lottspeich F, Nasmyth K. 1999. Sister chromatid separation at anaphase onset is promoted by cleavage of the cohesin subunit Scc1p. *Nature* 400:37–42

225. Uhlmann F, Nasmyth K. 1998. Cohesion between sister chromatids must be established during DNA replication. *Curr. Biol.* 8:1095–101

226. Uhlmann F, Wernic D, Poupart MA, Koonin E, Nasmyth K. 2000. Cleavage of cohesin by the CD clan protease separin triggers anaphase in yeast. *Cell* 103:375–86

227. Uren GA, O'Rourke K, Aravind L, Pisabarro TM, Seshagiri S, et al. 2000. Identification of paracaspases and metacaspases: two ancient families of caspase-like proteins, one of whic plays a key role in MALT lymphoma. *Mol. Cell* 6:961–67

228. van Heemst D, James H, Pöggeler S, Berteaux-Lecellier V, Zickler D. 1999. Spo76p is a conserved chromosome morphogenesis protein that links the mitotic and meiotic programms. *Cell* 98:261–71

229. Waizenegger I, Hauf S, Meinke A, Peters JM. 2000. Two distinct pathways remove mammalian cohesin from chromosome arms in prophase and from centromeres in anaphase. *Cell* 103:399–410

230. Deleted in proof

231. Wakefield JG, Huang JY, Raff JW. 2000. Centrosomes have a role in regulating the destruction of cyclin B in early Drosophila embryos. *Curr. Biol.* 10:1367–70

232. Walker MY, Hawley RS. 2000. Hanging on to your homolog: the roles of pairing, synapsis and recombination in the maintenance of homolog adhesion. *Chromosoma* 109:3–9

233. Wang HT, Frackman S, Kowalisyn J, Esposito RE, Elder R. 1987. Developmental regulation of SPO13, a gene required for separation of homologous chromosomes at meiosis I. *Mol. Cell Biol.* 7:1425–35

234. Wang Z, Castano IB, Fitzhugh DJ, De Las Penas A, Adams C, Christman MF. 2000. Pol Kappa: a DNA polymerase required for sister chromatid cohesion. *Science* 289:774–79

235. Warren WD, Steffensen S, Lin E, Coelho P, Loupart M, et al. 2000. The Drosophila RAD21 cohesin persists at the centromere region in mitosis. *Curr. Biol.* 10:1463–66

236. Deleted in proof

237. Deleted in proof

238. Watanabe Y, Nurse P. 1999. Cohesin Rec8

is required for reductional chromosome segregation at meiosis. *Nature* 400:461–64

239. Watanabe Y, Yokobayashi S, Yamamoto M, Nurse PN. 2001. Pre-meiotic S phase is linked to reductional chromosome segregation and recombination. *Nature* 409:359–63

240. Webb CD, Teleman A, Gordon S, Straight A, Belmont A, et al. 1997. Bipolar localization of the replication origin regions of chromosomes in vegetative and sporulating cells of *B. subtilis*. *Cell* 88:667–74

241. Weinstein J. 1997. Cell cycle-regulated expression, phosphorylation, and degradation of p55Cdc. A mammalian homolog of CDC20/Fizzy/slp1. *J. Biol. Chem.* 272:28501–11

242. Whitehouse HLK. 1973. *Towards An Understanding of the Mechanism of Heredity.* London: Arnold

243. Wohlschlegel JA, Dwyer BT, Dhar SK, Cvetic C, Walter JC, Dutta A. 2000. Inhibition of eukaryotic DNA replication by geminin binding to Cdt1. *Science* 290:2309–12

244. Wolstenholme J, Angell RA. 2000. Maternal age and trisomy—a unifying mechanism of formation. *Chromosoma* 109:435–38

245. Yamamoto A, Guacci V, Koshland D. 1996. Pds1p is required for faithful execution of anaphase in the yeast, *Saccharomyces cerevisiae. J. Cell Biol.* 133:85–97

246. Yamamoto A, Guacci V, Koshland D. 1996. Pds1p, an inhibitor of anaphase in budding yeast, plays a critical role in the APC and checkpoint pathways. *J. Cell Biol.* 133:99–110

247. Yanagida M. 2000. Cell cycle mechanisms of sister chormatid separation; roles of Cut1/separin and Cut2/securin. *Genes Cells* 5:1–8

248. Zachariae W, Nasmyth K. 1999. Whose end is destruction: cell division and the anaphase-promoting complex. *Genes Dev.* 13:2039–58

249. Zeng X, Kahana JA, Silver PA, Morphew MK, McIntosh JR, et al. 1999. Slk19p is a centromere protein that functions to stabilize mitotic spindles. *J. Cell Biol.* 146:415–25

250. Zenvirth D, Loidl J, Klein S, Arbel A, Shemesh R, Simchen G. 1997. Switching yeast from meiosis to mitosis: double-strand break repair, recombination and synaptonemal complex. *Genes Cells* 2:487–98

251. Zhang P, Hawley RS. 1990. The genetic analysis of distributive segregation in *Drosophila melanogaster*. II. Further genetic analysis of the nod locus. *Genetics* 125:115–27

252. Zickler D, Kleckner N. 1999. Meiotic chromosomes: integrating structure and function. *Annu. Rev. Genet.* 33:603–754

253. Zou H, McGarry TJ, Bernal T, Kirschner MW. 1999. Identification of a vertebrate sister-chromatid separation inhibitor involved in transformation and tumorigenesis. *Science* 285:418–22

254. Zur A, Brandeis M. 2001. Securin degradation is mediated by fzy and fzr, and is required for complete chromatid separation but not for cytokinesis. *EMBO J.* 20:792–801

Annu. Rev. Genet. 2001. 35:747–84

EPITHELIAL CELL POLARITY AND CELL JUNCTIONS IN *DROSOPHILA*

Ulrich Tepass and Guy Tanentzapf

Department of Zoology, University of Toronto, 25 Harbord Street, Toronto, Ontario M5S3G5, Canada; e-mail: utepass@zoo.utoronto.ca, guy@zoo.utoronto.ca

Robert Ward and Richard Fehon

DCMB Group, Department of Biology, Duke University, B333 LSRC Research Drive, Durham, North Carolina 27708; e-mail: rward@howard.genetics.utah.edu; rfehon@duke.edu

Key Words epithelium, polarity, cellular junctions, cellularization, membrane domain

■ **Abstract** The polarized architecture of epithelial cells and tissues is a fundamental determinant of animal anatomy and physiology. Recent progress made in the genetic and molecular analysis of epithelial polarity and cellular junctions in *Drosophila* has led to the most detailed understanding of these processes in a whole animal model system to date. Asymmetry of the plasma membrane and the differentiation of membrane domains and cellular junctions are controlled by protein complexes that assemble around transmembrane proteins such as DE-cadherin, Crumbs, and Neurexin IV, or other cytoplasmic protein complexes that associate with the plasma membrane. Much remains to be learned of how these complexes assemble, establish their polarized distribution, and contribute to the asymmetric organization of epithelial cells.

CONTENTS

0066-4197/01/1215-0747$14.00

INTRODUCTION

Epithelial tissues have emerged early during animal evolution, and their ability to form different shapes and to subdivide the body into physiologically distinct compartments is fundamental for the evolution of complex animal body plans. The plasma membrane of epithelial cells is subdivided into regions or domains that fulfill specialized roles in cell organization and physiology. The main subdivisions of the plasma membrane are the apical domain, which faces the external environment and the basolateral domain, which is in contact with the interstitial space of the body. These domains are segregated by a circumferential junctional complex (CJC) that binds adjacent epithelial cells together and forms a semipermeable barrier to the diffusion of solutes through the intercellular space (38). The movement of ions and molecules across an epithelial layer therefore requires regulated transport mechanisms that shuttle solutes from apical to basolateral, or vice versa, and allow epithelia to control the physiological composition of body compartments. In addition to the apical/basolateral distinction, membrane domains of epithelial cells are further regionalized. The basolateral membrane, for example, is subdivided into a basal domain characterized by cell-substrate adhesion and a lateral domain distinguished by cell-cell adhesion. Further, the lateral domain is partitioned into the apical CJC and a region basal to it (121).

The mechanisms that establish and maintain an asymmetric distribution of lipid and protein components of the plasma membrane of epithelial cells have been intensively studied in mammalian cell culture (100). Early work in this system led to a model suggesting that the sorting of plasma membrane components in the Trans-Golgi Network (TGN) into apical and basolateral transport vesicles and the subsequent polarized delivery to the appropriate surface domain are the key mechanisms by which epithelial polarity is maintained (130). However, this model failed to explain how apical and basolateral domains are established initially, and how the two main surface domains are further regionalized. Moreover, it was recognized that apical and basolateral transport vesicles are also formed in nonpolarized cells in which the components of such vesicles show overlapping distributions in the plasma membrane (98, 186).

Analysis of the role of cell adhesion and its consequences on cellular organization led to the current conceptual framework of epithelial polarity (30, 100).

External cues mediated by cell-cell or cell-substrate adhesion generate asymmetries within the plasma membrane that are elaborated by the formation of a local specific membrane cytoskeleton. Retention of transmembrane and cytoplasmic proteins that associate with this local actin/spectrin cytoskeleton emphasize regional differences (92). These differences are further elaborated by adhesion-dependent reorganization of the microtubule cytoskeleton that is necessary for vesicle traffic, and the formation of targeting patches at the lateral membrane that preferentially attract basolateral transport vesicles (51, 185). This model of membrane domain formation integrates several interdependent polarization mechanisms, the concerted activity of which is triggered by adhesive interactions that provide positional cues for cell polarization. The formation of the apical domain is thus viewed as a default pathway in which the plasma membrane assumes apical character wherever no adhesive interactions take place.

Using *Drosophila* as a genetic model to study epithelial polarization offers the opportunity to complement and expand on the mammalian cell culture studies by placing the mechanisms that control epithelial differentiation into a developmental context. Genetic screens have identified a number of factors essential for epithelial polarity that are either integral to, or associated with, the plasma membrane, but did not reveal components of the TGN that contribute to the formation of apical or basolateral transport vesicles. These findings are consistent with a predominant role of extrinsic cues mediated by transmembrane adhesion receptors and cyto-cortical factors in epithelial polarization. Work using *Drosophila* also revealed a number of regulators of epithelial polarity that are associated with the apical membrane, suggesting that the formation of the apical domain is not a default pathway but instead requires a specific molecular machinery. Finally, studies on *Drosophila* are beginning to reveal an unforeseen complexity in the mechanisms controlling epithelial polarization that may vary from tissue to tissue and over time in the same tissue. This review gives an overview of epithelial development in *Drosophila*, with emphasis on recent studies that have provided novel insights into polarity and the differentiation and function of cellular junctions in epithelial cells.

EPITHELIAL DIFFERENTIATION IN *DROSOPHILA*: AN OVERVIEW

Development of Primary and Secondary Epithelia

The first epithelium that emerges during *Drosophila* development is the blastoderm. It forms by a process known as cellularization, a modified form of cytokinesis discussed in detail below. Many epithelia in *Drosophila* derive from the blastoderm epithelium without a non-epithelial intermediate. Such primary epithelia, all derived from the ectoderm, include the larval and adult epidermis as well as the foregut, hindgut, Malphigian tubules, tracheae, and salivary glands.

In contrast, secondary epithelia arise by mesenchymal-epithelial transitions later in development. Embryonic secondary epithelia are the midgut epithelium, glia sheets that form the blood-nerve barrier, and the dorsal vessel (heart). In addition to their mode of formation, primary and secondary epithelia differ in structure and mechanisms used for cell polarization (150, 154).

Epithelial development can be subdivided roughly into three phases in which (I) the initial establishment of polarity is (II) followed by the consolidation and elaboration of surface domains and cytoplasmic asymmetries, and finally, (III) the terminal differentiation and specialization of surface domains. For primary epithelia, phase I occurs at cellularization during which distinct membrane domains are established. Phase II extends from gastrulation throughout organogenesis and includes the formation of a CJC. Groups of epithelia behave uniformly during this phase; for example, the ectoderm and its epithelial derivatives such as the epidermis and the tracheae, all establish the same CJC. During phase III, individual epithelia undergo sometimes dramatic specializations to accommodate their anatomical or physiological function, and tissue-specific control mechanisms for epithelial differentiation become apparent. For example, the Zinc-finger transcription factor Hindsight is required to maintain the integrity of the tracheal epithelium and controls the differentiation of a specialized tracheal cuticle, which contains a characteristic spiral-shaped superstructure, the taenidium, that prevents the tracheal lumen from collapsing (177). The mechanisms that control the terminal differentiation of specialized epithelia in *Drosophila* remain largely elusive. In addition, because little progress has been made in understanding the polarization mechanisms of embryonic secondary epithelia since we last reviewed the subject (150), this topic is not covered here.

Models for epithelial differentiation in postembryonic development of *Drosophila* include the imaginal discs and the ovarian follicular epithelium (96). The larval imaginal discs are fully polarized epithelial sheets with a well-developed junctional complex comparable to the epidermis of mid-embryonic stages from which they derive. Imaginal disc epithelia lend themselves to the analysis of epithelial maintenance and terminal differentiation and specialization of epithelial surface domains. In contrast, the follicular epithelium renews itself constantly as follicle cells originate from stem cells, and allows the analysis of the full range of phases in epithelial differentiation, including epithelial formation by mesenchymal-epithelial transition and the successive assembly of a CJC. The simplicity and accessibility of the follicular epithelium together with the large number of genes known that effect its epithelial integrity make the follicular epithelium a favored genetic system to study epithelial development (96, 147).

Cellular Junctions in *Drosophila* Epithelia

The complement of cellular junctions in *Drosophila* epithelia comprises spot adherens junctions (SAJs), the zonula adherens (ZA), pleated and smooth septate junctions (SJs), gap junctions, and hemiadherens junctions (HAJs) (154). Early during development, epithelia acquire a ZA, which assembles from the coalescence of individual SAJs. HAJs and SJs form only later during epithelial differentiation. Basal HAJs (elsewhere called focal contacts) are integrin-based and connect epithelial cells to basement membranes or specialize into muscle-tendon junctions (20, 118). Apical HAJs are seen in cuticle-secreting epithelia connecting the apical membrane to the cuticle. SJs act as the trans-epithelial barrier in most epithelia of non-chordate animals, and thus functionally substitute for the chordate tight junction. The CJC in primary epithelia and in the follicular epithelium is composed of the ZA and the SJ. Embryonic secondary epithelia such as glia sheets and the midgut epithelium lack a ZA but contain the SJ. Desmosomes and hemidesmosomes as well as tight junctions are not seen in *Drosophila* epithelia. Absence of desmosomes and hemidesmosomes is corroborated by the lack of cytoplasmic intermediate filaments in *Drosophila* (3), and the phylogenic analysis of the cadherin superfamily that suggests that desmosomal cadherins have evolved from classic cadherins within the chordate lineage (158). Gap junctions in *Drosophila* and other invertebrates are formed by innexins that appear unrelated in sequence to vertebrate connexins but perform similar functions (116). From gastrulation onwards, gap junctions are ubiquitous components of epithelia (154) but their role in epithelial differentiation is currently not understood.

EPITHELIUM FORMATION: CELLULARIZATION

The formation of an epithelial sheet is typically achieved through the reorganization of a cluster of mesenchymal cells into a monolayer of tightly adhering polarized cells. Such mesenchymal-epithelial transitions are seen many times during development but are best studied in early vertebrate embryos in which a cluster of mesenchymal blastomeres forms the blastula, a hollow ball, that is bound by a blastoderm epithelium. In contrast, the formation of the blastoderm in *Drosophila* embryos, and most other insects takes a different route to establish an epithelium. The fertilized *Drosophila* egg undergoes 13 nuclear divisions that are not followed by cytokinesis. At the end of the 13th cell cycle ~5000 nuclei form a monolayer just beneath the egg membrane. Invaginations of the egg membrane surround each nucleus and associated cytoplasm during the 14th cell cycle, cellularizing the blastoderm and establishing an epithelium of highly columnar cells (42–44, 126). As cytokinesis and epithelium formation go hand in hand, the exploration of epithelial polarization of the *Drosophila* blastoderm provides some unique challenges.

Cellularization is initiated by the formation of the furrow canal that remains at the leading edge of the invaginating membrane (Figure 1). As cellularization proceeds, SAJs are assembled next to the furrow canal in the emerging lateral membrane (55, 97, 149, 154). These "basal junctions" (55) remain associated with the furrow canal during cellularization but resolve as cellularization is completed. Additional SAJs form as the lateral membrane grows, and concentrate apically where they will form the ZA during gastrulation (97, 149). A number of proteins

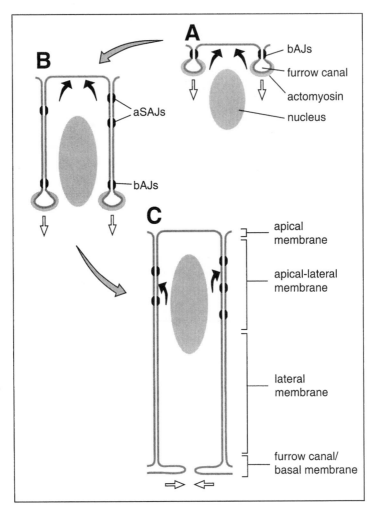

Figure 1   Cellularization forms the blastoderm epithelium in *Drosophila*. Three stages at early (*A*), mid- (*B*) and late- (*C*) cellularization are illustrated. The open arrows indicate the direction of plasma membrane movement and the black arrows point to the main membrane insertion sites as identified in (72). (*A*) Invaginations of the egg membrane surround each of ∼5000 nuclei and form the furrow canals. The basal adherens junctions (bAJs) remain closely linked with the furrow canals during cellularization. (*B*) Apical spot adherens junctions (aSAJs) form at midcellularization, increase in number and are retained in the apical one third of the lateral membrane. These SAJs will form the ZA during gastrulation. (*C*) Four membrane domains, indicated to the right, have formed at late-cellularization. The bAJs and the actomyosin ring resolve at this time and the furrow canals expand to form the basal membrane.

show an asymmetric distribution in the forming lateral membranes during cellularization (see below). These include the cadherin-catenin-complex (CCC) as part of the basal and the more apical SAJs (27, 55, 97, 149). Taken together, these observations suggest that a polarized lateral membrane domain is established during cellularization.

Cellularization is a modified form of cytokinesis, in which the furrow canal represents the leading edge of the cleavage furrows. As in other cells, the contractile ring of Actin and Myosin II at the furrow canal associates with Septins and Anillin during cytokinesis (2, 41, 43). The actomyosin rings of all 5000 blastoderm cells form an interlocking hexagonal array that plays an important role in cellularization (42, 189). Mutations in a number of genes have been described that disrupt the actomyosin array and compromise cellularization. These genes encode factors that co-localize with the actomyosin array such as Peanut, a septin (2), Bottleneck (125), Serendipity-α (127), Discontinuous Actin Hexagon (191, 192), the Formin-homology protein Diaphanous (4), and possibly the small GTPases Rho1 and Cdc42 (28). Others factors that contribute to the organization or function of the actomyosin array but act at a distance include Nullo, which localizes to the basal junction (55, 117, 131), the transcriptional regulator Lilliputian, which controls the expression of Serendipity-α (148), and Nuclear Fallout, a centrosome-associated protein required for recruitment of Actin and Discontinuous Actin Hexagon to the furrow canal (123, 124). Also Discs Lost (Dlt) localizes to the furrow canal during cellularization and, if disrupted, causes cuboidal rather than highly columnar cells to form (11).

Data from a number of systems point to a prominent role of forces generated by the insertion of new membrane in cytokinesis (53). In this model, contraction of the actomyosin ring is not the main driving force of cytokinesis but, instead, forms an elastic and tensile structure that orients and synchronizes membrane movement. For cellularization, it has now been established that at least the bulk of the membrane material required for the ingrowth of the plasma membrane is derived from the biosynthetic pathway (21, 72, 132) and does not come from unfolding of the egg membrane, as had been proposed previously (44, 166). The t-SNARE Syntaxin1 (21) and the Golgi-associated protein Lava Lamp (132) are required for cellularization, and the progression of the furrow canal is blocked in response to injection of Brefeldin A, an inhibitor of Golgi-derived vesicle transport (132). The movements of vesicles and Golgi bodies from a reservoir below the nuclei into the apical cytoplasm depends on (−) end-directed microtubule transport. This vesicle movement contributes to membrane growth in particular during early cellularization (42, 43, 132, 172).

One contentious issue is the location of the site of new membrane insertion. The close association of Golgi bodies with the furrow canal (132) and the alignment of vesicles in front of the progressing furrow canal (80) suggest that the furrow canal is the primary site of membrane insertion. This scenario implies that membrane turnover in the furrow canal is rapid, with new membrane being "exported" to the growing lateral membrane. However, a recent study that traces membrane flow during cellularization arrives at a different conclusion (72). Labeling of glycoproteins with fluorescent wheat germ agglutinin (WGA) in live embryos shortly before or early during cellularization revealed that the successive formation of membrane domains starts with the furrow canal. The main insertion site during early cellularization is the apical membrane. WGA remains with the furrow canals as they progress inwards, and apically inserted membrane moves basally to form lateral membrane. Late during cellularization the membrane insertion site shifts to the apical part of the lateral membrane. Membrane mixing appears rather limited as WGA-labeled membrane areas remain coherent and the label does not diffuse into other membrane domains. These findings suggest a sequential establishment of membrane domains during cellularization during which the furrow canal forms first followed by the basal part of lateral membrane, the apical membrane, and finally, the apical part of lateral membrane (Figure 1). The notion that the lateral membrane is subdivided into an apical and a basal region is further supported by the asymmetric distribution of molecular markers such as Neurotactin and Spectrin (72, 161).

We are only beginning to unravel the mechanisms that act to establish epithelial polarity during cellularization. Distinct apical and basolateral vesicle targeting mechanisms may not contribute to the formation of the blastoderm epithelium as bulk membrane insertion from the biosynthetic pathway appears to take place first at the apical membrane and later at the apical lateral membrane (72). Cadherin-based adhesive interactions may facilitate lateral membrane formation as adherens junctions are formed as soon as lateral membranes appear. Cellularization has so far not been studied in the complete absence of the CCC. However, recent analysis of Nullo indicates that it localizes to the basal adherens junctions and is required for their formation (55). This observation suggests that this junction plays an important role as cellularization in *nullo* mutant embryos is highly irregular (131). In wild-type embryos, Nullo degrades prior to the formation of the apical SAJs, and prolonged expression of Nullo blocks their assembly, causing abnormalities in epithelial morphology at gastrulation (55). Taken together, these findings suggest that the novel protein Nullo differentiates between the apical and the basal part of the lateral membrane and between the adherens junctions residing in these regions. These results also indicate that the adherens junctions that form during early cellularization play an important role in the formation of the blastoderm epithelium.

PROTEIN COMPLEXES INVOLVED IN SPECIFICATION AND REGIONALIZATION OF EPITHELIAL SURFACE DOMAINS

The exploration of epithelial differentiation in *Drosophila* has now led to the characterization of protein complexes that regulate polarity and junctional differentiation. Pioneering genetic work has identified several genes that are required for epithelial differentiation in *Drosophila*, including *crumbs (crb)*, *shotgun* (*shg*), *bazooka* (*baz*), *stardust* (*sdt*), *lethal giant larvae* (*lgl*), and *discs large* (*dlg*) (19, 45, 61, 102, 175). The products of these genes have now been characterized, and biochemical and/or genetic data suggest that these and other proteins form complexes that associate with the plasma membrane and show a polarized distribution. These complexes give essential cues that govern the polarized organization of epithelial cells. Aside from wondering about the molecular composition and mutual interactions within each complex, we need to ask how the polarized localization of these complexes is achieved, and how the activity of each complex affects cellular organization. Figure 2 illustrates the position of cellular junctions and the distribution of protein complexes important for epithelial differentiation.

Adherens Junctions and the Cadherin-Catenin Complex

The first event after cellularization that indicates the further elaboration of the epithelial cell surface is the formation of the ZA that occurs as cellularization nears completion and gastrulation proceeds. ZA formation has been characterized as a three-step process. First, SAJs form in the lateral membrane during cellularization. Second, at the onset of gastrulation these SAJs move toward the apicolateral edge of the cells. Third, SAJs fuse into a circumferential belt, the ZA, during gastrulation (97, 149, 150, 154). At late cellularization, apical markers such as Baz and β_{Heavy}-Spectrin (β_{H}-Spectrin), and lateral markers such as Arm and β-Spectrin are mixed in the apical part of the lateral membrane (97, 149, 161; A. Wodarz, personal communication). The segregation of these molecules into distinct plasma membrane domains leads to the formation of apical and basolateral domains that are separated by a ZA. The apical membrane domain is subdivided into two regions at this point, the free apical surface and the marginal zone, which represents a narrow region of cell-cell contact apical to the ZA (Figure 2) (149).

The formation of SAJs during cellularization presumably depends on the CCC, although direct evidence for such a requirement is still lacking. The CCC consists of DE-cadherin, the predominant epithelial cadherin in *Drosophila* encoded by the *shg* gene, Arm, the homolog of vertebrate α-catenin, Dα-catenin, and Dp120^{ctn} (103, 104, 107, 109, 152, 167; R. Cavallo & M. Peifer, personal communication). In addition to the pool of Arm molecules that are part of the CCC, Arm is also an effector of Wingless (Wg) signaling. The relation between cytosolic Arm that participates in Wg signaling and junctional Arm remains unresolved. The CCC is essential in the female germline and thus embryos that lack the maternal components of either DE-cadherin or Arm cannot be studied (27, 48, 105, 110, 152, 173) (mutations for Dα-catenin and Dp120^{ctn} are currently not available). However, if intermediate alleles of *shg* or *arm* are used, a limited number of fertilized eggs is recovered from females with a mutant germline. In such embryos, in which maternal and zygotic expression of *shg* or *arm* is strongly reduced, all epithelia that were examined lose integrity (27, 152). The development of *arm* mutant germline clone (*arm*^{GLC}) embryos was analyzed in detail (27). Here, plasma membrane-associated DE-cadherin is reduced and Dα-catenin is not membrane associated.

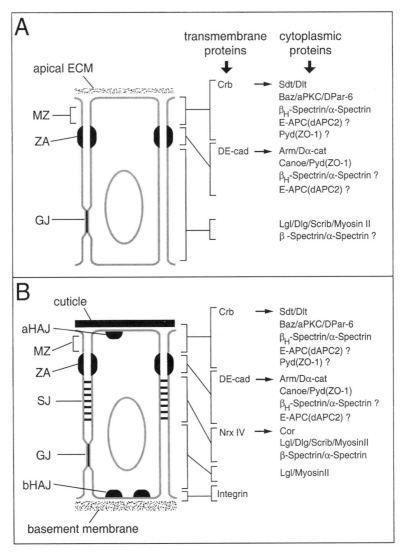

Figure 2   Schematic of epithelial cell structure in an ectodermal epithelial cell during gastrulation (*A*) and a late-embryonic/larval epidermal cell (*B*). Subdomains of the plasma membrane and cellular junctions are indicated to the left and the distribution of proteins discussed in this review are listed to the right. Cytoplasmic proteins that bind directly to a transmembrane protein are indicated by the small arrow. Synonyms of protein names are given in parenthesis. The question marks indicate that the localization of these proteins at the given position require confirmation, e.g., E-APC/dAPC2 might localize to the marginal zone, the zonula adherens or both regions. Abbreviations: aHAJ, apical hemi adherens junction; bHAJ, basal HAJ; DE-cad, DE-cadherin; ECM, extracellular matrix; GJ, gap junction; MZ, marginal zone; SJ, septate junction; ZA, zonula adherens.

Nevertheless, cellularization proceeds normally in these embryos, suggesting that very limited CCC activity is sufficient to promote cellularization. Alternatively, epithelium formation during cellularization might not require the CCC at all. This latter interpretation is supported by the radical shift in cell morphology seen in arm^{GLC} embryos at the onset of gastrulation (27). While the blastoderm forms normally in arm^{GLC} embryos, epithelia rapidly acquire a multilayered mesenchymal morphology at early gastrulation. At later stages of embryonic development, the CCC is needed to maintain integrity of all epithelial tissues that were studied (52, 151–153, 167). Thus, with the possible exception of the blastoderm, the CCC is essential to maintain adhesion and tissue architecture of *Drosophila* primary and secondary epithelia.

In the ovary, recent work indicates that the CCC is not required for the formation of the follicular epithelium but plays a role in its maintenance (147). Adherens junctions in the follicular epithelium contain DE-cadherin and DN-cadherin, which both disappear if follicle cells are rendered null for *arm*. *arm* mutant follicle cells are irregular in shape and sometimes form a multilayered epithelium. In most cases, however, *arm* mutant follicle cells remain within the epithelial layer and acquire a flat shape rather than being cuboidal or columnar, suggesting that the lateral membrane domain is reduced in size when the CCC is disrupted (96, 147). Interestingly, apical markers are lost (Crb and β_H-Spectrin) or mislocalized (Dlt) in *arm* mutant follicle cells although these cells retain a monolayered epithelial arrangement (147). These findings indicate that the loss of the CCC in the follicular epithelium disrupts the architecture of the apical domain without necessitating the breakdown of a monolayered epithelial tissue structure.

Adherens Junctions and Cell Signaling

Recent work in mammalian cell culture has established that the ZA, in addition to the CCC, contains a second complex composed of the immunoglobulin-like adhesion molecule Nectin and the cytoplasmic factors Afadin, a PDZ domain protein, and Ponsin, a SH3 domain protein. This complex interacts with both the CCC and the actin cytoskeleton. The knockout phenotype of Afadin in mice suggests that it has an essential role in maintaining epithelial integrity in the mouse ectoderm (56, 84, 85, 141, 144). *Drosophila* Canoe is the apparent ortholog of Afadin. Canoe was localized at the ZA in photoreceptor cells by immunoelectron microscopy (88), and appears to be a ubiquitous component of the ZA (143). Mutational analysis of *canoe* did not reveal a general requirement for epithelial or ZA integrity.

canoe mutant embryos show defects in dorsal closure, an epithelial migration process regulated by Jun N-terminal kinase (JNK) and Wg signaling (91, 101, 143). In fact, Canoe appears to be an upstream regulator of JNK signaling. In this process, Canoe colocalizes and interacts genetically and physically with *Drosophila* ZO-1, a MAGUK (membrane-associated guanylate kinase) protein encoded by the gene *polychaetoid* (previously also known as *tamou*) (143, 145, 170). Moreover, analysis of the role of Canoe in imaginal development suggests that it can physically interact with Ras1 and modulate Ras and Notch signaling (87, 95, 145). A close spatial association between the ZA and Ras and Notch signaling components has been described previously (22, 40, 163, 188). Taken together, next to Arm/β-catenin, the work on Canoe represents the best evidence that a bona fide component of the ZA regulates cell signaling. How Canoe modulates signaling remains to be elucidated. It will also be interesting to see whether and how Canoe and Pyd/ZO-1 interact with the CCC and whether Canoe interacts with Nectin and Ponsin-like molecules in *Drosophila*.

In addition to Arm/β-catenin, a second connection between the ZA and Wg signaling has now been made. E-APC/dAPC2, one of two *Drosophila* homologs of the adenomatous polyposis coli (APC) tumor suppressor, is predominantly expressed in epithelial cells and localizes to the apicolateral plasma membrane, a region that includes the ZA and the MZ (89, 187). Whether E-APC/dAPC2 is a component of the ZA or whether its association with the ZA is more peripheral remains to be established. Recent data from mammalian epithelial cells suggest that the majority of APC associates with the apical membrane domain and does not colocalize with the CCC at the lateral domain (120). The localization of E-APC/dAPC2 to the apicolateral region depends on the integrity of the ZA and the actin cytoskeleton (164, 187). The analysis of a hypomorphic E-APC/dAPC2 mutation and RNA interference experiments did not reveal overt effects on epithelial polarity, although junctional Armadillo is reduced in some tissues.

Both human APC and E-APC/dAPC2 act as part of a "destruction complex" that destabilizes Arm/β-catenin and thus negatively regulate Wg/Wnt signaling (111). A temperature-sensitive missense mutation in E-APC/dAPC2, dAPC$^{\Delta S}$, causes E-APC/dAPC2 to accumulate in the cytoplasm and compromises its role in Wg signaling, suggesting that the localization of E-APC/dAPC2 to the apicolateral region of the plasma membrane is essential for its signaling function (89). Interestingly, apical secretion of Wg, which is controlled by the apical localization of its mRNA, is important for effective signaling (128), and also the Wg receptor Frizzled is a component of the apical membrane (138). The concentration of both positive and negative elements of the Wg signaling cascade, in addition to the components of the Notch, Ras, and JNK signaling pathways, found at the marginal zone and/or ZA suggest that this region is at the crossroad of several signaling pathways.

The ZA and E-APC/dAPC2 have now been identified as sources of a cue that controls the orientation and symmetry of cell division in the ectodermal epithelium of *Drosophila* embryos (82). As epithelial cells divide they round off and rise to the apical surface of the epithelium, where they remain connected to adjacent cells via the ZA (154). The spindle is oriented in parallel to the planar axis of the epithelium. Two equally sized daughter cells form that receive a similar share of apical and basolateral membrane and an equal amount of polarity determinants that associate with these membrane domains from the mother cell (82). If the ZA or E-APC/dAPC2 activity is disrupted, spindle orientation is abnormal and cell division is asymmetric (82). Two daughter cells of unequal size form, with the smaller cell receiving only basal membrane. This type of division pattern is reminiscent of mesenchymal neural progenitor cells (neuroblasts) or epithelial cells that express

Inscutable, a key regulator of asymmetric cell division in the neuroectoderm (65). Thus, epithelial cells have the potential to divide asymmetrically. However, this potential is normally overridden by a ZA-associated cue that requires the activity of E-APC/dAPC2. Disruption of dEB1 causes similar defects as blocking of E-APC/dAPC2 activity (82). dEB1 is the *Drosophila* homolog of mammalian EB1 that binds to APC (139). Although dEB1 does not appear to interact physically with E-APC/dAPC2 as the latter lacks a dEB1 binding site, the phenotypic similarities suggest that both proteins interact functionally (82). EB1 is known to preferentially interact with the (+) end of microtubules (10). These findings raise the intriguing possibility that E-APC/dAPC2 and dEB1 may connect the ZA to the (+) end of astral microtubules during division, thereby orienting the spindle along the planar axis of the epithelium as a prerequisite of symmetric division.

Apical Polarization I: The Crumbs/Stardust/Discs Lost Complex

Aside from bona fide components of the adherens junction, the formation of the ZA depends on two protein complexes that associate with the apical membrane. The first complex is composed of Crb, Sdt, and Dlt (the Crb complex). *crb*, which encodes an apical transmembrane protein, was the first *Drosophila* gene characterized as a key regulator of epithelial polarization (155, 157). Currently, no interaction partners are known for the 30 EGF-like and 4 LG domains found in the extracellular part of Crb. The short cytoplasmic domain of Crb contains two functionally important motifs (64). One of these, the C-terminal amino acids ERLI, is a PDZ binding motif that interacts with Dlt and Sdt (7, 11, 54, 64). The physical interactions observed between Crb and Sdt are consistent with a previous genetic analysis suggesting that *sdt* acts downstream of *crb* (156). Dlt contains 4 PDZ domains whereas *sdt* gives rise to several splice forms, encoding either a MAGUK protein with a single PDZ, a SH3, and a GUK domain or a smaller protein containing only the GUK domain (7, 54). Crb, Dlt, and Sdt are conserved in *C. elegans* and mammalian species (7, 17, 29, 54, 112; own unpublished results). The current release of the human genome contains three *crb*-like genes (*CRB1*, *CRB2*, *CRB3*) (29, 112). *CRB1* was shown to correspond to the retinitis pigmentosa 12 (RP12) gene, mutations in which cause a degeneration of the retina (29).

As mentioned above, Dlt is found at the furrow canal during cellularization. In contrast, Crb and Sdt are first seen at the onset of gastrulation in association with the apical membrane, at which time Dlt is also recruited to the apical membrane (7, 11, 54, 149, 157). How the apical localization of the Crb complex is established initially is unclear. Recruitment of Sdt and Dlt to the apical membrane depends on interactions with the cytoplasmic tail of Crb (7, 11, 54, 64), whereas maintenance of apical Crb depends on Sdt and Dlt (11, 156). The distribution of the Crb complex within the apical domain is not uniform. Low concentrations are seen in the central region of the apical membrane, typically the "free" apical surface, whereas high levels of the Crb complex are found at the marginal zone. This accumulation of the

Crb complex may be driven by homophilic interactions between Crb molecules on opposing cell membranes, as the localization to the marginal zone depends on the presence of Crb in both contacting cells (112). Crb expression persists in all epithelia that have a ZA throughout development.

The ZA is not established in *crb* and *sdt* mutant embryos, and adherens junction material retains a spot-like distribution (50, 97, 149). Failure to assemble a ZA is likely to be a major contributor to the tissue breakdown seen in mutants that are affected for a component of the Crb complex. In addition, a number of apical markers disappear from the cell surface in these mutants, suggesting that the apical surface domain is lost (7, 11, 54, 156, 178). Failure to assemble adherens junctions may contribute to the loss of apical markers, as seen in the follicular epithelium (147). The Crb complex is presumably a component of a larger scaffold that controls the molecular composition of the apical membrane similar to the mutual dependencies seen between Crb complex components. In fact, Crb is sufficient to promote apical membrane differentiation as Crb overexpression results in an apicalization of the cell surface at the expense of the basolateral membrane and a complete disruption of the CJC (50, 179). Overexpression of the membrane-tethered cytoplasmic domain of Crb can rescue the *crb* mutant phenotype and cause membrane apicalization to a similar degree as overexpression of full-length Crb (179). The C-terminal PDZ binding motif in the cytoplasmic domain of Crb is essential for this activity (64). In contrast, overexpression of Dlt does not cause an apicalization phenotype (11; G.T. & U.T. unpublished data), whereas overexpression of Sdt still needs to be examined.

A functional overlap between Crb activity and other polarization mechanisms becomes apparent when the effects of *crb* mutations on different epithelial tissues that express Crb at similar levels and with similar subcellular distributions are compared. While the disruption of the ZA is uniform throughout ectodermal and endodermal epithelia during gastrulation (50, 149), tissue-specific responses to the lack of Crb or Sdt become apparent when individual organ primordia are established (155, 156). The epidermis is most strongly affected, and the great majority of cells die through programmed cell death. On the other extreme, there are tissues that appear more or less normal in late mutant embryos such as the Malphigian tubules or parts of the foregut and hindgut, implying that these cells were able to recover and establish normal polarity and a CJC although their ZA did not form during gastrulation (155–157). In fact, if epidermal cell death is prevented, surviving cells form small epithelial vesicles in which individual cells show normal polarity and possess a CJC (G.T. & U.T. unpublished data). Thus, it appears that Crb is required to establish normal polarity at early stages but is not needed to maintain epithelial polarity at later stages of development. Moreover, epithelial development in *crb* null mutants can be rescued to a large extent by increasing the wild-type gene copy number of *sdt* from two to three, suggesting that Sdt has a Crb independent apicalization activity (156). Further, in *crb* null mutant cells that do not die, Dlt is reduced but not lost entirely from the apical membrane, suggesting that a Crb-independent apical targeting mechanism must exist for Dlt that may partially or completely compensate for the loss of Crb (147). Taken together, these observations hint at complexities in the organization of protein scaffolds that define the apical surface domain of epithelial cells that remain largely elusive.

Apical Polarization II: The Bazooka/aPKC/DPar-6 Complex

A second complex that is important for the formation of the ZA but not a ZA component is composed of Bazooka (Baz), the *Drosophila* homolog of *C. elegans* Par-3 and vertebrate ASIP (66), *Drosophila* Par-6 (DPar-6) (115), and the *Drosophila* homolog of atypical Protein Kinase C (DaPKC) (180). This complex, called here the Baz/Par-3 complex, is conserved in *C. elegans* embryos where it contributes to the asymmetric division of the egg (122). In *Xenopus* oocytes this complex associates with the animal pole during maturation (99), and it localizes to tight junctions in mammalian epithelial cells where it is required for normal cell polarization (57–59, 78, 119, 140). Moreover, recent work in cell culture and *C. elegans* has shown that the small Rho family GTPases Cdc42 or Rac1 interact in their active, GTP-bound state with Par-6 and control localization and activity of the Par-3/Par-6/aPKC complex (49, 58, 59, 63, 78, 119).

Embryos that lack Baz function show defects at late cellularization/early gastrulation, at which time SAJs fail to concentrate at the apex of blastoderm cells and do not form a ZA (97). At early gastrulation in these mutant embryos, epithelial cells lose polarity, acquire mesenchymal characteristics, and gastrulation movements are compromised as a consequence. Later in development most cells die by programmed cell death (97, 174; U.T. unpublished data). This phenotype is very similar to the phenotype seen in arm^{GLC} mutants, suggesting that the failure of ZA assembly may be the major consequence of lack of Baz function. A similar phenotype is seen in embryos that lack DaPKC (180), whereas embryos that lack DPar-6 appear to undergo gastrulation movements normally and then subsequently lose epithelial integrity (115). Although the interactions between Cdc42 and the Baz/Par-3 complex have not been studied so far, embryos with reduced Cdc42 activity show defects in epithelial differentiation related to those in embryos with reduced activity of the Baz/Par-3 complex (46), raising the possibility that Cdc42 interacts with this complex as seen in other systems. The Baz/Par-3 complex associates with the entire apical membrane but is concentrated at the marginal zone similar to the Crb complex (180). It remains unclear as to how the Baz/Par-3 complex is linked to the plasma membrane, how it controls ZA assembly, and whether it has other important roles in epithelial polarity, in addition to ZA formation.

Apical Polarization III: The Lethal Giant Larvae/Discs Large/Scribble Complex

Molecular integrity and size of the apical domain also relies on the function of lateral protein complexes. Disruption of the CCC leads to loss of Crb and other factors from the apical membrane as mentioned above. A second interacting group of proteins that was recently implicated in the control of apical polarization is composed of Lgl, Dlg, and Scribble (Scrib) (14). Lgl is a Myosin II binding protein that contains WD40 repeats (93, 135), whereas Dlg and Scrib are multi-PDZ domain proteins; Dlg is a MAGUK (183) and Scrib belongs to the LAP subfamily of PDZ domain proteins that also contain leucine-rich repeats (13, 15). *lgl*, *dlg*, and *scrib* mutants display similar defects in the embryo, imaginal discs, and the follicular epithelium. In addition, colocalization and genetic interactions observed between these genes suggest that these proteins may form a biochemical complex, called here the Lgl complex (14). Lgl, Dlg, and Scrib homologs are found in *C. elegans* and vertebrates where they play a role in epithelial polarization as well (16, 17, 75, 91a). Also the Lgl homologs in yeast and humans have been shown to interact with Myosin II (62, 137), suggesting that Lgl may regulate Myosin II function. In fact, suppression of myosin II function by Lgl has recently been demonstarted in *Drosophila* neuroblast where Lgl and Dlg control neuroblast polarity during asymmetric division (106, 113).

lgl and *dlg* were identified as tumor suppressor genes that control proliferation and tissue integrity of imaginal discs (45, 93, 183). In imaginal discs and in late embryos, Dlg and Scrib are specific components of the SJ, whereas Lgl overlaps with the SJ but retains a broader distribution at the basolateral membrane (14, 15, 136, 183, 184). The role of the Lgl complex as a component of the SJ is discussed in more detail below. Defects in epithelial polarity in *lgl*, *dlg*, or *scrib* mutant embryos, which lack both maternal and zygotic expression of these genes, or mutant follicular epithelia appear long before SJ form (14, 15, 114). Cells of the ectodermal epithelium show a mislocalization of apical markers, such as Crb, and ZA markers, such as Arm. Both apical and ZA markers spread basally, suggesting that the marginal zone and the ZA do not form normally and the apical membrane expands basolaterally (14, 15). These defects are most prominent during gastrulation, whereas at later stages of development normal polarity is re-established and a normal CJC forms (G.T. & U.T. unpublished data). These findings suggest that the Lgl complex controls the segregation of apical and basolateral membrane domains at gastrulation and contributes to confinement of apical and apico-lateral markers to their normal position. How the Lgl complex acts to support normal differentiation of epithelial surface domains is unclear at present. As mentioned, the link to Myosin II may suggest that the regulation of Myosin activity is one target of the Lgl complex also in epithelial cells. Alternatively, the recent finding that the yeast homolog of Lgl interacts with a SNARE protein in polarized vesicle targeting (76) raises the possibility that the Lgl complex may regulate vesicle targeting to control epithelial polarity. This notion is also supported by the finding that a human Scrib homolog, ERBIN, restricts the ERBB2/HER2 receptor to the basolateral membrane (16).

THE SEPTATE JUNCTION

Structure and Functions of Septate Junctions

One of the most distinctive ultrastructural features of the CJC in invertebrate epithelial cells is the SJ. During *Drosophila* development, SJs first appear midway through embryogenesis, well after cellularization is completed, epithelial polarity has been established, and the ZA has formed. The SJ lies just basal to the ZA in epithelial cells, and within the SJ the membranes of adjacent cells maintain a constant distance of approximately 15 nm. In the pleated SJ (found in ectodermally derived epithelia and the glia sheets), regular arrays of electron-dense septae span the intermembranal space. In addition, freeze-fracture analysis reveals the presence of parallel rows of intramembranal particles (100a, 171) that presumably represent transmembrane proteins within the SJ. The septae form circumferential spirals around the cell much like the threads of a screw, and thereby greatly increase the distance that molecules must travel to pass between the apical and basolateral compartments of the epithelial sheet (25). Smooth SJs, which lack these ladder-like structures, are found only in the midgut and its derivatives. Relatively little is known of this variant of the SJ and for this reason we concentrate on the pleated SJ in this review.

Like other intercellular junctions, SJs have been proposed to play a role in formation of a trans-epithelial diffusion barrier, establishing and/or maintaining cell polarity, cell adhesion, and mediating interactions between cells. A variety of observations have led to the suggestion that SJs function in the formation of a trans-epithelial barrier. Morphological analysis, which revealed the existence of septae that fill the space between cells, led to the suggestion that SJs function to block direct paracellular flow between the apical and basolateral surfaces of epithelial sheets. This hypothesis was confirmed using injection of electron-dense dyes, which show restriction of dye diffusion at the SJ (25). More recently, mutational analysis of genes that encode SJ components has shown that disruption of the intercellular septae also results in disruption of the transepithelial seal (8, 68). A similar function has been ascribed to tight junctions in vertebrate epithelia, though SJs are quite different from tight junctions both morphologically and molecularly. Recent studies show that tight junctions regulate paracellular Na^+ and Mg^{++} ion flow by means of a selective "channel" function (129). Whether SJs display a similar ability to selectively regulate paracellular flow is a question that remains to be answered.

In addition to creating the paracellular barrier in ectodermally derived epithelia, pleated SJs have an essential role in the formation of the blood-nerve barrier in insect nervous systems (25). Although insect neurons are not myelinated as they are in vertebrates, neurons are typically surrounded by perineurial and glial sheath cells that form a diffusional barrier between the neurons of the central or peripheral nervous systems and the surrounding hemolymph. In some insects, though not in *Drosophila*, tight junction-like structures have been identified, in addition to SJs, in the surrounding cells (25, 70). Specific evidence regarding tight junctions in insects is discussed in a later section. All available evidence suggests that the SJs in epithelial cells and those in ensheathing perineurial and glia sheets are essentially indistinguishable.

Because SJs appear developmentally well after the time that epithelial polarity is established, they do not seem to be directly involved in this process, though they could be required to maintain that polarity once established. Thus the relationship between the SJ and apical-basal polarity is currently unclear. Mutations in some known SJ components disrupt the structure of the junction and the localization of other SJ components, but do not seem to affect the ZA, transmembrane proteins such as Notch, or apically localized components of the cytoskeleton (68). These observations suggest that SJs do not function as a fence that blocks diffusion of membrane components between the apical and basolateral surfaces. However, previous experiments have indicated that such a fence does exist in invertebrate epithelia that lack tight junctions, though the SJ was not directly shown to be the source of the fence function (181). In addition, mutations in other known SJ-associated proteins, notably *lgl*, *dlg* and *scrib*, do affect epithelial polarity (see earlier discussion). These seemingly contradictory results may indicate that the SJ has a selective fence function for particular proteins and that a mutation in one component may affect only a subset of apically or basally localized proteins. Alternatively, the early function of the Lgl complex in cell polarity may be distinct from its later role at the SJ.

Molecular Architecture of the Septate Junction

Although no systematic attempt has been made yet to characterize the molecular components of the SJ, molecular genetic analysis of several developmentally interesting genes has led to the discovery of SJ components. Of the SJ-associated proteins thus far identified, two, Coracle (Cor) and Neurexin-IV (Nrx-IV), appear to be most central to the morphologically defined SJ. Cor is a member of the Protein 4.1 superfamily of cytoplasmic proteins (39) that includes Protein 4.1, the Ezrin, Radixin, and Moesin (ERM) proteins, the NF2 tumor suppressor Merlin, Talin, several protein tyrosine phosphatases, unconventional myosins, and *Drosophila* Expanded (146). Cor is most similar to Protein 4.1, showing approximately 60% identity with Protein 4.1 in the amino-terminal 400 amino acids, a region of the molecule that is highly conserved in all members of the superfamily. This domain has been termed the FERM domain (26). In Cor, this domain appears to provide all functions that are required for localization to the SJ (168), as well as for SJ structure and function (169). Cor also shares a region of similarity with Protein 4.1 at the carboxy terminus that is not required for SJ function but is essential for viability. Phenotypic analysis of *cor* mutants revealed a role for *cor* during dorsal closure, salivary gland morphogenesis, and cuticle formation during embryonic development (39, 68, 168).

Examination of a null *cor* allele demonstrated a requirement for Cor in the formation of the SJ (68). *cor* mutant embryos lack the intercellular septae that are characteristic of the pleated SJ. The functional significance of this observed defect was tested by examining permeability of a 10-kD rhodamine-labeled dextran in living embryos. Dextran injected into the hemocoel of *cor* mutant embryos freely crosses the salivary gland epithelium, whereas in wild-type embryos injected in a similar manner dye cannot cross the epithelial barrier for at least one hour. Thus, *cor* function is clearly required for the trans-epithelial barrier function of the SJ, although, as noted previously, *cor* is not required for overall epithelial polarity or to restrict cell proliferation.

A significant step in understanding the role of Cor in SJ function was made with the discovery of the *Nrx-IV* gene (8). Nrx-IV is a *Drosophila* member of the Caspr (Contactin associated protein) family of neuronal receptors, which have a large extracellular domain with EGF and LG domains, and a single discoidin-like domain. *Drosophila* Nrx-IV possesses a single membrane-spanning region and, of particular interest, a short cytoplasmic domain that displays greater than 60% similarity to the cytoplasmic domain of glycophorin C, a transmembrane binding partner for Protein 4.1 in the erythrocyte (5). Like Cor, Nrx-IV is expressed in all cells that produce SJs, and its expression profile and subcellular localization are almost indistinguishable from that of Cor (8, 168).

Nrx-IV mutants display dorsal closure defects similar to those in *cor* and *dlg* mutant embryos (8, 168). Ultrastructural analysis revealed that the SJ is disrupted in *Nrx-IV* mutant embryos just as it is in *cor* mutants, identifying Nrx-IV as an important structural component of the SJ. *Nrx-IV* mutant embryos also display paralysis due to a breakdown of the blood-nerve barrier. Because this barrier is thought to be maintained by SJs, this result suggests that Nrx-IV, like Cor, is necessary for the barrier function of the SJ. Interestingly, mutations in *Drosophila gliotactin*, which encodes a neuroligin-like protein, also disrupt the blood-nerve barrier (6, 47). Neuroligins were originally identified as ligands for neurexins in neuronal synapses (23). Thus, it is possible that Gliotactin functions as a ligand for Nrx-IV in the SJ.

The similarity between the cytoplasmic tail of Nrx-IV and glycophorin C, the colocalization of Nrx-IV and Cor in the SJ, and the similarity of *cor* and *Nrx-IV* mutant phenotypes all suggest that *cor* and *Nrx-IV* may physically and functionally interact. Consistent with this notion, in *Nrx-IV* mutant embryos Cor fails to localize to the SJ and instead is distributed along the plasma membrane and in the cytoplasm (8, 168). Conversely, loss of *cor* function also affects Nrx-IV subcellular localization. Further studies have also shown that Cor and Nrx-IV can be co-immunoprecipitated from cell extracts, and that these proteins bind directly via the N-terminal conserved domain of Cor and the cytoplasmic tail of Nrx-IV (168). An unresolved question is how Cor and Nrx-IV initially target to the SJ, since they appear to show an interdependence that is incompatible with either one of them having this role. This observation suggests that at least one other protein whose identity is not yet known must be involved. By analogy with Protein 4.1 and glycophorin C, which interact with the PDZ domain containing proteins hDLG and p55 (83, 86), this third protein likely contains PDZ domains. The obvious candidate for this role is Dlg, but attempts to identify interactions between Dlg and either Cor or Nrx-IV have thus far produced negative results (168).

As indicated earlier, Dlg is another SJ-associated protein, and in fact was the first protein shown to localize preferentially to the SJ (183). Dlg is initially uniformly distributed along the lateral membrane and to a lesser extent throughout the cytoplasm. During mid-embryogenesis, this subcellular localization is refined to the presumptive SJ (183). Although the precise cellular function of Dlg in the SJ is not known, previous work has demonstrated a direct role for *dlg* in the ultrastructure and function of the SJ (184). Imaginal discs in *dlg* mutant larvae lack the septae that characterize the pleated SJ, whereas some ZA material is mislocalized to a more basal location. In addition, the apical-basal polarity of the imaginal tissues is disrupted in *dlg* mutants. Interestingly, these effects are less severe in the nondividing salivary gland epithelial cells, perhaps suggesting that the SJ can be maintained once established, at least in cells that are not mitotically active, and therefore are not disassembling and reassembling the CJC.

dlg mutant imaginal discs display loss of epithelial polarity, cellular apoptosis, and overproliferation that becomes apparent during the extended larval period (more than twice the normal length) that is a consequence of this mutation (1, 182). As described previously, Dlg is part of the Lgl complex that is essential for epithelial polarity in the early embryo (14). The core function of this complex is likely the same in early embryogenesis, before the SJ forms, and later when the complex is associated with the SJ. Less clear is how the function of these proteins relates to the structure of the SJ itself. One possibility is that Dlg, Scrib, and Lgl, cooperatively, establish a unique domain in the apical lateral membrane that serves as a scaffold upon which the later-acting SJ components, such as Cor and Nrx-IV, can assemble during the formation of the SJ.

In addition to their effects on epithelial polarity, mutations in *dlg*, *lgl*, and *scrib* also result in tumor-like overgrowth of imaginal epithelia. However, because null mutations in these genes also affect the formation of other junctions, it is unclear if the observed overproliferation effect results from a direct role in restricting cell proliferation, or instead from disruption of intercellular interactions. In contrast to the overproliferation phenotypes of *lgl*, *dlg*, and *scrib*, mutations in *cor* result in a decreased rate of cell proliferation and eventual loss of cells from the epithelium due to cell competition (68). In addition, *cor* mutations dominantly suppress the hypermorphic *Ellipse* allele of the *EGF receptor* gene (39). Unlike *dlg*, *scrib*, and *lgl*, *cor* mutations do not affect overall epithelial polarity or the ZA. This result could suggest that some aspect of SJ function is required to promote cell proliferation, whereas disruption of apical-basal polarity and CJC formation results in overproliferation, perhaps due to the loss of intercellular interactions that normally function to restrict proliferation.

Relationship between the Insect Epithelial and the Vertebrate Paranodal Septate Junction

Until recently, septate junctions were believed to be unique to invertebrates, unlike the ZA, which appears to be widespread throughout the metazoans. However, it is now apparent that within the vertebrate nervous system a structurally and functionally analogous SJ exists. Paranodal SJs are found in myelinated neurons at either end of each node of Ranvier, the region between adjacent sections of the myelin sheath in which components necessary for the action potential (primarily the voltage gated Na^+ channel) are clustered (Figure 3). These SJs form between the loops of myelinating oligodendrocytes and Schwann cells and the axons they ensheath. Morphologically, paranodal SJs are quite similar to SJs found in insect epithelial cells and glia, displaying a characteristic array of ladder-like cross-bridges (9, 108, 176). Functionally, paranodal SJs are thought to provide insulation between the nodal and internodal regions of the axon, thereby allowing the saltatory conduction that is essential for rapid transmission of electrical signals along myelinated nerve fibers. In addition, as the primary site of contact between axons and glia, they are almost certainly important in mediating signals between these very different but closely interlinked cell types.

Recent studies have made significant progress in understanding the molecular composition and genetic functions of the paranodal SJ. One of the first identified components of this junction, Caspr (Contactin-associated protein; also known as Paranodin), is the mammalian homologue of *Drosophila* Nrx-IV. Caspr is expressed only in neurons, and in mature myelinated neurons it is found exclusively in the paranodal SJs (36, 94). Caspr binds to a neuronal isoform of Protein 4.1 via its cytoplasmic tail (94), just as Nrx-IV binds to Cor (168). Thus two primary components of the invertebrate SJ have homologous counterparts in the vertebrate paranodal SJ. As its name implies, Caspr was isolated via its association with Contactin, a GPI-linked protein that is also expressed by neurons and localizes to the paranodal SJ. In addition to binding Caspr (in *cis* within the neuronal membrane), Contactin colocalizes with Neurofascin-155, an immunoglobulin superfamily adhesion molecule that is expressed on myelinating glia cells, although they do not appear to interact directly (142).

The functions of two components of the paranodal SJ, Caspr and Contactin, have recently been examined using knockout mutations in the mouse (12, 18). Mutation of either gene results in dramatic disruption of paranodal architecture and junctional function. As expected, both mutations alter the electrical properties of myelinated nerve fibers, resulting in reduced conduction velocity along the nerve. More surprising is the effect of these mutations on the organization of other proteins within the node of Ranvier, the paranodal regions, and in the myelinating cells. In *Caspr* knockout mice, the paranodal localizations of Contactin in the neuron and Neurofascin-155 in the myelinating cells are disrupted (12). Furthermore, Na^+ channels that are normally restricted to the node spread laterally along the axon into the paranodal region. Conversely, in *Caspr* mutant mice neuronal K^+ channels that normally are found just outside the paranodal region (in the juxtaparanode) redistribute into the paranodal region and into the node itself. Similar effects, including abnormal localizations of Neurofascin-155 and K^+ channels, are seen in the *contactin* knockout mutant (18). In addition, Caspr protein failed to be transported from the neuronal cell body to the axon in *contactin* mutant neurons, consistent with previous reports that these proteins form a complex prior to transport to the plasma membrane (37).

Taken together, these results provide strong evidence that the paranodal SJ not only provides a site of contact between the neuron and myelinating glia cell, but also serves as a molecular sieve that organizes the nodal, paranodal, and juxtaparanodal regions of myelinated neurons (108). This sieving effect appears to be analogous to the fence function within the plane of the plasma membrane that has

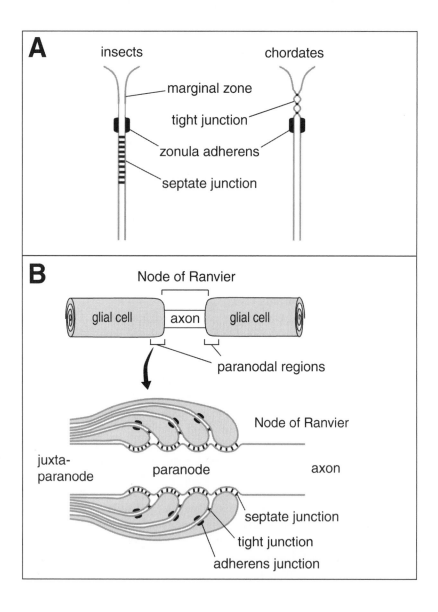

been shown for tight junctions in mammalian epithelial cells (133) and proposed for the SJ in invertebrate epithelia (181). Given their molecular and morphological similarities, the invertebrate SJ and the mammalian paranodal SJ probably both derive from a common ancestral junction and they are both structurally and functionally homologous. This observation has important implications for both SJs. By analogy with the paranodal SJ, epithelial SJ may have a selective fence function within the plane of the plasma membrane that has not yet been well characterized, and an as yet unidentified Contactin-like molecule may serve as a binding partner for Nrx-IV. Conversely, genetic studies of the epithelial SJ and epithelial polarity in *Drosophila* should provide new insights into the components and functions of the paranodal SJ in vertebrates. For example, we currently know little about the mammalian homologues of *lgl*, *scrib*, and *dlg* in neuronal development; however, the localization of these proteins to the epithelial SJ and their importance in epithelial polarity and SJ function suggest that their mammalian homologues are significant components of paranodal SJs and axonal cell polarity.

TIGHT JUNCTIONS IN DROSOPHILA?

As mentioned earlier, in vertebrate epithelia tight junctions are believed to form the principle paracellular barrier to transepithelial diffusion. Morphologically, tight junctions are characterized by strands of intramembranous particles in freeze fracture analysis. Based on this criterion, previous studies have reported the existence of tight junctions in a variety of invertebrate species, including insects (25, 69–71). However, careful morphological studies have so far failed to identify a tight junction-like structure in *Drosophila* (25, 154). For this reason, and because the SJ seems to provide at least some of the functions ascribed to tight junctions in vertebrate cells, there has not been a clear consensus on the existence of tight

----

Figure 3 Comparison of *Drosophila* and chordate apical junctional complexes (*A*) and the structure of the vertebrate paranodal junction (*B*). Insect and chordate epithelia are similar in that the junctional complex in both contains a zonula adherens. In insects the marginal zone is apical to the zonula adherens, whereas in chordates the tight junction is located at this position. Insect epithelial cells have in addition a septate junction that lies basal to the zonula adherens. In (*B*), the structure of the node of Ranvier in myelinated neurons is diagrammed above, with a higher magnification view of the paranodal region presented below. In the node of Ranvier, the myelin sheath is interrupted. At the edge of the sheath (the paranodal region) loops from the myelinating cell are closely apposed to the axon. At the point of contact between the neuron and the myelinating cell, a septate junction forms that is structurally and molecularly similar to the septate junction of insect epithelial and glial cells.

junctions in insects, nor have any functional studies been performed in *Drosophila*.

Despite the lack of evidence for tight junctions in *Drosophila*, molecular genetic analysis of developmentally important genes and the *Drosophila* genome project have identified apparent homologues of known components of the vertebrate tight junction. For example, the previously mentioned *pyd* gene encodes a protein that is similar to the mammalian ZO-1 protein, the first identified tight junction component (134, 145). Although PYD/ZO-1 was originally described as an SJ component, subsequent studies indicate that one isoform is localized apical to the SJ, while another seems more broadly distributed in the apical region of the cell (170). Analysis of vertebrate tight junctions has identified two other types of proteins that seem to be integral to the tight junction, the occludins and the claudins, although recent studies indicate that only the claudins are essential for tight junction function (165). The *Drosophila* genome does not contain any convincing occludin homologues (3). In contrast, there are at least two possible claudin-like genes in the genomic sequence (CG3770 and CG6982) that have four predicted transmembrane domains in a similar arrangement to the claudins (R. Fehon, unpublished observations). So far, neither of these predicted genes nor the proteins they encode have been studied.

Why then have tight junctions not been observed in *Drosophila*? Note that although PYD/ZO-1 is expressed apically in epithelial cells (170), we do not yet know how widely the claudin-like proteins are expressed. Thus, *Drosophila* tight junctions might be restricted to a particular developmental stage or tissue that has not been examined carefully enough to detect tight junctions (24). However, *Drosophila* epithelia might also retain some tight junctional structure, at least at the molecular level, but not have the occluding function of the mammalian tight junction (that is instead provided by the SJ). In mammalian epithelia, the tight junction is found at the apical-most point of contact between cells, just apical to the ZA. In *Drosophila*, the corresponding region is the marginal zone (149) (Figure 3), an area that lacks obvious junctional morphology but does seem to have an accumulation of transmembrane receptors and associated proteins (7, 11, 54, 66, 115, 149, 170, 180). Of particular interest in this regard is the recent demonstration that the Baz/Par-3 complex, which localizes to the marginal zone (66, 115, 180), has mammalian homologues that reside in the tight junction and is essential for tight junction assembly (57, 58; also see earlier discussion of these genes). Taken together, these results indicate that a number of tight junction proteins localize to the marginal zone in *Drosophila*, whereas currently no *Drosophila* homologues of tight junction proteins are known that associate with the SJ. These data suggest that the marginal zone in *Drosophila* epithelia may share some functions, in particular cell-cell signaling and perhaps the fence function within the plane of the plasma membrane, with the mammalian tight junction. In this regard it would be particularly interesting to know the subcellular localizations of the claudin-like proteins in *Drosophila* epithelial cells, if indeed they are expressed in these tissues.

THE SPECTRIN CYTOSKELETON
IN EPITHELIAL DIFFERENTIATION

One important aspect of epithelial polarity is the corresponding polarization of the underlying actin-based cytoskeleton that occurs via interactions between polarized transmembrane proteins, membrane-associated cytoplasmic proteins, and cytoskeletal proteins (100, 185). Among the many proteins that appear to be involved in this process, spectrin seems to play a crucial role. The spectrin protein is a tetrameric actin crosslinking protein comprised of two α and two β subunits. Epithelial cells contain a polarized spectrin cytoskeleton, in which distinct isoforms of spectrin associate with the apical or basolateral membrane. Spectrin contributes to polarized membrane organization by binding, and thus trapping membrane proteins at the basolateral surface (92, 100).

Drosophila has three different spectrin subunits, α, β, and β_H-Spectrin, which assemble into two different isoforms, $\alpha_2\beta_2$-Spectrin and $\alpha_2\beta_{H2}$-Spectrin. (32, 159). The two isoforms show non-overlapping polarized distributions in epithelial cells. $\alpha_2\beta_2$-Spectrin is found at the basolateral membrane where it forms a complex with Ankyrin (33, 35, 74). In contrast, the $\alpha_2\beta_{H2}$-Spectrin associates with the apical domain where it is enriched in the marginal zone and, possibly, the ZA (160, 162) (Figure 2). Before the onset of cellularization, $\alpha_2\beta_{H2}$-Spectrin associates with the egg membrane, and during cellularization it remains with the furrow canals, whereas $\alpha_2\beta_2$-Spectrin is added to the lateral membrane as it forms. At late cellularization, β-Spectrin and β_H-Spectrin overlap in the apical-lateral membrane, as mentioned above, before they segregate into their final distinct apical and basolateral positions (161).

Mutational analyses have been carried out for all three *Drosophila* spectrin genes but so far did not reveal a general role of the spectrin cytoskeleton in epithelial polarity. Lack of β_H-Spectrin does not cause defects in cellularization or epithelial polarity in early embryos (J.A. Williams & G.H. Thomas, personal communication), whereas the requirement of β- and α-Spectrin in early embryos remains to be analyzed. However, spectrin mutants exhibit a number of interesting defects, which suggest that spectrin has cell type-specific roles in epithelial differentiation. *α-spectrin* mutants die as larvae and exhibit loss of cell-cell contacts in the midgut whereas other epithelial tissues differentiate normally (74). *α-spectrin* mutations affect the cuprophilic cells of the midgut epithelium that are responsible for the acidification of the midgut content. In *α-spectrin* mutant larvae, β_H-Spectrin is lost from the apical membrane of cuprophilic cells. In addition, the actin cytoskeleton appears disorganized, and acid secretion is impaired (31, 74). In contrast, cuprophilic cells mutant for *β-spectrin* show a disrupted organization of the basolateral membrane that fails to accumulate the Na^+, K^+-ATPase, a defect not seen in *α-spectrin* mutants (34). These findings suggest that $\alpha_2\beta_{H2}$-Spectrin is required for the differentiation of the apical membrane in cuprophilic cells, and that β-Spectrin functions independently of α-Spectrin in the differentiation of the basolateral membrane domain.

β_H-Spectrin in encoded by the *karst* gene. *karst* mutations are semiviable and adult escapers exhibit bent wings, tracheal defects, sterility, and rough eyes (162). However, no obvious polarity defects were observed in *karst* mutant imaginal discs. A role for spectrin in the maintenance of epithelial polarity has been found in the ovarian follicular epithelium (73). Follicle cells that lack α-Spectrin form a normal follicular epithelium initially, but exhibit overproliferation, multilayering, and loss of the apical β_H-Spectrin at later stages. *α-spectrin* mutant follicle cells retain lateral β-Spectrin, suggesting that recruitment of β-Spectrin to the basolateral membrane is independent of α-Spectrin (73). Also, follicle cells that lack β_H-Spectrin form a follicular epithelium but lose apical α-Spectrin. No polarity defects are detected in *karst* mutant follicle cells. However, fragmentation of the ZA is observed in *karst* mutant follicle cells as they migrate posteriorly to cover the oocyte. At this time follicle cells in *karst* mutants fail to constrict apically, suggesting that β_H-Spectrin may stabilize the ZA during apical constriction (159, 190). These results suggest that the defects seen in *α-spectrin* mutants are largely independent of β_H-Spectrin, and that β_H-Spectrin is involved in maintaining the ZA during epithelial morphogenesis.

CONCLUSIONS AND PROSPECTS

The recent progress in our understanding of the mechanisms involved in epithelial polarization has focused our attention on a number of protein complexes that play essential roles in the formation of distinct plasma membrane domains. These protein complexes either congregate around transmembrane proteins (Cadherin, Crb, Nrx-IV) or represent cytocortical protein assemblies for which the mechanism of plasma membrane association remains obscure (Baz/Par-3 complex, Lgl complex). Additional components of these protein complexes and molecular interactions within these complexes remain to be characterized. Further, a remaining major challenge is to uncover how the activity of these complexes is integrated to generate a single polarized cellular architecture. How, for example, does the Crb complex control ZA formation and how, in turn, do adherens junctions control the stability of the Crb complex? Similarly, how does the Lgl complex, which localizes to the lateral membrane, confine the extent of the apical domain? These functional relationships suggest connections between these complexes, either in the form of physical linkages or perhaps via intracellular signaling pathways, which are currently not understood.

We have not discussed in detail a number of additional *Drosophila* genes that act in epithelial differentiation because their function and their relation to the larger themes elaborated in this review are not well understood. Among these genes is *bloated tubules*, which encodes a transmembrane protein related to vertebrate neurotransmitter symporters and controls the extent of the apical domain in Malphigian tubules (60). The *arc* gene encodes a PDZ domain protein that associates with the marginal zone and/or ZA of portions of embryonic and imaginal epithelia and controls the morphogenesis of imaginal discs (79). *faint sausage* encodes a GPI anchored adhesion molecule of the immunoglobulin superfamily and is required for maintenance of epithelial intergity from mid- to late-embryogenesis (77). Moesin and Merlin, members of the Protein 4.1 superfamily, are found in the apical region of epithelial cells (90). *Merlin* mutations do not appear to affect overall cell polarity (67), and the effects of *Moesin* mutations on polarity and epithelial integrity are currently being examined (O. Nikiforova & R. Fehon, unpublished results). A major challenge for the near future will be to explore the activity and molecular interactions of these proteins, and other yet unidentified genes that play a role in epithelial polarity. Moreover, one issue that plays a central role in the discussion of epithelial polarity in mammalian cell culture models, the contribution of protein and lipid sorting in the biosynthetic pathway, has so far not been vigorously pursued in *Drosophila*.

Intriguing parallels and differences become apparent when polarity in epithelial cells and non-epithelial neuroblasts are compared. Neuroblasts in *Drosophila* (similar to the one-cell *C. elegans* embryo and budding yeast) have two surface domains, an apical and a basal domain (anterior and posterior in *C. elegans* and bud site versus non-bud site in yeast) (30, 81, 122). In contrast, in differentiated epithelial cells we can distinguish at least six membrane domains, the free apical surface, the marginal zone, the ZA, the SJ, the lateral membrane basal to the SJ, and the basal membrane. Only the Baz/Par-3 complex and the Lgl complex act in both neuroblast and epithelial polarity, whereas the complexes that associate with Cadherin, Crb, and Nrx-IV are not needed for neuroblast polarity. This comparison emphasizes the central role of adhesive interactions mediated by transmembrane adhesion receptors in defining epithelial membrane domains. Moreover, it raises the question of how the polarized cortical localization of the Baz/Par-3 complex and the Lgl complex is generated in epithelial cells and neuroblasts in the absence of any known transmembrane components or other localization cues. Two inferences can be drawn from these observations. First, we may view polarity as seen in yeast, the *C. elegans* embryo, or *Drosophila* neuroblasts as a simpler form of cell polarization that is elaborated upon in epithelial cells by the impact of cell adhesion receptors and their associated protein complexes. Second, we are still far from fully understanding the mechanisms by which the intricate cellular architecture of epithelial cells, or even the relatively simpler polarized organization of other cells, is established and maintained in developing organisms.

ACKNOWLEDGMENTS

We are grateful to Y. Hong, Y. N. Jan, E. Knust, M. Peifer, G. Thomas and A. Wodarz for communicating results prior to publication. We thank D. Godt and J. Genova for critical comments on the manuscript. This work was supported by grants from the Canadian Institute of Heath Research and the Canadian Cancer Institute (to U. T.) and by grants from the American Cancer Society (RPG-97-026-04-DDC) and the National Institutes of Health (NS34783) (to R. F.).

Visit the Annual Reviews home page at www.AnnualReviews.org

LITERATURE CITED

1. Abbott LA, Natzle JE. 1992. Epithelial polarity and cell separation in the neoplastic *l(1)dlg-1* mutant of *Drosophila. Mech. Dev.* 37:43–56

2. Adam JC, Pringle JR, Peifer M. 2000. Evidence for functional differentiation among *Drosophila* septins in cytokinesis and cellularization. *Mol. Biol. Cell* 11: 3123–35

3. Adams MD, Celniker SE, Holt RA, Evans CA, Gocayne JD, et al. 2000. The genome sequence of *Drosophila melanogaster. Science* 287:2185–95

4. Afshar K, Stuart B, Wasserman SA. 2000. Functional analysis of the *Drosophila diaphanous* FH protein in early embryonic development. *Development* 127:1887–97

5. Anderson RA, Lovrien RE. 1984. Glycophorin is linked by band 4.1 protein to the human erythrocyte membrane skeleton. *Nature* 307:655–58

6. Auld VJ, Fetter RD, Broadie K, Goodman CS. 1995. Gliotactin, a novel transmembrane protein on peripheral glia, is required to form the blood-nerve barrier in *Drosophila. Cell* 81:757–67

7. Bachmann A, Schneider M, Theilenberg E, Grawe F, Knust E. 2001. Stardust, a novel *Drosophila* MAGUK, acts as a partner of Crumbs in the control of epithelial polarity. *Nature.* In press

8. Baumgartner S, Littleton JT, Broadie K, Bhat MA, Harbecke R, et al. 1996. A *Drosophila* neurexin is required for septate junction and blood-nerve barrier formation and function. *Cell* 87:1059–68

9. Bellen HJ, Lu Y, Beckstead R, Bhat MA. 1998. Neurexin IV, caspr and paranodin—novel members of the neurexin family: encounters of axons and glia. *Trends Neurosci.* 21:444–49

10. Berrueta L, Kraeft SK, Tirnauer JS, Schuyler SC, Chen LB, et al. 1998. The adenomatous polyposis coli-binding protein EB1 is associated with cytoplasmic and spindle microtubules. *Proc. Natl. Acad. Sci. USA* 95:10596–601

11. Bhat MA, Izaddoost S, Lu Y, Cho KO, Choi KW, Bellen H. 1999. Discs Lost, a novel multi-PDZ domain protein, establishes and maintains epithelial polarity. *Cell* 96:633–45

12. Bhat MA, Rios JC, Lu Y, Garcia-Fresco GP, Ching W, et al. 2001 Axon-glia interactions and the domain organization of myelinated axons requires neurexin IV/Caspr/Paranodin. *Neuron* 30:369–83

13. Bilder D, Birnbaum D, Borg JP, Bryant P, Huigbretse J, et al. 2000. Collective nomenclature for LAP proteins. *Nat. Cell Biol.* 2:E114

14. Bilder D, Li M, Perrimon N. 2000. Cooperative regulation of cell polarity and growth by *Drosophila* tumor suppressors. *Science* 289:113–16

15. Bilder D, Perrimon N. 2000. Localization of apical epithelial determinants by the basolateral PDZ protein Scribble. *Nature* 403:676–80

16. Borg JP, Marchetto S, Le Bivic A, Ollendorff V, Jaulin-Bastard F, et al. 2000. ERBIN: a basolateral PDZ protein that interacts with the mammalian ERBB2/HER2 receptor. *Nat. Cell Biol.* 2:407–14

17. Bossinger O, Klebes A, Segbert C, Theres C, Knust E. 2001. Zonula adherens formation in *Caenorhabditis elegans* requires *dlg-1*, the homologue of the *Drosophila* gene *discs large. Dev. Biol.* 230:29–42

18. Boyle ME, Berglund EO, Murai KK, Weber L, Peles E, Ranscht B. 2001. Contactin orchestrates assembly of the septate-like junctions at the paranode

in myelinated peripheral nerve. *Neuron* 30:385–97

19. Bridges CB, Brehme KS. 1944. *The Mutants of Drosophila melanogaster.* Carnegie Inst.

20. Brown NH, Gregory SL, Martin-Bermudo MD. 2000. Integrins as mediators of morphogenesis in *Drosophila. Dev. Biol.* 223:1–16

21. Burgess RW, Deitcher DL, Schwarz TL. 1997. The synaptic protein syntaxin1 is required for cellularization of *Drosophila* embryos. *J. Cell Biol.* 138:861–75

22. Cagan RL, Kramer H, Hart AC, Zipursky SL. 1992. The bride of sevenless and sevenless interaction: internalization of a transmembrane ligand. *Cell* 69:393–99

23. Cantallops I, Cline HT. 2000. Synapse formation: if it looks like a duck and quacks like a duck... . *Curr. Biol.* 10:R620–23

24. Carlson SD, Hilgers SL, Juang JL. 1997. Ultrastructure and blood-nerve barrier of chordotonal organs in the *Drosophila* embryo. *J. Neurocytol.* 26:377–88

25. Carlson SD, Juang JL, Hilgers SL, Garment MB. 2000. Blood barriers of the insect. *Annu. Rev. Entomol.* 45:151–74

26. Chishti AH, Kim AC, Marfatia SM, Lutchman M, Hanspal M, et al. 1998. The FERM domain: a unique module involved in the linkage of cytoplasmic proteins to the membrane. *Trends Biochem. Sci.* 23:281–82

27. Cox RT, Kirkpatrick C, Peifer M. 1996. Armadillo is required for adherens junction assembly, cell polarity, and morphogenesis during *Drosophila* embryogenesis. *J. Cell Biol.* 134:133–48

28. Crawford JM, Harden N, Leung T, Lim L, Kiehart DP. 1998. Cellularization in *Drosophila melanogaster* is disrupted by the inhibition of rho activity and the activation of Cdc42 function. *Dev. Biol.* 204:151–64

29. den Hollander AI, ten Brink JB, de Kok YJ, van Soest S, van den Born LI, et al. 1999. Mutations in a human homologue of *Drosophila crumbs* cause retinitis pigmentosa (RP12). *Nat. Genet.* 23:217–21

30. Drubin DG, Nelson WJ. 1996. Origins of cell polarity. *Cell* 84:335–44

31. Dubreuil RR, Frankel J, Wang P, Howrylak J, Kappil M, Grushko TA. 1998. Mutations of α-spectrin and labial block cuprophilic cell differentiation and acid secretion in the middle midgut of *Drosophila* larvae. *Dev. Biol.* 194:1–11

32. Dubreuil RR, Grushko T. 1998. Genetic studies of spectrin: new life for a ghost protein. *BioEssays* 20:875–78

33. Dubreuil RR, Maddux PB, Grushko TA, MacVicar GR. 1997. Segregation of two spectrin isoforms: polarized membrane-binding sites direct polarized membrane skeleton assembly. *Mol. Biol. Cell* 8:1933–42

34. Dubreuil RR, Wang P, Dahl S, Lee J, Goldstein LS. 2000. *Drosophila* β-spectrin functions independently of α-spectrin to polarize the Na,K ATPase in epithelial cells. *J Cell Biol.* 149:647–56

35. Dubreuil RR, Yu J. 1994. Ankyrin and β-spectrin accumulate independently of α-spectrin in *Drosophila. Proc. Natl. Acad. Sci. USA* 91:10285–89

36. Einheber S, Zanazzi G, Ching W, Scherer S, Milner TA, et al. 1997. The axonal membrane protein Caspr, a homologue of neurexin IV, is a component of the septate-like paranodal junctions that assemble during myelination. *J. Cell Biol.* 139:1495–506

37. Faivre-Sarrailh C, Gauthier F, Denisenko-Nehrbass N, Le Bivic A, Rougon G, Girault JA. 2000. The glycosylphosphatidyl inositol-anchored adhesion molecule F3/ contactin is required for surface transport of paranodin/contactin-associated protein (caspr). *J. Cell Biol.* 149:491–502

38. Farquhar MG, Palade GE. 1963. Junctional complexes in various epithelia. *J. Cell Biol.* 17:375–412

39. Fehon RG, Dawson IA, Artavanis-Tsakonas S. 1994. A *Drosophila* homologue of membrane-skeleton protein 4.1 is associated with septate junctions and is encoded by the *coracle* gene. *Development* 120:545–57

40. Fehon RG, Johansen K, Rebay I, Artavanis-Tsakonas S. 1991. Complex cellular and subcellular regulation of *Notch* expression during embryonic and imaginal development of implications for *Notch* function. *J. Cell Biol.* 113:657–69

41. Field CM, Alberts BM. 1995. Anillin, a contractile ring protein that cycles from the nucleus to the cell cortex. *J. Cell Biol.* 131:165–78

42. Foe VE, Alberts BM. 1983. Studies of nuclear and cytoplasmic behaviour during the five mitotic cycles that precede gastrulation in *Drosophila* embryogenesis. *J. Cell Sci.* 61:31–70

43. Foe VE, Odell GM, Edgar BA. 1993. Mitosis and morphogenesis in the *Drosophila* embryo: Point and counterpoint. In *The Development of Drosophila melanogaster*, ed. M Bate, A Martinez-Arias, pp. 149–300. Plainview, NY: CSHL Press

44. Fullilove SL, Jacobson AG. 1971. Nuclear elongation and cytokinesis in *Drosophila montana*. *Dev. Biol.* 26:560–77

45. Gateff E. 1978. Malignant neoplasms of genetic origin in *Drosophila melanogaster*. *Science* 200:1448–59

46. Genova JL, Jong S, Camp JT, Fehon RG. 2000. Functional analysis of Cdc42 in actin filament assembly, epithelial morphogenesis, and cell signaling during *Drosophila* development. *Dev. Biol.* 221: 181–94

47. Gilbert M, Smith J, Roskams AJ, Auld VJ. 2001. Neuroligin 3 is a vertebrate gliotactin expressed in the olfactory ensheathing glia, a growth-promoting class of macroglia. *Glia* 34:151–64

48. Godt D, Tepass U. 1998. *Drosophila* oocyte localization is mediated by differential cadherin-based adhesion. *Nature* 395:387–91

49. Gotta M, Abraham MC, Ahringer J. 2001. CDC-42 controls early cell polarity and spindle orientation in *C. elegans*. *Curr. Biol.* 11:482–88

50. Grawe F, Wodarz A, Lee B, Knust E, Skaer H. 1996. The *Drosophila* genes *crumbs* and *stardust* are involved in the biogenesis of adherens junctions. *Development* 122:951–59

51. Grindstaff KK, Yeaman C, Anandasabapathy N, Hsu SC, Rodriguez-Boulan E, et al. 1998. Sec6/8 complex is recruited to cell-cell contacts and specifies transport vesicle delivery to the basal-lateral membrane in polarized epithelial cells. *Cell* 93:731–40

52. Haag TA, Haag NP, Lekven AC, Hartenstein V. 1999. The role of cell adhesion molecules in *Drosophila* heart morphogenesis: faint sausage, shotgun/DE-cadherin, and laminin A are required for discrete stages in heart development. *Dev. Biol.* 208:56–69

53. Hales KG, Bi E, Wu JQ, Adam JC, Yu IC, Pringle JR. 1999. Cytokinesis: an emerging unified theory for eukaryotes? *Curr. Opin. Cell Biol.* 11:717–12

54. Hong Y, Stronach B, Perrimon N, Jan LY, Jan YN. 2001. Stardust interacts with Crumbs to control polarity of epithelia but neuroblasts in *Drosophila*. *Nature*. In press

55. Hunter C, Wieschaus E. 2000. Regulated expression of *nullo* is required for the formation of distinct apical and basal adherens junctions in the *Drosophila* blastoderm. *J. Cell Biol.* 150:391–401

56. Ikeda W, Nakanishi H, Miyoshi J, Mandai K, Ishizaki H, Tanaka M, et al. 1999. Afadin: a key molecule essential for structural organization of cell-cell junctions of polarized epithelia during embryogenesis. *J. Cell Biol.* 146:1117–32

57. Izumi Y, Hirose T, Tamai Y, Hirai S, Nagashima Y, et al. 1998. An atypical PKC directly associates and colocalizes at the epithelial tight junction with ASIP, a

mammalian homologue of *Caenorhabditis elegans* polarity protein PAR-3. *J. Cell Biol.* 143:95–106

58. Joberty G, Petersen C, Gao L, Macara IG. 2000. The cell-polarity protein Par6 links Par3 and atypical protein kinase C to Cdc42. *Nat. Cell Biol.* 2:531–39

59. Johansson A, Driessens M, Aspenstrom P. 2000. The mammalian homologue of the *Caenorhabditis elegans* polarity protein PAR-6 is a binding partner for the Rho GTPases Cdc42 and Rac1. *J. Cell Sci.* 113:3267–75

60. Johnson K, Knust E, Skaer H. 1999. *bloated tubules* (*blot*) encodes a *Drosophila* member of the neurotransmitter transporter family required for organisation of the apical cytocortex. *Dev. Biol.* 212:440–54

61. Jürgens G, Wieschaus E, Nüsslein-Volhard C, Kluding M. 1984. Mutations affecting the pattern of the larval cuticle in *Drosophila melanogaster*. II. Zygotic loci on the third chromosome. *Wilhelm Roux's Arch. Entwicklungsmech. Org.* 193:283–95

62. Kagami M, Toh-e A, Matsui Y. 1998. Sro7p, a *Saccharomyces cerevisiae* counterpart of the tumor suppressor l(2)gl protein, is related to myosins in function. *Genetics* 149:1717–27

63. Kay AJ, Hunter CP. 2001. CDC-42 regulates PAR protein localization and function to control cellular and embryonic polarity in C. *elegans. Curr. Biol.* 11:474–81

64. Klebes A, Knust E. 2000. A conserved motif in Crumbs is required for E-cadherin localisation and zonula adherens formation in *Drosophila. Curr. Biol.* 10:76–85

65. Kraut R, Chia W, Jan LY, Jan YN, Knoblich JA. 1996. Role of *inscuteable* in orienting asymmetric cell divisions in *Drosophila. Nature* 383:50–55

66. Kuchinke U, Grawe F, Knust E. 1998. Control of spindle orientation in *Drosophila* by the Par-3-related PDZ-domain protein Bazooka. *Curr. Biol.* 8:1357–65

67. LaJeunesse DR, McCartney BM, Fehon RG. 1998. Structural analysis of *Drosophila* merlin reveals functional domains important for growth control and subcellular localization. *J. Cell Biol.* 141:1589–99

68. Lamb RS, Ward RE, Schweizer L, Fehon RG. 1998. *Drosophila coracle*, a member of the protein 4.1 superfamily, has essential structural functions in the septate junctions and developmental functions in embryonic and adult epithelial cells. *Mol. Biol. Cell* 9:3505–19

69. Lane NJ. 1991. Morphology of glial blood-brain barriers. *Ann. NY Acad. Sci.* 633:348–62

70. Lane NJ, Chandler HJ. 1980. Definitive evidence for the existence of tight junctions in invertebrates. *J. Cell Biol.* 86:765–74

71. Lane NJ, Skaer HLB, Swales LS. 1977. Intercellular junctions in the central nervous system of insects. *J. Cell Sci.* 26:99–175

72. Lecuit T, Wieschaus E. 2000. Polarized insertion of new membrane from a cytoplasmic reservoir during cleavage of the *Drosophila* embryo. *J. Cell Biol.* 150:849–60

73. Lee JK, Brandin E, Branton D, Goldstein LS. 1997. α-Spectrin is required for ovarian follicle monolayer integrity in *Drosophila* melanogaster. *Development* 124:353–62

74. Lee JK, Coyne RS, Dubreuil RR, Goldstein LS, Branton D. 1993. Cell shape and interaction defects in α-*spectrin* mutants of *Drosophila melanogaster. J. Cell Biol.* 123:1797–809

75. Legouis R, Gansmuller A, Sookhareea S, Bosher JM, Baillie DL, Labouesse M. 2000. LET-413 is a basolateral protein required for the assembly of adherens junctions in *Caenorhabditis elegans. Nat. Cell Biol.* 2:415–22

76. Lehman K, Rossi G, Adamo JE, Brennwald P. 1999. Yeast homologues of tomosyn and lethal giant larvae function

in exocytosis and are associated with the plasma membrane SNARE, Sec9. *J. Cell Biol.* 146:125–40

77. Lekven AC, Tepass U, Keshmeshian M, Hartenstein V. 1998. *faint sausage* encodes a novel extracellular protein of the immunoglobulin superfamily required for cell migration and the establishment of normal axonal pathways in the *Drosophila* nervous system. *Development* 125:2747–58

78. Lin D, Edwards AS, Fawcett JP, Mbamalu G, Scott JD, Pawson T. 2000. A mammalian PAR-3-PAR-6 complex implicated in Cdc42/Rac1and aPKC signalling and cell polarity. *Nat. Cell Biol.* 2:540–47

79. Liu X, Lengyel JA. 2000. *Drosophila arc* encodes a novel adherens junction-associated PDZ domain protein required for wing and eye development. *Dev. Biol.* 221:419–34

80. Loncar D, Singer SJ. 1995. Cell membrane formation during the cellularization of the syncytial blastoderm of *Drosophila*. *Proc. Natl. Acad. Sci. USA* 92:2199–203

81. Lu B, Jan L, Jan YN. 2000. Control of cell divisions in the nervous system: symmetry and asymmetry. *Annu. Rev. Neurosci.* 23:531–56

82. Lu B, Roegiers F, Jan LY, Jan YN. 2001. Adherens junctions inhibit asymmetric division in the *Drosophila* epithelium. *Nature* 409:522–25

83. Lue RA, Brandin E, Chan EP, Branton D. 1996. Two independent domains of hDlg are sufficient for subcellular targeting: the PDZ1-2 conformational unit and an alternatively spliced domain. *J. Cell Biol.* 135:1125–37

84. Mandai K, Nakanishi H, Satoh A, Obaishi H, Wada M, et al. 1997. Afadin: a novel actin filament-binding protein with one PDZ domain localized at cadherin-based cell-to-cell adherens junction. *J. Cell Biol.* 139:517–28

85. Mandai K, Nakanishi H, Satoh A, Takahashi K, Satoh K, et al. 1999. An l-afadin-and vinculin-binding protein localized at cell-cell and cell-matrix adherens junctions. *J. Cell Biol.* 144:1001–17

86. Marfatia SM, Lue RA, Branton D, Chishti AH. 1994. In vitro binding studies suggest a membrane-associated complex between erythroid p55, protein 4.1, and glycophorin C. *J. Biol. Chem.* 269:8631–34

87. Matsuo T, Takahashi K, Kondo S, Kaibuchi K, Yamamoto D. 1997. Regulation of cone cell formation by Canoe and Ras in the developing *Drosophila* eye. *Development* 124:2671–80

88. Matsuo T, Takahashi K, Suzuki E, Yamamoto D. 1999. The Canoe protein is necessary in adherens junctions for development of ommatidial architecture in the *Drosophila* compound eye. *Cell Tissue Res.* 298:397–404

89. McCartney BM, Dierick HA, Kirkpatrick C, Moline MM, Baas A, et al. 1999. *Drosophila* APC2 is a cytoskeletally-associated protein that regulates *wingless* signaling in the embryonic epidermis. *J. Cell Biol.* 146:1303–18

90. McCartney BM, Fehon RG. 1996. Distinct cellular and subcellular patterns of expression imply distinct functions for the *Drosophila* homologues of moesin and the neurofibromatosis 2 tumor suppressor, merlin. *J. Cell Biol.* 133:843–52

91. McEwen DG, Cox RT, Peifer M. 2000. The canonical Wg and JNK signaling cascades collaborate to promote both dorsal closure and ventral patterning. *Development* 127:3607–17

91a. McMahon L, Legouis R, Vonesch JL, Labouesse M. 2001. Assembly of *C. elegans* apical junctions involves positioning and compaction by LET-413 and protein aggregation by the MAGUK protein DLG-1. *J. Cell Sci.* 114:2265–77

92. McNeill H, Ozawa M, Kemler R, Nelson WJ. 1990. Novel function of the cell adhesion molecule uvomorulin as an inducer of cell surface polarity. *Cell* 62:309–16

93. Mechler BM, McGinnis W, Gehring WJ. 1985. Molecular cloning of *lethal(2)giant larvae*, a recessive oncogene of *Drosophila melanogaster*. *EMBO J.* 4:1551–57

94. Menegoz M, Gaspar P, Le Bert M, Galvez T, Burgaya F, et al. 1997. Paranodin, a glycoprotein of neuronal paranodal membranes. *Neuron* 19:319–31

95. Miyamoto H, Nihonmatsu I, Kondo S, Ueda R, Togashi S, et al. 1995. *canoe* encodes a novel protein containing a GLGF/DHR motif and functions with Notch and scabrous in common developmental pathways in *Drosophila*. *Genes Dev.* 9:612–25

96. Müller HA. 2000. Genetic control of epithelial cell polarity: lessons from *Drosophila*. *Dev. Dyn.* 218:52–67

97. Müller HA, Wieschaus E. 1996. *armadillo*, *bazooka*, and *stardust* are critical for early stages in formation of the zonula adherens and maintenance of the polarized blastoderm epithelium in *Drosophila*. *J. Cell Biol.* 134:149–63

98. Müsch A, Xu H, Shields D, Rodriguez-Boulan E. 1996. Transport of vesicular stomatitis virus G protein to the cell surface is signal mediated in polarized and nonpolarized cells. *J. Cell Biol.* 133:543–58

99. Nakaya M, Fukui A, Izumi Y, Akimoto K, Asashima M, Ohno S. 2000. Meiotic maturation induces animal-vegetal asymmetric distribution of aPKC and ASIP/PAR-3 in *Xenopus* oocytes. *Development* 127:5021–31

100. Nelson WJ, Yeaman C, Grindstaff KK. 2000. Spatial cues for cellular asymmetry in polarized epithelia. In *Cell Polarity*, ed. DG Drubin, pp. 106–40. Oxford/New York: Oxford Univ. Press

100a. Noirot-Timothée C, Noirot C. 1980. Septate and scalariform junctions in arthropods. *Int. Rev. Cytol.* 63:97–141

101. Noselli S, Agnes F. 1999. Roles of the JNK signaling pathway in *Drosophila*

morphogenesis. *Curr. Opin. Genet. Dev.* 9:466–72

102. Nüsslein-Volhard C, Wieschaus E, Kluding M. 1984. Mutations affecting the pattern of the larval cuticle in *Drosophila melanogaster*. I. Zygotic loci on the second chromosome. *Wilhelm Roux's Arch. Entwicklungsmech. Org.* 193:267–282

103. Oda H, Uemura T, Harada Y, Iwai Y, Takeichi M. 1994. A *Drosophila* homolog of cadherin associated with Armadillo and essential for embryonic cell-cell adhesion. *Dev. Biol.* 165:716–26

104. Oda H, Uemura T, Shiomi T, Nagafuchi A, Tsukita S, Takeichi M. 1993. Identification of a *Drosophila* homologue of α-catenin and its association with the armadillo protein. *J. Cell Biol.* 121:1133–40

105. Oda H, Uemura T, Takeichi M. 1997. Phenotypic analysis of null mutants for DE-cadherin and Armadillo in *Drosophila* ovaries reveals distinct aspects of their functions in cell adhesion and cytoskeletal organization. *Genes Cells* 2:29–40

106. Ohshiro T, Yagami T, Zhang C, Matsuzaki F. 2000. Role of cortical tumour-suppressor proteins in asymmetric division of *Drosophila* neuroblast. *Nature* 408:593–96

107. Pai, L, Kirkpatrick C, Blanton, J, Oda, H, Takeichi M, Peifer M. 1996. α-catenin and DE-cadherin bind to distinct regions of *Drosophila* Armadillo. *J. Biol. Chem.* 271:32411–20

108. Pedraza L, Huang JK, Colman DR. 2001. Organizing principles of the axoglial apparatus. *Neuron* 30:335–44

109. Peifer M. 1993. The product of the *Drosophila* segment polarity gene *armadillo* is part of a protein complex resembling the vertebrate adherens junction. *J. Cell Sci.* 105:993–1000

110. Peifer M, Orsulic S, Sweeton D, Wieschaus E. 1993. A role for the *Drosophila* segment polarity gene *armadillo* in cell

adhesion and cytoskeletal integrity during oogenesis. *Development* 118:1191–207

111. Peifer M, Polakis P. 2000. Wnt signaling in oncogenesis and embryogenesis—a look outside the nucleus. *Science* 287:1606–9

112. Pellikka M, Tanentzapf G, Pinto M, Ready DF, Tepass U. 2001. Crumbs, the *Drosophila* homolog of human CRB1/RP12, is essential for photoreceptor morphogenesis. Submitted

113. Peng CY, Manning L, Albertson R, Doe CQ. 2000. The tumour-suppressor genes *lgl* and *dlg* regulate basal protein targeting in *Drosophila* neuroblasts. *Nature* 408:596–600

114. Perrimon N. 1988. The maternal effect of *lethal(1)discs-large-1*: a recessive oncogene of *Drosophila melanogaster*. *Dev. Biol.* 127:392–407

115. Petronczki M, Knoblich JA. 2001. DmPAR-6 directs epithelial polarity and asymmetric cell division of neuroblasts in *Drosophila. Nat. Cell Biol.* 3:43–49

116. Phelan P, Starich TA. 2001. Innexins get into the gap. *BioEssays* 23:388–96

117. Postner MA, Wieschaus EF. 1994. The Nullo protein is a component of the actin-myosin network that mediates cellularization in *Drosophila* melanogaster embryos. *J. Cell Sci.* 107:1863–73

118. Prokop A, Martin-Bermudo MD, Bate M, Brown NH. 1998. Absence of PS integrins or laminin A affects extracellular adhesion, but not intracellular assembly, of hemiadherens and neuromuscular junctions in *Drosophila* embryos. *Dev. Biol.* 196:58–76

119. Qiu RG, Abo A, Steven Martin G. 2000. A human homolog of the C. *elegans* polarity determinant Par-6 links Rac and Cdc42 to PKCζ signaling and cell transformation. *Curr. Biol.* 10:697–707

120. Reinacher-Schick A, Gumbiner BM. 2001. Apical membrane localization of the adenomatous polyposis coli tumor suppressor protein and subcellular distribution of the β-catenin destruction complex in polarized epithelial cells. *J. Cell Biol.* 152:491–502

121. Rodriguez-Boulan E, Nelson WJ. 1989. Morphogenesis of the polarized epithelial cell phenotype. *Science* 245:718–25

122. Rose LS, Kemphues KJ. 1998. Early patterning of the C. *elegans* embryo. *Annu. Rev. Genet.* 32:521–45

123. Rothwell WF, Fogarty P, Field CM, Sullivan W. 1998. Nuclear-fallout, a *Drosophila* protein that cycles from the cytoplasm to the centrosomes, regulates cortical microfilament organization. *Development* 125:1295–303

124. Rothwell WF, Zhang CX, Zelano C, Hsieh TS, Sullivan W. 1999. The *Drosophila* centrosomal protein Nuf is required for recruiting Dah, a membrane associated protein, to furrows in the early embryo. *J. Cell Sci.* 112:2885–93

125. Schejter ED, Wieschaus E. 1993. *bottleneck* acts as a regulator of the microfilament network governing cellularization of the *Drosophila* embryo. *Cell* 75:373–85

126. Schejter ED, Wieschaus E. 1993. Functional elements of the cytoskeleton in the early *Drosophila* embryo. *Annu. Rev. Cell Biol.* 9:67–99

127. Schweisguth F, Lepesant JA. Vincent A. 1990. The serendipity-α gene encodes a membrane-associated protein required for the cellularization of the *Drosophila* embryo. *Genes Dev.* 4:922–31

128. Simmonds AJ, dosSantos G, Livne-Bar I, Krause HM. 2001. Apical localization of *wingless* transcripts is required for *wingless* signaling. *Cell* 105:197–207

129. Simon DB, Lu Y, Choate KA, Velazquez H, Al-Sabban E, et al. 1999. Paracellin-1, a renal tight junction protein required for paracellular Mg^{2+} resorption. *Science* 285:103–6

130. Simons K, Wandinger-Ness A. 1990. Polarized sorting in epithelia. *Cell* 1990 62:207–10

131. Simpson L, Wieschaus E. 1990. Zygotic

activity of the *nullo* locus is required to stabilize the actin-myosin network during cellularization in *Drosophila. Development* 110:851–63

132. Sisson JC, Field C, Ventura R, Royou A, Sullivan W. 2000. Lava lamp, a novel peripheral golgi protein, is required for *Drosophila melanogaster* cellularization. *J. Cell Biol.* 151:905–18

133. Stevenson BR, Anderson JM, Bullivant S. 1988. The epithelial tight junction: structure, function and preliminary biochemical characterization. *Mol. Cell Biochem.* 83:129–45

134. Stevenson BR, Siliciano JD, Mooseker MS, Goodenough DA. 1986. Identification of ZO-1: a high molecular weight polypeptide associated with the tight junction (zonula occludens) in a variety of epithelia. *J. Cell Biol.* 103:755–66

135. Strand D, Jakobs R, Merdes G, Neumann B, Kalmes A, et al. 1994. The *Drosophila lethal(2)giant larvae* tumor suppressor protein forms homo-oligomers and is associated with nonmuscle myosin II heavy chain. *J. Cell Biol.* 127:1361–73

136. Strand D, Raska I, Mechler BM. 1994. The *Drosophila lethal(2)giant larvae* tumor suppressor protein is a component of the cytoskeleton. *J. Cell Biol.* 127:1345–60

137. Strand D, Unger S, Corvi R, Hartenstein K, Schenkel H, et al. 1995. A human homologue of the *Drosophila* tumour suppressor gene *l(2)gl* maps to 17p11.2-12 and codes for a cytoskeletal protein that associates with nonmuscle myosin II heavy chain. *Oncogene* 11:291–301

138. Strutt DI. 2001. Asymmetric localization of Frizzled and the establishment of cell polarity in the *Drosophila* wing. *Mol. Cell* 7:367–75

139. Su LK, Burrell M, Hill DE, Gyuris J, Brent R, et al. 1995. APC binds to the novel protein EB1. *Cancer Res.* 55:2972–77

140. Suzuki A, Yamanaka T, Hirose T, Manabe N, Mizuno K, et al. 2001. Atypical protein kinase C is involved in the evolutionarily conserved Par protein complex and plays a critical role in establishing epithelia-specific junctional structures. *J. Cell Biol.* 152:1183–96

141. Tachibana K, Nakanishi H, Mandai K0, Ozaki K, Ikeda W, et al. 2000. Two cell adhesion molecules, nectin and cadherin, interact through their cytoplasmic domain-associated proteins. *J. Cell Biol.* 150:1161–76

142. Tait S, Gunn-Moore F, Collinson JM, Huang J, Lubetzki C, et al. 2000. An oligodendrocyte cell adhesion molecule at the site of assembly of the paranodal axo-glial junction. *J. Cell Biol.* 150:657–66

143. Takahashi K, Matsuo T, Katsube T, Ueda R, Yamamoto D. 1998. Direct binding between two PDZ domain proteins Canoe and ZO-1 and their roles in regulation of the jun N-terminal kinase pathway in *Drosophila* morphogenesis. *Mech. Dev.* 78:97–111

144. Takahashi K, Nakanishi H, Miyahara M, Mandai K, Satoh K, et al. 1999. Nectin/PRR: an immunoglobulin-like cell adhesion molecule recruited to cadherin-based adherens junctions through interaction with Afadin, a PDZ domain containing protein. *J. Cell Biol.* 145:539–49

145. Takahisa M, Togashi S, Suzuki T, Kobayashi M, Murayama A, et al. 1996. The Drosophila tamou gene, a component of the activating pathway of extramacrochaetae expression, encodes a protein homologous to mammalian cell-cell junction-associated protein ZO-1. *Genes Dev.* 10:1783–95

146. Takeuchi K, Kawashima A, Nagafuchi A, Tsukita S. 1994. Structural diversity of band 4.1 superfamily members. *J. Cell Sci.* 107:1921–28

147. Tanentzapf G, Smith C, McGlade J, Tepass U. 2000. Apical, lateral, and basal

polarization cues contribute to the development of the follicular epithelium during *Drosophila* oogenesis. *J. Cell Biol.* 151:891–904

148. Tang AH, Neufeld TP, Rubin GM, Müller HA. 2001. Transcriptional regulation of cytoskeletal functions and segmentation by a novel maternal pair-rule gene, *lilliputian. Development* 128:801–13

149. Tepass U. 1996. Crumbs, a component of the apical membrane, is required for zonula adherens formation in primary epithelia of *Drosophila. Dev. Biol.* 177:217–25

150. Tepass U. 1997. Epithelial differentiation in *Drosophila. BioEssays.* 19:673–82

151. Tepass U. 1999. Genetic analysis of cadherin function in animal morphogenesis. *Curr. Opin. Cell Biol.* 11:540–48

152. Tepass U, Gruszynski-de Feo E, Haag TA, Omatyar L, Török T, Hartenstein V. 1996. *shotgun* encodes *Drosophila* E-cadherin and is preferentially required during cell rearrangement in the neuroectoderm and other morphogenetically active epithelia. *Genes Dev.* 10:672–85

153. Tepass U, Hartenstein V. 1994. Epithelium formation in the *Drosophila* midgut depends on the interaction of endoderm and mesoderm. *Development* 120:579–90

154. Tepass U, Hartenstein V. 1994. The development of cellular junctions in the *Drosophila* embryo. *Dev. Biol.* 161:563–96

155. Tepass U, Knust E. 1990. Phenotypic and developmental analysis of mutations at the *crumbs* locus, a gene required for the development of epithelia in *Drosophila melanogaster. Roux's Arch. Dev. Biol.* 199:189–206

156. Tepass U, Knust E. 1993. *crumbs* and *stardust* act in a genetic pathway that controls the organization of epithelia in *Drosophila melanogaster. Dev. Biol.* 158:311–26

157. Tepass U, Theres C, Knust E. 1990. *crumbs* encodes an EGF-like protein expressed on apical membranes of *Drosophila* epithelial cells and required for organization of epithelia. *Cell* 6:787–99

158. Tepass U, Truong K, Godt D, Ikura M, Peifer M. 2000. Cadherins in embryonic and neural morphogenesis. *Nat. Rev. Mol. Cell Biol.* 1:91–100

159. Thomas GH. 2001. Spectrin: the ghost in the machine. *BioEssays* 23:152–60

160. Thomas GH, Kiehart DP. 1994. β_{Heavy}-spectrin has a restricted tissue and subcellular distribution during *Drosophila* embryogenesis. *Development* 120:2039–50

161. Thomas GH, Williams JA 1999. Dynamic rearrangement of the spectrin membrane skeleton during the generation of epithelial polarity in *Drosophila. J. Cell Sci.* 112:2843–52

162. Thomas GH, Zarnescu DC, Juedes AE, Bales MA, Londergan A, et al. 1998. *Drosophila* β_{Heavy}-spectrin is essential for development and contributes to specific cell fates in the eye. *Development* 125:2125–34

163. Tomlinson A, Bowtell DD, Hafen E, Rubin GM. 1987. Localization of the sevenless protein, a putative receptor for positional information, in the eye imaginal disc of *Drosophila. Cell* 51:143–50

164. Townsley FM, Bienz M. 2000. Actin-dependent membrane association of a *Drosophila* epithelial APC protein and its effect on junctional Armadillo. *Curr Biol.* 10:1339–48

165. Tsukita S, Furuse M. 2000. Pores in the wall: claudins constitute tight junction strands containing aqueous pores. *J. Cell Biol.* 149:13–16

166. Turner FR, Mahowald AP. 1976. Scanning electron microscopy of *Drosophila* embryogenesis. 1. The structure of the egg envelopes and the formation of the cellular blastoderm. *Dev. Biol.* 50:95–108

167. Uemura T, Oda H, Kraut R, Hayashi

S, Takeichi M. 1996. Processes of dynamic epithelial cell rearrangement are the major targets of *cadE/shotgun* mutations in the *Drososphila* embryo. *Genes Dev.* 10:659–71

168. Ward RE, Lamb RS, Fehon RG. 1998. A conserved functional domain of *Drosophila* Coracle is required for localization at the septate junction and has membrane-organizing activity. *J. Cell Biol.* 140:1463–73

169. Ward RE, Schweizer L, Lamb RS, Fehon RG. 2001. The FERM domain of *Drosophila* Coracle, a cytoplasmic component of the septate junction, provides functions essential for embryonic development and imaginal cell proliferation. *Genetics.* In press

170. Wei X, Ellis HM. 2001. Localization of the *Drosophila* MAGUK protein Polychaetoid is controlled by alternative splicing. *Mech. Dev.* 100:217–31

171. Welsch U, Buchheim W. 1977. Freeze fracture studies on the annelid septate junction. *Cell Tissue Res.* 185:527–34

172. Welte MA, Gross SP, Postner M, Block SM, Wieschaus EF. 1998. Developmental regulation of vesicle transport in *Drosophila* embryos: forces and kinetics. *Cell* 92:547–57

173. White P, Aberle H, Vincent JP 1998. Signaling and adhesion activities of mammalian β-catenin and plakoglobin in *Drosophila. J. Cell Biol.* 140:183–95

174. Wieschaus E, Noell E. 1986. Specificity of embryonic lethal mutations in *Drosophila* analyzed in germ line clones. *Roux's Arch Dev. Biol.* 195:63–73

175. Wieschaus E, Nüsslein-Volhard C, Jürgens G. 1984. Mutations affecting the pattern of the larval cuticle in *Drosophila melanogaster*. III. Zygotic loci on the X-chromosome and fourth chromosome. *Wilhelm Roux's Arch. Entwicklungsmech. Org.* 193:296–307

176. Wiley CA, Ellisman MH. 1980. Rows of dimeric-particles within the axolemma and juxtaposed particles within glia, incorporated into a new model for the paranodal glial- axonal junction at the node of Ranvier. *J. Cell Biol.* 84:261–80

177. Wilk R, Reed BH, Tepass U, Lipshitz HD. 2000. The *hindsight* gene is required for epithelial maintenance and differentiation of the tracheal system in *Drosophila. Dev. Biol.* 219:183–96

178. Wodarz A, Grawe F, Knust E. 1993. Crumbs in involved in the control of apical protein targeting during *Drosophila* epithelial development. *Mech. Dev.* 44:175–87

179. Wodarz A, Hinz U, Engelbert M, Knust E. 1995. Expression of Crumbs confers apical character on plasma membrane domains of ectodermal epithelia of *Drosophila. Cell* 82:67–76

180. Wodarz A, Ramrath A, Grimm A, Knust E. 2000. *Drosophila* atypical protein kinase C associates with Bazooka and controls polarity of epithelia and neuroblasts. *J. Cell Biol.* 150:1361–74

181. Wood RL. 1990. The septate junction limits mobility of lipophilic markers in plasma membranes of *Hydra vulgaris* (attenuata). *Cell Tissue Res.* 259:61–66

182. Woods DF, Bryant PJ. 1989. Molecular cloning of the *lethal (1) discs large-1* oncogene of *Drosophila. Dev. Biol.* 134:222–35

183. Woods DF, Bryant PJ. 1991. The *discs-large* tumor suppressor gene of *Drosophila* encodes a guanylate kinase homolog localized at septate junctions. *Cell* 66:451–64

184. Woods DF, Hough C, Peel D, Callaini G, Bryant PJ. 1996. Dlg protein is required for junction structure, cell polarity, and proliferation control in *Drosophila* epithelia. *J. Cell Biol.* 134: 1469–82

185. Yeaman C, Grindstaff KK, Nelson WJ. 1999. New perspectives on mechanisms involved in generating epithelial cell polarity. *Physiol. Rev.* 79:73–98

186. Yoshimori T, Keller P, Roth MG, Simons K. 1996. Different biosynthetic

transport routes to the plasma membrane in BHK and CHO cells. *J. Cell Biol.* 133:247–56

187. Yu X, Waltzer L, Bienz M. 1999. A new *Drosophila* APC homologue associated with adhesive zones of epithelial cells. *Nat. Cell Biol.* 1:144–51

188. Zak NB, Shilo BZ. 1992. Localization of DER and the pattern of cell divisions in wild-type and Ellipse eye imaginal discs. *Dev. Biol.* 149:448–56

189. Zalokar M, Erk I. 1976. Division and migration of nuclei during early embryogenesis of *Drosophila melanogaster*. *J. Microbiol. Cell* 25:97–106

190. Zarnescu DC, Thomas GH. 1999. Apical spectrin is essential for epithelial morphogenesis but not apicobasal polarity in *Drosophila*. *J. Cell Biol.* 146:1075–86

191. Zhang CX, Lee MP, Chen AD, Brown SD, Hsieh T. 1996. Isolation and characterization of a *Drosophila* gene essential for early embryonic development and formation of cortical cleavage furrows. *J. Cell Biol.* 134:923–34

192. Zhang CX, Rothwell WF, Sullivan W, Hsieh TS. 2000. Discontinuous actin hexagon, a protein essential for cortical furrow formation in *Drosophila*, is membrane associated and hyperphosphorylated. *Mol. Biol. Cell* 11:1011–22

Annu. Rev. Genet. 2001. 35:785–800

INFORMED CONSENT AND OTHER ETHICAL ISSUES IN HUMAN POPULATION GENETICS

Henry T. Greely

Stanford Law School, Stanford University, Stanford, California 94305-8610;
e-mail: hgreely@stanford.edu

Key Words population genetics, genetics, ethics, law, society

■ **Abstract** Human population genetics has entered a new era of public interest, of controversy, and of ethical problems. Population genetics raises novel ethical problems because both the individuals and the populations being studied are, in effect, "subjects" of the research. Those populations are collectively subject to possible benefits and harms from the research and have interests, somewhat different from those of the individuals, that must be considered from both ethical and practical standpoints. The chapter first describes the new setting for research in human population genetics. It then examines the most controversial ethical issue in population genetics—whether researchers must obtain the informed consent of both the individual subjects and the group as a collectivity. Other vexing issues, including special problems caused by researchers' commercial interests, confidentiality, control over research uses and materials, and return of information to the population are also considered.

CONTENTS

0066-4197/01/1215-0785$14.00

INTRODUCTION

Although genetics has been a highly visible and controversial field for the past century, population genetics has been scarcely visible to the public eye. Neither its mathematical abstractions nor its hordes of collected fruit flies made much impression on public consciousness. In recent years, however, the increased ability of researchers to examine large numbers of genetic variations in humans and to draw conclusions of anthropological, historical, medical—and commercial— interest from them has increased public knowledge of this kind of research. This heightened ability of research in human population genetics to affect human societies has brought with it not only notoriety but also new ethical problems.

These ethical problems are in addition to the long-discussed ethical issues of more traditional individual or family-centered human genetics research (13). Each of the individual subjects of human population genetics research is an individual human subject, protected by governmental rules on informed consent and ethical review boards. What makes genetic research on populations different, and novel, is that the populations being studied are also, in effect, "subjects" of the research. Those populations are collectively subject to possible benefits and harms from the research. These collective risks are beginning to be discussed by researchers and ethicists, but have yet to be well addressed by regulators.

This review first describes the new setting for research in human population genetics. It then examines the ethical concerns affecting this research. It discusses first the most important, most controversial, and most debated issue—whether researchers must obtain the informed consent of the whole group. The chapter then discusses other vexing issues that have received much less attention. Those include special problems caused by researchers' commercial interests, confidentiality, control over the uses of the research and materials, and return of information to the population. All of these issues, including informed consent, are just beginning to be discussed. In the ethics of human population genetics, at this point hard questions are much more common than clear answers.

THE NEW HUMAN POPULATION GENETICS

The study of genetic differences between human populations dates back to the First World War and pioneering studies of the frequency of the different ABO blood groups among various ethnic groups (17, 31). But progress was slow. Phenotypic markers clearly linked to genetic variations were difficult to find and data were expensive to collect. Over the decades, a great deal of information did accumulate about particular classical markers in some populations (3). In the 1990s, however, the ability to test DNA directly for variations and the decreasing cost of that testing made it feasible to consider studying large numbers of markers in large samples.

Some researchers interested in human population genetics seized the scientific opportunity provided by the new technologies—and the apparent political opportunity provided by the Human Genome Project—to launch the Human Genome Diversity Project (HGDP) (2). While the Human Genome Project mapped and sequenced one reference human genome, the HGDP was to sample over 500 populations, preserve the samples both as cell lines and as extracted DNA, analyze the samples, and make their results publicly available through a computer database (18). Over the past decade, the HGDP has generated more controversy than samples (it remains largely in the planning process with very limited funding), but it did draw the attention of researchers, ethicists, and activists to the problems raised by population-wide genetic research (11, 15, 29).

Although the HGDP's interests were academic and largely non-medical, easier and cheaper analysis of genetic markers made population genetics of interest to biomedical researchers. Most research on genetic diseases had focused on extended families with a high rate of the specific disease. Researchers would examine the genomes of family members, both affected and unaffected, to see which genetic variations they shared. Some researchers argued that genes linked to disease would be easier to spot in populations that had high incidence of the disease and were relatively genetically homogenous. Academics and biotechnology firms both began "gene hunting" for disease-linked genes in these isolated populations. The involvement of the biotechnology firm Sequana in research on the genetics of asthma among the people of Tristan da Cuhna was well-known, and controversial (1, 30), but similar research was undertaken among a wide variety of populations.

Eventually, interest moved beyond research on particular diseases in small populations. Scientists and firms began to discuss the collection of large amounts of phenotypic data, usually from clinical medical information, and genetic data, from DNA samples, on entire populations. These population-wide "genotype/phenotype resources" became feasible as the costs of genetic analysis and computing power continued to decline. They were made attractive by the limited results from more traditional, family-focused research. Good evidence exists that many common diseases such as heart disease, asthma, non-insulin dependent diabetes, and schizophrenia, have substantial links to shared genetic variations. Although family studies have found the genetic variations linked to many rare diseases, they have largely failed to establish links to these more common disorders. Researchers hoped that statistical mining of a "genotype/phenotype resource" on a whole population might reveal weak, but still important, associations between variations in one or more genes and disease.

The first attempt to create such a population-wide resource has been made by deCODE Genetics in the Republic of Iceland. deCODE, founded by an Icelandic scientist, plans to create linked databases of clinical medical records, genotypes, and genealogical data on all 280,000 Icelanders (14). The company was founded in 1996; by 1998 it had convinced the government of Iceland to support legislation that would allow a private firm to create a "health sector database" made up of clinical medical records from all Icelanders. The law passed in December 1998, and deCODE was named the "licensee" under the Act and was empowered to create the database in January 2000. The firm has created the genealogical database and plans to create the genetic database without further government support (5). Although deCODE has not yet proven that its resource will be effective—the health sector database remains stalled in controversy—similar initiatives to create genotype/phenotype resources with large populations samples have been announced in Estonia, Sweden, Tonga, Newfoundland, and the United Kingdom (16). These resources are extremely expensive to create; most of the plans, like deCODE's, rely on commercial funding.

Thus, research on the genetics of human populations has become a high-profile business. These groups may have their own benefits, risks, and interests with respect to research that are somewhat different from those faced by individuals within the groups. But the regulation of research on human subjects has focused on individuals, not on groups. The existing federal regulations in the United States, the so-called Common Rule (35), do not recognize a role of groups; they deal only with individuals (24). Nongovernmental groups have made tentative proposals for guidelines that take group interests into consideration in research, notably in epidemiological standards promulgated by the Council for International Organizations of Medical Sciences and in the Model Ethical Protocol for Collecting DNA Samples of the North American Regional Committee of the HGDP (4, 26). The process of addressing the ethical issues of human population genetics is thus under way, but there is as yet no consensus—nor governmental regulation.

INFORMED CONSENT

Informed consent has raised the most controversy regarding the ethics of human population genetics. Although some have questioned the application of even individual informed consent in some parts of this research, the more interesting question, and the source of a burgeoning academic literature, is whether some form of consent should be required from the group as a whole.

Individual Informed Consent and Iceland

The individual informed consent of a human research subject is generally required by law in the United States and in most other countries and by relevant international guidelines, such as the Helsinki Declaration of the World Medical Association (34). Exceptions involve research with no significant risk; research involving children and the mentally incompetent (whose parents or guardians must give consent); and, in a very few cases, research that by its nature cannot be conducted with informed consent, such as research on patients brought unconscious to hospital emergency departments after heart attacks. The patient (or guardian) need not only consent, but must do so after being informed of the nature of the research and of the specific benefits and risks it might hold for him. The requirement for informed consent grew from both respect for the individual's autonomy and the belief that an individual's informed consent would be one check on overly dangerous research (7).

Iceland's Health Sector Database has been controversial in part because its authorizing legislation does not require that Icelanders give their informed consent before their clinical medical records are included in the database. Instead, the legislation gives living Icelanders a chance to opt out. They can file a form with the government stating that they do not wish to participate in the database. The form will prevent the addition of any new data on them to the database, although it will not lead to the removal of any data that has already been entered. There is no requirement that Icelanders be told of the specific research uses of their data.

The use of this "presumed consent" in place of informed consent has caused great unease. In Iceland, this has been a major objection by Mannvernd, an Icelandic organization on research ethics founded in response to the Health Sector Database law (22). It has also prompted denunciations from the World Medical Association and from various ethicists (6). Iceland and deCODE defend the lack of informed consent by comparing the planned uses of the Health Sector Database to epidemiological research, which typically proceeds without informed consent when it uses clinical medical data from which personal identifiers have been removed (5, 19). It notes that it plans to get individual informed consent for the genetic samples it collects. Critics of deCODE counter that the Health Sector Database should not be viewed in isolation from the genetics and genealogical databases; it is, they argue, a crucial part of this broader plan and should not follow the precedents of purely epidemiological research. Finally, deCODE points out that under ten percent of the Icelandic population has, in fact, opted out of the database. The Health Sector Database remains tied up in disputes in Iceland; it is not yet clear whether it will proceed as currently planned. Interestingly, the other proposed population-wide genotype/phenotype resources have made it clear that they will proceed only with full individual informed consent.

Group Informed Consent

The more interesting consent problem in human population genetics concerns the consent of the population. Should some kind of consent or at least consultation with a population as a whole be required, in addition to individual informed consent, before researchers study the genetic variations in that population? At least three positions have appeared—one in favor of a group consent requirement, one opposed to such a requirement, and one intermediate position that favors group consultation.

THE CASE FOR GROUP CONSENT   The case for a group consent requirement is based on both a principled and a pragmatic argument. The principled argument is that the population itself is, in effect, the research subject and should be treated as such. Research done on, for example, the Irish may have benefits and costs to everyone identified as Irish, whether or not that person ever consented. The risks may be quite concrete, such as an increased chance of discrimination in employment or insurance. They could be more stigmatic, if someone were to claim, for example, that the population had a high level of a genetic variation associated with alcoholism. Or risks could be more cultural in nature, as where the anthropological or historical uses of population genetics undercut a population's own history or even their political or legal claims to certain territories.

The pragmatic argument is that research in the midst of a population will often require the approval of the population's governance structure or leadership. This *de facto* group consent requirement often appears in some kinds of research. Without the cooperation of the local officials, the relevant religious figures, or the educational or medical professionals, much epidemiological research would not happen. Without the approval of the studied population, either formal or informal, ethnographic research would be impossible; without at least tacit consent, the ethnographer would never gain access to the population to study it. The shift from studying genetic variations in families with high rates of disease to studying genetic variations in communities will often force researchers to deal with some kind of collective consent. The pragmatic argument also has a political aspect; groups might not oppose certain types of research if they were confident that they would only be included in the research with their collective consent.

Some Native American tribes and indigenous organizations have asserted that group consent is required, and some organizations have, to a greater or lesser extent, agreed (33). The research organization that has been most active in endorsing group consent is the North American Regional Committee of the HGDP (on which the author serves). In 1996 that group completed a draft "Model Ethical Protocol for Collecting DNA Samples" (13, 26). That document aimed to give broad ethical and practical guidance to researchers and others in collecting DNA samples for the Project. Among other things, it stated that, in addition to individual informed consent:

> [T]he HGDP requires that researchers participating in the HGDP show that they have obtained the informed consent of the population, through its culturally appropriate authorities where such authorities exist, before they begin sampling. If, for example, the Navajo Nation decided that it would not participate in the HGDP, the HGDP would not accept samples taken from members of that population.

The Model Protocol recognized that there would be difficulties in implementing its group consent requirement. It foresaw two major problems in recognizing the "culturally appropriate authorities" and in defining the relevant group.

Who can provide consent? The Model Protocol singles out "culturally appropriate authorities." These are the people or groups whose authority the community recognizes. The Protocol notes that determining who the actual authorities are will often be difficult. For example, a federally recognized tribe will have a tribal government, which clearly would be at least one of the authorities. But it may also have people or organizations—elders, religious leaders, traditional leading families or clans—whom the members obey. In those cases, many different groups might be culturally appropriate authorities. Alternatively, in some cases, particularly in small populations, a community consensus, and not a discrete person or group, might be the relevant authority.

The problem of defining the group stems from the common lack of a formal or universally accepted definition of a group or its members. For ethnic groups or other communities, this has both a horizontal and a vertical component. Where does one draw the boundary between similar groups? Are two Cree bands living close together but on separate reservations one group or two groups? Do "the Irish" include only citizens of the Republic of Ireland or also the British subjects who live in Northern Ireland? The vertical component asks at what level a group is defined. Is the relevant group a particular village in the Navajo Nation (the proper term for the Navajo Reservation)? Is it the entire Navajo Nation? Is it all Southwestern Native Americans who speak languages related to Navajo, thus encompassing the Apache? Is it all Native Americans who speak a language in the NaDene language family, from inland Alaska to the Southwest? Or is the relevant group all Native Americans? Are Irish-Americans a group or is the relevant group all those with Irish ancestry?

The Model Protocol took an empirical approach to this issue. It recommended starting by assuming that the "group" was the community in which the research was being done. That population should then be asked its definition of its "group" for purposes of both the horizontal and the vertical definitions. The Model Protocol stated with respect to the vertical issue that

> Ultimately, the question of the levels at which consent should be sought is one that only the population can answer. Consent must be sought at higher levels if the population believes it is meaningfully part of such a higher level grouping and if there are entities operating at that level whose decisions on the population's participation in the research would be accepted by the local population as authoritative.

The definition of the group raises another dilemma not expressly addressed in the Model Protocol. What kinds of human organizations should count as "groups" (12)? The ethnic or religious communities that the Model Protocol describes are culturally meaningful entities whose members are expected to have genetic similarities through their genealogical connections. But other kinds of groups have also been the subjects of genetics research. Families have been the classic locus of genetics research. Should they count as "groups" for which group consent is needed? Similarly, many genetic (and nongenetic) diseases have spawned "disease organizations," entities advocating for those with the disease. Would the American Lung Association or the Cystic Fibrosis Foundation be a group whose consent would be required for research in their fields? The Model Ethical Protocol did not address these questions because the HGDP was not interested in those "populations," but, like ethnic groups, members of a family or victims of a particular disease all face some benefits and some risks from research in that family or on that disease. The logic behind group consent might extent to them, but the problems of implementation seem more daunting.

The Model Protocol recognized that neither identification of "culturally relevant authorities" nor definition of the relevant group would always be easy. It said that each question could only be answered "in the detailed factual context of a population," and it recommended that researchers come with the expertise, the funding, and the time to make informed determinations. The North American Regional Committee, in the Protocol and elsewhere, also recognized that there sometimes would not be any "culturally relevant authorities." Who, for example, would be a culturally appropriate authority for Irish-Americans or the Ashkenazim? The Model Protocol pointed out that

> In cases where communities do not have a culturally appropriate authority, there still may be institutions that provide a useful focus for community discussions and consensus. For example, in a Catholic parish in Seattle that served a largely Irish-American population, the parish priest would certainly not be a culturally appropriate authority to give permission to work with that population. But the priest might be a useful and knowledgeable figure with whom to discuss the community's participation and the parish might provide the best focus for the community. Through its auspices the researcher may be able to present information and seek approval from active members of the community.

In such situations, even the authors of the Model Protocol recognized that "group consent" was impossible. They required such consent only where it was feasible and settled for consultation and discussion in other circumstances.

THE CASE AGAINST GROUP CONSENT   The concept of group consent has not been widely adopted by researchers. Although adopted by the North American Regional Committee of the HGDP, it has not even been adopted by the entire HGDP. No doubt some of the reluctance is purely practical—why would a researcher take on, voluntarily, a new obligation that promises to be expensive and time consuming (27, 28)? Even the drafters of the Model Protocol acknowledge that its implementation will be difficult. But there are also more theoretical objections.

One common objection is that this concept favors group rights over individual rights. If a mentally competent adult wishes to participate in research, why should her right to do so be overruled by a collective decision, particularly where the group definition or the identification of culturally appropriate authorities is uncertain? Although, on occasion, a few U.S. courts have implied that the First Amendment includes a right to engage in scientific research, there is no precedent for finding a legal right to be a research subject. In fact much of the federal regulation of human subjects research is aimed at keeping people from being research subjects. The Common Rule, where it applies, requires not only informed consent but also review and approval of research protocols by ethics committees called Institutional Review

Boards (IRBs). IRBs are required to determine whether the research is appropriate, including whether its potential benefits justify its risks. If an IRB concludes that the proposed research does not meet that standard, the research may not take place no matter how much an individual wants to be a research subject. There seems little principled difference between an IRB making a decision, in opposition to a potential research subject's wishes, that the risks to her of the research are too high for it to proceed and a "culturally appropriate authority" determining that the risks to a population are too high for it to proceed.

Two other objections are less easily countered. Both were made by a National Research Council committee to study the HGDP (25) and by bioethicist Eric Juengst, who served on that committee (20, 21). The first objection is a version of the vertical problem of defining the relevant group. Juengst points out that all groups are nested within other groups. Ultimately, information about the genetic make-up of any human group, or human being, provides some information about every human group and every individual. This information then carries with it both potential benefits and potential risks. Indeed, given the strong similarity between many human and nonhuman genes, research with other animals could be said to trigger an obligation to seek group consent of all humans.

Juengst's other argument focuses on the possible effects of group consent. Although the scientists involved in the HGDP argue that genetic differences between ethnic groups are trivial and that human races have no genetic meaning, he points out that requiring "group consent" for genetic research implies that groups, including races, do have a genetic basis. Otherwise, why should genetic research in particular be subject to a group consent requirement? A group consent requirement could thus reinforce scientifically inaccurate and socially dangerous beliefs about genetics and race.

Juengst does recognize the reasons that have led some to argue for group consent. He proposes that, as a partial solution, members of the group be reminded, during their individual informed consent process, that the results of the research might have implications, good or bad, for all members of the group.

GROUP CONSULTATION—AN INTERMEDIATE POSITION   Other ethicists have opted for an intermediate position, endorsing group consultation but usually not requiring formal consent by the group. Morris Foster has been the leading proponent of this position, along with Richard Sharp (8, 9, 32). Drawing from his work with an Indian community in Oklahoma, Foster concludes that group consent is too difficult to implement. Researchers, he argues, should have an obligation to inform the community widely and to consult with it. This may have the same effect as a group consent requirement—if substantial community opposition develops, the research may not be able to proceed—but without the difficulties of formally identifying the group and its culturally appropriate authorities. Like both the North American Regional Committee and Juengst, Foster calls in addition for more attention to group considerations by IRBs that review population genetics proposals.

A Special Case of Group Consent—Federally Recognized Indian Tribes and Other Governments

In the United States, one set of groups does have a special status in consenting to research: the roughly 550 federally recognized Indian tribes and Alaska Native villages. These groups are political sovereignties, predating the United States and recognized by the United States as having political authority. Tribal governments have powers similar to those of states, including the power to make and enforce laws on the territories as long as they are not inconsistent with federal law. Some other countries recognize similar political powers in their indigenous groups.

At least within its territorial jurisdiction, therefore, a federally recognized tribal government has the power to make a collective decision whether to allow particular population genetics research in that jurisdiction. Some U.S. tribes have been exercising that jurisdiction, either through the regular mechanisms of their tribal governments, through tribal IRBs, or through use of an IRB established by the Indian Health Service (10). This application of group consent is not only legal, but it avoids some of the concept's implementation problems. The tribal government in this case is a culturally appropriate authority and its membership is clearly and legally defined. In the United States, Native American tribes, along with some tightly endogamous European religious groups such as the Amish or the Hutterites, are among the populations of most interest to researchers, for both biomedical and anthropological reasons. With respect to federally recognized tribes, group consent may be a fairly simple, and legally binding, requirement in the United States for much human population genetics research.

Similarly, in some cases a national government may be able to give group consent for its citizens. The deCODE project in Iceland, though it may lack individual informed consent, seems to meet the group consent requirement. A universally recognized (and democratic) government passed legislation approving the plan after substantial parliamentary and public discussion and debate. Many national governments, including Iceland's, accomplish similar ends through requiring approval of human subjects research by governmental bodies or the equivalents of IRBs. Those governments arguably are in a position to assess the overall risks and benefits to their people and to authorize, or veto, particular research projects.

The group consent concept does cause some difficulties even with national governments and federally recognized tribes. First, both national and tribal governments sometimes represent more than one population. Most countries have ethnic or cultural minorities. In some cases tribal governments encompass several different tribes, including sometimes communities with separate languages and cultures. Here the national or tribal governments may not be good representatives of, or "culturally appropriate authorities" for, their minorities. Consider, for example, the moral authority of the Nazi government to consent to research on German Jews.

Second, both tribal and national governments may have limited power over group members outside their territorial jurisdiction. The Navajo Nation may not learn of, or be able to control, research voluntarily undertaken by Navajos living in Los Angeles. The People's Republic of China, which recently adopted detailed regulations governing the export of "Chinese DNA," has little or no power over the vast Chinese diaspora worldwide. Unless the researchers themselves are bound to respect group consent, the territorially based power of governments may prove unable to enforce this principle.

OTHER ISSUES

Group consent has received the most attention, but the group nature of population genetics raises other special issues or, in some cases, raises conventional issues in unusual ways. At least four deserve mention: commercialism, confidentiality, control of research uses and materials, and return of information. These topics have only rarely been discussed in the scholarly literature (16) but are likely to be of increasing importance in human population genetics, at least with some groups. They have been the subject of much conversation at conferences with Native American leaders and researchers; the following analysis is drawn heavily from those discussions. Ethical questions are raised concerning a researcher's obligation to protect research subjects from unfair treatment or harm. There are also practical considerations that may well involve difficult negotiations between researchers and groups.

Commercialism

Biomedical research has become more commercial in the past two decades; even academic researchers often have connections with for-profit companies. For some research subjects, the increased commercialization strains the altruistic basis for their participation. When the researchers hope to make millions of dollars from stock options from the application of their findings, the subjects may find it hard to accept satisfaction at helping humanity as their only reward. This conflict is even more pronounced with associational research. When genetics research is based on disease families, research subjects have a strong motivation to participate to help themselves and their kin. Associational genetic research, because of the weakness of the connections its seeks, has no such direct link to the health of a research subject or his family. At the same time, the high cost of genotype/phenotype resources makes it very likely that commercial firms are performing the research.

In only a few instances have research subjects sought a share of the financial value of research, with the case of John Moore being the most prominent (23). Claims to share in future financial benefits may be more common with population-based research for several reasons. First, any one individual is unlikely to be crucial to an important genetic discovery. The discovery of BRCA1, for example, relied on thousands of people, affected and unaffected, with mutant genes and normal ones. Each was important to the collective result, but no single person was crucial. In research with a population at high risk for a disease, no individual member is crucial, but the population itself may be. To the extent that group consent or other forms of permission or consultation are involved, the population may have more awareness of the potential commercial value of the research than an individual would and may be better placed to negotiate for a share. This may be particularly true for governments. The Chinese regulations on export of DNA require some profit sharing with institutions in China. After much criticism, deCODE agreed to give the Icelandic health system a share of its profits (though capped at about $1 million per year). And recently a disease organization, the Canavan Foundation, has sued the hospital that holds the patent on the gene linked to the disease, claiming that its participation in the research gave it rights with respect to the gene's commercial use.

At the same time the group nature of population genetics makes financial claims more likely, it also makes financial sharing more feasible. Providing a benefit for participation in research is a tricky matter, as the benefit should not be inappropriate or so great as to be coercive, a matter of great concern to IRBs in reviewing research protocols. With population-based research, a share of the proceeds might be used in a way that benefits the entire relevant population. This would reward all those whose participation was important in the discovery rather than one individual whose DNA happened to be used to clone or sequence the relevant genetic variation. The dilution of the financial incentive this strategy would cause could have an added benefit. If an individual were promised a profit share for his participation, he might be unduly influenced to take part in the research, motivated by the speculative value of that share rather than a careful weighing of the possible medical benefits and the risks. The HGDP, both in North America and worldwide, committed itself to providing a fair share of any financial benefits to the participating populations. That example has not been widely followed, but organized populations may well demand such guarantees from researchers. Whether and on what conditions IRBs will permit such sharing of financial benefits remains unclear.

Confidentiality

Individual subjects usually want, and receive, assurances that their identities will be kept confidential. This restriction often involves using false names in publications, obscuring photographs, and sometimes even doctoring pedigrees so that individuals cannot be identified. Communities may have similar concerns. Thus, whereas a population might be willing to participate in research on a stigmatizing condition that is unusually common among its members, it might not want to be publicly identified. On the other hand, some populations might want their participation heralded.

Group identities, like individual identities, can be obscured. A geographical location can be referred to in broad terms; a population can be defined by language family rather than language. The Navajo, for example, could be described as a Southwestern tribe or as a tribe speaking a NaDene language. Either description would leave a reader uncertain just which tribe had participated. Such group confidentiality does have its costs, though. For example, obscuring the identity of the population may make it impossible to replicate the research. Or subsequent researchers may not be able to make good use of the initial publications for lack of an exact identification. Although the chances of a particular individual, or even family, being the subject of study by several research teams will usually be small, populations may well be studied by many researchers over many decades.

Again, little guidance exists on this issue. Researchers should discuss with the population and its leaders what kind of group identification they would like. At all costs, researchers should avoid unpleasant surprises to the research subject population. Unexpected publicity could well leave the group embittered, violating their expectations of the present researchers and making them less willing to participate in future research.

Control of Research Uses and Materials

Individual research subjects will have little knowledge of the possible uses of the results, data, and biological materials involved in the research. Populations may have more knowledge and greater ability to bargain about their uses and disposition. Based on conversations with members of Native American groups, three classes of safeguards emerge as particularly important to them: prior approval of publication, control over subsequent uses (and users) of data and materials, and the ultimate return of data and materials.

Some Indian tribes have demanded prior review and approval of all publications. They seek this in order to make sure that the publication accurately reflects tribal understandings and does not harm the tribe's interests. The tension caused by this demand is likely to grow. Researchers are understandably reluctant to put their findings into the hands of non-scientists for review. The potential for their results to be delayed, or entirely lost, certainly exists. On the other hand, tribes are becoming increasingly sensitive to unanticipated negative effects of research publications; the Navajo, in particular, were outraged that the Hanta virus outbreak of the late 1990s was initially termed the Navajo virus. One aspect of publication review is acutely susceptible to misunderstanding: the definition of what constitutes "publication." The researcher may be prepared to submit a draft article for review, but a tribal government might believe that all presentations of the data, including talks and posters, should be reviewed in advance. Prior review and approval may be contentious in many negotiations for group participation in research; the one universal goal should be to ensure that whatever agreement is reached is clearly and mutually understood.

Tribal governments have also come to understand that data and biological materials do not necessarily remain with one researcher. A tribe may have built a relationship of trust with a particular scientist but may be upset to discover that "their" data and materials have been transferred to other researchers. Apprehension may be particularly acute when biological materials are transferred to widely accessible repositories, such as the American Type Culture Collection or the repositories at the Coriell Institute. Tribes may increasingly demand the power to approve, or deny, any transfer of materials to different researchers, inside or outside of the lead researcher's institution. Such restriction on sharing data and materials does not fit easily into current research methods. To some extent, restricted access may even violate a researcher's perceived obligation to help colleagues in other laboratories try to verify or falsify his results. At the very least, transfer would become slower and more cumbersome, hence likely to be another charged subject of negotiation.

Finally, tribal governments may require the data and biological materials to be returned to their custody after a fixed period. Such a provision for return is some guarantee that the data and materials will not be transferred or used without their permission. For biological materials, return will also be consistent with traditions concerning the disposition of human materials, where relevant. Researchers, on the other hand, are likely to resist the return or destruction of data and materials. Once gone, the data or materials can no longer be used for confirming the initial results or for extending the earlier research.

Return of Information

One of the most difficult ethical issues concerning individual research subjects is whether, or to what extent, a researcher should return significant information gained in the research to the subject. Population-based research faces similar issues.

On the one hand, tribal leaders often complain that their people never learn the results, if any, from the research conducted with them. Mailing a reprint of a scientific article does not constitute effective communication, even where the population's first language is English. Some tribes are now asking for return visits to discuss results or for translations of articles resulting from research into colloquial English or into their own languages. On the other hand, populations may not want to receive some kinds of information. A group might be willing to participate in research that could lead to results inconsistent with its own histories and myths on the condition that such inconsistent information not be returned to the population. Once again, the only sure advice is to discuss these issues with the group in advance, reach a clear agreement, and keep that agreement.

CONCLUSION

Genetic variations are both individual and collective. Except for monozygotic twins, every human being has an entirely unique genome, a combination of genetic variations never before seen in the history of our species. But we share half of our genetic variations with our siblings, parents, and children. And as families extend into ethnic groups, the commonality, although diluted, continues. As genetic research moves increasingly into groups—communities, ethnic groups, tribes, nations—researchers and research subjects are confronting the uncomfortable reality that groups, like people, face potential benefits and harms from their participation. Serious ethical and practical challenges are involved in protecting the interests of groups—and of the people who constitute them. But the challenges must be met, for the protection both of the people whose participation makes research possible and, in the long run, for science itself. Embittered and angry research subjects are unlikely to participate themselves in future research or to provide political support for research. Given the potential of population genetics research for relieving human suffering, such an outcome would be tragic.

Visit the Annual Reviews home page at www.AnnualReviews.org

LITERATURE CITED

1. Andrews L, Nelkin D. 2001. *Body Bazaar: The Market for Human Tissue in the Biotechnology Age*. New York: Crown
2. Cavalli-Sforza LL, Wilson AC, Cantor CR, Cook-Deegan RM, King M-C. 1991. Call for a worldwide survey of human genetic diversity: a vanishing opportunity for the human genome project. *Genomics.* 11:490–91
3. Cavalli-Sforza LL, Menozzi P, Piazza A. 1994. *The History and Geography of Human Genes*. Princeton: Princeton Univ. Press
4. Counc. Int. Org. Med. Sci. 1991. *International Guidelines for Ethical Review of Epidemiological Studies*. CIOMS: Geneva.
5. deCODE. 2001. http://www.decode.com
6. Duncan N. 1999. World Medical Association opposes Icelandic gene database. *Br. Med. J.* 318:1096
7. Faden RR, Beauchamp TL. 1986. *A History and Theory of Informed Consent*. New York: Oxford Univ. Press.
8. Foster MW, Bernsten D, Carter TH. 1998. A model agreement for genetic research in socially identifiable populations. *Am. J. Hum. Genet.* 63:696–702
9. Foster MW, Sharp RR, Freeman WL, Chino M, Bernsten D, et al. 1999. The role of community review in evaluating the risks of human genetic variation research. *Am. J. Hum. Genet.* 64:1719–27
10. Freeman WL. 1998. The role of community in research with stored tissue samples. In *Stored Tissue Samples: Ethical, Le-*
gal, and Policy Implications, ed. RF Weir, pp. 267–301. Iowa City: Univ. Iowa Press
11. Greely HT. 1997. The ethics of the human genome diversity project: The North American Regional Committee's proposed model ethical protocol. *In Human DNA Sampling: Law and Policy—International and Comparative Perspectives*, ed. BM Knoppers, pp. 239–56. The Hague: Kluwer Law
12. Greely HT. 1997. The control of genetic research: involving the groups between. *Houston Law Rev.* 33:1397–430
13. Greely HT. 1998. Legal, ethical, and social issues in human genome research. *Annu. Rev. Anthropol.* 27:473–502
14. Greely HT. 2000. Iceland's plan for genomics research: facts and implications. *Jurimetrics* 40:153–91
15. Greely HT. 2000. Human genome diversity project. *In Encyclopedia of Ethical, Legal, and Policy Issues in Biotechnology,* ed. TJ Murray, MJ Mehlman, pp. 552–66. New York: Wiley
16. Greely HT. 2001. Human genomics research: new challenges for research ethics. *Perspect. Biol. Med.* 44:221–29
17. Hirschfeld L, Hirschfeld H. 1919. Serological differences between the blood of different races. *Lancet* II, pp. 675–95
18. Human Genome Diversity Project. 1994. *Summary Document.* http://www.stanford.edu/group/morrinst/HGDP.hmtl
19. Jonatansson H. 2000. Iceland's health sector database: a significant head start in the

search for the biological grail or an irreversible error? *Am. J. Law Med.* 26:31–67

20. Juengst ET. 1998. Groups as gatekeepers to genomic research: conceptually confusing, morally hazardous, and practically useless. *Kennedy Inst. Ethics. J.* 8:183–200

21. Juengst ET. 1998. Group identity and human diversity: keeping biology straight from culture. *Am. J. Hum. Genet.* 63:673–77

22. Mannvernd. 2000. Association of Icelanders for Ethics in Science and Medicine. http://www.mannvernd.is/English

23. Moore v. Regents of the Univ. Calif. 1990. 51 Cal. 3d 120, 793 P. 2d 479, 271 Cal. Report. 146

24. Natl. Bioethics Advis. Comm. 1999. *The Use of Human Biological Materials in Research*. Bethesda, MD: Natl. Bioethics Advis. Comm.

25. Natl. Res. Counc. 1997. *Evaluating Human Genetic Diversity*. Washington, DC: Natl. Acad. Press

26. North Am. Reg. Comm., Hum. Genome Divers. Proj. 1997. Proposed model ethical protocol for collecting DNA samples. *Houston Law Rev.* 33:1431–73; http://www.stanford.edu/group/morrinst/hgdp.html

27. Reilly PR. 1998. Rethinking risks to human subjects in genetic research. *Am. J. Hum. Genet.* 63:682–85

28. Reilly PR, Page DC. 1998. We're off to see the genome. *Nat. Genet.* 20:15–17

29. Rural Adv. Found. Int. 1993. *Communiqué: Patents, Indigenous People, and Human Genetic Diversity* (May). http://www.rafi.ca

30. Rural Adv. Found. Int. 1997. *Communiqué: The Human Tissue Trade* (Jan/Feb). http://www.rafi.ca

31. Schneider WH. 1996. The history of research on blood group genetics. *Hist. Philos. Life Sci.* 18:7–33

32. Sharp RR, Foster MW. 2000. Involving study populations in genetic research. *J. Law Med. Ethics* 28:41–51

33. Weijer C, Goldsand G, Emanuel EJ. 1999. Protecting communities in research: current guidelines and limits of extrapolation. *Nat. Genet.* 23:275–80

34. World Med. Assoc. 1964. *Declaration of Helsinki: Recommendations Guiding Physicians in Biomedical Research Involving Human Subjects*, Helsinki, as revis. Tokyo, Jpn. 1974; Venice, Italy, 1983; Hong Kong, China, 1989

35. 45 Code Fed. Reg. Part 46 (1998)

G. LEDYARD STEBBINS

Annu. Rev. Genet. 2001. 35:803–14

G. Ledyard Stebbins and the Evolutionary Synthesis

Vassiliki Betty Smocovitis

Department of History, University of Florida, Gainesville, Florida 32611;
e-mail: bsmocovi@history.ufl.edu

Key Words history of genetics, plant evolutionary biology, evolutionary synthesis, California botany, plant conservation

■ **Abstract**   More than any other individual, Stebbins synthesized knowledge from a disparate set of areas that included plant genetics, systematics, and evolution. This work culminated in 1950 with the appearance of his magnum opus, *Variation and Evolution in Plants*. This book gave plant evolution a coherent framework that was compatible with that emerging from the work of Theodosius Dobzhansky, Ernst Mayr, G. G. Simpson, and Julian Huxley, and others associated with establishing the synthetic theory of evolution. For this work he is regarded as the botanical "architect" of the evolutionary synthesis.

CONTENTS

G. LEDYARD STEBBINS

G. Ledyard Stebbins, one of the foremost scientists of the twentieth century, died in his home in Davis, California on January 19, 2000. His scientific career spanned most of the twentieth century and included three primary areas of research: botany, genetics, and evolution. He is especially well known for his masterful synthesis of these three areas his 1950 book *Variation and Evolution in Plants* (22). More than any other, this book formed the conceptual backbone of the new science of plant evolutionary biology and guided an entire generation of plant evolutionists. As a result of writing this book, Stebbins is generally regarded as the botanical "architect" of the evolutionary synthesis, ranking alongside other notable figures such as Theodosius Dobzhansky, Ernst Mayr, G. G. Simpson, and Julian Huxley (11, 14, 16). Much of his original research included studies into the evolutionary genetics of plant species like the genus *Crepis*. He is best known for his work with the Berkeley geneticist E. B. Babcock in articulating the idea of the polyploid complex, a complex of reproductive forms centering on sexual diploids surrounded by polyploids, which may be apomictic as in *Crepis*, and for being the master of the synthetic review article (3).

EARLY LIFE AND EDUCATION

Ledyard (as he preferred to be called) Stebbins began his career with a great love of plants. Like many botanists of his generation, he came from a privileged family who indulged their children's interests. He was born on January 6, 1906, in Lawrence, New York, but spent much of his early childhood going back and forth to the swank seaside resort of Seal Harbor, Maine, where his father was a successful real estate developer. His love of plants and natural history was apparent even as a very small child, a love that he shared with his father, mother, and his elder brother and sister. At the tender age of three, Ledyard had shown not only a marked tendency to enjoy time out of doors, but he had also begun to manifest a behavior pattern for which he later became renowned: a tendency to quick, sudden outbursts of anger, especially as a means to gain attention or win arguments rapidly. As an adult, his temper tantrums became the stuff of urban legends; the best-known story told of Ledyard Stebbins as an adult involved his throwing a typewriter out the window in a fit of anger. The other vestiges of his early childhood that remained with him until the end included a New England "preppie" accent and a schoolboy sense of humor, frequently seen in his love of rhyme and silly verse.

His early academic career was less than stellar, but he enjoyed botanizing and mountaineering. In 1924 he decided to attend Harvard University mostly because his family background dictated his choice of school and because his older brother, Henry, was already enrolled there. Initially without much focus or direction, Ledyard thought that he would pursue a career in law and planned to major in political science, but his love of plants prevailed. By his third year, he decided to major in botany and began courses with the celebrated Harvard faculty, renowned not just for their scientific standing, but also for their contentious and domineering personalites.

He began his graduate work in the botany department in 1928. While comedic with hindsight, his experience in graduate school shaped his subsequent research style. He had a life-long tendency to move from one vital area of research to another as the science demanded, even when this meant overcoming political obstacles and difficult personalities. He initially worked in floristic botany with the foremost eastern systematist, M. L. Fernald, in the Gray Herbarium, but quickly lost interest in Fernald's outdated taxonomic methods and inflexible personality. Instead of the static herbarium taxonomy practiced by a person whom he derided as one of the "eminent exsiccatae in the Gray Herbarium," Ledyard chose instead the newer and more exciting cytological study of reproductive processes in plants. For his doctoral research he began anatomical and cytological studies of megasporogenesis in the ovule and microsporogenesis in the pollen of the plant genus *Antennaria*. He was able to collect *Antennaria* easily in the nearby environs so that he could also study geographic variation in the genus. Unfortunately, he chose as his doctoral advisor the eminent morphologist and cytologist, E. C. Jeffrey. Jeffrey, also known as "the stormy petrel of botany," hated with a vengeance any of the work emerging from the school of genetics associated with Thomas Hunt Morgan. He preferred instead to contemplate the idiosyncratic hybridization theories made popular by J. P. Lotsy (9). Jeffrey actively campaigned against Morgan's genetics and vigorously discouraged Ledyard from pursuing it.

The growing interest in genetics took on a life of its own, however, especially since plant genetics, systematics, and evolution and the zones of contact between the three were becoming exciting new areas of research in the 1920s (6, 7). Stebbins traveled to the library of Harvard's Bussey Institution to keep up with the genetics journals like *Hereditas* and *Genetica*, which contained articles by A. Müntzing and C. Leonard Huskins. He sought out geneticists at the Bussey such as E. M. East and took courses with W. E. Castle but found that Castle's mammalian genetics was not immediately helpful. His growing interest in the genetical literature became serious after he began to work with the noted plant geneticist Karl Sax, then just appointed to the Arnold Arboretum. His growing collaboration and friendship with Sax, one of the leaders of plant genetics in his generation, did not, however, sit well with his thesis advisor, Jeffrey. When Sax located a serious error of interpretation in Stebbins's chromosomal studies in his doctoral research, Jeffrey threatened to resign from the thesis committee in retaliation for what he viewed as intrusion into his direction of a student. The thesis was eventually amended so many times to meet the demands of a squabbling committee that it bore numerous scissors and paste marks masking the "offending" passages. It still stands in the Harvard archives as a testament to the contentious Harvard personalities in the botany department. Ledyard's experiences there were nonetheless positive. He completed his dissertation in 1931 and published it as two papers in 1932 (17, 18). This was in addition to publications on the New England flora, especially that of Mt. Desert Island, Maine, that he had published earlier with the assistance of Fernald in his journal *Rhodora*.

THE SHIFT TO GENETICS

In 1931 Stebbins was appointed to Colgate University, where he was required to teach introductory biology. He wrote a textbook with his colleague in the psychology department, Clarence Young, for this course (30). It was an unusual mixture of biology and psychology that was adopted for use by the American armed forces in the late 1930s and 1940s. Despite a heavy teaching load, Ledyard found time for genetics, spending nearly all his spare time on chromosomal studies of plants. He worked in close collaboration with Percy Saunders (of the Canadian Saunders family of wheat-breeders), at nearby Hamilton College. Saunders was a keen collector and breeder of peonies; his backyard was a profusion of these beautiful plants, some of which could still be seen in 1987, when I visited the house. With Stebbins, Saunders engaged in chromosomal studies in the species hybrids of *Paeonia*, of both old world and new world forms and found numerous interesting deviations from the normal pairing relationships. Stebbins and Saunders spent many evenings working on these studies in a makeshift laboratory in the basement of the Saunders' house.

With Saunders, Ledyard took the important step of attending the 1932 meetings of the International Congress of Genetics in nearby Ithaca, New York. At those meetings Stebbins recalled seeing the famous exhibit that Sewall Wright had set up displaying his shifting balance theory of evolution, but not understanding what they represented. He attended one particularly memorable session that featured Sax and the English cytogeneticist C. D. Darlington. At the climax of an especially heated exchange between Sax and Darlington over the chiasmatype theory, two of the omnipresent stray Cornell dogs broke into the room and engaged in a fight precisely in front of Sax and Darlington. This sent the audience into a paroxysm of laughter over the simultaneity of the two dogfights.

Ledyard listened closely to Thomas Hunt Morgan's famous address on the future of genetics. He also studied John Belling's exhibit demonstrating the existence of chromomeres, which he had mistakenly identified as genes. Most exciting of all, however, was Barbara McClintock's presentation of some of her cytological studies in maize. Using her squashing technique, McClintock showed the linear pairing of parental chromosomes at mid-prophase or pachytene. The paired chromosomes clearly showed the effects of crossing over and demonstrated beautifully the effects of inversions and translocations in their characteristic configurations. Shortly thereafter, he replicated some of the same studies in his *Paeonia* material and was the first to detect ring formation in this genus. The work was hardly groundbreaking, but it did confirm what McClintock and others had been describing concerning chromosome behavior (29).

His interest in genetics was reinforced further when he began a life-long intense friendship with the botanist Edgar Anderson, who was a fellow at the John Innes Horticultural Institute at the time of their meeting. In scientific interests and even in a colorful personal style, Anderson most closely approximated Stebbins. The two had met at the Fifth International Botanical Congress in Cambridge, England, in 1930. Anderson was about to begin his work on detecting and measuring variation patterns in plants like *Iris*, which were frequent hybridizers. He eventually went on to pioneering work on hybridization and was the first to articulate the notion of introgressive hybridization, a phenomenon seen often in plants (8).

E. B. BABCOCK, *CREPIS,* AND BERKELEY GENETICS

In 1935, Stebbins took a giant step in his turn to genetics by accepting a position as junior geneticist to the noted Berkeley geneticist E. B. Babcock. With the aid of a Rockefeller grant, Babcock hired Stebbins to assist him in an enormous undertaking to understand the genetic basis of evolutionary change in the plant genus *Crepis.* Stebbins was recommended for the position by the Washington-based expert on the Compositae, Sidney F. Blake, and was hired even after Ledyard's father, who was a close friend of the Rockefellers (Ledyard had been a playmate of Nelson's), nearly undermined Ledyard's chances with Babcock by attempting to leverage a higher salary for the appointment. Babcock was not pleased by the attempted intervention using the Rockefeller connection.

Babcock was engaged in an ambitious team-oriented project to find a plant equivalent of *Drosophila.* Very much eclipsed by his contemporary Thomas Hunt Morgan at the California Institute of Technology, Babcock was one of the most important figures in establishing and institutionalizing genetics within the Agricultural College at Berkeley. It became one of the first departments of genetics in the country, thanks to the efforts of Babcock, who was convinced that genetics generally, and agricultural genetics in particular, was a vital part of the mission of the University of California. Babcock's vision for genetics at Berkeley was that it would rival the success of the Morgan school's project with *Drosophila melanogaster.* He chose the genus *Crepis* to be the plant equivalent of *Drosophila* even though it was a weed and not an important crop plant, mostly because he felt the genus with its diverse geographic variation patterns could be used to understand the genetic basis for evolutionary change, which could then form the basis for a taxonomic study (1). Preliminary work had begun as early as 1917–1918, but the project continued into the late 1940s and ended only with Babcock's retirement. Babcock considered his monograph on the genus *Crepis* to be the centerpiece of his life's work (2).

Stebbins's assignment assisting Babcock was in performing chromosome counts in some of the nearest relatives of *Crepis* in the tribe Cichorieae. He quickly developed an interest in Babcock's own research, which was in understanding some of the New World species of *Crepis,* because he recognized patterns of evolution that resembled those in *Antennaria* and *Paeonia.* Like these other genera, *Crepis* was a commonly hybridizing group that displayed polyploidy and could reproduce apomictically. In 1938, Babcock and Stebbins jointly published a monograph on the American species of *Crepis.* It laid the foundation for understanding polyploid complexes and the role of apomixis in the formation of some of them; for this reason, they first termed the American species of *Crepis* an agamic complex. They recognized clearly that certain plant genera consisted of a complex of reproductive forms that centered on sexual diploids and that had given rise to polyploids; sometimes as in *Crepis,* these were apomictic polyploids. Polyploids that combined the genetic patrimony of two species, they also showed, usually had the wider distribution pattern. Babcock and Stebbins's articulation of the polyploid complex, and their clear elucidation of its existence in the American species of *Crepis* was considered pathbreaking work at the time. Not only did it demonstrate in detail the complex interplay of apomixis, polyploidy, and hybridization in a geographic context, but it also offered insights into species formation, polymorphy in apomictic forms, and knowledge of how all these complex processes could inform an accurate phylogenetic history of the genus. Stebbins extended these ideas further with subsequent breeding studies in forage grasses and published a series of important articles in 1940, 1941, and 1947 (19–21). The latter article, entitled "Types of polyploids: their classification and significance," became a classic review that synthesized knowledge bearing on polyploidy in plants and constituted probably one of his most important contributions to understanding of plant evolution.

Stebbins worked closely with Babcock for six years. In 1939, he was successful in securing a position as assistant professor in the Berkeley genetics department. Babcock, who was impressed with Ledyard's energy and industry, was instrumental in making the appointment. Earlier, Stebbins had a significant disappointment in that he had failed to obtain the replacement position for Willis Linn Jepson in the botany department. Although he made himself at home with the botanists at Berkeley, Stebbins's interests were considered so heavily genetical that his colleagues in botany did not feel he was sufficiently focused on the curatorial work the position demanded. The position was offered to Lincoln Constance instead. The vacancy of a position in the genetics department, which required teaching of the general course on evolution, was opportune for Stebbins, whose interests were shifting to the exciting areas in evolutionary study opening in the late 1930s. He read voraciously in preparation of the course and quickly realized that there was a serious shortage of books in evolution helpful to him and to his course, which was taught out of the genetics department in the College of Agriculture. His growing interest in evolution was fueled by two additional factors: his interactions with a unique group of biologists all concerned with evolutionary approaches to systematics who called themselves "The Biosystematists," and his special relationship with the Russian émigré Theodosius Dobzhansky.

Beginning in the mid-1930s, the San Francisco Bay area became a hotbed for evolutionary activity. A new generation of systematists who incorporated insights from genetics and ecology had taken root in the Bay area at institutions like Stanford University, the Carnegie Institution at Stanford University, and the California Academy of Sciences, in addition to the University of California, Berkeley. Calling themselves "The Biosystematists," the group met at alternating locations every month to share in the new methods that were characterizing the "new systematics" as a whole. Ledyard was a prominent member of the group nearly from the start. He was active in inviting speakers, some of whom included visitors from other states like his close friend Edgar Anderson, from the Missouri Botanical Garden in St. Louis, and his fellow plant systematist at the University of California at Los Angeles, Carl Epling.

The critical players among the Biosystematists were the interdisciplinary Carnegie team that included the Danish genecologist Jens Clausen, the taxonomist David Keck, and the physiologist William Hiesey. By the mid-1930s, the team was engaged in series of long-term systematic studies that incorporated knowledge of genetics, ecology, and taxonomy to understand patterns of evolution in plants, initially to distinguish environmental from genetic factors in plant evolution. In particular, they studied patterns of variation of plants as they adapted along steep altitudes in the Californian landscape. Their work is considered pioneering in understanding the mechanisms responsible for plant adaptation along varying altitudinal gradients. Ledyard followed this work closely and visited the team in their experimental sites all through the 1940s.

Also in the mid-1930s, Stebbins began a close friendship with the evolutionary geneticist, Theodosius Dobzhansky. Stebbins met Dobzhansky on a visit to the California Institute of Technology in the spring of 1936 when Dobzhansky was just beginning to turn to his work on the genetics of natural populations using *Drosophila pseudoobscura*. The two interacted further when Dobzhansky frequented the Berkeley campus to see his close friend, the geneticist I. Michael Lerner, then in the Poultry Husbandry Department. Stebbins had interacted with Lerner in a fortnightly journal club called Genetics Associated. Even though Lerner and Dobzhansky frequently spoke to each other in Russian, Ledyard enjoyed listening to them discuss their mutual interests in evolutionary genetics. The friendship with Dobzhansky was to prove absolutely critical to Ledyard as his own interests were shifting more and more to evolutionary genetics, thanks to the teaching demands made by the evolution course. Dobzhansky, who published his own pathbreaking synthesis of evolutionary genetics under the title *Genetics and the Origin of Species* in 1937, began to foster Ledyard's evolutionary interests (4). Through the 1940s they came in closer contact when they met for field work at the Carnegie Institution's field site at Mather, California. Both were avid horseback riders and frequently collected hybrids from the back of a horse.

Dobzhansky played the single most important influence in Stebbins's career as an evolutionist. In 1945, thanks to Dobzhansky's recommendation, L. C. Dunn at Columbia University invited Stebbins to deliver the prestigious set of Morris K. Jesup Lectures at Columbia University. One reason why Stebbins had been selected was the need for a comprehensive synthesis of plant evolution. In 1941, Edgar Anderson had co-delivered the Jesup Lectures with the zoologist Ernst Mayr. While Mayr subsequently published his set of lectures under the title *Systematics and the Origin of Species from the Viewpoint of a Zoologist*, Anderson never completed the publication of his set of lectures (10). The viewpoint of the botanist was therefore needed in what was emerging as the new synthesis of evolution launched by Dobzhansky. In response to the invitation, Stebbins took the course notes he had been using for his evolution class and converted them to the Jesup Lectures. The published version of the lectures appeared with Columbia University Press in their Columbia Biological Series in 1950 under the title *Variation and Evolution in Plants*. It was published in the same series as Theodosius Dobzhansky's

1937 *Genetics and the Origin of Species*, Ernst Mayr's 1942 *Systematics and the Origin of Species*, and G. G. Simpson's 1944 *Tempo and Mode Evolution* (13). Taken as a whole, these books provided the backbone of the modern synthesis of evolution, which incorporated insights from a range of disciplines with evolutionary genetics. As far as botany went, Stebbins upheld the importance of most of the tenets emerging as part of the new consensus on evolution and followed his friend Dobzhansky, who had drawn more heavily from animal examples, closely. He stressed the centrality of natural selection but left plenty of room for random genetic drift and nonadaptive evolution. He also upheld Dobzhansky's and Mayr's notion of the biological species concept (BSC), though it took much explaining. (He subsequently backpedaled on the BSC'S application in botany.) The book also effectively killed any serious belief in alternative mechanisms of evolution like Lamarckian evolution or soft inheritance. At 643 pages in length and over 1250 citations, *Variation and Evolution in Plants* was the longest and the last of the books associated with the evolutionary synthesis. The book received instant recognition for its ambitious synthesis of a broad range of areas. It was so comprehensive that it opened a new field of research for younger scholars who recognized themselves as plant evolutionary biologists. Assessing the book, Peter Raven described it as "the most influential single book in plant systematics this century." It remains a heavily cited text (12).

THE DAVIS YEARS

With the publication of his magnum opus, Ledyard's life began to take different directions. He was already emerging as a leader in evolutionary biology. He was an active member of the first international society for the study of evolution, the Society for the Study of Evolution, and in 1948, he was elected as its third President. But his interests in evolution may have left him at odds with some of the members of his own department. By the 1940s, the Berkeley genetics department had become world-class and was leading the way in new areas of genetical research. Older areas like plant cytogenetics were becoming replaced with the newer physiological and biochemical genetics. With the retirement in 1947 of Babcock, who had served as chair of the department and was Ledyard's greatest supporter, Ledyard was increasingly becoming the odd one out. While publishing both original research and longer review articles in genetics journals and actively reading the literature, his own researches in genetics were always designed with the aim of understanding the mechanisms of evolutionary change. He was never really concerned with the mechanisms of gene action outside such an evolutionary framework. My sense is that Ledyard was increasingly beginning to feel uncomfortable in his own department and that there may have been tension building between him and R. E. Clausen, who succeeded Babcock as chair. Thus, in 1950 when the invitation came to move to the expanding Davis campus of the University of California and to organize a new department of genetics, Stebbins accepted the invitation enthusiastically. He explicitly told me several times that he liked the idea of being a "big fish in a small pond." In 1950, therefore, Ledyard moved to Davis, California, which became his home until the time of his death. He was instrumental in launching the genetics department there and stayed as its chair until 1963.

After his move to Davis, his research shifted once again to incorporate newer areas like developmental morphology and genetics in crop plants such as barley. He also became active in training graduate students, nearly all of whom were in developmental biology or plant developmental genetics. Ironically, the only students associated with Ledyard's primary area of research in plant evolutionary biology dated back to his Berkeley days. They were Verne Grant and Charles Heiser Jr. Ledyard had, however, only served as a committee member and not chair of their graduate committees. Other notable students included Ghurdev Khush and Michael Zohary.

From all indications, Ledyard was a popular and engaging teacher, especially at the undergraduate level. His undergraduate evaluations were consistently favorable. He frequently took students on day-trips that showed them California's unique flora. His claim to fame was that he could tell precisely the elevation of his whereabouts by identifying the plants closest to him. Students warmed to his energy, enthusiasm, and love of natural history. His graduate students speak favorably of working with Ledyard, but my sense is that he did not have the patient personality to supervise complex dissertations closely (see below). One recurring description of Ledyard as advisor concerns his legendary sloppiness and complete lack of dexterity in the laboratory: Graduate students and technicians usually prepared two sets of every important slide that they gave to him because he was likely to break it once it got into his hands.

Ledyard cared deeply about the teaching of evolution, and in the early and mid-1960s he worked as one of the faculty in the Biological Sciences and Curriculum Study to institute the teaching of evolution in American high schools. He echoed Dobzhansky closely in stating repeatedly that "nothing in biology makes sense except in the light of evolution." He actively fought the rise of "scientific creationist" groups in California and in the nation. Between 1960 and 1964 he served as secretary-general to the International Union of Biological Sciences. He was active in numerous societies and served as President for nearly all of them.

In the 1960s, his interest in evolution continued to grow. In 1965, he and Herbert Baker edited a collection of papers that came out of an Asilomar conference in a volume entitled *The Genetics of Colonizing Species* (28). His second most important book appeared in 1974 as *Flowering Plants: Evolution Above the Species Level*, following the Prather Lectures he gave at Harvard (26). His other books included a widely adopted textbook of evolution, *Processes of Organic Evolution*, which went through multiple editions (23). He also wrote *Chromosomal Evolution in Higher Plants*, which was also adopted as an advanced textbook, and *The Basis of Progressive Evolution* (24, 25). Along with Dobzhansky, Francisco Ayala, and James Valentine, he wrote the textbook *Evolution* in 1977 and in 1982 he completed his semipopular *Darwin to DNA: Molecules to Humanity* (5, 27).

After he moved to Davis, Ledyard became increasingly active in conservation work both with amateur and professional groups. He led innumerable public field trips to explore the California flora and effectively led a political campaign to prevent the destruction of a place he called "Evolution Hill," a strip of beach on the Monterey Peninsula that supported rare and endangered plants. He was instrumental in helping to form the California Native Plant Society and in contributing both scientific and popular articles on the subject of California native plants.

In the course of his long career, Ledyard Stebbins won numerous awards and honors in multiple areas of research. His greatest recognition was receiving the National Medal of Science from President Carter in 1979. In 1973 he became Emeritus Professor of Genetics.

BOTANIST, GENETICIST, OR EVOLUTIONIST: STEBBINS AS SYNTHESIZER

In science as in everything, small-scale synthesizers usually get credit from all constituent parties, but truly great synthesizers can fall between the cracks in the cycle of scientific credit. Ledyard Stebbins was in the latter category; neither fish nor fowl, he frequently failed to receive credit for work in some areas, usually at the hands of narrower colleagues. Systematists felt that he concentrated too heavily on genetics and too little on taxonomic studies, whereas geneticists felt that his work was too evolutionary, natural history–oriented and did not concentrate sufficiently on the mechanisms of gene action. Few, however, have challenged his contributions to plant evolutionary biology, nor questioned his ability to synthesize disparate literature into a coherent framework. This was a work style that he repeatedly demonstrated. His publication list at just over 260 articles bears a striking number of pieces that qualify as synthetic reviews. His ability to read quickly, recognize novel insights, digest new material, and then integrate the knowledge were the hallmarks of his scientific work style. He was a masterful synthesizer and master of the review essay or synthetic thought piece.

THE PERSONAL AND THE PROFESSIONAL: REFLECTIONS ON STEBBINS

Ledyard had such a strong personality, that it could not help but spill over into his professional life. He displayed a strangely predictable form of nonconformism. My sense is that this stemmed from a rebellious attitude toward his family. He suffered from the classic "latter-born son as rebel" syndrome that historian and psychologist Frank Sulloway described in his *Born to Rebel* (31). He considered himself a staunch liberal and a life-long supporter of the Democratic Party, but that was also a reaction against his father, who was a Republican. Though his family was Episcopalian (his father was an avid participant in the church at Seal Harbor),

Ledyard was never much of a believer, but in later years, after marriage to his second wife in the 1950s, he did become an active Unitarian.

Like many creative people, his strengths were simultaneously his weaknesses. He was industrious, intensely focused, and always enthusiastic. He seemed constantly excited by a some new insight that usually came out of his voracious reading. The new insight usually made its way into his latest project almost immediately. He loved following the work of younger people and supported them generously. At times, he seemed almost desperate to please people who mattered to him. At other times, however, he could be self-absorbed, petulant, and completely insensitive to the thoughts and wishes of the people around him. This latter behavior, combined with the tendency to "blow his top" (his expression), did not always make him an easy colleague. Many of his contemporaries dreaded collaborative work, committee work, or even spending an evening with him (he loved to speak in long, perfectly constructed paragraphs that gave little opportunity for conversations). Barbara Monaghan Stebbins, his second wife, best described Ledyard to me once as a "child." I think what she meant to convey with this description was his childlike wonder with the world, his fundamental belief in the goodness of people around him (he was never able to carry a grudge or remain angry for very long), but also his tendency to self-absorption and insensitivity to others. I think this may help explain the fact that no matter how much people may have been frustrated by Ledyard, nearly all admired his accomplishments and spoke of him with admiration, amusement, and affection.

Ledyard loved classical music passionately, frequented art shows with Barbara, and was fond of reading popular books of science. He especially loved to watch cultural programs on PBS. Toward the end of his life, with his eyesight failing, he would sit quietly listening to books on tape such as *Ancestral Passions*, the popular biography on the Leakey family. He absolutely loved having books read to him, and I was happy to read to him sections from David Quammen's *Song of the Dodo*. But though aged and infirm and in pain from a tumor growing on the side of his face, he was still capable of throwing a temper tantrum or two. Good liberal that he was, he did not discriminate against his target of momentary rage (15). To his end, he denied ever throwing that typewriter out the window.

Visit the Annual Reviews home page at www.AnnualReviews.org

LITERATURE CITED

1. Babcock EB. 1920. *Crepis*—a promising genus for genetic investigation. *Am. Nat.* 54:270–76

2. Babcock EB. 1947. *The Genus Crepis I and II*. Univ. Calif. Publ. Bot. 21, 22

3. Babcock EB, Stebbins GL Jr. 1938. The American species of *Crepis*: their interrelationships and distribution as affected by polyploidy. *Carnegie Inst. Washington Publ.* No. 504

4. Dobzhansky T. 1937. *Genetics and the Origin of Species*. New York: Columbia Univ. Press

5. Dobzhansky T, Ayala FJ, Stebbins GL, Valentine JW. 1977. *Evolution*. New York: WH Freeman

6. Hagen JB. 1982. *Experimental taxonomy, 1930–1950: The impact of cytology, ecology, and genetics on ideas of biological classification.* PhD thesis. *Oregon State Univ.*, Corvallis

7. Hagen JB. 1984. Experimentalists and naturalists in twentieth century botany, 1920–1950. *J. Hist. Biol.* 17:249–70

8. Kleinman K. 1999. His own synthesis: corn, Edgar Anderson, and evolutionary theory in the 1940s. *J. Hist. Biol.* 32:293–320

9. Lotsy JP. 1916. *Evolution by Means of Hybridization.* Hague: Martinus Nijhoff

10. Mayr E. 1942. *Systematics and the Origin of Species from the Viewpoint of a Zoologist.* New York: Columbia Univ. Press

11. Mayr E, Provine WB, eds. 1980. *The Evolutionary Synthesis.* Cambridge: Harvard Univ. Press

12. Raven P. 1974. Plant systematics 1947–1972. *Ann. Mo. Bot. Gard.* 61:166–78

13. Simpson GG. 1944. *Tempo and Mode in Evolution.* New York: Columbia Univ. Press

14. Smocovitis VB. 1997. G. Ledyard Stebbins and the evolutionary synthesis. *Am. J. Bot.* 84:1625–37

15. Smocovitis VB. 1999. Living with your biographical subject: problems of distance, privacy and trust in the writing of the biography of G. Ledyard Stebbins, Jr. *J. Hist. Biol.* 32:421–38

16. Solbrig O. 1979. George Ledyard Stebbins. In *Topics in Plant Population Biology*, ed. O Solbrig, S Jain, GB Johnson, PH Raven, pp. 1–17

17. Stebbins GL Jr. 1932. Cytology of *Antennaria.* I. Normal species. *Bot. Gaz.* 94:134–51

18. Stebbins GL Jr. 1932. Cytology of Antennaria. II. Parthenogenetic species. *Bot. Gaz.* 94:322–45

19. Stebbins GL Jr. 1940. The significance of polyploidy in plant evolution. *Am. Nat.* 74:54–66

20. Stebbins GL Jr. 1941. Apomixis in the angiosperms. *Bot. Rev.* 7:507–42

21. Stebbins GL Jr. 1947. Types of polyploids: their classification and significance. *Adv. Genet.* 1:403–29

22. Stebbins GL Jr. 1950. *Variation and Evolution in Plants*, New York: Columbia Univ. Press

23. Stebbins GL. 1966. *Processes of Organic Evolution.* Englewood Cliffs, NJ: Prentice-Hall

24. Stebbins GL. 1969. *The Basis of Progressive Evolution.* Chapel Hill: Univ. NC Press

25. Stebbins GL. 1971. *Chromosomal Evolution in Higher Plants.* London: Arnold

26. Stebbins GL. 1974. *Flowering Plants: Evolution above the Species Level.* Harvard: Harvard Univ. Press

27. Stebbins GL. 1982. *Darwin to DNA: Molecules to Humanity.* San Francisco: WH Freeman

28. Stebbins GL, Baker H. 1965. *The Genetics of Colonizing Species.* New York: Academic

29. Stebbins GL Jr, Ellerton S. 1939. Structural hybridity in *Paeonia Californica* and *P. Brownii. J. Genet.* 38:1–36

30. Stebbins GL, Young CW. 1938. *The Human Organism and the World of Life.* New York: Harper's

31. Sulloway F. 1996. *Born to Rebel: Birth Order, Family Dynamics, and Creative Lives.* New York: Pantheon

Subject Index

A

Abd genes
  chromatin insulators and
    boundaries, 197
Acetylcholine receptors
  ligand-gated
    epilepsy genes and,
    577–78
Actins
  hearing loss molecular
    genetics and, 595
  yeast calmodulin and its
    targets, 649, 661
"Action at a distance"
  homologous recombination
    near and far from DNA
    breaks, 243
Activation
  translational control in
    Drosophila and, 370,
    379–85, 388–89
Acyl-homoserine lactone
  quorum sensing
    acyl-HSLs
      accumulation, 449–50
      enzymology, 445
      membrane trafficking,
      449
      structural diversity,
      443–45
    cis-acting elements,
    453–54
    diffusible signals, 458–59
    DNA binding, 454–55
    eukaryotic signal
      molecules, 459
    fine tuning, 458–59
    intramolecular inhibition
      model, 452
    introduction, 440–42
    LuxI-type acyl-HSL

synthases
  ains family, 448–49
  structure-function
    studies, 445–48
  LuxR-type proteins,
    450–56
  membrane interactions,
    452
  molecular mechanisms,
    443–59
  multimerization, 452–53
  perspective, 460–61
  real-world environmental
    considerations, 442–43
  regulatory gene expression,
    456–57
  repression, 456
  transcriptional control,
    455–56
Adaptation
  fitness of hybrids and,
    31–47
  plant-enemy coevolution
    and, 476
ade genes
  homologous recombination
    near and far from DNA
    breaks, 265
Adenomatous polyposis coli
  tumor-suppressor gene
    models and, 228–29
Adenosine triphosphate
  (ATP)
  chaperones in early
    secretory pathway and,
    150–51, 155–56
  hearing loss molecular
    genetics and, 595, 598,
    622
  homologous recombination
    near and far from DNA

breaks, 250, 252–53,
    255–56
  recombination repair of
    E. coli replication forks
    and, 64, 72
  sister chromatid cohesion
    and, 686, 688
Adherens junctions
  epithelial cell polarity and
    cell junctions in
    Drosophila, 755–59
Adh gene
  genetic architecture of
    quantitative traits and, 323
  nonneutral evolution of
    mtDNA and, 551
A domains
  hearing loss molecular
    genetics and, 602, 620
Adriamycin
  tumor-suppressor gene
    models and, 216
Aedes albopictus
  nonneutral evolution of
    mtDNA and, 550
Agrobacterium tumefaciens
  acyl-homoserine lactone
    quorum sensing and, 441,
    444, 448, 450–56, 458
ains genes
  acyl-homoserine lactone
    quorum sensing and,
    448–49, 458
$\alpha2$ chain defect
  hearing loss molecular
    genetics and, 617–18
Alu elements
  L1 retrotransposons and,
    501, 503, 513–15
Aminoglycosides
  hearing loss molecular

815

CUMULATIVE INDEXES

CONTRIBUTING AUTHORS, VOLUMES 31–35

851

CHAPTER TITLES, VOLUMES 31–35

Introductory Chapters

Bacterial Genetics